Instructor's Solutions Manual

Volume II

Randy Gallaher • Kevin Bodden

Lewis and Clark Community College

Precalculus

Enhanced with Graphing Utilities

Fourth Edition

Sullivan
Sullivan

Upper Saddle River, NJ 07458

Editor-in-Chief: Sally Yagan
Acquisitions Editor: Adam Jaworski
Editorial Assistant: Christopher Truchan
Executive Managing Editor: Kathleen Schiaparelli
Assistant Managing Editor: Becca Richter
Production Editor: Elizabeth Klug
Supplement Cover Manager: Paul Gourhan
Supplement Cover Designer: Joanne Alexandris
Manufacturing Buyer: Ilene Kahn

Pearson Prentice Hall
Pearson Education, Inc.
Upper Saddle River, NJ 07458

Printed in the United States of America

10 9 8 7 6 5 4 3

ISBN 0-13-149101-6

Pearson Education Ltd., *London*
Pearson Education Australia Pty. Ltd., *Sydney*
Pearson Education Singapore, Pte. Ltd.
Pearson Education North Asia Ltd., *Hong Kong*
Pearson Education Canada, Inc., *Toronto*
Pearson Educación de Mexico, S.A. de C.V.
Pearson Education—Japan, *Tokyo*
Pearson Education Malaysia, Pte. Ltd.

Table of Contents

Chapter 8 Polar Coordinates; Vectors

Chapter 9 Analytic Geometry

Chapter 10 Systems of Equations and Inequalities

Chapter 11 Sequences; Induction; the Binomial Theorem

Preface

This solution manual accompanies *Precalculus: Enhanced with Graphing Utilities, 4th Edition* by Michael Sullivan and Michael Sullivan, III. The Instructor Solutions Manual (ISM) contains detailed solutions to all exercises in the text and the chapter projects (both in the text and those posted on the internet). The Student Solutions Manual (SSM) contains detailed solutions to all odd exercises in the text and all solutions to chapter tests. In both manuals, some TI-83 Plus graphing calculator screenshots have been included to demonstrate how technology can be used to solve problems and check solutions. A concerted effort has been made to make this manual as user-friendly and error free as possible. Please feel free to send us any suggestions or corrections.

We would like to extend our thanks to Dawn Murrin, Bob Walters, and Chris Truchan from Prentice Hall for all their help with manuscript pages and logistics. Thanks for everything!

We would also like to thank our wives (Angie and Karen) and our children (Annie, Ben, Ethan, Logan, Payton, and Shawn) for their patient support and for enduring many late evenings.

A special thanks to Bill Bodden for his help with error-checking solutions and for the many hours of great math discussions.

Kevin Bodden and Randy Gallaher
Department of Mathematics
Lewis and Clark Community College
5800 Godfrey Road
Godfrey, IL 62035
kbodden@lc.edu rgallahe@lc.edu

Chapter 7

Applications of Trigonometric Functions

Section 7.1

1. $a = \sqrt{5^2 - 3^2} = \sqrt{25-9} = \sqrt{16} = 4$

2. $\tan\theta = \frac{1}{2}, \quad 0 < \theta < 90°$

   $\theta = \tan^{-1}\left(\frac{1}{2}\right) \approx 26.6°$

3. False; $52° + 48° = 100° \neq 90°$

4. True

5. angle of elevation

6. angle of depression

7. True

8. False

9. opposite = 5; adjacent = 12; hypotenuse = ?
   $(\text{hypotenuse})^2 = 5^2 + 12^2 = 169$

   $\text{hypotenuse} = \sqrt{169} = 13$

   $\sin\theta = \frac{\text{opp}}{\text{hyp}} = \frac{5}{13} \qquad \csc\theta = \frac{\text{hyp}}{\text{opp}} = \frac{13}{5}$

   $\cos\theta = \frac{\text{adj}}{\text{hyp}} = \frac{12}{13} \qquad \sec\theta = \frac{\text{hyp}}{\text{adj}} = \frac{13}{12}$

   $\tan\theta = \frac{\text{opp}}{\text{adj}} = \frac{5}{12} \qquad \cot\theta = \frac{\text{adj}}{\text{opp}} = \frac{12}{5}$

10. opposite = 3; adjacent = 4, hypotenuse = ?
    $(\text{hypotenuse})^2 = 3^2 + 4^2 = 25$

    $\text{hypotenuse} = \sqrt{25} = 5$

    $\sin\theta = \frac{\text{opp}}{\text{hyp}} = \frac{3}{5} \qquad \csc\theta = \frac{\text{hyp}}{\text{opp}} = \frac{5}{3}$

    $\cos\theta = \frac{\text{adj}}{\text{hyp}} = \frac{4}{5} \qquad \sec\theta = \frac{\text{hyp}}{\text{adj}} = \frac{5}{4}$

    $\tan\theta = \frac{\text{opp}}{\text{adj}} = \frac{3}{4} \qquad \cot\theta = \frac{\text{adj}}{\text{opp}} = \frac{4}{3}$

11. opposite = 2; adjacent = 3; hypotenuse = ?
    $(\text{hypotenuse})^2 = 2^2 + 3^2 = 13$

    $\text{hypotenuse} = \sqrt{13}$

    $\sin\theta = \frac{\text{opp}}{\text{hyp}} = \frac{2}{\sqrt{13}} = \frac{2}{\sqrt{13}} \cdot \frac{\sqrt{13}}{\sqrt{13}} = \frac{2\sqrt{13}}{13}$

    $\cos\theta = \frac{\text{adj}}{\text{hyp}} = \frac{3}{\sqrt{13}} = \frac{3}{\sqrt{13}} \cdot \frac{\sqrt{13}}{\sqrt{13}} = \frac{3\sqrt{13}}{13}$

    $\tan\theta = \frac{\text{opp}}{\text{adj}} = \frac{2}{3}$

    $\csc\theta = \frac{\text{hyp}}{\text{opp}} = \frac{\sqrt{13}}{2}$

    $\sec\theta = \frac{\text{hyp}}{\text{adj}} = \frac{\sqrt{13}}{3}$

    $\cot\theta = \frac{\text{adj}}{\text{opp}} = \frac{3}{2}$

12. opposite = 3; adjacent = 3; hypotenuse = ?
    $(\text{hypotenuse})^2 = 3^2 + 3^2 = 18$

    $\text{hypotenuse} = \sqrt{18} = 3\sqrt{2}$

    $\sin\theta = \frac{\text{opp}}{\text{hyp}} = \frac{3}{3\sqrt{2}} = \frac{3}{3\sqrt{2}} \cdot \frac{\sqrt{2}}{\sqrt{2}} = \frac{\sqrt{2}}{2}$

    $\cos\theta = \frac{\text{adj}}{\text{hyp}} = \frac{3}{3\sqrt{2}} = \frac{3}{3\sqrt{2}} \cdot \frac{\sqrt{2}}{\sqrt{2}} = \frac{\sqrt{2}}{2}$

    $\tan\theta = \frac{\text{opp}}{\text{adj}} = \frac{3}{3} = 1$

    $\csc\theta = \frac{\text{hyp}}{\text{opp}} = \frac{3\sqrt{2}}{3} = \sqrt{2}$

    $\sec\theta = \frac{\text{hyp}}{\text{adj}} = \frac{3\sqrt{2}}{3} = \sqrt{2}$

    $\cot\theta = \frac{\text{adj}}{\text{opp}} = \frac{3}{3} = 1$

13. adjacent = 2; hypotenuse = 4; opposite = ?

$(\text{opposite})^2 + 2^2 = 4^2$

$(\text{opposite})^2 = 16 - 4 = 12$

$\text{opposite} = \sqrt{12} = 2\sqrt{3}$

$\sin\theta = \dfrac{\text{opp}}{\text{hyp}} = \dfrac{2\sqrt{3}}{4} = \dfrac{\sqrt{3}}{2}$

$\cos\theta = \dfrac{\text{adj}}{\text{hyp}} = \dfrac{2}{4} = \dfrac{1}{2}$

$\tan\theta = \dfrac{\text{opp}}{\text{adj}} = \dfrac{2\sqrt{3}}{2} = \sqrt{3}$

$\csc\theta = \dfrac{\text{hyp}}{\text{opp}} = \dfrac{4}{2\sqrt{3}} = \dfrac{4}{2\sqrt{3}} \cdot \dfrac{\sqrt{3}}{\sqrt{3}} = \dfrac{2\sqrt{3}}{3}$

$\sec\theta = \dfrac{\text{hyp}}{\text{adj}} = \dfrac{4}{2} = 2$

$\cot\theta = \dfrac{\text{adj}}{\text{opp}} = \dfrac{2}{2\sqrt{3}} = \dfrac{2}{2\sqrt{3}} \cdot \dfrac{\sqrt{3}}{\sqrt{3}} = \dfrac{\sqrt{3}}{3}$

14. opposite = 3; hypotenuse = 4; adjacent = ?

$3^2 + (\text{adjacent})^2 = 4^2$

$(\text{adjacent})^2 = 16 - 9 = 7$

$\text{adjacent} = \sqrt{7}$

$\sin\theta = \dfrac{\text{opp}}{\text{hyp}} = \dfrac{3}{4}$

$\cos\theta = \dfrac{\text{adj}}{\text{hyp}} = \dfrac{\sqrt{7}}{4}$

$\tan\theta = \dfrac{\text{opp}}{\text{adj}} = \dfrac{3}{\sqrt{7}} = \dfrac{3}{\sqrt{7}} \cdot \dfrac{\sqrt{7}}{\sqrt{7}} = \dfrac{3\sqrt{7}}{7}$

$\csc\theta = \dfrac{\text{hyp}}{\text{opp}} = \dfrac{4}{3}$

$\sec\theta = \dfrac{\text{hyp}}{\text{adj}} = \dfrac{4}{\sqrt{7}} = \dfrac{4}{\sqrt{7}} \cdot \dfrac{\sqrt{7}}{\sqrt{7}} = \dfrac{4\sqrt{7}}{7}$

$\cot\theta = \dfrac{\text{adj}}{\text{opp}} = \dfrac{\sqrt{7}}{3}$

15. opposite = $\sqrt{2}$; adjacent = 1; hypotenuse = ?

$(\text{hypotenuse})^2 = \left(\sqrt{2}\right)^2 + 1^2 = 3$

$\text{hypotenuse} = \sqrt{3}$

$\sin\theta = \dfrac{\text{opp}}{\text{hyp}} = \dfrac{\sqrt{2}}{\sqrt{3}} = \dfrac{\sqrt{2}}{\sqrt{3}} \cdot \dfrac{\sqrt{3}}{\sqrt{3}} = \dfrac{\sqrt{6}}{3}$

$\cos\theta = \dfrac{\text{adj}}{\text{hyp}} = \dfrac{1}{\sqrt{3}} = \dfrac{1}{\sqrt{3}} \cdot \dfrac{\sqrt{3}}{\sqrt{3}} = \dfrac{\sqrt{3}}{3}$

$\tan\theta = \dfrac{\text{opp}}{\text{adj}} = \dfrac{\sqrt{2}}{1} = \sqrt{2}$

$\csc\theta = \dfrac{\text{hyp}}{\text{opp}} = \dfrac{\sqrt{3}}{\sqrt{2}} = \dfrac{\sqrt{3}}{\sqrt{2}} \cdot \dfrac{\sqrt{2}}{\sqrt{2}} = \dfrac{\sqrt{6}}{2}$

$\sec\theta = \dfrac{\text{hyp}}{\text{adj}} = \dfrac{\sqrt{3}}{1} = \sqrt{3}$

$\cot\theta = \dfrac{\text{adj}}{\text{opp}} = \dfrac{1}{\sqrt{2}} = \dfrac{1}{\sqrt{2}} \cdot \dfrac{\sqrt{2}}{\sqrt{2}} = \dfrac{\sqrt{2}}{2}$

16. opposite = 2; adjacent = $\sqrt{3}$; hypotenuse = ?

$(\text{hypotenuse})^2 = 2^2 + \left(\sqrt{3}\right)^2 = 7$

$\text{hypotenuse} = \sqrt{7}$

$\sin\theta = \dfrac{\text{opp}}{\text{hyp}} = \dfrac{2}{\sqrt{7}} = \dfrac{2}{\sqrt{7}} \cdot \dfrac{\sqrt{7}}{\sqrt{7}} = \dfrac{2\sqrt{7}}{7}$

$\cos\theta = \dfrac{\text{adj}}{\text{hyp}} = \dfrac{\sqrt{3}}{\sqrt{7}} = \dfrac{\sqrt{3}}{\sqrt{7}} \cdot \dfrac{\sqrt{7}}{\sqrt{7}} = \dfrac{\sqrt{21}}{7}$

$\tan\theta = \dfrac{\text{opp}}{\text{adj}} = \dfrac{2}{\sqrt{3}} = \dfrac{2}{\sqrt{3}} \cdot \dfrac{\sqrt{3}}{\sqrt{3}} = \dfrac{2\sqrt{3}}{3}$

$\csc\theta = \dfrac{\text{hyp}}{\text{opp}} = \dfrac{\sqrt{7}}{2}$

$\sec\theta = \dfrac{\text{hyp}}{\text{adj}} = \dfrac{\sqrt{7}}{\sqrt{3}} = \dfrac{\sqrt{7}}{\sqrt{3}} \cdot \dfrac{\sqrt{3}}{\sqrt{3}} = \dfrac{\sqrt{21}}{3}$

$\cot\theta = \dfrac{\text{adj}}{\text{opp}} = \dfrac{\sqrt{3}}{2}$

17. opposite = 1; hypotenuse = $\sqrt{5}$; adjacent = ?

$1^2 + (\text{adjacent})^2 = \left(\sqrt{5}\right)^2$

$(\text{adjacent})^2 = 5 - 1 = 4$

$\text{adjacent} = \sqrt{4} = 2$

$\sin\theta = \dfrac{\text{opp}}{\text{hyp}} = \dfrac{1}{\sqrt{5}} = \dfrac{1}{\sqrt{5}} \cdot \dfrac{\sqrt{5}}{\sqrt{5}} = \dfrac{\sqrt{5}}{5}$

$\cos\theta = \dfrac{\text{adj}}{\text{hyp}} = \dfrac{2}{\sqrt{5}} = \dfrac{2}{\sqrt{5}} \cdot \dfrac{\sqrt{5}}{\sqrt{5}} = \dfrac{2\sqrt{5}}{5}$

$\tan\theta = \dfrac{\text{opp}}{\text{adj}} = \dfrac{1}{2}$

$\csc\theta = \dfrac{\text{hyp}}{\text{opp}} = \dfrac{\sqrt{5}}{1} = \sqrt{5}$

$\sec\theta = \dfrac{\text{hyp}}{\text{adj}} = \dfrac{\sqrt{5}}{2}$

$\cot\theta = \dfrac{\text{adj}}{\text{opp}} = \dfrac{2}{1} = 2$

18. adjacent = 2; hypotenuse = $\sqrt{5}$; opposite = ?

$$(\text{opposite})^2 + 2^2 = \left(\sqrt{5}\right)^2$$
$$(\text{opposite})^2 = 5 - 4 = 1$$
$$\text{opposite} = \sqrt{1} = 1$$

$$\sin\theta = \frac{\text{opp}}{\text{hyp}} = \frac{1}{\sqrt{5}} = \frac{1}{\sqrt{5}}\cdot\frac{\sqrt{5}}{\sqrt{5}} = \frac{\sqrt{5}}{5}$$

$$\cos\theta = \frac{\text{adj}}{\text{hyp}} = \frac{2}{\sqrt{5}} = \frac{2}{\sqrt{5}}\cdot\frac{\sqrt{5}}{\sqrt{5}} = \frac{2\sqrt{5}}{5}$$

$$\tan\theta = \frac{\text{opp}}{\text{adj}} = \frac{1}{2}$$

$$\csc\theta = \frac{\text{hyp}}{\text{opp}} = \frac{\sqrt{5}}{1} = \sqrt{5}$$

$$\sec\theta = \frac{\text{hyp}}{\text{adj}} = \frac{\sqrt{5}}{2}$$

$$\cot\theta = \frac{\text{adj}}{\text{opp}} = \frac{2}{1} = 2$$

19. $\sin 38° - \cos 52° = \sin 38° - \sin(90° - 52°)$
$= \sin 38° - \sin 38°$
$= 0$

20. $\tan 12° - \cot 78° = \tan 12° - \tan(90° - 78°)$
$= \tan 12° - \tan 12°$
$= 0$

21. $\dfrac{\cos 10°}{\sin 80°} = \dfrac{\sin(90° - 10°)}{\sin 80°} = \dfrac{\sin 80°}{\sin 80°} = 1$

22. $\dfrac{\cos 40°}{\sin 50°} = \dfrac{\sin(90° - 40°)}{\sin 50°} = \dfrac{\sin 50°}{\sin 50°} = 1$

23. $1 - \cos^2 20° - \cos^2 70° = 1 - \cos^2 20° - \sin^2(90° - 70°)$
$= 1 - \cos^2 20° - \sin^2(20°)$
$= 1 - \left(\cos^2 20° + \sin^2(20°)\right)$
$= 1 - 1$
$= 0$

24. $1 + \tan^2 5° - \csc^2 85° = \sec^2 5° - \csc^2 85°$
$= \sec^2 5° - \sec^2(90° - 85°)$
$= \sec^2 5° - \sec^2 5°$
$= 0$

25. $\tan 20° - \dfrac{\cos 70°}{\cos 20°} = \tan 20° - \dfrac{\sin(90° - 70°)}{\cos 20°}$
$= \tan 20° - \dfrac{\sin 20°}{\cos 20°}$
$= \tan 20° - \tan 20°$
$= 0$

26. $\cot 40° - \dfrac{\sin 50°}{\sin 40°} = \cot 40° - \dfrac{\cos(90° - 50°)}{\sin 40°}$
$= \cot 40° - \dfrac{\cos 40°}{\sin 40°}$
$= \cot 40° - \cot 40°$
$= 0$

27. $\cos 35°\cdot\sin 55° + \cos 55°\cdot\sin 35°$
$= \cos 35°\cdot\cos(90° - 55°) + \sin(90° - 55°)\cdot\sin 35°$
$= \cos 35°\cdot\cos 35° + \sin 35°\cdot\sin 35°$
$= \cos^2 35° + \sin^2 35°$
$= 1$

28. $\sec 35°\cdot\csc 55° - \tan 35°\cdot\cot 55°$
$= \sec 35°\cdot\sec(90° - 55°) - \tan 35°\cdot\tan(90° - 55°)$
$= \sec 35°\cdot\sec 35° - \tan 35°\cdot\tan 35°$
$= \sec^2 35° - \tan^2 35°$
$= (1 + \tan^2 35°) - \tan^2 35°$
$= 1$

29. $b = 5,\ \beta = 20°$

$$\tan\beta = \frac{b}{a}$$
$$\tan(20°) = \frac{5}{a}$$
$$a = \frac{5}{\tan(20°)} \approx \frac{5}{0.3640} \approx 13.74$$
$$\sin\beta = \frac{b}{c}$$
$$\sin(20°) = \frac{5}{c}$$
$$c = \frac{5}{\sin(20°)} \approx \frac{5}{0.3420} \approx 14.62$$
$$\alpha = 90° - \beta = 90° - 20° = 70°$$

30. $b=4,\ \beta=10°$

$$\tan\beta=\frac{b}{a}$$

$$\tan(10°)=\frac{4}{a}$$

$$a=\frac{4}{\tan(10°)}\approx\frac{4}{0.1763}\approx 22.69$$

$$\sin\beta=\frac{b}{c}$$

$$\sin(10°)=\frac{4}{c}$$

$$c=\frac{4}{\sin(10°)}\approx\frac{4}{0.1736}\approx 23.04$$

$$\alpha=90°-\beta=90°-10°=80°$$

31. $a=6,\ \beta=40°$

$$\tan\beta=\frac{b}{a}$$

$$\tan(40°)=\frac{b}{6}$$

$$b=6\tan(40°)\approx 6\cdot(0.8391)\approx 5.03$$

$$\cos\beta=\frac{a}{c}$$

$$\cos(40°)=\frac{6}{c}$$

$$c=\frac{6}{\cos(40°)}\approx\frac{6}{0.7660}\approx 7.83$$

$$\alpha=90°-\beta=90°-40°=50°$$

32. $a=7,\ \beta=50°$

$$\tan\beta=\frac{b}{a}$$

$$\tan(50°)=\frac{b}{7}$$

$$b=7\tan(50°)\approx 7\cdot(1.1918)\approx 8.34$$

$$\cos\beta=\frac{a}{c}$$

$$\cos(50°)=\frac{7}{c}$$

$$c=\frac{7}{\cos(50°)}\approx\frac{7}{0.6428}\approx 10.89$$

$$\alpha=90°-\beta=90°-50°=40°$$

33. $b=4,\ \alpha=10°$

$$\tan\alpha=\frac{a}{b}$$

$$\tan(10°)=\frac{a}{4}$$

$$a=4\tan(10°)\approx 4\cdot(0.1763)\approx 0.71$$

$$\cos\alpha=\frac{b}{c}$$

$$\cos(10°)=\frac{4}{c}$$

$$c=\frac{4}{\cos(10°)}\approx\frac{4}{0.9848}\approx 4.06$$

$$\beta=90°-\alpha=90°-10°=80°$$

34. $b=6,\ \alpha=20°$

$$\tan\alpha=\frac{a}{b}$$

$$\tan(20°)=\frac{a}{6}$$

$$a=6\tan(20°)\approx 6\cdot(0.3640)\approx 2.18$$

$$\cos\alpha=\frac{b}{c}$$

$$\cos(20°)=\frac{6}{c}$$

$$c=\frac{6}{\cos(20°)}\approx\frac{6}{0.9397}\approx 6.39$$

$$\beta=90°-\alpha=90°-20°=70°$$

35. $a=5,\ \alpha=25°$

$$\cot\alpha=\frac{b}{a}$$

$$\cot(25°)=\frac{b}{5}$$

$$b=5\cot(25°)\approx 5\cdot(2.1445)\approx 10.72$$

$$\csc\alpha=\frac{c}{a}$$

$$\csc(25°)=\frac{c}{5}$$

$$c=5\csc(25°)\approx 5\cdot(2.3662)\approx 11.83$$

$$\beta=90°-\alpha=\beta=90°-\alpha=90°-25°=65°$$

36. $a = 6,\ \alpha = 40°$

$$\cot\alpha = \frac{a}{b}$$

$$\cot(40°) = \frac{b}{6}$$

$$b = 6\cot(40°) \approx 6\cdot(1.1918) \approx 7.15$$

$$\csc\alpha = \frac{c}{a}$$

$$\csc(45°) = \frac{c}{6}$$

$$c = 6\csc(40°) \approx 6\cdot(1.5557) \approx 9.33$$

$$\beta = 90° - \alpha = 90° - 40° = 50°$$

37. $c = 9,\ \beta = 20°$

$$\sin\beta = \frac{b}{c}$$

$$\sin(20°) = \frac{b}{9}$$

$$b = 9\sin(20°) \approx 9\cdot(0.3420) \approx 3.08$$

$$\cos\beta = \frac{a}{c}$$

$$\cos(20°) = \frac{a}{9}$$

$$a = 9\cos(20°) \approx 9\cdot(0.9397) \approx 8.46$$

$$\alpha = 90° - \alpha = 90° - 20° = 70°$$

38. $c = 10,\ \alpha = 40°$

$$\sin\alpha = \frac{a}{c}$$

$$\sin(40°) = \frac{a}{10}$$

$$a = 10\sin(40°) \approx 10\cdot(0.6428) \approx 6.43$$

$$\cos\alpha = \frac{b}{c}$$

$$\cos(40°) = \frac{b}{10}$$

$$b = 10\cos(40°) \approx 10\cdot(0.7660) \approx 7.66$$

$$\beta = 90° - \alpha = 90° - 40° = 50°$$

39. $a = 5,\ b = 3$

$$c^2 = a^2 + b^2 = 5^2 + 3^2 = 25 + 9 = 34$$

$$c = \sqrt{34} \approx 5.83$$

$$\tan\alpha = \frac{a}{b} = \frac{5}{3}$$

$$\alpha = \tan^{-1}\left(\frac{5}{3}\right) \approx 59.0°$$

$$\beta = 90° - \alpha = 90° - 59.0° = 31.0°$$

40. $a = 2,\ b = 8$

$$c^2 = a^2 + b^2 = 2^2 + 8^2 = 4 + 64 = 68$$

$$c = \sqrt{68} \approx 8.25$$

$$\tan\alpha = \frac{a}{b} = \frac{2}{8} = \frac{1}{4}$$

$$\alpha = \tan^{-1}\left(\frac{1}{4}\right) \approx 14.0°$$

$$\beta = 90° - \alpha = 90° - 14.0° = 76.0°$$

41. $a = 2,\ c = 5$

$$c^2 = a^2 + b^2$$

$$b^2 = c^2 - a^2 = 5^2 - 2^2 = 25 - 4 = 21$$

$$b = \sqrt{21} \approx 4.58$$

$$\sin\alpha = \frac{a}{c} = \frac{2}{5}$$

$$\alpha = \tan^{-1}\left(\frac{2}{5}\right) \approx 23.6°$$

$$\beta = 90° - \alpha = 90° - 23.6° = 66.4°$$

42. $b = 4,\ c = 6$

$$c^2 = a^2 + b^2$$

$$a^2 = c^2 - b^2 = 6^2 - 4^2 = 36 - 16 = 20$$

$$a = \sqrt{20} \approx 4.47$$

$$\cos\alpha = \frac{b}{c} = \frac{4}{6} = \frac{2}{3}$$

$$\alpha = \tan^{-1}\left(\frac{2}{3}\right) \approx 48.2°$$

$$\beta = 90° - \alpha = 90° - 48.2° = 41.8°$$

43. $c = 8,\ \theta = 35°$

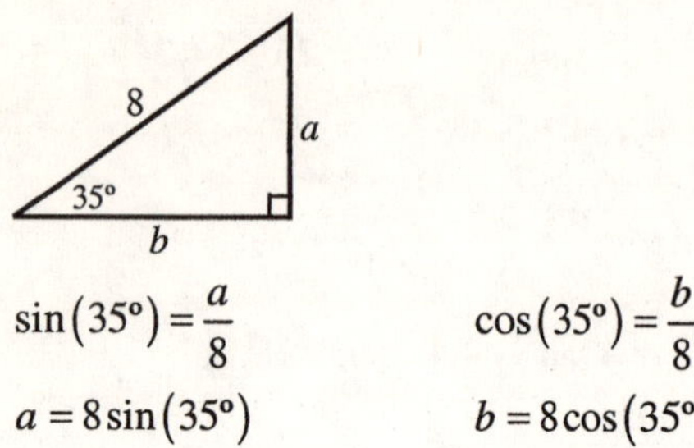

$\sin(35°) = \dfrac{a}{8}$

$a = 8\sin(35°)$
$\approx 8(0.5736)$
≈ 4.59 in.

$\cos(35°) = \dfrac{b}{8}$

$b = 8\cos(35°)$
$\approx 8(0.8192)$
≈ 6.55 in.

44. $c = 10,\ \theta = 40°$

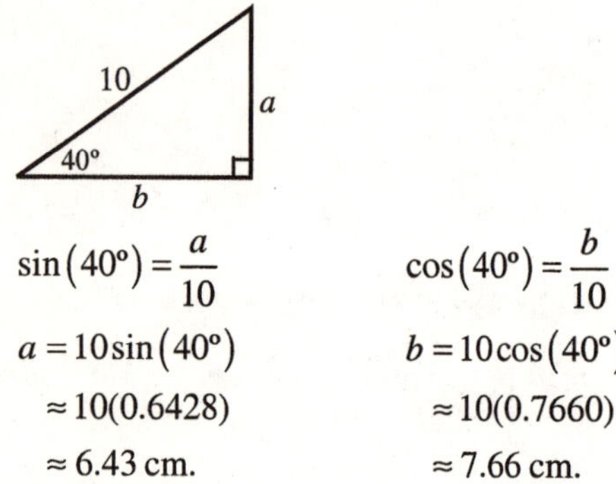

$\sin(40°) = \dfrac{a}{10}$

$a = 10\sin(40°)$
$\approx 10(0.6428)$
≈ 6.43 cm.

$\cos(40°) = \dfrac{b}{10}$

$b = 10\cos(40°)$
$\approx 10(0.7660)$
≈ 7.66 cm.

45. Case 1: $\theta = 25°,\ a = 5$

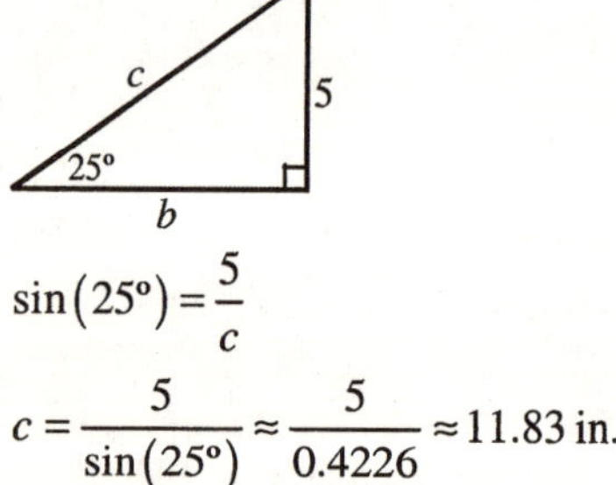

$\sin(25°) = \dfrac{5}{c}$

$c = \dfrac{5}{\sin(25°)} \approx \dfrac{5}{0.4226} \approx 11.83$ in.

Case 2: $\theta = 25°,\ b = 5$

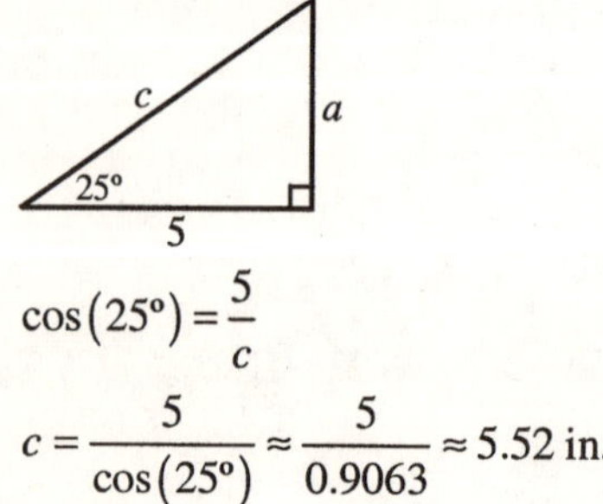

$\cos(25°) = \dfrac{5}{c}$

$c = \dfrac{5}{\cos(25°)} \approx \dfrac{5}{0.9063} \approx 5.52$ in.

46. Case 1: $\theta = \dfrac{\pi}{8},\ a = 3$

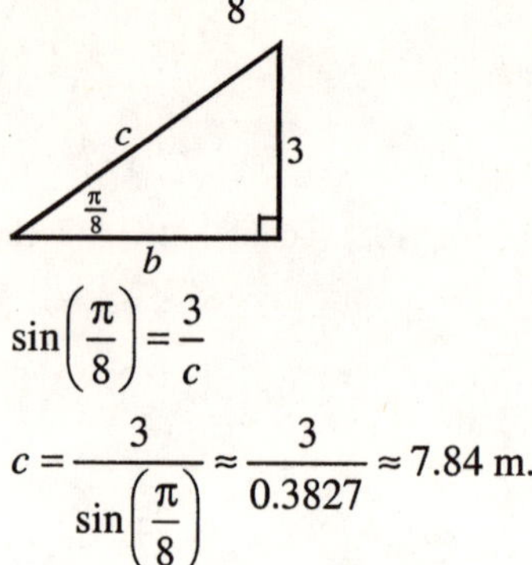

$\sin\left(\dfrac{\pi}{8}\right) = \dfrac{3}{c}$

$c = \dfrac{3}{\sin\left(\dfrac{\pi}{8}\right)} \approx \dfrac{3}{0.3827} \approx 7.84$ m.

Case 2: $\theta = \dfrac{\pi}{8},\ b = 3$

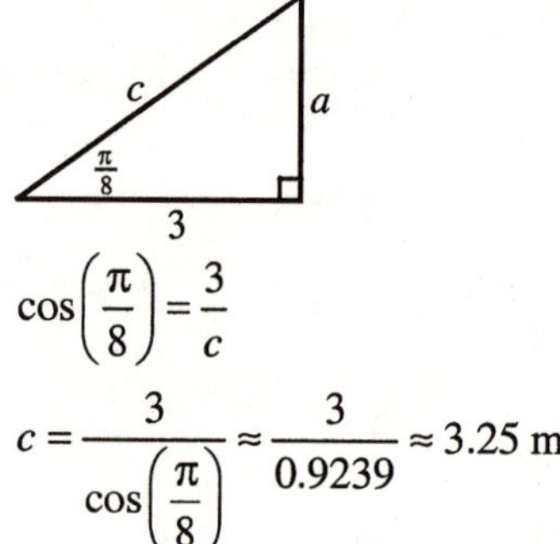

$\cos\left(\dfrac{\pi}{8}\right) = \dfrac{3}{c}$

$c = \dfrac{3}{\cos\left(\dfrac{\pi}{8}\right)} \approx \dfrac{3}{0.9239} \approx 3.25$ m.

47. $c = 5,\ a = 2$

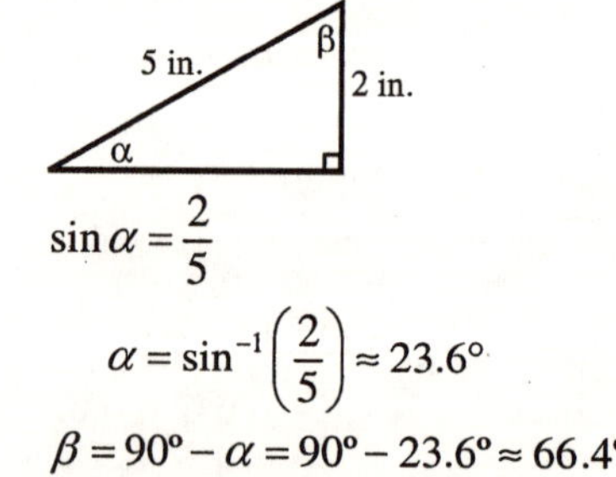

$\sin\alpha = \dfrac{2}{5}$

$\alpha = \sin^{-1}\left(\dfrac{2}{5}\right) \approx 23.6°$

$\beta = 90° - \alpha = 90° - 23.6° \approx 66.4°$

The two angles measure about $23.6°$ and $66.4°$.

48. $c = 3,\ a = 1$

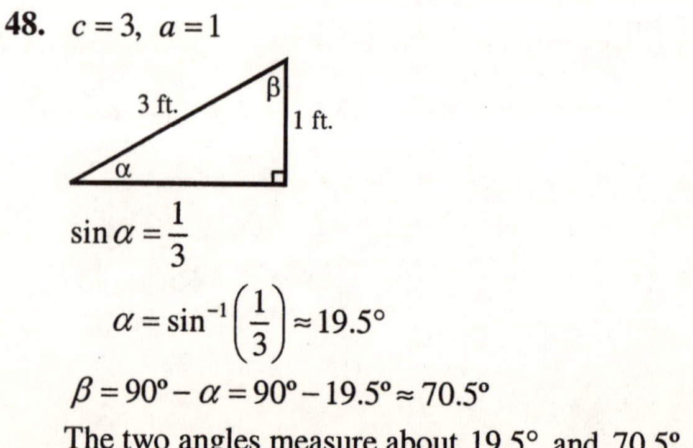

$\sin\alpha = \dfrac{1}{3}$

$\alpha = \sin^{-1}\left(\dfrac{1}{3}\right) \approx 19.5°$

$\beta = 90° - \alpha = 90° - 19.5° \approx 70.5°$

The two angles measure about $19.5°$ and $70.5°$.

49. $\tan(35^\circ) = \dfrac{|AC|}{100}$

$|AC| = 100\tan(35^\circ) \approx 100(0.7002) \approx 70.02$ feet

50. $\tan(40^\circ) = \dfrac{|AC|}{100}$

$|AC| = 100\tan(40^\circ) \approx 100(0.8391) \approx 83.91$ feet

51. Let x = the height of the Eiffel Tower.

$\tan(85.361^\circ) = \dfrac{x}{80}$

$x = 80\tan(85.361^\circ) \approx 80(12.3239) \approx 985.91$ feet

52. Let x = the distance to the shore.

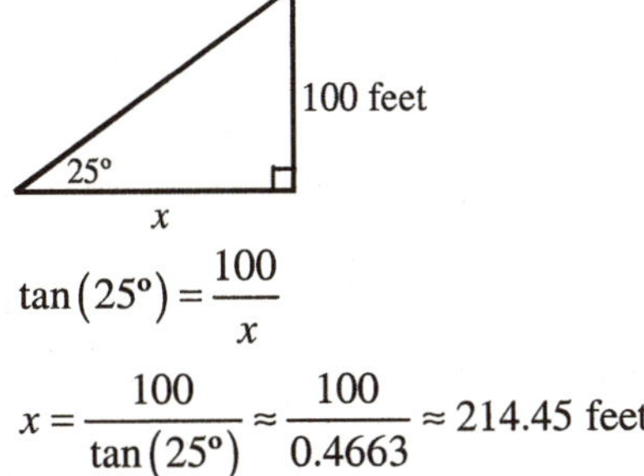

$\tan(25^\circ) = \dfrac{100}{x}$

$x = \dfrac{100}{\tan(25^\circ)} \approx \dfrac{100}{0.4663} \approx 214.45$ feet

53. Let x = the distance to the base of the plateau.

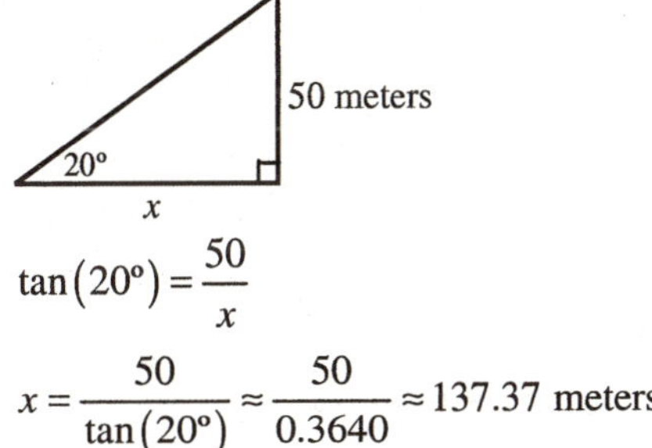

$\tan(20^\circ) = \dfrac{50}{x}$

$x = \dfrac{50}{\tan(20^\circ)} \approx \dfrac{50}{0.3640} \approx 137.37$ meters

54. Let x = the distance to the base of the statue.

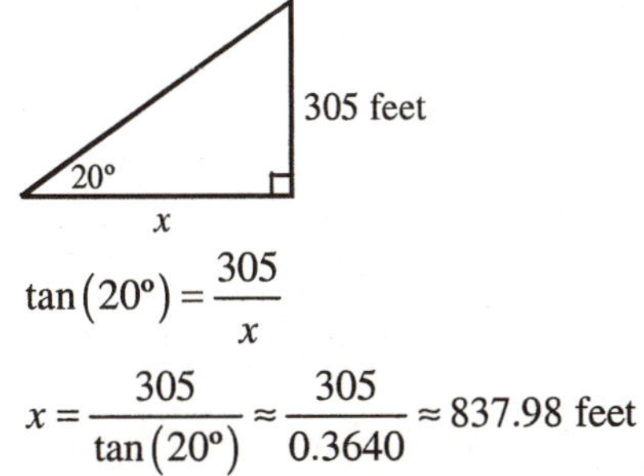

$\tan(20^\circ) = \dfrac{305}{x}$

$x = \dfrac{305}{\tan(20^\circ)} \approx \dfrac{305}{0.3640} \approx 837.98$ feet

55. Let x = the distance up the building

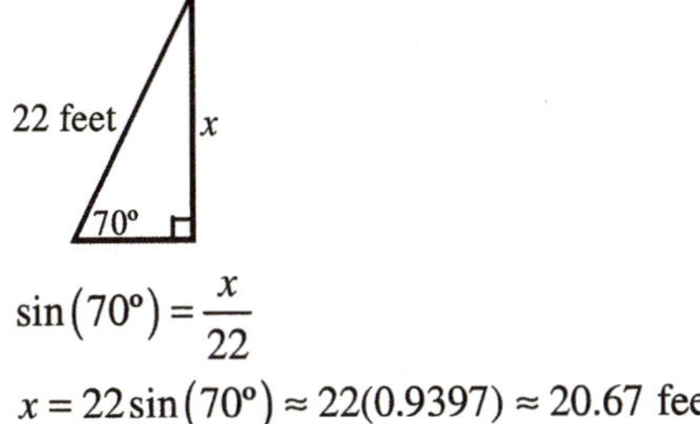

$\sin(70^\circ) = \dfrac{x}{22}$

$x = 22\sin(70^\circ) \approx 22(0.9397) \approx 20.67$ feet

56. Let h = the height of the building and let x = the distance from the building to the first sighting.

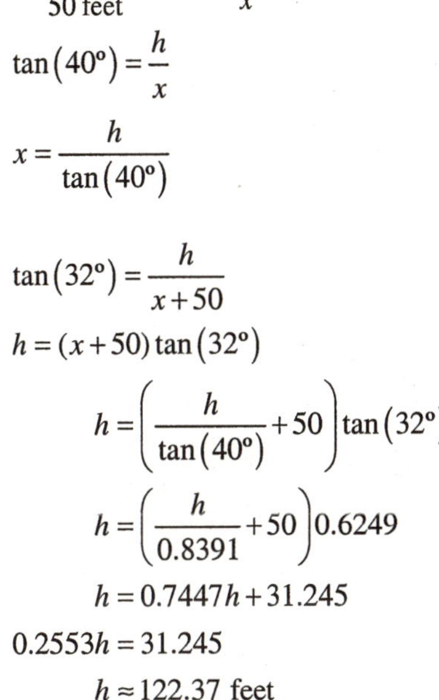

$\tan(40^\circ) = \dfrac{h}{x}$

$x = \dfrac{h}{\tan(40^\circ)}$

$\tan(32^\circ) = \dfrac{h}{x+50}$

$h = (x+50)\tan(32^\circ)$

$h = \left(\dfrac{h}{\tan(40^\circ)} + 50\right)\tan(32^\circ)$

$h = \left(\dfrac{h}{0.8391} + 50\right)0.6249$

$h = 0.7447h + 31.245$

$0.2553h = 31.245$

$h \approx 122.37$ feet

57. We construct the figure below:

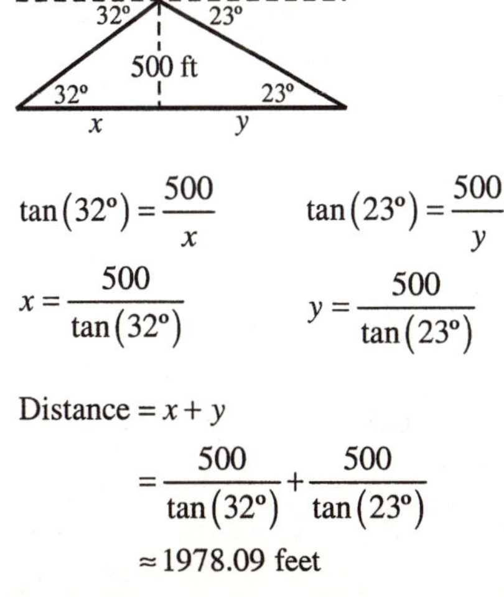

$\tan(32^\circ) = \dfrac{500}{x}$ $\qquad$ $\tan(23^\circ) = \dfrac{500}{y}$

$x = \dfrac{500}{\tan(32^\circ)}$ $\qquad$ $y = \dfrac{500}{\tan(23^\circ)}$

Distance $= x + y$

$= \dfrac{500}{\tan(32^\circ)} + \dfrac{500}{\tan(23^\circ)}$

≈ 1978.09 feet

58. Let α = the angle of elevation of the sun.

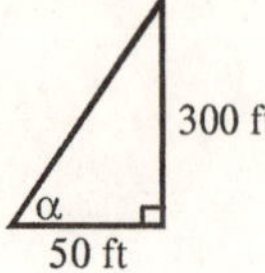

$$\tan\alpha = \frac{300}{50} = 6$$

$$\alpha = \tan^{-1} 6 \approx 80.5°$$

The angle of elevation of the sun is about $80.5°$.

59. Let h represent the height of Lincoln's face.

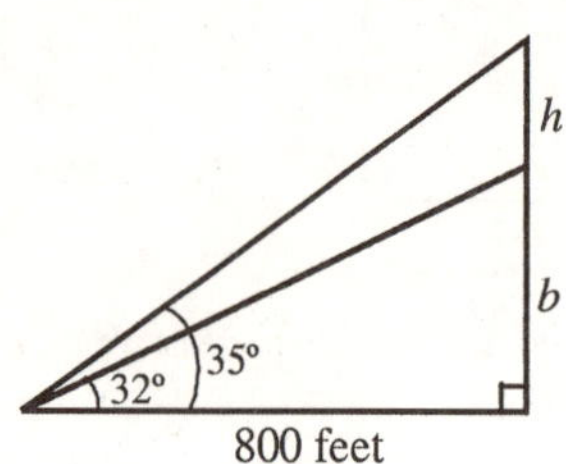

$$\tan(32°) = \frac{b}{800}$$

$$b = 800\tan(32°) \approx 499.90$$

$$\tan(35°) = \frac{b+h}{800}$$

$$b+h = 800\tan(35°) \approx 560.17$$

Thus, the height of Lincoln's face is:

$h = (b+h) - b = 560.17 - 499.90 \approx 60.27$ feet

60. opposite side = 10 feet, adjacent side = 35 feet

$$\tan\theta = \frac{10}{35}$$

$$\theta = \tan^{-1}\left(\frac{10}{35}\right) \approx 15.9°$$

61. Let x = the length of the guy wire.

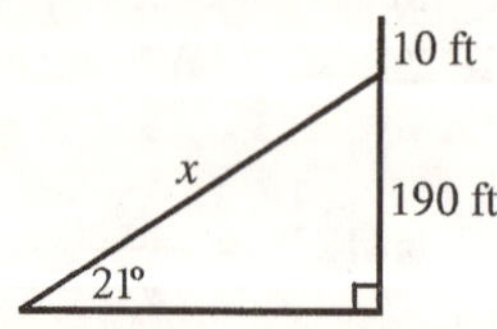

$$\sin(21°) = \frac{190}{x}$$

$$x = \frac{190}{\sin(21°)} \approx \frac{190}{0.3584} \approx 530.18 \text{ ft.}$$

62. Let h = the height of the tower.

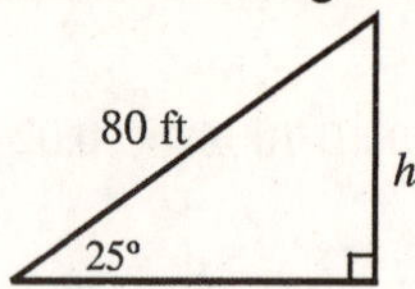

$$\sin(25°) = \frac{h}{80}$$

$$h = 80\sin(25°) \approx 80(0.4226) \approx 33.81 \text{ ft.}$$

63. Let h = the height of the monument.

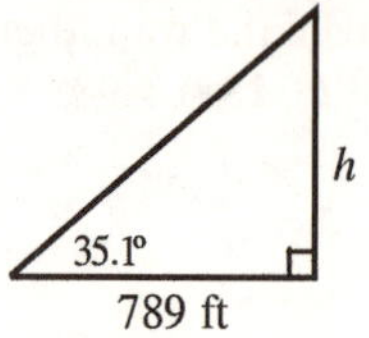

$$\tan(35.1°) = \frac{h}{789}$$

$$h = 789\tan(35.1°) \approx 789(0.7028) \approx 554.52 \text{ ft}$$

64. The elevation change is $11200 - 9000 = 2200$ ft.
Let x = the length of the trail.

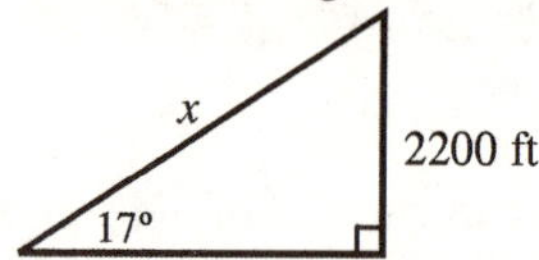

$$\sin 17° = \frac{2200}{x}$$

$$x = \frac{2200}{\sin(17°)} \approx \frac{2200}{0.2924} \approx 7524.67 \text{ ft.}$$

65. a. $\tan(15°) = \dfrac{30}{x}$

$$x = \frac{30}{\tan(15°)} \approx \frac{30}{0.2679} \approx 111.96 \text{ feet}$$

The truck is traveling at 111.96 ft/sec, or

$$\frac{111.96 \text{ ft}}{\text{sec}} \cdot \frac{1 \text{ mile}}{5280 \text{ ft}} \cdot \frac{3600 \text{ sec}}{\text{hr}} \approx 76.3 \text{ mi/hr}.$$

b. $\tan(20°) = \dfrac{30}{x}$

$$x = \frac{30}{\tan(20°)} \approx \frac{30}{0.3640} \approx 82.42 \text{ feet}$$

The truck is traveling at 82.42 ft/sec, or

$$\frac{82.42 \text{ ft}}{\text{sec}} \cdot \frac{1 \text{ mile}}{5280 \text{ ft}} \cdot \frac{3600 \text{ sec}}{\text{hr}} \approx 56.2 \text{ mi/hr}.$$

c. A ticket is issued for traveling at a speed of 60 mi/hr or more.

$$\frac{60\text{ mi}}{\text{hr}}\cdot\frac{5280\text{ ft}}{\text{mi}}\cdot\frac{1\text{hr}}{3600\text{ sec}}=88\text{ ft/sec.}$$

If $\tan\theta<\frac{30}{88}$, the trooper should issue a ticket. Now, $\tan^{-1}\left(\frac{30}{88}\right)\approx 18.8°$, so a ticket is issued if $\theta<18.8°$.

66. If the camera is to be directed to a spot 6 feet above the floor 12 feet from the wall, then the "side opposite" the angle of depression is 3 feet. (see figure)

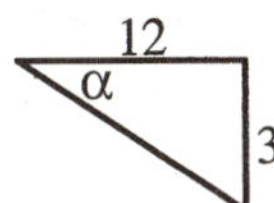

$$\tan\alpha=\frac{3}{12}=\frac{1}{4}$$

$$\alpha=\tan^{-1}\frac{1}{4}\approx 14.0°$$

The angle of depression should be about $14.0°$.

67. Begin by finding angle $\theta=\angle BAC$: (see figure)

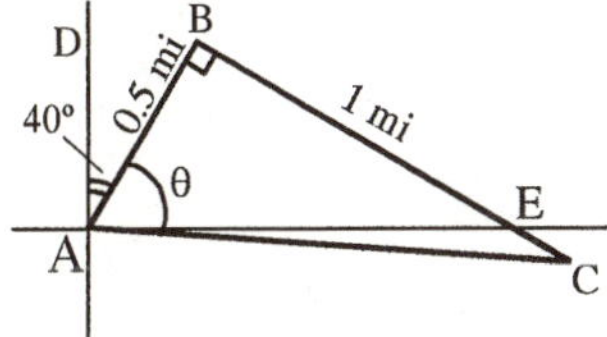

$$\tan\theta=\frac{1}{0.5}=2$$

$$\theta=63.4°$$

$$\angle DAC=40°+63.4°=103.4°$$

$$\angle EAC=103.4°-90°=13.4°$$

Now, $90°-13.4°=76.6°$

The control tower should use a bearing of S76.6˚E.

68. Find $\angle AMB$ and subtract from 80° to obtain θ (see figure).

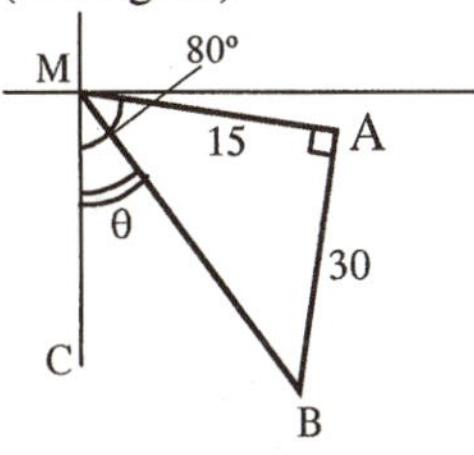

$$\angle CMA=80°$$

$$\tan\angle AMB=\frac{30}{15}=2$$

$$\angle AMB=\tan^{-1}2\approx 63.4°$$

$$\theta=80°-63.4°=16.6°$$

The bearing of the ship from port is S16.6ºE.

69. The height of the beam above the wall is $46-20=26$ feet.

$$\tan\theta=\frac{26}{10}=2.6$$

$$\theta=\tan^{-1}2.6\approx 69.0°$$

The pitch of the roof is about $69.0°$.

70. $\tan\alpha=\frac{10-6}{15}=\frac{4}{15}$

$$\alpha=\tan^{-1}\left(\frac{4}{15}\right)\approx 14.9°$$

The angle of elevation from the player's eyes to the center of the rim is about $14.9°$.

71. A line segment drawn from the vertex of the angle through the centers of the circles will bisect θ (see figure). Thus, the angle of the two right triangles formed is $\frac{\theta}{2}$.

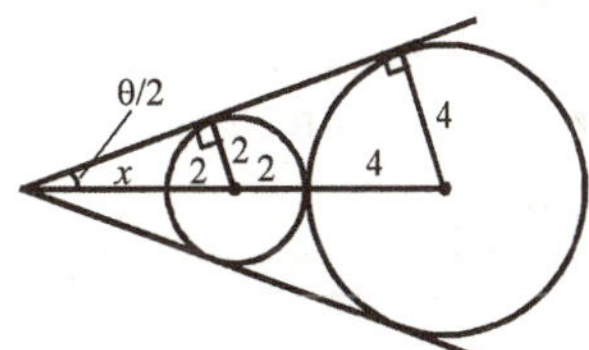

Let x = the length of the segment from the vertex of the angle to the smaller circle. Then the hypotenuse of the smaller right triangle is $x+2$, and the hypotenuse of the larger right triangle is $x+2+2+4=x+8$. Since the two right triangles are similar, we have that:

$$\frac{x+2}{2}=\frac{x+8}{4}$$

$$4(x+2)=2(x+8)$$

$$4x+8=2x+16$$

$$2x=8$$

$$x=4$$

Thus, the hypotenuse of the smaller triangle is $4+2=6$. Now, $\sin\left(\frac{\theta}{2}\right)=\frac{2}{6}=\frac{1}{3}$, so we have that: $\frac{\theta}{2}=\sin^{-1}\left(\frac{1}{3}\right)$

$$\theta=2\cdot\sin^{-1}\left(\frac{1}{3}\right)\approx 38.9°$$

72. **a.** $\cos\left(\frac{\theta}{2}\right) = \frac{3960}{3960+h}$

b. $d = 3960\theta$

c. $\cos\left(\frac{d}{7920}\right) = \frac{3960}{3960+h}$

d.
$$\cos\left(\frac{2500}{7920}\right) = \frac{3960}{3960+h}$$
$$0.9506 \approx \frac{3960}{3960+h}$$
$$0.9506(3960+h) \approx 3960$$
$$3764 + 0.9506h \approx 3960$$
$$0.9506h \approx 196$$
$$h \approx 206 \text{ miles}$$

e.
$$\cos\left(\frac{d}{7920}\right) = \frac{3960}{3960+300} = \frac{3960}{4260}$$
$$\frac{d}{7920} = \cos^{-1}\left(\frac{3960}{4260}\right)$$
$$d = 7970\cos^{-1}\left(\frac{3960}{4260}\right)$$
$$\approx 2990 \text{ miles}$$

73. Let x = the distance from George at which the camera must be set in order to see his head and feet.

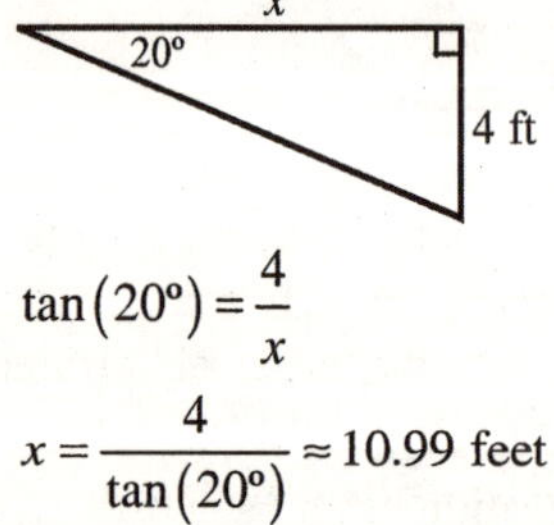

$$\tan(20°) = \frac{4}{x}$$
$$x = \frac{4}{\tan(20°)} \approx 10.99 \text{ feet}$$

If the camera is set at a distance of 10 feet from George, his feet will not be seen by the lens. The camera would need to be moved back about 1 additional foot (11 feet total).

74. Let r = the length of the ramp.

r
5 ft.
15°

$$\sin 15° = \frac{5}{r}$$
$$r = \frac{5}{\sin 15°} \approx 19.32$$

The ramp is about 19.32 feet long.

75. $\sin\theta = \frac{y}{1} = y$; $\cos\theta = \frac{x}{1} = x$

a. $A(\theta) = 2xy = 2\cos\theta\sin\theta$

b. $A(\theta) = 2\cos\theta\sin\theta = 2\sin\theta\cos\theta = \sin(2\theta)$ by the Double-angle Formula.

c. The largest value of the sine function is 1, so we solve:
$$\sin(2\theta) = 1$$
$$2\theta = \frac{\pi}{2}$$
$$\theta = \frac{\pi}{4} = 45°$$
The angle that results in the largest area is $\theta = \frac{\pi}{4} = 45°$.

d. $x = \cos\frac{\pi}{4} = \frac{\sqrt{2}}{2}$; $y = \sin\frac{\pi}{4} = \frac{\sqrt{2}}{2}$

The dimensions are $\sqrt{2}$ by $\frac{\sqrt{2}}{2}$.

76. Let b represent the base of the triangle.

$$\cos\frac{\theta}{2} = \frac{h}{s} \qquad \sin\frac{\theta}{2} = \frac{\frac{1}{2}b}{s}$$
$$h = s\cdot\cos\frac{\theta}{2} \qquad b = 2s\cdot\sin\frac{\theta}{2}$$

$$A = \frac{1}{2}b\cdot h$$
$$= \frac{1}{2}\left(2s\cdot\sin\frac{\theta}{2}\right)\left(s\cdot\cos\frac{\theta}{2}\right)$$
$$= s^2\cdot\sin\frac{\theta}{2}\cdot\cos\frac{\theta}{2}$$
$$= \frac{1}{2}\cdot s^2\cdot 2\sin\frac{\theta}{2}\cdot\cos\frac{\theta}{2}$$
$$= \frac{1}{2}s^2\sin\theta$$

77. Let θ = the central angle formed by the top of the lighthouse, the center of the Earth and the point P on the Earth's surface where the line of sight from the top of the lighthouse is tangent to the Earth. Note also that 362 feet $= \frac{362}{5280}$ miles.

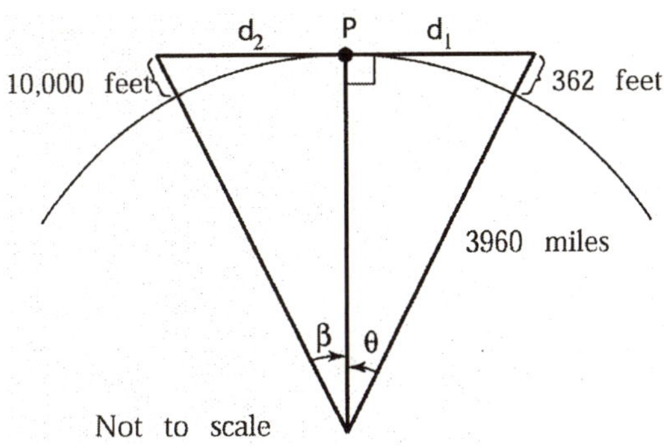

$$\theta = \cos^{-1}\left(\frac{3960}{3960 + 362/5280}\right) \approx 0.33715^\circ$$

Verify the airplane information:
Let β = the central angle formed by the plane, the center of the Earth and the point P.

$$\beta = \cos^{-1}\left(\frac{3960}{3960 + 10{,}000/5280}\right) \approx 1.77169^\circ$$

Note that

$$\tan\theta = \frac{d_1}{3690} \quad \text{and} \quad \tan\beta = \frac{d_2}{3690}$$
$$d_1 = 3960\tan\theta \qquad d_2 = 3960\tan\beta$$

So,

$$d_1 + d_2 = 3960\tan\theta + 3960\tan\beta$$
$$\approx 3960\tan(0.33715^\circ) + 3960\tan(1.77169^\circ)$$
$$\approx 146 \text{ miles}$$

To express this distance in nautical miles, we express the total angle $\theta + \beta$ in minutes. That is,

$$\theta + \beta \approx (0.33715^\circ + 1.77169^\circ) \cdot 60 \approx 126.5$$

nautical miles. Therefore, a plane flying at an altitude of 10,000 feet can see the lighthouse 120 miles away.

Verify the ship information:

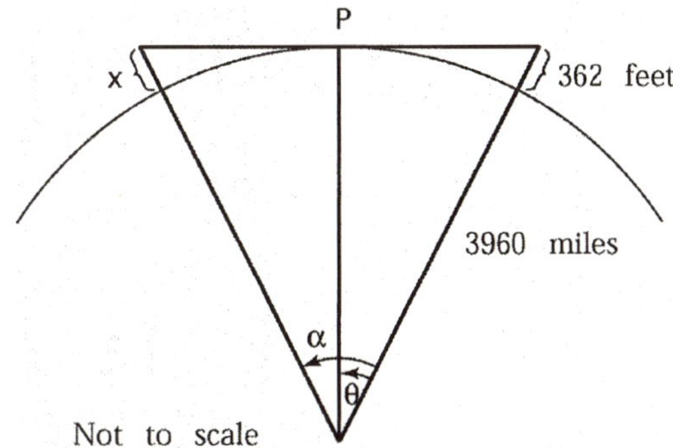

Let α = the central angle formed by 40 nautical miles then. $\alpha = \frac{40}{60} = \frac{2^\circ}{3}$

$$\cos(\alpha - \theta) = \frac{3960}{3960 + x}$$
$$\cos\left((2/3)^\circ - 0.33715^\circ\right) = \frac{3960}{3960 + x}$$
$$(3960 + x)\cos\left((2/3)^\circ - 0.33715^\circ\right) = 3960$$
$$3960 + x = \frac{3960}{\cos\left((2/3)^\circ - 0.33715^\circ\right)}$$
$$x = \frac{3960}{\cos\left((2/3)^\circ - 0.33715^\circ\right)} - 3960$$
$$\approx 0.06549 \text{ miles}$$
$$\approx (0.06549 \text{ mi})\left(5280\frac{\text{ft}}{\text{mi}}\right)$$
$$\approx 346 \text{ feet}$$

Therefore, a ship that is 346 feet above sea level can see the lighthouse from a distance of 40 nautical miles.

84 – 85. Answers will vary.

Section 7.2

1. $\sin\alpha\cos\beta - \cos\alpha\sin\beta$

2. $\sin\theta = \frac{1}{2}$

   $$\theta = \sin^{-1}\left(\frac{1}{2}\right) = \frac{\pi}{6} = 30^\circ$$

   The solution set is $\left\{\frac{\pi}{6} = 30^\circ\right\}$.

3. $\sin\theta = 2$
   There is no possible angle θ with $\sin\theta > 1$. Therefore, the equation has no solution.

4. oblique

5. $\frac{\sin\alpha}{a} = \frac{\sin\beta}{b} = \frac{\sin\gamma}{c}$

6. False

7. True

8. False

9. $c=5,\ \beta=45°,\ \gamma=95°$
$\alpha=180°-\beta-\gamma=180°-45°-95°=40°$

$$\frac{\sin\alpha}{a}=\frac{\sin\gamma}{c}$$
$$\frac{\sin 40°}{a}=\frac{\sin 95°}{5}$$
$$a=\frac{5\sin 40°}{\sin 95°}\approx 3.23$$
$$\frac{\sin\beta}{b}=\frac{\sin\gamma}{c}$$
$$\frac{\sin 45°}{b}=\frac{\sin 95°}{5}$$
$$b=\frac{5\sin 45°}{\sin 95°}\approx 3.55$$

10. $c=4,\ \alpha=45°,\ \beta=40°$
$\gamma=180°-\alpha-\beta=180°-45°-40°=95°$

$$\frac{\sin\alpha}{a}=\frac{\sin\gamma}{c}$$
$$\frac{\sin 45°}{a}=\frac{\sin 95°}{4}$$
$$a=\frac{4\sin 45°}{\sin 95°}\approx 2.84$$
$$\frac{\sin\beta}{b}=\frac{\sin\gamma}{c}$$
$$\frac{\sin 40°}{b}=\frac{\sin 95°}{4}$$
$$b=\frac{4\sin 40°}{\sin 95°}\approx 2.58$$

11. $b=3,\ \alpha=50°,\ \gamma=85°$
$\beta=180°-\alpha-\gamma=180°-50°-85°=45°$

$$\frac{\sin\alpha}{a}=\frac{\sin\beta}{b}$$
$$\frac{\sin 50°}{a}=\frac{\sin 45°}{3}$$
$$a=\frac{3\sin 50°}{\sin 45°}\approx 3.25$$
$$\frac{\sin\gamma}{c}=\frac{\sin\beta}{b}$$
$$\frac{\sin 85°}{c}=\frac{\sin 45°}{3}$$
$$c=\frac{3\sin 85°}{\sin 45°}\approx 4.23$$

12. $b=10,\ \beta=30°,\ \gamma=125°$
$\alpha=180°-\beta-\gamma=180°-30°-125°=25°$

$$\frac{\sin\alpha}{a}=\frac{\sin\beta}{b}$$
$$\frac{\sin 25°}{a}=\frac{\sin 30°}{10}$$
$$a=\frac{10\sin 25°}{\sin 30°}\approx 8.45$$
$$\frac{\sin\gamma}{c}=\frac{\sin\beta}{b}$$
$$\frac{\sin 125°}{c}=\frac{\sin 30°}{10}$$
$$c=\frac{10\sin 125°}{\sin 30°}\approx 16.38$$

13. $b=7,\ \alpha=40°,\ \beta=45°$
$\gamma=180°-\alpha-\beta=180°-40°-45°=95°$

$$\frac{\sin\alpha}{a}=\frac{\sin\beta}{b}$$
$$\frac{\sin 40°}{a}=\frac{\sin 45°}{7}$$
$$a=\frac{7\sin 40°}{\sin 45°}\approx 6.36$$
$$\frac{\sin\gamma}{c}=\frac{\sin\beta}{b}$$
$$\frac{\sin 95°}{c}=\frac{\sin 45°}{7}$$
$$c=\frac{7\sin 95°}{\sin 45°}\approx 9.86$$

14. $c=5,\ \alpha=10°,\ \beta=5°$
$\gamma=180°-\alpha-\beta=180°-10°-5°=165°$

$$\frac{\sin\alpha}{a}=\frac{\sin\gamma}{c}$$
$$\frac{\sin 10°}{a}=\frac{\sin 165°}{5}$$
$$a=\frac{5\sin 10°}{\sin 165°}\approx 3.35$$
$$\frac{\sin\beta}{b}=\frac{\sin\gamma}{c}$$
$$\frac{\sin 5°}{b}=\frac{\sin 165°}{5}$$
$$b=\frac{5\sin 5°}{\sin 165°}\approx 1.68$$

15. $b = 2,\ \beta = 40°,\ \gamma = 100°$
$\alpha = 180° - \beta - \gamma = 180° - 40° - 100° = 40°$

$$\frac{\sin\alpha}{a} = \frac{\sin\beta}{b}$$
$$\frac{\sin 40°}{a} = \frac{\sin 40°}{2}$$
$$a = \frac{2\sin 40°}{\sin 40°} = 2$$
$$\frac{\sin\gamma}{c} = \frac{\sin\beta}{b}$$
$$\frac{\sin 100°}{c} = \frac{\sin 40°}{2}$$
$$c = \frac{2\sin 100°}{\sin 40°} \approx 3.06$$

16. $b = 6,\ \alpha = 100°,\ \beta = 30°$
$\gamma = 180° - \alpha - \beta = 180° - 100° - 30° = 50°$

$$\frac{\sin\alpha}{a} = \frac{\sin\beta}{b}$$
$$\frac{\sin 100°}{a} = \frac{\sin 30°}{6}$$
$$a = \frac{6\sin 100°}{\sin 30°} \approx 11.82$$
$$\frac{\sin\gamma}{c} = \frac{\sin\beta}{b}$$
$$\frac{\sin 50°}{c} = \frac{\sin 30°}{6}$$
$$c = \frac{6\sin 50°}{\sin 30°} \approx 9.19$$

17. $\alpha = 40°,\ \beta = 20°,\ a = 2$
$\gamma = 180° - \alpha - \beta = 180° - 40° - 20° = 120°$

$$\frac{\sin\alpha}{a} = \frac{\sin\beta}{b}$$
$$\frac{\sin 40°}{2} = \frac{\sin 20°}{b}$$
$$b = \frac{2\sin 20°}{\sin 40°} \approx 1.06$$
$$\frac{\sin\gamma}{c} = \frac{\sin\alpha}{a}$$
$$\frac{\sin 120°}{c} = \frac{\sin 40°}{2}$$
$$c = \frac{2\sin 120°}{\sin 40°} \approx 2.69$$

18. $\alpha = 50°,\ \gamma = 20°,\ a = 3$
$\beta = 180° - \alpha - \gamma = 180° - 50° - 20° = 110°$

$$\frac{\sin\alpha}{a} = \frac{\sin\beta}{b}$$
$$\frac{\sin 50°}{3} = \frac{\sin 110°}{b}$$
$$b = \frac{3\sin 110°}{\sin 50°} \approx 3.68$$
$$\frac{\sin\gamma}{c} = \frac{\sin\alpha}{a}$$
$$\frac{\sin 20°}{c} = \frac{\sin 50°}{3}$$
$$c = \frac{3\sin 20°}{\sin 50°} \approx 1.34$$

19. $\beta = 70°,\ \gamma = 10°,\ b = 5$
$\alpha = 180° - \beta - \gamma = 180° - 70° - 10° = 100°$

$$\frac{\sin\alpha}{a} = \frac{\sin\beta}{b}$$
$$\frac{\sin 100°}{a} = \frac{\sin 70°}{5}$$
$$a = \frac{5\sin 100°}{\sin 70°} \approx 5.24$$
$$\frac{\sin\gamma}{c} = \frac{\sin\beta}{b}$$
$$\frac{\sin 10°}{c} = \frac{\sin 70°}{5}$$
$$c = \frac{5\sin 10°}{\sin 70°} \approx 0.92$$

20. $\alpha = 70°,\ \beta = 60°,\ c = 4$
$\gamma = 180° - \alpha - \beta = 180° - 70° - 60° = 50°$

$$\frac{\sin\alpha}{a} = \frac{\sin\gamma}{c}$$
$$\frac{\sin 70°}{a} = \frac{\sin 50°}{4}$$
$$a = \frac{4\sin 70°}{\sin 50°} \approx 4.91$$
$$\frac{\sin\beta}{b} = \frac{\sin\gamma}{c}$$
$$\frac{\sin 60°}{b} = \frac{\sin 50°}{4}$$
$$b = \frac{4\sin 60°}{\sin 50°} \approx 4.52$$

21. $\alpha = 110^\circ,\ \gamma = 30^\circ,\ c = 3$

$\beta = 180^\circ - \alpha - \gamma = 180^\circ - 110^\circ - 30^\circ = 40^\circ$

$$\frac{\sin\alpha}{a} = \frac{\sin\gamma}{c}$$
$$\frac{\sin 110^\circ}{a} = \frac{\sin 30^\circ}{3}$$
$$a = \frac{3\sin 110^\circ}{\sin 30^\circ} \approx 5.64$$
$$\frac{\sin\gamma}{c} = \frac{\sin\beta}{b}$$
$$\frac{\sin 30^\circ}{3} = \frac{\sin 40^\circ}{b}$$
$$b = \frac{3\sin 40^\circ}{\sin 30^\circ} \approx 3.86$$

22. $\beta = 10^\circ,\ \gamma = 100^\circ,\ b = 2$

$\alpha = 180^\circ - \beta - \gamma = 180^\circ - 10^\circ - 100^\circ = 70^\circ$

$$\frac{\sin\alpha}{a} = \frac{\sin\beta}{b}$$
$$\frac{\sin 70^\circ}{a} = \frac{\sin 10^\circ}{2}$$
$$a = \frac{2\sin 70^\circ}{\sin 10^\circ} \approx 10.82$$
$$\frac{\sin\gamma}{c} = \frac{\sin\beta}{b}$$
$$\frac{\sin 100^\circ}{c} = \frac{\sin 10^\circ}{2}$$
$$c = \frac{2\sin 100^\circ}{\sin 10^\circ} \approx 11.34$$

23. $\alpha = 40^\circ,\ \beta = 40^\circ,\ c = 2$

$\gamma = 180^\circ - \alpha - \beta = 180^\circ - 40^\circ - 40^\circ = 100^\circ$

$$\frac{\sin\alpha}{a} = \frac{\sin\gamma}{c}$$
$$\frac{\sin 40^\circ}{a} = \frac{\sin 100^\circ}{2}$$
$$a = \frac{2\sin 40^\circ}{\sin 100^\circ} \approx 1.31$$
$$\frac{\sin\beta}{b} = \frac{\sin\gamma}{c}$$
$$\frac{\sin 40^\circ}{b} = \frac{\sin 100^\circ}{2}$$
$$b = \frac{2\sin 40^\circ}{\sin 100^\circ} \approx 1.31$$

24. $\beta = 20^\circ,\ \gamma = 70^\circ,\ a = 1$

$\alpha = 180^\circ - \beta - \gamma = 180^\circ - 20^\circ - 70^\circ = 90^\circ$

$$\frac{\sin\alpha}{a} = \frac{\sin\beta}{b}$$
$$\frac{\sin 90^\circ}{1} = \frac{\sin 20^\circ}{b}$$
$$b = \frac{1\sin 20^\circ}{\sin 90^\circ} \approx 0.34$$
$$\frac{\sin\gamma}{c} = \frac{\sin\alpha}{a}$$
$$\frac{\sin 70^\circ}{c} = \frac{\sin 90^\circ}{1}$$
$$c = \frac{1\sin 70^\circ}{\sin 90^\circ} \approx 0.94$$

25. $a = 3,\ b = 2,\ \alpha = 50^\circ$

$$\frac{\sin\beta}{b} = \frac{\sin\alpha}{a}$$
$$\frac{\sin\beta}{2} = \frac{\sin(50^\circ)}{3}$$
$$\sin\beta = \frac{2\sin(50^\circ)}{3} \approx 0.5107$$

$\beta = \sin^{-1}(0.5107)$

$\beta = 30.7^\circ$ or $\beta = 149.3^\circ$

The second value is discarded because $\alpha + \beta > 180^\circ$.

$\gamma = 180^\circ - \alpha - \beta = 180^\circ - 50^\circ - 30.7^\circ = 99.3^\circ$

$$\frac{\sin\gamma}{c} = \frac{\sin\alpha}{a}$$
$$\frac{\sin 99.3^\circ}{c} = \frac{\sin 50^\circ}{3}$$
$$c = \frac{3\sin 99.3^\circ}{\sin 50^\circ} \approx 3.86$$

One triangle: $\beta \approx 30.7^\circ,\ \gamma \approx 99.3^\circ,\ c \approx 3.86$

26. $b = 4,\ c = 3,\ \beta = 40^\circ$

$$\frac{\sin\beta}{b} = \frac{\sin\gamma}{c}$$
$$\frac{\sin 40^\circ}{4} = \frac{\sin\gamma}{3}$$
$$\sin\gamma = \frac{3\sin 40^\circ}{4} \approx 0.4821$$

$\gamma = \sin^{-1}(0.4821)$

$\gamma = 28.8^\circ$ or $\gamma = 151.2^\circ$

The second value is discarded because $\beta + \gamma > 180^\circ$.

$\alpha = 180° - \beta - \gamma = 180° - 40° - 28.8° = 111.2°$

$\frac{\sin\beta}{b} = \frac{\sin\alpha}{a}$

$\frac{\sin 40°}{4} = \frac{\sin 111.2°}{a}$

$a = \frac{4\sin 111.2°}{\sin 40°} \approx 5.80$

One triangle: $\alpha \approx 111.2°$, $\gamma \approx 28.8°$, $a \approx 5.80$

27. $b = 5$, $c = 3$, $\beta = 100°$

$\frac{\sin\beta}{b} = \frac{\sin\gamma}{c}$

$\frac{\sin 100°}{5} = \frac{\sin\gamma}{3}$

$\sin\gamma = \frac{3\sin 100°}{5} \approx 0.5909$

$\gamma = \sin^{-1}(0.5909)$

$\gamma = 36.2°$ or $\gamma = 143.8°$

The second value is discarded because $\beta + \gamma > 180°$.

$\alpha = 180° - \beta - \gamma = 180° - 100° - 36.2° = 43.8°$

$\frac{\sin\beta}{b} = \frac{\sin\alpha}{a}$

$\frac{\sin 100°}{5} = \frac{\sin 43.8°}{a}$

$a = \frac{5\sin 43.8°}{\sin 100°} \approx 3.51$

One triangle: $\alpha \approx 43.8°$, $\gamma \approx 36.2°$, $a \approx 3.51$

28. $a = 2$, $c = 1$, $\alpha = 120°$

$\frac{\sin\gamma}{c} = \frac{\sin\alpha}{a}$

$\frac{\sin\gamma}{1} = \frac{\sin 120°}{2}$

$\sin\gamma = \frac{1\sin 120°}{2} \approx 0.4330$

$\gamma = \sin^{-1}(0.4330)$

$\gamma = 25.7°$ or $\gamma = 154.3°$

The second value is discarded because $\alpha + \gamma > 180°$.

$\beta = 180° - \alpha - \gamma = 180° - 120° - 25.7° = 34.3°$

$\frac{\sin\beta}{b} = \frac{\sin\alpha}{a}$

$\frac{\sin 34.3°}{b} = \frac{\sin 120°}{2}$

$b = \frac{2\sin 34.3°}{\sin 120°} \approx 1.30$

One triangle: $\beta \approx 34.3°$, $\gamma \approx 25.7°$, $b \approx 1.30$

29. $a = 4$, $b = 5$, $\alpha = 60°$

$\frac{\sin\beta}{b} = \frac{\sin\alpha}{a}$

$\frac{\sin\beta}{5} = \frac{\sin 60°}{4}$

$\sin\beta = \frac{5\sin 60°}{4} \approx 1.0825$

There is no angle β for which $\sin\beta > 1$. Therefore, there is no triangle with the given measurements.

30. $b = 2$, $c = 3$, $\beta = 40°$

$\frac{\sin\beta}{b} = \frac{\sin\gamma}{c}$

$\frac{\sin 40°}{2} = \frac{\sin\gamma}{3}$

$\sin\gamma = \frac{3\sin 40°}{2} \approx 0.9642$

$\gamma = \sin^{-1}(0.9642)$

$\gamma_1 = 74.6°$ or $\gamma_2 = 105.4°$

For both values, $\beta + \gamma < 180°$. Therefore, there are two triangles.

$\alpha_1 = 180° - \beta - \gamma_1 = 180° - 40° - 74.6° = 65.4°$

$\frac{\sin\beta}{b} = \frac{\sin\alpha_1}{a_1}$

$\frac{\sin 40°}{2} = \frac{\sin 65.4°}{a_1}$

$a_1 = \frac{2\sin 65.4°}{\sin 40°} \approx 2.83$

$\alpha_2 = 180° - \beta - \gamma_2 = 180° - 40° - 105.4° = 34.6°$

$\frac{\sin\beta}{b} = \frac{\sin\alpha_2}{a_2}$

$\frac{\sin 40°}{2} = \frac{\sin 34.6°}{a_2}$

$a_2 = \frac{2\sin 34.6°}{\sin 40°} \approx 1.77$

Two triangles:
$\alpha_1 \approx 65.4°$, $\gamma_1 \approx 74.6°$, $a_1 \approx 2.83$ or
$\alpha_2 \approx 34.6°$, $\gamma_2 \approx 105.4°$, $a_2 \approx 1.77$

31. $b=4,\ c=6,\ \beta=20^\circ$

$$\frac{\sin\beta}{b}=\frac{\sin\gamma}{c}$$
$$\frac{\sin 20^\circ}{4}=\frac{\sin\gamma}{6}$$
$$\sin\gamma=\frac{6\sin 20^\circ}{4}\approx 0.5130$$
$$\gamma=\sin^{-1}(0.5130)$$
$$\gamma_1=30.9^\circ \text{ or } \gamma_2=149.1^\circ$$

For both values, $\beta+\gamma<180^\circ$. Therefore, there are two triangles.

$$\alpha_1=180^\circ-\beta-\gamma_1=180^\circ-20^\circ-30.9^\circ=129.1^\circ$$
$$\frac{\sin\beta}{b}=\frac{\sin\alpha_1}{a_1}$$
$$\frac{\sin 20^\circ}{4}=\frac{\sin 129.1^\circ}{a_1}$$
$$a_1=\frac{4\sin 129.1^\circ}{\sin 20^\circ}\approx 9.08$$
$$\alpha_2=180^\circ-\beta-\gamma_2=180^\circ-20^\circ-149.1^\circ=10.9^\circ$$
$$\frac{\sin\beta}{b}=\frac{\sin\alpha_2}{a_2}$$
$$\frac{\sin 20^\circ}{4}=\frac{\sin 10.9^\circ}{a_2}$$
$$a_2=\frac{4\sin 10.9^\circ}{\sin 20^\circ}\approx 2.21$$

Two triangles:
$\alpha_1\approx 129.1^\circ,\ \gamma_1\approx 30.9^\circ,\ a_1\approx 9.08$ or
$\alpha_2\approx 10.9^\circ,\ \gamma_2\approx 149.1^\circ,\ a_2\approx 2.21$

32. $a=3,\ b=7,\ \alpha=70^\circ$

$$\frac{\sin\beta}{7}=\frac{\sin 70^\circ}{3}$$
$$\sin\beta=\frac{7\sin 70^\circ}{3}\approx 2.1926$$

There is no angle β for which $\sin\beta>1$. Thus, there is no triangle with the given measurements.

33. $a=2,\ c=1,\ \gamma=100^\circ$

$$\frac{\sin\gamma}{c}=\frac{\sin\alpha}{a}$$
$$\frac{\sin 100^\circ}{1}=\frac{\sin\alpha}{2}$$
$$\sin\alpha=2\sin 100^\circ\approx 1.9696$$

There is no angle α for which $\sin\alpha>1$. Thus, there is no triangle with the given measurements.

34. $b=4,\ c=5,\ \beta=95^\circ$

$$\frac{\sin\gamma}{c}=\frac{\sin\beta}{b}$$
$$\frac{\sin\gamma}{5}=\frac{\sin 95^\circ}{4}$$
$$\sin\gamma=\frac{5\sin 95^\circ}{4}\approx 1.2452$$

There is no angle γ for which $\sin\gamma>1$. Thus, there is no triangle with the given measurements.

35. $a=2,\ c=1,\ \gamma=25^\circ$

$$\frac{\sin\alpha}{a}=\frac{\sin\gamma}{c}$$
$$\frac{\sin\alpha}{2}=\frac{\sin 25^\circ}{1}$$
$$\sin\alpha=\frac{2\sin 25^\circ}{1}\approx 0.8452$$
$$\alpha=\sin^{-1}(0.8452)$$
$$\alpha_1=57.7^\circ \text{ or } \alpha_2=122.3^\circ$$

For both values, $\alpha+\gamma<180^\circ$. Therefore, there are two triangles.

$$\beta_1=180^\circ-\alpha_1-\gamma=180^\circ-57.7^\circ-25^\circ=97.3^\circ$$
$$\frac{\sin\beta_1}{b_1}=\frac{\sin\gamma}{c}$$
$$\frac{\sin 97.3^\circ}{b_1}=\frac{\sin 25^\circ}{1}$$
$$b_1=\frac{1\sin 97.3^\circ}{\sin 25^\circ}\approx 2.35$$
$$\beta_2=180^\circ-\alpha_2-\gamma=180^\circ-122.3^\circ-25^\circ=32.7^\circ$$
$$\frac{\sin\beta_2}{b_2}=\frac{\sin\gamma}{c}$$
$$\frac{\sin 32.7^\circ}{b_2}=\frac{\sin 25^\circ}{1}$$
$$b_2=\frac{1\sin 32.7^\circ}{\sin 25^\circ}\approx 1.28$$

Two triangles:
$\alpha_1\approx 57.7^\circ,\ \beta_1\approx 97.3^\circ,\ b_1\approx 2.35$ or
$\alpha_2\approx 122.3^\circ,\ \beta_2\approx 32.7^\circ,\ b_2\approx 1.28$

36. $b = 4,\ c = 5,\ \beta = 40°$

$$\frac{\sin\beta}{b} = \frac{\sin\gamma}{c}$$

$$\frac{\sin 40°}{4} = \frac{\sin\gamma}{5}$$

$$\sin\gamma = \frac{5\sin 40°}{4} \approx 0.8035$$

$$\gamma = \sin^{-1}(0.8035)$$

$$\gamma_1 = 53.5° \text{ or } \gamma_2 = 126.5°$$

For both values, $\beta + \gamma < 180°$. Therefore, there are two triangles.

$$\alpha_1 = 180° - \beta - \gamma_1 = 180° - 40° - 53.5° = 86.5°$$

$$\frac{\sin\beta}{b} = \frac{\sin\alpha_1}{a_1}$$

$$\frac{\sin 40°}{4} = \frac{\sin 86.5°}{a_1}$$

$$a_1 = \frac{4\sin 86.5°}{\sin 40°} \approx 6.21$$

$$\alpha_2 = 180° - \beta - \gamma_2 = 180° - 40° - 126.5° = 13.5°$$

$$\frac{\sin\beta}{b} = \frac{\sin\alpha_2}{a_2}$$

$$\frac{\sin 40°}{4} = \frac{\sin 13.5°}{a_2}$$

$$a_2 = \frac{4\sin 13.5°}{\sin 40°} \approx 1.45$$

Two triangles:

$\alpha_1 \approx 86.5°,\ \gamma_1 \approx 53.5°,\ a_1 \approx 6.21$ or

$\alpha_2 \approx 13.5°,\ \gamma_2 \approx 126.5°,\ a_2 \approx 1.45$

37. a. Find γ; then use the Law of Sines (see figure):

B, 60°, a, 150 mi, γ, S, 55°, b, A

$$\gamma = 180° - 60° - 55° = 65°$$

$$\frac{\sin 55°}{a} = \frac{\sin 65°}{150}$$

$$a = \frac{150\sin 55°}{\sin 65°} \approx 135.58 \text{ miles}$$

$$\frac{\sin 60°}{b} = \frac{\sin 65°}{150}$$

$$b = \frac{150\sin 60°}{\sin 65°} \approx 143.33 \text{ miles}$$

Station Able is about 143.33 miles from the ship, and Station Baker is about 135.58 miles from the ship.

b. $t = \dfrac{a}{r} = \dfrac{135.6}{200} \approx 0.68$ hours

$$0.68 \text{ hr} \cdot 60\ \frac{\text{min}}{\text{hr}} \approx 41 \text{ minutes}$$

38. Find β; then use the Law of Sines to find c (see figure):

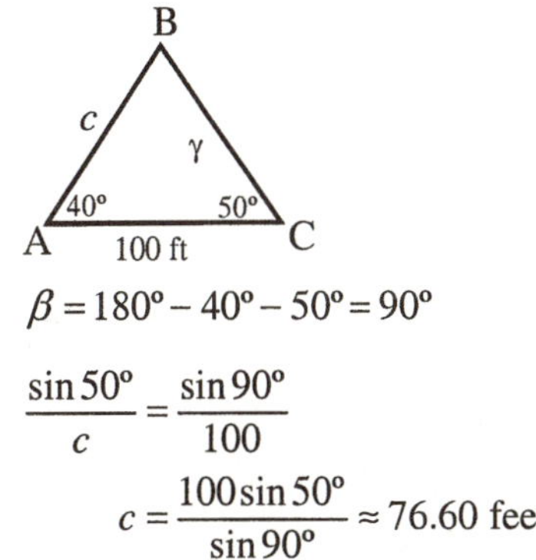

$$\beta = 180° - 40° - 50° = 90°$$

$$\frac{\sin 50°}{c} = \frac{\sin 90°}{100}$$

$$c = \frac{100\sin 50°}{\sin 90°} \approx 76.60 \text{ feet}$$

39. $\angle CAB = 180° - 25° = 155°$

$\angle ABC = 180° - 155° - 15° = 10°$

Let c represent the distance from A to B.

$$\frac{\sin 15°}{c} = \frac{\sin 10°}{1000}$$

$$c = \frac{1000\sin 15°}{\sin 10°} \approx 1490.48 \text{ feet}$$

40. From Problem 39, we have that the distance from A to B is 1490.48 feet. Let h represent the distance from B to D.

$$\sin 25° = \frac{h}{1490.48}$$

$$h = 1490.48\sin 25° \approx 629.90 \text{ feet}$$

41. Let h = the height of the plane, x = the distance from B to the plane, and $\gamma = \angle APB$ (see figure).

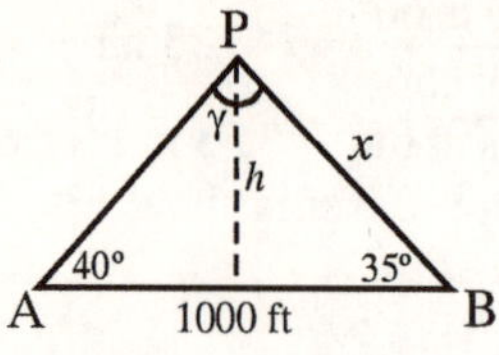

$\gamma = 180° - 40° - 35° = 105°$

$$\frac{\sin 40°}{x} = \frac{\sin 105°}{1000}$$

$$x = \frac{1000 \sin 40°}{\sin 105°} \approx 665.46 \text{ feet}$$

$$\sin 35° = \frac{h}{x} = \frac{h}{665.46}$$

$$h = (665.46)\sin 35° \approx 381.69 \text{ feet}$$

The plane is about 381.69 feet high.

42. Let h = the height of the bridge, x = the distance from C to point A, and $\gamma = \angle ACB$ (see figure).

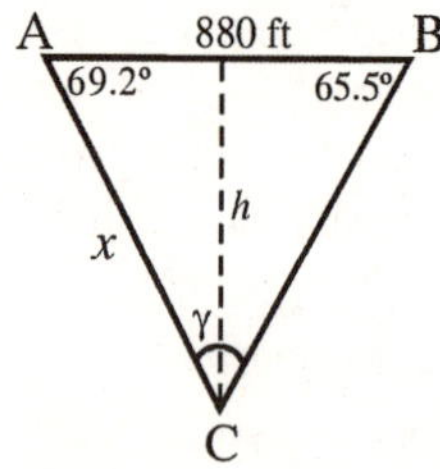

$\gamma = 180° - 69.2° - 65.5° = 45.3°$

$$\frac{\sin 65.5°}{x} = \frac{\sin 45.3°}{880}$$

$$x = \frac{880 \sin 65.5°}{\sin 45.3°} \approx 1126.57 \text{ feet}$$

$$\sin 69.2° = \frac{h}{x} = \frac{h}{1126.57}$$

$$h = (1126.57)\sin 69.2° \approx 1053.15 \text{ feet}$$

The bridge is about 1053.15 feet high.

43. a. $\angle ABC = 180° - 40° = 140°$

Find the angle at city C:

$$\frac{\sin C}{150} = \frac{\sin(140°)}{300}$$

$$\sin C = \frac{150 \sin(140°)}{300} \approx 0.3214$$

$$C = \sin^{-1}(0.3214) \approx 18.75°$$

Let y be the distance from city A to city C.

Find the angle at city A:
$A = 180° - 140° - 18.75° = 21.25°$

$$\frac{\sin 21.25°}{y} = \frac{\sin 140°}{300}$$

$$y = \frac{300 \sin 21.25°}{\sin 140°} \approx 169.16 \text{ miles}$$

The distance from city B to city C is approximately 169.16 miles.

b. To find the angle to turn, subtract angle C from 180°:
$180° - 18.75° = 161.25°$

The pilot needs to turn through an angle of 161.3° to return to city A.

44. The time of the actual trip was:

$$t = \frac{50+70}{250} = \frac{120}{250} = 0.48 \text{ hour}$$

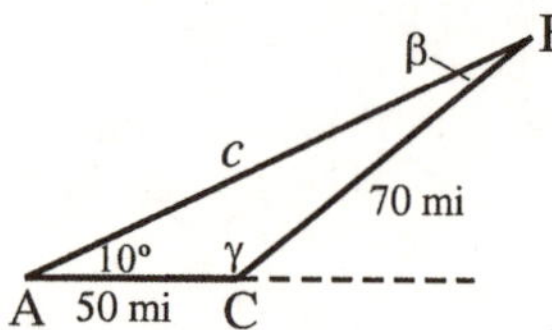

$a = 70,\ b = 50,\ \alpha = 10°$

Solve the triangle:

$$\frac{\sin 10°}{70} = \frac{\sin \beta}{50}$$

$$\sin \beta = \frac{50 \sin 10°}{70} \approx 0.1240$$

$$\beta = \sin^{-1}(0.1240) \approx 7.12°$$

$\gamma = 180° - 10° - 7.12° = 162.88°$

$$\frac{\sin 10°}{70} = \frac{\sin 162.88°}{c}$$

$$c = \frac{70 \sin 162.88°}{\sin 10°} \approx 118.67$$

$$t = \frac{118.67}{250} \approx 0.4747 \text{ hour}$$

The trip should have taken 0.4747 hour but, because of the incorrect course, took 0.48 hour. Thus, the trip took 0.0053 hour, or about 19 seconds, longer.

45. Let h = the perpendicular distance from C to AB, and let $\gamma = \angle ACB$ (see figure).

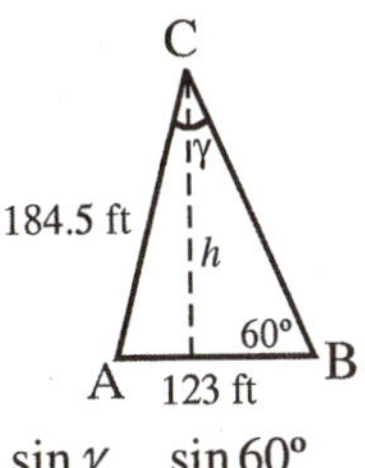

$$\frac{\sin\gamma}{123} = \frac{\sin 60^\circ}{184.5}$$

$$\sin\gamma = \frac{123\sin 60^\circ}{184.5} \approx 0.5774$$

$$\gamma = \sin^{-1}(0.5774) \approx 35.3^\circ$$

$$\angle CAB = 180^\circ - 60^\circ - 35.3^\circ \approx 84.7^\circ$$

$$\sin 84.7^\circ = \frac{h}{184.5}$$

$$h = 184.5\sin 84.7^\circ = 183.71 \text{ feet}$$

46. Let $\theta = \angle AOP$

$$\frac{\sin\theta}{9} = \frac{\sin 15^\circ}{3}$$

$$\sin\theta = \frac{9\sin 15^\circ}{3} \approx 0.7765$$

$$\theta = \sin^{-1}(0.7765) \approx 50.94^\circ$$

or $\theta \approx 180^\circ - 50.94^\circ = 129.06^\circ$

$A = 114.06^\circ$ or $A = 35.94^\circ$

$$\frac{\sin 114.06^\circ}{a} = \frac{\sin 15^\circ}{3}$$

$$a = \frac{3\sin 114.06^\circ}{\sin 15^\circ} \approx 10.58 \text{ inches}$$

or $\dfrac{\sin 35.94^\circ}{a} = \dfrac{\sin 15^\circ}{3}$

$$a = \frac{3\sin 35.94^\circ}{\sin 15^\circ} \approx 6.80 \text{ inches}$$

The approximate distance from the piston to the center of the crankshaft is either 6.80 inches or 10.58 inches.

47. $\alpha = 180^\circ - 140^\circ = 40^\circ$; $\beta = 180^\circ - 135^\circ = 45^\circ$; $\gamma = 180^\circ - 40^\circ - 45^\circ = 95^\circ$

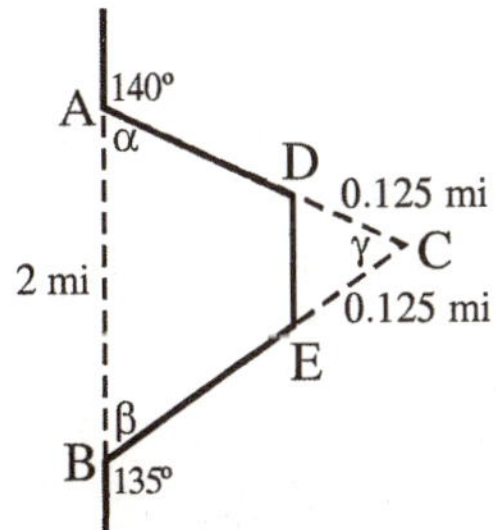

$$\frac{\sin 40^\circ}{BC} = \frac{\sin 95^\circ}{2}$$

$$BC = \frac{2\sin 40^\circ}{\sin 95^\circ} \approx 1.290 \text{ mi}$$

$$\frac{\sin 45^\circ}{AC} = \frac{\sin 95^\circ}{2}$$

$$AC = \frac{2\sin 45^\circ}{\sin 95^\circ} \approx 1.420 \text{ mi}$$

$$BE = 1.290 - 0.125 = 1.165 \text{ mi}$$

$$AD = 1.420 - 0.125 = 1.295 \text{ mi}$$

For the isosceles triangle,

$$\angle CDE = \angle CED = \frac{180^\circ - 95^\circ}{2} = 42.5^\circ$$

$$\frac{\sin 95^\circ}{DE} = \frac{\sin 42.5^\circ}{0.125}$$

$$DE = \frac{0.125\sin 95^\circ}{\sin 42.5^\circ} \approx 0.184 \text{ miles}$$

The approximate length of the highway is

$$AD + DE + BE = 1.295 + 0.184 + 1.165 \approx 2.64 \text{ mi.}$$

48. Let a = the distance from lighthouse A to the ship, b = the distance from lighthouse B to the ship, and d = the distance from the ship to the shore. From the diagram, $\angle ABC = 55^\circ$ and $\angle BAC = 75^\circ$, where point C is the ship.

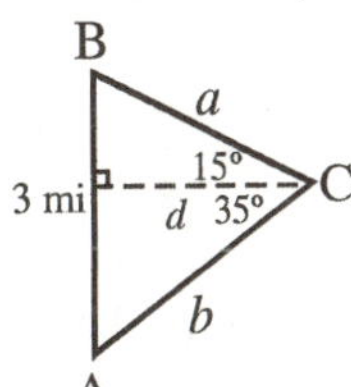

a. Use the Law of Sines:

$$\frac{\sin 50^\circ}{3} = \frac{\sin 55^\circ}{b}$$

$$b = \frac{3\sin 55^\circ}{\sin 50^\circ} \approx 3.21 \text{ miles}$$

The ship is about 3.21 miles from lighthouse A.

b. Use the Law of Sines:

$$\frac{\sin 50^\circ}{3} = \frac{\sin 75^\circ}{a}$$

$$a = \frac{3\sin 75^\circ}{\sin 50^\circ} \approx 3.78 \text{ miles}$$

The ship is about 3.78 miles from lighthouse B.

c. Use the Law of Sines:

$$\frac{\sin 90^\circ}{3.2} = \frac{\sin 75^\circ}{d}$$

$$d = \frac{3.2\sin 75^\circ}{\sin 90^\circ} \approx 3.10 \text{ miles}$$

The ship is about 3.10 miles from the shore.

49. Determine other angles in the figure:

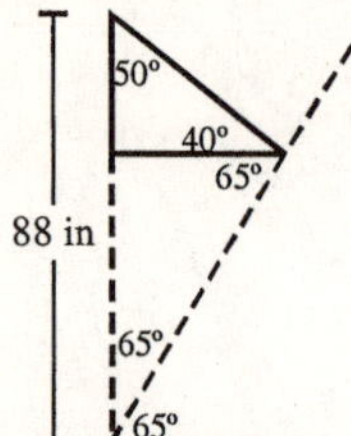

Using the Law of Sines:

$$\frac{\sin(65+40)°}{88}=\frac{\sin 25°}{L}$$

$$L=\frac{88\sin 25°}{\sin 105°}\approx 38.5 \text{ inches}$$

50. The tower forms an angle of 95° with the ground. Let x be the distance from the ranger to the tower.

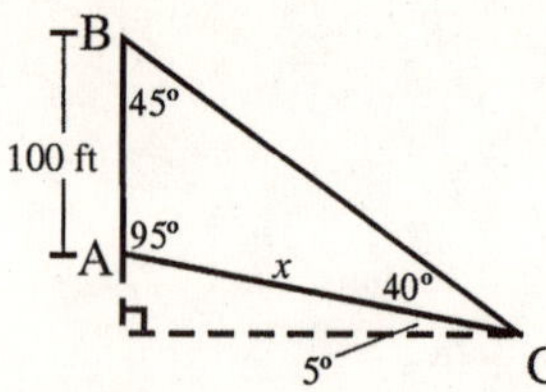

$$\angle ABC = 180° - 95° - 40° = 45°$$

$$\frac{\sin 40°}{100}=\frac{\sin 45°}{x}$$

$$x=\frac{100\sin 45°}{\sin 40°}\approx 110.01 \text{ feet}$$

The ranger is about 110.01 feet from the tower.

51. Let h = height of the pyramid, and let x = distance from the edge of the pyramid to the point beneath the tip of the pyramid (see figure).

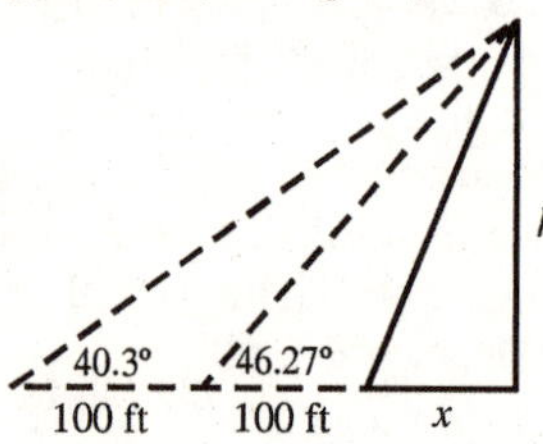

Using the Law of Sines twice yields two equations relating x and y:

Equation 1:

$$\frac{\sin 46.27°}{h}=\frac{\sin(90°-46.27°)}{x+100}$$

$$(x+100)\sin 46.27° = h\sin 43.73°$$

$$x\sin 46.27° + 100\sin 46.27° = h\sin 43.73°$$

$$x=\frac{h\sin 43.73° - 100\sin 46.27°}{\sin 46.27°}$$

Equation 2:

$$\frac{\sin 40.3°}{h}=\frac{\sin(90°-40.3°)}{x+200}$$

$$(x+200)\sin 40.3° = h\sin 49.7°$$

$$x\sin 40.3° + 200\sin 40.3° = h\sin 49.7°$$

$$x=\frac{h\sin 49.7° - 200\sin 40.3°}{\sin 40.3°}$$

Set the two equations equal to each other and solve for h.

$$\frac{h\sin 43.73° - 100\sin 46.27°}{\sin 46.27°}=\frac{h\sin 49.7° - 200\sin 40.3°}{\sin 40.3°}$$

$$h\sin 43.73°\cdot\sin 40.3° - 100\sin 46.27°\cdot\sin 40.3° = h\sin 49.7°\cdot\sin 46.27° - 200\sin 40.3°\cdot\sin 46.27°$$

$$h\sin 43.73°\cdot\sin 40.3° - h\sin 49.7°\cdot\sin 46.27° = 100\sin 46.27°\cdot\sin 40.3° - 200\sin 40.3°\cdot\sin 46.27°$$

$$h=\frac{100\sin 46.27°\cdot\sin 40.3° - 200\sin 40.3°\cdot\sin 46.27°}{\sin 43.73°\cdot\sin 40.3° - \sin 49.7°\cdot\sin 46.27°}$$

$$\approx 449.36 \text{ feet}$$

The current height of the pyramid is about 449.36 feet.

52. Let h = the height of the aircraft, and let x = the distance from the first sensor to a point on the ground beneath the airplane (see figure).

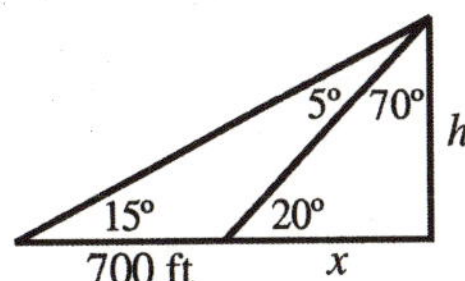

Using the Law of Sines twice yields two equations relating x and h.

Equation 1: $\dfrac{\sin 20^\circ}{h} = \dfrac{\sin 70^\circ}{x}$

$$x = \frac{h\sin 70^\circ}{\sin 20^\circ}$$

Equation 2: $\dfrac{\sin 15^\circ}{h} = \dfrac{\sin 75^\circ}{x+700}$

$$x = \frac{h\sin 75^\circ}{\sin 15^\circ} - 700$$

Set the two equations equal to each other and solve for h.

$$\frac{h\sin 70^\circ}{\sin 20^\circ} = \frac{h\sin 75^\circ}{\sin 15^\circ} - 700$$

$$h\left(\frac{\sin 70^\circ}{\sin 20^\circ} - \frac{\sin 75^\circ}{\sin 15^\circ}\right) = -700$$

$$h = \frac{-700}{\dfrac{\sin 70^\circ}{\sin 20^\circ} - \dfrac{\sin 75^\circ}{\sin 15^\circ}}$$

$$\approx 710.97 \text{ feet}$$

The height of the aircraft is about 710.97 feet.

53. Using the Law of Sines:

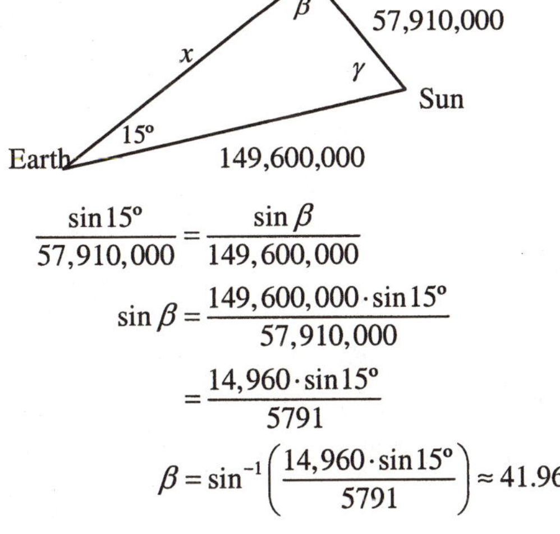

$$\frac{\sin 15^\circ}{57{,}910{,}000} = \frac{\sin\beta}{149{,}600{,}000}$$

$$\sin\beta = \frac{149{,}600{,}000\cdot\sin 15^\circ}{57{,}910{,}000}$$

$$= \frac{14{,}960\cdot\sin 15^\circ}{5791}$$

$$\beta = \sin^{-1}\left(\frac{14{,}960\cdot\sin 15^\circ}{5791}\right) \approx 41.96^\circ$$

or

$$\beta \approx 138.04^\circ$$

$\gamma \approx 180^\circ - 41.96^\circ - 15^\circ = 123.04^\circ$ or

$\gamma \approx 180^\circ - 138.04^\circ - 15^\circ = 26.96^\circ$

$$\frac{\sin 15^\circ}{57{,}910{,}000} = \frac{\sin\gamma}{x}$$

$$x = \frac{57{,}910{,}000\cdot\sin\gamma}{\sin 15^\circ}$$

$$= \frac{57{,}910{,}000\cdot\sin 123.04^\circ}{\sin 15^\circ}$$

$$\approx 187{,}564{,}951.5 \text{ km}$$

or

$$x = \frac{57{,}910{,}000\cdot\sin 26.96^\circ}{\sin 15^\circ}$$

$$\approx 101{,}439{,}834.5 \text{ km}$$

So the possible distances between Earth and Mercury are approximately 101,440,000 km and 187,600,000 km.

54. Using the Law of Sines:

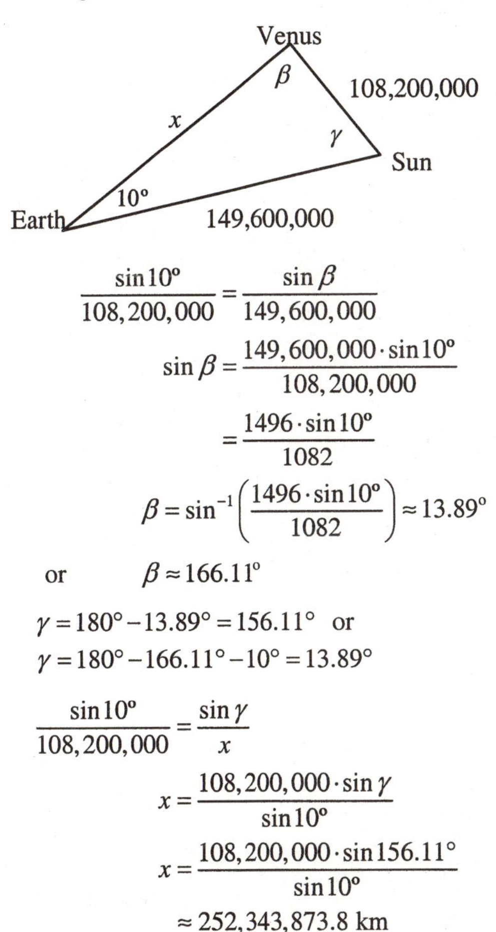

$$\frac{\sin 10^\circ}{108{,}200{,}000} = \frac{\sin\beta}{149{,}600{,}000}$$

$$\sin\beta = \frac{149{,}600{,}000\cdot\sin 10^\circ}{108{,}200{,}000}$$

$$= \frac{1496\cdot\sin 10^\circ}{1082}$$

$$\beta = \sin^{-1}\left(\frac{1496\cdot\sin 10^\circ}{1082}\right) \approx 13.89^\circ$$

or $\beta \approx 166.11^\circ$

$\gamma = 180^\circ - 13.89^\circ = 156.11^\circ$ or

$\gamma = 180^\circ - 166.11^\circ - 10^\circ = 13.89^\circ$

$$\frac{\sin 10^\circ}{108{,}200{,}000} = \frac{\sin\gamma}{x}$$

$$x = \frac{108{,}200{,}000\cdot\sin\gamma}{\sin 10^\circ}$$

$$x = \frac{108{,}200{,}000\cdot\sin 156.11^\circ}{\sin 10^\circ}$$

$$\approx 252{,}343{,}873.8 \text{ km}$$

or

$$x = \frac{108,200,000 \cdot \sin 3.89°}{\sin 10°}$$
$$\approx 42,271,757.2 \text{ km}$$

So the approximate possible distances between Earth and Venus are 42,300,000 km and 252,300,000 km.

55. Let h = the height of the tree, and let x = the distance from the first position to the center of the tree (see figure).

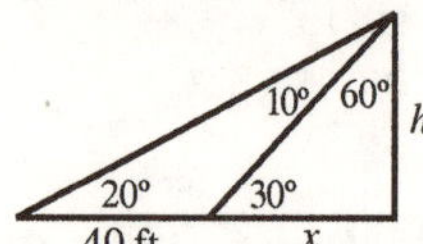

Using the Law of Sines twice yields two equations relating x and h.

Equation 1: $\frac{\sin 30°}{h} = \frac{\sin 60°}{x}$

$$x = \frac{h \sin 60°}{\sin 30°}$$

Equation 2: $\frac{\sin 20°}{h} = \frac{\sin 70°}{x+40}$

$$x = \frac{h \sin 70°}{\sin 20°} - 40$$

Set the two equations equal to each other and solve for h.

$$\frac{h \sin 60°}{\sin 30°} = \frac{h \sin 70°}{\sin 20°} - 40$$

$$h\left(\frac{\sin 60°}{\sin 30°} - \frac{\sin 70°}{\sin 20°}\right) = -40$$

$$h = \frac{-40}{\frac{\sin 60°}{\sin 30°} - \frac{\sin 70°}{\sin 20°}}$$
$$\approx 39.39 \text{ feet}$$

The height of the tree is about 39.39 feet.

56. Let x = the length of the new ramp (see figure).

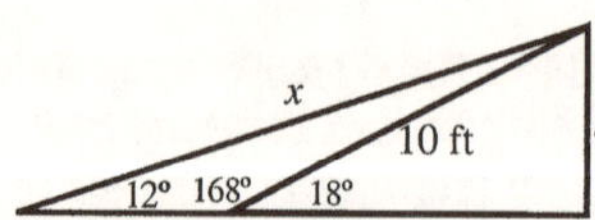

Using the Law of Sines:

$$\frac{\sin 162°}{x} = \frac{\sin 12°}{10}$$

$$x = \frac{10 \sin 162°}{\sin 12°} \approx 14.86 \text{ feet}$$

57. Let h = the height of the helicopter, x = the distance from observer A to the helicopter, and $\gamma = \angle AHB$ (see figure).

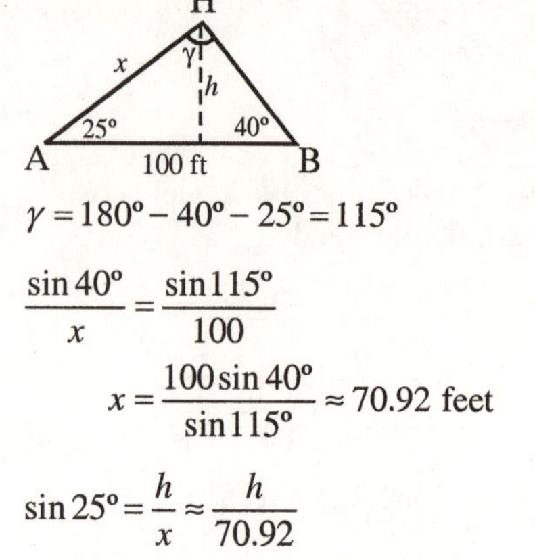

$$\gamma = 180° - 40° - 25° = 115°$$

$$\frac{\sin 40°}{x} = \frac{\sin 115°}{100}$$

$$x = \frac{100 \sin 40°}{\sin 115°} \approx 70.92 \text{ feet}$$

$$\sin 25° = \frac{h}{x} \approx \frac{h}{70.92}$$

$$h \approx 70.92 \sin 25° \approx 29.97 \text{ feet}$$

The helicopter is about 29.97 feet high.

58. $\frac{a+b}{c} = \frac{a}{c} + \frac{b}{c}$

$$= \frac{\sin\alpha}{\sin\gamma} + \frac{\sin\beta}{\sin\gamma}$$

$$= \frac{\sin\alpha + \sin\beta}{\sin\gamma}$$

$$= \frac{2\sin\left(\frac{\alpha+\beta}{2}\right)\cos\left(\frac{\alpha-\beta}{2}\right)}{2\sin\left(\frac{\gamma}{2}\right)\cos\left(\frac{\gamma}{2}\right)}$$

$$= \frac{\sin\left(\frac{\pi}{2} - \frac{\gamma}{2}\right)\cos\left(\frac{\alpha-\beta}{2}\right)}{\sin\left(\frac{\gamma}{2}\right)\cos\left(\frac{\gamma}{2}\right)}$$

$$= \frac{\cos\left(\frac{\gamma}{2}\right)\cos\left(\frac{\alpha-\beta}{2}\right)}{\sin\left(\frac{\gamma}{2}\right)\cos\left(\frac{\gamma}{2}\right)}$$

$$= \frac{\cos\left(\frac{\alpha-\beta}{2}\right)}{\sin\left(\frac{\gamma}{2}\right)}$$

$$= \frac{\cos\left[\frac{1}{2}(\alpha-\beta)\right]}{\sin\left(\frac{1}{2}\gamma\right)}$$

59. $\frac{a-b}{c}=\frac{a}{c}-\frac{b}{c}$
$=\frac{\sin\alpha}{\sin\gamma}-\frac{\sin\beta}{\sin\gamma}$
$=\frac{\sin\alpha-\sin\beta}{\sin\gamma}$
$=\frac{2\sin\left(\frac{\alpha-\beta}{2}\right)\cos\left(\frac{\alpha+\beta}{2}\right)}{\sin\left(2\cdot\frac{\gamma}{2}\right)}$
$=\frac{2\sin\left(\frac{\alpha-\beta}{2}\right)\cos\left(\frac{\alpha+\beta}{2}\right)}{2\sin\left(\frac{\gamma}{2}\right)\cos\left(\frac{\gamma}{2}\right)}$
$=\frac{\sin\left(\frac{\alpha-\beta}{2}\right)\cos\left(\frac{\pi}{2}-\frac{\gamma}{2}\right)}{\sin\left(\frac{\gamma}{2}\right)\cos\left(\frac{\gamma}{2}\right)}$
$=\frac{\sin\left(\frac{\alpha-\beta}{2}\right)\sin\left(\frac{\gamma}{2}\right)}{\sin\left(\frac{\gamma}{2}\right)\cos\left(\frac{\gamma}{2}\right)}$
$=\frac{\sin\left(\frac{\alpha-\beta}{2}\right)}{\cos\left(\frac{\gamma}{2}\right)}$
$=\frac{\sin\left[\frac{1}{2}(\alpha-\beta)\right]}{\cos\left(\frac{1}{2}\gamma\right)}$

60. $a=\frac{b\sin\alpha}{\sin\beta}$
$=\frac{b\sin[180°-(\beta+\gamma)]}{\sin\beta}$
$=\frac{b}{\sin\beta}\sin(\beta+\gamma)$
$=\frac{b}{\sin\beta}(\sin\beta\cos\gamma+\cos\beta\sin\gamma)$
$=b\cos\gamma+\frac{b\sin\gamma}{\sin\beta}\cos\beta$
$=b\cos\gamma+c\cos\beta$

61. $\frac{a-b}{a+b}=\frac{\frac{a-b}{c}}{\frac{a+b}{c}}$
$=\frac{\frac{\sin\left[\frac{1}{2}(\alpha-\beta)\right]}{\cos\left(\frac{1}{2}\gamma\right)}}{\frac{\cos\left[\frac{1}{2}(\alpha-\beta)\right]}{\sin\left(\frac{1}{2}\gamma\right)}}$
$=\frac{\sin\left[\frac{1}{2}(\alpha-\beta)\right]}{\cos\left(\frac{1}{2}\gamma\right)}\cdot\frac{\sin\left(\frac{1}{2}\gamma\right)}{\cos\left[\frac{1}{2}(\alpha-\beta)\right]}$
$=\tan\left[\frac{1}{2}(\alpha-\beta)\right]\tan\left(\frac{1}{2}\gamma\right)$
$=\tan\left[\frac{1}{2}(\alpha-\beta)\right]\tan\left[\frac{1}{2}(\pi-(\alpha+\beta))\right]$
$=\tan\left[\frac{1}{2}(\alpha-\beta)\right]\tan\left[\frac{\pi}{2}-\left(\frac{\alpha+\beta}{2}\right)\right]$
$=\tan\left[\frac{1}{2}(\alpha-\beta)\right]\cot\left(\frac{\alpha+\beta}{2}\right)$
$=\frac{\tan\left[\frac{1}{2}(\alpha-\beta)\right]}{\tan\left[\frac{1}{2}(\alpha+\beta)\right]}$

62. $\sin\beta=\sin(\angle ABC)=\sin(\angle AB'C)=\frac{b}{2r}$
$\frac{\sin\beta}{b}=\frac{1}{2r}$
The result follows from the Law of Sines.

63–65. Answers will vary.

Section 7.3

1. $d=\sqrt{(x_2-x_1)^2+(y_2-y_1)^2}$

2. $\cos\theta=\frac{\sqrt{2}}{2}$
$\theta=45°$
The solution set is $\{45°\}$.

3. Cosines

4. Sines

5. Cosines

6. False

7. False

8. True

9. $a=2,\ c=4,\ \beta=45°$

$b^2=a^2+c^2-2ac\cos\beta$

$b^2=2^2+4^2-2\cdot2\cdot4\cos45°$

$=20-16\cdot\frac{\sqrt{2}}{2}$

$=20-8\sqrt{2}$

$b=\sqrt{20-8\sqrt{2}}\approx2.95$

$a^2=b^2+c^2-2bc\cos\alpha$

$2bc\cos\alpha=b^2+c^2-a^2$

$\cos\alpha=\frac{b^2+c^2-a^2}{2bc}$

$\cos\alpha=\frac{2.95^2+4^2-2^2}{2(2.95)(4)}=\frac{20.7025}{23.6}$

$\alpha=\cos^{-1}\left(\frac{20.7025}{23.6}\right)\approx28.7°$

$\gamma=180°-\alpha-\beta\approx180°-28.7°-45°\approx106.3°$

10. $b=3,\ c=4,\ \alpha=30°$

$a^2=b^2+c^2-2bc\cos\alpha$

$a^2=3^2+4^2-2\cdot3\cdot4\cos30°$

$=25-24\left(\frac{\sqrt{3}}{2}\right)$

$=25-12\sqrt{3}$

$a=\sqrt{25-12\sqrt{3}}\approx2.05$

$c^2=a^2+b^2-2ab\cos\gamma$

$\cos\gamma=\frac{a^2+b^2-c^2}{2ab}$

$\cos\gamma=\frac{2.05^2+3^2-4^2}{2(2.05)(3)}=\frac{-2.7975}{12.3}$

$\gamma=\cos^{-1}\left(\frac{-2.7975}{12.3}\right)\approx103.1°$

$\beta=180°-\alpha-\gamma\approx180°-30°-103.1°=46.9°$

11. $a=2,\ b=3,\ \gamma=95°$

$c^2=a^2+b^2-2ab\cos\gamma$

$c^2=2^2+3^2-2\cdot2\cdot3\cos95°=13-12\cos95°$

$c=\sqrt{13-12\cos95°}\approx3.75$

$a^2=b^2+c^2-2bc\cos\alpha$

$\cos\alpha=\frac{b^2+c^2-a^2}{2bc}$

$\cos\alpha=\frac{3^2+3.75^2-2^2}{2(3)(3.75)}=\frac{19.0625}{22.5}$

$\alpha=\cos^{-1}\left(\frac{19.0625}{22.5}\right)\approx32.1°$

$\beta=180°-\alpha-\gamma\approx180°-32.1°-95°=52.9°$

12. $a=2,\ c=5,\ \beta=20°$

$b^2=a^2+c^2-2ac\cos\beta$

$b^2=2^2+5^2-2\cdot2\cdot5\cos20°=29-20\cos20°$

$b=\sqrt{29-20\cos20°}\approx3.19$

$a^2=b^2+c^2-2bc\cos\alpha$

$\cos\alpha=\frac{b^2+c^2-a^2}{2bc}$

$\cos\alpha=\frac{3.19^2+5^2-2^2}{2(3.19)(5)}=\frac{31.1761}{31.9}$

$\alpha=\cos^{-1}\left(\frac{31.1761}{31.9}\right)\approx12.2°$

$\gamma=180°-\alpha-\beta\approx180°-12.2°-20°\approx147.8°$

13. $a=6,\ b=5,\ c=8$

$a^2=b^2+c^2-2bc\cos\alpha$

$\cos\alpha=\frac{b^2+c^2-a^2}{2bc}=\frac{5^2+8^2-6^2}{2(5)(8)}=\frac{53}{80}$

$\alpha=\cos^{-1}\left(\frac{53}{80}\right)\approx48.5°$

$b^2=a^2+c^2-2ac\cos\beta$

$\cos\beta=\frac{a^2+c^2-b^2}{2ac}$

$\cos\beta=\frac{6^2+8^2-5^2}{2(6)(8)}=\frac{75}{96}$

$\beta=\cos^{-1}\left(\frac{75}{96}\right)\approx38.6°$

$\gamma=180°-\alpha-\beta\approx180°-48.5°-38.6°\approx92.9°$

14. $a=8,\ b=5,\ c=4$

$$a^2=b^2+c^2-2bc\cos\alpha$$

$$\cos\alpha=\frac{b^2+c^2-a^2}{2bc}=\frac{5^2+4^2-8^2}{2(5)(4)}=-\frac{23}{80}$$

$$\alpha=\cos^{-1}\left(-\frac{23}{80}\right)\approx 125.1^\circ$$

$$b^2=a^2+c^2-2ac\cos\beta$$

$$\cos\beta=\frac{a^2+c^2-b^2}{2ac}=\frac{8^2+4^2-5^2}{2(8)(4)}=\frac{55}{64}$$

$$\beta=\cos^{-1}\left(\frac{55}{64}\right)\approx 30.8^\circ$$

$$\gamma=180^\circ-\alpha-\beta\approx 180^\circ-125.1^\circ-30.8^\circ=24.1^\circ$$

15. $a=9,\ b=6,\ c=4$

$$a^2=b^2+c^2-2bc\cos\alpha$$

$$\cos\alpha=\frac{b^2+c^2-a^2}{2bc}=\frac{6^2+4^2-9^2}{2(6)(4)}=-\frac{29}{48}$$

$$\alpha=\cos^{-1}\left(-\frac{29}{48}\right)\approx 127.2^\circ$$

$$b^2=a^2+c^2-2ac\cos\beta$$

$$\cos\beta=\frac{a^2+c^2-b^2}{2ac}=\frac{9^2+4^2-6^2}{2(9)(4)}=\frac{61}{72}$$

$$\beta=\cos^{-1}\left(\frac{61}{72}\right)\approx 32.1^\circ$$

$$\gamma=180^\circ-\alpha-\beta\approx 180^\circ-127.2^\circ-32.1^\circ=20.7^\circ$$

16. $a=4,\ b=3,\ c=4$

$$a^2=b^2+c^2-2bc\cos\alpha$$

$$\cos\alpha=\frac{b^2+c^2-a^2}{2bc}=\frac{3^2+4^2-4^2}{2(3)(4)}=\frac{9}{24}$$

$$\alpha=\cos^{-1}\left(\frac{9}{24}\right)\approx 68.0^\circ$$

$$b^2=a^2+c^2-2ac\cos\beta$$

$$\cos\beta=\frac{a^2+c^2-b^2}{2ac}=\frac{4^2+4^2-3^2}{2(4)(4)}=\frac{23}{32}$$

$$\beta=\cos^{-1}\left(\frac{23}{32}\right)\approx 44.0^\circ$$

$$\gamma=180^\circ-\alpha-\beta\approx 180^\circ-68.0^\circ-44.0^\circ=68.0^\circ$$

17. $a=3,\ b=4,\ \gamma=40^\circ$

$$c^2=a^2+b^2-2ab\cos\gamma$$

$$c^2=3^2+4^2-2\cdot 3\cdot 4\cos 40^\circ=25-24\cos 40^\circ$$

$$c=\sqrt{25-24\cos 40^\circ}\approx 2.57$$

$$a^2=b^2+c^2-2bc\cos\alpha$$

$$\cos\alpha=\frac{b^2+c^2-a^2}{2bc}=\frac{4^2+2.57^2-3^2}{2(4)(2.57)}=\frac{13.6049}{20.56}$$

$$\alpha=\cos^{-1}\left(\frac{13.6049}{20.56}\right)\approx 48.6^\circ$$

$$\beta=180^\circ-\alpha-\gamma\approx 180^\circ-48.6^\circ-40^\circ=91.4^\circ$$

18. $a=2,\ c=1,\ \beta=10^\circ$

$$b^2=a^2+c^2-2ac\cos\beta$$

$$b^2=2^2+1^2-2\cdot 2\cdot 1\cos 10^\circ=5-4\cos 10^\circ$$

$$b=\sqrt{5-4\cos 10^\circ}\approx 1.03$$

$$a^2=b^2+c^2-2bc\cos\alpha$$

$$\cos\alpha=\frac{b^2+c^2-a^2}{2bc}=\frac{1.03^2+1^2-2^2}{2(1.03)(1)}=-\frac{1.9391}{2.06}$$

$$\alpha=\cos^{-1}\left(-\frac{1.9391}{2.06}\right)\approx 160.3^\circ$$

$$\gamma=180^\circ-\alpha-\beta\approx 180^\circ-160.3^\circ-10^\circ=9.7^\circ$$

19. $b=1,\ c=3,\ \alpha=80^\circ$

$$a^2=b^2+c^2-2bc\cos\alpha$$

$$a^2=1^2+3^2-2\cdot 1\cdot 3\cos 80^\circ=10-6\cos 80^\circ$$

$$a=\sqrt{10-6\cos 80^\circ}\approx 2.99$$

$$b^2=a^2+c^2-2ac\cos\beta$$

$$\cos\beta=\frac{a^2+c^2-b^2}{2ac}=\frac{2.99^2+3^2-1^2}{2(2.99)(3)}=\frac{16.9401}{17.94}$$

$$\beta=\cos^{-1}\left(\frac{16.9401}{17.94}\right)\approx 19.2^\circ$$

$$\gamma=180^\circ-\alpha-\beta\approx 180^\circ-80^\circ-19.2^\circ=80.8^\circ$$

20. $a=6,\ b=4,\ \gamma=60^\circ$

$c^2=a^2+b^2-2ab\cos\gamma$

$c^2=6^2+4^2-2\cdot 6\cdot 4\cos 60^\circ=28$

$c=\sqrt{28}\approx 5.29$

$a^2=b^2+c^2-2bc\cos\alpha$

$$\cos\alpha=\frac{b^2+c^2-a^2}{2bc}=\frac{4^2+5.29^2-6^2}{2(4)(5.29)}=\frac{7.9841}{42.32}$$

$$\alpha=\cos^{-1}\left(\frac{7.9841}{42.32}\right)\approx 79.1^\circ$$

$\beta=180^\circ-\alpha-\gamma\approx 180^\circ-79.1^\circ-60^\circ=40.9^\circ$

21. $a=3,\ c=2,\ \beta=110^\circ$

$b^2=a^2+c^2-2ac\cos\beta$

$b^2=3^2+2^2-2\cdot 3\cdot 2\cos 110^\circ=13-12\cos 110^\circ$

$b=\sqrt{13-12\cos 110^\circ}\approx 4.14$

$c^2=a^2+b^2-2ab\cos\gamma$

$$\cos\gamma=\frac{a^2+b^2-c^2}{2ab}=\frac{3^2+4.14^2-2^2}{2(3)(4.14)}=\frac{22.1396}{24.84}$$

$$\gamma=\cos^{-1}\left(\frac{22.1396}{24.84}\right)\approx 27.0^\circ$$

$\alpha=180^\circ-\beta-\gamma\approx 180^\circ-110^\circ-27.0^\circ=43.0^\circ$

22. $b=4,\ c=1,\ \alpha=120^\circ$

$a^2=b^2+c^2-2bc\cos\alpha$

$a^2=4^2+1^2-2\cdot 4\cdot 1\cos 120^\circ=21$

$a=\sqrt{21}\approx 4.58$

$c^2=a^2+b^2-2ab\cos\gamma$

$$\cos\gamma=\frac{a^2+b^2-c^2}{2ab}=\frac{4.58^2+4^2-1^2}{2(4.58)(4)}=\frac{35.9764}{36.64}$$

$$\gamma=\cos^{-1}\left(\frac{35.9764}{36.64}\right)\approx 10.9^\circ$$

$\beta=180^\circ-\alpha-\gamma\approx 180^\circ-120^\circ-10.9^\circ=49.1^\circ$

23. $a=2,\ b=2,\ \gamma=50^\circ$

$c^2=a^2+b^2-2ab\cos\gamma$

$c^2=2^2+2^2-2\cdot 2\cdot 2\cos 50^\circ=8-8\cos 50^\circ$

$c=\sqrt{8-8\cos 50^\circ}\approx 1.69$

$a^2=b^2+c^2-2bc\cos\alpha$

$$\cos\alpha=\frac{b^2+c^2-a^2}{2bc}=\frac{2^2+1.69^2-2^2}{2(2)(1.69)}=\frac{2.8561}{6.76}$$

$$\alpha=\cos^{-1}\left(\frac{2.8561}{6.76}\right)\approx 65.0^\circ$$

$\beta=180^\circ-\alpha-\gamma\approx 180^\circ-65.0^\circ-50^\circ=65.0^\circ$

24. $a=3,\ c=2,\ \beta=90^\circ$

$b^2=a^2+c^2-2ac\cos\beta$

$b^2=3^2+2^2-2\cdot 3\cdot 2\cos 90^\circ=13$

$b=\sqrt{13}\approx 3.61$

$a^2=b^2+c^2-2bc\cos\alpha$

$$\cos\alpha=\frac{b^2+c^2-a^2}{2bc}=\frac{3.61^2+2^2-3^2}{2(3.61)(2)}=\frac{8.0321}{14.44}$$

$$\alpha=\cos^{-1}\left(\frac{8.0321}{14.44}\right)\approx 56.2^\circ$$

$\gamma=180^\circ-\alpha-\beta\approx 10^\circ-56.2^\circ-90^\circ=33.8^\circ$

25. $a=12,\ b=13,\ c=5$

$a^2=b^2+c^2-2bc\cos\alpha$

$$\cos\alpha=\frac{b^2+c^2-a^2}{2bc}=\frac{13^2+5^2-12^2}{2(13)(5)}=\frac{50}{130}$$

$$\alpha=\cos^{-1}\left(\frac{50}{130}\right)\approx 67.4^\circ$$

$b^2=a^2+c^2-2ac\cos\beta$

$$\cos\beta=\frac{a^2+c^2-b^2}{2ac}=\frac{12^2+5^2-13^2}{2(12)(5)}=0$$

$\beta=\cos^{-1}0=90^\circ$

$\gamma=180^\circ-\alpha-\beta\approx 180^\circ-67.4^\circ-90^\circ=22.6^\circ$

26. $a=4,\ b=5,\ c=3$

$a^2=b^2+c^2-2bc\cos\alpha$

$$\cos\alpha=\frac{b^2+c^2-a^2}{2bc}=\frac{5^2+3^2-4^2}{2(5)(3)}=0.6$$

$\alpha=\cos^{-1}0.6\approx 53.1^\circ$

$b^2=a^2+c^2-2ac\cos\beta$

$$\cos\beta=\frac{a^2+c^2-b^2}{2ac}=\frac{4^2+3^2-5^2}{2(4)(3)}=0$$

$\beta=\cos^{-1}0=90^\circ$

$\gamma=180^\circ-\alpha-\beta\approx 180^\circ-53.1^\circ-90^\circ=36.9^\circ$

27. $a=2,\ b=2,\ c=2$

$$a^2=b^2+c^2-2bc\cos\alpha$$

$$\cos\alpha=\frac{b^2+c^2-a^2}{2bc}=\frac{2^2+2^2-2^2}{2(2)(2)}=0.5$$

$$\alpha=\cos^{-1}0.5=60°$$

$$b^2=a^2+c^2-2ac\cos\beta$$

$$\cos\beta=\frac{a^2+c^2-b^2}{2ac}=\frac{2^2+2^2-2^2}{2(2)(2)}=0.5$$

$$\beta=\cos^{-1}0.5=60°$$

$$\gamma=180°-\alpha-\beta\approx180°-60°-60°=60°$$

28. $a=3,\ b=3,\ c=2$

$$a^2=b^2+c^2-2bc\cos\alpha$$

$$\cos\alpha=\frac{b^2+c^2-a^2}{2bc}=\frac{3^2+2^2-3^2}{2(3)(2)}=\frac{1}{3}$$

$$\alpha=\cos^{-1}\left(\frac{1}{3}\right)\approx70.5°$$

$$b^2=a^2+c^2-2ac\cos\beta$$

$$\cos\beta=\frac{a^2+c^2-b^2}{2ac}=\frac{3^2+2^2-3^2}{2(3)(2)}=\frac{1}{3}$$

$$\beta=\cos^{-1}\left(\frac{1}{3}\right)\approx70.5°$$

$$\gamma=180°-\alpha-\beta\approx180°-70.5°-70.5°=39.0°$$

29. $a=5,\ b=8,\ c=9$

$$a^2=b^2+c^2-2bc\cos\alpha$$

$$\cos\alpha=\frac{b^2+c^2-a^2}{2bc}=\frac{8^2+9^2-5^2}{2(8)(9)}=\frac{120}{144}$$

$$\alpha=\cos^{-1}\left(\frac{120}{144}\right)\approx33.6°$$

$$b^2=a^2+c^2-2ac\cos\beta$$

$$\cos\beta=\frac{a^2+c^2-b^2}{2ac}=\frac{5^2+9^2-8^2}{2(5)(9)}=\frac{42}{90}$$

$$\beta=\cos^{-1}\left(\frac{42}{90}\right)\approx62.2°$$

$$\gamma=180°-\alpha-\beta\approx180°-33.6°-62.2°=84.2°$$

30. $a=4,\ b=3,\ c=6$

$$a^2=b^2+c^2-2bc\cos\alpha$$

$$\cos\alpha=\frac{b^2+c^2-a^2}{2bc}=\frac{3^2+6^2-4^2}{2(3)(6)}=\frac{29}{36}$$

$$\alpha=\cos^{-1}\left(\frac{29}{36}\right)\approx36.3°$$

$$b^2=a^2+c^2-2ac\cos\beta$$

$$\cos\beta=\frac{a^2+c^2-b^2}{2ac}=\frac{4^2+6^2-3^2}{2(4)(6)}=\frac{43}{48}$$

$$\beta=\cos^{-1}\left(\frac{43}{48}\right)\approx26.4°$$

$$\gamma=180°-\alpha-\beta\approx180°-36.3°-26.4°=117.3°$$

31. $a=10,\ b=8,\ c=5$

$$a^2=b^2+c^2-2bc\cos\alpha$$

$$\cos\alpha=\frac{b^2+c^2-a^2}{2bc}=\frac{8^2+5^2-10^2}{2(8)(5)}=-\frac{11}{80}$$

$$\alpha=\cos^{-1}\left(-\frac{11}{80}\right)\approx97.9°$$

$$b^2=a^2+c^2-2ac\cos\beta$$

$$\cos\beta=\frac{a^2+c^2-b^2}{2ac}=\frac{10^2+5^2-8^2}{2(10)(5)}=\frac{61}{100}$$

$$\beta=\cos^{-1}\left(\frac{61}{100}\right)\approx52.4°$$

$$\gamma=180°-\alpha-\beta\approx180°-97.9°-52.4°=29.7°$$

32. $a=9,\ b=7,\ c=10$

$$a^2=b^2+c^2-2bc\cos\alpha$$

$$\cos\alpha=\frac{b^2+c^2-a^2}{2bc}=\frac{7^2+10^2-9^2}{2(7)(10)}=\frac{68}{140}$$

$$\alpha=\cos^{-1}\left(\frac{68}{140}\right)\approx60.9°$$

$$b^2=a^2+c^2-2ac\cos\beta$$

$$\cos\beta=\frac{a^2+c^2-b^2}{2ac}=\frac{9^2+10^2-7^2}{2(9)(10)}=\frac{132}{180}$$

$$\beta=\cos^{-1}\left(\frac{132}{180}\right)\approx42.8°$$

$$\gamma=180°-\alpha-\beta\approx180°-60.9°-42.8°=76.3°$$

33. Find the third side of the triangle using the Law of Cosines: $a=150,\ b=35,\ \gamma=110°$

$$c^2=a^2+b^2-2ab\cos\gamma$$
$$=150^2+35^2-2\cdot150\cdot35\cos110°$$
$$=23{,}725-10{,}500\cos110°$$
$$c=\sqrt{23{,}725-10{,}500\cos110°}\approx165$$

The ball is approximately 165 yards from the center of the green.

34. a. The angle inside the triangle at Sarasota is $180°-50°=130°$. Use the Law of Cosines to find the third side:

$a=150,\ b=100,\ \gamma=130°$

$$c^2=a^2+b^2-2ab\cos\gamma$$
$$=150^2+100^2-2\cdot150\cdot100\cos130°$$
$$=32{,}500-30{,}000\cos130°$$
$$c=\sqrt{32{,}500-30{,}000\cos130°}\approx227.56\text{ mi}$$

b. Use the Law of Sines to find the angle inside the triangle at Orlando:

$$\frac{\sin\alpha}{150}=\frac{\sin130°}{227.6}$$
$$\sin\alpha=\frac{150\sin130°}{227.6}$$
$$\alpha=\sin^{-1}\left(\frac{150\sin130°}{227.6}\right)\approx30.3°$$

Since the angle of the triangle is 30.3°, the angle through which the pilot must turn is $180°-30.3°=149.7°$.

35. After 10 hours the ship will have traveled 150 nautical miles along its altered course. Use the Law of Cosines to find the distance from Barbados on the new course.

$a=600,\ b=150,\ \gamma=20°$

$$c^2=a^2+b^2-2ab\cos\gamma$$
$$=600^2+150^2-2\cdot600\cdot150\cos20°$$
$$=382{,}500-180{,}000\cos20°$$
$$c=\sqrt{382{,}500-180{,}000\cos20°}$$
$$\approx461.9\text{ nautical miles}$$

a. Use the Law of Cosines to find the angle opposite the side of 600:

$$\cos\alpha=\frac{b^2+c^2-a^2}{2bc}$$
$$\cos\alpha=\frac{150^2+461.9^2-600^2}{2(150)(461.9)}=-\frac{124{,}148.39}{138{,}570}$$
$$\alpha=\cos^{-1}\left(-\frac{124{,}148.39}{138{,}570}\right)\approx153.6°$$

The captain needs to turn the ship through an angle of $180°-153.6°=26.4°$.

b. $t=\dfrac{461.9\text{ nautical miles}}{15\text{ knots}}\approx30.8$ hours are required for the second leg of the trip. The total time for the trip will be about 40.8 hours.

36. a. After 15 minutes, the plane would have flown 220(0.25) = 55 miles. Find the third side of the triangle:

$a=55,\ b=330,\ \gamma=10°$

$$c^2=a^2+b^2-2ab\cos\gamma$$
$$=55^2+330^2-2\cdot55\cdot330\cos10°$$
$$=111{,}925-36{,}300\cos10°$$
$$c=\sqrt{111{,}925-36{,}300\cos10°}\approx276$$

Find the measure of the angle opposite the 330-mile side:

$$\cos\beta=\frac{a^2+c^2-b^2}{2ac}$$
$$=\frac{55^2+276^2-330^2}{2(55)(276)}=-\frac{29{,}699}{30{,}360}$$
$$\beta=\cos^{-1}\left(-\frac{29{,}699}{30{,}360}\right)\approx168.0°$$

The pilot should turn through an angle of $180°-168.0°=12.0°$.

b. If the total trip is to be done in 90 minutes, and 15 minutes were used already, then there are 75 minutes or 1.25 hours to complete the trip. The plane must travel 276 miles in 1.25 hours:

$$r=\frac{276}{1.25}=220.8\text{ miles/hour}$$

The pilot must maintain a speed of 220.8 mi/hr to complete the trip in 90 minutes.

37. a. Find x in the figure:

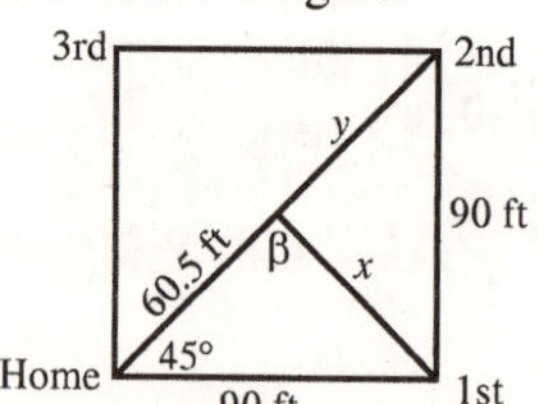

$$x^2 = 60.5^2 + 90^2 - 2(60.5)(90)\cos 45°$$
$$= 11,760.25 - 10,980\left(\frac{\sqrt{2}}{2}\right)$$
$$= 11,760.25 - 5445\sqrt{2}$$
$$x = \sqrt{11,760.25 - 5445\sqrt{2}} \approx 63.7 \text{ feet}$$

It is about 63.7 feet from the pitching rubber to first base.

b. Use the Pythagorean Theorem to find y in the figure:

$$90^2 + 90^2 = (60.5 + y)^2$$
$$16,200 = (60.5 + y)^2$$
$$60.5 + y = \sqrt{16,200} \approx 127.3$$
$$y \approx 66.8 \text{ feet}$$

It is about 66.8 feet from the pitching rubber to second base.

c. Find β in the figure by using the Law of Cosines:

$$\cos\beta = \frac{60.5^2 + 63.7^2 - 90^2}{2(60.5)(63.7)} = -\frac{382.06}{7707.7}$$
$$\beta = \cos^{-1}\left(-\frac{382.06}{7707.7}\right) \approx 92.8°$$

The pitcher needs to turn through an angle of about 92.8° to face first base.

38. a. Find x in the figure:

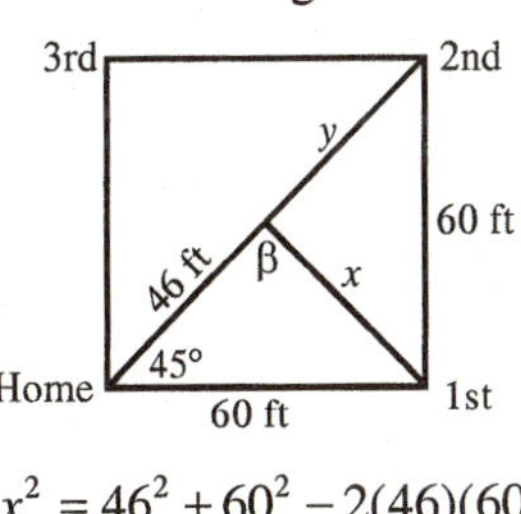

$$x^2 = 46^2 + 60^2 - 2(46)(60)\cos 45°$$
$$= 5716 - 5520\left(\frac{\sqrt{2}}{2}\right)$$
$$= 5716 - 2760\sqrt{2}$$
$$x = \sqrt{5716 - 2760\sqrt{2}} \approx 42.58 \text{ feet}$$

It is about 42.58 feet from the pitching rubber to first base.

b. Use the Pythagorean Theorem to find y in the figure:

$$60^2 + 60^2 = (46 + y)^2$$
$$7200 = (46 + y)^2$$
$$46 + y = \sqrt{7200} \approx 84.85$$
$$y \approx 38.85 \text{ feet}$$

It is about 38.85 feet from the pitching rubber to second base.

c. Find β in the figure by using the Law of Cosines:

$$\cos\beta = \frac{46^2 + 42.58^2 - 60^2}{2(46)(42.58)} = \frac{329.0564}{3917.36}$$
$$\beta = \cos^{-1}\left(\frac{329.0564}{3917.36}\right) \approx 85.2°$$

The pitcher needs to turn through an angle of 85.2° to face first base.

39. a. Find x by using the Law of Cosines:

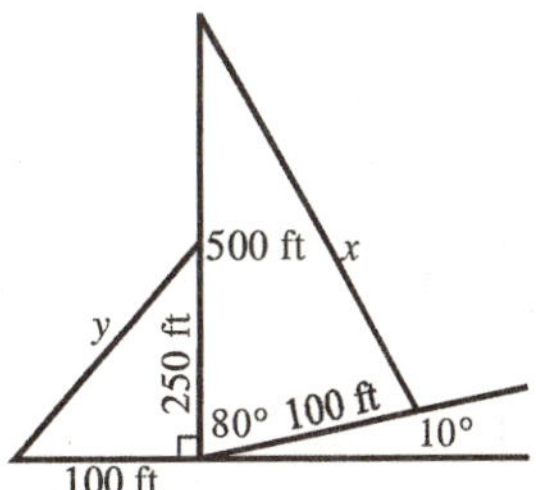

$$x^2 = 500^2 + 100^2 - 2(500)(100)\cos 80°$$
$$= 260,000 - 100,000\cos 80°$$
$$x = \sqrt{260,000 - 100,000\cos 80°} \approx 492.6 \text{ ft}$$

The guy wire needs to be about 492.6 feet long.

b. Use the Pythagorean Theorem to find the value of y:

$$y^2 = 100^2 + 250^2 = 72,500$$
$$y = 269.3 \text{ feet}$$

The guy wire needs to be about 269.3 feet long.

40. Find x by using the Law of Cosines:

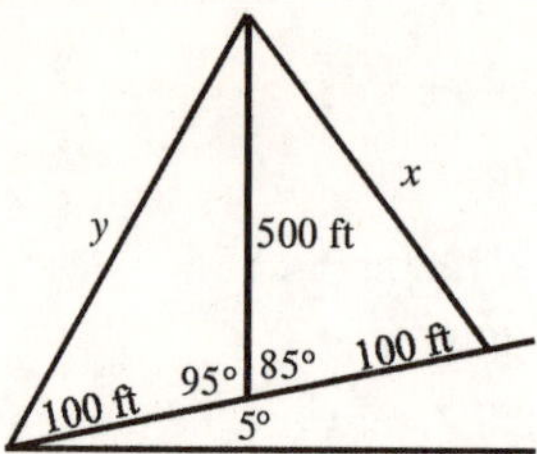

$$x^2 = 500^2 + 100^2 - 2(500)(100)\cos 85°$$
$$= 260,000 - 100,000\cos 85°$$
$$x = \sqrt{260,000 - 100,000\cos 85°} \approx 501.28 \text{ feet}$$

The guy wire needs to be about 501.28 feet long.

Find y by using the Law of Cosines:

$$y^2 = 500^2 + 100^2 - 2(500)100\cos 95°$$
$$= 260,000 - 100,000\cos 95°$$
$$y = \sqrt{260,000 - 100,000\cos 95°} \approx 518.38 \text{ feet}$$

The guy wire needs to be about 518.38 feet long.

41. Find x by using the Law of Cosines:

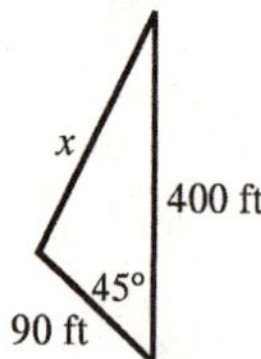

$$x^2 = 400^2 + 90^2 - 2(400)(90)\cos(45°)$$
$$= 168,100 - 36,000\sqrt{2}$$
$$x = \sqrt{168,100 - 36,000\sqrt{2}} \approx 342.33 \text{ feet}$$

It is approximately 342.33 feet from dead center to third base.

42. Find x by using the Law of Cosines:

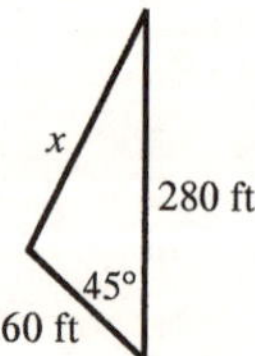

$$x^2 = 280^2 + 60^2 - 2(280)(60)\cos(45°)$$
$$= 82,000 - 16,800\sqrt{2}$$
$$x = \sqrt{82,000 - 16,800\sqrt{2}} \approx 241.33 \text{ feet}$$

It is approximately 241.33 feet from dead center to third base.

43. Use the Law of Cosines:

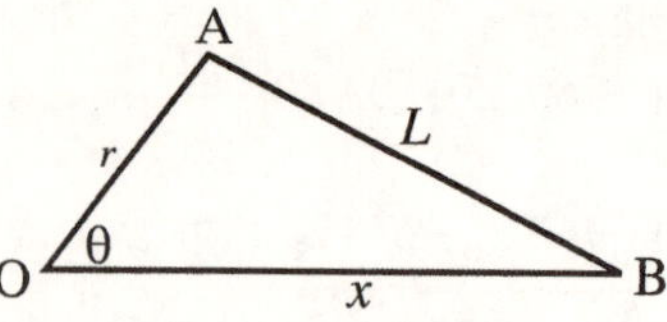

$$L^2 = x^2 + r^2 - 2xr\cos\theta$$
$$x^2 - 2xr\cos\theta + r^2 - L^2 = 0$$

Using the quadratic formula:

$$x = \frac{2r\cos\theta + \sqrt{(2r\cos\theta)^2 - 4(1)(r^2 - L^2)}}{2(1)}$$
$$x = \frac{2r\cos\theta + \sqrt{4r^2\cos^2\theta - 4\left(r^2 - L^2\right)}}{2}$$
$$x = \frac{2r\cos\theta + \sqrt{4\left(r^2\cos^2\theta - r^2 + L^2\right)}}{2}$$
$$x = \frac{2r\cos\theta + 2\sqrt{r^2\cos^2\theta - r^2 + L^2}}{2}$$
$$x = r\cos\theta + \sqrt{r^2\cos^2\theta + L^2 - r^2}$$

44. Use the Law of Cosines to find the length of side d:

$$d^2 = r^2 + r^2 - 2\cdot r\cdot r\cdot\cos\theta$$
$$= 2r^2 - 2r^2\cos\theta$$
$$= 2r^2(1 - \cos\theta)$$
$$= 4r^2\left(\frac{1-\cos\theta}{2}\right)$$
$$d = 2r\sqrt{\frac{1-\cos\theta}{2}} = 2r\sin\left(\frac{\theta}{2}\right)$$

If $s = r\theta$ is the length of the arc subtended by θ, then $d < s$, and we have $2r\sin\left(\frac{\theta}{2}\right) < r\theta$ or $\sin\left(\frac{\theta}{2}\right) < \frac{\theta}{2}$. Therefore, $\sin\theta < \theta$ for any angle $\theta > 0$.

45. $\cos\frac{\gamma}{2} = \sqrt{\frac{1+\cos\gamma}{2}}$

$$= \sqrt{\frac{1+\frac{a^2+b^2-c^2}{2ab}}{2}}$$

$$= \sqrt{\frac{2ab+a^2+b^2-c^2}{4ab}}$$

$$= \sqrt{\frac{(a+b)^2-c^2}{4ab}}$$

$$= \sqrt{\frac{(a+b+c)(a+b-c)}{4ab}}$$

$$= \sqrt{\frac{2s(2s-c-c)}{4ab}}$$

$$= \sqrt{\frac{4s(s-c)}{4ab}}$$

$$= \sqrt{\frac{s(s-c)}{ab}}$$

46. $\sin\frac{\gamma}{2} = \sqrt{\frac{1-\cos\gamma}{2}}$

$$= \sqrt{\frac{1-\frac{a^2+b^2-c^2}{2ab}}{2}}$$

$$= \sqrt{\frac{2ab-a^2-b^2+c^2}{4ab}}$$

$$= \sqrt{\frac{-(a^2-2ab+b^2-c^2)}{4ab}}$$

$$= \sqrt{\frac{-\left((a-b)^2-c^2\right)}{4ab}}$$

$$= \sqrt{\frac{-(a-b+c)(a-b-c)}{4ab}}$$

$$= \sqrt{\frac{(a-b+c)(b+c-a)}{4ab}}$$

$$= \sqrt{\frac{(2s-2b)(2s-2a)}{4ab}}$$

$$= \sqrt{\frac{4(s-b)(s-a)}{4ab}}$$

$$= \sqrt{\frac{(s-a)(s-b)}{ab}}$$

47. $\frac{\cos\alpha}{a}+\frac{\cos\beta}{b}+\frac{\cos\gamma}{c}$

$$= \frac{b^2+c^2-a^2}{2bca}+\frac{a^2+c^2-b^2}{2acb}+\frac{a^2+b^2-c^2}{2abc}$$

$$= \frac{b^2+c^2-a^2+a^2+c^2-b^2+a^2+b^2-c^2}{2abc}$$

$$= \frac{a^2+b^2+c^2}{2abc}$$

48–51. Answers will vary.

Section 7.4

1. $A=\frac{1}{2}bh$

2. Heron's

3. False

4. True

5. $a=2,\ c=4,\ \beta=45°$

$$A=\frac{1}{2}ac\sin\beta=\frac{1}{2}(2)(4)\sin 45°\approx 2.83$$

6. $b=3,\ c=4,\ \alpha=30°$

$$A=\frac{1}{2}bc\sin\alpha=\frac{1}{2}(3)(4)\sin 30°=3$$

7. $a=2,\ b=3,\ \gamma=95°$

$$A=\frac{1}{2}ab\sin\gamma=\frac{1}{2}(2)(3)\sin 95°\approx 2.99$$

8. $a=2,\ c=5,\ \beta=20°$

$$A=\frac{1}{2}ac\sin\beta=\frac{1}{2}(2)(5)\sin 20°\approx 1.71$$

9. $a=6,\ b=5,\ c=8$

$$s=\frac{1}{2}(a+b+c)=\frac{1}{2}(6+5+8)=\frac{19}{2}$$

$$A=\sqrt{s(s-a)(s-b)(s-c)}$$

$$=\sqrt{\left(\frac{19}{2}\right)\left(\frac{7}{2}\right)\left(\frac{9}{2}\right)\left(\frac{3}{2}\right)}=\sqrt{\frac{3591}{16}}\approx 14.98$$

10. $a=8,\ b=5,\ c=4$

$$s=\frac{1}{2}(a+b+c)=\frac{1}{2}(8+5+4)=\frac{17}{2}$$
$$A=\sqrt{s(s-a)(s-b)(s-c)}$$
$$=\sqrt{\left(\frac{17}{2}\right)\left(\frac{1}{2}\right)\left(\frac{7}{2}\right)\left(\frac{9}{2}\right)}=\sqrt{\frac{1071}{16}}\approx 8.18$$

11. $a=9,\ b=6,\ c=4$

$$s=\frac{1}{2}(a+b+c)=\frac{1}{2}(9+6+4)=\frac{19}{2}$$
$$A=\sqrt{s(s-a)(s-b)(s-c)}$$
$$=\sqrt{\left(\frac{19}{2}\right)\left(\frac{1}{2}\right)\left(\frac{7}{2}\right)\left(\frac{11}{2}\right)}=\sqrt{\frac{1463}{16}}\approx 9.56$$

12. $a=4,\ b=3,\ c=4$

$$s=\frac{1}{2}(a+b+c)=\frac{1}{2}(4+3+4)=\frac{11}{2}$$
$$A=\sqrt{s(s-a)(s-b)(s-c)}$$
$$=\sqrt{\left(\frac{11}{2}\right)\left(\frac{3}{2}\right)\left(\frac{5}{2}\right)\left(\frac{3}{2}\right)}=\sqrt{\frac{495}{16}}\approx 5.56$$

13. $a=3,\ b=4,\ \gamma=40°$

$$A=\frac{1}{2}ab\sin\gamma=\frac{1}{2}(3)(4)\sin 40°\approx 3.86$$

14. $a=2,\ c=1,\ \beta=10°$

$$A=\frac{1}{2}ac\sin\beta=\frac{1}{2}(2)(1)\sin 10°\approx 0.17$$

15. $b=1,\ c=3,\ \alpha=80°$

$$A=\frac{1}{2}bc\sin\alpha=\frac{1}{2}(1)(3)\sin 80°\approx 1.48$$

16. $a=6,\ b=4,\ \gamma=60°$

$$A=\frac{1}{2}ab\sin\gamma=\frac{1}{2}(6)(4)\sin 60°\approx 10.39$$

17. $a=3,\ c=2,\ \beta=110°$

$$A=\frac{1}{2}ac\sin\beta=\frac{1}{2}(3)(2)\sin 110°\approx 2.82$$

18. $b=4,\ c=1,\ \alpha=120°$

$$A=\frac{1}{2}bc\sin\alpha=\frac{1}{2}(4)(1)\sin 120°\approx 1.73$$

19. $a=12,\ b=13,\ c=5$

$$s=\frac{1}{2}(a+b+c)=\frac{1}{2}(12+13+5)=15$$
$$A=\sqrt{s(s-a)(s-b)(s-c)}$$
$$=\sqrt{(15)(3)(2)(10)}=\sqrt{900}=30$$

20. $a=4,\ b=5,\ c=3$

$$s=\frac{1}{2}(a+b+c)=\frac{1}{2}(4+5+3)=6$$
$$A=\sqrt{s(s-a)(s-b)(s-c)}$$
$$=\sqrt{(6)(2)(1)(3)}=\sqrt{36}=6$$

21. $a=2,\ b=2,\ c=2$

$$s=\frac{1}{2}(a+b+c)=\frac{1}{2}(2+2+2)=3$$
$$A=\sqrt{s(s-a)(s-b)(s-c)}$$
$$=\sqrt{(3)(1)(1)(1)}=\sqrt{3}\approx 1.73$$

22. $a=3,\ b=3,\ c=2$

$$s=\frac{1}{2}(a+b+c)=\frac{1}{2}(3+3+2)=4$$
$$A=\sqrt{s(s-a)(s-b)(s-c)}$$
$$=\sqrt{(4)(1)(1)(2)}=\sqrt{8}\approx 2.83$$

23. $a=5,\ b=8,\ c=9$

$$s=\frac{1}{2}(a+b+c)=\frac{1}{2}(5+8+9)=11$$
$$A=\sqrt{s(s-a)(s-b)(s-c)}$$
$$=\sqrt{(11)(6)(3)(2)}=\sqrt{396}\approx 19.90$$

24. $a=4,\ b=3,\ c=6$

$$s=\frac{1}{2}(a+b+c)=\frac{1}{2}(4+3+6)=\frac{13}{2}$$
$$A=\sqrt{s(s-a)(s-b)(s-c)}$$
$$=\sqrt{\left(\frac{13}{2}\right)\left(\frac{5}{2}\right)\left(\frac{7}{2}\right)\left(\frac{1}{2}\right)}=\sqrt{\frac{455}{16}}\approx 5.33$$

25. From the Law of Sines we know $\frac{\sin\alpha}{a}=\frac{\sin\beta}{b}$.

Solving for b, so we have that $b=\frac{a\sin\beta}{\sin\alpha}$. Thus,

$$A=\frac{1}{2}ab\sin\gamma=\frac{1}{2}a\left(\frac{a\sin\beta}{\sin\alpha}\right)\sin\gamma=\frac{a^2\sin\beta\sin\gamma}{2\sin\alpha}$$

26. From $\frac{\sin\alpha}{a}=\frac{\sin\beta}{b}=\frac{\sin\gamma}{c}$, we have that

$c=\frac{b\sin\gamma}{\sin\beta}$ and $a=\frac{c\sin\alpha}{\sin\gamma}$. Thus,

$$A=\frac{1}{2}bc\sin\alpha=\frac{1}{2}b\left(\frac{b\sin\gamma}{\sin\beta}\right)\sin\alpha=\frac{b^2\sin\alpha\sin\gamma}{2\sin\beta}$$

$$A=\frac{1}{2}ac\sin\beta=\frac{1}{2}\left(\frac{c\sin\alpha}{\sin\gamma}\right)c\sin\beta=\frac{c^2\sin\alpha\sin\beta}{2\sin\gamma}$$

27. $\alpha=40°,\ \beta=20°,\ a=2$

$\gamma=180°-\alpha-\beta=180°-40°-20°=120°$

$$A=\frac{a^2\sin\beta\sin\gamma}{2\sin\alpha}=\frac{2^2\sin 20°\cdot\sin 120°}{2\sin 40°}\approx 0.92$$

28. $\alpha=50°,\ \gamma=20°,\ a=3$

$\beta=180°-\alpha-\gamma=180°-50°-20°=110°$

$$A=\frac{a^2\sin\beta\sin\gamma}{2\sin\alpha}=\frac{3^2\sin 110°\cdot\sin 20°}{2\sin 50°}\approx 1.89$$

29. $\beta=70°,\ \gamma=10°,\ b=5$

$\alpha=180°-\beta-\gamma=180°-70°-10°=100°$

$$A=\frac{b^2\sin\alpha\sin\gamma}{2\sin\beta}=\frac{5^2\sin 100°\cdot\sin 10°}{2\sin 70°}\approx 2.27$$

30. $\alpha=70°,\ \beta=60°,\ c=4$

$\gamma=180°-\alpha-\beta=180°-70°-60°=50°$

$$A=\frac{c^2\sin\alpha\sin\beta}{2\sin\gamma}=\frac{4^2\sin 70°\cdot\sin 60°}{2\sin 50°}\approx 8.50$$

31. $\alpha=110°,\ \gamma=30°,\ c=3$

$\beta=180°-\alpha-\gamma=180°-110°-30°=40°$

$$A=\frac{c^2\sin\alpha\sin\beta}{2\sin\gamma}=\frac{3^2\sin 110°\cdot\sin 40°}{2\sin 30°}\approx 5.44$$

32. $\beta=10°,\ \gamma=100°,\ b=2$

$\alpha=180°-\beta-\gamma=180°-10°-100°=70°$

$$A=\frac{b^2\sin\alpha\sin\gamma}{2\sin\beta}=\frac{2^2\sin 70°\cdot\sin 100°}{2\sin 10°}\approx 10.66$$

33. Area of a sector $=\frac{1}{2}r^2\theta$ where θ is in radians.

$\theta=70\cdot\frac{\pi}{180}=\frac{7\pi}{18}$

$A_{\text{Sector}}=\frac{1}{2}\cdot 8^2\cdot\frac{7\pi}{18}=\frac{112\pi}{9}\text{ ft}^2$

$A_{\text{Triangle}}=\frac{1}{2}\cdot 8\cdot 8\sin 70°=32\sin 70°\text{ ft}^2$

$A_{\text{Segment}}=\frac{112\pi}{9}-32\sin 70°\approx 9.03\text{ ft}^2$

34. Area of a sector $=\frac{1}{2}r^2\theta$ where θ is in radians.

$\theta=40\cdot\frac{\pi}{180}=\frac{2\pi}{9}$

$A_{\text{Sector}}=\frac{1}{2}\cdot 5^2\cdot\frac{2\pi}{9}=\frac{25\pi}{9}\text{ in}^2$

$A_{\text{Triangle}}=\frac{1}{2}\cdot 5\cdot 5\sin 40°=\frac{25}{2}\sin 40°\text{ in}^2$

$A_{\text{Segment}}=\frac{25\pi}{9}-\frac{25}{2}\sin 40°\approx 0.69\text{ in}^2$

35. Find the area of the lot using Heron's Formula:
$a=100,\ b=50,\ c=75$

$s=\frac{1}{2}(a+b+c)=\frac{1}{2}(100+50+75)=\frac{225}{2}$

$$\begin{aligned}A&=\sqrt{s(s-a)(s-b)(s-c)}\\&=\sqrt{\left(\frac{225}{2}\right)\left(\frac{25}{2}\right)\left(\frac{125}{2}\right)\left(\frac{75}{2}\right)}\\&=\sqrt{\frac{52{,}734{,}375}{16}}\\&\approx 1815.46\end{aligned}$$

$\text{Cost}=(\$3)(1815.46)=\$5446.38$

36. Diameter of canvas is 24 feet; radius of canvas is 12 feet; angle is 260°.

Area of a sector $=\frac{1}{2}r^2\theta$ where θ is in radians.

$\theta=260\cdot\frac{\pi}{180}=\frac{13\pi}{9}$

$A_{\text{Sector}}=\frac{1}{2}\cdot 12^2\cdot\frac{13\pi}{9}=\frac{936\pi}{9}=104\pi\approx 326.73\text{ ft}^2$

37. Find the area of the shaded region by subtracting the area of the triangle from the area of the semicircle.

$$A_{\text{Semicircle}} = \frac{1}{2}\pi r^2 = \frac{1}{2}\pi(4)^2 = 8\pi \text{ cm}^2$$

The triangle is a right triangle. Find the other leg:

$$6^2 + b^2 = 8^2$$
$$b^2 = 64 - 36 = 28$$
$$b = \sqrt{28} = 2\sqrt{7}$$
$$A_{\text{Triangle}} = \frac{1}{2}\cdot 6\cdot 2\sqrt{7} = 6\sqrt{7} \text{ cm}^2$$
$$A_{\text{Shaded region}} = 8\pi - 6\sqrt{7} \approx 9.26 \text{ cm}^2$$

38. Find the area of the shaded region by subtracting the area of the triangle from the area of the semicircle.
Area of the semicircle

$$A_{\text{Semicircle}} = \frac{1}{2}\pi r^2 = \frac{1}{2}\pi(5)^2 = \frac{25}{2}\pi \text{ in}^2$$

The triangle is a right triangle. Find the other leg:

$$8^2 + b^2 = 10^2$$
$$b^2 = 100 - 64 = 36$$
$$b = \sqrt{36} = 6$$
$$A_{\text{Triangle}} = \frac{1}{2}\cdot 8\cdot 6 = 24 \text{ in}^2$$
$$A_{\text{Shaded region}} = 12.5\pi - 24 \approx 15.27 \text{ in}^2$$

39. The area is the sum of the area of a triangle and a sector.

$$A_{\text{Triangle}} = \frac{1}{2} r\cdot r\sin(\pi-\theta) = \frac{1}{2}r^2\sin(\pi-\theta)$$
$$A_{\text{Sector}} = \frac{1}{2}r^2\theta$$
$$\begin{aligned} A_{\text{Shaded region}} &= \frac{1}{2}r^2\sin(\pi-\theta) + \frac{1}{2}r^2\theta \\ &= \frac{1}{2}r^2\left(\sin(\pi-\theta)+\theta\right) \\ &= \frac{1}{2}r^2\left(\sin\pi\cos\theta - \cos\pi\sin\theta + \theta\right) \\ &= \frac{1}{2}r^2\left(0\cdot\cos\theta - (-1)\sin\theta + \theta\right) \\ &= \frac{1}{2}r^2\left(\theta + \sin\theta\right) \end{aligned}$$

40. Use the Law of Cosines to find the lengths of the diagonals of the polygon.

$$\begin{aligned} x^2 &= 35^2 + 80^2 - 2\cdot 35\cdot 80\cos 15^\circ \\ &= 7625 - 5600\cos 15^\circ \end{aligned}$$
$$x = \sqrt{7625 - 5600\cos 15^\circ} \approx 47.072 \text{ feet}$$

The interior angle of the third triangle is: $180^\circ - 100^\circ = 80^\circ$.

$$\begin{aligned} y^2 &= 45^2 + 20^2 - 2\cdot 45\cdot 20\cos 80^\circ \\ &= 2425 - 1800\cos 80^\circ \end{aligned}$$
$$y = \sqrt{2425 - 1800\cos 80^\circ} \approx 45.961 \text{ feet}$$

Find the area of the three triangles:

$$s_1 = \frac{1}{2}(35 + 80 + 47.072) = 81.036$$
$$s_1 - a_1 = 81.036 - 35 = 46.036$$
$$s_1 - b_1 = 81.036 - 80 = 1.036$$
$$s_1 - c_1 = 81.036 - 47.072 = 33.964$$
$$A_1 = \sqrt{81.036(46.036)(1.036)(33.964)} \approx 362.31 \text{ ft}^2$$

$$s_2 = \frac{1}{2}(40 + 45.961 + 47.072) = 66.5165$$
$$s_2 - a_2 = 66.5165 - 40 = 26.5165$$
$$s_2 - b_2 = 66.5165 - 45.961 = 20.5555$$
$$s_2 - c_2 = 66.5165 - 47.072 = 19.4445$$
$$A_2 = \sqrt{66.5165(26.5165)(20.5555)(19.4445)}$$
$$\approx 839.62 \text{ ft}^2$$

$$s_3 = \frac{1}{2}(45 + 20 + 45.961) = 55.4805$$
$$s_3 - a_3 = 55.4805 - 45 = 10.4805$$
$$s_3 - b_3 = 55.4805 - 20 = 35.4805$$
$$s_3 - c_3 = 55.4805 - 45.961 = 9.5195$$
$$A_3 = \sqrt{55.4805(10.4805)(35.4805)(9.5195)}$$
$$\approx 443.16 \text{ ft}^2$$

The approximate area of the lake is
$362.31 + 839.62 + 443.16 = 1645.09 \text{ ft}^2$

41. a.

$$\begin{aligned} \text{Area } \Delta OAC &= \frac{1}{2}|OC|\cdot|AC| \\ &= \frac{1}{2}\cdot\frac{|OC|}{1}\cdot\frac{|AC|}{1} \\ &= \frac{1}{2}\cos\alpha\sin\alpha \\ &= \frac{1}{2}\sin\alpha\cos\alpha \end{aligned}$$

b. $\text{Area } \Delta OCB = \frac{1}{2}|OC|\cdot|BC|$

$$= \frac{1}{2}\cdot|OB|^2\cdot\frac{|OC|}{|OB|}\cdot\frac{|BC|}{|OB|}$$

$$= \frac{1}{2}|OB|^2\cos\beta\sin\beta$$

$$= \frac{1}{2}|OB|^2\sin\beta\cos\beta$$

c. $\text{Area } \Delta OAB = \frac{1}{2}|BD|\cdot|OA|$

$$= \frac{1}{2}|BD|\cdot 1$$

$$= \frac{1}{2}\cdot|OB|\cdot\frac{|BD|}{|OB|}$$

$$= \frac{1}{2}|OB|\sin(\alpha+\beta)$$

d. $$\frac{\cos\alpha}{\cos\beta} = \frac{\frac{|OC|}{|OA|}}{\frac{|OC|}{|OB|}} = \frac{|OC|}{1}\cdot\frac{|OB|}{|OC|} = |OB|$$

e. $\text{Area } \Delta OAB = \text{Area } \Delta OAC + \text{Area } \Delta OCB$

$$\frac{1}{2}|OB|\sin(\alpha+\beta) = \frac{1}{2}\sin\alpha\cos\alpha + \frac{1}{2}|OB|^2\sin\beta\cos\beta$$

$$\frac{\cos\alpha}{\cos\beta}\sin(\alpha+\beta) = \sin\alpha\cos\alpha + \frac{\cos^2\alpha}{\cos^2\beta}\sin\beta\cos\beta$$

$$\sin(\alpha+\beta) = \frac{\cos\beta}{\cos\alpha}\sin\alpha\cos\alpha + \frac{\cos\alpha}{\cos\beta}\sin\beta\cos\beta$$

$$\sin(\alpha+\beta) = \sin\alpha\cos\beta + \cos\alpha\sin\beta$$

42. a. $\text{Area of } \Delta OBC = \frac{1}{2}\cdot 1\cdot 1\cdot\sin\theta = \frac{\sin\theta}{2}$

b. $\text{Area of } \Delta OBD = \frac{1}{2}\cdot 1\cdot\tan\theta = \frac{\tan\theta}{2} = \frac{\sin\theta}{2\cos\theta}$

c. $\text{Area } \Delta OBC < \text{Area } \widehat{OBC} < \text{Area } \Delta OBD$

$$\frac{1}{2}\sin\theta < \frac{1}{2}\theta < \frac{\sin\theta}{2\cos\theta}$$

$$\sin\theta < \theta < \frac{\sin\theta}{\cos\theta}$$

$$\frac{\sin\theta}{\sin\theta} < \frac{\theta}{\sin\theta} < \frac{\sin\theta}{\sin\theta\cos\theta}$$

$$1 < \frac{\theta}{\sin\theta} < \frac{1}{1\cos\theta}$$

43. The grazing area must be considered in sections. Region A_1 represents three-fourth of a circle with radius 100 feet. Thus,

$$A_1 = \frac{3}{4}\pi(100)^2 = 7500\pi \approx 23{,}561.94 \text{ ft}^2$$

Angles are needed to find regions A_2 and A_3: (see the figure)

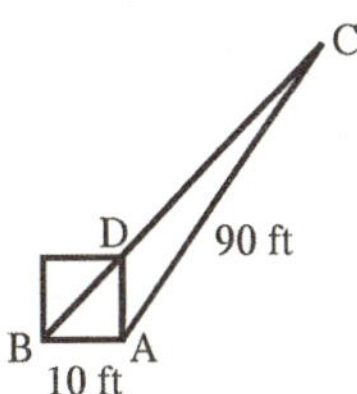

In ΔABC, $\angle CBA = 45°$, $AB = 10$, $AC = 90$.

Find $\angle BCA$:

$$\frac{\sin\angle CBA}{90} = \frac{\sin\angle BCA}{10}$$

$$\frac{\sin 45°}{90} = \frac{\sin\angle BCA}{10}$$

$$\sin\angle BCA = \frac{10\sin 45°}{90} \approx 0.0786$$

$$\angle BCA = \sin^{-1}\left(\frac{10\sin 45°}{90}\right) \approx 4.51°$$

$$\angle BAC = 180° - 45° - 4.51° = 130.49°$$

$$\angle DAC = 130.49° - 90° = 40.49°$$

$$A_3 = \frac{1}{2}(10)(90)\sin 40.49° \approx 292.19 \text{ ft}^2$$

The angle for the sector A_2 is $90° - 40.49° = 49.51°$.

$$A_2 = \frac{1}{2}(90)^2\left(49.51\cdot\frac{\pi}{180}\right) \approx 3499.66 \text{ ft}^2$$

Since the cow can go in either direction around the barn, both A_2 and A_3 must be doubled. Thus, the total grazing area is:

$$23{,}561.94 + 2(3499.66) + 2(292.19)$$

$$\approx 31{,}146 \text{ ft}^2$$

44. We begin by dividing the grazing area into five regions: three sectors and two triangles (see figure). Region A_1 is a sector representing three-fourths of a circle with radius 100 feet:

Thus, $A_1 = \frac{3}{4}\pi(100)^2 = 7500\pi \approx 23{,}561.9 \text{ ft}^2$.

To find the areas of regions A_2, A_3, A_4, and A_5, we first position the rectangular barn on a rectangular coordinate system so that the lower right corner is at the origin. The coordinates of the corners of the barn must then be $O(0,0)$, $P(-20,0)$, $Q(-20,10)$, and $R(0,10)$.

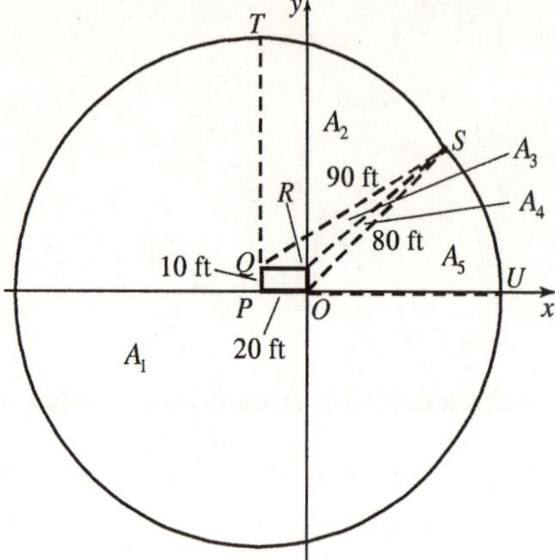

Now, region A_2 is a sector of a circle with center $Q(-20,10)$ and radius 90 feet. The equation of the circle then is $(x+20)^2+(y-10)^2=90^2$. Likewise, region A_5 is a sector of a circle with center $O(0,0)$ and radius 80 feet. The equation of this circle then is $x^2+y^2=80^2$. We use a graphing calculator to find the intersection point S of the two sectors. Let $Y_1=\sqrt{80^2-x^2}$ and $Y_2=\sqrt{90^2-(x+20)^2}+10$.

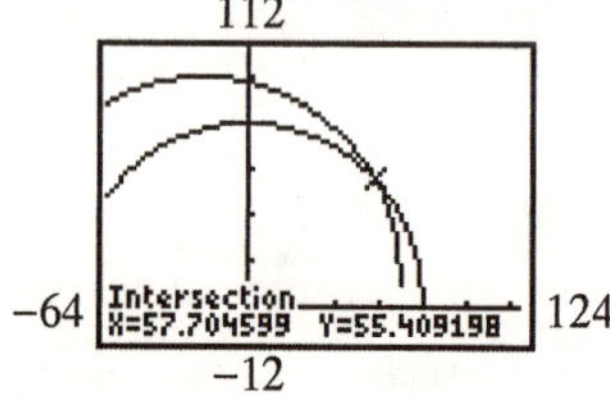

The approximate coordinates are $S(57.7,55.4)$. Now, consider ΔQRS (i.e. region A_3). The "base" of this triangle is 20 feet and the "height" is approximately $55.4-10=45.4$ feet (the y-coordinate of the intersection point minus the side of the barn). Thus, the area of region A_3 is $A_3 \approx \frac{1}{2}\cdot 20\cdot 45.4=454\text{ ft}^2$. Likewise, consider ΔORS (i.e. region A_4). The "base" of this triangle is 10 feet and the "height" is about 57.7 feet (the x-coordinate of the intersection point). Thus, $A_4 \approx \frac{1}{2}\cdot 10\cdot 57.7=288.5\text{ ft}^2$.

To find the area of sectors A_2 and A_5, we must determine their angles: $\angle TQS$ and $\angle SOU$, respectively. Now, we know $A_3 \approx 454\text{ ft}^2$. Also, we know $A_3=\frac{1}{2}\cdot 90\cdot 20\sin(\angle SQR)$.

Thus, $\frac{1}{2}\cdot 90\cdot 20\sin(\angle SQR)\approx 454$

$$\sin(\angle SQR)\approx 0.5044$$
$$\angle SQR \approx 0.5287\text{ rad.}$$

Since $\angle TQR$ is a right angle, we have

$\angle TQS \approx \frac{\pi}{2}-.5287\approx 1.0421$ rad. So,

$A_2=\frac{1}{2}r^2\theta \approx \frac{1}{2}\cdot 90^2\cdot 1.0421\approx 4220.5\text{ ft}^2$.

Similarly, $\frac{1}{2}\cdot 80\cdot 10\sin(\angle SRO)\approx 288.5$

$$\sin(\angle SQR)\approx 0.72125$$
$$\angle SQR \approx 0.8056\text{ rad.}$$

So, $\angle SOU \approx \frac{\pi}{2}-.8056\approx 0.7652$ rad. and

$A_5 \approx \frac{1}{2}\cdot 80^2\cdot 0.7652\approx 2448.6\text{ ft}^2$

Thus, the total grazing area is:
$23{,}561.9+4220.5+454+288.5+2448.6$
$=30{,}973\text{ ft}^2$

45. $K=\frac{1}{2}h_1 a$, so $h_1=\frac{2K}{a}$. Similarly, $h_2=\frac{2K}{b}$ and $h_3=\frac{2K}{c}$. Thus,

$$\frac{1}{h_1}+\frac{1}{h_2}+\frac{1}{h_3}=\frac{a}{2K}+\frac{b}{2K}+\frac{c}{2K}=\frac{a+b+c}{2K}=\frac{2s}{2K}=\frac{s}{K}$$

46. We know $A=\frac{1}{2}ah$ and $A=\frac{1}{2}ab\sin\gamma$, which means $h=b\sin\gamma$. From the Law of Sines, we know $\frac{\sin\alpha}{a}=\frac{\sin\beta}{b}$, so $b=\frac{a\sin\beta}{\sin\alpha}$. Therefore, $h=\left(\frac{a\sin\beta}{\sin\alpha}\right)\sin\gamma=\frac{a\sin\beta\sin\gamma}{\sin\alpha}$.

47. $h=\dfrac{a\sin\beta\sin\gamma}{\sin\alpha}$ where h is the altitude to side a.

In ΔOAB, c is opposite $\angle AOB$. The two adjacent angles are $\dfrac{\alpha}{2}$ and $\dfrac{\beta}{2}$. Then $r=\dfrac{c\sin\frac{\alpha}{2}\sin\frac{\beta}{2}}{\sin(\angle AOB)}$.

Now, $\angle AOB=\pi-\left(\dfrac{\alpha}{2}+\dfrac{\beta}{2}\right)$, so

$$\begin{aligned}\sin(\angle AOB)&=\sin\left[\pi-\left(\frac{\alpha}{2}+\frac{\beta}{2}\right)\right]\\&=\sin\left(\frac{\alpha}{2}+\frac{\beta}{2}\right)\\&=\sin\left(\frac{\alpha+\beta}{2}\right)\\&=\cos\left[\frac{\pi}{2}-\left(\frac{\alpha+\beta}{2}\right)\right]\\&=\cos\left[\frac{\pi-(\alpha+\beta)}{2}\right]\\&=\cos\frac{\gamma}{2}\end{aligned}$$

Thus, $r=\dfrac{c\sin\frac{\alpha}{2}\sin\frac{\beta}{2}}{\cos\frac{\gamma}{2}}$.

48.
$$\begin{aligned}\cot\frac{\gamma}{2}&=\frac{\cos\frac{\gamma}{2}}{\sin\frac{\gamma}{2}}\\&=\frac{\dfrac{c\cdot\sin\frac{\alpha}{2}\cdot\sin\frac{\beta}{2}}{r}}{\sqrt{\dfrac{(s-a)(s-b)}{ab}}}\\&=\frac{c\cdot\sqrt{\dfrac{(s-b)(s-c)}{bc}}\sqrt{\dfrac{(s-a)(s-c)}{ac}}}{r\sqrt{\dfrac{(s-a)(s-b)}{ab}}}\\&=\frac{c}{r}\sqrt{\frac{(s-b)(s-c)}{bc}}\sqrt{\frac{(s-a)(s-c)}{ac}}\sqrt{\frac{ab}{(s-a)(s-b)}}\\&=\frac{c}{r}\sqrt{\frac{ab(s-a)(s-b)(s-c)^2}{abc^2(s-a)(s-b)}}\\&=\frac{c}{r}\sqrt{\frac{(s-c)^2}{c^2}}\\&=\frac{c}{r}\cdot\frac{s-c}{c}\\&=\frac{s-c}{r}\end{aligned}$$

49. Use the result of Problem 48:
$$\begin{aligned}\cot\frac{\alpha}{2}+\cot\frac{\beta}{2}+\cot\frac{\gamma}{2}&=\frac{s-a}{r}+\frac{s-b}{r}+\frac{s-c}{r}\\&=\frac{s-a+s-b+s-c}{r}\\&=\frac{3s-(a+b+c)}{r}\\&=\frac{3s-2s}{r}\\&=\frac{s}{r}\end{aligned}$$

50.
$$\begin{aligned}K&=\text{Area }\Delta AOB+\text{ Area }\Delta AOC+\text{ Area }\Delta BOC\\&=\frac{1}{2}rc+\frac{1}{2}rb+\frac{1}{2}ra\\&=\frac{1}{2}r(a+b+c)\\&=rs\end{aligned}$$

Now, $K=\sqrt{s(s-a)(s-b)(s-c)}$, so
$$\begin{aligned}rs&=\sqrt{s(s-a)(s-b)(s-c)}\\r&=\frac{\sqrt{s(s-a)(s-b)(s-c)}}{s}\\&=\sqrt{\frac{(s-a)(s-b)(s-c)}{s}}\end{aligned}$$

51–52. Answers will vary.

Section 7.5

1. $|5|=5$; $\dfrac{2\pi}{4}=\dfrac{\pi}{2}$
2. simple harmonic; amplitude
3. simple harmonic motion; damped motion
4. True
5. $d=-5\cos(\pi t)$
6. $d=-10\cos\left(\dfrac{2\pi}{3}t\right)$
7. $d=-6\cos(2t)$
8. $d=-4\cos(4t)$
9. $d=-5\sin(\pi t)$

10. $d=-10\sin\left(\frac{2\pi}{3}t\right)$

11. $d=-6\sin(2t)$

12. $d=-4\sin(4t)$

13. $d=5\sin(3t)$

a. Simple harmonic

b. 5 meters

c. $\frac{2\pi}{3}$ seconds

d. $\frac{3}{2\pi}$ oscillation/second

14. $d=4\sin(2t)$

a. Simple harmonic

b. 4 meters

c. π seconds

d. $\frac{1}{\pi}$ oscillation/second

15. $d=6\cos(\pi t)$

a. Simple harmonic

b. 6 meters

c. 2 seconds

d. $\frac{1}{2}$ oscillation/second

16. $d=5\cos\left(\frac{\pi}{2}t\right)$

a. Simple harmonic

b. 5 meters

c. 4 seconds

d. $\frac{1}{4}$ oscillation/second

17. $d=-3\sin\left(\frac{1}{2}t\right)$

a. Simple harmonic

b. 3 meters

c. 4π seconds

d. $\frac{1}{4\pi}$ oscillation/second

18. $d=-2\cos(2t)$

a. Simple harmonic

b. 2 meters

c. π seconds

d. $\frac{1}{\pi}$ oscillation/second

19. $d=6+2\cos(2\pi t)$

a. Simple harmonic

b. 2 meters

c. 1 second

d. 1 oscillation/second

20. $d=4+3\sin(\pi t)$

a. Simple harmonic

b. 3 meters

c. 2 seconds

d. $\frac{1}{2}$ oscillation/second

21. a. $d=-10e^{-0.7t/2(25)}\cos\left(\sqrt{\left(\frac{2\pi}{5}\right)^2-\frac{(0.7)^2}{4(25)^2}}\,t\right)$

$d=-10e^{-0.7t/50}\cos\left(\sqrt{\frac{4\pi^2}{25}-\frac{0.49}{2500}}\,t\right)$

b.

10

0 25

−10

22. a. $d=-15e^{-0.75t/2(20)}\cos\left(\sqrt{\left(\frac{2\pi}{6}\right)^2-\frac{(0.75)^2}{4(20)^2}}\,t\right)$

$d=-15e^{-0.75t/40}\cos\left(\sqrt{\frac{\pi^2}{9}-\frac{0.5625}{1600}}\,t\right)$

b.

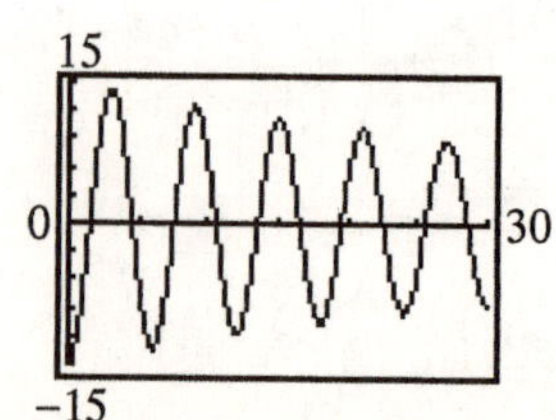

23. a. $d = -18e^{-0.6\,t/2(30)}\cos\left(\sqrt{\left(\frac{\pi}{2}\right)^2 - \frac{(0.6)^2}{4(30)^2}}\,t\right)$

$d = -18e^{-0.6\,t/60}\cos\left(\sqrt{\frac{\pi^2}{4} - \frac{0.36}{3600}}\,t\right)$

b. 18, 0, 20, −18

24. a. $d = -16e^{-0.65\,t/2(15)}\cos\left(\sqrt{\left(\frac{2\pi}{5}\right)^2 - \frac{(0.65)^2}{4(15)^2}}\,t\right)$

$d = -16e^{-0.65\,t/30}\cos\left(\sqrt{\frac{4\pi^2}{25} - \frac{0.4225}{900}}\,t\right)$

b. 16, 0, 25, −16

25. a. $d = -5e^{-0.8\,t/2(10)}\cos\left(\sqrt{\left(\frac{2\pi}{3}\right)^2 - \frac{(0.8)^2}{4(10)^2}}\,t\right)$

$d = -5e^{-0.8\,t/20}\cos\left(\sqrt{\frac{4\pi^2}{9} - \frac{0.64}{400}}\,t\right)$

b. 5, 0, 15, −5

26. a. $d = -5e^{-0.7\,t/2(10)}\cos\left(\sqrt{\left(\frac{2\pi}{3}\right)^2 - \frac{(0.7)^2}{4(10)^2}}\,t\right)$

$d = -5e^{-0.7\,t/20}\cos\left(\sqrt{\frac{4\pi^2}{9} - \frac{0.49}{400}}\,t\right)$

b.

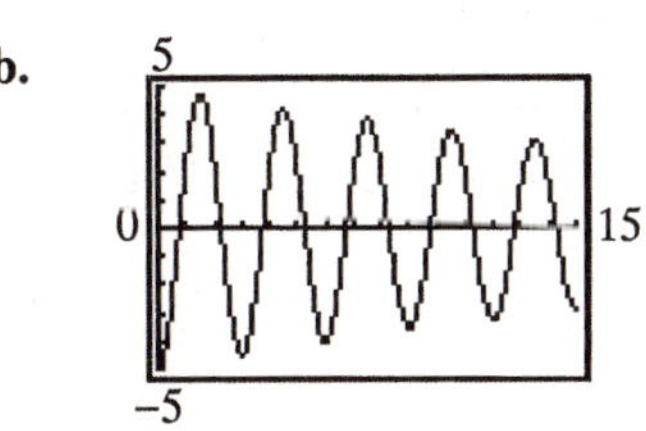

27. a. Damped motion with a bob of mass 20 kg and a damping factor of 0.7 kg/sec.

b. 20 meters leftward

c.

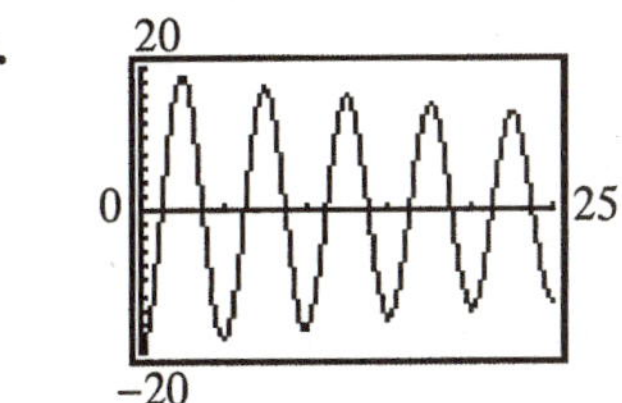

d. The displacement of the bob at the start of the second oscillation is about 18.33 meters.

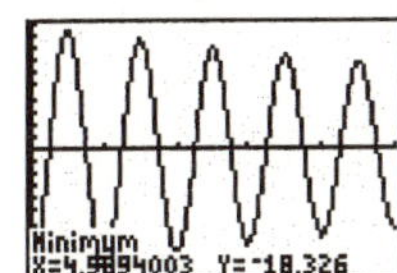

e. The displacement of the bob approaches zero, since $e^{-0.7t/40} \to 0$ as $t \to \infty$.

28. a. Damped motion with a bob of mass 20 kg and a damping factor of 0.8 kg/sec.

b. 20 meters leftward

c.

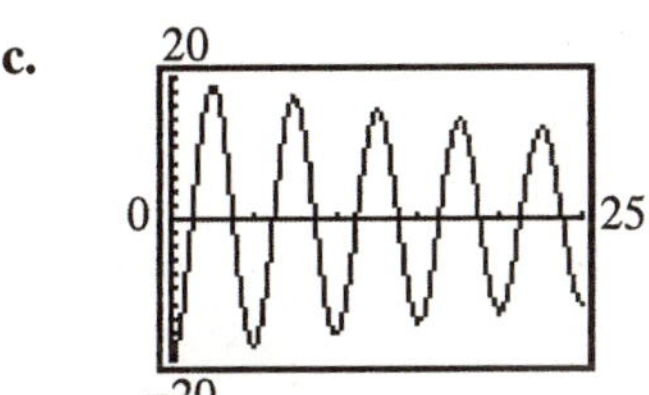

d. The displacement of the bob at the start of the second oscillation is about 18.10 meters.

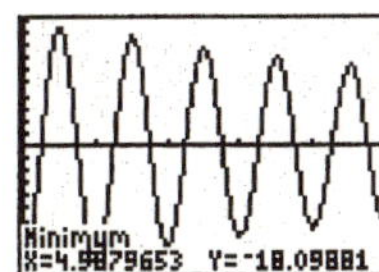

e. The displacement of the bob approaches zero, since $e^{-0.8t/40} \to 0$ as $t \to \infty$.

29. a. Damped motion with a bob of mass 40 kg and a damping factor of 0.6 kg/sec.

b. 30 meters leftward

c. 30, 0, 35, −30

d. The displacement of the bob at the start of the second oscillation is about 28.47 meters.

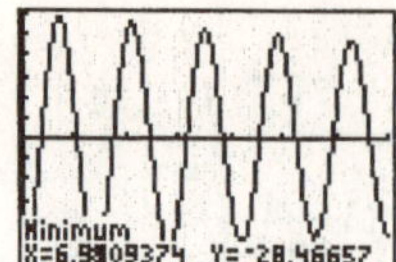

e. The displacement of the bob approaches zero, since $e^{-0.6t/80} \to 0$ as $t \to \infty$.

30. a. Damped motion with a bob of mass 35 kg and a damping factor of 0.5 kg/sec.

b. 30 meters leftward

c.

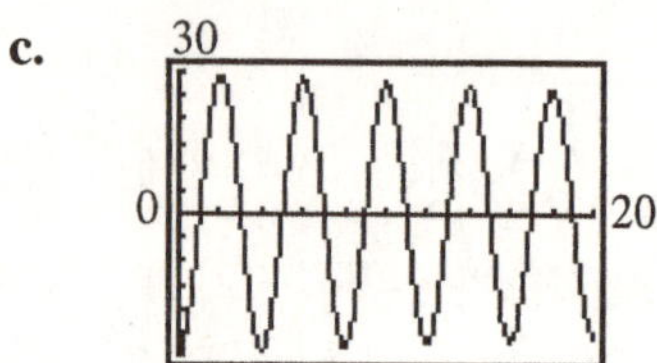

d. The displacement of the bob at the start of the second oscillation is about 29.15 meters.

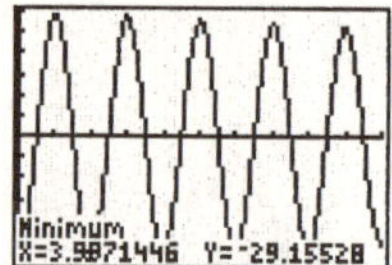

e. The displacement of the bob approaches zero, since $e^{-0.5t/70} \to 0$ as $t \to \infty$.

31. a. Damped motion with a bob of mass 15 kg and a damping factor of 0.9 kg/sec.

b. 15 meters leftward

c.

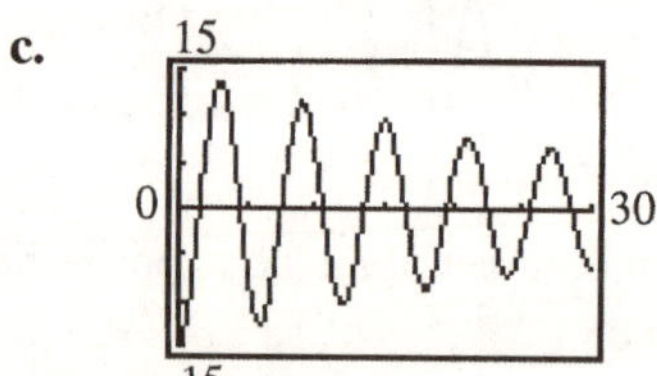

d. The displacement of the bob at the start of the second oscillation is about 12.53 meters.

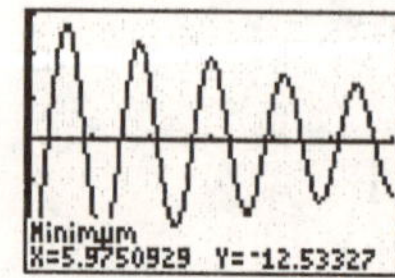

e. The displacement of the bob approaches zero, since $e^{-0.9t/30} \to 0$ as $t \to \infty$.

32. a. Damped motion with a bob of mass 25 kg and a damping factor of 0.8 kg/sec.

b. 10 meters leftward

c.

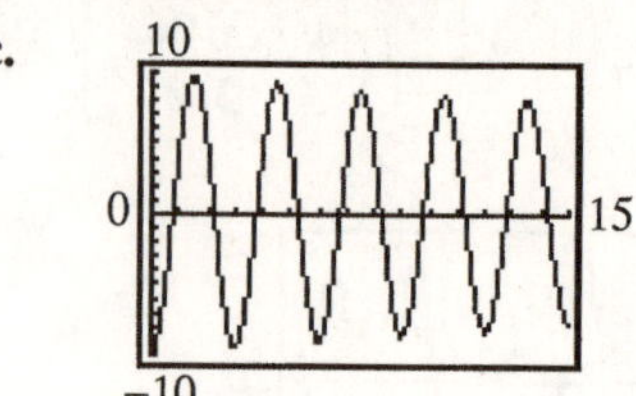

d. The displacement of the bob at the start of the second oscillation is 9.53 meters.

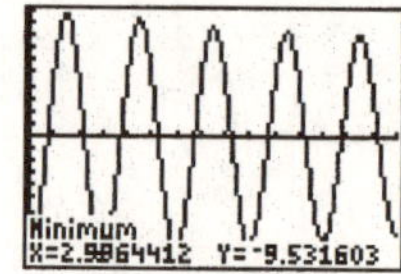

e. The displacement of the bob approaches zero, since $e^{-0.8t/50} \to 0$ as $t \to \infty$.

33. $f(x) = x + \cos x$

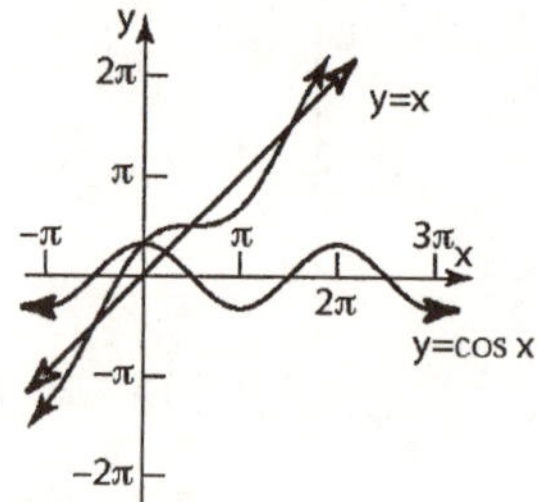

34. $f(x) = x + \cos(2x)$

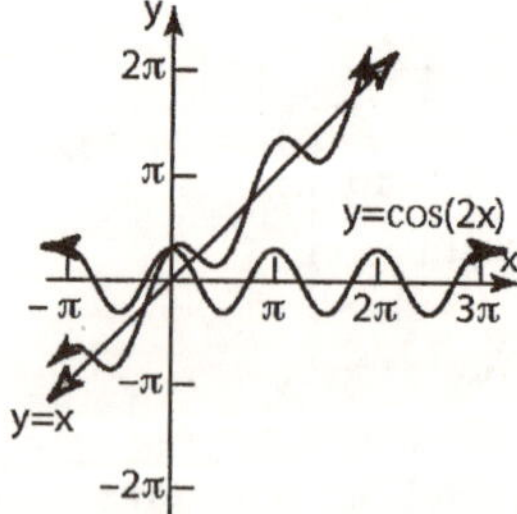

35. $f(x) = x - \sin x$

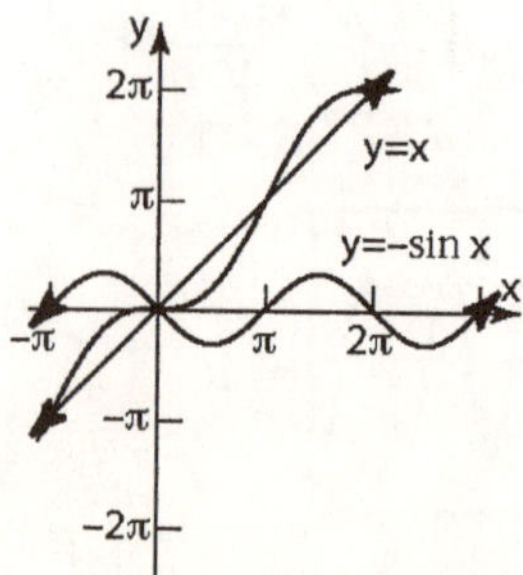

36. $f(x) = x - \cos x$

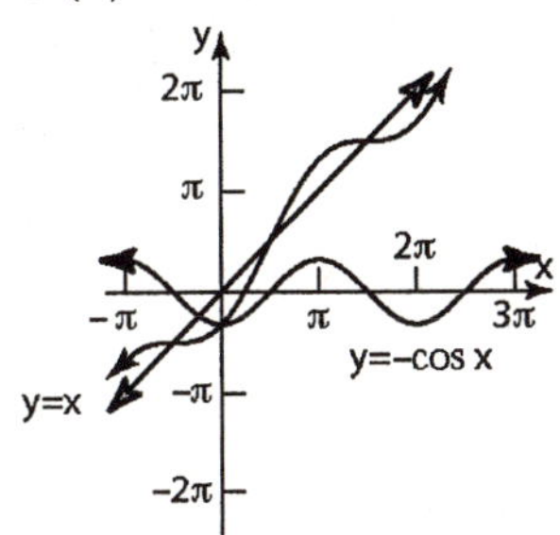

37. $f(x) = \sin x + \cos x$

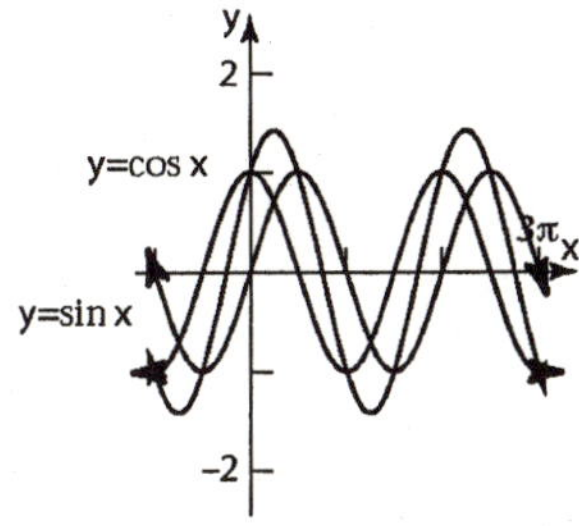

38. $f(x) = \sin(2x) + \cos x$

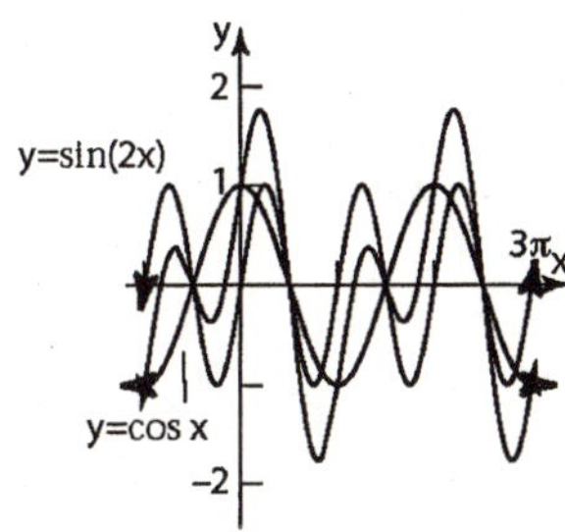

39. $f(x) = \sin x + \sin(2x)$

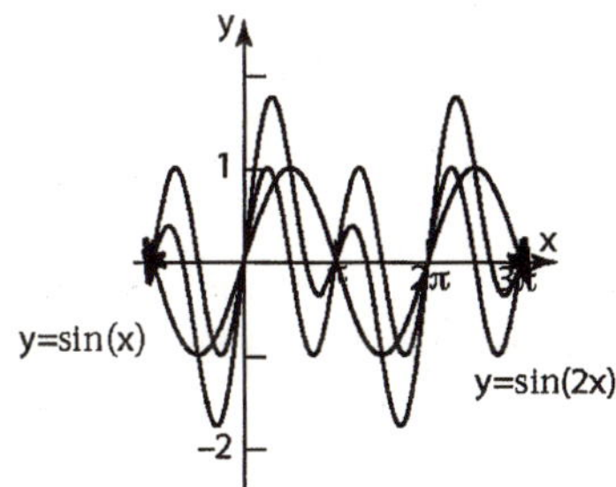

40. $f(x) = \cos(2x) + \cos x$

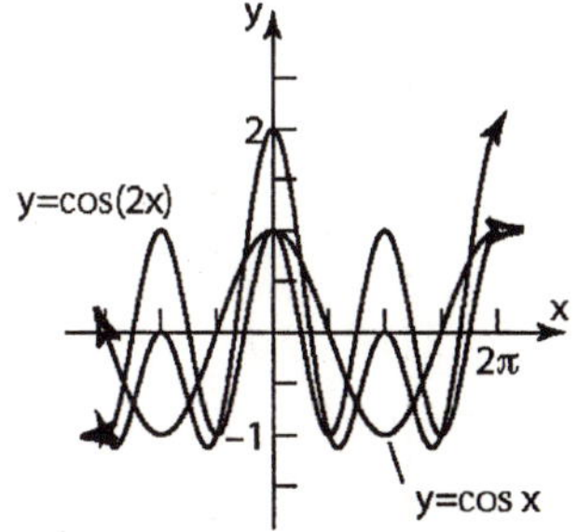

41. a.

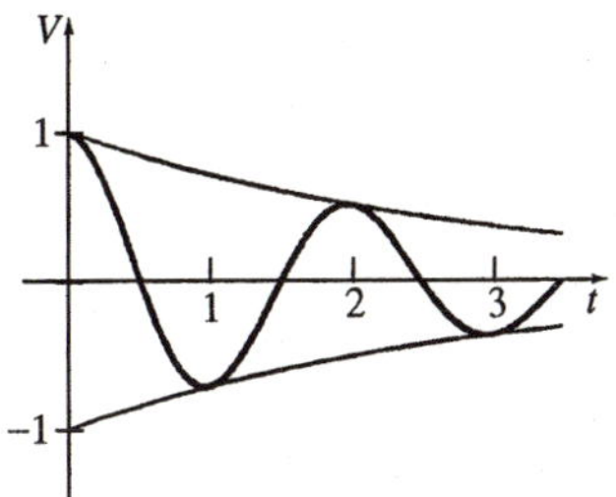

b. On the interval $0 \le t \le 3$, the graph of V touches the graph of $y = e^{-t/3}$ when $t = 0, 2$. The graph of V touches the graph of $y = -e^{-t/3}$ when $t = 1, 3$.

c. We need to solve the inequality $-0.4 < e^{-t/3}\cos(\pi t) < 0.4$ on the interval $0 \le t \le 3$. To do so, we consider the graphs of $y = -0.4$, $y = e^{-t/3}\cos(\pi t)$, and $y = 0.4$. On the interval $0 \le t \le 3$, we can use the INTERSECT feature on a calculator to determine that $y = e^{-t/3}\cos(\pi t)$ intersects $y = 0.4$ when $t \approx 0.35$, $t \approx 1.75$, and $t \approx 2.19$, $y = e^{-t/3}\cos(\pi t)$ intersects $y = -0.4$ when $t \approx 0.67$ and $t \approx 1.29$ and the graph shows that $-0.4 < e^{-t/3}\cos(\pi t) < 0.4$ when $t = 3$. Therefore, the voltage V is between -0.4 and 0.4 on the intervals $0.35 < t < 0.67$, $1.29 < t < 1.75$, and $2.19 < t \le 3$.

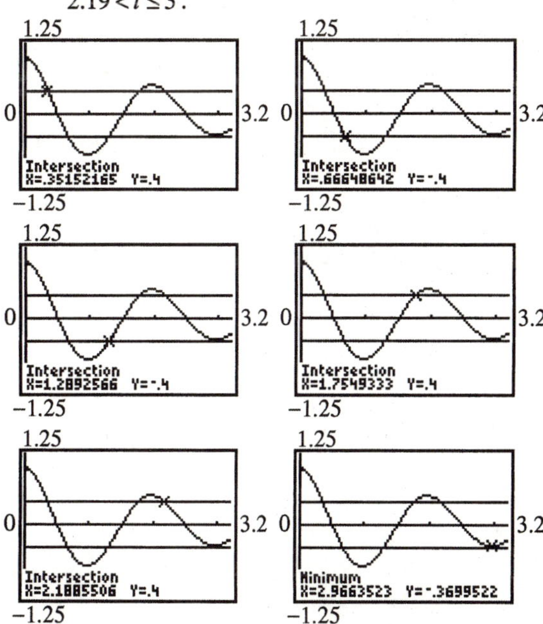

42. **a.** Let $Y_1 = \frac{1}{2}\sin(2\pi) + \frac{1}{4}\sin(4\pi x)$.

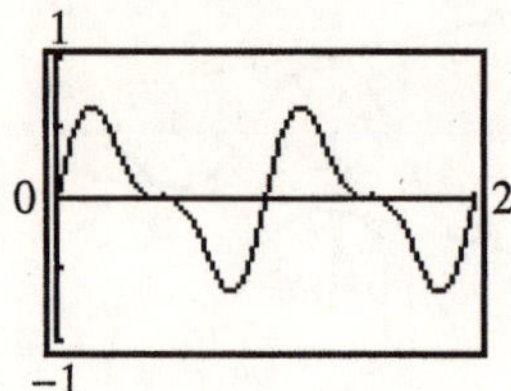

b. Let $Y_1 = \frac{1}{2}\sin(2\pi) + \frac{1}{4}\sin(4\pi x) + \frac{1}{8}\sin(8\pi x)$.

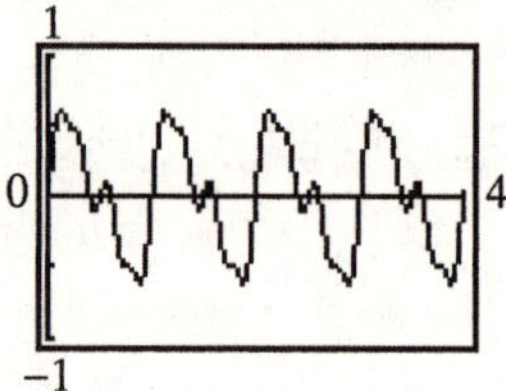

c. Let $Y_1 = \frac{1}{2}\sin(2\pi) + \frac{1}{4}\sin(4\pi x)$
$+ \frac{1}{8}\sin(8\pi x) + \frac{1}{16}\sin(16\pi x)$

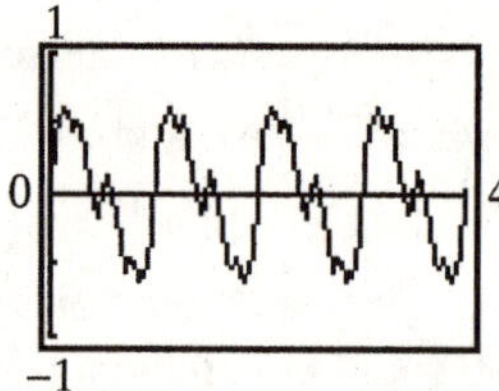

d. $f(x) = \frac{1}{2}\sin(2\pi x) + \frac{1}{4}\sin(4\pi x) + \frac{1}{8}\sin(8\pi x)$
$+ \frac{1}{16}\sin(16\pi x) + \frac{1}{32}\sin(32\pi x)$

43. Let $Y_1 = \sin(2\pi(852)x) + \sin(2\pi(1209)x)$.

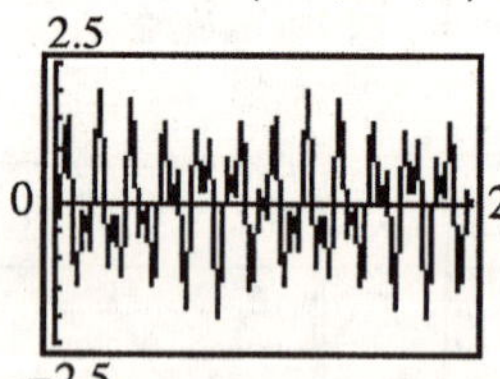

44. The sound emitted by touching * is
$y = \sin(2\pi(941)t) + \sin(2\pi(1209)t)$.
Let $Y_1 = \sin(2\pi(941)x) + \sin(2\pi(1209)x)$.

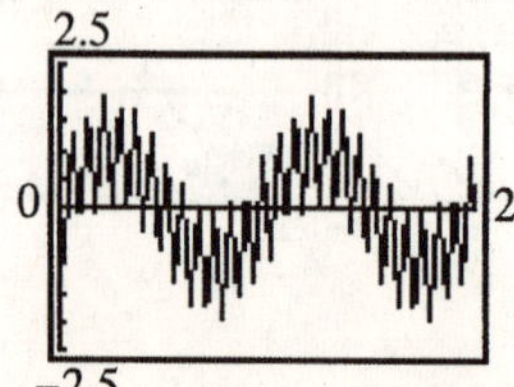

45 – 46. CBL Experiments

47. Let $Y_1 = \frac{\sin x}{x}$.

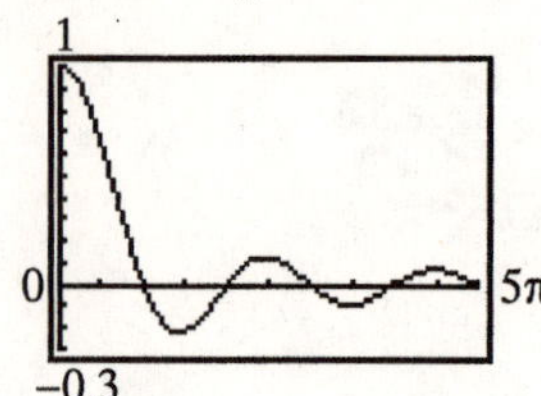

As x approaches 0, $\frac{\sin x}{x}$ approaches 1.

48. $y = x\sin x$

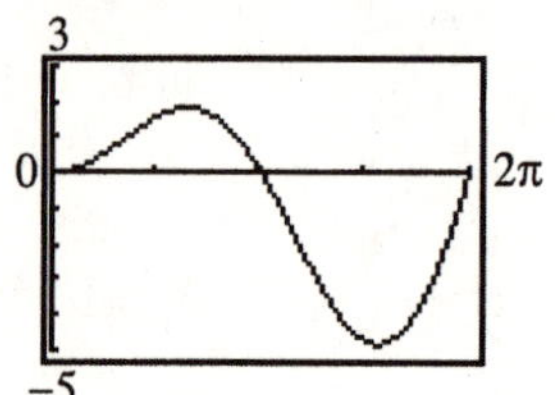

$y = x^2 \sin x$

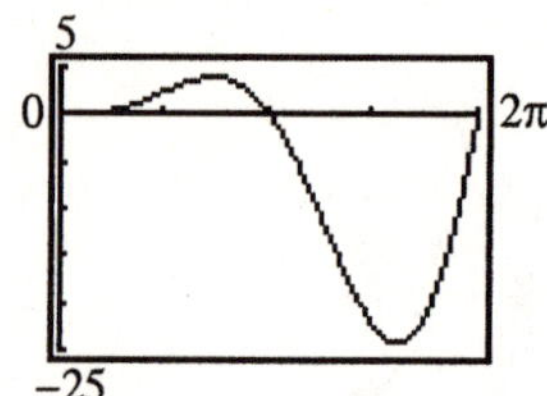

$y = x^3 \sin x$

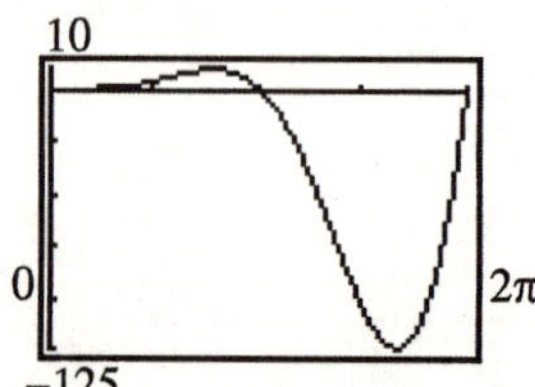

Possible observations: The graph lies between the bounding curves $y = \pm x,\ y = \pm x^2,\ y = \pm x^3$, respectively, touching them at odd multiples of $\frac{\pi}{2}$. The x-intercepts of each graph are the multiples of π.

49. $y = \left(\frac{1}{x}\right)\sin x$

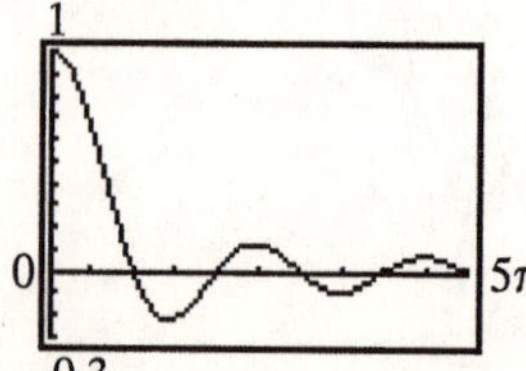

$y=\left(\frac{1}{x^2}\right)\sin x$

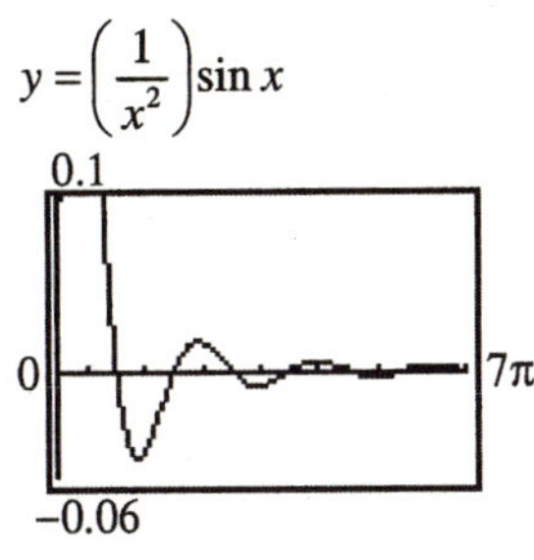

$y=\left(\frac{1}{x^3}\right)\sin x$

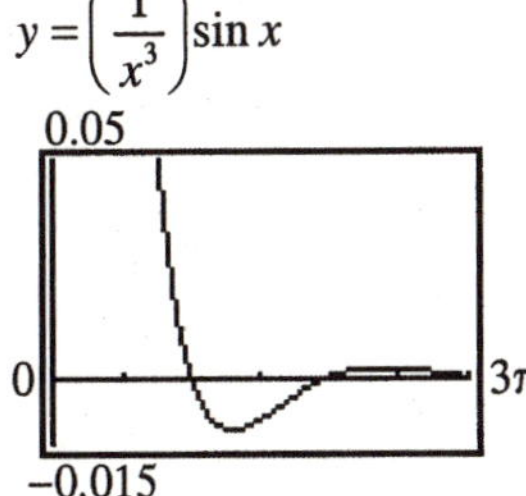

Possible observation: As x gets larger, the graph of $y=\left(\frac{1}{x^n}\right)\sin x$ gets closer to $y=0$.

50. Answers will vary.

Chapter 7 Review

1. opposite = 4; adjacent = 3; hypotenuse = ?

$(\text{hypotenuse})^2 = 4^2+3^2 = 25$

$\text{hypotenuse} = \sqrt{25} = 5$

$\sin\theta = \frac{\text{opp}}{\text{hyp}} = \frac{4}{5}$ $\quad \csc\theta = \frac{\text{hyp}}{\text{opp}} = \frac{5}{4}$

$\cos\theta = \frac{\text{adj}}{\text{hyp}} = \frac{3}{5}$ $\quad \sec\theta = \frac{\text{hyp}}{\text{adj}} = \frac{5}{3}$

$\tan\theta = \frac{\text{opp}}{\text{adj}} = \frac{4}{3}$ $\quad \cot\theta = \frac{\text{adj}}{\text{opp}} = \frac{3}{4}$

2. opposite = 3; adjacent = 5; hypotenuse = ?

$(\text{hypotenuse})^2 = 3^2+5^2 = 34$

$\text{hypotenuse} = \sqrt{34}$

$\sin\theta = \frac{\text{opp}}{\text{hyp}} = \frac{3}{\sqrt{34}} = \frac{3}{\sqrt{34}}\cdot\frac{\sqrt{34}}{\sqrt{34}} = \frac{3\sqrt{34}}{34}$

$\cos\theta = \frac{\text{adj}}{\text{hyp}} = \frac{5}{\sqrt{34}} = \frac{5}{\sqrt{34}}\cdot\frac{\sqrt{34}}{\sqrt{34}} = \frac{5\sqrt{34}}{34}$

$\tan\theta = \frac{\text{opp}}{\text{adj}} = \frac{3}{5};$ $\quad \cot\theta = \frac{\text{adj}}{\text{opp}} = \frac{5}{3}$

$\csc\theta = \frac{\text{hyp}}{\text{opp}} = \frac{\sqrt{34}}{3};$ $\quad \sec\theta = \frac{\text{hyp}}{\text{adj}} = \frac{\sqrt{34}}{5}$

3. adjacent = 2; hypotenuse = 4; opposite = ?

$(\text{opposite})^2 + 2^2 = 4^2$

$(\text{opposite})^2 = 16-4 = 12$

$\text{opposite} = \sqrt{12} = 2\sqrt{3}$

$\sin\theta = \frac{\text{opp}}{\text{hyp}} = \frac{2\sqrt{3}}{4} = \frac{\sqrt{3}}{2}$

$\cos\theta = \frac{\text{adj}}{\text{hyp}} = \frac{2}{4} = \frac{1}{2}$

$\tan\theta = \frac{\text{opp}}{\text{adj}} = \frac{2\sqrt{3}}{2} = \sqrt{3}$

$\csc\theta = \frac{\text{hyp}}{\text{opp}} = \frac{4}{2\sqrt{3}} = \frac{4}{2\sqrt{3}}\cdot\frac{\sqrt{3}}{\sqrt{3}} = \frac{2\sqrt{3}}{3}$

$\sec\theta = \frac{\text{hyp}}{\text{adj}} = \frac{4}{2} = 2$

$\cot\theta = \frac{\text{adj}}{\text{opp}} = \frac{2}{2\sqrt{3}} = \frac{2}{2\sqrt{3}}\cdot\frac{\sqrt{3}}{\sqrt{3}} = \frac{\sqrt{3}}{3}$

4. opposite = $\sqrt{2}$; hypotenuse = 2; adjacent = ?

$\left(\sqrt{2}\right)^2 + (\text{adjacent})^2 = 2^2$

$(\text{adjacent})^2 = 4-2 = 2$

$\text{opposite} = \sqrt{2}$

$\sin\theta = \frac{\text{opp}}{\text{hyp}} = \frac{\sqrt{2}}{2}$

$\cos\theta = \frac{\text{adj}}{\text{hyp}} = \frac{\sqrt{2}}{2}$

$\tan\theta = \frac{\text{opp}}{\text{adj}} = \frac{\sqrt{2}}{\sqrt{2}} = 1$

$\csc\theta = \frac{\text{hyp}}{\text{opp}} = \frac{2}{\sqrt{2}} = \frac{2}{\sqrt{2}}\cdot\frac{\sqrt{2}}{\sqrt{2}} = \frac{2\sqrt{2}}{2} = \sqrt{2}$

$\sec\theta = \frac{\text{hyp}}{\text{adj}} = \frac{2}{\sqrt{2}} = \frac{2}{\sqrt{2}}\cdot\frac{\sqrt{2}}{\sqrt{2}} = \frac{2\sqrt{2}}{2} = \sqrt{2}$

$\cot\theta = \frac{\text{adj}}{\text{opp}} = \frac{\sqrt{2}}{\sqrt{2}} = 1$

5. $\cos 62° - \sin 28° = \cos 62° - \cos(90° - 28°)$

$= \cos 62° - \cos 62°$

$= 0$

6. $\tan 15° - \cot 75° = \tan 15° - \tan(90° - 75°)$

$= \tan 15° - \tan 15°$

-0

7. $\dfrac{\sec 55°}{\csc 35°} = \dfrac{\csc(90° - 55°)}{\csc 35°} = \dfrac{\csc 35°}{\csc 35°} = 1$

8. $\dfrac{\tan 40°}{\cot 50°} = \dfrac{\tan 40°}{\tan(90° - 50°)} = \dfrac{\tan 40°}{\tan 40°} = 1$

9. $\cos^2 40° + \cos^2 50° = \sin^2(90° - 40°) + \cos^2 50°$
$= \sin^2 50° + \cos^2 50°$
$= 1$

10. $\tan^2 40° - \csc^2 50° = \left(-1 + \sec^2 40°\right) - \csc^2 50°$
$= -1 + \sec^2 40° - \sec^2(90° - 50°)$
$= -1 + \sec^2 40° - \sec^2 40°$
$= -1$

11. $c = 10,\ \beta = 20°$

$\sin\beta = \dfrac{b}{c}$

$\sin 20° = \dfrac{b}{10}$

$b = 10\sin 20° \approx 3.42$

$\cos\beta = \dfrac{a}{c}$

$\cos 20° = \dfrac{a}{10}$

$a = 10\cos 20° \approx 9.40$

$\alpha = 90° - \beta = 90° - 20° = 70°$

12. $a = 5,\ \alpha = 35°$

$\sin\alpha = \dfrac{a}{c}$

$\sin 35° = \dfrac{5}{c}$

$c = \dfrac{5}{\sin 35°} \approx 8.72$

$\tan\alpha = \dfrac{a}{b}$

$\tan 35° = \dfrac{5}{b}$

$b = \dfrac{5}{\tan 35°} \approx 7.14$

$\beta = 90° - \alpha = 90° - 35° = 55°$

13. $b = 2,\ c = 5$

$c^2 = a^2 + b^2$

$a^2 = c^2 - b^2 = 5^2 - 2^2 = 25 - 4 = 21$

$a = \sqrt{21} \approx 4.58$

$\sin\beta = \dfrac{b}{c} = \dfrac{2}{5}$

$\beta = \sin^{-1}\left(\dfrac{2}{5}\right) \approx 23.6°$

$\alpha = 90° - \beta \approx 90° - 23.6° = 66.4°$

14. $b = 1,\ a = 3$

$c^2 = a^2 + b^2$

$c^2 = 3^2 + 1^2 = 9 + 1 = 10$

$c = \sqrt{10} \approx 3.16$

$\tan\beta = \dfrac{b}{a} = \dfrac{1}{3}$

$\beta = \tan^{-1}\left(\dfrac{1}{3}\right) \approx 18.4°$

$\alpha = 90° - \beta \approx 90° - 18.4° = 71.6°$

15. $\alpha = 50°,\ \beta = 30°,\ a = 1$

$\gamma = 180° - \alpha - \beta = 180° - 50° - 30° = 100°$

$\dfrac{\sin\alpha}{a} = \dfrac{\sin\beta}{b}$

$\dfrac{\sin 50°}{1} = \dfrac{\sin 30°}{b}$

$b = \dfrac{1\sin 30°}{\sin 50°} \approx 0.65$

$\dfrac{\sin\gamma}{c} = \dfrac{\sin\alpha}{a}$

$\dfrac{\sin 100°}{c} = \dfrac{\sin 50°}{1}$

$c = \dfrac{1\sin 100°}{\sin 50°} \approx 1.29$

16. $\alpha = 10°,\ \gamma = 40°,\ c = 2$

$\dfrac{\sin\gamma}{c} = \dfrac{\sin\alpha}{a}$

$\dfrac{\sin 40°}{2} = \dfrac{\sin 10°}{a}$

$a = \dfrac{2\sin 10°}{\sin 40°} \approx 0.54$

$$\frac{\sin\gamma}{c} = \frac{\sin\beta}{b}$$
$$\frac{\sin 40^\circ}{2} = \frac{\sin 130^\circ}{b}$$
$$b = \frac{2\sin 130^\circ}{\sin 40^\circ} \approx 2.38$$
$$\beta = 180^\circ - \alpha - \gamma = 180^\circ - 10^\circ - 40^\circ = 130^\circ$$

17. $\alpha = 100^\circ,\ c = 2,\ a = 5$
$$\frac{\sin\gamma}{c} = \frac{\sin\alpha}{a}$$
$$\frac{\sin\gamma}{2} = \frac{\sin 100^\circ}{5}$$
$$\sin\gamma = \frac{2\sin 100^\circ}{5} \approx 0.3939$$
$$\gamma = \sin^{-1}\left(\frac{2\sin 100^\circ}{5}\right)$$
$\gamma \approx 23.2^\circ$ or $\gamma \approx 156.8^\circ$

The value 156.8° is discarded because $\alpha + \gamma > 180^\circ$. Thus, $\alpha \approx 23.2^\circ$.
$$\beta = 180^\circ - \alpha - \gamma \approx 180^\circ - 100^\circ - 23.2^\circ = 56.8^\circ$$
$$\frac{\sin\beta}{b} = \frac{\sin\alpha}{a}$$
$$\frac{\sin 56.8^\circ}{b} = \frac{\sin 100^\circ}{5}$$
$$b = \frac{5\sin 56.8^\circ}{\sin 100^\circ} \approx 4.25$$

18. $a = 2,\ c = 5,\ \alpha = 60^\circ$
$$\frac{\sin\gamma}{c} = \frac{\sin\alpha}{a}$$
$$\frac{\sin\gamma}{5} = \frac{\sin 60^\circ}{2}$$
$$\sin\gamma = \frac{5\sin 60^\circ}{2} \approx 2.1651$$
No angle γ exists for which $\sin\gamma > 1$. Therefore, there is not triangle with the given measurements.

19. $a = 3,\ c = 1,\ \gamma = 110^\circ$
$$\frac{\sin\gamma}{c} = \frac{\sin\alpha}{a}$$
$$\frac{\sin 110^\circ}{1} = \frac{\sin\alpha}{3}$$
$$\sin\alpha = \frac{3\sin 110^\circ}{1} \approx 2.8191$$
No angle α exists for which $\sin\alpha > 1$. Thus, there is no triangle with the given measurements.

20. $a = 3,\ c = 1,\ \gamma = 20^\circ$
$$\frac{\sin\gamma}{c} = \frac{\sin\alpha}{a}$$
$$\frac{\sin 20^\circ}{1} = \frac{\sin\alpha}{3}$$
$$\sin\alpha = \frac{3\sin 20^\circ}{1} \approx 1.0260$$
There is no angle α for which $\sin\alpha > 1$. Thus, there is no triangle with the given measurements.

21. $a = 3,\ c = 1,\ \beta = 100^\circ$
$$b^2 = a^2 + c^2 - 2ac\cos\beta$$
$$= 3^2 + 1^2 - 2\cdot 3\cdot 1\cos 100^\circ$$
$$= 10 - 6\cos 100^\circ$$
$$b = \sqrt{10 - 6\cos 100^\circ} \approx 3.32$$
$$a^2 = b^2 + c^2 - 2bc\cos\alpha$$
$$\cos\alpha = \frac{b^2 + c^2 - a^2}{2bc} = \frac{3.32^2 + 1^2 - 3^2}{2(3.32)(1)} = \frac{3.0224}{6.64}$$
$$\alpha = \cos^{-1}\left(\frac{3.0224}{6.64}\right) \approx 62.9^\circ$$
$$\gamma = 180^\circ - \alpha - \beta \approx 180^\circ - 62.9^\circ - 100^\circ = 17.1^\circ$$

22. $a = 3,\ b = 5,\ \beta = 80^\circ$
$$\frac{\sin\alpha}{a} = \frac{\sin\beta}{b}$$
$$\frac{\sin\alpha}{3} = \frac{\sin 80^\circ}{5}$$
$$\sin\alpha = \frac{3\sin 80^\circ}{5} \approx 0.5909$$
$$\alpha = \sin^{-1}\left(\frac{3\sin 80^\circ}{5}\right)$$
$\alpha \approx 36.2^\circ$ or $\alpha \approx 143.8^\circ$

The value 143.8° is discarded because $\alpha + \beta > 180^\circ$. Thus, $\alpha \approx 36.2^\circ$.
$$\gamma = 180^\circ - \alpha - \beta \approx 180^\circ - 36.2^\circ - 80^\circ = 63.8^\circ$$
$$\frac{\sin\gamma}{c} = \frac{\sin\beta}{b}$$
$$\frac{\sin 63.8^\circ}{c} = \frac{\sin 80^\circ}{5}$$
$$c = \frac{5\sin 63.8^\circ}{\sin 80^\circ} \approx 4.56$$

23. $a=2,\ b=3,\ c=1$

$$a^2=b^2+c^2-2bc\cos\alpha$$

$$\cos\alpha=\frac{b^2+c^2-a^2}{2bc}=\frac{3^2+1^2-2^2}{2(3)(1)}=1$$

$$\alpha=\cos^{-1}1=0°$$

No triangle exists with an angle of 0°.

24. $a=10,\ b=7,\ c=8$

$$a^2=b^2+c^2-2bc\cos\alpha$$

$$\cos\alpha=\frac{b^2+c^2-a^2}{2bc}=\frac{7^2+8^2-10^2}{2(7)(8)}=\frac{13}{112}$$

$$\alpha=\cos^{-1}\left(\frac{13}{112}\right)\approx 83.3°$$

$$b^2=a^2+c^2-2ac\cos\beta$$

$$\cos\beta=\frac{a^2+c^2-b^2}{2ac}=\frac{10^2+8^2-7^2}{2(10)(8)}=\frac{115}{160}$$

$$\beta=\cos^{-1}\left(\frac{115}{160}\right)\approx 44.0°$$

$$\gamma=180°-\alpha-\beta\approx 180°-83.3°-44.0°\approx 52.7°$$

25. $a=1,\ b=3,\ \gamma=40°$

$$c^2=a^2+b^2-2ab\cos\gamma$$
$$=1^2+3^2-2\cdot1\cdot3\cos 40°$$
$$=10-6\cos 40°$$
$$c=\sqrt{10-6\cos 40°}\approx 2.32$$

$$a^2=b^2+c^2-2bc\cos\alpha$$

$$\cos\alpha=\frac{b^2+c^2-a^2}{2bc}=\frac{3^2+2.32^2-1^2}{2(3)(2.32)}=\frac{13.3824}{13.92}$$

$$\alpha=\cos^{-1}\left(\frac{13.3824}{13.92}\right)\approx 16.0°$$

$$\beta=180°-\alpha-\gamma\approx 180°-16.0°-40°=124.0°$$

26. $a=4,\ b=1,\ \gamma=100°$

$$c^2=a^2+b^2-2ab\cos\gamma$$
$$c^2=4^2+1^2-2\cdot4\cdot1\cos 100°=17-8\cos 100°$$
$$c=\sqrt{17-8\cos 100°}\approx 4.29$$

$$a^2=b^2+c^2-2bc\cos\alpha$$

$$\cos\alpha=\frac{b^2+c^2-a^2}{2bc}=\frac{1^2+4.29^2-4^2}{2(1)(4.29)}=\frac{3.4041}{8.58}$$

$$\alpha=\cos^{-1}\left(\frac{3.4041}{8.58}\right)\approx 66.6°$$

$$\beta=180°-\alpha-\gamma\approx 180°-66.6°-100°=13.4°$$

27. $a=5,\ b=3,\ \alpha=80°$

$$\frac{\sin\beta}{b}=\frac{\sin\alpha}{a}$$
$$\frac{\sin\beta}{3}=\frac{\sin 80°}{5}$$
$$\sin\beta=\frac{3\sin 80°}{5}\approx 0.5909$$
$$\beta=\sin^{-1}\left(\frac{3\sin 80°}{5}\right)$$
$$\beta\approx 36.2° \text{ or } \beta\approx 143.8°$$

The value 143.8° is discarded because $\alpha+\beta>180°$. Thus, $\beta\approx 36.2°$.

$$\gamma=180°-\alpha-\beta\approx 180°-80°-36.2°=63.8°$$

$$\frac{\sin\gamma}{c}=\frac{\sin\alpha}{a}$$
$$\frac{\sin 63.8°}{c}=\frac{\sin 80°}{5}$$
$$c=\frac{5\sin 63.8°}{\sin 80°}\approx 4.56$$

28. $a=2,\ b=3,\ \alpha=20°$

$$\frac{\sin\beta}{b}=\frac{\sin\alpha}{a}$$
$$\frac{\sin\beta}{3}=\frac{\sin 20°}{2}$$
$$\sin\beta=\frac{3\sin 20°}{2}\approx 0.5130$$
$$\beta=\sin^{-1}\left(\frac{3\sin 20°}{2}\right)$$
$$\beta_1=30.9° \text{ or } \beta_2=149.1°$$

For both values, $\alpha+\beta<180°$. Therefore, there are two triangles.

$$\gamma_1=180°-\alpha-\beta_1\approx 180°-20°-30.9°\approx 129.1°$$

$$\frac{\sin\alpha}{a} = \frac{\sin\gamma_1}{c_1}$$

$$\frac{\sin 20^\circ}{2} = \frac{\sin 129.1^\circ}{c_1}$$

$$c_1 = \frac{2\sin 129.1^\circ}{\sin 20^\circ} \approx 4.54$$

$$\gamma_2 = 180^\circ - \alpha - \beta_2 \approx 180^\circ - 20^\circ - 149.1^\circ \approx 10.9^\circ$$

$$\frac{\sin\alpha}{a} = \frac{\sin\gamma_2}{c_2}$$

$$\frac{\sin 20^\circ}{2} = \frac{\sin 10.9^\circ}{c_2}$$

$$c_2 = \frac{2\sin 10.9^\circ}{\sin 20^\circ} \approx 1.11$$

Two triangles: $\beta_1 \approx 30.9^\circ, \gamma_1 \approx 129.1^\circ, c_1 \approx 4.54$
or $\beta_2 \approx 149.1^\circ, \gamma_2 \approx 10.9^\circ, c_2 \approx 1.11$

29. $a = 1,\ b = \frac{1}{2},\ c = \frac{4}{3}$

$$a^2 = b^2 + c^2 - 2bc\cos\alpha$$

$$\cos\alpha = \frac{b^2+c^2-a^2}{2bc} = \frac{\left(\frac{1}{2}\right)^2 + \left(\frac{4}{3}\right)^2 - 1^2}{2\left(\frac{1}{2}\right)\left(\frac{4}{3}\right)} = \frac{27}{37}$$

$$\alpha = \cos^{-1}\left(\frac{27}{37}\right) \approx 39.6^\circ$$

$$b^2 = a^2 + c^2 - 2ac\cos\beta$$

$$\cos\beta = \frac{a^2+c^2-b^2}{2ac} = \frac{1^2 + \left(\frac{4}{3}\right)^2 - \left(\frac{1}{2}\right)^2}{2(1)\left(\frac{4}{3}\right)} = \frac{91}{96}$$

$$\beta = \cos^{-1}\left(\frac{91}{96}\right) \approx 18.6^\circ$$

$$\gamma = 180^\circ - \alpha - \beta \approx 180^\circ - 39.6^\circ - 18.6^\circ \approx 121.8^\circ$$

30. $a = 3,\ b = 2,\ c = 2$

$$a^2 = b^2 + c^2 - 2bc\cos\alpha$$

$$\cos\alpha = \frac{b^2+c^2-a^2}{2bc} = \frac{2^2+2^2-3^2}{2(2)(2)} = -\frac{1}{8}$$

$$\alpha = \cos^{-1}\left(-\frac{1}{8}\right) \approx 97.2^\circ$$

$$b^2 = a^2 + c^2 - 2ac\cos\beta$$

$$\cos\beta = \frac{a^2+c^2-b^2}{2ac} = \frac{3^2+2^2-2^2}{2(3)(2)} = \frac{9}{12}$$

$$\beta = \cos^{-1}\left(\frac{9}{12}\right) \approx 41.4^\circ$$

$$\gamma = 180^\circ - \alpha - \beta \approx 180^\circ - 97.2^\circ - 41.4^\circ \approx 41.4^\circ$$

31. $a = 3,\ \alpha = 10^\circ,\ b = 4$

$$\frac{\sin\beta}{b} = \frac{\sin\alpha}{a}$$

$$\frac{\sin\beta}{4} = \frac{\sin 10^\circ}{3}$$

$$\sin\beta = \frac{4\sin 10^\circ}{3} \approx 0.2315$$

$$\beta = \sin^{-1}\left(\frac{4\sin 10^\circ}{3}\right)$$

$\beta_1 \approx 13.4^\circ$ or $\beta_2 \approx 166.6^\circ$

For both values, $\alpha + \beta < 180^\circ$. Therefore, there are two triangles.

$$\gamma_1 = 180^\circ - \alpha - \beta_1 \approx 180^\circ - 10^\circ - 13.4^\circ \approx 156.6^\circ$$

$$\frac{\sin\alpha}{a} = \frac{\sin\gamma_1}{c_1}$$

$$\frac{\sin 10^\circ}{3} = \frac{\sin 156.6^\circ}{c_1}$$

$$c_1 = \frac{3\sin 156.6^\circ}{\sin 10^\circ} \approx 6.86$$

$$\gamma_2 = 180^\circ - \alpha - \beta_2 = 180^\circ - 10^\circ - 166.6^\circ \approx 3.4^\circ$$

$$\frac{\sin\alpha}{a} = \frac{\sin\gamma_2}{c_2}$$

$$\frac{\sin 10^\circ}{3} = \frac{\sin 3.4^\circ}{c_2}$$

$$c_2 = \frac{3\sin 3.4^\circ}{\sin 10^\circ} \approx 1.02$$

Two triangles: $\beta_1 \approx 13.4^\circ, \gamma_1 \approx 156.6^\circ, c_1 \approx 6.86$
or $\beta_2 \approx 166.6^\circ, \gamma_2 \approx 3.4^\circ, c_2 \approx 1.02$

32. $a=4,\ \alpha=20^\circ,\ \beta=100^\circ$

$\gamma=180^\circ-\alpha-\beta=180^\circ-20^\circ-100^\circ=60^\circ$

$$\frac{\sin\alpha}{a}=\frac{\sin\beta}{b}$$
$$\frac{\sin 20^\circ}{4}=\frac{\sin 100^\circ}{b}$$
$$b=\frac{4\sin 100^\circ}{\sin 20^\circ}\approx 11.52$$
$$\frac{\sin\gamma}{c}=\frac{\sin\alpha}{a}$$
$$\frac{\sin 60^\circ}{c}=\frac{\sin 20^\circ}{4}$$
$$c=\frac{4\sin 60^\circ}{\sin 20^\circ}\approx 10.13$$

33. $c=5,\ b=4,\ \alpha=70^\circ$

$a^2=b^2+c^2-2bc\cos\alpha$

$a^2=4^2+5^2-2\cdot 4\cdot 5\cos 70^\circ=41-40\cos 70^\circ$

$a=\sqrt{41-40\cos 70^\circ}\approx 5.23$

$c^2=a^2+b^2-2ab\cos\gamma$

$$\cos\gamma=\frac{a^2+b^2-c^2}{2ab}$$
$$=\frac{5.23^2+4^2-5^2}{2(5.23)(4)}=\frac{18.3529}{41.48}$$
$$\gamma=\cos^{-1}\left(\frac{18.3529}{41.48}\right)\approx 64.0^\circ$$

$\beta=180^\circ-\alpha-\gamma\approx 180^\circ-70^\circ-64.0^\circ\approx 46.0^\circ$

34. $a=1,\ b=2,\ \gamma=60^\circ$

$c^2=a^2+b^2-2ab\cos\gamma$

$=1^2+2^2-2\cdot 1\cdot 2\cos 60^\circ=3$

$c=\sqrt{3}\approx 1.73$

$a^2=b^2+c^2-2bc\cos\alpha$

$$\cos\alpha=\frac{b^2+c^2-a^2}{2bc}=\frac{2^2+1.73^2-1^2}{2(2)(1.73)}=\frac{5.9929}{6.92}$$
$$\alpha=\cos^{-1}\left(\frac{5.9929}{6.92}\right)\approx 30^\circ$$

$\beta=180^\circ-\alpha-\gamma\approx 180^\circ-30^\circ-60^\circ=90^\circ$

35. $a=2,\ b=3,\ \gamma=40^\circ$

$$A=\frac{1}{2}ab\sin\gamma=\frac{1}{2}(2)(3)\sin 40^\circ\approx 1.93$$

36. $b=5,\ c=5,\ \alpha=20^\circ$

$$A=\frac{1}{2}bc\sin\alpha=\frac{1}{2}(5)(5)\sin 20^\circ\approx 4.28$$

37. $b=4,\ c=10,\ \alpha=70^\circ$

$$A=\frac{1}{2}bc\sin\alpha=\frac{1}{2}(4)(10)\sin 70^\circ\approx 18.79$$

38. $a=2,\ b=1,\ \gamma=100^\circ$

$$A=\frac{1}{2}ab\sin\gamma=\frac{1}{2}(2)(1)\sin 100^\circ\approx 0.98$$

39. $a=4,\ b=3,\ c=5$

$$s=\frac{1}{2}(a+b+c)=\frac{1}{2}(4+3+5)=6$$
$$A=\sqrt{s(s-a)(s-b)(s-c)}$$
$$=\sqrt{(6)(2)(3)(1)}=\sqrt{36}=6$$

40. $a=10,\ b=7,\ c=8$

$$s=\frac{1}{2}(a+b+c)=\frac{1}{2}(10+7+8)=12.5$$
$$A=\sqrt{s(s-a)(s-b)(s-c)}$$
$$=\sqrt{(12.5)(2.5)(5.5)(4.5)}$$
$$=\sqrt{773.4375}\approx 27.81$$

41. $a=4,\ b=2,\ c=5$

$$s=\frac{1}{2}(a+b+c)=\frac{1}{2}(4+2+5)=5.5$$
$$A=\sqrt{s(s-a)(s-b)(s-c)}$$
$$=\sqrt{(5.5)(1.5)(3.5)(0.5)}=\sqrt{14.4375}\approx 3.80$$

42. $a=3,\ b=2,\ c=2$

$$s=\frac{1}{2}(a+b+c)=\frac{1}{2}(3+2+2)=3.5$$
$$A=\sqrt{s(s-a)(s-b)(s-c)}$$
$$=\sqrt{(3.5)(0.5)(1.5)(1.5)}=\sqrt{3.9375}\approx 1.98$$

43. $\alpha=50^\circ, \beta=30^\circ, a=1$

$\gamma=180^\circ-\alpha-\beta=180^\circ-50^\circ-30^\circ=100^\circ$

$$A=\frac{a^2\sin\beta\sin\gamma}{2\sin\alpha}=\frac{1^2\sin 30^\circ\cdot\sin 100^\circ}{2\sin 50^\circ}\approx 0.32$$

44. $\alpha = 10°, \gamma = 40°, c = 3$
$\beta = 180° - \alpha - \gamma = 180° - 10° - 40° = 130°$
$$A = \frac{c^2 \sin\alpha \sin\beta}{2\sin\gamma} = \frac{3^2 \sin 10° \cdot \sin 130°}{2\sin 40°} \approx 0.93$$

45. Let θ = the inclination (grade) of the trail. The "rise" of the trail is $4100 - 5000 = -900$ feet (see figure).

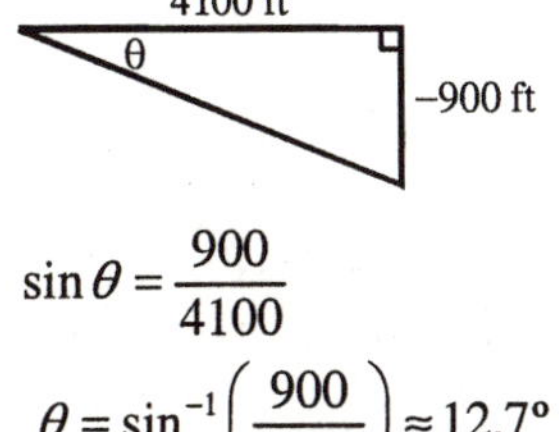

$$\sin\theta = \frac{900}{4100}$$
$$\theta = \sin^{-1}\left(\frac{900}{4100}\right) \approx 12.7°$$
The trail is inclined about $-12.7°$ from the hotel to the lake.

46. $c = 12$ feet, $a = 8$ feet. We need to find α and β (see figure).

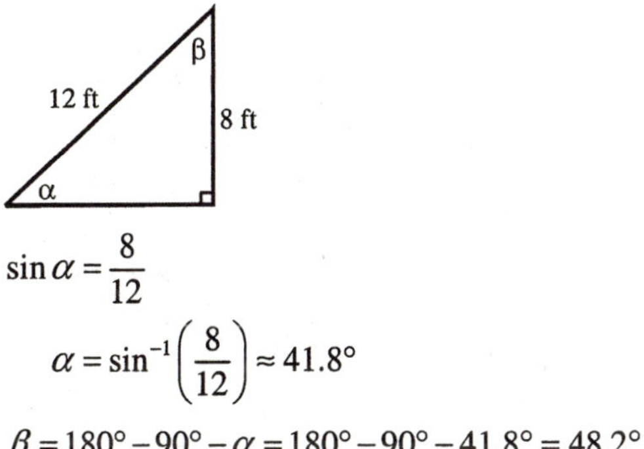

$$\sin\alpha = \frac{8}{12}$$
$$\alpha = \sin^{-1}\left(\frac{8}{12}\right) \approx 41.8°$$
$$\beta = 180° - 90° - \alpha = 180° - 90° - 41.8° = 48.2°$$

47. $\angle ABC = 180° - 20° = 160°$
Find the angle at city C:
$$\frac{\sin C}{100} = \frac{\sin 160°}{300}$$
$$\sin C = \frac{100\sin 160°}{300}$$
$$C = \sin^{-1}\left(\frac{100\sin 160°}{300}\right) \approx 6.55°$$
Find the angle at city A:
$A \approx 180° - 160° - 6.55° = 13.45°$
$$\frac{\sin 13.45°}{y} = \frac{\sin 160°}{300}$$
$$y = \frac{300\sin 13.45°}{\sin 160°} \approx 204.07 \text{ miles}$$
The distance from city B to city C is approximately 204.07 miles.

48. **a.** Note that 10 min $= \frac{1}{6}$ hr. The distance traveled prior to the discovery of the error is $420 \cdot \frac{1}{6} = 70$ miles.
Find the remaining distance to city B:
$$a^2 = 300^2 + 70^2 - 2(300)(70)\cos 5°$$
$$= 94{,}900 - 42{,}000\cos 5°$$
$$a = \sqrt{94{,}900 - 42{,}000\cos 5°} \approx 230.35 \text{ miles}$$
Find angle B: $\frac{\sin B}{70} = \frac{\sin 5°}{230.35}$
$$\sin B = \frac{70\sin 5°}{230.35} \approx 0.0265$$
$$B = \sin^{-1}\left(\frac{70\sin 5°}{230.35}\right) \approx 1.5°$$
Find angle C: $C = 180° - 5° - 1.5° = 173.5°$
Now, $180° - 173.5° = 6.5°$.
The pilot should turn through an angle of $6.5°$ to correct the course.

b. The time from city A to city B should be $\frac{300 \text{ mi}}{420 \text{ mi/hr}} = \frac{5}{7}$ hour. The time from city A to the point of the error is 10 minutes = $\frac{1}{6}$ hour. The remaining part of the trip must be completed in $\frac{5}{7} - \frac{1}{6} = \frac{23}{42}$ hour.
Calculate the rate:
$$r = \frac{230.35}{\frac{23}{42}} \approx 420.64 \text{ mi/hr}.$$

49. $\angle ACB = 12° + 30° = 42°$
$\angle ABC = 90° - 30° = 60°$
$\angle CAB = 90° - 12° = 78°$

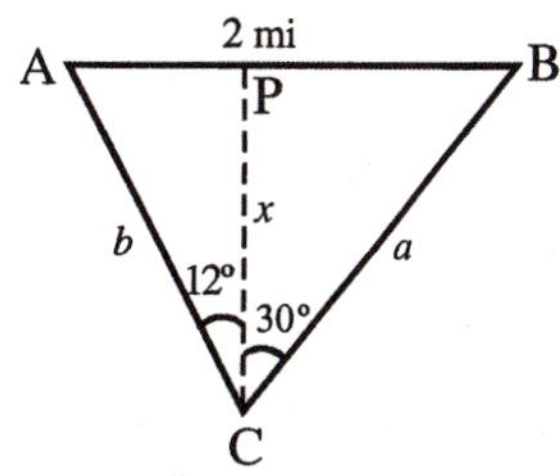

a. $$\frac{\sin 60°}{b} = \frac{\sin 42°}{2}$$
$$b = \frac{2\sin 60°}{\sin 42°} \approx 2.59 \text{ miles}$$

b. $\dfrac{\sin 78^\circ}{a}=\dfrac{\sin 42^\circ}{2}$

$a=\dfrac{2\sin 78^\circ}{\sin 42^\circ}\approx 2.92$ miles

c. $\cos 12^\circ=\dfrac{x}{2.59}$

$x=2.59\cos 12^\circ\approx 2.53$ miles

50. $\alpha=180^\circ-120^\circ=60^\circ$; $\beta=180^\circ-115^\circ=65^\circ$; $\gamma=180^\circ-60^\circ-65^\circ=55^\circ$

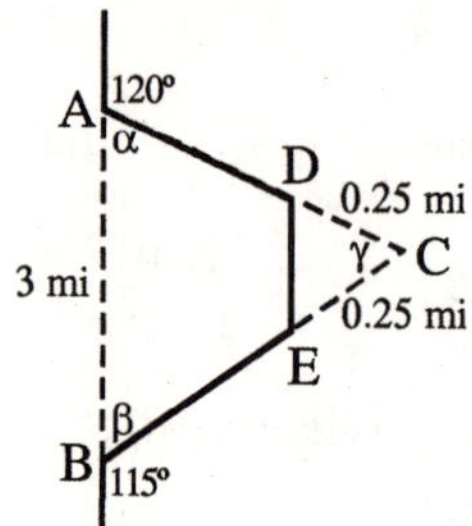

$\dfrac{\sin 60^\circ}{BC}=\dfrac{\sin 55^\circ}{3}$

$BC=\dfrac{3\sin 60^\circ}{\sin 55^\circ}\approx 3.17$ mi

$\dfrac{\sin 65^\circ}{AC}=\dfrac{\sin 55^\circ}{3}$

$AC=\dfrac{3\sin 65^\circ}{\sin 55^\circ}\approx 3.32$ mi

$BE=3.17-0.25=2.92$ mi

$AD=3.32-0.25=3.07$ mi

For the isosceles triangle,

$\angle CDE=\angle CED=\dfrac{180^\circ-55^\circ}{2}=62.5^\circ$

$\dfrac{\sin 55^\circ}{DE}=\dfrac{\sin 62.5^\circ}{0.25}$

$DE=\dfrac{0.25\sin 55^\circ}{\sin 62.5^\circ}\approx 0.23$ miles

The length of the highway is $2.92+3.07+0.23=6.22$ miles.

51. a. After 4 hours, the sailboat would have sailed $18(4)=72$ miles.

Find the third side of the triangle to determine the distance from the island:

ST 200 mi BWI 15° 72 mi β c

$a=72,\ b=200,\ \gamma=15^\circ$

$c^2=a^2+b^2-2ab\cos\gamma$

$c^2=72^2+200^2-2\cdot 72\cdot 200\cos 15^\circ$

$=45{,}184-28{,}800\cos 15^\circ$

$c\approx 131.8$ miles

The sailboat is about 131.8 miles from the island.

b. Find the measure of the angle opposite the 200 side:

$\cos\beta=\dfrac{a^2+c^2-b^2}{2ac}$

$\cos\beta=\dfrac{72^2+131.8^2-200^2}{2(72)(131.8)}=-\dfrac{17{,}444.76}{18{,}979.2}$

$\beta=\cos^{-1}\left(-\dfrac{17{,}444.76}{18{,}979.2}\right)\approx 156.8^\circ$

The sailboat should turn through an angle of $180^\circ-156.8^\circ=23.2^\circ$ to correct its course.

c. The original trip would have taken: $t=\dfrac{200}{18}\approx 11.11$ hours. The actual trip takes: $t=4+\dfrac{131.8}{18}\approx 11.32$ hours. The trip takes about 0.21 hour, or about 12.6 minutes longer.

52. Find the third side of the triangle using the Law of Cosines:

$a=50,\ b=60,\ \gamma=80^\circ$

$c^2=a^2+b^2-2ab\cos\gamma$

$=50^2+60^2-2\cdot 50\cdot 60\cos 80^\circ$

$=6100-6000\cos 80^\circ$

$c=\sqrt{6100-6000\cos 80^\circ}\approx 71.12$

The houses are approximately 71.12 feet apart.

53. Find the lengths of the two unknown sides of the middle triangle:

$x^2=100^2+125^2-2(100)(125)\cos 50^\circ$

$=25{,}625-25{,}000\cos 50^\circ$

$x=\sqrt{25{,}625-25{,}000\cos 50^\circ}\approx 97.75$ feet

$y^2=70^2+50^2-2(70)(50)\cos 100^\circ$

$=7400-7000\cos 100^\circ$

$y=\sqrt{7400-7000\cos 100^\circ}\approx 92.82$ feet

Find the areas of the three triangles:

$A_1=\dfrac{1}{2}(100)(125)\sin 50^\circ\approx 4787.78\text{ ft}^2$

$A_2 = \frac{1}{2}(50)(70)\sin 100° \approx 1723.41 \text{ ft}^2$

$s = \frac{1}{2}(50 + 97.75 + 92.82) = 120.285$

$A_3 = \sqrt{(120.285)(70.285)(22.535)(27.465)}$

$\approx 2287.47 \text{ ft}^2$

The approximate area of the lake is $4787.78 + 1723.41 + 2287.47 = 8798.67 \text{ ft}^2$.

54. Construct a diagonal. Find the area of the first triangle and the length of the diagonal:

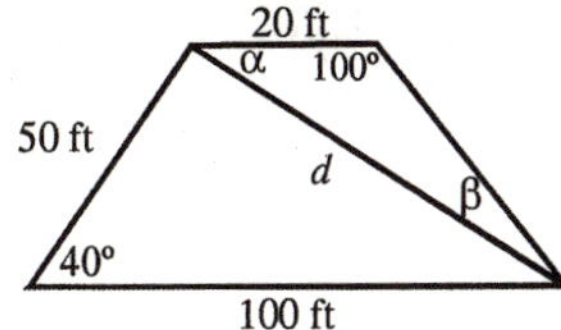

$A_1 = \frac{1}{2} \cdot 50 \cdot 100 \cdot \sin 40° \approx 1606.969 \text{ ft}^2$

$d^2 = 50^2 + 100^2 - 2 \cdot 50 \cdot 100 \cos 40°$

$= 12{,}500 - 10{,}000 \cos 40°$

$d = \sqrt{12{,}500 - 10{,}000 \cos 40°} \approx 69.566914$ feet

Use the Law of Sines to find β:

$\frac{\sin \beta}{20} = \frac{\sin 100°}{69.566914}$

$\sin \beta = \frac{20 \sin 100°}{69.566914}$

$\beta = \sin^{-1}\left(\frac{20 \sin 100°}{69.566914}\right) \approx 16.447°$

$\alpha \approx 180° - 100° - 16.447° = 63.553°$

Find the area of the second triangle:

$A_2 = \frac{1}{2} \cdot 20 \cdot 69.566914 \cdot \sin 63.553° \approx 622.865 \text{ ft}^2$

The cost of the parcel is approximately $100(1606.969 + 622.865) = \$222{,}983.40$.

55. To find the area of the segment, we subtract the area of the triangle from the area of the sector.

$A_{\text{Sector}} = \frac{1}{2}r^2\theta = \frac{1}{2} \cdot 6^2\left(50 \cdot \frac{\pi}{180}\right) \approx 15.708 \text{ in}^2$

$A_{\text{Triangle}} = \frac{1}{2}ab\sin\theta = \frac{1}{2} \cdot 6 \cdot 6 \sin 50° \approx 13.789 \text{ in}^2$

$A_{\text{Segment}} = 15.708 - 13.789 \approx 1.92 \text{ in}^2$

56. Find angle AMB and subtract from 80° to obtain θ.

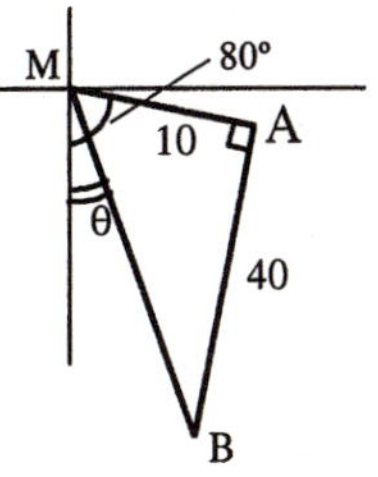

$\tan \angle AMB = \frac{40}{10} = 4$

$\angle AMB = \tan^{-1} 4 \approx 76.0°$

$\theta \approx 80° - 76.0° \approx 4.0°$

The bearing is S4.0°E.

57. Extend the tangent line until it meets a line extended through the centers of the pulleys. Label these extensions x and y. The distance between the points of tangency is z. Two similar triangles are formed.

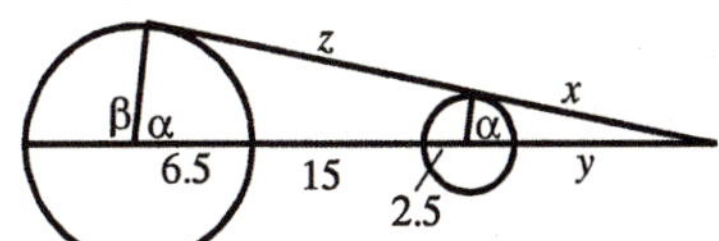

Thus, $\frac{24 + y}{y} = \frac{6.5}{2.5}$ where $24 + y$ is the hypotenuse of the larger triangle and y is the hypotenuse of the smaller triangle. Solve for y:

$6.5y = 2.5(24 + y)$

$6.5y = 60 + 2.5y$

$4y = 60$

$y = 15$

Use the Pythagorean Theorem to find x:

$x^2 + 2.5^2 = 15^2$

$x^2 = 225 - 6.25 = 218.75$

$x = \sqrt{218.75} \approx 14.79$

Use the Pythagorean Theorem to find z:

$(z + 14.79)^2 + 6.5^2 = (24 + 15)^2$

$(z + 14.79)^2 = 1521 - 42.25 = 1478.75$

$z + 14.79 = \sqrt{1478.75} \approx 38.45$

$z \approx 23.66$

Find α: $\cos\alpha = \frac{2.5}{15} \approx 0.1667$

$\alpha = \cos^{-1}\left(\frac{2.5}{15}\right) \approx 1.4033$ radians

$\beta \approx \pi - 1.4033 \approx 1.7383$ radians

The arc length on the top of the larger pulley is about: 6.5(1.7383) = 11.30 inches.

The arc length on the top of the smaller pulley is about: 2.5(1.4033) = 3.51 inches.

The distance between the points of tangency is about 23.66 inches.

The length of the belt is about:
2(11.30 + 3.51 + 23.66) = 76.94 inches.

58. Let x = the hypotenuse of the larger right triangle in the figure below. Then $24-x$ is the hypotenuse of the smaller triangle. The two triangles are similar.

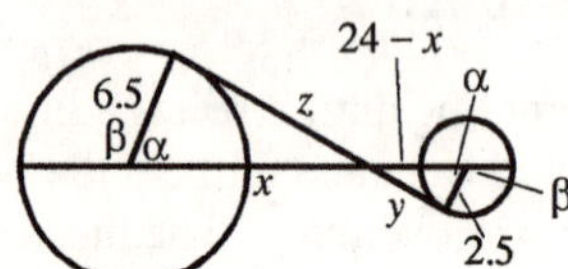

$$\frac{6.5}{2.5}=\frac{x}{24-x}$$
$$2.5x=6.5(24-x)$$
$$2.5x=156-6.5x$$
$$9x=156$$
$$x=\frac{52}{3}$$
$$24-x=\frac{20}{3}$$

$$\cos\alpha=\frac{6.5}{\frac{52}{3}}=\frac{3}{8}$$
$$\alpha=\cos^{-1}\left(\frac{3}{8}\right)\approx 67.98°$$

$$\beta=180°-67.98°=112.02°$$

$$z=6.5\tan 67.98°\approx 16.07\text{ in}$$
$$y=2.5\tan 67.98°\approx 6.18\text{ in}$$

The arc length on the top of the larger pulley is:
$6.5\left(112.02\cdot\frac{\pi}{180}\right)\approx 12.71$ inches.

The arc length on the top of the smaller pulley is
$2.5\left(112.02\cdot\frac{\pi}{180}\right)\approx 4.89$ inches.

The distance between the points of tangency is
$z+y\approx 16.07+6.18=22.25$ inches.

The length of the belt is about:
2(12.71 + 4.89 + 22.25) = 79.70 inches.

59. $d=-3\cos\left(\frac{\pi}{2}t\right)$

60. $d=-5\cos\left(\frac{\pi}{3}t\right)$

61. $d=6\sin(2t)$

a. Simple harmonic

b. 6 feet

c. π seconds

d. $\frac{1}{\pi}$ oscillation/second

62. $d=2\cos(4t)$

a. Simple harmonic

b. 2 feet

c. $\frac{\pi}{2}$ seconds

d. $\frac{2}{\pi}$ oscillation/second

63. $d=-2\cos(\pi t)$

a. Simple harmonic

b. 2 feet

c. 2 seconds

d. $\frac{1}{2}$ oscillation/second

64. $d=-3\sin\left(\frac{\pi}{2}t\right)$

a. Simple harmonic

b. 3 feet

c. 4 seconds

d. $\frac{1}{4}$ oscillation/second

65. a. $d=-15e^{-0.75t/[2(40)]}\cos\left(\sqrt{\left(\frac{2\pi}{5}\right)^2-\frac{(0.75)^2}{4(40)^2}}\,t\right)$

$d=-15e^{-0.75t/80}\cos\left(\sqrt{\frac{4\pi^2}{25}-\frac{0.5625}{6400}}\,t\right)$

b.

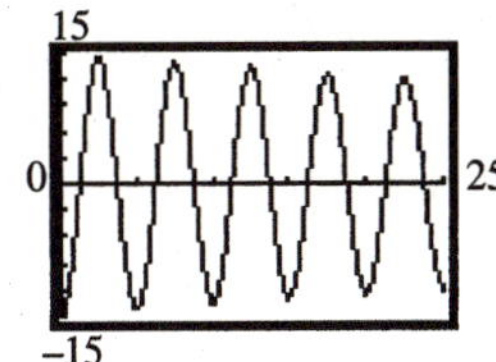

66. a. $d=-13e^{-0.65t/[2(25)]}\cos\left(\sqrt{\left(\frac{\pi}{2}\right)^2-\frac{(0.65)^2}{4(25)^2}}\,t\right)$

$d=-13e^{-0.65t/50}\cos\left(\sqrt{\frac{\pi^2}{4}-\frac{0.4225}{2500}}\,t\right)$

b.

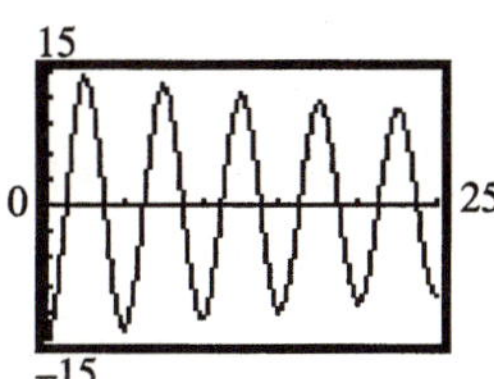

67. a. Damped motion with a bob of mass 20 kg and a damping factor of 0.6 kg/sec.

b. 15 meters leftward

c.

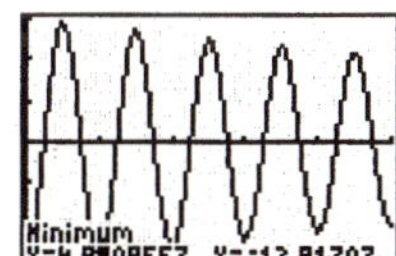

d. The displacement of the bob at the start of the second oscillation is about 13.92 meters.

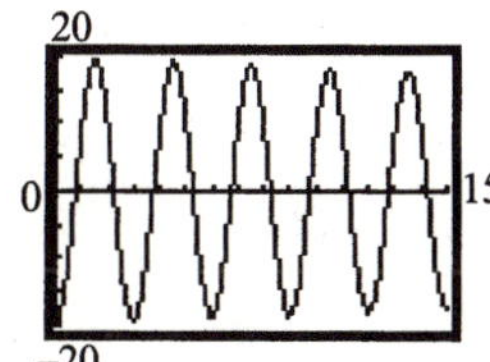

e. It approaches zero, since $e^{-0.6t/40}\to 0$ as $t\to\infty$.

68. a. Damped motion with a bob of mass 30 kg and a damping factor of 0.5 kg/sec.

b. 20 meters leftward

c.

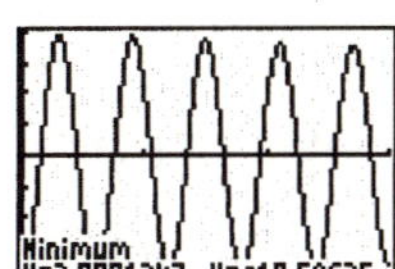

d. The displacement of the bob at the start of the second oscillation is about 19.51 meters.

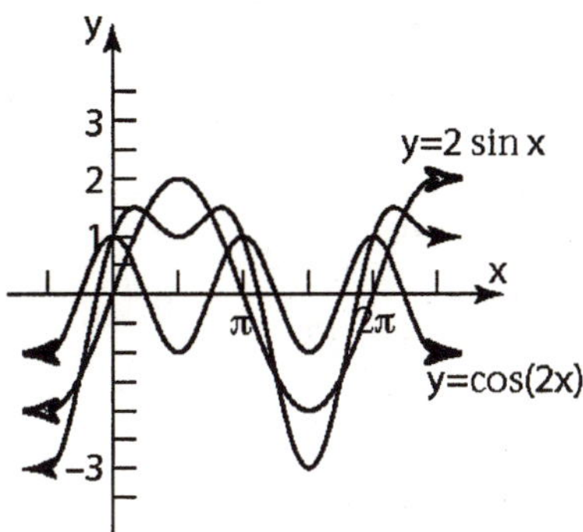

e. It approaches zero, since $e^{-0.5t/60}\to 0$ as $t\to\infty$.

69. $y=2\sin(x)+\cos(2x),\ 0\le x\le 2\pi$

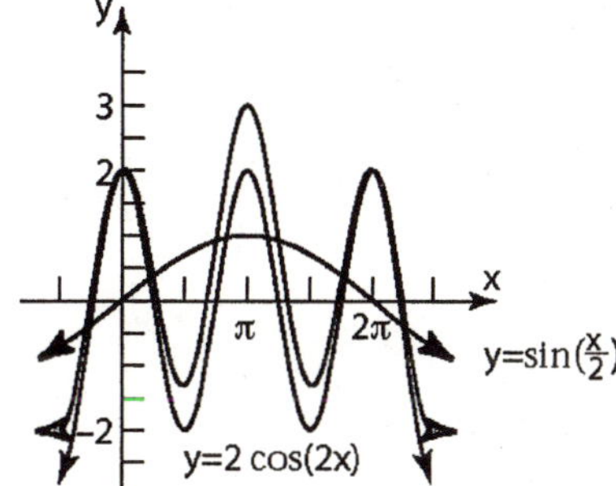

70. $y=2\cos(2x)+\sin\left(\frac{x}{2}\right),\ 0\le x\le 2\pi$

Chapter 7 Test

1. opposite = 3; adjacent = 6; hypotenuse = ?

$(\text{hypotenuse})^2=3^2+6^2=45$

$\text{hypotenuse}=\sqrt{45}=3\sqrt{5}$

$\sin\theta=\frac{\text{opp}}{\text{hyp}}=\frac{3}{3\sqrt{5}}=\frac{1}{\sqrt{5}}\cdot\frac{\sqrt{5}}{\sqrt{5}}=\frac{\sqrt{5}}{5}$

$\cos\theta=\frac{\text{adj}}{\text{hyp}}=\frac{6}{3\sqrt{5}}=\frac{2}{\sqrt{5}}\cdot\frac{\sqrt{5}}{\sqrt{5}}=\frac{2\sqrt{5}}{5}$

$\tan\theta=\frac{\text{opp}}{\text{adj}}=\frac{3}{6}=\frac{1}{2};\qquad \cot\theta=\frac{\text{adj}}{\text{opp}}=\frac{6}{3}=2$

$\csc\theta=\frac{\text{hyp}}{\text{opp}}=\frac{3\sqrt{5}}{3}=\sqrt{5};\ \sec\theta=\frac{\text{hyp}}{\text{adj}}=\frac{3\sqrt{5}}{6}=\frac{\sqrt{5}}{2}$

2. $\sin 40° - \cos 50° = \sin 40° - \sin(90° - 50°)$
$= \sin 40° - \sin 40°$
$= 0$

3. Let α = the angle formed by the ground and the ladder.

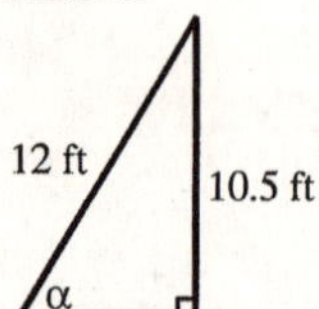

Then $\sin\alpha = \dfrac{10.5}{12}$

$$\alpha = \sin^{-1}\left(\frac{10.5}{12}\right) \approx 61.0°$$

The angle formed by the ladder and ground is about $61.0°$.

4. Let α = the angle of depression from the balloon to the airport.

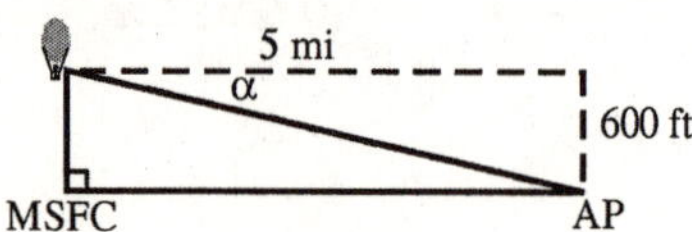

Note that 5 miles = 26,400 feet.

$$\tan\alpha = \frac{600}{26400} = \frac{1}{44}$$

$$a = \tan^{-1}\left(\frac{1}{44}\right) = 1.3°$$

The angle of depression from the balloon to the airport is about $1.3°$.

5. Use the law of cosines to find a:

$$a^2 = b^2 + c^2 - 2bc\cos\alpha$$
$$= (17)^2 + (19)^2 - 2(17)(19)\cos 52°$$
$$\approx 289 + 361 - 646(0.616)$$
$$= 252.064$$

$$a = \sqrt{252.064} \approx 15.88$$

Use the law of sines to find β.

$$\frac{a}{\sin\alpha} = \frac{b}{\sin\beta}$$

$$\frac{15.88}{\sin 52°} = \frac{17}{\sin\beta}$$

$$\sin\beta = \frac{17}{15.88}(\sin 52°) \approx 0.844$$

Since b is not the longest side of the triangle, we know that $\beta < 90°$. Therefore,

$$\beta = \sin^{-1}(0.844) \approx 57.56°$$

$$\lambda = 180° - \alpha - \beta = 180° - 52° - 57.56° = 70.44°$$

6. Use the law of sines to find b:

$$\frac{a}{\sin\alpha} = \frac{b}{\sin\beta}$$

$$\frac{12}{\sin 41°} = \frac{b}{\sin 22°}$$

$$b = \frac{12 \cdot \sin 22°}{\sin 41°} \approx 6.85$$

$$\gamma = 180° - \alpha - \beta = 180° - 41° - 22° = 117°$$

Use the law of sines to find c:

$$\frac{a}{\sin\alpha} = \frac{c}{\sin\gamma}$$

$$\frac{12}{\sin 41°} = \frac{c}{\sin 117°}$$

$$c = \frac{12 \cdot \sin 117°}{\sin 41°} \approx 16.30$$

7. Use the law of cosines to find α.

$$a^2 = b^2 + c^2 - 2bc\cos\alpha$$
$$8^2 = (5)^2 + (10)^2 - 2(5)(10)\cos\alpha$$
$$64 = 25 + 100 - 100\cos\alpha$$
$$100\cos\alpha = 61$$
$$\cos\alpha = 0.61$$
$$\alpha = \cos^{-1}(0.61) \approx 52.41°$$

Use the law of sines to find β.

$$\frac{a}{\sin\alpha} = \frac{b}{\sin\beta}$$

$$\frac{8}{\sin 52.41°} = \frac{5}{\sin\beta}$$

$$\sin\beta = \frac{5}{8}(\sin 52.41°)$$

$$\sin\beta \approx 0.495$$

Since b is not the longest side of the triangle, we have that $\beta < 90°$. Therefore

$$\beta = \sin^{-1}(0.495) \approx 29.67°$$

$$\gamma = 180° - \alpha - \beta$$
$$= 180° - 52.41° - 29.67°$$
$$= 97.92°$$

8. $\alpha = 55°,\ \gamma = 20°,\ a = 4$

$\beta = 180° - \alpha - \gamma = 180° - 55° - 20° = 105°$

Use the law of sines to find b.

$$\frac{\sin\alpha}{a} = \frac{\sin\beta}{b}$$

$$\frac{\sin 55°}{4} = \frac{\sin 105°}{b}$$

$$b = \frac{4\sin 105°}{\sin 55°} \approx 4.72$$

Use the law of sines to find c.

$$\frac{\sin\gamma}{c} = \frac{\sin\alpha}{a}$$

$$\frac{\sin 20°}{c} = \frac{\sin 55°}{4}$$

$$c = \frac{4\sin 20°}{\sin 55°} \approx 1.67$$

9. $a = 3,\ b = 7,\ \alpha = 40°$

Use the law of sines to find β

$$\frac{\sin\beta}{b} = \frac{\sin\alpha}{a}$$

$$\frac{\sin\beta}{7} = \frac{\sin 40°}{3}$$

$$\sin\beta = \frac{7\sin 40°}{3} \approx 1.4998$$

There is no angle β for which $\sin\beta > 1$.
Therefore, there is no triangle with the given measurements.

10. $a = 8,\ b = 4,\ \gamma = 70°$

$$c^2 = a^2 + b^2 - 2ab\cos\gamma$$

$$c^2 = 8^2 + 4^2 - 2\cdot 8\cdot 4\cos 70°$$

$$= 80 - 64\cos 70°$$

$$c = \sqrt{80 - 64\cos 70°} \approx 7.62$$

$$a^2 = b^2 + c^2 - 2bc\cos\alpha$$

$$\cos\alpha = \frac{b^2 + c^2 - a^2}{2bc} = \frac{4^2 + 7.62^2 - 8^2}{2(4)(7.62)} = \frac{10.0644}{60.96}$$

$$\alpha = \cos^{-1}\left(\frac{10.0644}{60.96}\right) \approx 80.5°$$

$$\beta = 180° - \alpha - \gamma \approx 180° - 80.5° - 70° = 29.5°$$

11. $a = 8,\ b = 4,\ \gamma = 70°$

$$A = \frac{1}{2}ab\sin\lambda$$

$$= \frac{1}{2}(8)(4)\sin 70° \approx 15.04 \text{ square units}$$

12. $a = 8,\ b = 5,\ c = 10$

$$s = \frac{1}{2}(a+b+c) = \frac{1}{2}(8+5+10) = 11.5$$

$$A = \sqrt{s(s-a)(s-b)(s-c)}$$

$$= \sqrt{11.5(11.5-8)(11.5-5)(11.5-10)}$$

$$= \sqrt{11.5(3.5)(6.5)(1.5)}$$

$$= \sqrt{392.4375}$$

$$\approx 19.81 \text{ square units}$$

13. Divide home plate into a rectangle and a triangle.

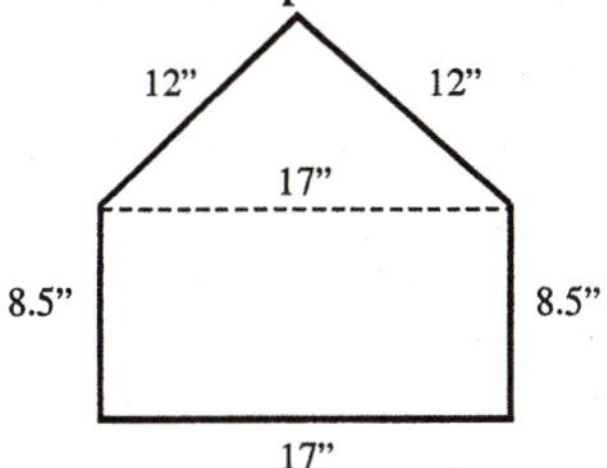

$A_{\text{Rectangle}} = lw = (17)(8.5) = 144.5 \text{ in}^2$

Using Heron's formula we get

$$A_{\text{Triangle}} = \sqrt{s(s-a)(s-b)(s-c)}$$

$$s = \frac{1}{2}(a+b+c) = \frac{1}{2}(12+12+17) = 20.5$$

Thus,

$$A_{\text{Triangle}} = \sqrt{(20.5)(20.5-12)(20.5-12)(20.5-17)}$$

$$= \sqrt{(20.5)(8.5)(8.5)(3.5)}$$

$$= \sqrt{5183.9375}$$

$$\approx 72.0 \text{ sq. in.}$$

So,

$$A_{\text{Total}} = A_{\text{Rectangle}} + A_{\text{Triangle}}$$

$$= 144.5 + 72.0$$

$$= 216.5 \text{ in}^2$$

The area of home plate is about 216.5 in^2.

14. Begin by adding a diagonal to the diagram.

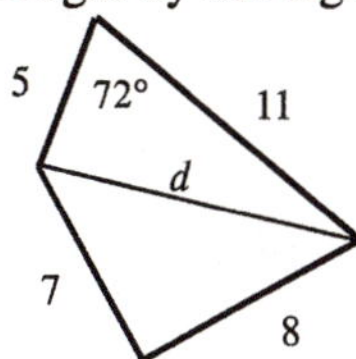

$$A_{\text{upper}\,\Delta} = \frac{1}{2}(5)(11)(\sin 72°) \approx 26.15 \text{ sq. units}$$

By the law of cosines,

$$d^2 = (5)^2 + (11)^2 - 2(5)(11)(\cos 72°)$$
$$= 25 + 121 - 110(0.309)$$
$$= 112.008$$
$$d = \sqrt{112.008} \approx 10.58$$

Using Heron's formula for the lower triangle,

$$s = \frac{7+8+10.58}{2} = 12.79$$
$$A_{\text{lower }\triangle} = \sqrt{12.79(5.79)(4.79)(2.21)}$$
$$= \sqrt{783.9293}$$
$$\approx 28.00 \text{ sq. units}$$

Total Area $= 26.15 + 28.00 = 54.15$ sq. units

15. Use the law of cosines to find c:

$$c^2 = a^2 + b^2 - 2ab\cos C$$
$$= (4.2)^2 + (3.5)^2 - 2(4.2)(3.5)\cos 32°$$
$$\approx 17.64 + 12.25 - 29.4(0.848)$$
$$= 4.9588$$
$$c = \sqrt{4.9588} \approx 2.23$$

Madison will have to swim about 2.23 miles.

16. Since $\triangle OAB$ is isosceles, we know that

$$\angle A = \angle B = \frac{180° - 40°}{2} = 70°$$

Then,

$$\frac{\sin A}{OB} = \frac{\sin 40°}{AB}$$
$$\frac{\sin 70°}{5} = \frac{\sin 40°}{AB}$$
$$AB = \frac{5\sin 40°}{\sin 70°} \approx 3.420$$

Now, AB is the diameter of the semicircle, so the radius is $\frac{3.420}{2} = 1.710$.

$$A_{\text{Semicircle}} = \frac{1}{2}\pi r^2 = \frac{1}{2}\pi(1.710)^2 \approx 4.593 \text{ sq. units}$$
$$A_{\text{Triangle}} = \frac{1}{2}ab\sin(\measuredangle O)$$
$$= \frac{1}{2}(5)(5)(\sin 40°) \approx 8.035 \text{ sq. units}$$
$$A_{\text{Total}} = A_{\text{Semicircle}} + A_{\text{Triangle}}$$
$$\approx 4.593 + 8.035 \approx 12.63 \text{ sq. units}$$

17. Using Heron's formula:

$$s = \frac{5x+6x+7x}{2} = \frac{18x}{2} = 9x$$
$$A = \sqrt{9x(9x-5x)(9x-6x)(9x-7x)}$$
$$= \sqrt{9x \cdot 4x \cdot 3x \cdot 2x}$$
$$= \sqrt{216x^4}$$
$$= \left(6\sqrt{6}\right)x^2$$

Thus,

$$(6\sqrt{6})x^2 = 54\sqrt{6}$$
$$x^2 = 9$$
$$x = 3$$

The sides are 15, 18, and 21.

18. Since we ignore all resistive forces, this is simple harmonic motion. Since the rest position $(t = 0)$ is the vertical position $(d = 0)$, the equation will have the form $d = a\sin(\omega t)$.

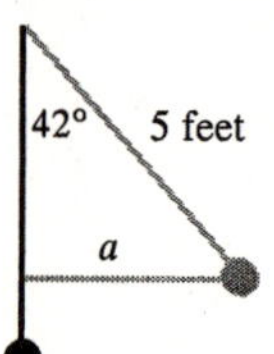

Now, the period is 6 seconds, so

$$6 = \frac{2\pi}{\omega}$$
$$\omega = \frac{2\pi}{6}$$
$$\omega = \frac{\pi}{3} \text{ radians/sec}$$

From the diagram we see

$$\frac{a}{5} = \sin 42°$$
$$a = 5(\sin 42°)$$

Thus, $d = 5(\sin 42°)\left(\sin\frac{\pi t}{3}\right)$ or

$$d \approx 3.346 \cdot \sin\left(\frac{\pi t}{3}\right).$$

Chapter 7 Projects

Project 1

A. a.

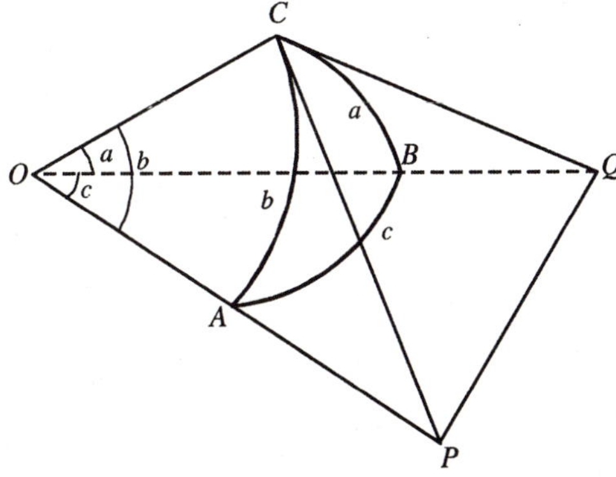

Triangles ΔOCP and ΔOCQ are plane right triangles.

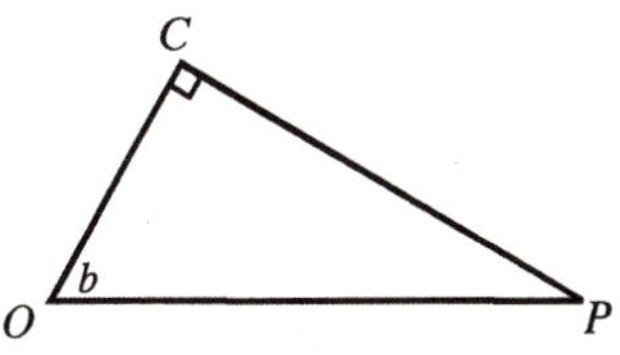

$$\sin b = \frac{CP}{OP},\quad OC^2 + CP^2 = OP^2$$

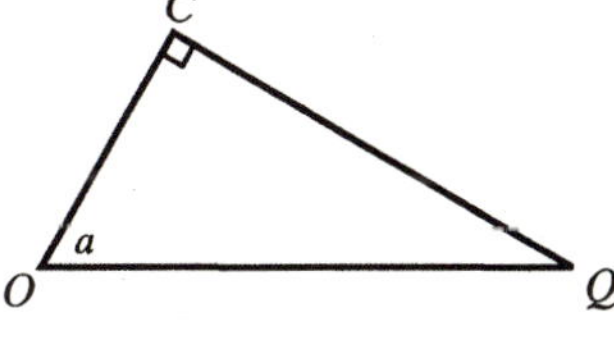

$$\sin a = \frac{CQ}{OQ},\quad OC^2 + CQ^2 = OQ^2$$

b. For ΔOPQ, we have that

$$(PQ)^2 = (OP)^2 + (OQ)^2 - 2(OQ)(OP)\cos c$$

For ΔCPQ, we have that

$$(PQ)^2 = (CQ)^2 + (CP)^2 - 2(CQ)(CP)\cos C$$

c.

$$0 = (OP)^2 + (OQ)^2 - 2(OQ)(OP)\cos c - \left[(CQ)^2 + (CP)^2 - 2(CQ)(CP)\cos C\right]$$

$$2(OQ)(OP)\cos c = (OQ)^2 - (CQ)^2 + (OP)^2 - (CQ)^2 + 2(CQ)(CP)\cos C$$

d. From part a: $(OC)^2 = (OQ)^2 - (CQ)^2$

$(OC)^2 = (OP)^2 - (CP)^2$

Thus,

$$2(OQ)(OP)\cos c = OQ^2 - CQ^2 + OP^2 - CQ^2 + 2(CQ)(CP)\cos C$$

$$2(OQ)(OP)\cos c = OC^2 + OC^2 + 2(CQ)(CP)\cos C$$

$$\cos c = \frac{2OC^2}{2(OQ)(OP)} + \frac{2(CQ)(CP)\cos C}{2(OQ)(OP)}$$

$$\cos c = \frac{OC^2}{(OQ)(OP)} + \frac{(CQ)(CP)\cos C}{(OQ)(OP)}$$

e.

$$\cos c = \frac{OC^2}{(OQ)(OP)} + \frac{(CQ)(CP)\cos C}{(OQ)(OP)}$$

$$= \frac{OC}{OQ}\cdot\frac{OC}{OP} + \frac{CQ}{OQ}\cdot\frac{CP}{OP}\cdot\cos C$$

$$= \cos a\cos b + \sin a\sin b\cos C$$

B. a. Putting this on a circle of radius 1 in order to apply the Law of Cosines from A:

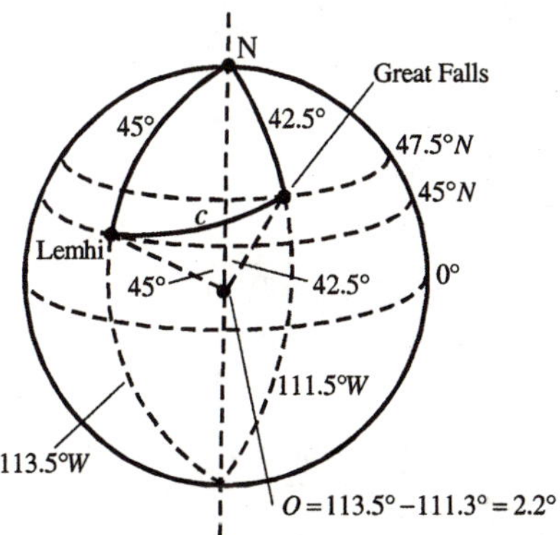

$$\cos c = \cos 45°\cos 42.5° + \sin 45°\sin 42.5°\cos 2.2°$$

$$\cos c = 0.99870$$

$$c = 2.93°$$

$$s = rc = 3960\left(2.93°\cdot\frac{\pi}{180°}\right) \approx 202.5$$

It is 202.5 miles from Great Falls to Lemhi.

b.

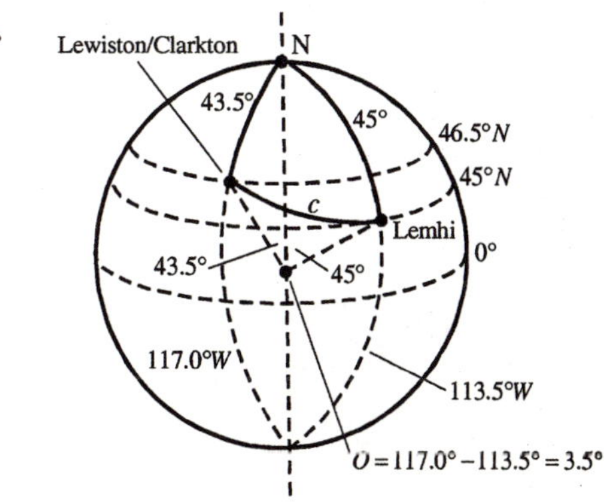

$\cos c = \cos 45°\cos 43.5° + \sin 45°\sin 43.5°\cos 3.5°$

$\cos c = 0.99875$

$c = 2.87°$

$$s = rc = 3960\left(2.87°\cdot\frac{\pi}{180°}\right) \approx 198.4$$

It is about 198.4 miles from Lemhi to Lewiston and Clarkston.

c. They traveled $202.5 + 198.4$ miles just to go from Great Falls to Lewiston and Clarkston.

d.

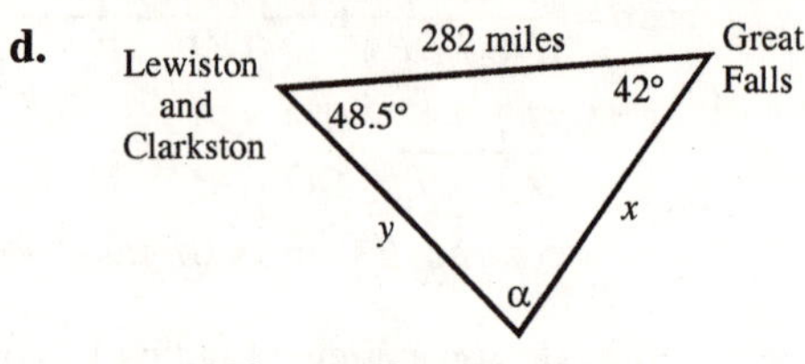

$\alpha = 180° - 48.5° - 42° = 89.5°$

$$\frac{x}{\sin 48.5°} = \frac{282}{\sin 89.5°}$$

$$x = \frac{282\sin 48.5°}{\sin 89.5°} \approx 211.2$$

$$\frac{y}{\sin 42°} = \frac{282}{\sin 89.5°}$$

$$y = \frac{282\sin 42°}{\sin 89.5°} \approx 188.8$$

Using a plane triangle, they traveled $211.2 + 188.7 = 399.9$ miles.

The mileage by using spherical triangles and that by using a plane triangle are relatively close. The total mileage is basically the same in each case. This is because compared to the surface of the Earth, these three towns are very close to each other and the surface can be approximated very closely by a plane.

Project 2 (web)

1. $f_1 = \sin(\pi t)$

$f_3 = \frac{1}{3}\sin(3\pi t)$

$f_5 = \frac{1}{5}\sin(5\pi t)$

$f_7 = \frac{1}{7}\sin(7\pi t)$

$f_9 = \frac{1}{9}\sin(9\pi t)$

2. $f_1 = \sin(\pi t)$

$f_1 + f_3 = \sin(\pi t) + \frac{1}{3}\sin(3\pi t)$

$f_1 + f_3 + f_5 = \sin(\pi t) + \frac{1}{3}\sin(3\pi t) + \frac{1}{5}\sin(5\pi t)$

$f_1 + f_3 + f_5 + f_7$

$= \sin(\pi t) + \frac{1}{3}\sin(3\pi t) + \frac{1}{5}\sin(5\pi t) + \frac{1}{7}\sin(7\pi t)$

$f_1 + f_3 + f_5 + f_7 + f_9$

$= \sin(\pi t) + \frac{1}{3}\sin(3\pi t) + \frac{1}{5}\sin(5\pi t)$

$+ \frac{1}{7}\sin(7\pi t) + \frac{1}{9}\sin(9\pi t)$

3. If one graphs each of these functions, one observes that with each iteration, the function becomes more square.

4. $f_1 + f_{13} + f_3 = \sin(\pi t) + \frac{1}{2}\cos(2\pi t) + \sin(3\pi t)$

By adding in the cosine term, the curve does not become as flat. The waves at the "tops" and the "bottoms" become deeper.

Project 3 (web)

a.

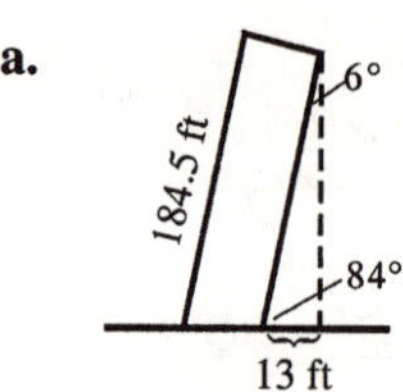

b. Let h = the height of the tower with a lean of $6°$.

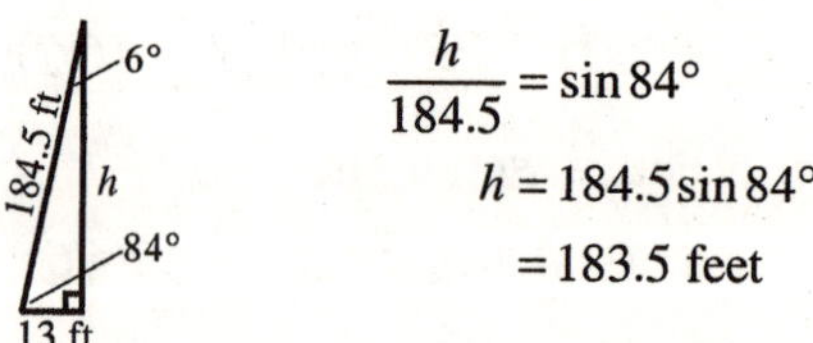

$$\frac{h}{184.5} = \sin 84°$$

$h = 184.5\sin 84°$

$= 183.5$ feet

c. $13 \text{ ft} - 16 \text{ in} = 11 \text{ ft } 8 \text{ in} = 11\frac{2}{3}$ ft

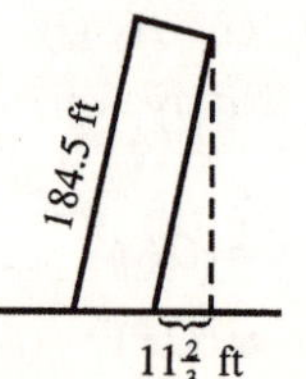

d. $\sin\alpha = \dfrac{11.67}{184.5}$

$$\alpha = \sin^{-1}\left(\frac{11.67}{184.5}\right) \approx 4°$$

e. $\dfrac{h}{184.5} = \sin 86°$

$$h = 184.5\sin 86° \approx 184.1 \text{ ft}$$

f. The angles are relatively small and for part (d), where 4° was acquired, it was arrived at by rounding.

g. Answers will vary.

Project 4 (web)

a. Answers will vary.

b.

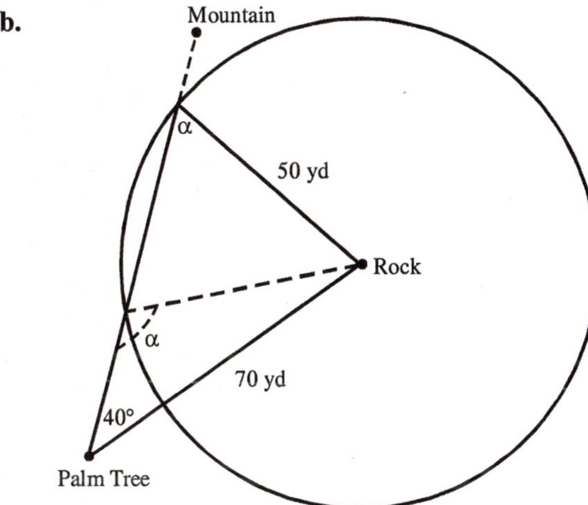

$$\frac{\sin\alpha}{70} = \frac{\sin 40°}{50}$$

$$\sin\alpha = \frac{70\sin 40°}{50} \approx 0.8999$$

$$\alpha = \sin^{-1}\left(\frac{70\sin 40°}{50}\right) \approx 64°$$

c. $\alpha = 180° - 64° = 116°$

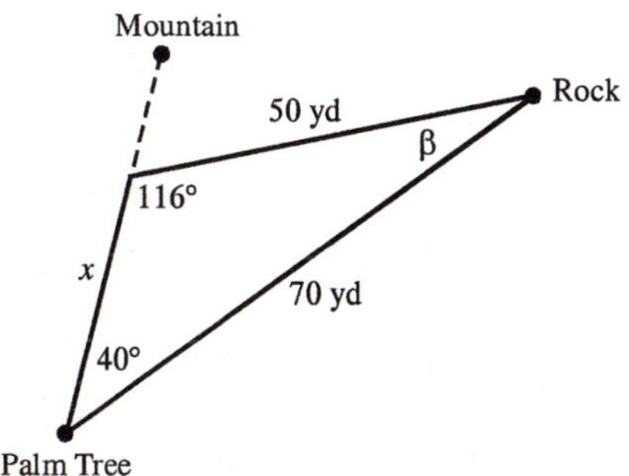

The third angle (i.e., at the rock) is 24°.

$$\frac{x}{\sin 24°} = \frac{70}{\sin 116°}$$

$$x = \frac{70\sin 24°}{\sin 116°} \approx 32$$

The treasure is about 32 yards away from the palm tree.

Cumulative Review R – 7

1.

$$3x^2 + 1 = 4x$$
$$3x^2 - 4x + 1 = 0$$
$$(3x-1)(x-1) = 0$$
$$x = \frac{1}{3} \text{ or } x = 1$$

The solution set is $\left\{\frac{1}{3}, 1\right\}$.

2. Center $(-5, 1)$; Radius 3

$$(x-h)^2 + (y-k)^2 = r^2$$
$$(x-(-5))^2 + (y-1)^2 = 3^2$$
$$(x+5)^2 + (y-1)^2 = 9$$

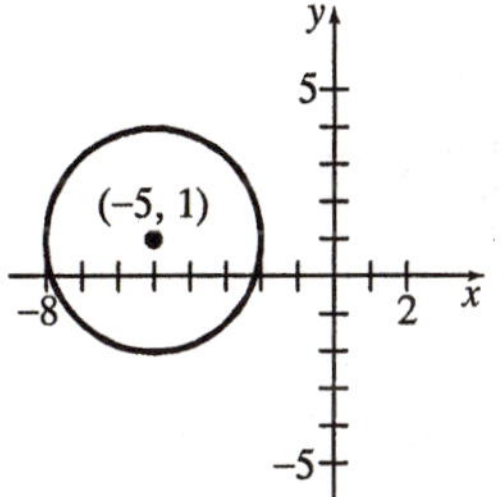

3. $f(x) = \sqrt{x^2 - 3x - 4}$

f will be defined provided

$g(x) = x^2 - 3x - 4 \geq 0$.

$$x^2 - 3x - 4 \geq 0$$
$$(x-4)(x+1) \geq 0$$

$x = 4$, $x = -1$ are the zeros.

Interval	Test Number	$g(x)$	Pos./Neg.
$-\infty < x < -1$	−2	6	Positive
$-1 < x < 4$	0	−4	Negative
$4 < x < \infty$	5	6	Positive

The domain of $f(x) = \sqrt{x^2 - 3x - 4}$ is $\{x \mid x \leq -1 \text{ or } x \geq 4\}$.

4. $y = 3\sin(\pi x)$

Amplitude: $|A| = |3| = 3$

Period: $T = \frac{2\pi}{\pi} = 2$

Phase Shift: $\frac{\phi}{\omega} = \frac{0}{\pi} = 0$

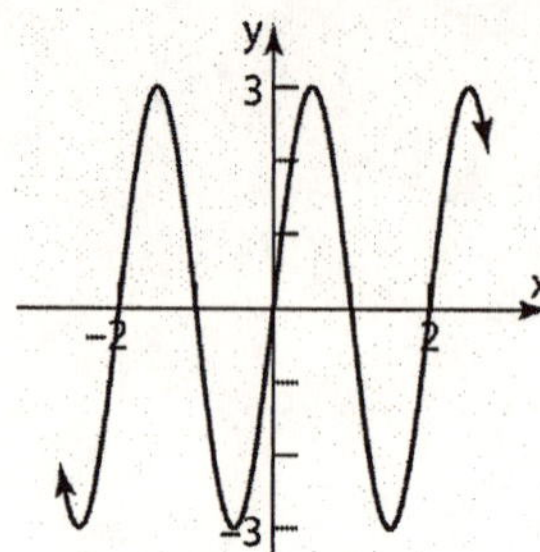

5. $y = -2\cos(2x - \pi) = -2\cos\left[2\left(x - \frac{\pi}{2}\right)\right]$

Amplitude: $|A| = |-2| = 2$

Period: $T = \frac{2\pi}{2} = \pi$

Phase Shift: $\frac{\phi}{\omega} = \frac{\pi}{2}$

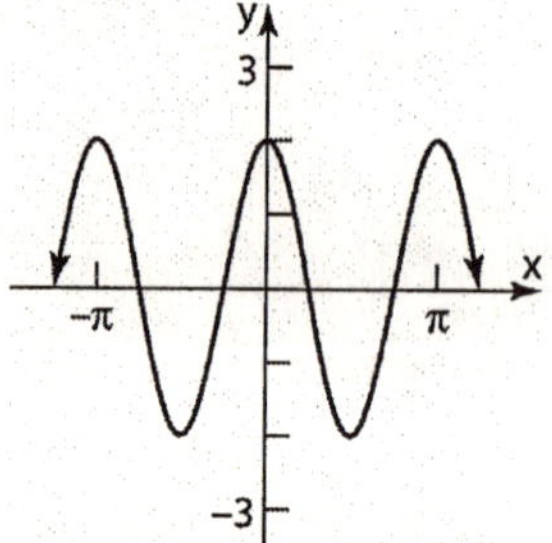

6. $\tan\theta = -2$, $\frac{3\pi}{2} < \theta < 2\pi$, so θ lies in quadrant IV.

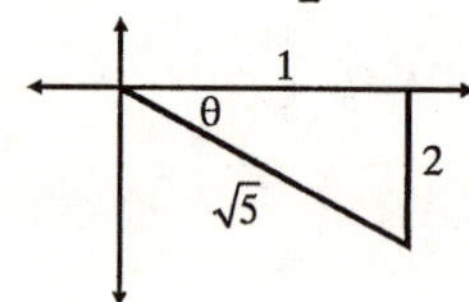

a. $\frac{3\pi}{2} < \theta < 2\pi$, so $\sin\theta < 0$.

$\sin\theta = -\frac{2}{\sqrt{5}} \cdot \frac{\sqrt{5}}{\sqrt{5}} = -\frac{2\sqrt{5}}{5}$

b. $\frac{3\pi}{2} < \theta < 2\pi$, so $\cos\theta > 0$

$\cos\theta = \frac{1}{\sqrt{5}} \cdot \frac{\sqrt{5}}{\sqrt{5}} = \frac{\sqrt{5}}{5}$

c. $\sin(2\theta) = 2\sin\theta\cos\theta$

$= 2\left(-\frac{2\sqrt{5}}{5}\right)\left(\frac{\sqrt{5}}{5}\right)$

$= -\frac{20}{25} = -\frac{4}{5}$

d. $\cos(2\theta) = \cos^2\theta - \sin^2\theta$

$= \left(\frac{\sqrt{5}}{5}\right)^2 - \left(-\frac{2\sqrt{5}}{5}\right)^2$

$= \frac{5}{25} - \frac{20}{25}$

$= -\frac{15}{25} = -\frac{3}{5}$

e. $\frac{3\pi}{2} < \theta < 2\pi$

$\frac{3\pi}{4} < \frac{\theta}{2} < \pi$

Since $\frac{\theta}{2}$ lies in Quadrant II, $\sin\frac{\theta}{2} > 0$.

$\sin\frac{\theta}{2} = \sqrt{\frac{1-\cos\theta}{2}}$

$= \sqrt{\frac{1-\frac{\sqrt{5}}{5}}{2}}$

$= \sqrt{\frac{\frac{5-\sqrt{5}}{5}}{2}}$

$= \sqrt{\frac{5-\sqrt{5}}{10}}$

f. $\frac{3\pi}{2} < \theta < 2\pi$

$\frac{3\pi}{4} < \frac{\theta}{2} < \pi$

Since $\frac{\theta}{2}$ lies in Quadrant II, $\cos\frac{\theta}{2} < 0$.

$\cos\frac{\theta}{2} = -\sqrt{\frac{1+\cos\theta}{2}}$

$= -\sqrt{\frac{1+\frac{\sqrt{5}}{5}}{2}}$

$= -\sqrt{\frac{\frac{5+\sqrt{5}}{5}}{2}}$

$= -\sqrt{\frac{5+\sqrt{5}}{10}}$

7. a. $y = e^x, \; 0 \le x \le 4$

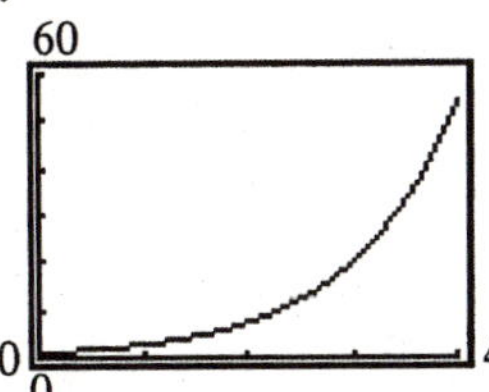

b. $y = \sin x, \; 0 \le x \le 4$

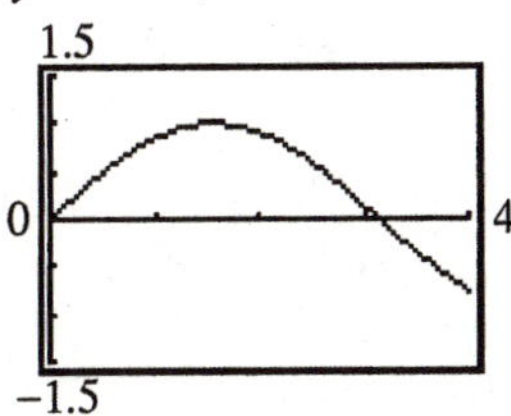

c. $y = e^x \sin x, \; 0 \le x \le 4$

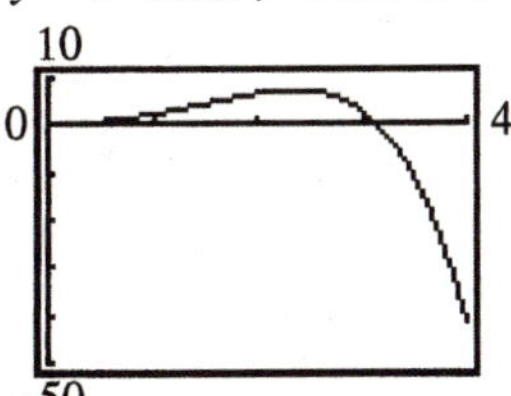

d. $y = 2x + \sin x, \; 0 \le x \le 4$

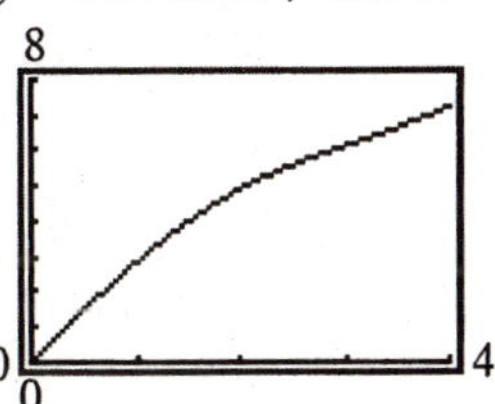

8. a. $y = x$

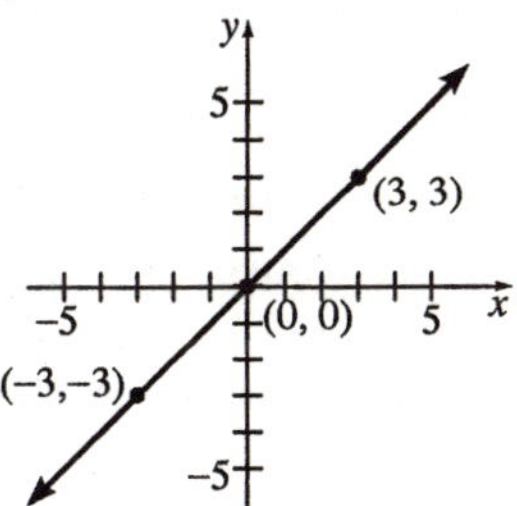

b. $y = x^2$

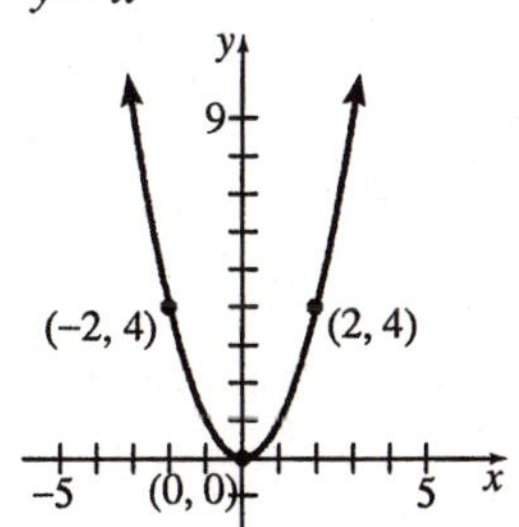

c. $y = \sqrt{x}$

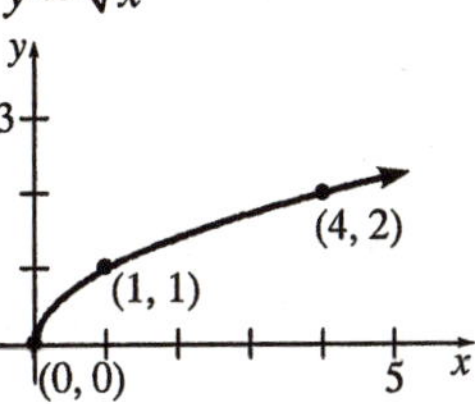

d. $y = x^3$

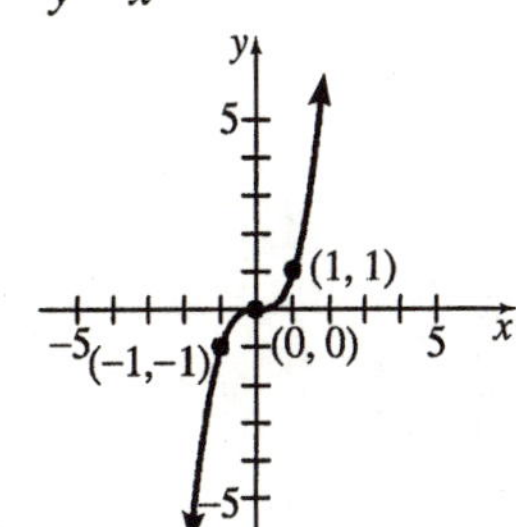

e. $y = e^x$

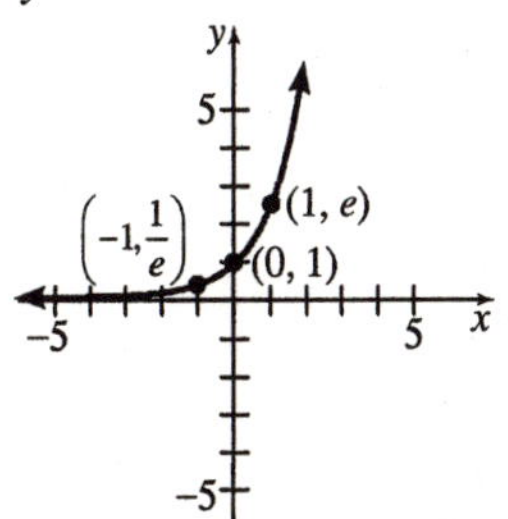

f. $y = \ln x$

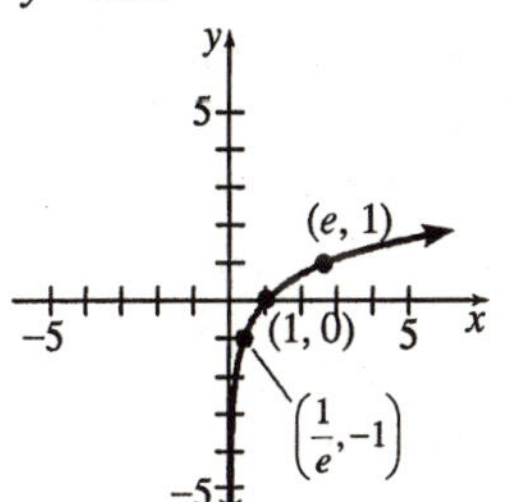

g. $y = \sin x$

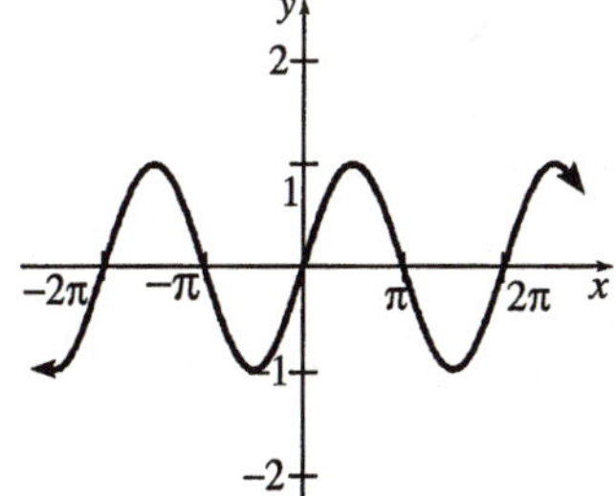

h. $y = \cos x$

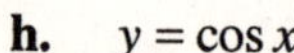

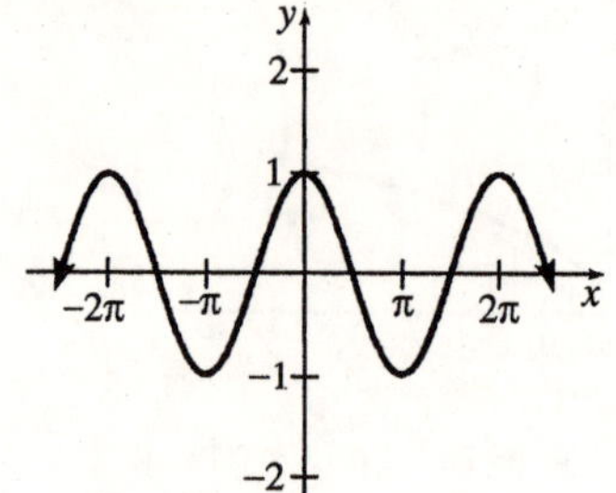

i. $y = \tan x$

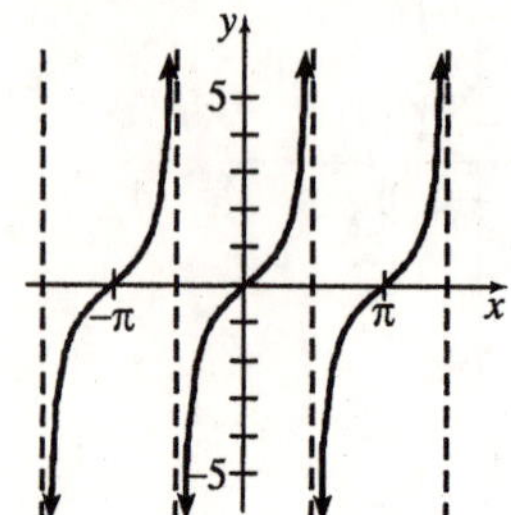

9. $a = 20,\ c = 15,\ \gamma = 40^\circ$

$$\frac{\sin\gamma}{c} = \frac{\sin\alpha}{a}$$

$$\frac{\sin 40^\circ}{15} = \frac{\sin\alpha}{20}$$

$$\sin\alpha = \frac{20\sin 40^\circ}{15}$$

$$\alpha = \sin^{-1}\left(\frac{20\sin 40^\circ}{15}\right)$$

$\alpha_1 \approx 59.0^\circ$ or $\alpha_2 \approx 121.0^\circ$

For both values, $\alpha + \gamma < 180^\circ$. Therefore, there are two triangles.

$$\beta_1 = 180^\circ - \alpha_1 - \gamma \approx 180^\circ - 40^\circ - 59.0^\circ = 81.0^\circ$$

$$\frac{\sin\gamma}{c} = \frac{\sin\beta_1}{b_1}$$

$$\frac{\sin 40^\circ}{15} = \frac{\sin 81.0^\circ}{b_1}$$

$$b_1 \approx \frac{15\sin 81.0^\circ}{\sin 40^\circ} \approx 23.05$$

$$\beta_2 = 180^\circ - \alpha_2 - \gamma \approx 180^\circ - 40^\circ - 121.0^\circ = 19.0^\circ$$

$$\frac{\sin\gamma}{c} = \frac{\sin\beta_2}{b_2}$$

$$\frac{\sin 40^\circ}{15} = \frac{\sin 19.0^\circ}{b_2}$$

$$b_2 \approx \frac{15\sin 19.0^\circ}{\sin 40^\circ} \approx 7.60$$

Two triangles: $\alpha_1 \approx 59.0^\circ$, $\beta_1 \approx 81.0^\circ$, $b_1 \approx 23.05$
or $\alpha_2 \approx 121.0^\circ$, $\beta_2 \approx 19.0^\circ$, $b_2 \approx 7.60$.

10. $3x^5 - 10x^4 + 21x^3 - 42x^2 + 36x - 8 = 0$

Let $f(x) = 3x^5 - 10x^4 + 21x^3 - 42x^2 + 36x - 8 = 0$

Step 1: $f(x)$ has at most 5 real zeros.

Step 2: Possible rational zeros:

$p = \pm1, \pm2, \pm4, \pm8;\quad q = \pm1, \pm2, \pm3;$

$$\frac{p}{q} = \pm1, \pm\frac{1}{2}, \pm\frac{1}{3}, \pm2, \pm\frac{2}{3}, \pm4, \pm\frac{4}{3}, \pm8, \pm\frac{8}{3}$$

Step 3: Using the Bounds on Zeros Theorem:

$$f(x) = 3\left(x^5 - \frac{10}{3}x^4 + 7x^3 - 14x^2 + 12x - \frac{8}{3}\right)$$

$$a_4 = -\frac{10}{3},\ a_3 = 7,\ a_2 = -14,\ a_1 = 12,\ a_0 = -\frac{8}{3}$$

$$\text{Max}\left\{1, \left|-\frac{8}{3}\right| + |\ 12\ | + |\ -14\ | + |7| + \left|-\frac{10}{3}\right|\right\}$$

$$= \text{Max}\{1, 39\} = 39$$

$$1 + \text{Max}\left\{\left|-\frac{8}{3}\right|, |\ 12\ |, |-14|, |7|, \left|-\frac{10}{3}\right|\right\}$$

$$= 1 + 14 = 15$$

The smaller of the two numbers is 15. Thus, every zero of f lies between −15 and 15.

Graphing using the bounds: (Second graph has a better window.)

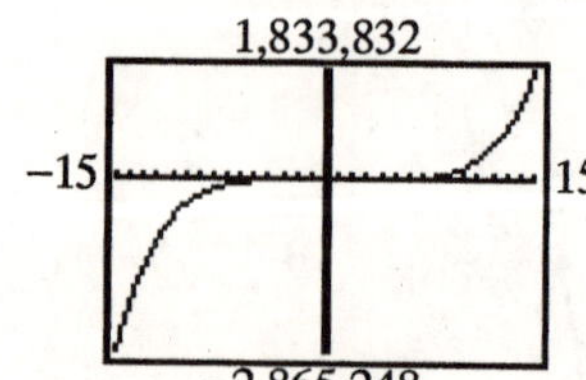

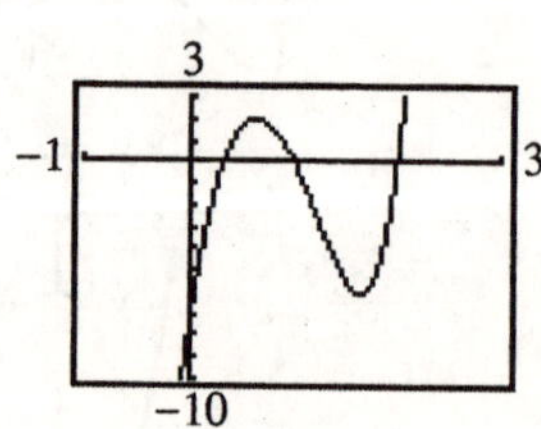

Step 4: From the graph it appears that there are x-intercepts at $\frac{1}{3}$, 1, and 2.

Using synthetic division with 1:

$$\begin{array}{r|rrrrrr} 1 & 3 & -10 & 21 & -42 & 36 & -8 \\ & & 3 & -7 & 14 & -28 & 8 \\ \hline & 3 & -7 & 14 & -28 & 8 & 0 \end{array}$$

Since the remainder is 0, $x-1$ is a factor. The other factor is the quotient: $3x^4-7x^3+14x^2-28x+8$.

Using synthetic division with 2 on the quotient:

$$\begin{array}{r|rrrrr} 2 & 3 & -7 & 14 & -28 & 8 \\ & & 6 & -2 & 24 & -8 \\ \hline & 3 & -1 & 12 & -4 & 0 \end{array}$$

Since the remainder is 0, $x-2$ is a factor. The other factor is the quotient: $3x^3-x^2+12x-4$.

Using synthetic division with $\frac{1}{3}$ on the quotient:

$$\begin{array}{r|rrrr} \frac{1}{3} & 3 & -1 & 12 & 4 \\ & & 1 & 0 & 4 \\ \hline & 3 & 0 & 12 & 0 \end{array}$$

Since the remainder is 0, $x-\frac{1}{3}$ is a factor. The other factor is the quotient:

$3x^2+12=3(x^2+4)=3(x+2i)(x-2i)$.

Factoring,

$$f(x)=3(x-1)(x-2)\left(x-\frac{1}{3}\right)(x+2i)(x-2i)$$

The real zeros are 1,2, and $\frac{1}{3}$. The imaginary zeros are $-2i$ and $2i$.

Therefore, over the complex numbers, the equation $3x^5-10x^4+21x^3-42x^2+36x-8=0$ has solution set $\left\{-2i,\ 2i,\ \frac{1}{3},\ 1,\ 2\right\}$.

11. $R(x)=\dfrac{2x^2-7x-4}{x^2+2x-15}=\dfrac{(2x+1)(x-4)}{(x-3)(x+5)}$

$p(x)=2x^2-7x-4;\ q(x)=x^2+2x-15;$
$n=2;\ m=2$

Step 1: Domain: $\{x \mid x\neq -5, x\neq 3\}$

Step 2: R is in lowest terms.

Step 3: The x-intercepts are the zeros of $p(x)$:

$$2x^2-7x-4=0$$
$$(2x+1)(x-4)=0$$
$$2x+1=0 \quad \text{or} \quad x-4=0$$
$$x=-\frac{1}{2} \qquad\qquad x=4$$

The y-intercept is

$$R(0)=\frac{2\cdot 0^2-7\cdot 0-4}{0^2+2\cdot 0-15}=\frac{-4}{-15}=\frac{4}{15}.$$

Step 4: $R(-x)=\dfrac{2(-x)^2-7(-x)-4}{(-x)^2+2(-x)-15}=\dfrac{2x^2+7x-4}{x^2-2x-15}$

which is neither $R(x)$ nor $-R(x)$, so there is no symmetry.

Step 5: The vertical asymptotes are the zeros of $q(x)$: $\quad x^2+2x-15=0$

$$(x+5)(x-3)=0$$
$$x+5=0 \quad \text{or} \quad x-3=0$$
$$x=-5 \qquad\qquad x=3$$

Step 6: Since $n=m$, the line $y=2$ is the horizontal asymptote.

$R(x)$ intersects $y=2$ at $\left(\frac{26}{11},2\right)$, since:

$$\frac{2x^2-7x-4}{x^2+2x-15}=2$$
$$2x^2-7x-4=2(x^2+2x-15)$$
$$2x^2-7x-4=2x^2+4x-30$$
$$-11x=-26$$
$$x=\frac{26}{11}$$

Step 7: Graphing utility:

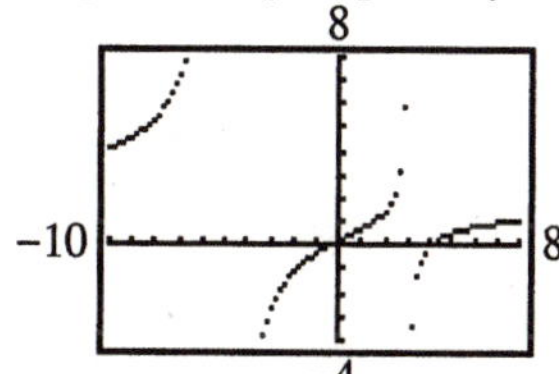

Step 8: Graph by hand:

Interval	Test number	Value of f	Location	Point
$(-\infty,-5)$	-6	≈ 12.22	Above x-axis	$(-6,12.22)$
$(-5,-0.5)$	-1	-0.3125	Below x-axis	$(-1,-0.3125)$
$(-0.5,3)$	0	≈ 0.27	Above x-axis	$(0,0.27)$
$(3,4)$	3.5	≈ -0.94	Below x-axis	$(3.5,-0.94)$
$(4,\infty)$	5	0.55	Above x-axis	$(5,0.55)$

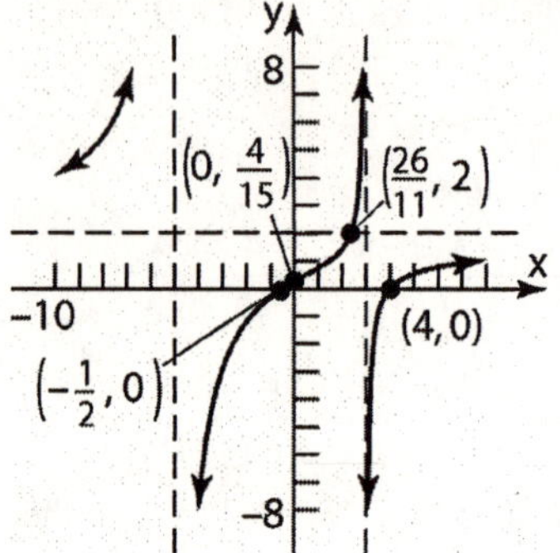

12.
$$3^x = 12$$
$$\ln(3^x) = \ln(12)$$
$$x\ln(3) = \ln(12)$$
$$x = \frac{\ln(12)}{\ln(3)} \approx 2.26$$

13. $\log_3(x+8) + \log_3 x = 2$
$$\log_3[(x+8)(x)] = 2$$
$$(x+8)(x) = 3^2$$
$$x^2 + 8x = 9$$
$$x^2 + 8x - 9 = 0$$
$$(x+9)(x-1) = 0$$
$$x = -9 \text{ or } x = 1$$
$x=-9$ is extraneous because it makes the original logarithms undefined. The solution set is $\{1\}$.

14. $f(x) = 4x+5$; $g(x) = x^2 + 5x - 24$

a.
$$f(x) = 0$$
$$4x+5 = 0$$
$$4x = -5$$
$$x = -\frac{5}{4}$$
The solution set is $\left\{-\frac{5}{4}\right\}$.

b.
$$f(x) = 13$$
$$4x+5 = 13$$
$$4x = 8$$
$$x = 2$$
The solution set is $\{2\}$.

c.
$$f(x) = g(x)$$
$$4x+4 = x^2 + 5x - 24$$
$$0 = x^2 + x - 28$$
$$x = \frac{-1 \pm \sqrt{1^2 - 4(1)(-29)}}{2(1)}$$
$$= \frac{-1 \pm \sqrt{117}}{2}$$
$$= \frac{-1 \pm 3\sqrt{13}}{2}$$
The solution set is $\left\{\frac{-1-3\sqrt{13}}{2}, \frac{-1+3\sqrt{13}}{2}\right\}$.

d.
$$f(x) > 0$$
$$4x+5 > 0$$
$$4x > -5$$
$$x > -\frac{5}{4}$$
The solution set is $\left\{x \middle| x > -\frac{5}{4}\right\}$ or $\left(-\frac{5}{4}, \infty\right)$.

e.
$$g(x) \le 0$$
$$x^2 + 5x - 24 \le 0$$
$$(x+8)(x-3) \le 0$$
$x = -8, x = 3$ are the zeros.

Interval	Test number	$g(x)$	Pos./Neg.
$(-\infty,-8)$	-9	12	Positive
$(-8,3)$	0	-24	Negative
$(3,\infty)$	4	12	Positive

The solution set is $\{x \mid -8 \le x \le 3\}$ or $[-8,3]$.

f. $y = f(x) = 4x + 5$

The graph of f is a line with slope 4 and y-intercept 5.

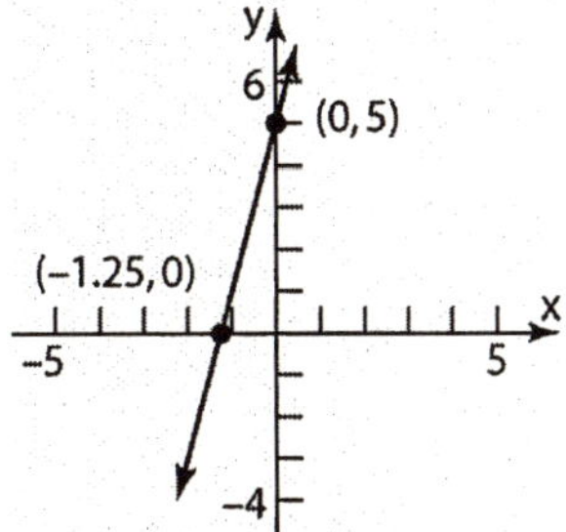

g. $y = g(x) = x^2 + 5x - 24$

The graph of g is a parabola with y-intercept -24 and x-intercepts -8 and 3. The x-coordinate of the vertex is

$$x = -\frac{b}{2a} = -\frac{5}{2(1)} = -\frac{5}{2} = -2.5.$$

The y-coordinate of the vertex is

$$y = f\left(-\frac{b}{2a}\right)$$

$$= f(-2.5) = (-2.5)^2 + 5(-2.5) - 24 = -30.25$$

The vertex is $(-2.5, -30.25)$.

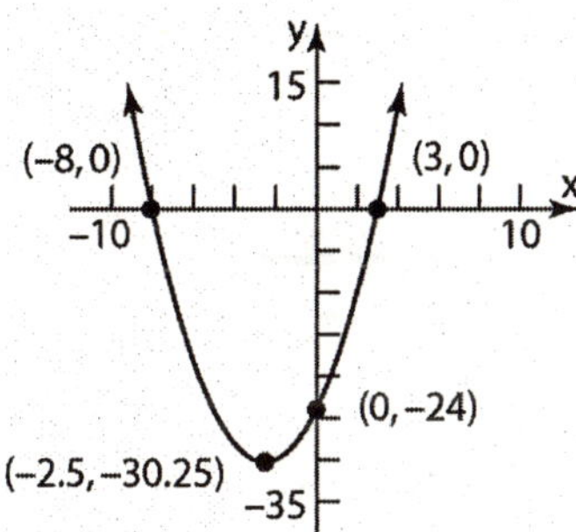

Chapter 8

Polar Coordinates; Vectors

Section 8.1

1. 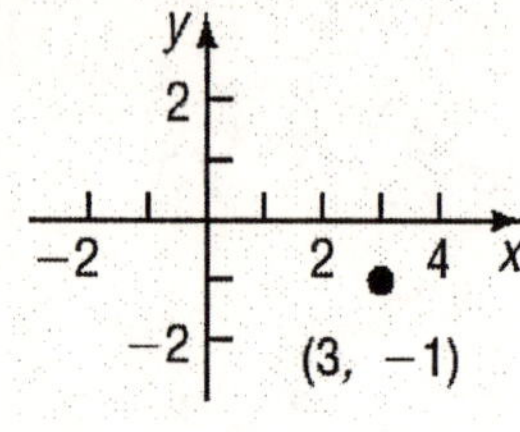

2. $\left(\frac{6}{2}\right)^2 = 9$

3. $\frac{b}{\sqrt{a^2+b^2}}$

4. $-\frac{\pi}{4}$

5. pole, polar axis

6. −2

7. $\left(-\sqrt{3},-1\right)$

8. False

9. True

10. True

11. *A*

12. *B*

13. *C*

14. *C*

15. *B*

16. *D*

17. *A*

18. *D*

19. 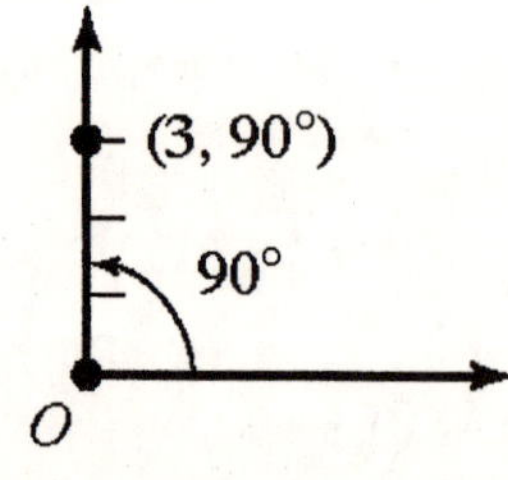

20. 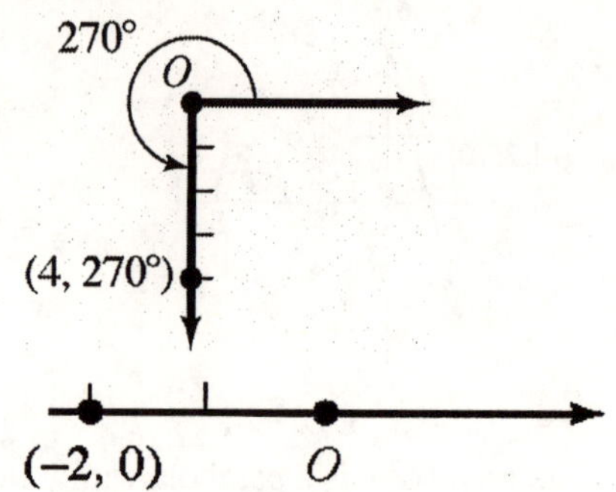

21. (−2, 0) *O*

22. π *O* (−3, π)

23. 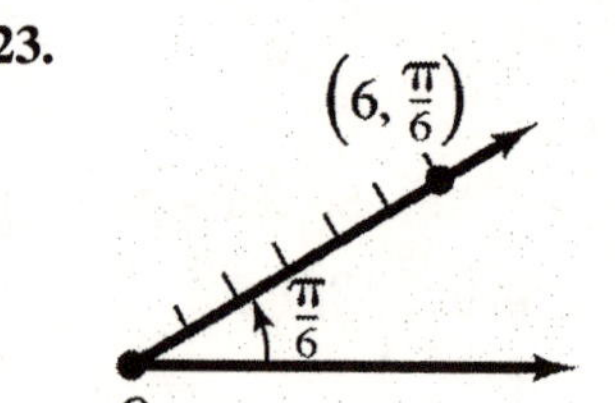

24. $\frac{5\pi}{3}$ *O* $\left(5, \frac{5\pi}{3}\right)$

25. 135° *O* (−2, 135°)

26. 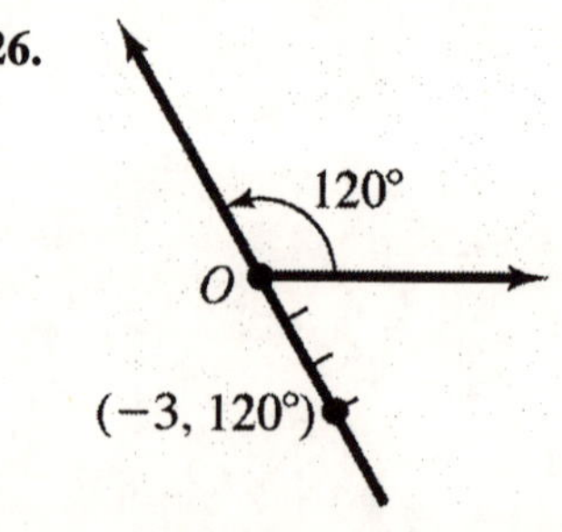

27.

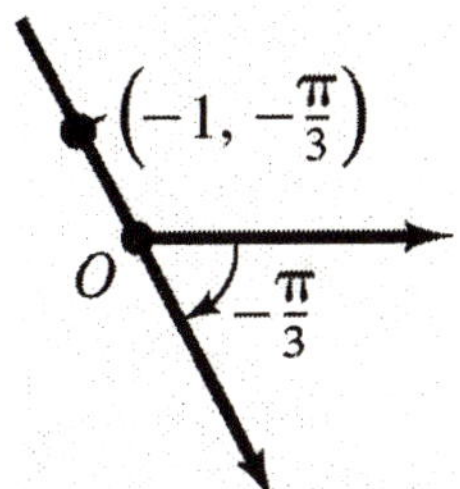

28.

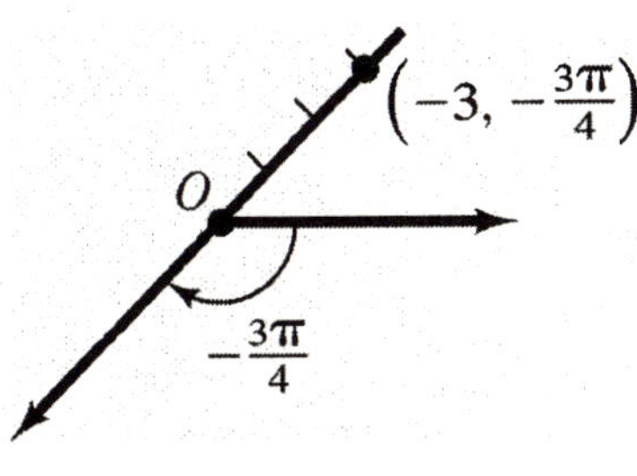

29.

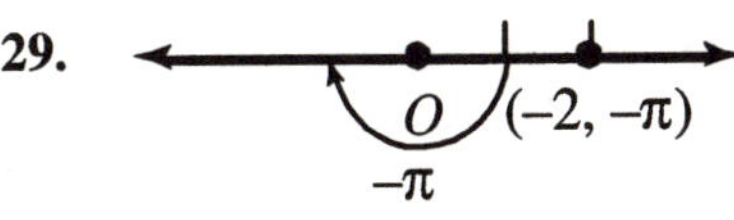

30.

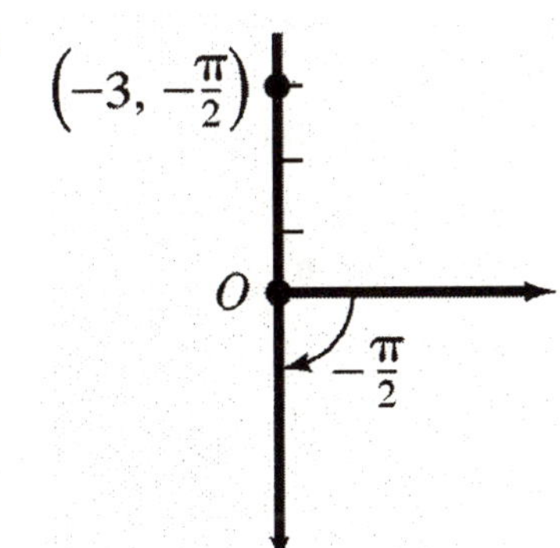

31.

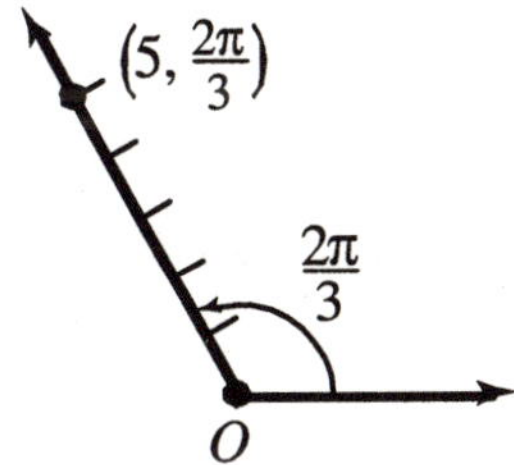

a. $r > 0,\ -2\pi \le \theta < 0$ $\left(5, -\frac{4\pi}{3}\right)$

b. $r < 0,\ 0 \le \theta < 2\pi$ $\left(-5, \frac{5\pi}{3}\right)$

c. $r > 0,\ 2\pi \le \theta < 4\pi$ $\left(5, \frac{8\pi}{3}\right)$

32.

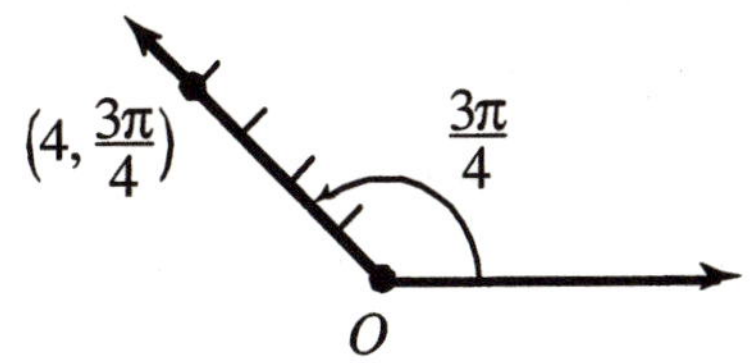

a. $r > 0,\ -2\pi \le \theta < 0$ $\left(4, -\frac{5\pi}{4}\right)$

b. $r < 0,\ 0 \le \theta < 2\pi$ $\left(-4, \frac{7\pi}{4}\right)$

c. $r > 0,\ 2\pi \le \theta < 4\pi$ $\left(4, \frac{11\pi}{4}\right)$

33.

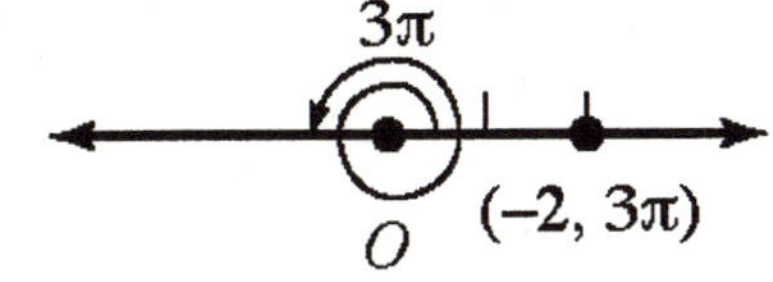

a. $r > 0,\ -2\pi \le \theta < 0$ $(2, -2\pi)$

b. $r < 0,\ 0 \le \theta < 2\pi$ $(-2, \pi)$

c. $r > 0,\ 2\pi \le \theta < 4\pi$ $(2, 2\pi)$

34.

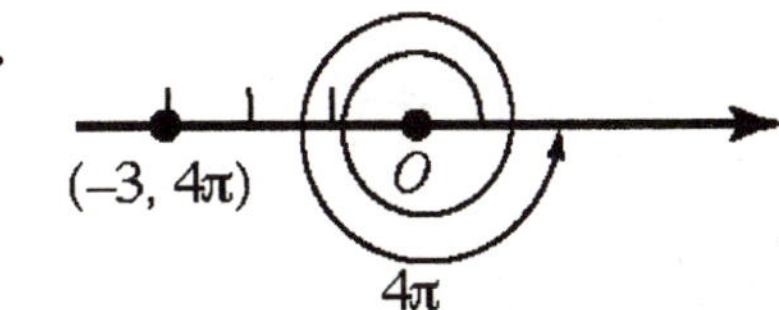

a. $r > 0,\ -2\pi \le \theta < 0$ $(3, -\pi)$

b. $r < 0,\ 0 \le \theta < 2\pi$ $(-3, 0)$

c. $r > 0,\ 2\pi \le \theta < 4\pi$ $(3, 3\pi)$

35.

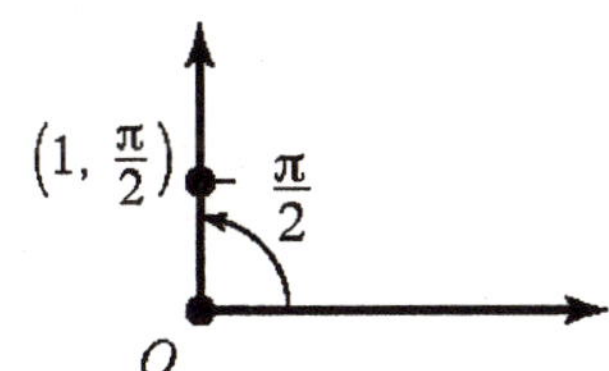

a. $r > 0,\ -2\pi \le \theta < 0$ $\left(1, -\frac{3\pi}{2}\right)$

b. $r < 0,\ 0 \le \theta < 2\pi$ $\left(-1, \frac{3\pi}{2}\right)$

c. $r > 0,\ 2\pi \le \theta < 4\pi$ $\left(1, \frac{5\pi}{2}\right)$

36.

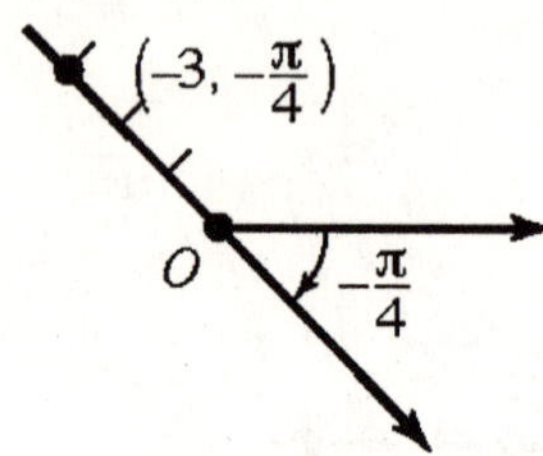

a. $r > 0,\ -2\pi \le \theta < 0 \quad (2, -\pi)$

b. $r < 0,\ 0 \le \theta < 2\pi \quad (-2, 0)$

c. $r > 0,\ 2\pi \le \theta < 4\pi \quad (2, 3\pi)$

37.

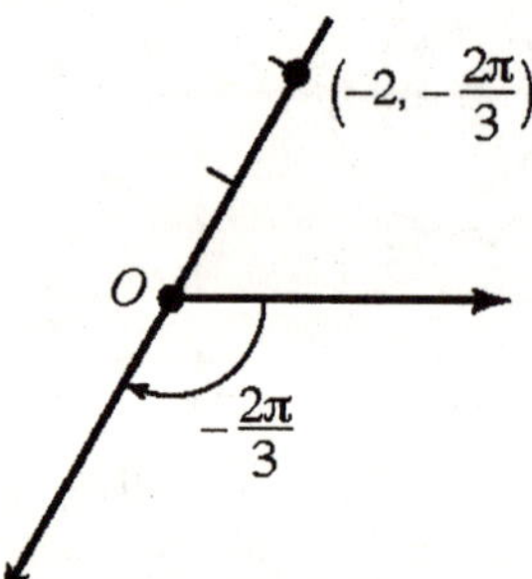

a. $r > 0,\ -2\pi \le \theta < 0 \quad \left(3, -\dfrac{5\pi}{4}\right)$

b. $r < 0,\ 0 \le \theta < 2\pi \quad \left(-3, \dfrac{7\pi}{4}\right)$

c. $(r > 0,\ 2\pi \le \theta < 4\pi \quad \left(3, \dfrac{11\pi}{4}\right)$

38.

a. $r > 0,\ -2\pi \le \theta < 0 \quad \left(2, -\dfrac{5\pi}{3}\right)$

b. $r < 0,\ 0 \le \theta < 2\pi \quad \left(-2, \dfrac{4\pi}{3}\right)$

c. $r > 0,\ 2\pi \le \theta < 4\pi \quad \left(2, \dfrac{7\pi}{3}\right)$

39. $x = r\cos\theta = 3\cos\dfrac{\pi}{2} = 3\cdot 0 = 0$

$y = r\sin\theta = 3\sin\dfrac{\pi}{2} = 3\cdot 1 = 3$

Rectangular coordinates of the point $\left(3, \dfrac{\pi}{2}\right)$ are $(0, 3)$.

40. $x = r\cos\theta = 4\cos\dfrac{3\pi}{2} = 4\cdot 0 = 0$

$y = r\sin\theta = 4\sin\dfrac{3\pi}{2} = 4\cdot(-1) = -4$

Rectangular coordinates of the point $\left(4, \dfrac{3\pi}{2}\right)$ are $(0, -4)$.

41. $x = r\cos\theta = -2\cos 0 = -2\cdot 1 = -2$

$y = r\sin\theta = -2\sin 0 = -2\cdot 0 = 0$

Rectangular coordinates of the point $(-2, 0)$ are $(-2, 0)$.

42. $x = r\cos\theta = -3\cos\pi = -3(-1) = 3$

$y = r\sin\theta = -3\sin\pi = -3\cdot 0 = 0$

Rectangular coordinates of the point $(-3, \pi)$ are $(3, 0)$.

43. $x = r\cos\theta = 6\cos 150° = 6\left(-\dfrac{\sqrt{3}}{2}\right) = -3\sqrt{3}$

$y = r\sin\theta = 6\sin 150° = 6\cdot\dfrac{1}{2} = 3$

Rectangular coordinates of the point $(6, 150°)$ are $\left(-3\sqrt{3}, 3\right)$.

44. $x = r\cos\theta = 5\cos 300° = 5\cdot\dfrac{1}{2} = \dfrac{5}{2}$

$y = r\sin\theta = 5\sin 300° = 5\left(-\dfrac{\sqrt{3}}{2}\right) = -\dfrac{5\sqrt{3}}{2}$

Rectangular coordinates of the point $(5, 300°)$ are $\left(\dfrac{5}{2}, -\dfrac{5\sqrt{3}}{2}\right)$.

45. $x = r\cos\theta = -2\cos\dfrac{3\pi}{4} = -2\left(-\dfrac{\sqrt{2}}{2}\right) = \sqrt{2}$

$y = r\sin\theta = -2\sin\dfrac{3\pi}{4} = -2\cdot\dfrac{\sqrt{2}}{2} = -\sqrt{2}$

Rectangular coordinates of the point $\left(-2, \dfrac{3\pi}{4}\right)$ are $\left(\sqrt{2}, -\sqrt{2}\right)$.

46. $x = r\cos\theta = -2\cos\frac{2\pi}{3} = -2\left(-\frac{1}{2}\right) = 1$

$y = r\sin\theta = -2\sin\frac{2\pi}{3} = -2\cdot\frac{\sqrt{3}}{2} = -\sqrt{3}$

Rectangular coordinates of the point $\left(-2, \frac{2\pi}{3}\right)$ are $\left(1, -\sqrt{3}\right)$.

47. $x = r\cos\theta = -1\cos\left(-\frac{\pi}{3}\right) = -1\cdot\frac{1}{2} = -\frac{1}{2}$

$y = r\sin\theta = -1\sin\left(-\frac{\pi}{3}\right) = -1\left(-\frac{\sqrt{3}}{2}\right) = \frac{\sqrt{3}}{2}$

Rectangular coordinates of the point $\left(-1, -\frac{\pi}{3}\right)$ are $\left(-\frac{1}{2}, \frac{\sqrt{3}}{2}\right)$.

48. $x = r\cos\theta = -3\cos\left(-\frac{3\pi}{4}\right) = -3\left(-\frac{\sqrt{2}}{2}\right) = \frac{3\sqrt{2}}{2}$

$y = r\sin\theta = -3\sin\left(-\frac{3\pi}{4}\right) = -3\left(-\frac{\sqrt{2}}{2}\right) = \frac{3\sqrt{2}}{2}$

Rectangular coordinates of the point $\left(-3, -\frac{3\pi}{4}\right)$ are $\left(\frac{3\sqrt{2}}{2}, \frac{3\sqrt{2}}{2}\right)$.

49. $x = r\cos\theta = -2\cos(-180°) = -2(-1) = 2$

$y = r\sin\theta = -2\sin(-180°) = -2\cdot 0 = 0$

Rectangular coordinates of the point $(-2, -180°)$ are $(2, 0)$.

50. $x = r\cos\theta = -3\cos(-90°) = -3\cdot 0 = 0$

$y = r\sin\theta = -3\sin(-90°) = -3(-1) = 3$

Rectangular coordinates of the point $(-3, -90°)$ are $(0, 3)$.

51. $x = r\cos\theta = 7.5\cos 110° \approx 7.5(-0.3420) \approx -2.57$

$y = r\sin\theta = 7.5\sin 110° \approx 7.5(0.9397) \approx 7.05$

Rectangular coordinates of the point $(7.5, 110°)$ are about $(-2.57, 7.05)$.

52. $x = r\cos\theta = -3.1\cos 182° \approx -3.1(-0.9994) \approx 3.10$

$y = r\sin\theta = -3.1\sin 182° \approx -3.1(-0.0349) \approx 0.11$

Rectangular coordinates of the point $(-3.1, 182°)$ are about $(3.10, 0.11)$.

53. $x = r\cos\theta = 6.3\cos(3.8) \approx 6.3(-0.7910) \approx -4.98$

$y = r\sin\theta = 6.3\sin(3.8) \approx 6.3(-0.6119) \approx -3.85$

Rectangular coordinates of the point $(6.3, 3.8)$ are about $(-4.98, -3.85)$.

54. $x = r\cos\theta = 8.1\cos(5.2) \approx 8.1(0.4685) \approx 3.79$

$y = r\sin\theta = 8.1\sin(5.2) \approx 8.1(-0.8835) \approx -7.16$

Rectangular coordinates of the point $(8.1, 5.2)$ are about $(3.79, -7.16)$.

55. $r = \sqrt{x^2 + y^2} = \sqrt{3^2 + 0^2} = \sqrt{9} = 3$

$\theta = \tan^{-1}\left(\frac{y}{x}\right) = \tan^{-1}\left(\frac{0}{3}\right) = \tan^{-1} 0 = 0$

Polar coordinates of the point $(3, 0)$ are $(3, 0)$.

56. $r = \sqrt{x^2 + y^2} = \sqrt{0^2 + 2^2} = \sqrt{4} = 2$

$\theta = \tan^{-1}\left(\frac{y}{x}\right) = \tan^{-1}\left(\frac{2}{0}\right)$

Since $\frac{2}{0}$ is undefined, $\theta = \frac{\pi}{2}$

Polar coordinates of the point $(0, 2)$ are $\left(2, \frac{\pi}{2}\right)$.

57. $r = \sqrt{x^2 + y^2} = \sqrt{(-1)^2 + 0^2} = \sqrt{1} = 1$

$\theta = \tan^{-1}\left(\frac{y}{x}\right) = \tan^{-1}\left(\frac{0}{-1}\right) = \tan^{-1} 0 = 0$

The point lies on the negative x-axis, so $\theta = \pi$.
Polar coordinates of the point $(-1, 0)$ are $(1, \pi)$.

58. $r = \sqrt{x^2 + y^2} = \sqrt{0^2 + (-2)^2} = \sqrt{4} = 2$

$\theta = \tan^{-1}\left(\frac{y}{x}\right) = \tan^{-1}\left(\frac{-2}{0}\right)$

Since $\frac{-2}{0}$ is undefined, $\theta = \frac{\pi}{2}$.

The point lies on the negative y-axis, so $\theta = -\frac{\pi}{2}$.

Polar coordinates of the point $(0, -2)$ are $\left(2, -\frac{\pi}{2}\right)$.

59. The point $(1,-1)$ lies in quadrant IV.

$r=\sqrt{x^2+y^2}=\sqrt{1^2+(-1)^2}=\sqrt{2}$

$\theta=\tan^{-1}\left(\frac{y}{x}\right)=\tan^{-1}\left(\frac{-1}{1}\right)=\tan^{-1}(-1)=-\frac{\pi}{4}$

Polar coordinates of the point $(1,-1)$ are $\left(\sqrt{2},-\frac{\pi}{4}\right)$.

60. The point $(-3, 3)$ lies in quadrant II.

$r=\sqrt{x^2+y^2}=\sqrt{(-3)^2+3^2}=3\sqrt{2}$

$\theta=\tan^{-1}\left(\frac{y}{x}\right)=\tan^{-1}\left(\frac{3}{-3}\right)=\tan^{-1}(-1)=-\frac{\pi}{4}$

Polar coordinates of the point $(-3,3)$ are $\left(3\sqrt{2},\frac{3\pi}{4}\right)$.

61. The point $\left(\sqrt{3},1\right)$ lies in quadrant I.

$r=\sqrt{x^2+y^2}=\sqrt{\left(\sqrt{3}\right)^2+1^2}=\sqrt{4}=2$

$\theta=\tan^{-1}\left(\frac{y}{x}\right)=\tan^{-1}\frac{1}{\sqrt{3}}=\frac{\pi}{6}$

Polar coordinates of the point $\left(\sqrt{3},1\right)$ are $\left(2,\frac{\pi}{6}\right)$.

62. The point $\left(-2,-2\sqrt{3}\right)$ lies in quadrant III.

$r=\sqrt{x^2+y^2}=\sqrt{(-2)^2+\left(-2\sqrt{3}\right)^2}=\sqrt{16}=4$

$\theta=\tan^{-1}\left(\frac{y}{x}\right)=\tan^{-1}\left(\frac{-2\sqrt{3}}{-2}\right)=\tan^{-1}\sqrt{3}=\frac{\pi}{3}$

The point lies in quadrant III, so $\theta=\frac{\pi}{3}-\pi=-\frac{2\pi}{3}$

Polar coordinates of the point $\left(-2,-2\sqrt{3}\right)$ are $\left(4,-\frac{2\pi}{3}\right)$.

63. The point $(1.3,-2.1)$ lies in quadrant IV.

$r=\sqrt{x^2+y^2}=\sqrt{1.3^2+(-2.1)^2}=\sqrt{6.1}\approx 2.47$

$\theta=\tan^{-1}\left(\frac{y}{x}\right)=\tan^{-1}\left(\frac{-2.1}{1.3}\right)\approx -1.02$

The polar coordinates of the point $(1.3,-2.1)$ are $(2.47,-1.02)$.

64. The point $(-0.8,-2.1)$ lies in quadrant III.

$r=\sqrt{x^2+y^2}=\sqrt{(-0.8)^2+(-2.1)^2}=\sqrt{5.05}=2.25$

$\theta=\tan^{-1}\left(\frac{y}{x}\right)=\tan^{-1}\left(\frac{-2.1}{-0.8}\right)\approx 1.21$

Since the point lies in quadrant III, $\theta=1.21-\pi\approx -1.93$.

The polar coordinates of the point $(-0.8,-2.1)$ are $(2.25,-1.93)$.

65. The point $(8.3, 4.2)$ lies in quadrant I.

$r=\sqrt{x^2+y^2}=\sqrt{8.3^2+4.2^2}=\sqrt{86.53}\approx 9.30$

$\theta=\tan^{-1}\left(\frac{y}{x}\right)=\tan^{-1}\left(\frac{4.2}{8.3}\right)\approx 0.47$

The polar coordinates of the point $(8.3, 4.2)$ are $(9.30, 0.47)$.

66. The point $(-2.3, 0.2)$ lies in quadrant II.

$r=\sqrt{x^2+y^2}=\sqrt{(-2.3)^2+0.2^2}=\sqrt{5.33}\approx 2.31$

$\theta=\tan^{-1}\left(\frac{y}{x}\right)=\tan^{-1}\left(\frac{0.2}{-2.3}\right)\approx -0.09$

Since the point lies in quadrant II, $\theta=\pi-0.09\approx 3.05$.

The polar coordinates of the point $(-2.3, 0.2)$ are $(2.31, 3.05)$.

67.

$$2x^2+2y^2=3$$
$$2\left(x^2+y^2\right)=3$$
$$2r^2=3$$
$$r^2=\frac{3}{2}$$

68.

$$x^2+y^2=x$$
$$r^2=r\cos\theta$$
$$r=\cos\theta$$

69.

$$x^2=4y$$
$$(r\cos\theta)^2=4r\sin\theta$$
$$r^2\cos^2\theta-4r\sin\theta=0$$

70.

$$y^2=2x$$
$$(r\sin\theta)^2=2r\cos\theta$$
$$r^2\sin^2\theta-2r\cos\theta=0$$

71.
$$2xy = 1$$
$$2(r\cos\theta)(r\sin\theta) = 1$$
$$r^2\left(2\sin\theta\cos\theta\right) = 1$$
$$r^2\sin 2\theta = 1$$

72.
$$4x^2y = 1$$
$$4(r\cos\theta)^2 r\sin\theta = 1$$
$$4r^2\cos^2\theta\, r\sin\theta = 1$$
$$r^3\cos^2\theta\sin\theta = \frac{1}{4}$$

73.
$$x = 4$$
$$r\cos\theta = 4$$

74.
$$y = -3$$
$$r\sin\theta = -3$$

75.
$$r = \cos\theta$$
$$r^2 = r\cos\theta$$
$$x^2 + y^2 = x$$
$$x^2 - x + y^2 = 0$$
$$x^2 - x + \frac{1}{4} + y^2 = \frac{1}{4}$$
$$\left(x - \frac{1}{2}\right)^2 + y^2 = \frac{1}{4}$$

76.
$$r = \sin\theta + 1$$
$$r^2 = r\sin\theta + r$$
$$x^2 + y^2 = y + \sqrt{x^2 + y^2}$$

77.
$$r^2 = \cos\theta$$
$$r^3 = r\cos\theta$$
$$\left(r^2\right)^{3/2} = r\cos\theta$$
$$\left(x^2 + y^2\right)^{3/2} = x$$
$$\left(x^2 + y^2\right)^{3/2} - x = 0$$

78.
$$r = \sin\theta - \cos\theta$$
$$r^2 = r\sin\theta - r\cos\theta$$
$$x^2 + y^2 = y - x$$
$$x^2 + x + y^2 - y = 0$$
$$x^2 + x + \frac{1}{4} + y^2 - y + \frac{1}{4} = \frac{1}{4} + \frac{1}{4}$$
$$\left(x + \frac{1}{2}\right)^2 + \left(y - \frac{1}{2}\right)^2 = \frac{1}{2}$$

79.
$$r = 2$$
$$r^2 = 4$$
$$x^2 + y^2 = 4$$

80.
$$r = 4$$
$$r^2 = 16$$
$$x^2 + y^2 = 16$$

81.
$$r = \frac{4}{1 - \cos\theta}$$
$$r(1 - \cos\theta) = 4$$
$$r - r\cos\theta = 4$$
$$\sqrt{x^2 + y^2} - x = 4$$
$$\sqrt{x^2 + y^2} = x + 4$$
$$x^2 + y^2 = x^2 + 8x + 16$$
$$y^2 = 8(x + 2)$$

82.
$$r = \frac{3}{3 - \cos\theta}$$
$$r(3 - \cos\theta) = 3$$
$$3r - r\cos\theta = 3$$
$$3\sqrt{x^2 + y^2} - x = 3$$
$$3\sqrt{x^2 + y^2} = x + 3$$
$$9\left(x^2 + y^2\right) = x^2 + 6x + 9$$
$$9x^2 + 9y^2 = x^2 + 6x + 9$$
$$8x^2 - 6x + 9y^2 - 9 = 0$$
$$64x^2 - 48x + 72y^2 - 72 = 0$$
$$64\left(x^2 - \frac{3}{4}x\right) + 72y^2 = 72$$
$$64\left(x^2 - \frac{3}{4}x + \frac{9}{64}\right) + 72y^2 = 72 + 64\left(\frac{9}{64}\right)$$
$$64\left(x - \frac{3}{8}\right)^2 + 72y^2 = 81$$

83. Rewrite the polar coordinates in rectangular form:
Since $P_1 = (r_1, \theta_1)$ and $P_2 = (r_2, \theta_2)$, we have that $(x_1, y_1) = (r_1 \cos\theta_1, r_1 \sin\theta_1)$ and $(x_1, y_1) = (r_2 \cos\theta_2, r_2 \sin\theta_2)$.

$$
\begin{aligned}
d &= \sqrt{(x_2 - x_1)^2 + (y_2 - y_1)^2} \\
&= \sqrt{(r_2 \cos\theta_2 - r_1 \cos\theta_1)^2 + (r_2 \sin\theta_2 - r_1 \sin\theta_1)^2} \\
&= \sqrt{r_2^2 \cos^2\theta_2 - 2r_1r_2 \cos\theta_2 \cos\theta_1 + r_1^2 \cos^2\theta_1 + r_2^2 \sin^2\theta_2 - 2r_1r_2 \sin\theta_2 \sin\theta_1 + r_1^2 \sin^2\theta_1} \\
&= \sqrt{r_2^2\left(\cos^2\theta_2 + \sin^2\theta_2\right) + r_1^2\left(\cos^2\theta_1 + \sin^2\theta_1\right) - 2r_1r_2\left(\cos\theta_2 \cos\theta_1 + \sin\theta_2 \sin\theta_1\right)} \\
&= \sqrt{r_2^2 + r_1^2 - 2r_1r_2 \cos(\theta_2 - \theta_1)}
\end{aligned}
$$

84. $x = r\cos\theta$ and $y = r\sin\theta$

85 – 86. Answers will vary.

Section 8.2

1. $(-4,6)$

2. $\cos\alpha\cos\beta + \sin\alpha\sin\beta$

3. $(x-(-2))^2 + (y-5)^2 = 3^2$

$(x+2)^2 + (y-5)^2 = 9$

4. odd, since $\sin(-x) = -\sin x$.

5. $-\dfrac{\sqrt{2}}{2}$

6. $-\dfrac{1}{2}$

7. polar equation

8. $r = 2\cos\theta$

9. $-r$

10. False

11. False

12. False

13. $r = 4$
The equation is of the form $r = a,\ a > 0$. It is a circle, center at the pole and radius 4. Transform to rectangular form:

$$
\begin{aligned}
r &= 4 \\
r^2 &= 16 \\
x^2 + y^2 &= 16
\end{aligned}
$$

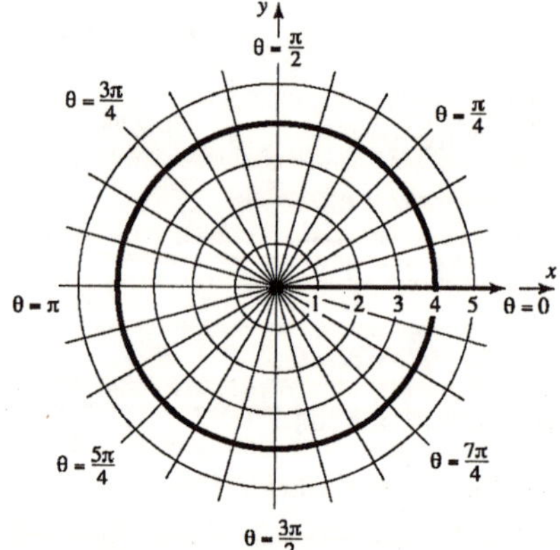

14. $r = 2$
The equation is of the form $r = a,\ a > 0$. It is a circle, center at the pole and radius 2. Transform to rectangular form:

$$
\begin{aligned}
r &= 2 \\
r^2 &= 4 \\
x^2 + y^2 &= 4
\end{aligned}
$$

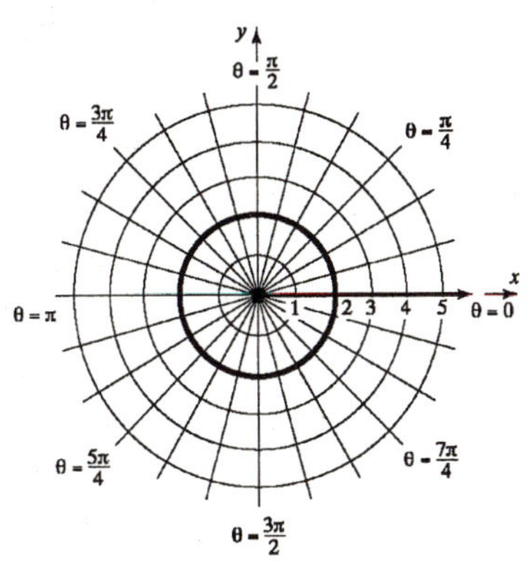

15. $\theta = \frac{\pi}{3}$
The equation is of the form $\theta = \alpha$. It is a line, passing through the pole at an angle of $\frac{\pi}{3}$ from the polar axis. Transform to rectangular form:
$$\theta = \frac{\pi}{3}$$
$$\tan\theta = \tan\frac{\pi}{3}$$
$$\frac{y}{x} = \sqrt{3}$$
$$y = \sqrt{3}x$$

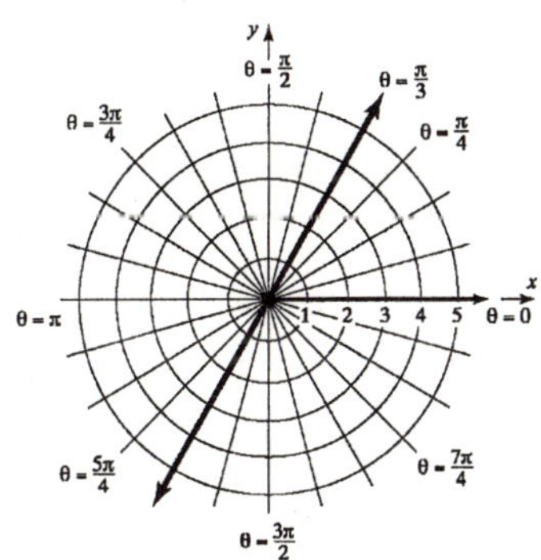

16. $\theta = -\frac{\pi}{4}$
The equation is of the form $\theta = \alpha$. It is a line, passing through the pole at an angle of $-\frac{\pi}{4}$ from the polar axis. Transform to rectangular form:
$$\theta = -\frac{\pi}{4}$$
$$\tan\theta = \tan\left(-\frac{\pi}{4}\right)$$
$$\frac{y}{x} = -1$$
$$y = -x$$

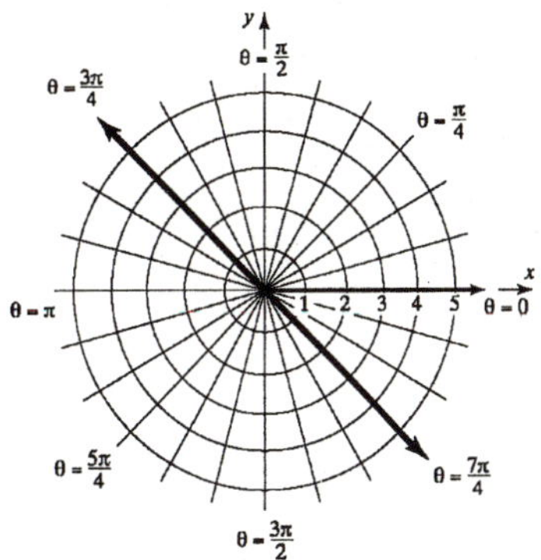

17. $r\sin\theta = 4$
The equation is of the form $r\sin\theta = b$. It is a horizontal line, 4 units above the pole.
Transform to rectangular form:
$$r\sin\theta = 4$$
$$y = 4$$

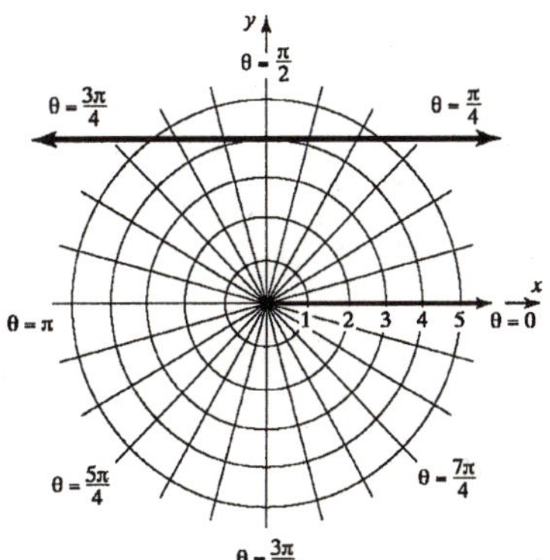

18. $r\cos\theta = 4$
The equation is of the form $r\cos\theta = a$. It is a vertical line, 4 units to the right of the pole.
Transform to rectangular form:
$$r\cos\theta = 4$$
$$x = 4$$

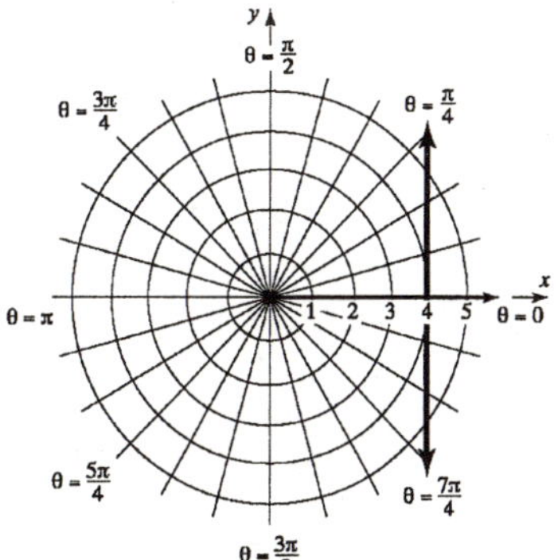

19. $r\cos\theta = -2$

The equation is of the form $r\cos\theta = a$. It is a vertical line, 2 units to the left of the pole. Transform to rectangular form:

$r\cos\theta = -2$

$x = -2$

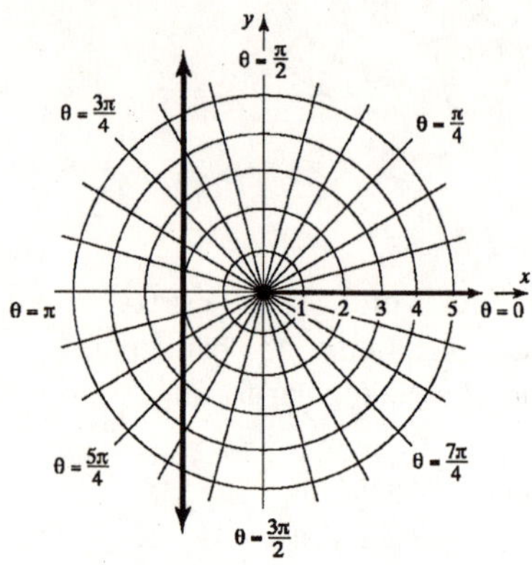

20. $r\sin\theta = -2$

The equation is of the form $r\sin\theta = b$. It is a horizontal line, 2 units below the pole. Transform to rectangular form:

$r\sin\theta = -2$

$y = -2$

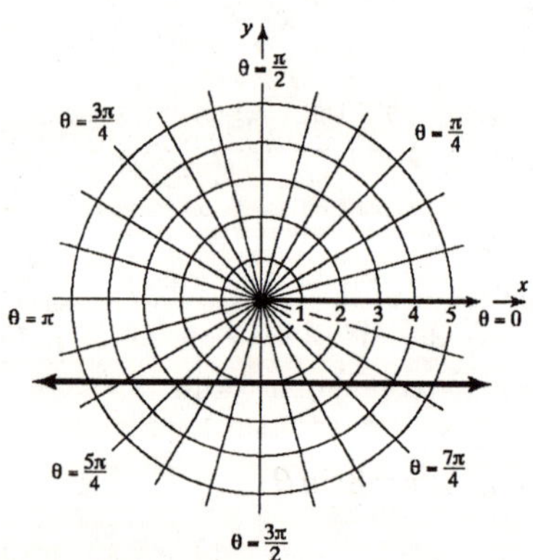

21. $r = 2\cos\theta$

The equation is of the form $r = \pm 2a\cos\theta,\ a > 0$. It is a circle, passing through the pole, and center on the polar axis. Transform to rectangular form:

$$r = 2\cos\theta$$
$$r^2 = 2r\cos\theta$$
$$x^2 + y^2 = 2x$$
$$x^2 - 2x + y^2 = 0$$
$$(x-1)^2 + y^2 = 1$$

center $(1,\ 0)$; radius 1

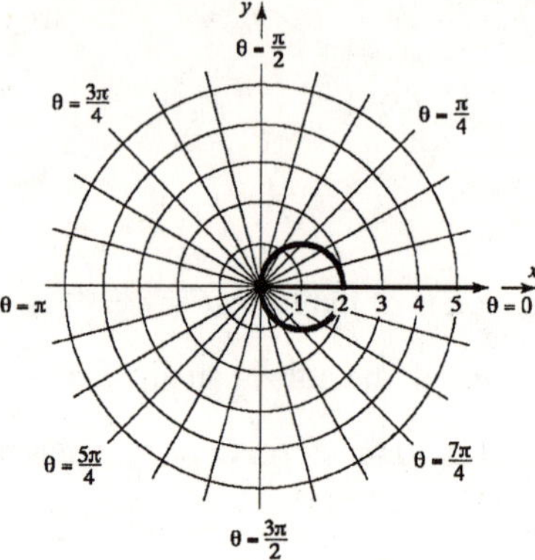

22. $r = 2\sin\theta$

The equation is of the form $r = \pm 2a\sin\theta,\ a > 0$. It is a circle, passing through the pole, and center on the line $\theta = \dfrac{\pi}{2}$. Transform to rectangular form:

$$r = 2\sin\theta$$
$$r^2 = 2r\sin\theta$$
$$x^2 + y^2 = 2y$$
$$x^2 + y^2 - 2y = 0$$
$$x^2 + (y-1)^2 = 1$$

center $(0,\ 1)$; radius 1

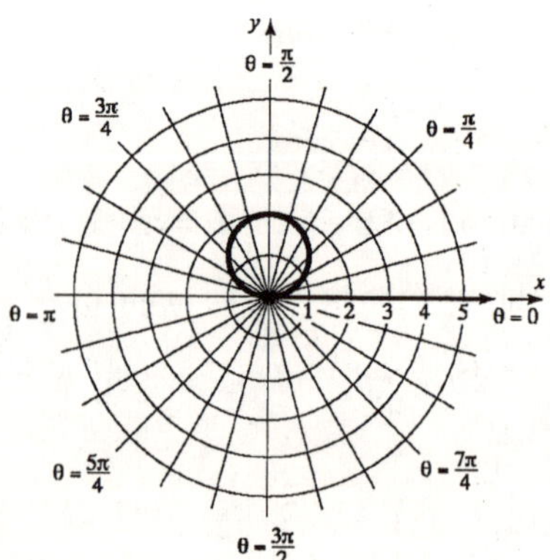

23. $r = -4\sin\theta$

The equation is of the form $r = \pm 2a\sin\theta,\ a > 0$. It is a circle, passing through the pole, and center on the line $\theta = \frac{\pi}{2}$. Transform to rectangular form:

$$r = -4\sin\theta$$
$$r^2 = -4r\sin\theta$$
$$x^2 + y^2 = -4y$$
$$x^2 + y^2 + 4y = 0$$
$$x^2 + (y+2)^2 = 4$$

center $(0, -2)$; radius 2

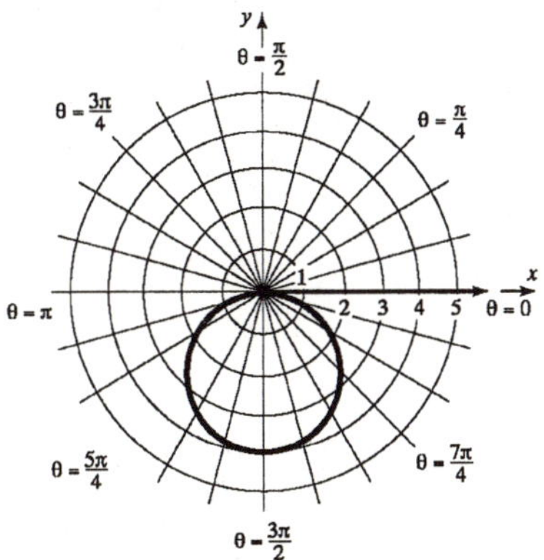

24. $r = -4\cos\theta$

The equation is of the form $r = 2a\cos\theta,\ a > 0$. It is a circle, passing through the pole, and center on the polar axis. Transform to rectangular form:

$$r = -4\cos\theta$$
$$r^2 = -4r\cos\theta$$
$$x^2 + y^2 = -4x$$
$$x^2 + 4x + y^2 = 0$$
$$(x+2)^2 + y^2 = 4$$

center $(-2, 0)$; radius 2

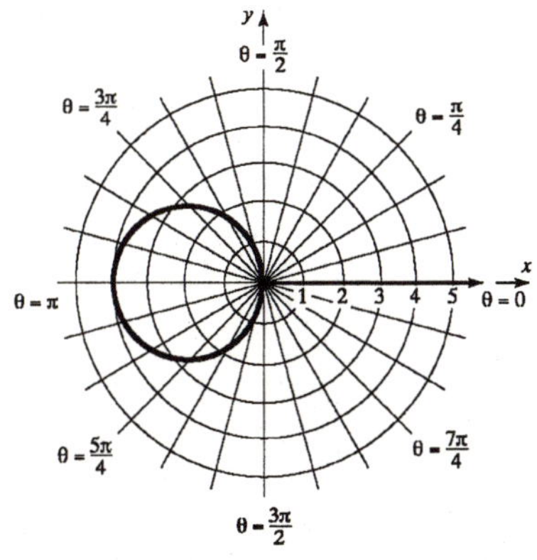

25. $r\sec\theta = 4$

The equation is a circle, passing through the pole, center on the polar axis and radius 2. Transform to rectangular form:

$$r\sec\theta = 4$$
$$r \cdot \frac{1}{\cos\theta} = 4$$
$$r = 4\cos\theta$$
$$r^2 = 4r\cos\theta$$
$$x^2 + y^2 = 4x$$
$$x^2 - 4x + y^2 = 0$$
$$(x-2)^2 + y^2 = 4$$

center $(2, 0)$; radius 2

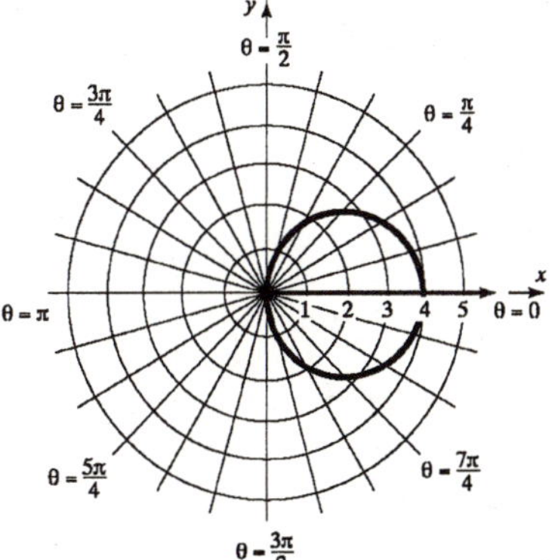

26. $r\csc\theta = 8$

The equation is a circle, passing through the pole, center on the line $\theta = \frac{\pi}{2}$ and radius 4. Transform to rectangular form:

$$r\csc\theta = 8$$
$$r \cdot \frac{1}{\sin\theta} = 8$$
$$r = 8\sin\theta$$
$$r^2 = 8r\sin\theta$$
$$x^2 + y^2 = 8y$$
$$x^2 + y^2 - 8y = 0$$
$$x^2 + (y-4)^2 = 16$$

center $(0, 4)$; radius 4

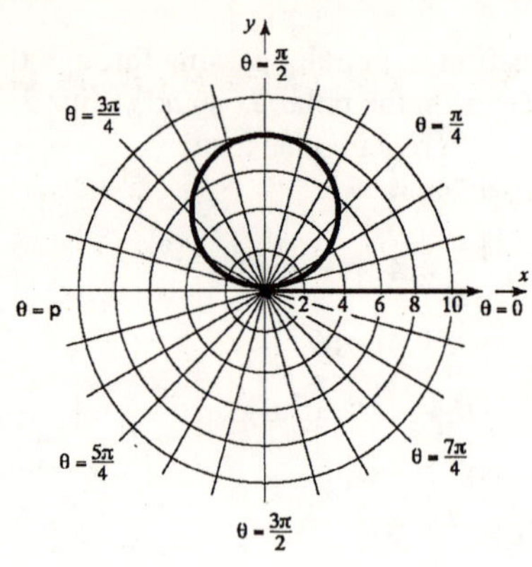

27. $r\csc\theta = -2$

The equation is a circle, passing through the pole, center on the line $\theta = \dfrac{\pi}{2}$ and radius 1.

Transform to rectangular form:

$$r\csc\theta = -2$$
$$r \cdot \frac{1}{\sin\theta} = -2$$
$$r = -2\sin\theta$$
$$r^2 = -2r\sin\theta$$
$$x^2 + y^2 = -2y$$
$$x^2 + y^2 + 2y = 0$$
$$x^2 + (y+1)^2 = 1$$

center $(0,-1)$; radius 1

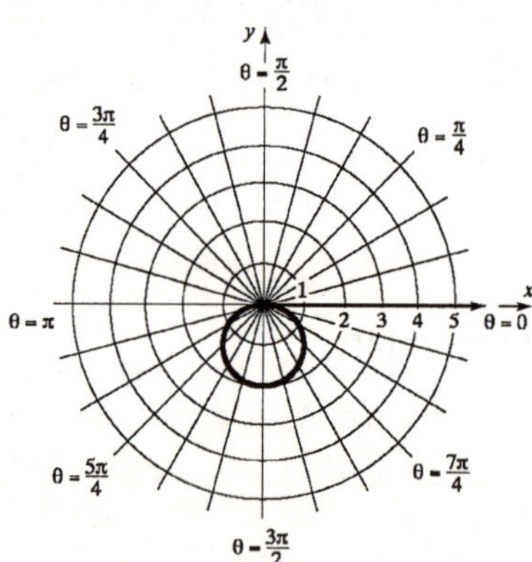

28. $r\sec\theta = -4$

The equation is a circle, passing through the pole, center on the polar axis and radius 2.

Transform to rectangular form:

$$r\sec\theta = -4$$
$$r \cdot \frac{1}{\cos\theta} = -4$$
$$r = -4\cos\theta$$
$$r^2 = -4r\cos\theta$$
$$x^2 + y^2 = -4x$$
$$x^2 + 4x + y^2 = 0$$
$$(x+2)^2 + y^2 = 4$$

center $(-2,\ 0)$; radius 2

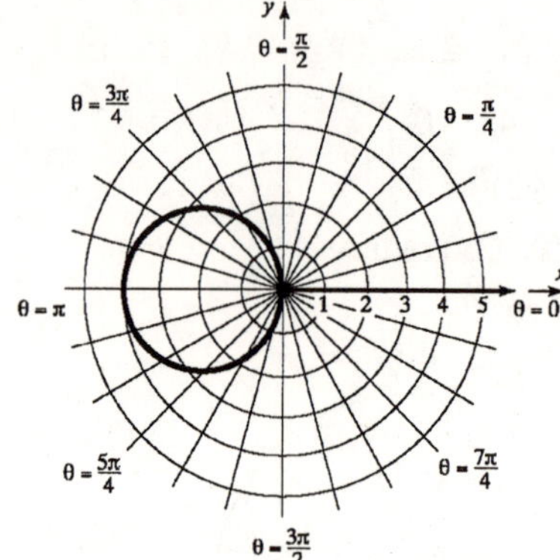

29. E

30. A

31. F

32. B

33. H

34. G

35. D

36. C

37. D

38. B

39. F

40. E

41. A

42. C

43. $r = 2 + 2\cos\theta$

The graph will be a cardioid. Check for symmetry:

Polar axis: Replace θ by $-\theta$. The result is $r = 2 + 2\cos(-\theta) = 2 + 2\cos\theta$.

The graph is symmetric with respect to the polar axis.

The line $\theta = \frac{\pi}{2}$: Replace θ by $\pi - \theta$.

$$\begin{aligned} r &= 2 + 2\cos(\pi - \theta) \\ &= 2 + 2[\cos(\pi)\cos\theta + \sin(\pi)\sin\theta] \\ &= 2 + 2(-\cos\theta + 0) \\ &= 2 - 2\cos\theta \end{aligned}$$

The test fails.

The pole: Replace r by $-r$.

$-r = 2 + 2\cos\theta$. The test fails.

Due to symmetry with respect to the polar axis, assign values to θ from 0 to π.

θ	$r = 2 + 2\cos\theta$
0	4
$\frac{\pi}{6}$	$2 + \sqrt{3} \approx 3.7$
$\frac{\pi}{3}$	3
$\frac{\pi}{2}$	2
$\frac{2\pi}{3}$	1
$\frac{5\pi}{6}$	$2 - \sqrt{3} \approx 0.3$
π	0

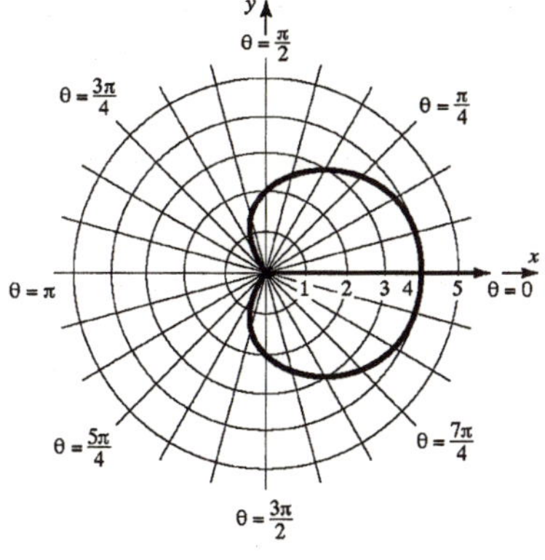

44. $r = 1 + \sin\theta$

The graph will be a cardioid. Check for symmetry:

Polar axis: Replace θ by $-\theta$. The result is $r = 1 + \sin(-\theta) = 1 - \sin\theta$. The test fails.

The line $\theta = \frac{\pi}{2}$: Replace θ by $\pi - \theta$.

$$\begin{aligned} r &= 1 + \sin(\pi - \theta) \\ &= 1 + [\sin(\pi)\cos\theta - \cos(\pi)\sin\theta] \\ &= 1 + (0 + \sin\theta) \\ &= 1 + \sin\theta \end{aligned}$$

The graph is symmetric with respect to the line $\theta = \frac{\pi}{2}$.

The pole: Replace r by $-r$. $-r = 1 + \sin\theta$. The test fails.

Due to symmetry with respect to the line $\theta = \frac{\pi}{2}$, assign values to θ from $-\frac{\pi}{2}$ to $\frac{\pi}{2}$.

θ	$r = 1 + \sin\theta$
$-\frac{\pi}{2}$	0
$-\frac{\pi}{3}$	$1 - \frac{\sqrt{3}}{2} \approx 0.1$
$-\frac{\pi}{6}$	$\frac{1}{2}$
0	1
$\frac{\pi}{6}$	$\frac{3}{2}$
$\frac{\pi}{3}$	$1 + \frac{\sqrt{3}}{2} \approx 1.9$
$\frac{\pi}{2}$	2

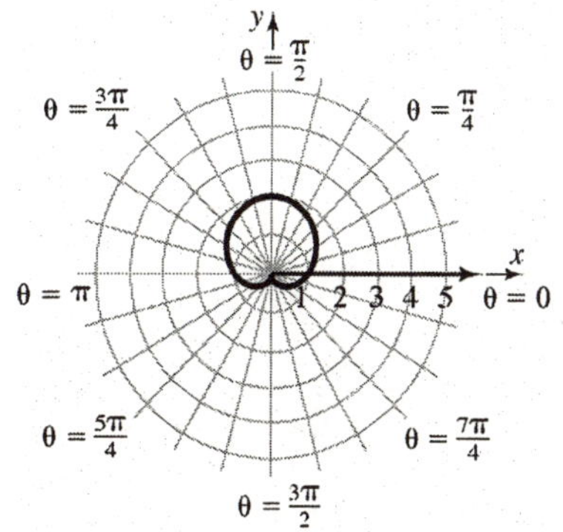

45. $r = 3 - 3\sin\theta$
The graph will be a cardioid. Check for symmetry:

Polar axis: Replace θ by $-\theta$. The result is $r = 3 - 3\sin(-\theta) = 3 + 3\sin\theta$. The test fails.

The line $\theta = \frac{\pi}{2}$: Replace θ by $\pi - \theta$.

$$\begin{aligned} r &= 3 - 3\sin(\pi - \theta) \\ &= 3 - 3\left[\sin(\pi)\cos\theta - \cos(\pi)\sin\theta\right] \\ &= 3 - 3(0 + \sin\theta) \\ &= 3 - 3\sin\theta \end{aligned}$$

The graph is symmetric with respect to the line $\theta = \frac{\pi}{2}$.

The pole: Replace r by $-r$. $-r = 3 - 3\sin\theta$. The test fails.

Due to symmetry with respect to the line $\theta = \frac{\pi}{2}$, assign values to θ from $-\frac{\pi}{2}$ to $\frac{\pi}{2}$.

θ	$r = 3 - 3\sin\theta$
$-\frac{\pi}{2}$	6
$-\frac{\pi}{3}$	$3 + \frac{3\sqrt{3}}{2} \approx 5.6$
$-\frac{\pi}{6}$	$\frac{9}{2}$
0	3
$\frac{\pi}{6}$	$\frac{3}{2}$
$\frac{\pi}{3}$	$3 - \frac{3\sqrt{3}}{2} \approx 0.4$
$\frac{\pi}{2}$	0

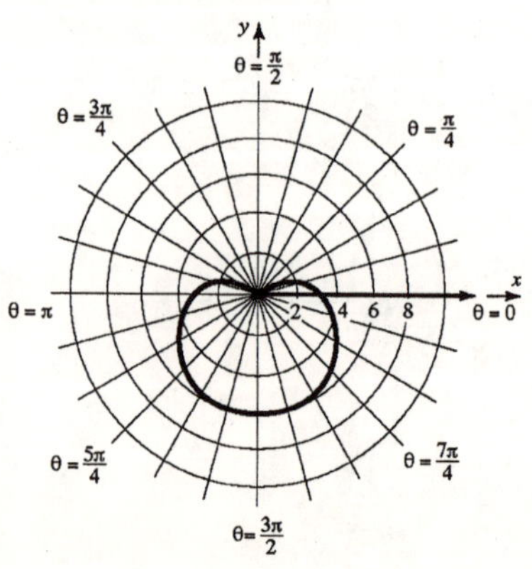

46. $r = 2 - 2\cos\theta$
The graph will be a cardioid. Check for symmetry:

Polar axis: Replace θ by $-\theta$. The result is $r = 2 - 2\cos(-\theta) = 2 - 2\cos\theta$. The graph is symmetric with respect to the polar axis.

The line $\theta = \frac{\pi}{2}$: Replace θ by $\pi - \theta$.

$$\begin{aligned} r &= 2 - 2\cos(\pi - \theta) \\ &= 2 - 2\left[\cos(\pi)\cos\theta + \sin(\pi)\sin\theta\right] \\ &= 2 - 2(-\cos\theta + 0) \\ &= 2 + 2\cos\theta \end{aligned}$$

The test fails.

The pole: Replace r by $-r$. $-r = 2 - 2\cos\theta$. The test fails.
Due to symmetry with respect to the polar axis, assign values to θ from 0 to π.

θ	$r = 2 - 2\cos\theta$
0	0
$\frac{\pi}{6}$	$2 - \sqrt{3} \approx 0.3$
$\frac{\pi}{3}$	1
$\frac{\pi}{2}$	2
$\frac{2\pi}{3}$	3
$\frac{5\pi}{6}$	$2 + \sqrt{3} \approx 3.7$
π	4

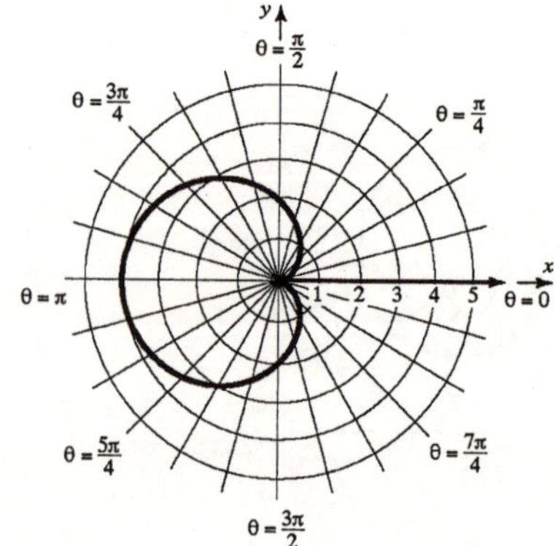

47. $r = 2 + \sin\theta$
The graph will be a limaçon without an inner loop. Check for symmetry:
Polar axis: Replace θ by $-\theta$. The result is $r = 2 + \sin(-\theta) = 2 - \sin\theta$. The test fails.

The line $\theta = \frac{\pi}{2}$: Replace θ by $\pi - \theta$.

$$\begin{aligned} r &= 2 + \sin(\pi - \theta) \\ &= 2 + \left[\sin(\pi)\cos\theta - \cos(\pi)\sin\theta\right] \\ &= 2 + (0 + \sin\theta) = 2 + \sin\theta \end{aligned}$$

The graph is symmetric with respect to the line $\theta = \frac{\pi}{2}$.

The pole: Replace r by $-r$. $-r = 2 + \sin\theta$. The test fails.

Due to symmetry with respect to the line $\theta = \frac{\pi}{2}$, assign values to θ from $-\frac{\pi}{2}$ to $\frac{\pi}{2}$.

θ	$r = 2 + \sin\theta$
$-\frac{\pi}{2}$	1
$-\frac{\pi}{3}$	$2 - \frac{\sqrt{3}}{2} \approx 1.1$
$-\frac{\pi}{6}$	$\frac{3}{2}$
0	2
$\frac{\pi}{6}$	$\frac{5}{2}$
$\frac{\pi}{3}$	$2 + \frac{\sqrt{3}}{2} \approx 2.9$
$\frac{\pi}{2}$	3

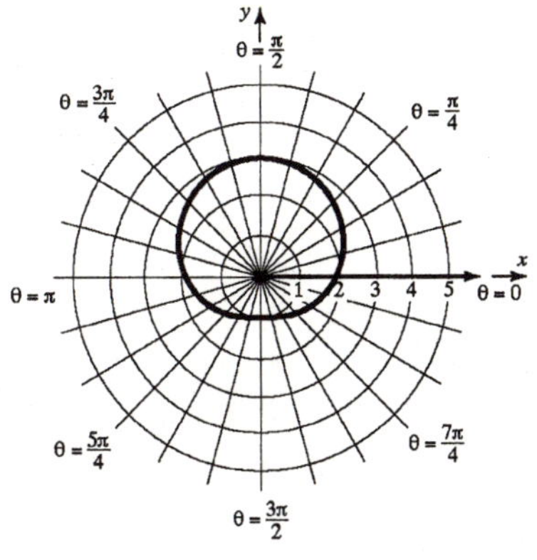

48. $r = 2 - \cos\theta$
The graph will be a limaçon without an inner loop. Check for symmetry:

Polar axis: Replace θ by $-\theta$. The result is $r = 2 - \cos(-\theta) = 2 - \cos\theta$. The graph is symmetric with respect to the polar axis.

The line $\theta = \frac{\pi}{2}$: Replace θ by $\pi - \theta$.

$$\begin{aligned} r &= 2 - \cos(\pi - \theta) \\ &= 2 - \left[\cos(\pi)\cos\theta + \sin(\pi)\sin\theta\right] \\ &= 2 - (-\cos\theta + 0) \\ &= 2 + \cos\theta \end{aligned}$$

The test fails.

The pole: Replace r by $-r$. $-r = 2 - \cos\theta$. The test fails.

Due to symmetry with respect to the polar axis, assign values to θ from 0 to π.

θ	$r = 2 - \cos\theta$
0	1
$\frac{\pi}{6}$	$2 - \frac{\sqrt{3}}{2} \approx 1.1$
$\frac{\pi}{3}$	$\frac{3}{2}$
$\frac{\pi}{2}$	2
$\frac{2\pi}{3}$	$\frac{5}{2}$
$\frac{5\pi}{6}$	$2 + \frac{\sqrt{3}}{2} \approx 2.9$
π	3

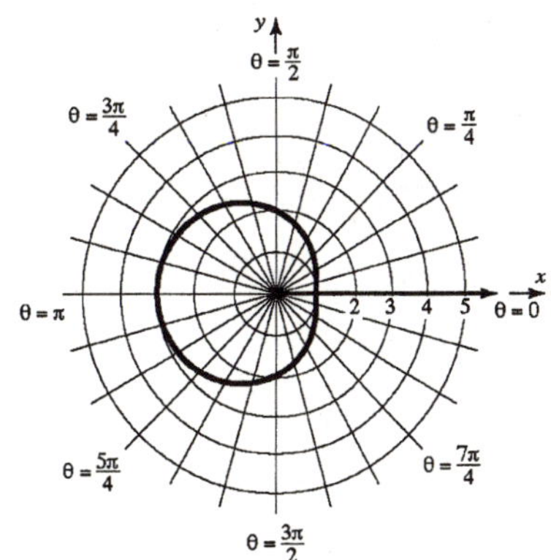

49. $r = 4 - 2\cos\theta$
The graph will be a limaçon without an inner loop. Check for symmetry:

Polar axis: Replace θ by $-\theta$. The result is $r = 4 - 2\cos(-\theta) = 4 - 2\cos\theta$. The graph is symmetric with respect to the polar axis.

The line $\theta = \frac{\pi}{2}$: Replace θ by $\pi - \theta$.

$$\begin{aligned} r &= 4 - 2\cos(\pi - \theta) \\ &= 4 - 2\left[\cos(\pi)\cos\theta + \sin(\pi)\sin\theta\right] \\ &= 4 - 2(-\cos\theta + 0) \\ &= 4 + 2\cos\theta \end{aligned}$$

The test fails.

The pole: Replace r by $-r$. $-r = 4 - 2\cos\theta$. The test fails.

Due to symmetry with respect to the polar axis, assign values to θ from 0 to π.

θ	$r = 4 - 2\cos\theta$
0	2
$\frac{\pi}{6}$	$4 - \sqrt{3} \approx 2.3$
$\frac{\pi}{3}$	3
$\frac{\pi}{2}$	4
$\frac{2\pi}{3}$	5
$\frac{5\pi}{6}$	$4 + \sqrt{3} \approx 5.7$
π	6

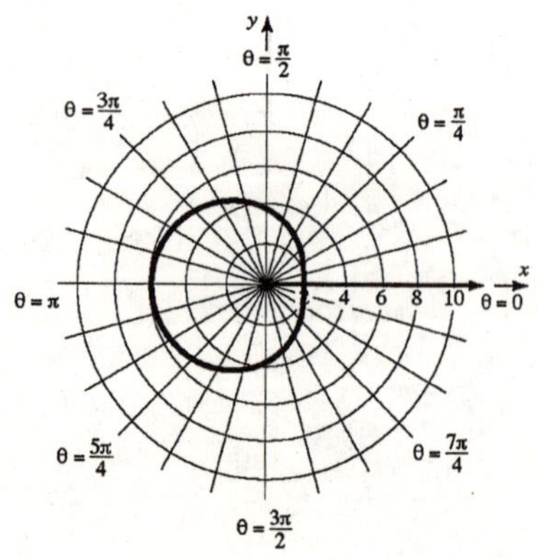

50. $r = 4 + 2\sin\theta$
The graph will be a limaçon without an inner loop. Check for symmetry:

Polar axis: Replace θ by $-\theta$. The result is $r = 4 + 2\sin(-\theta) = 4 - 2\sin\theta$. The test fails.

The line $\theta = \frac{\pi}{2}$: Replace θ by $\pi - \theta$.

$$\begin{aligned} r &= 4 + 2\sin(\pi - \theta) \\ &= 4 + 2\left[\sin(\pi)\cos\theta - \cos(\pi)\sin\theta\right] \\ &= 4 + 2(0 + \sin\theta) \\ &= 4 + 2\sin\theta \end{aligned}$$

The graph is symmetric with respect to the line $\theta = \frac{\pi}{2}$.

The pole: Replace r by $-r$. $-r = 4 + 2\sin\theta$. The test fails.

Due to symmetry with respect to the line $\theta = \frac{\pi}{2}$, assign values to θ from $-\frac{\pi}{2}$ to $\frac{\pi}{2}$.

θ	$r = 4 + 2\sin\theta$
$-\frac{\pi}{2}$	2
$-\frac{\pi}{3}$	$4 - \sqrt{3} \approx 2.3$
$-\frac{\pi}{6}$	3
0	4
$\frac{\pi}{6}$	5
$\frac{\pi}{3}$	$4 + \sqrt{3} \approx 5.7$
$\frac{\pi}{2}$	6

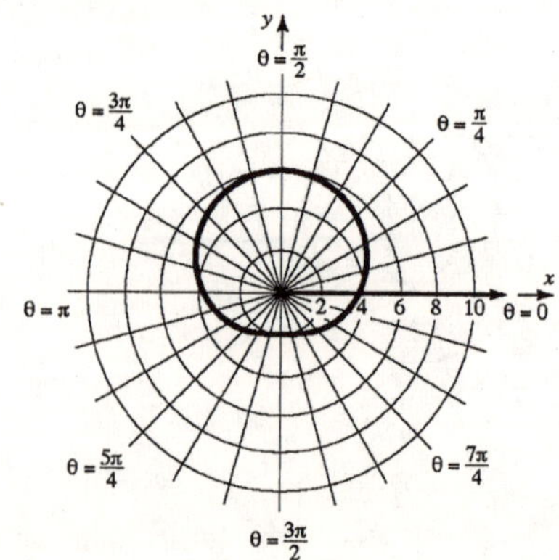

51. $r = 1 + 2\sin\theta$
The graph will be a limaçon with an inner loop.
Check for symmetry:

Polar axis: Replace θ by $-\theta$. The result is $r = 1 + 2\sin(-\theta) = 1 - 2\sin\theta$. The test fails.

The line $\theta = \frac{\pi}{2}$: Replace θ by $\pi - \theta$.

$$\begin{aligned} r &= 1 + 2\sin(\pi - \theta) \\ &= 1 + 2\left[\sin(\pi)\cos\theta - \cos(\pi)\sin\theta\right] \\ &= 1 + 2(0 + \sin\theta) \\ &= 1 + 2\sin\theta \end{aligned}$$

The graph is symmetric with respect to the line $\theta = \frac{\pi}{2}$.

The pole: Replace r by $-r$. $-r = 1 + 2\sin\theta$. The test fails.

Due to symmetry with respect to the line $\theta = \frac{\pi}{2}$, assign values to θ from $-\frac{\pi}{2}$ to $\frac{\pi}{2}$.

θ	$r = 1 + 2\sin\theta$
$-\frac{\pi}{2}$	-1
$-\frac{\pi}{3}$	$1 - \sqrt{3} \approx -0.7$
$-\frac{\pi}{6}$	0
0	1
$\frac{\pi}{6}$	2
$\frac{\pi}{3}$	$1 + \sqrt{3} \approx 2.7$
$\frac{\pi}{2}$	3

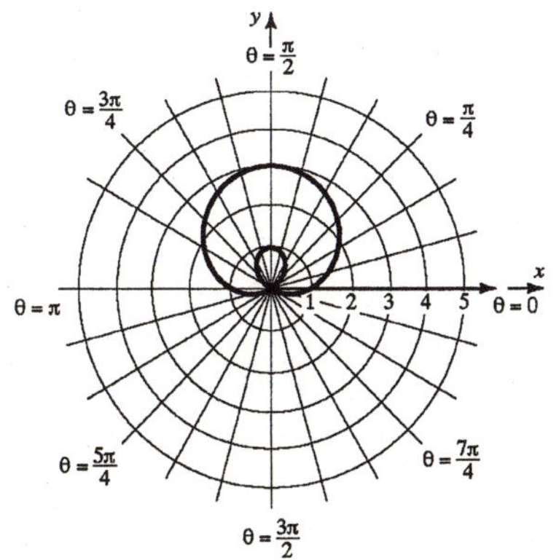

52. $r = 1 - 2\sin\theta$
The graph will be a limaçon with an inner loop.
Check for symmetry:

Polar axis: Replace θ by $-\theta$. The result is $r = 1 - 2\sin(-\theta) = 1 + 2\sin\theta$. The test fails.

The line $\theta = \frac{\pi}{2}$: Replace θ by $\pi - \theta$.

$$\begin{aligned} r &= 1 - 2\sin(\pi - \theta) \\ &= 1 - 2\left[\sin(\pi)\cos\theta - \cos(\pi)\sin\theta\right] \\ &= 1 - 2(0 + \sin\theta) \\ &= 1 - 2\sin\theta \end{aligned}$$

The graph is symmetric with respect to the line $\theta = \frac{\pi}{2}$.

The pole: Replace r by $-r$. $-r = 1 - 2\sin\theta$. The test fails.

Due to symmetry with respect to the line $\theta = \frac{\pi}{2}$, assign values to θ from $-\frac{\pi}{2}$ to $\frac{\pi}{2}$.

θ	$r = 1 - 2\sin\theta$
$-\frac{\pi}{2}$	3
$-\frac{\pi}{3}$	$1 + \sqrt{3} \approx 2.7$
$-\frac{\pi}{6}$	2
0	1
$\frac{\pi}{6}$	0
$\frac{\pi}{3}$	$1 - \sqrt{3} \approx -0.7$
$\frac{\pi}{2}$	-1

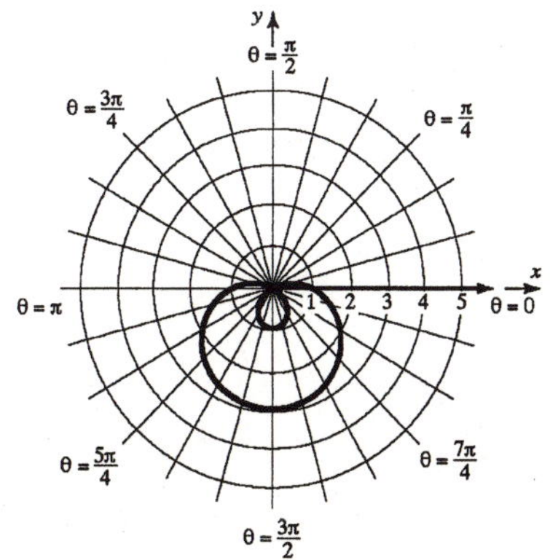

53. $r = 2 - 3\cos\theta$

The graph will be a limaçon with an inner loop. Check for symmetry:

Polar axis: Replace θ by $-\theta$. The result is $r = 2 - 3\cos(-\theta) = 2 - 3\cos\theta$.

The graph is symmetric with respect to the polar axis.

The line $\theta = \frac{\pi}{2}$: Replace θ by $\pi - \theta$.

$$\begin{aligned} r &= 2 - 3\cos(\pi - \theta) \\ &= 2 - 3\left[\cos(\pi)\cos\theta + \sin(\pi)\sin\theta\right] \\ &= 2 - 3(-\cos\theta + 0) \\ &= 2 + 3\cos\theta \end{aligned}$$

The test fails.

The pole: Replace r by $-r$. $-r = 2 - 3\cos\theta$. The test fails.

Due to symmetry with respect to the polar axis, assign values to θ from 0 to π.

θ	$r = 2 - 3\cos\theta$
0	−1
$\frac{\pi}{6}$	$2 - \frac{3\sqrt{3}}{2} \approx -0.6$
$\frac{\pi}{3}$	$\frac{1}{2}$
$\frac{\pi}{2}$	2
$\frac{2\pi}{3}$	$\frac{7}{2}$
$\frac{5\pi}{6}$	$2 + \frac{3\sqrt{3}}{2} \approx 4.6$
π	5

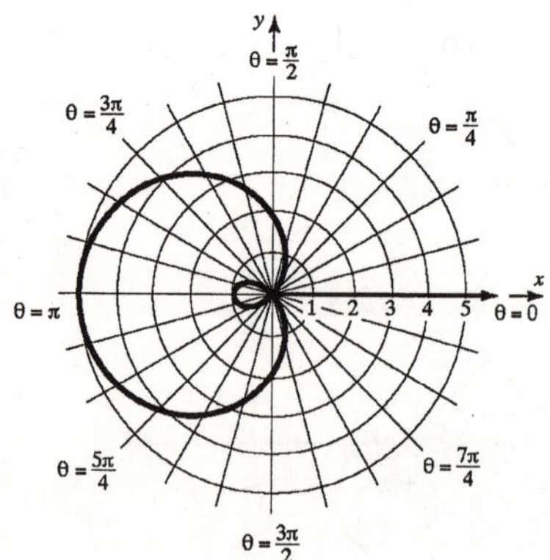

54. $r = 2 + 4\cos\theta$

The graph will be a limaçon with an inner loop. Check for symmetry:

Polar axis: Replace θ by $-\theta$. The result is $r = 2 + 4\cos(-\theta) = 2 + 4\cos\theta$.

The graph is symmetric with respect to the polar axis.

The line $\theta = \frac{\pi}{2}$: Replace θ by $\pi - \theta$.

$$\begin{aligned} r &= 2 + 4\cos(\pi - \theta) \\ &= 2 + 4\left[\cos(\pi)\cos\theta + \sin(\pi)\sin\theta\right] \\ &= 2 + 4(-\cos\theta + 0) \\ &= 2 - 4\cos\theta \end{aligned}$$

The test fails.

The pole: Replace r by $-r$. $-r = 2 + 4\cos\theta$. The test fails.

Due to symmetry with respect to the polar axis, assign values to θ from 0 to π.

θ	$r = 2 + 4\cos\theta$
0	6
$\frac{\pi}{6}$	$2 + 2\sqrt{3} \approx 5.5$
$\frac{\pi}{3}$	4
$\frac{\pi}{2}$	2
$\frac{2\pi}{3}$	0
$\frac{5\pi}{6}$	$2 - 2\sqrt{3} \approx -1.5$
π	−2

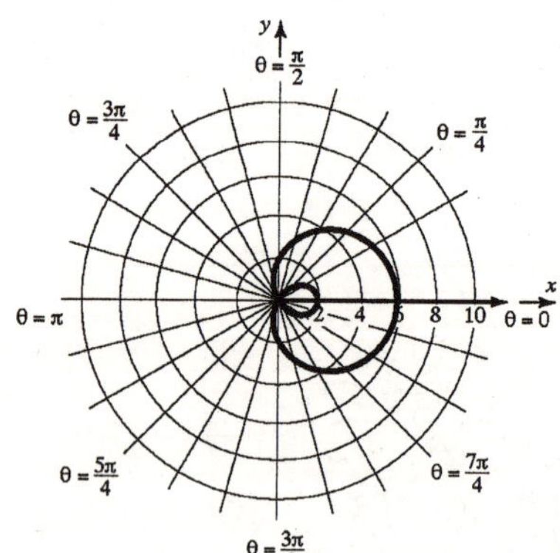

55. $r = 3\cos(2\theta)$

The graph will be a rose with four petals. Check for symmetry:

Polar axis: Replace θ by $-\theta$.
$r = 3\cos(2(-\theta)) = 3\cos(-2\theta) = 3\cos(2\theta)$. The graph is symmetric with respect to the polar axis.

The line $\theta = \frac{\pi}{2}$: Replace θ by $\pi - \theta$.

$$\begin{aligned} r &= 3\cos\left[2(\pi-\theta)\right] \\ &= 3\cos(2\pi - 2\theta) \\ &= 3\left[\cos(2\pi)\cos(2\theta) + \sin(2\pi)\sin(2\theta)\right] \\ &= 3(\cos 2\theta + 0) \\ &= 3\cos(2\theta) \end{aligned}$$

The graph is symmetric with respect to the line $\theta = \frac{\pi}{2}$.

The pole: Since the graph is symmetric with respect to both the polar axis and the line $\theta = \frac{\pi}{2}$, it is also symmetric with respect to the pole.

Due to symmetry, assign values to θ from 0 to $\frac{\pi}{2}$.

θ	$r = 3\cos(2\theta)$
	3
$\frac{\pi}{6}$	$\frac{3}{2}$
$\frac{\pi}{4}$	0
$\frac{\pi}{3}$	$-\frac{3}{2}$
$\frac{\pi}{2}$	-3

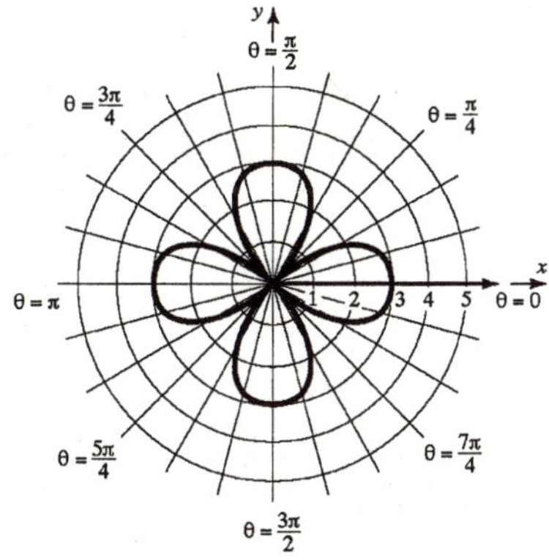

56. $r = 2\sin(3\theta)$
The graph will be a rose with three petals.
Check for symmetry:

Polar axis: Replace θ by $-\theta$.
$r = 2\sin\left[3(-\theta)\right] = 2\sin(-3\theta) = -2\sin(3\theta)$. The test fails.

The line $\theta = \frac{\pi}{2}$: Replace θ by $\pi - \theta$.

$$\begin{aligned} r &= 2\sin\left[3(\pi-\theta)\right] \\ &= 2\sin(3\pi - 3\theta) \\ &= 2\left[\sin(3\pi)\cos(3\theta) - \cos(3\pi)\sin(3\theta)\right] \\ &= 2\left[0 + \sin(3\theta)\right] \\ &= 2\sin(3\theta) \end{aligned}$$

The graph is symmetric with respect to the line $\theta = \frac{\pi}{2}$.

The pole: Replace r by $-r$. $-r = 2\sin(3\theta)$. The test fails.

Due to symmetry with respect to the line $\theta = \frac{\pi}{2}$, assign values to θ from $-\frac{\pi}{2}$ to $\frac{\pi}{2}$.

θ	$r = 2\sin(3\theta)$
$-\frac{\pi}{2}$	2
$-\frac{\pi}{3}$	0
$-\frac{\pi}{4}$	$-\sqrt{2} \approx -1.4$
$-\frac{\pi}{6}$	-2
0	0
$\frac{\pi}{6}$	2
$\frac{\pi}{4}$	$\sqrt{2} \approx 1.4$
$\frac{\pi}{3}$	0
$\frac{\pi}{2}$	-2

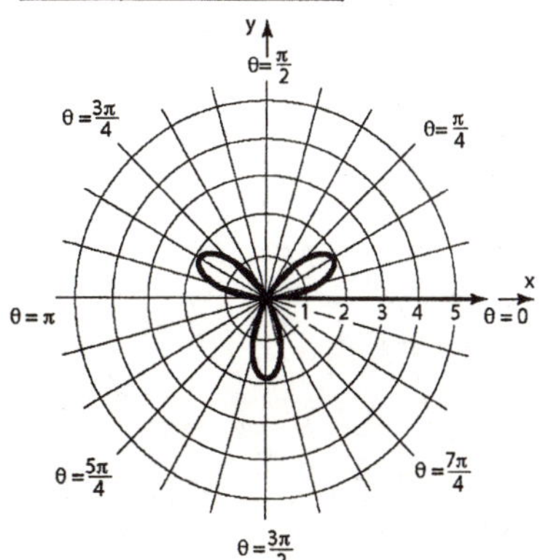

57. $r = 4\sin(5\theta)$

The graph will be a rose with five petals. Check for symmetry:

Polar axis: Replace θ by $-\theta$.

$r = 4\sin[5(-\theta)] = 4\sin(-5\theta) = -4\sin(5\theta)$.

The test fails.

The line $\theta = \frac{\pi}{2}$: Replace θ by $\pi - \theta$.

$$\begin{aligned} r &= 4\sin[5(\pi-\theta)] = 4\sin(5\pi - 5\theta) \\ &= 4[\sin(5\pi)\cos(5\theta) - \cos(5\pi)\sin(5\theta)] \\ &= 4[0 + \sin(5\theta)] \\ &= 4\sin(5\theta) \end{aligned}$$

The graph is symmetric with respect to the line $\theta = \frac{\pi}{2}$.

The pole: Replace r by $-r$. $-r = 4\sin(5\theta)$. The test fails.

Due to symmetry with respect to the line $\theta = \frac{\pi}{2}$, assign values to θ from $-\frac{\pi}{2}$ to $\frac{\pi}{2}$.

θ	$r = 4\sin(5\theta)$
$-\frac{\pi}{2}$	-4
$-\frac{\pi}{3}$	$2\sqrt{3} \approx 3.5$
$-\frac{\pi}{4}$	$2\sqrt{2} \approx 2.8$
$-\frac{\pi}{6}$	-2
0	0
$\frac{\pi}{6}$	2
$\frac{\pi}{4}$	$-2\sqrt{2} \approx -2.8$
$\frac{\pi}{3}$	$-2\sqrt{3} \approx -3.5$
$\frac{\pi}{2}$	4

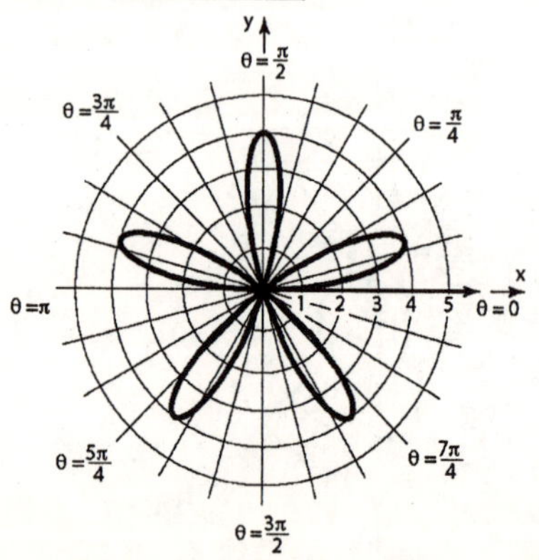

58. $r = 3\cos(4\theta)$

The graph will be a rose with eight petals. Check for symmetry:

Polar axis: Replace θ by $-\theta$.

$r = 3\cos(4(-\theta)) = 3\cos(-4\theta) = 3\cos(4\theta)$. The graph is symmetric with respect to the polar axis.

The line $\theta = \frac{\pi}{2}$: Replace θ by $\pi - \theta$.

$$\begin{aligned} r &= 3\cos[4(\pi-\theta)] \\ &= 3\cos(4\pi - 4\theta) \\ &= 3[\cos(4\pi)\cos(4\theta) + \sin(4\pi)\sin(4\theta)] \\ &= 3(\cos 4\theta + 0) \\ &= 3\cos(4\theta) \end{aligned}$$

The graph is symmetric with respect to the line $\theta = \frac{\pi}{2}$.

The pole: Since the graph is symmetric with respect to both the polar axis and the line $\theta = \frac{\pi}{2}$, it is also symmetric with respect to the pole.

Due to symmetry, assign values to θ from 0 to $\frac{\pi}{2}$.

θ	$r = 3\cos(4\theta)$
0	2
$\frac{\pi}{6}$	$-\frac{3}{2}$
$\frac{\pi}{4}$	-3
$\frac{\pi}{3}$	$-\frac{3}{2}$
$\frac{\pi}{2}$	3

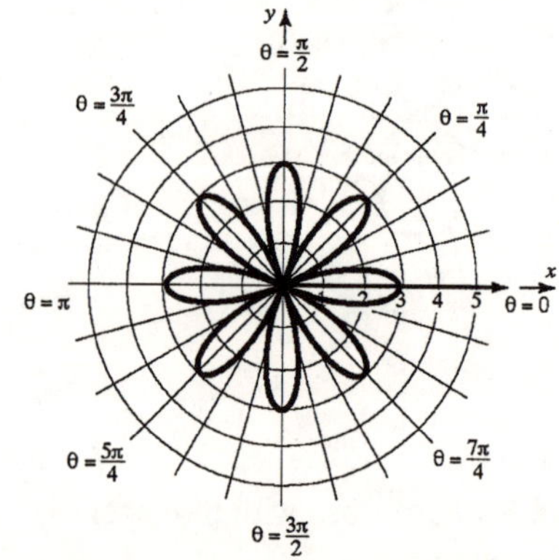

59. $r^2 = 9\cos(2\theta)$

The graph will be a lemniscate. Check for symmetry:

Polar axis: Replace θ by $-\theta$.

$r^2 = 9\cos(2(-\theta)) = 9\cos(-2\theta) = 9\cos(2\theta)$. The graph is symmetric with respect to the polar axis.

The line $\theta = \frac{\pi}{2}$: Replace θ by $\pi - \theta$.

$$\begin{aligned} r^2 &= 9\cos\left[2(\pi-\theta)\right] \\ &= 9\cos(2\pi - 2\theta) \\ &= 9\left[\cos(2\pi)\cos 2\theta + \sin(2\pi)\sin 2\theta\right] \\ &= 9(\cos 2\theta + 0) \\ &= 9\cos(2\theta) \end{aligned}$$

The graph is symmetric with respect to the line $\theta = \frac{\pi}{2}$.

The pole: Since the graph is symmetric with respect to both the polar axis and the line $\theta = \frac{\pi}{2}$, it is also symmetric with respect to the pole.

Due to symmetry, assign values to θ from 0 to $\frac{\pi}{2}$.

θ	$r = \pm\sqrt{9\cos(2\theta)}$
0	± 3
$\frac{\pi}{6}$	$\pm\frac{3\sqrt{2}}{2} \approx \pm 2.1$
$\frac{\pi}{4}$	0
$\frac{\pi}{3}$	undefined
$\frac{\pi}{2}$	undefined

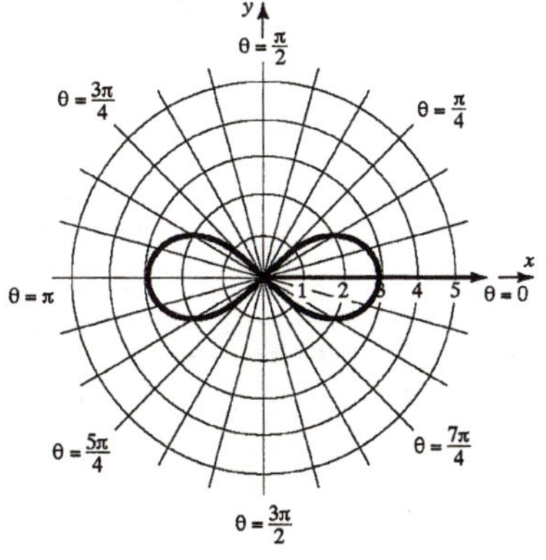

60. $r^2 = \sin(2\theta)$

The graph will be a lemniscate. Check for symmetry:

Polar axis: Replace θ by $-\theta$.

$r^2 = \sin(2(-\theta)) = \sin(-2\theta) = -\sin(2\theta)$. The test fails.

The line $\theta = \frac{\pi}{2}$: Replace θ by $\pi - \theta$.

$$\begin{aligned} r^2 &= \sin\left[2(\pi-\theta)\right] \\ &= \sin(2\pi - 2\theta) \\ &= \sin(2\pi)\cos 2\theta - \cos(2\pi)\sin(2\theta) \\ &= 0 - \sin(2\theta) \\ &= -\sin(2\theta) \end{aligned}$$

The test fails.

The pole: Replace r by $-r$.

$$\begin{aligned} (-r)^2 &= \sin(2\theta) \\ r^2 &= \sin(2\theta) \end{aligned}$$

The graph is symmetric with respect to the pole.

Due to symmetry, assign values to θ from 0 to π.

θ	$r = \pm\sqrt{\sin(2\theta)}$
0	0
$\frac{\pi}{6}$	$\pm\sqrt{\frac{\sqrt{3}}{2}}$
$\frac{\pi}{3}$	$\pm\sqrt{\frac{\sqrt{3}}{2}}$
$\frac{\pi}{2}$	0
$\frac{2\pi}{3}$	undefined
$\frac{5\pi}{6}$	undefined
π	0

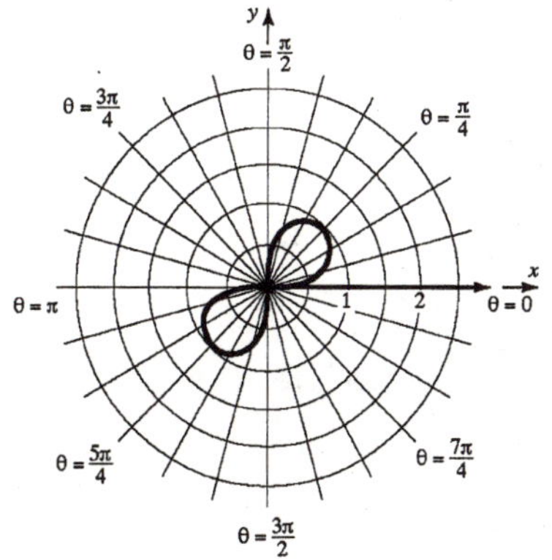

61. $r = 2^{\theta}$

The graph will be a spiral. Check for symmetry:

Polar axis: Replace θ by $-\theta$. $r = 2^{-\theta}$. The test fails.

The line $\theta = \frac{\pi}{2}$: Replace θ by $\pi - \theta$. $r = 2^{\pi-\theta}$. The test fails.

The pole: Replace r by $-r$. $-r = 2^{\theta}$. The test fails.

θ	$r = 2^{\theta}$
$-\pi$	0.1
$-\frac{\pi}{2}$	0.3
$-\frac{\pi}{4}$	0.6
0	1
$\frac{\pi}{4}$	1.7
$\frac{\pi}{2}$	3.0
π	8.8
$\frac{3\pi}{2}$	26.2
2π	77.9

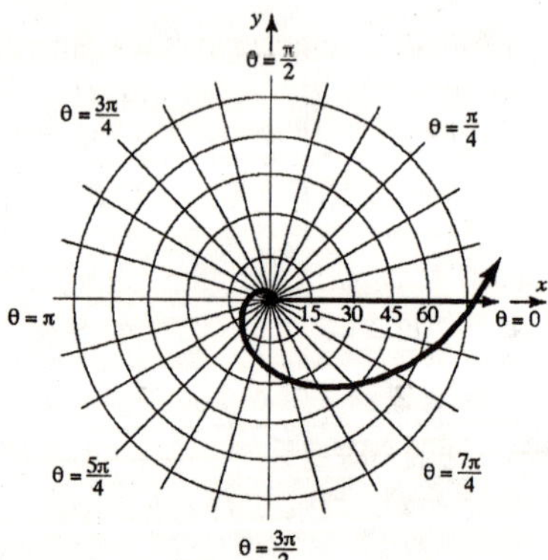

62. $r = 3^{\theta}$

The graph will be a spiral. Check for symmetry:

Polar axis: Replace θ by $-\theta$. $r = 3^{-\theta}$. The test fails.

The line $\theta = \frac{\pi}{2}$: Replace θ by $\pi - \theta$. $r = 3^{\pi-\theta}$. The test fails.

The pole: Replace r by $-r$. $-r = 3^{\theta}$. The test fails.

θ	$r = 3^{\theta}$
$-\pi$	0.03
$-\frac{\pi}{2}$	0.2
$-\frac{\pi}{4}$	0.4
0	1
$\frac{\pi}{4}$	2.4
$\frac{\pi}{2}$	5.6
π	31.5
$\frac{3\pi}{2}$	117.2
2π	995

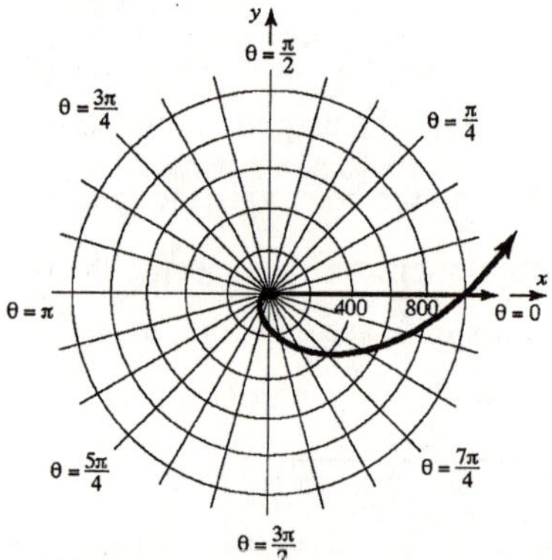

63. $r = 1 - \cos\theta$

The graph will be a cardioid. Check for symmetry:

Polar axis: Replace θ by $-\theta$. The result is $r = 1 - \cos(-\theta) = 1 - \cos\theta$.

The graph is symmetric with respect to the polar axis.

The line $\theta = \frac{\pi}{2}$: Replace θ by $\pi - \theta$.

$$\begin{aligned} r &= 1 - \cos(\pi - \theta) \\ &= 1 - (\cos(\pi)\cos\theta + \sin(\pi)\sin\theta) \\ &= 1 - (-\cos\theta + 0) \\ &= 1 + \cos\theta \end{aligned}$$

The test fails.

The pole: Replace r by $-r$. $-r = 1 - \cos\theta$. The test fails.

Due to symmetry, assign values to θ from 0 to π.

θ	$r=1-\cos\theta$
0	0
$\frac{\pi}{6}$	$1-\frac{\sqrt{3}}{2}\approx 0.1$
$\frac{\pi}{3}$	$\frac{1}{2}$
$\frac{\pi}{2}$	1
$\frac{2\pi}{3}$	$\frac{3}{2}$
$\frac{5\pi}{6}$	$1+\frac{\sqrt{3}}{2}\approx 1.9$
π	2

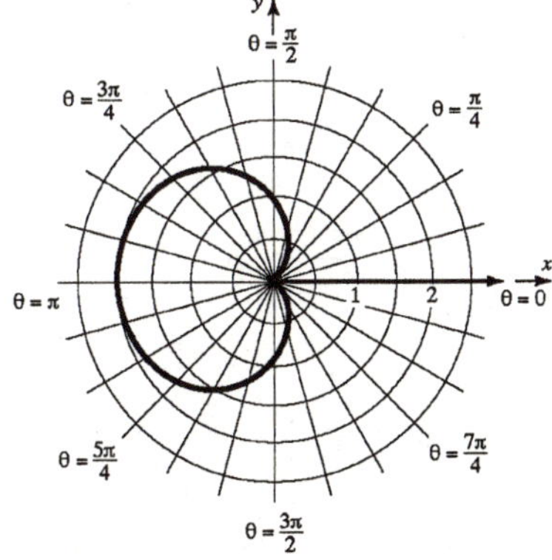

64. $r=3+\cos\theta$
The graph will be a limaçon without an inner loop. Check for symmetry:

Polar axis: Replace θ by $-\theta$. The result is $r=3+\cos(-\theta)=3+\cos\theta$. The graph is symmetric with respect to the polar axis.

The line $\theta=\frac{\pi}{2}$: Replace θ by $\pi-\theta$.

$$\begin{aligned} r&=3+\cos(\pi-\theta)\\ &=3+\left[\cos(\pi)\cos\theta+\sin(\pi)\sin\theta\right]\\ &=3+(-\cos\theta+0)\\ &=3-\cos\theta \end{aligned}$$

The test fails.

The pole: Replace r by $-r$. $-r=3+\cos\theta$. The test fails.

Due to symmetry, assign values to θ from 0 to π.

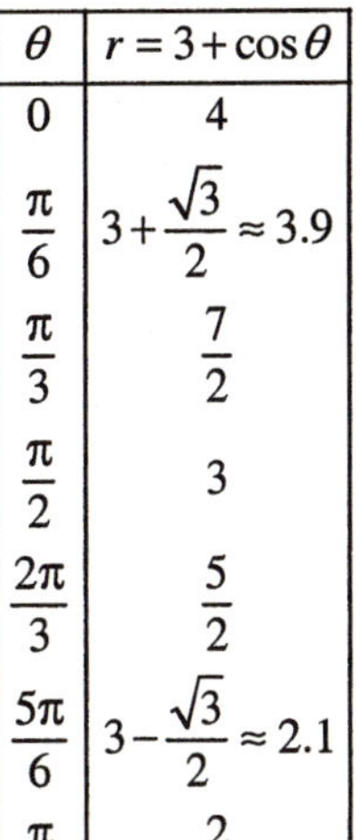

θ	$r=3+\cos\theta$
0	4
$\frac{\pi}{6}$	$3+\frac{\sqrt{3}}{2}\approx 3.9$
$\frac{\pi}{3}$	$\frac{7}{2}$
$\frac{\pi}{2}$	3
$\frac{2\pi}{3}$	$\frac{5}{2}$
$\frac{5\pi}{6}$	$3-\frac{\sqrt{3}}{2}\approx 2.1$
π	2

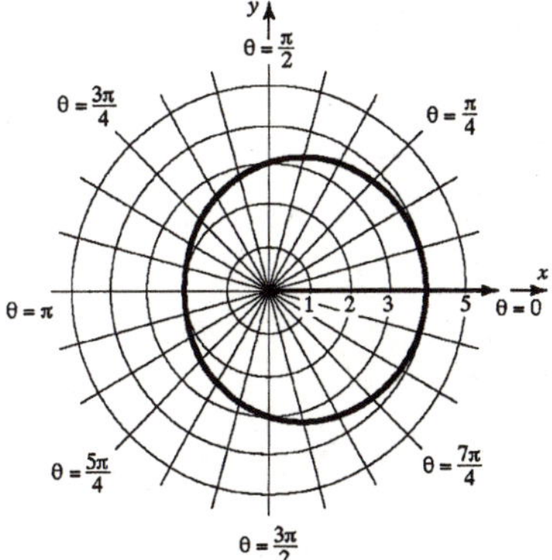

65. $r=1-3\cos\theta$
The graph will be a limaçon with an inner loop. Check for symmetry:

Polar axis: Replace θ by $-\theta$. The result is $r=1-3\cos(-\theta)=1-3\cos\theta$. The graph is symmetric with respect to the polar axis.

The line $\theta=\frac{\pi}{2}$: Replace θ by $\pi-\theta$.

$$\begin{aligned} r&=1-3\cos(\pi-\theta)\\ &=1-3\left[\cos(\pi)\cos\theta+\sin(\pi)\sin\theta\right]\\ &=1-3(-\cos\theta+0)\\ &=1+3\cos\theta \end{aligned}$$

The test fails.

The pole: Replace r by $-r$. $-r=1-3\cos\theta$. The test fails.

Due to symmetry, assign values to θ from 0 to π.

θ	$r=1-3\cos\theta$
0	-2
$\frac{\pi}{6}$	$1-\frac{3\sqrt{3}}{2}\approx-1.6$
$\frac{\pi}{3}$	$-\frac{1}{2}$
$\frac{\pi}{2}$	1
$\frac{2\pi}{3}$	$\frac{5}{2}$
$\frac{5\pi}{6}$	$1+\frac{3\sqrt{3}}{2}\approx 3.6$
π	4

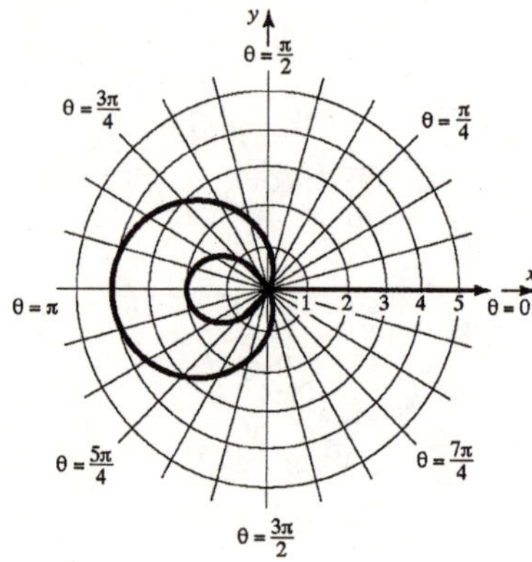

66. $r=4\cos(3\theta)$

The graph will be a rose with three petals.
Check for symmetry:

Polar axis: Replace θ by $-\theta$.
$r=4\cos(3(-\theta))=4\cos(-3\theta)=4\cos(3\theta)$. The graph is symmetric with respect to the polar axis.

The line $\theta=\frac{\pi}{2}$: Replace θ by $\pi-\theta$.

$$\begin{aligned} r&=4\cos\left[3(\pi-\theta)\right]\\ &=4\cos(3\pi-3\theta)\\ &=4\left[\cos(3\pi)\cos 3\theta+\sin(3\pi)\sin(3\theta)\right]\\ &=4(-\cos 3\theta+0)\\ &=-4\cos(3\theta)\end{aligned}$$

The test fails.

The pole: Replace r by $-r$. $-r=4\cos(3\theta)$.
The test fails.

Due to symmetry, assign values to θ from 0 to π.

θ	$r=4\cos(3\theta)$
0	4
$\frac{\pi}{6}$	0
$\frac{\pi}{3}$	-4
$\frac{\pi}{2}$	0
$\frac{2\pi}{3}$	4
$\frac{5\pi}{6}$	0
π	-4

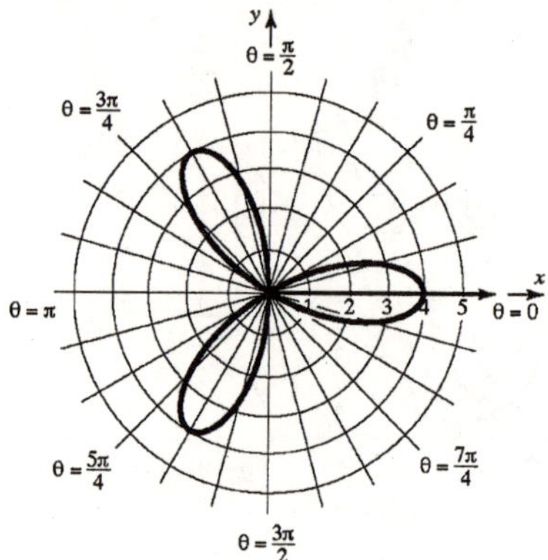

67. The graph is a cardioid whose equation is of the form $r=a+b\cos\theta$. The graph contains the point $(6,0)$, so we have

$6=a+b\cos 0$

$6=a+b(1)$

$6=a+b$

The graph contains the point $\left(3,\frac{\pi}{2}\right)$, so we have

$3=a+b\cos\frac{\pi}{2}$

$3=a+b(0)$

$3=a$

Substituting $a=3$ into the first equation yields:

$6=a+b$.

$6=3+b$

$3=b$

Therefore, the graph has equation $r=3+3\cos\theta$.

68. The graph is a cardioid whose equation is of the form $r = a + b\cos\theta$. The graph contains the point $(6, \pi)$, so we have

$6 = a + b\cos\pi$

$6 = a + b(-1)$

$6 = a - b$

The graph contains the point $\left(3, \frac{\pi}{2}\right)$, so we have

$3 = a + b\cos\frac{\pi}{2}$

$3 = a + b(0)$

$3 = a$

Substituting $a = 3$ into the first equation yields:

$6 = a - b$

$6 = 3 - b$

$b = -3$

Therefore, the graph has equation $r = 3 - 3\cos\theta$.

69. The graph is a limaçon without inner loop whose equation is of the form $r = a + b\sin\theta$, where $0 < b < a$. The graph contains the point $(4, 0)$, so we have

$4 = a + b\sin 0$

$4 = a + b(0)$

$4 = a$

The graph contains the point $\left(5, \frac{\pi}{2}\right)$, so we have

$5 = a + b\sin\frac{\pi}{2}$

$5 = a + b(1)$

$5 = a + b$

Substituting $a = 4$ into the second equation yields:

$5 = a + b$

$5 = 4 + b$

$1 = b$

Therefore, the graph has equation $r = 4 + \sin\theta$.

70. The graph is a limaçon with inner loop whose equation is of the form $r = a + b\sin\theta$, where $0 < a < b$. The graph contains the point $(1, 0)$, so we have

$1 = a + b\sin 0$

$1 = a + b(0)$

$1 = a$

The graph contains the point $\left(5, \frac{\pi}{2}\right)$, so we have

$5 = a + b\sin\frac{\pi}{2}$

$5 = a + b(1)$

$5 = a + b$

Substituting $a = 1$ into the second equation yields:

$5 = a + b$

$5 = 1 + b$

$4 = b$

Therefore, the graph has equation $r = 1 + 4\sin\theta$.

71. $r = \dfrac{2}{1 - \cos\theta}$ Check for symmetry:

Polar axis: Replace θ by $-\theta$. The result is

$r = \dfrac{2}{1 - \cos(-\theta)} = \dfrac{2}{1 - \cos\theta}$.

The graph is symmetric with respect to the polar axis.

The line $\theta = \frac{\pi}{2}$: Replace θ by $\pi - \theta$.

$$\begin{aligned} r &= \frac{2}{1 - \cos(\pi - \theta)} \\ &= \frac{2}{1 - (\cos\pi\cos\theta + \sin\pi\sin\theta)} \\ &= \frac{2}{1 - (-\cos\theta + 0)} \\ &= \frac{2}{1 + \cos\theta} \end{aligned}$$

The test fails.

The pole: Replace r by $-r$. $-r = \dfrac{2}{1 - \cos\theta}$.

The test fails.

Due to symmetry, assign values to θ from 0 to π.

θ	$r = \frac{2}{1 - \cos\theta}$
0	undefined
$\frac{\pi}{6}$	$\frac{2}{1 - \sqrt{3}/2} \approx 14.9$
$\frac{\pi}{3}$	4
$\frac{\pi}{2}$	2
$\frac{2\pi}{3}$	$\frac{4}{3}$
$\frac{5\pi}{6}$	$\frac{2}{1 + \sqrt{3}/2} \approx 1.1$
π	1

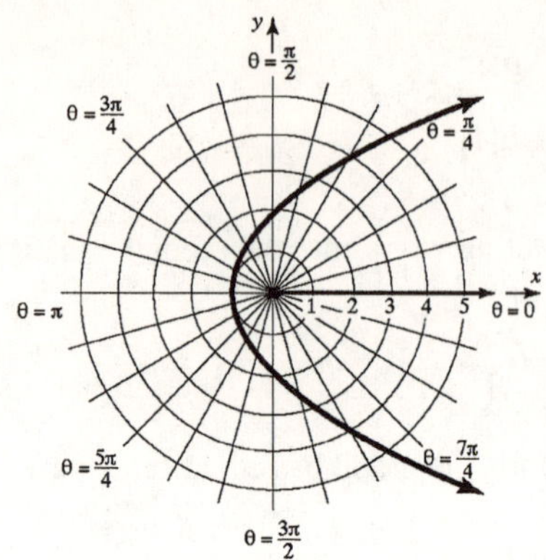

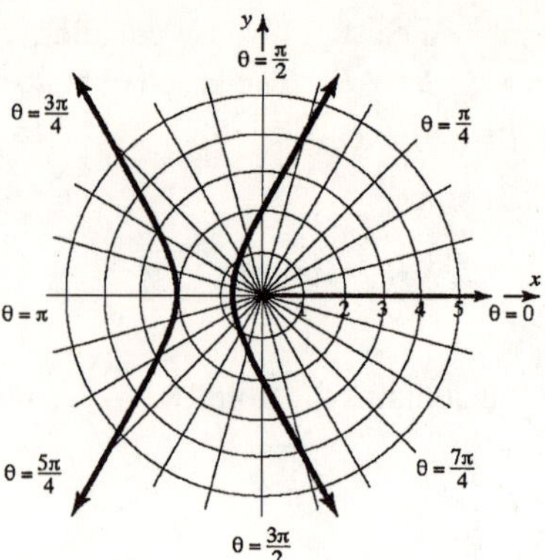

72. $r=\dfrac{2}{1-2\cos\theta}$ Check for symmetry:

Polar axis: Replace θ by $-\theta$. The result is $r=\dfrac{2}{1-2\cos(-\theta)}=\dfrac{2}{1-2\cos\theta}$. The graph is symmetric with respect to the polar axis.

The line $\theta=\dfrac{\pi}{2}$: Replace θ by $\pi-\theta$.

$$r=\frac{2}{1-2\cos(\pi-\theta)}$$
$$=\frac{2}{1-2(\cos\pi\cos\theta+\sin\pi\sin\theta)}$$
$$=\frac{2}{1-2(-\cos\theta+0)}$$
$$=\frac{2}{1+2\cos\theta}$$

The test fails.

The pole: Replace r by $-r$. $-r=\dfrac{2}{1-2\cos\theta}$.

The test fails.

Due to symmetry, assign values to θ from 0 to π.

θ	$r=\dfrac{2}{1-2\cos\theta}$
0	-2
$\frac{\pi}{6}$	$\frac{2}{1-\sqrt{3}}\approx-2.7$
$\frac{\pi}{3}$	undefined
$\frac{\pi}{2}$	2
$\frac{2\pi}{3}$	1
$\frac{5\pi}{6}$	$\frac{2}{1+\sqrt{3}}\approx 0.7$
π	$\frac{2}{3}$

73. $r=\dfrac{1}{3-2\cos\theta}$ Check for symmetry:

Polar axis: Replace θ by $-\theta$. The result is $r=\dfrac{1}{3-2\cos(-\theta)}=\dfrac{1}{3-2\cos\theta}$. The graph is symmetric with respect to the polar axis.

The line $\theta=\dfrac{\pi}{2}$: Replace θ by $\pi-\theta$.

$$r=\frac{1}{3-2\cos(\pi-\theta)}$$
$$=\frac{1}{3-2(\cos\pi\cos\theta+\sin\pi\sin\theta)}$$
$$=\frac{1}{3-2(-\cos\theta+0)}$$
$$=\frac{1}{3+2\cos\theta}$$

The test fails.

The pole: Replace r by $-r$. $-r=\dfrac{1}{3-2\cos\theta}$.

The test fails.

Due to symmetry, assign values to θ from 0 to π.

θ	$r=\dfrac{1}{3-2\cos\theta}$
0	1
$\frac{\pi}{6}$	$\frac{1}{3-\sqrt{3}}\approx 0.8$
$\frac{\pi}{3}$	$\frac{1}{2}$
$\frac{\pi}{2}$	$\frac{1}{3}$
$\frac{2\pi}{3}$	$\frac{1}{4}$
$\frac{5\pi}{6}$	$\frac{1}{3+\sqrt{3}}\approx 0.2$
π	$\frac{1}{5}$

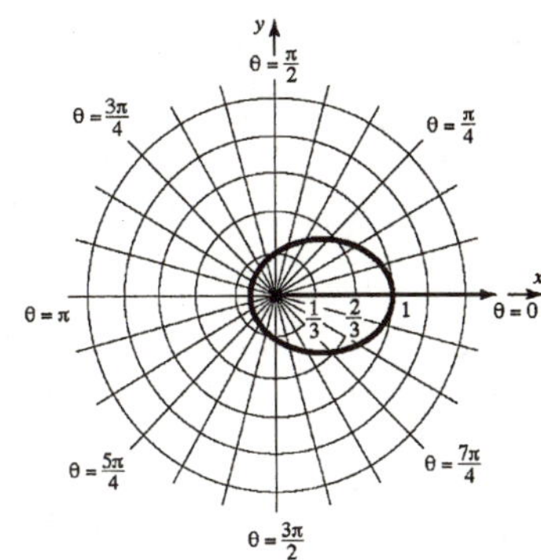

74. $r = \dfrac{1}{1-\cos\theta}$ Check for symmetry:

Polar axis: Replace θ by $-\theta$. The result is $r = \dfrac{1}{1-\cos(-\theta)} = \dfrac{1}{1-\cos\theta}$. The graph is symmetric with respect to the polar axis.

The line $\theta = \dfrac{\pi}{2}$: Replace θ by $\pi - \theta$.

$$r = \frac{1}{1-\cos(\pi-\theta)} = \frac{1}{1-(\cos\pi\cos\theta + \sin\pi\sin\theta)} = \frac{1}{1-(-\cos\theta+0)} = \frac{1}{1+\cos\theta}$$

The test fails.

The pole: Replace r by $-r$. $-r = \dfrac{1}{1-\cos\theta}$.

The test fails.

Due to symmetry, assign values to θ from 0 to π.

θ	$r = \dfrac{1}{1-\cos\theta}$
0	undefined
$\frac{\pi}{6}$	$\frac{1}{1-\sqrt{3}/2} \approx 7.5$
$\frac{\pi}{3}$	2
$\frac{\pi}{2}$	2
$\frac{2\pi}{3}$	$\frac{2}{3}$
$\frac{5\pi}{6}$	$\frac{1}{1+\sqrt{3}/2} \approx 0.5$
π	$\frac{1}{2}$

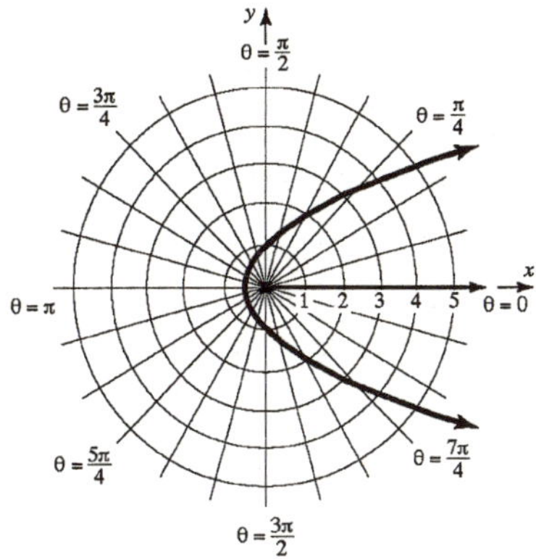

75. $r = \theta$, $\theta \ge 0$

Check for symmetry:

Polar axis: Replace θ by $-\theta$. $r = -\theta$. The test fails.

The line $\theta = \dfrac{\pi}{2}$: Replace θ by $\pi - \theta$. $r = \pi - \theta$. The test fails.

The pole: Replace r by $-r$. $-r = \theta$. The test fails.

θ	$r = \theta$
0	0
$\frac{\pi}{6}$	$\frac{\pi}{6} \approx 0.5$
$\frac{\pi}{3}$	$\frac{\pi}{3} \approx 1.0$
$\frac{\pi}{2}$	$\frac{\pi}{2} \approx 1.6$
π	$\pi \approx 3.1$
$\frac{3\pi}{2}$	$\frac{3\pi}{2} \approx 4.7$
2π	$2\pi \approx 6.3$

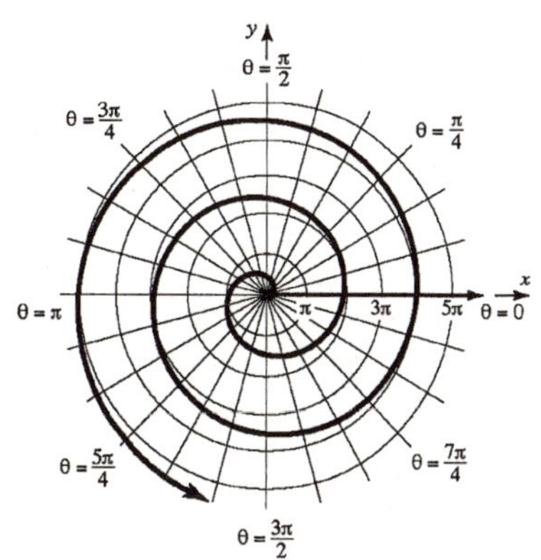

76. $r = \dfrac{3}{\theta}$ Check for symmetry:

Polar axis: Replace θ by $-\theta$. $r = \dfrac{3}{-\theta}$. The test fails.

The line $\theta = \dfrac{\pi}{2}$: Replace θ by $\pi - \theta$. $r = \dfrac{3}{\pi - \theta}$. The test fails.

The pole: Replace r by $-r$. $-r = \dfrac{3}{\theta}$. The test fails.

θ	$r = \dfrac{3}{\theta}$
0	undefined
$\dfrac{\pi}{6}$	$\dfrac{18}{\pi} \approx 5.7$
$\dfrac{\pi}{3}$	$\dfrac{9}{\pi} \approx 2.9$
$\dfrac{\pi}{2}$	$\dfrac{6}{\pi} \approx 1.9$
π	$\dfrac{3}{\pi} \approx 1.0$
$\dfrac{3\pi}{2}$	$\dfrac{2}{\pi} \approx 0.6$
2π	$\dfrac{3}{2\pi} \approx 0.5$

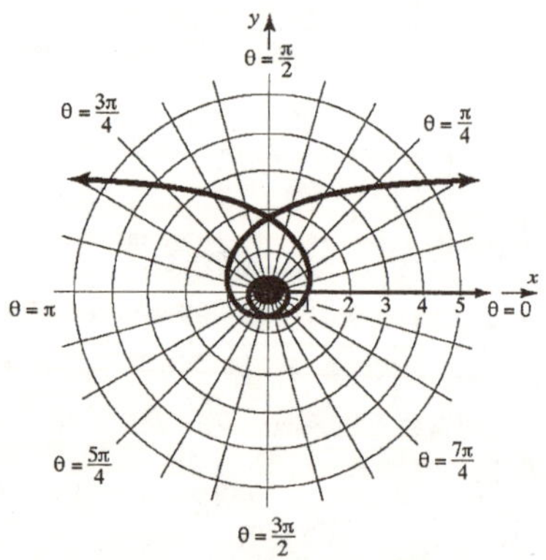

77. $r = \csc\theta - 2 = \dfrac{1}{\sin\theta} - 2, \quad 0 < \theta < \pi$

Check for symmetry:

Polar axis: Replace θ by $-\theta$.

$r = \csc(-\theta) - 2 = -\csc\theta - 2$. The test fails.

The line $\theta = \dfrac{\pi}{2}$: Replace θ by $\pi - \theta$.

$$
\begin{aligned}
r &= \csc(\pi - \theta) - 2 \\
&= \frac{1}{\sin(\pi - \theta)} - 2 \\
&= \frac{1}{\sin\pi\cos\theta - \cos\pi\sin\theta} - 2 \\
&= \frac{1}{0 \cdot \cos\theta - 1 \cdot \sin\theta} - 2 \\
&= \frac{1}{\sin\theta} - 2 \\
&= \csc\theta - 2
\end{aligned}
$$

The graph is symmetric with respect to the line $\theta = \dfrac{\pi}{2}$.

The pole: Replace r by $-r$. $-r = \csc\theta - 2$. The test fails.

Due to symmetry, assign values to θ from 0 to $\dfrac{\pi}{2}$.

θ	$r = \csc\theta - 2$
0	undefined
$\dfrac{\pi}{6}$	0
$\dfrac{\pi}{4}$	$\sqrt{2} - 2 \approx -0.6$
$\dfrac{\pi}{3}$	$\dfrac{2\sqrt{3}}{3} - 2 \approx -0.8$
$\dfrac{\pi}{2}$	-1

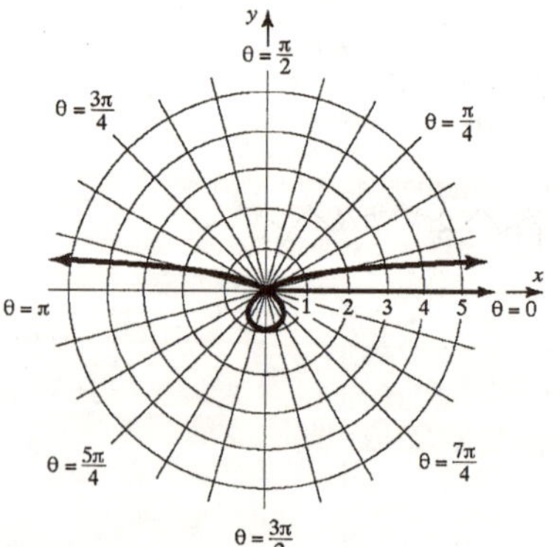

78. $r = \sin\theta\tan\theta$

Check for symmetry:

Polar axis: Replace θ by $-\theta$.

$$
\begin{aligned}
r &= \sin(-\theta)\tan(-\theta) \\
&= (-\sin\theta)(-\tan\theta) \\
&= \sin\theta\tan\theta
\end{aligned}
$$

The graph is symmetric with respect to the polar axis.

The line $\theta = \frac{\pi}{2}$: Replace θ by $\pi - \theta$.

$$r = \sin(\pi - \theta)\tan(\pi - \theta)$$
$$= (\sin\pi\cos\theta - \cos\pi\sin\theta)\left(\frac{\tan\pi - \tan\theta}{1 + \tan\pi\tan\theta}\right)$$
$$= \sin\theta \cdot \frac{-\tan\theta}{1}$$
$$= -\sin\theta\tan\theta$$

The test fails.

The pole: Replace r by $-r$. $-r = \sin\theta\tan\theta$. The test fails.

Due to symmetry, assign values to θ from 0 to π.

θ	$r = \sin\theta\tan\theta$
0	0
$\frac{\pi}{6}$	$\frac{1}{2} \cdot \frac{\sqrt{3}}{3} \approx 0.3$
$\frac{\pi}{3}$	$\frac{3}{2}$
$\frac{\pi}{2}$	undefined
$\frac{2\pi}{3}$	$-\frac{3}{2}$
$\frac{5\pi}{6}$	$\frac{1}{2} \cdot \left(-\frac{\sqrt{3}}{3}\right) \approx -0.3$
π	0

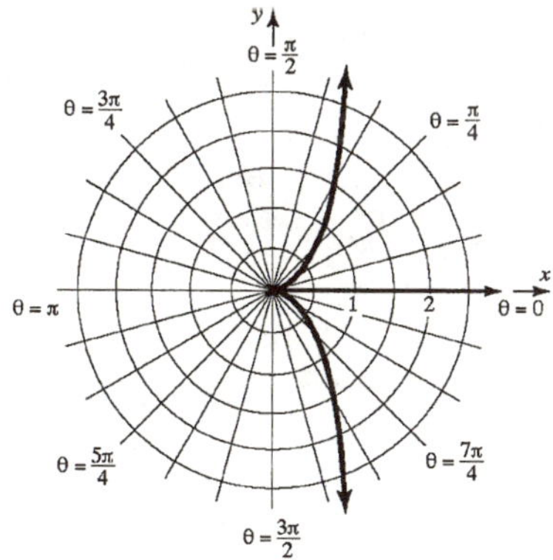

79. $r = \tan\theta, \quad -\frac{\pi}{2} < \theta < \frac{\pi}{2}$

Check for symmetry:

Polar axis: Replace θ by $-\theta$.

$r = \tan(-\theta) = -\tan\theta$. The test fails.

The line $\theta = \frac{\pi}{2}$: Replace θ by $\pi - \theta$.

$$r = \tan(\pi - \theta) = \frac{\tan(\pi) - \tan\theta}{1 + \tan(\pi)\tan\theta} = \frac{-\tan\theta}{1} = -\tan\theta$$

The test fails.

The pole: Replace r by $-r$. $-r = \tan\theta$. The test fails.

θ	$r = \tan\theta$
$-\frac{\pi}{3}$	$-\sqrt{3} \approx -1.7$
$-\frac{\pi}{4}$	-1
$-\frac{\pi}{6}$	$-\frac{\sqrt{3}}{3} \approx -0.6$
0	0
$\frac{\pi}{6}$	$\frac{\sqrt{3}}{3} \approx 0.6$
$\frac{\pi}{4}$	1
$\frac{\pi}{3}$	$\sqrt{3} \approx 1.7$

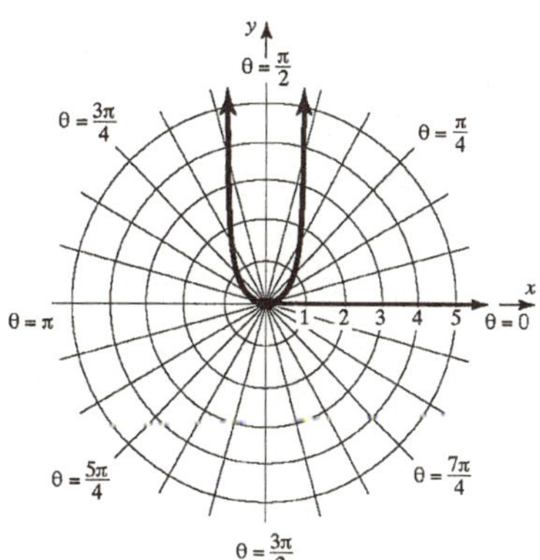

80. $r = \cos\frac{\theta}{2}$

Check for symmetry:

Polar axis: Replace θ by $-\theta$.

$r = \cos\left(-\frac{\theta}{2}\right) = \cos\frac{\theta}{2}$. The graph is symmetric with respect to the polar axis.

The line $\theta = \frac{\pi}{2}$: Replace θ by $\pi - \theta$.

$$r = \cos\left(\frac{\pi - \theta}{2}\right) = \cos\left(\frac{\pi}{2} - \frac{\theta}{2}\right)$$
$$= \cos\frac{\pi}{2} \cdot \cos\frac{\theta}{2} + \sin\frac{\pi}{2} \cdot \sin\frac{\theta}{2}$$
$$= \sin\frac{\theta}{2}$$

The test fails.

The pole: Replace r by $-r$. $-r=\cos\frac{\theta}{2}$. The test fails.

Due to symmetry, assign values to θ from 0 to π.

θ	$r=\cos\frac{\theta}{2}$
0	1
$\frac{\pi}{6}$	0.97
$\frac{\pi}{3}$	$\frac{\sqrt{3}}{2}\approx 0.87$
$\frac{\pi}{2}$	$\frac{\sqrt{2}}{2}\approx 0.71$
$\frac{2\pi}{3}$	$\frac{1}{2}$
$\frac{5\pi}{6}$	0.26
π	0

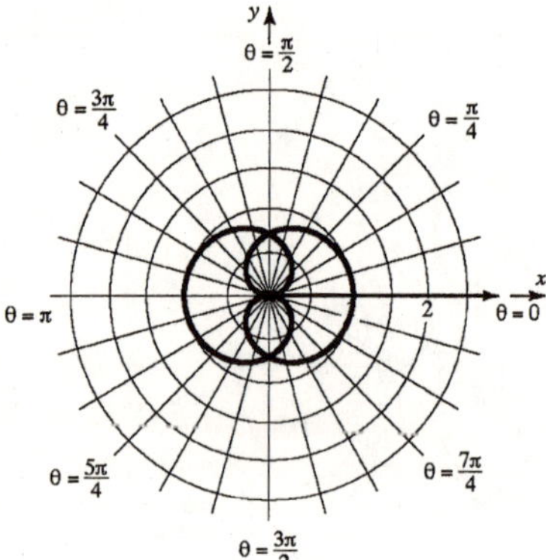

81. Convert the equation to rectangular form:
$r\sin\theta=a$

$y=a$

The graph of $r\sin\theta=a$ is a horizontal line a units above the pole if $a>0$, and $|a|$ units below the pole if $a<0$.

82. Convert the equation to rectangular form:
$r\cos\theta=a$

$x=a$

The graph of $r\cos\theta=a$ is a vertical line a units to the right of the pole if $a>0$, and $|a|$ units to the left of the pole if $a<0$.

83. Convert the equation to rectangular form:

$$r=2a\sin\theta,\ a>0$$
$$r^2=2ar\sin\theta$$
$$x^2+y^2=2ay$$
$$x^2+y^2-2ay=0$$
$$x^2+(y-a)^2=a^2$$

Circle: radius a, center at rectangular coordinates $(0, a)$.

84. Convert the equation to rectangular form:

$$r=-2a\sin\theta,\ a>0$$
$$r^2=-2ar\sin\theta$$
$$x^2+y^2=-2ay$$
$$x^2+y^2+2ay=0$$
$$x^2+(y+a)^2=a^2$$

Circle: radius a, center at rectangular coordinates $(0,-a)$.

85. Convert the equation to rectangular form:

$$r=2a\cos\theta,\ a>0$$
$$r^2=2ar\cos\theta$$
$$x^2+y^2=2ax$$
$$x^2-2ax+y^2=0$$
$$(x-a)^2+y^2=a^2$$

Circle: radius a, center at rectangular coordinates $(a, 0)$.

86. Convert the equation to rectangular form:

$$r=-2a\cos\theta,\ a>0$$
$$r^2=-2ar\cos\theta$$
$$x^2+y^2=-2ax$$
$$x^2+2ax+y^2=0$$
$$(x+a)^2+y^2=a^2$$

Circle: radius a, center at rectangular coordinates $(-a, 0)$.

87. a. $r^2=\cos\theta$: $r^2=\cos(\pi-\theta)$

$r^2=-\cos\theta$

Not equivalent; test fails.

$(-r)^2=\cos(-\theta)$

$r^2=\cos\theta$

New test works.

b. $r^2 = \sin\theta$: $r^2 = \sin(\pi-\theta)$

$r^2 = \sin\theta$

Test works.

$(-r)^2 = \sin(-\theta)$

$r^2 = -\sin\theta$

Not equivalent; new test fails.

88. Symmetry with respect to the pole: In a polar equation, replace θ by $\theta+\pi$. If an equivalent equation results, the graph is symmetric with respect to the pole.

a. $r^2 = \sin\theta$: $r^2 = \sin(\pi+\theta)$

$r^2 = -\sin\theta$

New test fails.

$(-r)^2 = \sin\theta$

$r^2 = \sin\theta$

Test works.

b. $r = \cos^2\theta$: $-r = \cos^2\theta$

Test fails.

$r = \cos^2(\pi+\theta) = (-\cos\theta)^2 = \cos^2\theta$

New test works.

89. Answers will vary.

Section 8.3

1. $-4+3i$

2. $\sin\alpha\cos\beta+\cos\alpha\sin\beta$

3. $\cos\alpha\cos\beta-\sin\alpha\sin\beta$

4. $\frac{\sqrt{3}}{2}, -\frac{1}{2}$

5. magnitude, modulus, argument

6. DeMoivre's

7. three

8. True

9. False

10. True

11. $r = \sqrt{x^2+y^2} = \sqrt{1^2+1^2} = \sqrt{2}$

$\tan\theta = \frac{y}{x} = 1$

$\theta = 45°$

The polar form of $z = 1+i$ is

$z = r(\cos\theta + i\sin\theta) = \sqrt{2}(\cos 45° + i\sin 45°)$

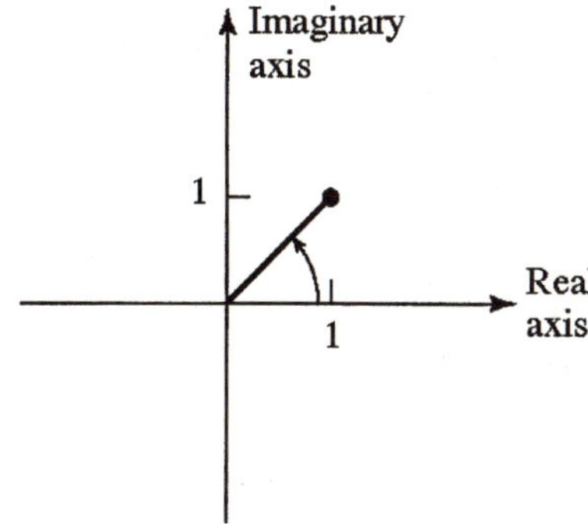

12. $r = \sqrt{x^2+y^2} = \sqrt{(-1)^2+1^2} = \sqrt{2}$

$\tan\theta = \frac{y}{x} = -1$

$\theta = 135°$

The polar form of $z = -1+i$ is

$z = r(\cos\theta + i\sin\theta) = \sqrt{2}(\cos 135° + i\sin 135°)$

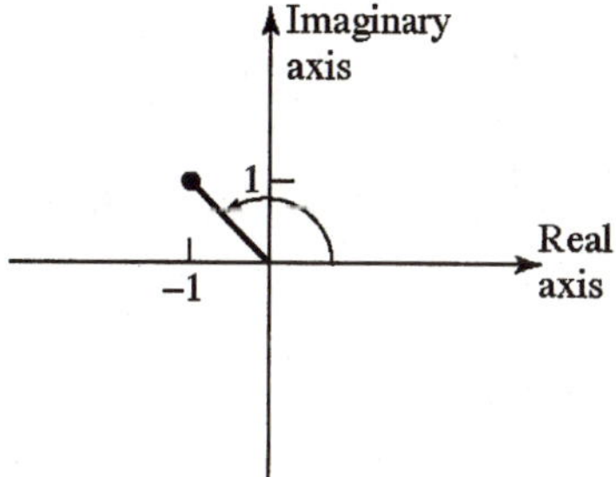

13. $r = \sqrt{x^2+y^2} = \sqrt{(\sqrt{3})^2+(-1)^2} = \sqrt{4} = 2$

$\tan\theta = \frac{y}{x} = \frac{-1}{\sqrt{3}} = -\frac{\sqrt{3}}{3}$

$\theta = 330°$

The polar form of $z = \sqrt{3} - i$ is

$z = r(\cos\theta + i\sin\theta) = 2(\cos 330° + i\sin 330°)$

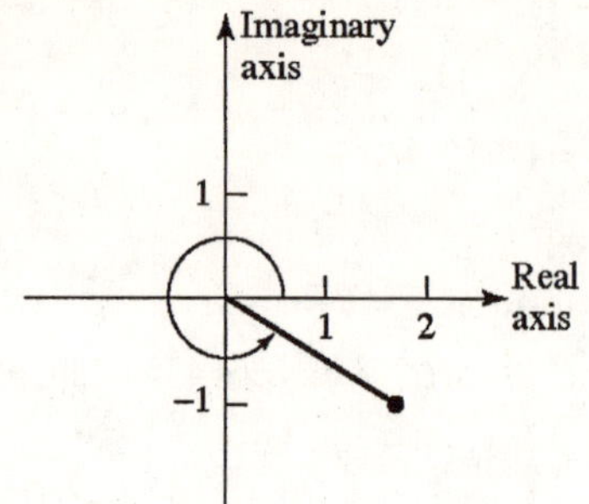

14. $r=\sqrt{x^2+y^2}=\sqrt{1^2+\left(-\sqrt{3}\right)^2}=\sqrt{4}=2$

$\tan\theta=\dfrac{y}{x}=\dfrac{-\sqrt{3}}{1}=-\sqrt{3}$

$\theta=300°$

The polar form of $z=1-\sqrt{3}i$ is

$z=r\left(\cos\theta+i\sin\theta\right)=2\left(\cos 300°+i\sin 300°\right)$

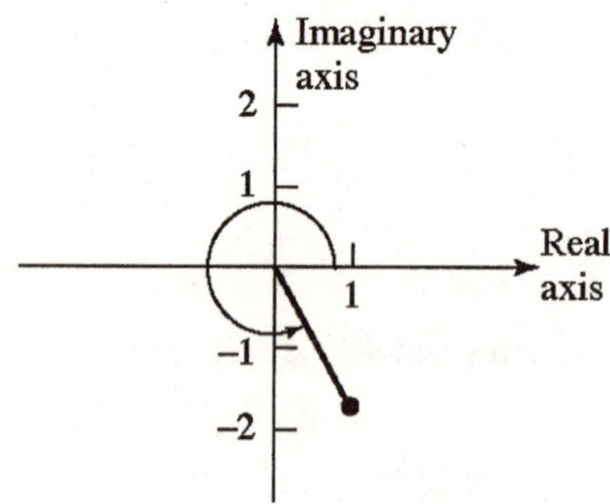

15. $r=\sqrt{x^2+y^2}=\sqrt{0^2+(-3)^2}=\sqrt{9}=3$

$\tan\theta=\dfrac{y}{x}=\dfrac{-3}{0}$

$\theta=270°$

The polar form of $z=-3i$ is

$z=r\left(\cos\theta+i\sin\theta\right)=3\left(\cos 270°+i\sin 270°\right)$

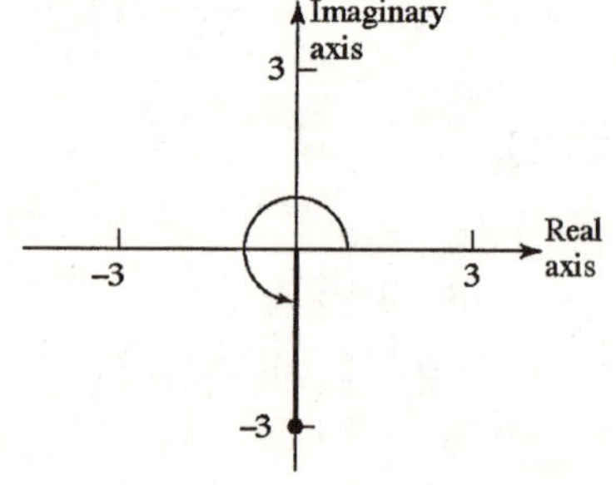

16. $r=\sqrt{x^2+y^2}=\sqrt{(-2)^2+0^2}=\sqrt{4}=2$

$\tan\theta=\dfrac{y}{x}=\dfrac{0}{-2}=0$

$\theta=180°$

The polar form of $z=-2$ is

$z=r\left(\cos\theta+i\sin\theta\right)=2\left(\cos 180°+i\sin 180°\right)$

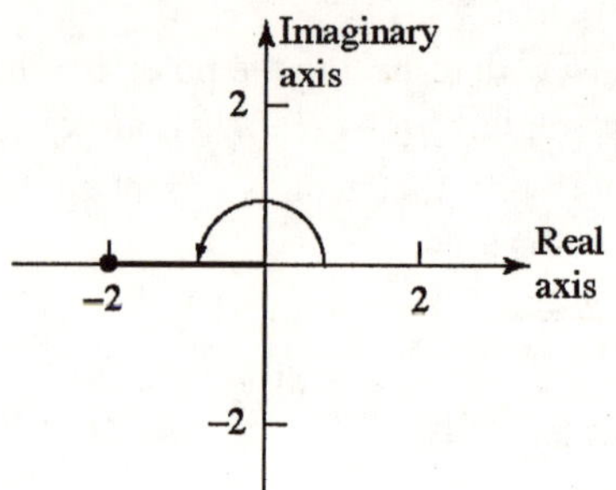

17. $r=\sqrt{x^2+y^2}=\sqrt{4^2+(-4)^2}=\sqrt{32}=4\sqrt{2}$

$\tan\theta=\dfrac{y}{x}=\dfrac{-4}{4}=-1$

$\theta=315°$

The polar form of $z=4-4i$ is

$z=r\left(\cos\theta+i\sin\theta\right)=4\sqrt{2}\left(\cos 315°+i\sin 315°\right)$

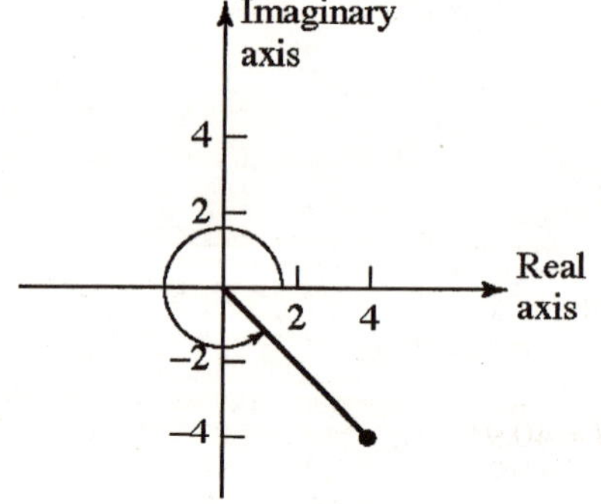

18. $r=\sqrt{x^2+y^2}=\sqrt{\left(9\sqrt{3}\right)^2+9^2}=\sqrt{324}=18$

$\tan\theta=\dfrac{y}{x}=\dfrac{9}{9\sqrt{3}}=\dfrac{\sqrt{3}}{3}$

$\theta=30°$

The polar form of $z=9\sqrt{3}+9i$ is

$z=r\left(\cos\theta+i\sin\theta\right)=18\left(\cos 30°+i\sin 30°\right).$

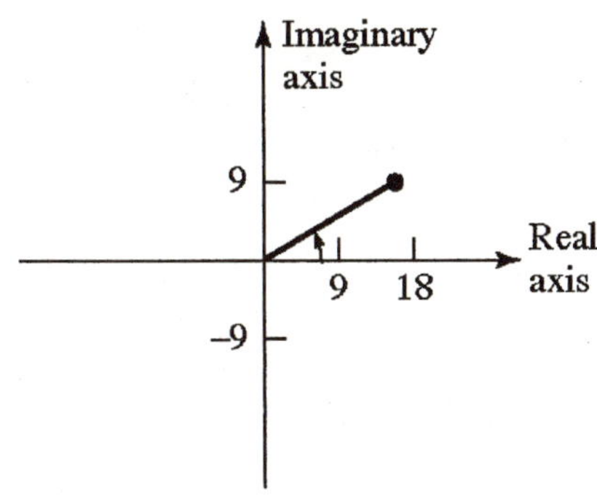

19. $r=\sqrt{x^2+y^2}=\sqrt{3^2+(-4)^2}=\sqrt{25}=5$

$\tan\theta=\dfrac{y}{x}=\dfrac{-4}{3}$

$\theta\approx 306.9°$

The polar form of $z=3-4i$ is

$z=r(\cos\theta+i\sin\theta)=5(\cos 306.9°+i\sin 306.9°)$

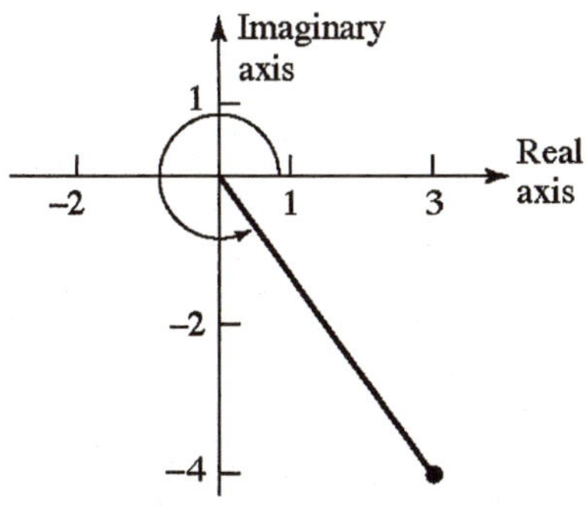

20. $r=\sqrt{x^2+y^2}=\sqrt{2^2+\left(\sqrt{3}\right)^2}=\sqrt{7}$

$\tan\theta=\dfrac{y}{x}=\dfrac{\sqrt{3}}{2}$

$\theta\approx 40.9°$

The polar form of $z=2+\sqrt{3}i$ is

$z=r(\cos\theta+i\sin\theta)$

$=\sqrt{7}(\cos 40.9°+i\sin 40.9°)$

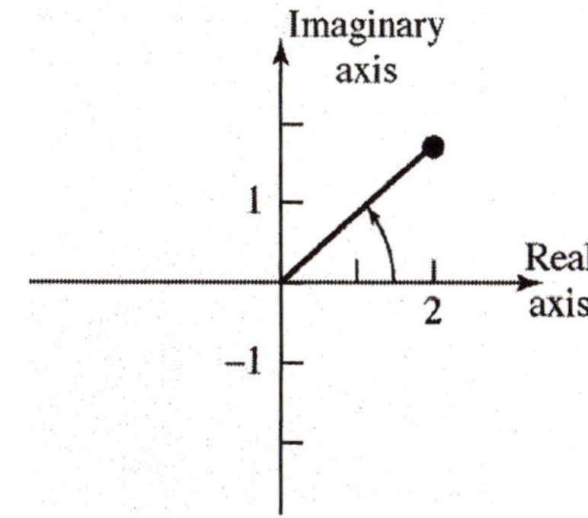

21. $r=\sqrt{x^2+y^2}=\sqrt{(-2)^2+3^2}=\sqrt{13}$

$\tan\theta=\dfrac{y}{x}=\dfrac{3}{-2}=-\dfrac{3}{2}$

$\theta\approx 123.7°$

The polar form of $z=-2+3i$ is

$z=r(\cos\theta+i\sin\theta)$

$=\sqrt{13}(\cos 123.7°+i\sin 123.7°)$

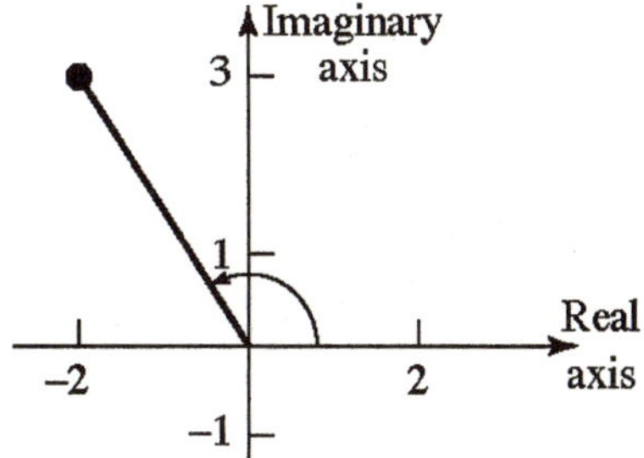

22. $r=\sqrt{x^2+y^2}=\sqrt{\left(\sqrt{5}\right)^2+(-1)^2}=\sqrt{6}$

$\tan\theta=\dfrac{y}{x}=\dfrac{-1}{\sqrt{5}}=-\dfrac{\sqrt{5}}{5}$

$\theta\approx 335.9°$

The polar form of $z=\sqrt{5}-i$ is

$z=r(\cos\theta+i\sin\theta)$

$=\sqrt{6}(\cos 335.9°+i\sin 335.9°)$

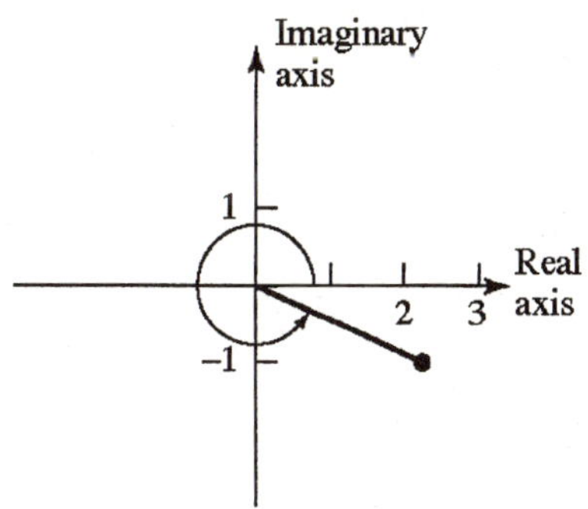

23. $2(\cos 120°+i\sin 120°)=2\left(-\dfrac{1}{2}+\dfrac{\sqrt{3}}{2}i\right)$

$=-1+\sqrt{3}\,i$

24. $3(\cos 210°+i\sin 210°)=3\left(-\dfrac{\sqrt{3}}{2}-\dfrac{1}{2}i\right)$

$=-\dfrac{3\sqrt{3}}{2}-\dfrac{3}{2}i$

25. $4\left(\cos\frac{7\pi}{4}+i\sin\frac{7\pi}{4}\right)=4\left(\frac{\sqrt{2}}{2}-\frac{\sqrt{2}}{2}i\right)$
$=2\sqrt{2}-2\sqrt{2}\,i$

26. $2\left(\cos\frac{5\pi}{6}+i\sin\frac{5\pi}{6}\right)=2\left(-\frac{\sqrt{3}}{2}+\frac{1}{2}i\right)=-\sqrt{3}+i$

27. $3\left(\cos\frac{3\pi}{2}+i\sin\frac{3\pi}{2}\right)=3(0-1i)=-3i$

28. $4\left(\cos\frac{\pi}{2}+i\sin\frac{\pi}{2}\right)=4(0+1i)=4i$

29. $0.2(\cos 100°+i\sin 100°)\approx 0.2(-0.1736+0.9848i)$
$\approx -0.035+0.197i$

30. $0.4(\cos 200°+i\sin 200°)\approx 0.4(-0.9397-0.3420i)$
$\approx -0.376-0.137i$

31. $2\left(\cos\frac{\pi}{18}+i\sin\frac{\pi}{18}\right)\approx 2(0.9848+0.1736i)$
$\approx 1.970+0.347i$

32. $3\left(\cos\frac{\pi}{10}+i\sin\frac{\pi}{10}\right)\approx 3(0.9511+0.3090i)$
$\approx 2.853+0.927i$

33. $z\cdot w=2(\cos 40°+i\sin 40°)\cdot 4(\cos 20°+i\sin 20°)$
$=2\cdot 4[\cos(40°+20°)+i\sin(40°+20°)]$
$=8(\cos 60°+i\sin 60°)$

$\frac{z}{w}=\frac{2(\cos 40°+i\sin 40°)}{4(\cos 20°+i\sin 20°)}$
$=\frac{2}{4}[\cos(40°-20°)+i\sin(40°-20°)]$
$=\frac{1}{2}(\cos 20°+i\sin 20°)$

34. $z\cdot w=(\cos 120°+i\sin 120°)\cdot(\cos 100°+i\sin 100°)$
$=\cos(120°+100°)+i\sin(120°+100°)$
$=\cos 220°+i\sin 220°$

$\frac{z}{w}=\frac{(\cos 120°+i\sin 120°)}{(\cos 100°+i\sin 100°)}$
$=\cos(120°-100°)+i\sin(120°-100°)$
$=\cos 20°+i\sin 20°$

35. $z\cdot w=3(\cos 130°+i\sin 130°)\cdot 4(\cos 270°+i\sin 270°)$
$=3\cdot 4[\cos(130°+270°)+i\sin(130°+270°)]$
$=12(\cos 400°+i\sin 400°)$
$=12[\cos(400°-360°)+i\sin(400°-360°)]$
$=12(\cos 40°+i\sin 40°)$

$\frac{z}{w}=\frac{3(\cos 130°+i\sin 130°)}{4(\cos 270°+i\sin 270°)}$
$=\frac{3}{4}[\cos(130°-270°)+i\sin(130°-270°)]$
$=\frac{3}{4}(\cos(-140°)+i\sin(-140°))$
$=\frac{3}{4}[\cos(360°-140°)+i\sin(360°-140°)]$
$=\frac{3}{4}(\cos 220°+i\sin 220°)$

36. $z\cdot w=2(\cos 80°+i\sin 80°)\cdot 6(\cos 200°+i\sin 200°)$
$=2\cdot 6[\cos(80°+200°)+i\sin(80°+200°)]$
$=12(\cos 280°+i\sin 280°)$

$\frac{z}{w}=\frac{2(\cos 80°+i\sin 80°)}{6(\cos 200°+i\sin 200°)}$
$=\frac{2}{6}[\cos(80°-200°)+i\sin(80°-200°)]$
$=\frac{1}{3}[\cos(-120°)+i\sin(-120°)]$
$=\frac{1}{3}[\cos(360°-120°)+i\sin(360°-120°)]$
$=\frac{1}{3}(\cos 240°+i\sin 240°)$

37. $z\cdot w=2\left(\cos\frac{\pi}{8}+i\sin\frac{\pi}{8}\right)\cdot 2\left(\cos\frac{\pi}{10}+i\sin\frac{\pi}{10}\right)$
$=2\cdot 2\left[\cos\left(\frac{\pi}{8}+\frac{\pi}{10}\right)+i\sin\left(\frac{\pi}{8}+\frac{\pi}{10}\right)\right]$
$=4\left(\cos\frac{9\pi}{40}+i\sin\frac{9\pi}{40}\right)$

$\frac{z}{w}=\frac{2\left(\cos\frac{\pi}{8}+i\sin\frac{\pi}{8}\right)}{2\left(\cos\frac{\pi}{10}+i\sin\frac{\pi}{10}\right)}$
$=\frac{2}{2}\left[\cos\left(\frac{\pi}{8}-\frac{\pi}{10}\right)+i\sin\left(\frac{\pi}{8}-\frac{\pi}{10}\right)\right]$
$=\cos\frac{\pi}{40}+i\sin\frac{\pi}{40}$

38. $z \cdot w = 4\left(\cos\frac{3\pi}{8} + i\sin\frac{3\pi}{8}\right) \cdot 2\left(\cos\frac{9\pi}{16} + i\sin\frac{9\pi}{16}\right)$

$= 4 \cdot 2\left[\cos\left(\frac{3\pi}{8} + \frac{9\pi}{16}\right) + i\sin\left(\frac{3\pi}{8} + \frac{9\pi}{16}\right)\right]$

$= 8\left(\cos\frac{15\pi}{16} + i\sin\frac{15\pi}{16}\right)$

$\frac{z}{w} = \frac{4\left(\cos\frac{3\pi}{8} + i\sin\frac{3\pi}{8}\right)}{2\left(\cos\frac{9\pi}{16} + i\sin\frac{9\pi}{16}\right)}$

$= \frac{4}{2}\left[\cos\left(\frac{3\pi}{8} - \frac{9\pi}{16}\right) + i\sin\left(\frac{3\pi}{8} - \frac{9\pi}{16}\right)\right]$

$= 2\left[\cos\left(-\frac{3\pi}{16}\right) + i\sin\left(-\frac{3\pi}{16}\right)\right]$

$= 2\left[\cos\left(2\pi - \frac{3\pi}{16}\right) + i\sin\left(2\pi - \frac{3\pi}{16}\right)\right]$

$= 2\left(\cos\frac{29\pi}{16} + i\sin\frac{29\pi}{16}\right)$

39. $z = 2 + 2i$

$r = \sqrt{2^2 + 2^2} = \sqrt{8} = 2\sqrt{2}$

$\tan\theta = \frac{2}{2} = 1$

$\theta = 45°$

$z = 2\sqrt{2}\left(\cos 45° + i\sin 45°\right)$

$w = \sqrt{3} - i$

$r = \sqrt{\left(\sqrt{3}\right)^2 + (-1)^2} = \sqrt{4} = 2$

$\tan\theta = \frac{-1}{\sqrt{3}} = -\frac{\sqrt{3}}{3}$

$\theta = 330°$

$w = 2\left(\cos 330° + i\sin 330°\right)$

$z \cdot w = 2\sqrt{2}\left(\cos 45° + i\sin 45°\right) \cdot 2\left(\cos 330° + i\sin 330°\right)$

$= 2\sqrt{2} \cdot 2\left[\cos\left(45° + 330°\right) + i\sin\left(45° + 330°\right)\right]$

$= 4\sqrt{2}\left(\cos 375° + i\sin 375°\right)$

$= 4\sqrt{2}\left[\cos\left(375° - 360°\right) + i\sin\left(375° - 360°\right)\right]$

$= 4\sqrt{2}\left(\cos 15° + i\sin 15°\right)$

$\frac{z}{w} = \frac{2\sqrt{2}\left(\cos 45° + i\sin 45°\right)}{2\left(\cos 330° + i\sin 330°\right)}$

$= \frac{2\sqrt{2}}{2}\left[\cos\left(45° - 330°\right) + i\sin\left(45° - 330°\right)\right]$

$= \sqrt{2}\left[\cos\left(-285°\right) + i\sin\left(-285°\right)\right]$

$= \sqrt{2}\left[\cos\left(360° - 285°\right) + i\sin\left(360° - 285°\right)\right]$

$= \sqrt{2}\left(\cos 75° + i\sin 75°\right)$

40. $z = 1 - i$

$r = \sqrt{1^2 + (-1)^2} = \sqrt{2}$

$\tan\theta = \frac{-1}{1} = -1$

$\theta = 315°$

$z = \sqrt{2}\left(\cos 315° + i\sin 315°\right)$

$w = 1 - \sqrt{3}i$

$r = \sqrt{1^2 + \left(-\sqrt{3}\right)^2} = \sqrt{4} = 2$

$\tan\theta = \frac{-\sqrt{3}}{1} = -\sqrt{3}$

$\theta = 300°$

$w = 2\left(\cos 300° + i\sin 300°\right)$

$z \cdot w = \sqrt{2}\left(\cos 315° + i\sin 315°\right) \cdot 2\left(\cos 300° + i\sin 300°\right)$

$= \sqrt{2} \cdot 2\left[\cos\left(315° + 300°\right) + i\sin\left(315° + 300°\right)\right]$

$= 2\sqrt{2}\left(\cos 615° + i\sin 615°\right)$

$= 2\sqrt{2}\left[\cos\left(615° - 360°\right) + i\sin\left(615° - 360°\right)\right]$

$= 2\sqrt{2}\left(\cos 255° + i\sin 255°\right)$

$\frac{z}{w} = \frac{\sqrt{2}\left(\cos 315° + i\sin 315°\right)}{2\left(\cos 300° + i\sin 300°\right)}$

$= \frac{\sqrt{2}}{2}\left[\cos\left(315° - 300°\right) + i\sin\left(315° - 300°\right)\right]$

$= \frac{\sqrt{2}}{2}\left(\cos 15° + i\sin 15°\right)$

41. $\left[4\left(\cos 40° + i\sin 40°\right)\right]^3$

$= 4^3\left[\cos\left(3 \cdot 40°\right) + i\sin\left(3 \cdot 40°\right)\right]$

$= 64\left(\cos 120° + i\sin 120°\right)$

$= 64\left(-\frac{1}{2} + \frac{\sqrt{3}}{2}i\right)$

$= -32 + 32\sqrt{3}\,i$

42. $\left[3(\cos 80^\circ + i\sin 80^\circ)\right]^3$
$= 3^3\left[\cos(3\cdot 80^\circ) + i\sin(3\cdot 80^\circ)\right]$
$= 27(\cos 240^\circ + i\sin 240^\circ)$
$= 27\left(-\frac{1}{2} - \frac{\sqrt{3}}{2}i\right)$
$= -\frac{27}{2} - \frac{27\sqrt{3}}{2}i$

43. $\left[2\left(\cos\frac{\pi}{10} + i\sin\frac{\pi}{10}\right)\right]^5$
$= 2^5\left[\cos\left(5\cdot\frac{\pi}{10}\right) + i\sin\left(5\cdot\frac{\pi}{10}\right)\right]$
$= 32\left(\cos\frac{\pi}{2} + i\sin\frac{\pi}{2}\right)$
$= 32(0 + 1i)$
$= 32i$

44. $\left[\sqrt{2}\left(\cos\frac{5\pi}{16} + i\sin\frac{5\pi}{16}\right)\right]^4$
$= \left(\sqrt{2}\right)^4\left[\cos\left(4\cdot\frac{5\pi}{16}\right) + i\sin\left(4\cdot\frac{5\pi}{16}\right)\right]$
$= 4\left(\cos\frac{5\pi}{4} + i\sin\frac{5\pi}{4}\right)$
$= 4\left(-\frac{\sqrt{2}}{2} - \frac{\sqrt{2}}{2}i\right)$
$= -2\sqrt{2} - 2\sqrt{2}\,i$

45. $\left[\sqrt{3}(\cos 10^\circ + i\sin 10^\circ)\right]^6$
$= \left(\sqrt{3}\right)^6\left[\cos(6\cdot 10^\circ) + i\sin(6\cdot 10^\circ)\right]$
$= 27(\cos 60^\circ + i\sin 60^\circ)$
$= 27\left(\frac{1}{2} + \frac{\sqrt{3}}{2}i\right)$
$= \frac{27}{2} + \frac{27\sqrt{3}}{2}i$

46. $\left[\frac{1}{2}\cdot(\cos 72^\circ + i\sin 72^\circ)\right]^5$
$= \left(\frac{1}{2}\right)^5\left[\cos(5\cdot 72^\circ) + i\sin(5\cdot 72^\circ)\right]$
$= \frac{1}{32}\cdot(\cos 360^\circ + i\sin 360^\circ)$
$= \frac{1}{32}\cdot(1 + 0i)$
$= \frac{1}{32}$

47. $\left[\sqrt{5}\left(\cos\frac{3\pi}{16} + i\sin\frac{3\pi}{16}\right)\right]^4$
$= \left(\sqrt{5}\right)^4\left[\cos\left(4\cdot\frac{3\pi}{16}\right) + i\sin\left(4\cdot\frac{3\pi}{16}\right)\right]$
$= 25\left(\cos\frac{3\pi}{4} + i\sin\frac{3\pi}{4}\right)$
$= 25\left(-\frac{\sqrt{2}}{2} + \frac{\sqrt{2}}{2}i\right)$
$= -\frac{25\sqrt{2}}{2} + \frac{25\sqrt{2}}{2}i$

48. $\left[\sqrt{3}\left(\cos\frac{5\pi}{18} + i\sin\frac{5\pi}{18}\right)\right]^6$
$= \left(\sqrt{3}\right)^6\left[\cos\left(6\cdot\frac{5\pi}{18}\right) + i\sin\left(6\cdot\frac{5\pi}{18}\right)\right]$
$= 27\left(\cos\frac{5\pi}{3} + i\sin\frac{5\pi}{3}\right)$
$= 27\left(\frac{1}{2} - \frac{\sqrt{3}}{2}i\right)$
$= \frac{27}{2} - \frac{27\sqrt{3}}{2}i$

49. $1 - i$
$r = \sqrt{1^2 + (-1)^2} = \sqrt{2}$
$\tan\theta = \frac{-1}{1} = -1$
$\theta = \frac{7\pi}{4}$
$1 - i = \sqrt{2}\left(\cos\frac{7\pi}{4} + i\sin\frac{7\pi}{4}\right)$

$$(1-i)^5 = \left[\sqrt{2}\left(\cos\frac{7\pi}{4} + i\sin\frac{7\pi}{4}\right)\right]^5$$
$$= \left(\sqrt{2}\right)^5\left[\cos\left(5\cdot\frac{7\pi}{4}\right) + i\sin\left(5\cdot\frac{7\pi}{4}\right)\right]$$
$$= 4\sqrt{2}\left(\cos\frac{35\pi}{4} + i\sin\frac{35\pi}{4}\right)$$
$$= 4\sqrt{2}\left(-\frac{\sqrt{2}}{2} + \frac{\sqrt{2}}{2}i\right)$$
$$= -4 + 4i$$

50. $\sqrt{3} - i$

$$r = \sqrt{\left(\sqrt{3}\right)^2 + (-1)^2} = \sqrt{4} = 2$$
$$\tan\theta = \frac{-1}{\sqrt{3}} = -\frac{\sqrt{3}}{3}$$
$$\theta = 330°$$
$$\sqrt{3} - i = 2(\cos 330° + i\sin 330°)$$
$$\left(\sqrt{3} - i\right)^6 = \left[2(\cos 330° + i\sin 330°)\right]^6$$
$$= 2^6\left[\cos(6\cdot 330°) + i\sin(6\cdot 330°)\right]$$
$$= 64(\cos 1980° + i\sin 1980°)$$
$$= 64(-1 + 0i)$$
$$= -64$$

51. $\sqrt{2} - i$

$$r = \sqrt{\left(\sqrt{2}\right)^2 + (-1)^2} = \sqrt{3}$$
$$\tan\theta = \frac{-1}{\sqrt{2}} = -\frac{\sqrt{2}}{2}$$
$$\theta \approx 324.736°$$
$$\sqrt{2} - i \approx \sqrt{3}(\cos 324.736° + i\sin 324.736°)$$
$$\left(\sqrt{2} - i\right)^6 \approx \left[\sqrt{3}(\cos 324.736° + i\sin 324.736°)\right]^6$$
$$= \left(\sqrt{3}\right)^6\left[\cos(6\cdot 324.736°) + i\sin(6\cdot 324.736°)\right]$$
$$= 27(\cos 1948.416° + i\sin 1948.416°)$$
$$\approx 27(-0.8519 + 0.5237i)$$
$$\approx -23 + 14.142i$$

52. $1 - \sqrt{5}i$

$$r = \sqrt{1^2 + \left(-\sqrt{5}\right)^2} = \sqrt{6}$$
$$\tan\theta = \frac{-\sqrt{5}}{1} = -\sqrt{5}$$
$$\theta \approx 294.095°$$
$$1 - \sqrt{5}\, i \approx \sqrt{6}(\cos 294.095° + i\sin 294.095°)$$
$$\left(1 - \sqrt{5}\, i\right)^8$$
$$\approx \left(\sqrt{6}\right)^8\left[\cos(8\cdot 294.095°) + i\sin(8\cdot 294.095°)\right]$$
$$= 1296[\cos 2352.76° + i\sin 2352.76°]$$
$$\approx 1296(-0.9753 - 0.2208i)$$
$$\approx -1294 - 286.217i$$

53. $1 + i$

$$r = \sqrt{1^2 + 1^2} = \sqrt{2}$$
$$\tan\theta = \frac{1}{1} = 1$$
$$\theta = 45°$$
$$1 + i = \sqrt{2}(\cos 45° + i\sin 45°)$$

The three complex cube roots of $1 + i = \sqrt{2}(\cos 45° + i\sin 45°)$ are:

$$z_k = \sqrt[3]{\sqrt{2}}\left[\cos\left(\frac{45°}{3} + \frac{360°k}{3}\right) + i\sin\left(\frac{45°}{3} + \frac{360°k}{3}\right)\right]$$
$$= \sqrt[6]{2}\left[\cos(15° + 120°k) + i\sin(15° + 120°k)\right]$$
$$z_0 = \sqrt[6]{2}\left[\cos(15° + 120°\cdot 0) + i\sin(15° + 120°\cdot 0)\right]$$
$$= \sqrt[6]{2}(\cos 15° + i\sin 15°)$$
$$z_1 = \sqrt[6]{2}\left[\cos(15° + 120°\cdot 1) + i\sin(15° + 120°\cdot 1)\right]$$
$$= \sqrt[6]{2}(\cos 135° + i\sin 135°)$$
$$z_2 = \sqrt[6]{2}\left[\cos(15° + 120°\cdot 2) + i\sin(15° + 120°\cdot 2)\right]$$
$$= \sqrt[6]{2}(\cos 255° + i\sin 255°)$$

54. $\sqrt{3} - i$

$$r = \sqrt{\left(\sqrt{3}\right)^2 + (-1)^2} = \sqrt{4} = 2$$
$$\tan\theta = \frac{-1}{\sqrt{3}} = -\frac{\sqrt{3}}{3}$$
$$\theta = 330°$$
$$\sqrt{3} - i = 2(\cos 330° + i\sin 330°)$$

The four complex fourth roots of
$\sqrt{3}-i=2(\cos 330°+i\sin 330°)$ are:

$$z_k=\sqrt[4]{2}\left[\cos\left(\frac{330°}{4}+\frac{360°k}{4}\right)+i\sin\left(\frac{330°}{4}+\frac{360°k}{4}\right)\right]$$
$$=\sqrt[4]{2}\left[\cos(82.5°+90°k)+i\sin(82.5°+90°k)\right]$$
$$z_0=\sqrt[4]{2}\left[\cos(82.5°+90°\cdot 0)+i\sin(82.5°+90°\cdot 0)\right]$$
$$=\sqrt[4]{2}\left[\cos(82.5°)+i\sin(82.5°)\right]$$
$$z_1=\sqrt[4]{2}\left[\cos(82.5°+90°\cdot 1)+i\sin(82.5°+90°\cdot 1)\right]$$
$$=\sqrt[4]{2}\left[\cos(172.5°)+i\sin(172.5°)\right]$$
$$z_2=\sqrt[4]{2}\left[\cos(82.5°+90°\cdot 2)+i\sin(82.5°+90°\cdot 2)\right]$$
$$=\sqrt[4]{2}\left[\cos(262.5°)+i\sin(262.5°)\right]$$
$$z_3=\sqrt[4]{2}\left[\cos(82.5°+90°\cdot 3)+i\sin(82.5°+90°\cdot 3)\right]$$
$$=\sqrt[4]{2}\left[\cos(352.5°)+i\sin(352.5°)\right]$$

55. $4-4\sqrt{3}i$

$$r=\sqrt{4^2+\left(-4\sqrt{3}\right)^2}=\sqrt{64}=8$$
$$\tan\theta=\frac{-4\sqrt{3}}{4}=-\sqrt{3}$$
$$\theta=300°$$
$$4-4\sqrt{3}i=8(\cos 300°+i\sin 300°)$$

The four complex fourth roots of
$4-4\sqrt{3}i=8(\cos 300°+i\sin 300°)$ are:

$$z_k=\sqrt[4]{8}\left[\cos\left(\frac{300°}{4}+\frac{360°k}{4}\right)+i\sin\left(\frac{300°}{4}+\frac{360°k}{4}\right)\right]$$
$$=\sqrt[4]{8}\left[\cos(75°+90°k)+i\sin(75°+90°k)\right]$$
$$z_0=\sqrt[4]{8}\left[\cos(75°+90°\cdot 0)+i\sin(75°+90°\cdot 0)\right]$$
$$=\sqrt[4]{8}(\cos 75°+i\sin 75°)$$
$$z_1=\sqrt[4]{8}\left[\cos(75°+90°\cdot 1)+i\sin(75°+90°\cdot 1)\right]$$
$$=\sqrt[4]{8}(\cos 165°+i\sin 165°)$$
$$z_2=\sqrt[4]{8}\left[\cos(75°+90°\cdot 2)+i\sin(75°+90°\cdot 2)\right]$$
$$=\sqrt[4]{8}(\cos 255°+i\sin 255°)$$
$$z_3=\sqrt[4]{8}\left[\cos(75°+90°\cdot 3)+i\sin(75°+90°\cdot 3)\right]$$
$$=\sqrt[4]{8}(\cos 345°+i\sin 345°)$$

56. $-8-8i$

$$r=\sqrt{(-8)^2+(-8)^2}=8\sqrt{2}$$
$$\tan\theta=\frac{-8}{-8}=1$$
$$\theta=225°$$
$$-8-8i=8\sqrt{2}(\cos 225°+i\sin 225°)$$

The three complex cube roots of
$-8-8i=8\sqrt{2}(\cos 225°+i\sin 225°)$ are:

$$z_k=\sqrt[3]{8\sqrt{2}}\left[\cos\left(\frac{225°}{3}+\frac{360°k}{3}\right)+i\sin\left(\frac{225°}{3}+\frac{360°k}{3}\right)\right]$$
$$=2\sqrt[6]{2}\left[\cos(75°+120°k)+i\sin(75°+120°k)\right]$$
$$z_0=2\sqrt[6]{2}\left[\cos(75°+120°\cdot 0)+i\sin(75°+120°\cdot 0)\right]$$
$$=2\sqrt[6]{2}(\cos 75°+i\sin 75°)$$
$$z_1=2\sqrt[6]{2}\left[\cos(75°+120°\cdot 1)+i\sin(75°+120°\cdot 1)\right]$$
$$=2\sqrt[6]{2}(\cos 195°+i\sin 195°)$$
$$z_2=2\sqrt[6]{2}\left[\cos(75°+120°\cdot 2)+i\sin(75°+120°\cdot 2)\right]$$
$$=2\sqrt[6]{2}(\cos 315°+i\sin 315°)$$

57. $-16i$

$$r=\sqrt{0^2+(-16)^2}=\sqrt{256}=16$$
$$\tan\theta=\frac{-16}{0}$$
$$\theta=270°$$
$$-16i=16(\cos 270°+i\sin 270°)$$

The four complex fourth roots of
$-16i=16(\cos 270°+i\sin 270°)$ are:

$$z_k=\sqrt[4]{16}\left[\cos\left(\frac{270°}{4}+\frac{360°k}{4}\right)+i\sin\left(\frac{270°}{4}+\frac{360°k}{4}\right)\right]$$
$$=2\left[\cos(67.5°+90°k)+i\sin(67.5°+90°k)\right]$$
$$z_0=2\left[\cos(67.5°+90°\cdot 0)+i\sin(67.5°+90°\cdot 0)\right]$$
$$=2(\cos 67.5°+i\sin 67.5°)$$
$$z_1=2\left[\cos(67.5°+90°\cdot 1)+i\sin(67.5°+90°\cdot 1)\right]$$
$$=2(\cos 157.5°+i\sin 157.5°)$$
$$z_2=2\left[\cos(67.5°+90°\cdot 2)+i\sin(67.5°+90°\cdot 2)\right]$$
$$=2(\cos 247.5°+i\sin 247.5°)$$
$$z_3=2\left[\cos(67.5°+90°\cdot 3)+i\sin(67.5°+90°\cdot 3)\right]$$
$$=2(\cos 337.5°+i\sin 337.5°)$$

58. -8

$r=\sqrt{(-8)^2+0^2}=8$

$\tan\theta=\frac{0}{-8}=0$

$\theta=180°$

$-8=8(\cos 180°+i\sin 180°)$

The three complex cube roots of $-8=8(\cos 180°+i\sin 180°)$ are:

$$z_k=\sqrt[3]{8}\left[\cos\left(\frac{180°}{3}+\frac{360°k}{3}\right)+i\sin\left(\frac{180°}{3}+\frac{360°k}{3}\right)\right]$$
$$=2\left[\cos(60°+120°k)+i\sin(60°+120°k)\right]$$

$$z_0=2\left[\cos(60°+120°\cdot 0)+i\sin(60°+120°\cdot 0)\right]$$
$$=2(\cos 60°+i\sin 60°)$$

$$z_1=2\left[\cos(60°+120°\cdot 1)+i\sin(60°+120°\cdot 1)\right]$$
$$=2(\cos 180°+i\sin 180°)$$

$$z_2=2\left[\cos(60°+120°\cdot 2)+i\sin(60°+120°\cdot 2)\right]$$
$$=2(\cos 300°+i\sin 300°)$$

59. i

$r=\sqrt{0^2+1^2}=\sqrt{1}=1$

$\tan\theta=\frac{1}{0}$

$\theta=90°$

$i=1(\cos 90°+i\sin 90°)$

The five complex fifth roots of $i=1(\cos 90°+i\sin 90°)$ are:

$$z_k=\sqrt[5]{1}\left[\cos\left(\frac{90°}{5}+\frac{360°k}{5}\right)+i\sin\left(\frac{90°}{5}+\frac{360°k}{5}\right)\right]$$
$$=1\left[\cos(18°+72°k)+i\sin(18°+72°k)\right]$$

$$z_0=1\left[\cos(18°+72°\cdot 0)+i\sin(18°+72°\cdot 0)\right]$$
$$=\cos 18°+i\sin 18°$$

$$z_1=1\left[\cos(18°+72°\cdot 1)+i\sin(18°+72°\cdot 1)\right]$$
$$=\cos 90°+i\sin 90°$$

$$z_2=1\left[\cos(18°+72°\cdot 2)+i\sin(18°+72°\cdot 2)\right]$$
$$=\cos 162°+i\sin 162°$$

$$z_3=1\left[\cos(18°+72°\cdot 3)+i\sin(18°+72°\cdot 3)\right]$$
$$=\cos 234°+i\sin 234°$$

$$z_4=1\left[\cos(18°+72°\cdot 4)+i\sin(18°+72°\cdot 4)\right]$$
$$=\cos 306°+i\sin 306°$$

60. $-i$

$r=\sqrt{0^2+(-1)^2}=\sqrt{1}=1$

$\tan\theta=\frac{-1}{0}$

$\theta=270°$

$-i=1(\cos 270°+i\sin 270°)$

The five complex fifth roots of $-i=1(\cos 270°+i\sin 270°)$ are:

$$z_k=\sqrt[5]{1}\left[\cos\left(\frac{270°}{5}+\frac{360°k}{5}\right)+i\sin\left(\frac{270°}{5}+\frac{360°k}{5}\right)\right]$$
$$=1\left[\cos(54°+72°k)+i\sin(54°+72°k)\right]$$

$$z_0=1\left[\cos(54°+72°\cdot 0)+i\sin(54°+72°\cdot 0)\right]$$
$$=\cos 54°+i\sin 54°$$

$$z_1=1\left[\cos(54°+72°\cdot 1)+i\sin(54°+72°\cdot 1)\right]$$
$$=\cos 126°+i\sin 126°$$

$$z_2=1\left[\cos(54°+72°\cdot 2)+i\sin(54°+72°\cdot 2)\right]$$
$$=\cos 198°+i\sin 198°$$

$$z_3=1\left[\cos(54°+72°\cdot 3)+i\sin(54°+72°\cdot 3)\right]$$
$$=\cos 270°+i\sin 270°$$

$$z_4=1\left[\cos(54°+72°\cdot 4)+i\sin(54°+72°\cdot 4)\right]$$
$$=\cos 342°+i\sin 342°$$

61. $1=1+0i$

$r=\sqrt{1^2+0^2}=\sqrt{1}=1$

$\tan\theta=\frac{0}{1}=0$

$\theta=0°$

$1+0i=1(\cos 0°+i\sin 0°)$

The four complex fourth roots of unity are:

$$z_k=\sqrt[4]{1}\left[\cos\left(\frac{0°}{4}+\frac{360°k}{4}\right)+i\sin\left(\frac{0°}{4}+\frac{360°k}{4}\right)\right]$$
$$=1\left[\cos(90°k)+i\sin(90°k)\right]$$

$$z_0=\cos(90°\cdot 0)+i\sin(90°\cdot 0)$$
$$=\cos 0°+i\sin 0°=1+0i=1$$

$$z_1=\cos(90°\cdot 1)+i\sin(90°\cdot 1)$$
$$=\cos 90°+i\sin 90°=0+1i=i$$

$$z_2=\cos(90°\cdot 2)+i\sin(90°\cdot 2)$$
$$=\cos 180°+i\sin 180°=-1+0i=-1$$

$$z_3=\cos(90°\cdot 3)+i\sin(90°\cdot 3)$$
$$=\cos 270°+i\sin 270°=0-1i=-i$$

The complex fourth roots of unity are:
$1, i, -1, -i$.

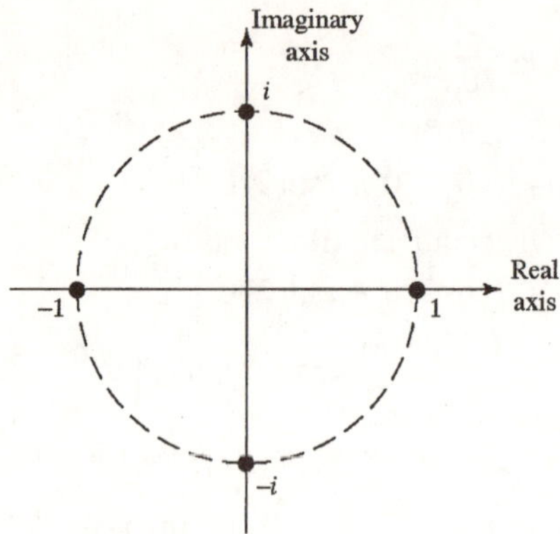

62. $1 = 1 + 0i$

$r = \sqrt{1^2 + 0^2} = \sqrt{1} = 1$

$\tan\theta = \dfrac{0}{1} = 0$

$\theta = 0°$

$1 + 0i = 1(\cos 0° + i\sin 0°)$

The six complex sixth roots of unity are:

$$z_k = \sqrt[6]{1}\left[\cos\left(\frac{0°}{6} + \frac{360° k}{6}\right) + i\sin\left(\frac{0°}{6} + \frac{360° k}{6}\right)\right]$$
$$= 1\left[\cos(60° k) + i\sin(60° k)\right]$$

$$z_0 = \cos(60°\cdot 0) + i\sin(60°\cdot 0)$$
$$= \cos 0° + i\sin 0° = 1 + 0i = 1$$

$$z_1 = \cos(60°\cdot 1) + i\sin(60°\cdot 1)$$
$$= \cos 60° + i\sin 60° = \frac{1}{2} + \frac{\sqrt{3}}{2}i$$

$$z_2 = \cos(60°\cdot 2) + i\sin(60°\cdot 2)$$
$$= \cos 120° + i\sin 120° = -\frac{1}{2} + \frac{\sqrt{3}}{2}i$$

$$z_3 = \cos(60°\cdot 3) + i\sin(60°\cdot 3)$$
$$= \cos 180° + i\sin 180° = -1 + 0i = -1$$

$$z_4 = \cos(60°\cdot 4) + i\sin(60°\cdot 4)$$
$$= \cos 240° + i\sin 240° = -\frac{1}{2} - \frac{\sqrt{3}}{2}i$$

$$z_5 = \cos(60°\cdot 5) + i\sin(60°\cdot 5)$$
$$= \cos 300° + i\sin 300° = \frac{1}{2} - \frac{\sqrt{3}}{2}i$$

The complex sixth roots of unity are:

$1, \frac{1}{2} + \frac{\sqrt{3}}{2}i, -\frac{1}{2} + \frac{\sqrt{3}}{2}i, -1, -\frac{1}{2} - \frac{\sqrt{3}}{2}i, \frac{1}{2} - \frac{\sqrt{3}}{2}i$.

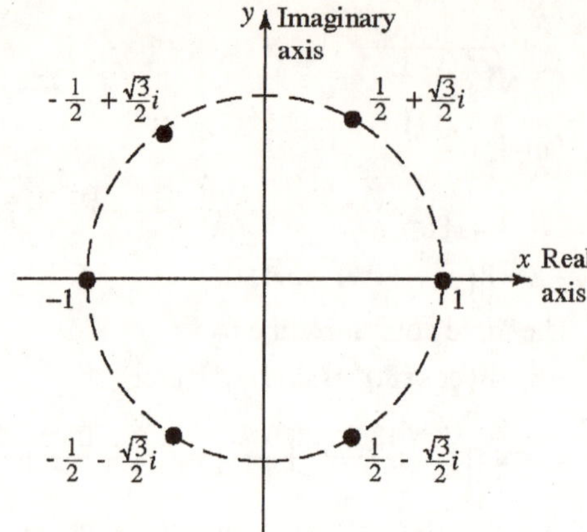

63. Let $w = r(\cos\theta + i\sin\theta)$ be a complex number. If $w \neq 0$, there are n distinct nth roots of w, given by the formula:

$$z_k = \sqrt[n]{r}\left[\cos\left(\frac{\theta}{n} + \frac{2k\pi}{n}\right) + i\sin\left(\frac{\theta}{n} + \frac{2k\pi}{n}\right)\right],$$

where $k = 0, 1, 2, \ldots, n-1$

$|z_k| = \sqrt[n]{r}$ for all k

64. Since $|z_k| = \sqrt[n]{r}$ for all k, each of the complex nth roots lies on a circle with center at the origin and radius $\sqrt[n]{|w|} = \sqrt[n]{r}$, where w is the original complex number.

65. Examining the formula for the distinct complex nth roots of the complex number $w = r(\cos\theta + i\sin\theta)$,

$$z_k = \sqrt[n]{r}\left[\cos\left(\frac{\theta}{n} + \frac{2k\pi}{n}\right) + i\sin\left(\frac{\theta}{n} + \frac{2k\pi}{n}\right)\right],$$

where $k = 0, 1, 2, \ldots, n-1$,

we see that the z_k are spaced apart by an angle of $\dfrac{2\pi}{n}$.

66. Let $z_1 = r_1\left(\cos\theta_1 + i\sin\theta_1\right)$ and $z_2 = r_2\left(\cos\theta_2 + i\sin\theta_2\right)$. Then

$$\frac{z_1}{z_2} = \frac{r_1\left(\cos\theta_1 + i\sin\theta_1\right)}{r_2\left(\cos\theta_2 + i\sin\theta_2\right)}$$

$$= \frac{r_1\left(\cos\theta_1 + i\sin\theta_1\right)}{r_2\left(\cos\theta_2 + i\sin\theta_2\right)} \cdot \frac{\left(\cos\theta_2 - i\sin\theta_2\right)}{\left(\cos\theta_2 - i\sin\theta_2\right)}$$

$$= \frac{r_1}{r_2} \cdot \frac{\cos\theta_1 \cdot \cos\theta_2 - i\cos\theta_1 \cdot \sin\theta_2 + i\sin\theta_1 \cdot \cos\theta_2 + \sin\theta_1 \cdot \sin\theta_2}{\cos^2\theta_2 + \sin^2\theta_2}$$

$$= \frac{r_1}{r_2} \cdot \frac{\cos\theta_1 \cdot \cos\theta_2 + \sin\theta_1 \cdot \sin\theta_2 + i\left(\sin\theta_1 \cdot \cos\theta_2 - \cos\theta_1 \cdot \sin\theta_2\right)}{1}$$

$$= \frac{r_1}{r_2}\left[\cos\left(\theta_1 - \theta_2\right) + i\sin\left(\theta_1 - \theta_2\right)\right]$$

Section 8.4

1. unit
2. scalar
3. horizontal, vertical
4. True
5. True
6. False
7. $\mathbf{v} + \mathbf{w}$

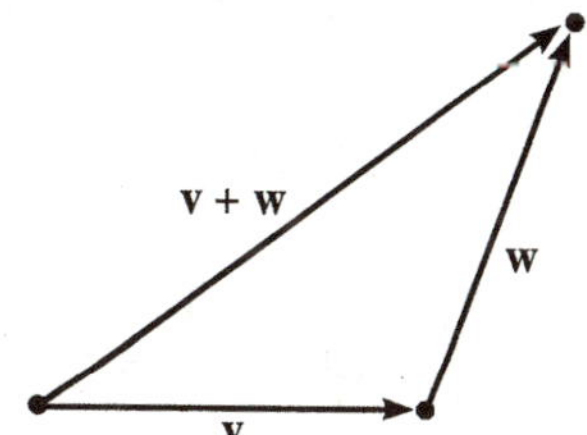

8. $\mathbf{u} + \mathbf{v}$

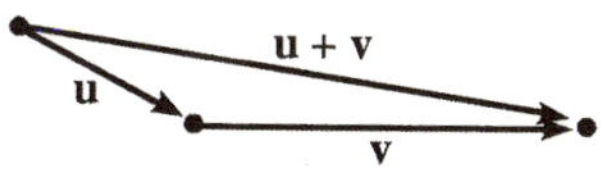

9. $3\mathbf{v}$

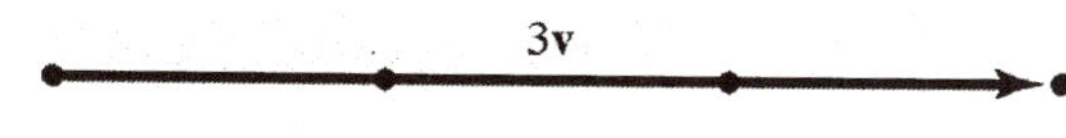

10. $4\mathbf{w}$

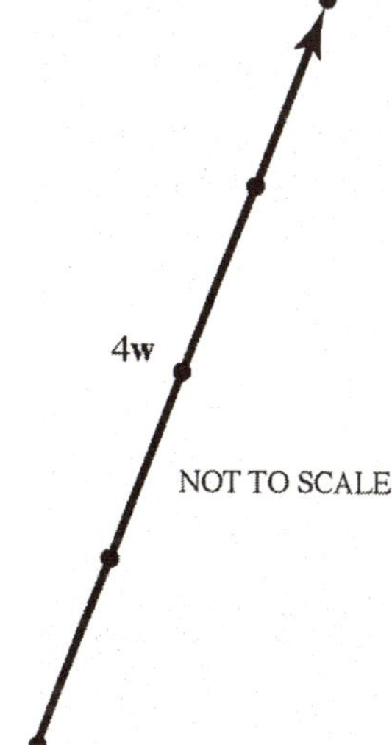

11. $\mathbf{v} - \mathbf{w}$

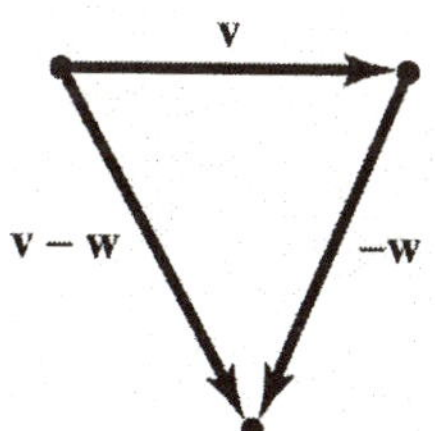

12. $\mathbf{u} - \mathbf{v}$

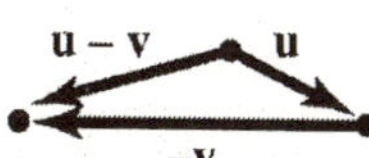

13. $3\mathbf{v}+\mathbf{u}-2\mathbf{w}$

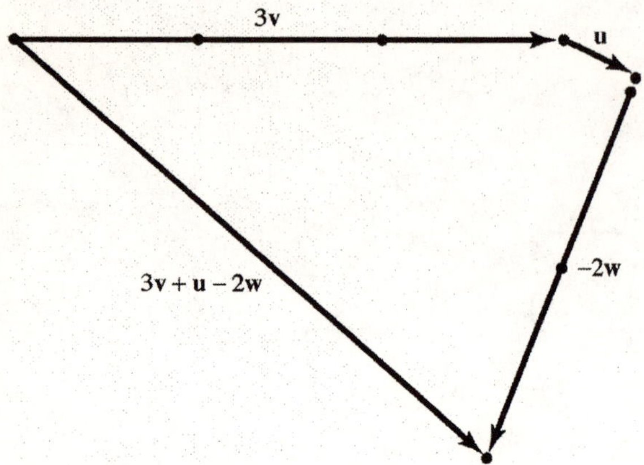

14. $2\mathbf{u}-3\mathbf{v}+\mathbf{w}$

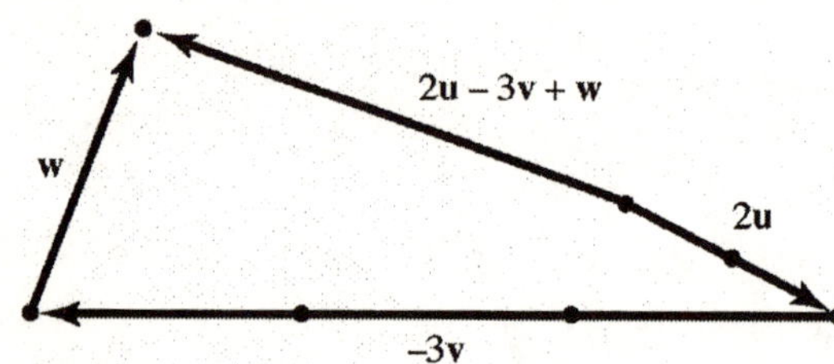

15. True

16. False $\mathbf{K}+\mathbf{G}=-\mathbf{F}$

17. False $\mathbf{C}=-\mathbf{F}+\mathbf{E}-\mathbf{D}$

18. True

19. False $\mathbf{D}-\mathbf{E}=\mathbf{H}+\mathbf{G}$

20. False $\mathbf{C}+\mathbf{H}=-\mathbf{G}-\mathbf{F}$

21. True

22. True

23. If $\|\mathbf{v}\|=4$, then $\|3\mathbf{v}\|=|3|\cdot\|\mathbf{v}\|=3\cdot 4=12$.

24. If $\|\mathbf{v}\|=2$, then $\|-4\mathbf{v}\|=|-4|\cdot\|\mathbf{v}\|=4\cdot 2=8$

25. $P=(0,0), Q=(3,4)$
$\mathbf{v}=(3-0)\mathbf{i}+(4-0)\mathbf{j}=3\mathbf{i}+4\mathbf{j}$

26. $P=(0,0), Q=(-3,-5)$
$\mathbf{v}=(-3-0)\mathbf{i}+(-5-0)\mathbf{j}=-3\mathbf{i}-5\mathbf{j}$

27. $P=(3,2), Q=(5,6)$
$\mathbf{v}=(5-3)\mathbf{i}+(6-2)\mathbf{j}=2\mathbf{i}+4\mathbf{j}$

28. $P=(-3,2), Q=(6,5)$
$\mathbf{v}=[6-(-3)]\mathbf{i}+(5-2)\mathbf{j}=9\mathbf{i}+3\mathbf{j}$

29. $P=(-2,-1), Q=(6,-2)$
$\mathbf{v}=[6-(-2)]\mathbf{i}+[-2-(-1)]\mathbf{j}=8\mathbf{i}-\mathbf{j}$

30. $P=(-1,4), Q=(6,2)$
$\mathbf{v}=[6-(-1)]\mathbf{i}+(2-4)\mathbf{j}=7\mathbf{i}-2\mathbf{j}$

31. $P=(1,0), Q=(0,1)$
$\mathbf{v}=(0-1)\mathbf{i}+(1-0)\mathbf{j}=-\mathbf{i}+\mathbf{j}$

32. $P=(1,1), Q=(2,2)$
$\mathbf{v}=(2-1)\mathbf{i}+(2-1)\mathbf{j}=\mathbf{i}+\mathbf{j}$

33. For $\mathbf{v}=3\mathbf{i}-4\mathbf{j}$, $\|\mathbf{v}\|=\sqrt{3^2+(-4)^2}=\sqrt{25}=5$.

34. For $\mathbf{v}=-5\mathbf{i}+12\mathbf{j}$,
$\|\mathbf{v}\|=\sqrt{(-5)^2+12^2}=\sqrt{169}=13$.

35. For $\mathbf{v}=\mathbf{i}-\mathbf{j}$, $\|\mathbf{v}\|=\sqrt{1^2+(-1)^2}=\sqrt{2}$.

36. For $\mathbf{v}=-\mathbf{i}-\mathbf{j}$, $\|\mathbf{v}\|=\sqrt{(-1)^2+(-1)^2}=\sqrt{2}$.

37. For $\mathbf{v}=-2\mathbf{i}+3\mathbf{j}$, $\|\mathbf{v}\|=\sqrt{(-2)^2+3^2}=\sqrt{13}$.

38. For $\mathbf{v}=6\mathbf{i}+2\mathbf{j}$, $\|\mathbf{v}\|=\sqrt{6^2+2^2}=\sqrt{40}=2\sqrt{10}$.

39.
$$\begin{aligned}2\mathbf{v}+3\mathbf{w}&=2(3\mathbf{i}-5\mathbf{j})+3(-2\mathbf{i}+3\mathbf{j})\\&=6\mathbf{i}-10\mathbf{j}-6\mathbf{i}+9\mathbf{j}\\&=-\mathbf{j}\end{aligned}$$

40.
$$\begin{aligned}3\mathbf{v}-2\mathbf{w}&=3(3\mathbf{i}-5\mathbf{j})-2(-2\mathbf{i}+3\mathbf{j})\\&=9\mathbf{i}-15\mathbf{j}+4\mathbf{i}-6\mathbf{j}\\&=13\mathbf{i}-21\mathbf{j}\end{aligned}$$

41.
$$\begin{aligned}\|\mathbf{v}-\mathbf{w}\|&=\|(3\mathbf{i}-5\mathbf{j})-(-2\mathbf{i}+3\mathbf{j})\|\\&=\|5\mathbf{i}-8\mathbf{j}\|\\&=\sqrt{5^2+(-8)^2}\\&=\sqrt{89}\end{aligned}$$

42. $\|\mathbf{v}+\mathbf{w}\| = \|(3\mathbf{i}-5\mathbf{j})+(-2\mathbf{i}+3\mathbf{j})\|$
$= \|\mathbf{i}-2\mathbf{j}\|$
$= \sqrt{1^2+(-2)^2}$
$= \sqrt{5}$

43. $\|\mathbf{v}\|-\|\mathbf{w}\| = \|3\mathbf{i}-5\mathbf{j}\|-\|-2\mathbf{i}+3\mathbf{j}\|$
$= \sqrt{3^2+(-5)^2}-\sqrt{(-2)^2+3^2}$
$= \sqrt{34}-\sqrt{13}$

44. $\|\mathbf{v}\|+\|\mathbf{w}\| = \|3\mathbf{i}-5\mathbf{j}\|+\|-2\mathbf{i}+3\mathbf{j}\|$
$= \sqrt{3^2+(-5)^2}+\sqrt{(-2)^2+3^2}$
$= \sqrt{34}+\sqrt{13}$

45. $\mathbf{u} = \dfrac{\mathbf{v}}{\|\mathbf{v}\|} = \dfrac{5\mathbf{i}}{\|5\mathbf{i}\|} = \dfrac{5\mathbf{i}}{\sqrt{25+0}} = \dfrac{5\mathbf{i}}{5} = \mathbf{i}$

46. $\mathbf{u} = \dfrac{\mathbf{v}}{\|\mathbf{v}\|} = \dfrac{-3\mathbf{j}}{\|-3\mathbf{j}\|} = \dfrac{-3\mathbf{j}}{\sqrt{0+9}} = \dfrac{-3\mathbf{j}}{3} = -\mathbf{j}$

47. $\mathbf{u} = \dfrac{\mathbf{v}}{\|\mathbf{v}\|} = \dfrac{3\mathbf{i}-4\mathbf{j}}{\|3\mathbf{i}-4\mathbf{j}\|} = \dfrac{3\mathbf{i}-4\mathbf{j}}{\sqrt{3^2+(-4)^2}}$
$= \dfrac{3\mathbf{i}-4\mathbf{j}}{\sqrt{25}}$
$= \dfrac{3\mathbf{i}-4\mathbf{j}}{5}$
$= \dfrac{3}{5}\mathbf{i}-\dfrac{4}{5}\mathbf{j}$

48. $\mathbf{u} = \dfrac{\mathbf{v}}{\|\mathbf{v}\|} = \dfrac{-5\mathbf{i}+12\mathbf{j}}{\|-5\mathbf{i}+12\mathbf{j}\|} = \dfrac{-5\mathbf{i}+12\mathbf{j}}{\sqrt{(-5)^2+12^2}}$
$= \dfrac{-5\mathbf{i}+12\mathbf{j}}{\sqrt{169}}$
$= \dfrac{-5\mathbf{i}+12\mathbf{j}}{13}$
$= -\dfrac{5}{13}\mathbf{i}+\dfrac{12}{13}\mathbf{j}$

49. $\mathbf{u} = \dfrac{\mathbf{v}}{\|\mathbf{v}\|} = \dfrac{\mathbf{i}-\mathbf{j}}{\|\mathbf{i}-\mathbf{j}\|} = \dfrac{\mathbf{i}-\mathbf{j}}{\sqrt{1^2+(-1)^2}} = \dfrac{\mathbf{i}-\mathbf{j}}{\sqrt{2}}$
$= \dfrac{1}{\sqrt{2}}\mathbf{i}-\dfrac{1}{\sqrt{2}}\mathbf{j}$
$= \dfrac{\sqrt{2}}{2}\mathbf{i}-\dfrac{\sqrt{2}}{2}\mathbf{j}$

50. $\mathbf{u} = \dfrac{\mathbf{v}}{\|\mathbf{v}\|} = \dfrac{2\mathbf{i}-\mathbf{j}}{\|2\mathbf{i}-\mathbf{j}\|} = \dfrac{2\mathbf{i}-\mathbf{j}}{\sqrt{2^2+(-1)^2}} = \dfrac{2\mathbf{i}-\mathbf{j}}{\sqrt{5}}$
$= \dfrac{2}{\sqrt{5}}\mathbf{i}-\dfrac{1}{\sqrt{5}}\mathbf{j}$
$= \dfrac{2\sqrt{5}}{5}\mathbf{i}-\dfrac{\sqrt{5}}{5}\mathbf{j}$

51. Let $\mathbf{v} = a\mathbf{i}+b\mathbf{j}$. We want $\|\mathbf{v}\| = 4$ and $a = 2b$.
$\|\mathbf{v}\| = \sqrt{a^2+b^2} = \sqrt{(2b)^2+b^2} = \sqrt{5b^2}$
$\sqrt{5b^2} = 4$
$5b^2 = 16$
$b^2 = \dfrac{16}{5}$
$b = \pm\sqrt{\dfrac{16}{5}} = \pm\dfrac{4}{\sqrt{5}} = \pm\dfrac{4\sqrt{5}}{5}$
$a = 2b = 2\left(\pm\dfrac{4\sqrt{5}}{5}\right) = \pm\dfrac{8\sqrt{5}}{5}$
$\mathbf{v} = \dfrac{8\sqrt{5}}{5}\mathbf{i}+\dfrac{4\sqrt{5}}{5}\mathbf{j}$ or $\mathbf{v} = -\dfrac{8\sqrt{5}}{5}\mathbf{i}-\dfrac{4\sqrt{5}}{5}\mathbf{j}$

52. Let $\mathbf{v} = a\mathbf{i}+b\mathbf{j}$. We want $\|\mathbf{v}\| = 3$ and $a = b$.
$\|\mathbf{v}\| = \sqrt{a^2+b^2} = \sqrt{b^2+b^2} = \sqrt{2b^2}$
$\sqrt{2b^2} = 3$
$2b^2 = 9$
$b^2 = \dfrac{9}{2}$
$b = \pm\sqrt{\dfrac{9}{2}} = \pm\dfrac{3}{\sqrt{2}} = \pm\dfrac{3\sqrt{2}}{2}$
$a = b = \pm\dfrac{3\sqrt{2}}{2}$
$\mathbf{v} = \dfrac{3\sqrt{2}}{2}\mathbf{i}+\dfrac{3\sqrt{2}}{2}\mathbf{j}$ or $\mathbf{v} = -\dfrac{3\sqrt{2}}{2}\mathbf{i}-\dfrac{3\sqrt{2}}{2}\mathbf{j}$

53. $\mathbf{v} = 2\mathbf{i}-\mathbf{j}$, $\mathbf{w} = x\mathbf{i}+3\mathbf{j}$, $\|\mathbf{v}+\mathbf{w}\| = 5$
$\|\mathbf{v}+\mathbf{w}\| = \|2\mathbf{i}-\mathbf{j}+x\mathbf{i}+3\mathbf{j}\|$
$= \|(2+x)\mathbf{i}+2\mathbf{j}\|$
$= \sqrt{(2+x)^2+2^2}$
$= \sqrt{x^2+4x+4+4}$
$= \sqrt{x^2+4x+8}$

Solve for x:

$$\sqrt{x^2+4x+8}=5$$
$$x^2+4x+8=25$$
$$x^2+4x-17=0$$
$$x=\frac{-4\pm\sqrt{16-4(1)(-17)}}{2(1)}$$
$$=\frac{-4\pm\sqrt{84}}{2}$$
$$=\frac{-4\pm 2\sqrt{21}}{2}$$
$$=-2\pm\sqrt{21}$$
$$x=-2+\sqrt{21} \quad \text{or} \quad x=-2-\sqrt{21}$$

54. $P=(-3,1), Q=(x,4),$

$$\mathbf{v}=[x-(-3)]\mathbf{i}+(4-1)\mathbf{j}=(x+3)\mathbf{i}+3\mathbf{j}$$
$$\|\mathbf{v}\|=\sqrt{(x+3)^2+3^2}$$
$$=\sqrt{x^2+6x+9+9}$$
$$=\sqrt{x^2+6x+18}$$

Solve for x:

$$\sqrt{x^2+6x+18}=5$$
$$x^2+6x+18=25$$
$$x^2+6x-7=0$$
$$(x+7)(x-1)=0$$
$$x=-7 \quad \text{or} \quad x=1$$

55. $\|\mathbf{v}\|=5, \ \alpha=60^\circ$

$$\mathbf{v}=\|\mathbf{v}\|(\cos\alpha\mathbf{i}+\sin\alpha\mathbf{j})$$
$$=5(\cos(60^\circ)\mathbf{i}+\sin(60^\circ)\mathbf{j})$$
$$=5\left(\frac{1}{2}\mathbf{i}+\frac{\sqrt{3}}{2}\mathbf{j}\right)$$
$$=\frac{5}{2}\mathbf{i}+\frac{5\sqrt{3}}{2}\mathbf{j}$$

56. $\|\mathbf{v}\|=8, \ \alpha=45^\circ$

$$\mathbf{v}=\|\mathbf{v}\|(\cos\alpha\mathbf{i}+\sin\alpha\mathbf{j})$$
$$=8[\cos(45^\circ)\mathbf{i}+\sin(45^\circ)\mathbf{j}]$$
$$=8\left(\frac{\sqrt{2}}{2}\mathbf{i}+\frac{\sqrt{2}}{2}\mathbf{j}\right)$$
$$=4\sqrt{2}\mathbf{i}+4\sqrt{2}\mathbf{j}$$

57. $\|\mathbf{v}\|=14, \ \alpha=120^\circ$

$$\mathbf{v}=\|\mathbf{v}\|(\cos\alpha\mathbf{i}+\sin\alpha\mathbf{j})$$
$$=14(\cos(120^\circ)\mathbf{i}+\sin(120^\circ)\mathbf{j})$$
$$=14\left(-\frac{1}{2}\mathbf{i}+\frac{\sqrt{3}}{2}\mathbf{j}\right)$$
$$=-7\mathbf{i}+7\sqrt{3}\mathbf{j}$$

58. $\|\mathbf{v}\|=3, \ \alpha=240^\circ$

$$\mathbf{v}=\|\mathbf{v}\|(\cos\alpha\mathbf{i}+\sin\alpha\mathbf{j})$$
$$=3[\cos(240^\circ)\mathbf{i}+\sin(240^\circ)\mathbf{j}]$$
$$=3\left(-\frac{1}{2}\mathbf{i}-\frac{\sqrt{3}}{2}\mathbf{j}\right)$$
$$=-\frac{3}{2}\mathbf{i}-\frac{3\sqrt{3}}{2}\mathbf{j}$$

59. $\|\mathbf{v}\|=25, \ \alpha=330^\circ$

$$\mathbf{v}=\|\mathbf{v}\|(\cos\alpha\mathbf{i}+\sin\alpha\mathbf{j})$$
$$=25[\cos(330^\circ)\mathbf{i}+\sin(330^\circ)\mathbf{j}]$$
$$=25\left(\frac{\sqrt{3}}{2}\mathbf{i}-\frac{1}{2}\mathbf{j}\right)$$
$$=\frac{25\sqrt{3}}{2}\mathbf{i}-\frac{25}{2}\mathbf{j}$$

60. $\|\mathbf{v}\|=15, \ \alpha=315^\circ$

$$\mathbf{v}=\|\mathbf{v}\|(\cos\alpha\mathbf{i}+\sin\alpha\mathbf{j})$$
$$=15(\cos 315^\circ\mathbf{i}+\sin 315^\circ\mathbf{j})$$
$$=15\left(\frac{\sqrt{2}}{2}\mathbf{i}-\frac{\sqrt{2}}{2}\mathbf{j}\right)$$
$$=\frac{15\sqrt{2}}{2}\mathbf{i}-\frac{15\sqrt{2}}{2}\mathbf{j}$$

61. $\mathbf{F}=40[\cos(30^\circ)\mathbf{i}+\sin(30^\circ)\mathbf{j}]$

$$=40\left(\frac{\sqrt{3}}{2}\mathbf{i}+\frac{1}{2}\mathbf{j}\right)$$
$$=20\sqrt{3}\mathbf{i}+20\mathbf{j}$$
$$=20(\sqrt{3}\mathbf{i}+\mathbf{j})$$

62. $\mathbf{F}=100\left[\cos(20°)\mathbf{i}+\sin(20°)\mathbf{j}\right]$
$\approx 100(0.9397\mathbf{i}+0.3420\mathbf{j})$
$=93.97\mathbf{i}+34.20\mathbf{j}$

63. $\mathbf{F}_1=40\left[\cos(30°)\mathbf{i}+\sin(30°)\mathbf{j}\right]$
$=40\left(\frac{\sqrt{3}}{2}\mathbf{i}+\frac{1}{2}\mathbf{j}\right)$
$=20\sqrt{3}\mathbf{i}+20\mathbf{j}$

$\mathbf{F}_2=60[\cos(-45°)\mathbf{i}+\sin(-45°)\mathbf{j}]$
$=60\left(\frac{\sqrt{2}}{2}\mathbf{i}-\frac{\sqrt{2}}{2}\mathbf{j}\right)$
$=30\sqrt{2}\mathbf{i}-30\sqrt{2}\mathbf{j}$

$\mathbf{F}_1+\mathbf{F}_2=20\sqrt{3}\mathbf{i}+20\mathbf{j}+30\sqrt{2}\mathbf{i}-30\sqrt{2}\mathbf{j}$
$=\left(20\sqrt{3}+30\sqrt{2}\right)\mathbf{i}+\left(20-30\sqrt{2}\right)\mathbf{j}$

64. $\mathbf{F}_1=30\left[\cos(45°)\mathbf{i}+\sin(45°)\mathbf{j}\right]$
$=30\left(\frac{\sqrt{2}}{2}\mathbf{i}+\frac{\sqrt{2}}{2}\mathbf{j}\right)$
$=15\sqrt{2}\mathbf{i}+15\sqrt{2}\mathbf{j}$

$\mathbf{F}_2=70\left[\cos(120°)\mathbf{i}+\sin(120°)\mathbf{j}\right]$
$=70\left(-\frac{1}{2}\mathbf{i}+\frac{\sqrt{3}}{2}\mathbf{j}\right)$
$=-35\mathbf{i}+35\sqrt{3}\mathbf{j}$

$\mathbf{F}_1+\mathbf{F}_2=15\sqrt{2}\mathbf{i}+15\sqrt{2}\mathbf{j}+(-35)\mathbf{i}+35\sqrt{3}\mathbf{j}$
$=\left(15\sqrt{2}-35\right)\mathbf{i}+\left(15\sqrt{2}+35\sqrt{3}\right)\mathbf{j}$

65. Let $\mathbf{F}_1$ be the tension on the left cable and $\mathbf{F}_2$ be the tension on the right cable. Let $\mathbf{F}_3$ represent the force of the weight of the box.
$\mathbf{F}_1=\|\mathbf{F}_1\|\left[\cos(155°)\mathbf{i}+\sin(155°)\mathbf{j}\right]$
$\approx\|\mathbf{F}_1\|(-0.9063\mathbf{i}+0.4226\mathbf{j})$
$\mathbf{F}_2=\|\mathbf{F}_2\|\left[\cos(40°)\mathbf{i}+\sin(40°)\mathbf{j}\right]$
$\approx\|\mathbf{F}_2\|(0.7660\mathbf{i}+0.6428\mathbf{j})$
$\mathbf{F}_3=-1000\mathbf{j}$
For equilibrium, the sum of the force vectors must be zero.
$\mathbf{F}_1+\mathbf{F}_2+\mathbf{F}_3$
$=-0.9063\|\mathbf{F}_1\|\mathbf{i}+0.4226\|\mathbf{F}_1\|\mathbf{j}$
$\quad+0.7660\|\mathbf{F}_2\|\mathbf{i}+0.6428\|\mathbf{F}_2\|\mathbf{j}-1000\mathbf{j}$
$=\left(-0.9063\|\mathbf{F}_1\|+0.7660\|\mathbf{F}_2\|\right)\mathbf{i}$
$\quad+\left(0.4226\|\mathbf{F}_1\|+0.6428\|\mathbf{F}_2\|-1000\right)\mathbf{j}$
$=0$
Set the $\mathbf{i}$ and $\mathbf{j}$ components equal to zero and solve:
$$\begin{cases}-0.9063\|\mathbf{F}_1\|+0.7660\|\mathbf{F}_2\|=0\\0.4226\|\mathbf{F}_1\|+0.6428\|\mathbf{F}_2\|-1000=0\end{cases}$$
Solve the first equation for $\|\mathbf{F}_2\|$ and substitute the result into the second equation to solve the system:
$\|\mathbf{F}_2\|=\frac{0.9063}{0.7660}\|\mathbf{F}_1\|\approx 1.1832\|\mathbf{F}_1\|$
$0.4226\|\mathbf{F}_1\|+0.6428\left(1.1832\|\mathbf{F}_1\|\right)-1000=0$
$1.1832\|\mathbf{F}_1\|=1000$
$\|\mathbf{F}_1\|\approx 845.2$
$\|\mathbf{F}_2\|\approx 1.1832(845.2)\approx 1000$
The tension in the left cable is about 845.2 pounds and the tension in the right cable is about 1000 pounds.

66. Let $\mathbf{F}_1$ be the tension on the left cable and $\mathbf{F}_2$ be the tension on the right cable. Let $\mathbf{F}_3$ represent the force of the weight of the box.
$\mathbf{F}_1=\|\mathbf{F}_1\|\left[\cos(145°)\mathbf{i}+\sin(145°)\mathbf{j}\right]$
$\approx\|\mathbf{F}_1\|(-0.8192\mathbf{i}+0.5736\mathbf{j})$
$\mathbf{F}_2=\|\mathbf{F}_2\|\left[\cos(50°)\mathbf{i}+\sin(50°)\mathbf{j}\right]$
$\approx\|\mathbf{F}_2\|(0.6428\mathbf{i}+0.7660\mathbf{j})$
$\mathbf{F}_3=-800\mathbf{j}$
For equilibrium, the sum of the force vectors must be zero.
$\mathbf{F}_1+\mathbf{F}_2+\mathbf{F}_3$
$=-0.8192\|\mathbf{F}_1\|\mathbf{i}+0.5736\|\mathbf{F}_1\|\mathbf{j}$
$\quad+0.6428\|\mathbf{F}_2\|\mathbf{i}+0.7660\|\mathbf{F}_2\|\mathbf{j}-800\mathbf{j}$
$=\left(-0.8192\|\mathbf{F}_1\|+0.6428\|\mathbf{F}_2\|\right)\mathbf{i}$
$\quad+\left(0.5736\|\mathbf{F}_1\|+0.7660\|\mathbf{F}_2\|-800\right)\mathbf{j}$
$=0$
Set the $\mathbf{i}$ and $\mathbf{j}$ components equal to zero and solve:

$$\begin{cases} -0.8192\|\mathbf{F}_1\|+0.6428\|\mathbf{F}_2\|=0 \\ 0.5736\|\mathbf{F}_1\|+0.7660\|\mathbf{F}_2\|-800=0 \end{cases}$$

Solve the first equation for $\|\mathbf{F}_2\|$ and substitute the result into the second equation to solve the system:

$$\|\mathbf{F}_2\|=\frac{0.8192}{0.6428}\|\mathbf{F}_1\|\approx 1.2744\|\mathbf{F}_1\|$$

$$0.5736\|\mathbf{F}_1\|+0.7660(1.2744\|\mathbf{F}_1\|)-800=0$$
$$1.5498\|\mathbf{F}_1\|=800$$
$$\|\mathbf{F}_1\|\approx 516.2$$

$$\|\mathbf{F}_2\|\approx 1.2744(516.2)\approx 657.8$$

The tension in the left cable is about 516.2 pounds and the tension in the right cable is about 657.8 pounds.

67. Let $\mathbf{F}_1$ be the tension on the left end of the rope and $\mathbf{F}_2$ be the tension on the right end of the rope. Let $\mathbf{F}_3$ represent the force of the weight of the tightrope walker.

$$\mathbf{F}_1=\|\mathbf{F}_1\|\left[\cos(175.8°)\mathbf{i}+\sin(175.8°)\mathbf{j}\right]$$
$$\approx\|\mathbf{F}_1\|(-0.9973\mathbf{i}+0.0732\mathbf{j})$$
$$\mathbf{F}_2=\|\mathbf{F}_2\|\left[\cos(3.7°)\mathbf{i}+\sin(3.7°)\mathbf{j}\right]$$
$$\approx\|\mathbf{F}_2\|(0.9979\mathbf{i}+0.0645\mathbf{j})$$
$$\mathbf{F}_3=-150\mathbf{j}$$

For equilibrium, the sum of the force vectors must be zero.

$$\mathbf{F}_1+\mathbf{F}_2+\mathbf{F}_3$$
$$=-0.9973\|\mathbf{F}_1\|\mathbf{i}+0.0732\|\mathbf{F}_1\|\mathbf{j}$$
$$+0.9979\|\mathbf{F}_2\|\mathbf{i}+0.0645\|\mathbf{F}_2\|\mathbf{j}-150\mathbf{j}$$
$$=(-0.9973\|\mathbf{F}_1\|+0.9979\|\mathbf{F}_2\|)\mathbf{i}$$
$$+(0.0732\|\mathbf{F}_1\|+0.0645\|\mathbf{F}_2\|-150)\mathbf{j}$$
$$=0$$

Set the **i** and **j** components equal to zero and solve:

$$\begin{cases} -0.9973\|\mathbf{F}_1\|+0.9979\|\mathbf{F}_2\|=0 \\ 0.0732\|\mathbf{F}_1\|+0.0645\|\mathbf{F}_2\|-150=0 \end{cases}$$

Solve the first equation for $\|\mathbf{F}_2\|$ and substitute the result into the second equation to solve the system:

$$\|\mathbf{F}_2\|=\frac{0.9973}{0.9979}\|\mathbf{F}_1\|\approx 0.9994\|\mathbf{F}_1\|$$

$$0.0732\|\mathbf{F}_1\|+0.0645(0.9994\|\mathbf{F}_1\|)-150=0$$
$$0.1377\|\mathbf{F}_1\|=150$$
$$\|\mathbf{F}_1\|=1089.3$$

$$\|\mathbf{F}_2\|=0.9994(1089.3)\approx 1088.6$$

The tension in the left end of the rope is about 1089.3 pounds and the tension in the right end of the rope is about 1088.6 pounds.

68. Let $\mathbf{F}_1$ be the tension on the left end of the rope and $\mathbf{F}_2$ be the tension on the right end of the rope. Let $\mathbf{F}_3$ represent the force of the weight of the tightrope walker.

$$\mathbf{F}_1=\|\mathbf{F}_1\|\left[\cos(176.2°)\mathbf{i}+\sin(176.2°)\mathbf{j}\right]$$
$$\approx\|\mathbf{F}_1\|(-0.9978\mathbf{i}+0.0663\mathbf{j})$$
$$\mathbf{F}_2=\|\mathbf{F}_2\|\left[\cos(2.6°)\mathbf{i}+\sin(2.6°)\mathbf{j}\right]$$
$$\approx\|\mathbf{F}_2\|(0.9990\mathbf{i}+0.0454\mathbf{j})$$
$$\mathbf{F}_3=-135\mathbf{j}$$

For equilibrium, the sum of the force vectors must be zero.

$$\mathbf{F}_1+\mathbf{F}_2+\mathbf{F}_3$$
$$=-0.9978\|\mathbf{F}_1\|\mathbf{i}+0.0663\|\mathbf{F}_1\|\mathbf{j}$$
$$+0.9990\|\mathbf{F}_2\|\mathbf{i}+0.0454\|\mathbf{F}_2\|\mathbf{j}-135\mathbf{j}$$
$$=(-0.9978\|\mathbf{F}_1\|+0.9990\|\mathbf{F}_2\|)\mathbf{i}$$
$$+(0.0663\|\mathbf{F}_1\|+0.0454\|\mathbf{F}_2\|-135)\mathbf{j}$$
$$=0$$

Set the **i** and **j** components equal to zero and solve:

$$\begin{cases} -0.9978\|\mathbf{F}_1\|+0.9990\|\mathbf{F}_2\|=0 \\ 0.0663\|\mathbf{F}_1\|+0.0454\|\mathbf{F}_2\|-135=0 \end{cases}$$

Solve the first equation for $\|\mathbf{F}_2\|$ and substitute the result into the second equation to solve the system:

$$\|\mathbf{F}_2\|=\frac{0.9978}{0.9990}\|\mathbf{F}_1\|\approx 0.9988\|\mathbf{F}_1\|$$

$$0.0663\|\mathbf{F}_1\|+0.0454(0.9988\|\mathbf{F}_1\|)-135=0$$
$$0.1116\|\mathbf{F}_1\|=135$$
$$\|\mathbf{F}_1\|\approx 1209.7$$

$$\|\mathbf{F}_2\|=0.9988(1209.7)=1208.2$$

The tension in the left end of the rope is about 1209.7 pounds and the tension in the right end of the rope is about 1208.2 pounds.

69. The given forces are:
$\mathbf{F}_1 = -3\mathbf{i}$; $\mathbf{F}_2 = -\mathbf{i} + 4\mathbf{j}$; $\mathbf{F}_3 = 4\mathbf{i} - 2\mathbf{j}$; $\mathbf{F}_4 = -4\mathbf{j}$
A vector $\mathbf{v} = a\mathbf{i} + b\mathbf{j}$ needs to be added for equilibrium. Find vector $\mathbf{v} = a\mathbf{i} + b\mathbf{j}$:

$$\mathbf{F}_1 + \mathbf{F}_2 + \mathbf{F}_3 + \mathbf{F}_4 + \mathbf{v} = 0$$
$$-3\mathbf{i} + (-\mathbf{i} + 4\mathbf{j}) + (4\mathbf{i} - 2\mathbf{j}) + (-4\mathbf{j}) + (a\mathbf{i} + b\mathbf{j}) = 0$$
$$0\mathbf{i} + -2\mathbf{j} + (a\mathbf{i} + b\mathbf{j}) = 0$$
$$a\mathbf{i} + (-2 + b)\mathbf{j} = 0$$
$$a = 0; \quad -2 + b = 0$$
$$b = 2$$

Therefore, $\mathbf{v} = 2\mathbf{j}$.

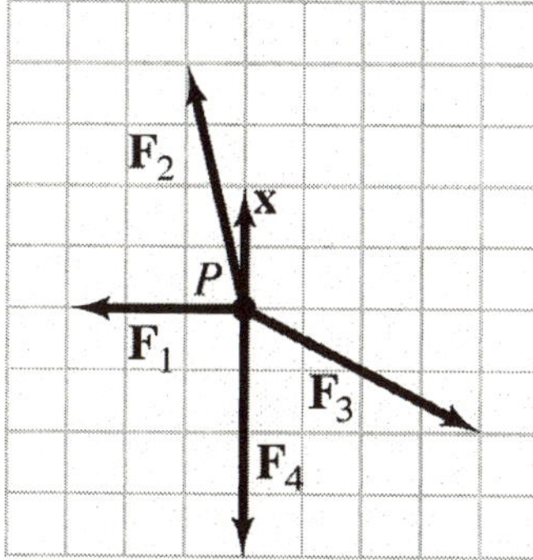

70–71. Answers will vary.

Section 8.5

1. $c^2 = a^2 + b^2 - 2ab\cos\gamma$

2. orthogonal

3. parallel

4. False

5. True

6. False

7. $\mathbf{v} = \mathbf{i} - \mathbf{j}, \quad \mathbf{w} = \mathbf{i} + \mathbf{j}$

a. $\mathbf{v} \cdot \mathbf{w} = 1(1) + (-1)(1) = 1 - 1 = 0$

b. $\cos\theta = \dfrac{\mathbf{v} \cdot \mathbf{w}}{\|\mathbf{v}\|\|\mathbf{w}\|} = \dfrac{0}{\sqrt{1^2 + (-1)^2}\sqrt{1^2 + 1^2}} = 0$

$\theta = 90°$

c. The vectors are orthogonal.

8. $\mathbf{v} = \mathbf{i} + \mathbf{j}, \quad \mathbf{w} = -\mathbf{i} + \mathbf{j}$

a. $\mathbf{v} \cdot \mathbf{w} = 1(-1) + 1(1) = -1 + 1 = 0$

b. $\cos\theta = \dfrac{\mathbf{v} \cdot \mathbf{w}}{\|\mathbf{v}\|\|\mathbf{w}\|} = \dfrac{0}{\sqrt{1^2 + 1^2}\sqrt{(-1)^2 + 1^2}} = 0$

$\theta = 90°$

c. The vectors are orthogonal.

9. $\mathbf{v} = 2\mathbf{i} + \mathbf{j}, \quad \mathbf{w} = \mathbf{i} + 2\mathbf{j}$

a. $\mathbf{v} \cdot \mathbf{w} = 2(1) + 1(2) = 2 + 2 = 4$

b. $\cos\theta = \dfrac{\mathbf{v} \cdot \mathbf{w}}{\|\mathbf{v}\|\|\mathbf{w}\|}$

$= \dfrac{4}{\sqrt{2^2 + 1^2}\sqrt{1^2 + 2^2}} = \dfrac{4}{\sqrt{5}\sqrt{5}} = \dfrac{4}{5}$

$\theta \approx 36.9°$

c. The vectors are neither parallel nor orthogonal.

10. $\mathbf{v} = 2\mathbf{i} + 2\mathbf{j}, \quad \mathbf{w} = \mathbf{i} + 2\mathbf{j}$

a. $\mathbf{v} \cdot \mathbf{w} = 2(1) + 2(2) = 2 + 4 = 6$

b. $\cos\theta = \dfrac{\mathbf{v} \cdot \mathbf{w}}{\|\mathbf{v}\|\|\mathbf{w}\|}$

$= \dfrac{6}{\sqrt{2^2 + 2^2}\sqrt{1^2 + 2^2}}$

$= \dfrac{6}{2\sqrt{2}\sqrt{5}} = \dfrac{3}{\sqrt{10}} = \dfrac{3\sqrt{10}}{10}$

$\theta \approx 18.4°$

c. The vectors are neither parallel nor orthogonal.

11. $\mathbf{v} = \sqrt{3}\mathbf{i} - \mathbf{j}, \quad \mathbf{w} = \mathbf{i} + \mathbf{j}$

a. $\mathbf{v} \cdot \mathbf{w} = \sqrt{3}(1) + (-1)(1) = \sqrt{3} - 1$

b. $\cos\theta = \dfrac{\mathbf{v} \cdot \mathbf{w}}{\|\mathbf{v}\|\|\mathbf{w}\|}$

$= \dfrac{\sqrt{3} - 1}{\sqrt{\left(\sqrt{3}\right)^2 + (-1)^2}\sqrt{1^2 + 1^2}}$

$= \dfrac{\sqrt{3} - 1}{\sqrt{4}\sqrt{2}} = \dfrac{\sqrt{3} - 1}{2\sqrt{2}} = \dfrac{\sqrt{6} - \sqrt{2}}{4}$

$\theta = 75°$

c. The vectors are neither parallel nor orthogonal.

12. $\mathbf{v}=\mathbf{i}+\sqrt{3}\mathbf{j},\quad \mathbf{w}=\mathbf{i}-\mathbf{j}$

a. $\mathbf{v}\cdot\mathbf{w}=1(1)+\sqrt{3}(-1)=1-\sqrt{3}$

b. $$\cos\theta=\frac{\mathbf{v}\cdot\mathbf{w}}{\|\mathbf{v}\|\|\mathbf{w}\|}=\frac{1-\sqrt{3}}{\sqrt{1^2+\left(\sqrt{3}\right)^2}\sqrt{1^2+(-1)^2}}=\frac{1-\sqrt{3}}{\sqrt{4}\sqrt{2}}=\frac{1-\sqrt{3}}{2\sqrt{2}}=\frac{\sqrt{2}-\sqrt{6}}{4}$$
$\theta=105°$

c. The vectors are neither parallel nor orthogonal.

13. $\mathbf{v}=3\mathbf{i}+4\mathbf{j},\quad \mathbf{w}=4\mathbf{i}+3\mathbf{j}$

a. $\mathbf{v}\cdot\mathbf{w}=3(4)+4(3)=12+12=24$

b. $$\cos\theta=\frac{\mathbf{v}\cdot\mathbf{w}}{\|\mathbf{v}\|\|\mathbf{w}\|}=\frac{24}{\sqrt{3^2+4^2}\sqrt{4^2+3^2}}=\frac{24}{\sqrt{25}\sqrt{25}}=\frac{24}{25}$$
$\theta\approx 16.3°$

c. The vectors are neither parallel nor orthogonal.

14. $\mathbf{v}=3\mathbf{i}-4\mathbf{j},\quad \mathbf{w}=4\mathbf{i}-3\mathbf{j}$

a. $\mathbf{v}\cdot\mathbf{w}=3(4)+(-4)(-3)=12+12=24$

b. $$\cos\theta=\frac{\mathbf{v}\cdot\mathbf{w}}{\|\mathbf{v}\|\|\mathbf{w}\|}=\frac{24}{\sqrt{3^2+(-4)^2}\sqrt{4^2+(-3)^2}}=\frac{24}{\sqrt{25}\sqrt{25}}=\frac{24}{25}$$
$\theta\approx 16.3°$

c. The vectors are neither parallel nor orthogonal.

15. $\mathbf{v}=4\mathbf{i},\quad \mathbf{w}=\mathbf{j}$

a. $\mathbf{v}\cdot\mathbf{w}=4(0)+0(1)=0+0=0$

b. $$\cos\theta=\frac{\mathbf{v}\cdot\mathbf{w}}{\|\mathbf{v}\|\|\mathbf{w}\|}=\frac{0}{\sqrt{4^2+0^2}\sqrt{0^2+1^2}}=\frac{0}{4\cdot 1}=\frac{0}{4}=0$$
$\theta=90°$

c. The vectors are orthogonal.

16. $\mathbf{v}=\mathbf{i},\quad \mathbf{w}=-3\mathbf{j}$

a. $\mathbf{v}\cdot\mathbf{w}=1(0)+0(-3)=0+0=0$

b. $$\cos\theta=\frac{\mathbf{v}\cdot\mathbf{w}}{\|\mathbf{v}\|\|\mathbf{w}\|}=\frac{0}{\sqrt{1^2+0^2}\sqrt{0^2+(-3)^2}}=\frac{0}{1\cdot 3}=\frac{0}{3}=0$$
$\theta=90°$

c. The vectors are orthogonal.

17. $\mathbf{v}=\mathbf{i}-a\mathbf{j},\quad \mathbf{w}=2\mathbf{i}+3\mathbf{j}$

Two vectors are orthogonal if the dot product is zero. Solve for a:
$$\mathbf{v}\cdot\mathbf{w}=0$$
$$1(2)+(-a)(3)=0$$
$$2-3a=0$$
$$3a=2$$
$$a=\frac{2}{3}$$

18. $\mathbf{v}=\mathbf{i}+\mathbf{j},\quad \mathbf{w}=\mathbf{i}+b\mathbf{j}$

Two vectors are orthogonal if the dot product is zero. Solve for b:
$$\mathbf{v}\cdot\mathbf{w}=0$$
$$1(1)+1(b)=0$$
$$1+b=0$$
$$b=-1$$

19. $\mathbf{v}=2\mathbf{i}-3\mathbf{j},\quad \mathbf{w}=\mathbf{i}-\mathbf{j}$
$$\mathbf{v}_1=\frac{\mathbf{v}\cdot\mathbf{w}}{(\|\mathbf{w}\|)^2}\mathbf{w}=\frac{2(1)+(-3)(-1)}{\left(\sqrt{1^2+(-1)^2}\right)^2}(\mathbf{i}-\mathbf{j})=\frac{5}{2}(\mathbf{i}-\mathbf{j})=\frac{5}{2}\mathbf{i}-\frac{5}{2}\mathbf{j}$$
$$\mathbf{v}_2=\mathbf{v}-\mathbf{v}_1=(2\mathbf{i}-3\mathbf{j})-\left(\frac{5}{2}\mathbf{i}-\frac{5}{2}\mathbf{j}\right)=-\frac{1}{2}\mathbf{i}-\frac{1}{2}\mathbf{j}$$

20. $\mathbf{v}=-3\mathbf{i}+2\mathbf{j},\quad \mathbf{w}=2\mathbf{i}+\mathbf{j}$
$$\mathbf{v}_1=\frac{\mathbf{v}\cdot\mathbf{w}}{(\|\mathbf{w}\|)^2}\mathbf{w}=\frac{-3(2)+2(1)}{\left(\sqrt{2^2+1^2}\right)^2}(2\mathbf{i}+\mathbf{j})=-\frac{4}{5}(2\mathbf{i}+\mathbf{j})=-\frac{8}{5}\mathbf{i}-\frac{4}{5}\mathbf{j}$$
$$\mathbf{v}_2=\mathbf{v}-\mathbf{v}_1=(-3\mathbf{i}+2\mathbf{j})-\left(-\frac{8}{5}\mathbf{i}-\frac{4}{5}\mathbf{j}\right)=-\frac{7}{5}\mathbf{i}+\frac{14}{5}\mathbf{j}$$

21. $\mathbf{v} = \mathbf{i} - \mathbf{j}, \quad \mathbf{w} = \mathbf{i} + 2\mathbf{j}$

$$\mathbf{v}_1 = \frac{\mathbf{v} \cdot \mathbf{w}}{(\|\mathbf{w}\|)^2}\mathbf{w} = \frac{1(1) + (-1)(2)}{\left(\sqrt{1^2 + 2^2}\right)^2}(\mathbf{i} + 2\mathbf{j}) = -\frac{1}{5}(\mathbf{i} + 2\mathbf{j}) = -\frac{1}{5}\mathbf{i} - \frac{2}{5}\mathbf{j}$$

$$\mathbf{v}_2 = \mathbf{v} - \mathbf{v}_1 = (\mathbf{i} - \mathbf{j}) - \left(-\frac{1}{5}\mathbf{i} - \frac{2}{5}\mathbf{j}\right) = \frac{6}{5}\mathbf{i} - \frac{3}{5}\mathbf{j}$$

22. $\mathbf{v} = 2\mathbf{i} - \mathbf{j}, \quad \mathbf{w} = \mathbf{i} - 2\mathbf{j}$

$$\mathbf{v}_1 = \frac{\mathbf{v} \cdot \mathbf{w}}{(\|\mathbf{w}\|)^2}\mathbf{w} = \frac{2(1) + (-1)(-2)}{\left(\sqrt{1^2 + (-2)^2}\right)^2}(\mathbf{i} - 2\mathbf{j}) = \frac{4}{5}(\mathbf{i} - 2\mathbf{j}) = \frac{4}{5}\mathbf{i} - \frac{8}{5}\mathbf{j}$$

$$\mathbf{v}_2 = \mathbf{v} - \mathbf{v}_1 = (2\mathbf{i} - \mathbf{j}) - \left(\frac{4}{5}\mathbf{i} - \frac{8}{5}\mathbf{j}\right) = \frac{6}{5}\mathbf{i} + \frac{3}{5}\mathbf{j}$$

23. $\mathbf{v} = 3\mathbf{i} + \mathbf{j}, \quad \mathbf{w} = -2\mathbf{i} - \mathbf{j}$

$$\mathbf{v}_1 = \frac{\mathbf{v} \cdot \mathbf{w}}{(\|\mathbf{w}\|)^2}\mathbf{w} = \frac{3(-2) + 1(-1)}{\left(\sqrt{(-2)^2 + (-1)^2}\right)^2}(-2\mathbf{i} - \mathbf{j}) = -\frac{7}{5}(-2\mathbf{i} - \mathbf{j}) = \frac{14}{5}\mathbf{i} + \frac{7}{5}\mathbf{j}$$

$$\mathbf{v}_2 = \mathbf{v} - \mathbf{v}_1 = (3\mathbf{i} + \mathbf{j}) - \left(\frac{14}{5}\mathbf{i} + \frac{7}{5}\mathbf{j}\right) = \frac{1}{5}\mathbf{i} - \frac{2}{5}\mathbf{j}$$

24. $\mathbf{v} = \mathbf{i} - 3\mathbf{j}, \quad \mathbf{w} = 4\mathbf{i} - \mathbf{j}$

$$\mathbf{v}_1 = \frac{\mathbf{v} \cdot \mathbf{w}}{(\|\mathbf{w}\|)^2}\mathbf{w} = \frac{1(4) + (-3)(-1)}{\left(\sqrt{4^2 + (-1)^2}\right)^2}(4\mathbf{i} - \mathbf{j}) = \frac{7}{17}(4\mathbf{i} - \mathbf{j}) = \frac{28}{17}\mathbf{i} - \frac{7}{17}\mathbf{j}$$

$$\mathbf{v}_2 = \mathbf{v} - \mathbf{v}_1 = (\mathbf{i} - 3\mathbf{j}) - \left(\frac{28}{17}\mathbf{i} - \frac{7}{17}\mathbf{j}\right) = -\frac{11}{17}\mathbf{i} - \frac{44}{17}\mathbf{j}$$

25. Let $\mathbf{v}_a$ = the velocity of the plane in still air, $\mathbf{v}_w$ = the velocity of the wind, and $\mathbf{v}_g$ = the velocity of the plane relative to the ground.

$$\mathbf{v}_g = \mathbf{v}_a + \mathbf{v}_w$$

$$\mathbf{v}_a = 550\left[\cos(225°)\mathbf{i} + \sin(225°)\mathbf{j}\right] = 550\left(-\frac{\sqrt{2}}{2}\mathbf{i} - \frac{\sqrt{2}}{2}\mathbf{j}\right) = -275\sqrt{2}\mathbf{i} - 275\sqrt{2}\mathbf{j}$$

$$\mathbf{v}_w = 80\mathbf{i}$$

$$\mathbf{v}_g = \mathbf{v}_a + \mathbf{v}_w = -275\sqrt{2}\mathbf{i} - 275\sqrt{2}\mathbf{j} + 80\mathbf{i} = \left(80 - 275\sqrt{2}\right)\mathbf{i} - 275\sqrt{2}\mathbf{j}$$

The speed of the plane relative to the ground is:

$$\|\mathbf{v}_g\| = \sqrt{\left(80 - 275\sqrt{2}\right)^2 + \left(-275\sqrt{2}\right)^2} = \sqrt{6400 - 44{,}000\sqrt{2} + 151{,}250 + 151{,}250} \approx \sqrt{246{,}674.6} \approx 496.7 \text{ miles per hour}$$

To find the direction, find the angle between $\mathbf{v}_g$ and a convenient vector such as due south, $-\mathbf{j}$.

$$\cos\theta = \frac{\mathbf{v}_g \cdot (-\mathbf{j})}{\|\mathbf{v}_g\|\|-\mathbf{j}\|} = \frac{\left(80 - 275\sqrt{2}\right)(0) + \left(-275\sqrt{2}\right)(-1)}{496.7\sqrt{0^2 + (-1)^2}} = \frac{275\sqrt{2}}{496.7} \approx 0.7830$$

$$\theta \approx 38.5°$$

The plane is traveling with a ground speed of about 496.7 miles per hour in an approximate direction of 38.5° degrees west of south.

26. Let $\mathbf{v_a}$ = the velocity of the plane in still air, $\mathbf{v_w}$ = the velocity of the wind, and $\mathbf{v_g}$ = the velocity of the plane relative to the ground.

$\mathbf{v_g} = \mathbf{v_a} + \mathbf{v_w}$

$\mathbf{v_a} = 250(\cos\alpha\,\mathbf{i} + \sin\alpha\,\mathbf{j})$

$$\mathbf{v_w} = 40\left(\frac{\mathbf{i}-\mathbf{j}}{\|\mathbf{i}-\mathbf{j}\|}\right)$$

$$= 40\left(\frac{\mathbf{i}-\mathbf{j}}{\sqrt{1+1}}\right) = \frac{40}{\sqrt{2}}(\mathbf{i}-\mathbf{j}) = 20\sqrt{2}(\mathbf{i}-\mathbf{j})$$

$\mathbf{v_g} = \mathbf{v_a} + \mathbf{v_w}$

$= 250\cos\alpha\,\mathbf{i} + 250\sin\alpha\,\mathbf{j} + 20\sqrt{2}\,\mathbf{i} - 20\sqrt{2}\,\mathbf{j}$

$= a\mathbf{i}$

Examining the $\mathbf{j}$ components:

$250\sin\alpha - 20\sqrt{2} = 0$

$250\sin\alpha = 20\sqrt{2}$

$$\sin\alpha = \frac{20\sqrt{2}}{250} \approx 0.11314$$

$\alpha \approx 6.5°$

The heading of the plane should be about N83.5°E, that is, about 6.5° north of east.

Examining the $\mathbf{i}$ components:

$250\cos(6.5°)\mathbf{i} + 20\sqrt{2}\,\mathbf{i} = a\mathbf{i}$

$276.7 \approx a$

The speed of the plane relative to the ground is about 276.7 miles per hour.

27. Let the positive x-axis point downstream, so that the velocity of the current is $\mathbf{v_c} = 3\mathbf{i}$. Let $\mathbf{v_w}$ = the velocity of the boat in the water, and $\mathbf{v_g}$ = the velocity of the boat relative to the land. Then $\mathbf{v_g} = \mathbf{v_w} + \mathbf{v_c}$ and $\mathbf{v_g} = \mathbf{k}$ since the boat is going directly across the river. The speed of the boat is $\|\mathbf{v_w}\| = 20$; we need to find the direction. Let $\mathbf{v_w} = a\mathbf{i} + b\mathbf{j}$, so

$\|\mathbf{v_w}\| = \sqrt{a^2+b^2} = 20$

$a^2 + b^2 = 400$

Since $\mathbf{v_g} = \mathbf{v_w} + \mathbf{v_c}$,

$k\mathbf{j} = a\mathbf{i} + b\mathbf{j} + 3\mathbf{i} = (a+3)\mathbf{i} + b\mathbf{j}$

$a + 3 = 0$

$a = -3$

$k = b$

$a^2 + b^2 = 400$

$9 + b^2 = 400$

$b^2 = 391$

$k = b = \sqrt{391} \approx 19.8$

$\mathbf{v_w} = -3\mathbf{i} + \sqrt{391}\mathbf{j}$ and $\mathbf{v_g} = \sqrt{391}\mathbf{j}$

Find the angle between $\mathbf{v}_w$ and $\mathbf{j}$:

$$\cos\theta = \frac{\mathbf{v_w}\cdot\mathbf{j}}{\|\mathbf{v_w}\|\|\mathbf{j}\|}$$

$$= \frac{-3(0) + \sqrt{391}(1)}{20\sqrt{0^2+1^2}} = \frac{\sqrt{391}}{20} = 0.9887$$

$\theta \approx 8.6°$

The heading of the boat needs to be about 8.6° upstream. The velocity of the boat directly across the river is about 19.8 kilometers per hour. The time to cross the river is:

$$t = \frac{0.5}{19.8} \approx 0.025 \text{ hours or}$$

$$t = \frac{0.5}{19.8}\cdot 60 \approx 1.52 \text{ minutes}.$$

28. Let the positive x-axis point downstream, so that the velocity of the current is $\mathbf{v_c} = 5\mathbf{i}$. Let $\mathbf{v_w}$ = the velocity of the boat in the water, and $\mathbf{v_g}$ = the velocity of the boat relative to the land. Then $\mathbf{v_g} = \mathbf{v_w} + \mathbf{v_c}$ and $\mathbf{v_g} = \mathbf{k}$ since the boat is going directly across the river. The speed of the boat is $\|\mathbf{v_w}\| = 20$; we need to find the direction. Let $\mathbf{v_w} = a\mathbf{i} + b\mathbf{j}$, so $\|\mathbf{v_w}\| = \sqrt{a^2+b^2} = 20$ or $a^2 + b^2 = 400$.

Since $\mathbf{v_g} = \mathbf{v_w} + \mathbf{v_c}$,

$k\mathbf{j} = a\mathbf{i} + b\mathbf{j} + 5\mathbf{i} = (a+5)\mathbf{i} + b\mathbf{j}$

$a + 5 = 0$

$a = -5$

$k = b$

$a^2 + b^2 = 400$

$25 + b^2 = 400$

$b^2 = 375$

$k = b = \sqrt{375} \approx 19.4$

$\mathbf{v_w} = -5\mathbf{i} + \sqrt{375}\mathbf{j}$ and $\mathbf{v_g} = \sqrt{375}\mathbf{j}$

Find the angle between $\mathbf{v}_w$ and $\mathbf{j}$:

$$\cos\theta = \frac{\mathbf{v}_w \cdot \mathbf{j}}{\|\mathbf{v}_w\|\|\mathbf{j}\|} = \frac{-5(0)+\sqrt{375}(1)}{20\sqrt{0^2+1^2}} = \frac{\sqrt{375}}{20} \approx 0.9682$$

$\theta \approx 14.5°$

The heading of the boat needs to be about 14.5° upstream. The velocity of the boat directly across the river is about 19.4 kilometers per hour. The time to cross the river is:

$t = \frac{0.5}{19.4} \approx 0.026$ hours or

$t = \frac{0.5}{19.4} \cdot 60 \approx 1.55$ minutes.

29. Split the force into the components going down the hill and perpendicular to the hill.

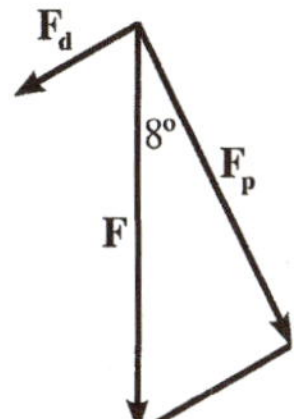

$\mathbf{F}_d = \mathbf{F}\sin 8° = 5300\sin 8° \approx 737.6$

$\mathbf{F}_p = \mathbf{F}\cos 8° = 5300\cos 8° \approx 5248.4$

The force required to keep the car from rolling down the hill is about 737.6 pounds. The force perpendicular to the hill is approximately 5248.4 pounds.

30. Split the force into the components going down the hill and perpendicular to the hill.

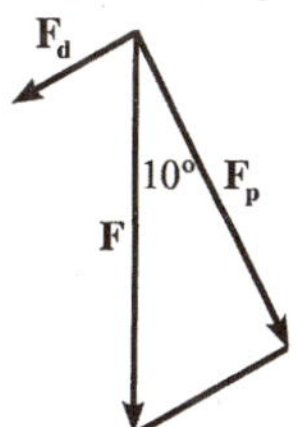

$\mathbf{F}_d = \mathbf{F}\sin 10° = 4500\sin 10° \approx 781.4$

$\mathbf{F}_p = \mathbf{F}\cos 10° = 4500\cos 10° \approx 4431.6$

The force required to keep the car from rolling down the hill is about 781.4 pounds. The force perpendicular to the hill is approximately 4431.6 pounds.

31. Let $\mathbf{v}_a$ = the velocity of the plane in still air, $\mathbf{v}_w$ = the velocity of the wind, and $\mathbf{v}_g$ = the velocity of the plane relative to the ground.

$\mathbf{v}_g = \mathbf{v}_a + \mathbf{v}_w$

$$\mathbf{v}_a = 500[\cos(45°)\mathbf{i} + \sin(45°)\mathbf{j}]$$
$$= 500\left(\frac{\sqrt{2}}{2}\mathbf{i} + \frac{\sqrt{2}}{2}\mathbf{j}\right) = 250\sqrt{2}\,\mathbf{i} + 250\sqrt{2}\,\mathbf{j}$$

$$\mathbf{v}_w = 60[\cos(120°)\mathbf{i} + \sin(120°)\mathbf{j}]$$
$$= 60\left(-\frac{1}{2}\mathbf{i} + \frac{\sqrt{3}}{2}\mathbf{j}\right) = -30\mathbf{i} + 30\sqrt{3}\,\mathbf{j}$$

$$\mathbf{v}_g = \mathbf{v}_a + \mathbf{v}_w$$
$$= 250\sqrt{2}\,\mathbf{i} + 250\sqrt{2}\,\mathbf{j} - 30\mathbf{i} + 30\sqrt{3}\,\mathbf{j}$$
$$= (250\sqrt{2} - 30)\mathbf{i} + (250\sqrt{2} + 30\sqrt{3})\mathbf{j}$$

The speed of the plane relative to the ground is:

$$\|\mathbf{v}_g\| = \sqrt{(-30+250\sqrt{2})^2 + (250\sqrt{2}+30\sqrt{3})^2}$$
$$\approx \sqrt{269{,}129.1} \approx 518.8 \text{ km/hr}$$

To find the direction, find the angle between $\mathbf{v}_g$ and a convenient vector such as due north, $\mathbf{j}$.

$$\cos\theta = \frac{\mathbf{v}_g \cdot \mathbf{j}}{\|\mathbf{v}_g\|\|\mathbf{j}\|}$$
$$= \frac{(-30+250\sqrt{2})(0) + (250\sqrt{2}+30\sqrt{3})(1)}{518.8\sqrt{0^2+1^2}}$$
$$= \frac{250\sqrt{2}+30\sqrt{3}}{518.8} \approx 0.7816$$

$\theta \approx 38.6°$

The plane is traveling with a ground speed of about 518.8 km/hr in a direction of 38.6° east of north.

32. Let $\mathbf{v}_a$ = the velocity of the plane in still air, $\mathbf{v}_w$ = the velocity of the wind, and $\mathbf{v}_g$ = the velocity of the plane relative to the ground.

$\mathbf{v}_g = \mathbf{v}_a + \mathbf{v}_w$

$$\mathbf{v}_a = 600[\cos(60°)\mathbf{i} - \sin(60°)\mathbf{j}]$$
$$= 600\left(\frac{1}{2}\mathbf{i} - \frac{\sqrt{3}}{2}\mathbf{j}\right) = 300\mathbf{i} - 300\sqrt{3}\,\mathbf{j}$$

$$\mathbf{v}_w = 40[\cos(45°)\mathbf{i} - \sin(45°)\mathbf{j}]$$
$$= 40\left(\frac{\sqrt{2}}{2}\mathbf{i} - \frac{\sqrt{2}}{2}\mathbf{j}\right) = 20\sqrt{2}\,\mathbf{i} - 20\sqrt{2}\,\mathbf{j}$$

$$\mathbf{v}_g = \mathbf{v}_a + \mathbf{v}_w$$
$$= 300\mathbf{i} - 300\sqrt{3}\,\mathbf{j} + 20\sqrt{2}\,\mathbf{i} - 20\sqrt{2}\,\mathbf{j}$$
$$= (300 + 20\sqrt{2})\mathbf{i} - (300\sqrt{3} + 20\sqrt{2})\mathbf{j}$$

The speed of the plane relative to the ground is:

$$\|\mathbf{v}_g\| = \sqrt{\left(300+20\sqrt{2}\right)^2 + \left(-300\sqrt{3}-20\sqrt{2}\right)^2}$$
$$\approx \sqrt{407{,}964} \approx 638.7 \text{ km/hr}$$

To find the direction, find the angle between $\mathbf{v}_g$ and a convenient vector such as due north, $\mathbf{j}$.

$$\cos\theta = \frac{\mathbf{v}_g \cdot \mathbf{j}}{\|\mathbf{v}_g\|\|\mathbf{j}\|}$$
$$= \frac{\left(300+20\sqrt{2}\right)(0) + \left(-300\sqrt{3}-20\sqrt{2}\right)(1)}{638.7\sqrt{0^2+1^2}}$$
$$= \frac{-300\sqrt{3}-20\sqrt{2}}{638.7} \approx -0.8574$$
$$\theta \approx 149.1^\circ$$

The plane is traveling with a ground speed of about 638.7 km/hr in a direction of about 30.9° degrees east of south (S30.9°E).

33. Let the positive x-axis point downstream, so that the velocity of the current is $\mathbf{v}_c = 3\mathbf{i}$. Let $\mathbf{v}_w$ = the velocity of the boat in the water and $\mathbf{v}_g$ = the velocity of the boat relative to the land. Then $\mathbf{v}_g = \mathbf{v}_w + \mathbf{v}_c$. The speed of the boat is $\|\mathbf{v}_w\| = 20$,and its heading is directly across the river, so $\mathbf{v}_w = 20\mathbf{j}$. Then

$$\mathbf{v}_g = \mathbf{v}_w + \mathbf{v}_c = 20\mathbf{j} + 3\mathbf{i} = 3\mathbf{i} + 20\mathbf{j}$$
$$\|\mathbf{v}_g\| = \sqrt{3^2+20^2} = \sqrt{409} \approx 20.2 \text{ mi/hr}$$

Find the angle between $\mathbf{v}_g$ and $\mathbf{j}$:

$$\cos\theta = \frac{\mathbf{v}_g \cdot \mathbf{j}}{\|\mathbf{v}_g\|\|\mathbf{j}\|}$$
$$= \frac{3(0)+20(1)}{\sqrt{409}\cdot\sqrt{0^2+1^2}} = \frac{20}{\sqrt{409}} \approx 0.9889$$
$$\theta \approx 8.5^\circ$$

The approximate direction of the boat will be 8.5° downstream (N8.5°E assuming the boat is traveling north and the current is traveling east).

34. Let the positive x-axis point downstream, so that the velocity of the current is $\mathbf{v}_c = 4\mathbf{i}$. Let $\mathbf{v}_w$ = the velocity of the boat in the water and $\mathbf{v}_g$ = the velocity of the boat relative to the land. Then $\mathbf{v}_g = \mathbf{v}_w + \mathbf{v}_c$. The speed of the boat is $\|\mathbf{v}_w\| = 10$, and its heading is directly across the river, so $\mathbf{v}_w = 10\mathbf{j}$. Then

$$\mathbf{v}_g = \mathbf{v}_w + \mathbf{v}_c = 10\mathbf{j} + 4\mathbf{i} = 4\mathbf{i} + 10\mathbf{j}$$
$$\|\mathbf{v}_g\| = \sqrt{4^2+10^2} = \sqrt{116} \approx 10.8 \text{ mi/hr}$$

Find the angle between $\mathbf{v}_g$ and $\mathbf{j}$:

$$\cos\theta = \frac{\mathbf{v}_g \cdot \mathbf{j}}{\|\mathbf{v}_g\|\|\mathbf{j}\|}$$
$$= \frac{4(0)+10(1)}{\sqrt{116}\cdot\sqrt{0^2+1^2}} = \frac{10}{\sqrt{116}} \approx 0.9285$$
$$\theta \approx 21.8^\circ$$

The approximate direction of the boat will be 21.8° downstream (N21.8°E assuming the boat is traveling north and the current is traveling east).

35. $\mathbf{F} = 3\left[\cos(60^\circ)\mathbf{i} + \sin(60^\circ)\mathbf{j}\right]$

$$= 3\left(\frac{1}{2}\mathbf{i} + \frac{\sqrt{3}}{2}\mathbf{j}\right) = \frac{3}{2}\mathbf{i} + \frac{3\sqrt{3}}{2}\mathbf{j}$$
$$W = \mathbf{F}\cdot\overrightarrow{AB} = \left(\frac{3}{2}\mathbf{i} + \frac{3\sqrt{3}}{2}\mathbf{j}\right)\cdot 2\mathbf{i}$$
$$= \frac{3}{2}(2) + \frac{3\sqrt{3}}{2}(0)$$
$$= 3 \text{ ft-lb}$$

36. $\mathbf{F} = 1\left[\cos(45^\circ)\mathbf{i} + \sin(45^\circ)\mathbf{j}\right]$

$$= 1\left(\frac{\sqrt{2}}{2}\mathbf{i} + \frac{\sqrt{2}}{2}\mathbf{j}\right) = \frac{\sqrt{2}}{2}\mathbf{i} + \frac{\sqrt{2}}{2}\mathbf{j}$$
$$W = \mathbf{F}\cdot\overrightarrow{AB} = \left(\frac{\sqrt{2}}{2}\mathbf{i} + \frac{\sqrt{2}}{2}\mathbf{j}\right)\cdot 5\mathbf{i}$$
$$= \frac{\sqrt{2}}{2}(5) + \frac{\sqrt{2}}{2}(0)$$
$$= \frac{5\sqrt{2}}{2} \approx 3.54 \text{ ft-lb}$$

37. $\mathbf{F} = 20\left[\cos(30^\circ)\mathbf{i} + \sin(30^\circ)\mathbf{j}\right]$

$$= 20\left(\frac{\sqrt{3}}{2}\mathbf{i} + \frac{1}{2}\mathbf{j}\right) = 10\sqrt{3}\mathbf{i} + 10\mathbf{j}$$
$$W = \mathbf{F}\cdot\overrightarrow{AB} = \left(10\sqrt{3}\mathbf{i} + 10\mathbf{j}\right)\cdot 100\mathbf{i}$$
$$= 10\sqrt{3}(100) + 10(0)$$
$$= 1000\sqrt{3} \approx 1732 \text{ ft-lb}$$

38. $W = \mathbf{F}\cdot\overrightarrow{AB} \qquad W = 2,\ \overrightarrow{AB} = 4\mathbf{i}$

$\mathbf{F} = \cos\alpha\mathbf{i} - \sin\alpha\mathbf{j}$

$2 = (\cos\alpha\mathbf{i} - \sin\alpha\mathbf{j})\cdot 4\mathbf{i}$

$2 = 4\cos\alpha$

$\frac{1}{2} = \cos\alpha$

$\alpha = 60°$

39. Let $\mathbf{u} = a_1\mathbf{i} + b_1\mathbf{j}$, $\mathbf{v} = a_2\mathbf{i} + b_2\mathbf{j}$, and $\mathbf{w} = a_3\mathbf{i} + b_3\mathbf{j}$.

$\mathbf{u}\cdot(\mathbf{v}+\mathbf{w})$

$= (a_1\mathbf{i} + b_1\mathbf{j})(a_2\mathbf{i} + b_2\mathbf{j} + a_3\mathbf{i} + b_3\mathbf{j})$

$= (a_1\mathbf{i} + b_1\mathbf{j})(a_2\mathbf{i} + a_3\mathbf{i} + b_2\mathbf{j} + b_3\mathbf{j})$

$= (a_1\mathbf{i} + b_1\mathbf{j})[(a_2 + a_3)\mathbf{i} + (b_2 + b_3)\mathbf{j}]$

$= a_1(a_2 + a_3) + b_1(b_2 + b_3)$

$= a_1a_2 + a_1a_3 + b_1b_2 + b_1b_3$

$= a_1a_2 + b_1b_2 + a_1a_3 + b_1b_3$

$= (a_1\mathbf{i} + b_1\mathbf{j})(a_2\mathbf{i} + b_2\mathbf{j}) + (a_1\mathbf{i} + b_1\mathbf{j})(a_3\mathbf{i} + b_3\mathbf{j})$

$= \mathbf{u}\cdot\mathbf{v} + \mathbf{u}\cdot\mathbf{w}$

40. Since $\mathbf{0} = 0\mathbf{i} + 0\mathbf{j}$ and $\mathbf{v}=a\mathbf{i}+b\mathbf{j}$, we have that $\mathbf{0}\cdot\mathbf{v} = 0\cdot a + 0\cdot b = 0$.

41. Let $\mathbf{v} = x\mathbf{i} + y\mathbf{j}$. Since $\mathbf{v}$ is a unit vector, we have that $\|\mathbf{v}\| = \sqrt{x^2 + y^2} = 1$, or $x^2 + y^2 = 1$. If α is the angle between $\mathbf{v}$ and $\mathbf{i}$, then

$\cos\alpha = \frac{\mathbf{v}\cdot\mathbf{i}}{\|\mathbf{v}\|\|\mathbf{i}\|} = \frac{(x\mathbf{i} + y\mathbf{j})\cdot\mathbf{i}}{1\cdot 1} = x$. Now,

$x^2 + y^2 = 1$

$\cos^2\alpha + y^2 = 1$

$y^2 = 1 - \cos^2\alpha$

$y^2 = \sin^2\alpha$

$y = \sin\alpha$

Thus, $\mathbf{v} = \cos\alpha\,\mathbf{i} + \sin\alpha\,\mathbf{j}$.

42. If $\mathbf{v} = a_1\mathbf{i} + b_1\mathbf{j} = \cos\alpha\mathbf{i} + \sin\alpha\mathbf{j}$ and $\mathbf{w} = a_2\mathbf{i} + b_2\mathbf{j} = \cos\beta\mathbf{i} + \sin\beta\mathbf{j}$, then

$\cos(\alpha - \beta) = \mathbf{v}\cdot\mathbf{w}$

$= a_1a_2 + b_1b_2$

$= \cos\alpha\cos\beta + \sin\alpha\sin\beta$

43. Let $\mathbf{v} = a\mathbf{i} + b\mathbf{j}$. The projection of $\mathbf{v}$ onto $\mathbf{i}$ is

$\mathbf{v}_1$ = the projection of $\mathbf{v}$ onto $\mathbf{i}$

$= \frac{\mathbf{v}_1\cdot\mathbf{i}}{(\|\mathbf{i}\|)^2}\mathbf{i} = \frac{(a\mathbf{i} + b\mathbf{i})\cdot\mathbf{i}}{\left(\sqrt{1^2 + 0^2}\right)^2}\mathbf{i} = \frac{a(1) + b(0)}{1^2}\mathbf{i} = a\mathbf{i}$

$\mathbf{v}_2 = \mathbf{v} - \mathbf{v}_1 = a\mathbf{i} + b\mathbf{j} - a\mathbf{i} = b\mathbf{j}$

Since $\mathbf{v}\cdot\mathbf{i} = (a\mathbf{i} + b\mathbf{j})\cdot\mathbf{i} = (a)(1) + (b)(0) = a$ and

$\mathbf{v}\cdot\mathbf{j} = (a\mathbf{i} + b\mathbf{j})\cdot\mathbf{j} = (a)(0) + (b)(1) = b$,

$\mathbf{v} = \mathbf{v}_1 + \mathbf{v}_2 = (\mathbf{v}\cdot\mathbf{i})\mathbf{i} + (\mathbf{v}\cdot\mathbf{j})\mathbf{j}$.

44. a. Let $\mathbf{u} = a_1\mathbf{i} + b_1\mathbf{j}$ and $\mathbf{v} = a_2\mathbf{i} + b_2\mathbf{j}$. Then,

$(\mathbf{u} + \mathbf{v})\cdot(\mathbf{u} - \mathbf{v})$

$= [(a_1 + a_2)\mathbf{i} + (b_1 + b_2)\mathbf{j}][(a_1 - a_2)\mathbf{i} + (b_1 - b_2)\mathbf{j}]$

$= (a_1 + a_2)(a_1 - a_2) + (b_1 + b_2)(b_1 - b_2)$

$= a_1^2 - a_2^2 + b_1^2 - b_2^2$

$= (a_1^2 + b_1^2) - (a_2^2 + b_2^2)$

$= \|\mathbf{u}\|^2 - \|\mathbf{v}\|^2$

Since the vectors have the same magnitudes and $(\mathbf{u} + \mathbf{v})(\mathbf{u} - \mathbf{v}) = 0$, the vectors are orthogonal.

b. Because the vectors $\mathbf{u}$ and $\mathbf{v}$ are radii of the circle, we know they have the same magnitude. Since $\mathbf{u} + \mathbf{v}$ and $\mathbf{u} - \mathbf{v}$ are sides of the angle inscribed in the semicircle and since we know that they are orthogonal by part (a), this angle must be a right angle.

45. $(\mathbf{v} - \alpha\mathbf{w})\cdot\mathbf{w} = \mathbf{v}\cdot\mathbf{w} - \alpha\mathbf{w}\cdot\mathbf{w}$

$= \mathbf{v}\cdot\mathbf{w} - \alpha(\|\mathbf{w}\|)^2$

$= \mathbf{v}\cdot\mathbf{w} - \frac{\mathbf{v}\cdot\mathbf{w}}{(\|\mathbf{w}\|)^2}(\|\mathbf{w}\|)^2$

$= 0$

Therefore, the vectors are orthogonal.

46. $(\|\mathbf{w}\|\mathbf{v} + \|\mathbf{v}\|\mathbf{w})\cdot(\|\mathbf{w}\|\mathbf{v} - \|\mathbf{v}\|\mathbf{w})$

$= (\|\mathbf{w}\|)^2\,\mathbf{v}\cdot\mathbf{v} - \|\mathbf{w}\|\|\mathbf{v}\|\mathbf{v}\cdot\mathbf{w}$

$\quad + \|\mathbf{w}\|\|\mathbf{v}\|\mathbf{v}\cdot\mathbf{w} - (\|\mathbf{v}\|)^2\,\mathbf{w}\cdot\mathbf{w}$

$= (\|\mathbf{w}\|)^2\,\mathbf{v}\cdot\mathbf{v} - (\|\mathbf{v}\|)^2\,\mathbf{w}\cdot\mathbf{w}$

$= (\|\mathbf{w}\|)^2(\|\mathbf{v}\|)^2 - (\|\mathbf{v}\|)^2(\|\mathbf{w}\|)^2$

$= 0$

Therefore, the vectors are orthogonal.

47. If $\mathbf{F}$ is orthogonal to $\overrightarrow{AB}$, then $\mathbf{F}\cdot\overrightarrow{AB}=0$. So, $W=\mathbf{F}\cdot\overrightarrow{AB}=0$.

48. $$\begin{aligned}&(\|\mathbf{u}+\mathbf{v}\|)^2-(\|\mathbf{u}-\mathbf{v}\|)^2\\&=(\mathbf{u}+\mathbf{v})\cdot(\mathbf{u}+\mathbf{v})-(\mathbf{u}-\mathbf{v})\cdot(\mathbf{u}-\mathbf{v})\\&=(\mathbf{u}\cdot\mathbf{u}+\mathbf{u}\cdot\mathbf{v}+\mathbf{v}\cdot\mathbf{u}+\mathbf{v}\cdot\mathbf{v})\\&\qquad-(\mathbf{u}\cdot\mathbf{u}-\mathbf{u}\cdot\mathbf{v}-\mathbf{v}\cdot\mathbf{u}+\mathbf{v}\cdot\mathbf{v})\\&=2(\mathbf{u}\cdot\mathbf{v})+2(\mathbf{u}\cdot\mathbf{v})\\&=4(\mathbf{u}\cdot\mathbf{v})\end{aligned}$$

49. Answers will vary.

Section 8.6

1. $\sqrt{(x_2-x_1)^2+(y_2-y_1)^2}$

2. *xy*-plane

3. components

4. 1

5. False

6. True

7. $y=0$ is the set of all points of the form $(x, 0, z)$, the set of all points in the *xz*-plane.

8. $x=0$ the set of all points of the form $(0, y, z)$, the set of all points in the *yz*-plane.

9. $z=2$ is the set of all points of the form $(x, y, 2)$, the plain two units above the *xy*-plane.

10. $y=3$ is the set of all points of the form $(x, 3, z)$, the plane three units to the right of the *xz*-plane.

11. $x=-4$ is the set of all points of the form $(-4, y, z)$, the plain four units to the left of the *yz*-plane.

12. $z=-3$ is the set of all points of the form $(x, y, -3)$, the plane three units below the *xy*-plane.

13. $x=1$ and $y=2$ is the set of all points of the form $(1, 2, z)$, a line parallel to the z-axis.

14. $x=3$ and $z=1$ is the set of all points of the form $(3, y, 1)$, a line parallel to the y-axis.

15. $$\begin{aligned}d&=\sqrt{(4-0)^2+(1-0)^2+(2-0)^2}\\&=\sqrt{16+1+4}\\&=\sqrt{21}\end{aligned}$$

16. $$\begin{aligned}d&=\sqrt{(1-0)^2+(-2-0)^2+(3-0)^2}\\&=\sqrt{1+4+9}\\&=\sqrt{14}\end{aligned}$$

17. $$\begin{aligned}d&=\sqrt{(0-(-1))^2+(-2-2)^2+(1-(-3))^2}\\&=\sqrt{1+16+16}\\&=\sqrt{33}\end{aligned}$$

18. $$\begin{aligned}d&=\sqrt{(4-(-2))^2+(0-2)^2+(-3-3)^2}\\&=\sqrt{36+4+36}\\&=\sqrt{76}\\&=2\sqrt{19}\end{aligned}$$

19. $$\begin{aligned}d&=\sqrt{(3-4)^2+(2-(-2))^2+(1-(-2))^2}\\&=\sqrt{1+16+9}\\&=\sqrt{26}\end{aligned}$$

20. $$\begin{aligned}d&=\sqrt{(4-2)^2+(1-(-3))^2+(-1-(-3))^2}\\&=\sqrt{4+16+4}\\&=\sqrt{24}\\&=2\sqrt{6}\end{aligned}$$

21. The bottom of the box is formed by the vertices (0, 0, 0), (2, 0, 0), (0, 1, 0), and (2, 1, 0). The top of the box is formed by the vertices (0, 0, 3), (2, 0, 3), (0, 1, 3), and (2, 1, 3).

22. The bottom of the box is formed by the vertices (0, 0, 0), (4, 0, 0), (4, 2, 0), and (0, 2, 0). The top of the box is formed by the vertices (0, 0, 2), (4, 0, 2), (0, 2, 2), and (4, 2, 2).

23. The bottom of the box is formed by the vertices (1, 2, 3), (3, 2, 3), (3, 4, 3), and (1, 4, 3). The top of the box is formed by the vertices (3, 4, 5), (1, 2, 5), (3, 2, 5), and (1, 4, 5).

24. The bottom of the box is formed by the vertices (5, 6, 1), (3, 6, 1), (5, 8, 1), and (3, 8, 1). The top of the box is formed by the vertices (3, 8, 2), (5, 6, 2), (3, 6, 2), and (5, 8, 2).

25. The bottom of the box is formed by the vertices (–1, 0, 2), (4, 0, 2), (–1, 2, 2), and (4, 2, 2). The top of the box is formed by the vertices (4, 2, 5), (–1, 0, 5), (4, 0, 5), and (–1, 2, 5).

26. The bottom of the box is formed by the vertices (–2, –3, 0), (–6, –3, 0), (–2, 7, 0), and (–6, 7, 0). The top of the box is formed by the vertices (–6, 7, 1), (–2, –3, 1), (–2, 7, 1), and (–6, –3, 1).

27. $\mathbf{v}=(3-0)\mathbf{i}+(4-0)\mathbf{j}+(-1-0)\mathbf{k}=3\mathbf{i}+4\mathbf{j}-\mathbf{k}$

28. $\mathbf{v}=(-3-0)\mathbf{i}+(-5-0)\mathbf{j}+(4-0)\mathbf{k}=-3\mathbf{i}-5\mathbf{j}+4\mathbf{k}$

29. $\mathbf{v}=(5-3)\mathbf{i}+(6-2)\mathbf{j}+(0-(-1))\mathbf{k}=2\mathbf{i}+4\mathbf{j}+\mathbf{k}$

30. $\mathbf{v}=(6-(-3))\mathbf{i}+(5-2)\mathbf{j}+(-1-0)\mathbf{k}=9\mathbf{i}+3\mathbf{j}-\mathbf{k}$

31. $\mathbf{v}=(6-(-2))\mathbf{i}+(-2-(-1))\mathbf{j}+(4-4)\mathbf{k}=8\mathbf{i}-\mathbf{j}$

32. $$\begin{aligned}\mathbf{v}&=(6-(-1))\mathbf{i}+(2-4)\mathbf{j}+(2-(-2))\mathbf{k}\\&=7\mathbf{i}-2\mathbf{j}+4\mathbf{k}\end{aligned}$$

33. $$\begin{aligned}\|\mathbf{v}\|&=\sqrt{3^2+(-6)^2+(-2)^2}\\&=\sqrt{9+36+4}\\&=\sqrt{49}\\&=7\end{aligned}$$

34. $$\begin{aligned}\|\mathbf{v}\|&=\sqrt{(-6)^2+12^2+4^2}\\&=\sqrt{36+144+16}\\&=\sqrt{196}\\&=14\end{aligned}$$

35. $\|\mathbf{v}\|=\sqrt{1^2+(-1)^2+1^2}=\sqrt{1+1+1}=\sqrt{3}$

36. $\|\mathbf{v}\|=\sqrt{(-1)^2+(-1)^2+1^2}=\sqrt{1+1+1}=\sqrt{3}$

37. $\|\mathbf{v}\|=\sqrt{(-2)^2+3^2+(-3)^2}=\sqrt{4+9+9}=\sqrt{22}$

38. $$\begin{aligned}\|\mathbf{v}\|&=\sqrt{6^2+2^2+(-2)^2}\\&=\sqrt{36+4+4}\\&=\sqrt{44}\\&=2\sqrt{11}\end{aligned}$$

39. $$\begin{aligned}2\mathbf{v}+3\mathbf{w}&=2(3\mathbf{i}-5\mathbf{j}+2\mathbf{k})+3(-2\mathbf{i}+3\mathbf{j}-2\mathbf{k})\\&=6\mathbf{i}-10\mathbf{j}+4\mathbf{k}-6\mathbf{i}+9\mathbf{j}-6\mathbf{k}\\&=-\mathbf{j}-2\mathbf{k}\end{aligned}$$

40. $$\begin{aligned}3\mathbf{v}-2\mathbf{w}&=3(3\mathbf{i}-5\mathbf{j}+2\mathbf{k})-2(-2\mathbf{i}+3\mathbf{j}-2\mathbf{k})\\&=9\mathbf{i}-15\mathbf{j}+6\mathbf{k}+4\mathbf{i}-6\mathbf{j}+4\mathbf{k}\\&=13\mathbf{i}-21\mathbf{j}+10\mathbf{k}\end{aligned}$$

41. $$\begin{aligned}\|\mathbf{v}-\mathbf{w}\|&=\|(3\mathbf{i}-5\mathbf{j}+2\mathbf{k})-(-2\mathbf{i}+3\mathbf{j}-2\mathbf{k})\|\\&=\|3\mathbf{i}-5\mathbf{j}+2\mathbf{k}+2\mathbf{i}-3\mathbf{j}+2\mathbf{k}\|\\&=\|5\mathbf{i}-8\mathbf{j}+4\mathbf{k}\|\\&=\sqrt{5^2+(-8)^2+4^2}\\&=\sqrt{25+64+16}\\&=\sqrt{105}\end{aligned}$$

42. $$\begin{aligned}\|\mathbf{v}+\mathbf{w}\|&=\|(3\mathbf{i}-5\mathbf{j}+2\mathbf{k})+(-2\mathbf{i}+3\mathbf{j}-2\mathbf{k})\|\\&=\|\mathbf{i}-2\mathbf{j}+0\mathbf{k}\|\\&=\sqrt{1^2+(-2)^2+0^2}\\&=\sqrt{1+4+0}\\&=\sqrt{5}\end{aligned}$$

43. $$\begin{aligned}&\|\mathbf{v}\|-\|\mathbf{w}\|\\&=\|3\mathbf{i}-5\mathbf{j}+2\mathbf{k}\|-\|-2\mathbf{i}+3\mathbf{j}-2\mathbf{k}\|\\&=\sqrt{3^2+(-5)^2+2^2}-\sqrt{(-2)^2+3^2+(-2)^2}\\&=\sqrt{38}-\sqrt{17}\end{aligned}$$

44. $$\begin{aligned}&\|\mathbf{v}\|+\|\mathbf{w}\|\\&=\|3\mathbf{i}-5\mathbf{j}+2\mathbf{k}\|+\|-2\mathbf{i}+3\mathbf{j}-2\mathbf{k}\|\\&=\sqrt{3^2+(-5)^2+2^2}+\sqrt{(-2)^2+3^2+(-2)^2}\\&=\sqrt{38}+\sqrt{17}\end{aligned}$$

45. $\mathbf{u}=\dfrac{\mathbf{v}}{\|\mathbf{v}\|}=\dfrac{5\mathbf{i}}{\sqrt{5^2+0^2+0^2}}=\dfrac{5\mathbf{i}}{5}=\mathbf{i}$

46. $\mathbf{u}=\dfrac{\mathbf{v}}{\|\mathbf{v}\|}=\dfrac{-3\mathbf{j}}{\sqrt{0^2+(-3)^2+0^2}}=\dfrac{-3\mathbf{j}}{3}=-\mathbf{j}$

47. $$\begin{aligned}\mathbf{u}&=\frac{\mathbf{v}}{\|\mathbf{v}\|}\\&=\frac{3\mathbf{i}-6\mathbf{j}-2\mathbf{k}}{\sqrt{3^2+(-6)^2+(-2)^2}}\\&=\frac{3\mathbf{i}-6\mathbf{j}-2\mathbf{k}}{7}\\&=\frac{3}{7}\mathbf{i}-\frac{6}{7}\mathbf{j}-\frac{2}{7}\mathbf{k}\end{aligned}$$

48. $\mathbf{u}=\dfrac{\mathbf{v}}{\|\mathbf{v}\|}$
$=\dfrac{-6\mathbf{i}+12\mathbf{j}+4\mathbf{k}}{\sqrt{(-6)^2+12^2+4^2}}$
$=\dfrac{-6\mathbf{i}+12\mathbf{j}+4\mathbf{k}}{14}$
$=-\dfrac{3}{7}\mathbf{i}+\dfrac{6}{7}\mathbf{j}+\dfrac{2}{7}\mathbf{k}$

49. $\mathbf{u}=\dfrac{\mathbf{v}}{\|\mathbf{v}\|}$
$=\dfrac{\mathbf{i}+\mathbf{j}+\mathbf{k}}{\sqrt{1^2+1^2+1^2}}$
$=\dfrac{\mathbf{i}+\mathbf{j}+\mathbf{k}}{\sqrt{3}}$
$=\dfrac{1}{\sqrt{3}}\mathbf{i}+\dfrac{1}{\sqrt{3}}\mathbf{j}+\dfrac{1}{\sqrt{3}}\mathbf{k}$
$=\dfrac{\sqrt{3}}{3}\mathbf{i}+\dfrac{\sqrt{3}}{3}\mathbf{j}+\dfrac{\sqrt{3}}{3}\mathbf{k}$

50. $\mathbf{u}=\dfrac{\mathbf{v}}{\|\mathbf{v}\|}$
$=\dfrac{2\mathbf{i}-\mathbf{j}+\mathbf{k}}{\sqrt{2^2+(-1)^2+1^2}}$
$=\dfrac{2\mathbf{i}-\mathbf{j}+\mathbf{k}}{\sqrt{6}}$
$=\dfrac{2}{\sqrt{6}}\mathbf{i}-\dfrac{1}{\sqrt{6}}\mathbf{j}+\dfrac{1}{\sqrt{6}}\mathbf{k}$
$=\dfrac{\sqrt{6}}{3}\mathbf{i}-\dfrac{\sqrt{6}}{6}\mathbf{j}+\dfrac{\sqrt{6}}{6}\mathbf{k}$

51. $\mathbf{v}\bullet\mathbf{w}=(\mathbf{i}-\mathbf{j})\bullet(\mathbf{i}+\mathbf{j}+\mathbf{k})$
$=1\cdot1+(-1)(1)+0\cdot1$
$=1-1+0$
$=0$

$\cos\theta=\dfrac{\mathbf{v}\bullet\mathbf{w}}{\|\mathbf{v}\|\|\mathbf{w}\|}$
$=\dfrac{0}{\sqrt{1^2+(-1)^2+0^2}\sqrt{1^2+1^2+1^2}}$
$=\dfrac{0}{\sqrt{2}\sqrt{3}}$
$=0$
$\theta=\dfrac{\pi}{2}$ radians $=90^\circ$

52. $\mathbf{v}\bullet\mathbf{w}=(\mathbf{i}+\mathbf{j})\bullet(-\mathbf{i}+\mathbf{j}-\mathbf{k})$
$=1\cdot(-1)+1(1)+0\cdot(-1)$
$=-1+1+0$
$=0$

$\cos\theta=\dfrac{\mathbf{v}\bullet\mathbf{w}}{\|\mathbf{v}\|\|\mathbf{w}\|}$
$=\dfrac{0}{\sqrt{1^2+1^2+0^2}\sqrt{(-1)^2+1^2+(-1)^2}}$
$=\dfrac{0}{\sqrt{2}\sqrt{3}}$
$=0$
$\theta=\dfrac{\pi}{2}$ radians $=90^\circ$

53. $\mathbf{v}\bullet\mathbf{w}=(2\mathbf{i}+\mathbf{j}-3\mathbf{k})\bullet(\mathbf{i}+2\mathbf{j}+2\mathbf{k})$
$=2\cdot1+1(2)+(-3)(2)$
$=2+2-6$
$=-2$

$\cos\theta=\dfrac{\mathbf{v}\bullet\mathbf{w}}{\|\mathbf{v}\|\|\mathbf{w}\|}$
$=\dfrac{-2}{\sqrt{2^2+1^2+(-3)^2}\sqrt{1^2+2^2+2^2}}$
$=\dfrac{-2}{\sqrt{14}\sqrt{9}}$
$=\dfrac{-2}{3\sqrt{14}}\approx-0.1782$
$\theta\approx1.75$ radians $\approx100.3^\circ$

54. $\mathbf{v}\bullet\mathbf{w}=(2\mathbf{i}+2\mathbf{j}-\mathbf{k})\bullet(\mathbf{i}+2\mathbf{j}+3\mathbf{k})$
$=2\cdot1+2(2)+(-1)(3)$
$=2+4-3$
$=3$

$\cos\theta=\dfrac{\mathbf{v}\bullet\mathbf{w}}{\|\mathbf{v}\|\|\mathbf{w}\|}$
$=\dfrac{3}{\sqrt{2^2+2^2+(-1)^2}\sqrt{1^2+2^2+3^2}}$
$=\dfrac{3}{\sqrt{9}\sqrt{14}}$
$=\dfrac{3}{3\sqrt{14}}\approx0.2673$
$\theta\approx1.30$ radians $\approx74.5^\circ$

55. $\mathbf{v}\bullet\mathbf{w}=(3\mathbf{i}-\mathbf{j}+2\mathbf{k})\bullet(\mathbf{i}+\mathbf{j}-\mathbf{k})$
$=3\cdot 1+(-1)(1)+2(-1)$
$=3-1-2$
$=0$

$$\cos\theta=\frac{\mathbf{v}\bullet\mathbf{w}}{\|\mathbf{v}\|\|\mathbf{w}\|}=\frac{0}{\sqrt{3^2+(-1)^2+2^2}\sqrt{1^2+1^2+(-1)^2}}=\frac{0}{\sqrt{14}\sqrt{3}}=0$$

$\theta=\frac{\pi}{2}$ radians $=90°$

56. $\mathbf{v}\bullet\mathbf{w}=(\mathbf{i}+3\mathbf{j}+2\mathbf{k})\bullet(\mathbf{i}-\mathbf{j}+\mathbf{k})$
$=1\cdot 1+3(-1)+2(1)$
$=1-3+2$
$=0$

$$\cos\theta=\frac{\mathbf{v}\bullet\mathbf{w}}{\|\mathbf{v}\|\|\mathbf{w}\|}=\frac{0}{\sqrt{1^2+3^2+2^2}\sqrt{1^2+(-1)^2+1^2}}=\frac{0}{\sqrt{14}\sqrt{3}}=0$$

$\theta=\frac{\pi}{2}$ radians $=90°$

57. $\mathbf{v}\bullet\mathbf{w}=(3\mathbf{i}+4\mathbf{j}+\mathbf{k})\bullet(6\mathbf{i}+8\mathbf{j}+2\mathbf{k})$
$=3\cdot 6+4\cdot 8+1\cdot 2$
$=18+32+2$
$=52$

$$\cos\theta=\frac{\mathbf{v}\bullet\mathbf{w}}{\|\mathbf{v}\|\|\mathbf{w}\|}=\frac{52}{\sqrt{3^2+4^2+1^2}\sqrt{6^2+8^2+2^2}}=\frac{52}{\sqrt{26}\sqrt{104}}=\frac{52}{52}=1$$

$\theta=0$ radians $=0°$

58. $\mathbf{v}\bullet\mathbf{w}=(3\mathbf{i}-4\mathbf{j}+\mathbf{k})\bullet(6\mathbf{i}-8\mathbf{j}+2\mathbf{k})$
$=3\cdot 6+(-4)\cdot(-8)+1\cdot 2$
$=18+32+2$
$=52$

$$\cos\theta=\frac{\mathbf{v}\bullet\mathbf{w}}{\|\mathbf{v}\|\|\mathbf{w}\|}=\frac{52}{\sqrt{3^2+(-4)^2+1^2}\sqrt{6^2+(-8)^2+2^2}}=\frac{52}{\sqrt{26}\sqrt{104}}=\frac{52}{52}=1$$

$\theta=0$ radians $=0°$

59. $\cos\alpha=\frac{a}{\|\mathbf{v}\|}=\frac{3}{\sqrt{3^2+(-6)^2+(-2)^2}}=\frac{3}{\sqrt{49}}=\frac{3}{7}$
$\alpha\approx 64.6°$

$\cos\beta=\frac{b}{\|\mathbf{v}\|}=\frac{-6}{\sqrt{3^2+(-6)^2+(-2)^2}}=\frac{-6}{\sqrt{49}}=-\frac{6}{7}$
$\beta\approx 149.0°$

$\cos\gamma=\frac{c}{\|\mathbf{v}\|}=\frac{-2}{\sqrt{3^2+(-6)^2+(-2)^2}}=\frac{-2}{\sqrt{49}}=-\frac{2}{7}$
$\gamma\approx 106.6°$

$\mathbf{v}=7(\cos 64.6°\mathbf{i}+\cos 149.0°\mathbf{j}+\cos 106.6°\mathbf{k})$

60. $\cos\alpha=\frac{a}{\|\mathbf{v}\|}=\frac{-6}{\sqrt{(-6)^2+12^2+4^2}}=\frac{-6}{\sqrt{196}}=-\frac{3}{7}$
$\alpha\approx 115.4°$

$\cos\beta=\frac{b}{\|\mathbf{v}\|}=\frac{12}{\sqrt{(-6)^2+12^2+4^2}}=\frac{12}{\sqrt{196}}=\frac{6}{7}$
$\beta\approx 31.0°$

$\cos\gamma=\frac{c}{\|\mathbf{v}\|}=\frac{4}{\sqrt{(-6)^2+12^2+4^2}}=\frac{4}{\sqrt{196}}=\frac{2}{7}$
$\gamma\approx 73.4°$

$\mathbf{v}=14(\cos 115.4°\mathbf{i}+\cos 31.0°\mathbf{j}+\cos 73.4°\mathbf{k})$

61. $\cos\alpha = \frac{a}{\|\mathbf{v}\|} = \frac{1}{\sqrt{1^2+1^2+1^2}} = \frac{1}{\sqrt{3}} = \frac{\sqrt{3}}{3}$
$\alpha \approx 54.7°$

$\cos\beta = \frac{b}{\|\mathbf{v}\|} = \frac{1}{\sqrt{1^2+1^2+1^2}} = \frac{1}{\sqrt{3}} = \frac{\sqrt{3}}{3}$
$\beta \approx 54.7°$

$\cos\gamma = \frac{c}{\|\mathbf{v}\|} = \frac{1}{\sqrt{1^2+1^2+1^2}} = \frac{1}{\sqrt{3}} = \frac{\sqrt{3}}{3}$
$\gamma \approx 54.7°$

$\mathbf{v} = \sqrt{3}\left(\cos 54.7°\mathbf{i} + \cos 54.7°\mathbf{j} + \cos 54.7°\mathbf{k}\right)$

62. $\cos\alpha = \frac{a}{\|\mathbf{v}\|} = \frac{1}{\sqrt{1^2+(-1)^2+(-1)^2}} = \frac{1}{\sqrt{3}} = \frac{\sqrt{3}}{3}$
$\alpha \approx 54.7°$

$\cos\beta = \frac{b}{\|\mathbf{v}\|} = \frac{-1}{\sqrt{1^2+(-1)^2+(-1)^2}} = \frac{-1}{\sqrt{3}} = -\frac{\sqrt{3}}{3}$
$\beta \approx 125.3°$

$\cos\gamma = \frac{c}{\|\mathbf{v}\|} = \frac{-1}{\sqrt{1^2+(-1)^2+(-1)^2}} = \frac{-1}{\sqrt{3}} = -\frac{\sqrt{3}}{3}$
$\gamma \approx 125.3°$

$\mathbf{v} = \sqrt{3}\left(\cos 54.7°\mathbf{i} + \cos 125.3°\mathbf{j} + \cos 125.3°\mathbf{k}\right)$

63. $\cos\alpha = \frac{a}{\|\mathbf{v}\|} = \frac{1}{\sqrt{1^2+1^2+0^2}} = \frac{1}{\sqrt{2}} = \frac{\sqrt{2}}{2}$
$\alpha = 45°$

$\cos\beta = \frac{b}{\|\mathbf{v}\|} = \frac{1}{\sqrt{1^2+1^2+0^2}} = \frac{1}{\sqrt{2}} = \frac{\sqrt{2}}{2}$
$\beta = 45°$

$\cos\gamma = \frac{c}{\|\mathbf{v}\|} = \frac{0}{\sqrt{1^2+1^2+0^2}} = \frac{0}{\sqrt{2}} = 0$
$\gamma = 90°$

$\mathbf{v} = \sqrt{2}\left(\cos 45°\mathbf{i} + \cos 45°\mathbf{j} + \cos 90°\mathbf{k}\right)$

64. $\cos\alpha = \frac{a}{\|\mathbf{v}\|} = \frac{0}{\sqrt{0^2+1^2+1^2}} = \frac{0}{\sqrt{2}} = 0$
$\alpha = 90°$

$\cos\beta = \frac{b}{\|\mathbf{v}\|} = \frac{1}{\sqrt{0^2+1^2+1^2}} = \frac{1}{\sqrt{2}} = \frac{\sqrt{2}}{2}$
$\beta = 45°$

$\cos\gamma = \frac{c}{\|\mathbf{v}\|} = \frac{1}{\sqrt{0^2+1^2+1^2}} = \frac{1}{\sqrt{2}} = \frac{\sqrt{2}}{2}$
$\gamma = 45°$

$\mathbf{v} = \sqrt{2}\left(\cos 90°\mathbf{i} + \cos 45°\mathbf{j} + \cos 45°\mathbf{k}\right)$

65. $\cos\alpha = \frac{a}{\|\mathbf{v}\|} = \frac{3}{\sqrt{3^2+(-5)^2+2^2}} = \frac{3}{\sqrt{38}}$
$\alpha \approx 60.9°$

$\cos\beta = \frac{b}{\|\mathbf{v}\|} = \frac{-5}{\sqrt{3^2+(-5)^2+2^2}} = -\frac{5}{\sqrt{38}}$
$\beta \approx 144.2°$

$\cos\gamma = \frac{c}{\|\mathbf{v}\|} = \frac{2}{\sqrt{3^2+(-5)^2+2^2}} = \frac{2}{\sqrt{38}}$
$\gamma \approx 71.1°$

$\mathbf{v} = \sqrt{38}\left(\cos 60.9°\mathbf{i} + \cos 144.2°\mathbf{j} + \cos 71.1°\mathbf{k}\right)$

66. $\cos\alpha = \frac{a}{\|\mathbf{v}\|} = \frac{2}{\sqrt{2^2+3^2+(-4)^2}} = \frac{2}{\sqrt{29}}$
$\alpha \approx 68.2°$

$\cos\beta = \frac{b}{\|\mathbf{v}\|} = \frac{3}{\sqrt{2^2+3^2+(-4)^2}} = \frac{3}{\sqrt{29}}$
$\beta \approx 56.1°$

$\cos\gamma = \frac{c}{\|\mathbf{v}\|} = \frac{-4}{\sqrt{2^2+3^2+(-4)^2}} = -\frac{4}{\sqrt{29}}$
$\gamma \approx 138.0°$

$\mathbf{v} = \sqrt{29}\left(\cos 68.2°\mathbf{i} + \cos 56.1°\mathbf{j} + \cos 138.0°\mathbf{k}\right)$

67. If the point $P = (x, y, z)$ is on the sphere with center $C = (x_0, y_0, z_0)$ and radius *r*, then the distance between *P* and *C* is
$d(P_0, P) = \sqrt{(x-x_0)^2 + (y-y_0)^2 + (z-z_0)^2} = r$
. Therefore, the equation of the sphere is
$(x-x_0)^2 + (y-y_0)^2 + (z-z_0)^2 = r^2$.

68. $(x-3)^2 + (y-1)^2 + (z-1)^2 = 1^2$
$(x-3)^2 + (y-1)^2 + (z-1)^2 = 1$

69. $(x-1)^2 + (y-2)^2 + (z-2)^2 = 2^2$
$(x-1)^2 + (y-2)^2 + (z-2)^2 = 4$

70. $(x+1)^2+(y-1)^2+(z-2)^2=3^2$
$(x+1)^2+(y-1)^2+(z-2)^2=9$

71. $x^2+y^2+z^2+2x-2y=2$
$(x^2+2x)+(y^2-2y)+z^2=2$
$(x^2+2x+1)+(y^2-2y+1)+z^2=2+1+1$
$(x+1)^2+(y-1)^2+(z-0)^2=4$
Center: (–1, 1, 0); Radius: 2

72. $x^2+y^2+z^2+2x-2z=-1$
$(x^2+2x)+y^2+(z^2-2z)=-1$
$(x^2+2x+1)+y^2+(z^2-2z+1)=-1+1+1$
$(x+1)^2+(y-0)^2+(z-1)^2=1$
Center: (–1, 0, 1); Radius: 1

73. $x^2+y^2+z^2-4x+4y+2z=0$
$(x^2-4x)+(y^2+4y)+(z^2+2z)=0$
$(x^2-4x+4)+(y^2+4y+4)+(z^2+2z+1)=4+4+1$
$(x-2)^2+(y+2)^2+(z+1)^2=9$
Center: (2, –2, –1); Radius: 3

74. $x^2+y^2+z^2-4x=0$
$(x^2-4x)+y^2+z^2=0$
$(x^2-4x+4)+y^2+z^2=4$
$(x-2)^2+(y-0)^2+(z-0)^2=4$
Center: (2, 0, 0); Radius: 2

75. $2x^2+2y^2+2z^2-8x+4z=-1$
$(x^2-4x)+y^2+(z^2+2z)=-\frac{1}{2}$
$(x^2-4x+4)+y^2+(z^2+2z+1)=-\frac{1}{2}+4+1$
$(x-2)^2+(y-0)^2+(z+1)^2=\frac{9}{2}$
Center: (2, 0, –1); Radius: $\frac{3\sqrt{2}}{2}$

76. $3x^2+3y^2+3z^2+6x-6y=3$
$(x^2+2x)+(y^2-2y)+z^2=1$
$(x^2+2x+1)+(y^2-2y+1)+z^2=1+1+1$
$(x+1)^2+(y-1)^2+(z-0)^2=3$
Center: (–1, 1, 0); Radius: $\sqrt{3}$

77. Write the force as a vector:
$\cos\alpha=\frac{2}{\sqrt{2^2+1^2+2^2}}=\frac{2}{\sqrt{9}}=\frac{2}{3}$
$\cos\beta=\frac{1}{3}$
$\cos\gamma=\frac{2}{3}$
$\mathbf{F}=3\left(\frac{2}{3}\mathbf{i}+\frac{1}{3}\mathbf{j}+\frac{2}{3}\mathbf{k}\right)$
$W=3\left(\frac{2}{3}\mathbf{i}+\frac{1}{3}\mathbf{j}+\frac{2}{3}\mathbf{k}\right)\bullet 2\mathbf{j}=3\left(\frac{1}{3}\cdot 2\right)=2$ joules

78. Write the force as a vector:
$\cos\alpha=\frac{2}{\sqrt{2^2+2^2+1^2}}=\frac{2}{\sqrt{9}}=\frac{2}{3}$
$\cos\beta=\frac{2}{3}$
$\cos\gamma=\frac{1}{3}$
$\mathbf{F}=1\left(\frac{2}{3}\mathbf{i}+\frac{2}{3}\mathbf{j}+\frac{1}{3}\mathbf{k}\right)$
$W=1\left(\frac{2}{3}\mathbf{i}+\frac{2}{3}\mathbf{j}+\frac{1}{3}\mathbf{k}\right)\bullet(1\mathbf{i}+2\mathbf{j}+2\mathbf{k})$
$=1\left(\frac{2}{3}\cdot 1+\frac{2}{3}\cdot 2+\frac{1}{3}\cdot 2\right)$
$=\frac{8}{3}$ joules

79. $W=\mathbf{F}\bullet\mathbf{AB}$
$=(2\mathbf{i}-\mathbf{j}-\mathbf{k})\bullet(3\mathbf{i}+2\mathbf{j}-5\mathbf{k})$
$=2\cdot 3+(-1)(2)+(-1)(-5)$
$=9$

Section 8.7

1. True
2. True
3. True
4. False
5. False
6. True
7. $\begin{vmatrix}3 & 4\\ 1 & 2\end{vmatrix}=3\cdot 2-1\cdot 4=6-4=2$

8. $\begin{vmatrix} -2 & 5 \\ 2 & -3 \end{vmatrix} = -2(-3) - 2\cdot 5 = 6 - 10 = -4$

9. $\begin{vmatrix} 6 & 5 \\ -2 & -1 \end{vmatrix} = 6(-1) - (-2)(5) = -6 + 10 = 4$

10. $\begin{vmatrix} -4 & 0 \\ 5 & 3 \end{vmatrix} = -4\cdot 3 - 5\cdot 0 = -12 - 0 = -12$

11. $\begin{vmatrix} A & B & C \\ 2 & 1 & 4 \\ 1 & 3 & 1 \end{vmatrix} = \begin{vmatrix} 1 & 4 \\ 3 & 1 \end{vmatrix} A - \begin{vmatrix} 2 & 4 \\ 1 & 1 \end{vmatrix} B + \begin{vmatrix} 2 & 1 \\ 1 & 3 \end{vmatrix} C$

$= (1-12)A - (2-4)B + (6-1)C$

$= -11A + 2B + 5C$

12. $\begin{vmatrix} A & B & C \\ 0 & 2 & 4 \\ 3 & 1 & 3 \end{vmatrix} = \begin{vmatrix} 2 & 4 \\ 1 & 3 \end{vmatrix} A - \begin{vmatrix} 0 & 4 \\ 3 & 3 \end{vmatrix} B + \begin{vmatrix} 0 & 2 \\ 3 & 1 \end{vmatrix} C$

$= (6-4)A - (0-12)B + (0-6)C$

$= 2A + 12B - 6C$

13. $\begin{vmatrix} A & B & C \\ -1 & 3 & 5 \\ 5 & 0 & -2 \end{vmatrix}$

$= \begin{vmatrix} 3 & 5 \\ 0 & -2 \end{vmatrix} A - \begin{vmatrix} -1 & 5 \\ 5 & -2 \end{vmatrix} B + \begin{vmatrix} -1 & 3 \\ 5 & 0 \end{vmatrix} C$

$= (-6-0)A - (2-25)B + (0-15)C$

$= -6A + 23B - 15C$

14. $\begin{vmatrix} A & B & C \\ 1 & -2 & -3 \\ 0 & 2 & -2 \end{vmatrix}$

$= \begin{vmatrix} -2 & -3 \\ 2 & -2 \end{vmatrix} A - \begin{vmatrix} 1 & -3 \\ 0 & -2 \end{vmatrix} B + \begin{vmatrix} 1 & -2 \\ 0 & 2 \end{vmatrix} C$

$= (4+6)A - (-2-0)B + (2-0)C$

$= 10A + 2B + 2C$

15. a. $\mathbf{v}\times\mathbf{w} = \begin{vmatrix} \mathbf{i} & \mathbf{j} & \mathbf{k} \\ 2 & -3 & 1 \\ 3 & -2 & -1 \end{vmatrix}$

$= \begin{vmatrix} -3 & 1 \\ -2 & -1 \end{vmatrix} \mathbf{i} - \begin{vmatrix} 2 & 1 \\ 3 & -1 \end{vmatrix} \mathbf{j} + \begin{vmatrix} 2 & -3 \\ 3 & -2 \end{vmatrix} \mathbf{k}$

$= 5\mathbf{i} + 5\mathbf{j} + 5\mathbf{k}$

b. $\mathbf{w}\times\mathbf{v} = \begin{vmatrix} \mathbf{i} & \mathbf{j} & \mathbf{k} \\ 3 & -2 & -1 \\ 2 & -3 & 1 \end{vmatrix}$

$= \begin{vmatrix} -2 & -1 \\ -3 & 1 \end{vmatrix} \mathbf{i} - \begin{vmatrix} 3 & -1 \\ 2 & 1 \end{vmatrix} \mathbf{j} + \begin{vmatrix} 3 & -2 \\ 2 & -3 \end{vmatrix} \mathbf{k}$

$= -5\mathbf{i} - 5\mathbf{j} - 5\mathbf{k}$

c. $\mathbf{w}\times\mathbf{w} = \begin{vmatrix} \mathbf{i} & \mathbf{j} & \mathbf{k} \\ 3 & -2 & -1 \\ 3 & -2 & -1 \end{vmatrix}$

$= \begin{vmatrix} -2 & -1 \\ -2 & -1 \end{vmatrix} \mathbf{i} - \begin{vmatrix} 3 & -1 \\ 3 & -1 \end{vmatrix} \mathbf{j} + \begin{vmatrix} 3 & -2 \\ 3 & -2 \end{vmatrix} \mathbf{k}$

$= 0\mathbf{i} + 0\mathbf{j} + 0\mathbf{k}$

$= \mathbf{0}$

d. $\mathbf{v}\times\mathbf{v} = \begin{vmatrix} \mathbf{i} & \mathbf{j} & \mathbf{k} \\ 2 & -3 & 1 \\ 2 & -3 & 1 \end{vmatrix}$

$= \begin{vmatrix} -3 & 1 \\ -3 & 1 \end{vmatrix} \mathbf{i} - \begin{vmatrix} 2 & 1 \\ 2 & 1 \end{vmatrix} \mathbf{j} + \begin{vmatrix} 2 & -3 \\ 2 & -3 \end{vmatrix} \mathbf{k}$

$= 0\mathbf{i} + 0\mathbf{j} + 0\mathbf{k}$

$= \mathbf{0}$

16. a. $\mathbf{v}\times\mathbf{w} = \begin{vmatrix} \mathbf{i} & \mathbf{j} & \mathbf{k} \\ -1 & 3 & 2 \\ 3 & -2 & -1 \end{vmatrix}$

$= \begin{vmatrix} 3 & 2 \\ -2 & -1 \end{vmatrix} \mathbf{i} - \begin{vmatrix} -1 & 2 \\ 3 & -1 \end{vmatrix} \mathbf{j} + \begin{vmatrix} -1 & 3 \\ 3 & -2 \end{vmatrix} \mathbf{k}$

$= \mathbf{i} + 5\mathbf{j} - 7\mathbf{k}$

b. $\mathbf{w}\times\mathbf{v} = \begin{vmatrix} \mathbf{i} & \mathbf{j} & \mathbf{k} \\ 3 & -2 & -1 \\ -1 & 3 & 2 \end{vmatrix}$

$= \begin{vmatrix} -2 & -1 \\ 3 & 2 \end{vmatrix} \mathbf{i} - \begin{vmatrix} 3 & -1 \\ -1 & 2 \end{vmatrix} \mathbf{j} + \begin{vmatrix} 3 & -2 \\ -1 & 3 \end{vmatrix} \mathbf{k}$

$= -1\mathbf{i} - 5\mathbf{j} + 7\mathbf{k}$

c. $\mathbf{w}\times\mathbf{w} = \begin{vmatrix} \mathbf{i} & \mathbf{j} & \mathbf{k} \\ 3 & -2 & -1 \\ 3 & -2 & -1 \end{vmatrix}$

$= \begin{vmatrix} -2 & -1 \\ -2 & -1 \end{vmatrix} \mathbf{i} - \begin{vmatrix} 3 & -1 \\ 3 & -1 \end{vmatrix} \mathbf{j} + \begin{vmatrix} 3 & -2 \\ 3 & -2 \end{vmatrix} \mathbf{k}$

$= 0\mathbf{i} + 0\mathbf{j} + 0\mathbf{k}$

$= \mathbf{0}$

d. $\mathbf{v}\times\mathbf{v} = \begin{vmatrix} \mathbf{i} & \mathbf{j} & \mathbf{k} \\ -1 & 3 & 2 \\ -1 & 3 & 2 \end{vmatrix}$

$= \begin{vmatrix} 3 & 2 \\ 3 & 2 \end{vmatrix}\mathbf{i} - \begin{vmatrix} -1 & 2 \\ -1 & 2 \end{vmatrix}\mathbf{j} + \begin{vmatrix} -1 & 3 \\ -1 & 3 \end{vmatrix}\mathbf{k}$

$= 0\mathbf{i} + 0\mathbf{j} + 0\mathbf{k}$

$= \mathbf{0}$

17. a. $\mathbf{v}\times\mathbf{w} = \begin{vmatrix} \mathbf{i} & \mathbf{j} & \mathbf{k} \\ 1 & 1 & 0 \\ 2 & 1 & 1 \end{vmatrix}$

$= \begin{vmatrix} 1 & 0 \\ 1 & 1 \end{vmatrix}\mathbf{i} - \begin{vmatrix} 1 & 0 \\ 2 & 1 \end{vmatrix}\mathbf{j} + \begin{vmatrix} 1 & 1 \\ 2 & 1 \end{vmatrix}\mathbf{k}$

$= 1\mathbf{i} - 1\mathbf{j} - 1\mathbf{k}$

$= \mathbf{i} - \mathbf{j} - \mathbf{k}$

b. $\mathbf{w}\times\mathbf{v} = \begin{vmatrix} \mathbf{i} & \mathbf{j} & \mathbf{k} \\ 2 & 1 & 1 \\ 1 & 1 & 0 \end{vmatrix}$

$= \begin{vmatrix} 1 & 1 \\ 1 & 0 \end{vmatrix}\mathbf{i} - \begin{vmatrix} 2 & 1 \\ 1 & 0 \end{vmatrix}\mathbf{j} + \begin{vmatrix} 2 & 1 \\ 1 & 1 \end{vmatrix}\mathbf{k}$

$= -1\mathbf{i} + 1\mathbf{j} + 1\mathbf{k}$

$= -\mathbf{i} + \mathbf{j} + \mathbf{k}$

c. $\mathbf{w}\times\mathbf{w} = \begin{vmatrix} \mathbf{i} & \mathbf{j} & \mathbf{k} \\ 2 & 1 & 1 \\ 2 & 1 & 1 \end{vmatrix}$

$= \begin{vmatrix} 1 & 1 \\ 1 & 1 \end{vmatrix}\mathbf{i} - \begin{vmatrix} 2 & 1 \\ 2 & 1 \end{vmatrix}\mathbf{j} + \begin{vmatrix} 2 & 1 \\ 2 & 1 \end{vmatrix}\mathbf{k}$

$= 0\mathbf{i} + 0\mathbf{j} + 0\mathbf{k}$

$= \mathbf{0}$

d. $\mathbf{v}\times\mathbf{v} = \begin{vmatrix} \mathbf{i} & \mathbf{j} & \mathbf{k} \\ 1 & 1 & 0 \\ 1 & 1 & 0 \end{vmatrix}$

$= \begin{vmatrix} 1 & 0 \\ 1 & 0 \end{vmatrix}\mathbf{i} - \begin{vmatrix} 1 & 0 \\ 1 & 0 \end{vmatrix}\mathbf{j} + \begin{vmatrix} 1 & 1 \\ 1 & 1 \end{vmatrix}\mathbf{k}$

$= 0\mathbf{i} + 0\mathbf{j} + 0\mathbf{k}$

$= \mathbf{0}$

18. a. $\mathbf{v}\times\mathbf{w} = \begin{vmatrix} \mathbf{i} & \mathbf{j} & \mathbf{k} \\ 1 & -4 & 2 \\ 3 & 2 & 1 \end{vmatrix}$

$= \begin{vmatrix} -4 & 2 \\ 2 & 1 \end{vmatrix}\mathbf{i} - \begin{vmatrix} 1 & 2 \\ 3 & 1 \end{vmatrix}\mathbf{j} + \begin{vmatrix} 1 & -4 \\ 3 & 2 \end{vmatrix}\mathbf{k}$

$= -8\mathbf{i} + 5\mathbf{j} + 14\mathbf{k}$

b. $\mathbf{w}\times\mathbf{v} = \begin{vmatrix} \mathbf{i} & \mathbf{j} & \mathbf{k} \\ 3 & 2 & 1 \\ 1 & -4 & 2 \end{vmatrix}$

$= \begin{vmatrix} 2 & 1 \\ -4 & 2 \end{vmatrix}\mathbf{i} - \begin{vmatrix} 3 & 1 \\ 1 & 2 \end{vmatrix}\mathbf{j} + \begin{vmatrix} 3 & 2 \\ 1 & -4 \end{vmatrix}\mathbf{k}$

$= 8\mathbf{i} - 5\mathbf{j} - 14\mathbf{k}$

c. $\mathbf{w}\times\mathbf{w} = \begin{vmatrix} \mathbf{i} & \mathbf{j} & \mathbf{k} \\ 3 & 2 & 1 \\ 3 & 2 & 1 \end{vmatrix}$

$= \begin{vmatrix} 2 & 1 \\ 2 & 1 \end{vmatrix}\mathbf{i} - \begin{vmatrix} 3 & 1 \\ 3 & 1 \end{vmatrix}\mathbf{j} + \begin{vmatrix} 3 & 2 \\ 3 & 2 \end{vmatrix}\mathbf{k}$

$= 0\mathbf{i} + 0\mathbf{j} + 0\mathbf{k}$

$= \mathbf{0}$

d. $\mathbf{v}\times\mathbf{v} = \begin{vmatrix} \mathbf{i} & \mathbf{j} & \mathbf{k} \\ 1 & -4 & 2 \\ 1 & -4 & 2 \end{vmatrix}$

$= \begin{vmatrix} -4 & 2 \\ -4 & 2 \end{vmatrix}\mathbf{i} - \begin{vmatrix} 1 & 2 \\ 1 & 2 \end{vmatrix}\mathbf{j} + \begin{vmatrix} 1 & -4 \\ 1 & -4 \end{vmatrix}\mathbf{k}$

$= 0\mathbf{i} + 0\mathbf{j} + 0\mathbf{k}$

$= \mathbf{0}$

19. a. $\mathbf{v}\times\mathbf{w} = \begin{vmatrix} \mathbf{i} & \mathbf{j} & \mathbf{k} \\ 2 & -1 & 2 \\ 0 & 1 & -1 \end{vmatrix}$

$= \begin{vmatrix} -1 & 2 \\ 1 & -1 \end{vmatrix}\mathbf{i} - \begin{vmatrix} 2 & 2 \\ 0 & -1 \end{vmatrix}\mathbf{j} + \begin{vmatrix} 2 & -1 \\ 0 & 1 \end{vmatrix}\mathbf{k}$

$= -1\mathbf{i} + 2\mathbf{j} + 2\mathbf{k}$

b. $\mathbf{w}\times\mathbf{v} = \begin{vmatrix} \mathbf{i} & \mathbf{j} & \mathbf{k} \\ 0 & 1 & -1 \\ 2 & -1 & 2 \end{vmatrix}$

$= \begin{vmatrix} 1 & -1 \\ -1 & 2 \end{vmatrix}\mathbf{i} - \begin{vmatrix} 0 & -1 \\ 2 & 2 \end{vmatrix}\mathbf{j} + \begin{vmatrix} 0 & 1 \\ 2 & -1 \end{vmatrix}\mathbf{k}$

$= 1\mathbf{i} - 2\mathbf{j} - 2\mathbf{k}$

c. $\mathbf{w}\times\mathbf{w}=\begin{vmatrix}\mathbf{i} & \mathbf{j} & \mathbf{k}\\ 0 & 1 & -1\\ 0 & 1 & -1\end{vmatrix}$

$=\begin{vmatrix}1 & -1\\ 1 & -1\end{vmatrix}\mathbf{i}-\begin{vmatrix}0 & -1\\ 0 & -1\end{vmatrix}\mathbf{j}+\begin{vmatrix}0 & 1\\ 0 & 1\end{vmatrix}\mathbf{k}$

$=0\mathbf{i}+0\mathbf{j}+0\mathbf{k}$

$=\mathbf{0}$

d. $\mathbf{v}\times\mathbf{v}=\begin{vmatrix}\mathbf{i} & \mathbf{j} & \mathbf{k}\\ 2 & -1 & 2\\ 2 & -1 & 2\end{vmatrix}$

$=\begin{vmatrix}-1 & 2\\ -1 & 2\end{vmatrix}\mathbf{i}-\begin{vmatrix}2 & 2\\ 2 & 2\end{vmatrix}\mathbf{j}+\begin{vmatrix}2 & -1\\ 2 & -1\end{vmatrix}\mathbf{k}$

$=0\mathbf{i}+0\mathbf{j}+0\mathbf{k}$

$=\mathbf{0}$

20. a. $\mathbf{v}\times\mathbf{w}=\begin{vmatrix}\mathbf{i} & \mathbf{j} & \mathbf{k}\\ 3 & 1 & 3\\ 1 & 0 & -1\end{vmatrix}$

$=\begin{vmatrix}1 & 3\\ 0 & -1\end{vmatrix}\mathbf{i}-\begin{vmatrix}3 & 3\\ 1 & -1\end{vmatrix}\mathbf{j}+\begin{vmatrix}3 & 1\\ 1 & 0\end{vmatrix}\mathbf{k}$

$=-\mathbf{i}+6\mathbf{j}-\mathbf{k}$

b. $\mathbf{w}\times\mathbf{v}=\begin{vmatrix}\mathbf{i} & \mathbf{j} & \mathbf{k}\\ 1 & 0 & -1\\ 3 & 1 & 3\end{vmatrix}$

$=\begin{vmatrix}0 & -1\\ 1 & 3\end{vmatrix}\mathbf{i}-\begin{vmatrix}1 & -1\\ 3 & 3\end{vmatrix}\mathbf{j}+\begin{vmatrix}1 & 0\\ 3 & 1\end{vmatrix}\mathbf{k}$

$=\mathbf{i}-6\mathbf{j}+\mathbf{k}$

c. $\mathbf{w}\times\mathbf{w}=\begin{vmatrix}\mathbf{i} & \mathbf{j} & \mathbf{k}\\ 1 & 0 & -1\\ 1 & 0 & -1\end{vmatrix}$

$=\begin{vmatrix}0 & -1\\ 0 & -1\end{vmatrix}\mathbf{i}-\begin{vmatrix}1 & -1\\ 1 & -1\end{vmatrix}\mathbf{j}+\begin{vmatrix}1 & 0\\ 1 & 0\end{vmatrix}\mathbf{k}$

$=0\mathbf{i}+0\mathbf{j}+0\mathbf{k}=\mathbf{0}$

d. $\mathbf{v}\times\mathbf{v}=\begin{vmatrix}\mathbf{i} & \mathbf{j} & \mathbf{k}\\ 3 & 1 & 3\\ 3 & 1 & 3\end{vmatrix}$

$=\begin{vmatrix}1 & 3\\ 1 & 3\end{vmatrix}\mathbf{i}-\begin{vmatrix}3 & 3\\ 3 & 3\end{vmatrix}\mathbf{j}+\begin{vmatrix}3 & 1\\ 3 & 1\end{vmatrix}\mathbf{k}$

$=0\mathbf{i}+0\mathbf{j}+0\mathbf{k}=\mathbf{0}$

21. a. $\mathbf{v}\times\mathbf{w}=\begin{vmatrix}\mathbf{i} & \mathbf{j} & \mathbf{k}\\ 1 & -1 & -1\\ 4 & 0 & -3\end{vmatrix}$

$=\begin{vmatrix}-1 & -1\\ 0 & -3\end{vmatrix}\mathbf{i}-\begin{vmatrix}1 & -1\\ 4 & -3\end{vmatrix}\mathbf{j}+\begin{vmatrix}1 & -1\\ 4 & 0\end{vmatrix}\mathbf{k}$

$=3\mathbf{i}-1\mathbf{j}+4\mathbf{k}$

b. $\mathbf{w}\times\mathbf{v}=\begin{vmatrix}\mathbf{i} & \mathbf{j} & \mathbf{k}\\ 4 & 0 & -3\\ 1 & -1 & -1\end{vmatrix}$

$=\begin{vmatrix}0 & -3\\ -1 & -1\end{vmatrix}\mathbf{i}-\begin{vmatrix}4 & -3\\ 1 & -1\end{vmatrix}\mathbf{j}+\begin{vmatrix}4 & 0\\ 1 & -1\end{vmatrix}\mathbf{k}$

$=-3\mathbf{i}+1\mathbf{j}-4\mathbf{k}$

c. $\mathbf{w}\times\mathbf{w}=\begin{vmatrix}\mathbf{i} & \mathbf{j} & \mathbf{k}\\ 4 & 0 & -3\\ 4 & 0 & -3\end{vmatrix}$

$=\begin{vmatrix}0 & -3\\ 0 & -3\end{vmatrix}\mathbf{i}-\begin{vmatrix}4 & -3\\ 4 & -3\end{vmatrix}\mathbf{j}+\begin{vmatrix}4 & 0\\ 4 & 0\end{vmatrix}\mathbf{k}$

$=0\mathbf{i}+0\mathbf{j}+0\mathbf{k}$

$=\mathbf{0}$

d. $\mathbf{v}\times\mathbf{v}=\begin{vmatrix}\mathbf{i} & \mathbf{j} & \mathbf{k}\\ 1 & -1 & -1\\ 1 & -1 & -1\end{vmatrix}$

$=\begin{vmatrix}-1 & -1\\ -1 & -1\end{vmatrix}\mathbf{i}-\begin{vmatrix}1 & -1\\ 1 & -1\end{vmatrix}\mathbf{j}+\begin{vmatrix}1 & -1\\ 1 & -1\end{vmatrix}\mathbf{k}$

$=0\mathbf{i}+0\mathbf{j}+0\mathbf{k}$

$=\mathbf{0}$

22. a. $\mathbf{v}\times\mathbf{w}=\begin{vmatrix}\mathbf{i} & \mathbf{j} & \mathbf{k}\\ 2 & -3 & 0\\ 0 & 3 & -2\end{vmatrix}$

$=\begin{vmatrix}-3 & 0\\ 3 & -2\end{vmatrix}\mathbf{i}-\begin{vmatrix}2 & 0\\ 0 & -2\end{vmatrix}\mathbf{j}+\begin{vmatrix}2 & -3\\ 0 & 3\end{vmatrix}\mathbf{k}$

$=6\mathbf{i}+4\mathbf{j}+6\mathbf{k}$

b. $\mathbf{w}\times\mathbf{v}=\begin{vmatrix}\mathbf{i} & \mathbf{j} & \mathbf{k}\\ 0 & 3 & -2\\ 2 & -3 & 0\end{vmatrix}$

$=\begin{vmatrix}3 & -2\\ -3 & 0\end{vmatrix}\mathbf{i}-\begin{vmatrix}0 & -2\\ 2 & 0\end{vmatrix}\mathbf{j}+\begin{vmatrix}0 & 3\\ 2 & -3\end{vmatrix}\mathbf{k}$

$=-6\mathbf{i}-4\mathbf{j}-6\mathbf{k}$

c. $\mathbf{w}\times\mathbf{w}=\begin{vmatrix}\mathbf{i} & \mathbf{j} & \mathbf{k}\\ 0 & 3 & -2\\ 0 & 3 & -2\end{vmatrix}$

$=\begin{vmatrix}3 & -2\\ 3 & -2\end{vmatrix}\mathbf{i}-\begin{vmatrix}0 & -2\\ 0 & -2\end{vmatrix}\mathbf{j}+\begin{vmatrix}0 & 3\\ 0 & 3\end{vmatrix}\mathbf{k}$

$=0\mathbf{i}+0\mathbf{j}+0\mathbf{k}$

$=\mathbf{0}$

d. $\mathbf{v}\times\mathbf{v}=\begin{vmatrix}\mathbf{i} & \mathbf{j} & \mathbf{k}\\ 2 & -3 & 0\\ 2 & -3 & 0\end{vmatrix}$

$=\begin{vmatrix}-3 & 0\\ -3 & 0\end{vmatrix}\mathbf{i}-\begin{vmatrix}2 & 0\\ 2 & 0\end{vmatrix}\mathbf{j}+\begin{vmatrix}2 & -3\\ 2 & -3\end{vmatrix}\mathbf{k}$

$=0\mathbf{i}+0\mathbf{j}+0\mathbf{k}$

$=\mathbf{0}$

23. $\mathbf{u}\times\mathbf{v}=\begin{vmatrix}\mathbf{i} & \mathbf{j} & \mathbf{k}\\ 2 & -3 & 1\\ -3 & 3 & 2\end{vmatrix}$

$=\begin{vmatrix}-3 & 1\\ 3 & 2\end{vmatrix}\mathbf{i}-\begin{vmatrix}2 & 1\\ -3 & 2\end{vmatrix}\mathbf{j}+\begin{vmatrix}2 & -3\\ -3 & 3\end{vmatrix}\mathbf{k}$

$=-9\mathbf{i}-7\mathbf{j}-3\mathbf{k}$

24. $\mathbf{v}\times\mathbf{w}=\begin{vmatrix}\mathbf{i} & \mathbf{j} & \mathbf{k}\\ -3 & 3 & 2\\ 1 & 1 & 3\end{vmatrix}$

$=\begin{vmatrix}3 & 2\\ 1 & 3\end{vmatrix}\mathbf{i}-\begin{vmatrix}-3 & 2\\ 1 & 3\end{vmatrix}\mathbf{j}+\begin{vmatrix}-3 & 3\\ 1 & 1\end{vmatrix}\mathbf{k}$

$=7\mathbf{i}+11\mathbf{j}-6\mathbf{k}$

25. $\mathbf{v}\times\mathbf{u}=\begin{vmatrix}\mathbf{i} & \mathbf{j} & \mathbf{k}\\ -3 & 3 & 2\\ 2 & -3 & 1\end{vmatrix}$

$=\begin{vmatrix}3 & 2\\ -3 & 1\end{vmatrix}\mathbf{i}-\begin{vmatrix}-3 & 2\\ 2 & 1\end{vmatrix}\mathbf{j}+\begin{vmatrix}-3 & 3\\ 2 & -3\end{vmatrix}\mathbf{k}$

$=9\mathbf{i}+7\mathbf{j}+3\mathbf{k}$

26. $\mathbf{w}\times\mathbf{v}=\begin{vmatrix}\mathbf{i} & \mathbf{j} & \mathbf{k}\\ 1 & 1 & 3\\ -3 & 3 & 2\end{vmatrix}$

$=\begin{vmatrix}1 & 3\\ 3 & 2\end{vmatrix}\mathbf{i}-\begin{vmatrix}1 & 3\\ -3 & 2\end{vmatrix}\mathbf{j}+\begin{vmatrix}1 & 1\\ -3 & 3\end{vmatrix}\mathbf{k}$

$=-7\mathbf{i}-11\mathbf{j}+6\mathbf{k}$

27. $\mathbf{v}\times\mathbf{v}=\begin{vmatrix}\mathbf{i} & \mathbf{j} & \mathbf{k}\\ -3 & 3 & 2\\ -3 & 3 & 2\end{vmatrix}$

$=\begin{vmatrix}3 & 2\\ 3 & 2\end{vmatrix}\mathbf{i}-\begin{vmatrix}-3 & 2\\ -3 & 2\end{vmatrix}\mathbf{j}+\begin{vmatrix}-3 & 3\\ -3 & 3\end{vmatrix}\mathbf{k}$

$=0\mathbf{i}+0\mathbf{j}+0\mathbf{k}$

$=\mathbf{0}$

28. $\mathbf{w}\times\mathbf{w}=\begin{vmatrix}\mathbf{i} & \mathbf{j} & \mathbf{k}\\ 1 & 1 & 3\\ 1 & 1 & 3\end{vmatrix}$

$=\begin{vmatrix}1 & 3\\ 1 & 3\end{vmatrix}\mathbf{i}-\begin{vmatrix}1 & 3\\ 1 & 3\end{vmatrix}\mathbf{j}+\begin{vmatrix}1 & 1\\ 1 & 1\end{vmatrix}\mathbf{k}$

$=0\mathbf{i}+0\mathbf{j}+0\mathbf{k}$

$=\mathbf{0}$

29. $(3\mathbf{u})\times\mathbf{v}=\begin{vmatrix}\mathbf{i} & \mathbf{j} & \mathbf{k}\\ 6 & -9 & 3\\ -3 & 3 & 2\end{vmatrix}$

$=\begin{vmatrix}-9 & 3\\ 3 & 2\end{vmatrix}\mathbf{i}-\begin{vmatrix}6 & 3\\ -3 & 2\end{vmatrix}\mathbf{j}+\begin{vmatrix}6 & -9\\ -3 & 3\end{vmatrix}\mathbf{k}$

$=-27\mathbf{i}-21\mathbf{j}-9\mathbf{k}$

30. $\mathbf{v}\times(4\mathbf{w})=\begin{vmatrix}\mathbf{i} & \mathbf{j} & \mathbf{k}\\ -3 & 3 & 2\\ 4 & 4 & 12\end{vmatrix}$

$=\begin{vmatrix}3 & 2\\ 4 & 12\end{vmatrix}\mathbf{i}-\begin{vmatrix}-3 & 2\\ 4 & 12\end{vmatrix}\mathbf{j}+\begin{vmatrix}-3 & 3\\ 4 & 4\end{vmatrix}\mathbf{k}$

$=28\mathbf{i}+44\mathbf{j}-24\mathbf{k}$

31. $\mathbf{u}\times(2\mathbf{v})=\begin{vmatrix}\mathbf{i} & \mathbf{j} & \mathbf{k}\\ 2 & -3 & 1\\ -6 & 6 & 4\end{vmatrix}$

$=\begin{vmatrix}-3 & 1\\ 6 & 4\end{vmatrix}\mathbf{i}-\begin{vmatrix}2 & 1\\ -6 & 4\end{vmatrix}\mathbf{j}+\begin{vmatrix}2 & -3\\ -6 & 6\end{vmatrix}\mathbf{k}$

$=-18\mathbf{i}-14\mathbf{j}-6\mathbf{k}$

32. $(-3\mathbf{v})\times\mathbf{w}=\begin{vmatrix}\mathbf{i} & \mathbf{j} & \mathbf{k}\\ 9 & -9 & -6\\ 1 & 1 & 3\end{vmatrix}$

$=\begin{vmatrix}-9 & -6\\ 1 & 3\end{vmatrix}\mathbf{i}-\begin{vmatrix}9 & -6\\ 1 & 3\end{vmatrix}\mathbf{j}+\begin{vmatrix}9 & -9\\ 1 & 1\end{vmatrix}\mathbf{k}$

$=-21\mathbf{i}-33\mathbf{j}+18\mathbf{k}$

33. $\mathbf{u}\bullet(\mathbf{u}\times\mathbf{v})$

$$= \mathbf{u}\bullet\begin{vmatrix}\mathbf{i} & \mathbf{j} & \mathbf{k}\\ 2 & -3 & 1\\ -3 & 3 & 2\end{vmatrix}$$

$$= \mathbf{u}\bullet\left(\begin{vmatrix}-3 & 1\\ 3 & 2\end{vmatrix}\mathbf{i}-\begin{vmatrix}2 & 1\\ -3 & 2\end{vmatrix}\mathbf{j}+\begin{vmatrix}2 & -3\\ -3 & 3\end{vmatrix}\mathbf{k}\right)$$

$$= (2\mathbf{i}-3\mathbf{j}+\mathbf{k})\bullet(-9\mathbf{i}-7\mathbf{j}-3\mathbf{k})$$
$$= 2(-9)+(-3)(-7)+1(-3)$$
$$= -18+21-3$$
$$= 0$$

34. $\mathbf{v}\bullet(\mathbf{v}\times\mathbf{w}) = \mathbf{v}\bullet\begin{vmatrix}\mathbf{i} & \mathbf{j} & \mathbf{k}\\ -3 & 3 & 2\\ 1 & 1 & 3\end{vmatrix}$

$$= \mathbf{v}\bullet\left(\begin{vmatrix}3 & 2\\ 1 & 3\end{vmatrix}\mathbf{i}-\begin{vmatrix}-3 & 2\\ 1 & 3\end{vmatrix}\mathbf{j}+\begin{vmatrix}-3 & 3\\ 1 & 1\end{vmatrix}\mathbf{k}\right)$$

$$= (-3\mathbf{i}+3\mathbf{j}+2\mathbf{k})\bullet(7\mathbf{i}+11\mathbf{j}-6\mathbf{k})$$
$$= -3\cdot 7+3(11)+2(-6)$$
$$= -21+33-12$$
$$= 0$$

35. $\mathbf{u}\bullet(\mathbf{v}\times\mathbf{w}) = \mathbf{u}\bullet\begin{vmatrix}\mathbf{i} & \mathbf{j} & \mathbf{k}\\ -3 & 3 & 2\\ 1 & 1 & 3\end{vmatrix}$

$$= \mathbf{u}\bullet\left(\begin{vmatrix}3 & 2\\ 1 & 3\end{vmatrix}\mathbf{i}-\begin{vmatrix}-3 & 2\\ 1 & 3\end{vmatrix}\mathbf{j}+\begin{vmatrix}-3 & 3\\ 1 & 1\end{vmatrix}\mathbf{k}\right)$$

$$= (2\mathbf{i}-3\mathbf{j}+\mathbf{k})\bullet(7\mathbf{i}+11\mathbf{j}-6\mathbf{k})$$
$$= 2\cdot 7+(-3)(11)+1(-6)$$
$$= 14-33-6$$
$$= -25$$

36. $(\mathbf{u}\times\mathbf{v})\bullet\mathbf{w}$

$$= \begin{vmatrix}\mathbf{i} & \mathbf{j} & \mathbf{k}\\ 2 & -3 & 1\\ -3 & 3 & 2\end{vmatrix}\bullet\mathbf{w}$$

$$= \left(\begin{vmatrix}-3 & 1\\ 3 & 2\end{vmatrix}\mathbf{i}-\begin{vmatrix}2 & 1\\ -3 & 2\end{vmatrix}\mathbf{j}+\begin{vmatrix}2 & -3\\ -3 & 3\end{vmatrix}\mathbf{k}\right)\bullet\mathbf{w}$$

$$= (-9\mathbf{i}-7\mathbf{j}-3\mathbf{k})\bullet(\mathbf{i}+\mathbf{j}+3\mathbf{k})$$
$$= -9\cdot 1+(-7)(1)+(-3)(3)$$
$$= -9-7-9$$
$$= -25$$

37. $\mathbf{v}\bullet(\mathbf{u}\times\mathbf{w}) = \mathbf{v}\bullet\begin{vmatrix}\mathbf{i} & \mathbf{j} & \mathbf{k}\\ 2 & -3 & 1\\ 1 & 1 & 3\end{vmatrix}$

$$= \mathbf{v}\bullet\left(\begin{vmatrix}-3 & 1\\ 1 & 3\end{vmatrix}\mathbf{i}-\begin{vmatrix}2 & 1\\ 1 & 3\end{vmatrix}\mathbf{j}+\begin{vmatrix}2 & -3\\ 1 & 1\end{vmatrix}\mathbf{k}\right)$$

$$= (-3\mathbf{i}+3\mathbf{j}+2\mathbf{k})\bullet(-10\mathbf{i}-5\mathbf{j}+5\mathbf{k})$$
$$= -3(-10)+3(-5)+2\cdot 5$$
$$= 30-15+10$$
$$= 25$$

38. $(\mathbf{v}\times\mathbf{u})\bullet\mathbf{w}$

$$= \begin{vmatrix}\mathbf{i} & \mathbf{j} & \mathbf{k}\\ -3 & 3 & 2\\ 2 & -3 & 1\end{vmatrix}\bullet\mathbf{w}$$

$$= \left(\begin{vmatrix}3 & 2\\ -3 & 1\end{vmatrix}\mathbf{i}-\begin{vmatrix}-3 & 2\\ 2 & 1\end{vmatrix}\mathbf{j}+\begin{vmatrix}-3 & 3\\ 2 & -3\end{vmatrix}\mathbf{k}\right)\bullet\mathbf{w}$$

$$= (9\mathbf{i}+7\mathbf{j}+3\mathbf{k})\bullet(\mathbf{i}+\mathbf{j}+3\mathbf{k})$$
$$= 9\cdot 1+7\cdot 1+3\cdot 3$$
$$= 9+7+9$$
$$= 25$$

39. $\mathbf{u}\times(\mathbf{v}\times\mathbf{v}) = \mathbf{u}\times\begin{vmatrix}\mathbf{i} & \mathbf{j} & \mathbf{k}\\ -3 & 3 & 2\\ -3 & 3 & 2\end{vmatrix}$

$$= \mathbf{u}\times\left(\begin{vmatrix}3 & 2\\ 3 & 2\end{vmatrix}\mathbf{i}-\begin{vmatrix}-3 & 2\\ -3 & 2\end{vmatrix}\mathbf{j}+\begin{vmatrix}-3 & 3\\ -3 & 3\end{vmatrix}\mathbf{k}\right)$$

$$= (2\mathbf{i}-3\mathbf{j}+\mathbf{k})\times(0\mathbf{i}+0\mathbf{j}+0\mathbf{k})$$

$$= \begin{vmatrix}\mathbf{i} & \mathbf{j} & \mathbf{k}\\ 2 & -3 & 1\\ 0 & 0 & 0\end{vmatrix}$$

$$= \begin{vmatrix}-3 & 1\\ 0 & 0\end{vmatrix}\mathbf{i}-\begin{vmatrix}2 & 1\\ 0 & 0\end{vmatrix}\mathbf{j}+\begin{vmatrix}2 & -3\\ 0 & 0\end{vmatrix}\mathbf{k}$$

$$= 0\mathbf{i}+0\mathbf{j}+0\mathbf{k}$$
$$= \mathbf{0}$$

40. $(\mathbf{w}\times\mathbf{w})\times\mathbf{v}=\begin{vmatrix}\mathbf{i} & \mathbf{j} & \mathbf{k}\\ 1 & 1 & 3\\ 1 & 1 & 3\end{vmatrix}\times\mathbf{v}$

$=\left(\begin{vmatrix}1 & 3\\ 1 & 3\end{vmatrix}\mathbf{i}-\begin{vmatrix}1 & 3\\ 1 & 3\end{vmatrix}\mathbf{j}+\begin{vmatrix}1 & 1\\ 1 & 1\end{vmatrix}\mathbf{k}\right)\times\mathbf{v}$

$=(0\mathbf{i}+0\mathbf{j}+0\mathbf{k})\times\mathbf{v}$

$=\begin{vmatrix}\mathbf{i} & \mathbf{j} & \mathbf{k}\\ 0 & 0 & 0\\ -3 & 3 & 2\end{vmatrix}$

$=\begin{vmatrix}0 & 0\\ 3 & 2\end{vmatrix}\mathbf{i}-\begin{vmatrix}0 & 0\\ -3 & 2\end{vmatrix}\mathbf{j}+\begin{vmatrix}0 & 0\\ -3 & 3\end{vmatrix}\mathbf{k}$

$=0\mathbf{i}+0\mathbf{j}+0\mathbf{k}$

$=\mathbf{0}$

41. $\mathbf{u}\times\mathbf{v}=\begin{vmatrix}\mathbf{i} & \mathbf{j} & \mathbf{k}\\ 2 & -3 & 1\\ -3 & 3 & 2\end{vmatrix}$

$=\begin{vmatrix}-3 & 1\\ 3 & 2\end{vmatrix}\mathbf{i}-\begin{vmatrix}2 & 1\\ -3 & 2\end{vmatrix}\mathbf{j}+\begin{vmatrix}2 & -3\\ -3 & 3\end{vmatrix}\mathbf{k}$

$=-9\mathbf{i}-7\mathbf{j}-3\mathbf{k}$

Actually, any vector of the form $c(-9\mathbf{i}-7\mathbf{j}-3\mathbf{k})$, where c is a nonzero scalar, is orthogonal to both $\mathbf{u}$ and $\mathbf{v}$.

42. $\mathbf{u}\times\mathbf{w}=\begin{vmatrix}\mathbf{i} & \mathbf{j} & \mathbf{k}\\ 2 & -3 & 1\\ 1 & 1 & 3\end{vmatrix}$

$=\begin{vmatrix}-3 & 1\\ 1 & 3\end{vmatrix}\mathbf{i}-\begin{vmatrix}2 & 1\\ 1 & 3\end{vmatrix}\mathbf{j}+\begin{vmatrix}2 & -3\\ 1 & 1\end{vmatrix}\mathbf{k}$

$=-10\mathbf{i}-5\mathbf{j}+5\mathbf{k}$

Actually, any vector of the form $c(-10\mathbf{i}-5\mathbf{j}+5\mathbf{k})$, where c is a nonzero scalar, is orthogonal to both $\mathbf{u}$ and $\mathbf{w}$.

43. A vector that is orthogonal to both $\mathbf{u}$ and $\mathbf{i}+\mathbf{j}$ is $\mathbf{u}\times(\mathbf{i}+\mathbf{j})$.

$\mathbf{u}\times(\mathbf{i}+\mathbf{j})=\begin{vmatrix}\mathbf{i} & \mathbf{j} & \mathbf{k}\\ 2 & -3 & 1\\ 1 & 1 & 0\end{vmatrix}$

$=\begin{vmatrix}-3 & 1\\ 1 & 0\end{vmatrix}\mathbf{i}-\begin{vmatrix}2 & 1\\ 1 & 0\end{vmatrix}\mathbf{j}+\begin{vmatrix}2 & -3\\ 1 & 1\end{vmatrix}\mathbf{k}$

$=-1\mathbf{i}+1\mathbf{j}+5\mathbf{k}$

Actually, any vector of the form $c(-\mathbf{i}+\mathbf{j}+5\mathbf{k})$, where c is a nonzero scalar, is orthogonal to both $\mathbf{u}$ and $\mathbf{i}+\mathbf{j}$

44. A vector that is orthogonal to both $\mathbf{u}$ and $\mathbf{j}+\mathbf{k}$ is $\mathbf{u}\times(\mathbf{j}+\mathbf{k})$.

$\mathbf{u}\times(\mathbf{j}+\mathbf{k})=\begin{vmatrix}\mathbf{i} & \mathbf{j} & \mathbf{k}\\ 2 & -3 & 1\\ 0 & 1 & 1\end{vmatrix}$

$=\begin{vmatrix}-3 & 1\\ 1 & 1\end{vmatrix}\mathbf{i}-\begin{vmatrix}2 & 1\\ 0 & 1\end{vmatrix}\mathbf{j}+\begin{vmatrix}2 & -3\\ 0 & 1\end{vmatrix}\mathbf{k}$

$=-4\mathbf{i}-2\mathbf{j}+2\mathbf{k}$

Actually, any vector of the form $c(-4\mathbf{i}-2\mathbf{j}+2\mathbf{k})$, where c is a nonzero scalar, is orthogonal to both $\mathbf{u}$ and $\mathbf{j}+\mathbf{k}$.

45. $\mathbf{u}=P_1P_2=1\mathbf{i}+2\mathbf{j}+3\mathbf{k}$

$\mathbf{v}=P_1P_3=-2\mathbf{i}+3\mathbf{j}+0\mathbf{k}$

$\mathbf{u}\times\mathbf{v}=\begin{vmatrix}\mathbf{i} & \mathbf{j} & \mathbf{k}\\ 1 & 2 & 3\\ -2 & 3 & 0\end{vmatrix}$

$=\begin{vmatrix}2 & 3\\ 3 & 0\end{vmatrix}\mathbf{i}-\begin{vmatrix}1 & 3\\ -2 & 0\end{vmatrix}\mathbf{j}+\begin{vmatrix}1 & 2\\ -2 & 3\end{vmatrix}\mathbf{k}$

$=-9\mathbf{i}-6\mathbf{j}+7\mathbf{k}$

$\text{Area}=\|\mathbf{u}\times\mathbf{v}\|=\sqrt{(-9)^2+(-6)^2+7^2}=\sqrt{166}$

46. $\mathbf{u}=P_1P_2=2\mathbf{i}+3\mathbf{j}+\mathbf{k}$

$\mathbf{v}=P_1P_3=-2\mathbf{i}+4\mathbf{j}+\mathbf{k}$

$\mathbf{u}\times\mathbf{v}=\begin{vmatrix}\mathbf{i} & \mathbf{j} & \mathbf{k}\\ 2 & 3 & 1\\ -2 & 4 & 1\end{vmatrix}$

$=\begin{vmatrix}3 & 1\\ 4 & 1\end{vmatrix}\mathbf{i}-\begin{vmatrix}2 & 1\\ -2 & 1\end{vmatrix}\mathbf{j}+\begin{vmatrix}2 & 3\\ -2 & 4\end{vmatrix}\mathbf{k}$

$=-\mathbf{i}-4\mathbf{j}+14\mathbf{k}$

$\text{Area}=\|\mathbf{u}\times\mathbf{v}\|=\sqrt{(-1)^2+(-4)^2+14^2}=\sqrt{213}$

47. $\mathbf{u} = P_1P_2 = -3\mathbf{i} + 1\mathbf{j} + 4\mathbf{k}$
$\mathbf{v} = P_1P_3 = -1\mathbf{i} - 4\mathbf{j} + 3\mathbf{k}$

$$\mathbf{u}\times\mathbf{v} = \begin{vmatrix} \mathbf{i} & \mathbf{j} & \mathbf{k} \\ -3 & 1 & 4 \\ -1 & -4 & 3 \end{vmatrix}$$
$$= \begin{vmatrix} 1 & 4 \\ -4 & 3 \end{vmatrix}\mathbf{i} - \begin{vmatrix} -3 & 4 \\ -1 & 3 \end{vmatrix}\mathbf{j} + \begin{vmatrix} -3 & 1 \\ -1 & -4 \end{vmatrix}\mathbf{k}$$
$$= 19\mathbf{i} + 5\mathbf{j} + 13\mathbf{k}$$
$$\text{Area} = \|\mathbf{u}\times\mathbf{v}\| = \sqrt{19^2 + 5^2 + 13^2} = \sqrt{555}$$

48. $\mathbf{u} = P_1P_2 = 4\mathbf{i} + \mathbf{j} - 3\mathbf{k}$
$\mathbf{v} = P_1P_3 = 4\mathbf{i} - 1\mathbf{j} + 0\mathbf{k}$

$$\mathbf{u}\times\mathbf{v} = \begin{vmatrix} \mathbf{i} & \mathbf{j} & \mathbf{k} \\ 4 & 1 & -3 \\ 4 & -1 & 0 \end{vmatrix}$$
$$= \begin{vmatrix} 1 & -3 \\ -1 & 0 \end{vmatrix}\mathbf{i} - \begin{vmatrix} 4 & -3 \\ 4 & 0 \end{vmatrix}\mathbf{j} + \begin{vmatrix} 4 & 1 \\ 4 & -1 \end{vmatrix}\mathbf{k}$$
$$= -3\mathbf{i} - 12\mathbf{j} - 8\mathbf{k}$$
$$\text{Area} = \|\mathbf{u}\times\mathbf{v}\|$$
$$= \sqrt{(-3)^2 + (-12)^2 + (-8)^2} = \sqrt{217}$$

49. $\mathbf{u} = P_1P_2 = 0\mathbf{i} + 1\mathbf{j} + 1\mathbf{k}$
$\mathbf{v} = P_1P_3 = -3\mathbf{i} + 2\mathbf{j} - 2\mathbf{k}$

$$\mathbf{u}\times\mathbf{v} = \begin{vmatrix} \mathbf{i} & \mathbf{j} & \mathbf{k} \\ 0 & 1 & 1 \\ -3 & 2 & -2 \end{vmatrix}$$
$$= \begin{vmatrix} 1 & 1 \\ 2 & -2 \end{vmatrix}\mathbf{i} - \begin{vmatrix} 0 & 1 \\ -3 & -2 \end{vmatrix}\mathbf{j} + \begin{vmatrix} 0 & 1 \\ -3 & 2 \end{vmatrix}\mathbf{k}$$
$$= -4\mathbf{i} - 3\mathbf{j} + 3\mathbf{k}$$
$$\text{Area} = \|\mathbf{u}\times\mathbf{v}\| = \sqrt{(-4)^2 + (-3)^2 + 3^2} = \sqrt{34}$$

50. $\mathbf{u} = P_1P_2 = 0\mathbf{i} + 2\mathbf{j} + 0\mathbf{k}$
$\mathbf{v} = P_1P_3 = -4\mathbf{i} + 3\mathbf{j} + 0\mathbf{k}$

$$\mathbf{u}\times\mathbf{v} = \begin{vmatrix} \mathbf{i} & \mathbf{j} & \mathbf{k} \\ 0 & 2 & 0 \\ -4 & 3 & 0 \end{vmatrix}$$
$$= \begin{vmatrix} 2 & 0 \\ 3 & 0 \end{vmatrix}\mathbf{i} - \begin{vmatrix} 0 & 0 \\ -4 & 0 \end{vmatrix}\mathbf{j} + \begin{vmatrix} 0 & 2 \\ -4 & 3 \end{vmatrix}\mathbf{k}$$
$$= 0\mathbf{i} + 0\mathbf{j} + 8\mathbf{k}$$
$$\text{Area} = \|\mathbf{u}\times\mathbf{v}\| = \sqrt{0^2 + 0^2 + 8^2} = \sqrt{64} = 8$$

51. $\mathbf{u} = P_1P_2 = 3\mathbf{i} + 0\mathbf{j} - 2\mathbf{k}$
$\mathbf{v} = P_1P_3 = 5\mathbf{i} - 7\mathbf{j} + 3\mathbf{k}$

$$\mathbf{u}\times\mathbf{v} = \begin{vmatrix} \mathbf{i} & \mathbf{j} & \mathbf{k} \\ 3 & 0 & -2 \\ 5 & -7 & 3 \end{vmatrix}$$
$$= \begin{vmatrix} 0 & -2 \\ -7 & 3 \end{vmatrix}\mathbf{i} - \begin{vmatrix} 3 & -2 \\ 5 & 3 \end{vmatrix}\mathbf{j} + \begin{vmatrix} 3 & 0 \\ 5 & -7 \end{vmatrix}\mathbf{k}$$
$$= -14\mathbf{i} - 19\mathbf{j} - 21\mathbf{k}$$
$$\text{Area} = \|\mathbf{u}\times\mathbf{v}\|$$
$$= \sqrt{(-14)^2 + (-19)^2 + (-21)^2} = \sqrt{998}$$

52. $\mathbf{u} = P_1P_2 = 0\mathbf{i} + 1\mathbf{j} + 1\mathbf{k}$
$\mathbf{v} = P_1P_3 = -2\mathbf{i} + 3\mathbf{j} - 6\mathbf{k}$

$$\mathbf{u}\times\mathbf{v} = \begin{vmatrix} \mathbf{i} & \mathbf{j} & \mathbf{k} \\ 0 & 1 & 1 \\ -2 & 3 & -6 \end{vmatrix}$$
$$= \begin{vmatrix} 1 & 1 \\ 3 & -6 \end{vmatrix}\mathbf{i} - \begin{vmatrix} 0 & 1 \\ -2 & -6 \end{vmatrix}\mathbf{j} + \begin{vmatrix} 0 & 1 \\ -2 & 3 \end{vmatrix}\mathbf{k}$$
$$= -9\mathbf{i} - 2\mathbf{j} + 2\mathbf{k}$$
$$\text{Area} = \|\mathbf{u}\times\mathbf{v}\| = \sqrt{(-9)^2 + (-2)^2 + 2^2} = \sqrt{89}$$

53. $$\mathbf{v}\times\mathbf{w} = \begin{vmatrix} \mathbf{i} & \mathbf{j} & \mathbf{k} \\ 1 & 3 & -2 \\ -2 & 1 & 3 \end{vmatrix}$$
$$= \begin{vmatrix} 3 & -2 \\ 1 & 3 \end{vmatrix}\mathbf{i} - \begin{vmatrix} 1 & -2 \\ -2 & 3 \end{vmatrix}\mathbf{j} + \begin{vmatrix} 1 & 3 \\ -2 & 1 \end{vmatrix}\mathbf{k}$$
$$= 11\mathbf{i} + 1\mathbf{j} + 7\mathbf{k}$$
$$\|\mathbf{v}\times\mathbf{w}\| = \sqrt{11^2 + 1^2 + 7^2} = \sqrt{171}$$
$$\mathbf{u} = \pm\frac{\mathbf{v}\times\mathbf{w}}{\|\mathbf{v}\times\mathbf{w}\|}$$
$$= \pm\frac{11\mathbf{i} + 1\mathbf{j} + 7\mathbf{k}}{\sqrt{171}}$$
$$= \pm\left(\frac{11}{\sqrt{171}}\mathbf{i} + \frac{1}{\sqrt{171}}\mathbf{j} + \frac{7}{\sqrt{171}}\mathbf{k}\right)$$
$$= \pm\left(\frac{11\sqrt{19}}{57}\mathbf{i} + \frac{\sqrt{19}}{57}\mathbf{j} + \frac{7\sqrt{19}}{57}\mathbf{k}\right)$$

So, $\frac{11\sqrt{19}}{57}\mathbf{i} + \frac{\sqrt{19}}{57}\mathbf{j} + \frac{7\sqrt{19}}{57}\mathbf{k}$ or
$-\frac{11\sqrt{19}}{57}\mathbf{i} - \frac{\sqrt{19}}{57}\mathbf{j} - \frac{7\sqrt{19}}{57}\mathbf{k}$.

54. $\mathbf{v}\times\mathbf{w}=\begin{vmatrix}\mathbf{i}&\mathbf{j}&\mathbf{k}\\2&3&-1\\-2&-4&-3\end{vmatrix}$

$=\begin{vmatrix}3&-1\\-4&-3\end{vmatrix}\mathbf{i}-\begin{vmatrix}2&-1\\-2&-3\end{vmatrix}\mathbf{j}+\begin{vmatrix}2&3\\-2&-4\end{vmatrix}\mathbf{k}$

$=-13\mathbf{i}+8\mathbf{j}-2\mathbf{k}$

$\|\mathbf{v}\times\mathbf{w}\|=\sqrt{(-13)^2+8^2+(-2)^2}=\sqrt{237}$

$\mathbf{u}=\pm\dfrac{\mathbf{v}\times\mathbf{w}}{\|\mathbf{v}\times\mathbf{w}\|}$

$=\pm\dfrac{-13\mathbf{i}+8\mathbf{j}-2\mathbf{k}}{\sqrt{237}}$

$=\pm\left(-\dfrac{13}{\sqrt{237}}\mathbf{i}+\dfrac{8}{\sqrt{237}}\mathbf{j}-\dfrac{2}{\sqrt{237}}\mathbf{k}\right)$

$=\pm\left(-\dfrac{13\sqrt{237}}{237}\mathbf{i}+\dfrac{8\sqrt{237}}{237}\mathbf{j}-\dfrac{2\sqrt{237}}{237}\mathbf{k}\right)$

So, $-\dfrac{13\sqrt{237}}{237}\mathbf{i}+\dfrac{8\sqrt{237}}{237}\mathbf{j}-\dfrac{2\sqrt{237}}{237}\mathbf{k}$ or $\dfrac{13\sqrt{237}}{237}\mathbf{i}-\dfrac{8\sqrt{237}}{237}\mathbf{j}+\dfrac{2\sqrt{237}}{237}\mathbf{k}$.

55. Prove: $\mathbf{u}\times\mathbf{v}=-(\mathbf{v}\times\mathbf{u})$

Let $\mathbf{u}=a_1\mathbf{i}+b_1\mathbf{j}+c_1\mathbf{k}$ and $\mathbf{v}=a_2\mathbf{i}+b_2\mathbf{j}+c_2\mathbf{k}$

$\mathbf{u}\times\mathbf{v}$

$=\begin{vmatrix}\mathbf{i}&\mathbf{j}&\mathbf{k}\\a_1&b_1&c_1\\a_2&b_2&c_2\end{vmatrix}$

$=\begin{vmatrix}b_1&c_1\\b_2&c_2\end{vmatrix}\mathbf{i}-\begin{vmatrix}a_1&c_1\\a_2&c_2\end{vmatrix}\mathbf{j}+\begin{vmatrix}a_1&b_1\\a_2&b_2\end{vmatrix}\mathbf{k}$

$=(b_1c_2-b_2c_1)\mathbf{i}-(a_1c_2-a_2c_1)\mathbf{j}+(a_1b_2-a_2b_1)\mathbf{k}$

$=-(b_2c_1-b_1c_2)\mathbf{i}+(a_2c_1-a_1c_2)\mathbf{j}-(a_2b_1-a_1b_2)\mathbf{k}$

$=-\left((b_2c_1-b_1c_2)\mathbf{i}-(a_2c_1-a_1c_2)\mathbf{j}+(a_2b_1-a_1b_2)\mathbf{k}\right)$

$=-\left(\begin{vmatrix}b_2&c_2\\b_1&c_1\end{vmatrix}\mathbf{i}-\begin{vmatrix}a_2&c_2\\a_1&c_1\end{vmatrix}\mathbf{j}+\begin{vmatrix}a_2&b_2\\a_1&b_1\end{vmatrix}\mathbf{k}\right)$

$=-\begin{vmatrix}\mathbf{i}&\mathbf{j}&\mathbf{k}\\a_2&b_2&c_2\\a_1&b_1&c_1\end{vmatrix}$

$=-(\mathbf{v}\times\mathbf{u})$

56. Prove: $\mathbf{u}\times(\mathbf{v}+\mathbf{w})=(\mathbf{u}\times\mathbf{v})+(\mathbf{u}\times\mathbf{w})$

Let $\mathbf{u}=a_1\mathbf{i}+b_1\mathbf{j}+c_1\mathbf{k}$, $\mathbf{v}=a_2\mathbf{i}+b_2\mathbf{j}+c_2\mathbf{k}$, and $\mathbf{w}=a_3\mathbf{i}+b_3\mathbf{j}+c_3\mathbf{k}$.

$\mathbf{u}\times(\mathbf{v}+\mathbf{w})$

$=\mathbf{u}\times\left((a_2+a_3)\mathbf{i}+(b_2+b_3)\mathbf{j}+(c_2+c_3)\mathbf{k}\right)$

$=\begin{vmatrix}\mathbf{i}&\mathbf{j}&\mathbf{k}\\a_1&b_1&c_1\\a_2+a_3&b_2+b_3&c_2+c_3\end{vmatrix}$

$=\begin{vmatrix}b_1&c_1\\b_2+b_3&c_2+c_3\end{vmatrix}\mathbf{i}-\begin{vmatrix}a_1&c_1\\a_2+a_3&c_2+c_3\end{vmatrix}\mathbf{j}+\begin{vmatrix}a_1&b_1\\a_2+a_3&b_2+b_3\end{vmatrix}\mathbf{k}$

$=(b_1c_2+b_1c_3-b_2c_1-b_3c_1)\mathbf{i}-(a_1c_2+a_1c_3-a_2c_1-a_3c_1)\mathbf{j}+(a_1b_2+a_1b_3-a_2b_1-a_3b_1)\mathbf{k}$

$=(b_1c_2-b_2c_1)\mathbf{i}+(b_1c_3-b_3c_1)\mathbf{i}-(a_1c_2-a_2c_1)\mathbf{j}-(a_1c_3-a_3c_1)\mathbf{j}+(a_1b_2-a_2b_1)\mathbf{k}+(a_1b_3-a_3b_1)\mathbf{k}$

$=(b_1c_2-b_2c_1)\mathbf{i}-(a_1c_2-a_2c_1)\mathbf{j}+(a_1b_2-a_2b_1)\mathbf{k}+(b_1c_3-b_3c_1)\mathbf{i}-(a_1c_3-a_3c_1)\mathbf{j}+(a_1b_3-a_3b_1)\mathbf{k}$

$=\left(\begin{vmatrix}b_1&c_1\\b_2&c_2\end{vmatrix}\mathbf{i}-\begin{vmatrix}a_1&c_1\\a_2&c_2\end{vmatrix}\mathbf{j}+\begin{vmatrix}a_1&b_1\\a_2&b_2\end{vmatrix}\mathbf{k}\right)+\left(\begin{vmatrix}b_1&c_1\\b_3&c_3\end{vmatrix}\mathbf{i}-\begin{vmatrix}a_1&c_1\\a_3&c_3\end{vmatrix}\mathbf{j}+\begin{vmatrix}a_1&b_1\\a_3&b_3\end{vmatrix}\mathbf{k}\right)$

$=\begin{vmatrix}\mathbf{i}&\mathbf{j}&\mathbf{k}\\a_1&b_1&c_1\\a_2&b_2&c_2\end{vmatrix}+\begin{vmatrix}\mathbf{i}&\mathbf{j}&\mathbf{k}\\a_1&b_1&c_1\\a_3&b_3&c_3\end{vmatrix}$

$=(\mathbf{u}\times\mathbf{v})+(\mathbf{u}\times\mathbf{w})$

57. Prove: $\|\mathbf{u}\times\mathbf{v}\|^2 = \|\mathbf{u}\|^2\|\mathbf{v}\|^2 - (\mathbf{u}\bullet\mathbf{v})^2$

Let $\mathbf{u} = a_1\mathbf{i} + b_1\mathbf{j} + c_1\mathbf{k}$ and $\mathbf{v} = a_2\mathbf{i} + b_2\mathbf{j} + c_2\mathbf{k}$.

$$\mathbf{u}\times\mathbf{v} = \begin{vmatrix} \mathbf{i} & \mathbf{j} & \mathbf{k} \\ a_1 & b_1 & c_1 \\ a_2 & b_2 & c_2 \end{vmatrix} = \begin{vmatrix} b_1 & c_1 \\ b_2 & c_2 \end{vmatrix}\mathbf{i} - \begin{vmatrix} a_1 & c_1 \\ a_2 & c_2 \end{vmatrix}\mathbf{j} + \begin{vmatrix} a_1 & b_1 \\ a_2 & b_2 \end{vmatrix}\mathbf{k} = (b_1c_2 - b_2c_1)\mathbf{i} - (a_1c_2 - a_2c_1)\mathbf{j} + (a_1b_2 - a_2b_1)\mathbf{k}$$

$$\begin{aligned}\|\mathbf{u}\times\mathbf{v}\|^2 &= (b_1c_2 - b_2c_1)^2 + (a_1c_2 - a_2c_1)^2 + (a_1b_2 - a_2b_1)^2 \\ &= b_1^2c_2^2 - 2b_1b_2c_1c_2 + b_2^2c_1^2 + a_1^2c_2^2 - 2a_1a_2c_1c_2 + a_2^2c_1^2 + a_1^2b_2^2 - 2a_1a_2b_1b_2 + a_2^2b_1^2 \\ &= a_1^2b_2^2 + a_1^2c_2^2 + a_2^2b_1^2 + a_2^2c_1^2 + b_1^2c_2^2 + b_2^2c_1^2 - 2a_1a_2b_1b_2 - 2a_1a_2c_1c_2 - 2b_1b_2c_1c_2\end{aligned}$$

$$\|\mathbf{u}\|^2 = a_1^2 + b_1^2 + c_1^2$$

$$\|\mathbf{v}\|^2 = a_2^2 + b_2^2 + c_2^2$$

$$(\mathbf{u}\bullet\mathbf{v})^2 = (a_1a_2 + b_1b_2 + c_1c_2)^2$$

$$\begin{aligned}\|\mathbf{u}\|^2\|\mathbf{v}\|^2 - (\mathbf{u}\bullet\mathbf{v})^2 &= (a_1^2 + b_1^2 + c_1^2)(a_2^2 + b_2^2 + c_2^2) - (a_1a_2 + b_1b_2 + c_1c_2)^2 \\ &= a_1^2a_2^2 + a_1^2b_2^2 + a_1^2c_2^2 + b_1^2a_2^2 + b_1^2b_2^2 + b_1^2c_2^2 + c_1^2a_2^2 + c_1^2b_2^2 + c_1^2c_2^2 \\ &\quad - (a_1^2a_2^2 + a_1a_2b_1b_2 + a_1a_2c_1c_2 + a_1a_2b_1b_2 + b_1^2b_2^2 + b_1b_2c_1c_2 + a_1a_2c_1c_2 + b_1b_2c_1c_2 + c_1^2c_2^2) \\ &= a_1^2b_2^2 + a_1^2c_2^2 + a_2^2b_1^2 + a_2^2c_1^2 + b_1^2c_2^2 + b_2^2c_1^2 - 2a_1a_2b_1b_2 - 2a_1a_2c_1c_2 - 2b_1b_2c_1c_2\end{aligned}$$

58. From Problem 57, we have:

$$\begin{aligned}\|\mathbf{u}\times\mathbf{v}\|^2 &= \|\mathbf{u}\|^2\|\mathbf{v}\|^2 - (\mathbf{u}\bullet\mathbf{v})^2 \\ &= \|\mathbf{u}\|^2\|\mathbf{v}\|^2 - (\|\mathbf{u}\|\|\mathbf{v}\|\cos\theta)^2 \\ &= \|\mathbf{u}\|^2\|\mathbf{v}\|^2 - \|\mathbf{u}\|^2\|\mathbf{v}\|^2\cos^2\theta \\ &= \|\mathbf{u}\|^2\|\mathbf{v}\|^2(1 - \cos^2\theta) \\ &= \|\mathbf{u}\|^2\|\mathbf{v}\|^2\sin^2\theta \\ \|\mathbf{u}\times\mathbf{v}\| &= \|\mathbf{u}\|\|\mathbf{v}\|\sin\theta\end{aligned}$$

59. If $\mathbf{u}$ and $\mathbf{v}$ are orthogonal, then $\mathbf{u}\bullet\mathbf{v} = 0$. From problem 57, then:

$$\begin{aligned}\|\mathbf{u}\times\mathbf{v}\|^2 &= \|\mathbf{u}\|^2\|\mathbf{v}\|^2 - (\mathbf{u}\bullet\mathbf{v})^2 \\ &= \|\mathbf{u}\|^2\|\mathbf{v}\|^2 - (0)^2 \\ &= \|\mathbf{u}\|^2\|\mathbf{v}\|^2 \\ \|\mathbf{u}\times\mathbf{v}\| &= \|\mathbf{u}\|\|\mathbf{v}\|\end{aligned}$$

60. $\mathbf{u}$ and $\mathbf{v}$ are unit vectors. If $\mathbf{u}$ and $\mathbf{v}$ are orthogonal, the area of the parallelogram is 1, thus $\|\mathbf{u}\times\mathbf{v}\| = 1$ and $\mathbf{u}\times\mathbf{v}$ is a unit vector. If $\mathbf{u}$ and $\mathbf{v}$ are not orthogonal, the area of the parallelogram is less than 1, thus $\|\mathbf{u}\times\mathbf{v}\| < 1$ and $\mathbf{u}\times\mathbf{v}$ is not a unit vector.

61. $\mathbf{u}\bullet\mathbf{v} = 0 \Rightarrow$ $\mathbf{u}$ and $\mathbf{v}$ are orthogonal.
$\mathbf{u}\times\mathbf{v} = \mathbf{0} \Rightarrow$ $\mathbf{u}$ and $\mathbf{v}$ are parallel.
Therefore, if $\mathbf{u}\bullet\mathbf{v} = 0$ and $\mathbf{u}\times\mathbf{v} = 0$, then either $\mathbf{u} = 0$ or $\mathbf{v} = 0$.

Chapter 8 Review

1. $\left(3, \frac{\pi}{6}\right)$

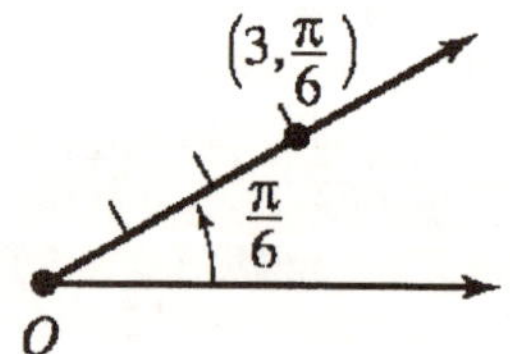

$x = 3\cos\frac{\pi}{6} = \frac{3\sqrt{3}}{2}$; $y = 3\sin\frac{\pi}{6} = \frac{3}{2}$

Rectangular coordinates of $\left(3, \frac{\pi}{6}\right)$ are $\left(\frac{3\sqrt{3}}{2}, \frac{3}{2}\right)$.

2. $\left(4, \frac{2\pi}{3}\right)$

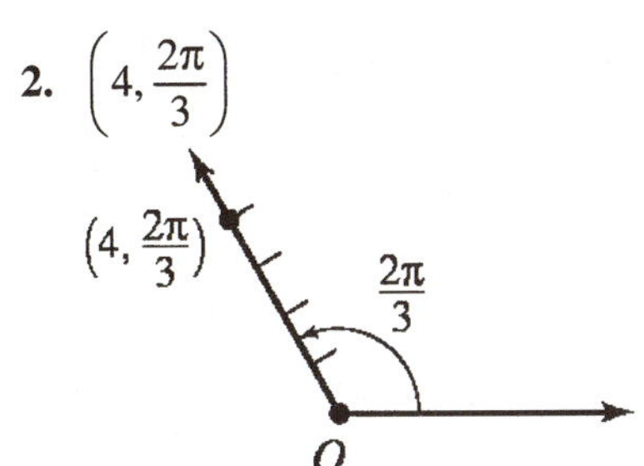

$x = 4\cos\frac{2\pi}{3} = -2;\ \ y = 4\sin\frac{2\pi}{3} = 2\sqrt{3}$

Rectangular coordinates of $\left(4, \frac{2\pi}{3}\right)$ are $\left(-2, 2\sqrt{3}\right)$.

3. $\left(-2, \frac{4\pi}{3}\right)$

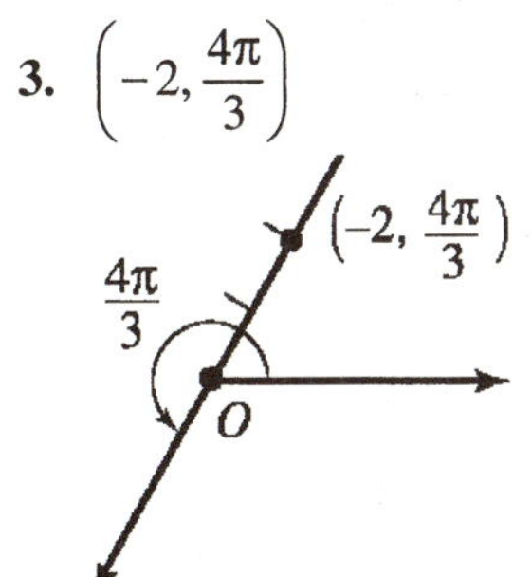

$x = -2\cos\frac{4\pi}{3} = 1;\ \ y = -2\sin\frac{4\pi}{3} = \sqrt{3}$

Rectangular coordinates of $\left(-2, \frac{4\pi}{3}\right)$ are $\left(1, \sqrt{3}\right)$.

4. $\left(-1, \frac{5\pi}{4}\right)$

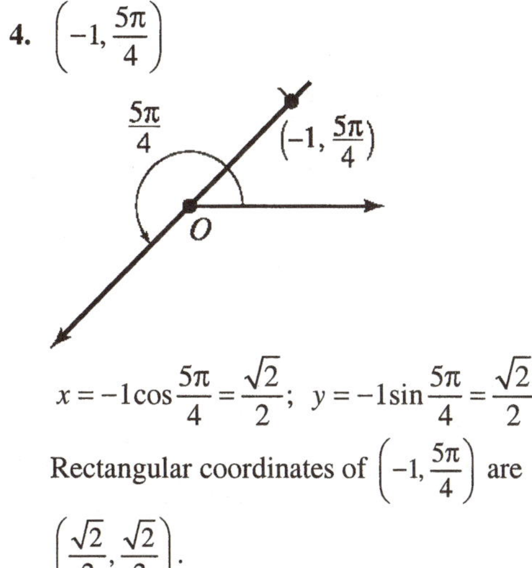

$x = -1\cos\frac{5\pi}{4} = \frac{\sqrt{2}}{2};\ \ y = -1\sin\frac{5\pi}{4} = \frac{\sqrt{2}}{2}$

Rectangular coordinates of $\left(-1, \frac{5\pi}{4}\right)$ are $\left(\frac{\sqrt{2}}{2}, \frac{\sqrt{2}}{2}\right)$.

5. $\left(-3, -\frac{\pi}{2}\right)$

$x = -3\cos\left(-\frac{\pi}{2}\right) = 0;\ \ y = -3\sin\left(-\frac{\pi}{2}\right) = 3$

Rectangular coordinates of $\left(-3, -\frac{\pi}{2}\right)$ are $(0, 3)$.

6. $\left(-4, -\frac{\pi}{4}\right)$

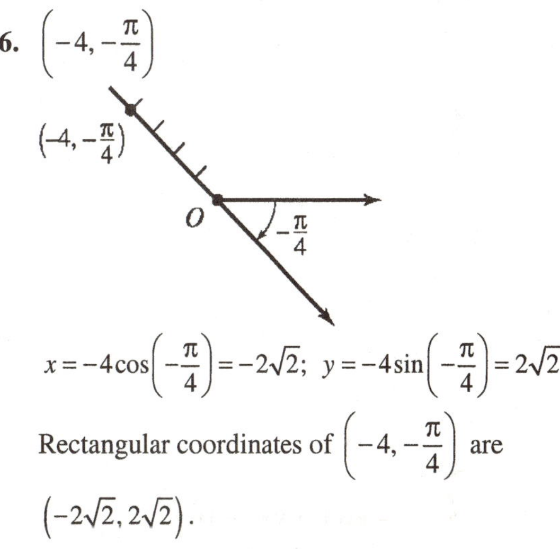

$x = -4\cos\left(-\frac{\pi}{4}\right) = -2\sqrt{2};\ \ y = -4\sin\left(-\frac{\pi}{4}\right) = 2\sqrt{2}$

Rectangular coordinates of $\left(-4, -\frac{\pi}{4}\right)$ are $\left(-2\sqrt{2}, 2\sqrt{2}\right)$.

7. The point $(-3, 3)$ lies in quadrant II.

$r = \sqrt{x^2 + y^2} = \sqrt{(-3)^2 + 3^2} = 3\sqrt{2}$

$\theta = \tan^{-1}\left(\frac{y}{x}\right) = \tan^{-1}\left(\frac{3}{-3}\right) = \tan^{-1}(-1) = -\frac{\pi}{4}$

Polar coordinates of the point $(-3, 3)$ are $\left(-3\sqrt{2}, -\frac{\pi}{4}\right)$ or $\left(3\sqrt{2}, \frac{3\pi}{4}\right)$.

8. The point $(1, -1)$ lies in quadrant IV.

$r = \sqrt{x^2 + y^2} = \sqrt{1^2 + (-1)^2} = \sqrt{2}$

$\theta = \tan^{-1}\left(\frac{y}{x}\right) = \tan^{-1}\left(\frac{-1}{1}\right) = \tan^{-1}(-1) = -\frac{\pi}{4}$

$-\frac{\pi}{4} + \pi = \frac{3\pi}{4}$

Polar coordinates of the point $(1, -1)$ are $\left(\sqrt{2}, -\frac{\pi}{4}\right)$ or $\left(-\sqrt{2}, \frac{3\pi}{4}\right)$.

9. The point $(0,-2)$ lies on the negative y-axis.

$r=\sqrt{x^2+y^2}=\sqrt{0^2+(-2)^2}=2$

$\theta=\tan^{-1}\left(\frac{y}{x}\right)=\tan^{-1}\left(\frac{-2}{0}\right)$

$\frac{-2}{0}$ is undefined, so $\theta=-\frac{\pi}{2}$; $-\frac{\pi}{2}+\pi=\frac{\pi}{2}$.

Polar coordinates of the point $(0,-2)$ are $\left(2,-\frac{\pi}{2}\right)$ or $\left(-2,\frac{\pi}{2}\right)$.

10. The point (2, 0) lies on the positive x-axis.

$r=\sqrt{x^2+y^2}=\sqrt{2^2+0^2}=\sqrt{4}=2$

$\theta=\tan^{-1}\left(\frac{y}{x}\right)=\tan^{-1}\left(\frac{0}{2}\right)=\tan^{-1}0=0$

$0+\pi=\pi$

Polar coordinates of (2, 0) are (2, 0) or $(-2, \pi)$.

11. The point $(3,4)$ lies in quadrant I.

$r=\sqrt{x^2+y^2}=\sqrt{3^2+4^2}=5$

$\theta=\tan^{-1}\left(\frac{y}{x}\right)=\tan^{-1}\left(\frac{4}{3}\right)\approx 0.93$

$\tan^{-1}\left(\frac{4}{3}\right)+\pi\approx 4.07$

Polar coordinates of the point $(3,4)$ are $(5, 0.93)$ or $(-5, 4.07)$.

12. The point $(-5,12)$ lies in quadrant II.

$r=\sqrt{x^2+y^2}=\sqrt{(-5)^2+12^2}=13$

$\theta=\tan^{-1}\left(\frac{y}{x}\right)=\tan^{-1}\left(\frac{12}{-5}\right)\approx -1.176$

$\tan^{-1}\left(\frac{12}{-5}\right)+\pi\approx 1.97$; $\tan^{-1}\left(\frac{12}{-5}\right)+2\pi\approx 5.11$

Polar coordinates of the point $(-5,12)$ are $(13, 1.97)$ or $(-13, 5.11)$.

13.

$$r=2\sin\theta$$
$$r^2=2r\sin\theta$$
$$x^2+y^2=2y$$
$$x^2+y^2-2y=0$$
$$x^2+y^2-2y+1=1$$
$$x^2+(y-1)^2=1^2$$

The graph is a circle with center $(0,1)$ and radius 1.

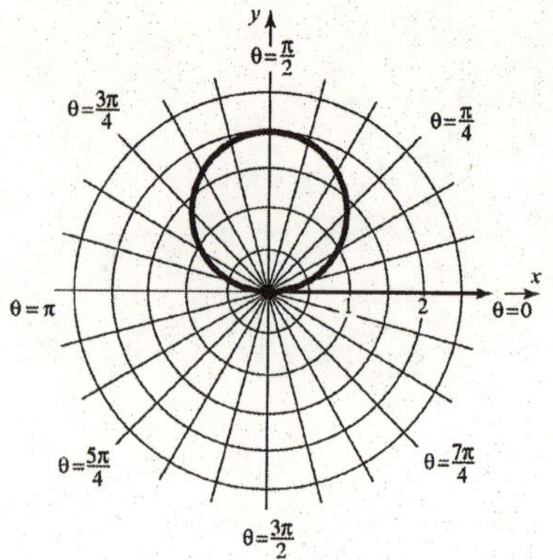

14.

$$3r=\sin\theta$$
$$3r^2=r\sin\theta$$
$$3(x^2+y^2)=y$$
$$3x^2+3y^2-y=0$$
$$x^2+y^2-\frac{1}{3}y=0$$
$$x^2+y^2-\frac{1}{3}y+\frac{1}{36}=\frac{1}{36}$$
$$x^2+\left(y-\frac{1}{6}\right)^2=\left(\frac{1}{6}\right)^2$$

The graph is a circle with center $\left(0,\frac{1}{6}\right)$ and radius $\frac{1}{6}$.

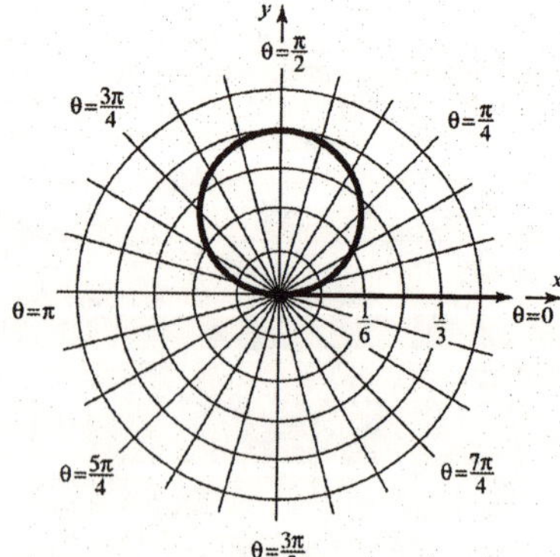

15.

$$r=5$$
$$r^2=25$$
$$x^2+y^2=5^2$$

The graph is a circle with center $(0,0)$ and radius 5.

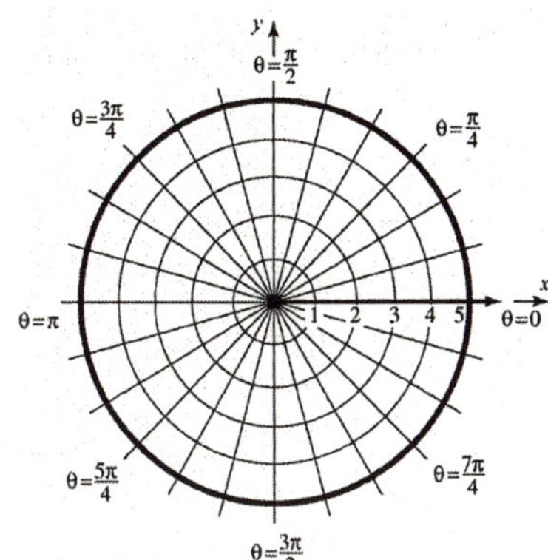

16. $\theta = \dfrac{\pi}{4}$

$\tan\theta = \tan\left(\dfrac{\pi}{4}\right)$

$\dfrac{y}{x} = 1$

$y = x$

$x - y = 0$

The graph is a line through the point $(0,0)$ with slope 1.

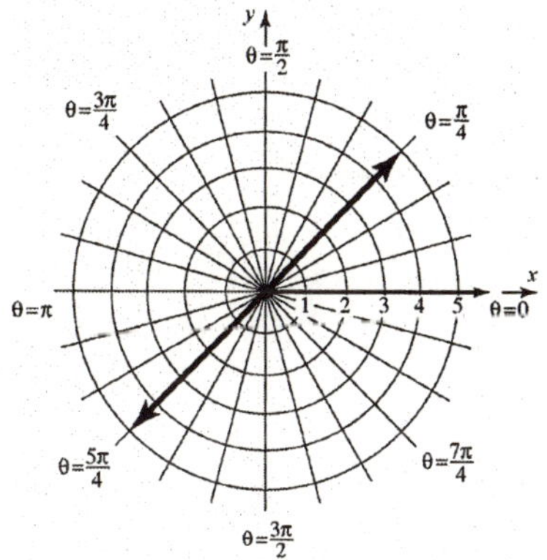

17. $r\cos\theta + 3r\sin\theta = 6$

$x + 3y = 6$

$3y = -x + 6$

$y = -\dfrac{1}{3}x + 2$

The graph is a line with y-intercept $(0,2)$ and slope $-\dfrac{1}{3}$.

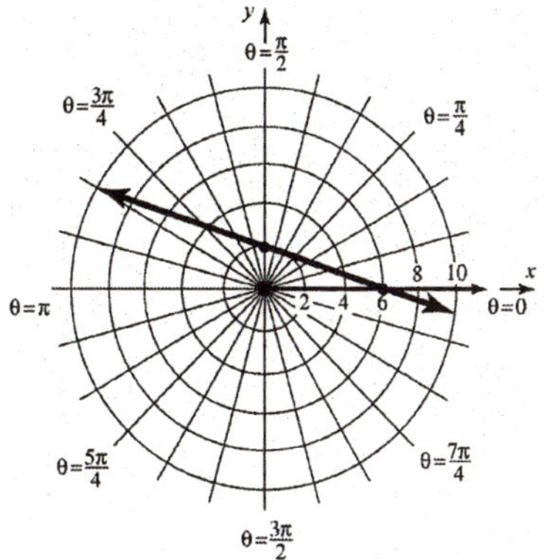

18. $r^2 + 4r\sin\theta - 8r\cos\theta = 5$

$x^2 + y^2 + 4y - 8x = 5$

$x^2 - 8x + 16 + y^2 + 4y + 4 = 5 + 16 + 4$

$(x-4)^2 + (y+2)^2 = 25$

The graph is a circle with center $(4,-2)$ and radius 5.

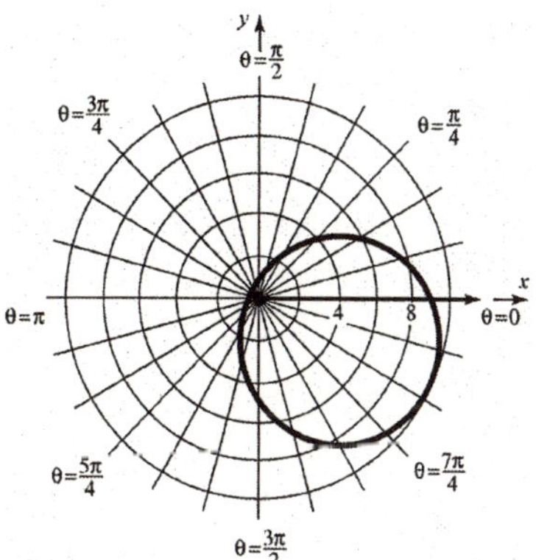

19. $r = 4\cos\theta$

The graph will be a circle with radius 2 and center (2, 0).

Check for symmetry:

Polar axis: Replace θ by $-\theta$. The result is $r = 4\cos(-\theta) = 4\cos\theta$.

The graph is symmetric with respect to the polar axis.

The line $\theta = \dfrac{\pi}{2}$: Replace θ by $\pi - \theta$.

$r = 4\cos(\pi - \theta)$

$= 4(\cos\pi\cos\theta + \sin\pi\sin\theta)$

$= 4(-\cos\theta + 0)$

$= -4\cos\theta$

The test fails.

The pole: Replace r by $-r$. $-r = 4\cos\theta$. The test fails.

Due to symmetry with respect to the polar axis, assign values to θ from 0 to π.

θ	$r = 4\cos\theta$
0	4
$\frac{\pi}{6}$	$2\sqrt{3} \approx 3.5$
$\frac{\pi}{3}$	2
$\frac{\pi}{2}$	0
$\frac{2\pi}{3}$	-2
$\frac{5\pi}{6}$	$-2\sqrt{3} \approx -3.5$
π	-4

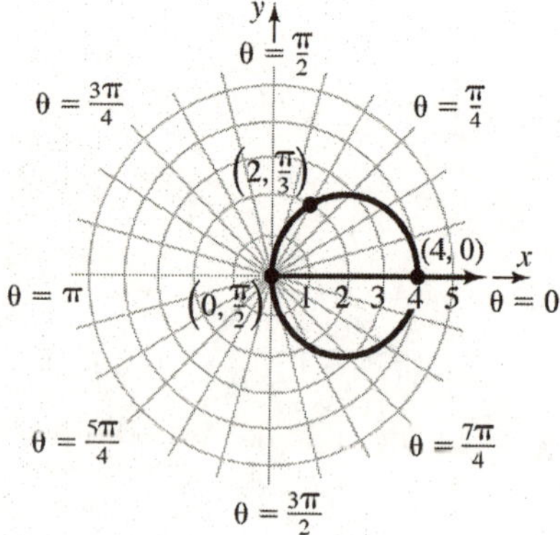

20. $r = 3\sin\theta$

The graph will be a circle with radius $\frac{3}{2}$ and center $\left(0, \frac{3}{2}\right)$.

Check for symmetry:

Polar axis: Replace θ by $-\theta$. The result is $r = 3\sin(-\theta) = -3\sin\theta$. The test fails.

The line $\theta = \frac{\pi}{2}$: Replace θ by $\pi - \theta$.

$r = 3\sin(\pi - \theta)$

$= 3(\sin\pi\cos\theta - \cos\pi\sin\theta)$

$= 3(0 + \sin\theta) = 3\sin\theta$

The graph is symmetric with respect to the line $\theta = \frac{\pi}{2}$.

The pole: Replace r by $-r$. $-r = 3\sin\theta$. The test fails.

Due to symmetry with respect to the line $\theta = \frac{\pi}{2}$, assign values to θ from $-\frac{\pi}{2}$ to $\frac{\pi}{2}$.

θ	$r = 3\sin\theta$
$-\frac{\pi}{2}$	-3
$-\frac{\pi}{3}$	$-\frac{3\sqrt{3}}{2} \approx -2.6$
$-\frac{\pi}{6}$	$-\frac{3}{2}$
0	0
$\frac{\pi}{6}$	$\frac{3}{2}$
$\frac{\pi}{3}$	$\frac{3\sqrt{3}}{2} \approx 2.6$
$\frac{\pi}{2}$	3

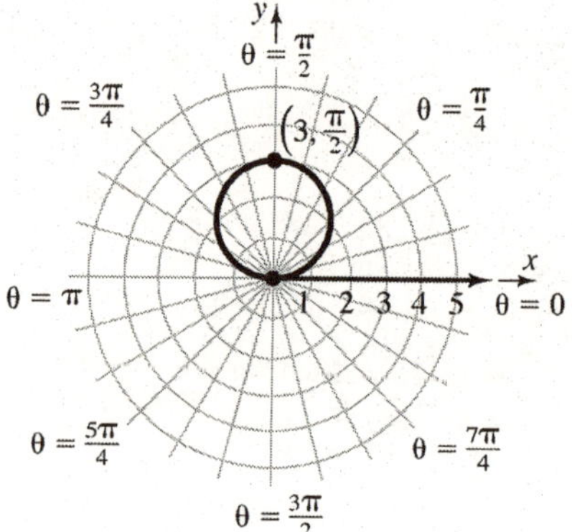

21. $r = 3 - 3\sin\theta$

The graph will be a cardioid. Check for symmetry:

Polar axis: Replace θ by $-\theta$. The result is $r = 3 - 3\sin(-\theta) = 3 + 3\sin\theta$. The test fails.

The line $\theta = \frac{\pi}{2}$: Replace θ by $\pi - \theta$.

$r = 3 - 3\sin(\pi - \theta)$

$= 3 - 3(\sin\pi\cos\theta - \cos\pi\sin\theta)$

$= 3 - 3(0 + \sin\theta)$

$= 3 - 3\sin\theta$

The graph is symmetric with respect to the line $\theta = \frac{\pi}{2}$.

The pole: Replace r by $-r$. $-r = 3 - 3\sin\theta$.
The test fails.

Due to symmetry with respect to the line $\theta = \frac{\pi}{2}$, assign values to θ from $-\frac{\pi}{2}$ to $\frac{\pi}{2}$.

θ	$r = 3 - 3\sin\theta$
$-\frac{\pi}{2}$	6
$-\frac{\pi}{3}$	$3 + \frac{3\sqrt{3}}{2} \approx 5.6$
$-\frac{\pi}{6}$	$\frac{9}{2}$
0	3
$\frac{\pi}{6}$	$\frac{3}{2}$
$\frac{\pi}{3}$	$3 - \frac{3\sqrt{3}}{2} \approx 0.4$
$\frac{\pi}{2}$	0

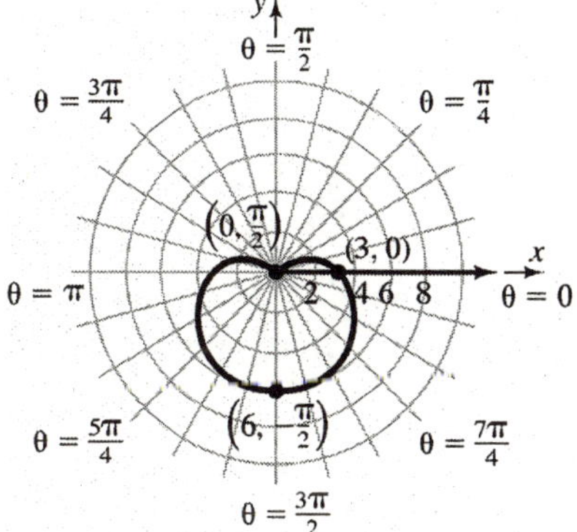

22. $r = 2 + \cos\theta$
The graph will be a limaçon without inner loop.
Check for symmetry:

Polar axis: Replace θ by $-\theta$. The result is $r = 2 + \cos(-\theta) = 2 + \cos\theta$.

The graph is symmetric with respect to the polar axis.

The line $\theta = \frac{\pi}{2}$: Replace θ by $\pi - \theta$.

$$\begin{aligned} r &= 2 + \cos(\pi - \theta) \\ &= 2 + (\cos\pi\cos\theta + \sin\pi\sin\theta) \\ &= 2 + (-\cos\theta + 0) \\ &= 2 - \cos\theta \end{aligned}$$

The test fails.

The pole: Replace r by $-r$. $-r = 2 + \cos\theta$.
The test fails.

Due to symmetry with respect to the polar axis, assign values to θ from 0 to π.

θ	$r = 2 + \cos\theta$
0	3
$\frac{\pi}{6}$	$2 + \frac{\sqrt{3}}{2} \approx 2.9$
$\frac{\pi}{3}$	$\frac{5}{2}$
$\frac{\pi}{2}$	2
$\frac{2\pi}{3}$	$\frac{3}{2}$
$\frac{5\pi}{6}$	$2 - \frac{\sqrt{3}}{2} \approx 1.1$
π	1

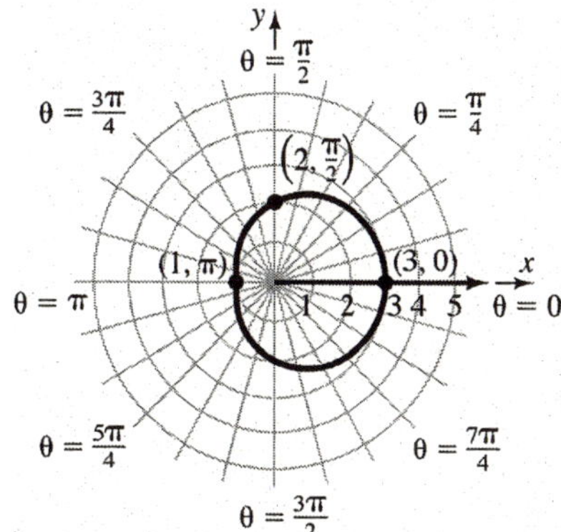

23. $r = 4 - \cos\theta$
The graph will be a limaçon without inner loop.
Check for symmetry:

Polar axis: Replace θ by $-\theta$. The result is $r = 4 - \cos(-\theta) = 4 - \cos\theta$.

The graph is symmetric with respect to the polar axis.

The line $\theta = \frac{\pi}{2}$: Replace θ by $\pi - \theta$.

$$\begin{aligned} r &= 4 - \cos(\pi - \theta) \\ &= 4 - (\cos\pi\cos\theta + \sin\pi\sin\theta) \\ &= 4 - (-\cos\theta + 0) \\ &= 4 + \cos\theta \end{aligned}$$

The test fails.

The pole: Replace r by $-r$. $-r = 4 - \cos\theta$.
The test fails.

Due to symmetry with respect to the polar axis, assign values to θ from 0 to π.

θ	$r=4-\cos\theta$
0	3
$\frac{\pi}{6}$	$4-\frac{\sqrt{3}}{2}\approx 3.1$
$\frac{\pi}{3}$	$\frac{7}{2}$
$\frac{\pi}{2}$	4
$\frac{2\pi}{3}$	$\frac{9}{2}$
$\frac{5\pi}{6}$	$4+\frac{\sqrt{3}}{2}\approx 4.9$
π	5

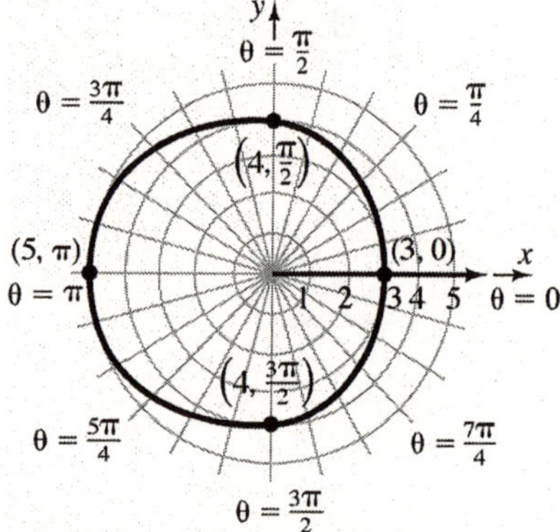

24. $r=1-2\sin\theta$
The graph will be a limaçon with inner loop.
Check for symmetry:

Polar axis: Replace θ by $-\theta$. The result is $r=1-2\sin(-\theta)=1+2\sin\theta$. The test fails.

The line $\theta=\frac{\pi}{2}$: Replace θ by $\pi-\theta$.

$$\begin{aligned} r&=1-2\sin(\pi-\theta)\\ &=1-2(\sin\pi\cos\theta-\cos\pi\sin\theta)\\ &=1-2(0+\sin\theta)\\ &=1-2\sin\theta \end{aligned}$$

The graph is symmetric with respect to the line $\theta=\frac{\pi}{2}$.

The pole: Replace r by $-r$. $-r=1-2\sin\theta$. The test fails.

Due to symmetry with respect to the line $\theta=\frac{\pi}{2}$, assign values to θ from $-\frac{\pi}{2}$ to $\frac{\pi}{2}$.

θ	$r=1-2\sin\theta$
$-\frac{\pi}{2}$	3
$-\frac{\pi}{3}$	$1+\sqrt{3}\approx 2.7$
$-\frac{\pi}{6}$	2
0	1
$\frac{\pi}{6}$	0
$\frac{\pi}{3}$	$1-\sqrt{3}\approx -0.7$
$\frac{\pi}{2}$	-1

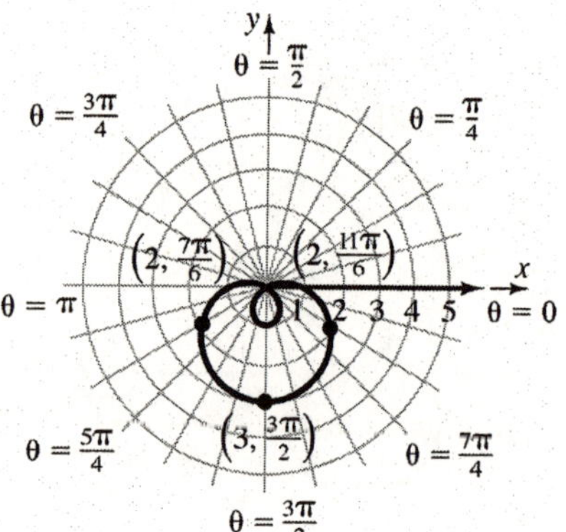

25. $r=\sqrt{x^2+y^2}=\sqrt{(-1)^2+(-1)^2}=\sqrt{2}$

$\tan\theta=\frac{y}{x}=\frac{-1}{-1}=1$

$\theta=225°$

The polar form of $z=-1-i$ is $z=r(\cos\theta+i\sin\theta)=\sqrt{2}(\cos 225°+i\sin 225°)$.

26. $r=\sqrt{x^2+y^2}=\sqrt{\left(-\sqrt{3}\right)^2+1^2}=\sqrt{4}=2$

$\tan\theta=\frac{y}{x}=\frac{1}{-\sqrt{3}}=-\frac{\sqrt{3}}{3}$

$\theta=150°$

The polar form of $z=-\sqrt{3}+i$ is $z=r(\cos\theta+i\sin\theta)=2(\cos 150°+i\sin 150°)$.

27. $r=\sqrt{x^2+y^2}=\sqrt{4^2+(-3)^2}=\sqrt{25}=5$

$\tan\theta=\frac{y}{x}=-\frac{3}{4}$

$\theta\approx 323.1°$

The polar form of $z=4-3i$ is

$z=r(\cos\theta+i\sin\theta)=5(\cos 323.1°+i\sin 323.1°)$.

28. $r=\sqrt{x^2+y^2}=\sqrt{3^2+(-2)^2}=\sqrt{13}$

$\tan\theta=\frac{y}{x}=-\frac{2}{3}$

$\theta\approx 326.3°$

The polar form of $z=3-2i$ is

$z=r(\cos\theta+i\sin\theta)$

$=\sqrt{13}(\cos 326.3°+i\sin 326.3°)$

29. $2(\cos 150°+i\sin 150°)=2\left(-\frac{\sqrt{3}}{2}+\frac{1}{2}i\right)$

$=-\sqrt{3}+i$

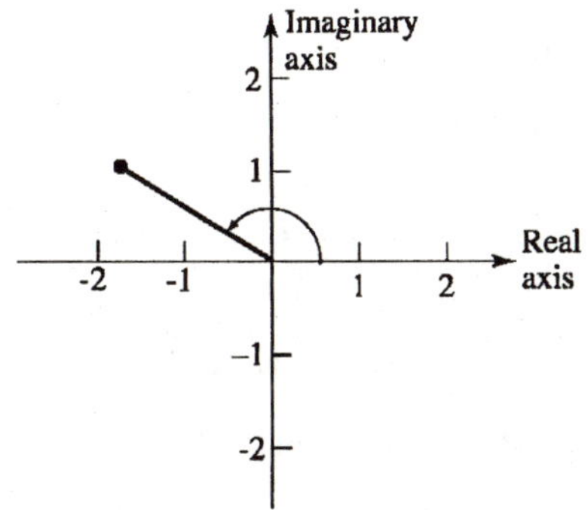

30. $3(\cos 60°+i\sin 60°)=3\left(\frac{1}{2}+\frac{\sqrt{3}}{2}i\right)=\frac{3}{2}+\frac{3\sqrt{3}}{2}i$

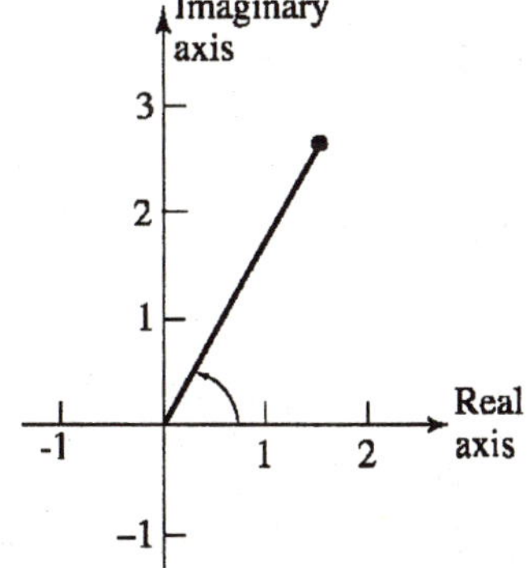

31. $3\left(\cos\frac{2\pi}{3}+i\sin\frac{2\pi}{3}\right)=3\left(-\frac{1}{2}+\frac{\sqrt{3}}{2}i\right)$

$=-\frac{3}{2}+\frac{3\sqrt{3}}{2}i$

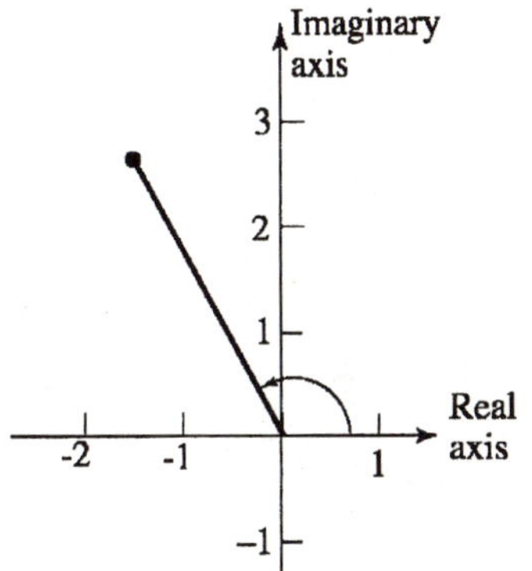

32. $4\left(\cos\frac{3\pi}{4}+i\sin\frac{3\pi}{4}\right)=4\left(-\frac{\sqrt{2}}{2}+\frac{\sqrt{2}}{2}i\right)$

$=-2\sqrt{2}+2\sqrt{2}i$

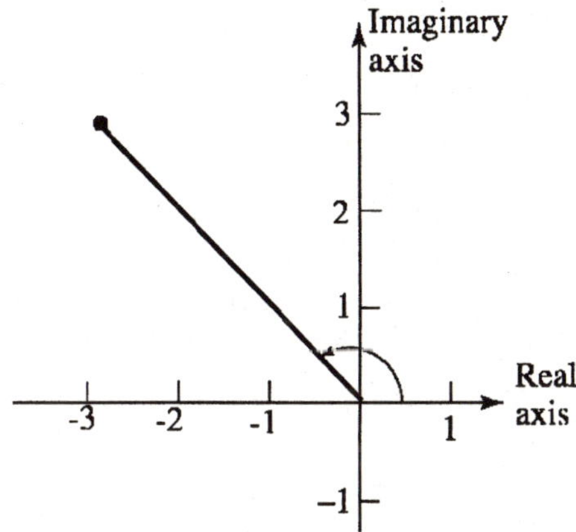

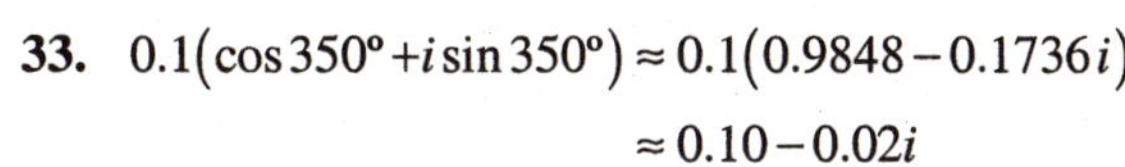

33. $0.1(\cos 350°+i\sin 350°)\approx 0.1(0.9848-0.1736i)$

$\approx 0.10-0.02i$

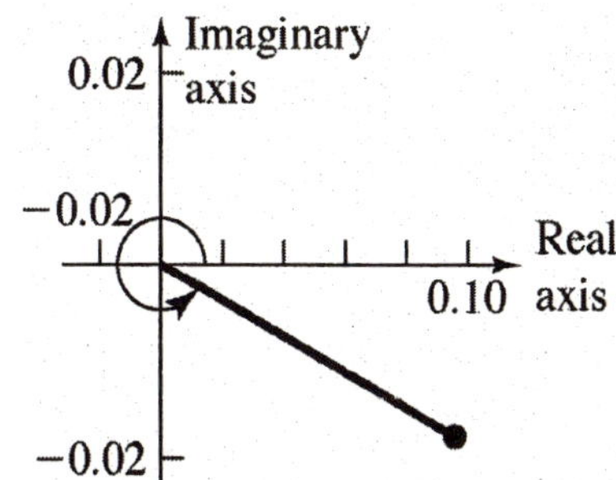

34. $0.5(\cos 160° + i\sin 160°) \approx 0.5(-0.9397 + 0.3420i)$
$\approx -0.47 + 0.17i$

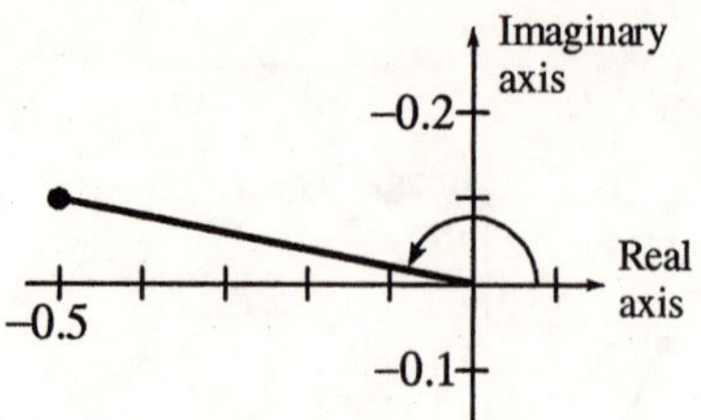

35. $z \cdot w = (\cos 80° + i\sin 80°) \cdot (\cos 50° + i\sin 50°)$
$= 1 \cdot 1[\cos(80° + 50°) + i\sin(80° + 50°)]$
$= \cos 130° + i\sin 130°$

$\frac{z}{w} = \frac{\cos 80° + i\sin 80°}{\cos 50° + i\sin 50°}$
$= \frac{1}{1}[\cos(80° - 50°) + i\sin(80° - 50°)]$
$= \cos 30° + i\sin 30°$

36. $z \cdot w = (\cos 205° + i\sin 205°) \cdot (\cos 85° + i\sin 85°)$
$= 1 \cdot 1[\cos(205° + 85°) + i\sin(205° + 85°)]$
$= \cos 290° + i\sin 290°$

$\frac{z}{w} = \frac{\cos 205° + i\sin 205°}{\cos 85° + i\sin 85°}$
$= \frac{1}{1}[\cos(205° - 85°) + i\sin(205° - 85°)]$
$= \cos 120° + i\sin 120°$

37. $z \cdot w = 3\left(\cos\frac{9\pi}{5} + i\sin\frac{9\pi}{5}\right) \cdot 2\left(\cos\frac{\pi}{5} + i\sin\frac{\pi}{5}\right)$
$= 3 \cdot 2\left[\cos\left(\frac{9\pi}{5} + \frac{\pi}{5}\right) + i\sin\left(\frac{9\pi}{5} + \frac{\pi}{5}\right)\right]$
$= 6 \cdot (\cos 2\pi + i\sin 2\pi)$
$= 6(\cos 0 + i\sin 0)$
$= 6$

$\frac{z}{w} = \frac{3 \cdot \left(\cos\frac{9\pi}{5} + i\sin\frac{9\pi}{5}\right)}{2 \cdot \left(\cos\frac{\pi}{5} + i\sin\frac{\pi}{5}\right)}$
$= \frac{3}{2}\left[\cos\left(\frac{9\pi}{5} - \frac{\pi}{5}\right) + i\sin\left(\frac{9\pi}{5} - \frac{\pi}{5}\right)\right]$
$= \frac{3}{2}\left(\cos\frac{8\pi}{5} + i\sin\frac{8\pi}{5}\right)$

38. $z \cdot w = 2 \cdot \left(\cos\frac{5\pi}{3} + i\sin\frac{5\pi}{3}\right) \cdot 3\left(\cos\frac{\pi}{3} + i\sin\frac{\pi}{3}\right)$
$= 2 \cdot 3\left[\cos\left(\frac{5\pi}{3} + \frac{\pi}{3}\right) + i\sin\left(\frac{5\pi}{3} + \frac{\pi}{3}\right)\right]$
$= 6 \cdot (\cos 2\pi + i\sin 2\pi)$
$= 6 \cdot (\cos 0 + i\sin 0)$
$= 6$

$\frac{z}{w} = \frac{2\left(\cos\frac{5\pi}{3} + i\sin\frac{5\pi}{3}\right)}{3\left(\cos\frac{\pi}{3} + i\sin\frac{\pi}{3}\right)}$
$= \frac{2}{3}\left[\cos\left(\frac{5\pi}{3} - \frac{\pi}{3}\right) + i\sin\left(\frac{5\pi}{3} - \frac{\pi}{3}\right)\right]$
$= \frac{2}{3}\left(\cos\frac{4\pi}{3} + i\sin\frac{4\pi}{3}\right)$

39. $z \cdot w = 5(\cos 10° + i\sin 10°) \cdot (\cos 355° + i\sin 355°)$
$= 5 \cdot 1[\cos(10° + 355°) + i\sin(10° + 355°)]$
$= 5(\cos 365° + i\sin 365°)$
$= 5(\cos 5° + i\sin 5°)$

$\frac{z}{w} = \frac{5(\cos 10° + i\sin 10°)}{\cos 355° + i\sin 355°}$
$= \frac{5}{1}[\cos(10° - 355°) + i\sin(10° - 355°)]$
$= 5[\cos(-345°) + i\sin(-345°)]$
$= 5(\cos 15° + i\sin 15°)$

40. $z \cdot w = 4 \cdot (\cos 50° + i\sin 50°) \cdot (\cos 340° + i\sin 340°)$
$= 4 \cdot 1[\cos(50° + 340°) + i\sin(50° + 340°)]$
$= 4 \cdot (\cos 390° + i\sin 390°)$
$= 4 \cdot (\cos 30° + i\sin 30°)$

$\frac{z}{w} = \frac{4(\cos 50° + i\sin 50°)}{\cos 340° + i\sin 340°}$
$= \frac{4}{1}[\cos(50° - 340°) + i\sin(50° - 340°)]$
$= 4 \cdot [\cos(-290°) + i\sin(-290°)]$
$= 4 \cdot (\cos 70° + i\sin 70°)$

41. $\left[3\left(\cos 20^\circ + i\sin 20^\circ\right)\right]^3$

$= 3^3\left[\cos(3\cdot 20^\circ) + i\sin(3\cdot 20^\circ)\right]$

$= 27\left(\cos 60^\circ + i\sin 60^\circ\right)$

$= 27\left(\frac{1}{2} + \frac{\sqrt{3}}{2}i\right) = \frac{27}{2} + \frac{27\sqrt{3}}{2}i$

42. $\left[2\left(\cos 50^\circ + i\sin 50^\circ\right)\right]^3$

$= 2^3\left[\cos(3\cdot 50^\circ) + i\sin(3\cdot 50^\circ)\right]$

$= 8\cdot\left(\cos 150^\circ + i\sin 150^\circ\right)$

$= 8\left(-\frac{\sqrt{3}}{2} + \frac{1}{2}i\right) = -4\sqrt{3} + 4i$

43. $\left[\sqrt{2}\left(\cos\frac{5\pi}{8} + i\sin\frac{5\pi}{8}\right)\right]^4$

$= \left(\sqrt{2}\right)^4\left[\cos\left(4\cdot\frac{5\pi}{8}\right) + i\sin\left(4\cdot\frac{5\pi}{8}\right)\right]$

$= 4\cdot\left(\cos\frac{5\pi}{2} + i\sin\frac{5\pi}{2}\right)$

$= 4\left(0 + 1i\right) = 4i$

44. $\left[2\left(\cos\frac{5\pi}{16} + i\sin\frac{5\pi}{16}\right)\right]^4$

$= 2^4\left[\cos\left(4\cdot\frac{5\pi}{16}\right) + i\sin\left(4\cdot\frac{5\pi}{16}\right)\right]$

$= 16\cdot\left(\cos\frac{5\pi}{4} + i\sin\frac{5\pi}{4}\right)$

$= 16\cdot\left(-\frac{\sqrt{2}}{2} - \frac{\sqrt{2}}{2}i\right) = -8\sqrt{2} - 8\sqrt{2}i$

45. $1 - \sqrt{3}i$

$r = \sqrt{1^2 + \left(-\sqrt{3}\right)^2} = 2$

$\tan\theta = \frac{-\sqrt{3}}{1} = -\sqrt{3}$

$\theta = 300^\circ$

$1 - \sqrt{3}i = 2\left(\cos 300^\circ + i\sin 300^\circ\right)$

$\left(1 - \sqrt{3}i\right)^6 = \left[2\left(\cos 300^\circ + i\sin 300^\circ\right)\right]^6$

$= 2^6\left[\cos\left(6\cdot 300^\circ\right) + i\sin\left(6\cdot 300^\circ\right)\right]$

$= 64\cdot\left(\cos 1800^\circ + i\sin 1800^\circ\right)$

$= 64\cdot\left(\cos 0^\circ + i\sin 0^\circ\right)$

$= 64 + 0\,i = 64$

46. $2 - 2i$

$r = \sqrt{2^2 + (-2)^2} = 2\sqrt{2}$

$\tan\theta = \frac{-2}{2} = -1$

$\theta = 315^\circ$

$2 - 2i = 2\sqrt{2}\cdot\left(\cos 315^\circ + i\sin 315^\circ\right)$

$(2 - 2i)^8 = \left[2\sqrt{2}\cdot\left(\cos 315^\circ + i\sin 315^\circ\right)\right]^8$

$= \left(2\sqrt{2}\right)^8\left[\cos\left(8\cdot 315^\circ\right) + i\sin\left(8\cdot 315^\circ\right)\right]$

$= 4096\cdot\left(\cos 2520^\circ + i\sin 2520^\circ\right)$

$= 4096\cdot\left(\cos 0^\circ + i\sin 0^\circ\right)$

$= 4096(1 + 0\,i) = 4096$

47. $3 + 4i$

$r = \sqrt{3^2 + 4^2} = 5$

$\tan\theta = \frac{4}{3}$

$\theta \approx 53.13^\circ$

$3 + 4i = 5\left[\cos\left(53.13^\circ\right) + i\sin\left(53.13^\circ\right)\right]$

$(3 + 4i)^4 = \left(5^4\left[\cos\left(4\cdot 53.13^\circ\right) + i\sin\left(4\cdot 53.13^\circ\right)\right]\right)^4$

$= 625\left[\cos\left(212.52^\circ\right) + i\sin\left(212.52^\circ\right)\right]$

$\approx 625\left(-0.8432 - 0.5376i\right)$

$= -527 - 336i$

48. $1 - 2i$

$r = \sqrt{1^2 + (-2)^2} = \sqrt{5}$

$\tan\theta = \frac{-2}{1} = -2$

$\theta \approx 296.57^\circ$

$1 - 2i = \sqrt{5}\cdot\left[\cos\left(296.57^\circ\right) + i\sin\left(296.57^\circ\right)\right]$

$(1 - 2i)^4$

$= \left(\sqrt{5}\right)^4\left[\cos\left(4\cdot 296.57^\circ\right) + i\sin\left(4\cdot 296.57^\circ\right)\right]$

$= 25\left[\cos\left(1186.26^\circ\right) + i\sin\left(1186.26^\circ\right)\right]$

$\approx 25\left(-0.2800 + .9600i\right)$

$= -7 + 24i$

49. $27+0i$

$r=\sqrt{27^2+0^2}=27$

$\tan\theta=\dfrac{0}{27}=0$

$\theta=0^\circ$

$27+0i=27\left(\cos 0^\circ+i\sin 0^\circ\right)$

The three complex cube roots of $27+0i=27\left(\cos 0^\circ+i\sin 0^\circ\right)$ are:

$$z_k=\sqrt[3]{27}\cdot\left[\cos\left(\frac{0^\circ}{3}+\frac{360^\circ\cdot k}{3}\right)+i\sin\left(\frac{0^\circ}{3}+\frac{360^\circ\cdot k}{3}\right)\right]$$
$$=3\left[\cos\left(120^\circ k\right)+i\sin\left(120^\circ\cdot k\right)\right]$$

$$z_0=3\left[\cos\left(120^\circ\cdot 0\right)+i\sin\left(120^\circ\cdot 0\right)\right]$$
$$=3\cdot\left(\cos 0^\circ+i\sin 0^\circ\right)=3$$

$$z_1=3\left[\cos\left(120^\circ\cdot 1\right)+i\sin\left(120^\circ\cdot 1\right)\right]$$
$$=3\cdot\left(\cos 120^\circ+i\sin 120^\circ\right)=-\frac{3}{2}+\frac{3\sqrt{3}}{2}i$$

$$z_2=3\left[\cos\left(120^\circ\cdot 2\right)+i\sin\left(120^\circ\cdot 2\right)\right]$$
$$=3\cdot\left(\cos 240^\circ+i\sin 240^\circ\right)=-\frac{3}{2}-\frac{3\sqrt{3}}{2}i$$

50. -16

$r=\sqrt{(-16)^2+0^2}=16$

$\tan\theta=\dfrac{0}{-16}=0$

$\theta=180^\circ$

$-16=16\left(\cos 180^\circ+i\sin 180^\circ\right)$

The four complex fourth roots of $-16=16\cdot\left(\cos 180^\circ+i\sin 180^\circ\right)$ are:

$$z_k=\sqrt[4]{16}\left[\cos\left(\frac{180^\circ}{4}+\frac{360^\circ\cdot k}{4}\right)+i\sin\left(\frac{180^\circ}{4}+\frac{360^\circ\cdot k}{4}\right)\right]$$
$$=2\left[\cos\left(45^\circ+90^\circ k\right)+i\sin\left(45^\circ+90^\circ\cdot k\right)\right]$$

$$z_0=2\left[\cos\left(45^\circ+90^\circ\cdot 0\right)+i\sin\left(45^\circ+90^\circ\cdot 0\right)\right]$$
$$=2\cdot\left(\cos 45^\circ+i\sin 45^\circ\right)=\sqrt{2}+\sqrt{2}i$$

$$z_1=2\left[\cos\left(45^\circ+90^\circ\cdot 1\right)+i\sin\left(45^\circ+90^\circ\cdot 1\right)\right]$$
$$=2\cdot\left(\cos 135^\circ+i\sin 135^\circ\right)=-\sqrt{2}+\sqrt{2}i$$

$$z_2=2\left[\cos\left(45^\circ+90^\circ\cdot 2\right)+i\sin\left(45^\circ+90^\circ\cdot 2\right)\right]$$
$$=2\cdot\left(\cos 225^\circ+i\sin 225^\circ\right)=-\sqrt{2}-\sqrt{2}i$$

$$z_3=2\left[\cos\left(45^\circ+90^\circ\cdot 3\right)+i\sin\left(45^\circ+90^\circ\cdot 3\right)\right]$$
$$=2\cdot\left(\cos 315^\circ+i\sin 315^\circ\right)=\sqrt{2}-\sqrt{2}i$$

51.

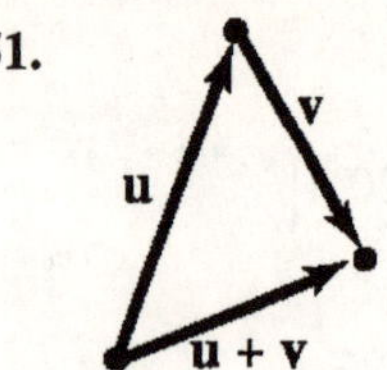

52.

v + w
v
w

53.

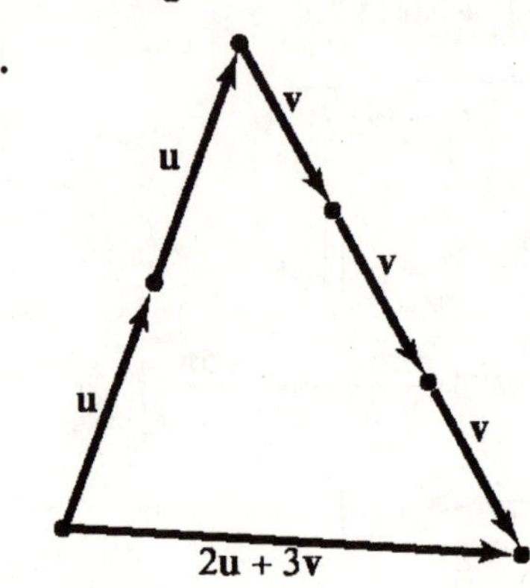

54.

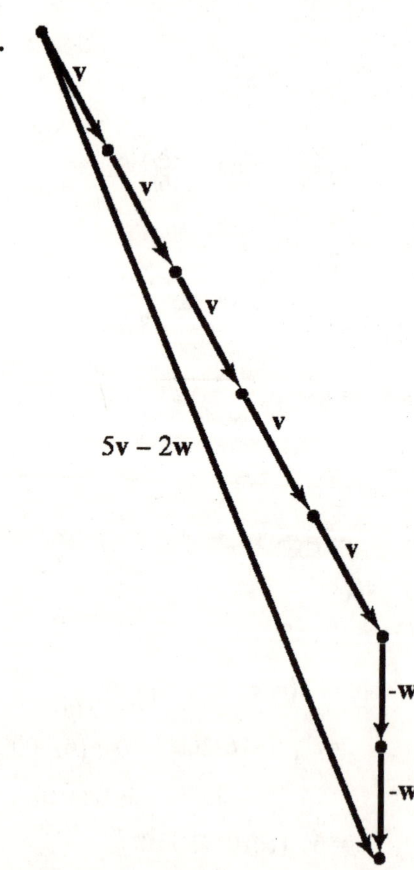

55. $P=(1,-2), Q=(3,-6)$
$\mathbf{v}=(3-1)\mathbf{i}+(-6-(-2))\mathbf{j}=2\mathbf{i}-4\mathbf{j}$
$\|\mathbf{v}\|=\sqrt{2^2+(-4)^2}=\sqrt{20}=2\sqrt{5}$

56. $P=(-3,1), Q=(4,-2)$
$\mathbf{v}=(4-(-3))\mathbf{i}+(-2-1)\mathbf{j}=7\mathbf{i}-3\mathbf{j}$
$\|\mathbf{v}\|=\sqrt{7^2+(-3)^2}=\sqrt{58}$

57. $P=(0,-2), Q=(-1,1)$
$\mathbf{v}=(-1-0)\mathbf{i}+(1-(-2))\mathbf{j}=-\mathbf{i}+3\mathbf{j}$
$\|\mathbf{v}\|=\sqrt{(-1)^2+3^2}=\sqrt{10}$

58. $P=(3,-4), Q=(-2,0)$
$\mathbf{v}=(-2-3)\mathbf{i}+(0-(-4))\mathbf{j}=-5\mathbf{i}+4\mathbf{j}$
$\|\mathbf{v}\|=\sqrt{(-5)^2+4^2}=\sqrt{41}$

59. $\mathbf{v}+\mathbf{w}=(-2\mathbf{i}+\mathbf{j})+(4\mathbf{i}-3\mathbf{j})=2\mathbf{i}-2\mathbf{j}$

60. $\mathbf{v}-\mathbf{w}=(-2\mathbf{i}+\mathbf{j})-(4\mathbf{i}-3\mathbf{j})=-6\mathbf{i}+4\mathbf{j}$

61. $4\mathbf{v}-3\mathbf{w}=4(-2\mathbf{i}+\mathbf{j})-3(4\mathbf{i}-3\mathbf{j})$
$=-8\mathbf{i}+4\mathbf{j}-12\mathbf{i}+9\mathbf{j}$
$=-20\mathbf{i}+13\mathbf{j}$

62. $-\mathbf{v}+2\mathbf{w}=-(-2\mathbf{i}+\mathbf{j})+2(4\mathbf{i}-3\mathbf{j})$
$=2\mathbf{i}-\mathbf{j}+8\mathbf{i}-6\mathbf{j}$
$=10\mathbf{i}-7\mathbf{j}$

63. $\|\mathbf{v}\|=\|-2\mathbf{i}+\mathbf{j}\|=\sqrt{(-2)^2+1^2}=\sqrt{5}$

64. $\|\mathbf{v}+\mathbf{w}\|=\|(-2\mathbf{i}+\mathbf{j})+(4\mathbf{i}-3\mathbf{j})\|$
$=\|2\mathbf{i}-2\mathbf{j}\|$
$=\sqrt{2^2+(-2)^2}=\sqrt{8}=2\sqrt{2}$

65. $\|\mathbf{v}\|+\|\mathbf{w}\|=\|-2\mathbf{i}+\mathbf{j}\|+\|4\mathbf{i}-3\mathbf{j}\|$
$=\sqrt{(-2)^2+1^2}+\sqrt{4^2+(-3)^2}$
$=\sqrt{5}+5\approx 7.24$

66. $\|2\mathbf{v}\|-3\|\mathbf{w}\|=\|2(-2\mathbf{i}+\mathbf{j})\|-3\|4\mathbf{i}-3\mathbf{j}\|$
$=\|-4\mathbf{i}+2\mathbf{j}\|-3\|4\mathbf{i}-3\mathbf{j}\|$
$=\sqrt{(-4)^2+2^2}-3\sqrt{4^2+(-3)^2}$
$=\sqrt{20}-3\cdot 5$
$=2\sqrt{5}-15\approx -10.53$

67. $\mathbf{u}=\dfrac{\mathbf{v}}{\|\mathbf{v}\|}$
$=\dfrac{-2\mathbf{i}+\mathbf{j}}{\|-2\mathbf{i}+\mathbf{j}\|}$
$=\dfrac{-2\mathbf{i}+\mathbf{j}}{\sqrt{(-2)^2+1^2}}$
$=\dfrac{-2\mathbf{i}+\mathbf{j}}{\sqrt{5}}=-\dfrac{2\sqrt{5}}{5}\mathbf{i}+\dfrac{\sqrt{5}}{5}\mathbf{j}$

68. $\mathbf{u}=\dfrac{-\mathbf{w}}{\|\mathbf{w}\|}$
$=\dfrac{-(4\mathbf{i}-3\mathbf{j})}{\|4\mathbf{i}-3\mathbf{j}\|}$
$=\dfrac{-4\mathbf{i}+3\mathbf{j}}{\sqrt{4^2+(-3)^2}}$
$=\dfrac{-4\mathbf{i}+3\mathbf{j}}{\sqrt{25}}=-\dfrac{4}{5}\mathbf{i}+\dfrac{3}{5}\mathbf{j}$

69. Let $\mathbf{v}=x\cdot\mathbf{i}+y\cdot\mathbf{j}$, $\|\mathbf{v}\|=3$, with the angle between $\mathbf{v}$ and $\mathbf{i}$ equal to $60°$.
$\|\mathbf{v}\|=3$
$\sqrt{x^2+y^2}=3$
$x^2+y^2=9$

The angle between $\mathbf{v}$ and $\mathbf{i}$ equals $60°$. Thus,
$\cos 60°=\dfrac{\mathbf{v}\cdot\mathbf{i}}{\|\mathbf{v}\|\cdot\|\mathbf{i}\|}=\dfrac{(x\mathbf{i}+y\mathbf{j})\cdot\mathbf{i}}{3\cdot 1}=\dfrac{x}{3\cdot 1}=\dfrac{x}{3}$.
We also conclude that $\mathbf{v}$ lies in Quadrant I.
$\cos 60°=\dfrac{x}{3}$
$\dfrac{1}{2}=\dfrac{x}{3}$
$x=\dfrac{3}{2}$

$$x^2 + y^2 = 9$$

$$\left(\frac{3}{2}\right)^2 + y^2 = 9$$

$$y^2 = 9 - \left(\frac{3}{2}\right)^2 = 9 - \frac{9}{4} = \frac{36-9}{4} = \frac{27}{4}$$

$$y = \pm\sqrt{\frac{27}{4}} = \pm\frac{3\sqrt{3}}{2}$$

Since $\mathbf{v}$ lies in Quadrant I, $y = \frac{3\sqrt{3}}{2}$. So,

$$\mathbf{v} = x\cdot\mathbf{i} + y\cdot\mathbf{j} = \frac{3}{2}\cdot\mathbf{i} + \frac{3\sqrt{3}}{2}\cdot\mathbf{j}.$$

70. Let $\mathbf{v} = x\cdot\mathbf{i} + y\cdot\mathbf{j}$, $\|\mathbf{v}\| = 5$, with the angle between $\mathbf{v}$ and $\mathbf{i}$ equal to $150°$.

$$\|\mathbf{v}\| = 5$$

$$\sqrt{x^2 + y^2} = 5$$

$$x^2 + y^2 = 25$$

The angle between $\mathbf{v}$ and $\mathbf{i}$ equals $150°$. Thus,

$$\cos 150° = \frac{\mathbf{v}\cdot\mathbf{i}}{\|\mathbf{v}\|\cdot\|\mathbf{i}\|} = \frac{(x\mathbf{i} + y\mathbf{j})\cdot\mathbf{i}}{5\cdot 1} = \frac{x}{5}.$$

We also conclude that $\mathbf{v}$ lies in Quadrant II.

$$\cos 150° = \frac{x}{5}$$

$$-\frac{\sqrt{3}}{2} = \frac{x}{5}$$

$$x = -\frac{5\sqrt{3}}{2}$$

$$x^2 + y^2 = 25$$

$$\left(-\frac{5\sqrt{3}}{2}\right)^2 + y^2 = 25$$

$$\frac{75}{4} + y^2 = 25$$

$$y^2 = 25 - \frac{75}{4} = \frac{100-75}{4} = \frac{25}{4}$$

$$y = \pm\sqrt{\frac{25}{4}} = \pm\frac{5}{2}$$

Since $\mathbf{v}$ lies in Quadrant II, $y = \frac{5}{2}$. So,

$$\mathbf{v} = x\cdot\mathbf{i} + y\cdot\mathbf{j} = -\frac{5\sqrt{3}}{2}\cdot\mathbf{i} + \frac{5}{2}\cdot\mathbf{j}.$$

71. $d(P_1, P_2) = \sqrt{(4-1)^2 + (-2-3)^2 + (1-(-2))^2}$
$= \sqrt{9 + 25 + 9}$
$= \sqrt{43} \approx 6.56$

72. $d(P_1, P_2) = \sqrt{(6-0)^2 + (-5-(-4))^2 + (-1-3)^2}$
$= \sqrt{36 + 1 + 16}$
$= \sqrt{53} \approx 7.28$

73. $\mathbf{v} = (4-1)\mathbf{i} + (-2-3)\mathbf{j} + (1-(-2))\mathbf{k}$
$= 3\mathbf{i} - 5\mathbf{j} + 3\mathbf{k}$

74. $\mathbf{v} = (6-0)\mathbf{i} + (-5-(-4))\mathbf{j} + (-1-3)\mathbf{k}$
$= 6\mathbf{i} - \mathbf{j} - 4\mathbf{k}$

75. $4\mathbf{v} - 3\mathbf{w} = 4(3\mathbf{i} + \mathbf{j} - 2\mathbf{k}) - 3(-3\mathbf{i} + 2\mathbf{j} - \mathbf{k})$
$= 12\mathbf{i} + 4\mathbf{j} - 8\mathbf{k} + 9\mathbf{i} - 6\mathbf{j} + 3\mathbf{k}$
$= 21\mathbf{i} - 2\mathbf{j} - 5\mathbf{k}$

76. $-\mathbf{v} + 2\mathbf{w} = -(3\mathbf{i} + \mathbf{j} - 2\mathbf{k}) + 2(-3\mathbf{i} + 2\mathbf{j} - \mathbf{k})$
$= -3\mathbf{i} - \mathbf{j} + 2\mathbf{k} - 6\mathbf{i} + 4\mathbf{j} - 2\mathbf{k}$
$= -9\mathbf{i} + 3\mathbf{j}$

77. $\|\mathbf{v} - \mathbf{w}\| = \|(3\mathbf{i} + \mathbf{j} - 2\mathbf{k}) - (-3\mathbf{i} + 2\mathbf{j} - \mathbf{k})\|$
$= \|3\mathbf{i} + \mathbf{j} - 2\mathbf{k} + 3\mathbf{i} - 2\mathbf{j} + \mathbf{k}\|$
$= \|6\mathbf{i} - \mathbf{j} - \mathbf{k}\|$
$= \sqrt{6^2 + (-1)^2 + (-1)^2}$
$= \sqrt{36 + 1 + 1}$
$= \sqrt{38}$

78. $\|\mathbf{v} + \mathbf{w}\| = \|(3\mathbf{i} + \mathbf{j} - 2\mathbf{k}) + (-3\mathbf{i} + 2\mathbf{j} - \mathbf{k})\|$
$= \|3\mathbf{j} - 3\mathbf{k}\|$
$= \sqrt{3^2 + (-3)^2}$
$= \sqrt{18}$
$= 3\sqrt{2}$

79. $\|\mathbf{v}\| - \|\mathbf{w}\|$
$= \|3\mathbf{i} + \mathbf{j} - 2\mathbf{k}\| - \|-3\mathbf{i} + 2\mathbf{j} - \mathbf{k}\|$
$= \sqrt{3^2 + 1^2 + (-2)^2} - \sqrt{(-3)^2 + 2^2 + (-1)^2}$
$= \sqrt{14} - \sqrt{14}$
$= 0$

80. $\|\mathbf{v}\|+\|\mathbf{w}\|=\|3\mathbf{i}+\mathbf{j}-2\mathbf{k}\|+\|-3\mathbf{i}+2\mathbf{j}-\mathbf{k}\|$
$=\sqrt{3^2+1^2+(-2)^2}+\sqrt{(-3)^2+2^2+(-1)^2}$
$=\sqrt{14}+\sqrt{14}$
$=2\sqrt{14}$

81. $\mathbf{v}\times\mathbf{w}=\begin{vmatrix}\mathbf{i} & \mathbf{j} & \mathbf{k}\\ 3 & 1 & -2\\ -3 & 2 & -1\end{vmatrix}$
$=\begin{vmatrix}1 & -2\\ 2 & -1\end{vmatrix}\mathbf{i}-\begin{vmatrix}3 & -2\\ -3 & -1\end{vmatrix}\mathbf{j}+\begin{vmatrix}3 & 1\\ -3 & 2\end{vmatrix}\mathbf{k}$
$=3\mathbf{i}+9\mathbf{j}+9\mathbf{k}$

82. $\mathbf{v}\bullet(\mathbf{v}\times\mathbf{w})$
$=\mathbf{v}\bullet\begin{vmatrix}\mathbf{i} & \mathbf{j} & \mathbf{k}\\ 3 & 1 & -2\\ -3 & 2 & -1\end{vmatrix}$
$=\mathbf{v}\bullet\left(\begin{vmatrix}1 & -2\\ 2 & -1\end{vmatrix}\mathbf{i}-\begin{vmatrix}3 & -2\\ -3 & -1\end{vmatrix}\mathbf{j}+\begin{vmatrix}3 & 1\\ -3 & 2\end{vmatrix}\mathbf{k}\right)$
$=(3\mathbf{i}+\mathbf{j}-2\mathbf{k})\bullet(3\mathbf{i}+9\mathbf{j}+9\mathbf{k})$
$=3\cdot 3+1\cdot 9+(-2)(9)$
$=0$

83. Same direction:
$\frac{\mathbf{v}}{\|\mathbf{v}\|}=\frac{3\mathbf{i}+\mathbf{j}-2\mathbf{k}}{\sqrt{3^2+1^2+(-2)^2}}$
$=\frac{3\mathbf{i}+\mathbf{j}-2\mathbf{k}}{\sqrt{14}}$
$=\frac{3\sqrt{14}}{14}\mathbf{i}+\frac{\sqrt{14}}{14}\mathbf{j}-\frac{\sqrt{14}}{7}\mathbf{k}$

Opposite direction:
$\frac{-\mathbf{v}}{\|\mathbf{v}\|}=-\frac{3\sqrt{14}}{14}\mathbf{i}-\frac{\sqrt{14}}{14}\mathbf{j}+\frac{\sqrt{14}}{7}\mathbf{k}$

84. $\pm\frac{\mathbf{v}\times\mathbf{w}}{\|\mathbf{v}\times\mathbf{w}\|}=\pm\frac{\begin{vmatrix}\mathbf{i} & \mathbf{j} & \mathbf{k}\\ 3 & 1 & -2\\ -3 & 2 & -1\end{vmatrix}}{\|\mathbf{v}\times\mathbf{w}\|}$
$=\pm\frac{\begin{vmatrix}1 & -2\\ 2 & -1\end{vmatrix}\mathbf{i}-\begin{vmatrix}3 & -2\\ -3 & -1\end{vmatrix}\mathbf{j}+\begin{vmatrix}3 & 1\\ -3 & 2\end{vmatrix}\mathbf{k}}{\|\mathbf{v}\times\mathbf{w}\|}$
$=\pm\left(\frac{3\mathbf{i}+9\mathbf{j}+9\mathbf{k}}{\sqrt{3^2+9^2+9^2}}\right)$
$=\pm\left(\frac{3}{3\sqrt{19}}\mathbf{i}+\frac{9}{3\sqrt{19}}\mathbf{j}+\frac{9}{3\sqrt{19}}\mathbf{k}\right)$
$=\pm\left(\frac{1}{\sqrt{19}}\mathbf{i}+\frac{3}{\sqrt{19}}\mathbf{j}+\frac{3}{\sqrt{19}}\mathbf{k}\right)$
$=\pm\left(\frac{\sqrt{19}}{19}\mathbf{i}+\frac{3\sqrt{19}}{19}\mathbf{j}+\frac{3\sqrt{19}}{19}\mathbf{k}\right)$

So, $\frac{\sqrt{19}}{19}\mathbf{i}+\frac{3\sqrt{19}}{19}\mathbf{j}+\frac{3\sqrt{19}}{19}\mathbf{k}$ or
$-\frac{\sqrt{19}}{19}\mathbf{i}-\frac{3\sqrt{19}}{19}\mathbf{j}-\frac{3\sqrt{19}}{19}\mathbf{k}$

85. $\mathbf{v}=-2\mathbf{i}+\mathbf{j},\quad \mathbf{w}=4\mathbf{i}-3\mathbf{j}$
$\mathbf{v}\cdot\mathbf{w}=-2(4)+1(-3)=-8-3=-11$
$\cos\theta=\frac{\mathbf{v}\cdot\mathbf{w}}{\|\mathbf{v}\|\|\mathbf{w}\|}$
$=\frac{-11}{\sqrt{(-2)^2+1^2}\sqrt{4^2+(-3)^2}}$
$=\frac{-11}{\sqrt{5}\cdot 5}=\frac{-11}{5\sqrt{5}}$
$\theta\approx 169.70°$

86. $\mathbf{v}=3\mathbf{i}-\mathbf{j},\quad \mathbf{w}=\mathbf{i}+\mathbf{j}$
$\mathbf{v}\cdot\mathbf{w}=3(1)+(-1)(1)=3-1=2$
$\cos\theta=\frac{\mathbf{v}\cdot\mathbf{w}}{\|\mathbf{v}\|\|\mathbf{w}\|}$
$=\frac{2}{\sqrt{3^2+(-1)^2}\sqrt{1^2+1^2}}$
$=\frac{2}{\sqrt{10}\sqrt{2}}=\frac{1}{\sqrt{5}}$
$\theta\approx 63.43°$

87. $\mathbf{v}=\mathbf{i}-3\mathbf{j},\quad \mathbf{w}=-\mathbf{i}+\mathbf{j}$
$\mathbf{v}\cdot\mathbf{w}=1(-1)+(-3)(1)=-1-3=-4$
$\cos\theta=\frac{\mathbf{v}\cdot\mathbf{w}}{\|\mathbf{v}\|\|\mathbf{w}\|}$
$=\frac{-4}{\sqrt{1^2+(-3)^2}\sqrt{(-1)^2+1^2}}$
$=\frac{-4}{\sqrt{10}\sqrt{2}}$
$=\frac{-2}{\sqrt{5}}=-\frac{2\sqrt{5}}{5}$
$\theta\approx 153.43°$

88. $\mathbf{v} = \mathbf{i} + 4\mathbf{j}, \quad \mathbf{w} = 3\mathbf{i} - 2\mathbf{j}$

$\mathbf{v} \cdot \mathbf{w} = 1(3) + (4)(-2) = 3 - 8 = -5$

$$\cos\theta = \frac{\mathbf{v} \cdot \mathbf{w}}{\|\mathbf{v}\|\|\mathbf{w}\|} = \frac{-5}{\sqrt{1^2 + 4^2}\sqrt{3^2 + (-2)^2}} = \frac{-5}{\sqrt{17}\sqrt{13}} = \frac{-5}{\sqrt{221}} = -\frac{5\sqrt{221}}{221}$$

$\theta \approx 109.65^\circ$

89. $\mathbf{v} \bullet \mathbf{w} = (\mathbf{i} + \mathbf{j} + \mathbf{k}) \bullet (\mathbf{i} - \mathbf{j} + \mathbf{k}) = 1 \cdot 1 + 1(-1) + 1 \cdot 1 = 1 - 1 + 1 = 1$

$$\cos\theta = \frac{\mathbf{v} \bullet \mathbf{w}}{\|\mathbf{v}\|\|\mathbf{w}\|} = \frac{1}{\sqrt{1^2 + 1^2 + 1^2}\sqrt{1^2 + (-1)^2 + 1^2}} = \frac{1}{\sqrt{3}\sqrt{3}} = \frac{1}{3}$$

$\theta \approx 70.5^\circ$

90. $\mathbf{v} \bullet \mathbf{w} = (\mathbf{i} - \mathbf{j} + \mathbf{k}) \bullet (2\mathbf{i} + \mathbf{j} + \mathbf{k}) = 1 \cdot 2 + (-1)(1) + 1 \cdot 1 = 2 - 1 + 1 = 2$

$$\cos\theta = \frac{\mathbf{v} \bullet \mathbf{w}}{\|\mathbf{v}\|\|\mathbf{w}\|} = \frac{2}{\sqrt{1^2 + (-1)^2 + 1^2}\sqrt{2^2 + 1^2 + 1^2}} = \frac{2}{\sqrt{3}\sqrt{6}} = \frac{2}{\sqrt{18}} = \frac{2}{3\sqrt{2}} = \frac{\sqrt{2}}{3}$$

$\theta \approx 61.9^\circ$

91. $\mathbf{v} \bullet \mathbf{w} = (4\mathbf{i} - \mathbf{j} + 2\mathbf{k}) \bullet (\mathbf{i} - 2\mathbf{j} - 3\mathbf{k}) = 4 \cdot 1 + (-1)(-2) + 2(-3) = 4 + 2 - 6 = 0$

$$\cos\theta = \frac{\mathbf{v} \bullet \mathbf{w}}{\|\mathbf{v}\|\|\mathbf{w}\|} = \frac{0}{\sqrt{4^2 + (-1)^2 + 2^2}\sqrt{1^2 + (-2)^2 + (-3)^2}} = 0$$

$\theta = 90^\circ$

92. $\mathbf{v} \bullet \mathbf{w} = (-\mathbf{i} - 2\mathbf{j} + 3\mathbf{k}) \bullet (5\mathbf{i} + \mathbf{j} + \mathbf{k}) = -1 \cdot 5 + (-2)(1) + 3 \cdot 1 = -5 - 2 + 3 = -4$

$$\cos\theta = \frac{\mathbf{v} \bullet \mathbf{w}}{\|\mathbf{v}\|\|\mathbf{w}\|} = \frac{-4}{\sqrt{(-1)^2 + (-2)^2 + 3^2}\sqrt{5^2 + 1^2 + 1^2}} = \frac{-4}{\sqrt{14}\sqrt{27}} = -\frac{4}{3\sqrt{42}} = -\frac{2\sqrt{42}}{63}$$

$\theta \approx 101.9^\circ$

93. $\mathbf{v} = 2\mathbf{i} + 3\mathbf{j}, \quad \mathbf{w} = -4\mathbf{i} - 6\mathbf{j}$

$\mathbf{v} \cdot \mathbf{w} = (2)(-4) + (3)(-6) = -8 - 18 = -26$

$$\cos\theta = \frac{\mathbf{v} \cdot \mathbf{w}}{\|\mathbf{v}\|\|\mathbf{w}\|} = \frac{-26}{\sqrt{2^2 + 3^2}\sqrt{(-4)^2 + (-6)^2}} = \frac{-26}{\sqrt{13}\sqrt{52}} = \frac{-26}{\sqrt{676}} = \frac{-26}{26} = -1$$

$\theta = \cos^{-1}(-1) = 180^\circ$

Thus, the vectors are parallel.

94. $\mathbf{v}=-2\mathbf{i}-\mathbf{j},\quad \mathbf{w}=2\mathbf{i}+\mathbf{j}$

$\mathbf{v}\cdot\mathbf{w}=(-2)(2)+(-1)(1)=-4-1=-5$

$$\cos\theta=\frac{\mathbf{v}\cdot\mathbf{w}}{\|\mathbf{v}\|\|\mathbf{w}\|}=\frac{-5}{\sqrt{(-2)^2+(-1)^2}\sqrt{2^2+1^2}}=\frac{-5}{\sqrt{5}\sqrt{5}}=\frac{-5}{5}=-1$$

$\theta=\cos^{-1}(-1)=180^\circ$

Thus, the vectors are parallel.

95. $\mathbf{v}=3\mathbf{i}-4\mathbf{j},\quad \mathbf{w}=-3\mathbf{i}+4\mathbf{j}$

$\mathbf{v}\cdot\mathbf{w}=(3)(-3)+(-4)(4)=-9-16=-25$

$$\cos\theta=\frac{\mathbf{v}\cdot\mathbf{w}}{\|\mathbf{v}\|\|\mathbf{w}\|}=\frac{-25}{\sqrt{3^2+(-4)^2}\sqrt{(-3)^2+4^2}}=\frac{-25}{\sqrt{25}\sqrt{25}}=\frac{-25}{25}=-1$$

$\theta=\cos^{-1}(-1)=180^\circ$

Thus, the vectors are parallel.

96. $\mathbf{v}=-2\mathbf{i}+2\mathbf{j},\quad \mathbf{w}=-3\mathbf{i}+2\mathbf{j}$

$\mathbf{v}\cdot\mathbf{w}=(-2)(-3)+(2)(2)=6+4=10$

$$\cos\theta=\frac{\mathbf{v}\cdot\mathbf{w}}{\|\mathbf{v}\|\|\mathbf{w}\|}=\frac{10}{\sqrt{(-2)^2+2^2}\sqrt{(-3)^2+2^2}}=\frac{10}{\sqrt{8}\cdot\sqrt{13}}=\frac{10}{\sqrt{104}}=0.9806$$

$$\theta=\cos^{-1}\left(\frac{10}{\sqrt{104}}\right)\approx 11.31^\circ$$

Thus, the vectors are neither parallel nor orthogonal.

97. $\mathbf{v}=3\mathbf{i}-2\mathbf{j},\quad \mathbf{w}=4\mathbf{i}+6\mathbf{j}$

$\mathbf{v}\cdot\mathbf{w}=(3)(4)+(-2)(6)=12-12=0$

Thus, the vectors are orthogonal.

98. $\mathbf{v}=-4\mathbf{i}+2\mathbf{j},\quad \mathbf{w}=2\mathbf{i}+4\mathbf{j}$

$\mathbf{v}\cdot\mathbf{w}=(-4)(2)+(2)(4)=-8+8=0$

Thus, the vectors are orthogonal.

99. $\mathbf{v}=2\mathbf{i}+\mathbf{j},\quad \mathbf{w}=-4\mathbf{i}+3\mathbf{j}$

The decomposition of $\mathbf{v}$ into 2 vectors $\mathbf{v}_1$ and $\mathbf{v}_2$ so that $\mathbf{v}_1$ is parallel to $\mathbf{w}$ and $\mathbf{v}_2$ is perpendicular to $\mathbf{w}$ is given by: $\mathbf{v}_1=\frac{\mathbf{v}\cdot\mathbf{w}}{\|\mathbf{w}\|^2}\mathbf{w}$ and $\mathbf{v}_2=\mathbf{v}-\mathbf{v}_1$

$$\mathbf{v}_1=\frac{\mathbf{v}\cdot\mathbf{w}}{\|\mathbf{w}\|^2}=\frac{(2\mathbf{i}+\mathbf{j})\cdot(-4\mathbf{i}+3\mathbf{j})}{\left(\sqrt{(-4)^2+3^2}\right)^2}(-4\mathbf{i}+3\mathbf{j})=\frac{(2)(-4)+(1)(3)}{25}(-4\mathbf{i}+3\mathbf{j})=\left(-\frac{1}{5}\right)(-4\mathbf{i}+3\mathbf{j})=\frac{4}{5}\mathbf{i}-\frac{3}{5}\mathbf{j}$$

$$\mathbf{v}_2=\mathbf{v}-\mathbf{v}_1=2\mathbf{i}+\mathbf{j}-\left(\frac{4}{5}\mathbf{i}-\frac{3}{5}\mathbf{j}\right)=\frac{6}{5}\mathbf{i}+\frac{8}{5}\mathbf{j}$$

100. $\mathbf{v}=-3\mathbf{i}+2\mathbf{j},\quad \mathbf{w}=-2\mathbf{i}+\mathbf{j}$

The decomposition of $\mathbf{v}$ into 2 vectors $\mathbf{v}_1$ and $\mathbf{v}_2$ so that $\mathbf{v}_1$ is parallel to $\mathbf{w}$ and $\mathbf{v}_2$ is perpendicular to $\mathbf{w}$ is given by: $\mathbf{v}_1=\frac{\mathbf{v}\cdot\mathbf{w}}{\|\mathbf{w}\|^2}\mathbf{w}$ and $\mathbf{v}_2=\mathbf{v}-\mathbf{v}_1$.

$$\mathbf{v}_1=\frac{\mathbf{v}\cdot\mathbf{w}}{\|\mathbf{w}\|^2}\mathbf{w}=\frac{(-3\mathbf{i}+2\mathbf{j})\cdot(-2\mathbf{i}+\mathbf{j})}{\left(\sqrt{(-2)^2+1^2}\right)^2}(-2\mathbf{i}+\mathbf{j})=\frac{(-3)(-2)+(2)(1)}{5}(-2\mathbf{i}+\mathbf{j})=\left(\frac{8}{5}\right)(-2\mathbf{i}+\mathbf{j})=-\frac{16}{5}\mathbf{i}+\frac{8}{5}\mathbf{j}$$

$$\mathbf{v}_2=\mathbf{v}-\mathbf{v}_1=-3\mathbf{i}+2\mathbf{j}-\left(-\frac{16}{5}\mathbf{i}+\frac{8}{5}\mathbf{j}\right)=\frac{1}{5}\mathbf{i}+\frac{2}{5}\mathbf{j}$$

101. $\mathbf{v} = 2\mathbf{i} + 3\mathbf{j}, \quad \mathbf{w} = 3\mathbf{i} + \mathbf{j}$

The projection of $\mathbf{v}$ onto $\mathbf{w}$ is given by:

$$\frac{\mathbf{v} \cdot \mathbf{w}}{\|\mathbf{w}\|^2}\mathbf{w} = \frac{(2\mathbf{i}+3\mathbf{j})\cdot(3\mathbf{i}+\mathbf{j})}{\left(\sqrt{3^2+1^2}\right)^2}(3\mathbf{i}+\mathbf{j})$$
$$= \frac{(2)(3)+(3)(1)}{10}(3\mathbf{i}+\mathbf{j})$$
$$= \left(\frac{9}{10}\right)(3\mathbf{i}+\mathbf{j}) = \frac{27}{10}\mathbf{i} + \frac{9}{10}\mathbf{j}$$

102. $\mathbf{v} = -\mathbf{i} + 2\mathbf{j}, \quad \mathbf{w} = 3\mathbf{i} - \mathbf{j}$

The projection of $\mathbf{v}$ onto $\mathbf{w}$ is given by:

$$\frac{\mathbf{v} \cdot \mathbf{w}}{\|\mathbf{w}\|^2}\mathbf{w} = \frac{(-\mathbf{i}+2\mathbf{j})\cdot(3\mathbf{i}-\mathbf{j})}{\left(\sqrt{3^2+(-1)^2}\right)^2}(3\mathbf{i}-\mathbf{j})$$
$$= \frac{(-1)(3)+(2)(-1)}{10}(3\mathbf{i}-\mathbf{j})$$
$$= \frac{-5}{10}(3\mathbf{i}-\mathbf{j})$$
$$= -\frac{1}{2}(3\mathbf{i}-\mathbf{j}) = -\frac{3}{2}\mathbf{i} + \frac{1}{2}\mathbf{j}$$

103. $$\cos\alpha = \frac{a}{\|\mathbf{v}\|} = \frac{3}{\sqrt{3^2+(-4)^2+2^2}} = \frac{3}{\sqrt{29}} = \frac{3\sqrt{29}}{29}$$

$\alpha \approx 56.1°$

$$\cos\beta = \frac{b}{\|\mathbf{v}\|} = \frac{-4}{\sqrt{3^2+(-4)^2+2^2}} = -\frac{4}{\sqrt{29}} = -\frac{4\sqrt{29}}{29}$$

$\beta \approx 138.0°$

$$\cos\gamma = \frac{c}{\|\mathbf{v}\|} = \frac{2}{\sqrt{3^2+(-4)^2+2^2}} = \frac{2}{\sqrt{29}} = \frac{2\sqrt{29}}{29}$$

$\gamma \approx 68.2°$

104. $$\cos\alpha = \frac{a}{\|\mathbf{v}\|} = \frac{1}{\sqrt{1^2+(-1)^2+2^2}} = \frac{1}{\sqrt{6}} = \frac{\sqrt{6}}{6}$$

$\alpha \approx 65.9°$

$$\cos\beta = \frac{b}{\|\mathbf{v}\|} = \frac{-1}{\sqrt{1^2+(-1)^2+2^2}} = -\frac{1}{\sqrt{6}} = -\frac{\sqrt{6}}{6}$$

$\beta \approx 114.1°$

$$\cos\gamma = \frac{c}{\|\mathbf{v}\|} = \frac{2}{\sqrt{1^2+(-1)^2+2^2}} = \frac{2}{\sqrt{6}} = \frac{\sqrt{6}}{3}$$

$\gamma \approx 35.3°$

105. $\mathbf{u} = \overrightarrow{P_1P_2} = \mathbf{i} + 2\mathbf{j} + 3\mathbf{k}$

$\mathbf{v} = \overrightarrow{P_1P_3} = 5\mathbf{i} + 4\mathbf{j} + \mathbf{k}$

$$\mathbf{u}\times\mathbf{v} = \begin{vmatrix} \mathbf{i} & \mathbf{j} & \mathbf{k} \\ 1 & 2 & 3 \\ 5 & 4 & 1 \end{vmatrix}$$
$$= \begin{vmatrix} 2 & 3 \\ 4 & 1 \end{vmatrix}\mathbf{i} - \begin{vmatrix} 1 & 3 \\ 5 & 1 \end{vmatrix}\mathbf{j} + \begin{vmatrix} 1 & 2 \\ 5 & 4 \end{vmatrix}\mathbf{k}$$
$$= -10\mathbf{i} + 14\mathbf{j} - 6\mathbf{k}$$

$$\text{Area} = \|\mathbf{u}\times\mathbf{v}\|$$
$$= \sqrt{(-10)^2 + 14^2 + (-6)^2}$$
$$= \sqrt{332} = 2\sqrt{83} \approx 18.2 \text{ sq. units}$$

106. $\mathbf{u} = \overrightarrow{P_1P_2} = 3\mathbf{i} + 2\mathbf{j} + 3\mathbf{k}$

$\mathbf{v} = \overrightarrow{P_1P_3} = -2\mathbf{i} + 2\mathbf{j} + 0\mathbf{k}$

$$\mathbf{u}\times\mathbf{v} = \begin{vmatrix} \mathbf{i} & \mathbf{j} & \mathbf{k} \\ 3 & 2 & 3 \\ -2 & 2 & 0 \end{vmatrix}$$
$$= \begin{vmatrix} 2 & 3 \\ 2 & 0 \end{vmatrix}\mathbf{i} - \begin{vmatrix} 3 & 3 \\ -2 & 0 \end{vmatrix}\mathbf{j} + \begin{vmatrix} 3 & 2 \\ -2 & 2 \end{vmatrix}\mathbf{k}$$
$$= -6\mathbf{i} - 6\mathbf{j} + 10\mathbf{k}$$

$$\text{Area} = \|\mathbf{u}\times\mathbf{v}\|$$
$$= \sqrt{(-6)^2 + (-6)^2 + 10^2}$$
$$= \sqrt{172} = 2\sqrt{43} \approx 13.1 \text{ sq. units}$$

107. $\mathbf{v}\times\mathbf{u} = -(\mathbf{u}\times\mathbf{v}) = -(2\mathbf{i} - 3\mathbf{j} + \mathbf{k}) = -2\mathbf{i} + 3\mathbf{j} - \mathbf{k}$

108. $\mathbf{u}\times\mathbf{v} = (3\mathbf{v})\times\mathbf{v} = 3(\mathbf{v}\times\mathbf{v}) = 3\cdot\mathbf{0} = \mathbf{0}$

109. Let the positive x-axis point downstream, so that the velocity of the current is $\mathbf{v_c} = 2\mathbf{i}$. Let $\mathbf{v_w}$ = the velocity of the swimmer in the water and $\mathbf{v_g}$ = the velocity of the swimmer relative to the land. Then $\mathbf{v_g} = \mathbf{v_w} + \mathbf{v_c}$. The speed of the swimmer is $\|\mathbf{v_w}\| = 5$, and the heading is directly across the river, so $\mathbf{v_w} = 5\mathbf{j}$. Then

$$\mathbf{v_g} = \mathbf{v_w} + \mathbf{v_c} = 5\mathbf{j} + 2\mathbf{i} = 2\mathbf{i} + 5\mathbf{j}$$

$$\|\mathbf{v_g}\| = \sqrt{2^2+5^2} = \sqrt{29} \approx 5.39 \text{ mi/hr}$$

Since the river is 1 mile wide, it takes the swimmer about 0.2 hour to cross the river. The swimmer will end up $(0.2)(2) = 0.4$ miles downstream.

110. Let $\mathbf{v}_a$ = the velocity of the plane in still air, $\mathbf{v}_w$ = the velocity of the wind, and $\mathbf{v}_g$ = the velocity of the plane relative to the ground.

$$\mathbf{v}_g = \mathbf{v}_a + \mathbf{v}_w$$

$$\mathbf{v}_a = 500\mathbf{j}$$

$$\mathbf{v}_w = 60\left(\frac{\sqrt{2}}{2}\mathbf{i} - \frac{\sqrt{2}}{2}\mathbf{j}\right) = 30\sqrt{2}\,\mathbf{i} - 30\sqrt{2}\,\mathbf{j}$$

$$\mathbf{v}_g = \mathbf{v}_a + \mathbf{v}_w = 500\mathbf{j} + 30\sqrt{2}\,\mathbf{i} - 30\sqrt{2}\,\mathbf{j} = \left(30\sqrt{2}\right)\mathbf{i} + \left(500 - 30\sqrt{2}\right)\mathbf{j}$$

The speed of the plane relative to the ground is:

$$\|\mathbf{v}_g\| = \sqrt{\left(30\sqrt{2}\right)^2 + \left(500 - 30\sqrt{2}\right)^2}$$
$$= \sqrt{1800 + 250{,}000 - 30{,}000\sqrt{2} + 1800}$$
$$\approx \sqrt{253{,}600 - 30{,}000\sqrt{2} + 1800}$$
$$\approx 459.54 \text{ km/hr}$$

To find the direction, find the angle between $\mathbf{v}_g$ and a convenient vector such as due north, $\mathbf{j}$.

$$\cos\theta = \frac{\mathbf{v}_g \cdot \mathbf{j}}{\|\mathbf{v}_g\|\|-\mathbf{j}\|}$$
$$= \frac{\left(30\sqrt{2}\right)(0) + \left(500 - 30\sqrt{2}\right)(1)}{459.54\sqrt{0^2 + 1^2}}$$
$$= \frac{500 - 30\sqrt{2}}{459.54} \approx 0.9957$$
$$\theta \approx 5.30^\circ$$

The plane is traveling with a ground speed of about 459.54 km/hr in a direction of about 5.30° east of north.

111. Let $\mathbf{F}_1$ = the tension on the left cable, $\mathbf{F}_2$ = the tension on the right cable, and $\mathbf{F}_3$ = the force of the weight of the box.

$$\mathbf{F}_1 = \|\mathbf{F}_1\|\left[\cos(140^\circ)\mathbf{i} + \sin(140^\circ)\mathbf{j}\right] \approx \|\mathbf{F}_1\|(-0.7660\mathbf{i} + 0.6428\mathbf{j})$$

$$\mathbf{F}_2 = \|\mathbf{F}_2\|\left[\cos(30^\circ)\mathbf{i} + \sin(30^\circ)\mathbf{j}\right] \approx \|\mathbf{F}_2\|(0.8660\mathbf{i} + 0.5000\mathbf{j})$$

$$\mathbf{F}_3 = -2000\mathbf{j}$$

For equilibrium, the sum of the force vectors must be zero.

$$\mathbf{F}_1 + \mathbf{F}_2 + \mathbf{F}_3$$
$$= -0.7660\|\mathbf{F}_1\|\mathbf{i} + 0.6428\|\mathbf{F}_1\|\mathbf{j} + 0.8660\|\mathbf{F}_2\|\mathbf{i} + 0.5000\|\mathbf{F}_2\|\mathbf{j} - 2000\mathbf{j}$$
$$= \left(-0.7660\|\mathbf{F}_1\| + 0.8660\|\mathbf{F}_2\|\right)\mathbf{i} + \left(0.6428\|\mathbf{F}_1\| + 0.5000\|\mathbf{F}_2\| - 2000\right)\mathbf{j}$$
$$= 0$$

Set the $\mathbf{i}$ and $\mathbf{j}$ components equal to zero and solve:

$$\begin{cases} -0.7660\|\mathbf{F}_1\| + 0.8660\|\mathbf{F}_2\| = 0 \\ 0.6428\|\mathbf{F}_1\| + 0.5000\|\mathbf{F}_2\| - 2000 = 0 \end{cases}$$

Solve the first equation for $\|\mathbf{F}_2\|$ and substitute the result into the second equation to solve the system:

$$\|\mathbf{F}_2\| \approx \frac{0.7660}{0.8660}\|\mathbf{F}_1\| \approx 0.8845\|\mathbf{F}_1\|$$

$$0.6428\|\mathbf{F}_1\| + 0.5000\left(0.8845\|\mathbf{F}_1\|\right) - 2000 = 0$$
$$1.08505\|\mathbf{F}_1\| = 2000$$
$$\|\mathbf{F}_1\| \approx 1843.23 \text{ lb}$$
$$\|\mathbf{F}_2\| = 0.8845(1843.23) = 1630.34 \text{ lb}$$

The tension in the left cable is about 1843.23 pounds and the tension in the right cable is about 1630.34 pounds.

112. Let $\mathbf{v}_t$ = the true velocity of the motorboat, $\mathbf{v}_b$ = the velocity of the boat in still water, and $\mathbf{v}_c$ = the velocity of the current.

$$\mathbf{v}_b = -11\mathbf{j}$$

$$\mathbf{v}_c = 3\left[-\cos(45^\circ)\mathbf{i} - \sin(45^\circ)\mathbf{j}\right] = -\frac{3\sqrt{2}}{2}\mathbf{i} - \frac{3\sqrt{2}}{2}\mathbf{j}$$

$\mathbf{v}_t = \mathbf{v}_b + \mathbf{v}_c$, so

$$\mathbf{v}_b = \mathbf{v}_t - \mathbf{v}_c$$
$$= -11\mathbf{j} - \left(-\frac{3\sqrt{2}}{2}\mathbf{i} - \frac{3\sqrt{2}}{2}\mathbf{j}\right)$$
$$= -11\mathbf{j} + \frac{3\sqrt{2}}{2}\mathbf{i} + \frac{3\sqrt{2}}{2}\mathbf{j}$$
$$= \frac{3\sqrt{2}}{2}\mathbf{i} + \left(-11 + \frac{3\sqrt{2}}{2}\right)\mathbf{j}$$

$$\|\mathbf{v_b}\| = \sqrt{\left(\frac{3\sqrt{2}}{2}\right)^2 + \left(-11+\frac{3\sqrt{2}}{2}\right)^2}$$

$$\approx \sqrt{83.33} \approx 9.13$$

To find the heading, find the angle between $\mathbf{v_b}$ and a convenient vector such as due south, $-\mathbf{j}$.

$$\cos\theta = \frac{\mathbf{v_b}\cdot(-\mathbf{j})}{\|\mathbf{v_b}\|\|-\mathbf{j}\|}$$

$$\approx \frac{\left(\frac{3\sqrt{2}}{2}\right)(0) + \left(-11+\frac{3\sqrt{2}}{2}\right)(-1)}{(9.13)\sqrt{0^2+1^2}}$$

$$= \frac{11-\frac{3\sqrt{2}}{2}}{9.13} \approx 0.9725$$

$$\theta \approx \cos^{-1}(0.9725) \approx 13.47^\circ$$

The speed of the motor boat relative to the water is about 9.13 mi/hr, and the compass indicates that the boat is headed in a direction of about 13.47° east of south.

113. $\mathbf{F} = 5\left[\cos(60^\circ)\mathbf{i} + \sin(60^\circ)\mathbf{j}\right]$

$$= 5\left(\frac{1}{2}\mathbf{i} + \frac{\sqrt{3}}{2}\mathbf{j}\right) = \frac{5}{2}\mathbf{i} + \frac{5\sqrt{3}}{2}\mathbf{j}$$

$$\overrightarrow{AB} = 20\mathbf{i}$$

$$W = \mathbf{F}\cdot\overrightarrow{AB} = \left(\frac{5}{2}\mathbf{i} + \frac{5\sqrt{3}}{2}\mathbf{j}\right)\cdot 20\mathbf{i}$$

$$= \left(\frac{5}{2}\right)(20) + \left(\frac{5\sqrt{3}}{2}\right)(0)$$

$$= 50 \text{ ft-lb}$$

Chapter 8 Test

1. – 3.

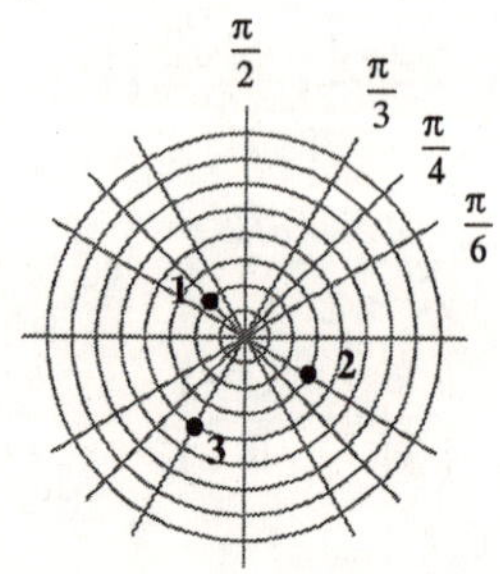

4. $x = 2$ and $y = 2\sqrt{3}$

$r = \sqrt{x^2+y^2} = \sqrt{(2)^2 + (2\sqrt{3})^2} = \sqrt{16} = 4$ and

$$\tan\theta = \frac{y}{x} = \frac{2\sqrt{3}}{2} = \sqrt{3}$$

Since (x, y) is in quadrant I, we have $\theta = \frac{\pi}{3}$. The point $(2, 2\sqrt{3})$ in rectangular coordinates is the point $\left(4, \frac{\pi}{3}\right)$ in polar coordinates.

5.

$$r = 7$$

$$\sqrt{x^2+y^2} = 7$$

$$\left(\sqrt{x^2+y^2}\right)^2 = 7^2$$

$$x^2 + y^2 = 49$$

Thus the equation is a circle with center (0, 0) and radius = 7.

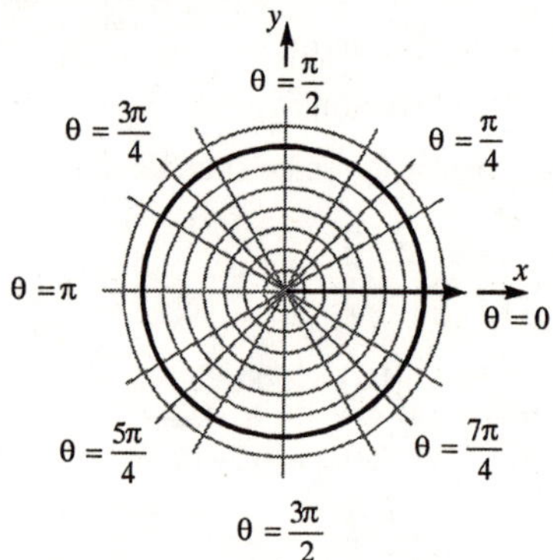

6. $\tan\theta = 3$

$$\frac{\sin\theta}{\cos\theta} = 3$$

$$\frac{r\sin\theta}{r\cos\theta} = 3$$

$$\frac{y}{x} = 3 \text{ or } y = 3x$$

Thus the equation is a straight line with $m = 3$ and $b = 0$.

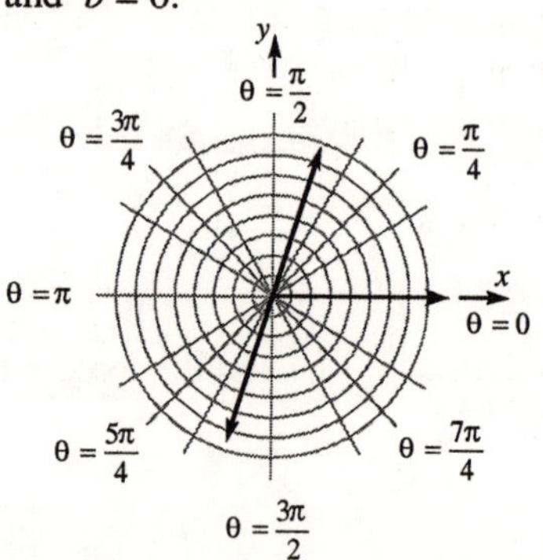

7. $r\sin^2\theta + 8\sin\theta = r$

$r^2\sin^2\theta + 8r\sin\theta = r^2$

$y^2 + 8y = x^2 + y^2$

$8y = x^2 \quad \text{or} \quad 4(2)y = x^2$

The graph is a parabola with vertex (0, 0) and focus (0, 2) in rectangular coordinates.

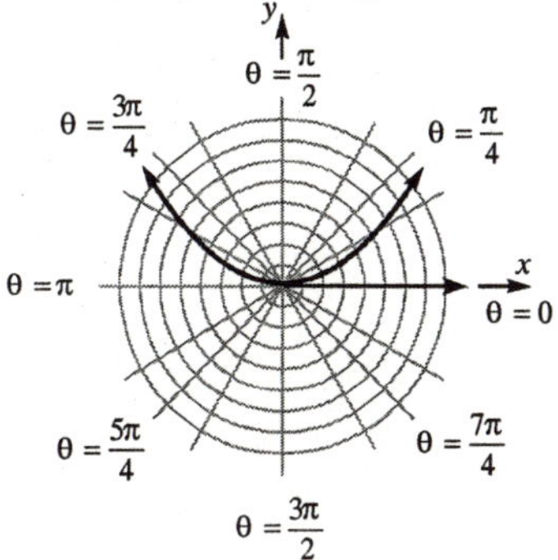

8. $r^2\cos\theta = 5$

Polar axis: Replace θ with $-\theta$:

$r^2\cos(-\theta) = 5$

$r^2\cos\theta = 5$

Since the resulting equation is the same as the original, the graph is symmetrical with respect to the polar axis.

The line $\theta = \frac{\pi}{2}$: Replace θ with $\pi - \theta$:

$r^2\cos(\pi - \theta) = 5$

$r^2(\cos\pi\cos\theta + \sin\pi\sin\theta) = 5$

$r^2(-\cos\theta) = 5$

$-r^2\cos\theta = 5$

Since the resulting equation is not the same as the original, the graph **may or may not be** symmetrical with respect to the line $\theta = \dfrac{\pi}{2}$.

The pole: Replacing r with $-r$:

$(-r)^2\cos\theta = 5$

$r^2\cos\theta = 5$

Since the resulting equation is the same as the original, the graph is symmetrical with respect to the pole.

Note: Since we have now established symmetry about the pole and the polar axis, it can be shown that the graph must also be symmetric about the line $\theta = \frac{\pi}{2}$.

9. $r = 5\sin\theta\cos^2\theta$

The polar axis: Replace θ with $-\theta$:

$r = 5\sin(-\theta)\cos^2(-\theta)$

$r = 5(-\sin\theta)\cos^2\theta$

$r = -5\sin\theta\cos^2\theta$

Since the resulting equation is the not same as the original, the graph **may or may not be** symmetrical with respect to the polar axis.

The line $\theta = \frac{\pi}{2}$: Replace θ with $\pi - \theta$:

$r = 5\sin(\pi - \theta)\cos^2(\pi - \theta)$

$r = 5(\sin\pi\cos\theta - \cos\pi\sin\theta)(-\cos\theta)^2$

$r = 5(0\cdot\cos\theta - (-1)\cdot\sin\theta)(\cos^2\theta)$

$r = 5\sin\theta\cos^2\theta$

Since the resulting equation is the same as the original, the graph is symmetrical with respect to the line $\theta = \frac{\pi}{2}$.

The pole: Replacing r with $-r$:

$-r = 5\sin\theta\cos^2\theta$

$r = -5\sin\theta\cos^2\theta$

Since the resulting equation is not the same as the original, the graph **may or may not be** symmetrical with respect to the pole.

10. $z\cdot w = [2(\cos 85° + i\sin 85°)][3(\cos 22° + i\sin 22°)]$

$= 2\cdot 3\cdot[\cos(85° + 22°) + i\sin(85° + 22°)]$

$= 6(\cos 107° + i\sin 107°)$

11. $\dfrac{w}{z} = \dfrac{3(\cos 22° + i\sin 22°)}{2(\cos 85° + i\sin 85°)}$

$= \dfrac{3}{2}[\cos(22° - 85°) + i\sin(22° - 85°)]$

$= \dfrac{3}{2}[\cos(-63°) + i\sin(-63°)]$

Since $-63°$ has the same terminal side as $297°$, we can write

$\dfrac{w}{z} = \dfrac{3}{2}(\cos 297° + i\sin 297°)$

12. $w^5 = [3(\cos 22° + i\sin 22°)]^5$

$= 3^5[\cos(5\cdot 22°) + i\sin(5\cdot 22°)]$

$= 243(\cos 110° + i\sin 110°)$

13. Let $w=-8+8\sqrt{3}i$; then

$$|w|=\sqrt{(-8)^2+(8\sqrt{3})^2}=\sqrt{64+192}=\sqrt{256}=16$$

so we can write w in polar form as

$$w=16\left(-\frac{1}{2}+\frac{\sqrt{3}}{2}i\right)=16(\cos 120^\circ+i\sin 120^\circ)$$

Using De Moivre's Theorem, the three distinct 3rd roots of w are

$$z_k=\sqrt[3]{16}\left[\cos\left(\frac{120^\circ}{3}+\frac{360^\circ k}{3}\right)+i\sin\left(\frac{120^\circ}{3}+\frac{360^\circ k}{3}\right)\right]$$
$$=2\sqrt[3]{2}\left[\cos(40^\circ+120^\circ k)+i\sin(40^\circ+120^\circ k)\right]$$

where $k=0$, 1, and 2 . So we have

$$z_0=2\sqrt[3]{2}\left[\cos(40^\circ+120^\circ\cdot 0)+i\sin(40^\circ+120^\circ\cdot 0)\right]$$
$$=2\sqrt[3]{2}\left(\cos 40^\circ+i\sin 40^\circ\right)$$
$$\approx 1.93+1.62i$$
$$z_1=2\sqrt[3]{2}\left[\cos(40^\circ+120^\circ\cdot 1)+i\sin(40^\circ+120^\circ\cdot 1)\right]$$
$$=2\sqrt[3]{2}\left(\cos 160^\circ+i\sin 160^\circ\right)$$
$$\approx -2.37+0.86i$$
$$z_2=2\sqrt[3]{2}\left[\cos(40^\circ+120^\circ\cdot 2)+i\sin(40^\circ+120^\circ\cdot 2)\right]$$
$$=2\sqrt[3]{2}\left(\cos 280^\circ+i\sin 280^\circ\right)$$
$$\approx 0.44-2.48i$$

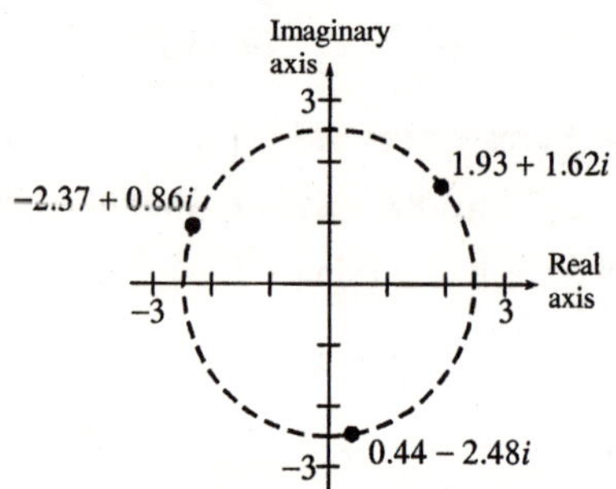

14. $v=\langle x_2-x_1, y_2-y_1\rangle$

$$=\langle 8\sqrt{2}-3\sqrt{2}\ ,\ 2\sqrt{2}-7\sqrt{2}\rangle$$
$$=\langle 5\sqrt{2},-5\sqrt{2}\rangle$$

15. $\|\mathbf{v}\|=\sqrt{a_1^{\,2}+b_1^{\,2}}$

$$=\sqrt{\left(5\sqrt{2}\right)^2+\left(-5\sqrt{2}\right)^2}$$
$$=\sqrt{100}=10$$

16. $\mathbf{u}=\dfrac{\mathbf{v}}{\|\mathbf{v}\|}=\dfrac{1}{10}\langle 5\sqrt{2},-5\sqrt{2}\rangle=\left\langle\dfrac{\sqrt{2}}{2},-\dfrac{\sqrt{2}}{2}\right\rangle$

17. From problem (16), we can write

$$\mathbf{v}=\|\mathbf{v}\|\cdot\mathbf{u}=10\left\langle\frac{\sqrt{2}}{2},-\frac{\sqrt{2}}{2}\right\rangle$$
$$=10\langle\cos 315^\circ, \sin 315^\circ\rangle$$
$$=10(\cos 315^\circ\mathbf{i}+\sin 315^\circ\mathbf{j})$$

Thus the angle between $\mathbf{v}$ and $\mathbf{i}$ is 315°.

18. From problem (17), we can write

$$\mathbf{v}=10\langle\cos 315^\circ\mathbf{i}+\sin 315^\circ\mathbf{j}\rangle$$
$$=10\left\langle\frac{\sqrt{2}}{2}\mathbf{i}+\left(-\frac{\sqrt{2}}{2}\right)\mathbf{j}\right\rangle$$
$$=5\sqrt{2}\,\mathbf{i}-5\sqrt{2}\,\mathbf{j}$$

19. $\mathbf{v}_1+2\mathbf{v}_2-\mathbf{v}_3=\langle 4,6\rangle+2\langle -3,-6\rangle-\langle -8,4\rangle$

$$=\langle 4+2(-3)-(-8),6+2(-6)-4\rangle$$
$$=\langle 4-6+8,6-12-4\rangle$$
$$=\langle 6,-10\rangle$$

20 – 21.

If θ_{ij} is the angle between $\mathbf{v}_i$ and $\mathbf{v}_j$, then

$\cos\theta_{ij}=\dfrac{\mathbf{v}_i\cdot\mathbf{v}_j}{\|\mathbf{v}_i\|\cdot\|\mathbf{v}_j\|}$. Thus,

$$\cos\theta_{12}=\frac{4(-3)+6(-6)}{\sqrt{52}\cdot\sqrt{45}}=\frac{-48}{48.4}=-0.992$$
$$\cos\theta_{13}=\frac{4(-8)+6(4)}{\sqrt{52}\cdot\sqrt{80}}=\frac{-8}{64.5}=-0.124$$
$$\cos\theta_{14}=\frac{4(10)+6(15)}{\sqrt{52}\cdot\sqrt{325}}=\frac{130}{130}=1$$
$$\cos\theta_{23}=\frac{(-3)(-8)+(-6)(4)}{\sqrt{45}\cdot\sqrt{80}}=\frac{0}{60}=0$$
$$\cos\theta_{24}=\frac{(-3)10+(-6)15}{\sqrt{45}\cdot\sqrt{325}}=\frac{-120}{120.9}=-0.992$$
$$\cos\theta_{34}=\frac{(-8)10+(4)15}{\sqrt{80}\cdot\sqrt{325}}=\frac{-20}{161.2}=-0.124$$

The vectors $\mathbf{v}_i$ and $\mathbf{v}_j$ are parallel if $\cos\theta_{ij}=1$ and are orthogonal if $\cos\theta_{ij}=0$. Thus, vectors $\mathbf{v}_1$ and $\mathbf{v}_4$ are parallel and $\mathbf{v}_2$ and $\mathbf{v}_3$ are orthogonal.

22. The angle between vectors $\mathbf{v}_1$ and $\mathbf{v}_2$ is

$$\cos^{-1}\left(\frac{\mathbf{v}_1\cdot\mathbf{v}_2}{\|\mathbf{v}_1\|\cdot\|\mathbf{v}_2\|}\right)\approx\cos^{-1}(-0.992)\approx 173^\circ.$$

23. $\mathbf{u}\times\mathbf{v}=\begin{vmatrix}\mathbf{i} & \mathbf{j} & \mathbf{k}\\ 2 & -3 & 1\\ -1 & 3 & 2\end{vmatrix}$

$=\begin{vmatrix}-3 & 1\\ 3 & 2\end{vmatrix}\mathbf{i}-\begin{vmatrix}2 & 1\\ -1 & 2\end{vmatrix}\mathbf{j}+\begin{vmatrix}2 & -3\\ -1 & 3\end{vmatrix}\mathbf{k}$

$=(-6-3)\mathbf{i}-(4-(-1))\mathbf{j}+(6-3)\mathbf{k}$

$=-9\mathbf{i}-5\mathbf{j}+3\mathbf{k}$

24. To find the direction angles for $\mathbf{u}$, we first evaluate $\|\mathbf{u}\|$.

$\|\mathbf{u}\|=\sqrt{(2)^2+(-3)^2+1^2}=\sqrt{4+9+1}=\sqrt{14}$

From the Theorem on Direction Angles, we have $\cos\alpha=\frac{2}{\sqrt{14}}$, $\cos\beta=\frac{-3}{\sqrt{14}}$, and $\cos\gamma=\frac{1}{\sqrt{14}}$.

Thus, the direction angles can be found by using the inverse cosine function. That is,

$\alpha=\cos^{-1}\frac{2}{\sqrt{14}}\approx 57.7°$

$\beta=\cos^{-1}\frac{-3}{\sqrt{14}}\approx 143.3°$

$\gamma=\cos^{-1}\frac{1}{\sqrt{14}}\approx 74.5°$

25. The area of a parallelogram with $\mathbf{u}$ and $\mathbf{v}$ as adjacent sides is given by $\|\mathbf{u}\times\mathbf{v}\|$.

From problem (1), we have that $\mathbf{u}\times\mathbf{v}=-9\mathbf{i}-5\mathbf{j}+3\mathbf{k}$.

Therefore, the area of the parallelogram is

$\|\mathbf{u}\times\mathbf{v}\|=\sqrt{(-9)^2+(-5)^2+3^2}$

$=\sqrt{81+25+9}$

$=\sqrt{115}\approx 10.724$ square units

26. We first calculate the angle α.

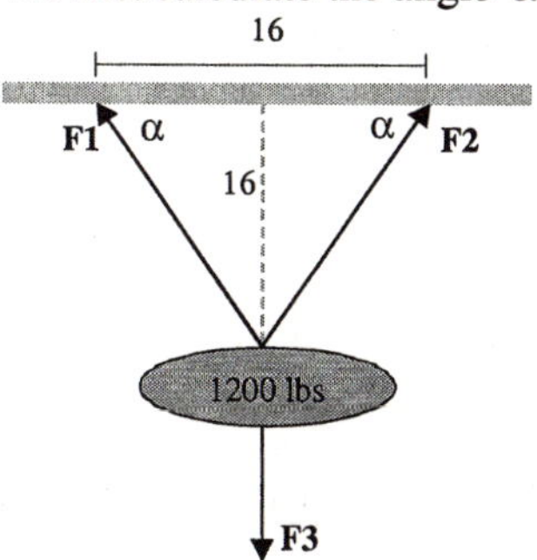

Using the right triangle in the sketch, we conclude $\tan\alpha=\frac{16}{8}=2$ so that

$\alpha=\tan^{-1}2\approx 63.43°$. Thus the three force vectors in the problem can be written as

$\mathbf{F1}=\|\mathbf{F1}\|(\cos 116.57°\,\mathbf{i}+\sin 116.57°\,\mathbf{j})$

$=\|\mathbf{F1}\|(-0.447\mathbf{i}+0.894\mathbf{j})$

$\mathbf{F2}=\|\mathbf{F2}\|(\cos 63.43°\,\mathbf{i}+\sin 63.43°\,\mathbf{j})$

$=\|\mathbf{F2}\|(0.447\mathbf{i}+0.894\mathbf{j})$

$\mathbf{F3}=-1200\mathbf{j}$

Since the system is in equilibrium, we need $\mathbf{F1}+\mathbf{F2}+\mathbf{F3}=0\mathbf{i}+0\mathbf{j}$.

This means

$\left[\|\mathbf{F1}\|(-0.447)+\|\mathbf{F2}\|(0.447)+0\right]=0$ and

$\left[\|\mathbf{F1}\|(0.894)+\|\mathbf{F2}\|(0.894)-1200\right]=0$.

The first equation gives $\|\mathbf{F1}\|=\|\mathbf{F2}\|$; if we call this common value $\|\mathbf{F}\|$ and substitute into the second equation we get

$2\cdot\|\mathbf{F}\|\cdot(0.894)=1200$

$$\|\mathbf{F}\|=\frac{1200}{2(0.894)}\approx 671.14$$

Thus the cable must be able to endure a tension of approximately 671.14 lbs.

Chapter 8 Projects

Project 1

a.

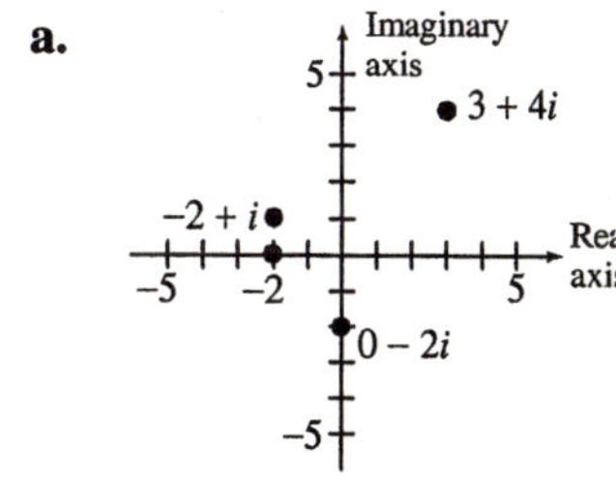

b.

$z = a_0$	a_1	a_2	a_3	a_4	a_5	a_6
$0.1-0.4i$	$-0.05-0.48i$	$-0.13-0.35i$	$-0.01-0.31i$	$0.01-0.35i$	$0.2-0.41i$	$0.07-0.42i$
$0.5+0.8i$	$0.11+1.6i$	$-2.05+1.15i$	$3.37-3.92i$	$-3.52-25.6i$	$-641.9+181.1i$	$379290-232509i$
$-0.9+0.7i$	$-0.58-0.56i$	$0.87+1.35i$	$-1.95-1.67i$	$0.13+7.21i$	$-52.88+2.56i$	$2788.5-269.6i$
$-1.1+0.1i$	$0.1-0.12i$	$-1.10+0.76i$	$0.11-0.068i$	$-1.09+0.085i$	$0.085-0.85i$	$-1.10+0.086i$
$0-1.3i$	$-1.69-1.3i$	$1.17+3.09i$	$-8.21+5.92i$	$32.46-98.47i$	$-8643.6-6393.7i$	$33833744+110529134.4i$
$1+1i$	$1+3i$	$-7+7i$	$1-97i$	$-9407-193i$	$88454401+3631103i$	$7.8\times10^{15}+6.4\times10^{14}i$

c. z_1 and z_4 are in the Mandlebrot set. a_6 for the complex numbers not in the set have very large components.

d.

z	$\|z\|$	$\|a_6\|$
$0.1-0.4i$	0.4	0.4
$0.5+0.8i$	0.9	444884
$-0.9+0.7i$	1.1	2802
$-1.1+0.1i$	1.1	1.1
$0-1.3i$	1.3	115591573
$1+1i$	1.4	7.8×10^{15}

The numbers which are not in the Mandlebrot set satisfy this condition. The numbers which are in the Mandlebrot set satisfy the condition $|a_n| \le 2$.

Project 2 (web)

1. If $h = 0$, then $\cos(2\pi h) = 1$ and $\sin(2\pi h) = 0$, then Magnitude $= |1 + 0.25(1)| = 1.25$ (max).

   If h is high enough such that $\cos(2\pi h) = \frac{\sqrt{2}}{2}$ and $\sin(2\pi h) = \frac{\sqrt{2}}{2}$, then

   $$\text{Magnitude} = \left| 1 + 0.25\left(\frac{\sqrt{2}}{2} + \frac{\sqrt{2}}{2}i\right) \right| = \sqrt{\left(1 + \frac{\sqrt{2}}{8}\right)^2 + \left(\frac{\sqrt{2}}{8}\right)^2} \approx 1.19$$

   If h is high enough such that $\cos(2\pi h) = 0$ and $\sin(2\pi h) = 1$, then

   $$\text{Magnitude} = \left| 1 + 0.25(0 + i) \right| = \sqrt{1 + (0.25)^2} \approx 1.03$$

   If h is high enough such that $\cos(2\pi h) = 0$ and $\sin(2\pi h) = -1$, then

   $$\text{Magnitude} = \left| 1 + 0.25(0 - i) \right| \approx 1.03$$

   If h is high enough such that $\cos(2\pi h) = -1$ and $\sin(2\pi h) = 0$, then

   $$\text{Magnitude} = \left| 1 + 0.25(-1) \right| \approx 0.75 \text{ (min).}$$

   Because sine and cosine oscillate between -1 and 1, the magnitude will oscillate between a maximum and a minimum.

   $$\text{SWR} = \frac{1.25}{.75} \approx 1.67$$

2. The distance between two consecutive minima will be the period of cosine (the value of cosine is -1 once per cycle): Period $= \frac{2\pi}{2\pi} = 1$.

3. The distance between two consecutive maxima will be the period of cosine (the value of cosine is 1 once per cycle): Period $= \frac{2\pi}{2\pi} = 1$.

4. The distance between a minimum and a maximum is half a period: $0.5(1) = 0.5$.

5. The pager must be sensitive enough to receive a minimum signal strength of 0.75.

6. 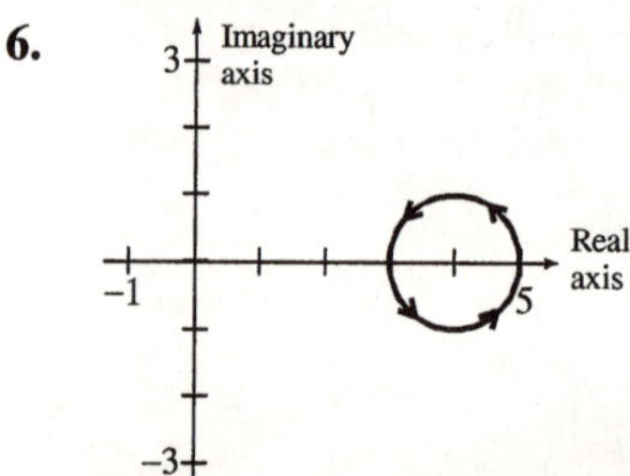

   Since the magnitude of a complex number is the distance from (0, 0) to the point, z, the minimum is 0.75 and the maximum is 1.25

Project 3 (web)

a. 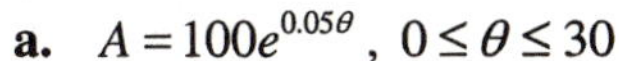
$A = 100e^{0.05\theta}$, $0 \le \theta \le 30$

400

-400 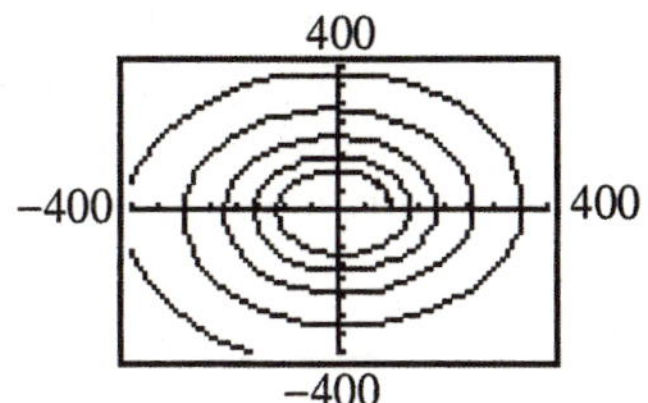 400

-400

b. Approximately 19.7 years.

400

-400 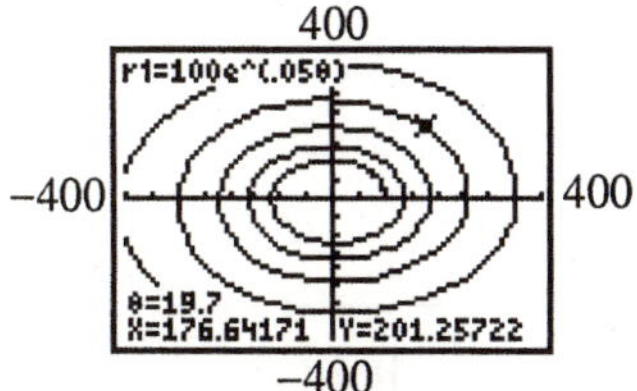 400

-400

c. Approximately 26.1 years.

400

-400 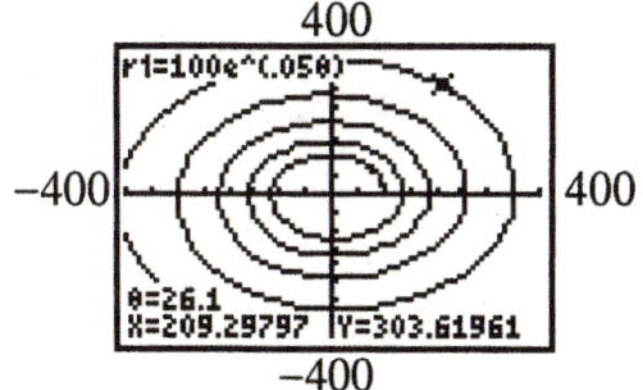 400

-400

d. $A = 100e^{0.10\theta}$, $0 \le \theta \le 30$

1000

-1000 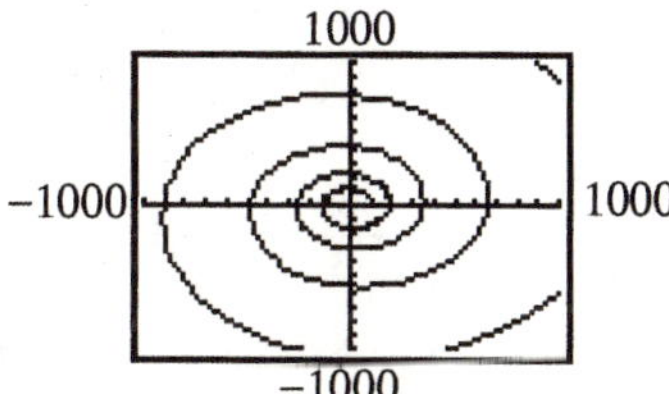 1000

-1000

e. Approximately 7.5 years

1000

-1000 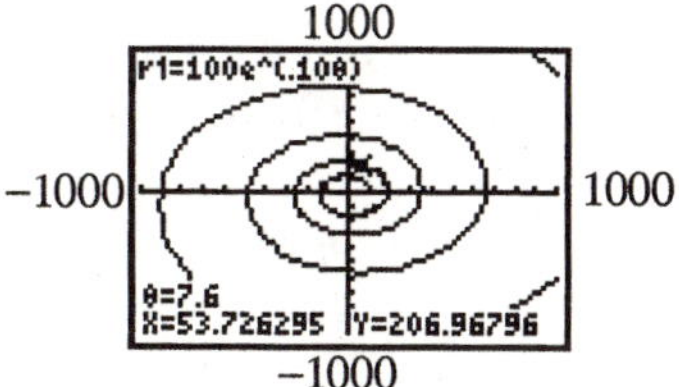 1000

-1000

f. Approximately 13.5 years.

1000

-1000 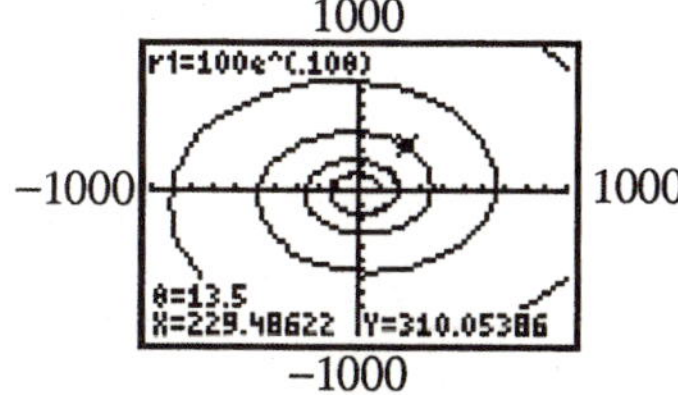 1000

-1000

g. Doubling the interest rate causes the time to double to be less than half. $(0.5 \times 19.7 - 9.85, \text{ not } 7.5)$

Project 4 (web)

A. a. $e^{i\pi} = \cos\pi + i\sin\pi = -1 + i(0) = -1$

Therefore, $e^{i\pi} + 1 = 0$.

b.
$$\frac{e^{ix} - e^{-ix}}{2i} = \frac{\cos x + i\sin x - [\cos(-x) + i\sin(-x)]}{2i}$$
$$= \frac{\cos x + i\sin x - \cos x + i\sin x}{2i}$$
$$= \frac{2i\sin x}{2i}$$
$$= \sin x$$

$$\frac{e^{ix} + e^{-ix}}{2} = \frac{\cos x + i\sin x + [\cos(-x) + i\sin(-x)]}{2}$$
$$= \frac{\cos x + i\sin x + \cos x - i\sin x}{2}$$
$$= \frac{2\cos x}{2}$$
$$= \cos x$$

c. $\sin(1+i)$
$$= \frac{e^{i(1+i)} - e^{-i(1+i)}}{2i}$$
$$= \frac{e^{-1+i} - e^{1-i}}{2i}$$
$$= \frac{e^{-1} \cdot e^{i(1)} - e \cdot e^{-i(1)}}{2i}$$
$$= \frac{e^{-1}(\cos 1 + i\sin 1) - e[\cos(-1) + i\sin(-1)]}{2i}$$
$$= \frac{0.1988 + i(0.3096) - 1.4687 + 2.2874i}{2i}$$
$$= \frac{-1.2699 + 2.597i}{2i} \cdot \frac{i}{i}$$
$$= 1.2985 + 0.6350i$$

d. $a + bi = r_1(\cos x + i\sin x) = r_1 e^{ix}$,

$r_1 = \sqrt{a^2 + b^2}$

$c + di = r_2(\cos y + i\sin y) = r_2 e^{iy}$,

$r_2 = \sqrt{c^2 + d^2}$

$$(a+bi)(c+di) = r_1 e^{ix} \cdot r_2 e^{iy}$$
$$= r_1 r_2 e^{ix} e^{iy}$$
$$= r_1 r_2 e^{ix+iy}$$
$$= r_1 r_2 e^{i(x+y)}$$
$$= r_1 r_2 [\cos(x+y) + i\sin(x+y)]$$

$$\frac{a+bi}{c+di}=\frac{r_1e^{ix}}{r_2e^{iy}}=\frac{r_1}{r_2}e^{ix-iy}=\frac{r_1}{r_2}e^{i(x-y)}$$
$$=\frac{r_1}{r_2}\left[\cos(x-y)+i\sin(x-y)\right]$$

B. a. 1. $x=u+iv$

$$i(u+iv)^3+8=0$$
$$i(u^3+3u^2(iv)+3u(iv)^2+(iv)^3)+8=0$$
$$i(u^3+3iu^2v-3uv^2-iv^3)+8=0$$
$$(v^3-3u^2v+8)+(u^3-3uv^2)i=0$$

2. $v^3-3u^2v+8=0$

$$u^2=\frac{8+v^3}{3v}$$

$$u^3-3uv^2=0$$
$$u(u^2-3v^2)=0$$
$$u=0 \quad \text{or} \quad u^2-3v^2=0$$

If $u=0$, then $0=\dfrac{8+v^3}{3v}$

$$8+v^3=0$$
$$v=-2$$

If $u^2-3v^2=0$, then $u^2-3v^2=0$

$$u^2=3v^2$$
$$3v^2=\frac{8+v^3}{3v}$$
$$9v^3=8+v^3$$
$$8v^3=8$$
$$v^3=1$$
$$v=1$$

If $u=0$ and $v=-2$, then $x=-2i$.

If $v=1$, then $u=\pm\sqrt{3}$. Thus, $x=\sqrt{3}+i$ or $x=-\sqrt{3}+i$

3. Graph $u=\pm\sqrt{\dfrac{8+v^3}{3v}}$ and $u=\pm\sqrt{3}v$

4. The points of intersection are (1, 1.73) and $(1,-1.73)$, which matches up to the second set of intersection points. There should be three points of intersection, however, the "missing" point is complex so it will not show up on the graph.

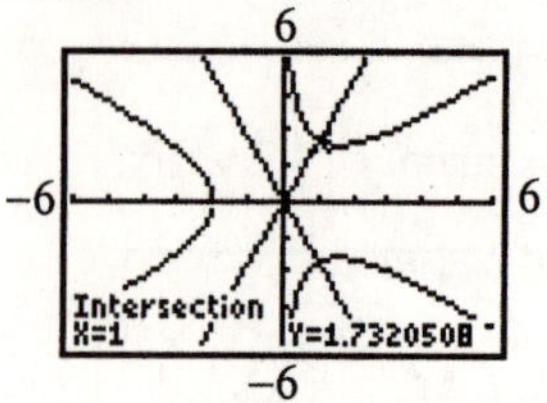

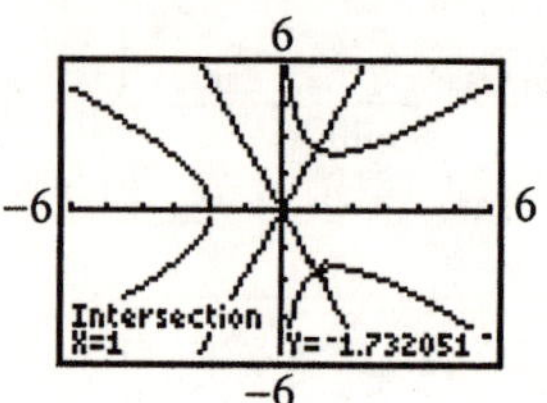

b. $ix^3+8=0$

$$-x^3+8i=0$$
$$x^3-8i=0$$
$$x^3=8i$$
$$8i=8(0+i)=8(\cos 90°+i\sin 90°)$$

$$z_k=\sqrt[3]{8}\left[\cos\left(\frac{90°}{3}+\frac{360°}{3}k\right)+i\sin\left(\frac{90°}{3}+\frac{360°}{3}k\right)\right]$$
$$=2\left[\cos(30°+120°k)+i\sin(30°+120°k)\right]$$

$$z_0=2\left[\cos(30°)+i\sin(30°)\right]$$
$$=2\left(\frac{\sqrt{3}}{2}+\frac{1}{2}i\right)=\sqrt{3}+i$$

$$z_1=2\left[\cos(30°+120°)+i\sin(30°+120°)\right]$$
$$=2\left(\frac{-\sqrt{3}}{2}+\frac{1}{2}i\right)=-\sqrt{3}+i$$

$$z_1=2\left[\cos(30°+240°)+i\sin(30°+240°)\right]$$
$$=2(0-1i)=-2i$$

These solutions do match those in part a.

Cumulative Review R – 8

1. $e^{x^2-9}=1$

$\ln\left(e^{x^2-9}\right)=\ln(1)$

$x^2-9=0$

$(x+3)(x-3)=0$

$x=-3$ or $x=3$

The solution set is $\{-3,3\}$.

2. The line containing point (0, 0), making an angle of 30° with the positive x-axis has polar equation $\theta=\frac{\pi}{6}$.

Using rectangular coordinates:

$\theta=\tan^{-1}\left(\frac{y}{x}\right)$

$\tan\theta=\frac{y}{x}$

$\theta=\frac{\pi}{6}$

$\tan\theta=\tan\left(\frac{\pi}{6}\right)=\frac{\sqrt{3}}{3}$

So, $\frac{\sqrt{3}}{3}=\frac{y}{x}$

$y=\frac{\sqrt{3}}{3}x$

3. The circle with center point (0,1) and radius has equation:

$(x-h)^2+(y-k)^2=r^2$

$(x-0)^2+(y-1)^2=3^2$

$x^2+(y-1)^2=9$

$x^2+y^2-2y+1=9$

$x^2+y^2-2y-8=0$

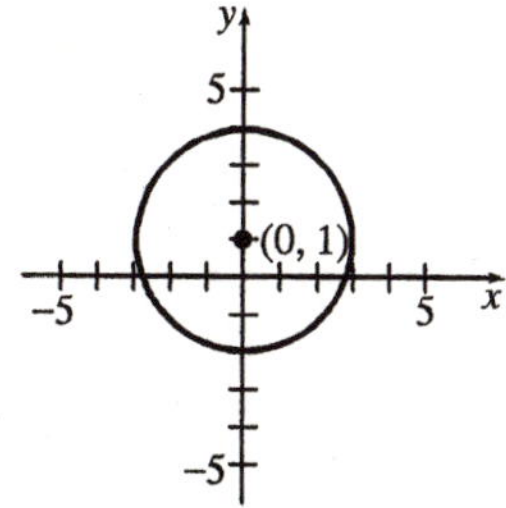

4. $f(x)=\ln(1-2x)$

f will be defined provided $1-2x>0$.

$1-2x>0$

$1>2x$

$\frac{1}{2}>x$

The domain of is $\left\{x \middle| x<\frac{1}{2}\right\}$ or $\left(-\infty,\frac{1}{2}\right)$.

5. $x^2+y^3=2x^4$

Test for symmetry:

x-axis: Replace y by $-y$:

$x^2+(-y)^3=2x^4$

$x^2-y^3=2x^4$

which is not equivalent to $x^2+y^3=2x^4$.

y-axis: Replace x by $-x$:

$(-x)^2+y^3=2(-x)^4$

$x^2+y^3=2x^4$

which is equivalent to

$x^2+y^3=2x^4$

Origin: Replace x by $-x$ and y by $-y$:

$(-x)^2+(-y)^3=2(-x)^4$

$x^2-y^3=2x^4$

which is not equivalent to $x^2+y^3=2x^4$.

Therefore, the graph is symmetric with respect to the y-axis.

6. $y=|\ln x|$

$=\begin{cases}\ln x, & \text{when } \ln x\ge 0\\ -\ln x, & \text{when } \ln x<0\end{cases}$

$=\begin{cases}\ln x, & \text{when } x\ge 1\\ -\ln x, & \text{when } 0<x<1\end{cases}$

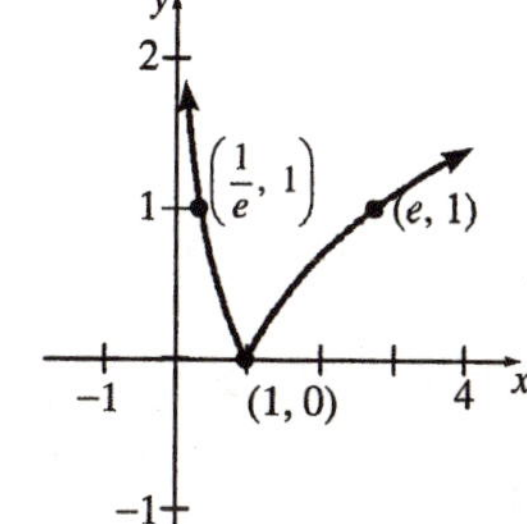

7. $y = |\sin x|$

$$= \begin{cases} \sin x, & \text{when } \sin x \geq 0 \\ -\sin x, & \text{when } \sin x < 0 \end{cases}$$

$$= \begin{cases} \sin x, & \text{when } 0 \leq x \leq \pi \\ -\sin x, & \text{when } \pi < x < 2\pi \end{cases}$$

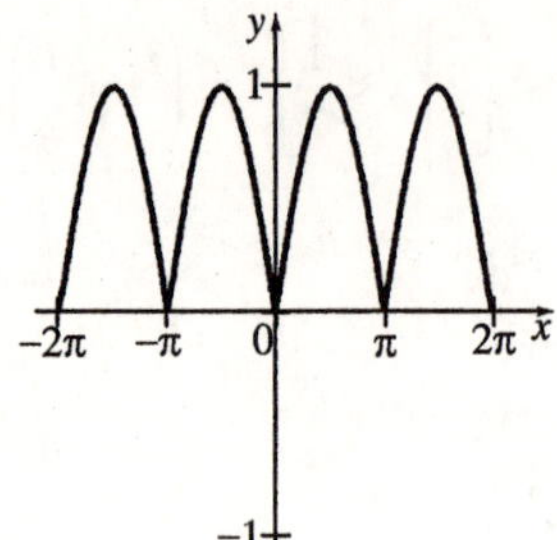

8. $y = \sin|x| = \begin{cases} \sin x, & \text{when } x \geq 0 \\ -\sin x, & \text{when } x < 0 \end{cases}$

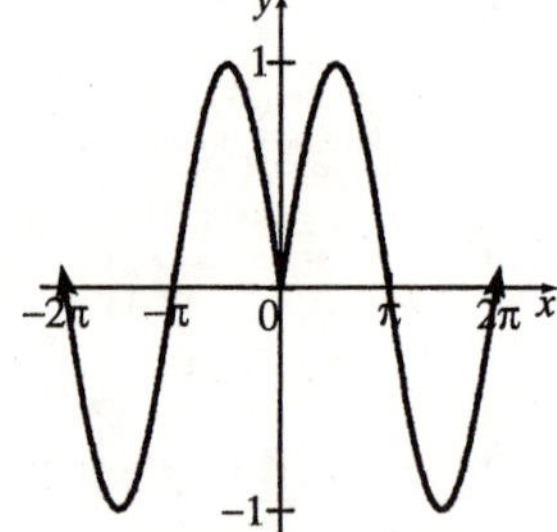

9. $\sin^{-1}\left(-\frac{1}{2}\right)$

We are finding the angle θ, $-\frac{\pi}{2} \leq \theta \leq \frac{\pi}{2}$, whose sine equals $-\frac{1}{2}$.

$$\sin\theta = -\frac{1}{2}, \quad -\frac{\pi}{2} \leq \theta \leq \frac{\pi}{2}$$

$$\theta = -\frac{\pi}{6}$$

$$\sin^{-1}\left(-\frac{1}{2}\right) = -\frac{\pi}{6}$$

10. Graphing $x = 3$ and $y = 4$ using rectangular coordinates:

$x = 3$ yields a vertical line passing through the point $(3, 0)$.

$y = 4$ yields a horizontal line passing through the point $(0, 4)$.

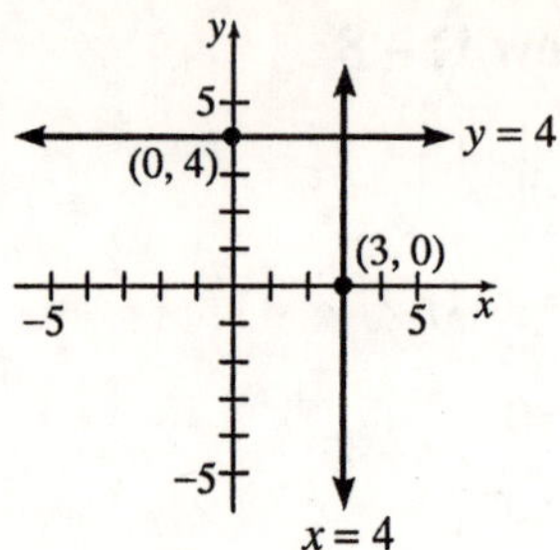

11. Graphing $r = 2$ and $\theta = \frac{\pi}{3}$ using polar coordinates:

$r = 2$ yields a circle, centered at $(0, 0)$, with radius = 2.

$\theta = \frac{\pi}{3}$ yields a line passing through the point $(0, 0)$, forming an angle of $\theta = \frac{\pi}{3}$ with the positive x-axis.

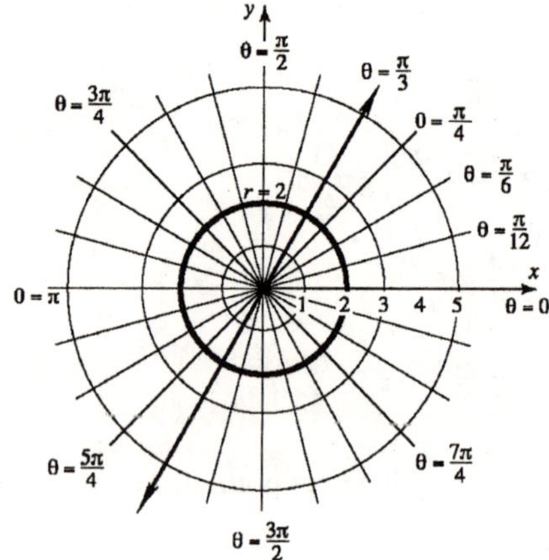

Chapter 9
Analytic Geometry

Section 9.1

Not applicable

Section 9.2

1. $\sqrt{(x_2 - x_1)^2 + (y_2 - y_1)^2}$

2. $\left(\frac{-4}{2}\right)^2 = 4$

3. $(x+4)^2 = 9$
   $x+4 = \pm 3$
   $x+4 = 3 \Rightarrow x = -1$
   or $x+4 = -3 \Rightarrow x = -7$
   The solution set is $\{-7, -1\}$

4. $(-2, -5)$

5. 3, up

6. parabola

7. paraboloid of revolution

8. True

9. True

10. True

11. B; the graph has a vertex $(h,k) = (0,0)$ and opens up. Therefore, the equation of the graph has the form $x^2 = 4ay$. The graph passes through the point $(2,1)$ so we have
    $(2)^2 = 4a(1)$
    $4 = 4a$
    $1 = a$
    Thus, the equation of the graph is $x^2 = 4y$.

12. G; the graph has vertex $(h,k) = (1,1)$ and opens to the left. Therefore, the equation of the graph has the form $(y-1)^2 = -4a(x-1)$.

13. E; the graph has vertex $(h,k) = (1,1)$ and opens to the right. Therefore, the equation of the graph has the form $(y-1)^2 = 4a(x-1)$.

14. D; the graph has vertex $(h,k) = (0,0)$ and opens down. Therefore, the equation of the graph has the form $x^2 = -4ay$. The graph passes through the point $(-2,-1)$ so we have
    $(-2)^2 = -4a(-1)$
    $4 = 4a$
    $a = 1$
    Thus, the equation of the graph is $x^2 = -4y$.

15. H; the graph has vertex $(h,k) = (-1,-1)$ and opens down. Therefore, the equation of the graph has the form $(x+1)^2 = -4a(y+1)$.

16. A; the graph has vertex $(h,k) = (0,0)$ and opens to the right. Therefore, the equation of the graph has the form $y^2 = 4ax$. The graph passes through the point $(1,2)$ so we have
    $(2)^2 = 4a(1)$
    $4 = 4a$
    $1 = a$
    Thus, the equation of the graph is $y^2 = 4x$.

17. C; the graph has vertex $(h,k) = (0,0)$ and opens to the left. Therefore, the equation of the graph has the form $y^2 = -4ax$. The graph passes through the point $(-1,-2)$ so we have
    $(-2)^2 = -4a(-1)$
    $4 = 4a$
    $1 = a$
    Thus, the equation of the graph is $y^2 = -4x$.

18. F; the graph has vertex $(h,k) = (-1,-1)$ and opens up. Therefore, the equation of the graph has the form $(x+1)^2 = 4a(y+1)$.

19. The focus is (4, 0) and the vertex is (0, 0). Both lie on the horizontal line $y = 0$. $a = 4$ and since (4, 0) is to the right of (0, 0), the parabola opens to the right. The equation of the parabola is:

$$y^2 = 4ax$$
$$y^2 = 4 \cdot 4 \cdot x$$
$$y^2 = 16x$$

Letting $x = 4$, we find $y^2 = 64$ or $y = \pm 8$. The points (4, 8) and (4, –8) define the latus rectum.

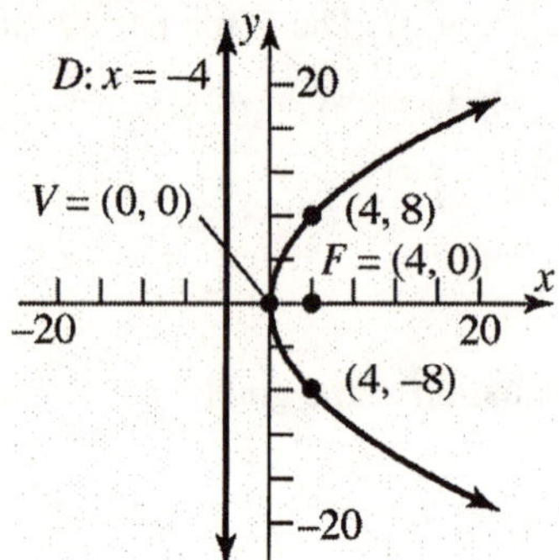

20. The focus is (0, 2) and the vertex is (0, 0). Both lie on the vertical line $x = 0$. $a = 2$ and since (0, 2) is above (0, 0), the parabola opens up. The equation of the parabola is:

$$x^2 = 4ay$$
$$x^2 = 4 \cdot 2 \cdot y$$
$$x^2 = 8y$$

Letting $y = 2$, we find $x^2 = 16$ or $x = \pm 4$. The points (–4, 2) and (4, 2) define the latus rectum.

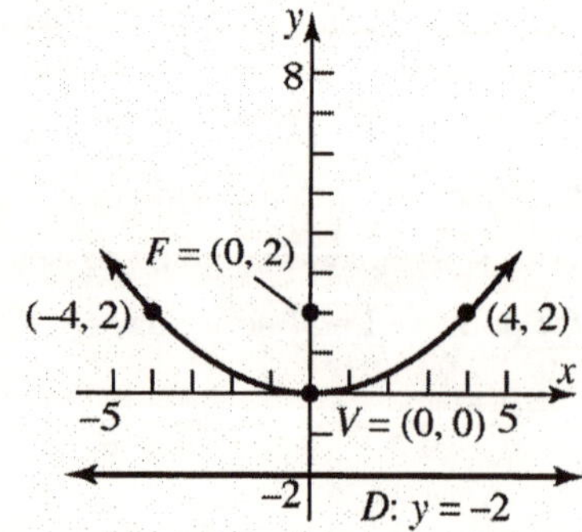

21. The focus is (0, –3) and the vertex is (0, 0). Both lie on the vertical line $x = 0$. $a = 3$ and since (0, –3) is below (0, 0), the parabola opens down. The equation of the parabola is:

$$x^2 = -4ay$$
$$x^2 = -4 \cdot 3 \cdot y$$
$$x^2 = -12y$$

Letting $y = -3$, we find $x^2 = 36$ or $x = \pm 6$. The points $(-6, -3)$ and $(6, -3)$ define the latus rectum.

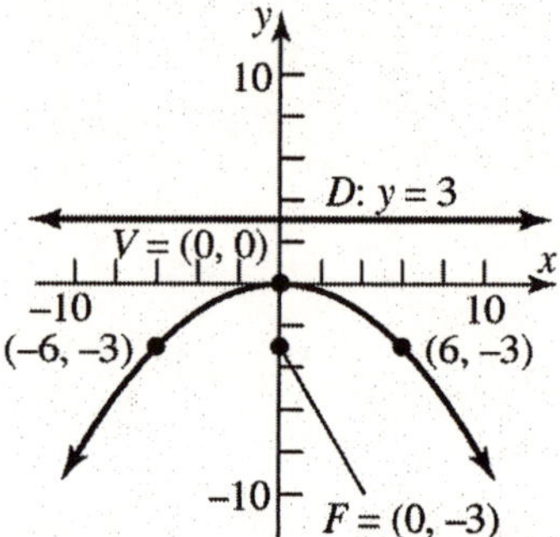

22. The focus is (–4, 0) and the vertex is (0, 0). Both lie on the horizontal line $y = 0$. $a = 4$ and since (–4, 0) is to the left of (0, 0), the parabola opens to the left. The equation of the parabola is:

$$y^2 = -4ax$$
$$y^2 = -4 \cdot 4 \cdot x$$
$$y^2 = -16x$$

Letting $x = -4$, we find $y^2 = 64$ or $y = \pm 8$. The points (–4, 8) and (–4, –8) define the latus rectum.

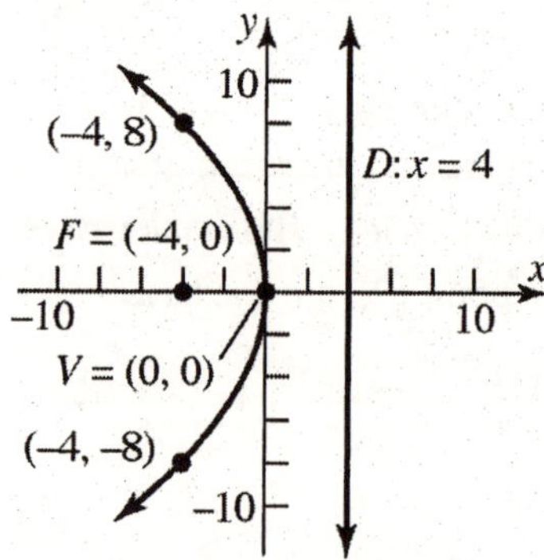

23. The focus is (–2, 0) and the directrix is $x = 2$. The vertex is (0, 0). $a = 2$ and since (–2, 0) is to the left of (0, 0), the parabola opens to the left. The equation of the parabola is:

$y^2 = -4ax$

$y^2 = -4 \cdot 2 \cdot x$

$y^2 = -8x$

Letting $x = -2$, we find $y^2 = 16$ or $y = \pm 4$. The points (–2, 4) and (–2, –4) define the latus rectum.

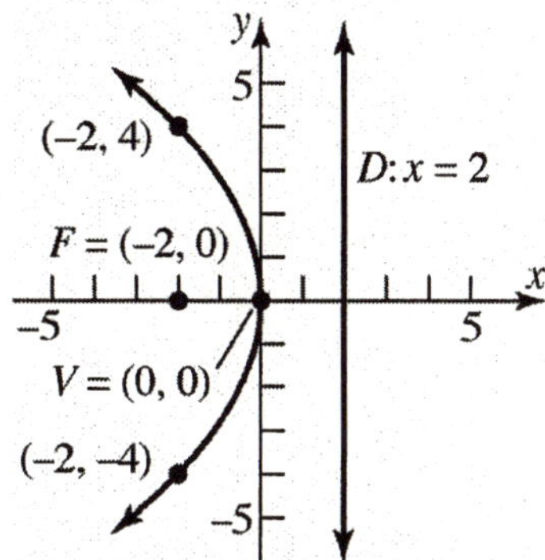

24. The focus is (0, –1) and the directrix is $y = 1$. The vertex is (0, 0). $a = 1$ and since (0, –1) is below (0, 0), the parabola opens down. The equation of the parabola is:

$x^2 = -4ay$

$x^2 = -4 \cdot 1 \cdot y$

$x^2 = -4y$

Letting $y = -1$, we find $x^2 = 4$ or $x = \pm 2$. The points (–2, –1) and (2, –1) define the latus rectum.

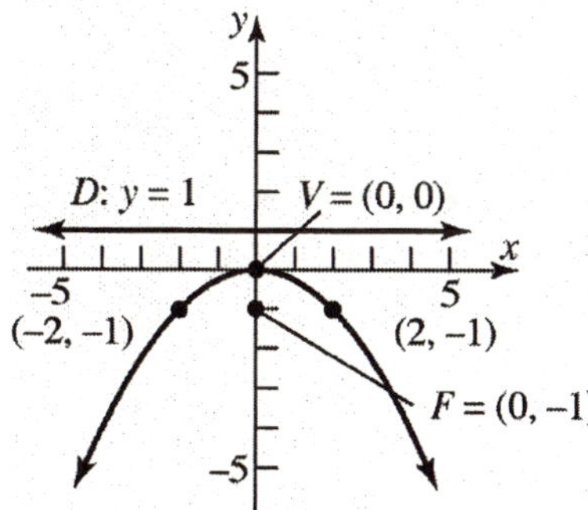

25. The directrix is $y = -\frac{1}{2}$ and the vertex is (0, 0). The focus is $\left(0, \frac{1}{2}\right)$. $a = \frac{1}{2}$ and since $\left(0, \frac{1}{2}\right)$ is above (0, 0), the parabola opens up. The equation of the parabola is:

$x^2 = 4ay$

$x^2 = 4 \cdot \frac{1}{2} \cdot y \Rightarrow x^2 = 2y$

Letting $y = \frac{1}{2}$, we find $x^2 = 1$ or $x = \pm 1$.

The points $\left(1, \frac{1}{2}\right)$ and $\left(-1, \frac{1}{2}\right)$ define the latus rectum.

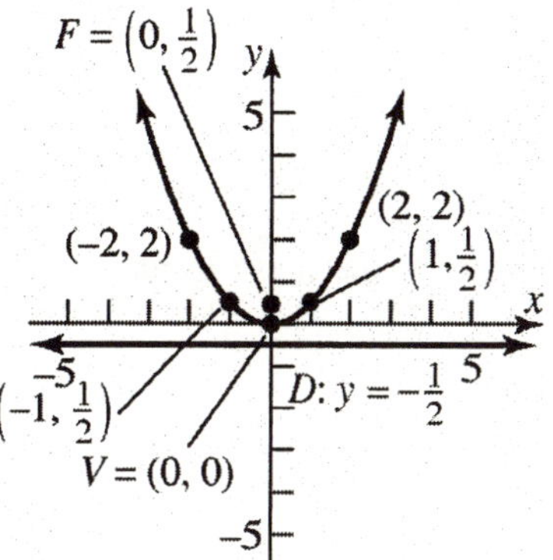

26. The directrix is $x = -\frac{1}{2}$ and the vertex is (0, 0). The focus is $\left(\frac{1}{2}, 0\right)$. $a = \frac{1}{2}$ and since $\left(\frac{1}{2}, 0\right)$ is to the right of (0, 0), the parabola opens to the right. The equation of the parabola is:

$y^2 = 4ax$

$y^2 = 4 \cdot \frac{1}{2} \cdot x \Rightarrow y^2 = 2x$

Letting $x = \frac{1}{2}$, we find $y^2 = 1$ or $y = \pm 1$. The points $\left(\frac{1}{2}, -1\right)$ and $\left(\frac{1}{2}, 1\right)$ define the latus rectum.

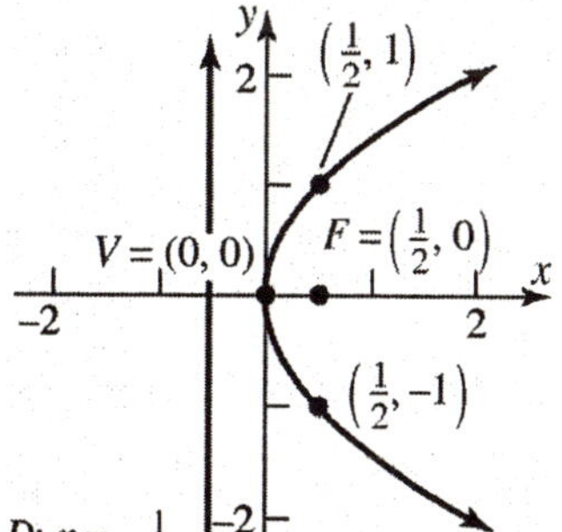

27. Vertex: (0, 0). Since the axis of symmetry is vertical, the parabola opens up or down. Since (2, 3) is above (0, 0), the parabola opens up. The equation has the form $x^2 = 4ay$. Substitute the coordinates of (2, 3) into the equation to find a:

$2^2 = 4a \cdot 3 \Rightarrow 4 = 12a \Rightarrow a = \frac{1}{3}$

The equation of the parabola is: $x^2 = \frac{4}{3}y$. The focus is $\left(0, \frac{1}{3}\right)$. Letting $y = \frac{1}{3}$, we find $x^2 = \frac{4}{9}$ or $x = \pm\frac{2}{3}$. The points $\left(\frac{2}{3}, \frac{1}{3}\right)$ and $\left(-\frac{2}{3}, \frac{1}{3}\right)$ define the latus rectum.

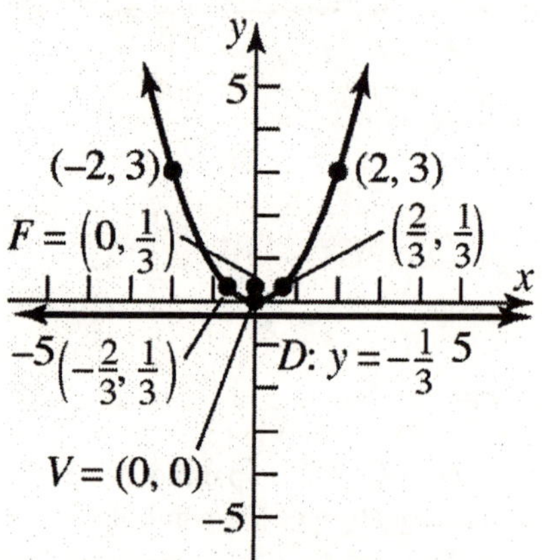

28. Vertex: (0, 0). Since the axis of symmetry is horizontal, the parabola opens left or right. Since (2, 3) is to the right of (0, 0), the parabola opens to the right. The equation has the form $y^2 = 4ax$. Substitute the coordinates of (2, 3) into the equation to find a:

$3^2 = 4a \cdot 2 \Rightarrow 9 = 8a \Rightarrow a = \frac{9}{8}$

The equation of the parabola is: $y^2 = \frac{9}{2}x$. The focus is $\left(\frac{9}{8}, 0\right)$. Letting $x = \frac{9}{8}$, we find $y^2 = \frac{81}{16}$ or $y = \pm\frac{9}{4}$. The points $\left(\frac{9}{8}, \frac{9}{4}\right)$ and $\left(\frac{9}{8}, -\frac{9}{4}\right)$ define the latus rectum.

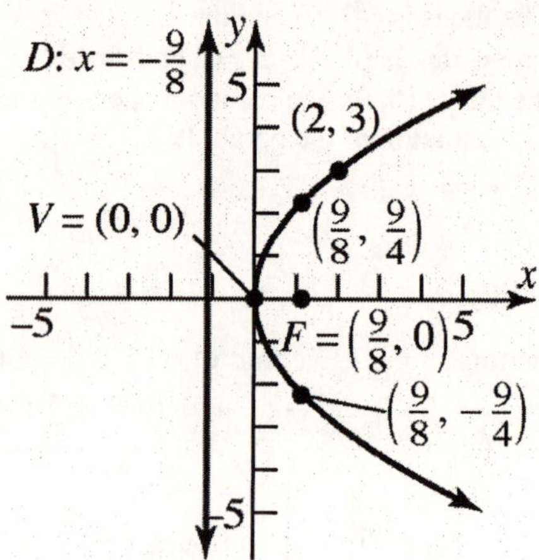

29. The vertex is (2, –3) and the focus is (2, –5). Both lie on the vertical line $x = 2$.
$a = |-5-(-3)| = 2$ and since (2, –5) is below (2, –3), the parabola opens down. The equation of the parabola is:

$(x-h)^2 = -4a(y-k)$

$(x-2)^2 = -4(2)(y-(-3))$

$(x-2)^2 = -8(y+3)$

Letting $y = -5$, we find

$(x-2)^2 = 16$

$x-2 = \pm 4 \Rightarrow x = -2$ or $x = 6$

The points (–2, –5) and (6, –5) define the latus rectum.

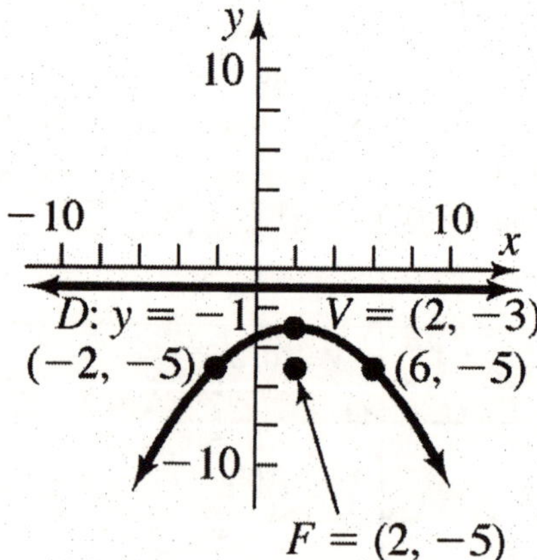

30. The vertex is (4, –2) and the focus is (6, –2). Both lie on the horizontal line $y = -2$.

$a = |4-6| = 2$ and since (6, –2) is to the right of (4, –2), the parabola opens to the right. The equation of the parabola is:

$(y-k)^2 = 4a(x-h)$

$(y-(-2))^2 = 4(2)(x-4)$

$(y+2)^2 = 8(x-4)$

Letting $x = 6$, we find

$(y+2)^2 = 16$

$y+2 = \pm 4 \Rightarrow y = -6 \text{ or } y = 2$

The points (6, –6) and (6, 2) define the latus rectum.

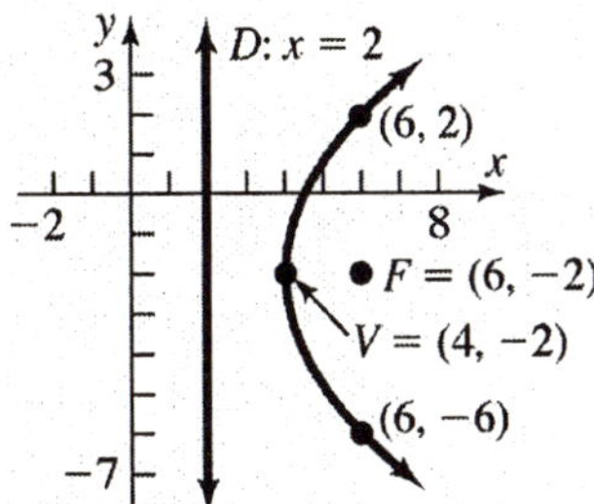

31. The vertex is (–1, –2) and the focus is (0, –2). Both lie on the horizontal line $y = -2$.

$a = |-1-0| = 1$ and since (0, –2) is to the right of (–1, –2), the parabola opens to the right. The equation of the parabola is:

$(y-k)^2 = 4a(x-h)$

$(y-(-2))^2 = 4(1)(x-(-1))$

$(y+2)^2 = 4(x+1)$

Letting $x = 0$, we find

$(y+2)^2 = 4$

$y+2 = \pm 2 \Rightarrow y = -4 \text{ or } y = 0$

The points (0, –4) and (0, 0) define the latus rectum.

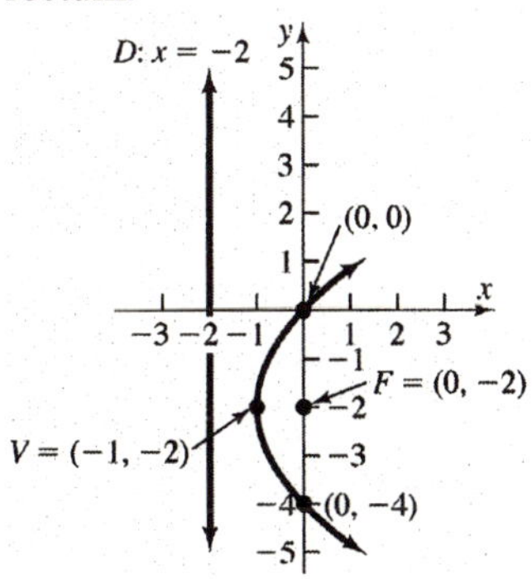

32. The vertex is (3, 0) and the focus is (3, –2). Both lie on the horizontal line $x = 3$. $a = |-2-0| = 2$ and since (3, –2) is below of (3, 0), the parabola opens down. The equation of the parabola is:

$(x-h)^2 = -4a(y-k)$

$(x-3)^2 = -4(2)(y-0)$

$(x-3)^2 = -8y$

Letting $y = -2$, we find

$(x-3)^2 = 16$

$x-3 = \pm 4 \Rightarrow x = -1 \text{ or } x = 7$

The points (–1, –2) and (7, –2) define the latus rectum.

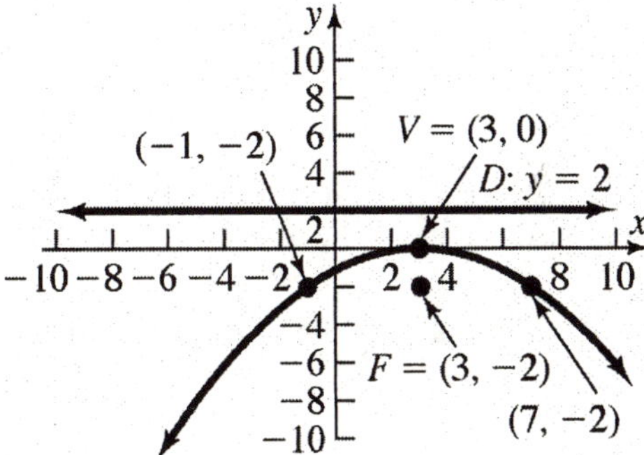

33. The directrix is $y = 2$ and the focus is (–3, 4). This is a vertical case, so the vertex is (–3, 3). $a = 1$ and since (–3, 4) is above $y = 2$, the parabola opens up. The equation of the parabola is: $(x-h)^2 = 4a(y-k)$

$(x-(-3))^2 = 4\cdot 1\cdot (y-3)$

$(x+3)^2 = 4(y-3)$

Letting $y = 4$, we find $(x+3)^2 = 4$ or $x+3 = \pm 2$. So, $x = -1$ or $x = -5$. The points (–1, 4) and (–5, 4) define the latus rectum.

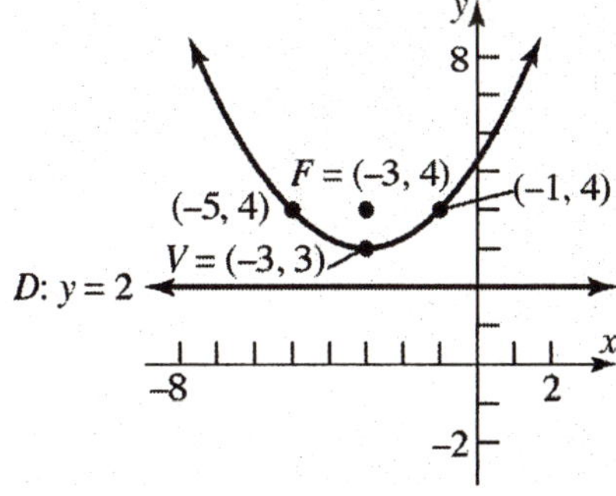

34. The directrix is $x=-4$ and the focus is (2, 4). This is a horizontal case, so the vertex is (–1, 4). $a=3$ and since (2, 4) is to the right of $x=-4$, the parabola opens to the right. The equation of the parabola is: $(y-k)^2=4a(x-h)$

$(y-4)^2=4\cdot3\cdot(x-(-1))$

$(y-4)^2=12(x+1)$

Letting $x=2$, we find $(y-4)^2=36$ or $y-4=\pm6$. So, $y=-2$ or $y=10$. The points (2, –2) and (2, 10) define the latus rectum.

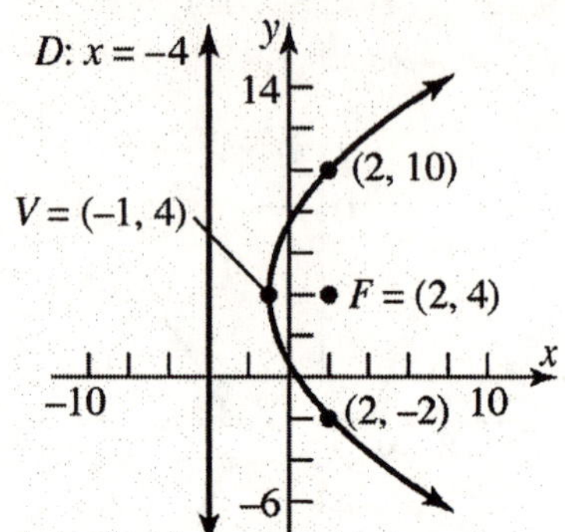

35. The directrix is $x=1$ and the focus is (–3, –2). This is a horizontal case, so the vertex is (–1, –2). $a=2$ and since (–3, –2) is to the left of $x=1$, the parabola opens to the left. The equation of the parabola is:

$(y-k)^2=-4a(x-h)$

$(y-(-2))^2=-4\cdot2\cdot(x-(-1))$

$(y+2)^2=-8(x+1)$

Letting $x=-3$, we find $(y+2)^2=16$ or $y+2=\pm4$. So, $y=2$ or $y=-6$. The points (–3, 2) and (–3, –6) define the latus rectum.

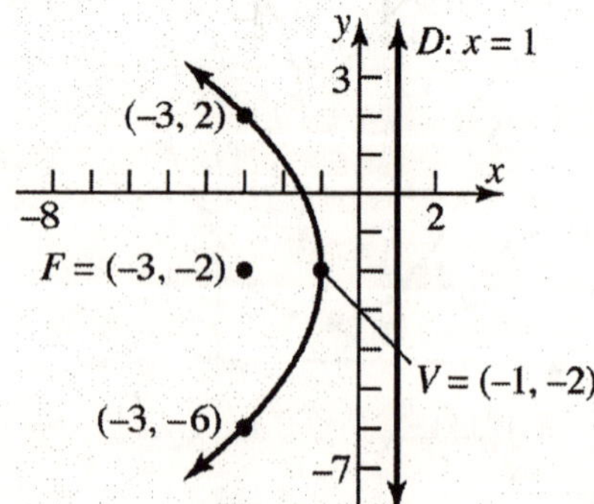

36. The directrix is $y=-2$ and the focus is (–4, 4). This is a vertical case, so the vertex is (–4, 1). $a=3$ and since (–4, 4) is above $y=-2$, the parabola opens up. The equation of the parabola is:

$(x-h)^2=4a(y-k)$

$(x-(-4))^2=4\cdot3\cdot(y-1)$

$(x+4)^2=12(y-1)$

Letting $y=4$, we find $(x+4)^2=36$ or $x+4=\pm6$. So, $x=-10$ or $x=2$. The points (–10, 4) and (2, 4) define the latus rectum.

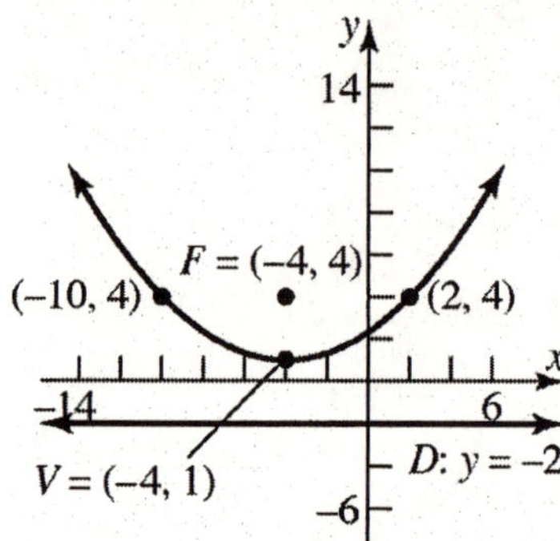

37. **a.** The equation $x^2=4y$ is in the form $x^2=4ay$ where $4a=4$ or $a=1$.

Thus, we have:
Vertex: (0, 0)
Focus: (0, 1)
Directrix: $y=-1$

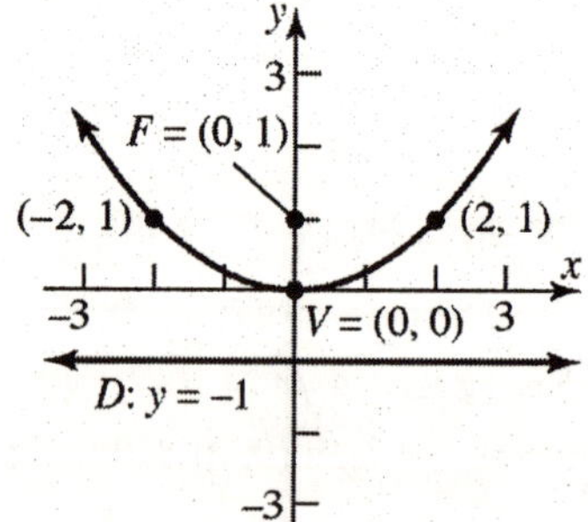

b.

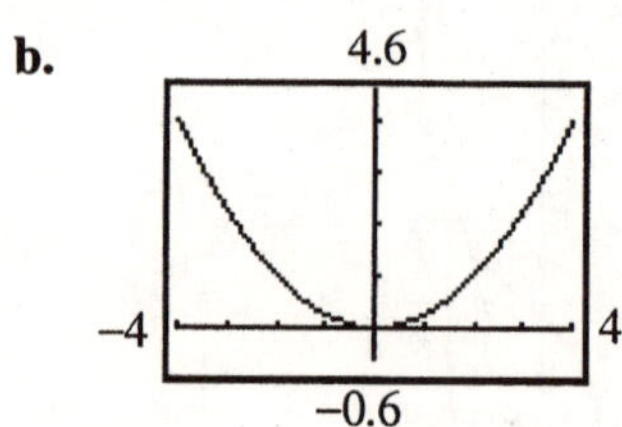

38. a. The equation $y^2 = 8x$ is in the form $y^2 = 4ax$ where $4a = 8$ or $a = 2$. Thus, we have:
Vertex: (0, 0)
Focus: (2, 0)
Directrix: $x = -2$

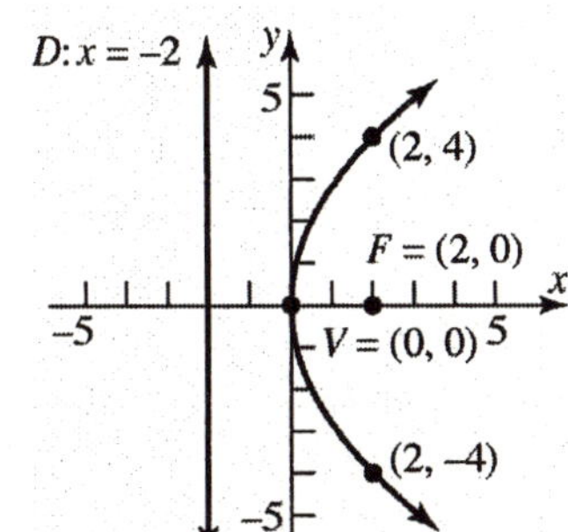

b.

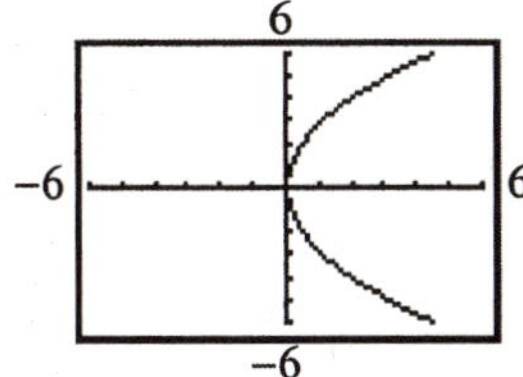

39. a. The equation $y^2 = -16x$ is in the form $y^2 = -4ax$ where $-4a = -16$ or $a = 4$. Thus, we have:
Vertex: (0, 0)
Focus: (−4, 0)
Directrix: $x = 4$

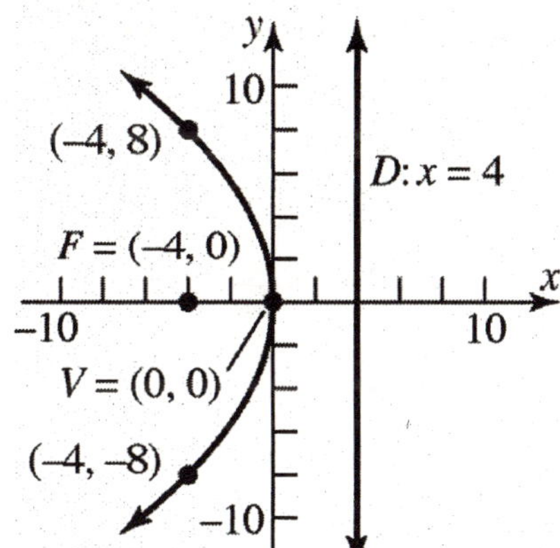

b.

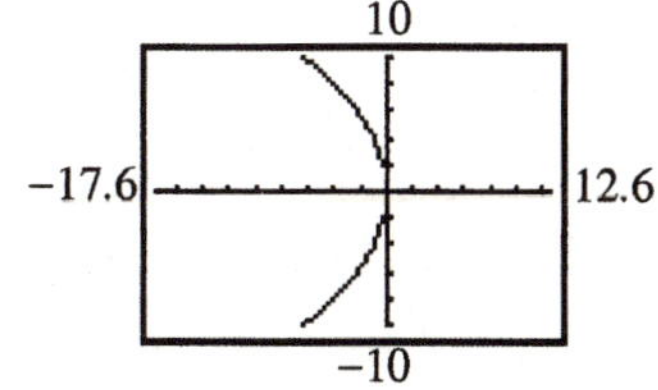

40. a. The equation $x^2 = -4y$ is in the form $x^2 = -4ay$ where $-4a = -4$ or $a = 1$. Thus, we have:
Vertex: (0, 0)
Focus: (0, −1)
Directrix: $y = 1$

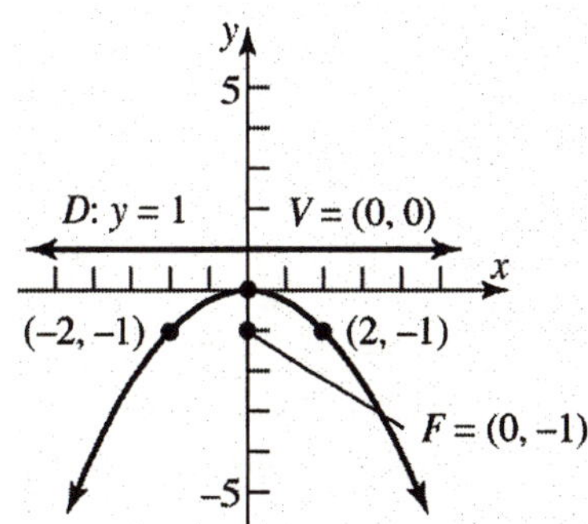

b.

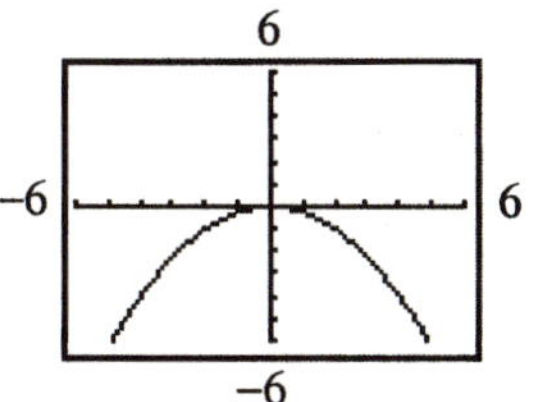

41. a. The equation $(y-2)^2 = 8(x+1)$ is in the form $(y-k)^2 = 4a(x-h)$ where $4a = 8$ or $a = 2$, $h = -1$, and $k = 2$. Thus, we have:
Vertex: (−1, 2)
Focus: (1, 2)
Directrix: $x = -3$

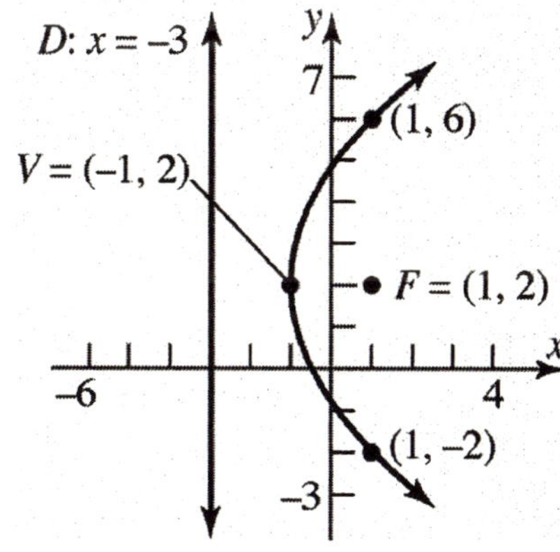

b.

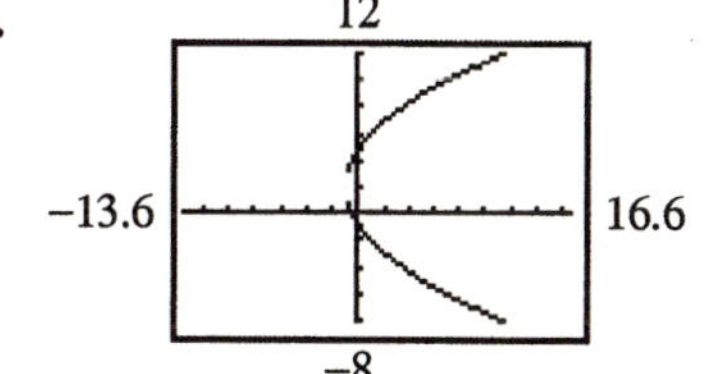

42. a. The equation $(x+4)^2 = 16(y+2)$ is in the form $(x-h)^2 = 4a(y-k)$ where $4a = 16$ or $a = 4$, $h = -4$, and $k = -2$.

Thus, we have:
Vertex: $(-4, -2)$
Focus: $(-4, 2)$
Directrix: $y = -6$

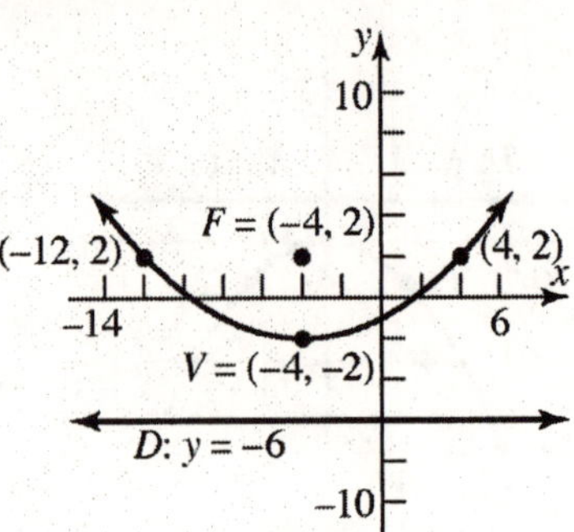

b.

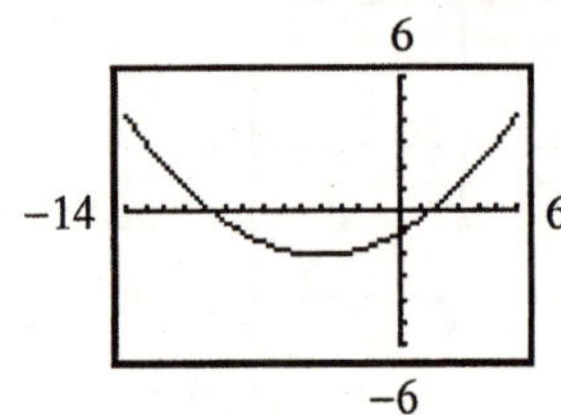

43. a. The equation $(x-3)^2 = -(y+1)$ is in the form $(x-h)^2 = -4a(y-k)$ where $-4a = -1$ or $a = \frac{1}{4}$, $h = 3$, and $k = -1$.

Thus, we have:

Vertex: $(3, -1)$; Focus: $\left(3, -\frac{5}{4}\right)$;

Directrix: $y = -\frac{3}{4}$

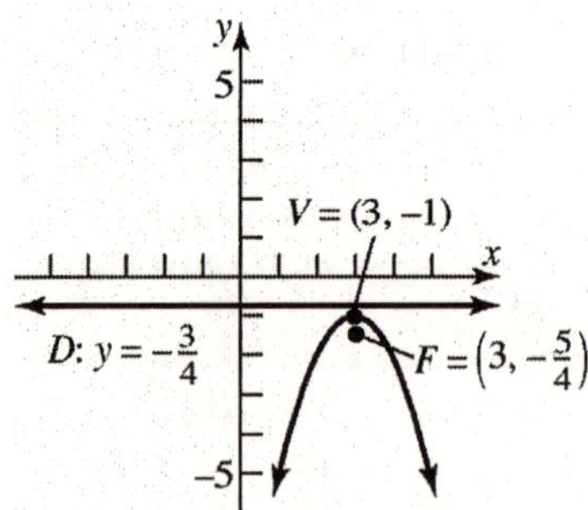

b.

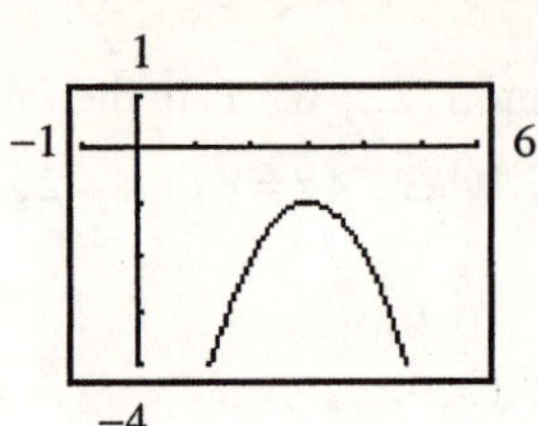

44. a. The equation $(y+1)^2 = -4(x-2)$ is in the form $(y-k)^2 = -4a(x-h)$ where $-4a = -4$ or $a = 1$, $h = 2$, and $k = -1$.

Thus, we have:
Vertex: $(2, -1)$
Focus: $(1, -1)$
Directrix: $x = 3$

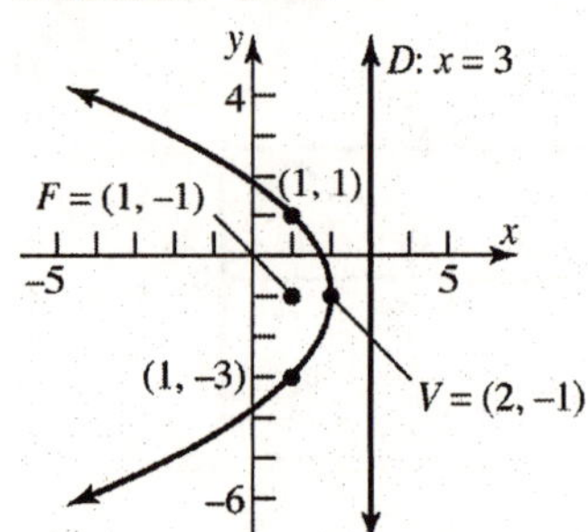

b.

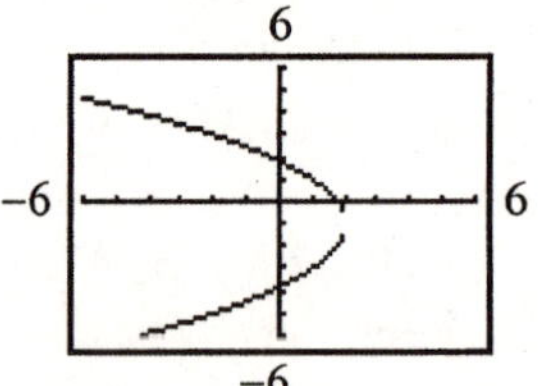

45. a. The equation $(y+3)^2 = 8(x-2)$ is in the form $(y-k)^2 = 4a(x-h)$ where $4a = 8$ or $a = 2$, $h = 2$, and $k = -3$. Thus, we have:
Vertex: $(2, -3)$
Focus: $(4, -3)$
Directrix: $x = 0$

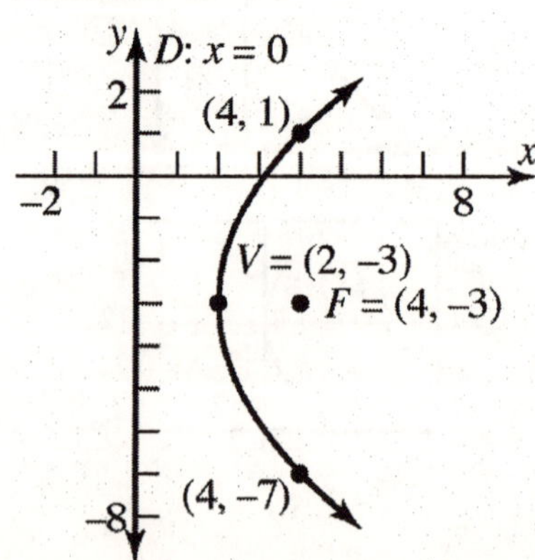

b.

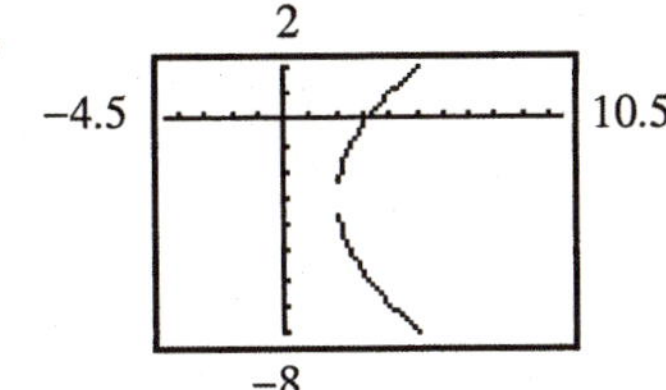

46. a. The equation $(x-2)^2 = 4(y-3)$ is in the form $(x-h)^2 = 4a(y-k)$ where $4a = 4$ or $a = 1$, $h = 2$, and $k = 3$. Thus, we have:
Vertex: (2, 3)
Focus: (2, 4)
Directrix: $y = 2$

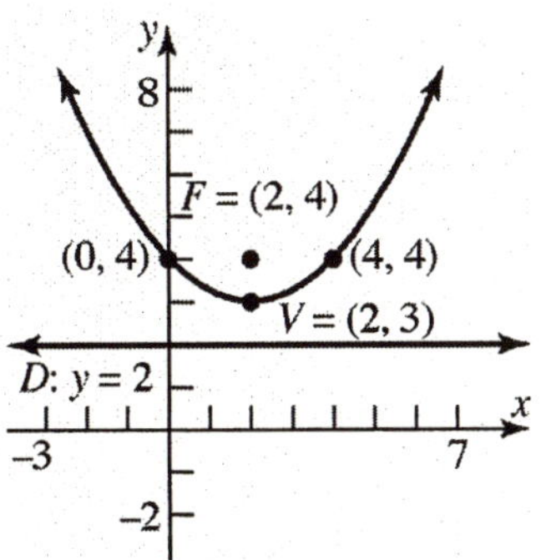

b.

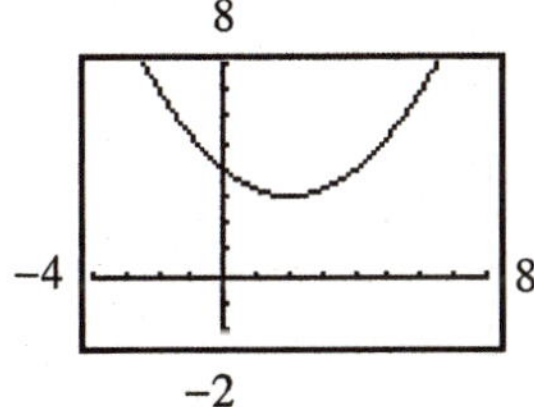

47. a. Complete the square to put in standard form:

$$y^2 - 4y + 4x + 4 = 0$$

$$y^2 - 4y + 4 = -4x$$

$$(y-2)^2 = -4x$$

The equation is in the form $(y-k)^2 = -4a(x-h)$ where $-4a = -4$ or $a = 1$, $h = 0$, and $k = 2$. Thus, we have:
Vertex: (0, 2)
Focus: (–1, 2)
Directrix: $x = 1$

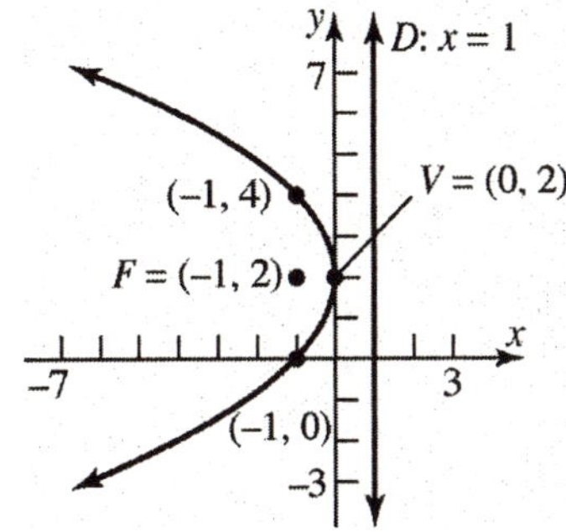

b.

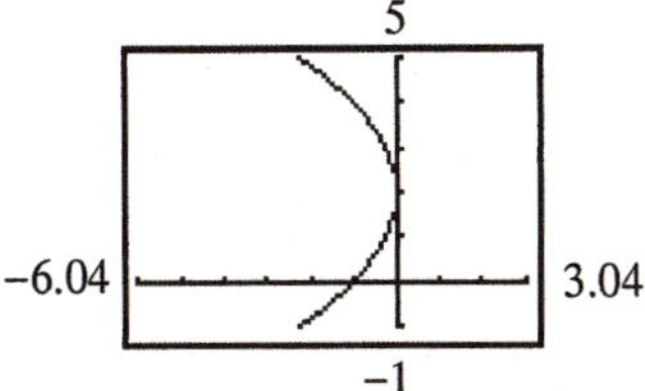

48. a. Complete the square to put in standard form:

$$x^2 + 6x - 4y + 1 = 0$$

$$x^2 + 6x + 9 = 4y - 1 + 9$$

$$(x+3)^2 = 4(y+2)$$

The equation is in the form $(x-h)^2 = 4a(y-k)$ where $4a = 4$ or $a = 1$, $h = -3$, and $k = -2$. Thus, we have:
Vertex: (–3, –2)
Focus: (–3, –1)
Directrix: $y = -3$

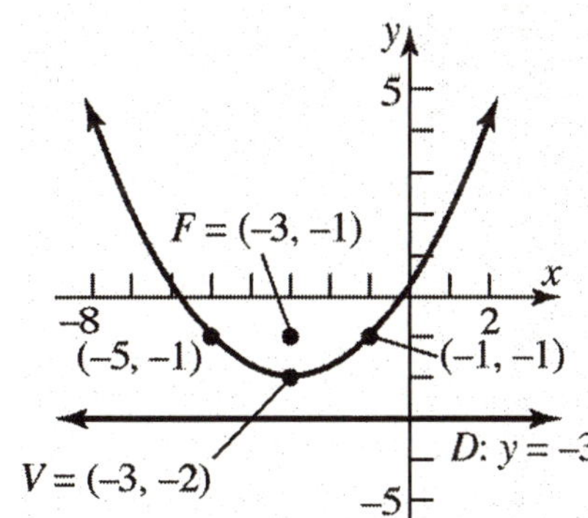

b.

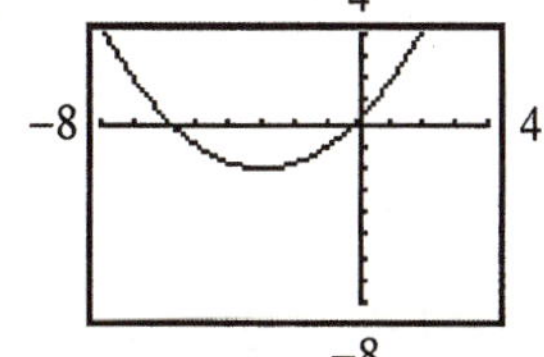

49. a. Complete the square to put in standard form:

$$x^2+8x=4y-8$$
$$x^2+8x+16=4y-8+16$$
$$(x+4)^2=4(y+2)$$

The equation is in the form $(x-h)^2=4a(y-k)$ where $4a=4$ or $a=1$, $h=-4$, and $k=-2$. Thus, we have:

Vertex: (–4, –2)
Focus: (–4, –1)
Directrix: $y=-3$

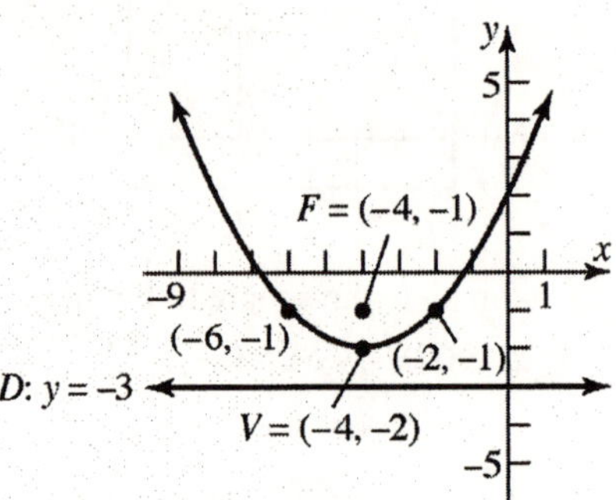

b.

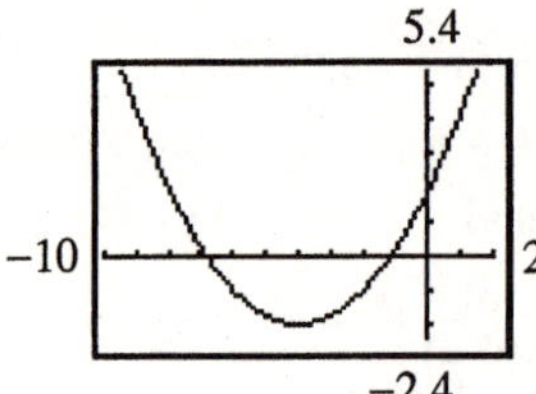

50. a. Complete the square to put in standard form:

$$y^2-2y=8x-1$$
$$y^2-2y+1=8x-1+1$$
$$(y-1)^2=8x$$

The equation is in the form $(y-k)^2=4a(x-h)$ where $4a=8$ or $a=2$, $h=0$, and $k=1$. Thus, we have:

Vertex: (0, 1)
Focus: (2, 1)
Directrix: $x=-2$

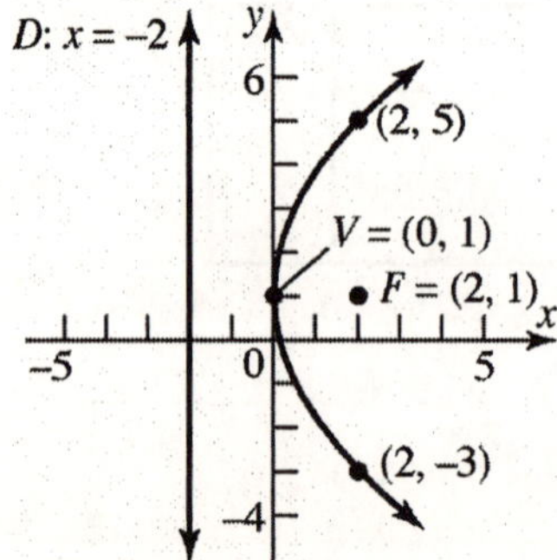

b.

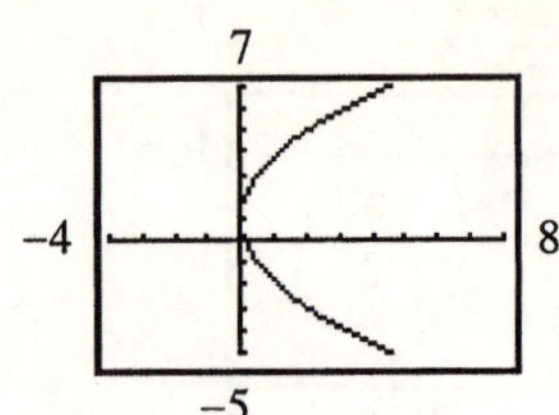

51. a. Complete the square to put in standard form:

$$y^2+2y-x=0$$
$$y^2+2y+1=x+1$$
$$(y+1)^2=x+1$$

The equation is in the form $(y-k)^2=4a(x-h)$ where $4a=1$ or $a=\frac{1}{4}$, $h=-1$, and $k=-1$. Thus, we have:

Vertex: (–1, –1)

Focus: $\left(-\frac{3}{4},-1\right)$

Directrix: $x=-\frac{5}{4}$

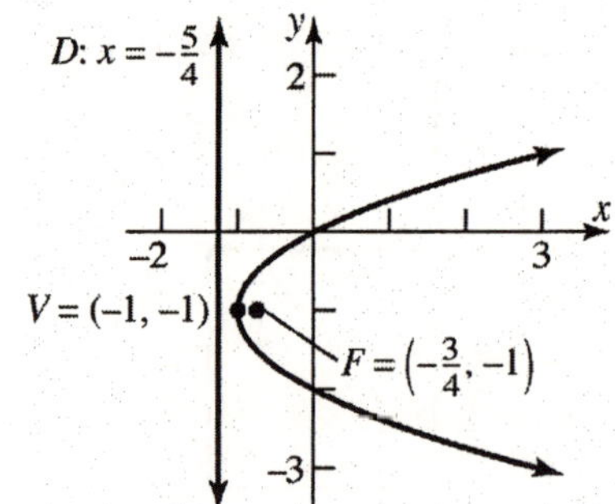

b.

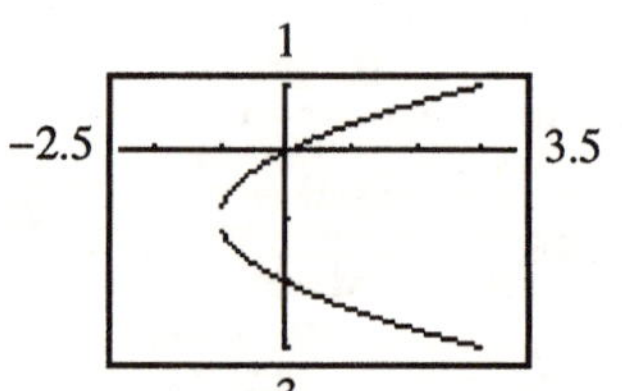

52. a. Complete the square to put in standard form:

$$x^2-4x=2y$$
$$x^2-4x+4=2y+4$$
$$(x-2)^2=2(y+2)$$

The equation is in the form $(x-h)^2=4a(y-k)$ where $4a=2$ or $a=\frac{1}{2}$, $h=2$, and $k=-2$. Thus, we have:

Vertex: (2, –2)

Focus: $\left(2, -\frac{3}{2}\right)$

Directrix: $y = -\frac{5}{2}$

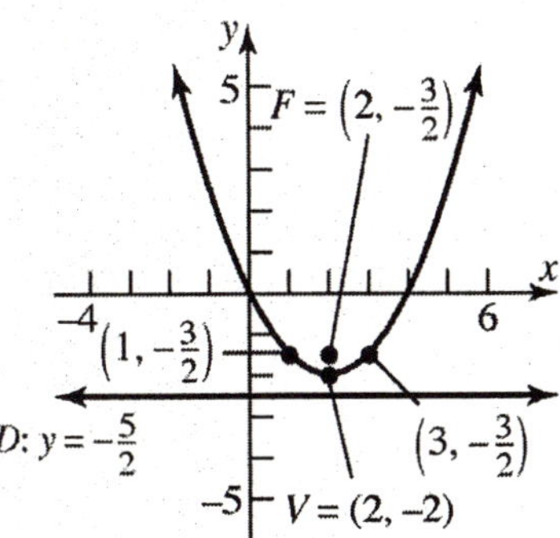

b.

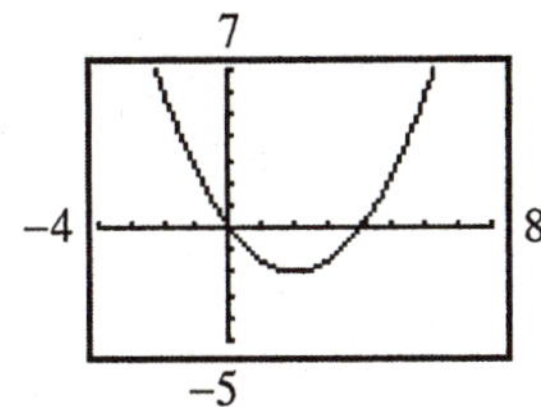

53. a. Complete the square to put in standard form:

$$x^2 - 4x = y + 4$$
$$x^2 - 4x + 4 = y + 4 + 4$$
$$(x-2)^2 = y + 8$$

The equation is in the form

$(x-h)^2 = 4a(y-k)$ where $4a = 1$ or $a = \frac{1}{4}$,

$h = 2$, and $k = -8$. Thus, we have:

Vertex: $(2, -8)$

Focus: $\left(2, -\frac{31}{4}\right)$

Directrix: $y = -\frac{33}{4}$

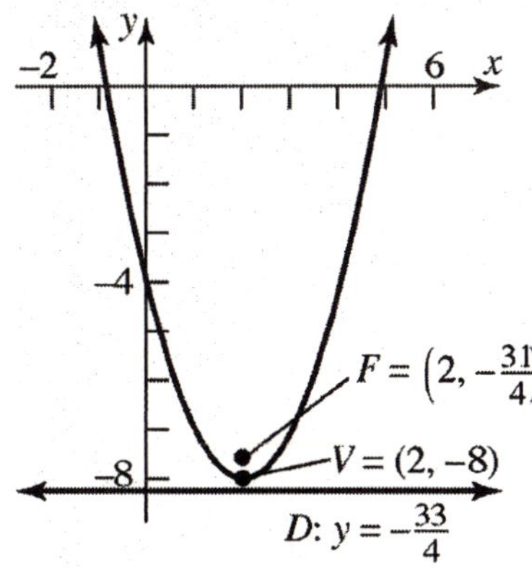

b.

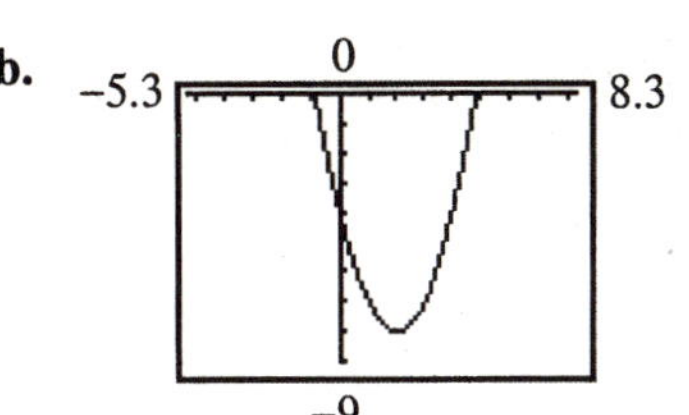

54. a. Complete the square to put in standard form:

$$y^2 + 12y = -x + 1$$
$$y^2 + 12y + 36 = -x + 1 + 36$$
$$(y+6)^2 = -(x-37)$$

The equation is in the form

$(y-k)^2 = -4a(x-h)$ where

$-4a = -1$ or $a = \frac{1}{4}$, $h = 37$, and $k = -6$.

Thus, we have:

Vertex: $(37, -6)$; Focus: $\left(\frac{147}{4}, -6\right)$;

Directrix: $x = \frac{149}{4}$

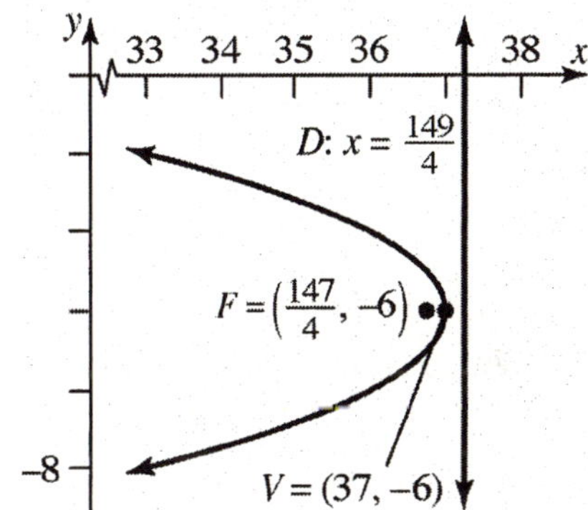

b.

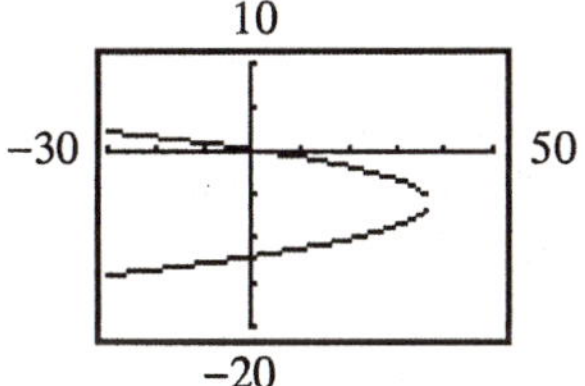

55. $(y-1)^2 = c(x-0)$

$$(y-1)^2 = cx$$
$$(2-1)^2 = c(1) \Rightarrow 1 = c$$
$$(y-1)^2 = x$$

56. $(x-1)^2 = c(y-2)$

$$(2-1)^2 = c(1-2)$$
$$1 = -c \Rightarrow c = -1$$
$$(x-1)^2 = -(y-2)$$

57. $(y-1)^2 = c(x-2)$

$(0-1)^2 = c(1-2)$

$1 = -c \Rightarrow c = -1$

$(y-1)^2 = -(x-2)$

58. $(x-0)^2 = c(y-(-1))$

$x^2 = c(y+1)$

$2^2 = c(0+1) \Rightarrow 4 = c$

$x^2 = 4(y+1)$

59. $(x-0)^2 = c(y-1)$

$x^2 = c(y-1)$

$2^2 = c(2-1)$

$4 = c$

$x^2 = 4(y-1)$

60. $(x-1)^2 = c(y-(-1))$

$(x-1)^2 = c(y+1)$

$(0-1)^2 = c(1+1) \Rightarrow 1 = 2c \Rightarrow c = \frac{1}{2}$

$(x-1)^2 = \frac{1}{2}(y+1)$

61. $(y-0)^2 = c(x-(-2))$

$y^2 = c(x+2)$

$1^2 = c(0+2) \Rightarrow 1 = 2c \Rightarrow c = \frac{1}{2}$

$y^2 = \frac{1}{2}(x+2)$

62. $(y-0)^2 = c(x-1)$

$y^2 = c(x-1)$

$1^2 = c(0-1)$

$1 = -c$

$c = -1$

$y^2 = -(x-1)$

63. Set up the problem so that the vertex of the parabola is at (0, 0) and it opens up. Then the equation of the parabola has the form: $x^2 = 4ay$. Since the parabola is 10 feet across and 4 feet deep, the points (5, 4) and (–5, 4) are on the parabola. Substitute and solve for a:

$5^2 = 4a(4) \Rightarrow 25 = 16a \Rightarrow a = \frac{25}{16}$

a is the distance from the vertex to the focus. Thus, the receiver (located at the focus) is $\frac{25}{16} = 1.5625$ feet, or 18.75 inches from the base of the dish, along the axis of the parabola.

64. Set up the problem so that the vertex of the parabola is at (0, 0) and it opens up. Then the equation of the parabola has the form: $x^2 = 4ay$. Since the parabola is 6 feet across and 2 feet deep, the points (3, 2) and (–3, 2) are on the parabola. Substitute and solve for a:

$3^2 = 4a(2) \Rightarrow 9 = 8a \Rightarrow a = \frac{9}{8}$

a is the distance from the vertex to the focus. Thus, the receiver (located at the focus) is $\frac{9}{8} = 1.125$ feet, or 13.5 inches from the base of the dish, along the axis of the parabola.

65. Set up the problem so that the vertex of the parabola is at (0, 0) and it opens up. Then the equation of the parabola has the form: $x^2 = 4ay$. Since the parabola is 4 inches across and 1 inch deep, the points (2, 1) and (–2, 1) are on the parabola. Substitute and solve for a:

$2^2 = 4a(1) \Rightarrow 4 = 4a \Rightarrow a = 1$

a is the distance from the vertex to the focus. Thus, the bulb (located at the focus) should be 1 inch from the vertex.

66. Set up the problem so that the vertex of the parabola is at (0, 0) and it opens up. Then the equation of the parabola has the form: $x^2 = 4ay$. Since the focus is 1 inch from the vertex and the depth is 2 inches, $a = 1$ and the points $(x, 2)$ and $(-x, 2)$ are on the parabola. Substitute and solve for x:

$x^2 = 4(1)(2) \Rightarrow x^2 = 8 \Rightarrow x = \pm 2\sqrt{2}$

The diameter of the headlight is $4\sqrt{2} \approx 5.66$ inches.

67. Set up the problem so that the vertex of the parabola is at (0, 0) and it opens up. Then the equation of the parabola has the form: $x^2 = cy$. The point (300, 80) is a point on the parabola. Solve for c and find the equation:

$300^2 = c(80) \Rightarrow c = 1125$

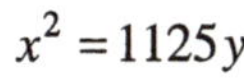

$x^2 = 1125y$

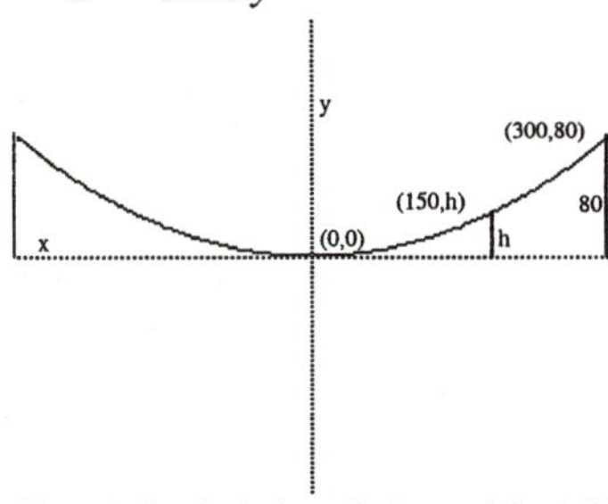

Since the height of the cable 150 feet from the center is to be found, the point (150, h) is a point on the parabola. Solve for h:

$$150^2 = 1125h$$
$$22{,}500 = 1125h$$
$$20 = h$$

The height of the cable 150 feet from the center is 20 feet.

68. Set up the problem so that the vertex of the parabola is at (0, 10) and it opens up. Then the equation of the parabola has the form:

$x^2 = c(y-10)$.

The point (200, 100) is a point on the parabola. Solve for c and find the equation:

$$200^2 = c(100-10)$$
$$40{,}000 = 90c$$
$$444.44 \approx c$$
$$x^2 = 444.44(y-10)$$

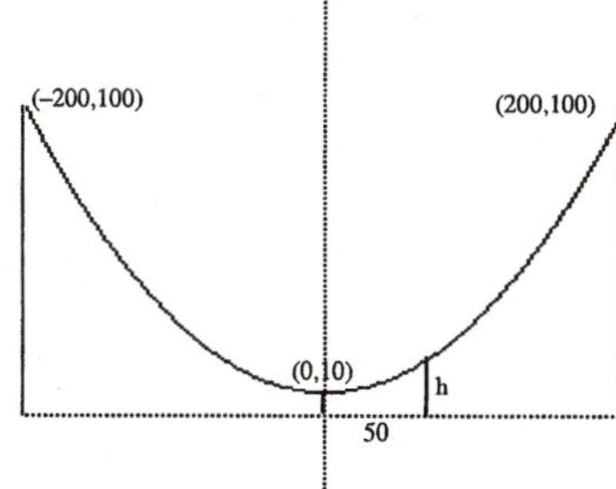

Since the height of the cable 50 feet from the center is to be found, the point (50, h) is a point on the parabola. Solve for h:

$$50^2 = 444.44(h-10)$$
$$2500 = 444.44h - 4444.4$$
$$6944.4 = 444.44h$$
$$15.625 \approx h$$

The height of the cable 50 feet from the center is about 15.625 feet.

69. Set up the problem so that the vertex of the parabola is at (0, 0) and it opens up. Then the equation of the parabola has the form: $x^2 = 4ay$. a is the distance from the vertex to the focus (where the source is located), so $a = 2$. Since the opening is 5 feet across, there is a point (2.5, y) on the parabola.

Solve for y: $x^2 = 8y$

$$2.5^2 = 8y$$
$$6.25 = 8y$$
$$y = 0.78125 \text{ feet}$$

The depth of the searchlight should be 0.78125 feet.

70. Set up the problem so that the vertex of the parabola is at (0, 0) and it opens up. Then the equation of the parabola has the form: $x^2 = 4ay$. a is the distance from the vertex to the focus (where the source is located), so $a = 2$. Since the depth is 4 feet, there is a point (x, 4) on the parabola. Solve for x:

$x^2 = 8y \Rightarrow x^2 = 8 \cdot 4 \Rightarrow x^2 = 32 \Rightarrow x = \pm 4\sqrt{2}$

The width of the opening of the searchlight should be $8\sqrt{2} \approx 11.31$ feet.

71. Set up the problem so that the vertex of the parabola is at (0, 0) and it opens up. Then the equation of the parabola has the form: $x^2 = 4ay$. Since the parabola is 20 feet across and 6 feet deep, the points (10, 6) and (–10, 6) are on the parabola. Substitute and solve for a:

$10^2 = 4a(6) \Rightarrow 100 = 24a \Rightarrow a \approx 4.17$ feet

The heat will be concentrated about 4.17 feet from the base, along the axis of symmetry.

72. Set up the problem so that the vertex of the parabola is at (0, 0) and it opens up. Then the equation of the parabola has the form: $x^2 = 4ay$. Since the parabola is 4 inches across and 3 feet deep, the points (2, 36) and (–2, 36) are on the parabola. Substitute and solve for a:

$2^2 = 4a(36)$

$4 = 144a$

$a = \frac{1}{36} \approx 0.0278$ inches

The collected light will be concentrated about 0.0278 inches from the base of the mirror.

73. Set up the problem so that the vertex of the parabola is at (0, 0) and it opens down. Then the equation of the parabola has the form: $x^2 = cy$. The point (60, –25) is a point on the parabola. Solve for c and find the equation:

$60^2 = c(-25) \Rightarrow c = -144$

$x^2 = -144y$

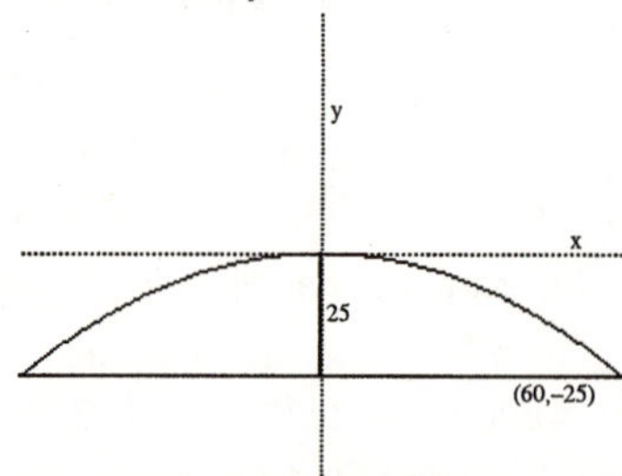

To find the height of the bridge 10 feet from the center the point (10, y) is a point on the parabola. Solve for y:

$10^2 = -144y$

$100 = -144y$

$-0.69 \approx y$

The height of the bridge 10 feet from the center is about 25 – 0.69 = 24.31 feet. To find the height of the bridge 30 feet from the center the point (30, y) is a point on the parabola. Solve for y:

$30^2 = -144y$

$900 = -144y$

$-6.25 = y$

The height of the bridge 30 feet from the center is 25 – 6.25 = 18.75 feet. To find the height of the bridge, 50 feet from the center, the point (50, y) is a point on the parabola. Solve for y:

$50^2 = -144y \Rightarrow 2500 = -144y \Rightarrow y = -17.36$

The height of the bridge 50 feet from the center is about 25 – 17.36 = 7.64 feet.

74. Set up the problem so that the vertex of the parabola is at (0, 0) and it opens down. Then the equation of the parabola has the form: $x^2 = cy$. The points (50, –h) and (40, –h+10) are points on the parabola. Substitute and solve for c and h:

$50^2 = c(-h)$ $\qquad 40^2 = c(-h+10)$

$ch = -2500$ $\qquad 1600 = -ch + 10c$

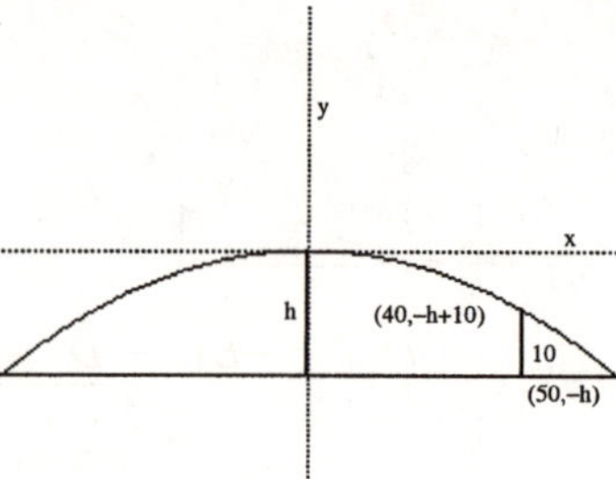

$1600 = -(-2500) + 10c$

$1600 = 2500 + 0c$

$-900 = 10c$

$-90 = c$

$-90h = -2500$

$h \approx 27.78$

The height of the bridge at the center is about 27.78 feet.

75. $Ax^2 + Ey = 0 \quad A \neq 0,\ E \neq 0$

$Ax^2 = -Ey \Rightarrow x^2 = -\frac{E}{A}y$

This is the equation of a parabola with vertex at (0, 0) and axis of symmetry being the y-axis. The focus is $\left(0, -\frac{E}{4A}\right)$. The directrix is $y = \frac{E}{4A}$.

The parabola opens up if $-\frac{E}{A} > 0$ and down if $-\frac{E}{A} < 0$.

76. $Cy^2 + Dx = 0 \quad C \neq 0,\ D \neq 0$

$Cy^2 = -Dx \Rightarrow y^2 = -\frac{D}{C}x$

This is the equation of a parabola with vertex at (0, 0) and whose axis of symmetry is the x-axis. The focus is $\left(-\frac{D}{4C}, 0\right)$. The directrix is $x = \frac{D}{4C}$.

The parabola opens to the right if $-\frac{D}{C} > 0$ and to the left if $-\frac{D}{C} < 0$.

77. $Ax^2 + Dx + Ey + F = 0 \quad A \neq 0$

a. If $E \neq 0$, then:

$$Ax^2 + Dx = -Ey - F$$

$$A\left(x^2 + \frac{D}{A}x + \frac{D^2}{4A^2}\right) = -Ey - F + \frac{D^2}{4A}$$

$$\left(x + \frac{D}{2A}\right)^2 = \frac{1}{A}\left(-Ey - F + \frac{D^2}{4A}\right)$$

$$\left(x + \frac{D}{2A}\right)^2 = \frac{-E}{A}\left(y + \frac{F}{E} - \frac{D^2}{4AE}\right)$$

$$\left(x + \frac{D}{2A}\right)^2 = \frac{-E}{A}\left(y - \frac{D^2 - 4AF}{4AE}\right)$$

This is the equation of a parabola whose vertex is $\left(-\frac{D}{2A}, \frac{D^2 - 4AF}{4AE}\right)$ and whose axis of symmetry is parallel to the y-axis.

b. If $E = 0$, then

$$Ax^2 + Dx + F = 0 \Rightarrow x = \frac{-D \pm \sqrt{D^2 - 4AF}}{2A}$$

If $D^2 - 4AF = 0$, then $x = -\frac{D}{2A}$ is a single vertical line.

c. If $E = 0$, then

$$Ax^2 + Dx + F = 0 \Rightarrow x = \frac{-D \pm \sqrt{D^2 - 4AF}}{2A}$$

If $D^2 - 4AF > 0$, then

$x = \frac{-D + \sqrt{D^2 - 4AF}}{2A}$ and

$x = \frac{-D - \sqrt{D^2 - 4AF}}{2A}$ are two vertical lines.

d. If $E = 0$, then

$$Ax^2 + Dx + F = 0 \Rightarrow x = \frac{-D \pm \sqrt{D^2 - 4AF}}{2A}$$

If $D^2 - 4AF < 0$, there is no real solution. The graph contains no points.

78. $Cy^2 + Dx + Ey + F = 0 \quad C \neq 0$

a. If $D \neq 0$, then:

$$Cy^2 + Ey = -Dx - F$$

$$C\left(y^2 + \frac{E}{C}y + \frac{E^2}{4C^2}\right) = -Dx - F + \frac{E^2}{4C}$$

$$\left(y + \frac{E}{2C}\right)^2 = \frac{1}{C}\left(-Dx - F + \frac{E^2}{4C}\right)$$

$$\left(y + \frac{E}{2C}\right)^2 = \frac{-D}{C}\left(x + \frac{F}{D} - \frac{E^2}{4CD}\right)$$

$$\left(y + \frac{E}{2C}\right)^2 = \frac{-D}{C}\left(x - \frac{E^2 - 4CF}{4CD}\right)$$

This is the equation of a parabola whose vertex is $\left(\frac{E^2 - 4CF}{4CD}, -\frac{E}{2C}\right)$, and whose axis of symmetry is parallel to the x-axis.

b. If $D = 0$, then

$$Cy^2 + Ey + F = 0 \Rightarrow y = \frac{-E \pm \sqrt{E^2 - 4CF}}{2C}$$

If $E^2 - 4CF = 0$, then $y = -\frac{E}{2C}$ is a single horizontal line.

c. If $D = 0$, then

$$Cy^2 + Ey + F = 0 \Rightarrow y = \frac{-E \pm \sqrt{E^2 - 4CF}}{2C}$$

If $E^2 - 4CF > 0$, then

$y = \frac{-E + \sqrt{E^2 - 4CF}}{2C}$ and

$y = \frac{-E - \sqrt{E^2 - 4CF}}{2C}$ are two horizontal lines.

d. If $D = 0$, then

$$Cy^2 + Ey + F = 0 \Rightarrow y = \frac{-E \pm \sqrt{E^2 - 4CF}}{2C}$$

If $E^2 - 4CF < 0$, then there is no real solution. The graph contains no points.

Section 9.3

1. $d=\sqrt{(4-2)^2+(-2-(-5))^2}=\sqrt{2^2+3^2}=\sqrt{13}$

2. $\left(-\frac{3}{2}\right)^2=\frac{9}{4}$

3. x-intercepts: $0^2=16-4x^2$
$4x^2=16$
$x^2=4$
$x=\pm 2\rightarrow(-2,0),(2,0)$

y-intercepts: $y^2=16-4(0)^2$
$y^2=16$
$y=\pm 4\rightarrow(0,-4),(0,4)$

The intercepts are $(-2,0)$, $(2,0)$, $(0,-4)$, and $(0,4)$.

4. $(2,5)$; change x to $-x$: $-2\rightarrow-(-2)=2$

5. left 1; down 4

6. $(x-2)^2+(y-(-3))^2=1^2$
$(x-2)^2+(y+3)^2=1$

7. ellipse

8. major

9. $(0,-5)$ and $(0,5)$

10. False; the foci, vertices, and center of an ellipse lie on a line called the major axis.

11. True

12. True

13. C; the major axis is along the x-axis and the vertices are at $(-4,0)$ and $(4,0)$.

14. D; the major axis is along the y-axis and the vertices are at $(0,-4)$ and $(0,4)$.

15. B; the major axis is along the y-axis and the vertices are at $(0,-2)$ and $(0,2)$.

16. A; the major axis is along the x-axis and the vertices are at $(-2,0)$ and $(2,0)$.

17. $\frac{x^2}{25}+\frac{y^2}{4}=1$
The center of the ellipse is at the origin.
$a=5,\ b=2$. The vertices are (5, 0) and (–5, 0).
Find the value of c:
$c^2=a^2-b^2=25-4=21\rightarrow c=\sqrt{21}$
The foci are $(\sqrt{21},0)$ and $(-\sqrt{21},0)$.

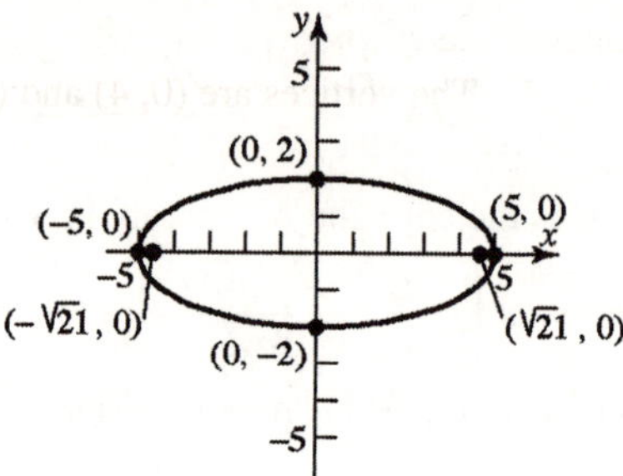

18. $\frac{x^2}{9}+\frac{y^2}{4}=1$
The center of the ellipse is at the origin.
$a=3,\ b=2$. The vertices are (3, 0) and (–3, 0).
Find the value of c:
$c^2=a^2-b^2=9-4=5\rightarrow c=\sqrt{5}$
The foci are $(\sqrt{5},0)$ and $(-\sqrt{5},0)$.

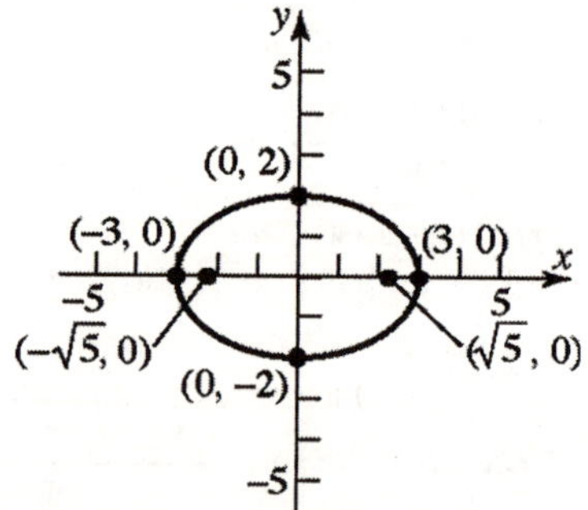

19. $\frac{x^2}{9}+\frac{y^2}{25}=1$
The center of the ellipse is at the origin.
$a=5,\ b=3$. The vertices are (0, 5) and (0, –5).
Find the value of c:
$c^2=a^2-b^2=25-9=16$
$c=4$

The foci are (0, 4) and (0, –4).

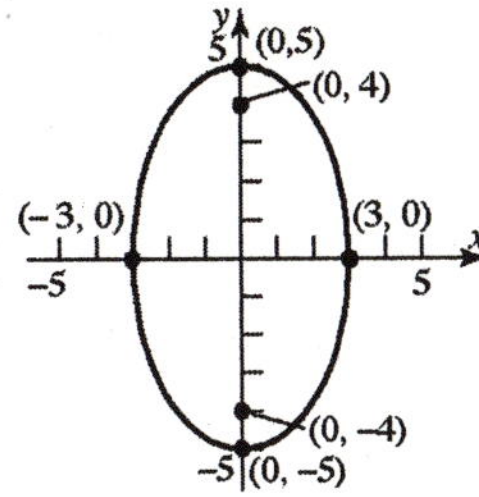

20. $x^2 + \frac{y^2}{16} = 1$

The center of the ellipse is at the origin.
$a = 4,\ b = 1$. The vertices are (0, 4) and (0, –4).
Find the value of c:
$c^2 = a^2 - b^2 = 16 - 1 = 15$
$c = \sqrt{15}$
The foci are $\left(0, \sqrt{15}\right)$ and $\left(0, -\sqrt{15}\right)$

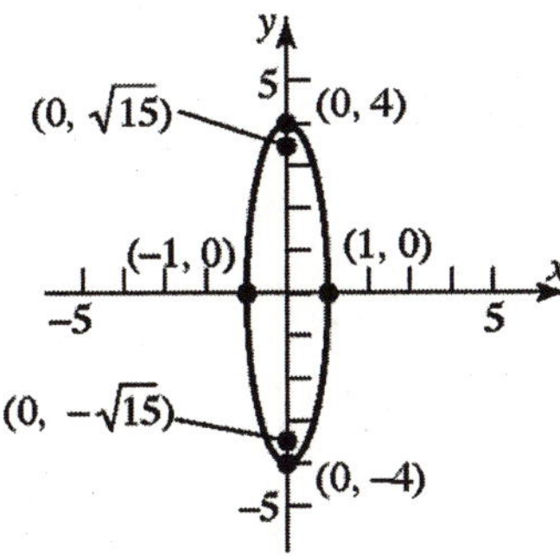

21. $4x^2 + y^2 = 16$

Divide by 16 to put in standard form:
$$\frac{4x^2}{16} + \frac{y^2}{16} = \frac{16}{16} \rightarrow \frac{x^2}{4} + \frac{y^2}{16} = 1$$
The center of the ellipse is at the origin.
$a = 4,\ b = 2$.
The vertices are (0, 4) and (0, –4). Find the value of c:
$c^2 = a^2 - b^2 = 16 - 4 = 12$
$c = \sqrt{12} = 2\sqrt{3}$
The foci are $\left(0, 2\sqrt{3}\right)$ and $\left(0, -2\sqrt{3}\right)$.

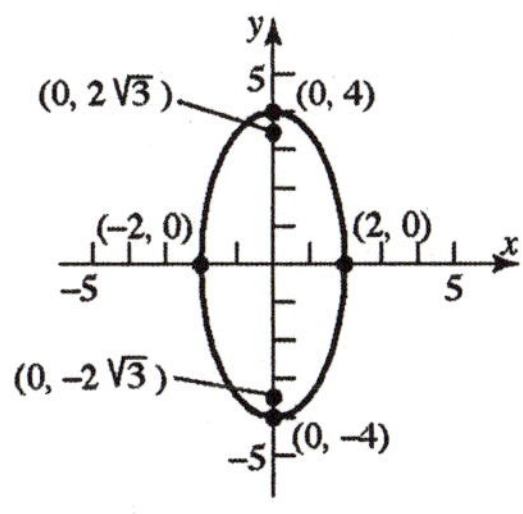

22. $x^2 + 9y^2 = 18$

Divide by 18 to put in standard form:
$$\frac{x^2}{18} + \frac{9y^2}{18} = \frac{18}{18} \rightarrow \frac{x^2}{18} + \frac{y^2}{2} = 1$$
The center of the ellipse is at the origin.
$a = 3\sqrt{2},\ b = \sqrt{2}$. The vertices are $\left(3\sqrt{2}, 0\right)$ and $\left(-3\sqrt{2}, 0\right)$. Find the value of c:
$c^2 = a^2 - b^2 = 18 - 2 = 16$
$c = 4$
The foci are (4, 0) and (–4, 0).

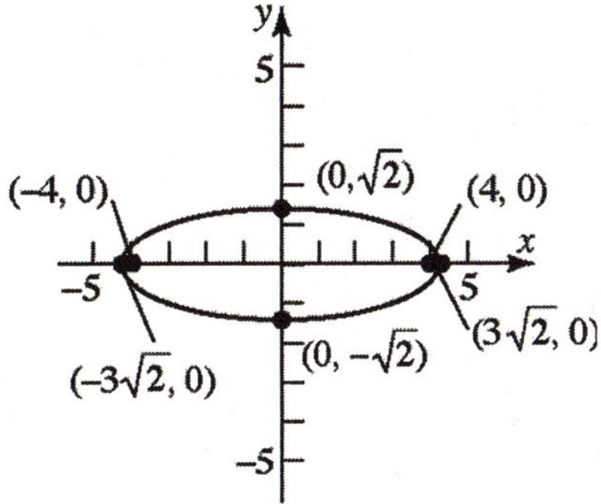

23. $4y^2 + x^2 = 8$

Divide by 8 to put in standard form:
$$\frac{4y^2}{8} + \frac{x^2}{8} = \frac{8}{8} \rightarrow \frac{x^2}{8} + \frac{y^2}{2} = 1$$
The center of the ellipse is at the origin.
$a = \sqrt{8} = 2\sqrt{2},\ b = \sqrt{2}$.
The vertices are $\left(2\sqrt{2}, 0\right)$ and $\left(-2\sqrt{2}, 0\right)$. Find the value of c:
$c^2 = a^2 - b^2 = 8 - 2 = 6$
$c = \sqrt{6}$
The foci are $\left(\sqrt{6}, 0\right)$ and $\left(-\sqrt{6}, 0\right)$.

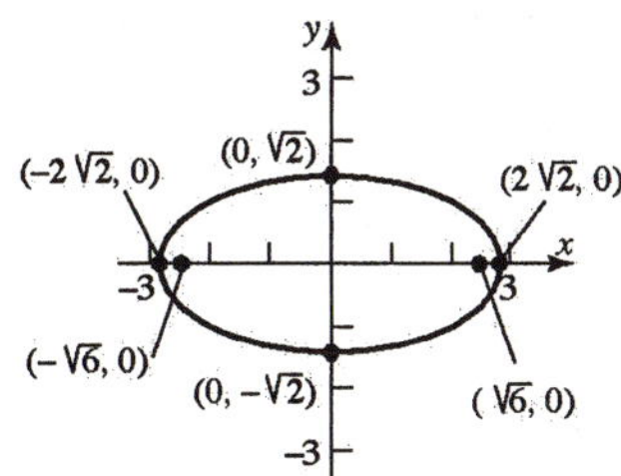

24. $4y^2 + 9x^2 = 36$

Divide by 36 to put in standard form:

$$\frac{4y^2}{36} + \frac{9x^2}{36} = \frac{36}{36} \rightarrow \frac{x^2}{4} + \frac{y^2}{9} = 1$$

The center of the ellipse is at the origin.

$a = 3$, $b = 2$. The vertices are (0, 3) and (0, –3).

Find the value of c:

$c^2 = a^2 - b^2 = 9 - 4 = 5$

$c = \sqrt{5}$

The foci are $(0, \sqrt{5})$ and $(0, -\sqrt{5})$.

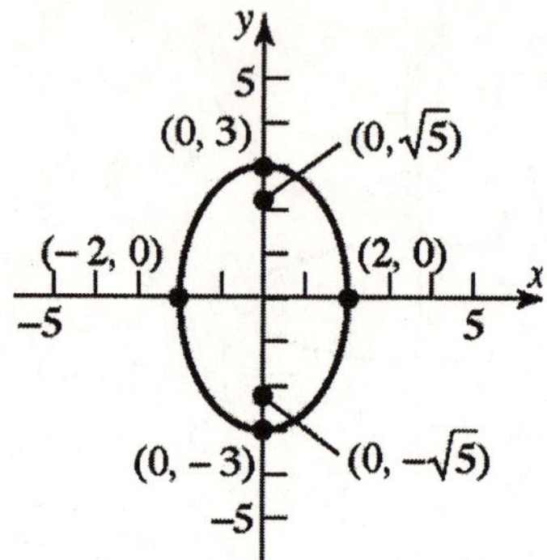

25. $x^2 + y^2 = 16$

This is the equation of a circle whose center is at (0, 0) and radius = 4.

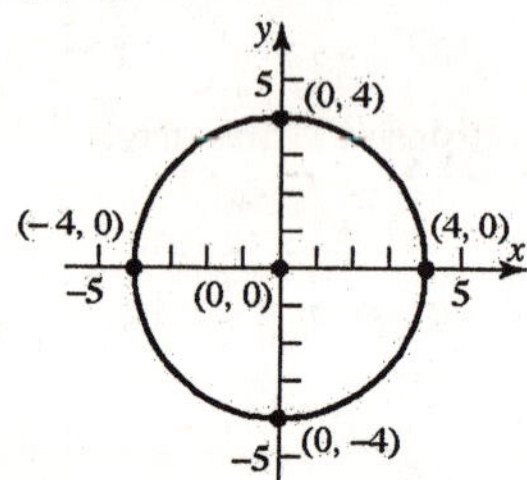

26. $x^2 + y^2 = 4$

This is the equation of a circle whose center is at (0, 0) and radius = 2.

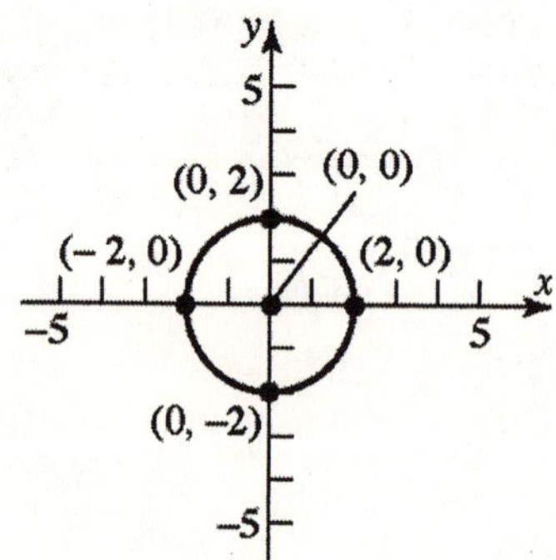

27. Center: (0, 0); Focus: (3, 0); Vertex: (5, 0); Major axis is the x-axis; $a = 5$; $c = 3$. Find b:

$b^2 = a^2 - c^2 = 25 - 9 = 16$

$b = 4$

Write the equation: $\dfrac{x^2}{25} + \dfrac{y^2}{16} = 1$

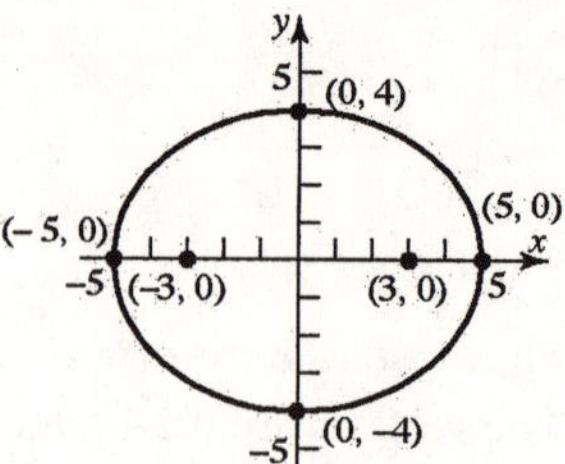

28. Center: (0, 0); Focus: (–1, 0); Vertex: (3, 0); Major axis is the x-axis; $a = 3$; $c = 1$. Find b:

$b^2 = a^2 - c^2 = 9 - 1 = 8$

$b = 2\sqrt{2}$

Write the equation: $\dfrac{x^2}{9} + \dfrac{y^2}{8} = 1$

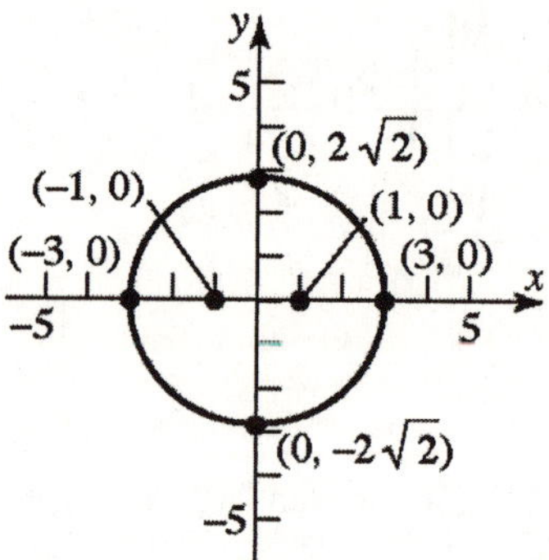

29. Center: (0, 0); Focus: (0, –4); Vertex: (0, 5); Major axis is the y-axis; $a = 5$; $c = 4$. Find b:

$b^2 = a^2 - c^2 = 25 - 16 = 9$

$b = 3$

Write the equation: $\dfrac{x^2}{9} + \dfrac{y^2}{25} = 1$

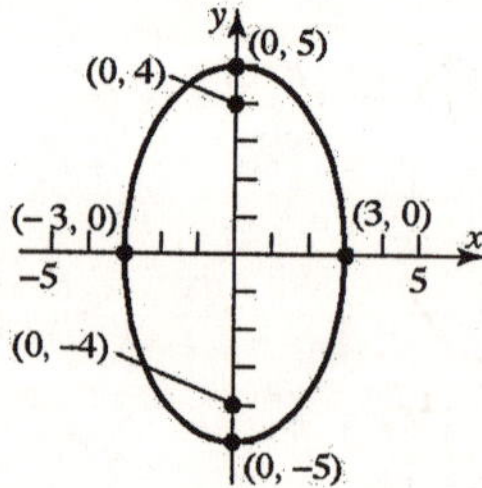

30. Center: (0, 0); Focus: (0, 1); Vertex: (0, –2); Major axis is the y-axis; $a = 2$; $c = 1$. Find b:

$b^2 = a^2 - c^2 = 4 - 1 = 3 \rightarrow b = \sqrt{3}$

Write the equation: $\frac{x^2}{3} + \frac{y^2}{4} = 1$

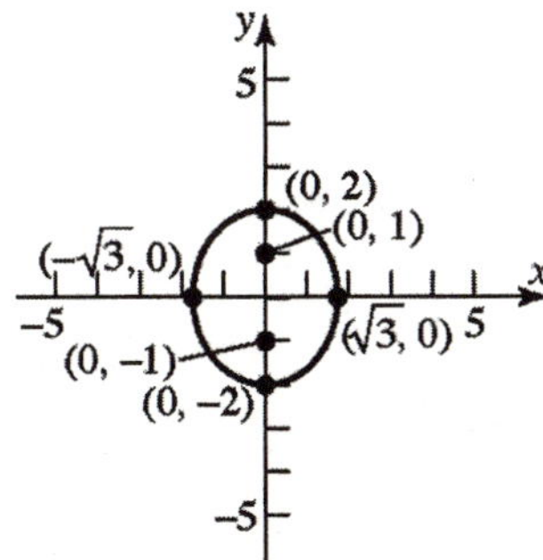

31. Foci: (±2, 0); Length of major axis is 6. Center: (0, 0); Major axis is the x-axis; $a = 3$; $c = 2$. Find b:

$b^2 = a^2 - c^2 = 9 - 4 = 5 \rightarrow b = \sqrt{5}$

Write the equation: $\frac{x^2}{9} + \frac{y^2}{5} = 1$

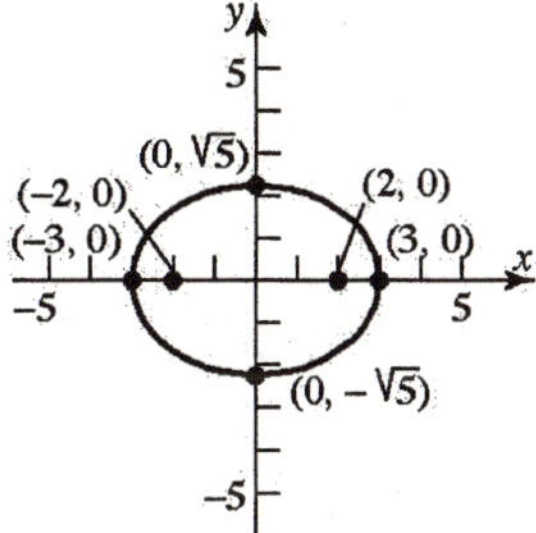

32. Foci: (0, ±2); length of the major axis is 8. Center: (0, 0); Major axis is the y-axis; $a = 4$; $c = 2$. Find b:

$b^2 = a^2 - c^2 = 16 - 4 = 12 \rightarrow b = 2\sqrt{3}$

Write the equation: $\frac{x^2}{12} + \frac{y^2}{16} = 1$

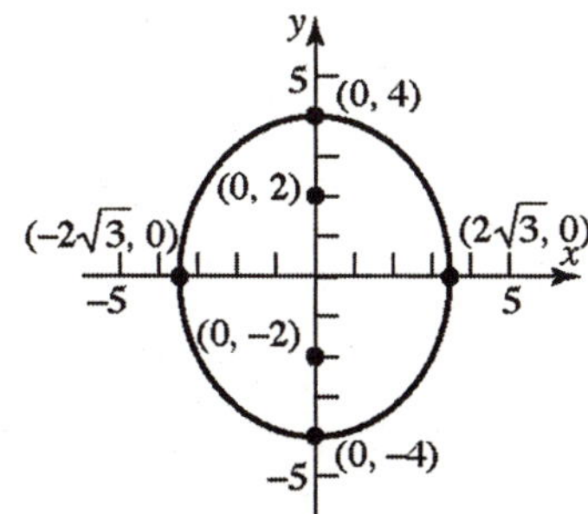

33. Focus: $(-4,0)$; Vertices: $(-5,0)$ and $(5,0)$; Center: $(0,0)$; Major axis is the x-axis. $a = 5$; $c = 4$. Find b:

$b^2 = a^2 - c^2 = 25 - 16 = 9 \rightarrow b = 3$

Write the equation: $\frac{x^2}{25} + \frac{y^2}{9} = 1$

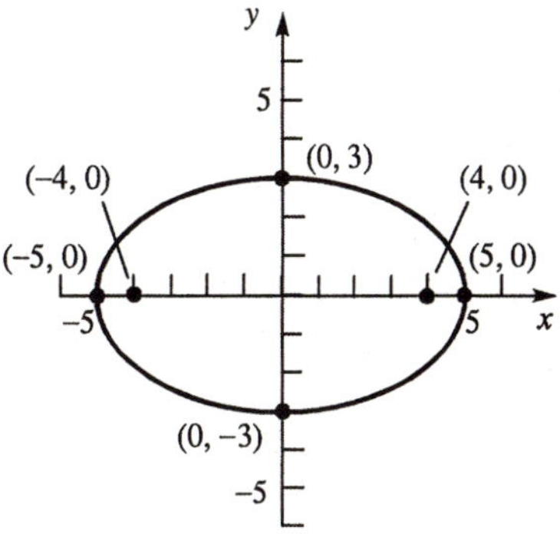

34. Focus: (0, –4); Vertices: (0, ±8). Center: (0, 0); Major axis is the y-axis; $a = 8$; $c = 4$. Find b:

$b^2 = a^2 - c^2 = 64 - 16 = 48 \rightarrow b = 4\sqrt{3}$

Write the equation: $\frac{x^2}{48} + \frac{y^2}{64} = 1$

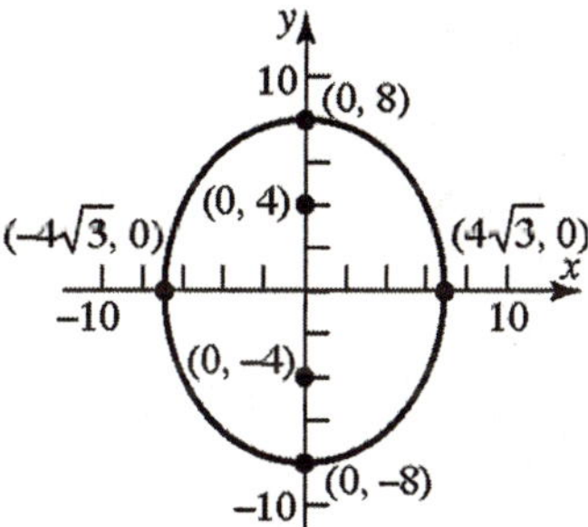

35. Foci: (0, ±3); x-intercepts are ±2. Center: (0, 0); Major axis is the y-axis; $c = 3$; $b = 2$. Find a:

$a^2 = b^2 + c^2 = 4 + 9 = 13 \rightarrow a = \sqrt{13}$

Write the equation: $\frac{x^2}{4} + \frac{y^2}{13} = 1$

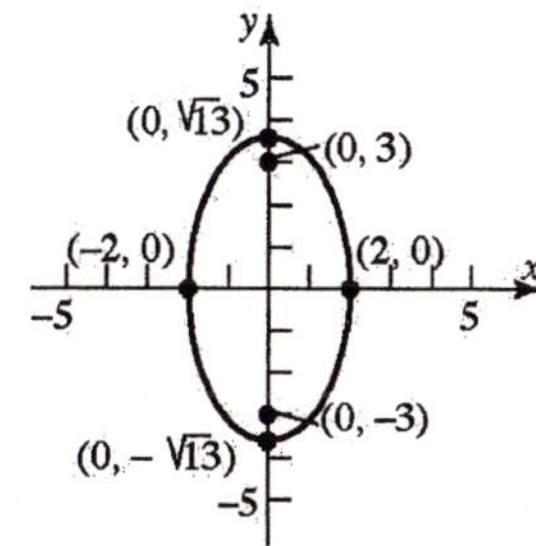

36. Vertices: $(\pm4, 0)$; y-intercepts are ±1. Center: (0, 0); Major axis is the x-axis; $a = 4$; $b = 1$. Find c:
$c^2 = a^2 - b^2 = 16 - 1 = 15$
$a = \sqrt{15}$

Write the equation: $\frac{x^2}{16} + y^2 = 1$

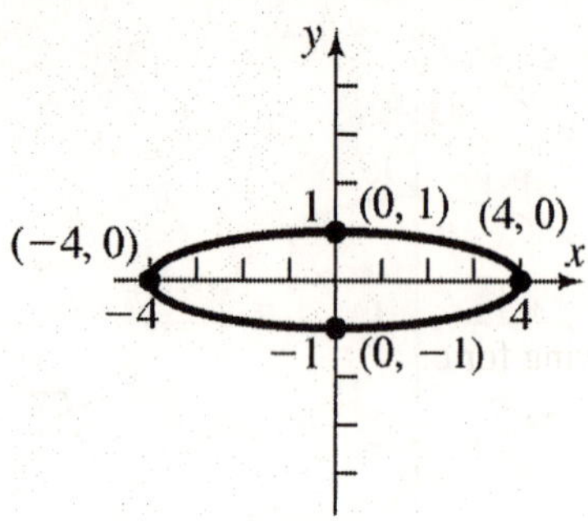

37. Center: (0, 0); Vertex: (0, 4); $b = 1$; Major axis is the y-axis; $a = 4$; $b = 1$.

Write the equation: $\frac{x^2}{1} + \frac{y^2}{16} = 1$

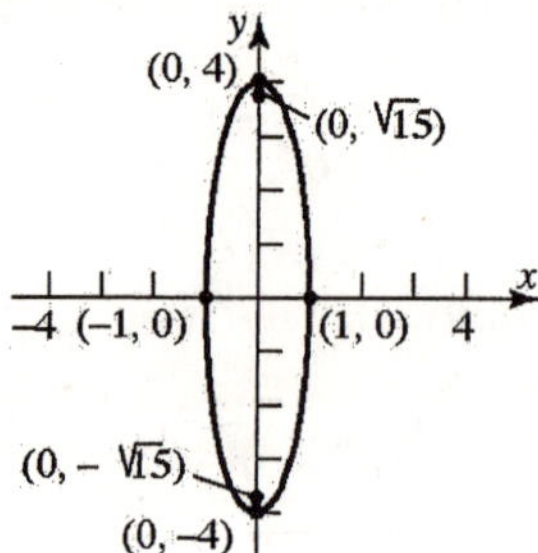

38. Vertices: $(\pm5, 0)$; $c = 2$; Major axis is the x-axis; $a = 5$; Find b:
$b^2 = a^2 - c^2 = 25 - 4 = 21$
$b = \sqrt{21}$

Write the equation: $\frac{x^2}{25} + \frac{y^2}{21} = 1$

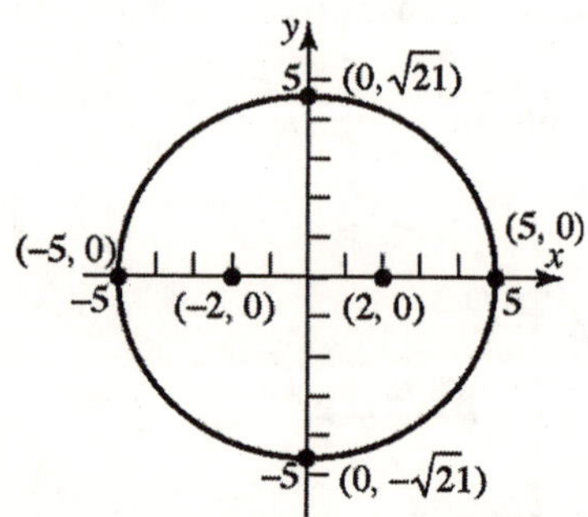

39. Center: $(-1, 1)$
Major axis: parallel to x-axis
Length of major axis: $4 = 2a \rightarrow a = 2$
Length of minor axis: $2 = 2b \rightarrow b = 1$
$$\frac{(x+1)^2}{4} + \frac{(y-1)^2}{1} = 1$$

40. Center: $(-1, -1)$
Major axis: parallel to y-axis
Length of major axis: $4 = 2a \rightarrow a = 2$
Length of minor axis: $2 = 2b \rightarrow b = 1$
$$\frac{(x+1)^2}{1} + \frac{(y+1)^2}{4} = 1$$

41. Center: $(1, 0)$
Major axis: parallel to y-axis
Length of major axis: $4 = 2a \rightarrow a = 2$
Length of minor axis: $2 = 2b \rightarrow b = 1$
$$\frac{(x-1)^2}{1} + \frac{y^2}{4} = 1$$

42. Center: $(0, 1)$
Major axis: parallel to x-axis
Length of major axis: $4 = 2a \rightarrow a = 2$
Length of minor axis: $2 = 2b \rightarrow b = 1$
$$\frac{x^2}{4} + \frac{(y-1)^2}{1} = 1$$

43. a. The equation $\frac{(x-3)^2}{4} + \frac{(y+1)^2}{9} = 1$ is in the form $\frac{(x-h)^2}{b^2} + \frac{(y-k)^2}{a^2} = 1$ (major axis parallel to the y-axis) where $a = 3$, $b = 2$, $h = 3$, and $k = -1$. Solving for c:
$c^2 = a^2 - b^2 = 9 - 4 = 5 \rightarrow c = \sqrt{5}$
Thus, we have:
Center: $(3, -1)$
Foci: $(3, -1+\sqrt{5})$, $(3, -1-\sqrt{5})$
Vertices: $(3, 2)$, $(3, -4)$

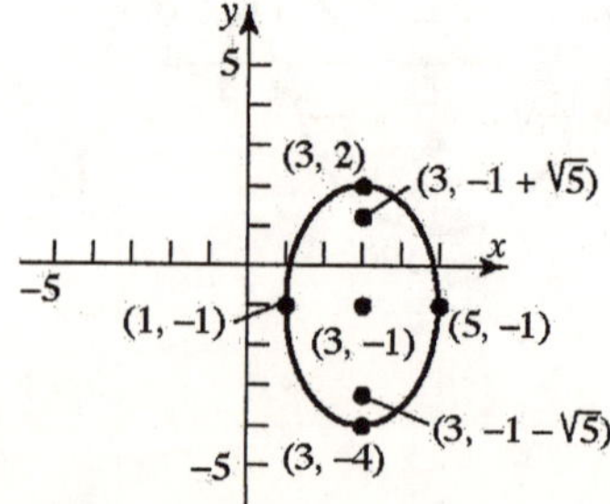

b. To graph, enter:

$y_1 = -1 + 3\sqrt{1-(x-3)^2/4}$;

$y_2 = -1 - 3\sqrt{1-(x-3)^2/4}$

2

−1.5 7.5

−4

44. a. The equation $\dfrac{(x+4)^2}{9} + \dfrac{(y+2)^2}{4} = 1$ is in the form $\dfrac{(x-h)^2}{a^2} + \dfrac{(y-k)^2}{b^2} = 1$ (major axis parallel to the x-axis) where $a = 3,\ b = 2,\ h = -4$, and $k = -2$.

Solving for c:

$c^2 = a^2 - b^2 = 9 - 4 = 5 \to c = \sqrt{5}$

Thus, we have:

Center: $(-4, -2)$

Foci: $\left(-4+\sqrt{5}, -2\right)$, $\left(-4-\sqrt{5}, -2\right)$

Vertices: $(-7, -2)$, $(-1, -2)$

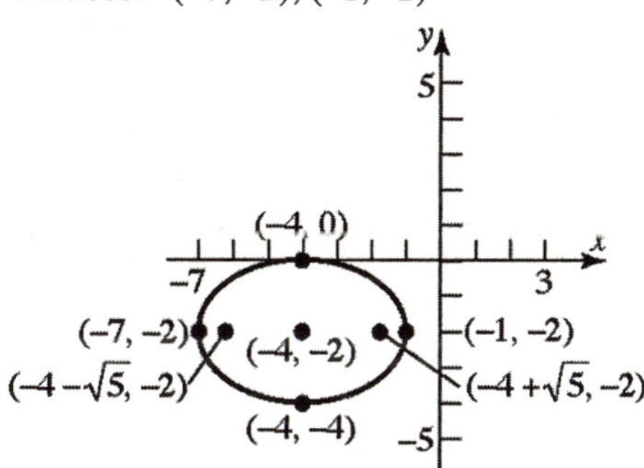

b. To graph, enter:

$y_1 = -2 + 2\sqrt{1-(x+4)^2/9}$;

$y_2 = -2 - 2\sqrt{1-(x+4)^2/9}$

1

−8 1

−5

45. a. Divide by 16 to put the equation in standard form:

$(x+5)^2 + 4(y-4)^2 = 16$

$\dfrac{(x+5)^2}{16} + \dfrac{4(y-4)^2}{16} = \dfrac{16}{16}$

$\dfrac{(x+5)^2}{16} + \dfrac{(y-4)^2}{4} = 1$

The equation is in the form

$\dfrac{(x-h)^2}{a^2} + \dfrac{(y-k)^2}{b^2} = 1$ (major axis parallel to the x-axis) where $a = 4,\ b = 2$, $h = -5$, and $k = 4$.

Solving for c:

$c^2 = a^2 - b^2 = 16 - 4 = 12 \to c = \sqrt{12} = 2\sqrt{3}$

Thus, we have:

Center: $(-5, 4)$

Foci: $\left(-5-2\sqrt{3}, 4\right)$, $\left(-5+2\sqrt{3}, 4\right)$

Vertices: $(-9, 4)$, $(-1, 4)$

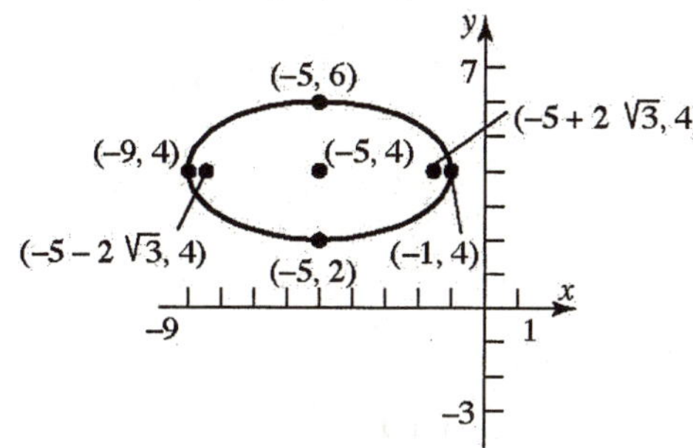

b. To graph, enter:

$y_1 = 4 + 2\sqrt{1-(x+5)^2/16}$;

$y_2 = 4 - 2\sqrt{1-(x+5)^2/16}$

6.6

−9 −1

1.3

46. a. Divide by 18 to put the equation in standard form:

$$9(x-3)^2+(y+2)^2=18$$

$$\frac{9(x-3)^2}{18}+\frac{(y+2)^2}{18}=\frac{18}{18}$$

$$\frac{(x-3)^2}{2}+\frac{(y+2)^2}{18}=1$$

The equation is in the form

$\frac{(x-h)^2}{b^2}+\frac{(y-k)^2}{a^2}=1$ (major axis parallel to the y-axis) where $a=3\sqrt{2},\ b=\sqrt{2},\ h=3,$ and $k=-2$.

Solving for c:

$c^2=a^2-b^2=18-2=16 \rightarrow c=4$

Thus, we have:

Center: (3, –2)

Foci: (3, 2), (3, –6)

Vertices: $\left(3,-2+3\sqrt{2}\right),\ \left(3,-2-3\sqrt{2}\right)$

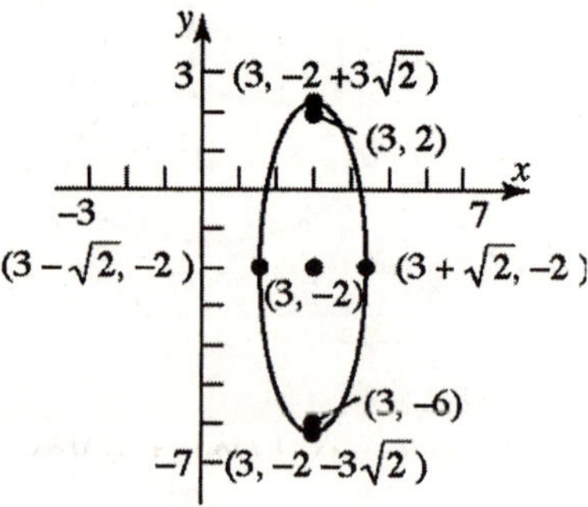

b. To graph, enter:

$y_1=-2+\sqrt{18-9(x-3)^2}$;

$y_2=-2-\sqrt{18-9(x-3)^2}$

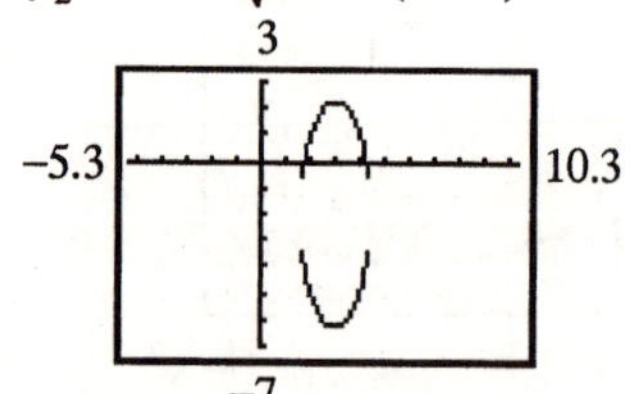

47. a. Complete the square to put the equation in standard form:

$$x^2+4x+4y^2-8y+4=0$$

$$(x^2+4x+4)+4(y^2-2y+1)=-4+4+4$$

$$(x+2)^2+4(y-1)^2=4$$

$$\frac{(x+2)^2}{4}+\frac{4(y-1)^2}{4}=\frac{4}{4}$$

$$\frac{(x+2)^2}{4}+\frac{(y-1)^2}{1}=1$$

The equation is in the form

$\frac{(x-h)^2}{a^2}+\frac{(y-k)^2}{b^2}=1$ (major axis parallel to the x-axis) where

$a=2,\ b=1,\ h=-2,$ and $k=1$.

Solving for c:

$c^2=a^2-b^2=4-1=3 \rightarrow c=\sqrt{3}$

Thus, we have:

Center: (–2, 1)

Foci: $\left(-2-\sqrt{3},1\right),\ \left(-2+\sqrt{3},1\right)$

Vertices: (–4, 1), (0, 1)

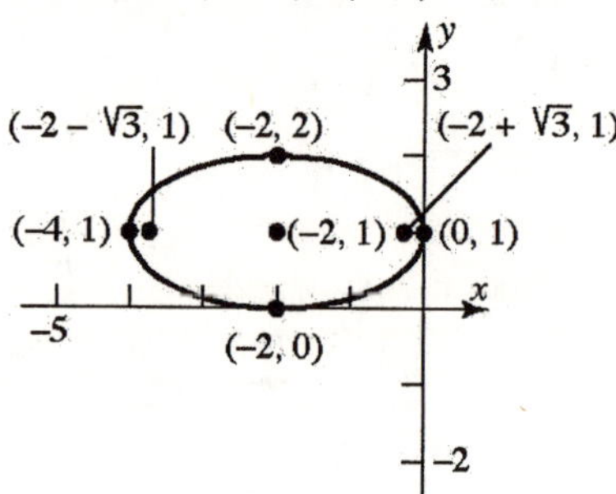

b. To graph, enter:

$y_1=1+\sqrt{1-(x+2)^2/4}$;

$y_2=1-\sqrt{1-(x+2)^2/4}$

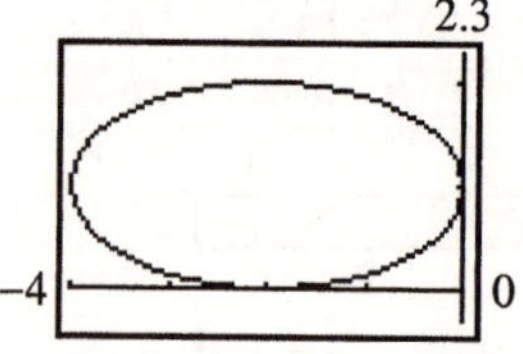

48. a. Complete the square to put the equation in standard form:

$$x^2+3y^2-12y+9=0$$
$$x^2+3(y^2-4y+4)=-9+12$$
$$x^2+3(y-2)^2=3$$
$$\frac{x^2}{3}+\frac{3(y-2)^2}{3}=\frac{3}{3}$$
$$\frac{x^2}{3}+\frac{(y-2)^2}{1}=1$$

The equation is in the form $\frac{(x-h)^2}{a^2}+\frac{(y-k)^2}{b^2}=1$ (major axis parallel to the x-axis) where $a=\sqrt{3},\ b=1,\ h=0,$ and $k=2$.

Solving for c:
$c^2=a^2-b^2=3-1=2 \rightarrow c=\sqrt{2}$

Thus, we have:
Center: (0, 2)
Foci: $\left(-\sqrt{2},2\right),\ \left(\sqrt{2},2\right)$
Vertices: $\left(-\sqrt{3},2\right),\ \left(\sqrt{3},2\right)$

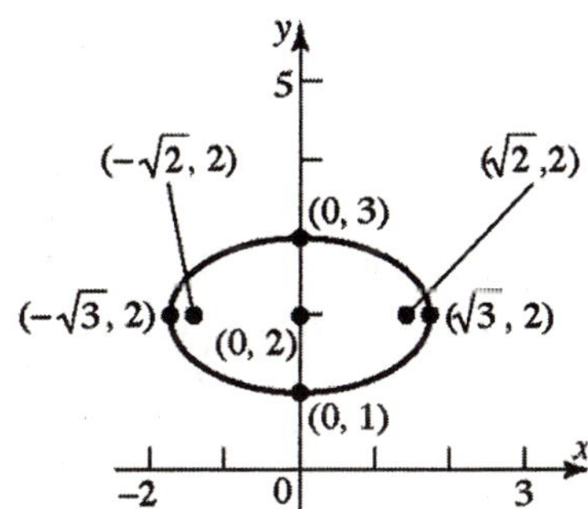

b. To graph, enter:
$y_1=2+\sqrt{1-x^2/3}$;
$y_2=2-\sqrt{1-x^2/3}$

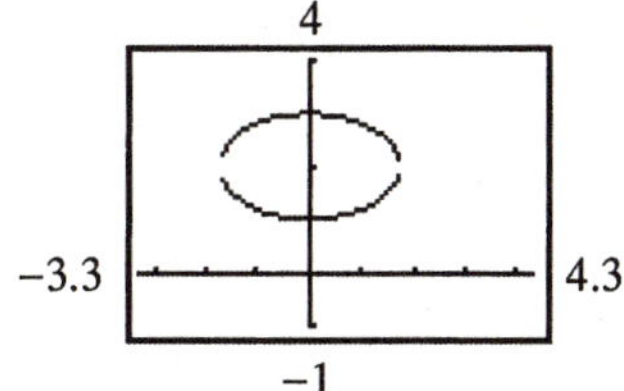

49. a. Complete the square to put the equation in standard form:

$$2x^2+3y^2-8x+6y+5=0$$
$$2(x^2-4x)+3(y^2+2y)=-5$$
$$2(x^2-4x+4)+3(y^2+2y+1)=-5+8+3$$
$$2(x-2)^2+3(y+1)^2=6$$
$$\frac{2(x-2)^2}{6}+\frac{3(y+1)^2}{6}=\frac{6}{6}$$
$$\frac{(x-2)^2}{3}+\frac{(y+1)^2}{2}=1$$

The equation is in the form $\frac{(x-h)^2}{a^2}+\frac{(y-k)^2}{b^2}=1$ (major axis parallel to the x-axis) where $a=\sqrt{3},\ b=\sqrt{2},\ h=2,$ and $k=-1$.

Solving for c:
$c^2=a^2-b^2=3-2=1 \ \rightarrow\ c=1$

Thus, we have:
Center: (2, –1)
Foci: (1, –1), (3, –1)
Vertices: $\left(2-\sqrt{3},-1\right),\ \left(2+\sqrt{3},-1\right)$

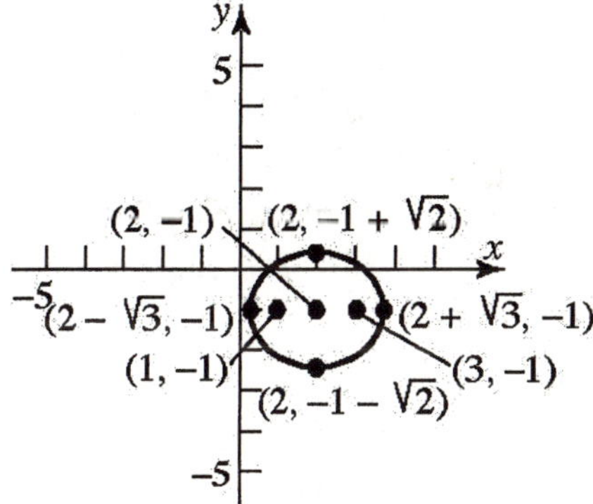

b. To graph, enter:
$y_1=-1+\sqrt{2-2(x-2)^2/3}$;
$y_2=-1-\sqrt{2-2(x-2)^2/3}$

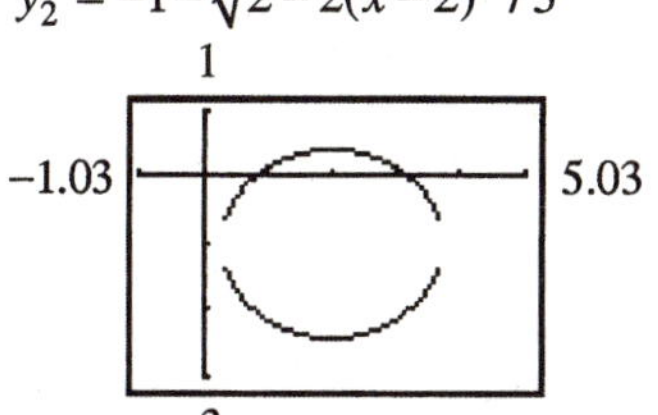

50. a. Complete the square to put the equation in standard form:

$$4x^2+3y^2+8x-6y=5$$
$$4(x^2+2x)+3(y^2-2y)=5$$
$$4(x^2+2x+1)+3(y^2-2y+1)=5+4+3$$
$$4(x+1)^2+3(y-1)^2=12$$
$$\frac{4(x+1)^2}{12}+\frac{3(y-1)^2}{12}=\frac{12}{12}$$
$$\frac{(x+1)^2}{3}+\frac{(y-1)^2}{4}=1$$

The equation is in the form $\frac{(x-h)^2}{b^2}+\frac{(y-k)^2}{a^2}=1$ (major axis parallel to the y-axis) where $a=2,\ b=\sqrt{3},\ h=-1,$ and $k=1$.

Solving for c:

$c^2=a^2-b^2=4-3=1\ \rightarrow\ c=1$

Thus, we have:

Center: $(-1, 1)$

Foci: $(-1, 0),\ (-1, 2)$

Vertices: $(-1,-1),\ (-1, 3)$

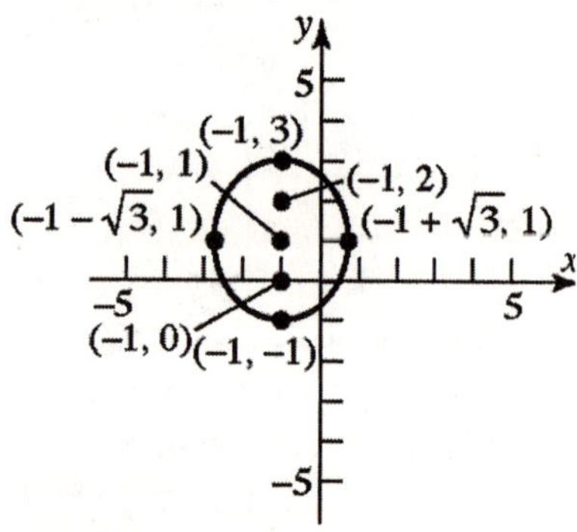

b. To graph, enter:

$y_1=1+2\sqrt{1-(x+1)^2/3};$

$y_2=1-2\sqrt{1-(x+1)^2/3}$

4

−4.5 4.5

−2

51. a. Complete the square to put the equation in standard form:

$$9x^2+4y^2-18x+16y-11=0$$
$$9(x^2-2x)+4(y^2+4y)=11$$
$$9(x^2-2x+1)+4(y^2+4y+4)=11+9+16$$
$$9(x-1)^2+4(y+2)^2=36$$
$$\frac{9(x-1)^2}{36}+\frac{4(y+2)^2}{36}=\frac{36}{36}$$
$$\frac{(x-1)^2}{4}+\frac{(y+2)^2}{9}=1$$

The equation is in the form $\frac{(x-h)^2}{b^2}+\frac{(y-k)^2}{a^2}=1$ (major axis parallel to the y-axis) where $a=3,\ b=2,\ h=1,$ and $k=-2$.

Solving for c:

$c^2=a^2-b^2=9-4=5\rightarrow c=\sqrt{5}$

Thus, we have:

Center: $(1,-2)$

Foci: $\left(1,-2+\sqrt{5}\right),\ \left(1,-2-\sqrt{5}\right)$

Vertices: $(1,1),\ (1,-5)$

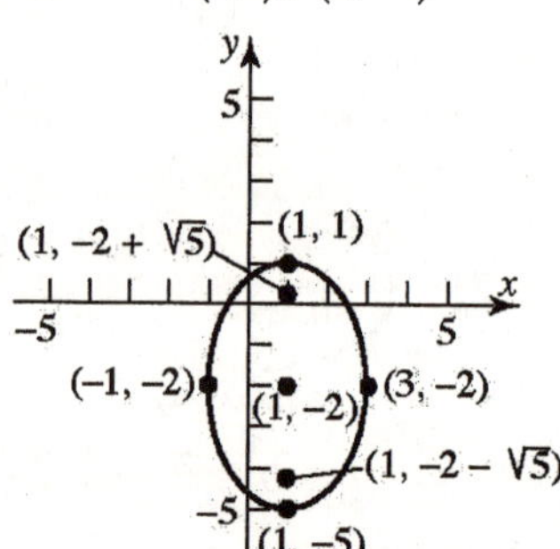

b. To graph, enter:

$y_1=-2+3\sqrt{1-(x-1)^2/4};$

$y_2=-2-3\sqrt{1-(x-1)^2/4}$

1

−3.5 3.5

−5

52. a. Complete the square to put the equation in standard form:

$$x^2+9y^2+6x-18y+9=0$$
$$(x^2+6x)+9(y^2-2y)=-9$$
$$(x^2+6x+9)+9(y^2-2y+1)=-9+9+9$$
$$(x+3)^2+9(y-1)^2=9$$
$$\frac{(x+3)^2}{9}+\frac{9(y-1)^2}{9}=\frac{9}{9}$$
$$\frac{(x+3)^2}{9}+\frac{(y-1)^2}{1}=1$$

The equation is in the form $\frac{(x-h)^2}{a^2}+\frac{(y-k)^2}{b^2}=1$ (major axis parallel to the x-axis) where $a=3,\ b=1,\ h=-3,$ and $k=1$.

Solving for c:
$c^2=a^2-b^2=9-1=8 \rightarrow c=2\sqrt{2}$

Thus, we have:
Center: (–3, 1))
Foci: $\left(-3+2\sqrt{2},1\right),\ \left(-3-2\sqrt{2},1\right)$
Vertices: (0, 1), (–6, 1)

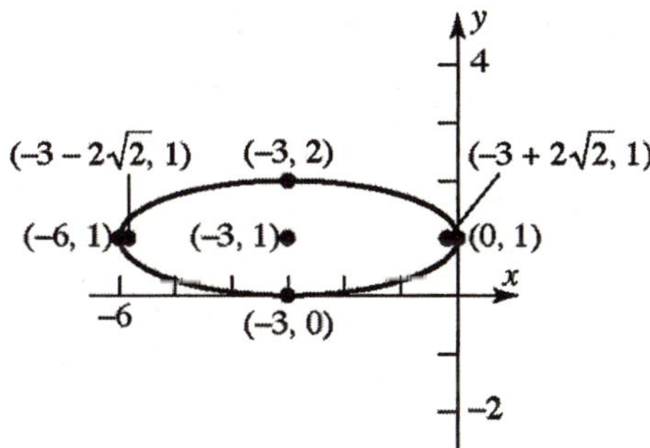

b. To graph, enter:
$y_1=1+\sqrt{1-(x+3)^2/9}$;
$y_2=1-\sqrt{1-(x+3)^2/9}$

3
–6 0
–1

53. a. Complete the square to put the equation in standard form:

$$4x^2+y^2+4y=0$$
$$4x^2+y^2+4y+4=4$$
$$4x^2+(y+2)^2=4$$
$$\frac{4x^2}{4}+\frac{(y+2)^2}{4}=\frac{4}{4}$$
$$\frac{x^2}{1}+\frac{(y+2)^2}{4}=1$$

The equation is in the form $\frac{(x-h)^2}{b^2}+\frac{(y-k)^2}{a^2}=1$ (major axis parallel to the y-axis) where $a=2,\ b=1,\ h=0,$ and $k=-2$.

Solving for c:
$c^2=a^2-b^2=4-1=3 \rightarrow c=\sqrt{3}$

Thus, we have:
Center: (0,–2)
Foci: $\left(0,-2+\sqrt{3}\right),\ \left(0,-2-\sqrt{3}\right)$
Vertices: (0, 0), (0, –4)

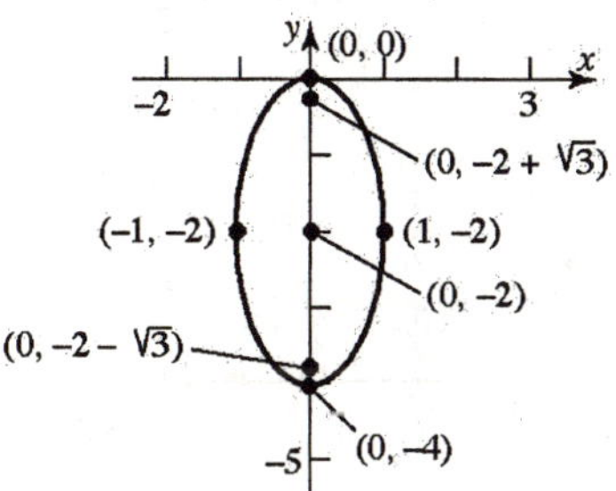

b. To graph, enter:
$y_1=-2+2\sqrt{1-x^2}$;
$y_2=-2-2\sqrt{1-x^2}$

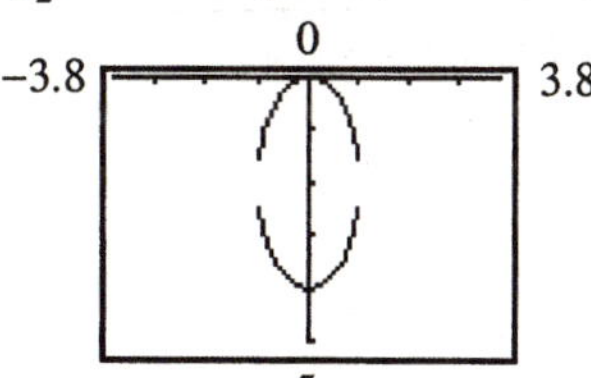

54. a. Complete the square to put the equation in standard form:

$$9x^2 + y^2 - 18x = 0$$
$$9(x^2 - 2x + 1) + y^2 = 9$$
$$9(x-1)^2 + y^2 = 9$$
$$\frac{9(x-1)^2}{9} + \frac{y^2}{9} = \frac{9}{9}$$
$$\frac{(x-1)^2}{1} + \frac{y^2}{9} = 1$$

The equation is in the form

$\frac{(x-h)^2}{b^2} + \frac{(y-k)^2}{a^2} = 1$ (major axis parallel to the y-axis) where

$a = 3,\ b = 1,\ h = 1,$ and $k = 0$.

Solving for c:

$c^2 = a^2 - b^2 = 9 - 1 = 8 \rightarrow c = 2\sqrt{2}$

Thus, we have:

Center: $(1, 0)$

Foci: $\left(1, 2\sqrt{2}\right),\ \left(1, -2\sqrt{2}\right)$

Vertices: $(1, 3),\ (1, -3)$

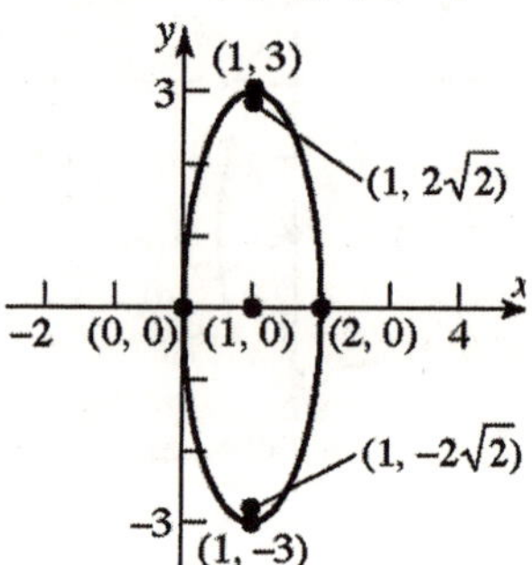

b. To graph, enter:

$y_1 = 3\sqrt{1-(x-1)^2}$;

$y_2 = -3\sqrt{1-(x-1)^2}$

3

–3.5 3.5

–3

55. Center: $(2, -2)$; Vertex: $(7, -2)$; Focus: $(4, -2)$; Major axis parallel to the x-axis; $a = 5$; $c = 2$.

Find b:

$b^2 = a^2 - c^2 = 25 - 4 = 21 \rightarrow b = \sqrt{21}$

Write the equation: $\frac{(x-2)^2}{25} + \frac{(y+2)^2}{21} = 1$

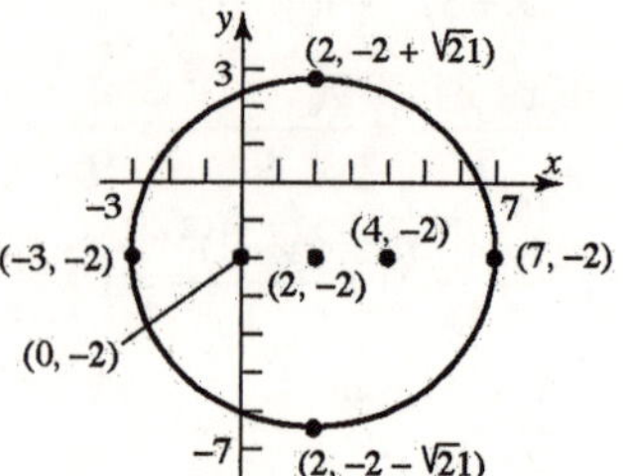

56. Center: $(-3, 1)$; Vertex: $(-3, 3)$; Focus: $(-3, 0)$; Major axis parallel to the y-axis; $a = 2$; $c = 1$.

Find b:

$b^2 = a^2 - c^2 = 4 - 1 = 3 \rightarrow b = \sqrt{3}$

Write the equation: $\frac{(x+3)^2}{3} + \frac{(y-1)^2}{4} = 1$

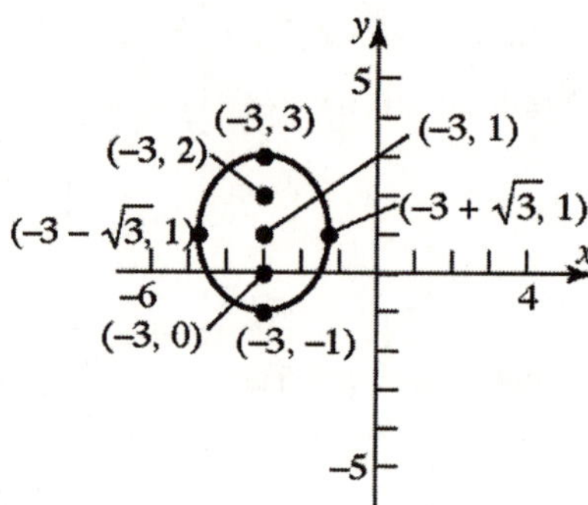

57. Vertices: $(4, 3), (4, 9)$; Focus: $(4, 8)$; Center: $(4, 6)$; Major axis parallel to the y-axis; $a = 3$; $c = 2$. Find b:

$b^2 = a^2 - c^2 = 9 - 4 = 5 \rightarrow b = \sqrt{5}$

Write the equation: $\frac{(x-4)^2}{5} + \frac{(y-6)^2}{9} = 1$

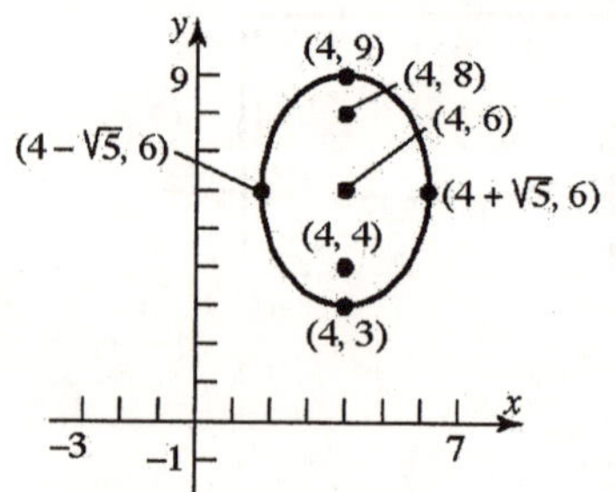

58. Foci: (1, 2), (–3, 2); Vertex: (–4, 2); Center: (–1, 2); Major axis parallel to the x-axis; $a = 3$; $c = 2$. Find b:

$b^2 = a^2 - c^2 = 9 - 4 = 5 \rightarrow b = \sqrt{5}$

Write the equation: $\dfrac{(x+1)^2}{9} + \dfrac{(y-2)^2}{5} = 1$

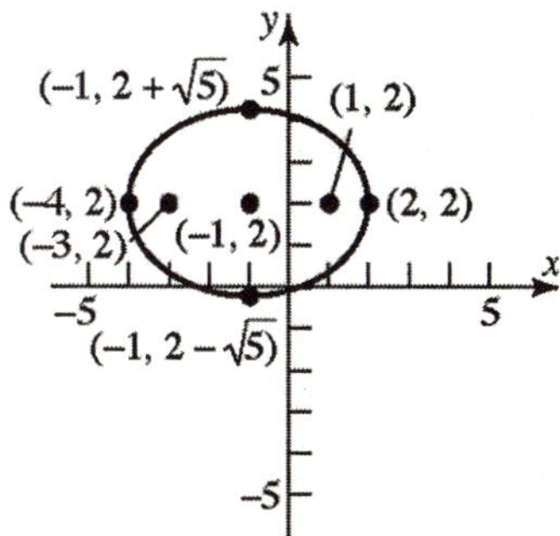

59. Foci: (5, 1), (–1, 1); Length of the major axis = 8; Center: (2, 1); Major axis parallel to the x-axis; $a = 4$; $c = 3$. Find b:

$b^2 = a^2 - c^2 = 16 - 9 = 7 \rightarrow b = \sqrt{7}$

Write the equation: $\dfrac{(x-2)^2}{16} + \dfrac{(y-1)^2}{7} = 1$

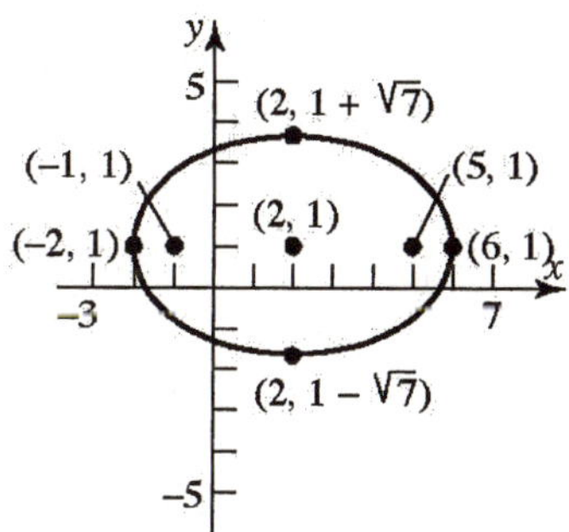

60. Vertices: (2, 5), (2, –1); $c = 2$; Center: (2, 2); Major axis parallel to the y-axis; $a = 3$; $c = 2$. Find b:

$b^2 = a^2 - c^2 = 9 - 4 = 5 \rightarrow b = \sqrt{5}$

Write the equation: $\dfrac{(x-2)^2}{5} + \dfrac{(y-2)^2}{9} = 1$

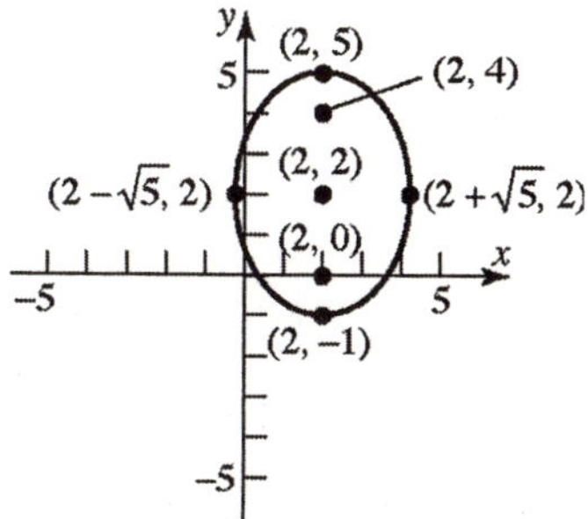

61. Center: (1, 2); Focus: (4, 2); contains the point (1, 3); Major axis parallel to the x-axis; $c = 3$. The equation has the form:

$\dfrac{(x-1)^2}{a^2} + \dfrac{(y-2)^2}{b^2} = 1$

Since the point (1, 3) is on the curve:

$\dfrac{0}{a^2} + \dfrac{1}{b^2} = 1$

$\dfrac{1}{b^2} = 1 \rightarrow b^2 = 1 \rightarrow b = 1$

Find a:

$a^2 = b^2 + c^2 = 1 + 9 = 10 \rightarrow a = \sqrt{10}$

Write the equation: $\dfrac{(x-1)^2}{10} + \dfrac{(y-2)^2}{1} = 1$

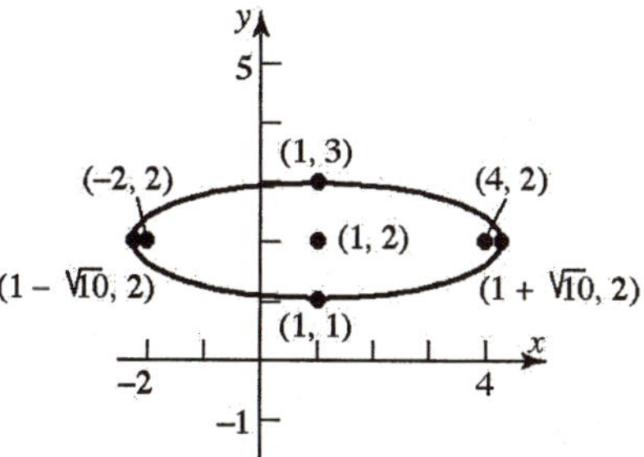

62. Center: (1, 2); Focus: (1, 4); contains the point (2, 2); Major axis parallel to the y-axis; $c = 2$. The equation has the form:

$\dfrac{(x-1)^2}{b^2} + \dfrac{(y-2)^2}{a^2} = 1$

Since the point (2, 2) is on the curve:

$\dfrac{1}{b^2} + \dfrac{0}{a^2} = 1$

$\dfrac{1}{b^2} = 1 \rightarrow b^2 = 1 \rightarrow b = 1$

Find a:

$a^2 = b^2 + c^2 = 1 + 4 = 5 \rightarrow a = \sqrt{5}$

Write the equation: $\dfrac{(x-1)^2}{1} + \dfrac{(y-2)^2}{5} = 1$

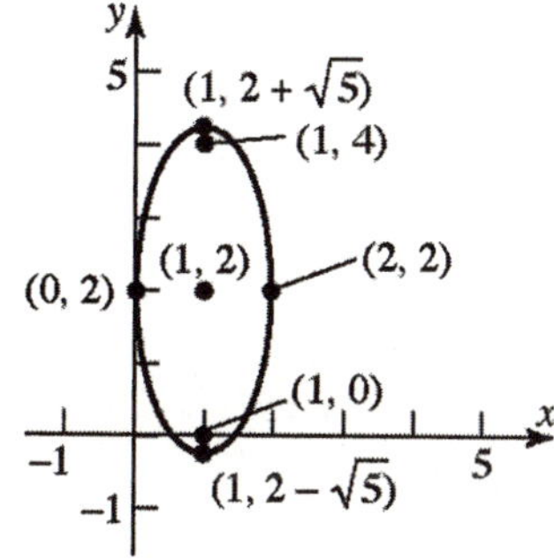

63. Center: $(1, 2)$; Vertex: $(4, 2)$; contains the point $(1, 3)$; Major axis parallel to the x-axis; $a = 3$.
The equation has the form:
$$\frac{(x-1)^2}{a^2}+\frac{(y-2)^2}{b^2}=1$$
Since the point (1, 3) is on the curve:
$$\frac{0}{9}+\frac{1}{b^2}=1$$
$$\frac{1}{b^2}=1 \rightarrow b^2=1 \rightarrow b=1$$
Solve for c:
$$c^2=a^2-b^2=9-1=8$$
Thus, $c=\pm 2\sqrt{2}$.

Write the equation: $\dfrac{(x-1)^2}{9}+\dfrac{(y-2)^2}{1}=1$

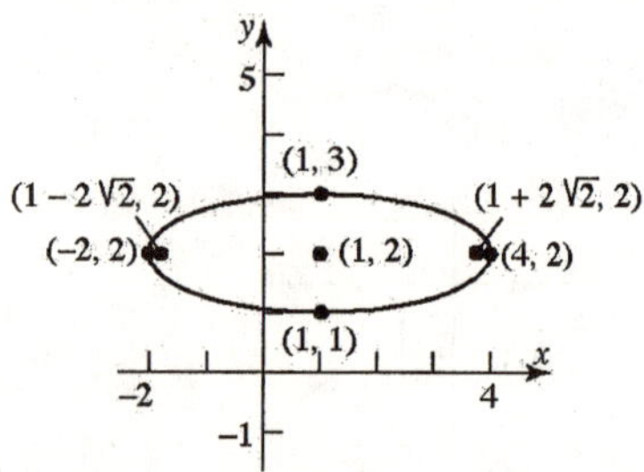

64. Center: $(1, 2)$; Vertex: $(1, 4)$; contains the point $(2, 2)$; Major axis parallel to the y-axis; $a = 2$.
The equation has the form:
$$\frac{(x-1)^2}{b^2}+\frac{(y-2)^2}{a^2}=1$$
Since the point (2, 2) is on the curve:
$$\frac{1}{b^2}+\frac{0}{a^2}=1$$
$$\frac{1}{b^2}=1 \rightarrow b^2=1 \rightarrow b=1$$

Write the equation: $\dfrac{(x-1)^2}{1}+\dfrac{(y-2)^2}{4}=1$

Solve for c:
$$c^2=a^2-b^2=4-1=3$$
Thus, $c=\pm\sqrt{3}$.

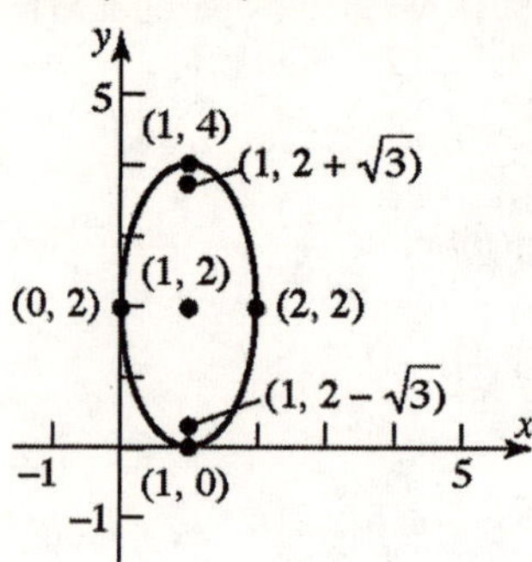

65. Rewrite the equation:
$$y=\sqrt{16-4x^2}$$
$$y^2=16-4x^2, \quad y\ge 0$$
$$4x^2+y^2=16, \quad y\ge 0$$
$$\frac{x^2}{4}+\frac{y^2}{16}=1, \quad y\ge 0$$

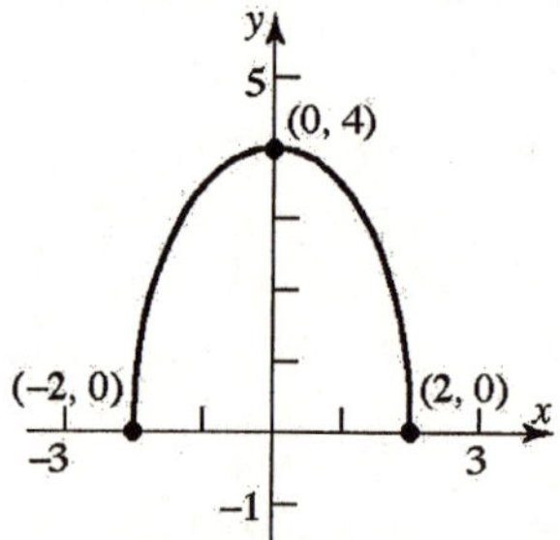

66. Rewrite the equation:
$$y=\sqrt{9-9x^2}$$
$$y^2=9-9x^2, \quad y\ge 0$$
$$9x^2+y^2=9, \quad y\ge 0$$
$$\frac{x^2}{1}+\frac{y^2}{9}=1, \quad y\ge 0$$

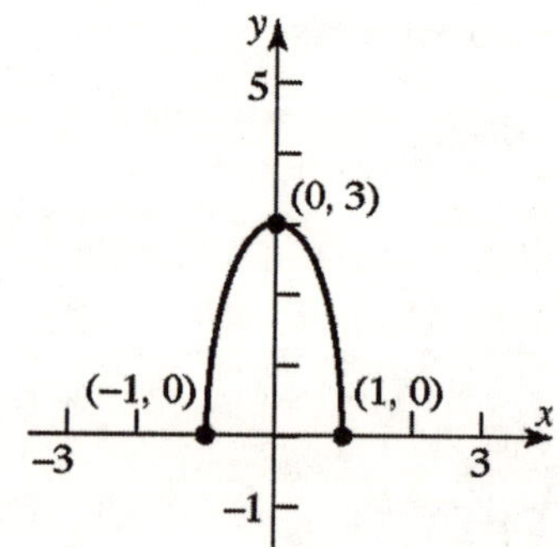

67. Rewrite the equation:

$$y = -\sqrt{64-16x^2}$$
$$y^2 = 64-16x^2, \quad y \le 0$$
$$16x^2 + y^2 = 64, \quad y \le 0$$
$$\frac{x^2}{4} + \frac{y^2}{64} = 1, \quad y \le 0$$

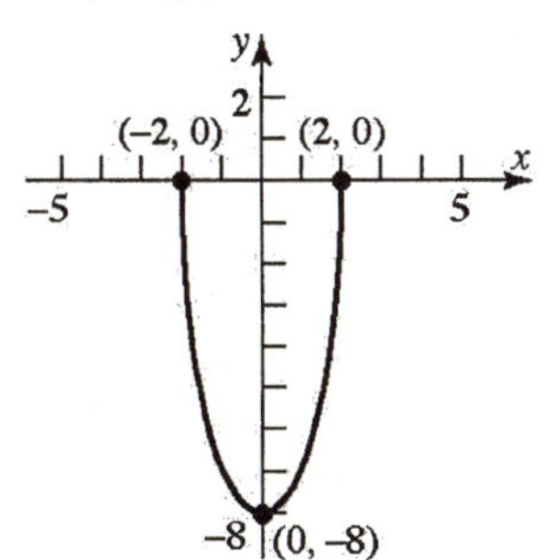

68. Rewrite the equation:

$$y = -\sqrt{4-4x^2}$$
$$y^2 = 4-4x^2, \quad y \le 0$$
$$4x^2 + y^2 = 4, \quad y \le 0$$
$$\frac{x^2}{1} + \frac{y^2}{4} = 1, \quad y \le 0$$

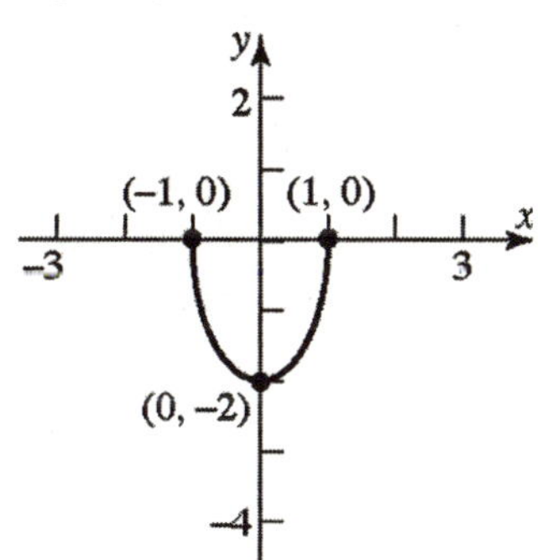

69. The center of the ellipse is (0, 0). The length of the major axis is 20, so $a = 10$. The length of half the minor axis is 6, so $b = 6$. The ellipse is situated with its major axis on the x-axis. The equation is: $\frac{x^2}{100} + \frac{y^2}{36} = 1$.

70. The center of the ellipse is (0, 0). The length of the major axis is 30, so $a = 15$. The length of half the minor axis is 10, so $b = 10$. The ellipse is situated with its major axis on the x-axis. The equation is: $\frac{x^2}{225} + \frac{y^2}{100} = 1$.

The roadway is 12 feet above the axis of the ellipse. At the center ($x = 0$), the roadway is 2 feet above the arch.

At a point 5 feet either side of the center, evaluate the equation at $x = 5$:

$$\frac{5^2}{225} + \frac{y^2}{100} = 1$$
$$\frac{y^2}{100} = 1 - \frac{25}{225} = \frac{200}{225}$$
$$y = 10\sqrt{\frac{200}{225}} \approx 9.43$$

The vertical distance from the roadway to the arch is $12 - 9.43 \approx 2.57$ feet.

At a point 10 feet either side of the center, evaluate the equation at $x = 10$:

$$\frac{10^2}{225} + \frac{y^2}{100} = 1$$
$$\frac{y^2}{100} = 1 - \frac{100}{225} = \frac{125}{225}$$
$$y = 10\sqrt{\frac{125}{225}} \approx 7.45$$

The vertical distance from the roadway to the arch is $12 - 7.45 \approx 4.55$ feet.

At a point 15 feet either side of the center, the roadway is 12 feet above the arch.

71. Assume that the half ellipse formed by the gallery is centered at (0, 0). Since the hall is 100 feet long, $2a = 100$ or $a = 50$. The distance from the center to the foci is 25 feet, so $c = 25$. Find the height of the gallery which is b:

$$b^2 = a^2 - c^2 = 2500 - 625 = 1875$$
$$b = \sqrt{1875} \approx 43.3$$

The ceiling will be 43.3 feet high in the center.

72. Assume that the half ellipse formed by the gallery is centered at (0, 0). Since the distance between the foci is 100 feet and Jim is 6 feet from the nearest wall, the length of the gallery is 112 feet. $2a = 112$ or $a = 56$. The distance from the center to the foci is 50 feet, so $c = 50$. Find the height of the gallery which is b:

$$b^2 = a^2 - c^2 = 3136 - 2500 = 636$$
$$b = \sqrt{636} \approx 25.2$$

The ceiling will be 25.2 feet high in the center.

73. Place the semi-elliptical arch so that the x-axis coincides with the water and the y-axis passes through the center of the arch. Since the bridge has a span of 120 feet, the length of the major axis is 120, or $2a = 120$ or $a = 60$. The maximum height of the bridge is 25 feet, so $b = 25$. The equation is: $\frac{x^2}{3600} + \frac{y^2}{625} = 1$.

The height 10 feet from the center:

$$\frac{10^2}{3600} + \frac{y^2}{625} = 1$$
$$\frac{y^2}{625} = 1 - \frac{100}{3600}$$
$$y^2 = 625 \cdot \frac{3500}{3600}$$
$$y \approx 24.65 \text{ feet}$$

The height 30 feet from the center:

$$\frac{30^2}{3600} + \frac{y^2}{625} = 1$$
$$\frac{y^2}{625} = 1 - \frac{900}{3600}$$
$$y^2 = 625 \cdot \frac{2700}{3600}$$
$$y \approx 21.65 \text{ feet}$$

The height 50 feet from the center:

$$\frac{50^2}{3600} + \frac{y^2}{625} = 1$$
$$\frac{y^2}{625} = 1 - \frac{2500}{3600}$$
$$y^2 = 625 \cdot \frac{1100}{3600}$$
$$y \approx 13.82 \text{ feet}$$

74. Place the semi-elliptical arch so that the x-axis coincides with the water and the y-axis passes through the center of the arch. Since the bridge has a span of 100 feet, the length of the major axis is 100, or $2a = 100$ or $a = 50$. Let h be the maximum height of the bridge. The equation is: $\frac{x^2}{2500} + \frac{y^2}{h^2} = 1$.

The height of the arch 40 feet from the center is 10 feet. So (40, 10) is a point on the ellipse. Substitute and solve for h:

$$\frac{40^2}{2500} + \frac{10^2}{h^2} = 1$$
$$\frac{10^2}{h^2} = 1 - \frac{1600}{2500} = \frac{9}{25}$$
$$9h^2 = 2500$$
$$h = \frac{50}{3} \approx 16.67$$

The height of the arch at its center is 16.67 feet.

75. Place the semi-elliptical arch so that the x-axis coincides with the major axis and the y-axis passes through the center of the arch. Since the ellipse is 40 feet wide, the length of the major axis is 40, or $2a = 40$ or $a = 20$. The height is 15 feet at the center, so $b = 15$. The equation is: $\frac{x^2}{400} + \frac{y^2}{225} = 1$.

The height a distance of 10 feet on either side of the center:

$$\frac{10^2}{400} + \frac{y^2}{225} = 1$$
$$\frac{y^2}{225} = 1 - \frac{100}{400}$$
$$y^2 = 225 \cdot \frac{3}{4}$$
$$y \approx 12.99 \text{ feet}$$

The height a distance of 20 feet on either side of the center:

$$\frac{20^2}{400} + \frac{y^2}{225} = 1$$
$$\frac{y^2}{225} = 1 - \frac{400}{400}$$
$$y^2 = 225 \cdot 0$$
$$y \approx 0 \text{ feet}$$

Heights: 0 ft, 12.99 ft, 15 ft, 12.99 ft, 0 ft.

76. Place the semi-elliptical arch so that the x-axis coincides with the major axis and the y-axis passes through the center of the arch. Since the height of the arch at the center is 20 feet, $b = 20$. The length of the major axis is to be found, so it is necessary to solve for a. The equation is: $\frac{x^2}{a^2} + \frac{y^2}{400} = 1$.

The height of the arch 28 feet from the center is to be 13 feet, so the point (28, 13) is on the ellipse. Substitute and solve for a:

$$\frac{28^2}{a^2}+\frac{13^2}{400}=1$$

$$\frac{784}{a^2}=1-\frac{169}{400}=\frac{231}{400}$$

$$231a^2=313600$$

$$a^2=1357.576$$

$$a=36.845$$

The span of the bridge is 73.69 feet.

77. Since the mean distance is 93 million miles, $a=93$ million. The length of the major axis is 186 million. The perihelion is 186 million – 94.5 million = 91.5 million miles.

The distance from the center of the ellipse to the sun (focus) is 93 million – 91.5 million = 1.5 million miles. Therefore, $c=1.5$ million. Find b:

$$b^2=a^2-c^2=\left(93\times10^6\right)^2-\left(1.5\times10^6\right)^2$$

$$=8.64675\times10^{15}=8646.75\times10^{12}$$

$$b=92.99\times10^6$$

The equation of the orbit is:

$$\frac{x^2}{\left(93\times10^6\right)^2}+\frac{y^2}{\left(92.99\times10^6\right)^2}=1$$

We can simplify the equation by letting our units for x and y be millions of miles. The equation then becomes:

$$\frac{x^2}{8649}+\frac{y^2}{8646.75}=1$$

78. Since the mean distance is 142 million miles, $a=142$ million. The length of the major axis is 284 million. The aphelion is 284 million – 128.5 million = 155.5 million miles.

The distance from the center of the ellipse to the sun (focus) is 142 million – 128.5 million = 13.5 million miles. Therefore, $c=13.5$ million. Find b:

$$b^2=a^2-c^2=\left(142\times10^6\right)^2-\left(13.5\times10^6\right)^2$$

$$=1.998175\times10^{16}$$

$$b=141.36\times10^6$$

The equation of the orbit is:

$$\frac{x^2}{\left(142\times10^6\right)^2}+\frac{y^2}{\left(141.36\times10^6\right)^2}=1$$

We can simplify the equation by letting our units for x and y be millions of miles. The equation then becomes:

$$\frac{x^2}{20{,}164}+\frac{y^2}{19{,}981.75}=1$$

79. The mean distance is 507 million – 23.2 million = 483.8 million miles.

The perihelion is 483.8 million – 23.2 million = 460.6 million miles.

Since $a=483.8\times10^6$ and $c=23.2\times10^6$, we can find b:

$$b^2=a^2-c^2=\left(483.8\times10^6\right)^2-\left(23.2\times10^6\right)^2$$

$$=2.335242\times10^{17}$$

$$b=483.2\times10^6$$

The equation of the orbit of Jupiter is:

$$\frac{x^2}{\left(483.8\times10^6\right)^2}+\frac{y^2}{\left(483.2\times10^6\right)^2}=1$$

We can simplify the equation by letting our units for x and y be millions of miles. The equation then becomes:

$$\frac{x^2}{234{,}062.44}+\frac{y^2}{233{,}524.2}=1$$

80. The mean distance is 4551 million + 897.5 million = 5448.5 million miles.

The aphelion is 5448.5 million + 897.5 million = 6346 million miles.

Since $a=5448.5\times10^6$ and $c=897.5\times10^6$, we can find b:

$$b^2=a^2-c^2=\left(5448.5\times10^6\right)^2-\left(897.5\times10^6\right)^2$$

$$=2.8880646\times10^{19}$$

$$b=5374.07\times10^6$$

The equation of the orbit of Pluto is:

$$\frac{x^2}{\left(5448.5\times10^6\right)^2}+\frac{y^2}{\left(5374.07\times10^6\right)^2}=1$$

We can simplify the equation by letting our units for x and y be millions of miles. The equation then becomes:

$$\frac{x^2}{29{,}686{,}152.25}+\frac{y^2}{28{,}880{,}646}=1$$

81. If the x-axis is placed along the 100 foot length and the y-axis is placed along the 50 foot length, the equation for the ellipse is: $\frac{x^2}{50^2}+\frac{y^2}{25^2}=1$.

Find y when x = 40:

$$\frac{40^2}{50^2}+\frac{y^2}{25^2}=1$$

$$\frac{y^2}{625}=1-\frac{1600}{2500}$$

$$y^2=625\cdot\frac{9}{25}$$

$$y\approx 15 \text{ feet}$$

To get the width of the ellipse at $x=40$, we need to double the y value. Thus, the width 10 feet from a vertex is 30 feet.

82. If the x-axis is placed along the 80 foot length, the y-axis is placed along the 40 foot width, and the center is at the origin, then the equation for the ellipse is: $\frac{x^2}{40^2}+\frac{y^2}{20^2}=1$.

Find y when x = 30 (that is, 10 feet from a vertex):

$$\frac{30^2}{40^2}+\frac{y^2}{20^2}=1$$

$$\frac{y^2}{400}=1-\frac{900}{1600}$$

$$y^2=400\cdot\frac{700}{1600}=175$$

$$y=\sqrt{175}=5\sqrt{7}$$

To get the width of the ellipse at $x=30$, we need to double the y value. Thus, the width 10 feet from the vertex is $2\left(5\sqrt{7}\right)=10\sqrt{7}\approx 26.5$ ft.

83. a. Put the equation in standard ellipse form:

$$Ax^2+Cy^2+F=0$$

$$Ax^2+Cy^2=-F$$

$$\frac{Ax^2}{-F}+\frac{Cy^2}{-F}=1$$

$$\frac{x^2}{(-F/A)}+\frac{y^2}{(-F/C)}=1$$

where $A\neq 0, C\neq 0, F\neq 0$, and $-F/A$ and $-F/C$ are positive.

If $A\neq C$, then $-\frac{F}{A}\neq-\frac{F}{C}$. So, this is the equation of an ellipse with center at (0, 0).

b. If $A=C$, the equation becomes:

$$Ax^2+Ay^2=-F \rightarrow x^2+y^2=\frac{-F}{A}$$

This is the equation of a circle with center at (0, 0) and radius of $\sqrt{\frac{-F}{A}}$.

84. Complete the square on the given equation:

$$Ax^2+Cy^2+Dx+Ey+F=0,$$

$$A\left(x^2+\frac{D}{A}x\right)+C\left(y^2+\frac{E}{C}y\right)=-F$$

$$A\left(x+\frac{D}{2A}\right)^2+C\left(y+\frac{E}{2C}\right)^2=\frac{D^2}{4A}+\frac{E^2}{4C}-F$$

where $A\cdot C>0$.

Let $U=\frac{D^2}{4A}+\frac{E^2}{4C}-F$.

a. If U is of the same sign as A (and C), then

$$\frac{\left(x+\frac{D}{2A}\right)^2}{\frac{U}{A}}+\frac{\left(y+\frac{E}{2C}\right)^2}{\frac{U}{C}}=1$$

This is the equation of an ellipse whose center is $\left(\frac{-D}{2A},\frac{-E}{2C}\right)$.

b. If $U=0$, the graph is the single point $\left(\frac{-D}{2A},\frac{-E}{2C}\right)$.

c. If U is of the opposite sign as A (and C), this graph contains no points since the left side always has the opposite sign of the right side.

85. Answers will vary.

Section 9.4

1. $d=\sqrt{(-2-3)^2+(1-(-4))^2}$

$$=\sqrt{(-5)^2+(5)^2}=\sqrt{25+25}$$

$$=\sqrt{50}=5\sqrt{2}$$

2. $\left(\frac{5}{2}\right)^2=\frac{25}{4}$

3. x-intercepts: $0^2 = 9 + 4x^2$

$$4x^2 = -9$$
$$x^2 = -\frac{9}{4} \text{ (no real solution)}$$

y-intercepts: $y^2 = 9 + 4(0)^2$

$$y^2 = 9$$
$$y = \pm 3 \rightarrow (0,-3),(0,3)$$

The intercepts are $(0,-3)$ and $(0,3)$.

4. True; the graph of $y^2 = 9 + x^2$ is a hyperbola with its center at the origin.

5. right 5 units; down 4 units

6. $y = \frac{x^2-9}{x^2-4}$; $p(x) = x^2 - 9$, $q(x) = x^2 - 4$

The vertical asymptotes are the zeros of q.

$$q(x) = 0$$
$$x^2 - 4 = 0$$
$$x^2 = 4 \Rightarrow x = -2, x = 2$$

The lines $x = -2$, and $x = 2$ are the vertical asymptotes. The degree of the numerator, $p(x) = x^2 - 9$, is $n = 2$. The degree of the denominator, $q(x) = x^2 - 4$, is $m = 2$.

Since $n = m$, the line $y = \frac{1}{1} = 1$ is a horizontal asymptote. Since there is a horizontal asymptote, there are no oblique asymptotes.

7. hyperbola

8. transverse axis

9. $y = \frac{3}{2}x$, $y = -\frac{3}{2}x$

10. False; the foci of a hyperbola lie on a line called the transverse axis.

11. True; hyperbolas always have two asymptotes.

12. False; a hyperbola intersects the transverse axis at the vertices.

13. *B*; the hyperbola opens to the left and right, and has vertices at $(\pm 1, 0)$. Thus, the graph has an equation of the form $x^2 - \frac{y^2}{b^2} = 1$.

14. *C*; the hyperbola opens up and down, and has vertices at $(0, \pm 2)$. Thus, the graph has an equation of the form $\frac{y^2}{4} - \frac{x^2}{b^2} = 1$.

15. *A*; the hyperbola opens to the left and right, and has vertices at $(\pm 2, 0)$. Thus, the graph has an equation of the form $\frac{x^2}{4} - \frac{y^2}{b^2} = 1$.

16. *D*; the hyperbola opens up and down, and has vertices at $(0, \pm 1)$. Thus, the graph has an equation of the form $y^2 - \frac{x^2}{b^2} = 1$.

17. Center: (0, 0); Focus: (3, 0); Vertex: (1, 0); Transverse axis is the *x*-axis; $a = 1$; $c = 3$. Find the value of *b*:

$$b^2 = c^2 - a^2 = 9 - 1 = 8$$
$$b = \sqrt{8} = 2\sqrt{2}$$

Write the equation: $\frac{x^2}{1} - \frac{y^2}{8} = 1$.

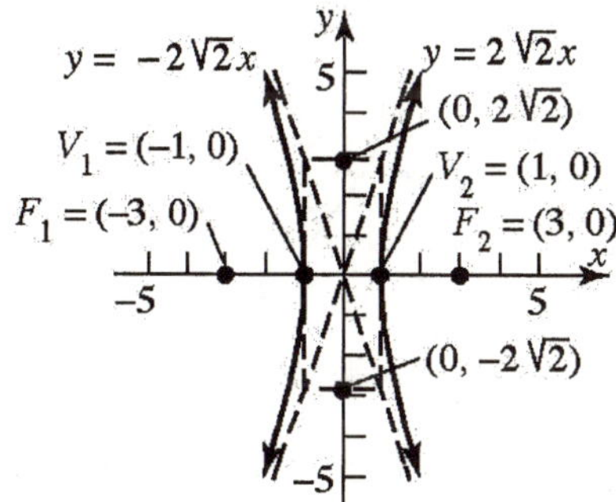

18. Center: (0, 0); Focus: (0, 5); Vertex: (0, 3); Transverse axis is the y-axis; $a=3$; $c=5$. Find the value of b:

$b^2=c^2-a^2=25-9=16$

$b=4$

Write the equation: $\frac{y^2}{9}-\frac{x^2}{16}=1$.

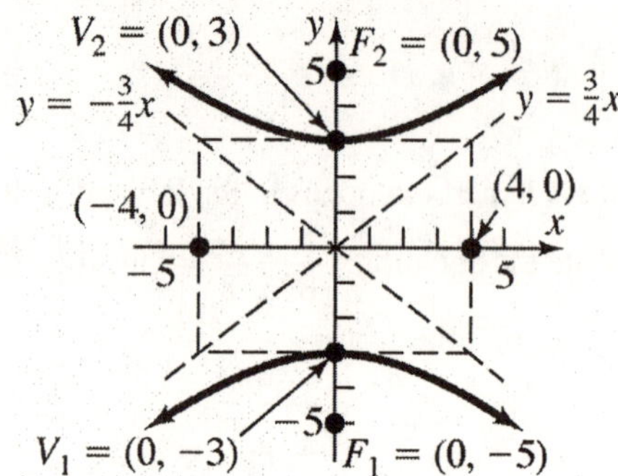

19. Center: (0, 0); Focus: (0, –6); Vertex: (0, 4) Transverse axis is the y-axis; $a=4$; $c=6$. Find the value of b:

$b^2=c^2-a^2=36-16=20$

$b=\sqrt{20}=2\sqrt{5}$

Write the equation: $\frac{y^2}{16}-\frac{x^2}{20}=1$.

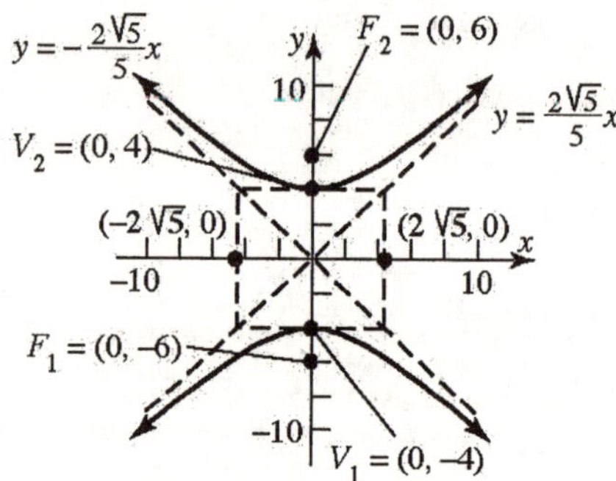

20. Center: (0, 0); Focus: (–3, 0); Vertex: (2, 0) Transverse axis is the x-axis; $a=2$; $c=3$. Find the value of b:

$b^2=c^2-a^2=9-4=5$

$b=\sqrt{5}$

Write the equation: $\frac{x^2}{4}-\frac{y^2}{5}=1$.

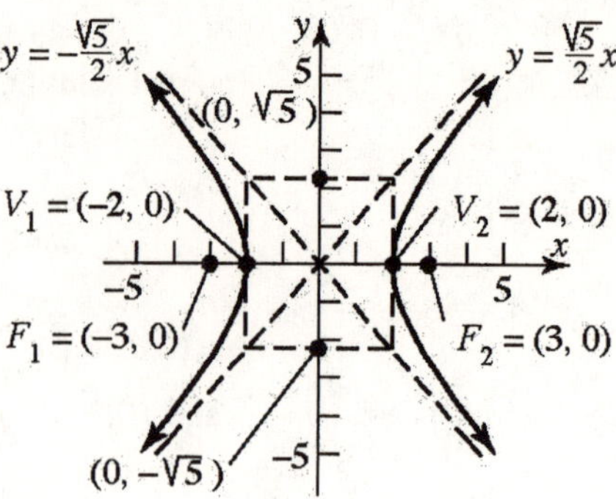

21. Foci: (–5, 0), (5, 0); Vertex: (3, 0) Center: (0, 0); Transverse axis is the x-axis; $a=3$; $c=5$.

Find the value of b:

$b^2=c^2-a^2=25-9=16 \Rightarrow b=4$

Write the equation: $\frac{x^2}{9}-\frac{y^2}{16}=1$.

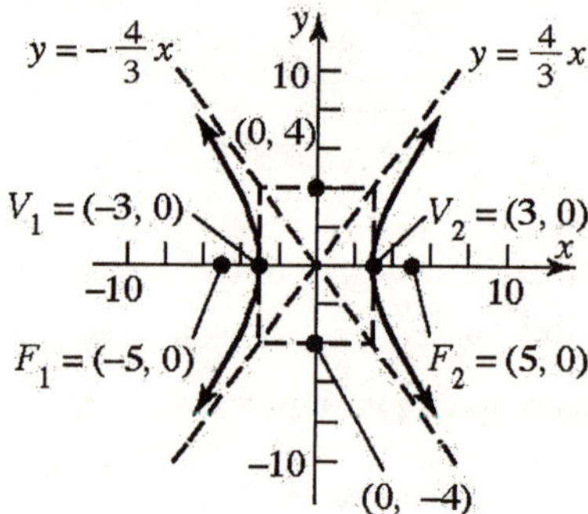

22. Focus: (0, 6); Vertices: (0, –2), (0, 2) Center: (0, 0); Transverse axis is the y-axis; $a=2$; $c=6$.

Find the value of b:

$b^2=c^2-a^2=36-4=32 \Rightarrow b=4\sqrt{2}$

Write the equation: $\frac{y^2}{4}-\frac{x^2}{32}=1$.

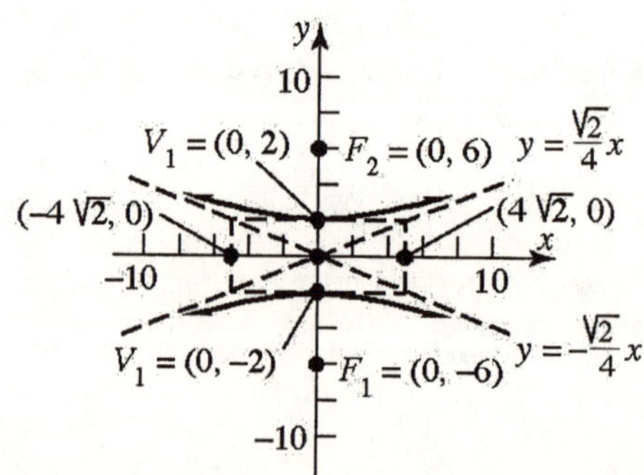

23. Vertices: (0, –6), (0, 6); asymptote: $y = 2x$;
Center: (0, 0); Transverse axis is the y-axis;
$a = 6$. Find the value of b using the slope of the asymptote: $\frac{a}{b} = \frac{6}{b} = 2 \Rightarrow 2b = 6 \Rightarrow b = 3$

Find the value of c:

$c^2 = a^2 + b^2 = 36 + 9 = 45$

$c = 3\sqrt{5}$

Write the equation: $\frac{y^2}{36} - \frac{x^2}{9} = 1$.

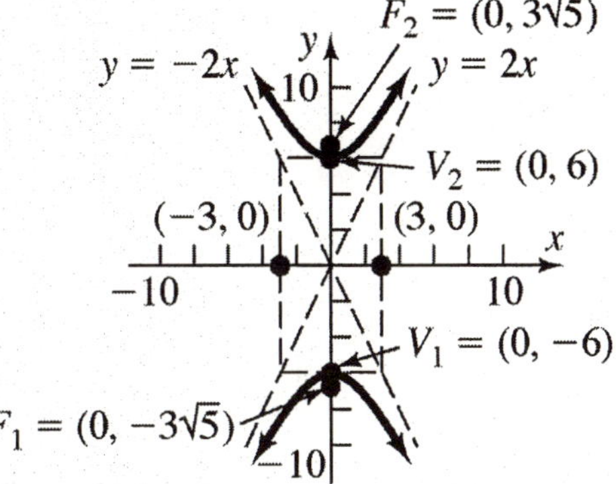

24. Vertices: (–4, 0), (4, 0); asymptote: $y = 2x$;
Center: (0, 0); Transverse axis is the x-axis;
$a = 4$. Find the value of b using the slope of the asymptote: $\frac{b}{a} = \frac{b}{4} = 2 \Rightarrow b = 8$

Find the value of c:

$c^2 = a^2 + b^2 = 16 + 64 = 80$

$c = 4\sqrt{5}$

Write the equation: $\frac{x^2}{16} - \frac{y^2}{64} = 1$.

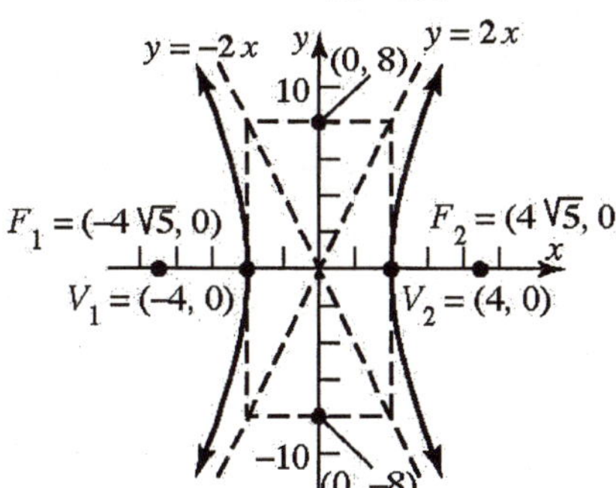

25. Foci: (–4, 0), (4, 0); asymptote: $y = -x$;
Center: (0, 0); Transverse axis is the x-axis;
$c = 4$. Using the slope of the asymptote:

$-\frac{b}{a} = -1 \Rightarrow -b = -a \Rightarrow b = a$.

Find the value of b:

$b^2 = c^2 - a^2 \Rightarrow a^2 + b^2 = c^2 \quad (c = 4)$

$b^2 + b^2 = 16 \Rightarrow 2b^2 = 16 \Rightarrow b^2 = 8$

$b = \sqrt{8} = 2\sqrt{2}$

$a = \sqrt{8} = 2\sqrt{2} \quad (a = b)$

Write the equation: $\frac{x^2}{8} - \frac{y^2}{8} = 1$.

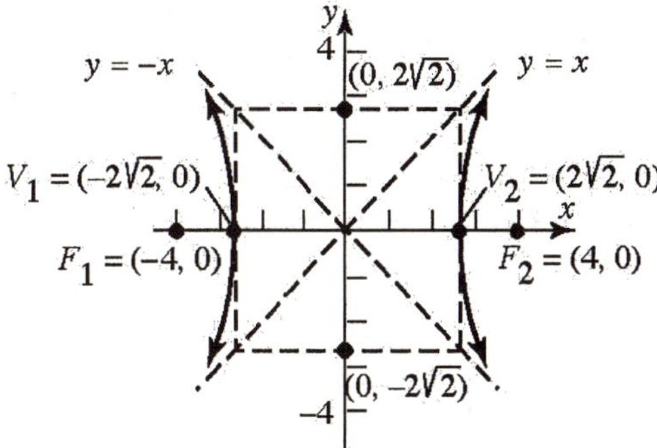

26. Foci: (0, –2), (0, 2); asymptote: $y = -x$;
Center: (0, 0); Transverse axis is the y-axis;
$c = 2$. Using the slope of the asymptote:

$-\frac{a}{b} = -1 \Rightarrow -b = -a \Rightarrow b = a$

Find the value of b:

$b^2 = c^2 - a^2$

$a^2 + b^2 = c^2 \quad (c = 2)$

$b^2 + b^2 = 4 \Rightarrow 2b^2 = 4$

$b^2 = 2 \Rightarrow b = \sqrt{2}$

$a = \sqrt{2} \quad (a = b)$

Write the equation: $\frac{y^2}{2} - \frac{x^2}{2} = 1$.

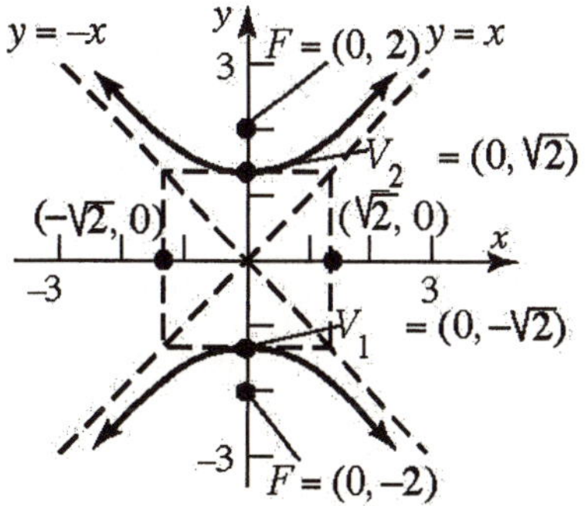

27. a. $\frac{x^2}{25}-\frac{y^2}{9}=1$

The center of the hyperbola is at (0, 0). $a=5,\ b=3$. The vertices are $(5,0)$ and $(-5,0)$. Find the value of c:

$c^2=a^2+b^2=25+9=34\Rightarrow c=\sqrt{34}$

The foci are $(\sqrt{34},0)$ and $(-\sqrt{34},0)$.

The transverse axis is the x-axis. The asymptotes are $y=\frac{3}{5}x;\ y=-\frac{3}{5}x$.

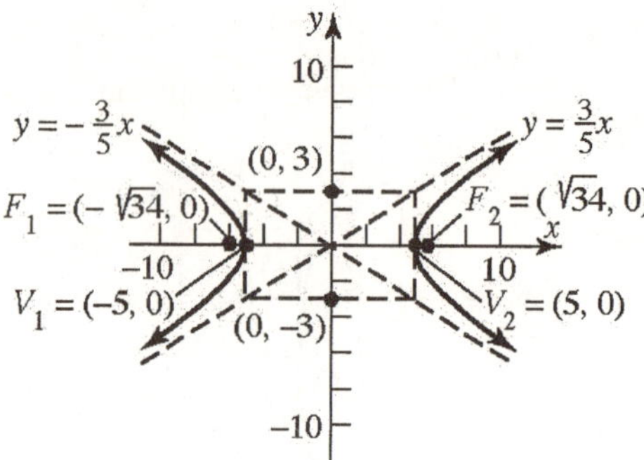

b. $\frac{x^2}{25}-\frac{y^2}{9}=1$

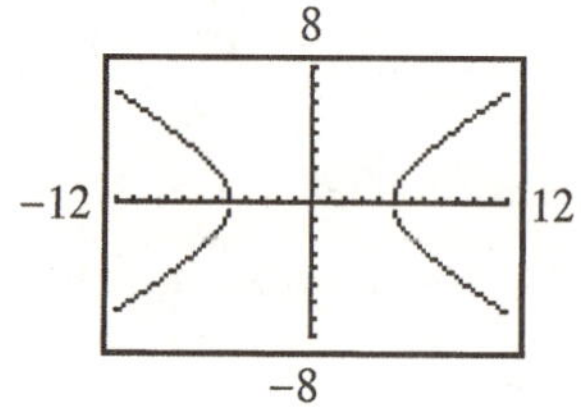

28. a. $\frac{y^2}{16}-\frac{x^2}{4}=1$

The center of the hyperbola is at (0, 0). $a=4,\ b=2$. The vertices are (0, 4) and (0, –4). Find the value of c:

$c^2=a^2+b^2=16+4=20$

$c=\sqrt{20}=2\sqrt{5}$

The foci are $(0,2\sqrt{5})$ and $(0,-2\sqrt{5})$.

The transverse axis is the y-axis. The asymptotes are $y=2x;\ y=-2x$.

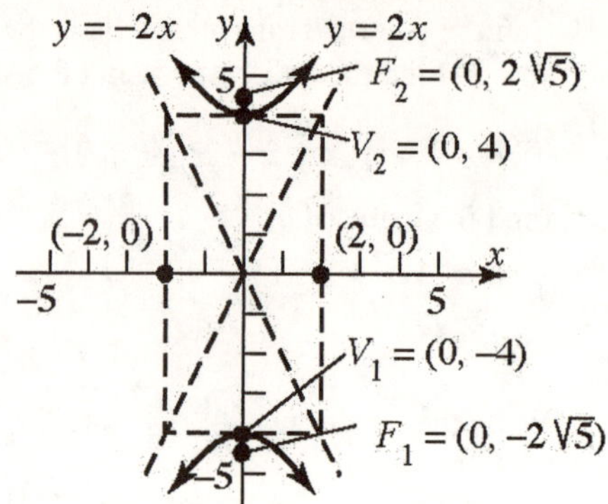

b. $\frac{y^2}{16}-\frac{x^2}{4}=1$

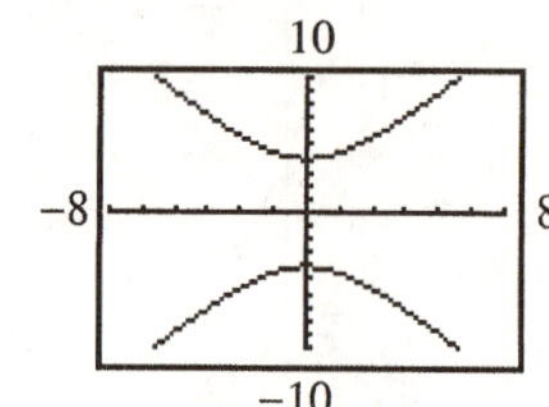

29. a. $4x^2-y^2=16$

Divide both sides by 16 to put in standard form: $\frac{4x^2}{16}-\frac{y^2}{16}=\frac{16}{16}\Rightarrow\frac{x^2}{4}-\frac{y^2}{16}=1$

. The center of the hyperbola is at (0, 0). $a=2,\ b=4$.

The vertices are (2, 0) and (–2, 0). Find the value of c:

$c^2=a^2+b^2=4+16=20$

$c=\sqrt{20}=2\sqrt{5}$

The foci are $(2\sqrt{5},0)$ and $(-2\sqrt{5},0)$.

The transverse axis is the x-axis. The asymptotes are $y=2x;\ y=-2x$.

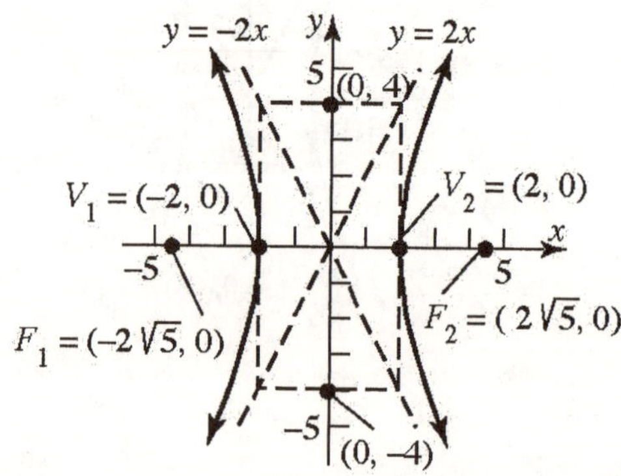

b. $4x^2 - y^2 = 16$

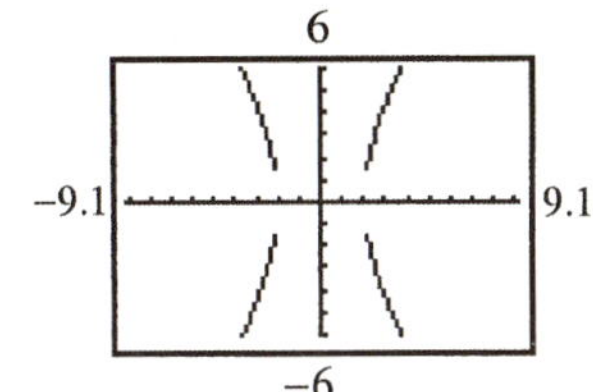

30. a. $4y^2 - x^2 = 16$

Divide both sides by 16 to put in standard form: $\frac{4y^2}{16} - \frac{x^2}{16} = \frac{16}{16} \Rightarrow \frac{y^2}{4} - \frac{x^2}{16} = 1$

The center of the hyperbola is at (0, 0). $a = 2,\ b = 4$. The vertices are $(0, 2)$ and $(0, -2)$. Find the value of c:

$c^2 = a^2 + b^2 = 4 + 16 = 20$

$c = \sqrt{20} = 2\sqrt{5}$

The foci are $(0, 2\sqrt{5})$ and $(0, -2\sqrt{5})$.

The transverse axis is the y-axis. The asymptotes are $y = \frac{1}{2}x$ and $y = -\frac{1}{2}x$.

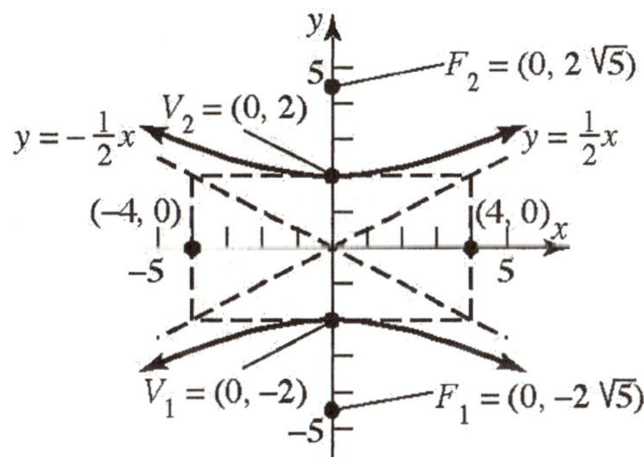

b. $4y^2 - x^2 = 16$

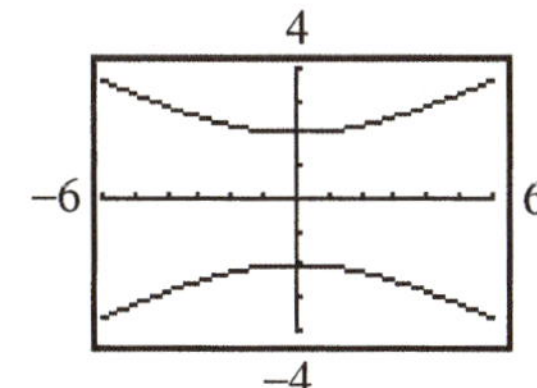

31. a. $y^2 - 9x^2 = 9$

Divide both sides by 9 to put in standard form: $\frac{y^2}{9} - \frac{9x^2}{9} = \frac{9}{9} \Rightarrow \frac{y^2}{9} - \frac{x^2}{1} = 1$

The center of the hyperbola is at (0, 0).

$a = 3,\ b = 1$.

The vertices are (0, 3) and (0, −3).

Find the value of c:

$c^2 = a^2 + b^2 = 9 + 1 = 10$

$c = \sqrt{10}$

The foci are $(0, \sqrt{10})$ and $(0, -\sqrt{10})$.

The transverse axis is the y-axis.

The asymptotes are $y = 3x;\ y = -3x$.

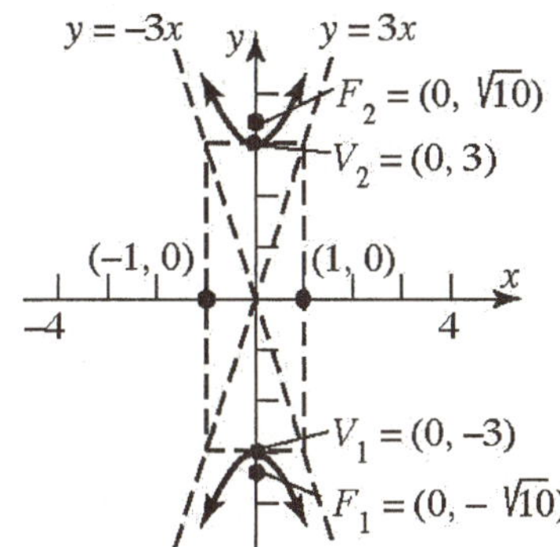

b. $y^2 - 9x^2 = 9$

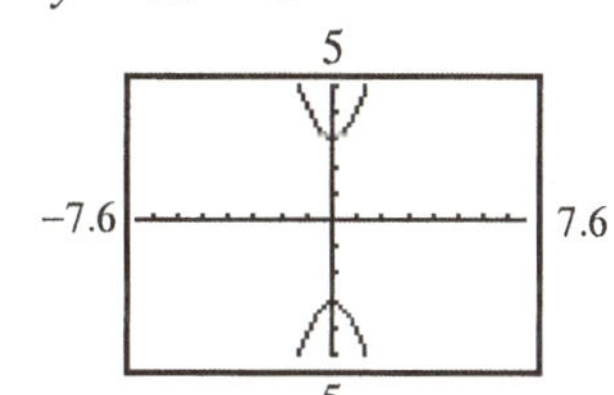

32. a. $x^2 - y^2 = 4$

Divide both sides by 4 to put in standard form: $\frac{x^2}{4} - \frac{y^2}{4} = \frac{4}{4} \Rightarrow \frac{x^2}{4} - \frac{y^2}{4} = 1$.

The center of the hyperbola is at (0, 0). $a = 2,\ b = 2$. The vertices are $(2, 0)$ and $(-2, 0)$. Find the value of c:

$c^2 = a^2 + b^2 = 4 + 4 = 8$

$c = \sqrt{8} = 2\sqrt{2}$

The foci are $(2\sqrt{2}, 0)$ and $(-2\sqrt{2}, 0)$.

The transverse axis is the x-axis.

The asymptotes are $y = x; y = -x$.

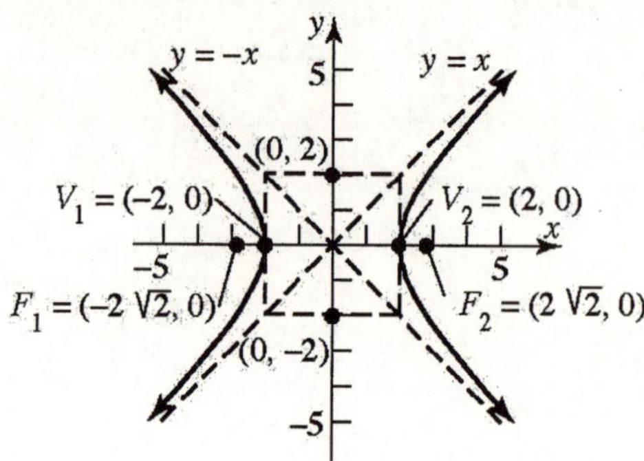

b. $x^2 - y^2 = 4$

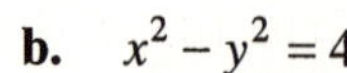

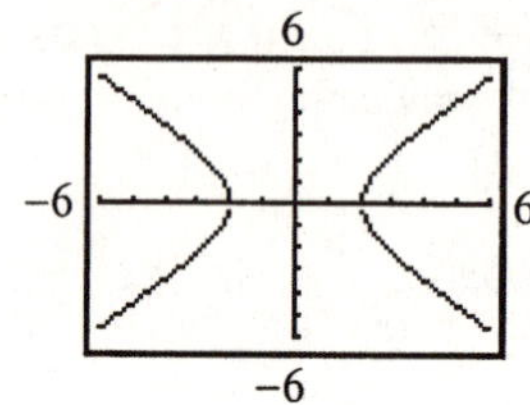

33. a. $y^2 - x^2 = 25$

Divide both sides by 25 to put in standard form: $\frac{y^2}{25} - \frac{x^2}{25} = 1$.

The center of the hyperbola is at (0, 0). $a = 5,\ b = 5$. The vertices are $(0,5)$ and $(0,-5)$. Find the value of c:

$c^2 = a^2 + b^2 = 25 + 25 = 50$

$c = \sqrt{50} = 5\sqrt{2}$

The foci are $\left(0,5\sqrt{2}\right)$ and $\left(0,-5\sqrt{2}\right)$.

The transverse axis is the y-axis.

The asymptotes are $y = x; y = -x$.

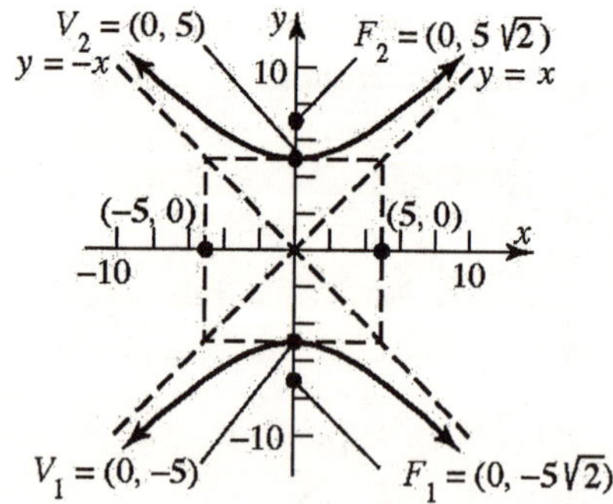

b. $y^2 - x^2 = 25$

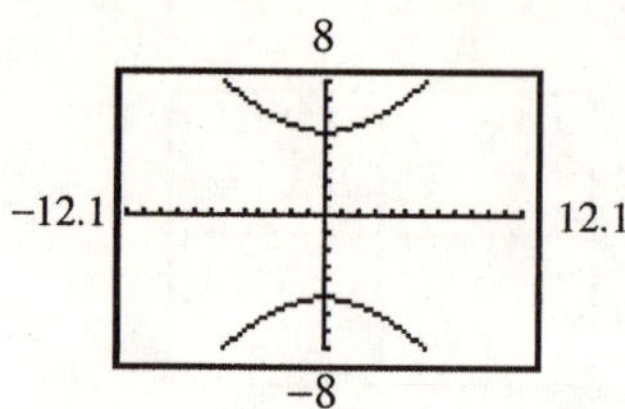

34. a. $2x^2 - y^2 = 4$

Divide both sides by 4 to put in standard form: $\frac{x^2}{2} - \frac{y^2}{4} = 1$.

The center of the hyperbola is at (0, 0).

$a = \sqrt{2},\ b = 2$.

The vertices are $\left(\sqrt{2},0\right)$ and $\left(-\sqrt{2},0\right)$.

Find the value of c:

$c^2 = a^2 + b^2 = 2 + 4 = 6$

$c = \sqrt{6}$

The foci are $\left(\sqrt{6},0\right)$ and $\left(-\sqrt{6},0\right)$.

The transverse axis is the x-axis.

The asymptotes are $y = \sqrt{2}x; y = -\sqrt{2}x$.

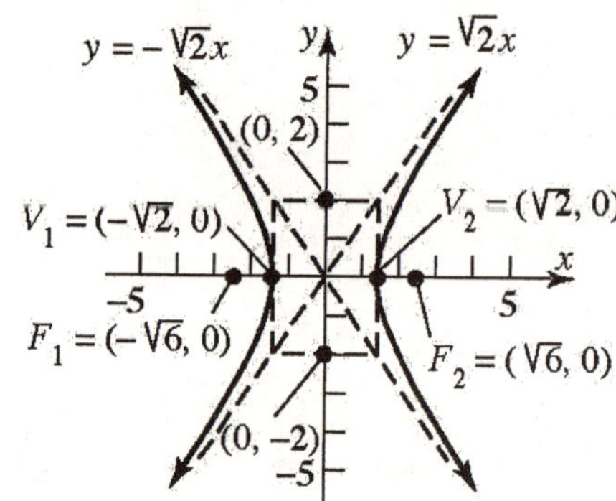

b. $2x^2 - y^2 = 4$

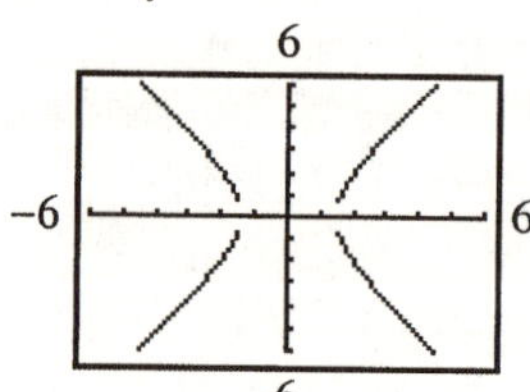

35. The center of the hyperbola is at (0, 0).
$a=1,\ b=1$. The vertices are $(1,0)$ and $(-1,0)$.
Find the value of c:
$c^2=a^2+b^2=1+1=2$
$c=\sqrt{2}$
The foci are $(\sqrt{2},0)$ and $(-\sqrt{2},0)$.
The transverse axis is the x-axis.
The asymptotes are $y=x;\ y=-x$.
The equation is: $x^2-y^2=1$.

36. The center of the hyperbola is at (0, 0).
$a=1,\ b=1$. The vertices are $(0,-1)$ and $(0,1)$.
Find the value of c:
$c^2=a^2+b^2=1+1=2$
$c=\sqrt{2}$
The foci are $(0,-\sqrt{2})$ and $(0,\sqrt{2})$.
The transverse axis is the y-axis.
The asymptotes are $y=x;\ y=-x$.
The equation is: $y^2-x^2=1$.

37. The center of the hyperbola is at (0, 0).
$a=6,\ b=3$.
The vertices are $(0,-6)$ and $(0,6)$. Find the value of c:
$c^2=a^2+b^2=36+9=45$
$c=\sqrt{45}=3\sqrt{5}$
The foci are $(0,-3\sqrt{5})$ and $(0,3\sqrt{5})$.
The transverse axis is the y-axis.
The asymptotes are $y=2x;\ y=-2x$. The equation is: $\dfrac{y^2}{36}-\dfrac{x^2}{9}=1$.

38. The center of the hyperbola is at (0, 0).
$a=2,\ b=4$.
The vertices are $(-2,0)$ and $(2,0)$.
Find the value of c:
$c^2=a^2+b^2=4+16=20$
$c=\sqrt{20}=2\sqrt{5}$
The foci are $(-2\sqrt{5},0)$ and $(2\sqrt{5},0)$.
The transverse axis is the x-axis.
The asymptotes are $y=2x;\ y=-2x$.
The equation is: $\dfrac{x^2}{4}-\dfrac{y^2}{16}=1$.

39. Center: (4, –1); Focus: (7, –1); Vertex: (6, –1);
Transverse axis is parallel to the x-axis;
$a=2;\ c=3$.
Find the value of b:
$b^2=c^2-a^2=9-4=5\Rightarrow b=\sqrt{5}$
Write the equation: $\dfrac{(x-4)^2}{4}-\dfrac{(y+1)^2}{5}=1$.

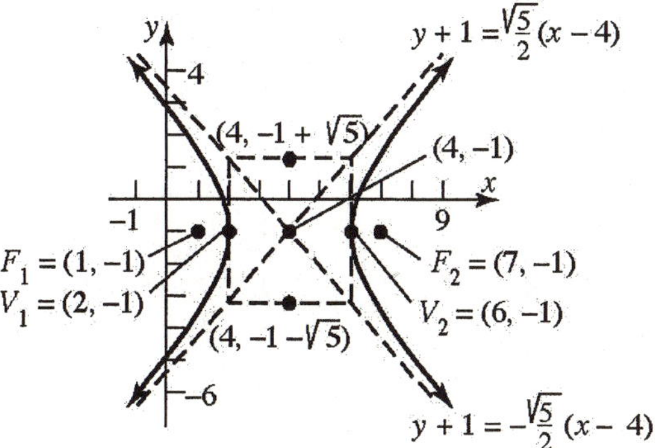

40. Center: (–3, 1); Focus: (–3, 6); Vertex: (–3, 4);
Transverse axis is parallel to the y-axis;
$a=3;\ c=5$. Find the value of b:
$b^2=c^2-a^2=25-9=16\Rightarrow b=4$
Write the equation: $\dfrac{(y-1)^2}{9}-\dfrac{(x+3)^2}{16}=1$.

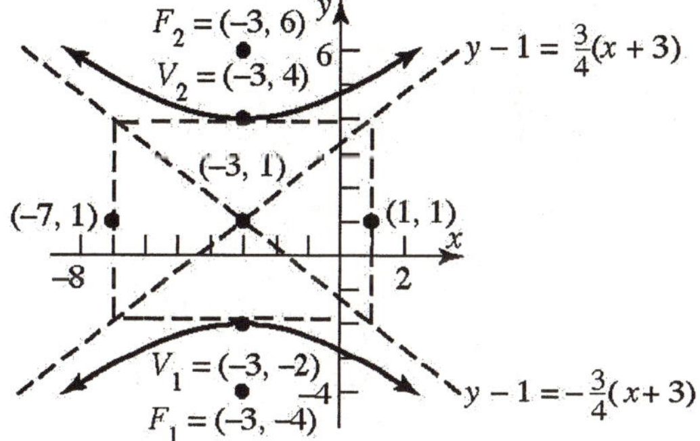

41. Center: (–3, –4); Focus: (–3, –8); Vertex: (–3, –2); Transverse axis is parallel to the y-axis; $a = 2$; $c = 4$.

Find the value of b:

$b^2 = c^2 - a^2 = 16 - 4 = 12$

$b = \sqrt{12} = 2\sqrt{3}$

Write the equation: $\dfrac{(y+4)^2}{4} - \dfrac{(x+3)^2}{12} = 1$.

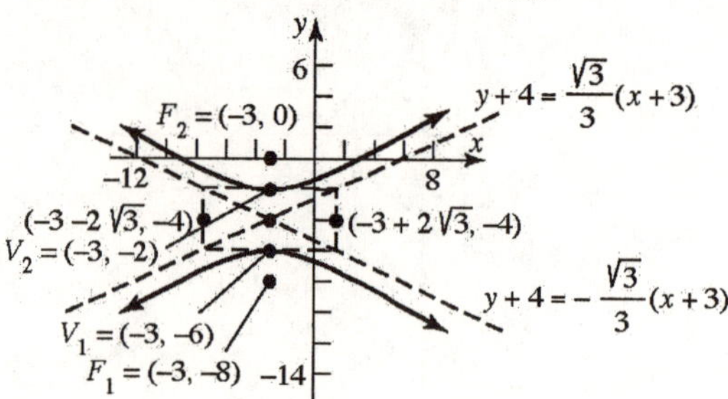

42. Center: (1, 4); Focus: (–2, 4); Vertex: (0, 4); Transverse axis is parallel to the x-axis; $a = 1$; $c = 3$. Find the value of b:

$b^2 = c^2 - a^2 = 9 - 1 = 8$

$b = \sqrt{8} = 2\sqrt{2}$

Write the equation: $\dfrac{(x-1)^2}{1} - \dfrac{(y-4)^2}{8} = 1$.

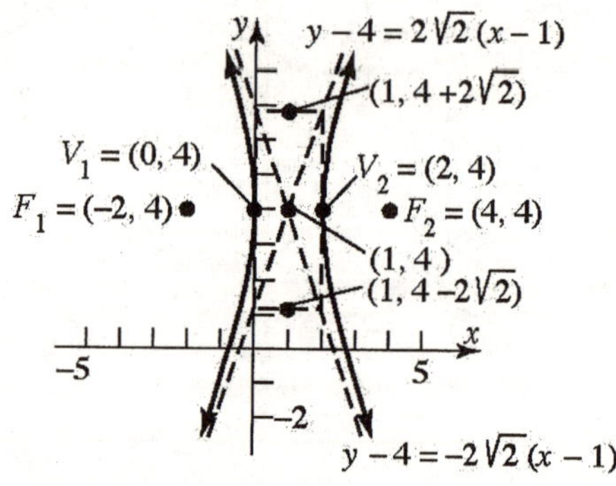

43. Foci: (3, 7), (7, 7); Vertex: (6, 7); Center: (5, 7); Transverse axis is parallel to the x-axis; $a = 1$; $c = 2$.

Find the value of b:

$b^2 = c^2 - a^2 = 4 - 1 = 3$

$b = \sqrt{3}$

Write the equation: $\dfrac{(x-5)^2}{1} - \dfrac{(y-7)^2}{3} = 1$.

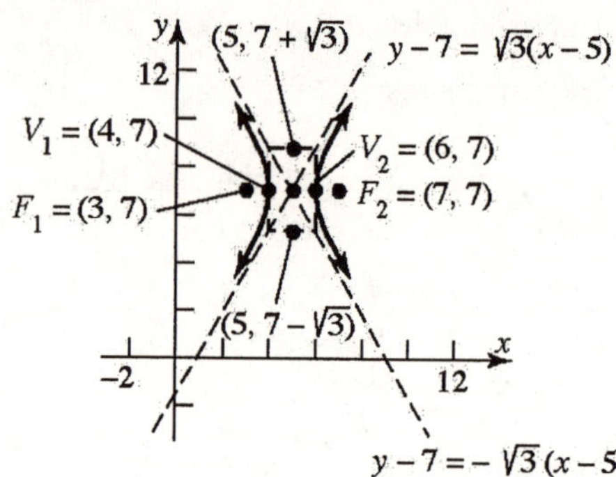

44. Focus: (–4, 0); Vertices: (–4, 4), (–4, 2); Center: (–4, 3); Transverse axis is parallel to the y-axis; $a = 1$; $c = 3$.

ind the value of b:

$b^2 = c^2 - a^2 = 9 - 1 = 8$

$b = \sqrt{8} = 2\sqrt{2}$

Write the equation: $\dfrac{(y-3)^2}{1} - \dfrac{(x+4)^2}{8} = 1$.

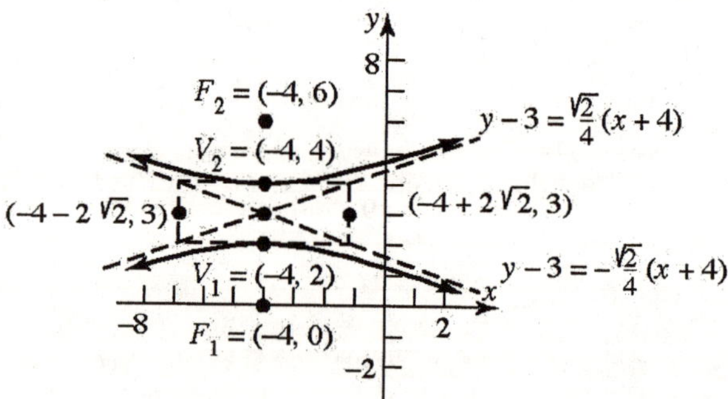

45. Vertices: (–1, –1), (3, –1); Center: (1, –1); Transverse axis is parallel to the x-axis; $a = 2$.

Asymptote: $y+1=\frac{3}{2}(x-1)$

Using the slope of the asymptote, find the value of b:

$\frac{b}{a}=\frac{b}{2}=\frac{3}{2} \Rightarrow b=3$

Find the value of c:

$c^2=a^2+b^2=4+9=13$

$c=\sqrt{13}$

Write the equation: $\frac{(x-1)^2}{4}-\frac{(y+1)^2}{9}=1$.

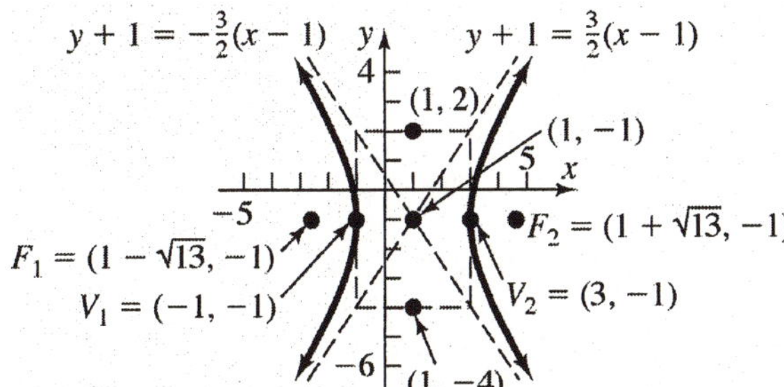

46. Vertices: (1, –3), (1, 1); Center: (1, –1); Transverse axis is parallel to the y-axis; $a = 2$.

Asymptote: $y+1=\frac{3}{2}(x-1)$

Using the slope of the asymptote, find the value of b:

$\frac{a}{b}=\frac{2}{b}=\frac{3}{2} \Rightarrow 3b=4 \Rightarrow b=\frac{4}{3}$

Find the value of c:

$c^2=a^2+b^2=4+\frac{16}{9}=\frac{52}{9} \Rightarrow c=\sqrt{\frac{52}{9}}=\frac{2\sqrt{13}}{3}$

Write the equation: $\frac{(y+1)^2}{4}-\frac{9(x-1)^2}{16}=1$.

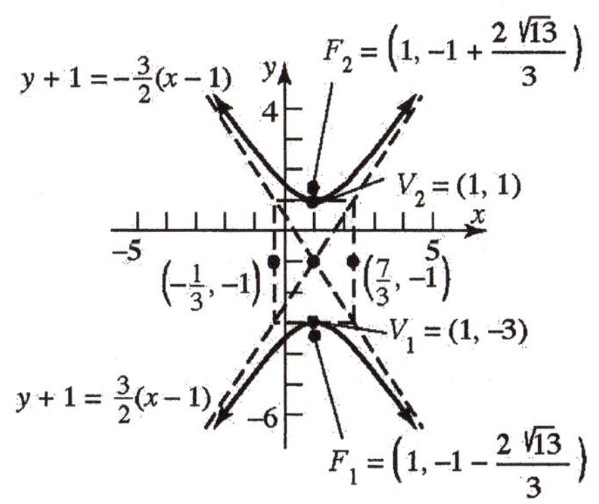

47. a. $\frac{(x-2)^2}{4}-\frac{(y+3)^2}{9}=1$

The center of the hyperbola is at (2, –3).

$a=2,\ b=3$.

The vertices are (0, –3) and (4, –3).

Find the value of c:

$c^2=a^2+b^2=4+9=13 \Rightarrow c=\sqrt{13}$

Foci: $\left(2-\sqrt{13},-3\right)$ and $\left(2+\sqrt{13},-3\right)$.

Transverse axis: $y=-3$, parallel to x-axis.

Asymptotes: $y+3=\frac{3}{2}(x-2)$;

$y+3=-\frac{3}{2}(x-2)$

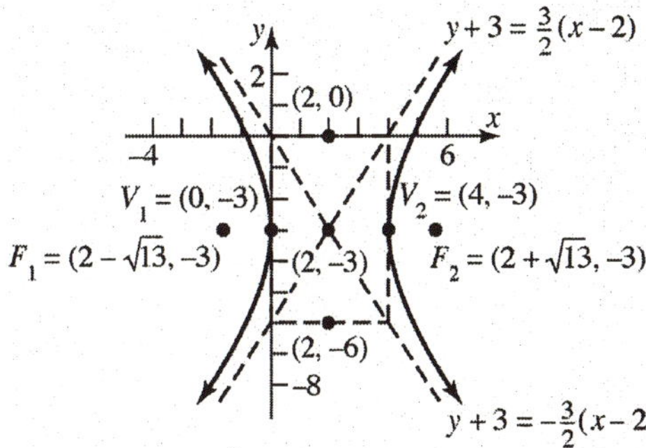

b. $\frac{(x-2)^2}{4}-\frac{(y+3)^2}{9}=1$

2

–5.5 9.5

–8

48. a. $\frac{(y+3)^2}{4}-\frac{(x-2)^2}{9}=1$

The center of the hyperbola is at (2, –3).

$a=2,\ b=3$. The vertices are (2, –1) and (2, –5). Find the value of c:

$c^2=a^2+b^2=4+9=13 \Rightarrow c=\sqrt{13}$

Foci: $\left(2,-3-\sqrt{13}\right)$ and $\left(2,-3+\sqrt{13}\right)$

Transverse axis: $x=2$, parallel to the y-axis

Asymptotes: $y+3=\frac{2}{3}(x-2)$;

$y+3=-\frac{2}{3}(x-2)$

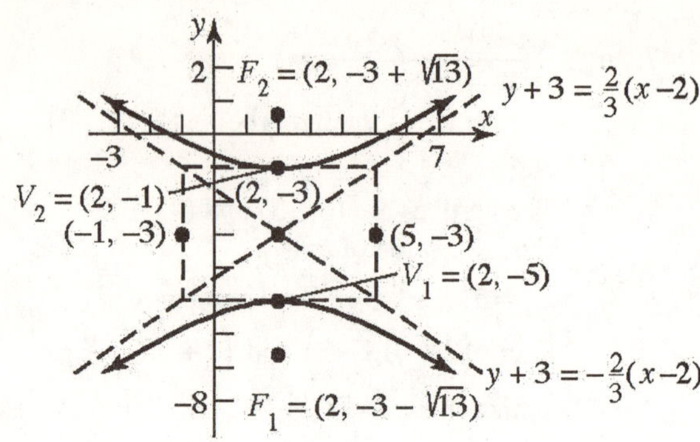

b. $\frac{(y+3)^2}{4} - \frac{(x-2)^2}{9} = 1$

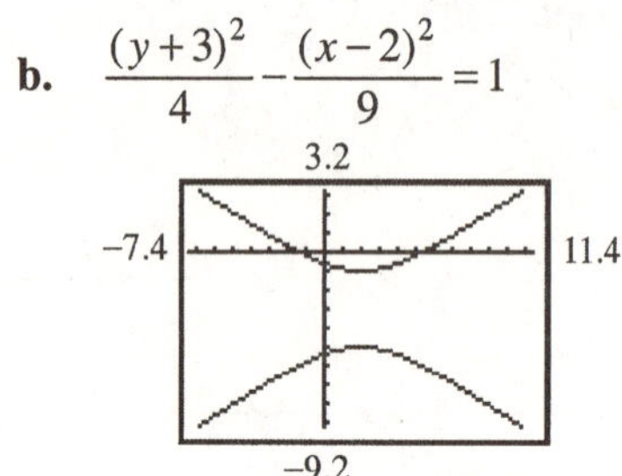

49. a. $(y-2)^2 - 4(x+2)^2 = 4$

Divide both sides by 4 to put in standard form: $\frac{(y-2)^2}{4} - \frac{(x+2)^2}{1} = 1$.

The center of the hyperbola is at (–2, 2).

$a = 2,\ b = 1$.

The vertices are (–2, 4) and (–2, 0). Find the value of c:

$c^2 = a^2 + b^2 = 4 + 1 = 5 \Rightarrow c = \sqrt{5}$

Foci: $\left(-2, 2-\sqrt{5}\right)$ and $\left(-2, 2+\sqrt{5}\right)$.

Transverse axis: $x = -2$, parallel to the y-axis.

Asymptotes:

$y - 2 = 2(x+2);\ y - 2 = -2(x+2)$.

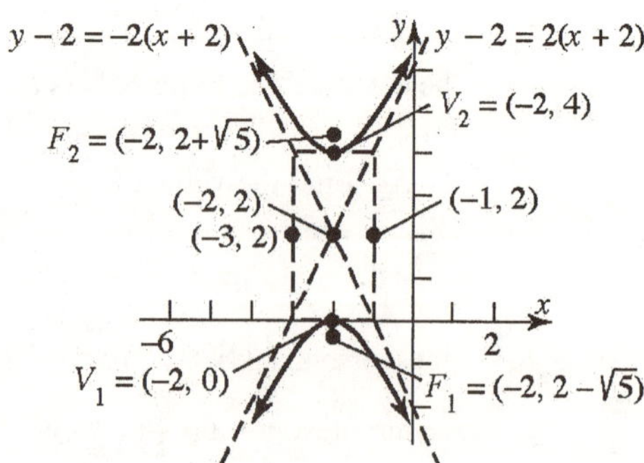

b. $(y-2)^2 - 4(x+2)^2 = 4$

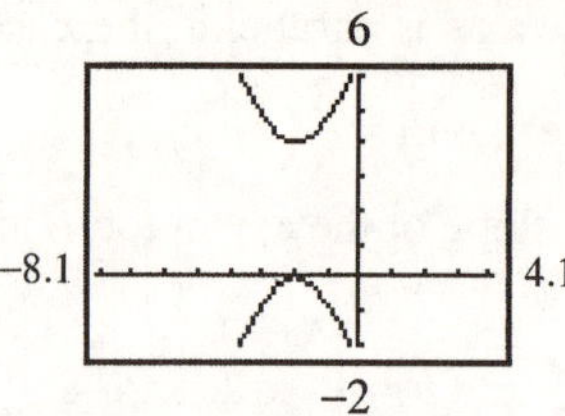

50. a. $(x+4)^2 - 9(y-3)^2 = 9$

Divide both sides by 9 to put in standard form: $\frac{(x+4)^2}{9} - \frac{(y-3)^2}{1} = 1$.

The center of the hyperbola is (–4, 3).

$a = 3,\ b = 1$.

The vertices are (–7, 3) and (–1, 3). Find the value of c:

$c^2 = a^2 + b^2 = 9 + 1 = 10$

$c = \sqrt{10}$

Foci: $\left(-4-\sqrt{10}, 3\right)$ and $\left(-4+\sqrt{10}, 3\right)$

Transverse axis: $y = 3$, parallel to the x-axis.

Asymptotes:

$y - 3 = \frac{1}{3}(x+4),\ y - 3 = -\frac{1}{3}(x+4)$

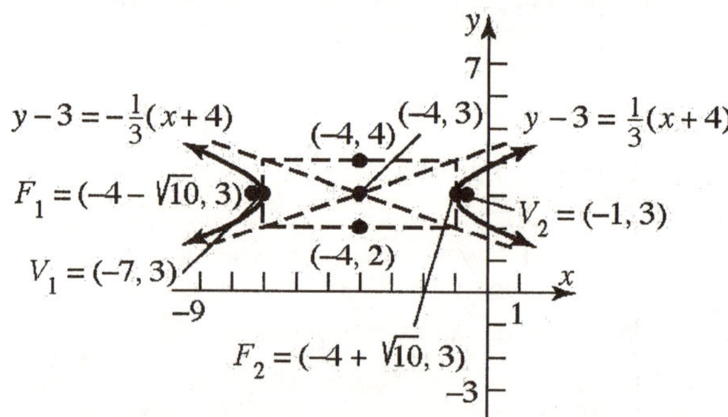

b. $(x+4)^2 - 9(y-3)^2 = 9$

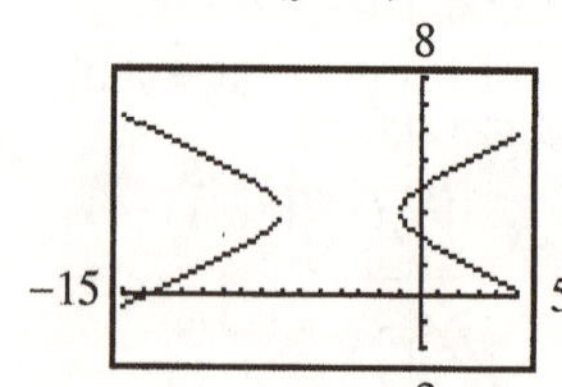

51. a. $(x+1)^2-(y+2)^2=4$

Divide both sides by 4 to put in standard form: $\frac{(x+1)^2}{4}-\frac{(y+2)^2}{4}=1$.

The center of the hyperbola is $(-1,-2)$.

$a=2,\ b=2$.

The vertices are $(-3,-2)$ and $(1,-2)$.

Find the value of c:

$c^2=a^2+b^2=4+4=8$

$c=\sqrt{8}=2\sqrt{2}$

Foci: $\left(-1-2\sqrt{2},-2\right)$ and $\left(-1+2\sqrt{2},-2\right)$

Transverse axis: $y=-2$, parallel to the x-axis.

Asymptotes: $y+2=x+1$;

$y+2=-(x+1)$

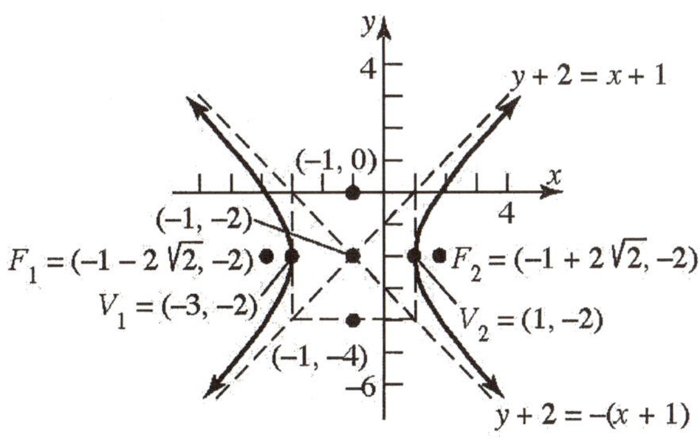

b. $(x+1)^2-(y+2)^2=4$

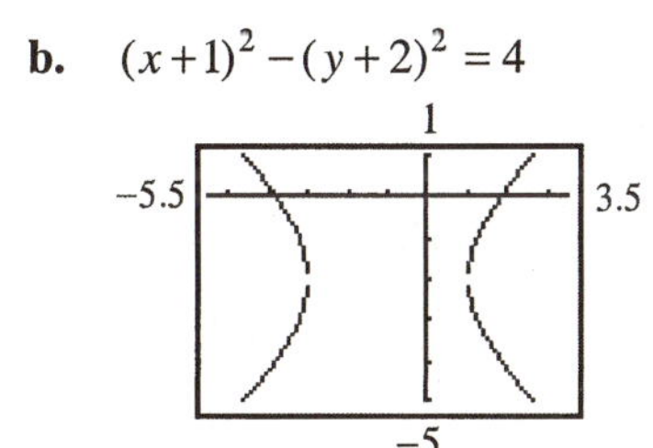

52. a. $(y-3)^2-(x+2)^2=4$

Divide both sides by 4 to put in standard form: $\frac{(y-3)^2}{4}-\frac{(x+2)^2}{4}=1$. The center of the hyperbola is at $(-2,3)$. $a=2,\ b=2$.

The vertices are $(-2,5)$ and $(-2,1)$.

Find the value of c:

$c^2=a^2+b^2=4+4=8$

$c=\sqrt{8}=2\sqrt{2}$

Foci: $\left(-2,3-2\sqrt{2}\right)$ and $\left(-2,3+2\sqrt{2}\right)$

Transverse axis: $x=-2$, parallel to the y-axis.

Asymptotes: $y-3=x+2$;

$y-3=-(x+2)$

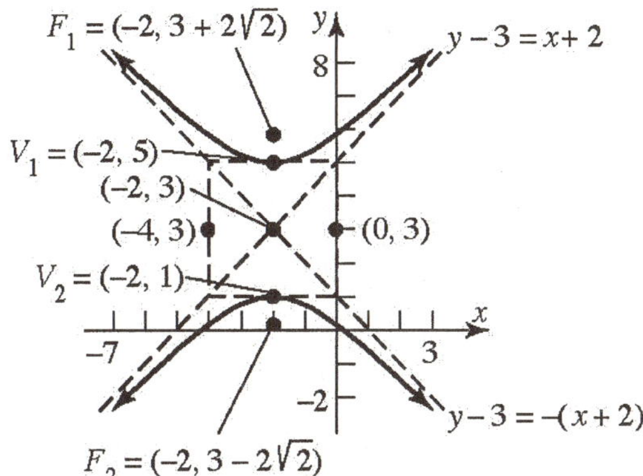

b. $(y-3)^2-(x+2)^2=4$

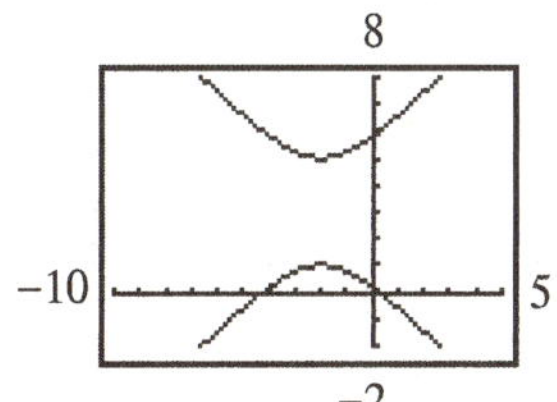

53. a. Complete the squares to put in standard form:

$$x^2-y^2-2x-2y-1=0$$

$$(x^2-2x+1)-(y^2+2y+1)=1+1-1$$

$$(x-1)^2-(y+1)^2=1$$

The center of the hyperbola is $(1,-1)$.

$a=1,\ b=1$. The vertices are $(0,-1)$ and $(2,-1)$. Find the value of c:

$c^2=a^2+b^2=1+1=2\Rightarrow c=\sqrt{2}$

Foci: $\left(1-\sqrt{2},-1\right)$ and $\left(1+\sqrt{2},-1\right)$.

Transverse axis: $y=-1$, parallel to x-axis.

Asymptotes: $y+1=x-1;\ y+1=-(x-1)$.

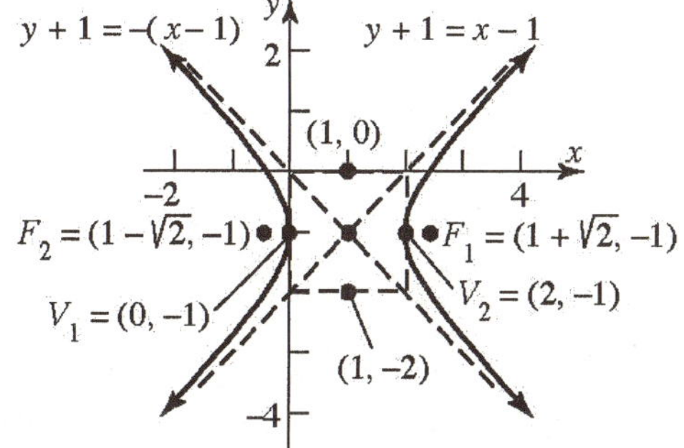

b. $(x-1)^2-(y+1)^2=1$

1

−2 4

−3

54. a. Complete the squares to put in standard form:

$$y^2-x^2-4y+4x-1=0$$

$$(y^2-4y+4)-(x^2-4x+4)=1+4-4$$

$$(y-2)^2-(x-2)^2=1$$

The center of the hyperbola is (2, 2).
$a=1,\ b=1$. The vertices are $(2,1)$ and $(2,3)$. Find the value of c:

$c^2=a^2+b^2=1+1=2\Rightarrow c=\sqrt{2}$

Foci: $\left(2,2-\sqrt{2}\right)$ and $\left(2,2+\sqrt{2}\right)$.

Transverse axis: $x=2$, parallel to the y-axis.

Asymptotes:
$y-2=x-2;\ y-2=-(x-2)$.

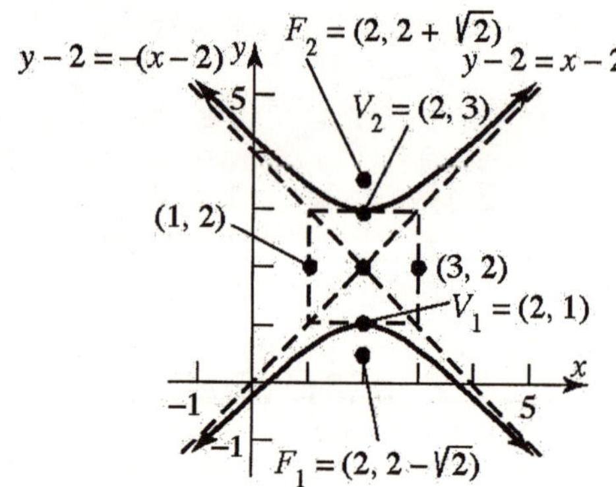

b. $(y-2)^2-(x-2)^2=1$

8

−5 10

−4

55. a. Complete the squares to put in standard form:

$$y^2-4x^2-4y-8x-4=0$$

$$(y^2-4y+4)-4(x^2+2x+1)=4+4-4$$

$$(y-2)^2-4(x+1)^2=4$$

$$\frac{(y-2)^2}{4}-\frac{(x+1)^2}{1}=1$$

The center of the hyperbola is (–1, 2).
$a=2,\ b=1$.

The vertices are (–1, 4) and (–1, 0).
Find the value of c:

$c^2=a^2+b^2=4+1=5\Rightarrow c=\sqrt{5}$

Foci: $\left(-1,2-\sqrt{5}\right)$ and $\left(-1,2+\sqrt{5}\right)$.

Transverse axis: $x=-1$, parallel to the y-axis.

Asymptotes:
$y-2=2(x+1);\ y-2=-2(x+1)$.

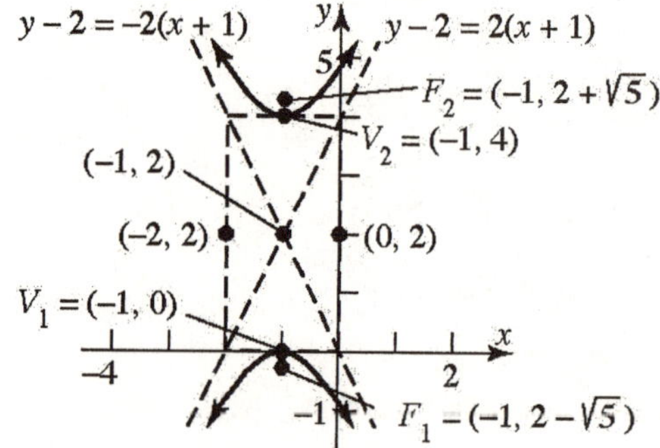

b. $\dfrac{(y-2)^2}{4}-(x+1)^2=1$

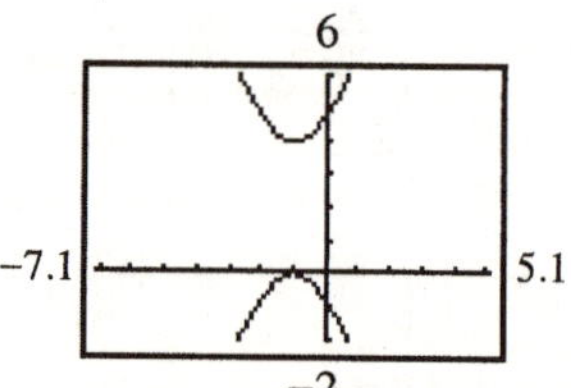

56. a. Complete the squares to put in standard form:

$$2x^2 - y^2 + 4x + 4y - 4 = 0$$
$$2(x^2 + 2x + 1) - (y^2 - 4y + 4) = 4 + 2 - 4$$
$$2(x+1)^2 - (y-2)^2 = 2$$
$$\frac{(x+1)^2}{1} - \frac{(y-2)^2}{2} = 1$$

The center of the hyperbola is (–1, 2).

$a = 1,\ b = \sqrt{2}$.

The vertices are (–2, 2) and (0, 2). Find the value of c:

$c^2 = a^2 + b^2 = 1 + 2 = 3 \Rightarrow c = \sqrt{3}$

Foci: $\left(-1-\sqrt{3}, 2\right)$ and $\left(-1+\sqrt{3}, 2\right)$.

Transverse axis: $y = 2$, parallel to the x-axis.

Asymptotes:

$y - 2 = \sqrt{2}(x+1);\ y - 2 = -\sqrt{2}(x+1)$.

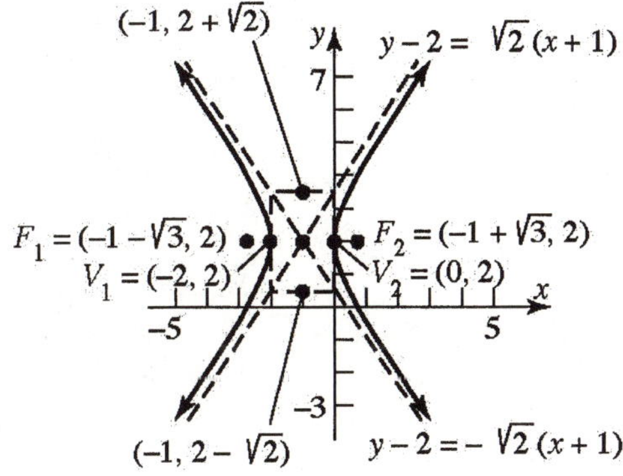

b. $(x+1)^2 - \frac{(y-2)^2}{2} = 1$

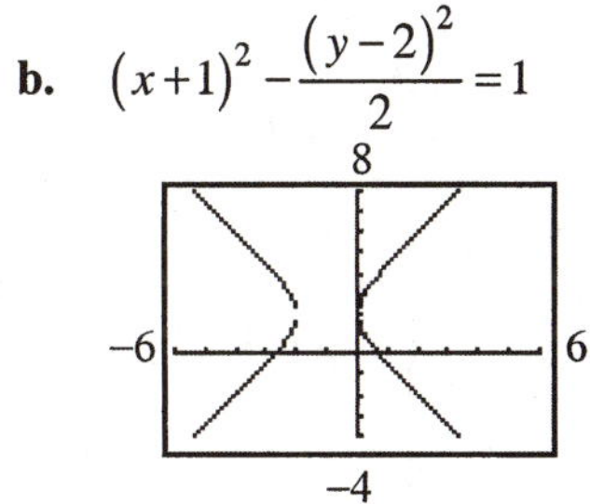

57. a. Complete the squares to put in standard form:

$$4x^2 - y^2 - 24x - 4y + 16 = 0$$
$$4(x^2 - 6x + 9) - (y^2 + 4y + 4) = -16 + 36 - 4$$
$$4(x-3)^2 - (y+2)^2 = 16$$
$$\frac{(x-3)^2}{4} - \frac{(y+2)^2}{16} = 1$$

The center of the hyperbola is (3, –2).

$a = 2,\ b = 4$.

The vertices are (1, –2) and (5, –2). Find the value of c:

$c^2 = a^2 + b^2 = 4 + 16 = 20$

$c = \sqrt{20} = 2\sqrt{5}$

Foci: $\left(3-2\sqrt{5}, -2\right)$ and $\left(3+2\sqrt{5}, -2\right)$.

Transverse axis: $y = -2$, parallel to x-axis.

Asymptotes: $y + 2 = 2(x-3)$;

$y + 2 = -2(x-3)$

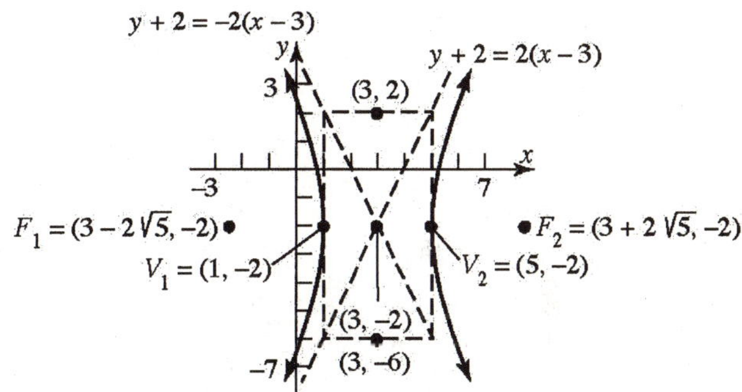

b. $\frac{(x-3)^2}{4} - \frac{(y+2)^2}{16} = 1$

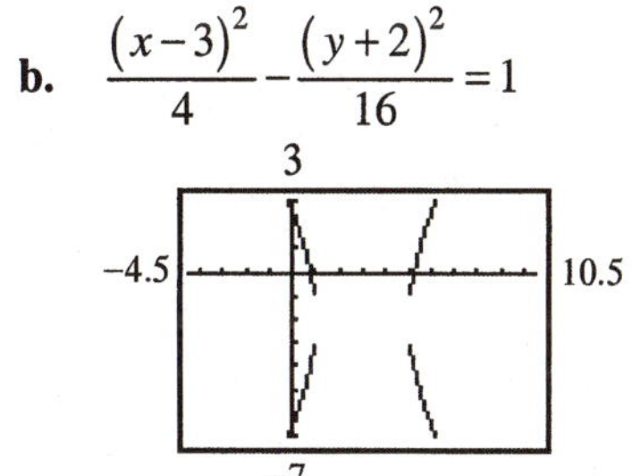

58. a. Complete the squares to put in standard form:

$$2y^2 - x^2 + 2x + 8y + 3 = 0$$
$$2(y^2 + 4y + 4) - (x^2 - 2x + 1) = -3 + 8 - 1$$
$$2(y+2)^2 - (x-1)^2 = 4$$
$$\frac{(y+2)^2}{2} - \frac{(x-1)^2}{4} = 1$$

The center of the hyperbola is (1,–2).
$a = \sqrt{2},\ b = 2$.

Vertices: $\left(1, -2-\sqrt{2}\right)$ and $\left(1, -2+\sqrt{2}\right)$

Find the value of c:
$c^2 = a^2 + b^2 = 2 + 4 = 6$
$c = \sqrt{6}$

Foci: $\left(1, -2-\sqrt{6}\right)$ and $\left(1, -2+\sqrt{6}\right)$.

Transverse axis: $x = 1$, parallel to the y-axis.

Asymptotes: $y + 2 = \frac{\sqrt{2}}{2}(x-1)$;
$y + 2 = -\frac{\sqrt{2}}{2}(x-1)$

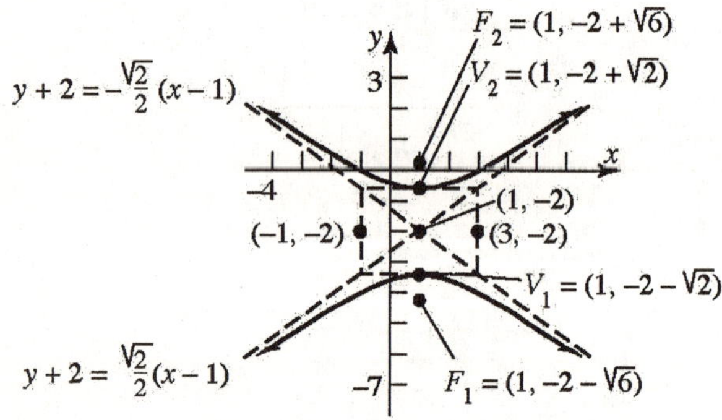

b. $\frac{(y+2)^2}{2} - \frac{(x-1)^2}{4} = 1$

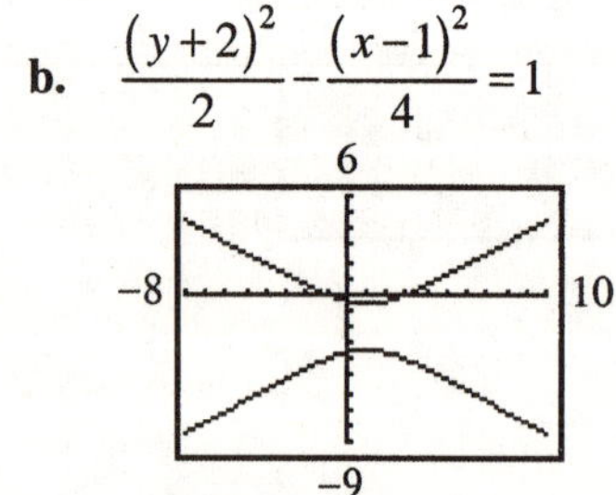

59. a. Complete the squares to put in standard form:

$$y^2 - 4x^2 - 16x - 2y - 19 = 0$$
$$(y^2 - 2y + 1) - 4(x^2 + 4x + 4) = 19 + 1 - 16$$
$$(y-1)^2 - 4(x+2)^2 = 4$$
$$\frac{(y-1)^2}{4} - \frac{(x+2)^2}{1} = 1$$

The center of the hyperbola is (–2, 1).
$a = 2,\ b = 1$.

The vertices are (–2, 3) and (–2, –1). Find the value of c:
$c^2 = a^2 + b^2 = 4 + 1 = 5$
$c = \sqrt{5}$

Foci: $\left(-2, 1-\sqrt{5}\right)$ and $\left(-2, 1+\sqrt{5}\right)$.

Transverse axis: $x = -2$, parallel to the y-axis.

Asymptotes:
$y - 1 = 2(x+2);\ y - 1 = -2(x+2)$.

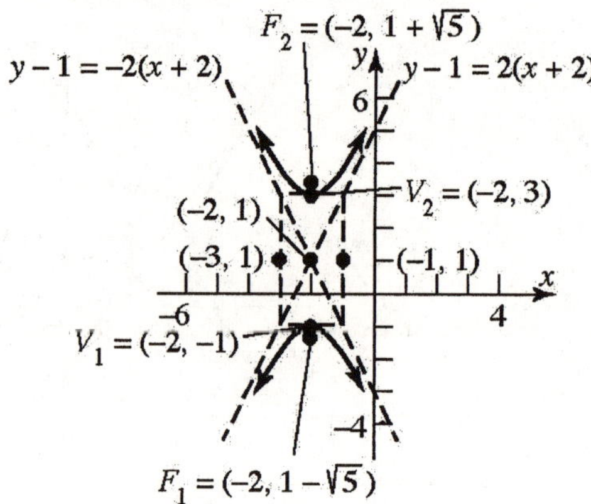

b. $\frac{(y-1)^2}{4} - \frac{(x+2)^2}{1} = 1$

5
–8.1 4.1
–3

60. a. Complete the squares to put in standard form:

$$x^2 - 3y^2 + 8x - 6y + 4 = 0$$
$$(x^2 + 8x + 16) - 3(y^2 + 2y + 1) = -4 + 16 - 3$$
$$(x+4)^2 - 3(y+1)^2 = 9$$
$$\frac{(x+4)^2}{9} - \frac{(y+1)^2}{3} = 1$$

The center of the hyperbola is $(-4, -1)$.

$a = 3,\ b = \sqrt{3}$. The vertices are $(-7, -1)$ and $(-1, -1)$. Find the value of c:

$$c^2 = a^2 + b^2 = 9 + 3 = 12$$
$$c = \sqrt{12} = 2\sqrt{3}$$

Foci: $\left(-4 - 2\sqrt{3}, -1\right)$ and $\left(-4 + 2\sqrt{3}, -1\right)$.

Transverse axis: $y = -1$, parallel to x-axis.

Asymptotes:

$$y + 1 = \frac{\sqrt{3}}{3}(x+4);\quad y + 1 = -\frac{\sqrt{3}}{3}(x+4)$$

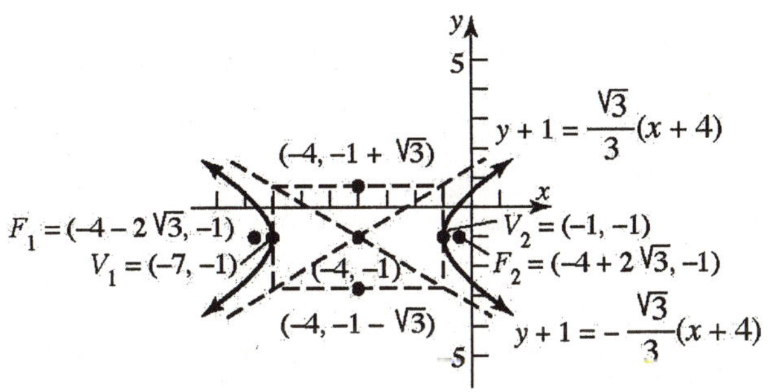

b. $\frac{(x+4)^2}{9} - \frac{(y+1)^2}{3} = 1$

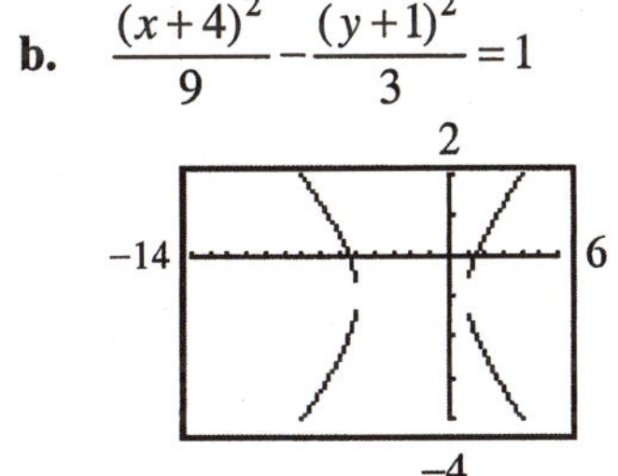

61. Rewrite the equation:

$$y = \sqrt{16 + 4x^2}$$
$$y^2 = 16 + 4x^2, \quad y \ge 0$$
$$y^2 - 4x^2 = 16, \quad y \ge 0$$
$$\frac{y^2}{16} - \frac{x^2}{4} = 1, \quad y \ge 0$$

62. Rewrite the equation:

$$y = -\sqrt{9 + 9x^2}$$
$$y^2 = 9 + 9x^2, \quad y \le 0$$
$$y^2 - 9x^2 = 9, \quad y \le 0$$
$$\frac{y^2}{9} - \frac{x^2}{1} = 1, \quad y \le 0$$

63. Rewrite the equation:

$$y = -\sqrt{-25 + x^2}$$
$$y^2 = -25 + x^2, \quad y \le 0$$
$$x^2 - y^2 = 25, \quad y \le 0$$
$$\frac{x^2}{25} - \frac{y^2}{25} = 1, \quad y \le 0$$

64. Rewrite the equation:

$$y = \sqrt{-1+x^2}$$
$$y^2 = -1+x^2, \quad y \ge 0$$
$$x^2 - y^2 = 1, \quad y \ge 0$$

65. First note that all points where a burst could take place, such that the time difference would be the same as that for the first burst, would form a hyperbola with A and B as the foci.
Start with a diagram:

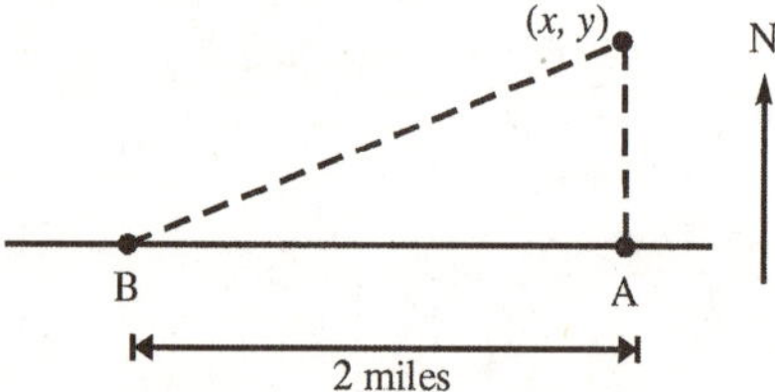

The ordered pair (x, y) represents the location of the fireworks. We know that sound travels at 1100 feet per second, so the person at point A is 1100 feet closer to the fireworks display than the person at point B. Since the difference of the distance from (x, y) to A and from (x, y) to B is the constant 1100, the point (x, y) lies on a hyperbola whose foci are at A and B. The hyperbola has the equation

$$\frac{x^2}{a^2} - \frac{y^2}{b^2} = 1$$

where $2a = 1100$, so $a = 550$. Because the distance between the two people is 2 miles (10,560 feet) and each person is at a focus of the hyperbola, we have

$$2c = 10{,}560$$
$$c = 5280$$
$$b^2 = c^2 - a^2 = 5280^2 - 550^2 = 27{,}575{,}900$$

The equation of the hyperbola that describes the location of the fireworks display is

$$\frac{x^2}{550^2} - \frac{y^2}{27{,}575{,}900} = 1$$

Since the fireworks display is due north of the individual at A, we let $x = 5280$ and solve the equation for y.

$$\frac{5280^2}{550^2} - \frac{y^2}{27{,}575{,}900} = 1$$
$$-\frac{y^2}{27{,}575{,}900} = -91.16$$
$$y^2 = 2{,}513{,}819{,}044$$
$$y = 50{,}138$$

Therefore, the fireworks display was 50,138 feet (approximately 9.5 miles) due north of the person at A.

66. First note that all points where the strike could take place, such that the time difference would be the same as that for the first strike, would form a hyperbola with A and B as the foci.
Start with a diagram:

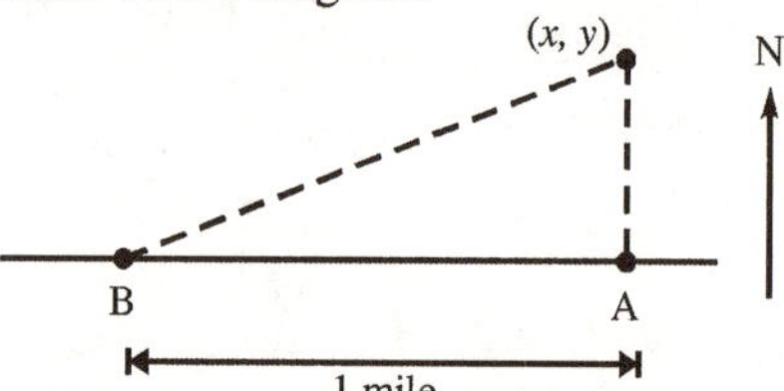

The ordered pair (x, y) represents the location of the lightning strike. We know that sound travels at 1100 feet per second, so the person at point A is 2200 feet closer to the lightning strike than the person at point B. Since the difference of the distance from (x, y) to A and from (x, y) to B is the constant 2200, the point (x, y) lies on a hyperbola whose foci are at A and B. The hyperbola has the equation

$$\frac{x^2}{a^2} - \frac{y^2}{b^2} = 1$$

where $2a = 2200$, so $a = 1100$. Because the distance between the two people is 1 mile (5,280 feet) and each person is at a focus of the hyperbola, we have

$$2c = 5{,}280$$
$$c = 2{,}640$$
$$b^2 = c^2 - a^2 = 2640^2 - 1100^2 = 5{,}759{,}600$$

The equation of the hyperbola that describes the location of the lightning strike is

$$\frac{x^2}{1100^2} - \frac{y^2}{5{,}759{,}600} = 1$$

Since the lightning strike is due north of the individual at A, we let $x = 2640$ and solve for y.

$$\frac{2640^2}{1100^2}-\frac{y^2}{5,759,600}=1$$
$$-\frac{y^2}{5,759,600}=-4.76$$
$$y^2=27,415,696$$
$$y=5236$$

The lightning strikes 5236 feet (approximately 0.99 miles) due north of the person at point A.

67. a. Since the particles are deflected at a 45° angle, the asymptotes will be $y=\pm x$.

b. Since the vertex is 10 cm from the center of the hyperbola, we know that $a=10$. The slope of the asymptotes is given by $\pm\frac{b}{a}$.

Therefore, we have

$$\frac{b}{a}=1$$
$$\frac{b}{10}=1$$
$$b=10$$

Using the origin as the center of the hyperbola, the equation of the particle path would be

$$\frac{x^2}{100}-\frac{y^2}{100}=1$$

68. First note that all points where an explosion could take place, such that the time difference would be the same as that for the first detonation, would form a hyperbola with A and B as the foci.

Start with a diagram:

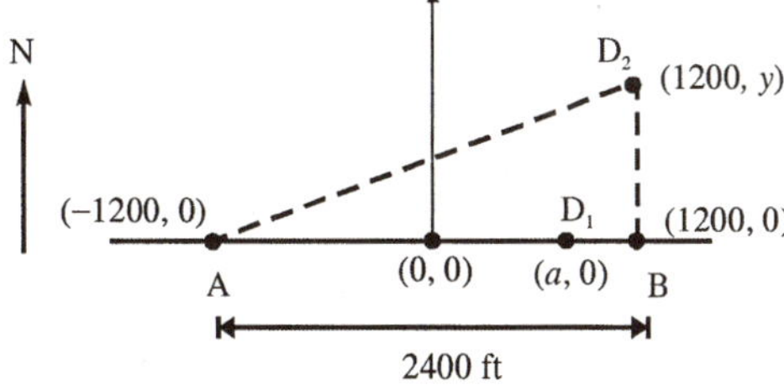

Since A and B are the foci, we have

$$2c=2400$$
$$c=1200$$

Since D_1 is on the transverse axis and is on the hyperbola, then it must be a vertex of the hyperbola. Since it is 300 feet from B, we have $a=900$. Finally,

$$b^2=c^2-a^2=1200^2-900^2=630,000$$

Thus, the equation of the hyperbola is

$$\frac{x^2}{810,000}-\frac{y^2}{630,000}=1$$

The point $(1200, y)$ needs to lie on the graph of the hyperbola. Thus, we have

$$\frac{(1200)^2}{810,000}-\frac{y^2}{630,000}=1$$
$$-\frac{y^2}{630,000}=-\frac{7}{9}$$
$$y^2=490,000$$
$$y=700$$

The second explosion should be set off 700 feet due north of point B.

69. Assume $\frac{x^2}{a^2}-\frac{y^2}{b^2}=1$.

If the eccentricity is close to 1, then $c\approx a$ and $b\approx 0$. When b is close to 0, the hyperbola is very narrow, because the slopes of the asymptotes are close to 0.

If the eccentricity is very large, then c is much larger than a and b is very large. The result is a hyperbola that is very wide, because the slopes of the asymptotes are very large.

70. If $a=b$, then $c^2=a^2+a^2=2a^2$. Thus, $\frac{c^2}{a^2}=2$ or $\frac{c}{a}=\sqrt{2}$. The eccentricity of an equilateral hyperbola is $\sqrt{2}$.

71. $\frac{x^2}{4}-y^2=1 \quad (a=2,\ b=1)$

This is a hyperbola with horizontal transverse axis, centered at (0, 0) and has asymptotes:

$$y=\pm\frac{1}{2}x$$

$y^2-\frac{x^2}{4}=1 \quad (a=1,\ b=2)$

This is a hyperbola with vertical transverse axis, centered at (0, 0) and has asymptotes:

$$y=\pm\frac{1}{2}x.$$

Since the two hyperbolas have the same asymptotes, they are conjugates.

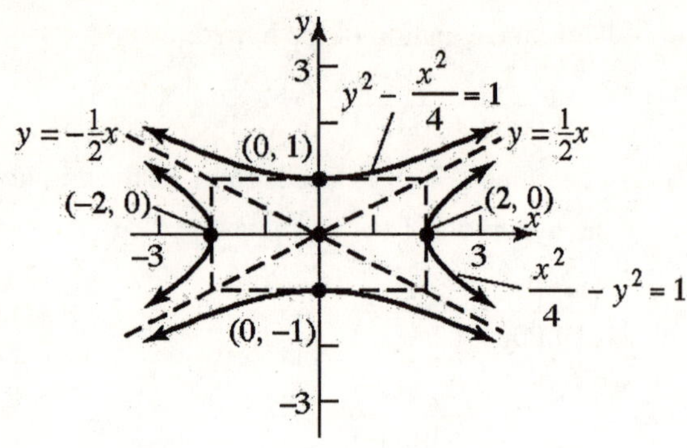

72. $\frac{y^2}{a^2} - \frac{x^2}{b^2} = 1$

Solve for y:

$$\frac{y^2}{a^2} = 1 + \frac{x^2}{b^2}$$

$$y^2 = a^2\left(1 + \frac{x^2}{b^2}\right)$$

$$y^2 = \frac{a^2x^2}{b^2}\left(\frac{b^2}{x^2} + 1\right)$$

$$y = \pm\frac{ax}{b}\sqrt{\frac{b^2}{x^2} + 1}$$

As $x \to -\infty$ or as $x \to \infty$, the term $\frac{b^2}{x^2}$ gets close to 0, so the expression under the radical gets closer to 1. Thus, the graph of the hyperbola gets closer to the lines $y = -\frac{a}{b}x$ and $y = \frac{a}{b}x$. These lines are the asymptotes of the hyperbola.

73. Put the equation in standard hyperbola form:

$$Ax^2 + Cy^2 + F = 0 \qquad A \cdot C < 0,\ F \neq 0$$

$$Ax^2 + Cy^2 = -F$$

$$\frac{Ax^2}{-F} + \frac{Cy^2}{-F} = 1$$

$$\frac{x^2}{\left(-\frac{F}{A}\right)} + \frac{y^2}{\left(-\frac{F}{C}\right)} = 1$$

Since $-F/A$ and $-F/C$ have opposite signs, this is a hyperbola with center (0, 0).
The transverse axis is the x-axis if $-F/A > 0$.
The transverse axis is the y-axis if $-F/A < 0$.

74. Complete the squares on the given equation:
$Ax^2 + Cy^2 + Dx + Ey + F = 0$, where $A \cdot C < 0$.

$$A\left(x^2 + \frac{D}{A}x\right) + C\left(y^2 + \frac{E}{C}y\right) = -F$$

To complete the square in x, we add $\left(\frac{D}{2A}\right)^2 = \frac{D^2}{4A^2}$ inside the first set of parentheses. Since we are multiplying by A, we need to add $A \cdot \frac{D^2}{4A^2} = \frac{D^2}{4A}$ to the right side of the equation.

To complete the square in y, we add $\left(\frac{E}{2C}\right)^2 = \frac{E^2}{4C^2}$ inside the second set of parentheses. Since we are multiplying by C, we add $C \cdot \frac{E^2}{4C^2} = \frac{E^2}{4C}$ to the right side of the equation. Therefore, we obtain the following:

$$A\left(x + \frac{D}{2A}\right)^2 + C\left(y + \frac{E}{2C}\right)^2 = \frac{D^2}{4A} + \frac{E^2}{4C} - F$$

Let $U = \frac{D^2}{4A} + \frac{E^2}{4C} - F$.

a. If $U \neq 0$, then

$$\frac{\left(x + \frac{D}{2A}\right)^2}{\frac{U}{A}} + \frac{\left(y + \frac{E}{2C}\right)^2}{\frac{U}{C}} = 1$$

with $\frac{U}{A}$ and $\frac{U}{C}$ having opposite signs. This is the equation of a hyperbola whose center is $\left(-\frac{D}{2A}, -\frac{E}{2C}\right)$.

b. If $U = 0$, then

$$A\left(x + \frac{D}{2A}\right)^2 + C\left(y + \frac{E}{2C}\right)^2 = 0$$

$$\left(y + \frac{E}{2C}\right)^2 = \frac{-A}{C}\left(x + \frac{D}{2A}\right)^2$$

$$y + \frac{E}{2C} = \pm\sqrt{\frac{-A}{C}}\left(x + \frac{D}{2A}\right)$$

which is the graph of two intersecting lines, through the point $\left(-\frac{D}{2A}, -\frac{E}{2C}\right)$, with slopes $\pm\sqrt{\left|\frac{A}{C}\right|}$.

Section 9.5

1. $\sin\alpha\cos\beta+\cos\alpha\sin\beta$

2. $2\sin\theta\cos\theta$

3. $\sqrt{\dfrac{1-\cos\theta}{2}}$

4. $\sqrt{\dfrac{1+\cos\theta}{2}}$

5. $\cot(2\theta)=\dfrac{A-C}{B}$

6. $x^2-2y^2-x-y-18=0$
   Since $B=0$, we consider only A and C.
   $A=1$ and $C=-2$; $AC=(1)(-2)=-2<0$.
   Since $AC<0$, the equation defines a hyperbola.

7. $x^2+2xy+3y^2-2x+4y+10=0$
   $A=1$, $B=2$, and $C=3$; .
   $B^2-4AC=2^2-4(1)(3)=-8$
   Since $B^2-4AC<0$, the equation defines an ellipse.

8. True

9. True

10. False; $\cot(2\theta)=\dfrac{A-C}{B}$

11. $x^2+4x+y+3=0$
    $A=1$ and $C=0$; $AC=(1)(0)=0$. Since $AC=0$, the equation defines a parabola.

12. $2y^2-3y+3x=0$
    $A=0$ and $C=2$; $AC=(0)(2)=0$. Since $AC=0$, the equation defines a parabola.

13. $6x^2+3y^2-12x+6y=0$
    $A=6$ and $C=3$; $AC=(6)(3)=18$. Since $AC>0$ and $A\neq C$, the equation defines an ellipse.

14. $2x^2+y^2-8x+4y+2=0$
    $A=2$ and $C=1$; $AC=(2)(1)=2$. Since $AC>0$ and $A\neq C$, the equation defines an ellipse.

15. $3x^2-2y^2+6x+4=0$
    $A=3$ and $C=-2$; $AC=(3)(-2)=-6$. Since $AC<0$, the equation defines a hyperbola.

16. $4x^2-3y^2-8x+6y+1=0$
    $A=4$ and $C=-3$; $AC=(4)(-3)=-12$. Since $AC<0$, the equation defines a hyperbola.

17. $2y^2-x^2-y+x=0$
    $A=-1$ and $C=2$; $AC=(-1)(2)=-2$. Since $AC<0$, the equation defines a hyperbola.

18. $y^2-8x^2-2x-y=0$
    $A=-8$ and $C=1$; $AC=(-8)(1)=-8$. Since $AC<0$, the equation defines a hyperbola.

19. $x^2+y^2-8x+4y=0$
    $A=1$ and $C=1$; $AC=(1)(1)=1$. Since $AC>0$ and $A=C$, the equation defines a circle.

20. $2x^2+2y^2-8x+8y=0$
    $A=2$ and $C=2$; $AC=(2)(2)=4$. Since $AC>0$ and $A=C$, the equation defines a circle.

21. $x^2+4xy+y^2-3=0$
    $A=1$, $B=4$, and $C=1$;
    $$\cot(2\theta)=\frac{A-C}{B}=\frac{1-1}{4}=\frac{0}{4}=0$$
    $$2\theta=\frac{\pi}{2}\Rightarrow\theta=\frac{\pi}{4}$$
    $$x=x'\cos\frac{\pi}{4}-y'\sin\frac{\pi}{4}$$
    $$=\frac{\sqrt{2}}{2}x'-\frac{\sqrt{2}}{2}y'=\frac{\sqrt{2}}{2}(x'-y')$$
    $$y=x'\sin\frac{\pi}{4}+y'\cos\frac{\pi}{4}=\frac{\sqrt{2}}{2}x'+\frac{\sqrt{2}}{2}y'$$
    $$=\frac{\sqrt{2}}{2}(x'+y')$$

22. $x^2 - 4xy + y^2 - 3 = 0$

$A = 1, B = -4, \text{ and } C = 1;$

$$\cot(2\theta) = \frac{A-C}{B} = \frac{1-1}{-4} = \frac{0}{-4} = 0$$

$$2\theta = \frac{\pi}{2} \Rightarrow \theta = \frac{\pi}{4}$$

$$x = x'\cos\frac{\pi}{4} - y'\sin\frac{\pi}{4} = \frac{\sqrt{2}}{2}x' - \frac{\sqrt{2}}{2}y' = \frac{\sqrt{2}}{2}(x' - y')$$

$$y = x'\sin\frac{\pi}{4} + y'\cos\frac{\pi}{4} = \frac{\sqrt{2}}{2}x' + \frac{\sqrt{2}}{2}y' = \frac{\sqrt{2}}{2}(x' + y')$$

23. $5x^2 + 6xy + 5y^2 - 8 = 0$

$A = 5, B = 6, \text{ and } C = 5;$

$$\cot(2\theta) = \frac{A-C}{B} = \frac{5-5}{6} = \frac{0}{6} = 0$$

$$2\theta = \frac{\pi}{2} \Rightarrow \theta = \frac{\pi}{4}$$

$$x = x'\cos\frac{\pi}{4} - y'\sin\frac{\pi}{4} = \frac{\sqrt{2}}{2}x' - \frac{\sqrt{2}}{2}y' = \frac{\sqrt{2}}{2}(x' - y')$$

$$y = x'\sin\frac{\pi}{4} + y'\cos\frac{\pi}{4} = \frac{\sqrt{2}}{2}x' + \frac{\sqrt{2}}{2}y' = \frac{\sqrt{2}}{2}(x' + y')$$

24. $3x^2 - 10xy + 3y^2 - 32 = 0$

$A = 3, B = -10, \text{ and } C = 3;$

$$\cot(2\theta) = \frac{A-C}{B} = \frac{3-3}{-10} = \frac{0}{-10} = 0 \Rightarrow$$

$$2\theta = \frac{\pi}{2} \Rightarrow \theta = \frac{\pi}{4}$$

$$x = x'\cos\frac{\pi}{4} - y'\sin\frac{\pi}{4} = \frac{\sqrt{2}}{2}x' - \frac{\sqrt{2}}{2}y' = \frac{\sqrt{2}}{2}(x' - y')$$

$$y = x'\sin\frac{\pi}{4} + y'\cos\frac{\pi}{4} = \frac{\sqrt{2}}{2}x' + \frac{\sqrt{2}}{2}y' = \frac{\sqrt{2}}{2}(x' + y')$$

25. $13x^2 - 6\sqrt{3}xy + 7y^2 - 16 = 0$

$A = 13, B = -6\sqrt{3}, \text{ and } C = 7;$

$$\cot(2\theta) = \frac{A-C}{B} = \frac{13-7}{-6\sqrt{3}} = \frac{6}{-6\sqrt{3}} = -\frac{\sqrt{3}}{3}$$

$$2\theta = \frac{2\pi}{3} \Rightarrow \theta = \frac{\pi}{3}$$

$$x = x'\cos\frac{\pi}{3} - y'\sin\frac{\pi}{3} = \frac{1}{2}x' - \frac{\sqrt{3}}{2}y' = \frac{1}{2}\left(x' - \sqrt{3}y'\right)$$

$$y = x'\sin\frac{\pi}{3} + y'\cos\frac{\pi}{3} = \frac{\sqrt{3}}{2}x' + \frac{1}{2}y' = \frac{1}{2}\left(\sqrt{3}x' + y'\right)$$

26. $11x^2 + 10\sqrt{3}xy + y^2 - 4 = 0$

$A = 11, B = 10\sqrt{3}, \text{ and } C = 1;$

$$\cot(2\theta) = \frac{A-C}{B} = \frac{11-1}{10\sqrt{3}} = \frac{10}{10\sqrt{3}} = \frac{\sqrt{3}}{3}$$

$$2\theta = \frac{\pi}{3} \Rightarrow \theta = \frac{\pi}{6}$$

$$x = x'\cos\frac{\pi}{6} - y'\sin\frac{\pi}{6} = \frac{\sqrt{3}}{2}x' - \frac{1}{2}y' = \frac{1}{2}\left(\sqrt{3}x' - y'\right)$$

$$y = x'\sin\frac{\pi}{6} + y'\cos\frac{\pi}{6} = \frac{1}{2}x' + \frac{\sqrt{3}}{2}y' = \frac{1}{2}\left(x' + \sqrt{3}y'\right)$$

27. $4x^2 - 4xy + y^2 - 8\sqrt{5}\,x - 16\sqrt{5}\,y = 0$

$A = 4$, $B = -4$, and $C = 1$;

$$\cot(2\theta) = \frac{A-C}{B} = \frac{4-1}{-4} = -\frac{3}{4};\quad \cos 2\theta = -\frac{3}{5}$$

$$\sin\theta = \sqrt{\frac{1-\left(-\frac{3}{5}\right)}{2}} = \sqrt{\frac{4}{5}} = \frac{2}{\sqrt{5}} = \frac{2\sqrt{5}}{5};$$

$$\cos\theta = \sqrt{\frac{1+\left(-\frac{3}{5}\right)}{2}} = \sqrt{\frac{1}{5}} = \frac{1}{\sqrt{5}} = \frac{\sqrt{5}}{5}$$

$$x = x'\cos\theta - y'\sin\theta = \frac{\sqrt{5}}{5}x' - \frac{2\sqrt{5}}{5}y' = \frac{\sqrt{5}}{5}(x' - 2y')$$

$$y = x'\sin\theta + y'\cos\theta = \frac{2\sqrt{5}}{5}x' + \frac{\sqrt{5}}{5}y' = \frac{\sqrt{5}}{5}(2x' + y')$$

28. $x^2 + 4xy + 4y^2 + 5\sqrt{5}\,y + 5 = 0$

$A = 1$, $B = 4$, and $C = 4$;

$$\cot(2\theta) = \frac{A-C}{B} = \frac{1-4}{4} = -\frac{3}{4};\quad \cos 2\theta = -\frac{3}{5}$$

$$\sin\theta = \sqrt{\frac{1-\left(-\frac{3}{5}\right)}{2}} = \sqrt{\frac{4}{5}} = \frac{2}{\sqrt{5}} = \frac{2\sqrt{5}}{5};$$

$$\cos\theta = \sqrt{\frac{1+\left(-\frac{3}{5}\right)}{2}} = \sqrt{\frac{1}{5}} = \frac{1}{\sqrt{5}} = \frac{\sqrt{5}}{5}$$

$$x = x'\cos\theta - y'\sin\theta = \frac{\sqrt{5}}{5}x' - \frac{2\sqrt{5}}{5}y' = \frac{\sqrt{5}}{5}(x' - 2y')$$

$$y = x'\sin\theta + y'\cos\theta = \frac{2\sqrt{5}}{5}x' + \frac{\sqrt{5}}{5}y' = \frac{\sqrt{5}}{5}(2x' + y')$$

29. $25x^2 - 36xy + 40y^2 - 12\sqrt{13}\,x - 8\sqrt{13}\,y = 0$

$A = 25$, $B = -36$, and $C = 40$;

$$\cot(2\theta) = \frac{A-C}{B} = \frac{25-40}{-36} = \frac{5}{12};\quad \cos 2\theta = \frac{5}{13}$$

$$\sin\theta = \sqrt{\frac{1-\frac{5}{13}}{2}} = \sqrt{\frac{4}{13}} = \frac{2}{\sqrt{13}} = \frac{2\sqrt{13}}{13};$$

$$\cos\theta = \sqrt{\frac{1+\frac{5}{13}}{2}} = \sqrt{\frac{9}{13}} = \frac{3}{\sqrt{13}} = \frac{3\sqrt{13}}{13}$$

$$x = x'\cos\theta - y'\sin\theta = \frac{3\sqrt{13}}{13}x' - \frac{2\sqrt{13}}{13}y' = \frac{\sqrt{13}}{13}(3x' - 2y')$$

$$y = x'\sin\theta + y'\cos\theta = \frac{2\sqrt{13}}{13}x' + \frac{3\sqrt{13}}{13}y' = \frac{\sqrt{13}}{13}(2x' + 3y')$$

30. $34x^2 - 24xy + 41y^2 - 25 = 0$

$A = 34,\ B = -24,$ and $C = 41;$

$$\cot(2\theta) = \frac{A-C}{B} = \frac{34-41}{-24} = \frac{7}{24};\quad \cos(2\theta) = \frac{7}{25}$$

$$\sin\theta = \sqrt{\frac{1-\frac{7}{25}}{2}} = \sqrt{\frac{9}{25}} = \frac{3}{5};\quad \cos\theta = \sqrt{\frac{1+\frac{7}{25}}{2}} = \sqrt{\frac{16}{25}} = \frac{4}{5}$$

$$x = x'\cos\theta - y'\sin\theta = \frac{4}{5}x' - \frac{3}{5}y' = \frac{1}{5}(4x' - 3y')$$

$$y = x'\sin\theta + y'\cos\theta = \frac{3}{5}x' + \frac{4}{5}y' = \frac{1}{5}(3x' + 4y')$$

31. $x^2 + 4xy + y^2 - 3 = 0;\ \theta = 45°$

$$\left(\frac{\sqrt{2}}{2}(x'-y')\right)^2 + 4\left(\frac{\sqrt{2}}{2}(x'-y')\right)\left(\frac{\sqrt{2}}{2}(x'+y')\right) + \left(\frac{\sqrt{2}}{2}(x'+y')\right)^2 - 3 = 0$$

$$\frac{1}{2}\left(x'^2 - 2x'y' + y'^2\right) + 2\left(x'^2 - y'^2\right) + \frac{1}{2}\left(x'^2 + 2x'y' + y'^2\right) - 3 = 0$$

$$\frac{1}{2}x'^2 - x'y' + \frac{1}{2}y'^2 + 2x'^2 - 2y'^2 + \frac{1}{2}x'^2 + x'y' + \frac{1}{2}y'^2 = 3$$

$$3x'^2 - y'^2 = 3$$

$$\frac{x'^2}{1} - \frac{y'^2}{3} = 1$$

Hyperbola; center at the origin, transverse axis is the x'-axis, vertices $(\pm 1, 0)$.

32. $x^2 - 4xy + y^2 - 3 = 0;\ \theta = 45°$

$$\left(\frac{\sqrt{2}}{2}(x'-y')\right)^2 - 4\left(\frac{\sqrt{2}}{2}(x'-y')\right)\left(\frac{\sqrt{2}}{2}(x'+y')\right) + \left(\frac{\sqrt{2}}{2}(x'+y')\right)^2 - 3 = 0$$

$$\frac{1}{2}\left(x'^2 - 2x'y' + y'^2\right) - 2\left(x'^2 - y'^2\right) + \frac{1}{2}\left(x'^2 + 2x'y' + y'^2\right) - 3 = 0$$

$$\frac{1}{2}x'^2 - x'y' + \frac{1}{2}y'^2 - 2x'^2 + 2y'^2 + \frac{1}{2}x'^2 + x'y' + \frac{1}{2}y'^2 = 3$$

$$-x'^2 + 3y'^2 = 3$$

$$\frac{y'^2}{1} - \frac{x'^2}{3} = 1$$

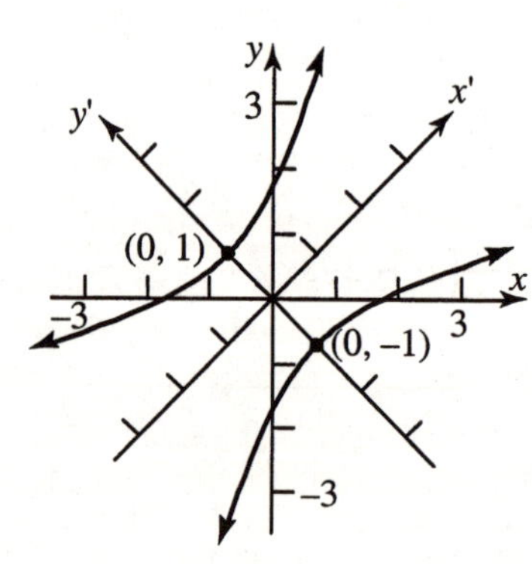

Hyperbola; center at the origin, transverse axis is the y'-axis, vertices $(0, \pm 1)$.

33. $5x^2+6xy+5y^2-8=0$; $\theta=45°$

$$5\left(\frac{\sqrt{2}}{2}(x'-y')\right)^2+6\left(\frac{\sqrt{2}}{2}(x'-y')\right)\left(\frac{\sqrt{2}}{2}(x'+y')\right)+5\left(\frac{\sqrt{2}}{2}(x'+y')\right)^2-8=0$$

$$\frac{5}{2}\left(x'^2-2x'y'+y'^2\right)+3\left(x'^2-y'^2\right)+\frac{5}{2}\left(x'^2+2x'y'+y'^2\right)-8=0$$

$$\frac{5}{2}x'^2-5x'y'+\frac{5}{2}y'^2+3x'^2-3y'^2+\frac{5}{2}x'^2+5x'y'+\frac{5}{2}y'^2=8$$

$$8x'^2+2y'^2=8$$

$$\frac{x'^2}{1}+\frac{y'^2}{4}=1$$

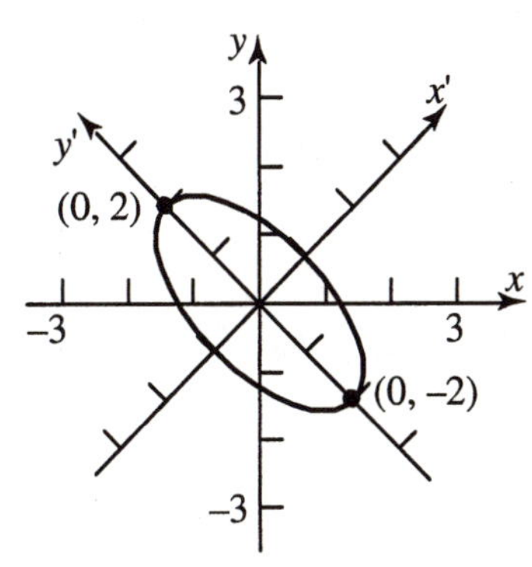

Ellipse; center at the origin, major axis is the y'-axis, vertices (0, ±2).

34. $3x^2-10xy+3y^2-32=0$; $\theta=45°$

$$3\left(\frac{\sqrt{2}}{2}(x'-y')\right)^2-10\left(\frac{\sqrt{2}}{2}(x'-y')\right)\left(\frac{\sqrt{2}}{2}(x'+y')\right)+3\left(\frac{\sqrt{2}}{2}(x'+y')\right)^2-32=0$$

$$\frac{3}{2}\left(x'^2-2x'y'+y'^2\right)-5\left(x'^2-y'^2\right)+\frac{3}{2}\left(x'^2+2x'y'+y'^2\right)-32=0$$

$$\frac{3}{2}x'^2-3x'y'+\frac{3}{2}y'^2-5x'^2+5y'^2+\frac{3}{2}x'^2+3x'y'+\frac{3}{2}y'^2=32$$

$$-2x'^2+8y'^2=32$$

$$\frac{y'^2}{4}-\frac{x'^2}{16}=1$$

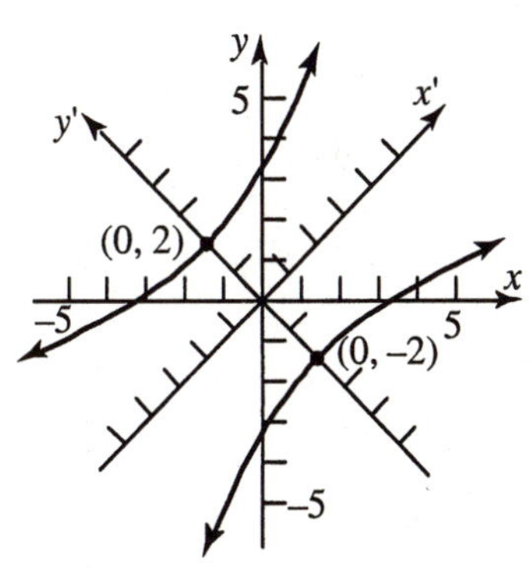

Hyperbola; center at the origin, transverse axis is the y'-axis, vertices (0, ±2).

35. $13x^2-6\sqrt{3}\,xy+7y^2-16=0$; $\theta=60°$

$$13\left(\frac{1}{2}\left(x'-\sqrt{3}y'\right)\right)^2-6\sqrt{3}\left(\frac{1}{2}\left(x'-\sqrt{3}y'\right)\right)\left(\frac{1}{2}\left(\sqrt{3}x'+y'\right)\right)+7\left(\frac{1}{2}\left(\sqrt{3}x'+y'\right)\right)^2-16=0$$

$$\frac{13}{4}\left(x'^2-2\sqrt{3}x'y'+3y'^2\right)-\frac{3\sqrt{3}}{2}\left(\sqrt{3}x'^2-2x'y'-\sqrt{3}y'^2\right)+\frac{7}{4}\left(3x'^2+2\sqrt{3}x'y'+y'^2\right)=16$$

$$\frac{13}{4}x'^2-\frac{13\sqrt{3}}{2}x'y'+\frac{39}{4}y'^2-\frac{9}{2}x'^2+3\sqrt{3}x'y'+\frac{9}{2}y'^2+\frac{21}{4}x'^2+\frac{7\sqrt{3}}{2}x'y'+\frac{7}{4}y'^2=16$$

$$4x'^2+16y'^2=16$$

$$\frac{x'^2}{4}+\frac{y'^2}{1}=1$$

Ellipse; center at the origin, major axis is the x'-axis, vertices (±2, 0).

36. $11x^2 + 10\sqrt{3}xy + y^2 - 4 = 0$; $\theta = 30°$

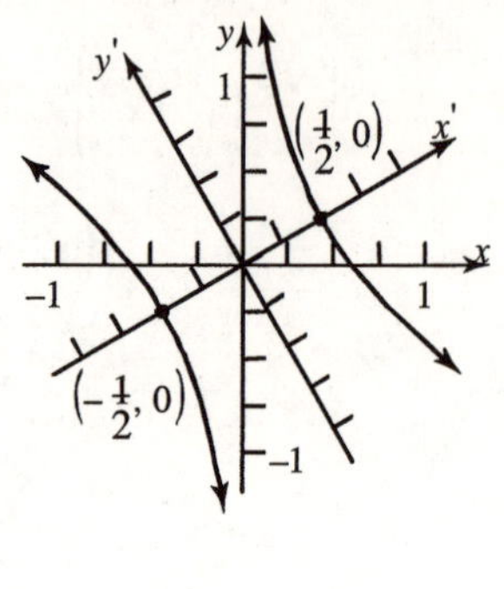

$$11\left(\frac{1}{2}\left(\sqrt{3}x' - y'\right)\right)^2 + 10\sqrt{3}\left(\frac{1}{2}\left(\sqrt{3}x' - y'\right)\right)\left(\frac{1}{2}\left(x' + \sqrt{3}y'\right)\right) + \left(\frac{1}{2}\left(x' + \sqrt{3}y'\right)\right)^2 - 4 = 0$$

$$\frac{11}{4}\left(3x'^2 - 2\sqrt{3}x'y' + y'^2\right) + \frac{5\sqrt{3}}{2}\left(\sqrt{3}x'^2 + 2x'y' - \sqrt{3}y'^2\right) + \frac{1}{4}\left(x'^2 + 2\sqrt{3}x'y' + 3y'^2\right) = 4$$

$$\frac{33}{4}x'^2 - \frac{11\sqrt{3}}{2}x'y' + \frac{11}{4}y'^2 + \frac{15}{2}x'^2 + 5\sqrt{3}x'y' - \frac{15}{2}y'^2 + \frac{1}{4}x'^2 + \frac{\sqrt{3}}{2}x'y' + \frac{3}{4}y'^2 = 4$$

$$16x'^2 - 4y'^2 = 4$$

$$4x'^2 - y'^2 = 1$$

$$\frac{x'^2}{\frac{1}{4}} - \frac{y'^2}{1} = 1$$

Hyperbola; center at the origin, transverse axis is the x'-axis, vertices $(\pm 0.5, 0)$.

37. $4x^2 - 4xy + y^2 - 8\sqrt{5}x - 16\sqrt{5}y = 0$; $\theta \approx 63.4°$

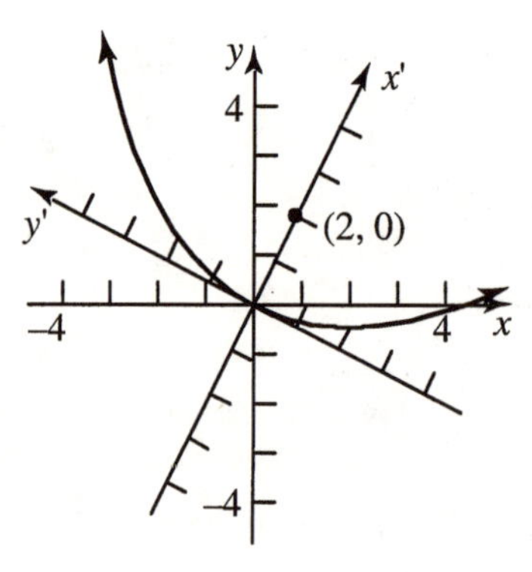

$$4\left(\frac{\sqrt{5}}{5}(x' - 2y')\right)^2 - 4\left(\frac{\sqrt{5}}{5}(x' - 2y')\right)\left(\frac{\sqrt{5}}{5}(2x' + y')\right) + \left(\frac{\sqrt{5}}{5}(2x' + y')\right)^2$$

$$-8\sqrt{5}\left(\frac{\sqrt{5}}{5}(x' - 2y')\right) - 16\sqrt{5}\left(\frac{\sqrt{5}}{5}(2x' + y')\right) = 0$$

$$\frac{4}{5}\left(x'^2 - 4x'y' + 4y'^2\right) - \frac{4}{5}\left(2x'^2 - 3x'y' - 2y'^2\right) + \frac{1}{5}\left(4x'^2 + 4x'y' + y'^2\right)$$

$$-8x' + 16y' - 32x' - 16y' = 0$$

$$\frac{4}{5}x'^2 - \frac{16}{5}x'y' + \frac{16}{5}y'^2 - \frac{8}{5}x'^2 + \frac{12}{5}x'y' + \frac{8}{5}y'^2 + \frac{4}{5}x'^2 + \frac{4}{5}x'y' + \frac{1}{5}y'^2 - 40x' = 0$$

$$5y'^2 - 40x' = 0$$

$$y'^2 = 8x'$$

Parabola; vertex at the origin, focus at (2, 0).

38. $x^2+4xy+4y^2+5\sqrt{5}\,y+5=0$; $\theta \approx 63.4°$

$$\left(\frac{\sqrt{5}}{5}(x'-2y')\right)^2+4\left(\frac{\sqrt{5}}{5}(x'-2y')\right)\left(\frac{\sqrt{5}}{5}(2x'+y')\right)+4\left(\frac{\sqrt{5}}{5}(2x'+y')\right)^2$$
$$+5\sqrt{5}\left(\frac{\sqrt{5}}{5}(2x'+y')\right)+5=0$$

$$\frac{1}{5}\left(x'^2-4x'y'+4y'^2\right)+\frac{4}{5}\left(2x'^2-3x'y'-2y'^2\right)+\frac{4}{5}\left(4x'^2+4x'y'+y'^2\right)$$
$$+10x'+5y'+5=0$$

$$\frac{1}{5}x'^2-\frac{4}{5}x'y'+\frac{4}{5}y'^2+\frac{8}{5}x'^2-\frac{12}{5}x'y'-\frac{8}{5}y'^2+\frac{16}{5}x'^2+\frac{16}{5}x'y'+\frac{4}{5}y'^2$$
$$+10x'+5y'+5=0$$
$$5x'^2+10x'+5y'+5=0$$
$$x'^2+2x'+1=-y'-1+1$$
$$(x'+1)^2=-y'$$

Parabola; vertex at (–1, 0), axis of symmetry parallel to the y'-axis; focus at $(x',y')=\left(-1,-\frac{1}{4}\right)$.

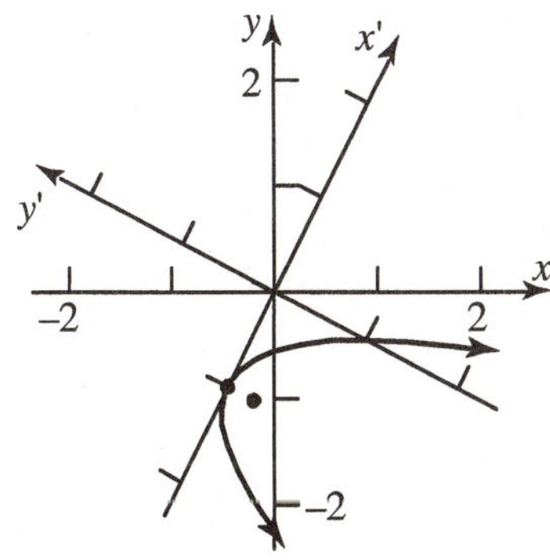

39. $25x^2-36xy+40y^2-12\sqrt{13}x-8\sqrt{13}y=0$; $\theta \approx 33.7°$

$$25\left(\frac{\sqrt{13}}{13}(3x'-2y')\right)^2-36\left(\frac{\sqrt{13}}{13}(3x'-2y')\right)\left(\frac{\sqrt{13}}{13}(2x'+3y')\right)+40\left(\frac{\sqrt{13}}{13}(2x'+3y')\right)^2$$
$$-12\sqrt{13}\left(\frac{\sqrt{13}}{13}(3x'-2y')\right)-8\sqrt{13}\left(\frac{\sqrt{13}}{13}(2x'+3y')\right)=0$$

$$\frac{25}{13}\left(9x'^2-12x'y'+4y'^2\right)-\frac{36}{13}\left(6x'^2+5x'y'-6y'^2\right)+\frac{40}{13}\left(4x'^2+12x'y'+9y'^2\right)$$
$$-36x'+24y'-16x'-24y'=0$$

$$\frac{225}{13}x'^2-\frac{300}{13}x'y'+\frac{100}{13}y'^2-\frac{216}{13}x'^2-\frac{180}{13}x'y'+\frac{216}{13}y'^2$$
$$+\frac{160}{13}x'^2+\frac{480}{13}x'y'+\frac{360}{13}y'^2-52x'=0$$

$$13x'^2 + 52y'^2 - 52x' = 0$$
$$x'^2 - 4x' + 4y'^2 = 0$$
$$(x'-2)^2 + 4y'^2 = 4$$
$$\frac{(x'-2)^2}{4} + \frac{y'^2}{1} = 1$$

Ellipse; center at (2, 0), major axis is the x'-axis, vertices (4, 0) and (0, 0).

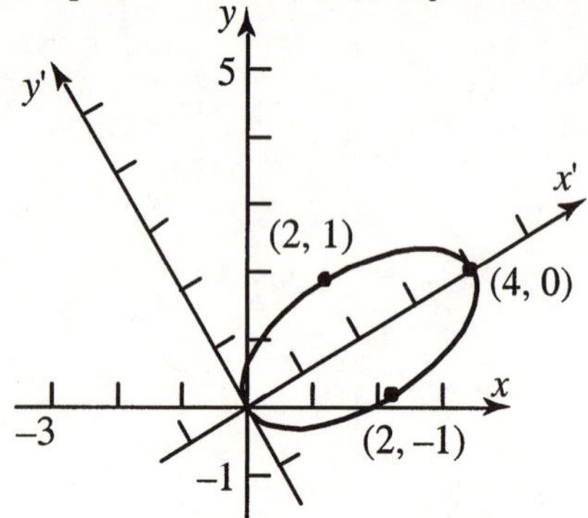

40. $34x^2 - 24xy + 41y^2 - 25 = 0$; $\theta \approx 36.9°$

$$34\left(\frac{1}{5}(4x'-3y')\right)^2 - 24\left(\frac{1}{5}(4x'-3y')\right)\left(\frac{1}{5}(3x'+4y')\right) + 41\left(\frac{1}{5}(3x'+4y')\right)^2 - 25 = 0$$

$$\frac{34}{25}\left(16x'^2 - 24x'y' + 9y'^2\right) - \frac{24}{25}\left(12x'^2 + 7x'y' - 12y'^2\right) + \frac{41}{25}\left(9x'^2 + 24x'y' + 16y'^2\right) = 25$$

$$\frac{544}{25}x'^2 - \frac{816}{25}x'y' + \frac{306}{25}y'^2 - \frac{288}{25}x'^2 - \frac{168}{25}x'y' + \frac{288}{25}y'^2 + \frac{369}{25}x'^2 + \frac{984}{25}x'y' + \frac{656}{25}y'^2 = 25$$

$$25x'^2 + 50y'^2 = 25$$
$$x'^2 + 2y'^2 = 1$$
$$\frac{x'^2}{1} + \frac{y'^2}{\frac{1}{2}} = 1$$

Ellipse; center at the origin, major axis is the x'-axis, vertices (±1, 0).

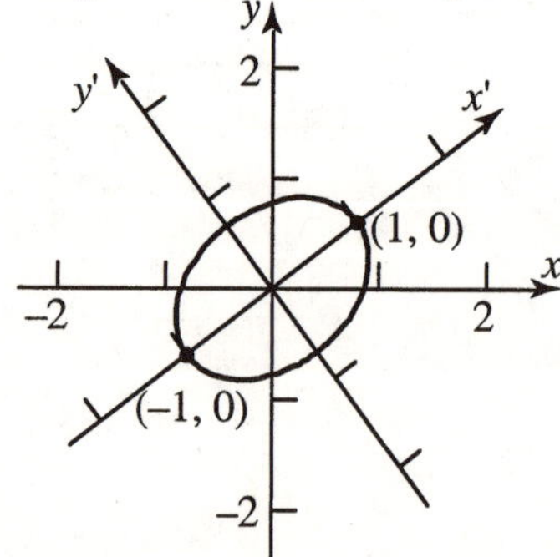

41. $16x^2+24xy+9y^2-130x+90y=0$

$$A=16,\ B=24,\ \text{and } C=9;\ \cot(2\theta)=\frac{A-C}{B}=\frac{16-9}{24}=\frac{7}{24}\Rightarrow\cos(2\theta)=\frac{7}{25}$$

$$\sin\theta=\sqrt{\frac{1-\frac{7}{25}}{2}}=\sqrt{\frac{9}{25}}=\frac{3}{5};\quad \cos\theta=\sqrt{\frac{1+\frac{7}{25}}{2}}=\sqrt{\frac{16}{25}}=\frac{4}{5}\Rightarrow\theta\approx 36.9^{\circ}$$

$$x=x'\cos\theta-y'\sin\theta=\frac{4}{5}x'-\frac{3}{5}y'=\frac{1}{5}(4x'-3y')$$

$$y=x'\sin\theta+y'\cos\theta=\frac{3}{5}x'+\frac{4}{5}y'=\frac{1}{5}(3x'+4y')$$

$$16\left(\frac{1}{5}(4x'-3y')\right)^2+24\left(\frac{1}{5}(4x'-3y')\right)\left(\frac{1}{5}(3x'+4y')\right)+9\left(\frac{1}{5}(3x'+4y')\right)^2$$
$$-130\left(\frac{1}{5}(4x'-3y')\right)+90\left(\frac{1}{5}(3x'+4y')\right)=0$$

$$\frac{16}{25}\left(16x'^2-24x'y'+9y'^2\right)+\frac{24}{25}\left(12x'^2+7x'y'-12y'^2\right)$$
$$+\frac{9}{25}\left(9x'^2+24x'y'+16y'^2\right)-104x'+78y'+54x'+72y'=0$$

$$\frac{256}{25}x'^2-\frac{384}{25}x'y'+\frac{144}{25}y'^2+\frac{288}{25}x'^2+\frac{168}{25}x'y'-\frac{288}{25}y'^2$$
$$+\frac{81}{25}x'^2+\frac{216}{25}x'y'+\frac{144}{25}y'^2-50x'+150y'=0$$

$$25x'^2-50x'+150y'=0$$
$$x'^2-2x'=-6y'$$
$$(x'-1)^2=-6y'+1$$
$$(x'-1)^2=-6\left(y'-\frac{1}{6}\right)$$

Parabola; vertex $\left(1,\frac{1}{6}\right)$, focus $\left(1,-\frac{4}{3}\right)$; axis of symmetry parallel to the y' axis.

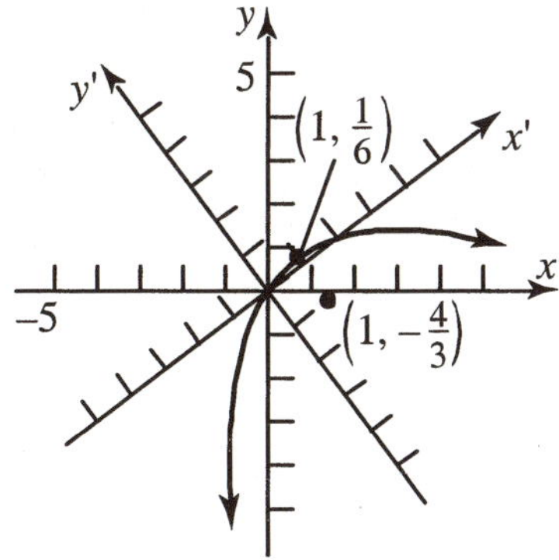

42. $16x^2+24xy+9y^2-60x+80y=0$

$A=16$, $B=24$, and $C=9$; $\cot(2\theta)=\frac{A-C}{B}=\frac{16-9}{24}=\frac{7}{24}\Rightarrow\cos(2\theta)=\frac{7}{25}$

$$\sin\theta=\sqrt{\frac{1-\frac{7}{25}}{2}}=\sqrt{\frac{9}{25}}=\frac{3}{5};\quad \cos\theta=\sqrt{\frac{1+\frac{7}{25}}{2}}=\sqrt{\frac{16}{25}}=\frac{4}{5}\Rightarrow\theta\approx 36.9^\circ$$

$$x=x'\cos\theta-y'\sin\theta=\frac{4}{5}x'-\frac{3}{5}y'=\frac{1}{5}(4x'-3y')$$

$$y=x'\sin\theta+y'\cos\theta=\frac{3}{5}x'+\frac{4}{5}y'=\frac{1}{5}(3x'+4y')$$

$$16\left(\frac{1}{5}(4x'-3y')\right)^2+24\left(\frac{1}{5}(4x'-3y')\right)\left(\frac{1}{5}(3x'+4y')\right)+9\left(\frac{1}{5}(3x'+4y')\right)^2$$
$$-60\left(\frac{1}{5}(4x'-3y')\right)+80\left(\frac{1}{5}(3x'+4y')\right)=0$$

$$\frac{16}{25}\left(16x'^2-24x'y'+9y'^2\right)+\frac{24}{25}\left(12x'^2+7x'y'-12y'^2\right)$$
$$+\frac{9}{25}\left(9x'^2+24x'y'+16y'^2\right)-48x'+36y'+48x'+64y'=0$$

$$\frac{256}{25}x'^2-\frac{384}{25}x'y'+\frac{144}{25}y'^2+\frac{288}{25}x'^2+\frac{168}{25}x'y'-\frac{288}{25}y'^2$$
$$+\frac{81}{25}x'^2+\frac{216}{25}x'y'+\frac{144}{25}y'^2+100y'=0$$
$$25x'^2+100y'=0$$
$$x'^2=-4y'$$

Parabola; vertex (0, 0), focus (0, –1).

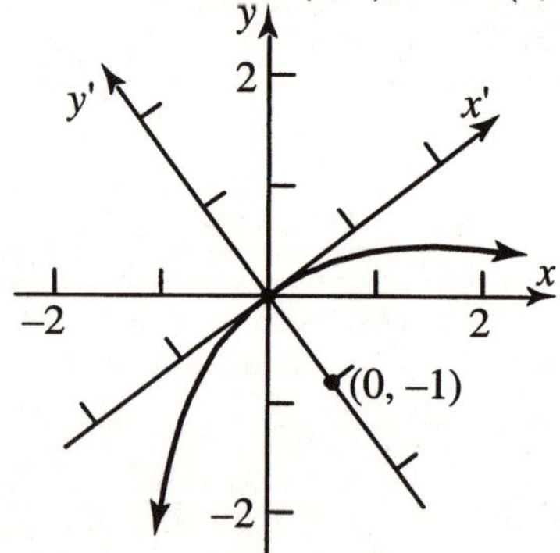

43. $A=1,\ B=3,\ C=-2\quad B^2-4AC=3^2-4(1)(-2)=17>0$; hyperbola

44. $A=2,\ B=-3,\ C=4\quad B^2-4AC=(-3)^2-4(2)(4)=-23<0$; ellipse

45. $A=1,\ B=-7,\ C=3\quad B^2-4AC=(-7)^2-4(1)(3)=37>0$; hyperbola

46. $A=2,\ B=-3,\ C=2\quad B^2-4AC=(-3)^2-4(2)(2)=-7<0$; ellipse

47. $A=9,\ B=12,\ C=4\quad B^2-4AC=12^2-4(9)(4)=0$; parabola

48. $A=10,\ B=12,\ C=4 \quad B^2-4AC=12^2-4(10)(4)=-16<0$; ellipse

49. $A=10,\ B=-12,\ C=4 \quad B^2-4AC=(-12)^2-4(10)(4)=-16<0$; ellipse

50. $A=4,\ B=12,\ C=9 \quad B^2-4AC=12^2-4(4)(9)=0$; parabola

51. $A=3,\ B=-2,\ C=1 \quad B^2-4AC=(-2)^2-4(3)(1)=-8<0$; ellipse

52. $A=3,\ B=2,\ C=1 \quad B^2-4AC=2^2-4(3)(1)=-8<0$; ellipse

53. $A'=A\cos^2\theta+B\sin\theta\cos\theta+C\sin^2\theta$

$B'=B(\cos^2\theta-\sin^2\theta)+2(C-A)(\sin\theta\cos\theta)$

$C'=A\sin^2\theta-B\sin\theta\cos\theta+C\cos^2\theta$

$D'=D\cos\theta+E\sin\theta$

$E'=-D\sin\theta+E\cos\theta$

$F'=F$

54. $A'+C'=\left(A\cos^2\theta+B\sin\theta\cos\theta+C\sin^2\theta\right)+\left(A\sin^2\theta-B\sin\theta\cos\theta+C\cos^2\theta\right)$

$=A\left(\cos^2\theta+\sin^2\theta\right)+C\left(\sin^2\theta+\cos^2\theta\right)=A(1)+C(1)=A+C$

55. $B'^2=[B(\cos^2\theta-\sin^2\theta)+2(C-A)\sin\theta\cos\theta]^2$

$=[B\cos 2\theta-(A-C)\sin 2\theta]^2$

$=B^2\cos^2 2\theta-2B(A-C)\sin 2\theta\cos 2\theta+(A-C)^2\sin^2 2\theta$

$4A'C'=4[A\cos^2\theta+B\sin\theta\cos\theta+C\sin^2\theta][A\sin^2\theta-B\sin\theta\cos\theta+C\cos^2\theta]$

$=4\left[A\left(\frac{1+\cos 2\theta}{2}\right)+B\left(\frac{\sin 2\theta}{2}\right)+C\left(\frac{1-\cos 2\theta}{2}\right)\right]\left[A\left(\frac{1-\cos 2\theta}{2}\right)-B\left(\frac{\sin 2\theta}{2}\right)+C\left(\frac{1+\cos 2\theta}{2}\right)\right]$

$=[A(1+\cos 2\theta)+B(\sin 2\theta)+C(1-\cos 2\theta)][A(1-\cos 2\theta)-B(\sin 2\theta)+C(1+\cos 2\theta)]$

$=[(A+C)+B\sin 2\theta+(A-C)\cos 2\theta][(A+C)-(B\sin 2\theta+(A-C)\cos 2\theta)]$

$=(A+C)^2-[B\sin 2\theta+(A-C)\cos 2\theta]^2$

$=(A+C)^2-[B^2\sin^2 2\theta+2B(A-C)\sin 2\theta\cos 2\theta+(A-C)^2\cos^2 2\theta]$

$B'^2-4A'C'=B^2\cos^2 2\theta-2B(A-C)\sin 2\theta\cos 2\theta+(A-C)^2\sin^2 2\theta$

$-(A+C)^2+B^2\sin^2 2\theta+2B(A-C)\sin 2\theta\cos 2\theta+(A-C)^2\cos^2 2\theta$

$=B^2(\cos^2 2\theta+\sin^2 2\theta)+(A-C)^2(\cos^2 2\theta+\sin^2 2\theta)-(A+C)^2$

$=B^2+(A-C)^2-(A+C)^2=B^2+(A^2-2AC+C^2)-(A^2+2AC+C^2)$

$=B^2-4AC$

56. Since $B^2-4AC=B'^2-4A'C'$ for any rotation θ (Problem 55), choose θ so that $B'=0$. Then $B^2-4AC=-4A'C'$.

a. If $B^2-4AC=-4A'C'=0$ then $A'C'=0$. Using the theorem for identifying conics without completing the square, the equation is a parabola.

b. If $B^2-4AC=-4A'C'<0$ then $A'C'>0$. Thus, the equation is an ellipse (or circle).

c. If $B^2-4AC=-4A'C'>0$ then $A'C'<0$. Thus, the equation is a hyperbola.

57.
$$\begin{aligned}
d^2 &= (y_2-y_1)^2+(x_2-x_1)^2\\
&= (x_2'\sin\theta+y_2'\cos\theta-x_1'\sin\theta-y_1'\cos\theta)^2+(x_2'\cos\theta-y_2'\sin\theta-x_1'\cos\theta+y_1'\sin\theta)^2\\
&= ((x_2'-x_1')\sin\theta+(y_2'-y_1')\cos\theta)^2+((x_2'-x_1')\cos\theta-(y_2'-y_1')\sin\theta)^2\\
&= (x_2'-x_1')^2\sin^2\theta+2(x_2'-x_1')(y_2'-y_1')\sin\theta\cos\theta+(y_2'-y_1')^2\cos^2\theta\\
&\qquad +(x_2'-x_1')^2\cos^2\theta-2(x_2'-x_1')(y_2'-y_1')\sin\theta\cos\theta+(y_2'-y_1')^2\sin^2\theta\\
&= (x_2'-x_1')^2\left(\sin^2\theta+\cos^2\theta\right)+(y_2'-y_1')^2\left(\cos^2\theta+\sin^2\theta\right)\\
&= \left[(x_2'-x_1')^2+(y_2'-y_1')^2\right]\left(\sin^2\theta+\cos^2\theta\right)\\
&= (x_2'-x_1')^2+(y_2'-y_1')^2
\end{aligned}$$

58.
$$\begin{aligned}
x^{1/2}+y^{1/2} &= a^{1/2}\\
y^{1/2} &= a^{1/2}-x^{1/2}\\
y &= \left(a^{1/2}-x^{1/2}\right)^2\\
y &= a-2a^{1/2}x^{1/2}+x\\
2a^{1/2}x^{1/2} &= (a+x)-y\\
4ax &= (a+x)^2-2y(a+x)+y^2\\
4ax &= a^2+2ax+x^2-2ay-2xy+y^2\\
0 &= x^2-2xy+y^2-2ax-2ay+a^2
\end{aligned}$$

$B^2-4AC=(-2)^2-4(1)(1)=4-4=0$

The graph of the equation is part of a parabola.

59. Answers will vary.

Section 9.6

1. $r\cos\theta$; $r\sin\theta$

2. $r = 6\cos\theta$
Begin by multiplying both sides of the equation by r to get $r^2 = 6r\cos\theta$. Since $r\cos\theta = x$ and $r^2 = x^2 + y^2$, we obtain $x^2 + y^2 = 6x$.
Move all the variable terms to the left side of the equation and complete the square in x.
$$x^2 - 6x + y^2 = 0$$
$$\left(x^2 - 6x + 9\right) + y^2 = 9$$
$$(x-3)^2 + y^2 = 9$$

3. $\frac{1}{2}$, ellipse, parallel, 4, below

4. 1; < 1; > 1

5. True

6. True

7. $e = 1$; $p = 1$; parabola; directrix is perpendicular to the polar axis and 1 unit to the right of the pole.

8. $e = 1$; $p = 3$; parabola; directrix is parallel to the polar axis and 3 units below the pole.

9. $r = \dfrac{4}{2 - 3\sin\theta} = \dfrac{4}{2\left(1 - \frac{3}{2}\sin\theta\right)}$
$= \dfrac{2}{1 - \frac{3}{2}\sin\theta}$
$ep = 2$, $e = \frac{3}{2}$; $p = \frac{4}{3}$
Hyperbola; directrix is parallel to the polar axis and $\frac{4}{3}$ units below the pole.

10. $r = \dfrac{2}{1 + 2\cos\theta}$; $ep = 2$, $e = 2$; $p = 1$
Hyperbola; directrix is perpendicular to the polar axis and 1 unit to the right of the pole.

11. $r = \dfrac{3}{4 - 2\cos\theta} = \dfrac{3}{4\left(1 - \frac{1}{2}\cos\theta\right)}$
$= \dfrac{\frac{3}{4}}{1 - \frac{1}{2}\cos\theta}$;
$ep = \frac{3}{4}$, $e = \frac{1}{2}$; $p = \frac{3}{2}$
Ellipse; directrix is perpendicular to the polar axis and $\frac{3}{2}$ units to the left of the pole.

12. $r = \dfrac{6}{8 + 2\sin\theta} = \dfrac{6}{8\left(1 + \frac{1}{4}\sin\theta\right)}$
$= \dfrac{\frac{3}{4}}{1 + \frac{1}{4}\sin\theta}$
$ep = \frac{3}{4}$, $e = \frac{1}{4}$; $p = 3$
Ellipse; directrix is parallel to the polar axis and 3 units above the pole.

13. $r = \dfrac{1}{1 + \cos\theta}$
$ep = 1$, $e = 1$, $p = 1$
Parabola; directrix is perpendicular to the polar axis 1 unit to the right of the pole; vertex is $\left(\frac{1}{2}, 0\right)$.

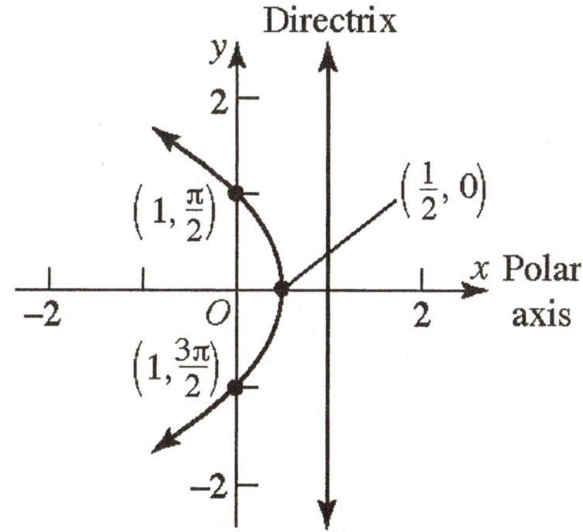

14. $r=\dfrac{3}{1-\sin\theta}$

$ep=3,\ e=1,\ p=3$

Parabola; directrix is parallel to the polar axis 3 units below the pole; vertex is $\left(\dfrac{3}{2},\dfrac{3\pi}{2}\right)$.

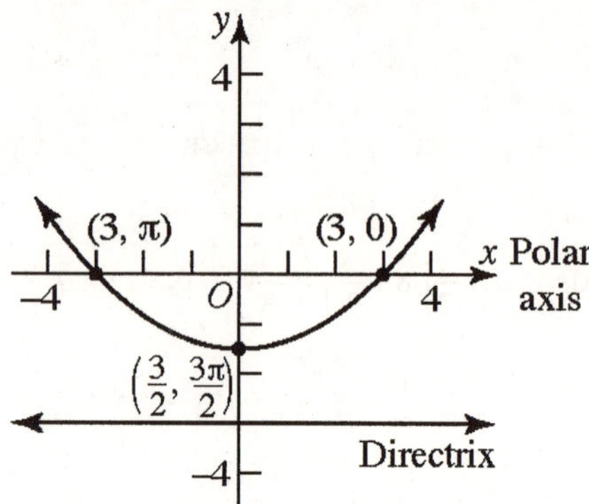

15. $r=\dfrac{8}{4+3\sin\theta}$

$r=\dfrac{8}{4\left(1+\dfrac{3}{4}\sin\theta\right)}=\dfrac{2}{1+\dfrac{3}{4}\sin\theta}$

$ep=2,\ e=\dfrac{3}{4},\ p=\dfrac{8}{3}$

Ellipse; directrix is parallel to the polar axis $\dfrac{8}{3}$ units above the pole; vertices are $\left(\dfrac{8}{7},\dfrac{\pi}{2}\right)$ and $\left(8,\dfrac{3\pi}{2}\right)$.

Also: $a=\dfrac{1}{2}\left(8+\dfrac{8}{7}\right)=\dfrac{32}{7}$ so the center is at $\left(8-\dfrac{32}{7},\dfrac{3\pi}{2}\right)=\left(\dfrac{24}{7},\dfrac{3\pi}{2}\right)$, and $c=\dfrac{24}{7}-0=\dfrac{24}{7}$ so that the second focus is at $\left(\dfrac{24}{7}+\dfrac{24}{7},\dfrac{3\pi}{2}\right)=\left(\dfrac{48}{7},\dfrac{3\pi}{2}\right)$

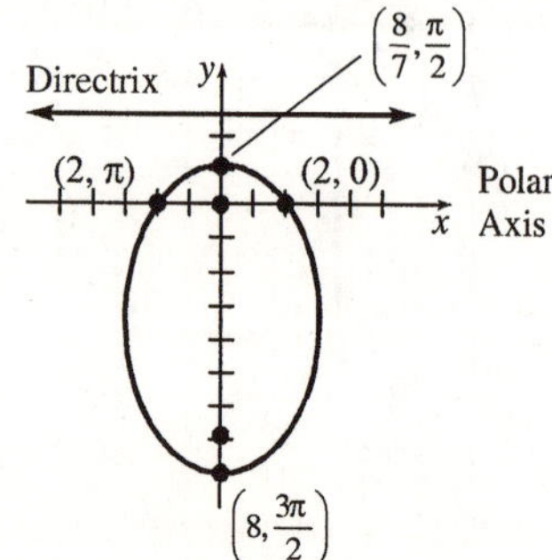

16. $r=\dfrac{10}{5+4\cos\theta}$

$r=\dfrac{10}{5\left(1+\dfrac{4}{5}\cos\theta\right)}=\dfrac{2}{1+\dfrac{4}{5}\cos\theta}$

$ep=2,\ e=\dfrac{4}{5},\ p=\dfrac{5}{2}$

Ellipse; directrix is perpendicular to the polar axis $\dfrac{5}{2}$ units to the right of the pole; vertices are $\left(\dfrac{10}{9},0\right)$ and $(10,\pi)$.

Also: $a=\dfrac{1}{2}\left(10+\dfrac{10}{9}\right)=\dfrac{50}{9}$ so the center is at $\left(10-\dfrac{50}{9},\pi\right)=\left(\dfrac{40}{9},\pi\right)$, and $c=\dfrac{40}{9}-0=\dfrac{40}{9}$ so that the second focus is at $\left(\dfrac{40}{9}+\dfrac{40}{9},\pi\right)=\left(\dfrac{80}{9},\pi\right)$.

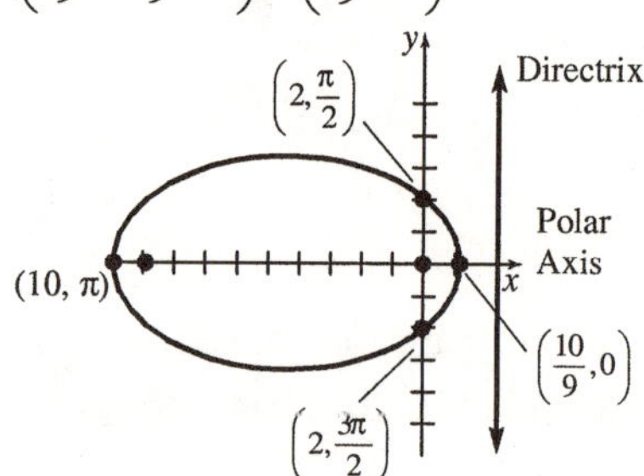

17. $r=\dfrac{9}{3-6\cos\theta}$

$r=\dfrac{9}{3(1-2\cos\theta)}=\dfrac{3}{1-2\cos\theta}$

$ep=3,\ e=2,\ p=\dfrac{3}{2}$

Hyperbola; directrix is perpendicular to the polar axis $\dfrac{3}{2}$ units to the left of the pole; vertices are $(-3,0)$ and $(1,\pi)$.

Also: $a=\dfrac{1}{2}(3-1)=1$ so the center is at $(1+1,\pi)=(2,\pi)$ [or $(-2,0)$], and $c=2-0=2$ so that the second focus is at $(2+2,\pi)=(4,\pi)$

[or $(-4,0)$].

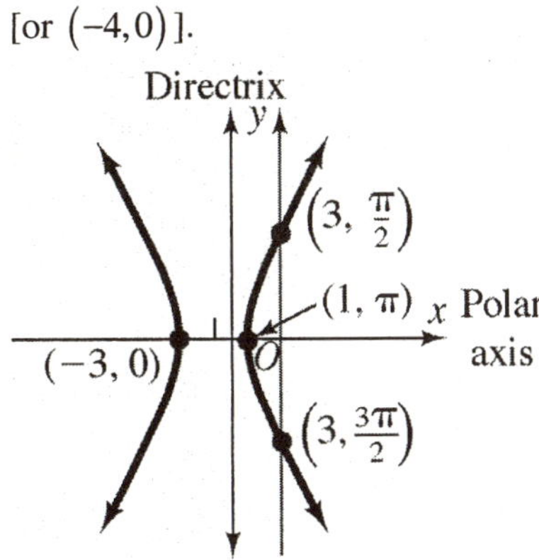

18. $r=\dfrac{12}{4+8\sin\theta}$

$r=\dfrac{12}{4(1+2\sin\theta)}=\dfrac{3}{1+2\sin\theta}$

$ep=3,\ e=2,\ p=\dfrac{3}{2}$

Hyperbola; directrix is parallel to the polar axis $\dfrac{3}{2}$ units above the pole; vertices are $\left(1,\dfrac{\pi}{2}\right)$ and $\left(-3,\dfrac{3\pi}{2}\right)$.

Also: $a=\dfrac{1}{2}(3-1)=1$ so the center is at $\left(1+1,\dfrac{\pi}{2}\right)=\left(2,\dfrac{\pi}{2}\right)$ [or $\left(-2,\dfrac{3\pi}{2}\right)$], and $c=2-0=2$ so that the second focus is at $\left(2+2,\dfrac{\pi}{2}\right)=\left(4,\dfrac{\pi}{2}\right)$ [or $\left(-4,\dfrac{3\pi}{2}\right)$].

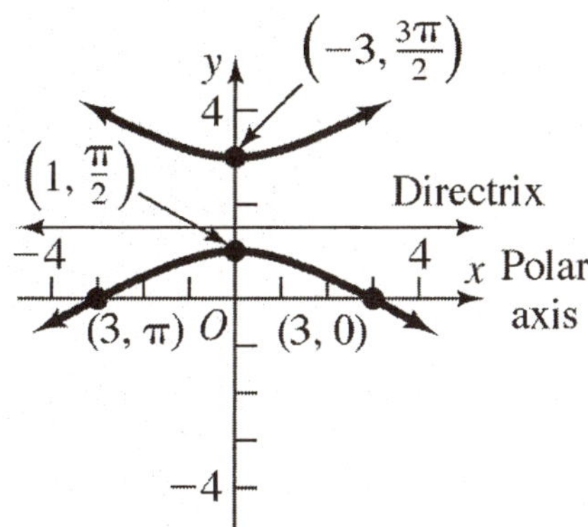

19. $r=\dfrac{8}{2-\sin\theta}$

$r=\dfrac{8}{2\left(1-\frac{1}{2}\sin\theta\right)}=\dfrac{4}{1-\frac{1}{2}\sin\theta}$

$ep=4,\ e=\dfrac{1}{2},\ p=8$

Ellipse; directrix is parallel to the polar axis 8 units below the pole; vertices are $\left(8,\dfrac{\pi}{2}\right)$ and $\left(\dfrac{8}{3},\dfrac{3\pi}{2}\right)$.

Also: $a=\dfrac{1}{2}\left(8+\dfrac{8}{3}\right)=\dfrac{16}{3}$ so the center is at $\left(8-\dfrac{16}{3},\dfrac{\pi}{2}\right)=\left(\dfrac{8}{3},\dfrac{\pi}{2}\right)$, and $c=\dfrac{8}{3}-0=\dfrac{8}{3}$ so that the second focus is at $\left(\dfrac{8}{3}+\dfrac{8}{3},\dfrac{\pi}{2}\right)=\left(\dfrac{16}{3},\dfrac{\pi}{2}\right)$.

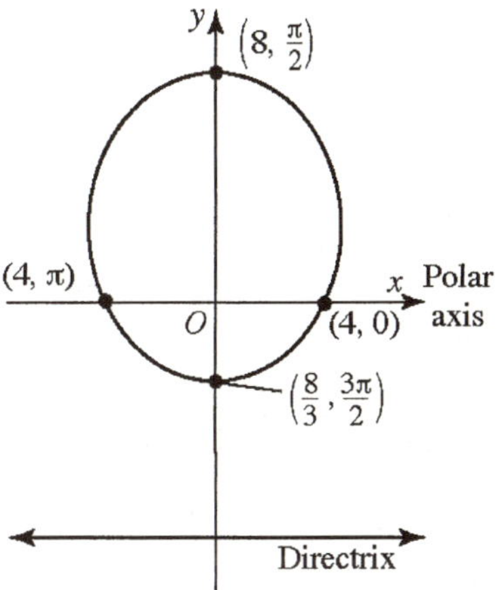

20. $r=\dfrac{8}{2+4\cos\theta}$

$r=\dfrac{8}{2(1+2\cos\theta)}=\dfrac{4}{1+2\cos\theta}$

$ep=4,\ e=2,\ p=2$

Hyperbola; directrix is perpendicular to the polar axis 2 units to the right of the pole; vertices are $\left(\dfrac{4}{3},0\right)$ and $(-4,\pi)$.

Also: $a=\dfrac{1}{2}\left(4-\dfrac{4}{3}\right)=\dfrac{4}{3}$ so the center is at $\left(\dfrac{4}{3}+\dfrac{4}{3},0\right)=\left(\dfrac{8}{3},0\right)$ [or $\left(-\dfrac{8}{3},\pi\right)$], and $c=\dfrac{8}{3}-0=\dfrac{8}{3}$ so that the second focus is at

$\left(\frac{8}{3}+\frac{8}{3},0\right)=\left(\frac{16}{3},0\right)$ [or $\left(-\frac{16}{3},\pi\right)$].

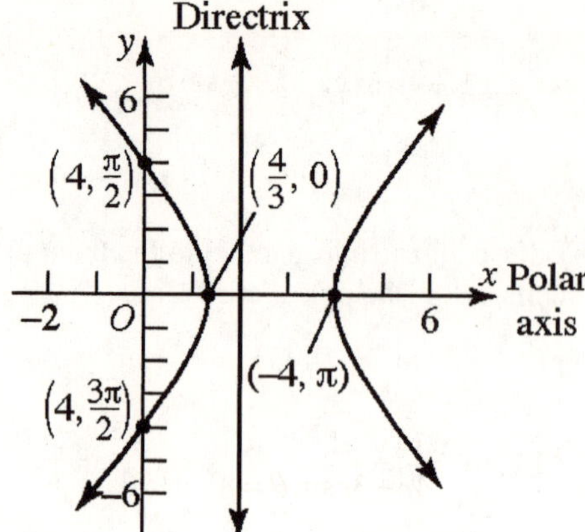

21. $r(3-2\sin\theta)=6 \Rightarrow r=\dfrac{6}{3-2\sin\theta}$

$$r=\frac{6}{3\left(1-\frac{2}{3}\sin\theta\right)}=\frac{2}{1-\frac{2}{3}\sin\theta}$$

$ep=2,\ e=\frac{2}{3},\ p=3$

Ellipse; directrix is parallel to the polar axis 3 units below the pole; vertices are $\left(6,\frac{\pi}{2}\right)$ and $\left(\frac{6}{5},\frac{3\pi}{2}\right)$.

Also: $a=\frac{1}{2}\left(6+\frac{6}{5}\right)=\frac{18}{5}$ so the center is at $\left(6-\frac{18}{5},\frac{\pi}{2}\right)=\left(\frac{12}{5},\frac{\pi}{2}\right)$, and $c=\frac{12}{5}-0=\frac{12}{5}$ so that the second focus is at $\left(\frac{12}{5}+\frac{12}{5},\frac{\pi}{2}\right)=\left(\frac{24}{5},\frac{\pi}{2}\right)$.

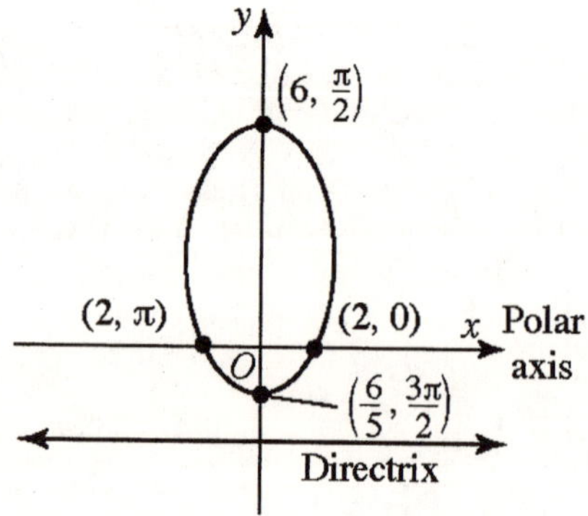

22. $r(2-\cos\theta)=2 \Rightarrow r=\dfrac{2}{2-\cos\theta}$

$$r=\frac{2}{2\left(1-\frac{1}{2}\cos\theta\right)}=\frac{1}{1-\frac{1}{2}\cos\theta}$$

$ep=1,\ e=\frac{1}{2},\ p=2$

Ellipse; directrix is perpendicular to the polar axis 2 units to the left of the pole; vertices are $(2,0)$ and $\left(\frac{2}{3},\pi\right)$.

Also: $a=\frac{1}{2}\left(2+\frac{2}{3}\right)=\frac{4}{3}$ so the center is at $\left(2-\frac{4}{3},0\right)=\left(\frac{2}{3},0\right)$, and $c=\frac{2}{3}-0=\frac{2}{3}$ so that the second focus is at $\left(\frac{2}{3}+\frac{2}{3},0\right)=\left(\frac{4}{3},0\right)$.

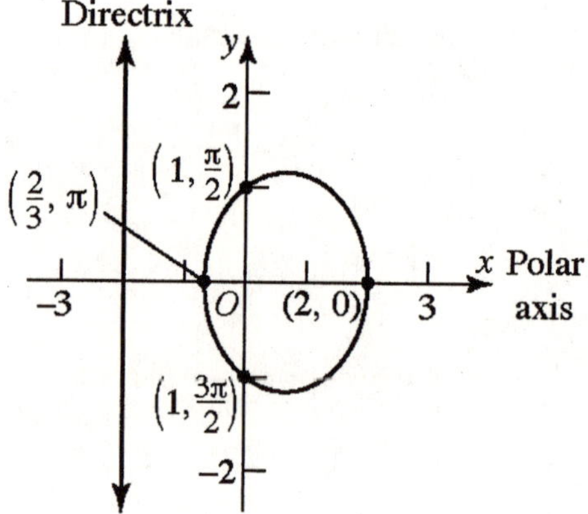

23. $r=\dfrac{6\sec\theta}{2\sec\theta-1}=\dfrac{6\left(\frac{1}{\cos\theta}\right)}{2\left(\frac{1}{\cos\theta}\right)-1}=\dfrac{\frac{6}{\cos\theta}}{\frac{2-\cos\theta}{\cos\theta}}$

$$=\left(\frac{6}{\cos\theta}\right)\left(\frac{\cos\theta}{2-\cos\theta}\right)=\frac{6}{2-\cos\theta}$$

$$r=\frac{6}{2\left(1-\frac{1}{2}\cos\theta\right)}=\frac{3}{1-\frac{1}{2}\cos\theta}$$

$ep=3,\ e=\frac{1}{2},\ p=6$

Ellipse; directrix is perpendicular to the polar axis 6 units to the left of the pole; vertices are $(6,0)$ and $(2,\pi)$.

Also: $a=\frac{1}{2}(6+2)=4$ so the center is at

$(6-4,0)=(2,0)$, and $c=2-0=2$ so that the second focus is at $(2+2,0)=(4,0)$.

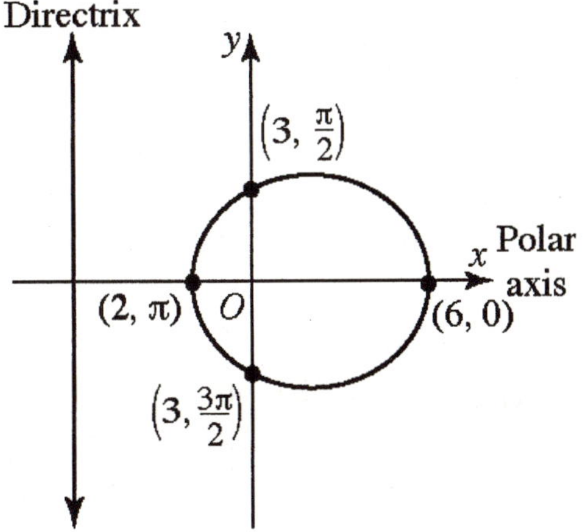

24. $r=\dfrac{3\csc\theta}{\csc\theta-1}=\dfrac{3\left(\frac{1}{\sin\theta}\right)}{\frac{1}{\sin\theta}-1}=\dfrac{\frac{3}{\sin\theta}}{\frac{1-\sin\theta}{\sin\theta}}$

$$=\left(\frac{3}{\sin\theta}\right)\left(\frac{\sin\theta}{1-\sin\theta}\right)=\frac{3}{1-\sin\theta}$$

$ep=3,\ e=1,\ p=3$

Parabola; directrix is parallel to the polar axis 3 units below the pole; vertex is $\left(\frac{3}{2},\frac{3\pi}{2}\right)$.

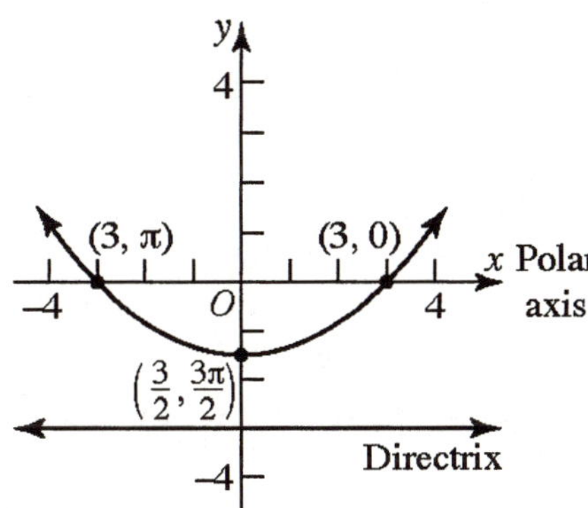

25. $r=\dfrac{1}{1+\cos\theta}$

$$r+r\cos\theta=1$$
$$r=1-r\cos\theta$$
$$r^2=(1-r\cos\theta)^2$$
$$x^2+y^2=(1-x)^2$$
$$x^2+y^2=1-2x+x^2$$
$$y^2+2x-1=0$$

26. $r=\dfrac{3}{1-\sin\theta}$

$$r-r\sin\theta=3$$
$$r=3+r\sin\theta$$
$$r^2=(3+r\sin\theta)^2$$
$$x^2+y^2=(3+y)^2$$
$$x^2+y^2=9+6y+y^2$$
$$x^2-6y-9=0$$

27. $r=\dfrac{8}{4+3\sin\theta}$

$$4r+3r\sin\theta=8$$
$$4r=8-3r\sin\theta$$
$$16r^2=(8-3r\sin\theta)^2$$
$$16(x^2+y^2)=(8-3y)^2$$
$$16x^2+16y^2=64-48y+9y^2$$
$$16x^2+7y^2+48y-64=0$$

28. $r=\dfrac{10}{5+4\cos\theta}$

$$5r+4r\cos\theta=10$$
$$5r=10-4r\cos\theta$$
$$25r^2=(10-4r\cos\theta)^2$$
$$25(x^2+y^2)=(10-4x)^2$$
$$25x^2+25y^2=100-80x+16x^2$$
$$9x^2+25y^2+80x-100=0$$

29. $r=\dfrac{9}{3-6\cos\theta}$

$$3r-6r\cos\theta=9$$
$$3r=9+6r\cos\theta$$
$$r=3+2r\cos\theta$$
$$r^2=(3+2r\cos\theta)^2$$
$$x^2+y^2=(3+2x)^2$$
$$x^2+y^2=9+12x+4x^2$$
$$3x^2-y^2+12x+9=0$$

30. $r=\dfrac{12}{4+8\sin\theta}$

$$4r+8r\sin\theta=12$$
$$4r=12-8r\sin\theta$$
$$r=3-2r\sin\theta$$
$$r^2=(3-2r\sin\theta)^2$$
$$x^2+y^2=(3-2y)^2$$
$$x^2+y^2=9-12y+4y^2$$
$$x^2-3y^2+12y-9=0$$

31. $r=\dfrac{8}{2-\sin\theta}$

$$2r-r\sin\theta=8$$
$$2r=8+r\sin\theta$$
$$4r^2=(8+r\sin\theta)^2$$
$$4(x^2+y^2)=(8+y)^2$$
$$4x^2+4y^2=64+16y+y^2$$
$$4x^2+3y^2-16y-64=0$$

32. $r=\dfrac{8}{2+4\cos\theta}$

$$2r+4r\cos\theta=8$$
$$2r=8-4r\cos\theta$$
$$r=4-2r\cos\theta$$
$$r^2=(4-2r\cos\theta)^2$$
$$x^2+y^2=(4-2x)^2$$
$$x^2+y^2=16-16x+4x^2$$
$$3x^2-y^2-16x+16=0$$

33. $r(3-2\sin\theta)=6$

$$3r-2r\sin\theta=6$$
$$3r=6+2r\sin\theta$$
$$9r^2=(6+2r\sin\theta)^2$$
$$9(x^2+y^2)=(6+2y)^2$$
$$9x^2+9y^2=36+24y+4y^2$$
$$9x^2+5y^2-24y-36=0$$

34. $r(2-\cos\theta)=2$

$$2r-r\cos\theta=2$$
$$2r=2+r\cos\theta$$
$$4r^2=(2+r\cos\theta)^2$$
$$4(x^2+y^2)=(2+x)^2$$
$$4x^2+4y^2=4+4x+x^2$$
$$3x^2+4y^2-4x-4=0$$

35. $r=\dfrac{6\sec\theta}{2\sec\theta-1}$

$$r=\frac{6}{2-\cos\theta}$$
$$2r-r\cos\theta=6$$
$$2r=6+r\cos\theta$$
$$4r^2=(6+r\cos\theta)^2$$
$$4(x^2+y^2)=(6+x)^2$$
$$4x^2+4y^2=36+12x+x^2$$
$$3x^2+4y^2-12x-36=0$$

36. $r=\dfrac{3\csc\theta}{\csc\theta-1}$

$$r=\frac{3}{1-\sin\theta}$$
$$r-r\sin\theta=3$$
$$r=3+r\sin\theta$$
$$r^2=(3+r\sin\theta)^2$$
$$x^2+y^2=(3+y)^2$$
$$x^2+y^2=9+6y+y^2$$
$$x^2-6y-9=0$$

37. $r=\dfrac{ep}{1+e\sin\theta}$

$e=1;\quad p=1$

$$r=\frac{1}{1+\sin\theta}$$

38. $r=\dfrac{ep}{1-e\sin\theta}$

$e=1;\quad p=2$

$$r=\frac{2}{1-\sin\theta}$$

39. $r = \dfrac{ep}{1 - e\cos\theta}$

$e = \dfrac{4}{5};\quad p = 3$

$r = \dfrac{\frac{12}{5}}{1 - \frac{4}{5}\cos\theta} = \dfrac{12}{5 - 4\cos\theta}$

40. $r = \dfrac{ep}{1 + e\sin\theta}$

$e = \dfrac{2}{3};\quad p = 3$

$r = \dfrac{2}{1 + \frac{2}{3}\sin\theta} = \dfrac{6}{3 + 2\sin\theta}$

41. $r = \dfrac{ep}{1 - e\sin\theta}$

$e = 6;\quad p = 2$

$r = \dfrac{12}{1 - 6\sin\theta}$

42. $r = \dfrac{ep}{1 + e\cos\theta}$

$e = 5;\quad p = 5$

$r = \dfrac{25}{1 + 5\cos\theta}$

43. Consider the following diagram:

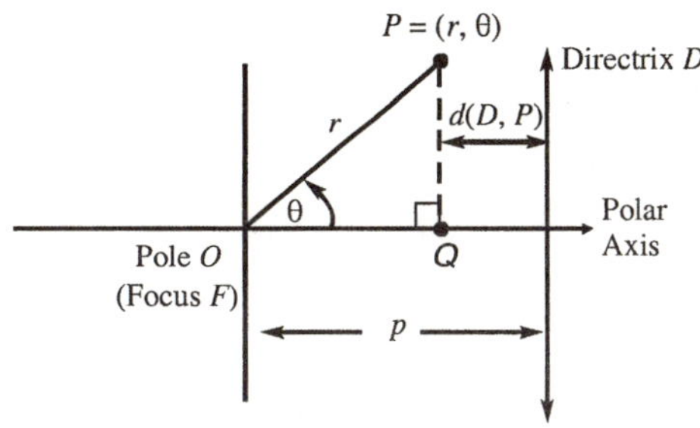

$$d(F, P) = e \cdot d(D, P)$$
$$d(D, P) = p - r\cos\theta$$
$$r = e(p - r\cos\theta)$$
$$r = ep - er\cos\theta$$
$$r + er\cos\theta = ep$$
$$r(1 + e\cos\theta) = ep$$
$$r = \frac{ep}{1 + e\cos\theta}$$

44. Consider the following diagram:

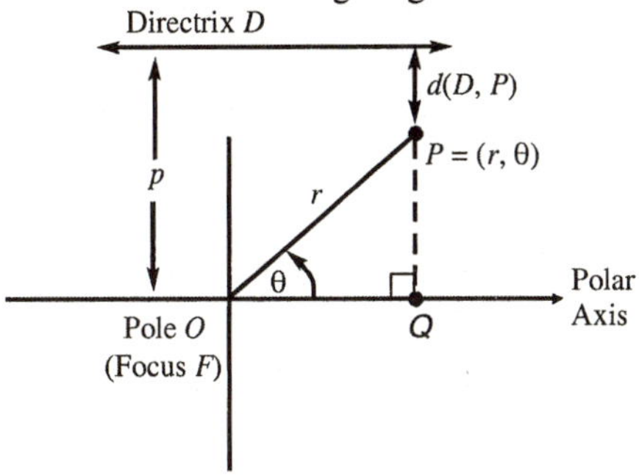

$$d(F, P) = e \cdot d(D, P)$$
$$d(D, P) = p - r\sin\theta$$
$$r = e(p - r\sin\theta)$$
$$r = ep - er\sin\theta$$
$$r + er\sin\theta = ep$$
$$r(1 + e\sin\theta) = ep$$
$$r = \frac{ep}{1 + e\sin\theta}$$

45. Consider the following diagram:

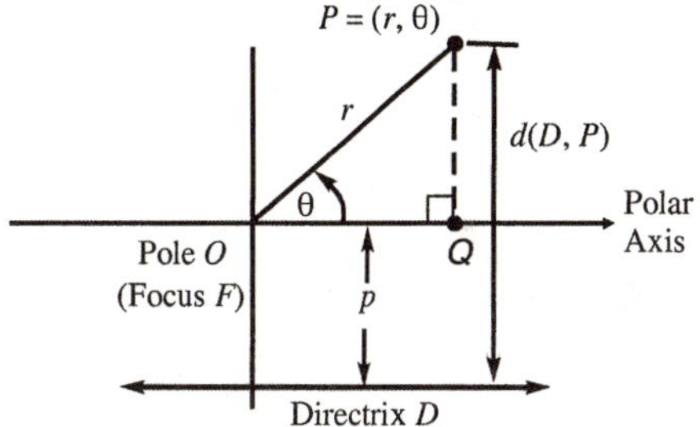

$$d(F, P) = e \cdot d(D, P)$$
$$d(D, P) = p + r\sin\theta$$
$$r = e(p + r\sin\theta)$$
$$r = ep + er\sin\theta$$
$$r - er\sin\theta = ep$$
$$r(1 - e\sin\theta) = ep$$
$$r = \frac{ep}{1 - e\sin\theta}$$

46. $r = \dfrac{(3.442)10^7}{1-0.206\cos\theta}$

At aphelion, the greatest distance from the sun, $\cos\theta = 1$.

$r = \dfrac{(3.442)10^7}{1-0.206(1)} = \dfrac{(3.442)10^7}{0.794}$

$\approx 4.335\times10^7$ miles

At perihelion, the shortest distance from the sun, $\cos\theta = -1$.

$r = \dfrac{(3.442)10^7}{1-0.206(-1)} = \dfrac{(3.442)10^7}{1.206}$

$\approx 2.854\times10^7$ miles

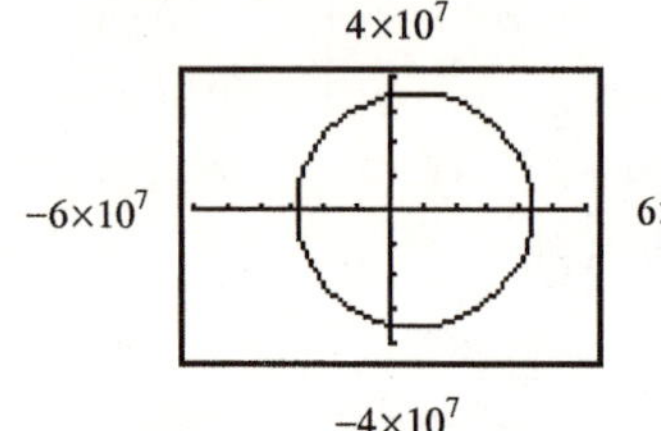

Section 9.7

1. $|3| = 3;\ \dfrac{2\pi}{4} = \dfrac{\pi}{2}$
2. plane curve; parameter
3. ellipse
4. cycloid
5. False; for example: $x = 2\cos t$, $y = 3\sin t$ define the same curve as in problem 3.
6. True
7. $x = 3t+2,\ y = t+1,\ 0 \le t \le 4$

   $x = 3(y-1)+2$

   $x = 3y-3+2$

   $x = 3y-1$

   $x-3y+1 = 0$

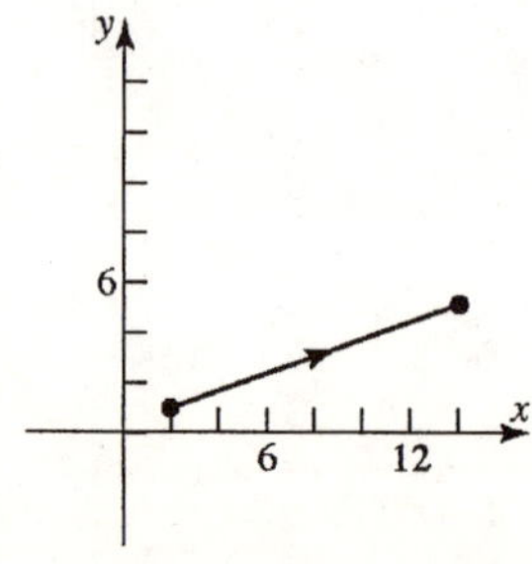

8. $x = t-3,\ y = 2t+4,\ 0 \le t \le 2$

   $y = 2(x+3)+4$

   $y = 2x+6+4$

   $y = 2x+10$

   $2x-y+10 = 0$

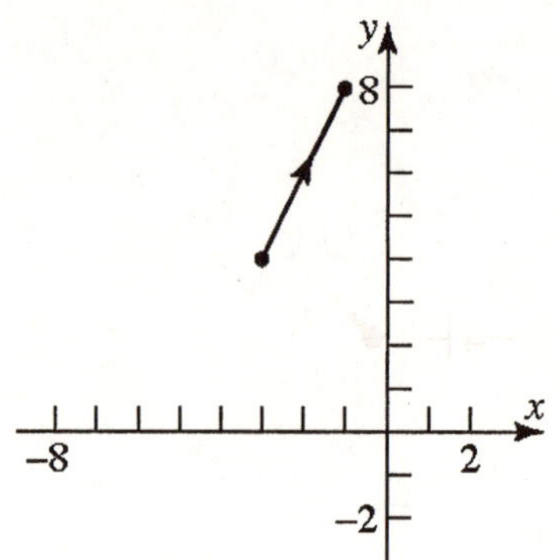

9. $x = t+2,\ y = \sqrt{t},\ t \ge 0$

   $y = \sqrt{x-2}$

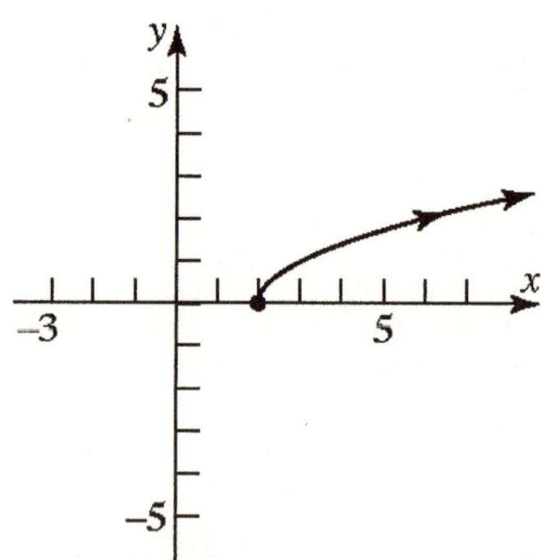

10. $x = \sqrt{2t},\ y = 4t,\ t \ge 0$

    $y = 4\left(\dfrac{x^2}{2}\right) = 2x^2$

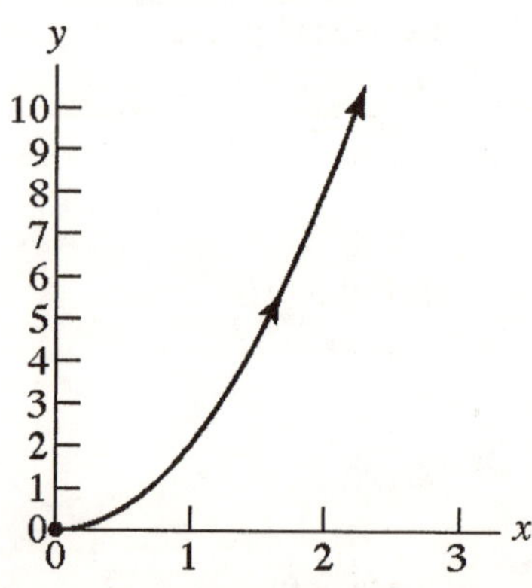

11. $x = t^2 + 4,\ y = t^2 - 4,\ -\infty < t < \infty$

$y = (x-4) - 4$

$y = x - 8$

For $-\infty < t < 0$ the movement is to the left. For $0 < t < \infty$ the movement is to the right.

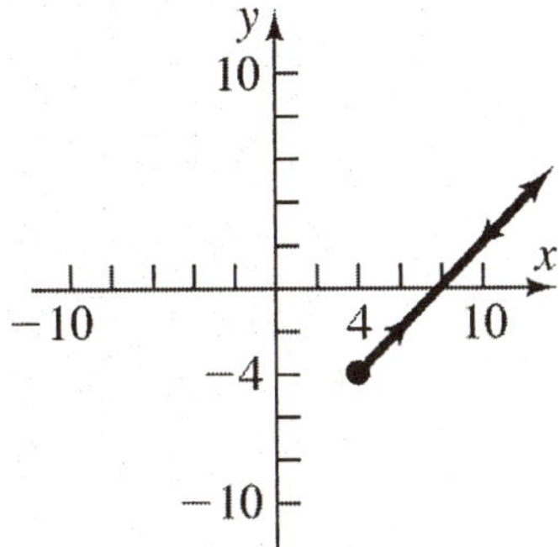

12. $x = \sqrt{t} + 4,\ y = \sqrt{t} - 4,\ t \ge 0$

$y = x - 4 - 4 = x - 8$

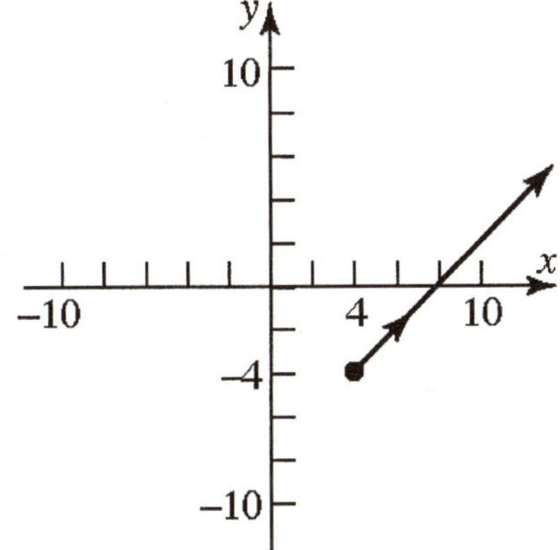

13. $x = 3t^2,\ y = t + 1,\ -\infty < t < \infty$

$x = 3(y-1)^2$

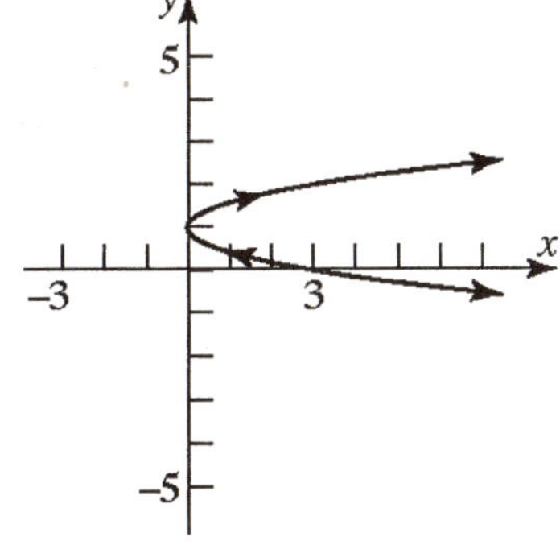

14. $x = 2t - 4,\ y = 4t^2,\ -\infty < t < \infty$

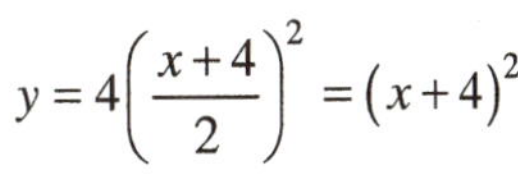

$y = 4\left(\frac{x+4}{2}\right)^2 = (x+4)^2$

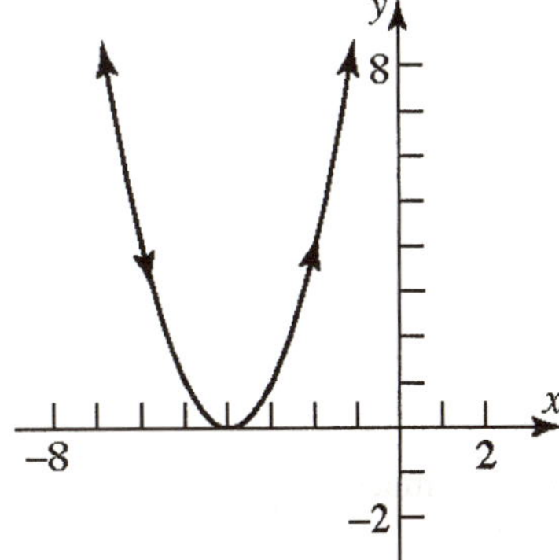

15. $x = 2e^t,\ y = 1 + e^t,\ t \ge 0$

$y = 1 + \frac{x}{2}$

$2y = 2 + x$

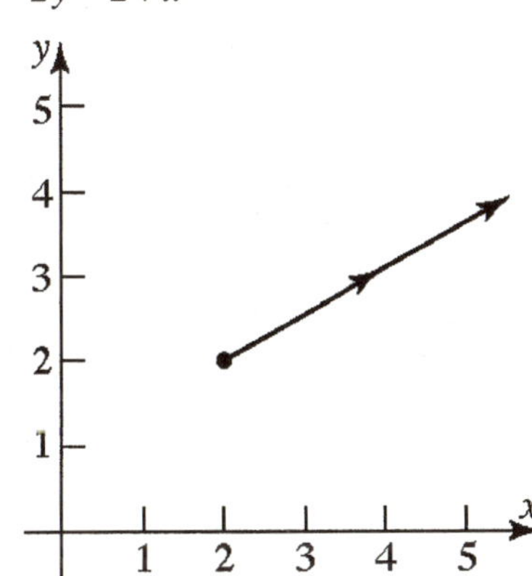

16. $x = e^t,\ y = e^{-t},\ t \ge 0$

$y = x^{-1} = \frac{1}{x}$

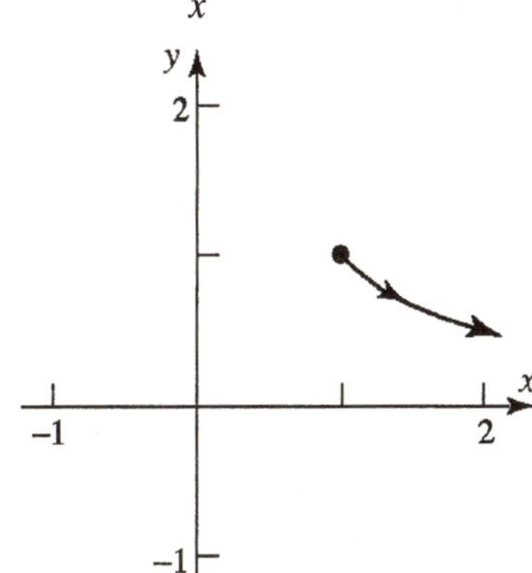

17. $x=\sqrt{t},\ y=t^{3/2}\ ,\ t\ge 0$

$y=\left(x^2\right)^{3/2}$

$y=x^3$

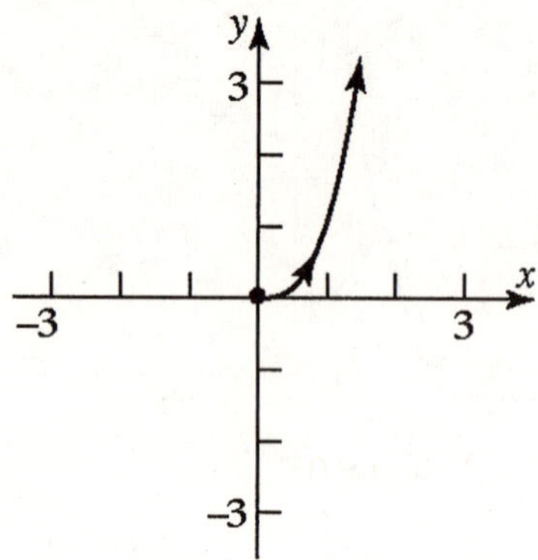

18. $x=t^{3/2}+1,\ y=\sqrt{t}\ ,\ t\ge 0$

$x=\left(y^2\right)^{3/2}+1$

$x=y^3+1$

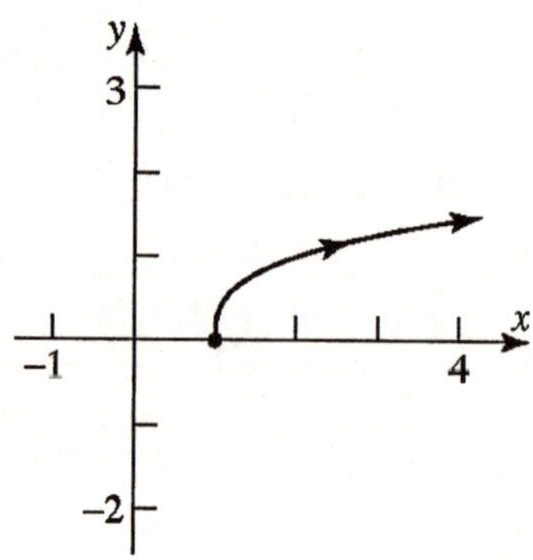

19. $x=2\cos t,\ y=3\sin t\ ,\ 0\le t\le 2\pi$

$$\frac{x}{2}=\cos t \qquad \frac{y}{3}=\sin t$$

$$\left(\frac{x}{2}\right)^2+\left(\frac{y}{3}\right)^2=\cos^2 t+\sin^2 t=1$$

$$\frac{x^2}{4}+\frac{y^2}{9}=1$$

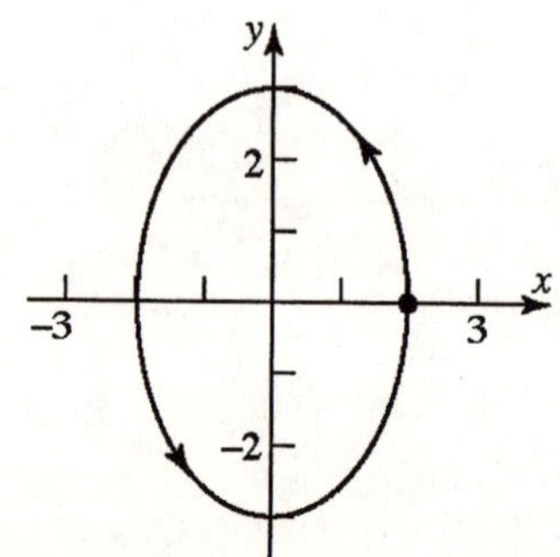

20. $x=2\cos t,\ y=3\sin t\ ,\ 0\le t\le \pi$

$$\frac{x}{2}=\cos t \qquad \frac{y}{3}=\sin t$$

$$\left(\frac{x}{2}\right)^2+\left(\frac{y}{3}\right)^2=\cos^2 t+\sin^2 t=1$$

$$\frac{x^2}{4}+\frac{y^2}{9}=1 \qquad y\ge 0$$

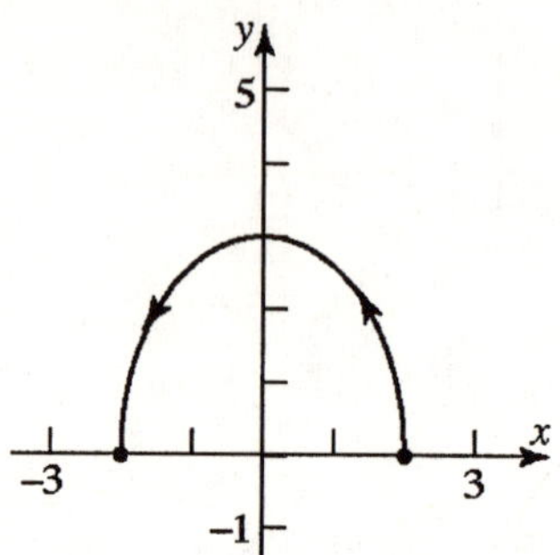

21. $x=2\cos t,\ y=3\sin t\ ,\ -\pi\le t\le 0$

$$\frac{x}{2}=\cos t \qquad \frac{y}{3}=\sin t$$

$$\left(\frac{x}{2}\right)^2+\left(\frac{y}{3}\right)^2=\cos^2 t+\sin^2 t=1$$

$$\frac{x^2}{4}+\frac{y^2}{9}=1 \qquad y\le 0$$

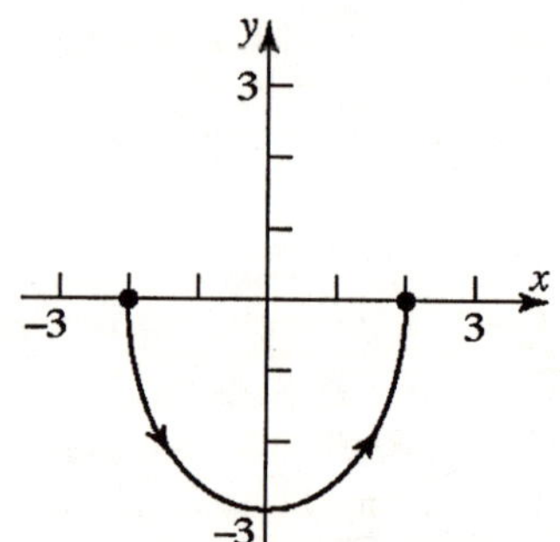

22. $x = 2\cos t,\ y = \sin t,\ 0 \le t \le \dfrac{\pi}{2}$

$$\frac{x}{2} = \cos t \qquad y = \sin t$$

$$\left(\frac{x}{2}\right)^2 + (y)^2 = \cos^2 t + \sin^2 t = 1$$

$$\frac{x^2}{4} + y^2 = 1 \quad x \ge 0,\ y \ge 0$$

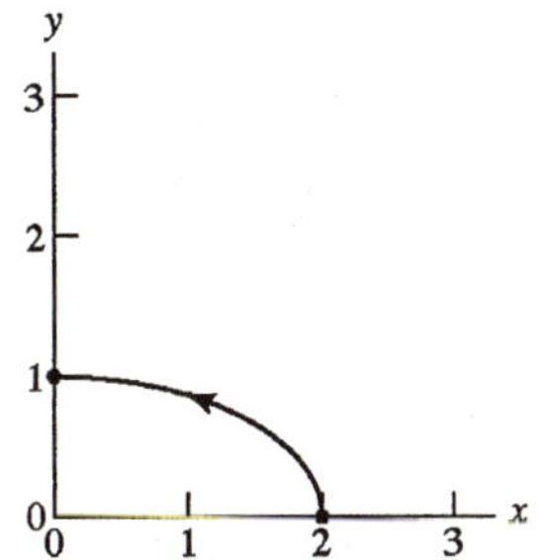

23. $x = \sec t,\ y = \tan t,\ 0 \le t \le \dfrac{\pi}{4}$

$$\sec^2 t = 1 + \tan^2 t$$

$$x^2 = 1 + y^2$$

$$x^2 - y^2 = 1 \quad 1 \le x \le \sqrt{2},\ 0 \le y \le 1$$

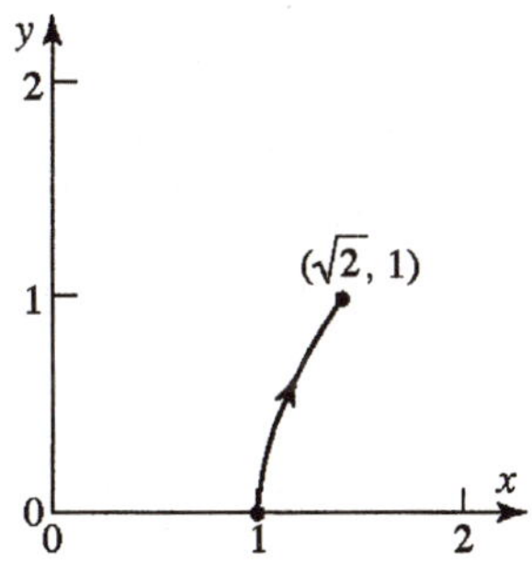

24. $x = \csc t,\ y = \cot t,\ \dfrac{\pi}{4} \le t \le \dfrac{\pi}{2}$

$$\csc^2 t = 1 + \cot^2 t$$

$$x^2 = 1 + y^2$$

$$x^2 - y^2 = 1 \quad 1 \le x \le \sqrt{2},\ 0 \le y \le 1$$

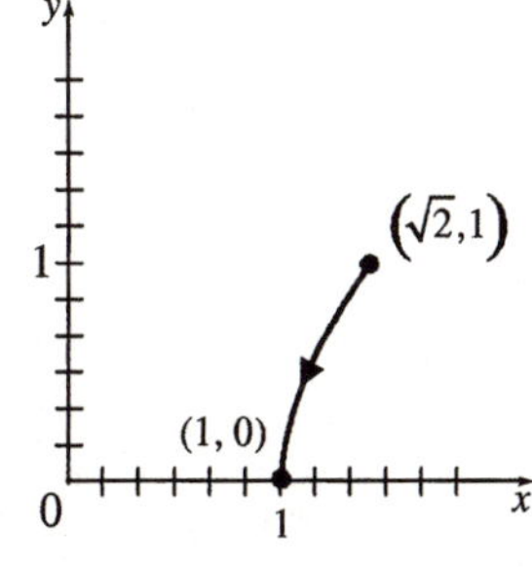

25. $x = \sin^2 t,\ y = \cos^2 t,\ 0 \le t \le 2\pi$

$$\sin^2 t + \cos^2 t = 1$$

$$x + y = 1$$

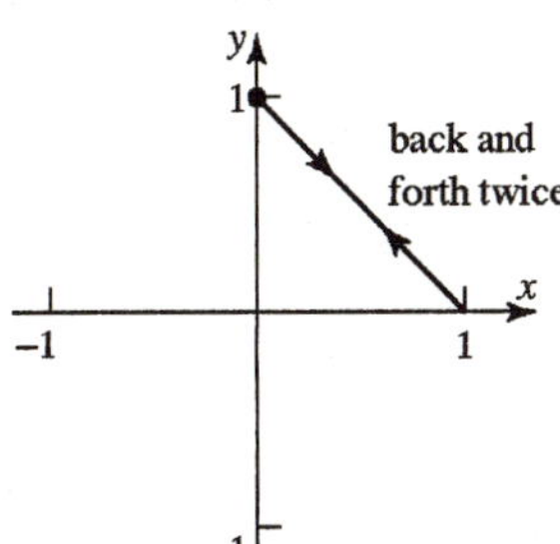

26. $x = t^2,\ y = \ln t,\ t > 0$

$$y = \ln\sqrt{x}$$

$$y = \frac{1}{2}\ln x$$

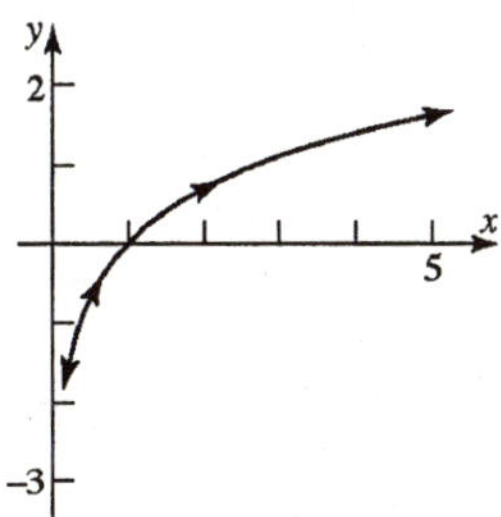

27. $x = t,\ y = 4t - 1 \qquad x = t + 1,\ y = 4t + 3$

28. $x = t,\ y = -8t + 3 \qquad x = t + 1,\ y = -8t - 5$

29. $x = t,\ y = t^2 + 1 \qquad x = t - 1,\ y = t^2 - 2t + 2$

30. $x = t,\ y = -2t^2 + 1$

$x = t - 1,\ y = -2t^2 + 4t - 1$

31. $x = t,\ y = t^3 \qquad x = \sqrt[3]{t},\ y = t$

32. $x = t,\ y = t^4 + 1 \qquad x = \sqrt[4]{t - 1},\ y = t$

33. $x = t^{3/2},\ y = t \qquad x = t,\ y = t^{2/3}$

34. $x = \sqrt{t},\ y = t \qquad x = t,\ y = t^2$

35. $x = t + 2,\ y = t;\ 0 \le t \le 5$

36. $x = t,\ y = -t + 1;\ -1 \le t \le 3$

37. $x = 3\cos t,\ y = 2\sin t;\ 0 \le t \le 2\pi$

38. $x = \cos t,\ y = 4\sin t;\ \ -\frac{\pi}{2} \le t \le \frac{\pi}{2}$

39. Since the motion begins at (2, 0), we want $x = 2$ and $y = 0$ when $t = 0$. For the motion to be clockwise, we must have x positive and y negative initially.

$x = 2\cos(\omega t),\ y = -3\sin(\omega t)$

$\frac{2\pi}{\omega} = 2 \Rightarrow \omega = \pi$

$x = 2\cos(\pi t),\ y = -3\sin(\pi t),\ \ 0 \le t \le 2$

40. Since the motion begins at (0, 3), we want $x = 0$ and $y = 3$ when $t = 0$. For the motion to be counter-clockwise, we need x negative and y positive initially.

$x = -2\sin(\omega t),\ y = 3\cos(\omega t)$

$\frac{2\pi}{\omega} = 1 \Rightarrow \omega = 2\pi$

$x = -2\sin(2\pi t),\ y = 3\cos(2\pi t),\ \ 0 \le t \le 1$

41. Since the motion begins at (0, 3), we want $x = 0$ and $y = 3$ when $t = 0$. For the motion to be clockwise, we need x positive and y positive initially.

$x = 2\sin(\omega t),\ y = 3\cos(\omega t)$

$\frac{2\pi}{\omega} = 1 \Rightarrow \omega = 2\pi$

$x = 2\sin(2\pi t),\ y = 3\cos(2\pi t),\ \ 0 \le t \le 1$

42. Since the motion begins at (2, 0), we want $x = 2$ and $y = 0$ when $t = 0$. For the motion to be counter-clockwise, we need x positive and y positive initially.

$x = 2\cos(\omega t),\ y = 3\sin(\omega t)$

$\frac{2\pi}{\omega} = 3 \Rightarrow \omega = \frac{2\pi}{3}$

$x = 2\cos\left(\frac{2\pi}{3}t\right),\ y = 3\sin\left(\frac{2\pi}{3}t\right),\ \ 0 \le t \le 3$

43. C_1

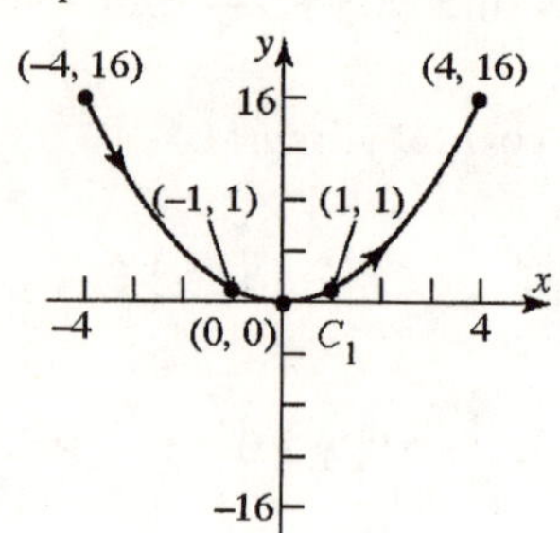

C_2

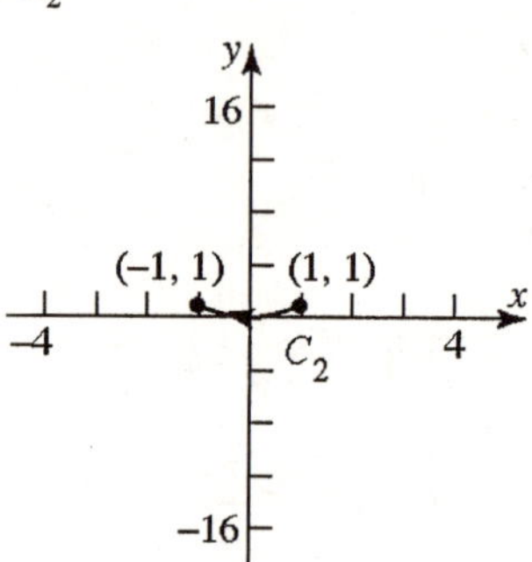

C_3

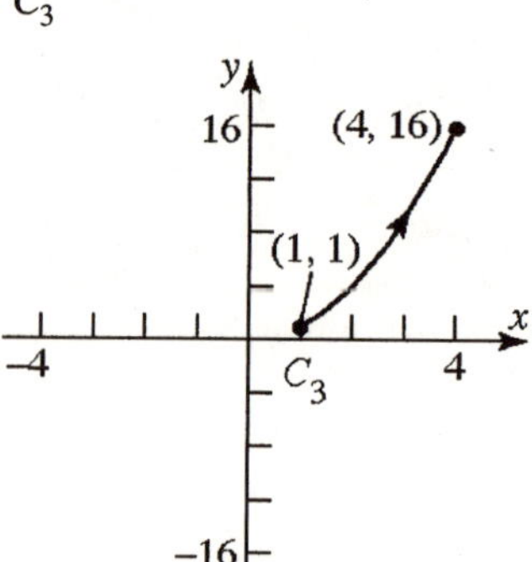

C_4

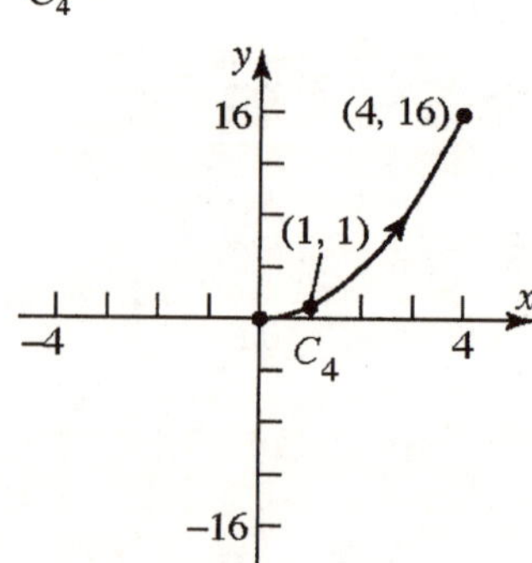

44. C_1

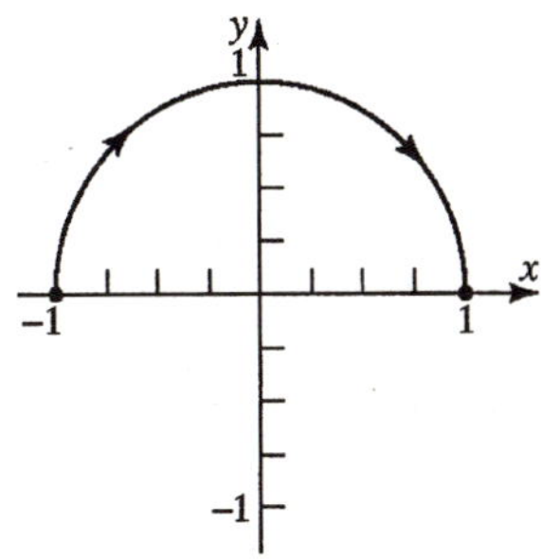

C_2

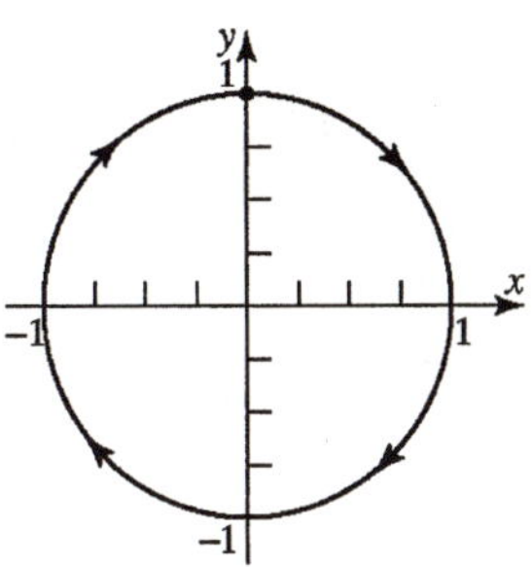

C_3

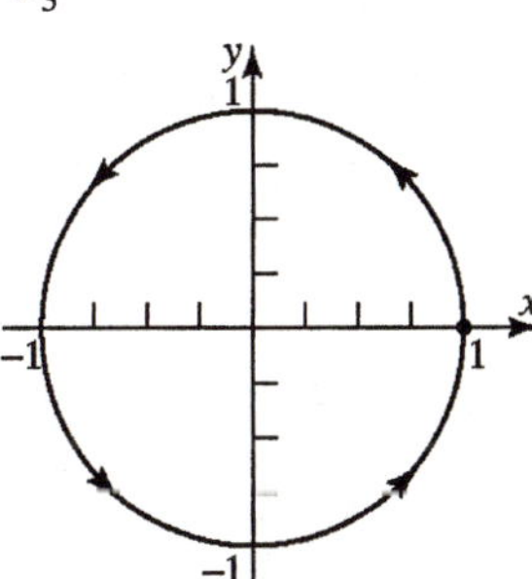

C_4

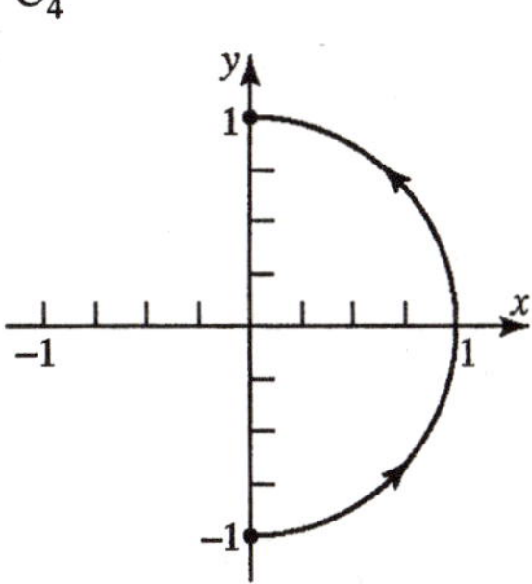

45. $x = t\sin t,\ y = t\cos t$

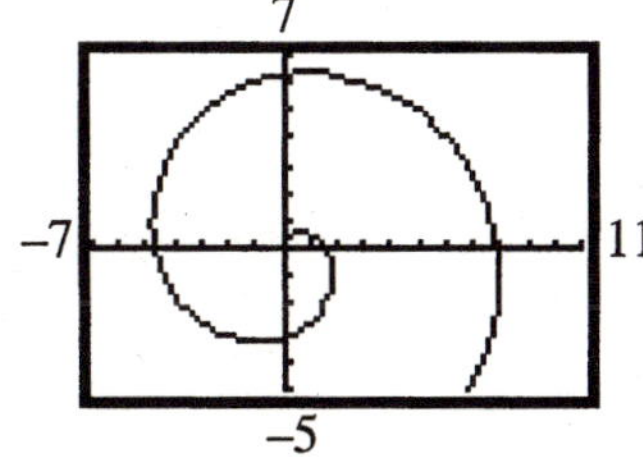

46. $x = \sin t + \cos t,\ y = \sin t - \cos t$

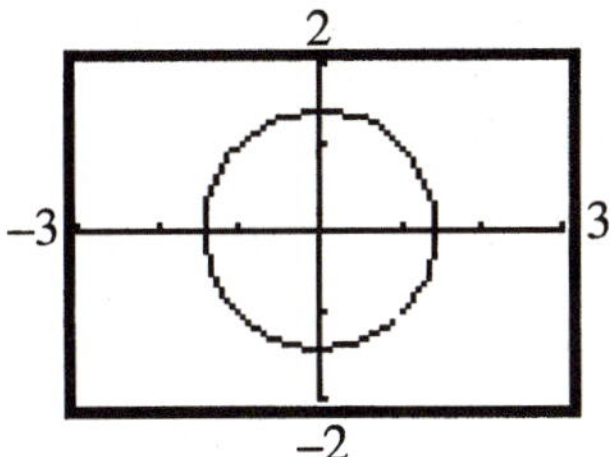

47. $x = 4\sin t - 2\sin(2t)$
$y = 4\cos t - 2\cos(2t)$

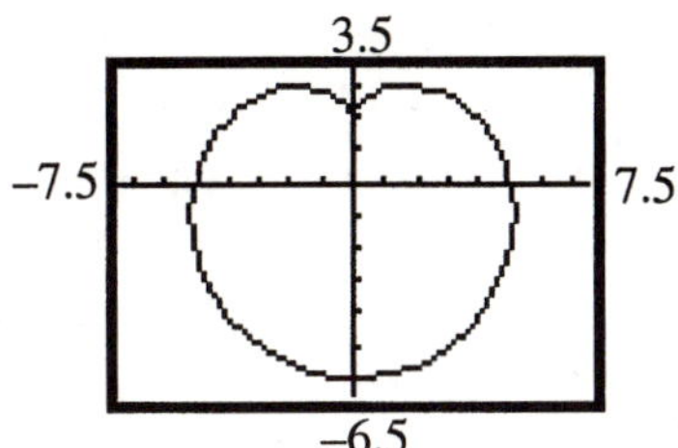

48. $x = 4\sin t + 2\sin(2t)$
$y = 4\cos t + 2\cos(2t)$

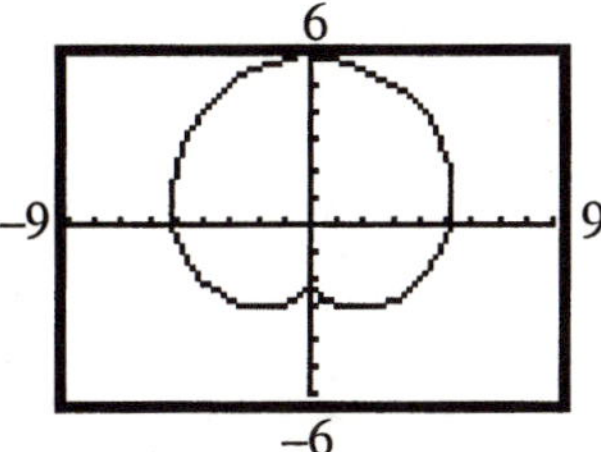

49. a. Use equations (2):

$$x = (50\cos 90°)t = 0$$

$$y = -\frac{1}{2}(32)t^2 + (50\sin 90°)t + 6$$
$$= -16t^2 + 50t + 6$$

b. The ball is in the air until $y = 0$. Solve:

$$-16t^2 + 50t + 6 = 0$$

$$t = \frac{-50 \pm \sqrt{50^2 - 4(-16)(6)}}{2(-16)}$$
$$= \frac{-50 \pm \sqrt{2884}}{-32}$$
$$\approx -0.12 \text{ or } 3.24$$

The ball is in the air for about 3.24 seconds. (The negative solution is extraneous.)

c. The maximum height occurs at the vertex of the quadratic function.

$$t = -\frac{b}{2a} = -\frac{50}{2(-16)} = 1.5625 \text{ seconds}$$

Evaluate the function to find the maximum height:

$-16(1.5625)^2 + 50(1.5625) + 6 = 45.0625$

The maximum height is 45.0625 feet.

d. We use $x = 3$ so that the line is not on top of the y-axis.

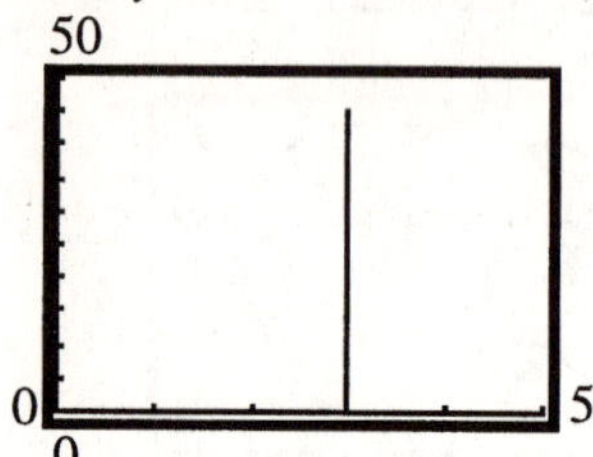

50. a. Use equations (2):

$x = (40\cos 90°)t = 0$

$$y = -\frac{1}{2}(32)t^2 + (40\sin 90°)t + 5$$
$$= -16t^2 + 40t + 5$$

b. The ball is in the air until $y = 0$. Solve:

$-16t^2 + 40t + 5 = 0$

$$t = \frac{-40 \pm \sqrt{40^2 - 4(-16)(5)}}{2(-16)}$$
$$= \frac{-40 \pm \sqrt{1920}}{-32} \approx -0.12 \text{ or } 2.62$$

The ball is in the air for about 2.62 seconds. (The negative solution is extraneous.)

c. The maximum height occurs at the vertex of the quadratic function.

$$t = -\frac{b}{2a} = -\frac{40}{2(-16)} = 1.25 \text{ seconds}$$

Evaluate the function to find the maximum height:

$-16(1.25)^2 + 40(1.25) + 5 = 30$

The maximum height is 30 feet.

d. We use $x = 3$ so that the line is not on top of the y-axis.

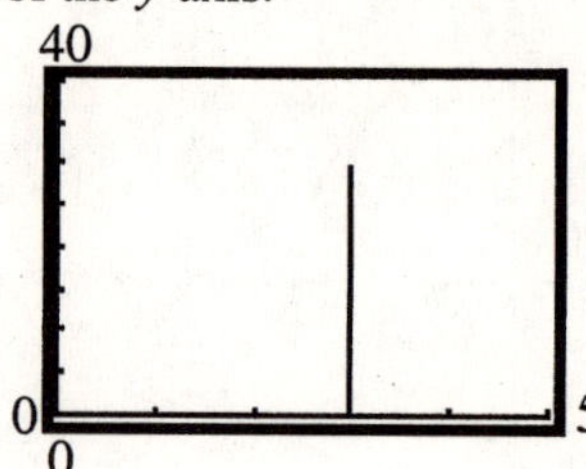

51. Let $y_1 = 1$ be the train's path and $y_2 = 5$ be Bill's path.

a. Train: Using the hint,

$$x_1 = \frac{1}{2}(2)t^2 = t^2$$

$y_1 = 1$

Bill:

$x_2 = 5(t-5)$

$y_2 = 5$

b. Bill will catch the train if $x_1 = x_2$.

$$t^2 = 5(t-5)$$
$$t^2 = 5t - 25$$

$t^2 - 5t + 25 = 0$

Since

$$b^2 - 4ac = (-5)^2 - 4(1)(25) = 25 - 100 = -75 < 0,$$

the equation has no real solution. Thus, Bill will not catch the train.

c.

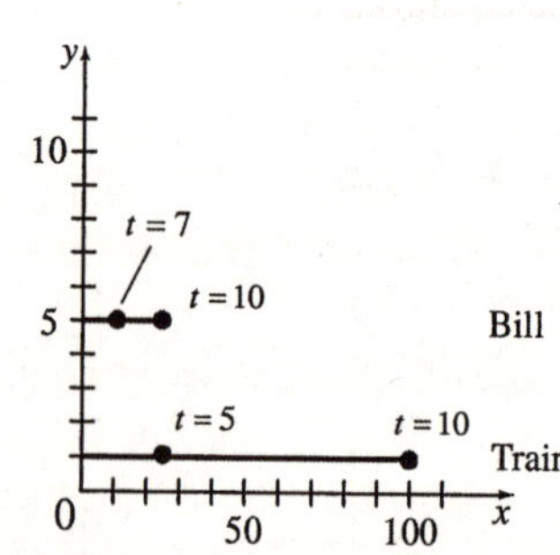

52. Let $y_1 = 1$ be the bus's path and $y_2 = 5$ be Jodi's path.

a. Bus: Using the hint,

$x_1 = \frac{1}{2}(3)t^2 = 1.5t^2$

$y_1 = 1$

Jodi:

$x_2 = 5(t-2)$

$y_2 = 5$

b. Jodi will catch the bus if $x_1 = x_2$.

$1.5t^2 = 5(t-2)$

$1.5t^2 = 5t - 10$

$1.5t^2 - 5t + 10 = 0$

Since

$b^2 - 4ac = (-5)^2 - 4(1.5)(10)$,

$= 25 - 60 = -35 < 0$

the equation has no real solution. Thus, Jodi will not catch the bus.

c.

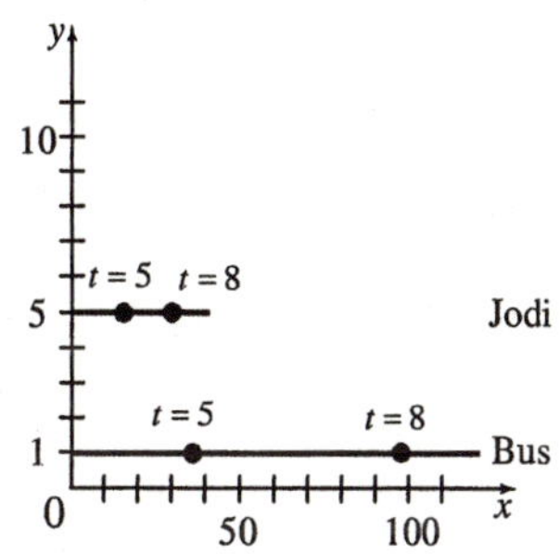

53. a. Use equations (2):

$x = (145\cos 20°)t$

$y = -\frac{1}{2}(32)t^2 + (145\sin 20°)t + 5$

b. The ball is in the air until $y = 0$. Solve:

$-16t^2 + (145\sin 20°)t + 5 = 0$

$$t = \frac{-145\sin 20° \pm \sqrt{(145\sin 20°)^2 - 4(-16)(5)}}{2(-16)}$$

≈ -0.10 or 3.20

The ball is in the air for about 3.20 seconds. (The negative solution is extraneous.)

c. The maximum height occurs at the vertex of the quadratic function.

$t = -\frac{b}{2a} = -\frac{145\sin 20°}{2(-16)} \approx 1.55$ seconds

Evaluate the function to find the maximum height:

$-16(1.55)^2 + (145\sin 20°)(1.55) + 5 \approx 43.43$

The maximum height is about 43.43 feet.

d. Find the horizontal displacement:

$x = (145\cos 20°)(3.20) \approx 436$ feet

e.

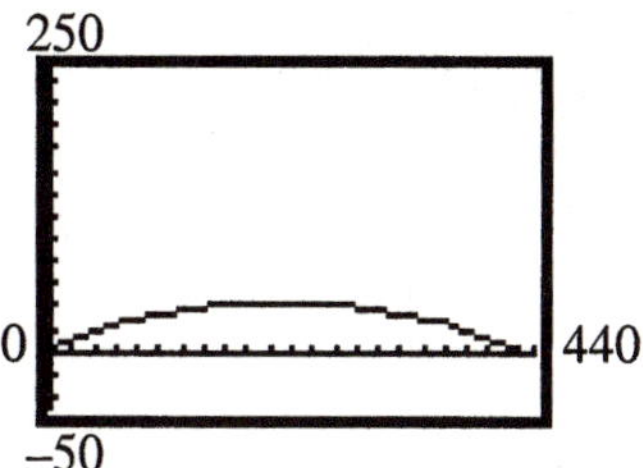

54. a. Use equations (2):

$x = (125\cos 40°)t$

$y = -16t^2 + (125\sin 40°)t + 3$

The ball is in the air until $y = 0$. Solve:

$-16t^2 + (125\sin 40°)t + 3 = 0$

$$t = \frac{-125\sin 40° \pm \sqrt{(125\sin 40°)^2 - 4(-16)(3)}}{2(-16)}$$

≈ -0.037 or 5.059

The ball is in the air for about 5.059 sec. (The negative solution is extraneous.)

c. The maximum height occurs at the vertex of the quadratic function.

$t = -\frac{b}{2a} = -\frac{125\sin 40°}{2(-16)} \approx 2.51$ seconds

Evaluate the function to find the maximum height:

$-16(2.51)^2 + (125\sin 40°)(2.51) + 3 \approx 103.87$

The maximum height is about 103.87 feet.

d. Find the horizontal displacement:

$x = (125\cos 40°)(5.059) \approx 484.43$ feet

e. $x = (125\cos 40°)t$

$y = -16t^2 + (125\sin 40°)t + 3$

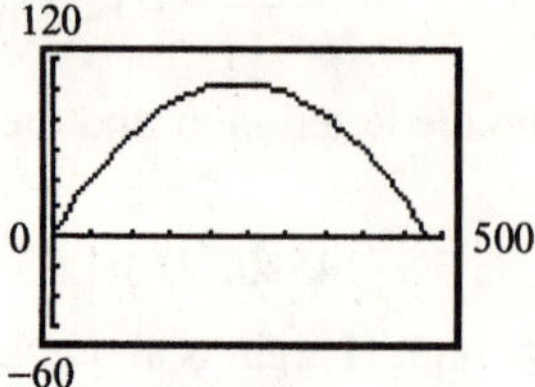

55. a. Use equations (2):

$x = (40\cos 45°)t = 20\sqrt{2}t$

$y = -\frac{1}{2}(9.8)t^2 + (40\sin 45°)t + 300$

$= -4.9t^2 + 20\sqrt{2}t + 300$

b. The ball is in the air until $y = 0$. Solve:

$-4.9t^2 + 20\sqrt{2}t + 300 = 0$

$$t = \frac{-20\sqrt{2} \pm \sqrt{\left(20\sqrt{2}\right)^2 - 4(-4.9)(300)}}{2(-4.9)}$$

≈ -5.45 or 11.23

The ball is in the air for about 11.23 seconds. (The negative solution is extraneous.)

c. The maximum height occurs at the vertex of the quadratic function.

$t = -\frac{b}{2a} = -\frac{20\sqrt{2}}{2(-4.9)} \approx 2.89$ seconds

Evaluate the function to find the maximum height:

$-4.9(2.89)^2 + 20\sqrt{2}(2.89) + 300$

$= 340.8$ meters

d. Find the horizontal displacement:

$x = \left(20\sqrt{2}\right)(11.23) \approx 317.6$ meters

e.

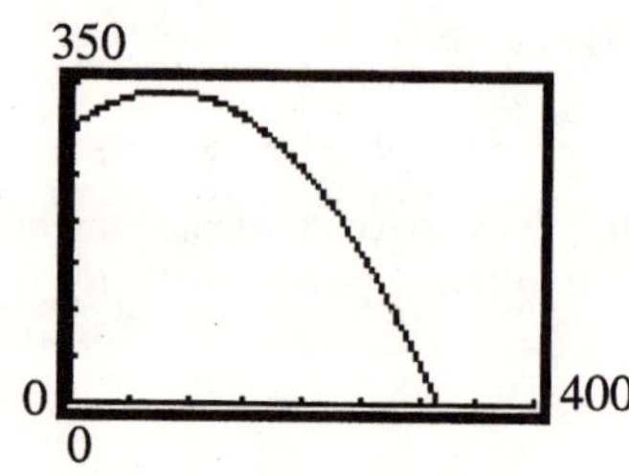

56. a. Use equations (2):

$x = (40\cos 45°)t = 20\sqrt{2}t$

$y = -\frac{1}{2}\cdot\frac{1}{6}(9.8)t^2 + (40\sin 45°)t + 300$

$= -\frac{4.9}{6}t^2 + 20\sqrt{2}t + 300$

b. The ball is in the air until $y = 0$. Solve:

$-\frac{4.9}{6}t^2 + 20\sqrt{2}t + 300 = 0$

$$t = \frac{-20\sqrt{2} \pm \sqrt{\left(20\sqrt{2}\right)^2 - 4\left(-\frac{4.9}{6}\right)(300)}}{2\left(\frac{-4.9}{6}\right)}$$

≈ -8.51 or 43.15

The ball is in the air for about 43.15 seconds. (The negative solution is extraneous.)

c. The maximum height occurs at the vertex of the quadratic function.

$t = -\frac{b}{2a} = -\frac{20\sqrt{2}}{2\left(\frac{-4.9}{6}\right)} \approx 17.32$ seconds

Evaluate the function to find the maximum height:

$-\frac{4.9}{6}(17.32)^2 + 20\sqrt{2}(17.32) + 300$

≈ 544.9 meters

d. Find the horizontal displacement:

$x = \left(20\sqrt{2}\right)(43.15) \approx 1220.5$ meters

e.

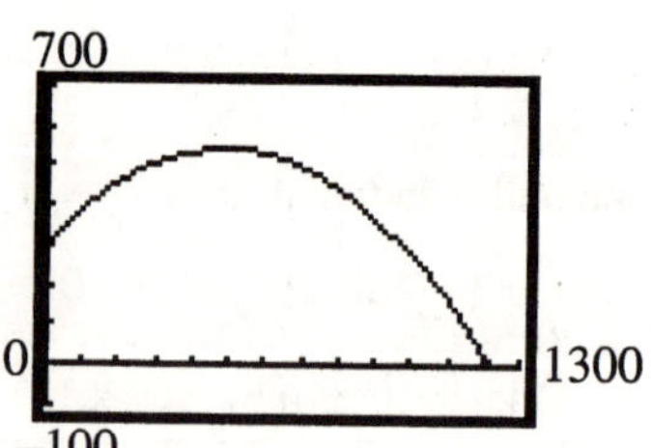

57. a. At $t = 0$, the Paseo is 5 miles from the intersection (at (0, 0)) traveling east (along the x-axis) at 40 mph. Thus, $x = 40t - 5$ describes the position of the Paseo as a function of time. The Bonneville, at $t = 0$, is 4 miles from the intersection traveling north (along the y-axis) at 30 mph. Thus, $y = 30t - 4$ describes the position of the Bonneville as a function of time.

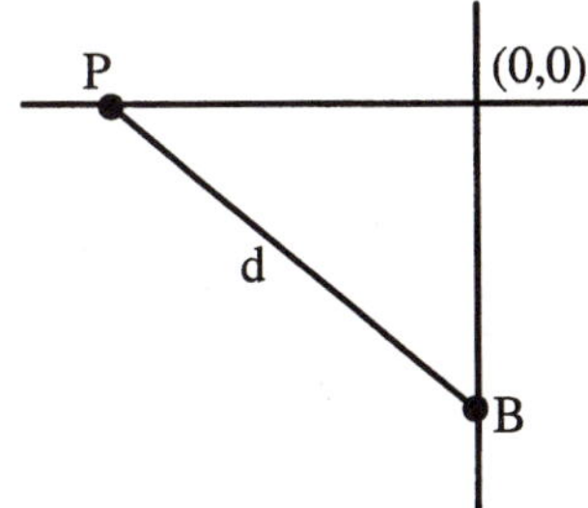

b. Let d represent the distance between the cars. Use the Pythagorean Theorem to find the distance: $d = \sqrt{(40t-5)^2 + (30t-4)^2}$.

c. Note this is a function graph not a parametric graph.

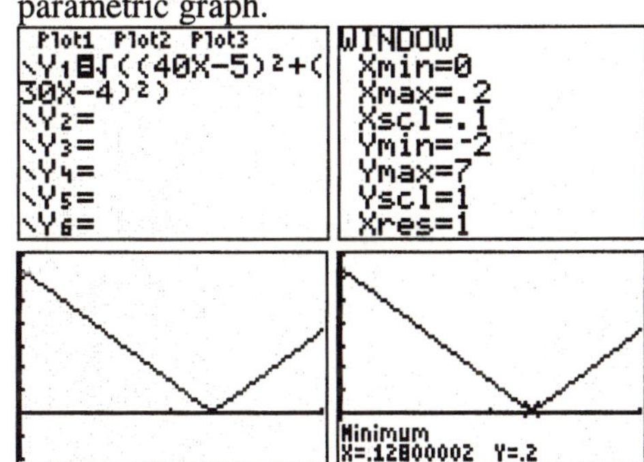

d. The minimum distance between the cars is 0.2 miles and occurs at 0.128 hours.

e.

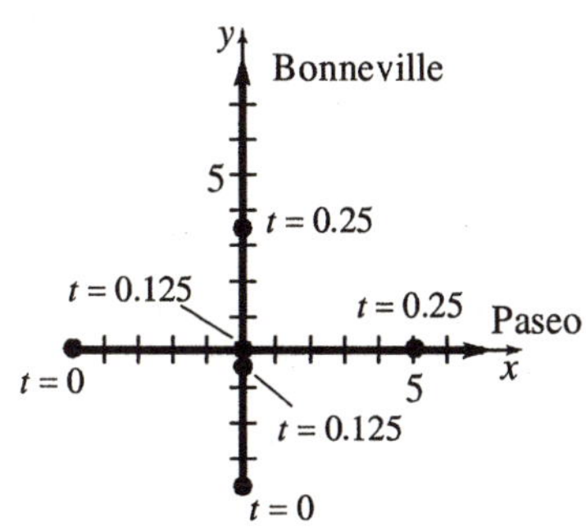

58. a. At $t = 0$, the Boeing 747 is 550 miles from the intersection (at (0, 0)) traveling west (along the x-axis) at 600 mph. Thus, $x = -600t + 550$ describes the position of the Boeing 747 as a function of time. The Cessna, at $t = 0$, is 100 miles from the intersection traveling south (along the y-axis) at 120 mph. Thus, $y = -120t + 100$ describes the position of the Cessna as a function of time.

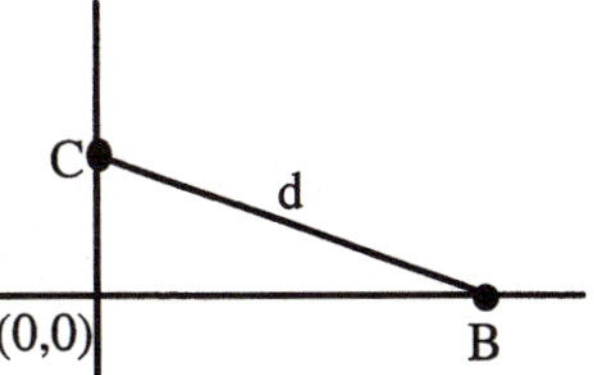

b. Let d represent the distance between the planes. Use the Pythagorean Theorem to find the distance:

$$d = \sqrt{(-600t+550)^2 + (-120t+100)^2}.$$

c. Note this is a function graph not a parametric graph.

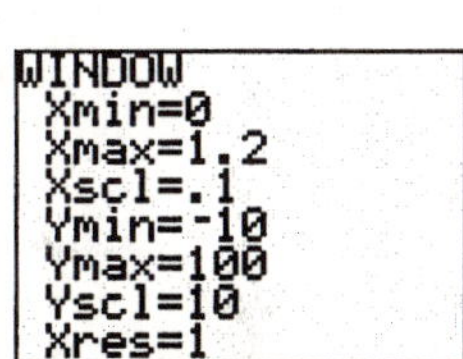

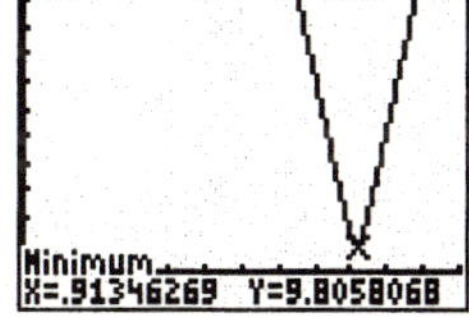

d. The minimum distance between the planes is 9.8 miles and occurs at 0.91 hours.

e.

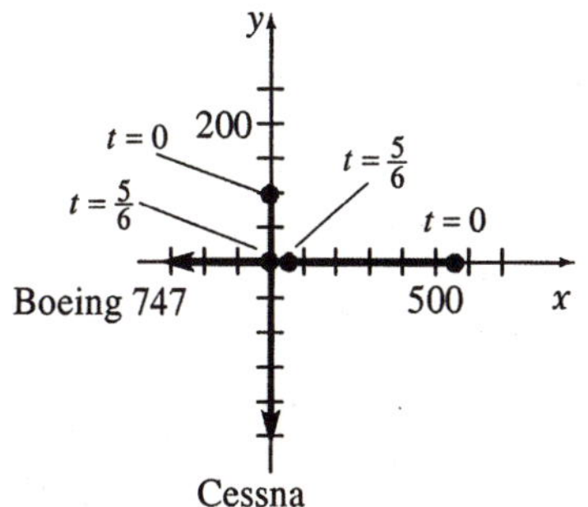

59. $x=(x_2-x_1)t+x_1$,
$y=(y_2-y_1)t+y_1$, $-\infty<t<\infty$

$$\frac{x-x_1}{x_2-x_1}=t$$

$$y=(y_2-y_1)\left(\frac{x-x_1}{x_2-x_1}\right)+y_1$$

$$y-y_1=\left(\frac{y_2-y_1}{x_2-x_1}\right)(x-x_1)$$

This is the two-point form for the equation of a line. Its orientation is from (x_1, y_1) to (x_2, y_2).

60. a. $x=(v_0\cos\theta)t$, $y=(v_0\sin\theta)t-16t^2$

$$t=\frac{x}{v_0\cos\theta}$$

$$y=v_0\sin\theta\left(\frac{x}{v_0\cos\theta}\right)-16\left(\frac{x}{v_0\cos\theta}\right)^2$$

$$y=(\tan\theta)x-\frac{16}{v_0^2\cos^2\theta}x^2$$

y is a quadratic function of x; its graph is a parabola with $a=\dfrac{-16}{v_0^2\cos^2\theta}$, $b=\tan\theta$, and $c=0$.

b.

$$y=0$$

$$(v_0\sin\theta)t-16t^2=0$$

$$t(v_0\sin\theta-16t)=0$$

$$t=0 \text{ or } v_0\sin\theta-16t=0$$

$$t=\frac{v_0\sin\theta}{16}$$

c. $x=(v_0\cos\theta)t$

$$=(v_0\cos\theta)\left(\frac{v_0\sin\theta}{16}\right)$$

$$=\frac{v_0^2\sin 2\theta}{32}\text{ feet}$$

d. $x=y$

$$(v_0\cos\theta)t=(v_0\sin\theta)t-16t^2$$

$$16t^2+(v_0\cos\theta)t-(v_0\sin\theta)t=0$$

$$t(16t+(v_0\cos\theta)-(v_0\sin\theta))=0$$

$$t=0 \text{ or } 16t+(v_0\cos\theta)-(v_0\sin\theta)=0$$

$$t=\frac{v_0\sin\theta-v_0\cos\theta}{16}=\frac{v_0}{16}(\sin\theta-\cos\theta)$$

At $t = \frac{v_0}{16}(\sin\theta - \cos\theta)$:

$$x = v_0 \cos\theta\left(\frac{v_0}{16}(\sin\theta - \cos\theta)\right) = \frac{v_0^2}{16}\cos\theta(\sin\theta - \cos\theta)$$

$$y = v_0 \sin\theta\left(\frac{v_0}{16}(\sin\theta - \cos\theta)\right) - 16\left(\frac{v_0}{16}(\sin\theta - \cos\theta)\right)^2$$

$$= \frac{v_0^2}{16}\sin\theta(\sin\theta - \cos\theta) - \frac{v_0^2}{16}(\sin^2\theta - 2\sin\theta\cos\theta + \cos^2\theta)$$

$$= \frac{v_0^2}{16}(\sin^2\theta - \sin\theta\cos\theta - \sin^2\theta + 2\sin\theta\cos\theta - \cos^2\theta)$$

$$= \frac{v_0^2}{16}(-\cos^2\theta + \sin\theta\cos\theta)$$

$$\sqrt{x^2 + y^2} = \sqrt{\left(\frac{v_0^2}{16}\cos\theta(\sin\theta - \cos\theta)\right)^2 + \left(\frac{v_0^2}{16}(-\cos^2\theta + \sin\theta\cos\theta)\right)^2}$$

$$= \sqrt{\left(\frac{v_0^2}{16}\cos\theta(\sin\theta - \cos\theta)\right)^2 + \left(\frac{v_0^2}{16}\cos\theta(\sin\theta - \cos\theta)\right)^2}$$

$$= \sqrt{2\left(\frac{v_0^2}{16}\cos\theta(\sin\theta - \cos\theta)\right)^2}$$

$$= \frac{v_0^2}{16}\sqrt{2}\cos\theta(\sin\theta - \cos\theta)$$

Note: since θ must be greater than 45° ($\sin\theta - \cos\theta > 0$), thus absolute value is not needed.

61. a. $x(t) = \cos^3 t,\ y(t) = \sin^3 t,\ 0 \le t \le 2\pi$

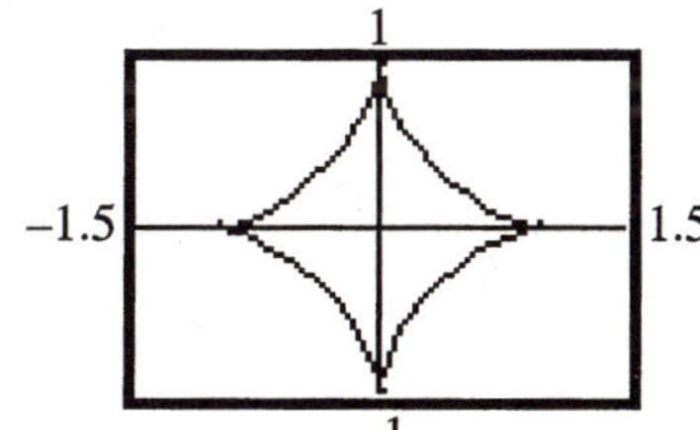

b. $\cos^2 t + \sin^2 t = (x^{1/3})^2 + (y^{1/3})^2$

$x^{2/3} + y^{2/3} = 1$

62. Answers will vary.

63. Answers will vary.

Chapter 9 Review

1. $y^2 = -16x$

This is a parabola.

$a = 4$

Vertex: (0, 0)

Focus: (–4, 0)

Directrix: $x = 4$

2. $16x^2 = y \rightarrow x^2 = \frac{1}{16}y$

This is a parabola.

$a = \frac{1}{64}$

Vertex: (0, 0)

Focus: $\left(0, \frac{1}{64}\right)$

Directrix: $y = -\frac{1}{64}$

3. $\frac{x^2}{25} - y^2 = 1$

This is a hyperbola.

$a = 5,\ b = 1.$

Find the value of c:

$c^2 = a^2 + b^2 = 25 + 1 = 26$

$c = \sqrt{26}$

Center: (0, 0)

Vertices: (5, 0), (–5, 0)

Foci: $\left(\sqrt{26}, 0\right), \left(-\sqrt{26}, 0\right)$

Asymptotes: $y = \frac{1}{5}x;\ \ y = -\frac{1}{5}x$

4. $\frac{y^2}{25} - x^2 = 1$

This is a hyperbola.

$a = 5,\ b = 1.$

Find the value of c:

$c^2 = a^2 + b^2 = 25 + 1 = 26$

$c = \sqrt{26}$

Center: (0, 0)

Vertices: (0, 5), (0, –5)

Foci: $\left(0, \sqrt{26}\right), \left(0, -\sqrt{26}\right)$

Asymptotes: $y = 5x;\ \ y = -5x$

5. $\frac{y^2}{25} + \frac{x^2}{16} = 1$

This is an ellipse.

$a = 5,\ b = 4.$

Find the value of c:

$c^2 = a^2 - b^2 = 25 - 16 = 9$

$c = 3$

Center: (0, 0)

Vertices: (0, 5), (0, –5)

Foci: (0, 3), (0, –3)

6. $\frac{x^2}{9} + \frac{y^2}{16} = 1$

This is an ellipse.

$a = 4,\ b = 3.$

Find the value of c:

$c^2 = a^2 - b^2 = 16 - 9 = 7$

$c = \sqrt{7}$

Center: (0, 0)

Vertices: (0, 4), (0, –4)

Foci: $\left(0, \sqrt{7}\right), \left(0, -\sqrt{7}\right)$

7. $x^2 + 4y = 4$

This is a parabola.

Write in standard form:

$x^2 = -4y + 4$

$x^2 = -4(y - 1)$

$a = 1$

Vertex: (0, 1)

Focus: (0, 0)

Directrix: $y = 2$

8. $3y^2 - x^2 = 9$

This is a hyperbola.

Write in standard form:

$\frac{y^2}{3} - \frac{x^2}{9} = 1$

$a = \sqrt{3},\ b = 3$

Find the value of c:

$c^2 = a^2 + b^2 = 3 + 9 = 12$

$c = \sqrt{12} = 2\sqrt{3}$

Center: (0, 0)

Vertices: $\left(0, \sqrt{3}\right), \left(0, -\sqrt{3}\right)$

Foci: $\left(0, 2\sqrt{3}\right), \left(0, -2\sqrt{3}\right)$

Asymptotes: $y = \frac{\sqrt{3}}{3}x;\ \ y = -\frac{\sqrt{3}}{3}x$

9. $4x^2 - y^2 = 8$

This is a hyperbola.

Write in standard form:

$\frac{x^2}{2} - \frac{y^2}{8} = 1$

$a = \sqrt{2},\ b = \sqrt{8} = 2\sqrt{2}.$

Find the value of c:

$c^2 = a^2 + b^2 = 2 + 8 = 10$

$c = \sqrt{10}$

Center: (0, 0)

Vertices: $\left(-\sqrt{2}, 0\right), \left(\sqrt{2}, 0\right)$

Foci: $\left(-\sqrt{10}, 0\right), \left(\sqrt{10}, 0\right)$

Asymptotes: $y = 2x;\ \ y = -2x$

10. $9x^2 + 4y^2 = 36$
This is an ellipse.
Write in standard form:
$$\frac{x^2}{4} + \frac{y^2}{9} = 1$$
$a = 3,\ b = 2$.
Find the value of c:
$$c^2 = a^2 - b^2 = 9 - 4 = 5$$
$$c = \sqrt{5}$$
Cnter: (0, 0)
Vertices: (0, 3), (0, –3)
Foci: $\left(0, \sqrt{5}\right), \left(0, -\sqrt{5}\right)$

11. $x^2 - 4x = 2y$
This is a parabola.
Write in standard form:
$$x^2 - 4x + 4 = 2y + 4$$
$$(x-2)^2 = 2(y+2)$$
$$a = \frac{1}{2}$$
Vertex: (2, –2)
Focus: $\left(2, -\frac{3}{2}\right)$
Directrix: $y = -\frac{5}{2}$

12. $2y^2 - 4y = x - 2$
This is a parabola.
Write in standard form:
$$2\left(y^2 - 2y + 1\right) = x - 2 + 2$$
$$(y-1)^2 = \frac{1}{2}x$$
$$a = \frac{1}{8}$$
Vertex: (0, 1)
Focus: $\left(\frac{1}{8}, 1\right)$
Directrix: $x = -\frac{1}{8}$

13. $y^2 - 4y - 4x^2 + 8x = 4$
This is a hyperbola.
Write in standard form:
$$(y^2 - 4y + 4) - 4(x^2 - 2x + 1) = 4 + 4 - 4$$
$$(y-2)^2 - 4(x-1)^2 = 4$$
$$\frac{(y-2)^2}{4} - \frac{(x-1)^2}{1} = 1$$
$a = 2,\ b = 1$.
Find the value of c:
$$c^2 = a^2 + b^2 = 4 + 1 = 5$$
$$c = \sqrt{5}$$
Center: (1, 2)
Vertices: (1, 0), (1, 4)
Foci: $\left(1, 2 - \sqrt{5}\right), \left(1, 2 + \sqrt{5}\right)$
Asymptotes: $y - 2 = 2(x-1);\quad y - 2 = -2(x-1)$

14. $4x^2 + y^2 + 8x - 4y + 4 = 0$
This is an ellipse.
Write in standard form:
$$4(x^2 + 2x + 1) + (y^2 - 4y + 4) = -4 + 4 + 4$$
$$4(x+1)^2 + (y-2)^2 = 4$$
$$\frac{(x+1)^2}{1} + \frac{(y-2)^2}{4} = 1$$
$a = 2,\ b = 1$
Find the value of c:
$$c^2 = a^2 - b^2 = 4 - 1 = 3 \quad \rightarrow \quad c = \sqrt{3}$$
Center: (–1, 2)
Vertices: (–1, 0), (–1, 4)
Foci: $\left(-1, 2 - \sqrt{3}\right), \left(-1, 2 + \sqrt{3}\right)$

15. $4x^2 + 9y^2 - 16x - 18y = 11$
This is an ellipse.
Write in standard form:
$$4x^2 + 9y^2 - 16x - 18y = 11$$
$$4(x^2 - 4x + 4) + 9(y^2 - 2y + 1) = 11 + 16 + 9$$
$$4(x-2)^2 + 9(y-1)^2 = 36$$
$$\frac{(x-2)^2}{9} + \frac{(y-1)^2}{4} = 1$$
$a = 3,\ b = 2$. Find the value of c:
$$c^2 = a^2 - b^2 = 9 - 4 = 5 \quad \rightarrow \quad c = \sqrt{5}$$
Center: (2, 1); Vertices: (–1, 1), (5, 1)
Foci: $\left(2 - \sqrt{5}, 1\right), \left(2 + \sqrt{5}, 1\right)$

16. $4x^2+9y^2-16x+18y=11$
This is an ellipse.
Write in standard form:
$$4x^2+9y^2-16x+18y=11$$
$$4(x^2-4x+4)+9(y^2+2y+1)=11+16+9$$
$$4(x-2)^2+9(y+1)^2=36$$
$$\frac{(x-2)^2}{9}+\frac{(y+1)^2}{4}=1$$
$a=3,\ b=2$.
Find the value of c:
$c^2=a^2-b^2=9-4=5$
$c=\sqrt{5}$
Center: (2, –1); Vertices: (–1, –1), (5, –1)
Foci: $\left(2-\sqrt{5},-1\right),\left(2+\sqrt{5},-1\right)$

17. $4x^2-16x+16y+32=0$
This is a parabola.
Write in standard form:
$$4(x^2-4x+4)=-16y-32+16$$
$$4(x-2)^2=-16(y+1)$$
$$(x-2)^2=-4(y+1)$$
$a=1$
Vertex: (2, –1); Focus: $(2,-2)$;
Directrix: $y=0$

18. $4y^2+3x-16y+19=0$
This is a parabola.
Write in standard form:
$$4(y^2-4y+4)=-3x-19+16$$
$$4(y-2)^2=-3(x+1)$$
$$(y-2)^2=-\frac{3}{4}(x+1)$$
$a=\frac{3}{16}$
Vertex: (–1, 2); Focus: $\left(-\frac{19}{16},2\right)$
Directrix: $x=-\frac{13}{16}$

19. $9x^2+4y^2-18x+8y=23$
This is an ellipse.
Write in standard form:
$$9(x^2-2x+1)+4(y^2+2y+1)=23+9+4$$
$$9(x-1)^2+4(y+1)^2=36$$
$$\frac{(x-1)^2}{4}+\frac{(y+1)^2}{9}=1$$
$a=3,\ b=2$.
Find the value of c:
$c^2=a^2-b^2=9-4=5$
$c=\sqrt{5}$
Center: (1, –1)
Vertices: (1, –4), (1, 2)
Foci: $\left(1,-1-\sqrt{5}\right),\left(1,-1+\sqrt{5}\right)$

20. $x^2-y^2-2x-2y=1$
This is a hyperbola.
Write in standard form:
$$(x^2-2x+1)-(y^2+2y+1)=1+1-1$$
$$(x-1)^2-(y+1)^2=1$$
$a=1,\ b=1$.
Find the value of c:
$c^2=a^2+b^2=1+1=2$
$c=\sqrt{2}$
Center: (1, –1)
Vertices: (0, –1), (2, –1)
Foci: $\left(1+\sqrt{2},-1\right),\left(1-\sqrt{2},-1\right)$
Asymptotes: $y+1=x-1;\ \ y+1=-(x-1)$

21. Parabola: The focus is (–2, 0) and the directrix is $x=2$. The vertex is (0, 0). $a=2$ and since (–2, 0) is to the left of (0, 0), the parabola opens to the left. The equation of the parabola is:
$$y^2=-4ax$$
$$y^2=-4\cdot 2\cdot x$$
$$y^2=-8x$$

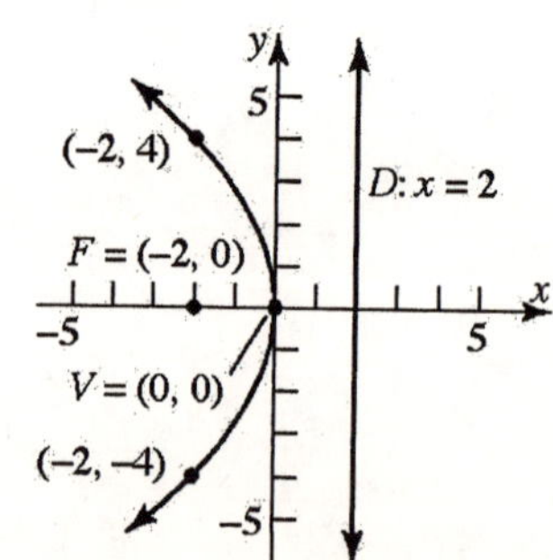

22. Ellipse: The center is (0, 0), a focus is (0, 3), and a vertex is (0, 5). The major axis is $x = 0$. $a = 5,\ c = 3$. Find b:

$b^2 = a^2 - c^2 = 25 - 9 = 16$. So, $b = 4$. The equation of the ellipse is:

$$\frac{x^2}{b^2} + \frac{y^2}{a^2} = 1$$

$$\frac{x^2}{4^2} + \frac{y^2}{5^2} = 1$$

$$\frac{x^2}{16} + \frac{y^2}{25} = 1$$

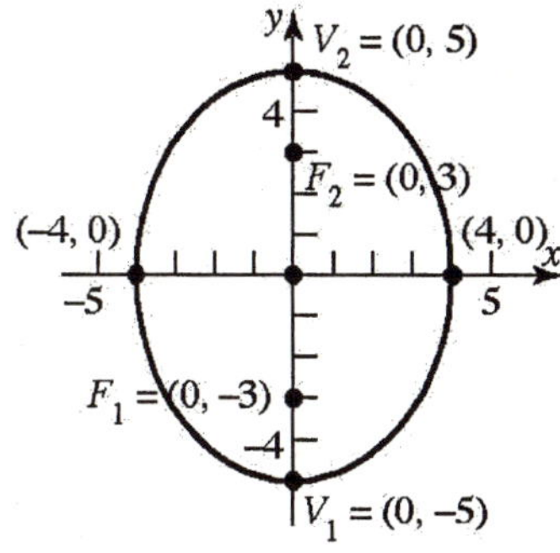

23. Hyperbola: Center: (0, 0);
Focus: (0, 4); Vertex: (0, –2);
Transverse axis is the y-axis; $a = 2$; $c = 4$.
Find b:

$b^2 = c^2 - a^2 = 16 - 4 = 12$

$b = \sqrt{12} = 2\sqrt{3}$

Write the equation: $\dfrac{y^2}{4} - \dfrac{x^2}{12} = 1$

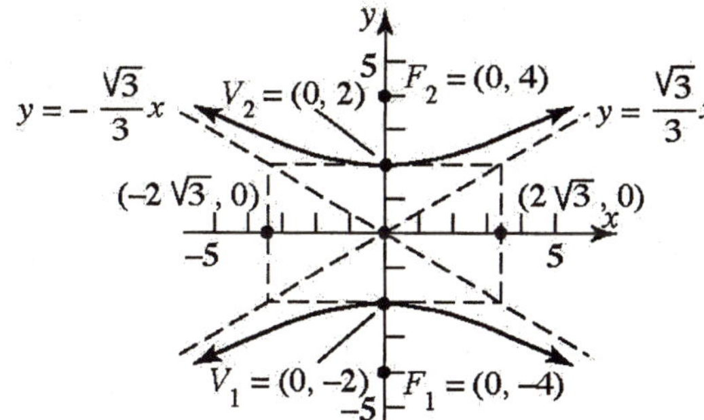

24. Parabola: Vertex: (0, 0); Directrix: $y = -3$; $a = 3$; the focus is the point $(0,3)$; the graph opens up. The equation of the parabola is:

$x^2 = 4ay$

$x^2 = 4(3)y$

$x^2 = 12y$

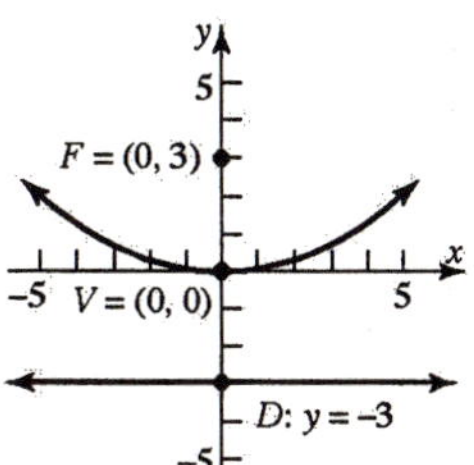

25. Ellipse: Foci: (–3, 0), (3, 0); Vertex: (4, 0);
Center: (0, 0); Major axis is the x-axis;
$a = 4$; $c = 3$. Find b:

$b^2 = a^2 - c^2 = 16 - 9 = 7$

$b = \sqrt{7}$

Write the equation: $\dfrac{x^2}{16} + \dfrac{y^2}{7} = 1$

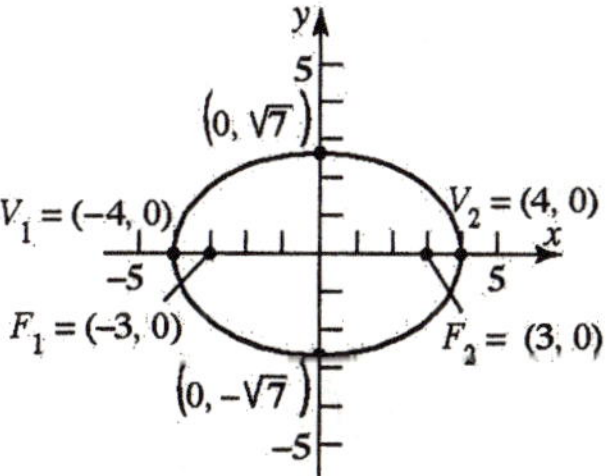

26. Hyperbola: Vertices: (–2, 0), (2, 0);
Focus: (4, 0); Center: (0, 0); Transverse axis is the x-axis; $a = 2$; $c = 4$.
Find b:

$b^2 = c^2 - a^2 = 16 - 4 = 12$

$b = \sqrt{12} = 2\sqrt{3}$

Write the equation: $\dfrac{x^2}{4} - \dfrac{y^2}{12} = 1$

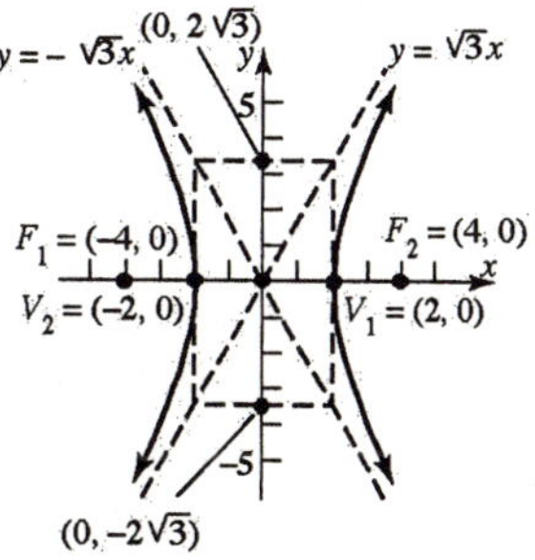

27. Parabola: The focus is (2, –4) and the vertex is (2, –3). Both lie on the vertical line $x = 2$. $a = 1$ and since (2, –4) is below (2, –3), the parabola opens down. The equation of the parabola is:

$$(x-h)^2 = -4a(y-k)$$
$$(x-2)^2 = -4\cdot 1\cdot(y-(-3))$$
$$(x-2)^2 = -4(y+3)$$

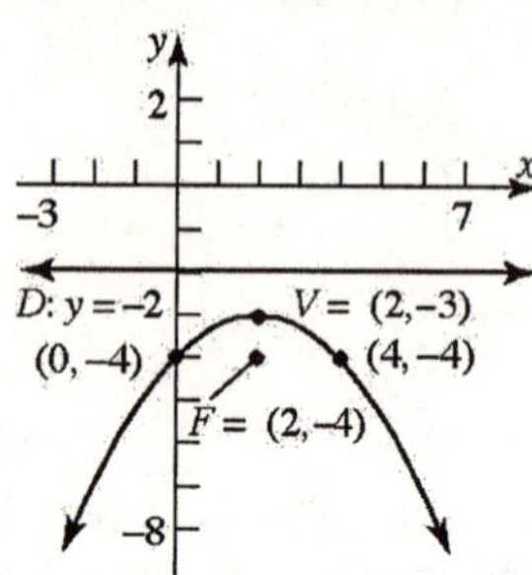

28. Ellipse: Center: (–1, 2); Focus: (0, 2); Vertex: (2, 2). Major axis: $y = 2$. $a = 3$; $c = 1$. Find b:

$$b^2 = a^2 - c^2 = 9 - 1 = 8$$
$$b = \sqrt{8} = 2\sqrt{2}$$

Write the equation: $\dfrac{(x+1)^2}{9} + \dfrac{(y-2)^2}{8} = 1$

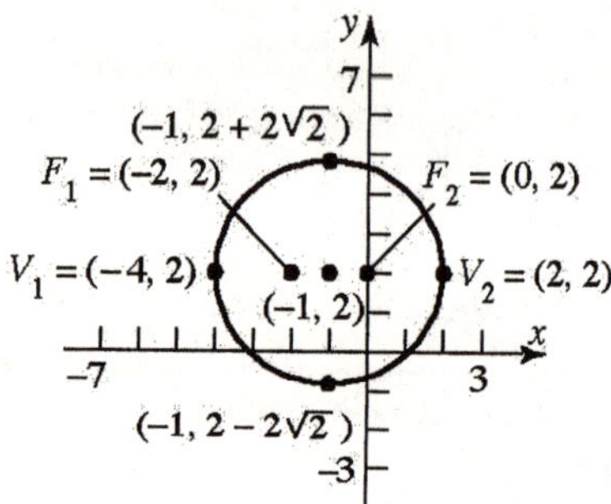

29. Hyperbola: Center: (–2, –3); Focus: (–4, –3); Vertex: (–3, –3); Transverse axis is parallel to the x-axis; $a = 1$; $c = 2$. Find b:

$$b^2 = c^2 - a^2 = 4 - 1 = 3$$
$$b = \sqrt{3}$$

Write the equation: $\dfrac{(x+2)^2}{1} - \dfrac{(y+3)^2}{3} = 1$

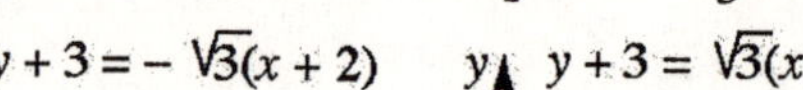

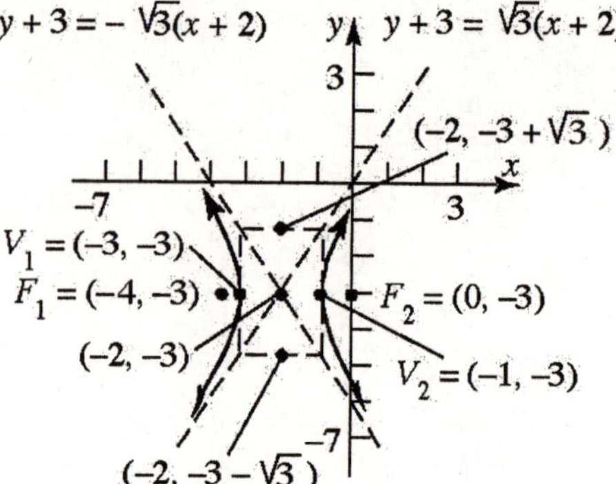

30. Parabola: Focus: (3, 6); Directrix: $y = 8$; Parabola opens down. Vertex: (3, 7) $a = 1$. The equation of the parabola is:

$$(x-h)^2 = -4a(y-k)$$
$$(x-3)^2 = -4(1)(y-7)$$
$$(x-3)^2 = -4(y-7)$$

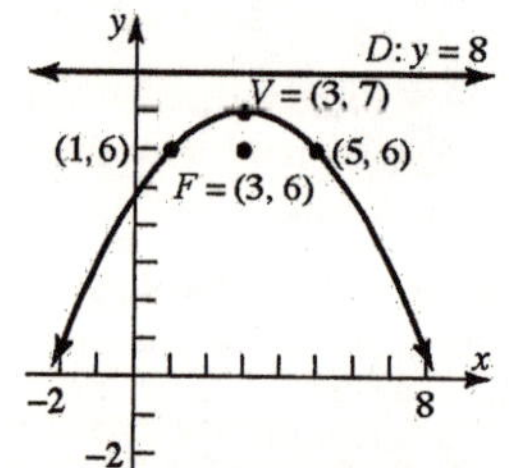

31. Ellipse: Foci: (–4, 2), (–4, 8); Vertex: (–4, 10); Center: (–4, 5); Major axis is parallel to the y-axis; $a = 5$; $c = 3$. Find b:

$$b^2 = a^2 - c^2 = 25 - 9 = 16 \quad \rightarrow \quad b = 4$$

Write the equation: $\dfrac{(x+4)^2}{16} + \dfrac{(y-5)^2}{25} = 1$

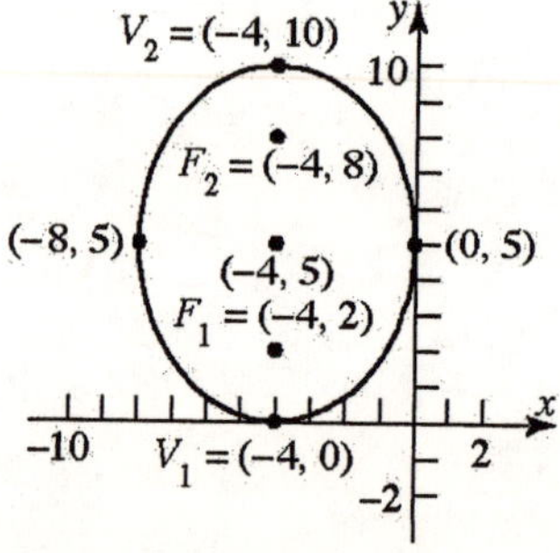

32. Hyperbola: Vertices: (–3, 3), (5, 3); Focus: (7, 3); Center: (1, 3); Major axis is parallel to the x-axis; $a = 4$; $c = 6$. Find b:

$b^2 = c^2 - a^2 = 36 - 16 = 20$

$b = \sqrt{20} = 2\sqrt{5}$

Write the equation: $\dfrac{(x-1)^2}{16} - \dfrac{(y-3)^2}{20} = 1$

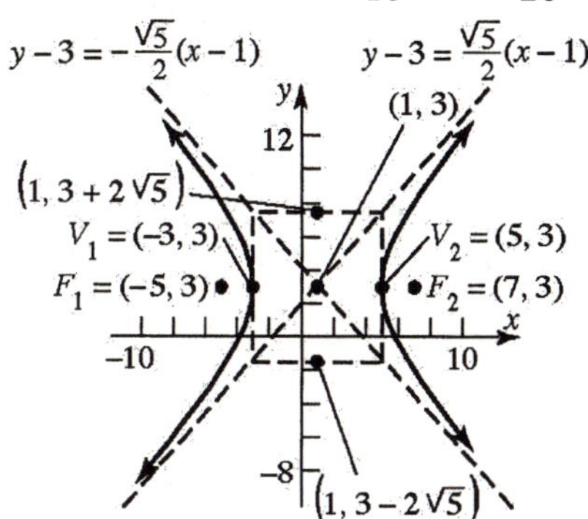

33. Hyperbola: Center: (–1, 2); $a = 3$; $c = 4$; Transverse axis parallel to the x-axis; Find b:

$b^2 = c^2 - a^2 = 16 - 9 = 7$

$b = \sqrt{7}$

Write the equation: $\dfrac{(x+1)^2}{9} - \dfrac{(y-2)^2}{7} = 1$

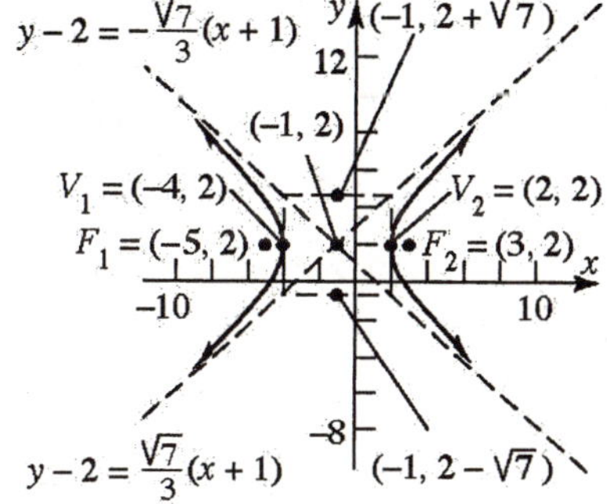

34. Hyperbola: Center: (4, –2); $a = 1$; $c = 4$; Transverse axis parallel to the y-axis. Fnd b:

$b^2 = c^2 - a^2 = 16 - 1 = 15 \rightarrow b = \sqrt{15}$

Write the equation: $\dfrac{(y+2)^2}{1} - \dfrac{(x-4)^2}{15} = 1$

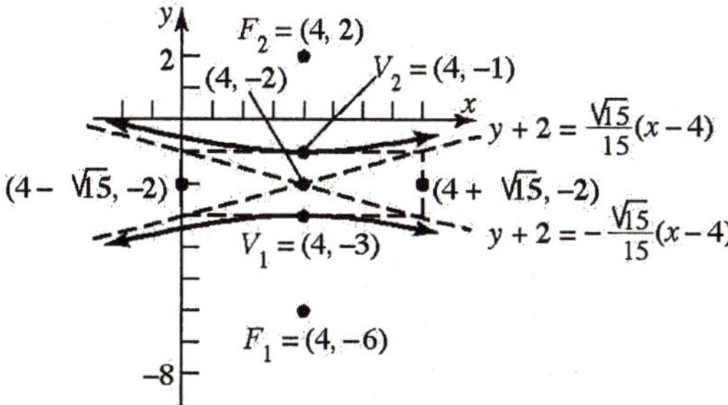

35. Hyperbola: Vertices: (0, 1), (6, 1); Asymptote: $3y + 2x - 9 = 0$; Center: (3, 1); Transverse axis is parallel to the x-axis; $a = 3$; The slope of the asymptote is $-\dfrac{2}{3}$; Find b:

$\dfrac{-b}{a} = \dfrac{-b}{3} = \dfrac{-2}{3} \rightarrow -3b = -6 \rightarrow b = 2$

Write the equation: $\dfrac{(x-3)^2}{9} - \dfrac{(y-1)^2}{4} = 1$

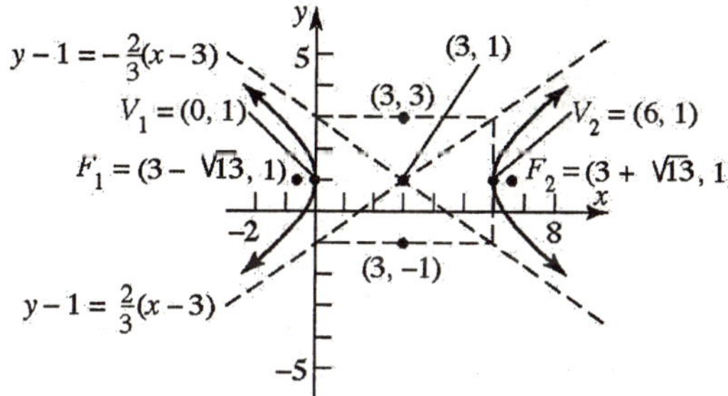

36. Hyperbola: Vertices: (4, 0), (4, 4);
Asymptote: $y+2x-10=0$; Center: (4, 2);
Transverse axis is parallel to the y-axis; $a=2$;
The slope of the asymptote is -2; Find b:

$$\frac{-a}{b}=\frac{-2}{b}=-2 \rightarrow -2b=-2 \rightarrow b=1$$

Write the equation: $\frac{(y-2)^2}{4}-\frac{(x-4)^2}{1}=1$

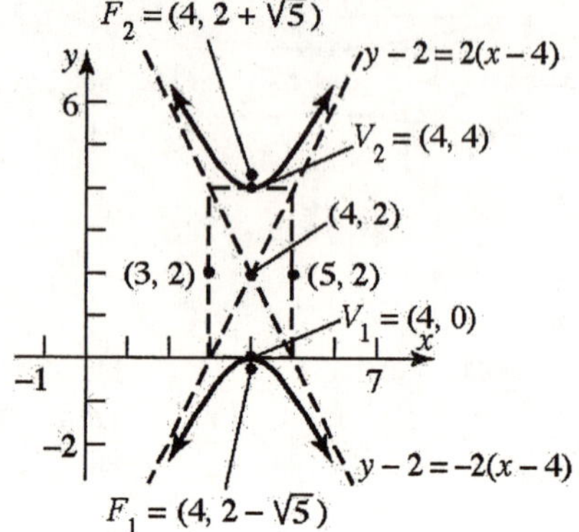

37. $y^2+4x+3y-8=0$
$A=0$ and $C=1$; $AC=(0)(1)=0$. Since $AC=0$, the equation defines a parabola.

38. $2x^2-y+8x=0$
$A=2$ and $C=0$; $AC=(2)(0)=0$. Since $AC=0$, the equation defines a parabola.

39. $x^2+2y^2+4x-8y+2=0$
$A=1$ and $C=2$; $AC=(1)(2)=2$. Since $AC>0$ and $A\neq C$, the equation defines an ellipse.

40. $x^2-8y^2-x-2y=0$
$A=1$ and $C=-8$; $AC=(1)(-8)=-8$. Since $AC<0$, the equation defines a hyperbola.

41. $9x^2-12xy+4y^2+8x+12y=0$
$A=9,\ B=-12,\ C=4$
$B^2-4AC=(-12)^2-4(9)(4)=0$
Parabola

42. $4x^2+4xy+y^2-8\sqrt{5}\,x+16\sqrt{5}\,y=0$
$A=4,\ B=4,\ C=1$
$B^2-4AC=4^2-4(4)(1)=0$
Parabola

43. $4x^2+10xy+4y^2-9=0$
$A=4,\ B=10,\ C=4$
$B^2-4AC=10^2-4(4)(4)=36>0$
Hyperbola

44. $4x^2-10xy+4y^2-9=0$
$A=4,\ B=-10,\ C=4$
$B^2-4AC=(-10)^2-4(4)(4)=36>0$
Hyperbola

45. $x^2-2xy+3y^2+2x+4y-1=0$
$A=1,\ B=-2,\ C=3$
$B^2-4AC=(-2)^2-4(1)(3)=-8<0$
Ellipse

46. $4x^2+12xy-10y^2+x+y-10=0$
$A=4,\ B=12,\ C=-10$
$B^2-4AC=12^2-4(4)(-10)=304>0$
Hyperbola

47. $2x^2+5xy+2y^2-\frac{9}{2}=0$

$A=2, B=5,$ and $C=2$; $\cot(2\theta)=\frac{A-C}{B}=\frac{2-2}{5}=0 \rightarrow 2\theta=\frac{\pi}{2} \rightarrow \theta=\frac{\pi}{4}$

$$x=x'\cos\theta-y'\sin\theta=\frac{\sqrt{2}}{2}x'-\frac{\sqrt{2}}{2}y'=\frac{\sqrt{2}}{2}(x'-y')$$

$$y=x'\sin\theta+y'\cos\theta=\frac{\sqrt{2}}{2}x'+\frac{\sqrt{2}}{2}y'=\frac{\sqrt{2}}{2}(x'+y')$$

$$2\left(\frac{\sqrt{2}}{2}(x'-y')\right)^2+5\left(\frac{\sqrt{2}}{2}(x'-y')\right)\left(\frac{\sqrt{2}}{2}(x'+y')\right)+2\left(\frac{\sqrt{2}}{2}(x'+y')\right)^2-\frac{9}{2}=0$$

$$\left(x'^2-2x'y'+y'^2\right)+\frac{5}{2}\left(x'^2-y'^2\right)+\left(x'^2+2x'y'+y'^2\right)-\frac{9}{2}=0$$

$$\frac{9}{2}x'^2-\frac{1}{2}y'^2=\frac{9}{2}\to 9x'^2-y'^2=9\to\frac{x'^2}{1}-\frac{y'^2}{9}=1$$

Hyperbola; center at (0, 0), transverse axis is the x'-axis, vertices at $(x',y')=(\pm1,0)$; foci at $(x',y')=(\pm\sqrt{10},0)$; asymptotes: $y'=\pm3x'$.

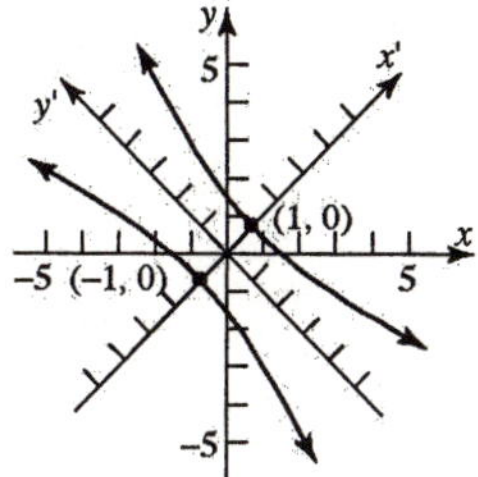

48. $2x^2-5xy+2y^2-\frac{9}{2}=0$

$A=2, B=-5,$ and $C=2$; $\cot(2\theta)=\frac{A-C}{B}=\frac{2-2}{-5}=0 \;\to\; 2\theta=\frac{\pi}{2} \;\to\; \theta=\frac{\pi}{4}$

$$x=x'\cos\theta-y'\sin\theta=\frac{\sqrt{2}}{2}x'-\frac{\sqrt{2}}{2}y'=\frac{\sqrt{2}}{2}(x'-y')$$

$$y=x'\sin\theta+y'\cos\theta=\frac{\sqrt{2}}{2}x'+\frac{\sqrt{2}}{2}y'=\frac{\sqrt{2}}{2}(x'+y')$$

$$2\left(\frac{\sqrt{2}}{2}(x'-y')\right)^2-5\left(\frac{\sqrt{2}}{2}(x'-y')\right)\left(\frac{\sqrt{2}}{2}(x'+y')\right)+2\left(\frac{\sqrt{2}}{2}(x'+y')\right)^2-\frac{9}{2}=0$$

$$\left(x'^2-2x'y'+y'^2\right)-\frac{5}{2}\left(x'^2-y'^2\right)+\left(x'^2+2x'y'+y'^2\right)-\frac{9}{2}=0$$

$$-\frac{1}{2}x'^2+\frac{9}{2}y'^2=\frac{9}{2}\to -x'^2+9y'^2=9\to\frac{y'^2}{1}-\frac{x'^2}{9}=1$$

Hyperbola; center at (0, 0), transverse axis is the y'-axis, vertices at $(x',y')=(0,\pm1)$; foci at $(x',y')=(0,\pm\sqrt{10})$; asymptotes: $y'=\pm\frac{1}{3}x'$.

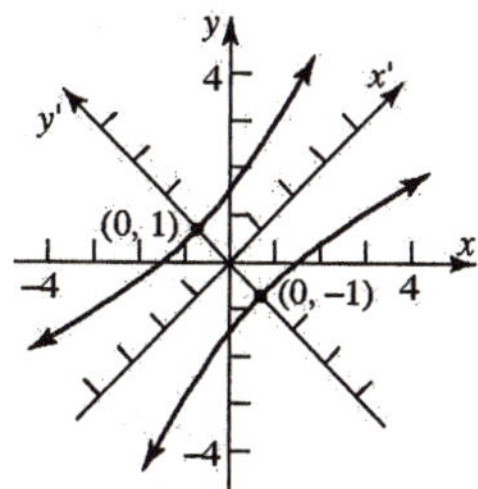

49. $6x^2+4xy+9y^2-20=0$

$A=6$, $B=4$, and $C=9$; $\cot(2\theta)=\dfrac{A-C}{B}=\dfrac{6-9}{4}=-\dfrac{3}{4} \rightarrow \cos(2\theta)=-\dfrac{3}{5}$

$\sin\theta=\sqrt{\dfrac{\left(1-\left(-\frac{3}{5}\right)\right)}{2}}=\sqrt{\dfrac{4}{5}}=\dfrac{2\sqrt{5}}{5}$; $\cos\theta=\sqrt{\dfrac{\left(1+\left(-\frac{3}{5}\right)\right)}{2}}=\sqrt{\dfrac{1}{5}}=\dfrac{\sqrt{5}}{5} \rightarrow \theta\approx 63.4°$

$x=x'\cos\theta-y'\sin\theta=\dfrac{\sqrt{5}}{5}x'-\dfrac{2\sqrt{5}}{5}y'=\dfrac{\sqrt{5}}{5}(x'-2y')$

$y=x'\sin\theta+y'\cos\theta=\dfrac{2\sqrt{5}}{5}x'+\dfrac{\sqrt{5}}{5}y'=\dfrac{\sqrt{5}}{5}(2x'+y')$

$$6\left(\frac{\sqrt{5}}{5}(x'-2y')\right)^2+4\left(\frac{\sqrt{5}}{5}(x'-2y')\right)\left(\frac{\sqrt{5}}{5}(2x'+y')\right)+9\left(\frac{\sqrt{5}}{5}(2x'+y')\right)^2-20=0$$

$$\frac{6}{5}\left(x'^2-4x'y'+4y'^2\right)+\frac{4}{5}\left(2x'^2-3x'y'-2y'^2\right)+\frac{9}{5}\left(4x'^2+4x'y'+y'^2\right)-20=0$$

$$\frac{6}{5}x'^2-\frac{24}{5}x'y'+\frac{24}{5}y'^2+\frac{8}{5}x'^2-\frac{12}{5}x'y'-\frac{8}{5}y'^2+\frac{36}{5}x'^2+\frac{36}{5}x'y'+\frac{9}{5}y'^2=20$$

$$10x'^2+5y'^2=20 \rightarrow \frac{x'^2}{2}+\frac{y'^2}{4}=1$$

Ellipse; center at the origin, major axis is the y'-axis, vertices at $(x',y')=(0,\pm 2)$; foci at $(x',y')=\left(0,\pm\sqrt{2}\right)$.

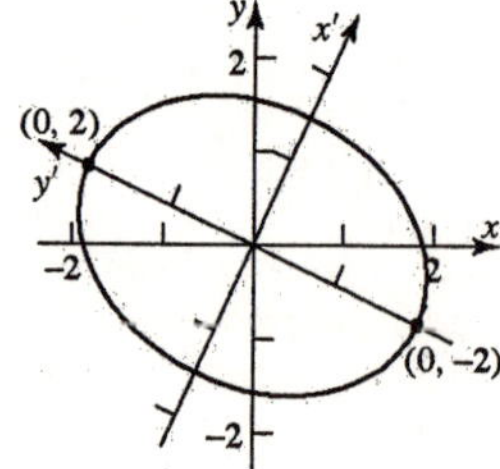

50. $x^2+4xy+4y^2+16\sqrt{5}\,x-8\sqrt{5}\,y=0$

$A=1$, $B=4$, and $C=4$; $\cot(2\theta)=\dfrac{A-C}{B}=\dfrac{1-4}{4}=-\dfrac{3}{4}$; $\cos(2\theta)=-\dfrac{3}{5}$

$\sin\theta=\sqrt{\dfrac{1-\left(-\frac{3}{5}\right)}{2}}=\sqrt{\dfrac{4}{5}}=\dfrac{2\sqrt{5}}{5}$; $\cos\theta=\sqrt{\dfrac{1+\left(-\frac{3}{5}\right)}{2}}=\sqrt{\dfrac{1}{5}}=\dfrac{\sqrt{5}}{5} \rightarrow \theta=63.4°$

$x=x'\cos\theta-y'\sin\theta=\dfrac{\sqrt{5}}{5}x'-\dfrac{2\sqrt{5}}{5}y'=\dfrac{\sqrt{5}}{5}(x'-2y')$

$y=x'\sin\theta+y'\cos\theta=\dfrac{2\sqrt{5}}{5}x'+\dfrac{\sqrt{5}}{5}y'=\dfrac{\sqrt{5}}{5}(2x'+y')$

$$\left(\frac{\sqrt{5}}{5}(x'-2y')\right)^2+4\left(\frac{\sqrt{5}}{5}(x'-2y')\right)\left(\frac{\sqrt{5}}{5}(2x'+y')\right)+4\left(\frac{\sqrt{5}}{5}(2x'+y')\right)^2$$
$$+16\sqrt{5}\left(\frac{\sqrt{5}}{5}(x'-2y')\right)-8\sqrt{5}\left(\frac{\sqrt{5}}{5}(2x'+y')\right)=0$$

$$\frac{1}{5}\left(x'^2-4x'y'+4y'^2\right)+\frac{4}{5}\left(2x'^2-3x'y'-2y'^2\right)+\frac{4}{5}\left(4x'^2+4x'y'+y'^2\right)+16x'-32y'-16x'-8y'=0$$

$$\frac{1}{5}x'^2-\frac{4}{5}x'y'+\frac{4}{5}y'^2+\frac{8}{5}x'^2-\frac{12}{5}x'y'-\frac{8}{5}y'^2+\frac{16}{5}x'^2+\frac{16}{5}x'y'+\frac{4}{5}y'^2-40y'=0$$

$$5x'^2-40y'=0 \rightarrow x'^2=8y'$$

Parabola; vertex at the origin, focus at $(x',y')=(0,2)$.

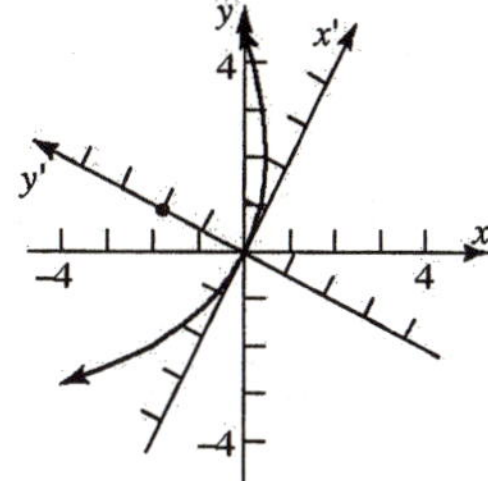

51. $4x^2-12xy+9y^2+12x+8y=0$

$A=4$, $B=-12$, and $C=9$; $\cot(2\theta)=\frac{A-C}{B}=\frac{4-9}{-12}=\frac{5}{12} \rightarrow \cos(2\theta)=\frac{5}{13}$

$$\sin\theta=\sqrt{\frac{\left(1-\frac{5}{13}\right)}{2}}=\sqrt{\frac{4}{13}}=\frac{2\sqrt{13}}{13};\quad \cos\theta=\sqrt{\frac{\left(1+\frac{5}{13}\right)}{2}}=\sqrt{\frac{9}{13}}=\frac{3\sqrt{13}}{13} \rightarrow \theta\approx 33.7^\circ$$

$$x=x'\cos\theta \quad y'\sin\theta=\frac{3\sqrt{13}}{13}x'-\frac{2\sqrt{13}}{13}y'-\frac{\sqrt{13}}{13}(3x'-2y')$$

$$y=x'\sin\theta+y'\cos\theta=\frac{2\sqrt{13}}{13}x'+\frac{3\sqrt{13}}{13}y'=\frac{\sqrt{13}}{13}(2x'+3y')$$

$$4\left(\frac{\sqrt{13}}{13}(3x'-2y')\right)^2-12\left(\frac{\sqrt{13}}{13}(3x'-2y')\right)\left(\frac{\sqrt{13}}{13}(2x'+3y')\right)+9\left(\frac{\sqrt{13}}{13}(2x'+3y')\right)^2$$
$$+12\left(\frac{\sqrt{13}}{13}(3x'-2y')\right)+8\left(\frac{\sqrt{13}}{13}(2x'+3y')\right)=0$$

$$\frac{4}{13}\left(9x'^2-12x'y'+4y'^2\right)-\frac{12}{13}\left(6x'^2+5x'y'-6y'^2\right)+\frac{9}{13}\left(4x'^2+12x'y'+9y'^2\right)$$
$$+\frac{36\sqrt{13}}{13}x'-\frac{24\sqrt{13}}{13}y'+\frac{16\sqrt{13}}{13}x'+\frac{24\sqrt{13}}{13}y'=0$$

$$\frac{36}{13}x'^2-\frac{48}{13}x'y'+\frac{16}{13}y'^2-\frac{72}{13}x'^2-\frac{60}{13}x'y'+\frac{72}{13}y'^2+\frac{36}{13}x'^2+\frac{108}{13}x'y'+\frac{81}{13}y'^2+4\sqrt{13}x'=0$$

$$13y'^2+4\sqrt{13}x'=0 \Rightarrow y'^2=-\frac{4\sqrt{13}}{13}x'$$

Parabola; vertex at the origin, focus at $(x', y') = \left(-\frac{\sqrt{13}}{13}, 0\right)$.

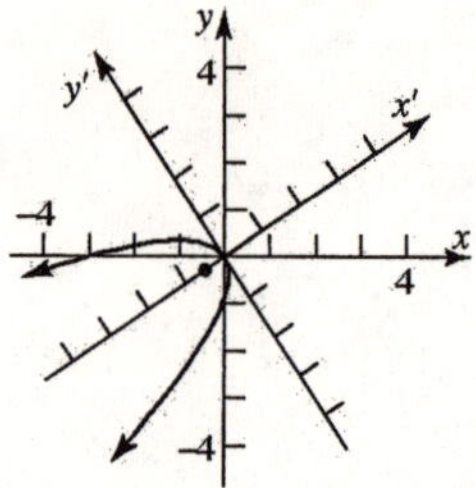

52. $9x^2 - 24xy + 16y^2 + 80x + 60y = 0$

$A = 9$, $B = -24$, and $C = 16$; $\cot(2\theta) = \frac{A-C}{B} = \frac{9-16}{-24} = \frac{7}{24} \rightarrow \cos(2\theta) = \frac{7}{25}$

$\sin\theta = \sqrt{\frac{\left(1-\frac{7}{25}\right)}{2}} = \sqrt{\frac{9}{25}} = \frac{3}{5}$; $\cos\theta = \sqrt{\frac{\left(1+\frac{7}{25}\right)}{2}} = \sqrt{\frac{16}{25}} = \frac{4}{5} \rightarrow \theta \approx 36.9$

$x = x'\cos\theta - y'\sin\theta = \frac{4}{5}x' - \frac{3}{5}y' = \frac{1}{5}(4x' - 3y')$

$y = x'\sin\theta + y'\cos\theta = \frac{3}{5}x' + \frac{4}{5}y' = \frac{1}{5}(3x' + 4y')$

$9\left(\frac{1}{5}(4x'-3y')\right)^2 - 24\left(\frac{1}{5}(4x'-3y')\right)\left(\frac{1}{5}(3x'+4y')\right) + 16\left(\frac{1}{5}(3x'+4y')\right)^2 + 80\left(\frac{1}{5}(4x'-3y')\right) + 60\left(\frac{1}{5}(3x'+4y')\right) = 0$

$\frac{9}{25}(16x'^2 - 24x'y' + 9y'^2) - \frac{24}{25}(12x'^2 + 7x'y' - 12y'^2) + \frac{16}{25}(9x'^2 + 24x'y' + 16y'^2) + 64x' - 48y' + 36x' + 48y' = 0$

$\frac{144}{25}x'^2 - \frac{216}{25}x'y' + \frac{81}{25}y'^2 - \frac{288}{25}x'^2 - \frac{168}{25}x'y' + \frac{288}{25}y'^2 + \frac{144}{25}x'^2 + \frac{384}{25}x'y' + \frac{256}{25}y'^2 + 100x' = 0$

$25y'^2 + 100x' = 0 \rightarrow y'^2 = -4x'$

Parabola; vertex at the origin; focus at $(x', y') = (-1, 0)$.

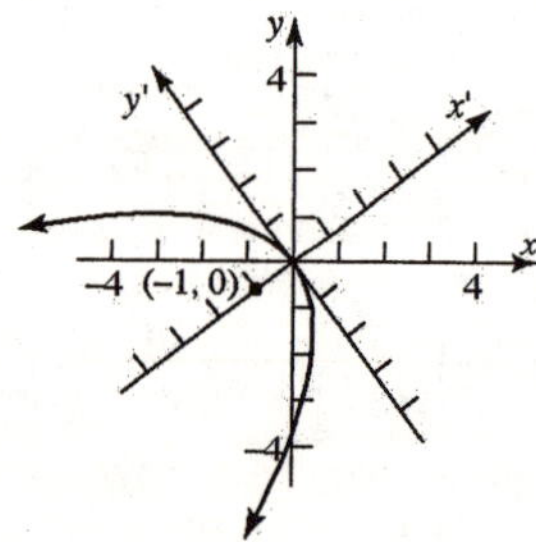

53. $r = \frac{4}{1-\cos\theta}$

$ep = 4$, $e = 1$, $p = 4$

Parabola; directrix is perpendicular to the polar axis 4 units to the left of the pole; vertex is $(2, \pi)$.

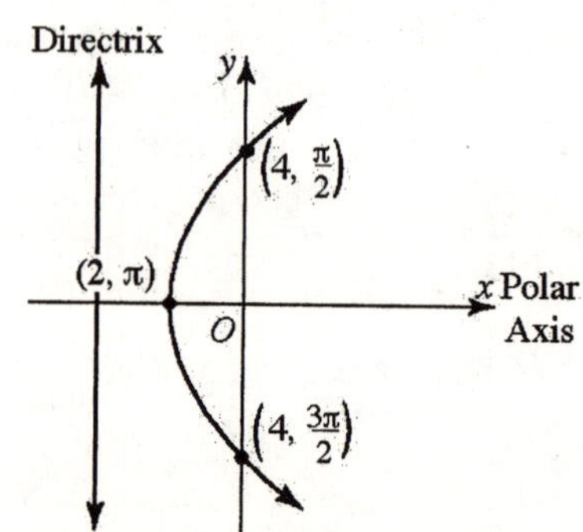

54. $r = \dfrac{6}{1+\sin\theta}$
$ep = 6,\ e = 1,\ p = 6$
Parabola; directrix is parallel to the polar axis 6 units above the pole; vertex is $\left(3, \dfrac{\pi}{2}\right)$.

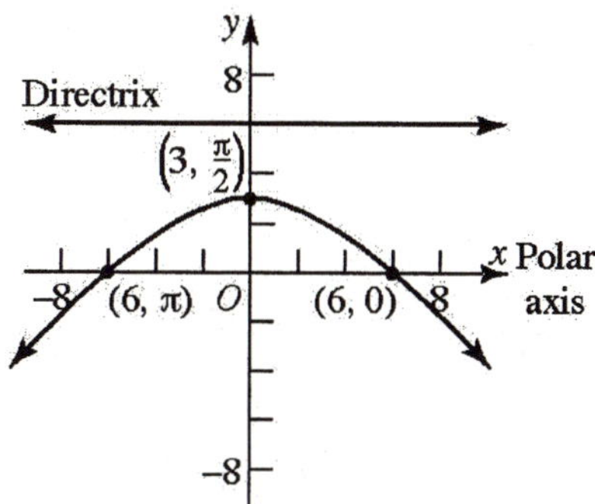

55. $r = \dfrac{6}{2-\sin\theta} = \dfrac{3}{\left(1-\frac{1}{2}\sin\theta\right)}$
$ep = 3,\ e = \dfrac{1}{2},\ p = 6$
Ellipse; directrix is parallel to the polar axis 6 units below the pole; vertices are $\left(6, \dfrac{\pi}{2}\right)$ and $\left(2, \dfrac{3\pi}{2}\right)$. Center at $\left(2, \dfrac{\pi}{2}\right)$; other focus at $\left(4, \dfrac{\pi}{2}\right)$.

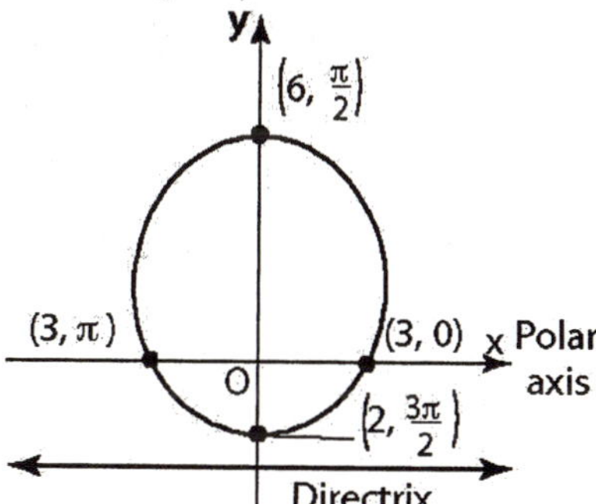

56. $r = \dfrac{2}{3+2\cos\theta} = \dfrac{\left(\frac{2}{3}\right)}{\left(1+\frac{2}{3}\cos\theta\right)}$
$ep = \dfrac{2}{3},\ e = \dfrac{2}{3},\ p = 1$
Ellipse; directrix is perpendicular to the polar axis 1 unit to the right of the pole; vertices are $\left(\dfrac{2}{5}, 0\right)$ and $(2, \pi)$. Center at $\left(\dfrac{1}{5}, \pi\right)$; other focus at $\left(\dfrac{8}{5}, \pi\right)$.

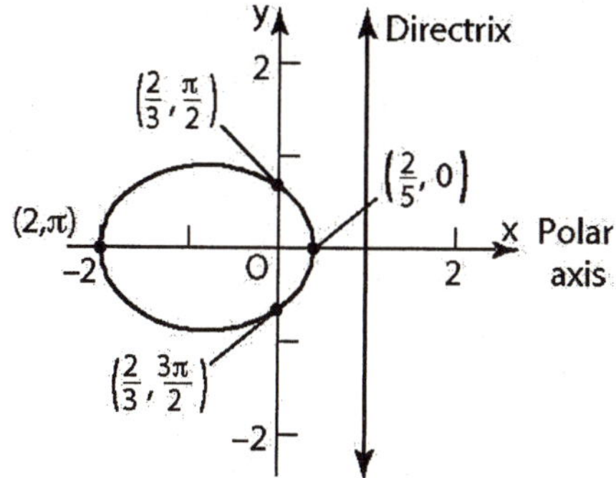

57. $r = \dfrac{8}{4+8\cos\theta} = \dfrac{2}{1+2\cos\theta}$
$ep = 2,\ e = 2,\ p = 1$
Hyperbola; directrix is perpendicular to the polar axis 1 unit to the right of the pole; vertices are $\left(\dfrac{2}{3}, 0\right)$ and $(-2, \pi)$. Center at $\left(\dfrac{4}{3}, 0\right)$; other focus at $\left(-\dfrac{8}{3}, \pi\right)$.

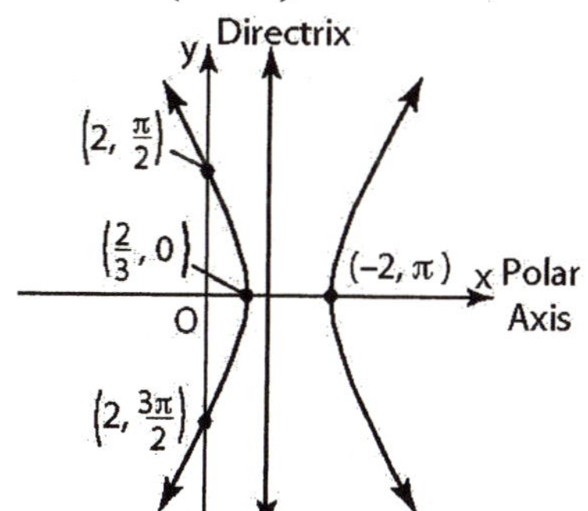

58. $r = \dfrac{10}{5+20\sin\theta} = \dfrac{2}{1+4\sin\theta}$

$ep = 2,\ e = 4,\ p = \dfrac{1}{2}$

Hyperbola; directrix is parallel to the polar axis $\dfrac{1}{2}$ unit above the pole; vertices are $\left(\dfrac{2}{5}, \dfrac{\pi}{2}\right)$ and $\left(-\dfrac{2}{3}, \dfrac{3\pi}{2}\right)$. Center at $\left(\dfrac{8}{15}, \dfrac{\pi}{2}\right)$; other focus at $\left(-\dfrac{16}{15}, \dfrac{3\pi}{2}\right)$.

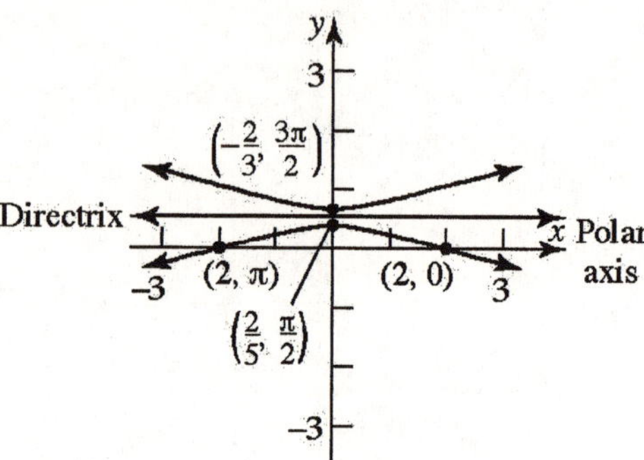

59. $r = \dfrac{4}{1-\cos\theta}$

$$r - r\cos\theta = 4$$
$$r = 4 + r\cos\theta$$
$$r^2 = (4 + r\cos\theta)^2$$
$$x^2 + y^2 = (4 + x)^2$$
$$x^2 + y^2 = 16 + 8x + x^2$$
$$y^2 - 8x - 16 = 0$$

60. $r = \dfrac{6}{2-\sin\theta}$

$$2r - r\sin\theta = 6$$
$$2r = 6 + r\sin\theta$$
$$4r^2 = (6 + r\sin\theta)^2$$
$$4\left(x^2 + y^2\right) = (6 + y)^2$$
$$4x^2 + 4y^2 = 36 + 12y + y^2$$
$$4x^2 + 3y^2 - 12y - 36 = 0$$

61. $r = \dfrac{8}{4+8\cos\theta}$

$$4r + 8r\cos\theta = 8$$
$$4r = 8 - 8r\cos\theta$$
$$r = 2 - 2r\cos\theta$$
$$r^2 = (2 - 2r\cos\theta)^2$$
$$x^2 + y^2 = (2 - 2x)^2$$
$$x^2 + y^2 = 4 - 8x + 4x^2$$
$$3x^2 - y^2 - 8x + 4 = 0$$

62. $r = \dfrac{2}{3+2\cos\theta}$

$$3r + 2r\cos\theta = 2$$
$$3r = 2 - 2r\cos\theta$$
$$9r^2 = (2 - 2r\cos\theta)^2$$
$$9\left(x^2 + y^2\right) = (2 - 2x)^2$$
$$9x^2 + 9y^2 = 4 - 8x + 4x^2$$
$$5x^2 + 9y^2 + 8x - 4 = 0$$

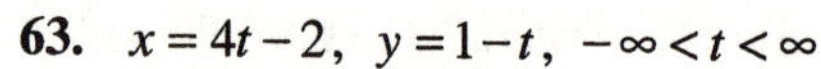

63. $x = 4t - 2,\ y = 1 - t,\ -\infty < t < \infty$

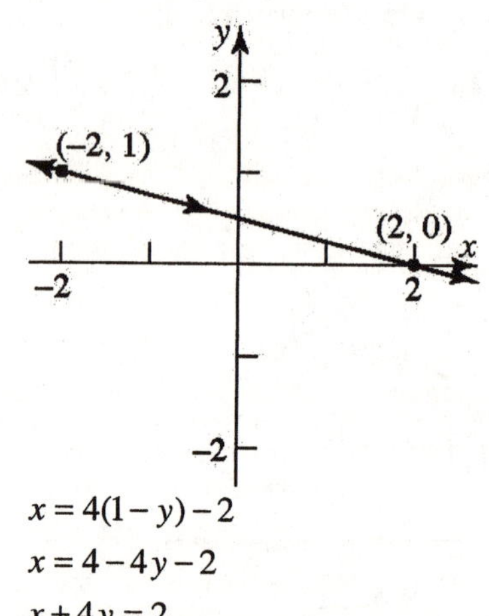

$$x = 4(1 - y) - 2$$
$$x = 4 - 4y - 2$$
$$x + 4y = 2$$

64. $x = 2t^2 + 6,\ y = 5 - t,\ -\infty < t < \infty$

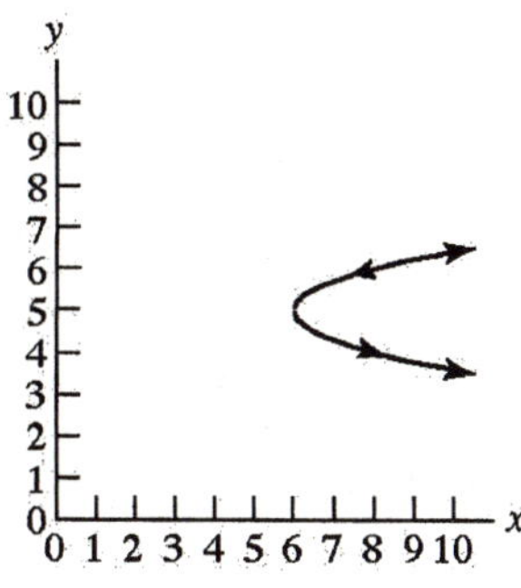

$$x = 2(5 - y)^2 + 6$$
$$x = 2\left(25 - 10y + y^2\right) + 6$$
$$x = 50 - 20y + 2y^2 + 6$$
$$x = 2y^2 - 20y + 56$$

65. $x = 3\sin t,\ y = 4\cos t + 2,\ 0 \le t \le 2\pi$

$$\frac{x}{3} = \sin t,\quad \frac{y-2}{4} = \cos t$$
$$\sin^2 t + \cos^2 t = 1$$
$$\left(\frac{x}{3}\right)^2 + \left(\frac{y-2}{4}\right)^2 = 1$$
$$\frac{x^2}{9} + \frac{(y-2)^2}{16} = 1$$

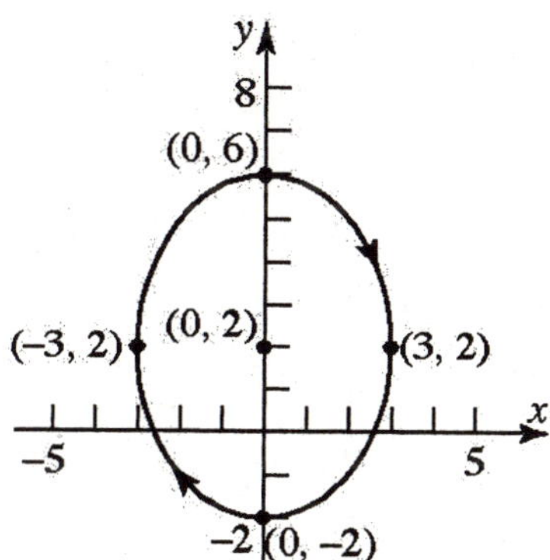

66. $x = \ln t,\ y = t^3,\ t > 0$

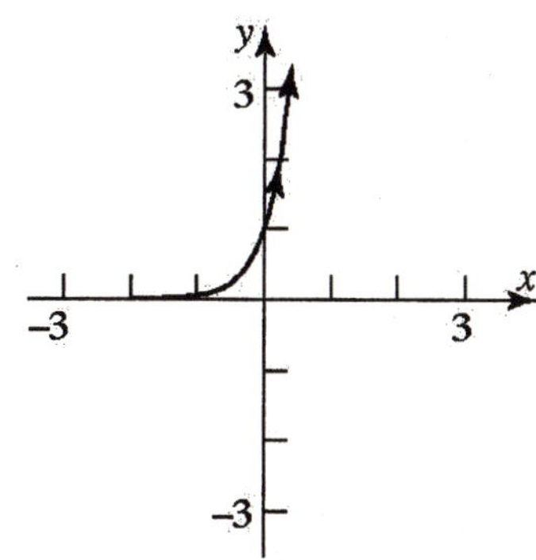

$$\ln y = \ln t^3$$
$$\ln y = 3\ln t$$
$$\ln y = 3x \quad \rightarrow \quad y = e^{3x}$$

67. $x = \sec^2 t,\ y = \tan^2 t,\ 0 \le t \le \frac{\pi}{4}$

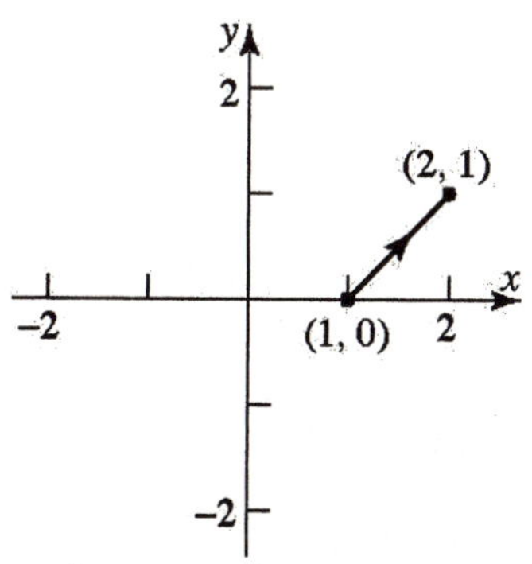

$$\tan^2 t + 1 = \sec^2 t \rightarrow y + 1 = x$$

68. $x = t^{3/2},\ y = 2t + 4,\ t \ge 0$

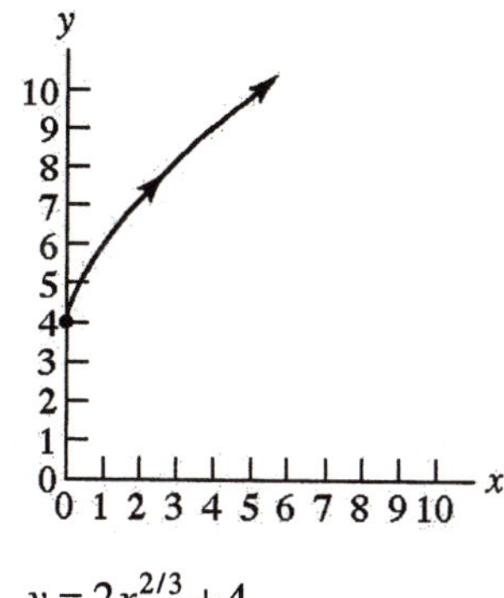

$$y = 2x^{2/3} + 4$$

69. Answers will vary. One example:

$$y = -2x + 4$$
$$x = t,\ y = -2t + 4$$
$$x = \frac{4-t}{2},\ y = t$$

70. Answers will vary. One example:

$y = 2x^2 - 8$

$x = t,\ y = 2t^2 - 8$

$x = 2t,\ y = 2(2t)^2 - 8 = 8t^2 - 8$

71. $\frac{x}{4} = \cos(\omega t);\ \frac{y}{3} = \sin(\omega t)$

$\frac{2\pi}{\omega} = 4 \Rightarrow 4\omega = 2\pi \Rightarrow \omega = \frac{2\pi}{4} = \frac{\pi}{2}$

$\therefore\ \frac{x}{4} = \cos\left(\frac{\pi}{2}t\right); \frac{y}{3} = \sin\left(\frac{\pi}{2}t\right)$

$\Rightarrow x = 4\cos\left(\frac{\pi}{2}t\right);\ y = 3\sin\left(\frac{\pi}{2}t\right),\quad 0 \le t \le 4$

72. $\frac{x}{4} = \sin(\omega t);\ \frac{y}{3} = \cos(\omega t)$

$\frac{2\pi}{\omega} = 5 \Rightarrow 5\omega = 2\pi \Rightarrow \omega = \frac{2\pi}{5}$

$\therefore\ \frac{x}{4} = \sin\left(\frac{2\pi}{5}t\right);\ \frac{y}{3} = \cos\left(\frac{2\pi}{5}t\right)$

$\Rightarrow x = 4\sin\left(\frac{2\pi}{5}t\right);\ y = 3\cos\left(\frac{2\pi}{5}t\right),\quad 0 \le t \le 5$

73. Write the equation in standard form:

$4x^2 + 9y^2 = 36 \rightarrow \frac{x^2}{9} + \frac{y^2}{4} = 1$

The center of the ellipse is (0, 0). The major axis is the x-axis.

$a = 3;\ b = 2;$

$c^2 = a^2 - b^2 = 9 - 4 = 5 \rightarrow c = \sqrt{5}.$

For the ellipse:

Vertices: (–3, 0), (3, 0);

Foci: $\left(-\sqrt{5}, 0\right), \left(\sqrt{5}, 0\right)$

For the hyperbola:

Foci: (–3, 0), (3, 0);

Vertices: $\left(-\sqrt{5}, 0\right), \left(\sqrt{5}, 0\right)$;

Center: (0, 0)

$a = \sqrt{5};\ c = 3;$

$b^2 = c^2 - a^2 = 9 - 5 = 4 \rightarrow b = 2$

The equation of the hyperbola is: $\frac{x^2}{5} - \frac{y^2}{4} = 1$

74. Write the equation in standard form:

$x^2 - 4y^2 = 16 \rightarrow \frac{x^2}{16} - \frac{y^2}{4} = 1$

The center of the hyperbola is (0, 0). The transverse axis is the x-axis.

$a = 4;\ b = 2;$

$c^2 = a^2 + b^2 = 16 + 4 = 20 \rightarrow c = \sqrt{20} = 2\sqrt{5}.$

For the hyperbola:

Vertices: (–4, 0), (4, 0);

Foci: $\left(-2\sqrt{5}, 0\right), \left(2\sqrt{5}, 0\right)$

For the ellipse:

Foci: (–4, 0), (4, 0);

Vertices: $\left(-2\sqrt{5}, 0\right), \left(2\sqrt{5}, 0\right)$;

Center: (0, 0)

$a = 2\sqrt{5};\ c = 4;$

$b^2 = a^2 - c^2 = 20 - 16 = 4 \rightarrow b = 2$

The equation of the ellipse is: $\frac{x^2}{20} + \frac{y^2}{4} = 1$

75. Let (*x*, *y*) be any point in the collection of points.

The distance from

(x, y) to $(3, 0) = \sqrt{(x-3)^2 + y^2}$.

The distance from

(x, y) to the line $x = \frac{16}{3}$ is $\left|x - \frac{16}{3}\right|$.

Relating the distances, we have:

$$\sqrt{(x-3)^2 + y^2} = \frac{3}{4}\left|x - \frac{16}{3}\right|$$

$$(x-3)^2 + y^2 = \frac{9}{16}\left(x - \frac{16}{3}\right)^2$$

$$x^2 - 6x + 9 + y^2 = \frac{9}{16}\left(x^2 - \frac{32}{3}x + \frac{256}{9}\right)$$

$$16x^2 - 96x + 144 + 16y^2 = 9x^2 - 96x + 256$$

$$7x^2 + 16y^2 = 112$$

$$\frac{7x^2}{112} + \frac{16y^2}{112} = 1$$

$$\frac{x^2}{16} + \frac{y^2}{7} = 1$$

The set of points is an ellipse.

76. Let (x, y) be any point in the collection of points. The distance from (x, y) to $(5, 0) = \sqrt{(x-5)^2 + y^2}$.
The distance from (x, y) to the line $x = \frac{16}{5}$ is $\left|x - \frac{16}{5}\right|$.
Relating the distances, we have:

$$\sqrt{(x-5)^2 + y^2} = \frac{5}{4}\left|x - \frac{16}{5}\right|$$

$$(x-5)^2 + y^2 = \frac{25}{16}\left(x - \frac{16}{5}\right)^2$$

$$x^2 - 10x + 25 + y^2 = \frac{25}{16}\left(x^2 - \frac{32}{5}x + \frac{256}{25}\right)$$

$$16x^2 - 160x + 400 + 16y^2 = 25x^2 - 160x + 256$$

$$9x^2 - 16y^2 = 144$$

$$\frac{9x^2}{144} - \frac{16y^2}{144} = 1$$

$$\frac{x^2}{16} - \frac{y^2}{9} = 1$$

The set of points is a hyperbola.

77. Locate the parabola so that the vertex is at (0, 0) and opens up. It then has the equation: $x^2 = 4ay$. Since the light source is located at the focus and is 1 foot from the base, $a = 1$. Thus, $x^2 = 4y$. The diameter is 2, so the point (1, y) is located on the parabola. Solve for y:
$1^2 = 4y \to 1 = 4y \to y = 0.25$ feet
The mirror is 0.25 feet, or 3 inches, deep.

78. Set up the problem so that the vertex of the parabola is at (0, 0) and it opens down. Then the equation of the parabola has the form: $x^2 = cy$. The point (30, –20) is a point on the parabola. Solve for c and find the equation:
$30^2 = c(-20) \to c = -45$
$x^2 = -45y$

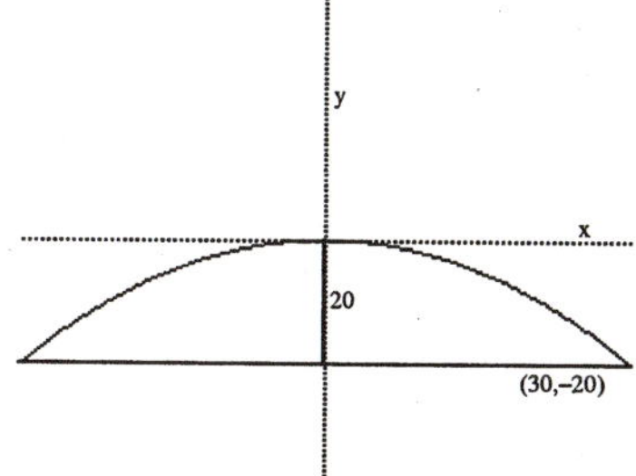

To find the height of the bridge, 5 feet from the center, the point (5, y) is a point on the parabola. Solve for y:
$5^2 = -45y \to 25 = -45y \to y \approx -0.56$
The height of the bridge, 5 feet from the center, is 20 – 0.56 = 19.44 feet.
To find the height of the bridge, 10 feet from the center, the point (10, y) is a point on the parabola. Solve for y:
$10^2 = -45y \to 100 = -45y \to y \approx -2.22$
The height of the bridge, 10 feet from the center, is 20 – 2.22 = 17.78 feet.
To find the height of the bridge, 20 feet from the center, the point (20, y) is a point on the parabola. Solve for y:
$20^2 = -45y \to 400 = -45y \to y \approx -8.89$
The height of the bridge, 20 feet from the center, is 20 – 8.89 = 11.11 feet.

79. Place the semi-elliptical arch so that the x-axis coincides with the water and the y-axis passes through the center of the arch. Since the bridge has a span of 60 feet, the length of the major axis is 60, or $2a = 60$ or $a = 30$. The maximum height of the bridge is 20 feet, so $b = 20$. The equation is: $\frac{x^2}{900} + \frac{y^2}{400} = 1$.
The height 5 feet from the center:

$$\frac{5^2}{900} + \frac{y^2}{400} = 1$$

$$\frac{y^2}{400} = 1 - \frac{25}{900}$$

$$y^2 = 400 \cdot \frac{875}{900} \to y \approx 19.72 \text{ feet}$$

The height 10 feet from the center:

$$\frac{10^2}{900} + \frac{y^2}{400} = 1$$

$$\frac{y^2}{400} = 1 - \frac{100}{900}$$

$$y^2 = 400 \cdot \frac{800}{900} \to y \approx 18.86 \text{ feet}$$

The height 20 feet from the center:

$$\frac{20^2}{900} + \frac{y^2}{400} = 1$$

$$\frac{y^2}{400} = 1 - \frac{400}{900}$$

$$y^2 = 400 \cdot \frac{500}{900} \to y \approx 14.91 \text{ feet}$$

80. The major axis is 80 feet; therefore, $2a = 80$ or $a = 40$. The maximum height is 25 feet, so $b = 25$. To locate the foci, find c:

$$c^2 = a^2 - b^2 = 1600 - 625 = 975 \quad \rightarrow \quad c \approx 31.22$$

The foci are 31.22 feet from the center of the room or 8.78 feet from each wall.

81. First note that all points where an explosion could take place, such that the time difference would be the same as that for the first detonation, would form a hyperbola with A and B as the foci.

Start with a diagram:

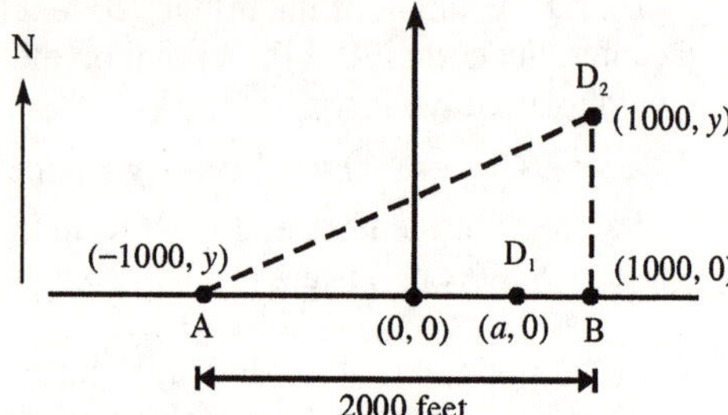

Since A and B are the foci, we have

$$2c = 2000$$
$$c = 1000$$

Since D_1 is on the transverse axis and is on the hyperbola, then it must be a vertex of the hyperbola. Since it is 200 feet from B, we have $a = 800$. Finally,

$$b^2 = c^2 - a^2 = 1000^2 - 800^2 = 360{,}000$$

Thus, the equation of the hyperbola is

$$\frac{x^2}{640{,}000} - \frac{y^2}{360{,}000} = 1$$

The point $(1000, y)$ needs to lie on the graph of the hyperbola. Thus, we have

$$\frac{(1000)^2}{640{,}000} - \frac{y^2}{360{,}000} = 1$$
$$-\frac{y^2}{360{,}000} = -\frac{9}{16}$$
$$y^2 = 202{,}500$$
$$y = 450$$

The second explosion should be set off 450 feet due north of point B.

82. a. Train:

$x_1 = \frac{3}{2}t^2$;

Let $y_1 = 1$ for plotting convenience.

Mary:

$x_2 = 6(t-2)$;

Let $y_2 = 3$ for plotting convenience.

b. Mary will catch the train if $x_1 = x_2$.

$$\frac{3}{2}t^2 = 6(t-2) \quad t = 2$$
$$\frac{3}{2}t^2 = 6t - 12 \quad t = 3$$
$$t = 4$$
$$t^2 - 4t + 8 = 0$$

Since $b^2 - 4ac = (-4)^2 - 4(1)(8)$
$= 16 - 32 = -16 < 0$

the equation has no real solution. Thus, Mary will not catch the train.

c.

y
5
t = 2
t = 3 t = 4
Mary
t = 2 t = 3 t = 4
Train
10 20 x

83. a. Use equation (2) in section 10.7.

$$x = \left(80\cos(35°)\right)t$$
$$y = -\frac{1}{2}(32)t^2 + \left(80\sin(35°)\right)t + 6$$

b. The ball is in the air until $y = 0$. Solve:

$$-16t^2 + \left(80\sin(35°)\right)t + 6 = 0$$
$$t = \frac{-80\sin(35°) \pm \sqrt{\left(80\sin(35°)\right)^2 - 4(-16)(6)}}{2(-16)}$$
$$\approx \frac{-45.89 \pm \sqrt{2489.54}}{-32}$$
$$\approx -0.13 \text{ or } 2.99$$

The ball is in the air for about 2.99 seconds. (The negative solution is extraneous.)

c. The maximum height occurs at the vertex of the quadratic function.

$$t = \frac{-b}{2a} = \frac{-80\sin(35°)}{2(-16)} \approx 1.43 \text{ seconds}$$

Evaluate the function to find the maximum height:

$$-16(1.43)^2 + (80\sin(35°))(1.43) + 6 \approx 38.9 \text{ ft}$$

d. Find the horizontal displacement:

$$x = (80\cos(35°))(2.99) \approx 196 \text{ feet}$$

e.

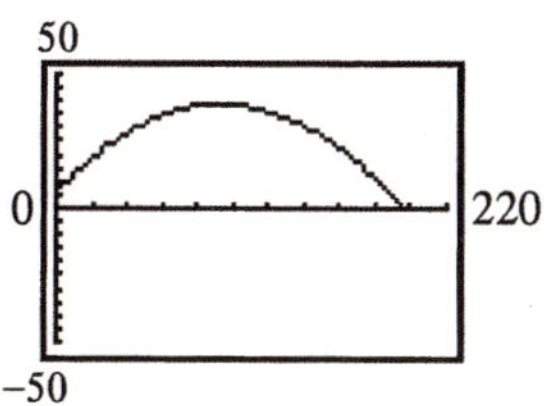

84. Answers may vary.

Chapter 9 Test

1. $\dfrac{(x+1)^2}{4} - \dfrac{y^2}{9} = 1$

Rewriting the equation as $\dfrac{(x-(-1))^2}{2^2} - \dfrac{(y-0)^2}{3^2} = 1$, we see that this is the equation of a hyperbola in the form $\dfrac{(x-h)^2}{a^2} - \dfrac{(y-k)^2}{b^2} = 1$. Therefore, we have $h = -1$, $k = 0$, $a = 2$, and $b = 3$. Since $a^2 = 4$ and $b^2 = 9$, we get $c^2 = a^2 + b^2 = 4 + 9 = 13$, or $c = \sqrt{13}$. The center is at $(-1, 0)$ and the transverse axis is the x-axis. The vertices are at $(h \pm a, k) = (-1 \pm 2, 0)$, or $(-3, 0)$ and $(1, 0)$. The foci are at $(h \pm c, k) = \left(-1 \pm \sqrt{13}, 0\right)$, or $\left(-1 - \sqrt{13}, 0\right)$ and $\left(-1 + \sqrt{13}, 0\right)$. The asymptotes are $y - 0 = \pm\frac{3}{2}(x - (-1))$, or $y = -\frac{3}{2}(x+1)$ and $y = \frac{3}{2}(x+1)$.

2. $8y = (x-1)^2 - 4$

Rewriting gives

$$(x-1)^2 = 8y + 4$$

$$(x-1)^2 = 8\left(y - \left(-\frac{1}{2}\right)\right)$$

$$(x-1)^2 = 4(2)\left(y - \left(-\frac{1}{2}\right)\right)$$

This is the equation of a parabola in the form $(x-h)^2 = 4a(y-k)$. Therefore, the axis of symmetry is parallel to the y-axis and we have $(h,k) = \left(1, -\frac{1}{2}\right)$ and $a = 2$. The vertex is at $(h,k) = \left(1, -\frac{1}{2}\right)$, the axis of symmetry is $x = 1$, the focus is at $(h, k+a) = \left(1, -\frac{1}{2} + 2\right) = \left(1, \frac{3}{2}\right)$, and the directrix is given by the line $y = k - a$, or $y = -\frac{5}{2}$.

3. $2x^2 + 3y^2 + 4x - 6y = 13$

Rewrite the equation by completing the square in x and y.

$$2x^2 + 3y^2 + 4x - 6y = 13$$

$$2x^2 + 4x + 3y^2 - 6y = 13$$

$$2\left(x^2 + 2x\right) + 3\left(y^2 - 2y\right) = 13$$

$$2\left(x^2 + 2x + 1\right) + 3\left(y^2 - 2y + 1\right) = 13 + 2 + 3$$

$$2(x+1)^2 + 3(y-1)^2 = 18$$

$$\frac{(x-(-1))^2}{9} + \frac{(y-1)^2}{6} = 1$$

This is the equation of an ellipse with center at $(-1, 1)$ and major axis parallel to the x-axis. Since $a^2 = 9$ and $b^2 = 6$, we have $c^2 = a^2 - b^2 = 9 - 6 = 3$, or $c = \sqrt{3}$. The foci are $(h \pm c, k) = \left(-1 \pm \sqrt{3}, 1\right)$ or $\left(-1 - \sqrt{3}, 1\right)$ and $\left(-1 + \sqrt{3}, 1\right)$. The vertices are at $(h \pm a, k) = (-1 \pm 3, 1)$, or $(-4, 1)$ and $(2, 1)$.

4. The vertex $(-1,3)$ and the focus $(-1,4.5)$ both lie on the vertical line $x=-1$ (the axis of symmetry). The distance a from the vertex to the focus is $a=1.5$. Because the focus lies above the vertex, we know the parabola opens upward. As a result, the form of the equation is
$(x-h)^2 = 4a(y-k)$
where $(h,k)=(-1,3)$ and $a=1.5$. Therefore, the equation is
$(x+1)^2 = 4(1.5)(y-3)$
$(x+1)^2 = 6(y-3)$
The points $(h\pm 2a,k)$, that is $(-4,4.5)$ and $(2,4.5)$, define the lattice rectum; the line $y=1.5$ is the directrix.

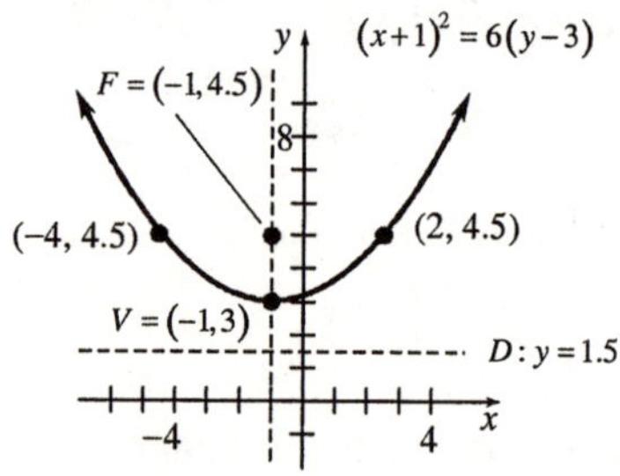

5. The center is $(h,k)=(0,0)$ so $h=0$ and $k=0$. Since the center, focus, and vertex all lie on the line $x=0$, the major axis is the y-axis. The distance from the center $(0,0)$ to a focus $(0,3)$ is $c=3$. The distance from the center $(0,0)$ to a vertex $(0,-4)$ is $a=4$. Then,
$b^2 = a^2 - c^2 = 4^2 - 3^2 = 16 - 9 = 7$
The form of the equation is
$$\frac{(x-h)^2}{b^2}+\frac{(y-k)^2}{a^2}=1$$
where $h=0$, $k=0$, $a=4$, and $b=\sqrt{7}$. Thus, we get
$$\frac{x^2}{7}+\frac{y^2}{16}=1$$
To graph the equation, we use the center $(h,k)=(0,0)$ to locate the vertices. The major axis is the y-axis, so the vertices are $a=4$ units above and below the center. Therefore, the vertices are $V_1=(0,4)$ and $V_2=(0,-4)$. Since $c=3$ and the major axis is the y-axis, the foci are 3 units above and below the center. Therefore, the foci are $F_1=(0,3)$ and $F_2=(0,-3)$. Finally, we use the value $b=\sqrt{7}$ to find the two points left and right of the center: $(-\sqrt{7},0)$ and $(\sqrt{7},0)$.

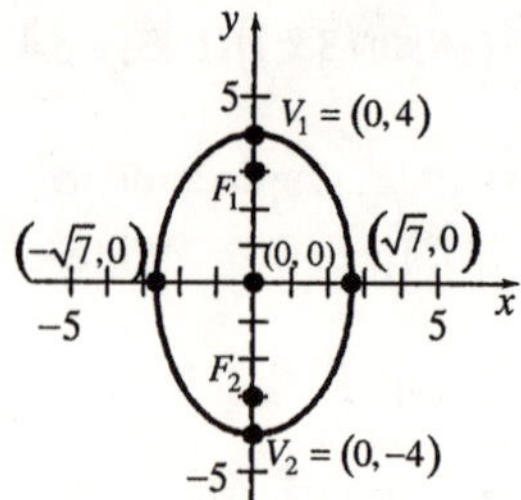

6. The center $(h,k)=(2,2)$ and vertex $(2,4)$ both lie on the line $x=2$, the transverse axis is parallel to the y-axis. The distance from the center $(2,2)$ to the vertex $(2,4)$ is $a=2$, so the other vertex must be $(2,0)$. The form of the equation is
$$\frac{(y-k)^2}{a^2}-\frac{(x-h)^2}{b^2}=1$$
where $h=2$, $k=2$, and $a=2$. This gives us
$$\frac{(y-2)^2}{4}-\frac{(x-2)^2}{b^2}=1$$
Since the graph contains the point $(x,y)=(2+\sqrt{10},5)$, we can use this point to determine the value for b.
$$\frac{(5-2)^2}{4}-\frac{(2+\sqrt{10}-2)^2}{b^2}=1$$
$$\frac{9}{4}-\frac{10}{b^2}=1$$
$$\frac{5}{4}=\frac{10}{b^2}$$
$$b^2=8$$
$$b=2\sqrt{2}$$
Therefore, the equation becomes
$$\frac{(y-2)^2}{4}-\frac{(x-2)^2}{8}=1$$
Since $c^2=a^2+b^2=4+8=12$, the distance from the center to either focus is $c=2\sqrt{3}$. Therefore, the foci are $c=2\sqrt{3}$ units above and below the center. The foci are $F_1=(2,2+2\sqrt{3})$

and $F_2 = \left(2, 2-2\sqrt{3}\right)$. The asymptotes are given by the lines

$y-k=\pm\frac{a}{b}(x-h)$. Therefore, the asymptotes are

$$y-2=\pm\frac{2}{2\sqrt{2}}(x-2)$$

$$y=\pm\frac{\sqrt{2}}{2}(x-2)+2$$

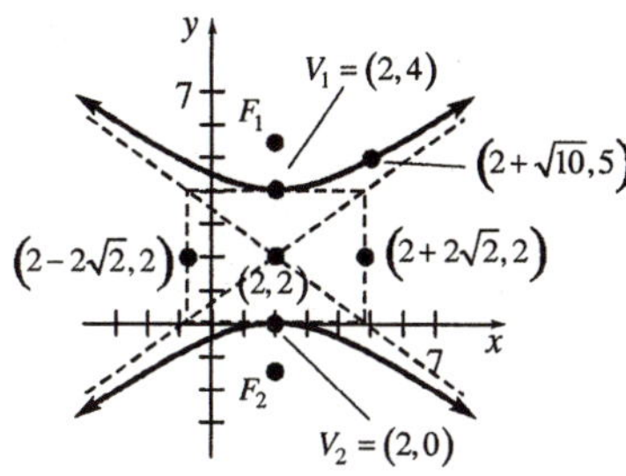

7. $2x^2+5xy+3y^2+3x-7=0$ is in the form $Ax^2+Bxy+Cy^2+Dx+Ey+F=0$ where $A=2$, $B=5$, and $C=3$.

$$\begin{aligned}B^2-4AC &= 5^2-4(2)(3)\\ &= 25-24\\ &= 1\end{aligned}$$

Since $B^2-4AC>0$, the equation defines a hyperbola.

8. $3x^2-xy+2y^2+3y+1=0$ is in the form $Ax^2+Bxy+Cy^2+Dx+Ey+F=0$ where $A=3$, $B=-1$, and $C=2$.

$$\begin{aligned}B^2-4AC &= (-1)^2-4(3)(2)\\ &= 1-24\\ &= -23\end{aligned}$$

Since $B^2-4AC<0$, the equation defines an ellipse.

9. $x^2-6xy+9y^2+2x-3y-2=0$ is in the form $Ax^2+Bxy+Cy^2+Dx+Ey+F=0$ where $A=1$, $B=-6$, and $C=9$.

$$\begin{aligned}B^2-4AC &= (-6)^2-4(1)(9)\\ &= 36-36\\ &= 0\end{aligned}$$

Since $B^2-4AC=0$, the equation defines a parabola.

10. $41x^2-24xy+34y^2-25=0$

Substituting $x=x'\cos\theta-y'\sin\theta$ and $y=x'\sin\theta+y'\cos\theta$ transforms the equation into one that represents a rotation through an angle θ. To eliminate the $x'y'$ term in the new equation, we need $\cot(2\theta)=\frac{A-C}{B}$. That is, we need to solve

$$\cot(2\theta)=\frac{41-34}{-24}$$

$$\cot(2\theta)=-\frac{7}{24}$$

Since $\cot(2\theta)<0$, we may choose $90°<2\theta<180°$, or $45°<\theta<90°$.

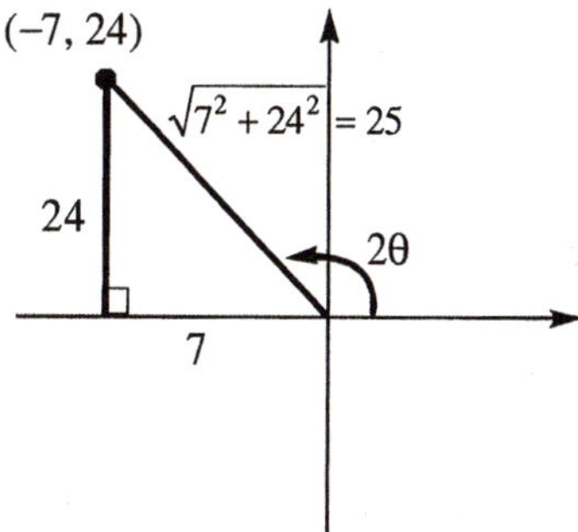

We have $\cot(2\theta)=-\frac{7}{24}$ so it follows that $\cos(2\theta)=-\frac{7}{25}$.

$$\sin\theta=\sqrt{\frac{1-\cos(2\theta)}{2}}=\sqrt{\frac{1-\left(-\frac{7}{25}\right)}{2}}=\sqrt{\frac{16}{25}}=\frac{4}{5}$$

$$\cos\theta=\sqrt{\frac{1+\cos(2\theta)}{2}}=\sqrt{\frac{1+\left(-\frac{7}{25}\right)}{2}}=\sqrt{\frac{9}{25}}=\frac{3}{5}$$

$$\theta=\cos^{-1}\left(\frac{3}{5}\right)\approx 53.13°$$

With these values, the rotation formulas are

$$x=\frac{3}{5}x'-\frac{4}{5}y' \quad\text{and}\quad y=\frac{4}{5}x'+\frac{3}{5}y'$$

Substituting these values into the original equation and simplifying, we obtain

$$41x^2-24xy+34y^2-25=0$$

$$41\left(\frac{3}{5}x'-\frac{4}{5}y'\right)^2-24\left(\frac{3}{5}x'-\frac{4}{5}y'\right)\left(\frac{4}{5}x'+\frac{3}{5}y'\right)$$
$$+34\left(\frac{4}{5}x'+\frac{3}{5}y'\right)^2=25$$

Multiply both sides by 25 and expand to obtain

$41\left(9x'^2-24x'y'+16y'^2\right)-24\left(12x'^2-7x'y'-12y'^2\right)$

$+34\left(16x'^2+24x'y'+9y'^2\right)=625$

$625x'^2+1250y'^2=625$

$x'^2+2y'^2=1$

$$\frac{x'^2}{1}+\frac{y'^2}{\frac{1}{2}}=1$$

Thus: $a=\sqrt{1}=1$ and $b=\sqrt{\frac{1}{2}}=\frac{\sqrt{2}}{2}$

This is the equation of an ellipse with center at $(0,0)$ in the $x'y'$–plane. The vertices are at $(-1,0)$ and $(1,0)$ in the $x'y'$–plane.

$c^2=a^2-b^2=1-\frac{1}{2}=\frac{1}{2} \quad \to \quad c=\frac{\sqrt{2}}{2}$

The foci are located at $\left(\pm\frac{\sqrt{2}}{2},0\right)$ in the $x'y'$–plane. In summary:

	x′y′ – plane	*xy – plane*
center	(0,0)	(0,0)
vertices	$(\pm 1,0)$	$\left(\frac{3}{5},\frac{4}{5}\right),\left(-\frac{3}{5},-\frac{4}{5}\right)$
minor axis intercepts	$\left(0,\pm\frac{\sqrt{2}}{2}\right)$	$\left(\frac{2\sqrt{2}}{5},\frac{3\sqrt{2}}{10}\right)$, $\left(-\frac{2\sqrt{2}}{5},-\frac{3\sqrt{2}}{10}\right)$
foci	$\left(\pm\frac{\sqrt{2}}{2},0\right)$	$\left(\frac{3\sqrt{2}}{10},\frac{2\sqrt{2}}{5}\right)$, $\left(-\frac{3\sqrt{2}}{10},-\frac{2\sqrt{2}}{5}\right)$

The graph is given below.

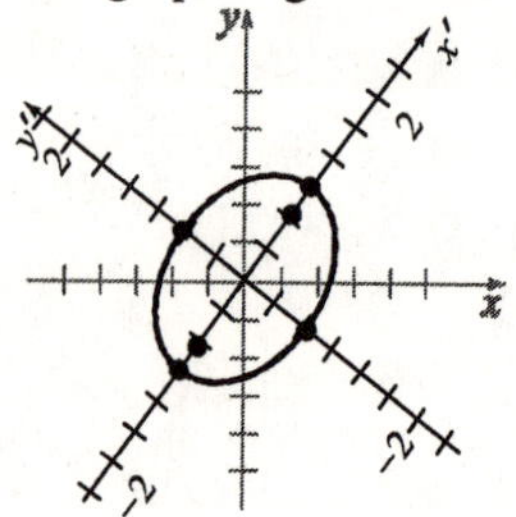

11. $r=\frac{ep}{1-e\cos\theta}=\frac{3}{1-2\cos\theta}$

Therefore, $e=2$ and $p=\frac{3}{2}$. Since $e>1$, this is the equation of a hyperbola.

$$r=\frac{3}{1-2\cos\theta}$$
$$r(1-2\cos\theta)=3$$
$$r-2r\cos\theta=3$$

Since $x=r\cos\theta$ and $x^2+y^2=r^2$, we get

$$r-2r\cos\theta=3$$
$$r=2r\cos\theta+3$$
$$r^2=(2r\cos\theta+3)^2$$
$$x^2+y^2=(2x+3)^2$$
$$x^2+y^2=4x^2+12x+9$$
$$3x^2+12x-y^2=-9$$
$$3(x^2+4x)-y^2=-9$$
$$3(x^2+4x+4)-y^2=-9+12$$
$$3(x+2)^2-y^2=3$$
$$\frac{(x+2)^2}{1}-\frac{y^2}{3}=1$$

12. $\begin{cases} x=3t-2 & (1) \\ y=1-\sqrt{t} & (2) \end{cases}$

t	x	y	(x,y)
0	$x=3(0)-2=-2$	$y=1-\sqrt{0}=1$	$(-2,1)$
1	$x=3(1)-2=1$	$y=1-\sqrt{1}=0$	$(1,0)$
4	$x=3(4)-2=10$	$y=1-\sqrt{4}=-1$	$(10,-1)$
9	$x=3(9)-2=25$	$y=1-\sqrt{9}=-2$	$(25,-2)$

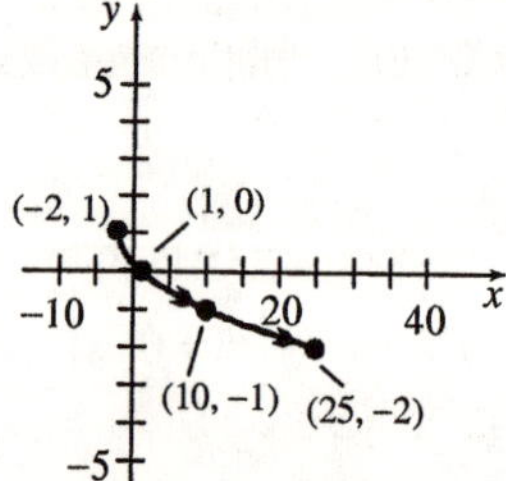

To find the rectangular equation for the curve, we need to eliminate the variable t from the equations.

We can start by solving equation (1) for t.

$$x = 3t - 2$$
$$3t = x + 2$$
$$t = \frac{x+2}{3}$$

Substituting this result for t into equation (2) gives

$$y = 1 - \sqrt{\frac{x+2}{3}}, \quad -2 \le x \le 25$$

13. We can draw the parabola used to form the reflector on a rectangular coordinate system so that the vertex of the parabola is at the origin and its focus is on the positive y-axis.

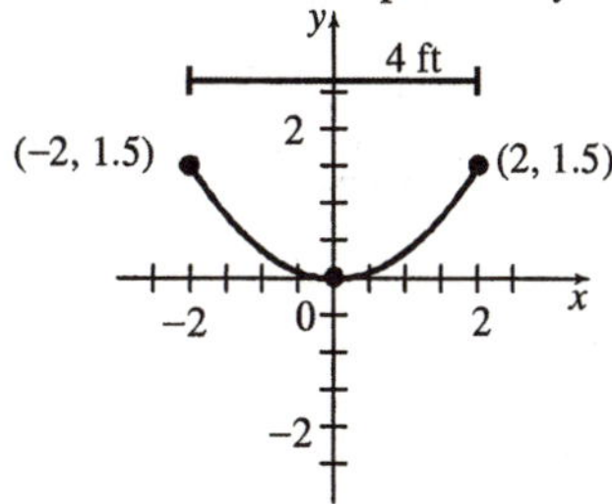

The form of the equation of the parabola is $x^2 = 4ay$ and its focus is at $(0, a)$. Since the point $(2, 1.5)$ is on the graph, we have

$$2^2 = 4a(1.5)$$
$$4 = 6a$$
$$a = \frac{2}{3}$$

The microphone should be located $\frac{2}{3}$ feet (or 8 inches) from the base of the reflector, along its axis of symmetry.

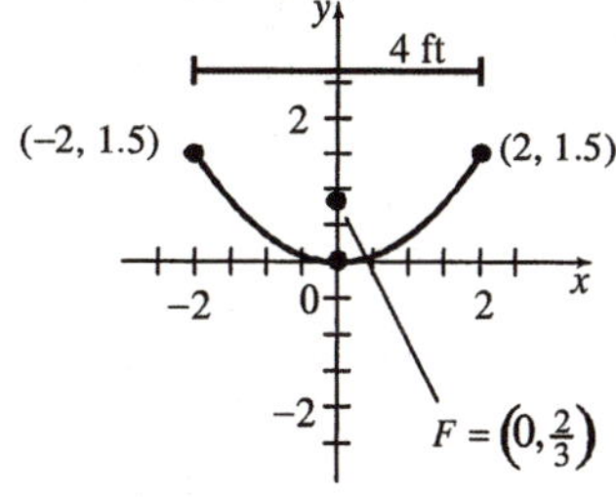

Chapter 9 Projects

Project 1

a. Figure:

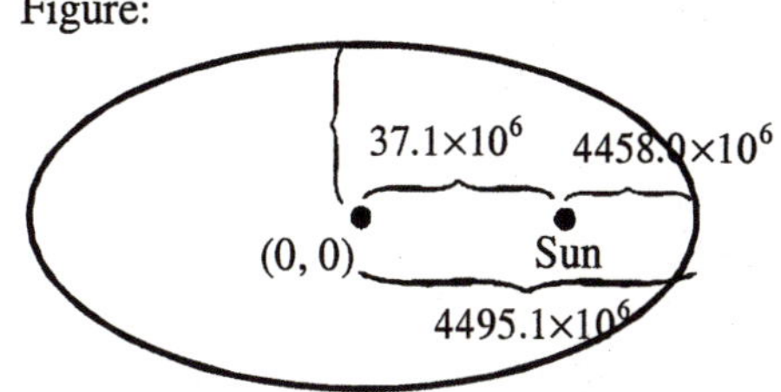

$$c = 37.2 \times 10^6$$
$$b^2 = a^2 - c^2$$
$$b^2 = \left(4495.1 \times 10^6\right)^2 - \left(37.1 \times 10^6\right)^2$$
$$b = 4494.9 \times 10^6$$
$$\frac{x^2}{(4495.1x10^6)^2} + \frac{y^2}{(4494.9x10^6)^2} = 1$$

b. Figure:

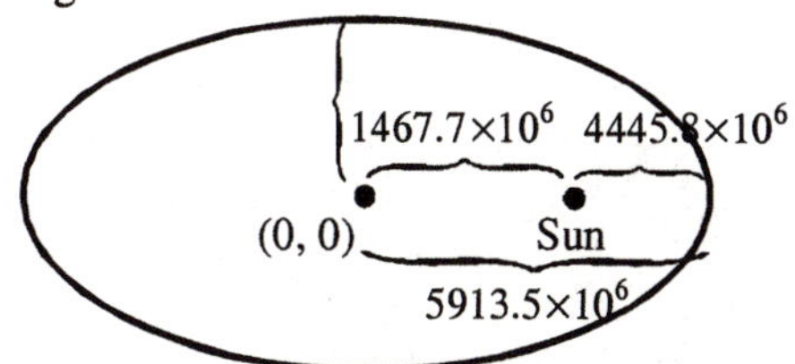

$$77381.2 \times 10^6 + 4445.8 \times 10^6 = 11827 \times 10^6$$
$$a = 0.5\left(11827 \times 10^6\right) = 5913.5 \times 10^6$$
$$c = 1467.7 \times 10^6$$
$$b^2 = \left(5913.5 \times 10^6\right)^2 - \left(1467.7 \times 10^6\right)^2$$
$$b = 5728.5 \times 10^6$$
$$\frac{x^2}{(5913.5x10^6)^2} + \frac{y^2}{(5728.5x10^6)^2} = 1$$

c. The two graphs are being graphed with the same center. Actually, the sun should remain in the same place for each graph. This means that the graph of Pluto needs to be adjusted.

d. Shift = Pluto's distance − Neptune's distance

$= 1467.7\times10^6 - 37.1\times10^6$

$= 1430.6\times10^6$

$$\frac{(x+1430.6x10^6)^2}{(5913.5x10^6)^2}+\frac{y^2}{(5728.5x10^6)^2}=1$$

e. Yes. One must adjust the scale accordingly to see it.

f. $\left(4431.6\times10^6, 752.6\times10^6\right)$,

$\left(4431.6\times10^6, 752.6\times10^6\right)$

g. No, The timing is different. They do not both pass through those points at the same time.

Project 2 (web)

1. As an example, T_1 will be used. (Note that any of the targets will yield the same result.)

$z = 4ax^2 + 4ay^2$

$0.5 = 4a(0)^2 + 4a(-2)^2$

$0.5 = 16a$

$a = \frac{1}{32}$

The focal length is 0.03125 m.

$z = \frac{1}{8}x^2 + \frac{1}{8}y^2$

2.

Target	x	R	θ	Z	y
T_1	0	9.551	−11.78	0.5	−1.950
T_2	0	9.948	−5.65	0.125	−0.979
T_3	0	9.928	5.90	0.125	1.021
T_4	0	9.708	11.89	0.5	2.000
T_5	9.510	9.722	−11.99	11.31	0
T_6	9.865	9.917	−5.85	12.165	0
T_7	9.875	9.925	5.78	12.189	0
T_8	9.350	9.551	11.78	10.928	0

$z = \frac{1}{8}x^2 + \frac{1}{8}y^2$, y = Rsinθ, x = Rcosθ

4. T_1 through T_4 do not need to be adjusted. T_5 must move 11.510 m toward the y-axis and the z coordinate must move down 10.81 m. T_6 must move 10.865 m toward the y-axis and the z coordinate must move down 12.04 m. T_7 must move 8.875 toward the y-axis and z must move down 12.064m. T_8 must move 7.35 m toward the y-axis and z must move down 10.425 m.

Project 3 (web)

Figure 1

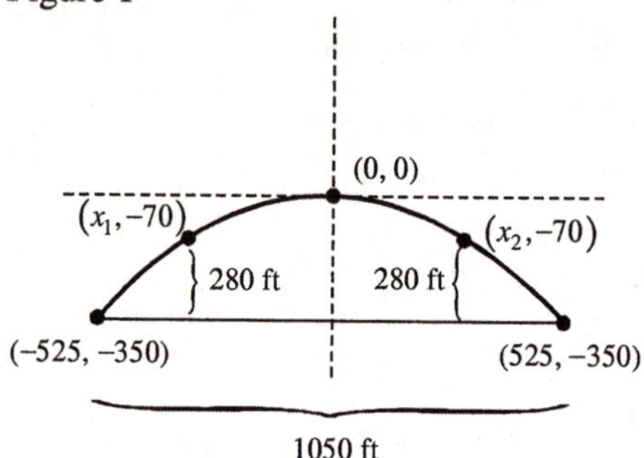

a. $x^2 = -4ay$

$(525)^2 = -4a(-350)$

$272625 = 1400a$

$a = 196.875$

$x^2 = -787.5y$

b. Let $y = -70$. (The arch needs to be 280 ft high. Remember the vertex is at (0, 0), so we must measure down to the arch from the x-axis at the point where the arch's height is 280 ft.)

$x^2 = -787.5(-70)$

$x^2 = 55125$

$x = \pm 234.8$

The channel will be 469.6 ft wide.

c. Figure 2

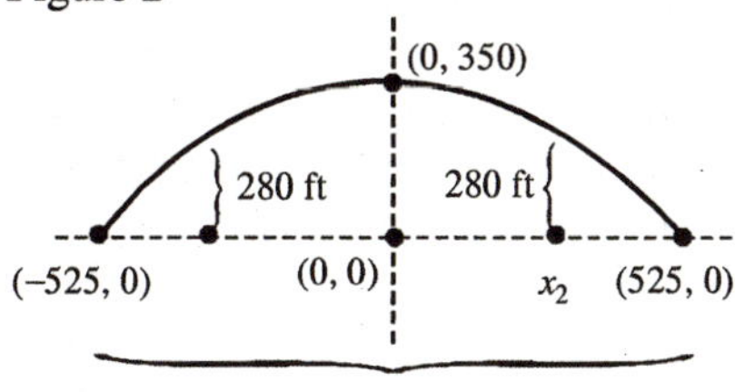

$$\frac{x^2}{a^2}+\frac{y^2}{b^2}=1$$

$$\frac{x^2}{275625}+\frac{y^2}{122500}=1$$

$$\frac{x^2}{275625}+\frac{(280)^2}{122500}=1$$

d. $x^2=\left(1-\frac{(280)^2}{122500}\right)275625$

$x^2=99225$

$x=\pm315$

The channel will be 630 feet wide.

e. If the river rises 10 feet, then we need to look for how wide the channel is when the height is 290 ft.

For the parabolic shape:

$x^2=-787.5(-60)$

$x^2=47250$

$x=\pm217.4$

There is still a 435 ft wide channel for the ship.

For the semi-ellipse:

$$\frac{x^2}{275625}+\frac{(290)^2}{122500}=1$$

$$x^2=\left(1-\frac{(290)^2}{122500}\right)275625$$

$x^2=86400$

$x=\pm293.9$

The ship has a 588-ft channel. A semi-ellipse would be more practical since the channel doesn't shrink in width in a flood as fast as a parabola.

Project 4 (web)

a. $4t-2=\sec^2 t, \qquad 0\le t\le\frac{\pi}{4}$

$1-t=\tan^2 t, \qquad 0\le t\le\frac{\pi}{4}$

For the x-values, t = 1.99, which is not in the domain [0, π/4]. Therefore there is no t-value that allows the two x values to be equal.

b. On the graphing utility, solve these in parametric form, using a t-step of π/32.
It appears that the two graphs intersect at about (1.14, 0.214). However, for the first pair, t = 0.785 at that point. That t-value gives the point (2, 1) on the second pair. There is no intersection point.

c. Since there were no solutions found for each method, the "solutions" matched.

d. $x_1=4t-2 \qquad y_1=1-t$ D=**R**; R = **R**

$1-y_1=t$

$x=4(1-y)-2$

$x+4y=2$

$x_2=\sec^2 t \qquad y_2=\tan^2 t$

$1+\tan^2 t=\sec^2 t$ D=[1,2], R=[0,1]

$1+y=x$

$x-y=1$

$$\begin{cases} x+4y=1 \\ x-y=1 \end{cases}$$

$5y=0$

$y=0$

$x=1$

The t-values that go with those x, y values are not the same for both pairs. Thus, again, there is no solution.

e. $x:\ t^{3/2}=\ln t$

$y:\ t^3=2t+4$

Graphing each of these and finding the intersection: There is no intersection for the x-values, so there is no intersection for the system.
Graphing the two parametric pairs: The parametric equations show an intersection point. However, the t-value that gives that point for each parametric pair is not the same. Thus there is no solution for the system.
Putting each parametric pair into rectangular coordinates:

$x_1 = \ln t \rightarrow t = e^x \qquad D = \mathbf{R},\ R = (0,\infty)$

$y_1 = t^3 = e^{3x}$

$x_2 = t^{3/2} \rightarrow t = x^{2/3} \qquad D = [0,\infty), \quad R = [4,\infty)$

$y_2 = 2t + 4 = 2x^{2/3} + 4$

Then solving that system:

$$\begin{cases} y = e^{3x} \\ y = 2x^{2/3} + 4 \end{cases}$$

This system has an intersection point at (0.56, 5.39).

However, ln t = 0.56, gives t = 1.75 and $t = (0.56)^{2/3} \approx 0.68$. Since the t-values are not the same, the point of intersection is false for the system.

f. x: 3 sin t = 2 cos t ➔ tan t = 2/3 ➔ t = 0.588

y: 4 cos t +2 = 4 sin t ➔ by graphing, the solution is t = 1.15 or t = 3.57.

Neither of these are the same as for the x-values, thus the system has no solution.

Graphing parametrically: If the graphs are done simultaneously on the graphing utility, the two graphs do not intersect at the same t-value. Tracing the graphs shows the same thing. This backs up the conclusion reached the first way.

$x_1 = 3\sin t \qquad y_1 = 4\cos t + 2$

$\sin t = \dfrac{x}{3} \qquad \cos t = \dfrac{y-2}{4}$

$\sin^2 t + \cos^2 t = 1$

$\dfrac{x^2}{9} + \dfrac{(y-2)^2}{16} = 1$

D=[-3, 3], R=[-2, 6]

$x_2 = 2\cos t \qquad y_2 = 4\sin t$

$\cos t = \dfrac{x}{2} \qquad \sin t = \dfrac{y}{4}$

$\cos^2 t + \sin^2 t = 1$

$\dfrac{x^2}{4} + \dfrac{y^2}{16} = 1$

D=[-2, 2], R=[-4,4]

Solving the system graphically: $x = 1.3$, $y = -3.05$. However, the t-values associated with these values are not the same. Thus there is no solution. (Similarly with the symmetric pair in the third quadrant.)

g. Efficiency depends upon the equations. Graphing the parametric pairs allows one to see immediately whether the t-values will be the same for each pair at any point of intersection. Sometimes, solving for t, as was done in the first method is easy and can be quicker. It leads straight to the t-values, so that allow the method to be efficient.

If the two graphs intersect, one must be careful to check that the t-values are the same for each curve at that point of intersection.

Cumulative Review 1-9

1. $\sin(2\theta) = 0.5$

$$2\theta = \frac{\pi}{6} + 2k\pi \quad \Rightarrow \quad \theta = \frac{\pi}{12} + k\pi,$$

$$\text{or} \quad 2\theta = \frac{5\pi}{6} + 2k\pi \quad \Rightarrow \quad \theta = \frac{5\pi}{12} + k\pi$$

where k is any integer.

2. The line containing point (0,0), making an angle of 30° with the positive x-axis has polar equation: $\theta = \dfrac{\pi}{6}$.

3. Using rectangular coordinates, the circle with center point (0, 4) and radius 4 has the equation:

$(x-h)^2 + (y-k)^2 = r^2$

$(x-0)^2 + (y-4)^2 = 4^2$

$x^2 + y^2 - 8y + 16 = 16$

$x^2 + y^2 - 8y = 0$

Converting to polar coordinates:

$r^2 - 8r\sin\theta = 0$

$r^2 = 8r\sin\theta$

$r = 8\sin\theta$

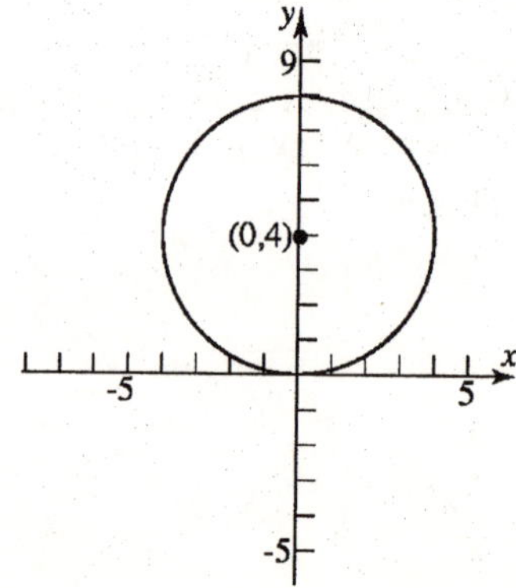

4. $f(x) = \dfrac{3}{\sin x + \cos x}$

f will be defined provided $\sin x + \cos x \neq 0$.

$\sin x + \cos x = 0$

$$\sin x = -\cos x$$

$$\frac{\sin x}{\cos x} = -1$$

$$\tan x = -1$$

$$x = \frac{3\pi}{4} + k\pi, \text{ } k \text{ is any integer}$$

The domain is

$$\left\{ x \,\middle|\, x \neq \frac{3\pi}{4} + k\pi, \text{ where } k \text{ is any integer} \right\}.$$

5. $\dfrac{f(x+h) - f(x)}{h}$

$$= \frac{-3(x+h)^2 + 5(x+h) - 2 - (-3x^2 + 5x - 2)}{h}$$

$$= \frac{-3(x^2 + 2xh + h^2) + 5x + 5h - 2 + 3x^2 - 5x + 2}{h}$$

$$= \frac{-3x^2 - 6xh - 3h^2 + 5x + 5h - 2 + 3x^2 - 5x + 2}{h}$$

$$= \frac{-6xh - 3h^2 + 5h}{h} = \frac{h(-6x - 3h + 5)}{h}$$

$$= -6x - 3h + 5$$

6. $f(x) = 3^x + 2$

a. Domain: $(-\infty, \infty)$.

Range: $(2, \infty)$.

b. $f(x) = 3^x + 2$

$y = 3^x + 2$

$x = 3^y + 2$ Inverse

$x - 2 = 3^y$

$\log_3(x-2) = y \Rightarrow f^{-1}(x) = \log_3(x-2)$

Domain of f^{-1} = range of f $= (2, \infty)$.

Range of f^{-1} = domain of f $= (-\infty, \infty)$.

7. $9x^4 + 33x^3 - 71x^2 - 57x - 10 = 0$

There are at most 4 real zeros.

Possible rational zeros:

$p = \pm 1, \pm 2, \pm 5, \pm 10;$ $q = \pm 1, \pm 3, \pm 9;$

$$\frac{p}{q} = \pm 1, \pm\frac{1}{3}, \pm\frac{1}{9}, \pm 2, \pm\frac{2}{3}, \pm\frac{2}{9}, \pm 5,$$

$$\pm\frac{5}{3}, \pm\frac{5}{9}, \pm 10, \pm\frac{10}{3}, \pm\frac{10}{9}$$

Graphing $y_1 = 9x^4 + 33x^3 - 71x^2 - 57x - 10$ indicates that there appear to be zeros at $x = -5$ and at $x = 2$.

Using synthetic division with $x = -5$:

−5	9	33	−71	−57	−10
		−45	60	55	10
	9	−12	−11	−2	0

Since the remainder is 0, –5 is a zero for *f*. So $x - (-5) = x + 5$ is a factor.

The other factor is the quotient:

$9x^3 - 12x^2 - 11x - 2$.

Thus, $f(x) = (x+5)(9x^3 - 12x^2 - 11x - 2)$.

Using synthetic division on the quotient and $x = 2$:

2	9	−12	−11	−2
		18	12	2
	9	6	1	0

Since the remainder is 0, 2 is a zero for *f*. So $x - 2$ is a factor; thus,

$$f(x) = (x+5)(x-2)(9x^2 + 6x + 1)$$

$$= (x+5)(x-2)(3x+1)(3x+1)$$

Therefore, $x = -\dfrac{1}{3}$ is also a zero for *f* (with multiplicity 2).

Solution set: $\left\{-5, -\frac{1}{3}, 2\right\}$.

8. $6-x\ge x^2$

$0\ge x^2+x-6$

$x^2+x-6\le 0$

$(x+3)(x-2)\le 0$

$f(x)=x^2+x-6$

$x=-3,\ x=2$ are the zeros of f.

Interval	$(-\infty,-3)$	$(-3,2)$	$(2,\infty)$
Test Value	–4	0	3
Value of f	6	–6	6
Conclusion	Positive	Negative	Positive

The solution set is $\{x\mid -3\le x\le 2\}$, or $[-3,2]$.

9. $\cot(2\theta)=1,$ where $0^\circ<\theta<90^\circ$

$2\theta=\frac{\pi}{4}+k\pi\Rightarrow\theta=\frac{\pi}{8}+\frac{k\pi}{2}$, where k is any integer. On the interval $0^\circ<\theta<90^\circ$, the solution is $\theta=\frac{\pi}{8}=22.5^\circ$.

10. a. This graph is a line containing points $(0,-2)$ and $(1,0)$.

$\text{slope}=\frac{\Delta y}{\Delta x}=\frac{0-(-2)}{1-0}=\frac{2}{1}=2$

using $y-y_1=m(x-x_1)$

$y-0=2(x-1)$

$y=2x-2$ or $2x-y-2=0$

b. This graph is a circle with center point (2, 0) and radius 2.

$(x-h)^2+(y-k)^2=r^2$

$(x-2)^2+(y-0)^2=2^2$

$(x-2)^2+y^2=4$

c. This graph is an ellipse with center point (0, 0); vertices $(\pm3,0)$ and y-intercepts $(0,\pm2)$.

$\frac{(x-h)^2}{a^2}+\frac{(y-k)^2}{b^2}=1$

$\frac{(x-0)^2}{3^2}+\frac{(y-0)^2}{2^2}=1\ \Rightarrow\ \frac{x^2}{9}+\frac{y^2}{4}=1$

d. This graph is a parabola with vertex $(1,0)$ and y-intercept $(0,2)$.

$(x-h)^2=4a(y-k)$

$(x-1)^2=4ay$

$(0-1)^2=4a(2)$

$1=8a$

$a=\frac{1}{8}$

$(x-1)^2=\frac{1}{2}y$ or $y=2(x-1)^2$

e. This graph is a hyperbola with center $(0,0)$ and vertices $(0,\pm1)$, containing the point $(3,2)$.

$\frac{(y-k)^2}{a^2}-\frac{(x-h)^2}{b^2}=1$

$\frac{y^2}{1}-\frac{x^2}{b^2}=1$

$\frac{(2)^2}{1}-\frac{(3)^2}{b^2}=1$

$\frac{4}{1}-\frac{9}{b^2}=1$

$4-\frac{9}{b^2}=1$

$3=\frac{9}{b^2}$

$3b^2=9$

$b^2=3$

The equation of the hyperbola is:

$\frac{y^2}{1}-\frac{x^2}{3}=1$

f. This is the graph of an exponential function with y-intercept $(0,1)$, containing the point $(1,4)$.

$y=A\cdot b^x$

y-intercept $(0,1)\Rightarrow 1=A\cdot b^0=A\cdot1\Rightarrow A=1$

point $(1,4)\Rightarrow 4=b^1=b$

Therefore, $y=4^x$.

11. $f(x) = \log_4(x-2)$

a. $f(x) = \log_4(x-2) = 2$

$x - 2 = 4^2$

$x - 2 = 16$

$x = 18$

The solution set is $\{18\}$.

b. $f(x) = \log_4(x-2) \le 2$

$x - 2 \le 4^2$ and $x - 2 > 0$

$x - 2 \le 16$ and $x > 2$

$x \le 18$ and $x > 2$

$2 < x \le 18$

$(2,18]$

Chapter 10
Systems of Equations and Inequalities

Section 10.1

1. $3x+4=8-x$

   $4x=4$

   $x=1$

   The solution set is $\{1\}$.

2. a. $3x+4y=12$

   x-intercept: $3x+4(0)=12$

   $3x=12$

   $x=4$

   y-intercept: $3(0)+4y=12$

   $4y=12$

   $y=3$

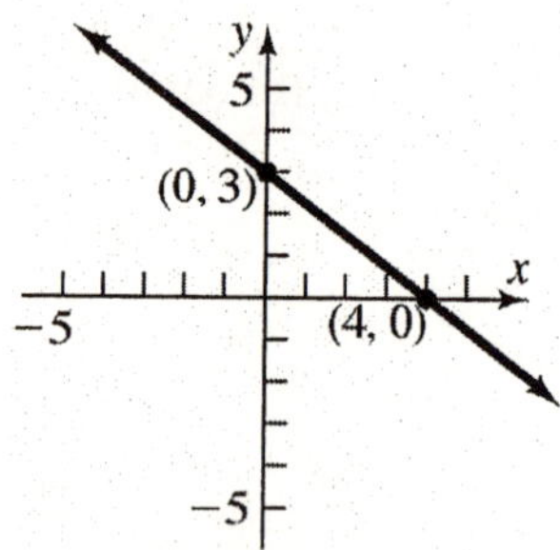

   b. $3x+4y=12$

   $4y=-3x+12$

   $y=-\frac{3}{4}x+3$

   A parallel line would have slope $-\frac{3}{4}$.

3. inconsistent

4. consistent

5. False

6. True

7. $\begin{cases} 2x-\ y=5 \\ 5x+2y=8 \end{cases}$

   Substituting the values of the variables:

   $\begin{cases} 2(2)-\ (-1)=4+1=5 \\ 5(2)+2(-1)=10-2=8 \end{cases}$

   Each equation is satisfied, so $x=2,\ y=-1$ is a solution of the system of equations.

8. $\begin{cases} 3x+2y=2 \\ x-7y=-30 \end{cases}$

   Substituting the values of the variables:

   $\begin{cases} 3(-2)+2(4)=-6+8=2 \\ (-2)-7(4)=-2-28=-30 \end{cases}$

   Each equation is satisfied, so $x=-2,\ y=4$ is a solution of the system of equations.

9. $\begin{cases} 3x-4y=\ 4 \\ \frac{1}{2}x-3y=-\frac{1}{2} \end{cases}$

   Substituting the values of the variables:

   $\begin{cases} 3(2)-4\left(\frac{1}{2}\right)=6-2=4 \\ \frac{1}{2}(2)-3\left(\frac{1}{2}\right)=1-\frac{3}{2}=-\frac{1}{2} \end{cases}$

   Each equation is satisfied, so $x=2,\ y=\frac{1}{2}$ is a solution of the system of equations.

10. $\begin{cases} 2x+\frac{1}{2}y=0 \\ 3x-4y=-\frac{19}{2} \end{cases}$

    Substituting the values of the variables, we obtain:

    $\begin{cases} 2\left(-\frac{1}{2}\right)+\frac{1}{2}(2)=-1+1=0 \\ 3\left(-\frac{1}{2}\right)-4(2)=-\frac{3}{2}-8=-\frac{19}{2} \end{cases}$

    Each equation is satisfied, so $x=-\frac{1}{2},\ y=2$ is a solution of the system of equations.

11. $\begin{cases} x-y=\ 3 \\ \frac{1}{2}x+y=3 \end{cases}$

    Substituting the values of the variables, we obtain:

    $\begin{cases} 4-1=3 \\ \frac{1}{2}(4)+1=2+1=3 \end{cases}$

    Each equation is satisfied, so $x=4,\ y=1$ is a solution of the system of equations.

12. $\begin{cases} x-y=3 \\ -3x+y=1 \end{cases}$

Substituting the values of the variables:

$\begin{cases} (-2)-(-5)=-2+5=3 \\ -3(-2)+-5=6-5=1 \end{cases}$

Each equation is satisfied, so $x=-2,\ y=-5$ is a solution of the system of equations.

13. $\begin{cases} 3x+3y+3z=4 \\ x-y-z=0 \\ 2y-3z=-8 \end{cases}$

Substituting the values of the variables:

$\begin{cases} 3(1)+3(-1)+2(2)=3-3+4=4 \\ 1-(-1)-2=1+1-2=0 \\ 2(-1)-3(2)=-2-6=-8 \end{cases}$

Each equation is satisfied, so $x=1,\ y=-1,\ z=2$ is a solution of the system of equations.

14. $\begin{cases} 4x-z=7 \\ 8x+5y-z=0 \\ -x-y+5z=6 \end{cases}$

Substituting the values of the variables:

$\begin{cases} 4(2)-1=8-1=7 \\ 8(2)+5(-3)-1=16-15-1=0 \\ -2-(-3)+5(1)=-2+3+5=6 \end{cases}$

Each equation is satisfied, so $x=2$, $y=-3$, $z=1$ is a solution of the system of equations.

15. $\begin{cases} 3x+3y+2z=4 \\ x-3y+z=10 \\ 5x-2y-3z=8 \end{cases}$

Substituting the values of the variables:

$\begin{cases} 3(2)+3(-2)+2(2)=6-6+4=4 \\ 2-3(-2)+2=2+6+2=10 \\ 5(2)-2(-2)-3(2)=10+4-6=8 \end{cases}$

Each equation is satisfied, so $x=2$, $y=-2$, $z=2$ is a solution of the system of equations.

16. $\begin{cases} 4x-5z=6 \\ 5y-z=-17 \\ -x-6y+5z=24 \end{cases}$

Substituting the values of the variables:

$\begin{cases} 4(4)-5(2)=16-10=6 \\ 5(-3)-(2)=-15-2=-17 \\ -(4)-6(-3)+5(2)=-4+18+10=24 \end{cases}$

Each equation is satisfied, so $x=4$, $y=-3$, $z=2$ is a solution of the system of equations.

17. $\begin{cases} x+y=8 \\ x-y=4 \end{cases}$

Solve the first equation for *y*, substitute into the second equation and solve:

$\begin{cases} y=8-x \\ x-y=4 \end{cases}$

$$x-(8-x)=4$$
$$x-8+x=4$$
$$2x=12$$
$$x=6$$

Since $x=6,\ y=8-6=2$. The solution of the system is $x=6,\ y=2$.

18. $\begin{cases} x+2y=5 \\ x+y=3 \end{cases}$

Solve the first equation for *x*, substitute into the second equation and solve:

$\begin{cases} x=5-2y \\ x+y=3 \end{cases}$

$$(5-2y)+y=3$$
$$5-y=3$$
$$2=y$$

Since $y=2,\ x=5-2(2)=1$. The solution of the system is $x=1,\ y=2$.

19. $\begin{cases} 5x - y = 13 \\ 2x + 3y = 12 \end{cases}$

Multiply each side of the first equation by 3 and add the equations to eliminate y:

$$\begin{cases} 15x - 3y = 39 \\ 2x + 3y = 12 \end{cases}$$

$$17x = 51$$
$$x = 3$$

Substitute and solve for y:

$$5(3) - y = 13$$
$$15 - y = 13$$
$$-y = -2$$
$$y = 2$$

The solution of the system is $x = 3,\ y = 2$.

20. $\begin{cases} x + 3y = 5 \\ 2x - 3y = -8 \end{cases}$

Add the equations:

$$\begin{cases} x + 3y = 5 \\ 2x - 3y = -8 \end{cases}$$

$$3x = -3$$
$$x = -1$$

Substitute and solve for y:

$$-1 + 3y = 5$$
$$3y = 6$$
$$y = 2$$

The solution of the system is $x = -1,\ y = 2$.

21. $\begin{cases} 3x = 24 \\ x + 2y = 0 \end{cases}$

Solve the first equation for x and substitute into the second equation:

$$\begin{cases} x = 8 \\ x + 2y = 0 \end{cases}$$

$$8 + 2y = 0$$
$$2y = -8$$
$$y = -4$$

The solution of the system is $x = 8,\ y = -4$.

22. $\begin{cases} 4x + 5y = -3 \\ -2y = -4 \end{cases}$

Solve the second equation for y and substitute into the first equation:

$$\begin{cases} 4x + 5y = -3 \\ y = 2 \end{cases}$$

$$4x + 5(2) = -3$$
$$4x + 10 = -3$$
$$4x = -13$$
$$x = -\frac{13}{4}$$

The solution of the system is $x = -\frac{13}{4},\ y = 2$.

23. $\begin{cases} 3x - 6y = 2 \\ 5x + 4y = 1 \end{cases}$

Multiply each side of the first equation by 2 and each side of the second equation by 3, then add to eliminate y:

$$\begin{cases} 6x - 12y = 4 \\ 15x + 12y = 3 \end{cases}$$

$$21x = 7$$
$$x = \frac{1}{3}$$

Substitute and solve for y:

$$3(1/3) - 6y = 2$$
$$1 - 6y = 2$$
$$-6y = 1$$
$$y = -\frac{1}{6}$$

The solution of the system is $x = \frac{1}{3},\ y = -\frac{1}{6}$.

24. $\begin{cases} 2x+4y=\dfrac{2}{3} \\ 3x-5y=-10 \end{cases}$

Multiply each side of the first equation by 5 and each side of the second equation by 4, then add to eliminate *y*:

$$\begin{cases} 10x+20y=\dfrac{10}{3} \\ 12x-20y=-40 \end{cases}$$

$$22x \quad =\frac{110}{3}$$

$$x=-\frac{5}{3}$$

Substitute and solve for *y*:

$$\begin{aligned} 3(-5/3)-5y&=-10 \\ -5-5y&=-10 \\ -5y&=-5 \\ y&=1 \end{aligned}$$

The solution of the system is $x=-\frac{5}{3},\ y=1$.

25. $\begin{cases} 2x+\ \ y=1 \\ 4x+2y=3 \end{cases}$

Solve the first equation for *y*, substitute into the second equation and solve:

$$\begin{cases} y=1-2x \\ 4x+2y-3 \end{cases}$$

$$\begin{aligned} 4x+2(1-2x)&=3 \\ 4x+2-4x&=3 \\ 0x&=1 \end{aligned}$$

This has no solution, so the system is inconsistent.

26. $\begin{cases} x-\ \ y=5 \\ -3x+3y=2 \end{cases}$

Solve the first equation for *x*, substitute into the second equation and solve:

$$\begin{cases} x=y+5 \\ -3x+3y=2 \end{cases}$$

$$\begin{aligned} -3(y+5)+3y&=2 \\ -3y-15+3y&=2 \\ 0y&=17 \end{aligned}$$

This has no solution, so the system is inconsistent.

27. $\begin{cases} 2x-\ \ y=0 \\ 3x+2y=7 \end{cases}$

Solve the first equation for *y*, substitute into the second equation and solve:

$$\begin{cases} y=2x \\ 3x+2y=7 \end{cases}$$

$$\begin{aligned} 3x+2(2x)&=7 \\ 3x+4x&=7 \\ 7x&=7 \\ x&=1 \end{aligned}$$

Since $x=1,\ y=2(1)=2$

The solution of the system is $x=1,\ y=2$.

28. $\begin{cases} 3x+3y=-1 \\ 4x+\ \ y=\dfrac{8}{3} \end{cases}$

Solve the second equation for *y*, substitute into the first equation and solve:

$$\begin{cases} 3x+3y=-1 \\ y=\dfrac{8}{3}-4x \end{cases}$$

$$\begin{aligned} 3x+3\left(\frac{8}{3}-4x\right)&=-1 \\ 3x+8-12x&=-1 \\ -9x&=-9 \\ x&=1 \end{aligned}$$

Since $x=1,\ y=\frac{8}{3}-4(1)=\frac{8}{3}-4=-\frac{4}{3}$.

The solution of the system is $x=1,\ y=-\frac{4}{3}$.

29. $\begin{cases} x+2y=4 \\ 2x+4y=8 \end{cases}$

Solve the first equation for *x*, substitute into the second equation and solve:

$$\begin{cases} x=4-2y \\ 2x+4y=8 \end{cases}$$

$$\begin{aligned} 2(4-2y)+4y&=8 \\ 8-4y+4y&=8 \\ 0y&=0 \end{aligned}$$

These equations are dependent. Any real number is a solution for *y*. The solution of the system is $x=4-2y$, where *y* is any real number.

30. $\begin{cases} 3x - y = 7 \\ 9x - 3y = 21 \end{cases}$

Solve the first equation for *y*, substitute into the second equation and solve:

$\begin{cases} y = 3x - 7 \\ 9x - 3y = 21 \end{cases}$

$$9x - 3(3x - 7) = 21$$
$$9x - 9x + 21 = 21$$
$$0x = 0$$

These equations are dependent. Any real number is a solution for *x*. The solution of the system is $y = 3x - 7$, where *x* is any real number.

31. $\begin{cases} 2x - 3y = -1 \\ 10x + y = 11 \end{cases}$

Multiply each side of the first equation by –5, and add the equations to eliminate *x*:

$\begin{cases} -10x + 15y = 5 \\ 10x + y = 11 \end{cases}$

$$16y = 16$$
$$y = 1$$

Substitute and solve for *x*:

$$2x - 3(1) = -1$$
$$2x - 3 = -1$$
$$2x = 2$$
$$x = 1$$

The solution of the system is $x = 1,\ y = 1$.

32. $\begin{cases} 3x - 2y = 0 \\ 5x + 10y = 4 \end{cases}$

Multiply each side of the first equation by 5, and add the equations to eliminate *y*:

$\begin{cases} 15x - 10y = 0 \\ 5x + 10y = 4 \end{cases}$

$$20x = 4$$
$$x = \frac{1}{5}$$

Substitute and solve for *y*:

$$5(1/5) + 10y = 4$$
$$1 + 10y = 4$$
$$10y = 3$$
$$y = \frac{3}{10}$$

The solution of the system is $x = \frac{1}{5},\ y = \frac{3}{10}$.

33. $\begin{cases} 2x + 3y = 6 \\ x - y = \frac{1}{2} \end{cases}$

Solve the second equation for *x*, substitute into the first equation and solve:

$\begin{cases} 2x + 3y = 6 \\ x = y + \frac{1}{2} \end{cases}$

$$2\left(y + \frac{1}{2}\right) + 3y = 6$$
$$2y + 1 + 3y = 6$$
$$5y = 5$$
$$y = 1$$

Since $y = 1,\ x = 1 + \frac{1}{2} = \frac{3}{2}$. The solution of the system is $x = \frac{3}{2},\ y = 1$.

34. $\begin{cases} \frac{1}{2}x + y = -2 \\ x - 2y = 8 \end{cases}$

Solve the second equation for *x*, substitute into the first equation and solve:

$\begin{cases} \frac{1}{2}x + y = -2 \\ x = 2y + 8 \end{cases}$

$$\frac{1}{2}(2y + 8) + y = -2$$
$$y + 4 + y = -2$$
$$2y = -6$$
$$y = -3$$

Since $y = -3,\ x = 2(-3) + 8 = -6 + 8 = 2$. The solution of the system is $x = 2,\ y = -3$.

35. $\begin{cases} \frac{1}{2}x + \frac{1}{3}y = 3 \\ \frac{1}{4}x - \frac{2}{3}y = -1 \end{cases}$

Multiply each side of the first equation by –6 and each side of the second equation by 12, then add to eliminate *x*:

$\begin{cases} -3x - 2y = -18 \\ 3x - 8y = -12 \end{cases}$

$$-10y = -30$$
$$y = 3$$

Substitute and solve for x:

$$\frac{1}{2}x+\frac{1}{3}(3)=3$$
$$\frac{1}{2}x+1=3$$
$$\frac{1}{2}x=2$$
$$x=4$$

The solution of the system is $x=4,\ y=3$.

36. $\begin{cases}\frac{1}{3}x-\frac{3}{2}y=-5\\ \frac{3}{4}x+\frac{1}{3}y=11\end{cases}$

Multiply each side of the first equation by –54 and each side of the second equation by 24, then add to eliminate x:

$$\begin{cases}-18x+81y=270\\ 18x+\ 8y=264\end{cases}$$
$$89y=534$$
$$y=6$$

Substitute and solve for x:

$$\frac{3}{4}x+\frac{1}{3}(6)=11$$
$$\frac{3}{4}x+2=11$$
$$\frac{3}{4}x=9$$
$$x=12$$

The solution of the system is $x=12,\ y=6$.

37. $\begin{cases}3x-5y=3\\ 15x+5y=21\end{cases}$

Add the equations to eliminate y:

$$\begin{cases}3x-5y=3\\ 15x+5y=21\end{cases}$$
$$18x\quad=24$$
$$x=\frac{4}{3}$$

Substitute and solve for y:

$$3(4/3)-5y=3$$
$$4-5y=3$$
$$-5y=-1$$
$$y=\frac{1}{5}$$

The solution of the system is $x=\frac{4}{3},\ y=\frac{1}{5}$.

38. $\begin{cases}2x-\ y=-1\\ x+\frac{1}{2}y=\ \frac{3}{2}\end{cases}$

Multiply each side of the second equation by 2, and add the equations to eliminate y:

$$\begin{cases}2x-y=-1\\ 2x+y=\ 3\end{cases}$$
$$4x\quad=2$$
$$x=\frac{1}{2}$$

Substitute and solve for y: $2\left(\frac{1}{2}\right)-y=-1$

$$1-y=-1$$
$$-y=-2$$
$$y=2$$

The solution of the system is $x=\frac{1}{2},\ y=2$.

39. $\begin{cases}\frac{1}{x}+\frac{1}{y}=8\\ \frac{3}{x}-\frac{5}{y}=0\end{cases}$

Rewrite letting $u=\frac{1}{x},\ v=\frac{1}{y}$:

$$\begin{cases}u+v=8\\ 3u-5v=0\end{cases}$$

Solve the first equation for u, substitute into the second equation and solve:

$$\begin{cases}u=8-v\\ 3u-5v=0\end{cases}$$

$$3(8-v)-5v=0$$
$$24-3v-5v=0$$
$$-8v=-24$$
$$v=3$$

Since $v=3,\ u=8-3=5$. Thus, $x=\frac{1}{u}=\frac{1}{5}$, $y=\frac{1}{v}=\frac{1}{3}$. The solution of the system is $x=\frac{1}{5},\ y=\frac{1}{3}$.

40. $\begin{cases} \frac{4}{x} - \frac{3}{y} = 0 \\ \frac{6}{x} + \frac{3}{2y} = 2 \end{cases}$

Rewrite letting $u = \frac{1}{x},\ v = \frac{1}{y}$:

$\begin{cases} 4u - 3v = 0 \\ 6u + \frac{3}{2}v = 2 \end{cases}$

Multiply each side of the second equation by 2, and add the equations to eliminate v:

$$\begin{cases} 4u - 3v = 0 \\ 12u + 3v = 4 \end{cases}$$
$$16u \quad = 4$$
$$u = \frac{4}{16} = \frac{1}{4}$$

Substitute and solve for v:

$$4\left(\frac{1}{4}\right) - 3v = 0$$
$$1 - 3v = 0$$
$$-3v = -1$$
$$v = \frac{1}{3}$$

Thus, $x = \frac{1}{u} = 4,\ y = \frac{1}{v} = 3$. The solution of the system is $x = 4,\ y = 3$.

41. $\begin{cases} x - y = 6 \\ 2x - 3z = 16 \\ 2y + z = 4 \end{cases}$

Multiply each side of the first equation by –2 and add to the second equation to eliminate x:

$$-2x + 2y \quad = -12$$
$$2x \quad - 3z = 16$$
$$2y - 3z = 4$$

Multiply each side of the result by –1 and add to the original third equation to eliminate y:

$$-2y + 3z = -4$$
$$2y + z = 4$$
$$4z = 0$$
$$z = 0$$

Substituting and solving for the other variables:

$$2y + 0 = 4 \qquad 2x - 3(0) = 16$$
$$2y = 4 \qquad 2x = 16$$
$$y = 2 \qquad x = 8$$

The solution is $x = 8,\ y = 2,\ z = 0$.

42. $\begin{cases} 2x + y = -4 \\ -2y + 4z = 0 \\ 3x - 2z = -11 \end{cases}$

Multiply each side of the first equation by 2 and add to the second equation to eliminate y:

$$4x + 2y \quad = -8$$
$$-2y + 4z = 0$$
$$4x \quad + 4z = -8$$

Multiply each side of the result by $\frac{1}{2}$ and add to the original third equation to eliminate z:

$$2x + 2z = -4$$
$$3x - 2z = -11$$
$$5x \quad = -15$$
$$x = -3$$

Substituting and solving for the other variables:

$$2(-3) + y = -4 \qquad 3(-3) - 2z = -11$$
$$-6 + y = -4 \qquad -9 - 2z = -11$$
$$y = 2 \qquad -2z = -2$$
$$z = 1$$

The solution is $x = -3,\ y = 2,\ z = 1$.

43. $\begin{cases} x - 2y + 3z = 7 \\ 2x + y + z = 4 \\ -3x + 2y - 2z = -10 \end{cases}$

Multiply each side of the first equation by –2 and add to the second equation to eliminate x; and multiply each side of the first equation by 3 and add to the third equation to eliminate x:

$$-2x + 4y - 6z = -14$$
$$2x + y + z = 4$$
$$5y - 5z = -10$$

$$3x - 6y + 9z = 21$$
$$-3x + 2y - 2z = -10$$
$$-4y + 7z = 11$$

Multiply each side of the first result by $\frac{4}{5}$ and add to the second result to eliminate y:

$$\begin{array}{r} 4y-4z=-8 \\ -4y+7z=11 \\ \hline 3z=3 \\ z=1 \end{array}$$

Substituting and solving for the other variables:

$$y-1=-2 \qquad y=-1$$

$$x-2(-1)+3(1)=7$$
$$x+2+3=7$$
$$x=2$$

The solution is $x=2,\ y=-1,\ z=1$.

44. $\begin{cases} 2x+y-3z=0 \\ -2x+2y+z=-7 \\ 3x-4y-3z=7 \end{cases}$

Multiply each side of the first equation by –2 and add to the second equation to eliminate *y*; and multiply each side of the first equation by 4 and add to the third equation to eliminate *y*:

$$\begin{array}{r} -4x-2y+6z=0 \\ -2x+2y+z=-7 \\ \hline -6x \quad +7z=-7 \end{array}$$

$$\begin{array}{r} 8x+4y-12z=0 \\ 3x-4y-3z=7 \\ \hline 11x \quad -15z=7 \end{array}$$

Multiply each side of the first result by 11 and multiply each side of the second result by 6 to eliminate *x*:

$$\begin{array}{r} -66x+77z=-77 \\ 66x-90z=42 \\ \hline -13z=-35 \\ z=\frac{35}{13} \end{array}$$

Substituting and solving for the other variables:

$$-6x+7\left(\frac{35}{13}\right)=-7$$
$$-6x+\frac{245}{13}=-7$$
$$-6x=-\frac{336}{13}$$
$$x=\frac{56}{13}$$

$$2\left(\frac{56}{13}\right)+y-3\left(\frac{35}{13}\right)=0$$
$$\frac{112}{13}+y-\frac{105}{13}=0$$
$$y=-\frac{7}{13}$$

The solution is $x=\frac{56}{13},\ y=-\frac{7}{13},\ z=\frac{35}{13}$.

45. $\begin{cases} x-y-z=1 \\ 2x+3y+z=2 \\ 3x+2y=0 \end{cases}$

Add the first and second equations to eliminate *z*:

$$\begin{array}{r} x-y-z=1 \\ 2x+3y+z=2 \\ \hline 3x+2y \quad =3 \end{array}$$

Multiply each side of the result by –1 and add to the original third equation to eliminate *y*:

$$\begin{array}{r} -3x-2y=-3 \\ 3x+2y=0 \\ \hline 0=-3 \end{array}$$

This result has no solution, so the system is inconsistent.

46. $\begin{cases} 2x-3y-z=0 \\ -x+2y+z=5 \\ 3x-4y-z=1 \end{cases}$

Add the first and second equations to eliminate *z*; then add the second and third equations to eliminate *z*:

$$\begin{array}{r} 2x-3y-z=0 \\ -x+2y+z=5 \\ \hline x-y \quad =5 \end{array}$$

$$\begin{array}{r} -x+2y+z=5 \\ 3x-4y-z=1 \\ \hline 2x-2y \quad =6 \end{array}$$

Multiply each side of the first result by –2 and add to the second result to eliminate *y*:

$$\begin{array}{r} -2x+2y=-10 \\ 2x-2y=6 \\ \hline 0=-2 \end{array}$$

This result has no solution, so the system is inconsistent.

47. $\begin{cases} x-y-z=1 \\ -x+2y-3z=-4 \\ 3x-2y-7z=0 \end{cases}$

Add the first and second equations to eliminate x; multiply the first equation by –3 and add to the third equation to eliminate x:

$$\begin{array}{r} x-y-z=1 \\ -x+2y-3z=-4 \\ \hline y-4z=-3 \end{array}$$

$$\begin{array}{r} -3x+3y+3z=-3 \\ 3x-2y-7z=0 \\ \hline y-4z=-3 \end{array}$$

Multiply each side of the first result by –1 and add to the second result to eliminate y:

$$\begin{array}{r} -y+4z=3 \\ y-4z=-3 \\ \hline 0=0 \end{array}$$

The system is dependent. If z is any real number, then $y=4z-3$.

Solving for x in terms of z in the first equation:

$$x-(4z-3)-z=1$$
$$x-4z+3-z=1$$
$$x-5z+3=1$$
$$x=5z-2$$

The solution is $x=5z-2$, $y=4z-3$, z is any real number.

48. $\begin{cases} 2x-3y-z=0 \\ 3x+2y+2z=2 \\ x+5y+3z=2 \end{cases}$

Multiply the first equation by 2 and add to the second equation to eliminate z; multiply the first equation by 3 and add to the third equation to eliminate z:

$$\begin{array}{r} 4x-6y-2z=0 \\ 3x+2y+2z=2 \\ \hline 7x-4y \quad =2 \end{array}$$

$$\begin{array}{r} 6x-9y-3z=0 \\ x+5y+3z=2 \\ \hline 7x-4y \quad =2 \end{array}$$

Multiply each side of the first result by –1 and add to the second result to eliminate y:

$$\begin{array}{r} -7x+4y=-2 \\ 7x-4y=2 \\ \hline 0=0 \end{array}$$

The system is dependent. If y is any real number, then $x=\frac{4}{7}y+\frac{2}{7}$.

Solving for z in terms of x in the first equation:

$$z=2x-3y$$
$$=2\left(\frac{4y+2}{7}\right)-3y$$
$$=\frac{8y+4-21y}{7}$$
$$=\frac{-13y+4}{7}$$

The solution is $x=\frac{4}{7}y+\frac{2}{7}$, $z=-\frac{13}{7}y+\frac{4}{7}$, y is any real number.

49. $\begin{cases} 2x-2y+3z=6 \\ 4x-3y+2z=0 \\ -2x+3y-7z=1 \end{cases}$

Multiply the first equation by –2 and add to the second equation to eliminate x; add the first and third equations to eliminate x:

$$\begin{array}{r} -4x+4y-6z=-12 \\ 4x-3y+2z=0 \\ \hline y-4z=-12 \end{array}$$

$$\begin{array}{r} 2x-2y+3z=6 \\ -2x+3y-7z=1 \\ \hline y-4z=7 \end{array}$$

Multiply each side of the first result by –1 and add to the second result to eliminate y:

$$\begin{array}{r} -y+4z=12 \\ y-4z=7 \\ \hline 0=19 \end{array}$$

This result has no solution, so the system is inconsistent.

50. $\begin{cases} 3x-2y+2z=6 \\ 7x-3y+2z=-1 \\ 2x-3y+4z=0 \end{cases}$

Multiply the first equation by –1 and add to the second equation to eliminate z; multiply the first equation by –2 and add to the third equation to eliminate z:

$$\begin{array}{r} -3x+2y-2z=-6 \\ \underline{7x-3y+2z=-1} \\ 4x-y=-7 \end{array}$$

$$\begin{array}{r} -6x+4y-4z=-12 \\ \underline{2x-3y+4z=0} \\ -4x+y=-12 \end{array}$$

Add the first result to the second result to eliminate y:

$$\begin{array}{r} 4x-y=-7 \\ \underline{-4x+y=-12} \\ 0=-19 \end{array}$$

This result has no solution, so the system is inconsistent.

51. $\begin{cases} x+y-z=6 \\ 3x-2y+z=-5 \\ x+3y-2z=14 \end{cases}$

Add the first and second equations to eliminate z; multiply the second equation by 2 and add to the third equation to eliminate z:

$$\begin{array}{r} x+y-z=6 \\ \underline{3x-2y+z=-5} \\ 4x-y=1 \end{array}$$

$$\begin{array}{r} 6x-4y+2z=-10 \\ \underline{x+3y-2z=14} \\ 7x-y=4 \end{array}$$

Multiply each side of the first result by –1 and add to the second result to eliminate y:

$$\begin{array}{r} -4x+y=-1 \\ \underline{7x-y=4} \\ 3x=3 \\ x=1 \end{array}$$

Substituting and solving for the other variables:

$$\begin{array}{rr} 4(1)-y=1 & 3(1)-2(3)+z=-5 \\ -y=-3 & 3-6+z=-5 \\ y=3 & z=-2 \end{array}$$

The solution is $x=1,\ y=3,\ z=-2$.

52. $\begin{cases} x-y+z=-4 \\ 2x-3y+4z=-15 \\ 5x+y-2z=12 \end{cases}$

Multiply the first equation by –3 and add to the second equation to eliminate y; add the first and third equations to eliminate y:

$$\begin{array}{r} -3x+3y-3z=12 \\ \underline{2x-3y+4z=-15} \\ -x+z=-3 \\ z=x-3 \end{array}$$

$$\begin{array}{r} x-y+z=-4 \\ \underline{5x+y-2z=12} \\ 6x-z=8 \end{array}$$

Substitute and solve:

$$\begin{array}{r} 6x-(x-3)=8 \\ 6x-x+3=8 \\ 5x=5 \\ x=1 \end{array}$$

$z=x-3=1-3=-2$

$y=12-5x+2z=12-5(1)+2(-2)=12-5-4=3$

The solution is $x=1,\ y=3,\ z=-2$.

53. $\begin{cases} x+2y-z=-3 \\ 2x-4y+z=-7 \\ -2x+2y-3z=4 \end{cases}$

Add the first and second equations to eliminate z; multiply the second equation by 3 and add to the third equation to eliminate z:

$$\begin{array}{r} x+2y-z=-3 \\ \underline{2x-4y+z=-7} \\ 3x-2y=-10 \end{array}$$

$$\begin{array}{r} 6x-12y+3z=-21 \\ \underline{-2x+2y-3z=4} \\ 4x-10y=-17 \end{array}$$

Multiply each side of the first result by –5 and add to the second result to eliminate y:

$$\begin{array}{r} -15x+10y=50 \\ \underline{4x-10y=-17} \\ -11x=33 \\ x=-3 \end{array}$$

Substituting and solving for the other variables:

$$3(-3)-2y=-10$$
$$-9-2y=-10$$
$$-2y=-1$$
$$y=\frac{1}{2}$$

$$-3+2\left(\frac{1}{2}\right)-z=-3$$
$$-3+1-z=-3$$
$$-z=-1$$
$$z=1$$

The solution is $x=-3,\ y=\frac{1}{2},\ z=1$.

54. $\begin{cases} x+4y-3z=-8 \\ 3x-\ y+3z=\ 12 \\ x+\ y+6z=\ 1 \end{cases}$

Add the first and second equations to eliminate z; multiply the first equation by 2 and add to the third equation to eliminate z:

$$\begin{array}{l} x+4y-3z=-8 \\ \underline{3x-\ y+3z=\ 12} \\ 4x+3y\qquad=\ 4 \end{array}$$

$$\begin{array}{l} 2x+8y-6z=-16 \\ \underline{x+\ y+6z=\ 1} \\ 3x+9y\qquad=-15 \end{array}$$

Multiply each side of the second result by $-1/3$ and add to the first result to eliminate y:

$$\begin{array}{l} 4x+3y=4 \\ \underline{-x-3y=5} \\ 3x\qquad=9 \end{array}$$
$$x=3$$

Substituting and solving for the other variables:

$$3+3y=-5$$
$$3y=-8$$
$$y=-\frac{8}{3}$$

$$3+\left(-\frac{8}{3}\right)+6z=1$$
$$6z=\frac{2}{3}$$
$$z=\frac{1}{9}$$

The solution is $x=3,\ y=-\frac{8}{3},\ z=\frac{1}{9}$.

55. Let l be the length of the rectangle and w be the width of the rectangle. Then:
$l=2w$ and $2l+2w=90$

Solve by substitution:

$$2(2w)+2w=90$$
$$4w+2w=90$$
$$6w=90$$
$$w=15\text{ feet}$$
$$l=2(15)=30\text{ feet}$$

The dimensions of the floor are 15 feet by 30 feet.

56. Let l be the length of the rectangle and w be the width of the rectangle. Then:
$l=w+50$ and $2l+2w=3000$

Solve by substitution:

$$2(w+50)+2w=3000$$
$$2w+100+2w=3000$$
$$4w=2900$$
$$w=725\text{ meters}$$
$$l=725+50=775\text{ meters}$$

The dimensions of the field are 775 meters by 725 meters.

57. Let x = the cost of one cheeseburger and y = the cost of one shake. Then:
$4x+2y=790$ and $2y=x+15$

Solve by substitution:

$$4x+x+15=790 \qquad 2y=155+15$$
$$5x=775 \qquad 2y=170$$
$$x=155 \qquad y=85$$

A cheeseburger cost \$1.55 and a shake costs \$0.85.

58. Let x = the number of adult tickets sold and y = the number of senior tickets sold. Then:

$\begin{cases} x+\ y=\ 325 \\ 9x+7y=2495 \end{cases}$

Solve the first equation for y: $y=325-x$

Solve by substitution:

$$9x+7(325-x)=2495$$
$$9x+2275-7x=2495$$
$$2x=220$$
$$x=110$$
$$y=325-110=215$$

There were 110 adult tickets sold and 215 senior citizen tickets sold.

59. Let x = the number of pounds of cashews.
Let y = is the number of pounds in the mixture.
The value of the cashews is $5x$.
The value of the peanuts is $1.50(30) = 45$.
The value of the mixture is $3y$.
Then $x+30=y$ represents the amount of mixture.
$5x+45=3y$ represents the value of the mixture.

Solve by substitution:

$$5x+45=3(x+30)$$
$$2x=45$$
$$x=22.5$$

So, 22.5 pounds of cashews should be used in the mixture.

60. Let x = the amount invested in AA bonds.
Let y = the amount invested in the Bank Certificate.

a. Then $x+y=150,000$ represents the total investment.
$0.10x+0.05y=12,000$ represents the earnings on the investment.
Solve by substitution:

$$0.10(150,000-y)+0.05y=12,000$$
$$15,000-0.10y+0.05y=12,000$$
$$-0.05y=-3000$$
$$y=60,000$$
$$x=150,000-60,000=90,000$$

Thus, \$90,000 should be invested in AA Bonds and \$60,000 in a Bank Certificate.

b. Then $x+y=150,000$ represents the total investment.
$0.10x+0.05y=14,000$ represents the earnings on the investment.
Solve by substitution:

$$0.10(150,000-y)+0.05y=14,000$$
$$15,000-0.10y+0.05y=14,000$$
$$-0.05y=-1000$$
$$y=20,000$$
$$x=150,000-20,000=130,000$$

Thus, \$130,000 should be invested in AA Bonds and \$20,000 in a Bank Certificate.

61. Let x = the plane's air speed and y = the wind speed.

	Rate	Time	Distance
With Wind	$x+y$	3	600
Against	$x-y$	4	600

$$\begin{cases}(x+y)(3)=600\\(x-y)(4)=600\end{cases}$$

Multiply each side of the first equation by $\frac{1}{3}$, multiply each side of the second equation by $\frac{1}{4}$, and add the result to eliminate y

$$x+y=200$$
$$x-y=150$$
$$2x=350$$
$$x=175$$

$$175+y=200$$
$$y=25$$

The airspeed of the plane is 175 mph, and the wind speed is 25 mph.

62. Let x = the wind speed.
Let y = the distance.

	Rate	Time	Distance
With Wind	$150+x$	2	y
Against	$150-x$	3	y

$$\begin{cases}(150+x)(2)=y\\(150-x)(3)=y\end{cases}$$

Solve by substitution:

$$(150+x)(2)=(150-x)(3)$$
$$300+2x=450-3x$$
$$5x=150$$
$$x=30$$

Thus, the wind speed is 30 mph.

63. Let x = the number of \$25-design.
Let y = the number of \$45-design.
Then $x+y$ = the total number of sets of dishes.
$25x+45y$ = the cost of the dishes.

Setting up the equations and solving by substitution:

$$\begin{cases} x+ \quad y = 200 \\ 25x+45y=7400 \end{cases}$$

Solve the first equation for y, the solve by substitution: $y=200-x$

$$25x+45(200-x)=7400$$
$$25x+9000-45x=7400$$
$$-20x=-1600$$
$$x=80$$
$$y=200-80=120$$

Thus, 80 sets of the \$25 dishes and 120 sets of the \$45 dishes should be ordered.

64. Let x = the cost of a hot dog.
Let y = the cost of a soft drink.

Setting up the equations and solving by substitution:

$$\begin{cases} 10x+5y=12.50 \\ 7x+4y=9.00 \end{cases}$$

$$10x+5y=12.50$$
$$2x+y=2.50$$
$$y=2.50-2x$$

$$7x+4(2.50-2x)=9.00$$
$$7x+10.00-8x=9.00$$
$$-x=-1$$
$$x=1$$
$$y=2.50-2(1)=0.50$$

A hot dog costs \$1.00 and a soft drink costs \$0.50.

65. Let x = the cost per package of bacon.
Let y = the cost of a carton of eggs.

Set up a system of equations for the problem:

$$\begin{cases} 3x+2y=7.45 \\ 2x+3y=6.45 \end{cases}$$

Multiply each side of the first equation by 3 and each side of the second equation by –2 and solve by elimination:

$$\begin{array}{r} 9x+6y= 22.35 \\ -4x-6y=-12.90 \\ \hline 5x \quad = 9.45 \end{array}$$
$$x=1.89$$

Substitute and solve for y:

$$3(1.89)+2y=7.45$$
$$5.67+2y=7.45$$
$$2y=1.78$$
$$y=0.89$$

A package of bacon costs \$1.89 and a carton of eggs cost \$0.89. The refund for 2 packages of bacon and 2 cartons of eggs will be 2(\$1.89) + 2(\$0.89) =\$5.56.

66. Let x = Pamela's speed in still water.
Let y = the speed of the current.

	Rate	Time	Distance
Downstream	$x+y$	3	15
Upstream	$x-y$	5	15

Set up a system of equations for the problem:

$$\begin{cases} 3(x+y)=15 \\ 5(x-y)=15 \end{cases}$$

Multiply each side of the first equation by $\frac{1}{3}$, multiply each side of the second equation by $\frac{1}{5}$, and add the result to eliminate y:

$$\begin{array}{r} x+y=5 \\ x-y=3 \\ \hline 2x=8 \end{array}$$
$$x=4$$
$$4+y=5$$
$$y=1$$

Pamela's speed is 4 miles per hour and the speed of the current is 1 mile per hour.

67. Let x = the # of mg of compound 1.
Let y = the # of mg of compound 2.
Setting up the equations and solving by substitution:

$$\begin{cases} 0.2x + 0.4y = 40 & \text{vitamin C} \\ 0.3x + 0.2y = 30 & \text{vitamin D} \end{cases}$$

Multiplying each equation by 10 yields

$$\begin{cases} 2x + 4y = 400 \\ 6x + 4y = 600 \end{cases}$$

Subtracting the bottom equation from the top equation yields

$$2x + 4y - (6x + 4y) = 400 - 600$$
$$2x - 6x = -200$$
$$-4x = -200$$
$$x = 50$$

$$2(50) + 4y = 400$$
$$100 + 4y = 400$$
$$4y = 300$$
$$y = \frac{300}{4} = 75$$

So 50 mg of compound 1 should be mixed with 75 mg of compound 2.

68. Let x = the # of units of powder 1.
Let y = the # of units of powder 2.
Setting up the equations and solving by substitution:

$$\begin{cases} 0.2x + 0.4y = 12 & \text{vitamin } B_{12} \\ 0.3x + 0.2y = 12 & \text{vitamin E} \end{cases}$$

Multiplying each equation by 10 yields

$$\begin{cases} 2x + 4y = 120 \\ 6x + 4y = 240 \end{cases}$$

Subtracting the bottom equation from the top equation yields

$$2x + 4y - (6x + 4y) = 120 - 240$$
$$-4x = -120$$
$$x = 30$$

$$2(30) + 4y = 120$$
$$60 + 4y = 120$$
$$4y = 60$$
$$y = \frac{60}{4} = 15$$

So 30 units of powder 1 should be mixed with 15 units of powder 2.

69. $y = ax^2 + bx + c$

At (–1, 4) the equation becomes:

$$4 = a(-1)^2 + b(-1) + c$$
$$4 = a - b + c$$

At (2, 3) the equation becomes:

$$3 = a(2)^2 + b(2) + c$$
$$3 = 4a + 2b + c$$

At (0, 1) the equation becomes:

$$1 = a(0)^2 + b(0) + c$$
$$1 = c$$

The system of equations is:

$$\begin{cases} a - b + c = 4 \\ 4a + 2b + c = 3 \\ c = 1 \end{cases}$$

Substitute $c = 1$ into the first and second equations and simplify:

$$a - b + 1 = 4 \qquad 4a + 2b + 1 = 3$$
$$a - b = 3 \qquad 4a + 2b = 2$$
$$a = b + 3$$

Solve the first result for a, substitute into the second result and solve:

$$4(b + 3) + 2b = 2$$
$$4b + 12 + 2b = 2$$
$$6b = -10$$
$$b = -\frac{5}{3}$$

$$a = -\frac{5}{3} + 3 = \frac{4}{3}$$

The solution is $a = \frac{4}{3}$, $b = -\frac{5}{3}$, $c = 1$. The equation is $y = \frac{4}{3}x^2 - \frac{5}{3}x + 1$.

70. $y = ax^2 + bx + c$

At (–1, –2) the equation becomes:

$$-2 = a(-1)^2 + b(-1) + c$$
$$a - b + c = -2$$

At (1, –4) the equation becomes:

$$-4 = a(1)^2 + b(1) + c$$
$$a + b + c = -4$$

At (2, 4) the equation becomes:

$$4 = a(2)^2 + b(2) + c$$
$$4a + 2b + c = 4$$

The system of equations is:

$$\begin{cases} a-b+c=-2 \\ a+b+c=-4 \\ 4a+2b+c=4 \end{cases}$$

Multiply the first equation by –1 and add to the second equation; multiply the first equation by –1 and add to the third equation to eliminate *c*:

$$\begin{cases} -a+b-c=2 \\ a+b+c=-4 \end{cases}$$

$$2b=-2$$
$$b=-1$$

$$-a+b-c=2$$
$$4a+2b+c=4$$
$$3a+3b=6$$
$$a+b=2$$

Substitute and solve:

$$a+(-1)=2$$
$$a=3$$

$$c=-a-b-4$$
$$=-3-(-1)-4$$
$$=-6$$

The solution is $a=3,\ b=-1,\ c=-6$. The equation is $y=3x^2-x-6$

71. $\begin{cases} 0.06Y-5000r=240 \\ 0.06Y+6000r=900 \end{cases}$

Multiply the first equation by -1, the add the result to the second equation to eliminate *Y*.

$$-0.06Y+5000r=-240$$
$$0.06Y+6000r=900$$
$$11000r=660$$
$$r=0.06$$

Substitute this result into the first equation to find *Y*.

$$0.06Y-5000(0.06)=240$$
$$0.06Y-300=240$$
$$0.06Y=540$$
$$Y=9000$$

The equilibrium level of income and interest rates is $9000 million and 6%.

72. $\begin{cases} 0.05Y-1000r=10 \\ 0.05Y+800r=100 \end{cases}$

Multiply the first equation by -1, the add the result to the second equation to eliminate *Y*.

$$-0.05Y+1000r=-10$$
$$0.05Y+800r=100$$
$$1800r=90$$
$$r=0.05$$

Substitute this result into the first equation to find *Y*.

$$0.05Y-1000(0.05)=10$$
$$0.05Y-50=10$$
$$0.05Y=60$$
$$Y=1200$$

The equilibrium level of income and interest rates is $1200 million and 5%.

73. $\begin{cases} I_2=I_1+I_3 \\ 5-3I_1-5I_2=0 \\ 10-5I_2-7I_3=0 \end{cases}$

Substitute the expression for I_2 into the second and third equations and simplify:

$$5-3I_1-5(I_1+I_3)=0$$
$$-8I_1-5I_3=-5$$

$$10-5(I_1+I_3)-7I_3=0$$
$$-5I_1-12I_3=-10$$

Multiply both sides of the first result by 5 and multiply both sides of the second result by –8 to eliminate I_1:

$$-40I_1-25I_3=-25$$
$$40I_1+96I_3=80$$
$$71I_3=55$$
$$I_3=\frac{55}{71}$$

Substituting and solving for the other variables:

$$-8I_1-5\left(\frac{55}{71}\right)=-5$$
$$-8I_1-\frac{275}{71}=-5$$
$$-8I_1=-\frac{80}{71}$$
$$I_1=\frac{10}{71}$$

$$I_2=\left(\frac{10}{71}\right)+\frac{55}{71}$$
$$I_2=\frac{65}{71}$$

The solution is $I_1=\frac{10}{71},\ I_2=\frac{65}{71},\ I_3=\frac{55}{71}$.

74. $\begin{cases} I_3 = I_1 + I_2 \\ 8 = 4I_3 + 6I_2 \\ 8I_1 = 4 + 6I_2 \end{cases}$

Substitute the expression for I_3 into the second equation and simplify:

$8 = 4(I_1 + I_2) + 6I_2$ $\quad 8I_1 = 4 + 6I_2$

$8 = 4I_1 + 10I_2$ $\quad 8I_1 - 6I_2 = 4$

$4I_1 + 10I_2 = 8$

Multiply both sides of the first result by –2 and add to the second result to eliminate I_1:

$$\begin{aligned} -8I_1 - 20I_2 &= -16 \\ 8I_1 - 6I_2 &= 4 \\ \hline -26I_2 &= -12 \\ I_2 &= \frac{-12}{-26} = \frac{6}{13} \end{aligned}$$

Substituting and solving for the other variables:

$$\begin{aligned} 4I_1 + 10\left(\frac{6}{13}\right) &= 8 \\ 4I_1 + \frac{60}{13} &= 8 \\ 4I_1 &= \frac{44}{13} \\ I_1 &= \frac{11}{13} \end{aligned}$$

$$I_3 = I_1 + I_2 = \frac{11}{13} + \frac{6}{13} = \frac{17}{13}$$

The solution is $I_1 = \frac{11}{13},\ I_2 = \frac{6}{13},\ I_3 = \frac{17}{13}$.

75. Let x = the number of orchestra seats.
Let y = the number of main seats.
Let z = the number of balcony seats.
Since the total number of seats is 500, $x + y + z = 500$.

Since the total revenue is \$17,100 if all seats are sold, $50x + 35y + 25z = 17{,}100$.

If only half of the orchestra seats are sold, the revenue is \$14,600.

So, $50\left(\frac{1}{2}x\right) + 35y + 25z = 14{,}600$.

Thus, we have the following system:

$$\begin{cases} x + y + z = 500 \\ 50x + 35y + 25z = 17{,}100 \\ 25x + 35y + 25z = 14{,}600 \end{cases}$$

Multiply each side of the first equation by –25 and add to the second equation to eliminate z; multiply each side of the third equation by –1 and add to the second equation to eliminate z:

$$\begin{aligned} -25x - 25y - 25z &= -12{,}500 \\ 50x + 35y + 25z &= 17{,}100 \\ \hline 25x + 10y &= 4600 \end{aligned}$$

$$\begin{aligned} 50x + 35y + 25z &= 17{,}100 \\ -25x - 35y - 25z &= -14{,}600 \\ \hline 25x &= 2500 \\ x &= 100 \end{aligned}$$

Substituting and solving for the other variables:

$$\begin{aligned} 25(100) + 10y &= 4600 \\ 2500 + 10y &= 4600 \\ 10y &= 2100 \\ y &= 210 \end{aligned} \qquad \begin{aligned} 100 + 210 + z &= 500 \\ 310 + z &= 500 \\ z &= 190 \end{aligned}$$

There are 100 orchestra seats, 210 main seats, and 190 balcony seats.

76. Let x = the number of adult tickets.
Let y = the number of child tickets.
Let z = the number of senior citizen tickets.
Since the total number of tickets is 405, $x + y + z = 405$.

Since the total revenue is \$2320, $8x + 4.50y + 6z = 2320$.

Twice as many children's tickets as adult tickets are sold. So, $y = 2x$.

Thus, we have the following system:

$$\begin{cases} x + y + z = 405 \\ 8x + 4.50y + 6z = 2320 \\ y = 2x \end{cases}$$

Substitute for y in the first two equations and simplify:

$$\begin{aligned} x + (2x) + z &= 405 \\ 3x + z &= 405 \end{aligned}$$

$$\begin{aligned} 8x + 4.50(2x) + 6z &= 2320 \\ 17x + 6z &= 2320 \end{aligned}$$

Multiply the first result by –6 and add to the second result to eliminate z:

$$\begin{aligned} -18x - 6z &= -2430 \\ 17x + 6z &= 2320 \\ \hline -x &= -110 \\ x &= 110 \end{aligned}$$

$$\begin{aligned} y &= 2x \\ &= 2(110) \\ &= 220 \end{aligned} \qquad \begin{aligned} 3x + z &= 405 \\ 3(110) + z &= 405 \\ 330 + z &= 405 \\ z &= 75 \end{aligned}$$

There were 110 adults, 220 children, and 75 senior citizens that bought tickets.

77. Let x = the number of servings of chicken.
Let y = the number of servings of corn.
Let z = the number of servings of 2% milk.

Protein equation: $30x + 3y + 9z = 66$

Carbohydrate equation: $35x + 16y + 13z = 94.5$

Calcium equation: $200x + 10y + 300z = 910$

Multiply each side of the first equation by –16 and multiply each side of the second equation by 3 and add them to eliminate *y*; multiply each side of the second equation by –5 and multiply each side of the third equation by 8 and add to eliminate *y*:

$$\begin{aligned} -480x - 48y - 144z &= -1056 \\ 105x + 48y + 39z &= 283.5 \\ \hline -375x \qquad - 105z &= -772.5 \end{aligned}$$

$$\begin{aligned} -175x - 80y - 65z &= -472.5 \\ 1600x + 80y + 2400z &= 7280 \\ \hline 1425x \qquad + 2335z &= 6807.5 \end{aligned}$$

Multiply each side of the first result by 19 and multiply each side of the second result by 5 to eliminate *x*:

$$\begin{aligned} -7125x - 1995z &= -14,677.5 \\ 7125x + 11,675z &= 34,037.5 \\ \hline 9680z &= 19,360 \\ z &= 2 \end{aligned}$$

Substituting and solving for the other variables:

$$\begin{aligned} -375x - 105(2) &= -772.5 \\ -375x - 210 &= -772.5 \\ -375x &= -562.5 \\ x &= 1.5 \end{aligned}$$

$$\begin{aligned} 30(1.5) + 3y + 9(2) &= 66 \\ 45 + 3y + 18 &= 66 \\ 3y &= 3 \\ y &= 1 \end{aligned}$$

The dietitian should serve 1.5 servings of chicken, 1 serving of corn, and 2 servings of 2% milk.

78. Let x = the amount in Treasury bills.
Let y = the amount in Treasury bonds.
Let z = the amount in corporate bonds.

Since the total investment is \$20,000,
$x + y + z = 20,000$

Since the total income is to be \$1390,
$0.05x + 0.07y + 0.10z = 1390$

The investment in Treasury bills is to be \$3000 more than the investment in corporate bonds.
So, $x = 3000 + z$

Substitute for *x* in the first two equations and simplify:

$$\begin{aligned} (3000 + z) + y + z &= 20,000 \\ y + 2z &= 17,000 \end{aligned}$$

$$\begin{aligned} 5(3000 + z) + 7y + 10z &= 139,000 \\ 7y + 15z &= 124,000 \end{aligned}$$

Multiply each side of the first result by -7 and add to the second result to eliminate *y*:

$$\begin{aligned} -7y - 14z &= -119,000 \\ 7y + 15z &= 124,000 \\ \hline z &= 5,000 \end{aligned}$$

$$x = 3000 + z = 3000 + 5000 = 8000$$

$$\begin{aligned} y + 2z &= 17,000 \\ y + 2(5000) &= 17,000 \\ y + 10,000 &= 17,000 \\ y &= 7000 \end{aligned}$$

Kelly should invest \$8000 in Treasury bills, \$7000 in Treasury bonds, and \$5000 in corporate bonds.

79. Let x = the price of 1 hamburger.
Let y = the price of 1 order of fries.
Let z = the price of 1 drink.

We can construct the system

$$\begin{cases} 8x + 6y + 6z = 26.10 \\ 10x + 6y + 8z = 31.60 \end{cases}$$

A system involving only 2 equations that contain 3 or more unknowns cannot be solved uniquely.

Multiply the first equation by $-\frac{1}{2}$ and the second equation by $\frac{1}{2}$, then add to eliminate *y*:

$$\begin{array}{r} -4x-3y-3z=-13.05 \\ \underline{5x+3y+4z=15.80} \\ x \quad\quad +z=2.75 \\ x=2.75-z \end{array}$$

Substitute and solve for y in terms of z:

$$5(2.75-z)+3y+4z=15.80$$
$$13.75+3y-z=15.80$$
$$3y=z+2.05$$
$$y=\frac{1}{3}z+\frac{41}{60}$$

Solutions of the system are: $x=2.75-z$, $y=\frac{1}{3}z+\frac{41}{60}$.

Since we are given that $0.60 \le z \le 0.90$, we choose values of z that give two-decimal-place values of x and y with $1.75 \le x \le 2.25$ and $0.75 \le y \le 1.00$.

The possible values of x, y, and z are shown in the table.

x	y	z
2.13	0.89	0.62
2.10	0.90	0.65
2.07	.091	0.68
2.04	.092	0.71
2.01	.093	0.74
1.98	.094	0.77
1.95	0.95	0.80
1.92	0.96	0.83
1.89	0.97	0.86
1.86	0.98	0.89
1.83	0.99	0.92
1.80	1.00	0.95

80. Let x = the price of 1 hamburger.
Let y = the price of 1 order of fries.
Let z = the price of 1 drink

We can construct the system

$$\begin{cases} 8x+6y+6z=26.10 \\ 10x+6y+8z=31.60 \\ 3x+2y+4z=10.95 \end{cases}$$

Subtract the second equation from the first equation to eliminate y:

$$\begin{array}{r} 8x+6y+6z=26.10 \\ \underline{10x+6y+8z=31.60} \\ -2x-2z=-5.5 \end{array}$$

Multiply the third equation by –3 and add it to the second equation to eliminate y:

$$\begin{array}{r} 10x+6y+8z=31.60 \\ \underline{-9x-6y-12z=-32.85} \\ x \quad\quad -4z=-1.25 \end{array}$$

Multiply the second result by 2 and add it to the first result to eliminate x:

$$\begin{array}{r} -2x-2z=-5.5 \\ \underline{2x-8z=-2.5} \\ -10z=-8 \\ z=0.8 \end{array}$$

Substitute for z to find the other variables:

$$x-4(0.8)=-1.25$$
$$x-3.2=-1.25$$
$$x=1.95$$

$$3(1.95)+2y+4(0.8)=10.95$$
$$5.85+2y+3.2=1.095$$
$$2y=1.9$$
$$y=0.95$$

Therefore, one hamburger costs \$1.95, one order of fries costs \$0.95, and one drink costs \$0.80.

81. Let x = Beth's time working alone.
Let y = Bill's time working alone.
Let z = Edie's time working alone.

We can use the following tables to organize our work:

	Beth	Bill	Edie
Hours to do job	x	y	z
Part of job done in 1 hour	$\frac{1}{x}$	$\frac{1}{y}$	$\frac{1}{z}$

In 10 hours they complete 1 entire job, so

$$10\left(\frac{1}{x}+\frac{1}{y}+\frac{1}{z}\right)=1$$

$$\frac{1}{x}+\frac{1}{y}+\frac{1}{z}=\frac{1}{10}$$

	Bill	Edie
Hours to do job	y	z
Part of job done in 1 hour	$\frac{1}{y}$	$\frac{1}{z}$

In 15 hours they complete 1 entire job, so

$$15\left(\frac{1}{y}+\frac{1}{z}\right)=1.$$

$$\frac{1}{y}+\frac{1}{z}=\frac{1}{15}$$

	Beth	Bill	Edie
Hours to do job	x	y	z
Part of job done in 1 hour	$\frac{1}{x}$	$\frac{1}{y}$	$\frac{1}{z}$

With all 3 working for 4 hours and Beth and Bill working for an additional 8 hours, they complete 1 entire job, so $4\left(\frac{1}{x}+\frac{1}{y}+\frac{1}{z}\right)+8\left(\frac{1}{x}+\frac{1}{y}\right)=1$

$$\frac{12}{x}+\frac{12}{y}+\frac{4}{z}=1$$

We have the system

$$\begin{cases}\frac{1}{x}+\frac{1}{y}+\frac{1}{z}=\frac{1}{10}\\ \frac{1}{y}+\frac{1}{z}=\frac{1}{15}\\ \frac{12}{x}+\frac{12}{y}+\frac{4}{z}=1\end{cases}$$

Subtract the second equation from the first equation:

$$\begin{array}{r} \frac{1}{x}+\frac{1}{y}+\frac{1}{z}=\frac{1}{10}\\ \frac{1}{y}+\frac{1}{z}=\frac{1}{15}\\ \hline \frac{1}{x}\qquad\quad=\frac{1}{30}\\ x=30\end{array}$$

Substitute $x = 30$ into the third equation:

$$\frac{12}{30}+\frac{12}{y}+\frac{4}{z}=1$$

$$\frac{12}{y}+\frac{4}{z}=\frac{3}{5}$$

Now consider the system consisting of the last result and the second original equation. Multiply the second original equation by -12 and add it to the last result to eliminate y:

$$\begin{array}{r} \frac{-12}{y}+\frac{-12}{z}=\frac{-12}{15}\\ \frac{12}{y}+\frac{4}{z}=\frac{3}{5}\\ \hline -\frac{8}{z}=-\frac{3}{15}\\ z=40\end{array}$$

Plugging $z = 40$ to find y:

$$\frac{12}{y}+\frac{4}{z}=\frac{3}{5}$$

$$\frac{12}{y}+\frac{4}{40}=\frac{3}{5}$$

$$\frac{12}{y}=\frac{1}{2}$$

$$y=24$$

Working alone, it would take Beth 30 hours, Bill 24 hours, and Edie 40 hours to complete the job.

82 – 84. Answers will vary.

Section 10.2

1. matrix
2. augmented
3. True
4. True
5. Writing the augmented matrix for the system of equations:
$$\begin{cases}x-5y=5\\ 4x+3y=6\end{cases}\rightarrow\left[\begin{array}{cc|c}1 & -5 & 5\\ 4 & 3 & 6\end{array}\right]$$
6. Writing the augmented matrix for the system of equations:
$$\begin{cases}3x+4y=7\\ 4x-2y=5\end{cases}\rightarrow\left[\begin{array}{cc|c}3 & 4 & 7\\ 4 & -2 & 5\end{array}\right]$$

7. $\begin{cases} 2x+3y-6=0 \\ 4x-6y+2=0 \end{cases}$

Write the system in standard form and then write the augmented matrix for the system of equations:

$$\begin{cases} 2x+3y=6 \\ 4x-6y=-2 \end{cases} \rightarrow \left[\begin{array}{cc|c} 2 & 3 & 6 \\ 4 & -6 & -2 \end{array}\right]$$

8. $\begin{cases} 9x-y=0 \\ 3x-y-4=0 \end{cases}$

Write the system in standard form and then write the augmented matrix for the system of equations:

$$\begin{cases} 9x-y=0 \\ 3x-y=4 \end{cases} \rightarrow \left[\begin{array}{cc|c} 9 & -1 & 0 \\ 3 & -1 & 4 \end{array}\right]$$

9. Writing the augmented matrix for the system of equations:

$$\begin{cases} 0.01x-0.03y=0.06 \\ 0.13x+0.10y=0.20 \end{cases} \rightarrow \left[\begin{array}{cc|c} 0.01 & -0.03 & 0.06 \\ 0.13 & 0.10 & 0.20 \end{array}\right]$$

10. Writing the augmented matrix for the system of equations:

$$\begin{cases} \frac{4}{3}x-\frac{3}{2}y=\frac{3}{4} \\ -\frac{1}{4}x+\frac{1}{3}y=\frac{2}{3} \end{cases} \rightarrow \left[\begin{array}{cc|c} \frac{4}{3} & -\frac{3}{2} & \frac{3}{4} \\ -\frac{1}{4} & \frac{1}{3} & \frac{2}{3} \end{array}\right]$$

11. Writing the augmented matrix for the system of equations:

$$\begin{cases} x-y+z=10 \\ 3x+3y=5 \\ x+y+2z=2 \end{cases} \rightarrow \left[\begin{array}{ccc|c} 1 & -1 & 1 & 10 \\ 3 & 3 & 0 & 5 \\ 1 & 1 & 2 & 2 \end{array}\right]$$

12. Writing the augmented matrix for the system of equations:

$$\begin{cases} 5x-y-z=0 \\ x+y=5 \\ 2x-3z=2 \end{cases} \rightarrow \left[\begin{array}{ccc|c} 5 & -1 & -1 & 0 \\ 1 & 1 & 0 & 5 \\ 2 & 0 & -3 & 2 \end{array}\right]$$

13. Writing the augmented matrix for the system of equations:

$$\begin{cases} x+y-z=2 \\ 3x-2y=2 \\ 5x+3y-z=1 \end{cases} \rightarrow \left[\begin{array}{ccc|c} 1 & 1 & -1 & 2 \\ 3 & -2 & 0 & 2 \\ 5 & 3 & -1 & 1 \end{array}\right]$$

14. $\begin{cases} 2x+3y-4z=0 \\ x-5z+2=0 \\ x+2y-3z=-2 \end{cases}$

Write the system in standard form and then write the augmented matrix for the system of equations:

$$\begin{cases} 2x+3y-4z=0 \\ x-5z=-2 \\ x+2y-3z=-2 \end{cases} \rightarrow \left[\begin{array}{ccc|c} 2 & 3 & -4 & 0 \\ 1 & 0 & -5 & -2 \\ 1 & 2 & -3 & -2 \end{array}\right]$$

15. Writing the augmented matrix for the system of equations:

$$\begin{cases} x-y-z=10 \\ 2x+y+2z=-1 \\ -3x+4y=5 \\ 4x-5y+z=0 \end{cases} \rightarrow \left[\begin{array}{ccc|c} 1 & -1 & -1 & 10 \\ 2 & 1 & 2 & -1 \\ -3 & 4 & 0 & 5 \\ 4 & -5 & 1 & 0 \end{array}\right]$$

16. Writing the augmented matrix for the system of equations:

$$\begin{cases} x-y+2z-w=5 \\ x+3y-4z+2w=2 \\ 3x-y-5z-w=-1 \end{cases} \rightarrow \left[\begin{array}{cccc|c} 1 & -1 & 2 & -1 & 5 \\ 1 & 3 & -4 & 2 & 2 \\ 3 & -1 & -5 & -1 & -1 \end{array}\right]$$

17. $R_2=-2r_1+r_2$

$$\left[\begin{array}{cc|c} 1 & -3 & -2 \\ 2 & -5 & 5 \end{array}\right] \rightarrow \left[\begin{array}{cc|c} 1 & -3 & -2 \\ -2(1)+2 & 2(-3)-5 & -2(-2)+5 \end{array}\right]$$

$$\rightarrow \left[\begin{array}{cc|c} 1 & 3 & 2 \\ 0 & 1 & 9 \end{array}\right]$$

18. $R_2=-2r_1+r_2$

$$\left[\begin{array}{cc|c} 1 & -3 & -3 \\ 2 & -5 & -4 \end{array}\right] \rightarrow \left[\begin{array}{cc|c} 1 & -3 & -3 \\ -2(1)+2 & -2(-3)-5 & -2(-3)-4 \end{array}\right]$$

$$\rightarrow \left[\begin{array}{cc|c} 1 & -3 & -3 \\ 0 & 1 & 2 \end{array}\right]$$

19. a. $R_2=-2r_1+r_2$

$$\left[\begin{array}{ccc|c} 1 & -3 & 4 & 3 \\ 2 & -5 & 6 & 6 \\ -3 & 3 & 4 & 6 \end{array}\right]$$

$$\rightarrow \left[\begin{array}{ccc|c} 1 & -3 & 4 & 3 \\ -2(1)+2 & -2(-3)-5 & -2(4)+6 & -2(3)+6 \\ -3 & 3 & 4 & 6 \end{array}\right]$$

$$\rightarrow \left[\begin{array}{ccc|c} 1 & -3 & 4 & 3 \\ 0 & 1 & -2 & 0 \\ -3 & 3 & 4 & 6 \end{array}\right]$$

b. $R_3 = 3r_1 + r_3$

$$\left[\begin{array}{ccc|c} 1 & -3 & 4 & 3 \\ 2 & -5 & 6 & 6 \\ -3 & 3 & 4 & 6 \end{array}\right]$$

$$\rightarrow \left[\begin{array}{ccc|c} 1 & -3 & 4 & 3 \\ 2 & -5 & 6 & 6 \\ 3(1)-3 & 3(-3)+3 & 3(4)+4 & 3(3)+6 \end{array}\right]$$

$$\rightarrow \left[\begin{array}{ccc|c} 1 & -3 & 4 & 3 \\ 2 & -5 & 6 & 6 \\ 0 & -6 & 16 & 15 \end{array}\right]$$

20. a. $R_2 = -2r_1 + r_2$

$$\left[\begin{array}{ccc|c} 1 & -3 & 3 & -5 \\ 2 & -5 & -3 & -5 \\ -3 & -2 & 4 & 6 \end{array}\right]$$

$$\rightarrow \left[\begin{array}{ccc|c} 1 & -3 & 4 & 3 \\ -2(1)+2 & -2(-3)-5 & -2(3)-3 & -2(-5)-5 \\ -3 & 3 & 4 & 6 \end{array}\right]$$

$$\rightarrow \left[\begin{array}{ccc|c} 1 & -3 & 3 & -5 \\ 0 & 1 & -9 & 5 \\ -3 & -2 & 4 & 6 \end{array}\right]$$

b. $R_3 = 3r_1 + r_3$

$$\left[\begin{array}{ccc|c} 1 & -3 & 3 & -5 \\ 2 & -5 & -3 & -5 \\ -3 & -2 & 4 & 6 \end{array}\right]$$

$$\rightarrow \left[\begin{array}{ccc|c} 1 & -3 & 3 & -5 \\ 2 & -5 & -3 & -5 \\ 3(1)-3 & 3(-3)-2 & 3(3)+4 & 3(-5)+6 \end{array}\right]$$

$$\rightarrow \left[\begin{array}{ccc|c} 1 & -3 & 3 & -5 \\ 2 & -5 & -3 & -5 \\ 0 & -11 & 13 & -9 \end{array}\right]$$

21. a. $R_2 = -2r_1 + r_2$

$$\left[\begin{array}{ccc|c} 1 & -3 & 2 & -6 \\ 2 & -5 & 3 & -4 \\ -3 & -6 & 4 & 6 \end{array}\right]$$

$$\rightarrow \left[\begin{array}{ccc|c} 1 & -3 & 2 & -6 \\ -2(1)+2 & -2(-3)-5 & -2(2)+3 & -2(-6)-4 \\ -3 & -6 & 4 & 6 \end{array}\right]$$

$$\rightarrow \left[\begin{array}{ccc|c} 1 & -3 & 2 & -6 \\ 0 & 1 & -1 & 8 \\ -3 & -6 & 4 & 6 \end{array}\right]$$

b. $R_3 = 3r_1 + r_3$

$$\left[\begin{array}{ccc|c} 1 & -3 & 2 & -6 \\ 2 & -5 & 3 & -4 \\ -3 & -6 & 4 & 6 \end{array}\right]$$

$$\rightarrow \left[\begin{array}{ccc|c} 1 & -3 & 2 & -6 \\ 2 & -5 & 3 & -4 \\ 3(1)-3 & 3(-3)-6 & 3(2)+4 & 3(-6)+6 \end{array}\right]$$

$$\rightarrow \left[\begin{array}{ccc|c} 1 & -3 & 2 & -6 \\ 2 & -5 & 3 & -4 \\ 0 & -15 & 10 & -12 \end{array}\right]$$

22. a. $R_2 = -2r_1 + r_2$

$$\left[\begin{array}{ccc|c} 1 & -3 & -4 & -6 \\ 2 & -5 & 6 & -6 \\ -3 & 1 & 4 & 6 \end{array}\right]$$

$$\rightarrow \left[\begin{array}{ccc|c} 1 & -3 & -4 & -6 \\ -2(1)+2 & -2(-3)-5 & -2(-4)+6 & -2(-6)-6 \\ -3 & 1 & 4 & 6 \end{array}\right]$$

$$\rightarrow \left[\begin{array}{ccc|c} 1 & -3 & -4 & -6 \\ 0 & 1 & 14 & 6 \\ -3 & 1 & 4 & 6 \end{array}\right]$$

b. $R_3 = 3r_1 + r_3$

$$\left[\begin{array}{ccc|c} 1 & -3 & -4 & -6 \\ 2 & -5 & 6 & -6 \\ -3 & 1 & 4 & 6 \end{array}\right]$$

$$\rightarrow \left[\begin{array}{ccc|c} 1 & -3 & -4 & -6 \\ 2 & -5 & 6 & -6 \\ 3(1)-3 & 3(-3)+1 & 3(-4)+4 & 3(-6)+6 \end{array}\right]$$

$$\rightarrow \left[\begin{array}{ccc|c} 1 & -3 & -4 & -6 \\ 2 & -5 & 6 & -6 \\ 0 & -8 & -8 & -12 \end{array}\right]$$

23. a. $R_2 = -2r_1 + r_2$

$$\left[\begin{array}{ccc|c} 1 & -3 & 1 & -2 \\ 2 & -5 & 6 & -2 \\ -3 & 1 & 4 & 6 \end{array}\right]$$

$$\rightarrow \left[\begin{array}{ccc|c} 1 & -3 & 1 & -2 \\ -2(1)+2 & -2(-3)-5 & -2(1)+6 & -2(-2)-2 \\ -3 & 1 & 4 & 6 \end{array}\right]$$

$$\rightarrow \left[\begin{array}{ccc|c} 1 & -3 & 1 & -2 \\ 0 & 1 & 4 & 2 \\ -3 & 1 & 4 & 6 \end{array}\right]$$

b. $R_3 = 3r_1 + r_3$

$$\left[\begin{array}{ccc|c} 1 & -3 & 1 & -2 \\ 2 & -5 & 6 & -2 \\ -3 & 1 & 4 & 6 \end{array}\right]$$

$$\to \left[\begin{array}{ccc|c} 1 & -3 & 1 & -2 \\ 2 & -5 & 6 & -2 \\ 3(1)-3 & 3(-3)+1 & 3(1)+4 & 3(-2)+6 \end{array}\right]$$

$$\to \left[\begin{array}{ccc|c} 1 & -3 & 1 & -2 \\ 2 & -5 & 6 & -2 \\ 0 & -8 & 7 & 0 \end{array}\right]$$

24. a. $R_2 = -2r_1 + r_2$

$$\left[\begin{array}{ccc|c} 1 & -3 & -1 & 2 \\ 2 & -5 & 2 & 6 \\ -3 & -6 & 4 & 6 \end{array}\right]$$

$$\to \left[\begin{array}{ccc|c} 1 & -3 & -1 & 2 \\ -2(1)+2 & -2(-3)-5 & -2(-1)+2 & -2(2)+6 \\ -3 & -6 & 4 & 6 \end{array}\right]$$

$$\to \left[\begin{array}{ccc|c} 1 & -3 & -1 & 2 \\ 0 & 1 & 4 & 2 \\ -3 & -6 & 4 & 6 \end{array}\right]$$

b. $R_3 = 3r_1 + r_3$

$$\left[\begin{array}{ccc|c} 1 & -3 & -1 & 2 \\ 2 & -5 & 2 & 6 \\ -3 & -6 & 4 & 6 \end{array}\right]$$

$$\to \left[\begin{array}{ccc|c} 1 & -3 & -1 & 2 \\ 2 & -5 & 2 & 6 \\ 3(1)-3 & 3(-3)-6 & 3(-1)+4 & 3(2)+6 \end{array}\right]$$

$$\to \left[\begin{array}{ccc|c} 1 & -3 & -1 & 2 \\ 2 & -5 & 2 & 6 \\ 0 & -15 & 1 & 12 \end{array}\right]$$

25. $\begin{cases} x = 5 \\ y = -1 \end{cases}$ consistent; $x = 5,\ y = -1$

26. $\begin{cases} x = -4 \\ y = 0 \end{cases}$ consistent; $x = -4,\ y = 0$

27. $\begin{cases} x = 1 \\ y = 2 \\ 0 = 3 \end{cases}$ inconsistent

28. $\begin{cases} x = 0 \\ y = 0 \\ 0 = 2 \end{cases}$ inconsistent

29. $\begin{cases} x + 2z = -1 \\ y - 4z = -2 \\ 0 = 0 \end{cases}$ consistent; $x = -1 - 2z,\ y = -2 + 4z$, z is any real number

30. $\begin{cases} x + 4z = 4 \\ y + 3z = 2 \\ 0 = 0 \end{cases}$ consistent; $x = 4 - 4z,\ y = 2 - 3z$, z is any real number

31. $\begin{cases} x_1 = 1 \\ x_2 + x_4 = 2 \\ x_3 + 2x_4 = 3 \end{cases}$ consistent; $x_1 = 1,\ x_2 = 2 - x_4,\ x_3 = 3 - 2x_4$, x_4 is any real number

32. $\begin{cases} x_1 = 1 \\ x_2 + 2x_4 = 2 \\ x_3 + 3x_4 = 0 \end{cases}$ consistent; $x_1 = 1,\ x_2 = 2 - 2x_4,\ x_3 = -3x_4$, x_4 is any real number

33. $\begin{cases} x_1 + 4x_4 = 2 \\ x_2 + x_3 + 3x_4 = 3 \\ 0 = 0 \end{cases}$ consistent; $x_1 = 2 - 4x_4$, $x_2 = 3 - x_3 - 3x_4$, x_3, x_4 are any real numbers

34. $\begin{cases} x_1 = 1 \\ x_2 = 2 \\ x_3 + 2x_4 = 3 \end{cases}$ consistent; $x_1 = 1, x_2 = 2,\ x_3 = 3 - 2x_4$, x_4 is any real number

35. $\begin{cases} x_1 + x_4 = -2 \\ x_2 + 2x_4 = 2 \\ x_3 - x_4 = 0 \\ 0 = 0 \end{cases}$ consistent; $x_1 = -2 - x_4$, $x_2 = 2 - 2x_4,\ x_3 = x_4$, x_4 is any real number

36. $\begin{cases} x_1 = 1 \\ x_2 = 2 \\ x_3 = 3 \\ x_4 = 0 \end{cases}$ consistent; $x_1 = 1_4,\ x_2 = 2$, $x_3 = 3,\ x_4 = 0$

37. $\begin{cases} x+y=8 \\ x-y=4 \end{cases}$

Write the augmented matrix:

$$\left[\begin{array}{cc|c} 1 & 1 & 8 \\ 1 & -1 & 4 \end{array}\right] \to \left[\begin{array}{cc|c} 1 & 1 & 8 \\ 0 & -2 & -4 \end{array}\right] \quad (R_2 = -r_1 + r_2)$$

$$\to \left[\begin{array}{cc|c} 1 & 1 & 8 \\ 0 & 1 & 2 \end{array}\right] \quad \left(R_2 = -\tfrac{1}{2} r_2\right)$$

$$\to \left[\begin{array}{cc|c} 1 & 0 & 6 \\ 0 & 1 & 2 \end{array}\right] \quad (R_1 = -r_2 + r_1)$$

The solution is $x = 6,\ y = 2$.

38. $\begin{cases} x+2y=5 \\ x+\ \ y=3 \end{cases}$

Write the augmented matrix:

$$\left[\begin{array}{cc|c} 1 & 2 & 5 \\ 1 & 1 & 3 \end{array}\right] \to \left[\begin{array}{cc|c} 1 & 2 & 5 \\ 0 & -1 & -2 \end{array}\right] \quad (R_2 = -r_1 + r_2)$$

$$\to \left[\begin{array}{cc|c} 1 & 2 & 5 \\ 0 & 1 & 2 \end{array}\right] \quad (R_2 = -r_2)$$

$$\to \left[\begin{array}{cc|c} 1 & 0 & 1 \\ 0 & 1 & 2 \end{array}\right] \quad (R_1 = -2r_2 + r_1)$$

The solution is $x = 1,\ y = 2$.

39. $\begin{cases} 2x-4y=-2 \\ 3x+2y=\ \ 3 \end{cases}$

Write the augmented matrix:

$$\left[\begin{array}{cc|c} 2 & -4 & -2 \\ 3 & 2 & 3 \end{array}\right] \to \left[\begin{array}{cc|c} 1 & -2 & -1 \\ 3 & 2 & 3 \end{array}\right] \quad \left(R_1 = \tfrac{1}{2} r_1\right)$$

$$\to \left[\begin{array}{cc|c} 1 & -2 & -1 \\ 0 & 8 & 6 \end{array}\right] \quad (R_2 = -3r_1 + r_2)$$

$$\to \left[\begin{array}{cc|c} 1 & -2 & -1 \\ 0 & 1 & \frac{3}{4} \end{array}\right] \quad \left(R_2 = \tfrac{1}{8} r_2\right)$$

$$\to \left[\begin{array}{cc|c} 1 & 0 & \frac{1}{2} \\ 0 & 1 & \frac{3}{4} \end{array}\right] \quad (R_1 = 2r_2 + r_1)$$

The solution is $x = \dfrac{1}{2},\ y = \dfrac{3}{4}$.

40. $\begin{cases} 3x+3y=3 \\ 4x+2y=\dfrac{8}{3} \end{cases}$

Write the augmented matrix:

$$\left[\begin{array}{cc|c} 3 & 3 & 3 \\ 4 & 2 & \frac{8}{3} \end{array}\right] \to \left[\begin{array}{cc|c} 1 & 1 & 1 \\ 4 & 2 & \frac{8}{3} \end{array}\right] \quad \left(R_1 = \tfrac{1}{3} r_1\right)$$

$$\to \left[\begin{array}{cc|c} 1 & 1 & 1 \\ 0 & -2 & -\frac{4}{3} \end{array}\right] \quad (R_2 = -4r_1 + r_2)$$

$$\to \left[\begin{array}{cc|c} 1 & 1 & 1 \\ 0 & 1 & \frac{2}{3} \end{array}\right] \quad \left(R_2 = -\tfrac{1}{2} r_2\right)$$

$$\to \left[\begin{array}{cc|c} 1 & 0 & \frac{1}{3} \\ 0 & 1 & \frac{2}{3} \end{array}\right] \quad (R_1 = -r_2 + r_1)$$

The solution is $x = \dfrac{1}{3},\ y = \dfrac{2}{3}$.

41. $\begin{cases} x+2y=4 \\ 2x+4y=8 \end{cases}$

Write the augmented matrix:

$$\left[\begin{array}{cc|c} 1 & 2 & 4 \\ 2 & 4 & 8 \end{array}\right] \to \left[\begin{array}{cc|c} 1 & 2 & 4 \\ 0 & 0 & 0 \end{array}\right] \quad (R_2 = -2r_1 + r_2)$$

This is a dependent system and the solution is $x = 4 - 2y$, y is any real number .

42. $\begin{cases} 3x-\ \ y=\ \ 7 \\ 9x-3y=21 \end{cases}$

Write the augmented matrix:

$$\left[\begin{array}{cc|c} 3 & -1 & 7 \\ 9 & -3 & 21 \end{array}\right] \to \left[\begin{array}{cc|c} 1 & -\frac{1}{3} & \frac{7}{3} \\ 9 & -3 & 21 \end{array}\right] \quad \left(R_1 = \tfrac{1}{3} r_1\right)$$

$$\to \left[\begin{array}{cc|c} 1 & -\frac{1}{3} & \frac{7}{3} \\ 0 & 0 & 0 \end{array}\right] \quad (R_2 = -9r_1 + r_2)$$

This is a dependent system and the solution is $y = 3x - 7$, x is any real number .

43. $\begin{cases} 2x+3y=6 \\ x-\ \ y=\dfrac{1}{2} \end{cases}$

Write the augmented matrix:

$$\left[\begin{array}{cc|c} 2 & 3 & 6 \\ 1 & -1 & \frac{1}{2} \end{array}\right] \to \left[\begin{array}{cc|c} 1 & \frac{3}{2} & 3 \\ 1 & -1 & \frac{1}{2} \end{array}\right] \quad \left(R_1 = \tfrac{1}{2} r_1\right)$$

$$\to \left[\begin{array}{cc|c} 1 & \frac{3}{2} & 3 \\ 0 & -\frac{5}{2} & -\frac{5}{2} \end{array}\right] \quad (R_2 = -r_1 + r_2)$$

$$\to \left[\begin{array}{cc|c} 1 & \frac{3}{2} & 3 \\ 0 & 1 & 1 \end{array}\right] \quad \left(R_2 = -\tfrac{2}{5} r_2\right)$$

$$\to \left[\begin{array}{cc|c} 1 & 0 & \frac{3}{2} \\ 0 & 1 & 1 \end{array}\right] \quad \left(R_1 = -\tfrac{3}{2} r_2 + r_1\right)$$

The solution is $x = \dfrac{3}{2},\ y = 1$.

44. $\begin{cases} \frac{1}{2}x + y = -2 \\ x - 2y = 8 \end{cases}$

Write the augmented matrix:

$$\left[\begin{array}{cc|c} \frac{1}{2} & 1 & -2 \\ 1 & -2 & 8 \end{array}\right] \to \left[\begin{array}{cc|c} 1 & 2 & -4 \\ 1 & -2 & 8 \end{array}\right] \quad (R_1 = 2r_1)$$

$$\to \left[\begin{array}{cc|c} 1 & 2 & -4 \\ 0 & -4 & 12 \end{array}\right] \quad (R_2 = -r_1 + r_2)$$

$$\to \left[\begin{array}{cc|c} 1 & 2 & -4 \\ 0 & 1 & -3 \end{array}\right] \quad (R_2 = -\tfrac{1}{4}r_2)$$

$$\to \left[\begin{array}{cc|c} 1 & 0 & 2 \\ 0 & 1 & -3 \end{array}\right] \quad (R_1 = -2r_2 + r_1)$$

The solution is $x = 2$, $y = -3$.

45. $\begin{cases} 3x - 5y = 3 \\ 15x + 5y = 21 \end{cases}$

Write the augmented matrix:

$$\left[\begin{array}{cc|c} 3 & -5 & 3 \\ 15 & 5 & 21 \end{array}\right] \to \left[\begin{array}{cc|c} 1 & -\frac{5}{3} & 1 \\ 15 & 5 & 21 \end{array}\right] \quad (R_1 = \tfrac{1}{3}r_1)$$

$$\to \left[\begin{array}{cc|c} 1 & -\frac{5}{3} & 1 \\ 0 & 30 & 6 \end{array}\right] \quad (R_2 = -15r_1 + r_2)$$

$$\to \left[\begin{array}{cc|c} 1 & -\frac{5}{3} & 1 \\ 0 & 1 & \frac{1}{5} \end{array}\right] \quad (R_2 = \tfrac{1}{30}r_2)$$

$$\to \left[\begin{array}{cc|c} 1 & 0 & \frac{4}{3} \\ 0 & 1 & \frac{1}{5} \end{array}\right] \quad (R_1 = \tfrac{5}{3}r_2 + r_1)$$

The solution is $x = \dfrac{4}{3}$, $y = \dfrac{1}{5}$.

46. $\begin{cases} 2x - y = -1 \\ x + \frac{1}{2}y = \frac{3}{2} \end{cases}$

$$\left[\begin{array}{cc|c} 2 & -1 & -1 \\ 1 & \frac{1}{2} & \frac{3}{2} \end{array}\right] \to \left[\begin{array}{cc|c} 1 & -\frac{1}{2} & -\frac{1}{2} \\ 1 & \frac{1}{2} & \frac{3}{2} \end{array}\right] \quad (R_1 = \tfrac{1}{2}r_1)$$

$$\to \left[\begin{array}{cc|c} 1 & -\frac{1}{2} & -\frac{1}{2} \\ 0 & 1 & 2 \end{array}\right] \quad (R_2 = -1r_1 + r_2)$$

$$\to \left[\begin{array}{cc|c} 1 & 0 & \frac{1}{2} \\ 0 & 1 & 2 \end{array}\right] \quad (R_1 = \tfrac{1}{2}r_2 + r_1)$$

The solution is $x = \dfrac{1}{2}$, $y = 2$.

47. $\begin{cases} x - y = 6 \\ 2x - 3z = 16 \\ 2y + z = 4 \end{cases}$

Write the augmented matrix:

$$\left[\begin{array}{ccc|c} 1 & -1 & 0 & 6 \\ 2 & 0 & -3 & 16 \\ 0 & 2 & 1 & 4 \end{array}\right]$$

$$\to \left[\begin{array}{ccc|c} 1 & -1 & 0 & 6 \\ 0 & 2 & -3 & 4 \\ 0 & 2 & 1 & 4 \end{array}\right] \quad (R_2 = -2r_1 + r_2)$$

$$\to \left[\begin{array}{ccc|c} 1 & -1 & 0 & 6 \\ 0 & 1 & -\frac{3}{2} & 2 \\ 0 & 2 & 1 & 4 \end{array}\right] \quad (R_2 = \tfrac{1}{2}r_2)$$

$$\to \left[\begin{array}{ccc|c} 1 & 0 & -\frac{3}{2} & 8 \\ 0 & 1 & -\frac{3}{2} & 2 \\ 0 & 0 & 4 & 0 \end{array}\right] \quad \begin{pmatrix} R_1 = r_2 + r_1 \\ R_3 = -2r_2 + r_3 \end{pmatrix}$$

$$\to \left[\begin{array}{ccc|c} 1 & 0 & -\frac{3}{2} & 8 \\ 0 & 1 & -\frac{3}{2} & 2 \\ 0 & 0 & 1 & 0 \end{array}\right] \quad (R_3 = \tfrac{1}{4}r_3)$$

$$\to \left[\begin{array}{ccc|c} 1 & 0 & 0 & 8 \\ 0 & 1 & 0 & 2 \\ 0 & 0 & 1 & 0 \end{array}\right] \quad \begin{pmatrix} R_1 = \frac{3}{2}r_3 + r_1 \\ R_2 = \frac{3}{2}r_3 + r_2 \end{pmatrix}$$

The solution is $x = 8$, $y = 2$, $z = 0$.

48. $\begin{cases} 2x + y = -4 \\ -2y + 4z = 0 \\ 3x - 2z = -11 \end{cases}$

Write the augmented matrix:

$$\left[\begin{array}{ccc|c} 2 & 1 & 0 & -4 \\ 0 & -2 & 4 & 0 \\ 3 & 0 & -2 & -11 \end{array}\right]$$

$$\to \left[\begin{array}{ccc|c} 1 & \frac{1}{2} & 0 & -2 \\ 0 & -2 & 4 & 0 \\ 3 & 0 & -2 & -11 \end{array}\right] \quad (R_1 = \tfrac{1}{2}r_1)$$

$$\rightarrow\left[\begin{array}{ccc|c} 1 & \frac{1}{2} & 0 & -2 \\ 0 & -2 & 4 & 0 \\ 0 & -\frac{3}{2} & -2 & -5 \end{array}\right] \quad (R_3 = -3r_1 + r_3)$$

$$\rightarrow\left[\begin{array}{ccc|c} 1 & \frac{1}{2} & 0 & -2 \\ 0 & 1 & -2 & 0 \\ 0 & -\frac{3}{2} & -2 & -5 \end{array}\right] \quad (R_2 = -\tfrac{1}{2}r_2)$$

$$\rightarrow\left[\begin{array}{ccc|c} 1 & 0 & 1 & -2 \\ 0 & 1 & -2 & 0 \\ 0 & 0 & -5 & -5 \end{array}\right] \quad \begin{pmatrix} R_1 = -\frac{1}{2}r_2 + r_1 \\ R_3 = \frac{3}{2}r_2 + r_3 \end{pmatrix}$$

$$\rightarrow\left[\begin{array}{ccc|c} 1 & 0 & 1 & -2 \\ 0 & 1 & -2 & 0 \\ 0 & 0 & 1 & 1 \end{array}\right] \quad (R_3 = -\tfrac{1}{5}r_3)$$

$$\rightarrow\left[\begin{array}{ccc|c} 1 & 0 & 0 & -3 \\ 0 & 1 & 0 & 2 \\ 0 & 0 & 1 & 1 \end{array}\right] \quad \begin{pmatrix} R_1 = -r_3 + r_1 \\ R_2 = 2r_3 + r_2 \end{pmatrix}$$

The solution is $x = -3,\ y = 2,\ z = 1$.

49. $\begin{cases} x - 2y + 3z = 7 \\ 2x + y + z = 4 \\ -3x + 2y - 2z = -10 \end{cases}$

Write the augmented matrix:

$$\left[\begin{array}{ccc|c} 1 & -2 & 3 & 7 \\ 2 & 1 & 1 & 4 \\ -3 & 2 & -2 & -10 \end{array}\right]$$

$$\rightarrow\left[\begin{array}{ccc|c} 1 & -2 & 3 & 7 \\ 0 & 5 & -5 & -10 \\ 0 & -4 & 7 & 11 \end{array}\right] \quad \begin{pmatrix} R_2 = -2r_1 + r_2 \\ R_3 = 3r_1 + r_3 \end{pmatrix}$$

$$\rightarrow\left[\begin{array}{ccc|c} 1 & -2 & 3 & 7 \\ 0 & 1 & -1 & -2 \\ 0 & -4 & 7 & 11 \end{array}\right] \quad (R_2 = \tfrac{1}{5}r_2)$$

$$\rightarrow\left[\begin{array}{ccc|c} 1 & 0 & 1 & 3 \\ 0 & 1 & -1 & -2 \\ 0 & 0 & 3 & 3 \end{array}\right] \quad \begin{pmatrix} R_1 = 2r_2 + r_1 \\ R_3 = 4r_2 + r_3 \end{pmatrix}$$

$$\rightarrow\left[\begin{array}{ccc|c} 1 & 0 & 1 & 3 \\ 0 & 1 & -1 & -2 \\ 0 & 0 & 1 & 1 \end{array}\right] \quad (R_3 = \tfrac{1}{3}r_3)$$

$$\rightarrow\left[\begin{array}{ccc|c} 1 & 0 & 0 & 2 \\ 0 & 1 & 0 & -1 \\ 0 & 0 & 1 & 1 \end{array}\right] \quad \begin{pmatrix} R_1 = -r_3 + r_1 \\ R_2 = r_3 + r_2 \end{pmatrix}$$

The solution is $x = 2,\ y = -1,\ z = 1$.

50. $\begin{cases} 2x + y - 3z = 0 \\ -2x + 2y + z = -7 \\ 3x - 4y - 3z = 7 \end{cases}$

Write the augmented matrix:

$$\left[\begin{array}{ccc|c} 2 & 1 & -3 & 0 \\ -2 & 2 & 1 & -7 \\ 3 & -4 & -3 & 7 \end{array}\right]$$

$$\rightarrow\left[\begin{array}{ccc|c} 1 & \frac{1}{2} & -\frac{3}{2} & 0 \\ -2 & 2 & 1 & -7 \\ 3 & -4 & -3 & 7 \end{array}\right] \quad (R_1 = \tfrac{1}{2}r_1)$$

$$\rightarrow\left[\begin{array}{ccc|c} 1 & \frac{1}{2} & -\frac{3}{2} & 0 \\ 0 & 3 & -2 & -7 \\ 0 & -\frac{11}{2} & \frac{3}{2} & 7 \end{array}\right] \quad \begin{pmatrix} R_2 = 2r_1 + r_2 \\ R_3 = -3r_1 + r_3 \end{pmatrix}$$

$$\rightarrow\left[\begin{array}{ccc|c} 1 & \frac{1}{2} & -\frac{3}{2} & 0 \\ 0 & 1 & -\frac{2}{3} & -\frac{7}{3} \\ 0 & -\frac{11}{2} & \frac{3}{2} & 7 \end{array}\right] \quad (R_2 = \tfrac{1}{3}r_2)$$

$$\rightarrow\left[\begin{array}{ccc|c} 1 & 0 & -\frac{7}{6} & \frac{7}{6} \\ 0 & 1 & -\frac{2}{3} & -\frac{7}{3} \\ 0 & 0 & -\frac{13}{6} & -\frac{35}{6} \end{array}\right] \quad \begin{pmatrix} R_1 = -\frac{1}{2}r_2 + r_1 \\ R_3 = \frac{11}{2}r_2 + r_3 \end{pmatrix}$$

$$\rightarrow\left[\begin{array}{ccc|c} 1 & 0 & -\frac{7}{6} & \frac{7}{6} \\ 0 & 1 & -\frac{2}{3} & -\frac{7}{3} \\ 0 & 0 & 1 & \frac{35}{13} \end{array}\right] \quad (R_3 = -\tfrac{6}{13}r_3)$$

$$\rightarrow\left[\begin{array}{ccc|c} 1 & 0 & 0 & \frac{56}{13} \\ 0 & 1 & 0 & -\frac{7}{13} \\ 0 & 0 & 1 & \frac{35}{13} \end{array}\right] \quad \begin{pmatrix} R_1 = \frac{7}{6}r_3 + r_1 \\ R_2 = \frac{2}{3}r_3 + r_2 \end{pmatrix}$$

The solution is $x = \dfrac{56}{13},\ y = -\dfrac{7}{13},\ z = \dfrac{35}{13}$.

51. $\begin{cases} 2x-2y-2z=2 \\ 2x+3y+z=2 \\ 3x+2y=0 \end{cases}$

Write the augmented matrix:

$$\left[\begin{array}{ccc|c} 2 & -2 & -2 & 2 \\ 2 & 3 & 1 & 2 \\ 3 & 2 & 0 & 0 \end{array}\right]$$

$$\rightarrow \left[\begin{array}{ccc|c} 1 & -1 & -1 & 1 \\ 2 & 3 & 1 & 2 \\ 3 & 2 & 0 & 0 \end{array}\right] \quad \left(R_1 = \tfrac{1}{2}r_1\right)$$

$$\rightarrow \left[\begin{array}{ccc|c} 1 & -1 & -1 & 1 \\ 0 & 5 & 3 & 0 \\ 0 & 5 & 3 & -3 \end{array}\right] \quad \begin{pmatrix} R_2 = -2r_1 + r_2 \\ R_3 = -3r_1 + r_3 \end{pmatrix}$$

$$\rightarrow \left[\begin{array}{ccc|c} 1 & -1 & -1 & 1 \\ 0 & 5 & 3 & 0 \\ 0 & 0 & 0 & -3 \end{array}\right] \quad \left(R_3 = -r_2 + r_3\right)$$

There is no solution. The system is inconsistent.

52. $\begin{cases} 2x-3y-z=0 \\ -x+2y+z=5 \\ 3x-4y-z=1 \end{cases}$

Write the augmented matrix:

$$\left[\begin{array}{ccc|c} 2 & -3 & -1 & 0 \\ -1 & 2 & 1 & 5 \\ 3 & -4 & -1 & 1 \end{array}\right]$$

$$\rightarrow \left[\begin{array}{ccc|c} 1 & -2 & -1 & -5 \\ 2 & -3 & -1 & 0 \\ 3 & -4 & -1 & 1 \end{array}\right] \quad \begin{pmatrix} \text{Interchange} \\ r_1 \text{ and } -r_2 \end{pmatrix}$$

$$\rightarrow \left[\begin{array}{ccc|c} 1 & -2 & -1 & -5 \\ 0 & 1 & 1 & 10 \\ 0 & 2 & 2 & 16 \end{array}\right] \quad \begin{pmatrix} R_2 = -2r_1 + r_2 \\ R_3 = -3r_1 + r_3 \end{pmatrix}$$

$$\rightarrow \left[\begin{array}{ccc|c} 1 & -2 & -1 & -5 \\ 0 & 1 & 1 & 10 \\ 0 & 0 & 0 & -4 \end{array}\right] \quad \left(R_3 = -2r_2 + r_3\right)$$

There is no solution. The system is inconsistent.

53. $\begin{cases} -x+y+z=-1 \\ -x+2y-3z=-4 \\ 3x-2y-7z=0 \end{cases}$

Write the augmented matrix:

$$\left[\begin{array}{ccc|c} -1 & 1 & 1 & -1 \\ -1 & 2 & -3 & -4 \\ 3 & -2 & -7 & 0 \end{array}\right]$$

$$\rightarrow \left[\begin{array}{ccc|c} 1 & -1 & -1 & 1 \\ -1 & 2 & -3 & -4 \\ 3 & -2 & -7 & 0 \end{array}\right] \quad \left(R_1 = -r_1\right)$$

$$\rightarrow \left[\begin{array}{ccc|c} 1 & -1 & -1 & 1 \\ 0 & 1 & -4 & -3 \\ 0 & 1 & -4 & -3 \end{array}\right] \quad \begin{pmatrix} R_2 = r_1 + r_2 \\ R_3 = -3r_1 + r_3 \end{pmatrix}$$

$$\rightarrow \left[\begin{array}{ccc|c} 1 & 0 & -5 & -2 \\ 0 & 1 & -4 & -3 \\ 0 & 0 & 0 & 0 \end{array}\right] \quad \begin{pmatrix} R_1 = r_2 + r_1 \\ R_3 = -r_2 + r_3 \end{pmatrix}$$

The matrix in the last step represents the system

$\begin{cases} x-5z=-2 \\ y-4z=-3 \\ 0=0 \end{cases}$

The solution is $x = 5z - 2$, $y = 4z - 3$, z is any real number.

54. $\begin{cases} 2x-3y-z=0 \\ 3x+2y+2z=2 \\ x+5y+3z=2 \end{cases}$

Write the augmented matrix:

$$\left[\begin{array}{ccc|c} 2 & -3 & -1 & 0 \\ 3 & 2 & 2 & 2 \\ 1 & 5 & 3 & 2 \end{array}\right]$$

$$\rightarrow \left[\begin{array}{ccc|c} 1 & 5 & 3 & 2 \\ 3 & 2 & 2 & 2 \\ 2 & -3 & -1 & 0 \end{array}\right] \quad \begin{pmatrix} \text{Interchange} \\ r_1 \text{ and } r_3 \end{pmatrix}$$

$$\rightarrow \left[\begin{array}{ccc|c} 1 & 5 & 3 & 2 \\ 0 & -13 & -7 & -4 \\ 0 & -13 & -7 & -4 \end{array}\right] \quad \begin{pmatrix} R_2 = -3r_1 + r_2 \\ R_3 = -2r_1 + r_3 \end{pmatrix}$$

$$\rightarrow \left[\begin{array}{ccc|c} 1 & 5 & 3 & 2 \\ 0 & 1 & \frac{7}{13} & \frac{4}{13} \\ 0 & 0 & 0 & 0 \end{array}\right] \quad \begin{pmatrix} R_3 = -r_2 + r_3 \\ R_2 = -\frac{1}{13}r_2 \end{pmatrix}$$

$$\rightarrow \left[\begin{array}{ccc|c} 1 & 0 & \frac{4}{13} & \frac{6}{13} \\ 0 & 1 & \frac{7}{13} & \frac{4}{13} \\ 0 & 0 & 0 & 0 \end{array}\right] \quad \left(R_1 = -5r_2 + r_1\right)$$

The matrix in the last step represents the system

$$\begin{cases} x+\dfrac{4}{13}z=\dfrac{6}{13} \\ y+\dfrac{7}{13}z=\dfrac{4}{13} \\ 0=0 \end{cases}$$

The solution is $x=-\dfrac{4}{13}z+\dfrac{6}{13}$, $y=-\dfrac{7}{13}z+\dfrac{4}{13}$, z is any real number.

55. $\begin{cases} 2x-2y+3z=6 \\ 4x-3y+2z=0 \\ -2x+3y-7z=1 \end{cases}$

Write the augmented matrix:

$$\left[\begin{array}{rrr|r} 2 & -2 & 3 & 6 \\ 4 & -3 & 2 & 0 \\ -2 & 3 & -7 & 1 \end{array}\right]$$

$$\rightarrow\left[\begin{array}{rrr|r} 1 & -1 & \frac{3}{2} & 3 \\ 4 & -3 & 2 & 0 \\ -2 & 3 & -7 & 1 \end{array}\right] \quad \left(R_1=\tfrac{1}{2}r_1\right)$$

$$\rightarrow\left[\begin{array}{rrr|r} 1 & -1 & \frac{3}{2} & 3 \\ 0 & 1 & -4 & -12 \\ 0 & 1 & -4 & 7 \end{array}\right] \quad \left(\begin{array}{l} R_2=-4r_1+r_2 \\ R_3=2r_1+r_3 \end{array}\right)$$

$$\rightarrow\left[\begin{array}{rrr|r} 1 & 0 & -\frac{5}{2} & -9 \\ 0 & 1 & -4 & -12 \\ 0 & 0 & 0 & 19 \end{array}\right] \quad \left(\begin{array}{l} R_1=r_2+r_1 \\ R_3=-r_2+r_3 \end{array}\right)$$

There is no solution. The system is inconsistent.

56. $\begin{cases} 3x-2y+2z=6 \\ 7x-3y+2z=-1 \\ 2x-3y+4z=0 \end{cases}$

Write the augmented matrix:

$$\left[\begin{array}{rrr|r} 3 & -2 & 2 & 6 \\ 7 & -3 & 2 & -1 \\ 2 & -3 & 4 & 0 \end{array}\right]$$

$$\rightarrow\left[\begin{array}{rrr|r} 1 & -\frac{2}{3} & \frac{2}{3} & 2 \\ 7 & -3 & 2 & -1 \\ 2 & -3 & 4 & 0 \end{array}\right] \quad \left(R_1=\tfrac{1}{3}r_1\right)$$

$$\rightarrow\left[\begin{array}{rrr|r} 1 & -\frac{2}{3} & \frac{2}{3} & 2 \\ 0 & \frac{5}{3} & -\frac{8}{3} & -15 \\ 0 & -\frac{5}{3} & \frac{8}{3} & -4 \end{array}\right] \quad \left(\begin{array}{l} R_2=-7r_1+r_2 \\ R_3=-2r_1+r_3 \end{array}\right)$$

$$\rightarrow\left[\begin{array}{rrr|r} 1 & -\frac{2}{3} & \frac{2}{3} & 2 \\ 0 & \frac{5}{3} & -\frac{8}{3} & -15 \\ 0 & 0 & 0 & -19 \end{array}\right] \quad \left(R_3=r_2+r_3\right)$$

There is no solution. The system is inconsistent.

57. $\begin{cases} x+y-z=6 \\ 3x-2y+z=-5 \\ x+3y-2z=14 \end{cases}$

Write the augmented matrix:

$$\left[\begin{array}{rrr|r} 1 & 1 & -1 & 6 \\ 3 & -2 & 1 & -5 \\ 1 & 3 & -2 & 14 \end{array}\right]$$

$$\rightarrow\left[\begin{array}{rrr|r} 1 & 1 & -1 & 6 \\ 0 & -5 & 4 & -23 \\ 0 & 2 & -1 & 8 \end{array}\right] \quad \left(\begin{array}{l} R_2=-3r_1+r_2 \\ R_3=-r_1+r_3 \end{array}\right)$$

$$\rightarrow\left[\begin{array}{rrr|r} 1 & 1 & -1 & 6 \\ 0 & 1 & -\frac{4}{5} & \frac{23}{5} \\ 0 & 2 & -1 & 8 \end{array}\right] \quad \left(R_2=-\tfrac{1}{5}r_2\right)$$

$$\rightarrow\left[\begin{array}{rrr|r} 1 & 0 & -\frac{1}{5} & \frac{7}{5} \\ 0 & 1 & -\frac{4}{5} & \frac{23}{5} \\ 0 & 0 & \frac{3}{5} & -\frac{6}{5} \end{array}\right] \quad \left(\begin{array}{l} R_1=-r_2+r_1 \\ R_3=-2r_2+r_3 \end{array}\right)$$

$$\rightarrow\left[\begin{array}{rrr|r} 1 & 0 & -\frac{1}{5} & \frac{7}{5} \\ 0 & 1 & -\frac{4}{5} & \frac{23}{5} \\ 0 & 0 & 1 & -2 \end{array}\right] \quad \left(R_3=\tfrac{5}{3}r_3\right)$$

$$\rightarrow\left[\begin{array}{rrr|r} 1 & 0 & 0 & 1 \\ 0 & 1 & 0 & 3 \\ 0 & 0 & 1 & -2 \end{array}\right] \quad \left(\begin{array}{l} R_1=\frac{1}{5}r_3+r_1 \\ R_2=\frac{4}{5}r_3+r_2 \end{array}\right)$$

The solution is $x=1$, $y=3$, $z=-2$.

58. $\begin{cases} x-y+z=-4 \\ 2x-3y+4z=-15 \\ 5x+y-2z=12 \end{cases}$

Write the augmented matrix:

$$\left[\begin{array}{ccc|c} 1 & -1 & 1 & -4 \\ 2 & -3 & 4 & -15 \\ 5 & 1 & -2 & 12 \end{array}\right]$$

$$\rightarrow\left[\begin{array}{ccc|c} 1 & -1 & 1 & -4 \\ 0 & -1 & 2 & -7 \\ 0 & 6 & -7 & 32 \end{array}\right] \quad \begin{pmatrix} R_2=-2r_1+r_2 \\ R_3=-5r_1+r_3 \end{pmatrix}$$

$$\rightarrow\left[\begin{array}{ccc|c} 1 & -1 & 1 & -4 \\ 0 & 1 & -2 & 7 \\ 0 & 6 & -7 & 32 \end{array}\right] \quad (R_2=-r_2)$$

$$\rightarrow\left[\begin{array}{ccc|c} 1 & 0 & -1 & 3 \\ 0 & 1 & -2 & 7 \\ 0 & 0 & 5 & -10 \end{array}\right] \quad \begin{pmatrix} R_1=r_2+r_1 \\ R_3=-6r_2+r_3 \end{pmatrix}$$

$$\rightarrow\left[\begin{array}{ccc|c} 1 & 0 & -1 & 3 \\ 0 & 1 & -2 & 7 \\ 0 & 0 & 1 & -2 \end{array}\right] \quad \left(R_3=\tfrac{1}{5}r_3\right)$$

$$\rightarrow\left[\begin{array}{ccc|c} 1 & 0 & 0 & 1 \\ 0 & 1 & 0 & 3 \\ 0 & 0 & 1 & -2 \end{array}\right] \quad \begin{pmatrix} R_1=r_3+r_1 \\ R_2=2r_3+r_2 \end{pmatrix}$$

The solution is $x=1,\ y=3,\ z=-2$.

59. $\begin{cases} x+2y-z=-3 \\ 2x-4y+z=-7 \\ -2x+2y-3z=4 \end{cases}$

Write the augmented matrix:

$$\left[\begin{array}{ccc|c} 1 & 2 & -1 & -3 \\ 2 & -4 & 1 & -7 \\ -2 & 2 & -3 & 4 \end{array}\right]$$

$$\rightarrow\left[\begin{array}{ccc|c} 1 & 2 & -1 & -3 \\ 0 & -8 & 3 & -1 \\ 0 & 6 & -5 & -2 \end{array}\right] \quad \begin{pmatrix} R_2=-2r_1+r_2 \\ R_3=2r_1+r_3 \end{pmatrix}$$

$$\rightarrow\left[\begin{array}{ccc|c} 1 & 2 & -1 & -3 \\ 0 & 1 & -\frac{3}{8} & \frac{1}{8} \\ 0 & 6 & -5 & -2 \end{array}\right] \quad \left(R_2=-\tfrac{1}{8}r_2\right)$$

$$\rightarrow\left[\begin{array}{ccc|c} 1 & 0 & -\frac{1}{4} & -\frac{13}{4} \\ 0 & 1 & -\frac{3}{8} & \frac{1}{8} \\ 0 & 0 & -\frac{11}{4} & -\frac{11}{4} \end{array}\right] \quad \begin{pmatrix} R_1=-2r_2+r_1 \\ R_3=-6r_2+r_3 \end{pmatrix}$$

$$\rightarrow\left[\begin{array}{ccc|c} 1 & 0 & -\frac{1}{4} & -\frac{13}{4} \\ 0 & 1 & -\frac{3}{8} & \frac{1}{8} \\ 0 & 0 & 1 & 1 \end{array}\right] \quad \left(R_3=-\tfrac{4}{11}r_3\right)$$

$$\rightarrow\left[\begin{array}{ccc|c} 1 & 0 & 0 & -3 \\ 0 & 1 & 0 & \frac{1}{2} \\ 0 & 0 & 1 & 1 \end{array}\right] \quad \begin{pmatrix} R_1=\frac{1}{4}r_3+r_1 \\ R_2=\frac{3}{8}r_3+r_2 \end{pmatrix}$$

The solution is $x=-3,\ y=\dfrac{1}{2},\ z=1$.

60. $\begin{cases} x+4y-3z=-8 \\ 3x-y+3z=12 \\ x+y+6z=1 \end{cases}$

Write the augmented matrix:

$$\left[\begin{array}{ccc|c} 1 & 4 & -3 & -8 \\ 3 & -1 & 3 & 12 \\ 1 & 1 & 6 & 1 \end{array}\right]$$

$$\rightarrow\left[\begin{array}{ccc|c} 1 & 4 & -3 & -8 \\ 0 & -13 & 12 & 36 \\ 0 & -3 & 9 & 9 \end{array}\right] \quad \begin{pmatrix} R_2=-3r_1+r_2 \\ R_3=-r_1+r_3 \end{pmatrix}$$

$$\rightarrow\left[\begin{array}{ccc|c} 1 & 4 & -3 & -8 \\ 0 & 1 & -\frac{12}{13} & -\frac{36}{13} \\ 0 & -3 & 9 & 9 \end{array}\right] \quad \left(R_2=-\tfrac{1}{13}r_2\right)$$

$$\rightarrow\left[\begin{array}{ccc|c} 1 & 0 & \frac{9}{13} & \frac{40}{13} \\ 0 & 1 & -\frac{12}{13} & -\frac{36}{13} \\ 0 & 0 & \frac{81}{13} & \frac{9}{13} \end{array}\right] \quad \begin{pmatrix} R_1=-4r_2+r_1 \\ R_3=3r_2+r_3 \end{pmatrix}$$

$$\rightarrow\left[\begin{array}{ccc|c} 1 & 0 & \frac{9}{13} & -\frac{40}{13} \\ 0 & 1 & -\frac{12}{13} & -\frac{36}{13} \\ 0 & 0 & 1 & \frac{1}{9} \end{array}\right] \quad \left(R_3=\tfrac{13}{81}r_3\right)$$

$$\rightarrow\left[\begin{array}{ccc|c} 1 & 0 & 0 & 3 \\ 0 & 1 & 0 & -\frac{8}{3} \\ 0 & 0 & 1 & \frac{1}{9} \end{array}\right] \quad \begin{pmatrix} R_1=-\frac{9}{13}r_3+r_1 \\ R_3=\frac{12}{13}r_3+r_2 \end{pmatrix}$$

The solution is $x=3,\ y=-\dfrac{8}{3},\ z=\dfrac{1}{9}$.

61. $\begin{cases} 3x+y-z=\dfrac{2}{3} \\ 2x-y+z=1 \\ 4x+2y=\dfrac{8}{3} \end{cases}$

Write the augmented matrix:

$$\left[\begin{array}{ccc|c} 3 & 1 & -1 & \frac{2}{3} \\ 2 & -1 & 1 & 1 \\ 4 & 2 & 0 & \frac{8}{3} \end{array}\right]$$

$$\rightarrow \left[\begin{array}{ccc|c} 1 & \frac{1}{3} & -\frac{1}{3} & \frac{2}{9} \\ 2 & -1 & 1 & 1 \\ 4 & 2 & 0 & \frac{8}{3} \end{array}\right] \quad \left(R_1 = \tfrac{1}{3}r_1\right)$$

$$\rightarrow \left[\begin{array}{ccc|c} 1 & \frac{1}{3} & -\frac{1}{3} & \frac{2}{9} \\ 0 & -\frac{5}{3} & \frac{5}{3} & \frac{5}{9} \\ 0 & \frac{2}{3} & \frac{4}{3} & \frac{16}{9} \end{array}\right] \quad \begin{pmatrix} R_2 = -2r_1 + r_2 \\ R_3 = -4r_1 + r_3 \end{pmatrix}$$

$$\rightarrow \left[\begin{array}{ccc|c} 1 & \frac{1}{3} & -\frac{1}{3} & \frac{2}{9} \\ 0 & 1 & -1 & -\frac{1}{3} \\ 0 & \frac{2}{3} & \frac{4}{3} & \frac{16}{9} \end{array}\right] \quad \left(R_2 = -\tfrac{3}{5}r_2\right)$$

$$\rightarrow \left[\begin{array}{ccc|c} 1 & 0 & 0 & \frac{1}{3} \\ 0 & 1 & -1 & -\frac{1}{3} \\ 0 & 0 & 2 & 2 \end{array}\right] \quad \begin{pmatrix} R_1 = -\frac{1}{3}r_2 + r_1 \\ R_3 = -\frac{2}{3}r_2 + r_3 \end{pmatrix}$$

$$\rightarrow \left[\begin{array}{ccc|c} 1 & 0 & 0 & \frac{1}{3} \\ 0 & 1 & -1 & -\frac{1}{3} \\ 0 & 0 & 1 & 1 \end{array}\right] \quad \left(R_3 = \tfrac{1}{2}r_3\right)$$

$$\rightarrow \left[\begin{array}{ccc|c} 1 & 0 & 0 & \frac{1}{3} \\ 0 & 1 & 0 & \frac{2}{3} \\ 0 & 0 & 1 & 1 \end{array}\right] \quad \left(R_2 = r_3 + r_2\right)$$

The solution is $x = \dfrac{1}{3}$, $y = \dfrac{2}{3}$, $z = 1$.

62. $\begin{cases} x+y=1 \\ 2x-y+z=1 \\ x+2y+z=\frac{8}{3} \end{cases}$

Write the augmented matrix:

$$\left[\begin{array}{ccc|c} 1 & 1 & 0 & 1 \\ 2 & -1 & 1 & 1 \\ 1 & 2 & 1 & \frac{8}{3} \end{array}\right]$$

$$\rightarrow \left[\begin{array}{ccc|c} 1 & 1 & 0 & 1 \\ 0 & -3 & 1 & -1 \\ 0 & 1 & 1 & \frac{5}{3} \end{array}\right] \quad \begin{pmatrix} R_2 = -2r_1 + r_2 \\ R_3 = -r_1 + r_3 \end{pmatrix}$$

$$\rightarrow \left[\begin{array}{ccc|c} 1 & 1 & 0 & 1 \\ 0 & 1 & 1 & \frac{5}{3} \\ 0 & -3 & 1 & -1 \end{array}\right] \quad \begin{pmatrix} \text{Interchange} \\ r_2 \text{ and } r_3 \end{pmatrix}$$

$$\rightarrow \left[\begin{array}{ccc|c} 1 & 1 & 0 & 1 \\ 0 & 1 & 1 & \frac{5}{3} \\ 0 & 0 & 4 & 4 \end{array}\right] \quad \left(R_3 = 3r_2 + r_3\right)$$

$$\rightarrow \left[\begin{array}{ccc|c} 1 & 1 & 0 & 1 \\ 0 & 1 & 1 & \frac{5}{3} \\ 0 & 0 & 1 & 1 \end{array}\right] \quad \left(R_3 = \tfrac{1}{4}r_3\right)$$

$$\rightarrow \left[\begin{array}{ccc|c} 1 & 1 & 0 & 1 \\ 0 & 1 & 0 & \frac{2}{3} \\ 0 & 0 & 1 & 1 \end{array}\right] \quad \left(R_2 = r_2 - r_3\right)$$

$$\rightarrow \left[\begin{array}{ccc|c} 1 & 0 & 0 & \frac{1}{3} \\ 0 & 1 & 0 & \frac{2}{3} \\ 0 & 0 & 1 & 1 \end{array}\right] \quad \left(R_1 = r_1 - r_2\right)$$

The solution is $x = \dfrac{1}{3}$, $y = \dfrac{2}{3}$, $z = 1$.

63. $\begin{cases} x+y+z+w=4 \\ 2x-y+z=0 \\ 3x+2y+z-w=6 \\ x-2y-2z+2w=-1 \end{cases}$

Write the augmented matrix:

$$\left[\begin{array}{cccc|c} 1 & 1 & 1 & 1 & 4 \\ 2 & -1 & 1 & 0 & 0 \\ 3 & 2 & 1 & -1 & 6 \\ 1 & -2 & -2 & 2 & -1 \end{array}\right]$$

$$\rightarrow \left[\begin{array}{cccc|c} 1 & 1 & 1 & 1 & 4 \\ 0 & -3 & -1 & -2 & -8 \\ 0 & -1 & -2 & -4 & -6 \\ 0 & -3 & -3 & 1 & -5 \end{array}\right] \quad \begin{pmatrix} R_2 = -2r_1 + r_2 \\ R_3 = -3r_1 + r_3 \\ R_4 = -r_1 + r_4 \end{pmatrix}$$

$$\rightarrow\left[\begin{array}{cccc|c} 1 & 1 & 1 & 1 & 4 \\ 0 & -1 & -2 & -4 & -6 \\ 0 & -3 & -1 & -2 & -8 \\ 0 & -3 & -3 & 1 & -5 \end{array}\right] \quad \left(\begin{array}{l}\text{Interchange} \\ r_2 \text{ and } r_3\end{array}\right)$$

$$\rightarrow\left[\begin{array}{cccc|c} 1 & 1 & 1 & 1 & 4 \\ 0 & 1 & 2 & 4 & 6 \\ 0 & -3 & -1 & -2 & -8 \\ 0 & -3 & -3 & 1 & -5 \end{array}\right] \quad (R_2 = -r_2)$$

$$\rightarrow\left[\begin{array}{cccc|c} 1 & 0 & -1 & -3 & -2 \\ 0 & 1 & 2 & 4 & 6 \\ 0 & 0 & 5 & 10 & 10 \\ 0 & 0 & 3 & 13 & 13 \end{array}\right] \quad \left(\begin{array}{l} R_1 = -r_2 + r_1 \\ R_3 = 3r_2 + r_3 \\ R_4 = 3r_2 + r_4 \end{array}\right)$$

$$\rightarrow\left[\begin{array}{cccc|c} 1 & 0 & -1 & -3 & -2 \\ 0 & 1 & 2 & 4 & 6 \\ 0 & 0 & 1 & 2 & 2 \\ 0 & 0 & 3 & 13 & 13 \end{array}\right] \quad \left(R_3 = \tfrac{1}{5} r_3\right)$$

$$\rightarrow\left[\begin{array}{cccc|c} 1 & 0 & 0 & -1 & 0 \\ 0 & 1 & 0 & 0 & 2 \\ 0 & 0 & 1 & 2 & 2 \\ 0 & 0 & 0 & 7 & 7 \end{array}\right] \quad \left(\begin{array}{l} R_1 = r_3 + r_1 \\ R_2 = -2r_3 + r_2 \\ R_4 = -3r_3 + r_4 \end{array}\right)$$

$$\rightarrow\left[\begin{array}{cccc|c} 1 & 0 & 0 & -1 & 0 \\ 0 & 1 & 0 & 0 & 2 \\ 0 & 0 & 1 & 2 & 2 \\ 0 & 0 & 0 & 1 & 1 \end{array}\right] \quad \left(R_4 = \tfrac{1}{7} r_4\right)$$

$$\rightarrow\left[\begin{array}{cccc|c} 1 & 0 & 0 & 0 & 1 \\ 0 & 1 & 0 & 0 & 2 \\ 0 & 0 & 1 & 0 & 0 \\ 0 & 0 & 0 & 1 & 1 \end{array}\right] \quad \left(\begin{array}{l} R_1 = r_4 + r_1 \\ R_3 = -2r_4 + r_3 \end{array}\right)$$

The solution is $x = 1,\ y = 2,\ z = 0,\ w = 1$.

64. $\begin{cases} x + y + z + w = 4 \\ -x + 2y + z = 0 \\ 2x + 3y + z - w = 6 \\ -2x + y - 2z + 2w = -1 \end{cases}$

Write the augmented matrix:

$$\left[\begin{array}{cccc|c} 1 & 1 & 1 & 1 & 4 \\ -1 & 2 & 1 & 0 & 0 \\ 2 & 3 & 1 & -1 & 6 \\ -2 & 1 & -2 & 2 & -1 \end{array}\right]$$

$$\rightarrow\left[\begin{array}{cccc|c} 1 & 1 & 1 & 1 & 4 \\ 0 & 3 & 2 & 1 & 4 \\ 0 & 1 & -1 & -3 & -2 \\ 0 & 3 & 0 & 4 & 7 \end{array}\right] \quad \left(\begin{array}{l} R_2 = r_1 + r_2 \\ R_3 = -2r_1 + r_3 \\ R_4 = 2r_1 + r_4 \end{array}\right)$$

$$\rightarrow\left[\begin{array}{cccc|c} 1 & 1 & 1 & 1 & 4 \\ 0 & 1 & -1 & -3 & -2 \\ 0 & 3 & 2 & 1 & 4 \\ 0 & 3 & 0 & 4 & 7 \end{array}\right] \quad \left(\begin{array}{l}\text{Interchange} \\ r_2 \text{ and } r_3\end{array}\right)$$

$$\rightarrow\left[\begin{array}{cccc|c} 1 & 0 & 2 & 4 & 6 \\ 0 & 1 & -1 & -3 & -2 \\ 0 & 0 & 5 & 10 & 10 \\ 0 & 0 & 3 & 13 & 13 \end{array}\right] \quad \left(\begin{array}{l} R_1 = -r_2 + r_1 \\ R_3 = -3r_2 + r_3 \\ R_4 = -3r_2 + r_4 \end{array}\right)$$

$$\rightarrow\left[\begin{array}{cccc|c} 1 & 0 & 2 & 4 & 6 \\ 0 & 1 & -1 & -3 & -2 \\ 0 & 0 & 5 & 10 & 10 \\ 0 & 0 & 0 & -35 & -35 \end{array}\right] \quad (R_4 = 3r_3 - 5r_4)$$

$$\rightarrow\left[\begin{array}{cccc|c} 1 & 0 & 2 & 4 & 6 \\ 0 & 1 & -1 & -3 & -2 \\ 0 & 0 & 1 & 2 & 2 \\ 0 & 0 & 0 & 1 & 1 \end{array}\right] \quad \left(\begin{array}{l} R_4 = \tfrac{1}{5} r_3 \\ R_4 = -\tfrac{1}{35} r_4 \end{array}\right)$$

Write the matrix as the corresponding system:

$$\begin{cases} x + 2z + 4w = 6 \\ y - z - 3w = -2 \\ z + 2w = 2 \\ w = 1 \end{cases}$$

Substitute and solve:

$$z + 2(1) = 2$$
$$z + 2 = 2$$
$$z = 0$$

$$y - 0 - 3(1) = -2$$
$$y - 3 = -2$$
$$y = 1$$

$$x + 2(0) + 4(1) = 6$$
$$x + 0 + 4 = 6$$
$$x = 2$$

The solution is $x = 2,\ y = 1,\ z = 0,\ w = 1$.

65. $\begin{cases} x+2y+\ \ z=1 \\ 2x-\ y+2z=2 \\ 3x+\ y+3z=3 \end{cases}$

Write the augmented matrix:

$$\left[\begin{array}{ccc|c} 1 & 2 & 1 & 1 \\ 2 & -1 & 2 & 2 \\ 3 & 1 & 3 & 3 \end{array}\right]$$

$$\to\left[\begin{array}{ccc|c} 1 & 2 & 1 & 1 \\ 0 & -5 & 0 & 0 \\ 0 & -5 & 0 & 0 \end{array}\right] \quad \begin{pmatrix} R_2=-2r_1+r_2 \\ R_3=-3r_1+r_3 \end{pmatrix}$$

$$\to\left[\begin{array}{ccc|c} 1 & 2 & 1 & 1 \\ 0 & -5 & 0 & 0 \\ 0 & 0 & 0 & 0 \end{array}\right] \quad (R_3=-r_2+r_3)$$

The matrix in the last step represents the system

$$\begin{cases} x+2y+z=1 \\ -5y\quad =0 \\ 0=0 \end{cases}$$

Substitute and solve:

$$\begin{aligned} -5y&=0 & x+2(0)+z&=1 \\ y&=0 & x+z&=1 \\ & & z&=1-x \end{aligned}$$

The solution is $y=0,\ z=1-x,\ x$ is any real number.

66. $\begin{cases} x+2y-\ z=3 \\ 2x-\ y+2z=6 \\ x-3y+3z=4 \end{cases}$

Write the augmented matrix:

$$\left[\begin{array}{ccc|c} 1 & 2 & -1 & 3 \\ 2 & -1 & 2 & 6 \\ 1 & -3 & 3 & 4 \end{array}\right]$$

$$\to\left[\begin{array}{ccc|c} 1 & 2 & -1 & 3 \\ 0 & -5 & 4 & 0 \\ 0 & -5 & 4 & 1 \end{array}\right] \quad \begin{pmatrix} R_2=-2r_1+r_2 \\ R_3=-r_1+r_3 \end{pmatrix}$$

$$\to\left[\begin{array}{ccc|c} 1 & 2 & -1 & 3 \\ 0 & -5 & 4 & 0 \\ 0 & 0 & 0 & 1 \end{array}\right] \quad (R_3=-r_2+r_3)$$

There is no solution. The system is inconsistent.

67. $\begin{cases} x-y+z=5 \\ 3x+2y-2z=0 \end{cases}$

Write the augmented matrix:

$$\left[\begin{array}{ccc|c} 1 & -1 & 1 & 5 \\ 3 & 2 & -2 & 0 \end{array}\right]$$

$$\to\left[\begin{array}{ccc|c} 1 & -1 & 1 & 5 \\ 0 & 5 & -5 & -15 \end{array}\right] \quad (R_2=-3r_1+r_2)$$

$$\to\left[\begin{array}{ccc|c} 1 & -1 & 1 & 5 \\ 0 & 1 & -1 & -3 \end{array}\right] \quad \left(R_2=\tfrac{1}{5}r_2\right)$$

$$\to\left[\begin{array}{ccc|c} 1 & 0 & 0 & 2 \\ 0 & 1 & -1 & -3 \end{array}\right] \quad (R_1=r_2+r_1)$$

The matrix in the last step represents the system

$$\begin{cases} x=2 \\ y-z=-3 \end{cases}$$

Thus, the solution is $x=2$, $y=z-3$, z is any real number.

68. $\begin{cases} 2x+y-z=4 \\ -x+y+3z=1 \end{cases}$

Write the augmented matrix:

$$\left[\begin{array}{ccc|c} 2 & 1 & -1 & 4 \\ -1 & 1 & 3 & 1 \end{array}\right]$$

$$\to\left[\begin{array}{ccc|c} 1 & -1 & -3 & -1 \\ 2 & 1 & -1 & 4 \end{array}\right] \quad \begin{pmatrix} \text{interchange} \\ r_1 \text{ and } -r_2 \end{pmatrix}$$

$$\to\left[\begin{array}{ccc|c} 1 & -1 & -3 & -1 \\ 0 & 3 & 5 & 6 \end{array}\right] \quad (R_2=-2r_1+r_2)$$

$$\to\left[\begin{array}{ccc|c} 1 & -1 & -3 & -1 \\ 0 & 1 & \frac{5}{3} & 2 \end{array}\right] \quad \left(R_2=\tfrac{1}{3}r_2\right)$$

$$\to\left[\begin{array}{ccc|c} 1 & 0 & -\frac{4}{3} & 1 \\ 0 & 1 & \frac{5}{3} & 2 \end{array}\right] \quad (R_1=r_2+r_1)$$

The matrix in the last step represents the system

$$\begin{cases} x-\dfrac{4}{3}z=1 \\ y+\dfrac{5}{3}z=2 \end{cases}$$

Thus, the solution is: $x=1+\dfrac{4}{3}z$, $y=2-\dfrac{5}{3}z$, z is any real number.

69. $\begin{cases} 2x+3y-z=3 \\ x-y-z=0 \\ -x+y+z=0 \\ x+y+3z=5 \end{cases}$

Write the augmented matrix:

$$\left[\begin{array}{ccc|c} 2 & 3 & -1 & 3 \\ 1 & -1 & -1 & 0 \\ -1 & 1 & 1 & 0 \\ 1 & 1 & 3 & 5 \end{array}\right]$$

$$\rightarrow \left[\begin{array}{ccc|c} 1 & -1 & -1 & 0 \\ 2 & 3 & -1 & 3 \\ -1 & 1 & 1 & 0 \\ 1 & 1 & 3 & 5 \end{array}\right] \quad \begin{pmatrix} \text{interchange} \\ r_1 \text{ and } r_2 \end{pmatrix}$$

$$\rightarrow \left[\begin{array}{ccc|c} 1 & -1 & -1 & 0 \\ 0 & 5 & 1 & 3 \\ 0 & 0 & 0 & 0 \\ 0 & 2 & 4 & 5 \end{array}\right] \quad \begin{pmatrix} R_2 = -2r_1 + r_2 \\ R_3 = r_1 + r_3 \\ R_4 = -r_1 + r_4 \end{pmatrix}$$

$$\rightarrow \left[\begin{array}{ccc|c} 1 & -1 & -1 & 0 \\ 0 & 5 & 1 & 3 \\ 0 & 2 & 4 & 5 \\ 0 & 0 & 0 & 0 \end{array}\right] \quad \begin{pmatrix} \text{interchange} \\ r_3 \text{ and } r_4 \end{pmatrix}$$

$$\rightarrow \left[\begin{array}{ccc|c} 1 & -1 & -1 & 0 \\ 0 & 1 & -7 & -7 \\ 0 & 2 & 4 & 5 \\ 0 & 0 & 0 & 0 \end{array}\right] \quad \left(R_2 = -2r_3 + r_2\right)$$

$$\rightarrow \left[\begin{array}{ccc|c} 1 & 0 & -8 & -7 \\ 0 & 1 & -7 & -7 \\ 0 & 0 & 18 & 19 \\ 0 & 0 & 0 & 0 \end{array}\right] \quad \begin{pmatrix} R_1 = r_2 + r_1 \\ R_3 = -2r_2 + r_3 \end{pmatrix}$$

$$\rightarrow \left[\begin{array}{ccc|c} 1 & 0 & -8 & -7 \\ 0 & 1 & -7 & -7 \\ 0 & 0 & 1 & \frac{19}{18} \\ 0 & 0 & 0 & 0 \end{array}\right] \quad \left(R_3 = \tfrac{1}{18} r_3\right)$$

The matrix in the last step represents the system

$$\begin{cases} x-8z=-7 \\ y-7z=-7 \\ z=\dfrac{19}{18} \end{cases}$$

Substitute and solve:

$$y-7\left(\frac{19}{18}\right)=7 \qquad x-8\left(\frac{19}{18}\right)=-7$$

$$y=\frac{7}{18} \qquad x=\frac{13}{9}$$

Thus, the solution is $x=\dfrac{13}{9}$, $y=\dfrac{7}{18}$, $z=\dfrac{19}{18}$.

70. $\begin{cases} x-3y+z=1 \\ 2x-y-4z=0 \\ x-3y+2z=1 \\ x-2y=5 \end{cases}$

Write the augmented matrix:

$$\left[\begin{array}{ccc|c} 1 & -3 & 1 & 1 \\ 2 & -1 & -4 & 0 \\ 1 & -3 & 2 & 1 \\ 1 & -2 & 0 & 5 \end{array}\right]$$

$$\rightarrow \left[\begin{array}{ccc|c} 1 & -3 & 1 & 1 \\ 0 & 5 & -6 & -2 \\ 0 & 0 & 1 & 0 \\ 0 & 1 & -1 & 4 \end{array}\right] \quad \begin{pmatrix} R_2 = 2r_1 + r_2 \\ R_3 = -r_1 + r_2 \\ R_4 = -r_1 + r_2 \end{pmatrix}$$

$$\rightarrow \left[\begin{array}{ccc|c} 1 & -3 & 1 & 1 \\ 0 & 1 & -1 & 4 \\ 0 & 0 & 1 & 0 \\ 0 & 5 & -6 & -2 \end{array}\right] \quad \begin{pmatrix} \text{interchange} \\ r_2 \text{ and } r_4 \end{pmatrix}$$

$$\rightarrow \left[\begin{array}{ccc|c} 1 & 0 & -2 & 13 \\ 0 & 1 & -1 & 4 \\ 0 & 0 & 1 & 0 \\ 0 & 0 & -1 & -22 \end{array}\right] \quad \begin{pmatrix} R_1 = 3r_2 + r_1 \\ R_4 = -5r_2 + r_4 \end{pmatrix}$$

$$\rightarrow \left[\begin{array}{ccc|c} 1 & 0 & 0 & 13 \\ 0 & 1 & 0 & 4 \\ 0 & 0 & 1 & 0 \\ 0 & 0 & 0 & -22 \end{array}\right] \quad \begin{pmatrix} R_1 = 2r_3 + r_1 \\ R_2 = r_3 + r_2 \\ R_4 = r_3 + r_4 \end{pmatrix}$$

There is no solution. The system is inconsistent.

71. $\begin{cases} 4x+y+z-w=4 \\ x-y+2z+3w=3 \end{cases}$

Write the augmented matrix:

$$\left[\begin{array}{cccc|c} 4 & 1 & 1 & -1 & 4 \\ 1 & -1 & 2 & 3 & 3 \end{array}\right]$$

$$\rightarrow \left[\begin{array}{cccc|c} 1 & -1 & 2 & 3 & 3 \\ 4 & 1 & 1 & -1 & 4 \end{array}\right] \quad \begin{pmatrix} \text{interchange} \\ r_1 \text{ and } r_2 \end{pmatrix}$$

$$\rightarrow \left[\begin{array}{cccc|c} 1 & -1 & 2 & 3 & 3 \\ 0 & 5 & -7 & -13 & -8 \end{array}\right] \quad (R_2 = -4r_1 + r_2)$$

The matrix in the last step represents the system

$$\begin{cases} x - y + 2z + 3w = 3 \\ 5y - 7z - 13w = -8 \end{cases}$$

The second equation yields

$$5y - 7z - 13w = -8$$
$$5y = 7z + 13w - 8$$
$$y = \frac{7}{5}z + \frac{13}{5}w - \frac{8}{5}$$

The first equation yields

$$x - y + 2z + 3w = 3$$
$$x = 3 + y - 2z - 3w$$

Substituting for *y:*

$$x = 3 + \left(-\frac{8}{5} + \frac{7}{5}z + \frac{13}{5}w\right) - 2z - 3w$$
$$x = -\frac{3}{5}z - \frac{2}{5}w + \frac{7}{5}$$

Thus, the solution is $x = -\frac{3}{5}z - \frac{2}{5}w + \frac{7}{5}$, $y = \frac{7}{5}z + \frac{13}{5}w - \frac{8}{5}$, z and w are any real numbers.

72. $\begin{cases} -4x + y = 5 \\ 2x - y + z - w = 5 \\ z + w = 4 \end{cases}$

Write the augmented matrix:

$$\left[\begin{array}{cccc|c} -4 & 1 & 0 & 0 & 5 \\ 2 & -1 & 1 & -1 & 5 \\ 0 & 0 & 1 & 1 & 4 \end{array}\right]$$

$$\rightarrow \left[\begin{array}{cccc|c} 1 & -\frac{1}{4} & 0 & 0 & -\frac{5}{4} \\ 2 & -1 & 1 & -1 & 5 \\ 0 & 0 & 1 & 1 & 4 \end{array}\right] \quad (R_1 = -\tfrac{1}{4}r_1)$$

$$\rightarrow \left[\begin{array}{cccc|c} 1 & -\frac{1}{4} & 0 & 0 & -\frac{5}{4} \\ 0 & -\frac{1}{2} & 1 & -1 & \frac{15}{2} \\ 0 & 0 & 1 & 1 & 4 \end{array}\right] \quad (R_2 = -2r_1 + r_2)$$

$$\rightarrow \left[\begin{array}{cccc|c} 1 & -\frac{1}{4} & 0 & 0 & -\frac{5}{4} \\ 0 & 1 & -2 & 2 & -15 \\ 0 & 0 & 1 & 1 & 4 \end{array}\right] \quad (R_2 = -2r_2)$$

$$\rightarrow \left[\begin{array}{cccc|c} 1 & 0 & -\frac{1}{2} & \frac{1}{2} & -5 \\ 0 & 1 & -2 & 2 & -15 \\ 0 & 0 & 1 & 1 & 4 \end{array}\right] \quad (R_1 = \tfrac{1}{4}r_2 + r_1)$$

$$\rightarrow \left[\begin{array}{cccc|c} 1 & 0 & 0 & 1 & -3 \\ 0 & 1 & 0 & 4 & -7 \\ 0 & 0 & 1 & 1 & 4 \end{array}\right] \quad \begin{pmatrix} R_1 = \frac{1}{2}r_3 + r_1 \\ R_2 = 2r_3 + r_2 \end{pmatrix}$$

The matrix in the last step represents the system

$$\begin{cases} x + w = -3 \\ y + 4w = -7 \\ z + w = 4 \end{cases}$$

Thus, the solution is $x = -3 - w$, $y = -7 - 4w$, $z = 4 - w$, w is any real number.

73. Each of the points must satisfy the equation $y = ax^2 + bx + c$.

$(1,2)$: $2 = a + b + c$

$(-2,-7)$: $-7 = 4a - 2b + c$

$(2,-3)$: $-3 = 4a + 2b + c$

Set up a matrix and solve:

$$\left[\begin{array}{ccc|c} 1 & 1 & 1 & 2 \\ 4 & -2 & 1 & -7 \\ 4 & 2 & 1 & -3 \end{array}\right]$$

$$\rightarrow \left[\begin{array}{ccc|c} 1 & 1 & 1 & 2 \\ 0 & -6 & -3 & -15 \\ 0 & -2 & -3 & -11 \end{array}\right] \quad \begin{pmatrix} R_2 = -4r_1 + r_2 \\ R_3 = -4r_1 + r_3 \end{pmatrix}$$

$$\rightarrow \left[\begin{array}{ccc|c} 1 & 1 & 1 & 2 \\ 0 & 1 & \frac{1}{2} & \frac{5}{2} \\ 0 & -2 & -3 & -11 \end{array}\right] \quad (R_2 = -\tfrac{1}{6}r_2)$$

$$\rightarrow \left[\begin{array}{ccc|c} 1 & 0 & \frac{1}{2} & -\frac{1}{2} \\ 0 & 1 & \frac{1}{2} & \frac{5}{2} \\ 0 & 0 & -2 & -6 \end{array}\right] \quad \begin{pmatrix} R_1 = -r_2 + r_1 \\ R_3 = 2r_2 + r_3 \end{pmatrix}$$

$$\rightarrow \left[\begin{array}{ccc|c} 1 & 0 & \frac{1}{2} & -\frac{1}{2} \\ 0 & 1 & \frac{1}{2} & \frac{5}{2} \\ 0 & 0 & 1 & 3 \end{array}\right] \rightarrow \quad (R_3 = -\tfrac{1}{2}r_3)$$

$$\rightarrow \left[\begin{array}{ccc|c} 1 & 0 & 0 & -2 \\ 0 & 1 & 0 & 1 \\ 0 & 0 & 1 & 3 \end{array}\right] \quad \begin{pmatrix} R_1 = -\frac{1}{2}r_3 + r_1 \\ R_2 = -\frac{1}{2}r_3 + r_2 \end{pmatrix}$$

The solution is $a = -2, b = 1, c = 3$; so the equation is $y = -2x^2 + x + 3$.

74. Each of the points must satisfy the equation $y = ax^2 + bx + c$.

$(1,-1)$: $-1 = a + b + c$

$(3,-1)$: $-1 = 9a + 3b + c$

$(-2,14)$: $14 = 4a - 2b + c$

Set up a matrix and solve:

$$\left[\begin{array}{ccc|c} 1 & 1 & 1 & -1 \\ 9 & 3 & 1 & -1 \\ 4 & -2 & 1 & 14 \end{array}\right] \rightarrow$$

$$\left[\begin{array}{ccc|c} 1 & 1 & 1 & -1 \\ 0 & -6 & -8 & 8 \\ 0 & -6 & -3 & 18 \end{array}\right] \begin{pmatrix} R_2 = -9r_1 + r_2 \\ R_3 = -4r_1 + r_3 \end{pmatrix}$$

$$\rightarrow \left[\begin{array}{ccc|c} 1 & 1 & 1 & -1 \\ 0 & 1 & \frac{4}{3} & -\frac{4}{3} \\ 0 & 0 & 5 & 10 \end{array}\right] \begin{pmatrix} R_2 = -\frac{1}{6}r_2 \\ R_3 = -r_2 + r_3 \end{pmatrix}$$

$$\rightarrow \left[\begin{array}{ccc|c} 1 & 0 & -\frac{1}{3} & \frac{1}{3} \\ 0 & 1 & \frac{4}{3} & -\frac{4}{3} \\ 0 & 0 & 1 & 2 \end{array}\right] \begin{pmatrix} R_1 = -r_2 + r_1 \\ R_3 = \frac{1}{5}r_3 \end{pmatrix}$$

$$\rightarrow \left[\begin{array}{ccc|c} 1 & 0 & 0 & 1 \\ 0 & 1 & 0 & -4 \\ 0 & 0 & 1 & 2 \end{array}\right] \begin{pmatrix} R_1 = \frac{1}{3}r_3 + r_1 \\ R_1 = -\frac{4}{3}r_3 + r_2 \end{pmatrix}$$

The solution is $a = 1, b = -4, c = 2$; so the equation is $y = x^2 - 4x + 2$.

75. Each of the points must satisfy the equation $f(x) = ax^3 + bx^2 + cx + d$.

$f(-3) = -12$: $-27a + 9b - 3c + d = -112$

$f(-1) = -2$: $-a + b - c + d = -2$

$f(1) = 4$: $a + b + c + d = 4$

$f(2) = 13$: $8a + 4b + 2c + d = 13$

Set up a matrix and solve:

$$\left[\begin{array}{cccc|c} -27 & 9 & -3 & 1 & -112 \\ -1 & 1 & -1 & 1 & -2 \\ 1 & 1 & 1 & 1 & 4 \\ 8 & 4 & 2 & 1 & 13 \end{array}\right]$$

$$\rightarrow \left[\begin{array}{cccc|c} 1 & 1 & 1 & 1 & 4 \\ -1 & 1 & -1 & 1 & -2 \\ -27 & 9 & -3 & 1 & -112 \\ 8 & 4 & 2 & 1 & 13 \end{array}\right] \begin{pmatrix} \text{Interchange} \\ r_3 \text{ and } r_1 \end{pmatrix}$$

$$\rightarrow \left[\begin{array}{cccc|c} 1 & 1 & 1 & 1 & 4 \\ 0 & 2 & 0 & 2 & 2 \\ 0 & 36 & 24 & 28 & -4 \\ 0 & -4 & -6 & -7 & -19 \end{array}\right] \begin{pmatrix} R_2 = r_1 + r_2 \\ R_3 = 27r_1 + r_3 \\ R_4 = -8r_1 + r_4 \end{pmatrix}$$

$$\rightarrow \left[\begin{array}{cccc|c} 1 & 1 & 1 & 1 & 4 \\ 0 & 1 & 0 & 1 & 1 \\ 0 & 36 & 24 & 28 & -4 \\ 0 & -4 & -6 & -7 & -19 \end{array}\right] \left(R_2 = \tfrac{1}{2}r_2\right)$$

$$\rightarrow \left[\begin{array}{cccc|c} 1 & 0 & 1 & 0 & 3 \\ 0 & 1 & 0 & 1 & 1 \\ 0 & 0 & 24 & -8 & -40 \\ 0 & 0 & -6 & -3 & -15 \end{array}\right] \begin{pmatrix} R_1 = -r_2 + r_1 \\ R_3 = -36r_2 + r_3 \\ R_4 = 4r_2 + r_4 \end{pmatrix}$$

$$\rightarrow \left[\begin{array}{cccc|c} 1 & 0 & 1 & 0 & 3 \\ 0 & 1 & 0 & 1 & 1 \\ 0 & 0 & 1 & -\frac{5}{3} & -\frac{5}{3} \\ 0 & 0 & -6 & -3 & -15 \end{array}\right] \left(R_3 = \tfrac{1}{24}r_3\right)$$

$$\rightarrow \left[\begin{array}{cccc|c} 1 & 0 & 0 & \frac{1}{3} & \frac{14}{3} \\ 0 & 1 & 0 & 1 & 1 \\ 0 & 0 & 1 & -\frac{1}{3} & -\frac{5}{3} \\ 0 & 0 & 0 & -5 & -25 \end{array}\right] \begin{pmatrix} R_1 = -r_3 + r_1 \\ R_4 = 6r_3 + r_4 \end{pmatrix}$$

$$\rightarrow \left[\begin{array}{cccc|c} 1 & 0 & 0 & \frac{1}{3} & \frac{14}{3} \\ 0 & 1 & 0 & 1 & 1 \\ 0 & 0 & 1 & -\frac{1}{3} & -\frac{5}{3} \\ 0 & 0 & 0 & 1 & 5 \end{array}\right] \left(R_4 = -\tfrac{1}{5}r_4\right)$$

$$\rightarrow \left[\begin{array}{cccc|c} 1 & 0 & 0 & 0 & 3 \\ 0 & 1 & 0 & 0 & -4 \\ 0 & 0 & 1 & 0 & 0 \\ 0 & 0 & 0 & 1 & 5 \end{array}\right] \begin{pmatrix} R_1 = -\frac{1}{3}r_4 + r_1 \\ R_2 = -r_4 + r_2 \\ R_3 = \frac{1}{3}r_4 + r_3 \end{pmatrix}$$

The solution is $a = 3, b = -4, c = 0, d = 5$; so the equation is $f(x) = 3x^3 - 4x^2 + 5$.

76. Each of the points must satisfy the equation $f(x) = ax^3 + bx^2 + cx + d$.

$f(-2) = -10$: $-8a + 4b - 2c + d = -10$

$f(-1) = 3$: $-a + b - c + d = 3$

$f(1) = 5$: $a + b + c + d = 5$

$f(3) = 15$: $27a + 9b + 3c + d = 15$

Set up a matrix and solve:

$$\left[\begin{array}{cccc|c} -8 & 4 & -2 & 1 & -10 \\ -1 & 1 & -1 & 1 & 3 \\ 1 & 1 & 1 & 1 & 5 \\ 27 & 9 & 3 & 1 & 15 \end{array}\right]$$

$$\rightarrow \left[\begin{array}{cccc|c} 1 & 1 & 1 & 1 & 5 \\ -1 & 1 & -1 & 1 & 3 \\ -8 & 4 & -2 & 1 & -10 \\ 27 & 9 & 3 & 1 & 15 \end{array}\right] \left(\begin{array}{l} \text{Interchange} \\ r_3 \text{ and } r_1 \end{array}\right)$$

$$\rightarrow \left[\begin{array}{cccc|c} 1 & 1 & 1 & 1 & 5 \\ 0 & 2 & 0 & 2 & 8 \\ 0 & 12 & 6 & 9 & 30 \\ 0 & -18 & -24 & -26 & -120 \end{array}\right] \left(\begin{array}{l} R_2 = r_1 + r_2 \\ R_3 = 8r_1 + r_3 \\ R_4 = -27r_1 + r_4 \end{array}\right)$$

$$\rightarrow \left[\begin{array}{cccc|c} 1 & 1 & 1 & 1 & 5 \\ 0 & 1 & 0 & 1 & 4 \\ 0 & 12 & 6 & 9 & 30 \\ 0 & -18 & -24 & -26 & -120 \end{array}\right] \left(R_2 = \tfrac{1}{2}r_2\right)$$

$$\rightarrow \left[\begin{array}{cccc|c} 1 & 0 & 1 & 0 & 1 \\ 0 & 1 & 0 & 1 & 4 \\ 0 & 0 & 6 & -3 & -18 \\ 0 & 0 & -24 & -8 & -48 \end{array}\right] \left(\begin{array}{l} R_1 = -r_2 + r_1 \\ R_3 = -12r_2 + r_3 \\ R_4 = 18r_2 + r_4 \end{array}\right)$$

$$\rightarrow \left[\begin{array}{cccc|c} 1 & 0 & 1 & 0 & 1 \\ 0 & 1 & 0 & 1 & 4 \\ 0 & 0 & 1 & -\frac{1}{2} & -3 \\ 0 & 0 & -24 & -8 & -48 \end{array}\right] \left(R_3 = \tfrac{1}{6}r_3\right)$$

$$\rightarrow \left[\begin{array}{cccc|c} 1 & 0 & 0 & \frac{1}{2} & 4 \\ 0 & 1 & 0 & 1 & 4 \\ 0 & 0 & 1 & -\frac{1}{2} & -3 \\ 0 & 0 & 0 & -20 & -120 \end{array}\right] \left(\begin{array}{l} R_1 = -r_3 + r_1 \\ R_4 = 24r_3 + r_4 \end{array}\right)$$

$$\rightarrow \left[\begin{array}{cccc|c} 1 & 0 & 0 & \frac{1}{2} & 4 \\ 0 & 1 & 0 & 1 & 4 \\ 0 & 0 & 1 & -\frac{1}{2} & -3 \\ 0 & 0 & 0 & 1 & 6 \end{array}\right] \left(R_4 = -\tfrac{1}{20}r_4\right)$$

$$\rightarrow \left[\begin{array}{cccc|c} 1 & 0 & 0 & 0 & 1 \\ 0 & 1 & 0 & 0 & -2 \\ 0 & 0 & 1 & 0 & 0 \\ 0 & 0 & 0 & 1 & 6 \end{array}\right] \left(\begin{array}{l} R_1 = -\frac{1}{2}r_4 + r_1 \\ R_2 = -r_4 + r_2 \\ R_3 = \frac{1}{2}r_4 + r_3 \end{array}\right)$$

The solution is $a = 1, b = -2, c = 0, d = 6$; so the equation is $f(x) = x^3 - 2x^2 + 6$.

77. Let x = the number of servings of salmon steak.
Let y = the number of servings of baked eggs.
Let z = the number of servings of acorn squash.
Protein equation: $30x + 15y + 3z = 78$
Carbohydrate equation: $20x + 2y + 25z = 59$
Vitamin A equation: $2x + 20y + 32z = 75$
Set up a matrix and solve:

$$\left[\begin{array}{ccc|c} 30 & 15 & 3 & 78 \\ 20 & 2 & 25 & 59 \\ 2 & 20 & 32 & 75 \end{array}\right]$$

$$\rightarrow \left[\begin{array}{ccc|c} 2 & 20 & 32 & 75 \\ 20 & 2 & 25 & 59 \\ 30 & 15 & 3 & 78 \end{array}\right] \left(\begin{array}{l} \text{Interchange} \\ r_3 \text{ and } r_1 \end{array}\right)$$

$$\rightarrow \left[\begin{array}{ccc|c} 1 & 10 & 16 & 37.5 \\ 20 & 2 & 25 & 59 \\ 30 & 15 & 3 & 78 \end{array}\right] \left(R_1 = \tfrac{1}{2}r_1\right)$$

$$\rightarrow \left[\begin{array}{ccc|c} 1 & 10 & 16 & 37.5 \\ 0 & -198 & -295 & -691 \\ 0 & -285 & -477 & -1047 \end{array}\right] \left(\begin{array}{l} R_2 = -20r_1 + r_2 \\ R_3 = -30r_1 + r_3 \end{array}\right)$$

$$\rightarrow \left[\begin{array}{ccc|c} 1 & 10 & 16 & 37.5 \\ 0 & -198 & -295 & -691 \\ 0 & 0 & -\frac{3457}{66} & -\frac{3457}{66} \end{array}\right] \left(R_3 = -\tfrac{95}{66}r_2 + r_3\right)$$

$$\rightarrow \left[\begin{array}{ccc|c} 1 & 10 & 16 & 37.5 \\ 0 & -198 & -295 & -691 \\ 0 & 0 & 1 & 1 \end{array}\right] \left(R_3 = -\tfrac{66}{3457}r_3\right)$$

Substitute $z = 1$ and solve:

$$-198y - 295(1) = -691$$
$$-198y = -396$$
$$y = 2$$

$$x + 10(2) + 16(1) = 37.5$$
$$x + 36 = 37.5$$
$$x = 1.5$$

The dietitian should serve 1.5 servings of salmon steak, 2 servings of baked eggs, and 1 serving of acorn squash.

78. Let x = the number of servings of pork chops.
Let y = the number of servings of corn on the cob.
Let z = the number of servings of 2% milk.
Protein equation: $23x+3y+9z=47$
Carbohydrate equation: $16y+13z=58$
Calcium equation: $10x+10y+300z=630$
Set up a matrix and solve:

$$\left[\begin{array}{ccc|c} 23 & 3 & 9 & 47 \\ 0 & 16 & 13 & 58 \\ 10 & 10 & 300 & 630 \end{array}\right]$$

$$\rightarrow\left[\begin{array}{ccc|c} 1 & 1 & 30 & 63 \\ 0 & 16 & 13 & 58 \\ 23 & 3 & 9 & 47 \end{array}\right] \quad \left(\begin{array}{c}\text{Interchange} \\ \frac{1}{10}r_3 \text{ and } r_1\end{array}\right)$$

$$\rightarrow\left[\begin{array}{ccc|c} 1 & 1 & 30 & 63 \\ 0 & 1 & \frac{13}{16} & \frac{29}{8} \\ 0 & -20 & -681 & -1402 \end{array}\right] \quad \left(\begin{array}{l} R_3=-23r_1+r_3 \\ R_2=\frac{1}{16}r_2 \end{array}\right)$$

$$\rightarrow\left[\begin{array}{ccc|c} 1 & 0 & \frac{467}{16} & \frac{475}{8} \\ 0 & 1 & \frac{13}{16} & \frac{29}{8} \\ 0 & 0 & -\frac{2659}{4} & -\frac{2659}{2} \end{array}\right] \quad \left(\begin{array}{l} R_1=-r_2+r_1 \\ R_3=20r_2+r_3 \end{array}\right)$$

$$\rightarrow\left[\begin{array}{ccc|c} 1 & 0 & \frac{467}{16} & \frac{475}{8} \\ 0 & 1 & \frac{13}{16} & \frac{29}{8} \\ 0 & 0 & 1 & 2 \end{array}\right] \quad \left(R_3=-\frac{4}{2659}r_3\right)$$

$$\rightarrow\left[\begin{array}{ccc|c} 1 & 0 & 0 & 1 \\ 0 & 1 & 0 & 2 \\ 0 & 0 & 1 & 2 \end{array}\right] \quad \left(\begin{array}{l} R_1=-\frac{467}{16}r_3+r_1 \\ R_2=-\frac{13}{16}r_3+r_2 \end{array}\right)$$

The dietitian should provide 1 serving of pork chops, 2 servings of corn on the cob, and 2 servings of 2% milk.

79. Let x = the amount invested in Treasury bills.
Let y = the amount invested in Treasury bonds.
Let z = the amount invested in corporate bonds.
Total investment equation: $x+y+z=10{,}000$
Annual income equation:
$0.06x+0.07y+0.08z=680$
Condition on investment equation:
$z=0.5x$
$x-2z=0$
Set up a matrix and solve:

$$\left[\begin{array}{ccc|c} 1 & 1 & 1 & 10{,}000 \\ 0.06 & 0.07 & 0.08 & 680 \\ 1 & 0 & -2 & 0 \end{array}\right]$$

$$\rightarrow\left[\begin{array}{ccc|c} 1 & 1 & 1 & 10{,}000 \\ 0 & 0.01 & 0.02 & 80 \\ 0 & -1 & -3 & -10{,}000 \end{array}\right] \quad \left(\begin{array}{l} R_2=-0.06r_1+r_2 \\ R_3=-r_1+r_3 \end{array}\right)$$

$$\rightarrow\left[\begin{array}{ccc|c} 1 & 1 & 1 & 10{,}000 \\ 0 & 1 & 2 & 8000 \\ 0 & -1 & -3 & -10{,}000 \end{array}\right] \quad (R_2=100r_2)$$

$$\rightarrow\left[\begin{array}{ccc|c} 1 & 0 & -1 & 2000 \\ 0 & 1 & 2 & 8000 \\ 0 & 0 & -1 & -2000 \end{array}\right] \quad \left(\begin{array}{l} R_1=-r_2+r_1 \\ R_3=r_2+r_3 \end{array}\right)$$

$$\rightarrow\left[\begin{array}{ccc|c} 1 & 0 & -1 & 2000 \\ 0 & 1 & 2 & 8000 \\ 0 & 0 & 1 & 2000 \end{array}\right] \quad (R_3=-r_3)$$

$$\rightarrow\left[\begin{array}{ccc|c} 1 & 0 & 0 & 4000 \\ 0 & 1 & 0 & 4000 \\ 0 & 0 & 1 & 2000 \end{array}\right] \quad \left(\begin{array}{l} R_1=r_3+r_1 \\ R_2=-2r_3+r_2 \end{array}\right)$$

Carletta should invest \$4000 in Treasury bills, \$4000 in Treasury bonds, and \$2000 in corporate bonds.

80. Let x = the amount invested in Treasury bills.
Let y = the amount invested in Treasury bonds.
Let z = the amount invested in corporate bonds.
Total investment equation: $x+y+z=20{,}000$
Annual income equation:
$0.05x+0.07y+0.09z=1280$
Condition on investment equation:
$x=2z \Rightarrow x-2z=0$
Set up a matrix and solve:

$$\left[\begin{array}{ccc|c} 1 & 1 & 1 & 20{,}000 \\ 0.05 & 0.07 & 0.09 & 1280 \\ 1 & 0 & -2 & 0 \end{array}\right]$$

$$\rightarrow\left[\begin{array}{ccc|c} 1 & 1 & 1 & 20{,}000 \\ 0 & 0.02 & 0.04 & 280 \\ 0 & -1 & -3 & -20{,}000 \end{array}\right] \quad \left(\begin{array}{l} R_2=-0.05r_1+r_2 \\ R_3=-r_1+r_3 \end{array}\right)$$

$$\rightarrow\left[\begin{array}{ccc|c} 1 & 1 & 1 & 20{,}000 \\ 0 & 1 & 2 & 14{,}000 \\ 0 & -1 & -3 & -20{,}000 \end{array}\right] \quad (R_2=50r_2)$$

$$\rightarrow\left[\begin{array}{ccc|c} 1 & 0 & -1 & 6000 \\ 0 & 1 & 2 & 14{,}000 \\ 0 & 0 & -1 & -6000 \end{array}\right] \quad \left(\begin{array}{l} R_1=-r_2+r_1 \\ R_3=r_2+r_3 \end{array}\right)$$

$$\rightarrow\left[\begin{array}{ccc|c} 1 & 0 & -1 & 6000 \\ 0 & 1 & 2 & 14{,}000 \\ 0 & 0 & 1 & 6000 \end{array}\right] \quad (R_3=-r_3)$$

$$\rightarrow\left[\begin{array}{ccc|c}1 & 0 & 0 & 12{,}000\\0 & 1 & 0 & 2000\\0 & 0 & 1 & 6000\end{array}\right]\quad\begin{pmatrix}R_1 = r_3 + r_1\\R_2 = -2r_3 + r_2\end{pmatrix}$$

John should invest \$12,000 in Treasury bills, \$2000 in Treasury bonds, and \$6000 in corporate bonds.

81. Let x = the number of Deltas produced.
Let y = the number of Betas produced.
Let z = the number of Sigmas produced.
Painting equation: $10x+16y+8z=240$
Drying equation: $3x+5y+2z=69$
Polishing equation: $2x+3y+z=41$
Set up a matrix and solve:

$$\left[\begin{array}{ccc|c}10 & 16 & 8 & 240\\3 & 5 & 2 & 69\\2 & 3 & 1 & 41\end{array}\right]$$

$$\rightarrow\left[\begin{array}{ccc|c}1 & 1 & 2 & 33\\3 & 5 & 2 & 69\\2 & 3 & 1 & 41\end{array}\right]\quad(R_1 = -3r_2 + r_1)$$

$$\rightarrow\left[\begin{array}{ccc|c}1 & 1 & 2 & 33\\0 & 2 & -4 & -30\\0 & 1 & -3 & -25\end{array}\right]\quad\begin{pmatrix}R_2 = -3r_1 + r_2\\R_3 = -2r_1 + r_3\end{pmatrix}$$

$$\rightarrow\left[\begin{array}{ccc|c}1 & 1 & 2 & 33\\0 & 1 & -2 & -15\\0 & 1 & -3 & -25\end{array}\right]\quad\left(R_2 = \tfrac{1}{2}r_2\right)$$

$$\rightarrow\left[\begin{array}{ccc|c}1 & 0 & 4 & 48\\0 & 1 & -2 & -15\\0 & 0 & -1 & -10\end{array}\right]\quad\begin{pmatrix}R_1 = r_1 - r_2\\R_3 = r_3 - r_2\end{pmatrix}$$

$$\rightarrow\left[\begin{array}{ccc|c}1 & 0 & 4 & 48\\0 & 1 & -2 & -15\\0 & 0 & 1 & 10\end{array}\right]\quad(R_3 = -r_3)$$

$$\rightarrow\left[\begin{array}{ccc|c}1 & 0 & 0 & 8\\0 & 1 & 0 & 5\\0 & 0 & 1 & 10\end{array}\right]\quad\begin{pmatrix}R_1 = -4r_3 + r_1\\R_2 = 2r_3 + r_2\end{pmatrix}$$

The company should produce 8 Deltas, 5 Betas, and 10 Sigmas.

82. Let x = the number of cases of orange juice produced.
Let y = the number of cases of grapefruit juice produced.
Let z = the number of cases of tomato juice produced.
Sterilizing equation: $9x+10y+12z=398$
Filling equation: $6x+4y+4z=164$
Labeling equation: $x+2y+z=58$
Set up a matrix and solve:

$$\left[\begin{array}{ccc|c}9 & 10 & 12 & 398\\6 & 4 & 4 & 164\\1 & 2 & 1 & 58\end{array}\right]$$

$$\rightarrow\left[\begin{array}{ccc|c}1 & 2 & 1 & 58\\6 & 4 & 4 & 164\\9 & 10 & 12 & 398\end{array}\right]\quad\begin{pmatrix}\text{Interchange}\\r_1 \text{ and } r_3\end{pmatrix}$$

$$\rightarrow\left[\begin{array}{ccc|c}1 & 2 & 1 & 58\\0 & -8 & -2 & -184\\0 & -8 & 3 & -124\end{array}\right]\quad\begin{pmatrix}R_2 = -6r_1 + r_2\\R_3 = -9r_1 + r_3\end{pmatrix}$$

$$\rightarrow\left[\begin{array}{ccc|c}1 & 2 & 1 & 58\\0 & 1 & \frac{1}{4} & 23\\0 & -8 & 3 & -124\end{array}\right]\quad\left(R_2 = -\tfrac{1}{8}r_2\right)$$

$$\rightarrow\left[\begin{array}{ccc|c}1 & 0 & \frac{1}{2} & 12\\0 & 1 & \frac{1}{4} & 23\\0 & 0 & 5 & 60\end{array}\right]\quad\begin{pmatrix}R_1 = -2r_2 + r_1\\R_3 = 8r_2 + r_3\end{pmatrix}$$

$$\rightarrow\left[\begin{array}{ccc|c}1 & 0 & \frac{1}{2} & 12\\0 & 1 & \frac{1}{4} & 23\\0 & 0 & 1 & 12\end{array}\right]\quad\left(R_3 = \tfrac{1}{5}r_3\right)$$

$$\rightarrow\left[\begin{array}{ccc|c}1 & 0 & 0 & 6\\0 & 1 & 0 & 20\\0 & 0 & 1 & 12\end{array}\right]\quad\begin{pmatrix}R_1 = -\frac{1}{2}r_3 + r_1\\R_2 = -\frac{1}{4}r_3 + r_2\end{pmatrix}$$

The company should prepare 6 cases of orange juice, 20 cases of grapefruit juice, and 12 cases of tomato juice.

83. Rewrite the system to set up the matrix and solve:

$$\begin{cases} -4+8-2I_2=0 \\ 8=5I_4+I_1 \\ 4=3I_3+I_1 \\ I_3+I_4=I_1 \end{cases} \rightarrow \begin{cases} 2I_2=4 \\ I_1+5I_4=8 \\ I_1+3I_3=4 \\ I_1-I_3-I_4=0 \end{cases}$$

$$\left[\begin{array}{cccc|c} 0 & 2 & 0 & 0 & 4 \\ 1 & 0 & 0 & 5 & 8 \\ 1 & 0 & 3 & 0 & 4 \\ 1 & 0 & -1 & -1 & 0 \end{array}\right]$$

$$\rightarrow \left[\begin{array}{cccc|c} 1 & 0 & 0 & 5 & 8 \\ 0 & 2 & 0 & 0 & 4 \\ 1 & 0 & 3 & 0 & 4 \\ 1 & 0 & -1 & -1 & 0 \end{array}\right] \quad \left(\begin{array}{l}\text{Interchange} \\ r_2 \text{ and } r_1\end{array}\right)$$

$$\rightarrow \left[\begin{array}{cccc|c} 1 & 0 & 0 & 5 & 8 \\ 0 & 1 & 0 & 0 & 2 \\ 0 & 0 & 3 & -5 & -4 \\ 0 & 0 & -1 & -6 & -8 \end{array}\right] \quad \left(\begin{array}{l} R_2=\frac{1}{2}r_2 \\ R_3=-r_1+r_3 \\ R_4=-r_1+r_4 \end{array}\right)$$

$$\rightarrow \left[\begin{array}{cccc|c} 1 & 0 & 0 & 5 & 8 \\ 0 & 1 & 0 & 0 & 2 \\ 0 & 0 & -1 & -6 & -8 \\ 0 & 0 & 3 & -5 & -4 \end{array}\right] \quad \left(\begin{array}{l}\text{Interchange} \\ r_3 \text{ and } r_4\end{array}\right)$$

$$\rightarrow \left[\begin{array}{cccc|c} 1 & 0 & 0 & 5 & 8 \\ 0 & 1 & 0 & 0 & 2 \\ 0 & 0 & 1 & 6 & 8 \\ 0 & 0 & 0 & -23 & -28 \end{array}\right] \quad \left(\begin{array}{l} R_3=-r_3 \\ R_4=-3r_3+r_4 \end{array}\right)$$

$$\rightarrow \left[\begin{array}{cccc|c} 1 & 0 & 0 & 5 & 8 \\ 0 & 1 & 0 & 0 & 2 \\ 0 & 0 & 1 & 6 & 8 \\ 0 & 0 & 0 & 1 & \frac{28}{23} \end{array}\right] \quad \left(R_4=-\frac{1}{23}r_4\right)$$

$$\rightarrow \left[\begin{array}{cccc|c} 1 & 0 & 0 & 0 & \frac{44}{23} \\ 0 & 1 & 0 & 0 & 2 \\ 0 & 0 & 1 & 0 & \frac{16}{23} \\ 0 & 0 & 0 & 1 & \frac{28}{23} \end{array}\right] \quad \left(\begin{array}{l} R_1=-5r_4+r_1 \\ R_3=-6r_4+r_3 \end{array}\right)$$

The solution is $I_1=\dfrac{44}{23}$, $I_2=2$, $I_3=\dfrac{16}{23}$, $I_4=\dfrac{28}{23}$.

84. Rewrite the system to set up the matrix and solve:

$$\begin{cases} I_1=I_3+I_2 \\ 24-6I_1-3I_3=0 \\ 12+24-6I_1-6I_2=0 \end{cases} \rightarrow \begin{cases} I_1-I_2-I_3=0 \\ -6I_1-3I_3=-24 \\ -6I_1-6I_2=-36 \end{cases}$$

$$\left[\begin{array}{ccc|c} 1 & -1 & -1 & 0 \\ -6 & 0 & -3 & -24 \\ -6 & -6 & 0 & -36 \end{array}\right]$$

$$\rightarrow \left[\begin{array}{ccc|c} 1 & -1 & -1 & 0 \\ 0 & -6 & -9 & -24 \\ 0 & -12 & -6 & -36 \end{array}\right] \quad \left(\begin{array}{l} R_2=6r_1+r_2 \\ R_3=6r_1+r_3 \end{array}\right)$$

$$\rightarrow \left[\begin{array}{ccc|c} 1 & -1 & -1 & 0 \\ 0 & 1 & \frac{3}{2} & 4 \\ 0 & -12 & -6 & -36 \end{array}\right] \quad \left(R_2=-\frac{1}{6}r_2\right)$$

$$\rightarrow \left[\begin{array}{ccc|c} 1 & 0 & \frac{1}{2} & 4 \\ 0 & 1 & \frac{3}{2} & 4 \\ 0 & 0 & 12 & 12 \end{array}\right] \quad \left(\begin{array}{l} R_1=r_2+r_1 \\ R_3=12r_2+r_3 \end{array}\right)$$

$$\rightarrow \left[\begin{array}{ccc|c} 1 & 0 & \frac{1}{2} & 4 \\ 0 & 1 & \frac{3}{2} & 4 \\ 0 & 0 & 1 & 1 \end{array}\right] \quad \left(R_3=\frac{1}{12}r_3\right)$$

$$\rightarrow \left[\begin{array}{ccc|c} 1 & 0 & 0 & 3.5 \\ 0 & 1 & 0 & 2.5 \\ 0 & 0 & 1 & 1 \end{array}\right] \quad \left(\begin{array}{l} R_1=-\frac{1}{2}r_3+r_1 \\ R_2=-\frac{3}{2}r_3+r_2 \end{array}\right)$$

The solution is $I_1-3.5$, $I_2=2.5$, $I_3=1$.

85. Let x = the amount invested in Treasury bills.
Let y = the amount invested in corporate bonds.
Let z = the amount invested in junk bonds.

a. Total investment equation:

$$x+y+z=20{,}000$$

Annual income equation:

$$0.07x+0.09y+0.11z=2000$$

Set up a matrix and solve:

$$\left[\begin{array}{ccc|c} 1 & 1 & 1 & 20{,}000 \\ 0.07 & 0.09 & 0.11 & 2000 \end{array}\right]$$

$$\rightarrow \left[\begin{array}{ccc|c} 1 & 1 & 1 & 20{,}000 \\ 7 & 9 & 11 & 200{,}000 \end{array}\right] \quad (R_2=100r_2)$$

$$\rightarrow \left[\begin{array}{ccc|c} 1 & 1 & 1 & 20{,}000 \\ 0 & 2 & 4 & 60{,}000 \end{array}\right] \quad (R_2=r_2-7r_1)$$

$$\rightarrow \left[\begin{array}{ccc|c} 1 & 1 & 1 & 20{,}000 \\ 0 & 1 & 2 & 30{,}000 \end{array}\right] \quad \left(R_2=\frac{1}{2}r_2\right)$$

$$\rightarrow\left[\begin{array}{ccc|c}1 & 0 & -1 & -10{,}000\\0 & 1 & 2 & 30{,}000\end{array}\right] \quad (R_1 = r_1 - r_2)$$

The matrix in the last step represents the system $\begin{cases} x - z = -10{,}000 \\ y + 2z = 30{,}000 \end{cases}$

Therefore the solution is $x = -10{,}000 + z$, $y = 30{,}000 - 2z$, z is any real number.

Possible investment strategies:

Amount Invested At

7%	9%	11%
0	10,000	10,000
1000	8000	11,000
2000	6000	12,000
3000	4000	13,000
4000	2000	14,000
5000	0	15,000

b. Total investment equation:
$x + y + z = 25{,}000$

Annual income equation:
$0.07x + 0.09y + 0.11z = 2000$

Set up a matrix and solve:

$$\left[\begin{array}{ccc|c}1 & 1 & 1 & 25{,}000\\0.07 & 0.09 & 0.11 & 2000\end{array}\right]$$

$$\rightarrow\left[\begin{array}{ccc|c}1 & 1 & 1 & 25{,}000\\7 & 9 & 11 & 200{,}000\end{array}\right] \quad (R_2 = 100r_2)$$

$$\rightarrow\left[\begin{array}{ccc|c}1 & 1 & 1 & 25{,}000\\0 & 2 & 4 & 25{,}000\end{array}\right] \quad (R_2 = r_2 - 7r_1)$$

$$\rightarrow\left[\begin{array}{ccc|c}1 & 1 & 1 & 25{,}000\\0 & 1 & 2 & 12{,}500\end{array}\right] \quad (R_2 = \tfrac{1}{2}r_2)$$

$$\rightarrow\left[\begin{array}{ccc|c}1 & 0 & -1 & 12{,}500\\0 & 1 & 2 & 12{,}500\end{array}\right] \quad (R_1 = r_1 - r_2)$$

The matrix in the last step represents the system $\begin{cases} x - z = 12{,}500 \\ y + 2z = 12{,}500 \end{cases}$

Thus, the solution is $x = z + 12{,}500$, $y = -2z + 12{,}500$, z is any real number.

Possible investment strategies:

Amount Invested At

7%	9%	11%
12,500	12,500	0
14,500	8500	2000
16,500	4500	4000
18,750	0	6250

c. Total investment equation:
$x + y + z = 30{,}000$

Annual income equation:
$0.07x + 0.09y + 0.11z = 2000$

Set up a matrix and solve:

$$\left[\begin{array}{ccc|c}1 & 1 & 1 & 30{,}000\\0.07 & 0.09 & 0.11 & 2000\end{array}\right]$$

$$\rightarrow\left[\begin{array}{ccc|c}1 & 1 & 1 & 30{,}000\\7 & 9 & 11 & 200{,}000\end{array}\right] \quad (R_2 = 100r_2)$$

$$\rightarrow\left[\begin{array}{ccc|c}1 & 1 & 1 & 30{,}000\\0 & 2 & 4 & -10{,}000\end{array}\right] \quad (R_1 = r_2 - 7r_1)$$

$$\rightarrow\left[\begin{array}{ccc|c}1 & 1 & 1 & 30{,}000\\0 & 1 & 2 & -5000\end{array}\right] \quad (R_2 = \tfrac{1}{2}r_2)$$

$$\rightarrow\left[\begin{array}{ccc|c}1 & 0 & -1 & 35{,}000\\0 & 1 & 2 & -5000\end{array}\right] \quad (R_1 = r_1 - r_2)$$

The matrix in the last step represents the system $\begin{cases} x - z = 35{,}000 \\ y + 2z = -5000 \end{cases}$

Thus, the solution is $x = z + 35{,}000$, $y = -2z - 5000$, z is any real number.
However, y and z cannot be negative. From $y = -2z - 5000$, we must have $y = z = 0$.

One possible investment strategy

Amount Invested At

7%	9%	11%
30,000	0	0

This will yield ($30,000)(0.07) = $2100, which is more than the required income.

d. Answers will vary.

86. Let x = the amount invested in Treasury bills.
Let y = the amount invested in corporate bonds.
Let z = the amount invested in junk bonds.
Let I = income
Total investment equation: $x+y+z=25{,}000$
Annual income equation:
$0.07x+0.09y+0.11z=I$
Set up a matrix and solve:

$$\left[\begin{array}{ccc|c} 1 & 1 & 1 & 25{,}000 \\ 0.07 & 0.09 & 0.11 & I \end{array}\right]$$

$$\to\left[\begin{array}{ccc|c} 1 & 1 & 1 & 25{,}000 \\ 7 & 9 & 11 & 100I \end{array}\right] \quad (R_2 = 100r_2)$$

$$\to\left[\begin{array}{ccc|c} 1 & 1 & 1 & 25{,}000 \\ 0 & 2 & 4 & 100I-175{,}000 \end{array}\right] \quad (R_1 = r_2 - 7r_1)$$

$$\to\left[\begin{array}{ccc|c} 1 & 1 & 1 & 25{,}000 \\ 0 & 1 & 2 & 50I-87{,}500 \end{array}\right] \quad (R_2 = \tfrac{1}{2}r_2)$$

$$\to\left[\begin{array}{ccc|c} 1 & 0 & -1 & 112{,}500-50I \\ 0 & 1 & 2 & 50I-87{,}500 \end{array}\right] \quad (R_1 = r_1 - r_2)$$

The matrix in the last step represents the system

$$\begin{cases} x-z=112{,}500-50I \\ y+2z=50I-87{,}500 \end{cases}$$

Thus, the solution is $x=112{,}500-50I+z$, $y=50I-87{,}500-2z$, z is any real number.

a. $I=1500$

$x=112{,}500-50I+z$
$=112{,}500-50(1500)+z$
$=37{,}500+z$
$y=50I-87{,}500-2z$
$=50(1500)-87{,}500-2z$
$=-12{,}500-2z$
z is any real number.
Since y and z cannot be negative, we must have $y=z=0$. Investing all of the money at 7% yields \$1750, which is more than what is required.

b. $I=2000$

$x=112{,}500-50T+z$
$=112{,}500-50(2000)+z$
$=12{,}500+z$
$y=50I-87{,}500-2z$
$=50(2000)-87{,}500-2z$
$=12{,}500-2z$
z is any real number.

Possible investment strategies:

Amount Invested At

7%	9%	11%
12,500	12,500	0
15,500	6500	3000
18,750	0	6250

c. $I=2500$

$x=112{,}500-50T+z$
$=112{,}500-50(2500)+z$
$=-12{,}500+z$
$y=50I-87{,}500-2z$
$=50(2500)-87{,}500-2z$
$=37{,}500-2z$
z is any real number.

Possible investment strategies:

Amount invested at

7%	9%	11%
0	12,500	12,500
1000	10,500	13,500
6250	0	18,750

87. Let x = the amount of supplement 1.
Let y = the amount of supplement 2.
Let z = the amount of supplement 3.

$$\begin{cases} 0.20x+0.40y+0.30z=40 & \text{Vitamin C} \\ 0.30x+0.20y+0.50z=30 & \text{Vitamin D} \end{cases}$$

Multiplying each equation by 10 yields

$$\begin{cases} 2x+4y+3z=400 \\ 3x+2y+5z=300 \end{cases}$$

Set up a matrix and solve:

$$\left[\begin{array}{ccc|c} 2 & 4 & 3 & 400 \\ 3 & 2 & 5 & 300 \end{array}\right]$$

$$\to\left[\begin{array}{ccc|c} 1 & 2 & \frac{3}{2} & 200 \\ 3 & 2 & 5 & 300 \end{array}\right] \quad (R_1 = \tfrac{1}{2}r_1)$$

$$\to\left[\begin{array}{ccc|c} 1 & 2 & \frac{3}{2} & 200 \\ 0 & -4 & \frac{1}{2} & -300 \end{array}\right] \quad (R_2 = r_2 - 3r_1)$$

$$\to\left[\begin{array}{ccc|c} 1 & 2 & \frac{3}{2} & 200 \\ 0 & 1 & -\frac{1}{8} & 75 \end{array}\right] \quad (R_2 = -\tfrac{1}{4}r_2)$$

$$\rightarrow \left[\begin{array}{ccc|c} 1 & 0 & \frac{7}{4} & 50 \\ 0 & 1 & -\frac{1}{8} & 75 \end{array}\right] \quad (R_1 = r_1 - 2r_2)$$

The matrix in the last step represents the system

$$\begin{cases} x + \frac{7}{4}z = 50 \\ y - \frac{1}{8}z = 75 \end{cases}$$

Therefore the solution is $x = 50 - \dfrac{7}{4}z$,

$y = 75 + \dfrac{1}{8}z$, z is any real number.

Possible combinations:

Supplement 1	Supplement 2	Supplement 3
50mg	75mg	0mg
36mg	76mg	8mg
22mg	77mg	16mg
8mg	78mg	24mg

88. Let x = the amount of powder 1.
Let y = the amount of powder 2.
Let z = the amount of powder 3.

$$\begin{cases} 0.20x + 0.40y + 0.30z = 12 & \text{Vitamin B}_{12} \\ 0.30x + 0.20y + 0.40z = 12 & \text{Vitamin E} \end{cases}$$

Multiplying each equation by 10 yields

$$\begin{cases} 2x + 4y + 3z = 120 \\ 3x + 2y + 4z = 120 \end{cases}$$

Set up a matrix and solve:

$$\left[\begin{array}{ccc|c} 2 & 4 & 3 & 120 \\ 3 & 2 & 4 & 120 \end{array}\right]$$

$$\rightarrow \left[\begin{array}{ccc|c} 2 & 4 & 3 & 120 \\ 0 & -4 & -0.5 & -60 \end{array}\right] \quad (R_2 = r_2 - \tfrac{3}{2}r_1)$$

$$\rightarrow \left[\begin{array}{ccc|c} 2 & 0 & 2.5 & 60 \\ 0 & -4 & -0.5 & -60 \end{array}\right] \quad (R_1 = r_1 + r_2)$$

The matrix in the last step represents the system

$$\begin{cases} 2x + 2.5z = 60 \\ -4y - 0.5z = -60 \end{cases}$$

Thus, the solution is $x = 30 - 1.25z$, $y = 15 - 0.125z$, z is any real number.

Possible combinations:

Powder 1	Powder 2	Powder 3
30 units	15 units	0 units
20 units	14 units	8 units
10 units	13 units	16 units
0 units	12 units	24 units

89–91. Answers will vary.

Section 10.3

1. determinants

2. $ad - bc$

3. False

4. False

5. $\begin{vmatrix} 3 & 1 \\ 4 & 2 \end{vmatrix} = 3(2) - 4(1) = 6 - 4 = 2$

6. $\begin{vmatrix} 6 & 1 \\ 5 & 2 \end{vmatrix} = 6(2) - 5(1) = 12 - 5 = 7$

7. $\begin{vmatrix} 6 & 4 \\ -1 & 3 \end{vmatrix} = 6(3) - (-1)(4) = 18 + 4 = 22$

8. $\begin{vmatrix} 8 & -3 \\ 4 & 2 \end{vmatrix} = 8(2) - 4(-3) = 16 + 12 = 28$

9. $\begin{vmatrix} -3 & -1 \\ 4 & 2 \end{vmatrix} = -3(2) - 4(-1) = -6 + 4 = -2$

10. $\begin{vmatrix} -4 & 2 \\ -5 & 3 \end{vmatrix} = -4(3) - (-5)(2) = -12 + 10 = -2$

11.
$$\begin{vmatrix} 3 & 4 & 2 \\ 1 & -1 & 5 \\ 1 & 2 & -2 \end{vmatrix} = 3\begin{vmatrix} -1 & 5 \\ 2 & -2 \end{vmatrix} - 4\begin{vmatrix} 1 & 5 \\ 1 & -2 \end{vmatrix} + 2\begin{vmatrix} 1 & -1 \\ 1 & 2 \end{vmatrix}$$

$$\begin{aligned} &= 3[(-1)(-2) - 2(5)] - 4[1(-2) - 1(5)] + 2[1(2) - 1(-1)] \\ &= 3(-8) - 4(-7) + 2(3) \\ &= -24 + 28 + 6 \\ &= 10 \end{aligned}$$

12. $\begin{vmatrix} 1 & 3 & -2 \\ 6 & 1 & -5 \\ 8 & 2 & 3 \end{vmatrix} = 1\begin{vmatrix} 1 & -5 \\ 2 & 3 \end{vmatrix} - 3\begin{vmatrix} 6 & -5 \\ 8 & 3 \end{vmatrix} + (-2)\begin{vmatrix} 6 & 1 \\ 8 & 2 \end{vmatrix}$

$$\begin{aligned} &= 1[1(3)-2(-5)]-3[6(3)-8(-5)] \\ &\quad -2[6(2)-8(1)] \\ &= 1(13)-3(58)-2(4) \\ &= 13-174-8 \\ &= -169 \end{aligned}$$

13. $\begin{vmatrix} 4 & -1 & 2 \\ 6 & -1 & 0 \\ 1 & -3 & 4 \end{vmatrix} = 4\begin{vmatrix} -1 & 0 \\ -3 & 4 \end{vmatrix} - (-1)\begin{vmatrix} 6 & 0 \\ 1 & 4 \end{vmatrix} + 2\begin{vmatrix} 6 & -1 \\ 1 & -3 \end{vmatrix}$

$$\begin{aligned} &= 4[-1(4)-0(-3)]+1[6(4)-1(0)] \\ &\quad +2[6(-3)-1(-1)] \\ &= 4(-4)+1(24)+2(-17) \\ &= -16+24-34 \\ &= -26 \end{aligned}$$

14. $\begin{vmatrix} 3 & -9 & 4 \\ 1 & 4 & 0 \\ 8 & -3 & 1 \end{vmatrix} = 3\begin{vmatrix} 4 & 0 \\ -3 & 1 \end{vmatrix} - (-9)\begin{vmatrix} 1 & 0 \\ 8 & 1 \end{vmatrix} + 4\begin{vmatrix} 1 & 4 \\ 8 & -3 \end{vmatrix}$

$$\begin{aligned} &= 3[4(1)-(-3)(0)]+9[1(1)-8(0)] \\ &\quad +4[1(-3)-8(4)] \\ &= 3(4)+9(1)+4(-35) \\ &= 12+9-140 \\ &= -119 \end{aligned}$$

15. $\begin{cases} x+y=8 \\ x-y=4 \end{cases}$

$D = \begin{vmatrix} 1 & 1 \\ 1 & -1 \end{vmatrix} = -1-1 = -2$

$D_x = \begin{vmatrix} 8 & 1 \\ 4 & -1 \end{vmatrix} = -8-4 = -12$

$D_y = \begin{vmatrix} 1 & 8 \\ 1 & 4 \end{vmatrix} = 4-8 = -4$

Find the solutions by Cramer's Rule:

$x = \frac{D_x}{D} = \frac{-12}{-2} = 6 \qquad y = \frac{D_y}{D} = \frac{-4}{-2} = 2$

16. $\begin{cases} x+2y=5 \\ x-\ y=3 \end{cases}$

$D = \begin{vmatrix} 1 & 2 \\ 1 & -1 \end{vmatrix} = -1-2 = -3$

$D_x = \begin{vmatrix} 5 & 2 \\ 3 & -1 \end{vmatrix} = -5-6 = -11$

$D_y = \begin{vmatrix} 1 & 5 \\ 1 & 3 \end{vmatrix} = 3-5 = -2$

Find the solutions by Cramer's Rule:

$x = \frac{D_x}{D} = \frac{-11}{-3} = \frac{11}{3} \qquad y = \frac{D_y}{D} = \frac{-2}{-3} = \frac{2}{3}$

17. $\begin{cases} 5x-\ y=13 \\ 2x+3y=12 \end{cases}$

$D = \begin{vmatrix} 5 & -1 \\ 2 & 3 \end{vmatrix} = 15+2 = 17$

$D_x = \begin{vmatrix} 13 & -1 \\ 12 & 3 \end{vmatrix} = 39+12 = 51$

$D_y = \begin{vmatrix} 5 & 13 \\ 2 & 12 \end{vmatrix} = 60-26 = 34$

Find the solutions by Cramer's Rule:

$x = \frac{D_x}{D} = \frac{51}{17} = 3 \qquad y = \frac{D_y}{D} = \frac{34}{17} = 2$

18. $\begin{cases} x+3y=\ 5 \\ 2x-3y=-8 \end{cases}$

$D = \begin{vmatrix} 1 & 3 \\ 2 & -3 \end{vmatrix} = -3-6 = -9$

$D_x = \begin{vmatrix} 5 & 3 \\ -8 & -3 \end{vmatrix} = -15-(-24) = 9$

$D_y = \begin{vmatrix} 1 & 5 \\ 2 & -8 \end{vmatrix} = -8-10 = -18$

Find the solutions by Cramer's Rule:

$x = \frac{D_x}{D} = \frac{9}{-9} = -1 \qquad y = \frac{D_y}{D} = \frac{-18}{-9} = 2$

19. $\begin{cases} 3x \quad\;\; = 24 \\ x+2y = 0 \end{cases}$

$$D = \begin{vmatrix} 3 & 0 \\ 1 & 2 \end{vmatrix} = 6-0 = 6$$

$$D_x = \begin{vmatrix} 24 & 0 \\ 0 & 2 \end{vmatrix} = 48-0 = 48$$

$$D_y = \begin{vmatrix} 3 & 24 \\ 1 & 0 \end{vmatrix} = 0-24 = -24$$

Find the solutions by Cramer's Rule:

$$x = \frac{D_x}{D} = \frac{48}{6} = 8 \qquad y = \frac{D_y}{D} = \frac{-24}{6} = -4$$

20. $\begin{cases} 4x+5y = -3 \\ \quad -2y = -4 \end{cases}$

$$D = \begin{vmatrix} 4 & 5 \\ 0 & -2 \end{vmatrix} = -8-0 = -8$$

$$D_x = \begin{vmatrix} -3 & 5 \\ -4 & -2 \end{vmatrix} = 6-(-20) = 26$$

$$D_y = \begin{vmatrix} 4 & -3 \\ 0 & -4 \end{vmatrix} = -16-0 = -16$$

Find the solutions by Cramer's Rule:

$$x = \frac{D_x}{D} = \frac{26}{-8} = -\frac{13}{4} \qquad y = \frac{D_y}{D} = \frac{-16}{-8} = 2$$

21. $\begin{cases} 3x-6y = 24 \\ 5x+4y = 12 \end{cases}$

$$D = \begin{vmatrix} 3 & -6 \\ 5 & 4 \end{vmatrix} = 12-(-30) = 42$$

$$D_x = \begin{vmatrix} 24 & -6 \\ 12 & 4 \end{vmatrix} = 96-(-72) = 168$$

$$D_y = \begin{vmatrix} 3 & 24 \\ 5 & 12 \end{vmatrix} = 36-120 = -84$$

Find the solutions by Cramer's Rule:

$$x = \frac{D_x}{D} = \frac{168}{42} = 4 \qquad y = \frac{D_y}{D} = \frac{-84}{42} = -2$$

22. $\begin{cases} 2x+4y = 16 \\ 3x-5y = -9 \end{cases}$

$$D = \begin{vmatrix} 2 & 4 \\ 3 & -5 \end{vmatrix} = -10-12 = -22$$

$$D_x = \begin{vmatrix} 16 & 4 \\ -9 & -5 \end{vmatrix} = -80+36 = -44$$

$$D_y = \begin{vmatrix} 2 & 16 \\ 3 & -9 \end{vmatrix} = -18-48 = -66$$

Find the solutions by Cramer's Rule:

$$x = \frac{D_x}{D} = \frac{-44}{-22} = 2 \qquad y = \frac{D_y}{D} = \frac{-66}{-22} = 3$$

23. $\begin{cases} 3x-2y = 4 \\ 6x-4y = 0 \end{cases}$

$$D = \begin{vmatrix} 3 & -2 \\ 6 & -4 \end{vmatrix} = -12-(-12) = 0$$

Since $D = 0$, Cramer's Rule does not apply.

24. $\begin{cases} -x+2y = 5 \\ 4x-8y = 6 \end{cases}$

$$D = \begin{vmatrix} -1 & 2 \\ 4 & -8 \end{vmatrix} = 8-8 = 0$$

Since $D = 0$, Cramer's Rule does not apply.

25. $\begin{cases} 2x-4y = -2 \\ 3x+2y = 3 \end{cases}$

$$D = \begin{vmatrix} 2 & -4 \\ 3 & 2 \end{vmatrix} = 4+12 = 16$$

$$D_x = \begin{vmatrix} -2 & -4 \\ 3 & 2 \end{vmatrix} = -4+12 = 8$$

$$D_y = \begin{vmatrix} 2 & -2 \\ 3 & 3 \end{vmatrix} = 6+6 = 12$$

Find the solutions by Cramer's Rule:

$$x = \frac{D_x}{D} = \frac{8}{16} = \frac{1}{2} \qquad y = \frac{D_y}{D} = \frac{12}{16} = \frac{3}{4}$$

26. $\begin{cases} 3x+3y = 3 \\ 4x+2y = \frac{8}{3} \end{cases}$

$$D = \begin{vmatrix} 3 & 3 \\ 4 & 2 \end{vmatrix} = 6-12 = -6$$

$$D_x = \begin{vmatrix} 3 & 3 \\ \frac{8}{3} & 2 \end{vmatrix} = 6-8 = -2$$

$$D_y = \begin{vmatrix} 3 & 3 \\ 4 & \frac{8}{3} \end{vmatrix} = 8-12 = -4$$

Find the solutions by Cramer's Rule:

$$x = \frac{D_x}{D} = \frac{-2}{-6} = \frac{1}{3} \qquad y = \frac{D_y}{D} = \frac{-4}{-6} = \frac{2}{3}$$

27. $\begin{cases} 2x - 3y = -1 \\ 10x + 10y = 5 \end{cases}$

$D = \begin{vmatrix} 2 & -3 \\ 10 & 10 \end{vmatrix} = 20 - (-30) = 50$

$D_x = \begin{vmatrix} -1 & -3 \\ 5 & 10 \end{vmatrix} = -10 - (-15) = 5$

$D_y = \begin{vmatrix} 2 & -1 \\ 10 & 5 \end{vmatrix} = 10 - (-10) = 20$

Find the solutions by Cramer's Rule:

$x = \frac{D_x}{D} = \frac{5}{50} = \frac{1}{10} \qquad y = \frac{D_y}{D} = \frac{20}{50} = \frac{2}{5}$

28. $\begin{cases} 3x - 2y = 0 \\ 5x + 10y = 4 \end{cases}$

$D = \begin{vmatrix} 3 & -2 \\ 5 & 10 \end{vmatrix} = 30 - (-10) = 40$

$D_x = \begin{vmatrix} 0 & -2 \\ 4 & 10 \end{vmatrix} = 0 - (-8) = 8$

$D_y = \begin{vmatrix} 3 & 0 \\ 5 & 4 \end{vmatrix} = 12 - 0 = 12$

Find the solutions by Cramer's Rule:

$x = \frac{D_x}{D} = \frac{8}{40} = \frac{1}{5} \qquad y = \frac{D_y}{D} = \frac{12}{40} = \frac{3}{10}$

29. $\begin{cases} 2x + 3y = 6 \\ x - y = \frac{1}{2} \end{cases}$

$D = \begin{vmatrix} 2 & 3 \\ 1 & -1 \end{vmatrix} = -2 - 3 = -5$

$D_x = \begin{vmatrix} 6 & 3 \\ \frac{1}{2} & -1 \end{vmatrix} = -6 - \frac{3}{2} = -\frac{15}{2}$

$D_y = \begin{vmatrix} 2 & 6 \\ 1 & \frac{1}{2} \end{vmatrix} = 1 - 6 = -5$

Find the solutions by Cramer's Rule:

$x = \frac{D_x}{D} = \frac{-\frac{15}{2}}{-5} = \frac{3}{2} \qquad y = \frac{D_y}{D} = \frac{-5}{-5} = 1$

30. $\begin{cases} \frac{1}{2}x + y = -2 \\ x - 2y = 8 \end{cases}$

$D = \begin{vmatrix} \frac{1}{2} & 1 \\ 1 & -2 \end{vmatrix} = -1 - 1 = -2$

$D_x = \begin{vmatrix} -2 & 1 \\ 8 & -2 \end{vmatrix} = 4 - 8 = -4$

$D_y = \begin{vmatrix} \frac{1}{2} & -2 \\ 1 & 8 \end{vmatrix} = 4 - (-2) = 6$

Find the solutions by Cramer's Rule:

$x = \frac{D_x}{D} = \frac{-4}{-2} = 2 \qquad y = \frac{D_y}{D} = \frac{6}{-2} = -3$

31. $\begin{cases} 3x - 5y = 3 \\ 15x + 5y = 21 \end{cases}$

$D = \begin{vmatrix} 3 & -5 \\ 15 & 5 \end{vmatrix} = 15 - (-75) = 90$

$D_x = \begin{vmatrix} 3 & -5 \\ 21 & 5 \end{vmatrix} = 15 - (-105) = 120$

$D_y = \begin{vmatrix} 3 & 3 \\ 15 & 21 \end{vmatrix} = 63 - 45 = 18$

Find the solutions by Cramer's Rule:

$x = \frac{D_x}{D} = \frac{120}{90} = \frac{4}{3} \qquad y = \frac{D_y}{D} = \frac{18}{90} = \frac{1}{5}$

32. $\begin{cases} 2x - y = -1 \\ x + \frac{1}{2}y = \frac{3}{2} \end{cases}$

$D = \begin{vmatrix} 2 & -1 \\ 1 & \frac{1}{2} \end{vmatrix} = 1 + 1 = 2$

$D_x = \begin{vmatrix} -1 & -1 \\ \frac{3}{2} & \frac{1}{2} \end{vmatrix} = -\frac{1}{2} + \frac{3}{2} = 1$

$D_y = \begin{vmatrix} 2 & -1 \\ 1 & \frac{3}{2} \end{vmatrix} = 3 + 1 = 4$

Find the solutions by Cramer's Rule:

$x = \frac{D_x}{D} = \frac{1}{2} \qquad y = \frac{D_y}{D} = \frac{4}{2} = 2$

33. $\begin{cases} x+\ y-\ z=\ 6 \\ 3x-2y+\ z=-5 \\ x+3y-2z=14 \end{cases}$

$$D=\begin{vmatrix} 1 & 1 & -1 \\ 3 & -2 & 1 \\ 1 & 3 & -2 \end{vmatrix}$$

$$=1\begin{vmatrix} -2 & 1 \\ 3 & -2 \end{vmatrix}-1\begin{vmatrix} 3 & 1 \\ 1 & -2 \end{vmatrix}+(-1)\begin{vmatrix} 3 & -2 \\ 1 & 3 \end{vmatrix}$$
$$=1(4-3)-1(-6-1)-1(9+2)$$
$$=1+7-11$$
$$=-3$$

$$D_x=\begin{vmatrix} 6 & 1 & -1 \\ -5 & -2 & 1 \\ 14 & 3 & -2 \end{vmatrix}$$

$$=6\begin{vmatrix} -2 & 1 \\ 3 & -2 \end{vmatrix}-1\begin{vmatrix} -5 & 1 \\ 14 & -2 \end{vmatrix}+(-1)\begin{vmatrix} -5 & -2 \\ 14 & 3 \end{vmatrix}$$
$$=6(4-3)-1(10-14)-1(-15+28)$$
$$=6+4-13$$
$$=-3$$

$$D_y=\begin{vmatrix} 1 & 6 & -1 \\ 3 & -5 & 1 \\ 1 & 14 & -2 \end{vmatrix}$$

$$=1\begin{vmatrix} -5 & 1 \\ 14 & -2 \end{vmatrix}-6\begin{vmatrix} 3 & 1 \\ 1 & -2 \end{vmatrix}+(-1)\begin{vmatrix} 3 & -5 \\ 1 & 14 \end{vmatrix}$$
$$=1(10-14)-6(-6-1)-1(42+5)$$
$$=-4+42-47$$
$$=-9$$

$$D_z=\begin{vmatrix} 1 & 1 & 6 \\ 3 & -2 & -5 \\ 1 & 3 & 14 \end{vmatrix}$$

$$=1\begin{vmatrix} -2 & -5 \\ 3 & 14 \end{vmatrix}-1\begin{vmatrix} 3 & -5 \\ 1 & 14 \end{vmatrix}+6\begin{vmatrix} 3 & -2 \\ 1 & 3 \end{vmatrix}$$
$$=1(-28+15)-1(42+5)+6(9+2)$$
$$=-13-47+66$$
$$=6$$

Find the solutions by Cramer's Rule:

$$x=\frac{D_x}{D}=\frac{-3}{-3}=1 \qquad y=\frac{D_y}{D}=\frac{-9}{-3}=3$$
$$z=\frac{D_z}{D}=\frac{6}{-3}=-2$$

34. $\begin{cases} x-\ y+\ z=-4 \\ 2x-3y+4z=-15 \\ 5x+\ y-2z=\ 12 \end{cases}$

$$D=\begin{vmatrix} 1 & -1 & 1 \\ 2 & -3 & 4 \\ 5 & 1 & -2 \end{vmatrix}$$

$$=1\begin{vmatrix} -3 & 4 \\ 1 & -2 \end{vmatrix}-(-1)\begin{vmatrix} 2 & 4 \\ 5 & -2 \end{vmatrix}+1\begin{vmatrix} 2 & -3 \\ 5 & 1 \end{vmatrix}$$
$$=1(6-4)+1(-4-20)+1(2+15)$$
$$=2-24+17$$
$$=-5$$

$$D_x=\begin{vmatrix} -4 & -1 & 1 \\ -15 & -3 & 4 \\ 12 & 1 & -2 \end{vmatrix}$$

$$=-4\begin{vmatrix} -3 & 4 \\ 1 & -2 \end{vmatrix}-(-1)\begin{vmatrix} -15 & 4 \\ 12 & -2 \end{vmatrix}+1\begin{vmatrix} -15 & -3 \\ 12 & 1 \end{vmatrix}$$
$$=-4(6-4)+1(30-48)+1(-15+36)$$
$$=-8-18+21$$
$$=-5$$

$$D_y=\begin{vmatrix} 1 & -4 & 1 \\ 2 & -15 & 4 \\ 5 & 12 & -2 \end{vmatrix}$$

$$=1\begin{vmatrix} -15 & 4 \\ 12 & -2 \end{vmatrix}-(-4)\begin{vmatrix} 2 & 4 \\ 5 & -2 \end{vmatrix}+1\begin{vmatrix} 2 & -15 \\ 5 & 12 \end{vmatrix}$$
$$=1(30-48)+4(-4-20)+1(24+75)$$
$$=-18-96+99$$
$$=-15$$

$$D_z=\begin{vmatrix} 1 & -1 & -4 \\ 2 & -3 & -15 \\ 5 & 1 & 12 \end{vmatrix}$$

$$=1\begin{vmatrix} -3 & -15 \\ 1 & 12 \end{vmatrix}-(-1)\begin{vmatrix} 2 & -15 \\ 5 & 12 \end{vmatrix}+(-4)\begin{vmatrix} 2 & -3 \\ 5 & 1 \end{vmatrix}$$
$$=1(-36+15)+1(24+75)-4(2+15)$$
$$=-21+99-68$$
$$=10$$

Find the solutions by Cramer's Rule:

$$x=\frac{D_x}{D}=\frac{-5}{-5}=1 \qquad y=\frac{D_y}{D}=\frac{-15}{-5}=3$$
$$z=\frac{D_z}{D}=\frac{10}{-5}=-2$$

35. $\begin{cases} x+2y-z=-3 \\ 2x-4y+z=-7 \\ -2x+2y-3z=4 \end{cases}$

$$D=\begin{vmatrix} 1 & 2 & -1 \\ 2 & -4 & 1 \\ -2 & 2 & -3 \end{vmatrix}$$

$$=1\begin{vmatrix} -4 & 1 \\ 2 & -3 \end{vmatrix}-2\begin{vmatrix} 2 & 1 \\ -2 & -3 \end{vmatrix}+(-1)\begin{vmatrix} 2 & -4 \\ -2 & 2 \end{vmatrix}$$

$$=1(12-2)-2(-6+2)-1(4-8)$$
$$=10+8+4$$
$$=22$$

$$D_x=\begin{vmatrix} -3 & 2 & -1 \\ -7 & -4 & 1 \\ 4 & 2 & -3 \end{vmatrix}$$

$$=-3\begin{vmatrix} -4 & 1 \\ 2 & -3 \end{vmatrix}-2\begin{vmatrix} -7 & 1 \\ 4 & -3 \end{vmatrix}+(-1)\begin{vmatrix} -7 & -4 \\ 4 & 2 \end{vmatrix}$$

$$=-3(12-2)-2(21-4)-1(-14+16)$$
$$=-30-34-2$$
$$=-66$$

$$D_y=\begin{vmatrix} 1 & -3 & -1 \\ 2 & -7 & 1 \\ -2 & 4 & -3 \end{vmatrix}$$

$$=1\begin{vmatrix} -7 & 1 \\ 4 & -3 \end{vmatrix}-(-3)\begin{vmatrix} 2 & 1 \\ -2 & -3 \end{vmatrix}+(-1)\begin{vmatrix} 2 & -7 \\ -2 & 4 \end{vmatrix}$$

$$=1(21-4)+3(-6+2)-1(8-14)$$
$$=17-12+6$$
$$=11$$

$$D_z=\begin{vmatrix} 1 & 2 & -3 \\ 2 & -4 & -7 \\ -2 & 2 & 4 \end{vmatrix}$$

$$=1\begin{vmatrix} -4 & -7 \\ 2 & 4 \end{vmatrix}-2\begin{vmatrix} 2 & -7 \\ -2 & 4 \end{vmatrix}+(-3)\begin{vmatrix} 2 & -4 \\ -2 & 2 \end{vmatrix}$$

$$=1(-16+14)-2(8-14)-3(4-8)$$
$$=-2+12+12$$
$$=22$$

Find the solutions by Cramer's Rule:

$$x=\frac{D_x}{D}=\frac{-66}{22}=-3 \qquad y=\frac{D_y}{D}=\frac{11}{22}=\frac{1}{2}$$

$$z=\frac{D_z}{D}=\frac{22}{22}=1$$

36. $\begin{cases} x+4y-3z=-8 \\ 3x-y+3z=12 \\ x+y+6z=1 \end{cases}$

$$D=\begin{vmatrix} 1 & 4 & -3 \\ 3 & -1 & 3 \\ 1 & 1 & 6 \end{vmatrix}$$

$$=1\begin{vmatrix} -1 & 3 \\ 1 & 6 \end{vmatrix}-4\begin{vmatrix} 3 & 3 \\ 1 & 6 \end{vmatrix}+(-3)\begin{vmatrix} 3 & -1 \\ 1 & 1 \end{vmatrix}$$

$$=1(-6-3)-4(18-3)-3(3+1)$$
$$=-9-60-12$$
$$=-81$$

$$D_x=\begin{vmatrix} -8 & 4 & -3 \\ 12 & -1 & 3 \\ 1 & 1 & 6 \end{vmatrix}$$

$$=-8\begin{vmatrix} -1 & 3 \\ 1 & 6 \end{vmatrix}-4\begin{vmatrix} 12 & 3 \\ 1 & 6 \end{vmatrix}+(-3)\begin{vmatrix} 12 & -1 \\ 1 & 1 \end{vmatrix}$$

$$=-8(-6-3)-4(72-3)-3(12+1)$$
$$=72-276-39$$
$$=-243$$

$$D_y=\begin{vmatrix} 1 & -8 & -3 \\ 3 & 12 & 3 \\ 1 & 1 & 6 \end{vmatrix}$$

$$=1\begin{vmatrix} 12 & 3 \\ 1 & 6 \end{vmatrix}-(-8)\begin{vmatrix} 3 & 3 \\ 1 & 6 \end{vmatrix}+(-3)\begin{vmatrix} 3 & 12 \\ 1 & 1 \end{vmatrix}$$

$$=1(72-3)+8(18-3)-3(3-12)$$
$$=69+120+27$$
$$=216$$

$$D_z=\begin{vmatrix} 1 & 4 & -8 \\ 3 & -1 & 12 \\ 1 & 1 & 1 \end{vmatrix}$$

$$=1\begin{vmatrix} -1 & 12 \\ 1 & 1 \end{vmatrix}-4\begin{vmatrix} 3 & 12 \\ 1 & 1 \end{vmatrix}+(-8)\begin{vmatrix} 3 & -1 \\ 1 & 1 \end{vmatrix}$$

$$=1(-1-12)-4(3-12)-8(3+1)$$
$$=-13+36-32$$
$$=-9$$

Find the solutions by Cramer's Rule:

$$x=\frac{D_x}{D}=\frac{-243}{-81}=3 \qquad y=\frac{D_y}{D}=\frac{216}{-81}=-\frac{8}{3}$$

$$z=\frac{D_z}{D}=\frac{-9}{-81}=\frac{1}{9}$$

37. $\begin{cases} x-2y+3z=1 \\ 3x+\ \ y-2z=0 \\ 2x-4y+6z=2 \end{cases}$

$$D=\begin{vmatrix} 1 & -2 & 3 \\ 3 & 1 & -2 \\ 2 & -4 & 6 \end{vmatrix}$$

$$=1\begin{vmatrix} 1 & -2 \\ -4 & 6 \end{vmatrix}-(-2)\begin{vmatrix} 3 & -2 \\ 2 & 6 \end{vmatrix}+3\begin{vmatrix} 3 & 1 \\ 2 & -4 \end{vmatrix}$$

$$=1(6-8)+2(18+4)+3(-12-2)$$

$$=-2+44-42$$

$$=0$$

Since $D=0$, Cramer's Rule does not apply.

38. $\begin{cases} x-\ \ y+2z=\ \ 5 \\ 3x+2y\qquad =\ \ 4 \\ -2x+2y-4z=-10 \end{cases}$

$$D=\begin{vmatrix} 1 & -1 & 2 \\ 3 & 2 & 0 \\ -2 & 2 & -4 \end{vmatrix}$$

$$=1\begin{vmatrix} 2 & 0 \\ 2 & -4 \end{vmatrix}-(-1)\begin{vmatrix} 3 & 0 \\ -2 & -4 \end{vmatrix}+2\begin{vmatrix} 3 & 2 \\ -2 & 2 \end{vmatrix}$$

$$=1(-8-0)+1(-12-0)+2(6+4)$$

$$=-8-12+20$$

$$=0$$

Since $D=0$, Cramer's Rule does not apply.

39. $\begin{cases} x+2y-\ \ z=0 \\ 2x-4y+\ \ z=0 \\ -2x+2y-3z=0 \end{cases}$

$$D=\begin{vmatrix} 1 & 2 & -1 \\ 2 & -4 & 1 \\ -2 & 2 & -3 \end{vmatrix}$$

$$=1\begin{vmatrix} -4 & 1 \\ 2 & -3 \end{vmatrix}-2\begin{vmatrix} 2 & 1 \\ -2 & -3 \end{vmatrix}+(-1)\begin{vmatrix} 2 & -4 \\ -2 & 2 \end{vmatrix}$$

$$=1(12-2)-2(-6+2)-1(4-8)$$

$$=10+8+4$$

$$=22$$

$$D_x=\begin{vmatrix} 0 & 2 & -1 \\ 0 & -4 & 1 \\ 0 & 2 & -3 \end{vmatrix}=0 \text{ (By Theorem 12)}$$

$$D_y=\begin{vmatrix} 1 & 0 & -1 \\ 2 & 0 & 1 \\ -2 & 0 & -3 \end{vmatrix}=0 \text{ (By Theorem 12)}$$

$$D_z=\begin{vmatrix} 1 & 2 & 0 \\ 2 & -4 & 0 \\ -2 & 2 & 0 \end{vmatrix}=0 \text{ (By Theorem 12)}$$

Find the solutions by Cramer's Rule:

$$x=\frac{D_x}{D}=\frac{0}{22}=0 \qquad y=\frac{D_y}{D}=\frac{0}{22}=0$$

$$z=\frac{D_z}{D}=\frac{0}{22}=0$$

40. $\begin{cases} x+4y-3z=0 \\ 3x-\ \ y+3z=0 \\ x+\ \ y+6z=0 \end{cases}$

$$D=\begin{vmatrix} 1 & 4 & -3 \\ 3 & -1 & 3 \\ 1 & 1 & 6 \end{vmatrix}$$

$$=1\begin{vmatrix} -1 & 3 \\ 1 & 6 \end{vmatrix}-4\begin{vmatrix} 3 & 3 \\ 1 & 6 \end{vmatrix}+(-3)\begin{vmatrix} 3 & -1 \\ 1 & 1 \end{vmatrix}$$

$$=1(-6-3)-4(18-3)-3(3+1)$$

$$=-9-60-12$$

$$=-81$$

$$D_x=\begin{vmatrix} 0 & 4 & -3 \\ 0 & -1 & 3 \\ 0 & 1 & 6 \end{vmatrix}=0 \text{ (By Theorem 12)}$$

$$D_y=\begin{vmatrix} 1 & 0 & -3 \\ 3 & 0 & 3 \\ 1 & 0 & 6 \end{vmatrix}=0 \text{ (By Theorem 12)}$$

$$D_z=\begin{vmatrix} 1 & 4 & 0 \\ 3 & -1 & 0 \\ 1 & 1 & 0 \end{vmatrix}=0 \text{ (By Theorem 12)}$$

Find the solutions by Cramer's Rule:

$$x=\frac{D_x}{D}=\frac{0}{-81}=0 \qquad y=\frac{D_y}{D}=\frac{0}{-81}=0$$

$$z=\frac{D_z}{D}=\frac{0}{-81}=0$$

41. $\begin{cases} x-2y+3z=0 \\ 3x+\ \ y-2z=0 \\ 2x-4y+6z=0 \end{cases}$

$$D=\begin{vmatrix} 1 & -2 & 3 \\ 3 & 1 & -2 \\ 2 & -4 & 6 \end{vmatrix}$$

$$=1\begin{vmatrix} 1 & -2 \\ -4 & 6 \end{vmatrix}-(-2)\begin{vmatrix} 3 & -2 \\ 2 & 6 \end{vmatrix}+3\begin{vmatrix} 3 & 1 \\ 2 & -4 \end{vmatrix}$$
$$=1(6-8)+2(18+4)+3(-12-2)$$
$$=-2+44-42$$
$$=0$$

Since $D=0$, Cramer's Rule does not apply.

42. $\begin{cases} x-\ y+2z=0 \\ 3x+2y\qquad=0 \\ -2x+2y-4z=0 \end{cases}$

$$D=\begin{vmatrix} 1 & -1 & 2 \\ 3 & 2 & 0 \\ -2 & 2 & -4 \end{vmatrix}$$

$$=1\begin{vmatrix} 2 & 0 \\ 2 & -4 \end{vmatrix}-(-1)\begin{vmatrix} 3 & 0 \\ -2 & -4 \end{vmatrix}+2\begin{vmatrix} 3 & 2 \\ -2 & 2 \end{vmatrix}$$
$$=1(-8-0)+1(-12-0)+2(6+4)$$
$$=-8-12+20$$
$$=0$$

Since $D=0$, Cramer's Rule does not apply.

43. Solve for x:

$$\begin{vmatrix} x & x \\ 4 & 3 \end{vmatrix}=5$$
$$3x-4x=5$$
$$-x=5$$
$$x=-5$$

44. Solve for x:

$$\begin{vmatrix} x & 1 \\ 3 & x \end{vmatrix}=-2$$
$$x^2-3=-2$$
$$x^2=1$$
$$x=\pm 1$$

45. Solve for x:

$$\begin{vmatrix} x & 1 & 1 \\ 4 & 3 & 2 \\ -1 & 2 & 5 \end{vmatrix}=2$$

$$x\begin{vmatrix} 3 & 2 \\ 2 & 5 \end{vmatrix}-1\begin{vmatrix} 4 & 2 \\ -1 & 5 \end{vmatrix}+1\begin{vmatrix} 4 & 3 \\ -1 & 2 \end{vmatrix}=2$$
$$x(15-4)-(20+2)+(8+3)=2$$
$$11x-22+11=2$$
$$11x=13$$
$$x=\frac{13}{11}$$

46. Solve for x:

$$\begin{vmatrix} 3 & 2 & 4 \\ 1 & x & 5 \\ 0 & 1 & -2 \end{vmatrix}=0$$

$$3\begin{vmatrix} x & 5 \\ 1 & -2 \end{vmatrix}-2\begin{vmatrix} 1 & 5 \\ 0 & -2 \end{vmatrix}+4\begin{vmatrix} 1 & x \\ 0 & 1 \end{vmatrix}=0$$
$$3(-2x-5)-2(-2)+4(1)=0$$
$$-6x-15+4+4=0$$
$$-6x-7=0$$
$$-6x=7$$
$$x=-\frac{7}{6}$$

47. Solve for x:

$$\begin{vmatrix} x & 2 & 3 \\ 1 & x & 0 \\ 6 & 1 & -2 \end{vmatrix}=7$$

$$x\begin{vmatrix} x & 0 \\ 1 & -2 \end{vmatrix}-2\begin{vmatrix} 1 & 0 \\ 6 & -2 \end{vmatrix}+3\begin{vmatrix} 1 & x \\ 6 & 1 \end{vmatrix}=7$$
$$x(-2x)-2(-2)+3(1-6x)=7$$
$$-2x^2+4+3-18x=7$$
$$-2x^2-18x=0$$
$$-2x(x+9)=0$$
$$x=0 \text{ or } x=-9$$

48. Solve for *x*:

$$\begin{vmatrix} x & 1 & 2 \\ 1 & x & 3 \\ 0 & 1 & 2 \end{vmatrix} = -4x$$

$$x\begin{vmatrix} x & 3 \\ 1 & 2 \end{vmatrix} - 1\begin{vmatrix} 1 & 3 \\ 0 & 2 \end{vmatrix} + 2\begin{vmatrix} 1 & x \\ 0 & 1 \end{vmatrix} = -4x$$

$$x(2x-3) - 1(2) + 2(1) = -4x$$

$$2x^2 - 3x - 2 + 2 = -4x$$

$$2x^2 + x = 0$$

$$x(2x+1) = 0$$

$$x = 0 \text{ or } x = -\frac{1}{2}$$

49. $\begin{vmatrix} x & y & z \\ u & v & w \\ 1 & 2 & 3 \end{vmatrix} = 4$

By Theorem (11), the value of a determinant changes sign if any two rows are interchanged.

Thus, $\begin{vmatrix} 1 & 2 & 3 \\ u & v & w \\ x & y & z \end{vmatrix} = -4$.

50. $\begin{vmatrix} x & y & z \\ u & v & w \\ 1 & 2 & 3 \end{vmatrix} = 4$

By Theorem (14), if any row of a determinant is multiplied by a nonzero number *k*, the value of the determinant is also changed by a factor of *k*.

Thus, $\begin{vmatrix} x & y & z \\ u & v & w \\ 2 & 4 & 6 \end{vmatrix} = 2\begin{vmatrix} x & y & z \\ u & v & w \\ 1 & 2 & 3 \end{vmatrix} = 2(4) = 8$.

51. Let $\begin{vmatrix} x & y & z \\ u & v & w \\ 1 & 2 & 3 \end{vmatrix} = 4$.

$$\begin{vmatrix} x & y & z \\ -3 & -6 & -9 \\ u & v & w \end{vmatrix} = -3\begin{vmatrix} x & y & z \\ 1 & 2 & 3 \\ u & v & w \end{vmatrix} \quad \text{Theorem (14)}$$

$$= -3(-1)\begin{vmatrix} x & y & z \\ u & v & w \\ 1 & 2 & 3 \end{vmatrix} \quad \text{Theorem (11)}$$

$$= 3(4)$$

$$= 12$$

52. Let $\begin{vmatrix} x & y & z \\ u & v & w \\ 1 & 2 & 3 \end{vmatrix} = 4$

$$\begin{vmatrix} 1 & 2 & 3 \\ x-u & y-v & z-w \\ u & v & w \end{vmatrix} = \begin{vmatrix} 1 & 2 & 3 \\ x & y & z \\ u & v & w \end{vmatrix} \quad \text{Theorem (15)} \ (R_2 = r_2 + r_3)$$

$$= (-1)\begin{vmatrix} x & y & z \\ 1 & 2 & 3 \\ u & v & w \end{vmatrix} \quad \text{Theorem (11)}$$

$$= (-1)(-1)\begin{vmatrix} x & y & z \\ u & v & w \\ 1 & 2 & 3 \end{vmatrix} \quad \text{Theorem (11)}$$

$$= \begin{vmatrix} x & y & z \\ u & v & w \\ 1 & 2 & 3 \end{vmatrix}$$

$$= 4$$

53. Let $\begin{vmatrix} x & y & z \\ u & v & w \\ 1 & 2 & 3 \end{vmatrix} = 4$

$$\begin{vmatrix} 1 & 2 & 3 \\ x-3 & y-6 & z-9 \\ 2u & 2v & 2w \end{vmatrix}$$

$$= 2\begin{vmatrix} 1 & 2 & 3 \\ x-3 & y-6 & z-9 \\ u & v & w \end{vmatrix} \quad \text{Theorem (14)}$$

$$= 2(-1)\begin{vmatrix} x-3 & y-6 & z-9 \\ 1 & 2 & 3 \\ u & v & w \end{vmatrix} \quad \text{Theorem (11)}$$

$$= 2(-1)(-1)\begin{vmatrix} x-3 & y-6 & z-9 \\ u & v & w \\ 1 & 2 & 3 \end{vmatrix} \quad \text{Theorem (11)}$$

$$= 2(-1)(-1)\begin{vmatrix} x & y & z \\ u & v & w \\ 1 & 2 & 3 \end{vmatrix} \quad \text{Theorem (15)} \ (R_1 = -3r_3 + r_1)$$

$$= 2(-1)(-1)(4)$$

$$= 8$$

54. Let $\begin{vmatrix} x & y & z \\ u & v & w \\ 1 & 2 & 3 \end{vmatrix} = 4$

$$\begin{vmatrix} x & y & z-x \\ u & v & w-u \\ 1 & 2 & 2 \end{vmatrix} = \begin{vmatrix} x & y & z \\ u & v & w \\ 1 & 2 & 3 \end{vmatrix} \quad \text{Theorem (15)} \ (C_3 = c_1 + c_3)$$

$$= 4$$

55. Let $\begin{vmatrix} x & y & z \\ u & v & w \\ 1 & 2 & 3 \end{vmatrix} = 4$

$$\begin{vmatrix} 1 & 2 & 3 \\ 2x & 2y & 2z \\ u-1 & v-2 & w-3 \end{vmatrix}$$

$$= 2\begin{vmatrix} 1 & 2 & 3 \\ x & y & z \\ u-1 & v-2 & w-3 \end{vmatrix} \quad \text{Theorem (14)}$$

$$= 2(-1)\begin{vmatrix} x & y & z \\ 1 & 2 & 3 \\ u-1 & v-2 & w-3 \end{vmatrix} \quad \text{Theorem (11)}$$

$$= 2(-1)(-1)\begin{vmatrix} x & y & z \\ u-1 & v-2 & w-3 \\ 1 & 2 & 3 \end{vmatrix} \quad \text{Theorem (11)}$$

$$= 2(-1)(-1)\begin{vmatrix} x & y & z \\ u & v & w \\ 1 & 2 & 3 \end{vmatrix} \quad \text{Theorem (15)} \ (R_2 = -r_3 + r_2)$$

$$= 2(-1)(-1)(4)$$

$$= 8$$

56. Let $\begin{vmatrix} x & y & z \\ u & v & w \\ 1 & 2 & 3 \end{vmatrix} = 4$

$$\begin{vmatrix} x+3 & y+6 & z+9 \\ 3u-1 & 3v-2 & 3w-3 \\ 1 & 2 & 3 \end{vmatrix}$$

$$= \begin{vmatrix} x & y & z \\ 3u-1 & 3v-2 & 3w-3 \\ 1 & 2 & 3 \end{vmatrix} \quad \text{Theorem (15)} \ (R_1 = 3r_3 + r_1)$$

$$= \begin{vmatrix} x & y & z \\ 3u & 3v & 3w \\ 1 & 2 & 3 \end{vmatrix} \quad \text{Theorem (15)} \ (R_2 = -r_3 + r_2)$$

$$= 3\begin{vmatrix} x & y & z \\ u & v & w \\ 1 & 2 & 3 \end{vmatrix} \quad \text{Theorem (14)}$$

$$= 3(4)$$

$$= 12$$

57. Expanding the determinant:

$$\begin{vmatrix} x & y & 1 \\ x_1 & y_1 & 1 \\ x_2 & y_2 & 1 \end{vmatrix} = 0$$

$$x\begin{vmatrix} y_1 & 1 \\ y_2 & 1 \end{vmatrix} - y\begin{vmatrix} x_1 & 1 \\ x_2 & 1 \end{vmatrix} + 1\begin{vmatrix} x_1 & y_1 \\ x_2 & y_2 \end{vmatrix} = 0$$

$$x(y_1 - y_2) - y(x_1 - x_2) + (x_1 y_2 - x_2 y_1) = 0$$

$$x(y_1 - y_2) + y(x_2 - x_1) = x_2 y_1 - x_1 y_2$$

$$y(x_2 - x_1) = x_2 y_1 - x_1 y_2 + x(y_2 - y_1)$$

$$y(x_2 - x_1) - y_1(x_2 - x_1) = x_2 y_1 - x_1 y_2 + x(y_2 - y_1) - y_1(x_2 - x_1)$$

$$(x_2 - x_1)(y - y_1) = x(y_2 - y_1) + x_2 y_1 - x_1 y_2 - y_1 x_2 + y_1 x_1$$

$$(x_2 - x_1)(y - y_1) = (y_2 - y_1)x - (y_2 - y_1)x_1$$

$$(x_2 - x_1)(y - y_1) = (y_2 - y_1)(x - x_1)$$

$$(y - y_1) = \frac{(y_2 - y_1)}{(x_2 - x_1)}(x - x_1)$$

This is the 2-point form of the equation for a line.

58. Any point (x, y) on the line containing (x_2, y_2) and (x_3, y_3) satisfies:

$$\begin{vmatrix} x & y & 1 \\ x_2 & y_2 & 1 \\ x_3 & y_3 & 1 \end{vmatrix} = 0$$

If the point (x_1, y_1) is on the line containing (x_2, y_2) and (x_3, y_3) (the points are collinear),

then $\begin{vmatrix} x_1 & y_1 & 1 \\ x_2 & y_2 & 1 \\ x_3 & y_3 & 1 \end{vmatrix} = 0$.

Conversely, if $\begin{vmatrix} x_1 & y_1 & 1 \\ x_2 & y_2 & 1 \\ x_3 & y_3 & 1 \end{vmatrix} = 0$, then (x_1, y_1) is on the line containing (x_2, y_2) and (x_3, y_3), and the points are collinear.

59. Expanding the determinant:

$$\begin{vmatrix} x^2 & x & 1 \\ y^2 & y & 1 \\ z^2 & z & 1 \end{vmatrix}$$

$$= x^2\begin{vmatrix} y & 1 \\ z & 1 \end{vmatrix} - x\begin{vmatrix} y^2 & 1 \\ z^2 & 1 \end{vmatrix} + 1\begin{vmatrix} y^2 & y \\ z^2 & z \end{vmatrix}$$

$$= x^2(y-z) - x(y^2-z^2) + 1(y^2z - z^2y)$$

$$= x^2(y-z) - x(y-z)(y+z) + yz(y-z)$$

$$= (y-z)\left[x^2 - xy - xz + yz\right]$$

$$= (y-z)\left[x(x-y) - z(x-y)\right]$$

$$= (y-z)(x-y)(x-z)$$

60. If $a = 0$, then $b \neq 0$ and $c \neq 0$ since $ad - bc \neq 0$, and the system is $\begin{cases} by = s \\ cx + dy = t \end{cases}$.

The solution of the system is $y = \frac{s}{b}$, $x = \frac{t - dy}{c} = \frac{t - d\left(\frac{s}{b}\right)}{c} = \frac{tb - sd}{bc}$. Using Cramer's Rule, we get $D = \begin{vmatrix} 0 & b \\ c & d \end{vmatrix} = -bc$,

$D_x = \begin{vmatrix} s & b \\ t & d \end{vmatrix} = sd - tb$,

$D_y = \begin{vmatrix} 0 & s \\ c & t \end{vmatrix} = 0 - sc = -sc$, so

$x = \frac{D_x}{D} = \frac{ds - tb}{-bc} = \frac{td - sd}{bc}$ and

$y = \frac{D_y}{D} = \frac{-sc}{-bc} = \frac{s}{b}$, which is the solution. Note that these solutions agree if $d = 0$.

If $b = 0$, then $a \neq 0$ and $d \neq 0$ since $ad - bc \neq 0$, and the system is $\begin{cases} ax = s \\ cx + dy = t \end{cases}$.

The solution of the system is $x = \frac{s}{a}$, $y = \frac{t - cx}{d} = \frac{at - cs}{ad}$. Using Cramer's Rule, we get $D = \begin{vmatrix} a & 0 \\ c & d \end{vmatrix} = ad$, $D_x = \begin{vmatrix} s & 0 \\ t & d \end{vmatrix} = sd$, and

$D_y = \begin{vmatrix} a & s \\ c & t \end{vmatrix} = at - cs$, so $x = \frac{D_x}{D} = \frac{sd}{ad} = \frac{s}{a}$

and $y = \frac{D_y}{D} = \frac{at - cs}{ad}$, which is the solution. Note that these solutions agree if $c = 0$.

If $c = 0$, then $a \neq 0$ and $d \neq 0$ since $ad - bc \neq 0$, and the system is $\begin{cases} ax + by = s \\ dy = t \end{cases}$.

The solution of the system is $y = \frac{t}{d}$, $x = \frac{s - by}{a} = \frac{sd - tb}{ad}$. Using Cramer's Rule, we get $D = \begin{vmatrix} a & b \\ 0 & d \end{vmatrix} = ad$, $D_x = \begin{vmatrix} s & b \\ t & d \end{vmatrix} = sd - tb$,

and $D_y = \begin{vmatrix} a & s \\ 0 & t \end{vmatrix} = at$, so $x = \frac{D_x}{D} = \frac{sd - tb}{ad}$ and $y = \frac{D_y}{D} = \frac{at}{ad} = \frac{t}{d}$, which is the solution. Note that these solutions agree if $b = 0$.

If $d = 0$, then $b \neq 0$ and $c \neq 0$ since $ad - bc \neq 0$, and the system is $\begin{cases} ax + by = s \\ cx = t \end{cases}$.

The solution of the system is $x = \frac{t}{c}$, $y = \frac{s - ax}{b} = \frac{cs - at}{bc}$. Using Cramer's Rule, we get $D = \begin{vmatrix} a & b \\ c & 0 \end{vmatrix} = 0 - bc = -bc$,

$D_x = \begin{vmatrix} s & b \\ t & 0 \end{vmatrix} = 0 - tb = -tb$, and

$D_y = \begin{vmatrix} a & s \\ c & t \end{vmatrix} = at - cs$, so $x = \frac{D_x}{D} = \frac{t}{c}$ and

$y = \frac{D_y}{D} = \frac{at - cs}{-bc} = \frac{cs - at}{bc}$, which is the solution. Note that these solutions agree if $a = 0$.

61. Evaluating the determinant to show the relationship:

$$\begin{vmatrix} a_{13} & a_{12} & a_{11} \\ a_{23} & a_{22} & a_{21} \\ a_{33} & a_{32} & a_{31} \end{vmatrix}$$

$$= a_{13}\begin{vmatrix} a_{22} & a_{21} \\ a_{32} & a_{31} \end{vmatrix} - a_{12}\begin{vmatrix} a_{23} & a_{21} \\ a_{33} & a_{31} \end{vmatrix} + a_{11}\begin{vmatrix} a_{23} & a_{22} \\ a_{33} & a_{32} \end{vmatrix}$$

$$= a_{13}(a_{22}a_{31} - a_{21}a_{32}) - a_{12}(a_{23}a_{31} - a_{21}a_{33}) + a_{11}(a_{23}a_{32} - a_{22}a_{33})$$

$$= a_{13}a_{22}a_{31} - a_{13}a_{21}a_{32} - a_{12}a_{23}a_{31} + a_{12}a_{21}a_{33} + a_{11}a_{23}a_{32} - a_{11}a_{22}a_{33}$$

$$= -a_{11}a_{22}a_{33} + a_{11}a_{23}a_{32} + a_{12}a_{21}a_{33} - a_{12}a_{23}a_{31} - a_{13}a_{21}a_{32} + a_{13}a_{22}a_{31}$$

$$= -a_{11}(a_{22}a_{33} - a_{23}a_{32}) + a_{12}(a_{21}a_{33} - a_{23}a_{31}) - a_{13}(a_{21}a_{32} - a_{22}a_{31})$$

$$= -a_{11}\begin{vmatrix} a_{22} & a_{23} \\ a_{32} & a_{33} \end{vmatrix} + a_{12}\begin{vmatrix} a_{21} & a_{23} \\ a_{31} & a_{33} \end{vmatrix} - a_{13}\begin{vmatrix} a_{21} & a_{22} \\ a_{31} & a_{32} \end{vmatrix}$$

$$= -\left(a_{11}\begin{vmatrix} a_{22} & a_{23} \\ a_{32} & a_{33} \end{vmatrix} - a_{12}\begin{vmatrix} a_{21} & a_{23} \\ a_{31} & a_{33} \end{vmatrix} + a_{13}\begin{vmatrix} a_{21} & a_{22} \\ a_{31} & a_{32} \end{vmatrix} \right)$$

$$= -\begin{vmatrix} a_{11} & a_{12} & a_{13} \\ a_{21} & a_{22} & a_{23} \\ a_{31} & a_{32} & a_{33} \end{vmatrix}$$

62. Evaluating the determinant to show the relationship:

$$\begin{vmatrix} a_{11} & a_{12} & a_{13} \\ ka_{21} & ka_{22} & ka_{23} \\ a_{31} & a_{32} & a_{33} \end{vmatrix}$$

$$= a_{11}\begin{vmatrix} ka_{22} & ka_{23} \\ a_{32} & a_{33} \end{vmatrix} - a_{12}\begin{vmatrix} ka_{21} & ka_{23} \\ a_{31} & a_{33} \end{vmatrix} + a_{13}\begin{vmatrix} ka_{21} & ka_{22} \\ a_{31} & a_{32} \end{vmatrix}$$

$$= a_{11}(ka_{22}a_{33} - ka_{23}a_{32}) - a_{12}(ka_{21}a_{33} - ka_{23}a_{31}) + a_{13}(ka_{21}a_{32} - ka_{22}a_{31})$$

$$= ka_{11}(a_{22}a_{33} - a_{23}a_{32}) - ka_{12}(a_{21}a_{33} - a_{23}a_{31}) + ka_{13}(a_{21}a_{32} - a_{22}a_{31})$$

$$= k\big(a_{11}(a_{22}a_{33} - a_{23}a_{32}) - a_{12}(a_{21}a_{33} - a_{23}a_{31}) + a_{13}(a_{21}a_{32} - a_{22}a_{31})\big)$$

$$= k\left(a_{11}\begin{vmatrix} a_{22} & a_{23} \\ a_{32} & a_{33} \end{vmatrix} - a_{12}\begin{vmatrix} a_{21} & a_{23} \\ a_{31} & a_{33} \end{vmatrix} + a_{13}\begin{vmatrix} a_{21} & a_{22} \\ a_{31} & a_{32} \end{vmatrix} \right)$$

$$= k\begin{vmatrix} a_{11} & a_{12} & a_{13} \\ a_{21} & a_{22} & a_{23} \\ a_{31} & a_{32} & a_{33} \end{vmatrix}$$

63. Set up a 3 by 3 determinant in which the first column and third column are the same and evaluate by expanding down column 2:

$$\begin{vmatrix} a & b & a \\ c & d & c \\ e & f & e \end{vmatrix} = -b\begin{vmatrix} c & c \\ e & e \end{vmatrix} + d\begin{vmatrix} a & a \\ e & e \end{vmatrix} - f\begin{vmatrix} a & a \\ c & c \end{vmatrix}$$

$$= -b(ce - ce) + d(ae - ae) - f(ac - ac)$$

$$= -b(0) + d(0) - f(0)$$

$$= 0$$

64. Evaluating the determinant to show the relationship:

$$\begin{vmatrix} a_{11} + ka_{21} & a_{12} + ka_{22} & a_{13} + ka_{23} \\ a_{21} & a_{22} & a_{23} \\ a_{31} & a_{32} & a_{33} \end{vmatrix}$$

$$= (a_{11} + ka_{21})\begin{vmatrix} a_{22} & a_{23} \\ a_{32} & a_{33} \end{vmatrix} - (a_{12} + ka_{22})\begin{vmatrix} a_{21} & a_{23} \\ a_{31} & a_{33} \end{vmatrix} + (a_{13} + ka_{23})\begin{vmatrix} a_{21} & a_{22} \\ a_{31} & a_{32} \end{vmatrix}$$

$$= (a_{11} + ka_{21})(a_{22}a_{33} - a_{23}a_{32}) - (a_{12} + ka_{22})(a_{21}a_{33} - a_{23}a_{31}) + (a_{13} + ka_{23})(a_{21}a_{32} - a_{22}a_{31})$$

$$= a_{11}(a_{22}a_{33} - a_{23}a_{32}) + ka_{21}(a_{22}a_{33} - a_{23}a_{32}) - a_{12}(a_{21}a_{33} - a_{23}a_{31}) - ka_{22}(a_{21}a_{33} - a_{23}a_{31}) + a_{13}(a_{21}a_{32} - a_{22}a_{31}) + ka_{23}(a_{21}a_{32} - a_{22}a_{31})$$

$$= a_{11}(a_{22}a_{33} - a_{23}a_{32}) + ka_{21}a_{22}a_{33} - ka_{21}a_{23}a_{32} - a_{12}(a_{21}a_{33} - a_{23}a_{31}) - ka_{22}a_{21}a_{33} + ka_{22}a_{23}a_{31} + a_{13}(a_{21}a_{32} - a_{22}a_{31}) + ka_{23}a_{21}a_{32} - ka_{23}a_{22}a_{31}$$

$$= a_{11}(a_{22}a_{33} - a_{23}a_{32}) - a_{12}(a_{21}a_{33} - a_{23}a_{31}) + a_{13}(a_{21}a_{32} - a_{22}a_{31})$$

$$= a_{11}\begin{vmatrix} a_{22} & a_{23} \\ a_{32} & a_{33} \end{vmatrix} - a_{12}\begin{vmatrix} a_{21} & a_{23} \\ a_{31} & a_{33} \end{vmatrix} + a_{13}\begin{vmatrix} a_{21} & a_{22} \\ a_{31} & a_{32} \end{vmatrix}$$

$$= \begin{vmatrix} a_{11} & a_{12} & a_{13} \\ a_{21} & a_{22} & a_{23} \\ a_{31} & a_{32} & a_{33} \end{vmatrix}$$

Section 10.4

1. inverse

2. square

3. identity

4. False

5. False

6. False

7. a. $A+B=\begin{bmatrix} 0 & 3 & -5 \\ 1 & 2 & 6 \end{bmatrix}+\begin{bmatrix} 4 & 1 & 0 \\ -2 & 3 & -2 \end{bmatrix}$
$=\begin{bmatrix} 0+4 & 3+1 & -5+0 \\ 1+(-2) & 2+3 & 6+(-2) \end{bmatrix}$
$=\begin{bmatrix} 4 & 4 & -5 \\ -1 & 5 & 4 \end{bmatrix}$

   b. Enter the matrices into a graphing utility. The result is shown below:

```
[A]+[B]
      [[4  4 -5]
       [-1 5 4 ]]
```

8. a. $A-B=\begin{bmatrix} 0 & 3 & -5 \\ 1 & 2 & 6 \end{bmatrix}-\begin{bmatrix} 4 & 1 & 0 \\ -2 & 3 & -2 \end{bmatrix}$
$=\begin{bmatrix} 0-4 & 3-1 & -5-0 \\ 1-(-2) & 2-3 & 6-(-2) \end{bmatrix}$
$=\begin{bmatrix} -4 & 2 & -5 \\ 3 & -1 & 8 \end{bmatrix}$

   b. Enter the matrices into a graphing utility. The result is shown below:

```
[A]-[B]
      [[-4 2  -5]
       [3  -1 8 ]]
```

9. a. $4A=4\begin{bmatrix} 0 & 3 & -5 \\ 1 & 2 & 6 \end{bmatrix}$
$=\begin{bmatrix} 4\cdot 0 & 4\cdot 3 & 4(-5) \\ 4\cdot 1 & 4\cdot 2 & 4\cdot 6 \end{bmatrix}$
$=\begin{bmatrix} 0 & 12 & -20 \\ 4 & 8 & 24 \end{bmatrix}$

   b. Enter the matrices into a graphing utility. The result is shown below:

```
4[A]
      [[0 12 -20]
       [4 8  24 ]]
```

10. a. $-3B=-3\begin{bmatrix} 4 & 1 & 0 \\ -2 & 3 & -2 \end{bmatrix}$
$=\begin{bmatrix} -3\cdot 4 & -3\cdot 1 & -3\cdot 0 \\ -3\cdot -2 & -3\cdot 3 & -3\cdot -2 \end{bmatrix}$
$=\begin{bmatrix} -12 & -3 & 0 \\ 6 & -9 & 6 \end{bmatrix}$

   b. Enter the matrices into a graphing utility. The result is shown below:

```
-3[B]
      [[-12 -3 0]
       [6   -9 6]]
```

11. a. $3A-2B=3\begin{bmatrix} 0 & 3 & -5 \\ 1 & 2 & 6 \end{bmatrix}-2\begin{bmatrix} 4 & 1 & 0 \\ -2 & 3 & -2 \end{bmatrix}$
$=\begin{bmatrix} 0 & 9 & -15 \\ 3 & 6 & 18 \end{bmatrix}-\begin{bmatrix} 8 & 2 & 0 \\ -4 & 6 & -4 \end{bmatrix}$
$=\begin{bmatrix} -8 & 7 & -15 \\ 7 & 0 & 22 \end{bmatrix}$

   b. Enter the matrices into a graphing utility. The result is shown below:

```
3[A]-2[B]
      [[-8 7 -15]
       [7  0 22 ]]
```

12. a. $2A+4B=2\begin{bmatrix} 0 & 3 & -5 \\ 1 & 2 & 6 \end{bmatrix}+4\begin{bmatrix} 4 & 1 & 0 \\ -2 & 3 & -2 \end{bmatrix}$
$=\begin{bmatrix} 0 & 6 & -10 \\ 2 & 4 & 12 \end{bmatrix}+\begin{bmatrix} 16 & 4 & 0 \\ -8 & 12 & -8 \end{bmatrix}$
$=\begin{bmatrix} 16 & 10 & -10 \\ -6 & 16 & 4 \end{bmatrix}$

   b. Enter the matrices into a graphing utility. The result is shown below:

```
2[A]+4[B]
      [[16 10 -10]
       [-6 16 4  ]]
```

13. a. $AC = \begin{bmatrix} 0 & 3 & -5 \\ 1 & 2 & 6 \end{bmatrix} \cdot \begin{bmatrix} 4 & 1 \\ 6 & 2 \\ -2 & 3 \end{bmatrix}$

$= \begin{bmatrix} 0(4)+3(6)+(-5)(-2) & 0(1)+3(2)+(-5)(3) \\ 1(4)+2(6)+6(-2) & 1(1)+2(2)+6(3) \end{bmatrix}$

$= \begin{bmatrix} 28 & -9 \\ 4 & 23 \end{bmatrix}$

b. Enter the matrices into a graphing utility. The result is shown below:

```
[A][C]
        [[28 -9]
         [4  23]]
```

14. a. $BC = \begin{bmatrix} 4 & 1 & 0 \\ -2 & 3 & -2 \end{bmatrix} \cdot \begin{bmatrix} 4 & 1 \\ 6 & 2 \\ -2 & 3 \end{bmatrix}$

$= \begin{bmatrix} 4(4)+1(6)+0(-2) & 4(1)+1(2)+0(3) \\ -2(4)+3(6)+(-2)(-2) & -2(1)+3(2)+(-2)(3) \end{bmatrix}$

$= \begin{bmatrix} 22 & 6 \\ 14 & -2 \end{bmatrix}$

b. Enter the matrices into a graphing utility. The result is shown below:

```
[B][C]
        [[22 6 ]
         [14 -2]]
```

15. a. $CA = \begin{bmatrix} 4 & 1 \\ 6 & 2 \\ -2 & 3 \end{bmatrix} \cdot \begin{bmatrix} 0 & 3 & -5 \\ 1 & 2 & 6 \end{bmatrix}$

$= \begin{bmatrix} 4(0)+1(1) & 4(3)+1(2) & 4(-5)+1(6) \\ 6(0)+2(1) & 6(3)+2(2) & 6(-5)+2(6) \\ -2(0)+3(1) & -2(3)+3(2) & -2(-5)+3(6) \end{bmatrix}$

$= \begin{bmatrix} 1 & 14 & -14 \\ 2 & 22 & -18 \\ 3 & 0 & 28 \end{bmatrix}$

b. Enter the matrices into a graphing utility. The result is shown below:

```
[C][A]
     [[1 14 -14]
      [2 22 -18]
      [3 0  28 ]]
```

16. a. $CB = \begin{bmatrix} 4 & 1 \\ 6 & 2 \\ -2 & 3 \end{bmatrix} \cdot \begin{bmatrix} 4 & 1 & 0 \\ -2 & 3 & -2 \end{bmatrix}$

$= \begin{bmatrix} 4(4)+1(-2) & 4(1)+1(3) & 4(0)+1(-2) \\ 6(4)+2(-2) & 6(1)+2(3) & 6(0)+2(-2) \\ -2(4)+3(-2) & -2(1)+3(3) & -2(0)+3(-2) \end{bmatrix}$

$= \begin{bmatrix} 14 & 7 & -2 \\ 20 & 12 & -4 \\ -14 & 7 & -6 \end{bmatrix}$

b. Enter the matrices into a graphing utility. The result is shown below:

```
[C][B]
 [[14  7  -2]
  [20  12 -4]
  [-14 7  -6]]
```

17. a. $C(A+B) = \begin{bmatrix} 4 & 1 \\ 6 & 2 \\ -2 & 3 \end{bmatrix} \left(\begin{bmatrix} 0 & 3 & -5 \\ 1 & 2 & 6 \end{bmatrix} + \begin{bmatrix} 4 & 1 & 0 \\ -2 & 3 & -2 \end{bmatrix} \right)$

$= \begin{bmatrix} 4 & 1 \\ 6 & 2 \\ -2 & 3 \end{bmatrix} \cdot \begin{bmatrix} 4 & 4 & -5 \\ -1 & 5 & 4 \end{bmatrix}$

$= \begin{bmatrix} 15 & 21 & -16 \\ 22 & 34 & -22 \\ -11 & 7 & 22 \end{bmatrix}$

b. Enter the matrices into a graphing utility. The result is shown below:

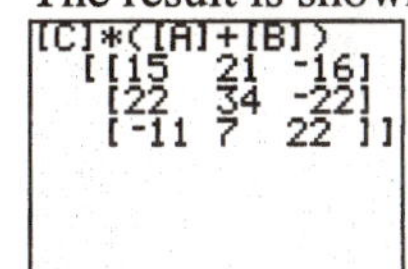

18. a. $(A+B)C = \left(\begin{bmatrix} 0 & 3 & -5 \\ 1 & 2 & 6 \end{bmatrix} + \begin{bmatrix} 4 & 1 & 0 \\ -2 & 3 & -2 \end{bmatrix} \right) \begin{bmatrix} 4 & 1 \\ 6 & 2 \\ -2 & 3 \end{bmatrix}$

$= \begin{bmatrix} 4 & 4 & -5 \\ -1 & 5 & 4 \end{bmatrix} \cdot \begin{bmatrix} 4 & 1 \\ 6 & 2 \\ -2 & 3 \end{bmatrix}$

$= \begin{bmatrix} 50 & -3 \\ 18 & 21 \end{bmatrix}$

b. Enter the matrices into a graphing utility. The result is shown below:

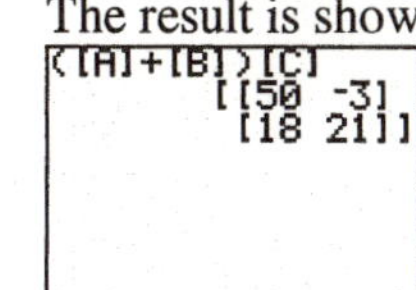

19. a. $AC - 3I_2 = \begin{bmatrix} 0 & 3 & -5 \\ 1 & 2 & 6 \end{bmatrix} \cdot \begin{bmatrix} 4 & 1 \\ 6 & 2 \\ -2 & 3 \end{bmatrix} - 3\begin{bmatrix} 1 & 0 \\ 0 & 1 \end{bmatrix}$

$= \begin{bmatrix} 28 & -9 \\ 4 & 23 \end{bmatrix} - \begin{bmatrix} 3 & 0 \\ 0 & 3 \end{bmatrix}$

$= \begin{bmatrix} 25 & -9 \\ 4 & 20 \end{bmatrix}$

b. Enter the matrices into a graphing utility. Use $I = \begin{bmatrix} 1 & 0 \\ 0 & 1 \end{bmatrix}$. The result is shown below:

```
[A][C]-3[I]
      [[25 -9]
       [4  20]]
```

20. a. $CA + 5I_3 = \begin{bmatrix} 4 & 1 \\ 6 & 2 \\ -2 & 3 \end{bmatrix} \cdot \begin{bmatrix} 0 & 3 & -5 \\ 1 & 2 & 6 \end{bmatrix} + 5\begin{bmatrix} 1 & 0 & 0 \\ 0 & 1 & 0 \\ 0 & 0 & 1 \end{bmatrix}$

$= \begin{bmatrix} 1 & 14 & -14 \\ 2 & 22 & -18 \\ 3 & 0 & 28 \end{bmatrix} + \begin{bmatrix} 5 & 0 & 0 \\ 0 & 5 & 0 \\ 0 & 0 & 5 \end{bmatrix}$

$= \begin{bmatrix} 6 & 14 & -14 \\ 2 & 27 & -18 \\ 3 & 0 & 33 \end{bmatrix}$

b. Enter the matrices into a graphing utility. Use $I = \begin{bmatrix} 1 & 0 & 0 \\ 0 & 1 & 0 \\ 0 & 0 & 1 \end{bmatrix}$. The result is shown below:

```
[C][A]+5[I]
      [[6 14 -14]
       [2 27 -18]
       [3 0  33 ]]
```

21. a. $CA - CB$

$= \begin{bmatrix} 4 & 1 \\ 6 & 2 \\ -2 & 3 \end{bmatrix} \cdot \begin{bmatrix} 0 & 3 & -5 \\ 1 & 2 & 6 \end{bmatrix} - \begin{bmatrix} 4 & 1 \\ 6 & 2 \\ -2 & 3 \end{bmatrix} \cdot \begin{bmatrix} 4 & 1 & 0 \\ -2 & 3 & -2 \end{bmatrix}$

$= \begin{bmatrix} 1 & 14 & -14 \\ 2 & 22 & -18 \\ 3 & 0 & 28 \end{bmatrix} - \begin{bmatrix} 14 & 7 & -2 \\ 20 & 12 & -4 \\ -14 & 7 & -6 \end{bmatrix}$

$= \begin{bmatrix} -13 & 7 & -12 \\ -18 & 10 & -14 \\ 17 & -7 & 34 \end{bmatrix}$

b. Enter the matrices into a graphing utility. The result is shown below:

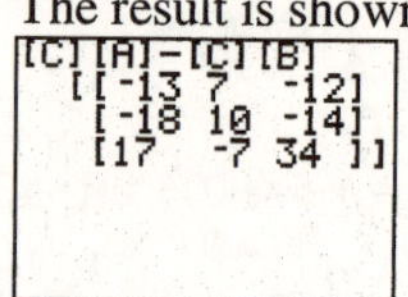

22. a. $AC + BC$

$= \begin{bmatrix} 0 & 3 & -5 \\ 1 & 2 & 6 \end{bmatrix} \cdot \begin{bmatrix} 4 & 1 \\ 6 & 2 \\ -2 & 3 \end{bmatrix} + \begin{bmatrix} 4 & 1 & 0 \\ -2 & 3 & -2 \end{bmatrix} \cdot \begin{bmatrix} 4 & 1 \\ 6 & 2 \\ -2 & 3 \end{bmatrix}$

$= \begin{bmatrix} 28 & -9 \\ 4 & 23 \end{bmatrix} + \begin{bmatrix} 22 & 6 \\ 14 & -2 \end{bmatrix}$

$= \begin{bmatrix} 50 & -3 \\ 18 & 21 \end{bmatrix}$

b. Enter the matrices into a graphing utility. The result is shown below:

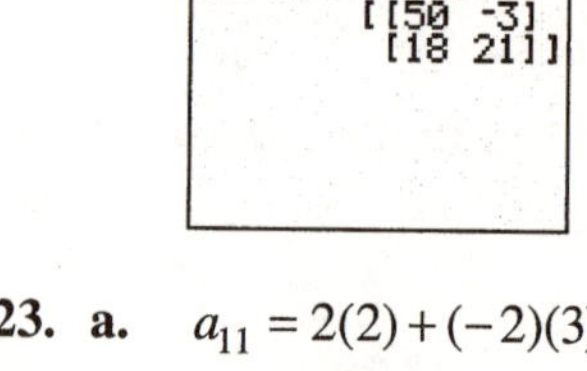

23. a. $a_{11} = 2(2) + (-2)(3) = -2$

$a_{12} = 2(1) + (-2)(-1) = 4$

$a_{13} = 2(4) + (-2)(3) = 2$

$a_{14} = 2(6) + (-2)(2) = 8$

$a_{21} = 1(2) + 0(3) = 2$

$a_{22} = 1(1) + 0(-1) = 1$

$a_{23} = 1(4) + 0(3) = 4$

$a_{24} = 1(6) + 0(2) = 6$

$\begin{bmatrix} 2 & -2 \\ 1 & 0 \end{bmatrix}\begin{bmatrix} 2 & 1 & 4 & 6 \\ 3 & -1 & 3 & 2 \end{bmatrix} = \begin{bmatrix} -2 & 4 & 2 & 8 \\ 2 & 1 & 4 & 6 \end{bmatrix}$

b. Enter the matrices into a graphing utility using $A = \begin{bmatrix} 2 & -2 \\ 1 & 0 \end{bmatrix}$ and $B = \begin{bmatrix} 2 & 1 & 4 & 6 \\ 3 & -1 & 3 & 2 \end{bmatrix}$.

The result is shown below:

```
[A][B]
      [[-2 4 2 8]
       [2  1 4 6]]
```

24. a. $a_{11} = 4(-6) + 1(2) = -22$
$a_{12} = 4(6) + 1(5) = 29$
$a_{13} = 4(1) + 1(4) = 8$
$a_{14} = 4(0) + 1(-1) = -1$
$a_{21} = 2(-6) + 1(2) = -10$
$a_{22} = 2(6) + 1(5) = 17$
$a_{23} = 2(1) + 1(4) = 6$
$a_{24} = 2(0) + 1(-1) = -1$

$$\begin{bmatrix} 4 & 1 \\ 2 & 1 \end{bmatrix}\begin{bmatrix} -6 & 6 & 1 & 0 \\ 2 & 5 & 4 & -1 \end{bmatrix} = \begin{bmatrix} -22 & 29 & 8 & -1 \\ -10 & 17 & 6 & -1 \end{bmatrix}$$

b. Enter the matrices into a graphing utility using $A = \begin{bmatrix} 4 & 1 \\ 2 & 1 \end{bmatrix}$ and $B = \begin{bmatrix} -6 & 6 & 1 & 0 \\ 2 & 5 & 4 & -1 \end{bmatrix}$.

The result is shown below:

```
[A][B]
 [[-22 29 8 -1]
  [-10 17 6 -1]]
```

25. a.
$$\begin{bmatrix} 1 & 2 & 3 \\ 0 & -1 & 4 \end{bmatrix}\begin{bmatrix} 1 & 2 \\ -1 & 0 \\ 2 & 4 \end{bmatrix}$$
$$= \begin{bmatrix} 1(1)+2(-1)+3(2) & 1(2)+2(0)+3(4) \\ 0(1)+(-1)(-1)+4(2) & 0(2)+(-1)(0)+4(4) \end{bmatrix}$$
$$= \begin{bmatrix} 5 & 14 \\ 9 & 16 \end{bmatrix}$$

b. Enter the matrices into a graphing utility using $A = \begin{bmatrix} 1 & 2 & 3 \\ 0 & -1 & 4 \end{bmatrix}$ and $B = \begin{bmatrix} 1 & 2 \\ -1 & 0 \\ 2 & 4 \end{bmatrix}$.

The result is shown below:

```
[A][B]
       [[5 14]
        [9 16]]
```

26. a.
$$\begin{bmatrix} 1 & -1 \\ -3 & 2 \\ 0 & 5 \end{bmatrix}\begin{bmatrix} 2 & 8 & -1 \\ 3 & 6 & 0 \end{bmatrix}$$
$$= \begin{bmatrix} 1(2)+(-1)(3) & 1(8)+(-1)(6) & 1(-1)+(-1)(0) \\ (-3)(2)+2(3) & (-3)(8)+2(6) & (-3)(-1)+2(0) \\ 0(2)+5(3) & 0(8)+5(6) & 0(-1)+5(0) \end{bmatrix}$$
$$= \begin{bmatrix} -1 & 2 & -1 \\ 0 & -12 & 3 \\ 15 & 30 & 0 \end{bmatrix}$$

b. Enter the matrices into a graphing utility using $A = \begin{bmatrix} 1 & -1 \\ -3 & 2 \\ 0 & 5 \end{bmatrix}$ and $B = \begin{bmatrix} 2 & 8 & -1 \\ 3 & 6 & 0 \end{bmatrix}$.

The result is shown below:

```
[A][B]
  [[-1 2   -1]
   [0  -12 3 ]
   [15 30  0 ]]
```

27. a.
$$\begin{bmatrix} 1 & 0 & 1 \\ 2 & 4 & 1 \\ 3 & 6 & 1 \end{bmatrix}\begin{bmatrix} 1 & 3 \\ 6 & 2 \\ 8 & -1 \end{bmatrix}$$
$$= \begin{bmatrix} 1(1)+0(6)+1(8) & 1(3)+0(2)+1(-1) \\ 2(1)+4(6)+1(8) & 2(3)+4(2)+1(-1) \\ 3(1)+6(6)+1(8) & 3(3)+6(2)+1(-1) \end{bmatrix}$$
$$= \begin{bmatrix} 9 & 2 \\ 34 & 13 \\ 47 & 20 \end{bmatrix}$$

b. Enter the matrices into a graphing utility using $A = \begin{bmatrix} 1 & 0 & 1 \\ 2 & 4 & 1 \\ 3 & 6 & 1 \end{bmatrix}$ and $B = \begin{bmatrix} 1 & 3 \\ 6 & 2 \\ 8 & -1 \end{bmatrix}$. The result is shown below:

```
[A][B]
      [[9  2 ]
       [34 13]
       [47 20]]
```

28. a.
$$\begin{bmatrix} 4 & -2 & 3 \\ 0 & 1 & 2 \\ -1 & 0 & 1 \end{bmatrix}\begin{bmatrix} 2 & 6 \\ 1 & -1 \\ 0 & 2 \end{bmatrix}$$
$$= \begin{bmatrix} 4(2)+(-2)(1)+3(0) & 4(6)+(-2)(-1)+3(2) \\ 0(2)+1(1)+2(0) & 0(6)+1(-1)+2(2) \\ -1(2)+0(1)+1(0) & -1(6)+0(-1)+1(2) \end{bmatrix}$$
$$= \begin{bmatrix} 6 & 32 \\ 1 & 3 \\ -2 & -4 \end{bmatrix}$$

b. Enter the matrices into a graphing utility using $A = \begin{bmatrix} 4 & -2 & 3 \\ 0 & 1 & 2 \\ -1 & 0 & 1 \end{bmatrix}$ and $B = \begin{bmatrix} 2 & 6 \\ 1 & -1 \\ 0 & 2 \end{bmatrix}$.

The result is shown below:

```
[A][B]
      [[6  28]
       [1  5 ]
       [-2 -4]]
```

29. $A=\begin{bmatrix}2 & 1\\1 & 1\end{bmatrix}$

Augment the matrix with the identity and use row operations to find the inverse:

$$\left[\begin{array}{cc|cc}2 & 1 & 1 & 0\\1 & 1 & 0 & 1\end{array}\right]$$

$$\to\left[\begin{array}{cc|cc}1 & 1 & 0 & 1\\2 & 1 & 1 & 0\end{array}\right]\begin{pmatrix}\text{Interchange}\\ r_1 \text{ and } r_2\end{pmatrix}$$

$$\to\left[\begin{array}{cc|cc}1 & 1 & 0 & 1\\0 & -1 & 1 & -2\end{array}\right]\ (R_2=-2r_1+r_2)$$

$$\to\left[\begin{array}{cc|cc}1 & 1 & 0 & 1\\0 & 1 & -1 & 2\end{array}\right]\ (R_2=-r_2)$$

$$\to\left[\begin{array}{cc|cc}1 & 0 & 1 & -1\\0 & 1 & -1 & 2\end{array}\right]\ (R_1=-r_2+r_1)$$

Thus, $A^{-1}=\begin{bmatrix}1 & -1\\-1 & 2\end{bmatrix}$.

```
[A]-1
      [[1  -1]
       [-1 2 ]]
```

30. $A=\begin{bmatrix}3 & -1\\-2 & 1\end{bmatrix}$

Augment the matrix with the identity and use row operations to find the inverse:

$$\left[\begin{array}{cc|cc}3 & -1 & 1 & 0\\-2 & 1 & 0 & 1\end{array}\right]$$

$$\to\left[\begin{array}{cc|cc}1 & 0 & 1 & 1\\-2 & 1 & 0 & 1\end{array}\right]\ (R_1=r_2+r_1)$$

$$\to\left[\begin{array}{cc|cc}1 & 0 & 1 & 1\\0 & 1 & 2 & 3\end{array}\right]\ (R_2=2r_1+r_2)$$

Thus, $A^{-1}=\begin{bmatrix}1 & 1\\2 & 3\end{bmatrix}$.

```
[A]-1
      [[1 1]
       [2 3]]
```

31. $A=\begin{bmatrix}6 & 5\\2 & 2\end{bmatrix}$

Augment the matrix with the identity and use row operations to find the inverse:

$$\left[\begin{array}{cc|cc}6 & 5 & 1 & 0\\2 & 2 & 0 & 1\end{array}\right]$$

$$\to\left[\begin{array}{cc|cc}2 & 2 & 0 & 1\\6 & 5 & 1 & 0\end{array}\right]\begin{pmatrix}\text{Interchange}\\ r_1 \text{ and } r_2\end{pmatrix}$$

$$\to\left[\begin{array}{cc|cc}2 & 2 & 0 & 1\\0 & -1 & 1 & -3\end{array}\right]\ (R_2=-3r_1+r_2)$$

$$\to\left[\begin{array}{cc|cc}1 & 1 & 0 & \frac{1}{2}\\0 & 1 & -1 & 3\end{array}\right]\begin{pmatrix}R_1=\frac{1}{2}r_1\\ R_2=-r_2\end{pmatrix}$$

$$\to\left[\begin{array}{cc|cc}1 & 0 & 1 & -\frac{5}{2}\\0 & 1 & -1 & 3\end{array}\right]\ (R_1=-r_2+r_1)$$

Thus, $A^{-1}=\begin{bmatrix}1 & -\frac{5}{2}\\-1 & 3\end{bmatrix}$.

```
[A]-1▸Frac
   [[1   -5/2]
    [-1 3    ]]
```

32. $A=\begin{bmatrix}-4 & 1\\6 & -2\end{bmatrix}$

Augment the matrix with the identity and use row operations to find the inverse:

$$\left[\begin{array}{cc|cc}-4 & 1 & 1 & 0\\6 & -2 & 0 & 1\end{array}\right]$$

$$\to\left[\begin{array}{cc|cc}1 & -\frac{1}{4} & -\frac{1}{4} & 0\\6 & -2 & 0 & 1\end{array}\right]\ (R_1=-\tfrac{1}{4}r_1)$$

$$\to\left[\begin{array}{cc|cc}1 & -\frac{1}{4} & -\frac{1}{4} & 0\\0 & -\frac{1}{2} & \frac{3}{2} & 1\end{array}\right]\ (R_2=-6r_1+r_2)$$

$$\to\left[\begin{array}{cc|cc}1 & -\frac{1}{4} & -\frac{1}{4} & 0\\0 & 1 & -3 & -2\end{array}\right]\ (R_2=-2r_2)$$

$$\to\left[\begin{array}{cc|cc}1 & 0 & -1 & -\frac{1}{2}\\0 & 1 & -3 & -2\end{array}\right]\ (R_1=\tfrac{1}{4}r_2+r)$$

Thus, $A^{-1}=\begin{bmatrix}-1 & -\frac{1}{2}\\-3 & -2\end{bmatrix}$.

```
[A]-1▸Frac
   [[-1  -1/2]
    [-3 -2   ]]
```

33. $A=\begin{bmatrix} 2 & 1 \\ a & a \end{bmatrix}$ where $a \neq 0$.

Augment the matrix with the identity and use row operations to find the inverse:

$$\left[\begin{array}{cc|cc} 2 & 1 & 1 & 0 \\ a & a & 0 & 1 \end{array}\right]$$

$$\to \left[\begin{array}{cc|cc} 1 & \frac{1}{2} & \frac{1}{2} & 0 \\ a & a & 0 & 1 \end{array}\right] \quad \left(R_1 = \tfrac{1}{2} r_1\right)$$

$$\to \left[\begin{array}{cc|cc} 1 & \frac{1}{2} & \frac{1}{2} & 0 \\ 0 & \frac{1}{2}a & -\frac{1}{2}a & 1 \end{array}\right] \quad \left(R_2 = -a r_1 + r_2\right)$$

$$\to \left[\begin{array}{cc|cc} 1 & \frac{1}{2} & \frac{1}{2} & 0 \\ 0 & 1 & -1 & \frac{2}{a} \end{array}\right] \quad \left(R_2 = \tfrac{2}{a} r_2\right)$$

$$\to \left[\begin{array}{cc|cc} 1 & 0 & 1 & -\frac{1}{a} \\ 0 & 1 & -1 & \frac{2}{a} \end{array}\right] \quad \left(R_1 = -\tfrac{1}{2} r_2 + r_1\right)$$

Thus, $A^{-1} = \begin{bmatrix} 1 & -\frac{1}{a} \\ -1 & \frac{2}{a} \end{bmatrix}$.

34. $A=\begin{bmatrix} b & 3 \\ b & 2 \end{bmatrix}$ where $b \neq 0$.

Augment the matrix with the identity and use row operations to find the inverse:

$$\left[\begin{array}{cc|cc} b & 3 & 1 & 0 \\ b & 2 & 0 & 1 \end{array}\right]$$

$$\to \left[\begin{array}{cc|cc} b & 3 & 1 & 0 \\ 0 & -1 & -1 & 1 \end{array}\right] \quad \left(R_2 = -r_1 + r_2\right)$$

$$\to \left[\begin{array}{cc|cc} 1 & \frac{3}{b} & \frac{1}{b} & 0 \\ 0 & -1 & -1 & 1 \end{array}\right] \quad \left(R_1 = \tfrac{1}{b} r_1\right)$$

$$\to \left[\begin{array}{cc|cc} 1 & \frac{3}{b} & \frac{1}{b} & 0 \\ 0 & 1 & 1 & -1 \end{array}\right] \quad \left(R_2 = -r_2\right)$$

$$\to \left[\begin{array}{cc|cc} 1 & 0 & -\frac{2}{b} & \frac{3}{b} \\ 0 & 1 & 1 & -1 \end{array}\right] \quad \left(R_1 = -\tfrac{3}{b} r_2 + r_1\right)$$

Thus, $A^{-1} = \begin{bmatrix} -\frac{2}{b} & \frac{3}{b} \\ 1 & -1 \end{bmatrix}$.

35. $A=\begin{bmatrix} 1 & -1 & 1 \\ 0 & -2 & 1 \\ -2 & -3 & 0 \end{bmatrix}$

Augment the matrix with the identity and use row operations to find the inverse:

$$\left[\begin{array}{ccc|ccc} 1 & -1 & 1 & 1 & 0 & 0 \\ 0 & -2 & 1 & 0 & 1 & 0 \\ -2 & -3 & 0 & 0 & 0 & 1 \end{array}\right]$$

$$\to \left[\begin{array}{ccc|ccc} 1 & -1 & 1 & 1 & 0 & 0 \\ 0 & -2 & 1 & 0 & 1 & 0 \\ 0 & -5 & 2 & 2 & 0 & 1 \end{array}\right] \quad \left(R_3 = 2r_1 + r_3\right)$$

$$\to \left[\begin{array}{ccc|ccc} 1 & -1 & 1 & 1 & 0 & 0 \\ 0 & 1 & -\frac{1}{2} & 0 & -\frac{1}{2} & 0 \\ 0 & -5 & 2 & 2 & 0 & 1 \end{array}\right] \quad \left(R_2 = -\tfrac{1}{2} r_2\right)$$

$$\to \left[\begin{array}{ccc|ccc} 1 & 0 & \frac{1}{2} & 1 & -\frac{1}{2} & 0 \\ 0 & 1 & -\frac{1}{2} & 0 & -\frac{1}{2} & 0 \\ 0 & 0 & -\frac{1}{2} & 2 & -\frac{5}{2} & 1 \end{array}\right] \quad \begin{pmatrix} R_1 = r_2 + r_1 \\ R_3 = 5r_2 + r_3 \end{pmatrix}$$

$$\to \left[\begin{array}{ccc|ccc} 1 & 0 & \frac{1}{2} & 1 & -\frac{1}{2} & 0 \\ 0 & 1 & -\frac{1}{2} & 0 & -\frac{1}{2} & 0 \\ 0 & 0 & 1 & -4 & 5 & -2 \end{array}\right] \quad \left(R_3 = -2r_3\right)$$

$$\to \left[\begin{array}{ccc|ccc} 1 & 0 & 0 & 3 & -3 & 1 \\ 0 & 1 & 0 & -2 & 2 & -1 \\ 0 & 0 & 1 & -4 & 5 & -2 \end{array}\right] \quad \begin{pmatrix} R_1 = -\frac{1}{2} r_3 + r_1 \\ R_2 = \frac{1}{2} r_3 + r_2 \end{pmatrix}$$

Thus, $A^{-1} = \begin{bmatrix} 3 & -3 & 1 \\ -2 & 2 & -1 \\ -4 & 5 & -2 \end{bmatrix}$.

```
[A]-1
 [[3  -3 1 ]
  [-2 2  -1]
  [-4 5  -2]]
```

36. $A=\begin{bmatrix} 1 & 0 & 2 \\ -1 & 2 & 3 \\ 1 & -1 & 0 \end{bmatrix}$

Augment the matrix with the identity and use row operations to find the inverse:

$$\left[\begin{array}{ccc|ccc} 1 & 0 & 2 & 1 & 0 & 0 \\ -1 & 2 & 3 & 0 & 1 & 0 \\ 1 & -1 & 0 & 0 & 0 & 1 \end{array}\right]$$

$$\to \left[\begin{array}{ccc|ccc} 1 & 0 & 2 & 1 & 0 & 0 \\ 0 & 2 & 5 & 1 & 1 & 0 \\ 0 & -1 & -2 & -1 & 0 & 1 \end{array}\right] \quad \begin{pmatrix} R_2 = r_1 + r_2 \\ R_3 = -r_1 + r_3 \end{pmatrix}$$

$$\to \left[\begin{array}{ccc|ccc} 1 & 0 & 2 & 1 & 0 & 0 \\ 0 & 1 & \frac{5}{2} & \frac{1}{2} & \frac{1}{2} & 0 \\ 0 & -1 & -2 & -1 & 0 & 1 \end{array}\right] \quad \left(R_2 = \tfrac{1}{2} r_2\right)$$

$$\rightarrow\left[\begin{array}{ccc|ccc}1 & 0 & 2 & 1 & 0 & 0\\ 0 & 1 & \frac{5}{2} & \frac{1}{2} & \frac{1}{2} & 0\\ 0 & 0 & \frac{1}{2} & -\frac{1}{2} & \frac{1}{2} & 1\end{array}\right] \quad (R_3 = r_2 + r_3)$$

$$\rightarrow\left[\begin{array}{ccc|ccc}1 & 0 & 2 & 1 & 0 & 0\\ 0 & 1 & \frac{5}{2} & \frac{1}{2} & \frac{1}{2} & 0\\ 0 & 0 & 1 & -1 & 1 & 2\end{array}\right] \quad (R_3 = 2r_3)$$

$$\rightarrow\left[\begin{array}{ccc|ccc}1 & 0 & 0 & 3 & -2 & -4\\ 0 & 1 & 0 & 3 & -2 & -5\\ 0 & 0 & 1 & -1 & 1 & 2\end{array}\right] \quad \begin{pmatrix}R_1 = -2r_3 + r_1\\ R_2 = -\frac{5}{2}r_3 + r_2\end{pmatrix}$$

Thus, $A^{-1} = \begin{bmatrix}3 & -2 & -4\\ 3 & -2 & -5\\ -1 & 1 & 2\end{bmatrix}$.

```
[A]-1
  [[3  -2 -4]
   [3  -2 -5]
   [-1 1  2 ]]
```

37. $A = \begin{bmatrix}1 & 1 & 1\\ 3 & 2 & -1\\ 3 & 1 & 2\end{bmatrix}$

Augment the matrix with the identity and use row operations to find the inverse:

$$\left[\begin{array}{ccc|ccc}1 & 1 & 1 & 1 & 0 & 0\\ 3 & 2 & -1 & 0 & 1 & 0\\ 3 & 1 & 2 & 0 & 0 & 1\end{array}\right]$$

$$\rightarrow\left[\begin{array}{ccc|ccc}1 & 1 & 1 & 1 & 0 & 0\\ 0 & -1 & -4 & -3 & 1 & 0\\ 0 & -2 & -1 & -3 & 0 & 1\end{array}\right] \quad \begin{pmatrix}R_2 = -3r_1 + r_2\\ R_3 = -3r_1 + r_3\end{pmatrix}$$

$$\rightarrow\left[\begin{array}{ccc|ccc}1 & 1 & 1 & 1 & 0 & 0\\ 0 & 1 & 4 & 3 & -1 & 0\\ 0 & -2 & -1 & -3 & 0 & 1\end{array}\right] \quad (R_2 = -r_2)$$

$$\rightarrow\left[\begin{array}{ccc|ccc}1 & 0 & -3 & -2 & 1 & 0\\ 0 & 1 & 4 & 3 & -1 & 0\\ 0 & 0 & 1 & \frac{3}{7} & -\frac{2}{7} & \frac{1}{7}\end{array}\right] \quad (R_3 = \tfrac{1}{7}r_3)$$

$$\rightarrow\left[\begin{array}{ccc|ccc}1 & 0 & 0 & -\frac{5}{7} & \frac{1}{7} & \frac{3}{7}\\ 0 & 1 & 0 & \frac{9}{7} & \frac{1}{7} & -\frac{4}{7}\\ 0 & 0 & 1 & \frac{3}{7} & -\frac{2}{7} & \frac{1}{7}\end{array}\right] \quad \begin{pmatrix}R_1 = 3r_3 + r_1\\ R_2 = -4r_3 + r_2\end{pmatrix}$$

Thus, $A^{-1} = \begin{bmatrix}-\frac{5}{7} & \frac{1}{7} & \frac{3}{7}\\ \frac{9}{7} & \frac{1}{7} & -\frac{4}{7}\\ \frac{3}{7} & -\frac{2}{7} & \frac{1}{7}\end{bmatrix}$.

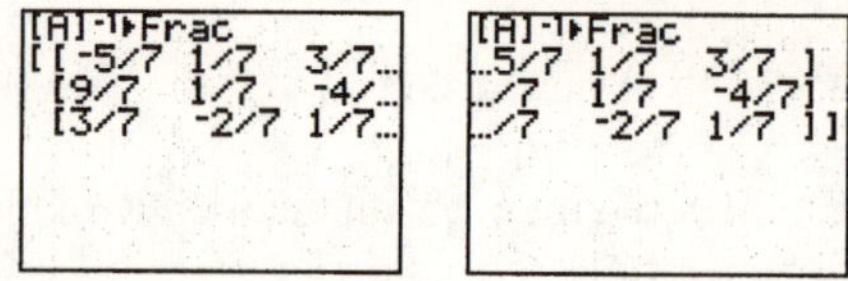

38. $A = \begin{bmatrix}3 & 3 & 1\\ 1 & 2 & 1\\ 2 & -1 & 1\end{bmatrix}$

Augment the matrix with the identity and use row operations to find the inverse:

$$\left[\begin{array}{ccc|ccc}3 & 3 & 1 & 1 & 0 & 0\\ 1 & 2 & 1 & 0 & 1 & 0\\ 2 & -1 & 1 & 0 & 0 & 1\end{array}\right]$$

$$\rightarrow\left[\begin{array}{ccc|ccc}1 & 2 & 1 & 0 & 1 & 0\\ 3 & 3 & 1 & 1 & 0 & 0\\ 2 & -1 & 1 & 0 & 0 & 1\end{array}\right] \quad \begin{pmatrix}\text{Interchange}\\ r_1 \text{ and } r_2\end{pmatrix}$$

$$\rightarrow\left[\begin{array}{ccc|ccc}1 & 2 & 1 & 0 & 1 & 0\\ 0 & -3 & -2 & 1 & -3 & 0\\ 0 & -5 & -1 & 0 & -2 & 1\end{array}\right] \quad \begin{pmatrix}R_2 = -3r_1 + r_2\\ R_3 = -2r_1 + r_3\end{pmatrix}$$

$$\rightarrow\left[\begin{array}{ccc|ccc}1 & 2 & 1 & 0 & 1 & 0\\ 0 & 1 & \frac{2}{3} & -\frac{1}{3} & 1 & 0\\ 0 & -5 & -1 & 0 & -2 & 1\end{array}\right] \quad (R_2 = -\tfrac{1}{3}r_2)$$

$$\rightarrow\left[\begin{array}{ccc|ccc}1 & 0 & -\frac{1}{3} & \frac{2}{3} & -1 & 0\\ 0 & 1 & \frac{2}{3} & -\frac{1}{3} & 1 & 0\\ 0 & 0 & \frac{7}{3} & -\frac{5}{3} & 3 & 1\end{array}\right] \quad \begin{pmatrix}R_1 = -2r_2 + r_1\\ R_3 = 5r_2 + r_3\end{pmatrix}$$

$$\rightarrow\left[\begin{array}{ccc|ccc}1 & 0 & -\frac{1}{3} & \frac{2}{3} & -1 & 0\\ 0 & 1 & \frac{2}{3} & -\frac{1}{3} & 1 & 0\\ 0 & 0 & 1 & -\frac{5}{7} & \frac{9}{7} & \frac{3}{7}\end{array}\right] \quad (R_3 = \tfrac{3}{7}r_3)$$

$$\rightarrow\left[\begin{array}{ccc|ccc}1 & 0 & 0 & \frac{3}{7} & -\frac{4}{7} & \frac{1}{7}\\ 0 & 1 & 0 & \frac{1}{7} & \frac{1}{7} & -\frac{2}{7}\\ 0 & 0 & 1 & -\frac{5}{7} & \frac{9}{7} & \frac{3}{7}\end{array}\right] \quad \begin{pmatrix}R_1 = \frac{1}{3}r_3 + r_1\\ R_2 = -\frac{2}{3}r_3 + r_2\end{pmatrix}$$

Thus, $A^{-1} = \begin{bmatrix}\frac{3}{7} & -\frac{4}{7} & \frac{1}{7}\\ \frac{1}{7} & \frac{1}{7} & -\frac{2}{7}\\ -\frac{5}{7} & \frac{9}{7} & \frac{3}{7}\end{bmatrix}$.

```
[A]-1▸Frac
[[3/7  -4/7 1/7...
 [1/7  1/7  -2/...
 [-5/7 9/7  3/7...
```

```
[A]-1▸Frac
...7   -4/7 1/7 ]
...7   1/7  -2/7]
...5/7 9/7  3/7 ]]
```

39. $\begin{cases} 2x + y = 8 \\ x + y = 5 \end{cases}$

Rewrite the system of equations in matrix form:

$$A = \begin{bmatrix} 2 & 1 \\ 1 & 1 \end{bmatrix}, \quad X = \begin{bmatrix} x \\ y \end{bmatrix}, \quad B = \begin{bmatrix} 8 \\ 5 \end{bmatrix}$$

Find the inverse of A and solve $X = A^{-1}B$:

From Problem 29, $A^{-1} = \begin{bmatrix} 1 & -1 \\ -1 & 2 \end{bmatrix}$, so

$$X = A^{-1}B = \begin{bmatrix} 1 & -1 \\ -1 & 2 \end{bmatrix}\begin{bmatrix} 8 \\ 5 \end{bmatrix} = \begin{bmatrix} 3 \\ 2 \end{bmatrix}.$$

The solution is $x = 3,\ y = 2$.

40. $\begin{cases} 3x - y = 8 \\ -2x + y = 4 \end{cases}$

Rewrite the system of equations in matrix form:

$$A = \begin{bmatrix} 3 & -1 \\ -2 & 1 \end{bmatrix}, \quad X = \begin{bmatrix} x \\ y \end{bmatrix}, \quad B = \begin{bmatrix} 8 \\ 4 \end{bmatrix}$$

Find the inverse of A and solve $X = A^{-1}B$:

From Problem 30, $A^{-1} = \begin{bmatrix} 1 & 1 \\ 2 & 3 \end{bmatrix}$, so

$$X = A^{-1}B = \begin{bmatrix} 1 & 1 \\ 2 & 3 \end{bmatrix}\begin{bmatrix} 8 \\ 4 \end{bmatrix} = \begin{bmatrix} 12 \\ 28 \end{bmatrix}.$$

The solution is $x = 12,\ y = 28$.

41. $\begin{cases} 2x + y = 0 \\ x + y = 5 \end{cases}$

Rewrite the system of equations in matrix form:

$$A = \begin{bmatrix} 2 & 1 \\ 1 & 1 \end{bmatrix}, \quad X = \begin{bmatrix} x \\ y \end{bmatrix}, \quad B = \begin{bmatrix} 0 \\ 5 \end{bmatrix}$$

Find the inverse of A and solve $X = A^{-1}B$:

From Problem 29, $A^{-1} = \begin{bmatrix} 1 & -1 \\ -1 & 2 \end{bmatrix}$, so

$$X = A^{-1}B = \begin{bmatrix} 1 & -1 \\ -1 & 2 \end{bmatrix}\begin{bmatrix} 0 \\ 5 \end{bmatrix} = \begin{bmatrix} -5 \\ 10 \end{bmatrix}.$$

The solution is $x = -5,\ y = 10$.

42. $\begin{cases} 3x - y = 4 \\ -2x + y = 5 \end{cases}$

Rewrite the system of equations in matrix form:

$$A = \begin{bmatrix} 3 & -1 \\ -2 & 1 \end{bmatrix}, \quad X = \begin{bmatrix} x \\ y \end{bmatrix}, \quad B = \begin{bmatrix} 4 \\ 5 \end{bmatrix}$$

Find the inverse of A and solve $X = A^{-1}B$:

From Problem 30, $A^{-1} = \begin{bmatrix} 1 & 1 \\ 2 & 3 \end{bmatrix}$, so

$$X = A^{-1}B = \begin{bmatrix} 1 & 1 \\ 2 & 3 \end{bmatrix}\begin{bmatrix} 4 \\ 5 \end{bmatrix} = \begin{bmatrix} 9 \\ 23 \end{bmatrix}.$$

The solution is $x = 9,\ y = 23$.

43. $\begin{cases} 6x + 5y = 7 \\ 2x + 2y = 2 \end{cases}$

Rewrite the system of equations in matrix form:

$$A = \begin{bmatrix} 6 & 5 \\ 2 & 2 \end{bmatrix}, \quad X = \begin{bmatrix} x \\ y \end{bmatrix}, \quad B = \begin{bmatrix} 7 \\ 2 \end{bmatrix}$$

Find the inverse of A and solve $X = A^{-1}B$:

From Problem 31, $A^{-1} = \begin{bmatrix} 1 & -\frac{5}{2} \\ -1 & 3 \end{bmatrix}$, so

$$X = A^{-1}B = \begin{bmatrix} 1 & -\frac{5}{2} \\ -1 & 3 \end{bmatrix}\begin{bmatrix} 7 \\ 2 \end{bmatrix} = \begin{bmatrix} 2 \\ -1 \end{bmatrix}.$$

The solution is $x = 2,\ y = -1$.

44. $\begin{cases} -4x + y = 0 \\ 6x - 2y = 14 \end{cases}$

Rewrite the system of equations in matrix form:

$$A = \begin{bmatrix} -4 & 1 \\ 6 & -2 \end{bmatrix}, \quad X = \begin{bmatrix} x \\ y \end{bmatrix}, \quad B = \begin{bmatrix} 0 \\ 14 \end{bmatrix}$$

Find the inverse of A and solve $X = A^{-1}B$:

From Problem 32, $A^{-1} = \begin{bmatrix} -1 & -\frac{1}{2} \\ -3 & -2 \end{bmatrix}$, so

$$X = A^{-1}B = \begin{bmatrix} -1 & -\frac{1}{2} \\ -3 & -2 \end{bmatrix}\begin{bmatrix} 0 \\ 14 \end{bmatrix} = \begin{bmatrix} -7 \\ -28 \end{bmatrix}.$$

The solution is $x = -7,\ y = -28$.

45. $\begin{cases} 6x + 5y = 13 \\ 2x + 2y = 5 \end{cases}$

Rewrite the system of equations in matrix form:

$$A = \begin{bmatrix} 6 & 5 \\ 2 & 2 \end{bmatrix}, \quad X = \begin{bmatrix} x \\ y \end{bmatrix}, \quad B = \begin{bmatrix} 13 \\ 5 \end{bmatrix}$$

Find the inverse of A and solve $X = A^{-1}B$:

From Problem 31, $A^{-1} = \begin{bmatrix} 1 & -\frac{5}{2} \\ -1 & 3 \end{bmatrix}$, so

$$X = A^{-1}B = \begin{bmatrix} 1 & -\frac{5}{2} \\ -1 & 3 \end{bmatrix}\begin{bmatrix} 13 \\ 5 \end{bmatrix} = \begin{bmatrix} \frac{1}{2} \\ 2 \end{bmatrix}.$$

The solution is $x = \frac{1}{2},\ y = 2$.

46. $\begin{cases} -4x + y = 5 \\ 6x - 2y = -9 \end{cases}$

Rewrite the system of equations in matrix form:

$$A = \begin{bmatrix} -4 & 1 \\ 6 & -2 \end{bmatrix}, \quad X = \begin{bmatrix} x \\ y \end{bmatrix}, \quad B = \begin{bmatrix} 5 \\ -9 \end{bmatrix}$$

Find the inverse of A and solve $X = A^{-1}B$:

From Problem 32, $A^{-1} = \begin{bmatrix} -1 & -\frac{1}{2} \\ -3 & -2 \end{bmatrix}$, so

$$X = A^{-1}B = \begin{bmatrix} -1 & -\frac{1}{2} \\ -3 & -2 \end{bmatrix} \begin{bmatrix} 5 \\ -9 \end{bmatrix} = \begin{bmatrix} -\frac{1}{2} \\ 3 \end{bmatrix}.$$

The solution is $x = -\frac{1}{2}$, $y = 3$.

47. $\begin{cases} 2x + y = -3 \\ ax + ay = -a \end{cases} \quad a \neq 0$

Rewrite the system of equations in matrix form:

$$A = \begin{bmatrix} 2 & 1 \\ a & a \end{bmatrix}, \quad X = \begin{bmatrix} x \\ y \end{bmatrix}, \quad B = \begin{bmatrix} -3 \\ -a \end{bmatrix}$$

Find the inverse of A and solve $X = A^{-1}B$:

From Problem 33, $A^{-1} = \begin{bmatrix} 1 & -\frac{1}{a} \\ -1 & \frac{2}{a} \end{bmatrix}$, so

$$X = A^{-1}B = \begin{bmatrix} 1 & -\frac{1}{a} \\ -1 & \frac{2}{a} \end{bmatrix} \begin{bmatrix} -3 \\ -a \end{bmatrix} = \begin{bmatrix} -2 \\ 1 \end{bmatrix}.$$

The solution is $x = -2$, $y = 1$.

48. $\begin{cases} bx + 3y = 2b + 3 \\ bx + 2y = 2b + 2 \end{cases} \quad b \neq 0$

Rewrite the system of equations in matrix form:

$$A = \begin{bmatrix} b & 3 \\ b & 2 \end{bmatrix}, \quad X = \begin{bmatrix} x \\ y \end{bmatrix}, \quad B = \begin{bmatrix} 2b+3 \\ 2b+2 \end{bmatrix}$$

Find the inverse of A and solve $X = A^{-1}B$:

From Problem 34, $A^{-1} = \begin{bmatrix} -\frac{2}{b} & \frac{3}{b} \\ 1 & -1 \end{bmatrix}$, so

$$X = A^{-1}B = \begin{bmatrix} -\frac{2}{b} & \frac{3}{b} \\ 1 & -1 \end{bmatrix} \begin{bmatrix} 2b+3 \\ 2b+2 \end{bmatrix} = \begin{bmatrix} 2 \\ 1 \end{bmatrix}.$$

The solution is $x = 2$, $y = 1$.

49. $\begin{cases} 2x + y = \frac{7}{a} \\ ax + ay = 5 \end{cases} \quad a \neq 0$

Rewrite the system of equations in matrix form:

$$A = \begin{bmatrix} 2 & 1 \\ a & a \end{bmatrix}, \quad X = \begin{bmatrix} x \\ y \end{bmatrix}, \quad B = \begin{bmatrix} \frac{7}{a} \\ 5 \end{bmatrix}$$

Find the inverse of A and solve $X = A^{-1}B$:

From Problem 33, $A^{-1} = \begin{bmatrix} 1 & -\frac{1}{a} \\ -1 & \frac{2}{a} \end{bmatrix}$, so

$$X = A^{-1}B = \begin{bmatrix} 1 & -\frac{1}{a} \\ -1 & \frac{2}{a} \end{bmatrix} \begin{bmatrix} \frac{7}{a} \\ 5 \end{bmatrix} = \begin{bmatrix} \frac{2}{a} \\ \frac{3}{a} \end{bmatrix}.$$

The solution is $x = \frac{2}{a}$, $y = \frac{3}{a}$.

50. $\begin{cases} bx + 3y = 14 \\ bx + 2y = 10 \end{cases} \quad b \neq 0$

Rewrite the system of equations in matrix form:

$$A = \begin{bmatrix} b & 3 \\ b & 2 \end{bmatrix}, \quad X = \begin{bmatrix} x \\ y \end{bmatrix}, \quad B = \begin{bmatrix} 14 \\ 10 \end{bmatrix}$$

Find the inverse of A and solve $X = A^{-1}B$:

From Problem 34, $A^{-1} = \begin{bmatrix} -\frac{2}{b} & \frac{3}{b} \\ 1 & -1 \end{bmatrix}$, so

$$X = A^{-1}B = \begin{bmatrix} -\frac{2}{b} & \frac{3}{b} \\ 1 & -1 \end{bmatrix} \begin{bmatrix} 14 \\ 10 \end{bmatrix} = \begin{bmatrix} \frac{2}{b} \\ 4 \end{bmatrix}.$$

The solution is $x = \frac{2}{b}$, $y = 4$.

51. $\begin{cases} x - y + z = 0 \\ -2y + z = -1 \\ -2x - 3y = -5 \end{cases}$

Rewrite the system of equations in matrix form:

$$A = \begin{bmatrix} 1 & -1 & 1 \\ 0 & -2 & 1 \\ -2 & -3 & 0 \end{bmatrix}, \quad X = \begin{bmatrix} x \\ y \\ z \end{bmatrix}, \quad B = \begin{bmatrix} 0 \\ -1 \\ -5 \end{bmatrix}$$

Find the inverse of A and solve $X = A^{-1}B$:

From Problem 35, $A^{-1} = \begin{bmatrix} 3 & -3 & 1 \\ -2 & 2 & -1 \\ -4 & 5 & -2 \end{bmatrix}$, so

$$X = A^{-1}B = \begin{bmatrix} 3 & -3 & 1 \\ -2 & 2 & -1 \\ -4 & 5 & -2 \end{bmatrix} \begin{bmatrix} 0 \\ -1 \\ -5 \end{bmatrix} = \begin{bmatrix} -2 \\ 3 \\ 5 \end{bmatrix}.$$

The solution is $x = -2$, $y = 3$, $z = 5$.

52. $\begin{cases} x \quad +2z = 6 \\ -x+2y+3z=-5 \\ x-y \quad = 6 \end{cases}$

Rewrite the system of equations in matrix form:

$$A=\begin{bmatrix}1&0&2\\-1&2&3\\1&-1&0\end{bmatrix},\quad X=\begin{bmatrix}x\\y\\z\end{bmatrix},\quad B=\begin{bmatrix}6\\-5\\6\end{bmatrix}$$

Find the inverse of A and solve $X = A^{-1}B$:

From Problem 36, $A^{-1}=\begin{bmatrix}3&-2&-4\\3&-2&-5\\-1&1&2\end{bmatrix}$, so

$$X=A^{-1}B=\begin{bmatrix}3&-2&-4\\3&-2&-5\\-1&1&2\end{bmatrix}\begin{bmatrix}6\\-5\\6\end{bmatrix}=\begin{bmatrix}4\\-2\\1\end{bmatrix}.$$

The solution is $x=4,\ y=-2,\ z=1$.

53. $\begin{cases} x-y+z=2 \\ -2y+z=2 \\ -2x-3y \quad =\frac{1}{2} \end{cases}$

Rewrite the system of equations in matrix form:

$$A=\begin{bmatrix}1&-1&1\\0&-2&1\\-2&-3&0\end{bmatrix},\quad X=\begin{bmatrix}x\\y\\z\end{bmatrix},\quad B=\begin{bmatrix}2\\2\\\frac{1}{2}\end{bmatrix}$$

Find the inverse of A and solve $X = A^{-1}B$:

From Problem 35, $A^{-1}=\begin{bmatrix}3&-3&1\\-2&2&-1\\-4&5&-2\end{bmatrix}$, so

$$X=A^{-1}B=\begin{bmatrix}3&-3&1\\-2&2&-1\\-4&5&-2\end{bmatrix}\begin{bmatrix}2\\2\\\frac{1}{2}\end{bmatrix}=\begin{bmatrix}\frac{1}{2}\\-\frac{1}{2}\\1\end{bmatrix}$$

The solution is $x=\frac{1}{2},\ y=-\frac{1}{2},\ z=1$.

54. $\begin{cases} x \quad +2z = 2 \\ -x+2y+3z=-\frac{3}{2} \\ x-y \quad = 2 \end{cases}$

Rewrite the system of equations in matrix form:

$$A=\begin{bmatrix}1&0&2\\-1&2&3\\1&-1&0\end{bmatrix},\quad X=\begin{bmatrix}x\\y\\z\end{bmatrix},\quad B=\begin{bmatrix}2\\-\frac{3}{2}\\2\end{bmatrix}$$

Find the inverse of A and solve $X = A^{-1}B$:

From Problem 36, $A^{-1}=\begin{bmatrix}3&-2&-4\\3&-2&-5\\-1&1&2\end{bmatrix}$, so

$$X=A^{-1}B=\begin{bmatrix}3&-2&-4\\3&-2&-5\\-1&1&2\end{bmatrix}\begin{bmatrix}2\\-\frac{3}{2}\\2\end{bmatrix}=\begin{bmatrix}1\\-1\\\frac{1}{2}\end{bmatrix}.$$

The solution is $x=1,\ y=-1,\ z=\frac{1}{2}$.

55. $\begin{cases} x+y+z=9 \\ 3x+2y-z=8 \\ 3x+y+2z=1 \end{cases}$

Rewrite the system of equations in matrix form:

$$A=\begin{bmatrix}1&1&1\\3&2&-1\\3&1&2\end{bmatrix},\quad X=\begin{bmatrix}x\\y\\z\end{bmatrix},\quad B=\begin{bmatrix}9\\8\\1\end{bmatrix}$$

Find the inverse of A and solve $X = A^{-1}B$:

From Problem 37, $A^{-1}=\begin{bmatrix}-\frac{5}{7}&\frac{1}{7}&\frac{3}{7}\\\frac{9}{7}&\frac{1}{7}&-\frac{4}{7}\\\frac{3}{7}&-\frac{2}{7}&\frac{1}{7}\end{bmatrix}$, so

$$X=A^{-1}B=\begin{bmatrix}-\frac{5}{7}&\frac{1}{7}&\frac{3}{7}\\\frac{9}{7}&\frac{1}{7}&-\frac{4}{7}\\\frac{3}{7}&-\frac{2}{7}&\frac{1}{7}\end{bmatrix}\begin{bmatrix}9\\8\\1\end{bmatrix}=\begin{bmatrix}-\frac{34}{7}\\\frac{85}{7}\\\frac{12}{7}\end{bmatrix}.$$

The solution is $x=-\frac{34}{7},\ y=\frac{85}{7},\ z=\frac{12}{7}$.

56. $\begin{cases} 3x+3y+z=8 \\ x+2y+z=5 \\ 2x-y+z=4 \end{cases}$

Rewrite the system of equations in matrix form:

$$A=\begin{bmatrix}3&3&1\\1&2&1\\2&-1&1\end{bmatrix},\quad X=\begin{bmatrix}x\\y\\z\end{bmatrix},\quad B=\begin{bmatrix}8\\5\\4\end{bmatrix}$$

Find the inverse of A and solve $X = A^{-1}B$:

From Problem 38, $A^{-1}=\begin{bmatrix}\frac{3}{7}&-\frac{4}{7}&\frac{1}{7}\\\frac{1}{7}&\frac{1}{7}&-\frac{2}{7}\\-\frac{5}{7}&\frac{9}{7}&\frac{3}{7}\end{bmatrix}$, so

$$X = A^{-1}B = \begin{bmatrix} \frac{3}{7} & -\frac{4}{7} & \frac{1}{7} \\ \frac{1}{7} & \frac{1}{7} & -\frac{2}{7} \\ -\frac{5}{7} & \frac{9}{7} & \frac{3}{7} \end{bmatrix}\begin{bmatrix} 8 \\ 5 \\ 4 \end{bmatrix} = \begin{bmatrix} \frac{8}{7} \\ \frac{5}{7} \\ \frac{17}{7} \end{bmatrix}.$$

The solution is $x = \frac{8}{7},\ y = \frac{5}{7},\ z = \frac{17}{7}$.

57. $\begin{cases} x + y + z = 2 \\ 3x + 2y - z = \frac{7}{3} \\ 3x + y + 2z = \frac{10}{3} \end{cases}$

Rewrite the system of equations in matrix form:

$$A = \begin{bmatrix} 1 & 1 & 1 \\ 3 & 2 & -1 \\ 3 & 1 & 2 \end{bmatrix},\quad X = \begin{bmatrix} x \\ y \\ z \end{bmatrix},\quad B = \begin{bmatrix} 2 \\ \frac{7}{3} \\ \frac{10}{3} \end{bmatrix}$$

Find the inverse of A and solve $X = A^{-1}B$:

From Problem 37, $A^{-1} = \begin{bmatrix} -\frac{5}{7} & \frac{1}{7} & \frac{3}{7} \\ \frac{9}{7} & \frac{1}{7} & -\frac{4}{7} \\ \frac{3}{7} & -\frac{2}{7} & \frac{1}{7} \end{bmatrix}$, so

$$X = A^{-1}B = \begin{bmatrix} -\frac{5}{7} & \frac{1}{7} & \frac{3}{7} \\ \frac{9}{7} & \frac{1}{7} & -\frac{4}{7} \\ \frac{3}{7} & -\frac{2}{7} & \frac{1}{7} \end{bmatrix}\begin{bmatrix} 2 \\ \frac{7}{3} \\ \frac{10}{3} \end{bmatrix} = \begin{bmatrix} \frac{1}{3} \\ 1 \\ \frac{2}{3} \end{bmatrix}.$$

The solution is $x = \frac{1}{3},\ y = 1,\ z = \frac{2}{3}$.

58. $\begin{cases} 3x + 3y + z = 1 \\ x + 2y + z = 0 \\ 2x - y + z = 4 \end{cases}$

Rewrite the system of equations in matrix form:

$$A = \begin{bmatrix} 3 & 3 & 1 \\ 1 & 2 & 1 \\ 2 & -1 & 1 \end{bmatrix},\quad X = \begin{bmatrix} x \\ y \\ z \end{bmatrix},\quad B = \begin{bmatrix} 1 \\ 0 \\ 4 \end{bmatrix}$$

Find the inverse of A and solve $X = A^{-1}B$:

From Problem 38, $A^{-1} = \begin{bmatrix} \frac{3}{7} & -\frac{4}{7} & \frac{1}{7} \\ \frac{1}{7} & \frac{1}{7} & -\frac{2}{7} \\ -\frac{5}{7} & \frac{9}{7} & \frac{3}{7} \end{bmatrix}$, so

$$X = A^{-1}B = \begin{bmatrix} \frac{3}{7} & -\frac{4}{7} & \frac{1}{7} \\ \frac{1}{7} & \frac{1}{7} & -\frac{2}{7} \\ -\frac{5}{7} & \frac{9}{7} & \frac{3}{7} \end{bmatrix}\begin{bmatrix} 1 \\ 0 \\ 4 \end{bmatrix} = \begin{bmatrix} 1 \\ -1 \\ 1 \end{bmatrix}.$$

The solution is $x = 1,\ y = -1,\ z = 1$.

59. $A = \begin{bmatrix} 4 & 2 \\ 2 & 1 \end{bmatrix}$

Augment the matrix with the identity and use row operations to find the inverse:

$$\left[\begin{array}{cc|cc} 4 & 2 & 1 & 0 \\ 2 & 1 & 0 & 1 \end{array}\right]$$

$$\rightarrow \left[\begin{array}{cc|cc} 4 & 2 & 1 & 0 \\ 0 & 0 & -\frac{1}{2} & 1 \end{array}\right] \quad \left(R_2 = -\tfrac{1}{2}r_1 + r_2\right)$$

$$\rightarrow \left[\begin{array}{cc|cc} 1 & \frac{1}{2} & \frac{1}{4} & 0 \\ 0 & 0 & -\frac{1}{2} & 1 \end{array}\right] \quad \left(R_1 = \tfrac{1}{4}r_1\right)$$

There is no way to obtain the identity matrix on the left. Thus, this matrix has no inverse.

60. $A = \begin{bmatrix} -3 & \frac{1}{2} \\ 6 & -1 \end{bmatrix}$

Augment the matrix with the identity and use row operations to find the inverse:

$$\left[\begin{array}{cc|cc} -3 & \frac{1}{2} & 1 & 0 \\ 6 & -1 & 0 & 1 \end{array}\right]$$

$$\rightarrow \left[\begin{array}{cc|cc} -3 & \frac{1}{2} & 1 & 0 \\ 0 & 0 & 2 & 1 \end{array}\right] \quad \left(R_2 = 2r_1 + r_2\right)$$

$$\rightarrow \left[\begin{array}{cc|cc} 1 & -\frac{1}{6} & -\frac{1}{3} & 0 \\ 0 & 0 & 2 & 1 \end{array}\right] \quad \left(R_1 = -\tfrac{1}{3}r_1\right)$$

There is no way to obtain the identity matrix on the left. Thus, this matrix has no inverse.

61. $A = \begin{bmatrix} 15 & 3 \\ 10 & 2 \end{bmatrix}$

Augment the matrix with the identity and use row operations to find the inverse:

$$\left[\begin{array}{cc|cc} 15 & 3 & 1 & 0 \\ 10 & 2 & 0 & 1 \end{array}\right]$$

$$\rightarrow \left[\begin{array}{cc|cc} 15 & 3 & 1 & 0 \\ 0 & 0 & -\frac{2}{3} & 1 \end{array}\right] \quad \left(R_2 = -\tfrac{2}{3}r_1 + r_2\right)$$

$$\rightarrow \left[\begin{array}{cc|cc} 1 & \frac{1}{5} & \frac{1}{15} & 0 \\ 0 & 0 & -\frac{2}{3} & 1 \end{array}\right] \quad \left(R_1 = \tfrac{1}{15}r_1\right)$$

There is no way to obtain the identity matrix on the left; thus, there is no inverse.

62. $A=\begin{bmatrix}-3 & 0\\ 4 & 0\end{bmatrix}$

Augment the matrix with the identity and use row operations to find the inverse:

$$\left[\begin{array}{cc|cc}-3 & 0 & 1 & 0\\ 4 & 0 & 0 & 1\end{array}\right]$$

$$\rightarrow\left[\begin{array}{cc|cc}-3 & 0 & 1 & 0\\ 0 & 0 & \frac{4}{3} & 1\end{array}\right]\quad\left(R_2=\tfrac{4}{3}r_1+r_2\right)$$

$$\rightarrow\left[\begin{array}{cc|cc}1 & 0 & -\frac{1}{3} & 0\\ 0 & 0 & \frac{4}{3} & 1\end{array}\right]\quad\left(R_1=-\tfrac{1}{3}r_1\right)$$

There is no way to obtain the identity matrix on the left; thus, there is no inverse.

63. $A=\begin{bmatrix}-3 & 1 & -1\\ 1 & -4 & -7\\ 1 & 2 & 5\end{bmatrix}$

Augment the matrix with the identity and use row operations to find the inverse:

$$\left[\begin{array}{ccc|ccc}-3 & 1 & -1 & 1 & 0 & 0\\ 1 & -4 & -7 & 0 & 1 & 0\\ 1 & 2 & 5 & 0 & 0 & 1\end{array}\right]$$

$$\rightarrow\left[\begin{array}{ccc|ccc}1 & 2 & 5 & 0 & 0 & 1\\ 1 & -4 & -7 & 0 & 1 & 0\\ -3 & 1 & -1 & 1 & 0 & 0\end{array}\right]\quad\left(\begin{array}{c}\text{Interchange}\\ r_1\text{ and }r_3\end{array}\right)$$

$$\rightarrow\left[\begin{array}{ccc|ccc}1 & 2 & 5 & 0 & 0 & 1\\ 0 & -6 & -12 & 0 & 1 & -1\\ 0 & 7 & 14 & 1 & 0 & 3\end{array}\right]\quad\left(\begin{array}{l}R_2=-r_1+r_2\\ R_3=3r_1+r_3\end{array}\right)$$

$$\rightarrow\left[\begin{array}{ccc|ccc}1 & 2 & 5 & 0 & 0 & 1\\ 0 & 1 & 2 & 0 & -\frac{1}{6} & \frac{1}{6}\\ 0 & 7 & 14 & 1 & 0 & 3\end{array}\right]\quad\left(R_2=-\tfrac{1}{6}r_2\right)$$

$$\rightarrow\left[\begin{array}{ccc|ccc}1 & 0 & 1 & 0 & \frac{1}{3} & \frac{2}{3}\\ 0 & 1 & 2 & 0 & -\frac{1}{6} & \frac{1}{6}\\ 0 & 0 & 0 & 1 & \frac{7}{6} & \frac{11}{6}\end{array}\right]\quad\left(\begin{array}{l}R_1=-2r_2+r_1\\ R_3=-7r_2+r_3\end{array}\right)$$

There is no way to obtain the identity matrix on the left; thus, there is no inverse.

64. $A=\begin{bmatrix}1 & 1 & -3\\ 2 & -4 & 1\\ -5 & 7 & 1\end{bmatrix}$

Augment the matrix with the identity and use row operations to find the inverse:

$$\left[\begin{array}{ccc|ccc}1 & 1 & -3 & 1 & 0 & 0\\ 2 & -4 & 1 & 0 & 1 & 0\\ -5 & 7 & 1 & 0 & 0 & 1\end{array}\right]$$

$$\rightarrow\left[\begin{array}{ccc|ccc}1 & 1 & -3 & 1 & 0 & 0\\ 0 & -6 & 7 & -2 & 1 & 0\\ 0 & 12 & -14 & 5 & 0 & 1\end{array}\right]\quad\left(\begin{array}{l}R_2=-2r_1+r_2\\ R_3=5r_1+r_3\end{array}\right)$$

$$\rightarrow\left[\begin{array}{ccc|ccc}1 & 1 & -3 & 1 & 0 & 0\\ 0 & 1 & -\frac{7}{6} & \frac{1}{3} & -\frac{1}{6} & 0\\ 0 & 12 & -14 & 5 & 0 & 1\end{array}\right]\quad\left(R_2=-\tfrac{1}{6}r_2\right)$$

$$\rightarrow\left[\begin{array}{ccc|ccc}1 & 0 & -\frac{11}{6} & \frac{2}{3} & \frac{1}{6} & 0\\ 0 & 1 & -\frac{7}{6} & \frac{1}{3} & -\frac{1}{6} & 0\\ 0 & 0 & 0 & 1 & 2 & 1\end{array}\right]\quad\left(\begin{array}{l}R_1=-r_2+r_1\\ R_3=-12r_2+r_3\end{array}\right)$$

There is no way to obtain the identity matrix on the left; thus, there is no inverse.

65. $A=\begin{bmatrix}25 & 61 & -12\\ 18 & -2 & 4\\ 8 & 35 & 21\end{bmatrix}$

```
[A]-1
[[.01  .05  -.0...
 [.01  -.02 .01...
 [-.02 .01  .03...
```

```
[A]-1
...01  .05  -.01]
...01  -.02 .01 ]
...02  .01  .03 ]]
```

Thus, $A^{-1}\approx\begin{bmatrix}0.01 & 0.05 & -0.01\\ 0.01 & -0.02 & 0.01\\ -0.02 & 0.01 & 0.03\end{bmatrix}$

66. $A=\begin{bmatrix}18 & -3 & 4\\ 6 & -20 & 14\\ 10 & 25 & -15\end{bmatrix}$

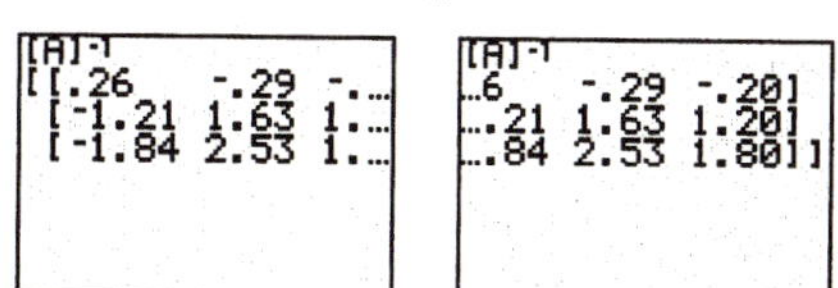

Thus, $A^{-1}\approx\begin{bmatrix}0.26 & -0.29 & -0.20\\ -1.21 & 1.63 & 1.20\\ -1.84 & 2.53 & 1.80\end{bmatrix}$

67. $A=\begin{bmatrix}44 & 21 & 18 & 6\\ -2 & 10 & 15 & 5\\ 21 & 12 & -12 & 4\\ -8 & -16 & 4 & 9\end{bmatrix}$

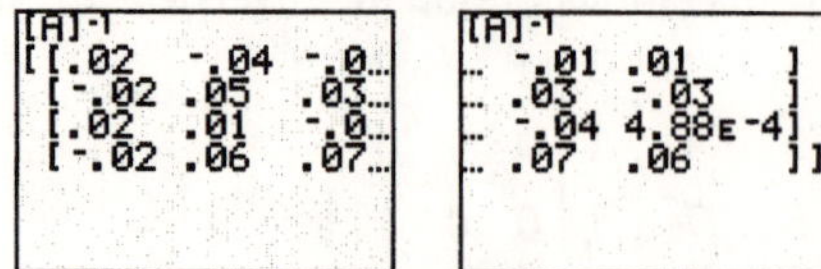

Thus, $A^{-1}\approx\begin{bmatrix}0.02 & -0.04 & -0.01 & 0.01\\ -0.02 & 0.05 & 0.03 & -0.03\\ 0.02 & 0.01 & -0.04 & 0.00\\ -0.02 & 0.06 & 0.07 & 0.06\end{bmatrix}$.

68. $A=\begin{bmatrix}16 & 22 & -3 & 5\\ 21 & -17 & 4 & 8\\ 2 & 8 & 27 & 20\\ 5 & 15 & -3 & -10\end{bmatrix}$

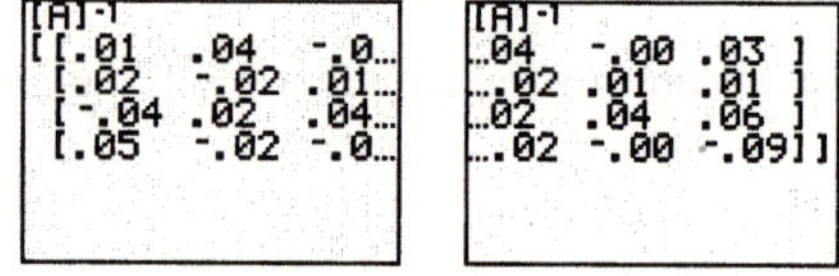

Thus, $A^{-1}=\begin{bmatrix}0.01 & 0.04 & 0.00 & 0.03\\ 0.02 & -0.02 & 0.01 & 0.01\\ -0.04 & 0.02 & 0.04 & 0.06\\ 0.05 & -0.02 & 0.00 & -0.09\end{bmatrix}$.

69. $A=\begin{bmatrix}25 & 61 & -12\\ 18 & -12 & 7\\ 3 & 4 & -1\end{bmatrix}$; $B=\begin{bmatrix}10\\ -9\\ 12\end{bmatrix}$

Enter the matrices into a graphing utility and use $A^{-1}B$ to solve the system. The result is shown below:

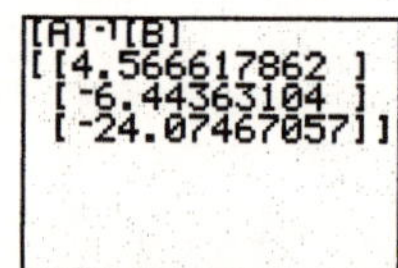

Thus, the solution to the system is $x\approx 4.57$, $y\approx -6.44$, $z\approx -24.07$.

70. $A=\begin{bmatrix}25 & 61 & -12\\ 18 & -12 & 7\\ 3 & 4 & -1\end{bmatrix}$; $B=\begin{bmatrix}15\\ -3\\ 12\end{bmatrix}$

Enter the matrices into a graphing utility and use $A^{-1}B$ to solve the system. The result is shown below:

Thus, the solution to the system is $x\approx 4.56$, $y\approx -6.06$, $z\approx -22.55$.

71. $A=\begin{bmatrix}25 & 61 & -12\\ 18 & -12 & 7\\ 3 & 4 & -1\end{bmatrix}$; $B=\begin{bmatrix}21\\ 7\\ -2\end{bmatrix}$

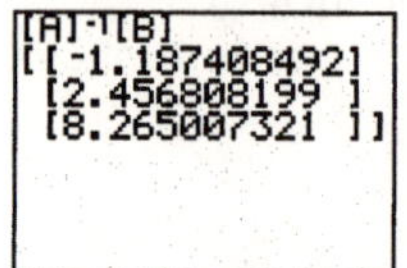

Thus, the solution to the system is $x\approx -1.19$, $y\approx 2.46$, $z\approx 8.27$.

72. $A=\begin{bmatrix}25 & 61 & -12\\ 18 & -12 & 7\\ 3 & 4 & -1\end{bmatrix}$; $B=\begin{bmatrix}25\\ 10\\ -4\end{bmatrix}$

Thus, the solution to the system is $x\approx -2.05$, $y\approx 3.88$, $z\approx 13.36$.

73. a. The rows of the 2 by 3 matrix represent stainless steel and aluminum. The columns represent 10-gallon, 5-gallon, and 1-gallon.

The 2 by 3 matrix is: $\begin{bmatrix}500 & 350 & 400\\ 700 & 500 & 850\end{bmatrix}$.

The 3 by 2 matrix is: $\begin{bmatrix}500 & 700\\ 350 & 500\\ 400 & 850\end{bmatrix}$.

b. The 3 by 1 matrix representing the amount of material is: $\begin{bmatrix}15\\ 8\\ 3\end{bmatrix}$.

c. The days usage of materials is:

$$\begin{bmatrix} 500 & 350 & 400 \\ 700 & 500 & 850 \end{bmatrix} \cdot \begin{bmatrix} 15 \\ 8 \\ 3 \end{bmatrix} = \begin{bmatrix} 11{,}500 \\ 17{,}050 \end{bmatrix}$$

Thus, 11,500 pounds of stainless steel and 17,050 pounds of aluminum were used that day.

d. The 1 by 2 matrix representing cost is: $[0.10 \quad 0.05]$.

e. The total cost of the day's production was:

$$[0.10 \quad 0.05] \cdot \begin{bmatrix} 11{,}500 \\ 17{,}050 \end{bmatrix} = [2002.50].$$

The total cost of the day's production was $2002.50.

74. a. The rows of the 2 by 3 matrix represent the location. The columns represent the type of car sold. The 2 by 3 matrix for January is: $\begin{bmatrix} 400 & 250 & 50 \\ 450 & 200 & 140 \end{bmatrix}$. The 2 by 3 matrix for February is: $\begin{bmatrix} 350 & 100 & 30 \\ 350 & 300 & 100 \end{bmatrix}$.

b. Adding the matrices:

$$\begin{bmatrix} 400 & 250 & 50 \\ 450 & 200 & 140 \end{bmatrix} + \begin{bmatrix} 350 & 100 & 30 \\ 350 & 300 & 100 \end{bmatrix} = \begin{bmatrix} 750 & 350 & 80 \\ 800 & 500 & 240 \end{bmatrix}$$

c. The 3 by 1 matrix representing profit: $\begin{bmatrix} 100 \\ 150 \\ 200 \end{bmatrix}$.

d. Multiplying to find the profit at each location:

$$\begin{bmatrix} 750 & 350 & 80 \\ 800 & 500 & 240 \end{bmatrix} \cdot \begin{bmatrix} 100 \\ 150 \\ 200 \end{bmatrix} = \begin{bmatrix} 143{,}500 \\ 203{,}000 \end{bmatrix}.$$

The city location has a two-month profit of $143,500. The suburban location has a two-month profit of $203,000.

75. $A = \begin{bmatrix} a & b \\ c & d \end{bmatrix}$

If $D = ad - bc \neq 0$, then $a \neq 0$ and $d \neq 0$, or $b \neq 0$ and $c \neq 0$. Assuming the former, then

$$\left[\begin{array}{cc|cc} a & b & 1 & 0 \\ c & d & 0 & 1 \end{array}\right]$$

$$\to \left[\begin{array}{cc|cc} 1 & \frac{b}{a} & \frac{1}{a} & 0 \\ c & d & 0 & 1 \end{array}\right] \quad \left(R_1 = \tfrac{1}{a} \cdot r_1\right)$$

$$\to \left[\begin{array}{cc|cc} 1 & \frac{b}{a} & \frac{1}{a} & 0 \\ 0 & d - \frac{bc}{a} & -\frac{c}{a} & 1 \end{array}\right] \quad \left(R_2 = -c \cdot r_1 + r_2\right)$$

$$\to \left[\begin{array}{cc|cc} 1 & \frac{b}{a} & \frac{1}{a} & 0 \\ 0 & \frac{ad-bc}{a} & -\frac{c}{a} & 1 \end{array}\right]$$

$$\to \left[\begin{array}{cc|cc} 1 & \frac{b}{a} & \frac{1}{a} & 0 \\ 0 & 1 & \frac{-c}{ad-bc} & \frac{a}{ad-bc} \end{array}\right] \quad \left(R_2 = \tfrac{a}{ad-bc} \cdot r_2\right)$$

$$\to \left[\begin{array}{cc|cc} 1 & 0 & \frac{1}{a} + \frac{bc}{a(ad-bc)} & \frac{-b}{ad-bc} \\ 0 & 1 & \frac{-c}{ad-bc} & \frac{a}{ad-bc} \end{array}\right] \quad \left(R_1 = -\tfrac{b}{a} \cdot r_2 + r_1\right)$$

$$\to \left[\begin{array}{cc|cc} 1 & 0 & \frac{d}{ad-bc} & \frac{-b}{ad-bc} \\ 0 & 1 & \frac{-c}{ad-bc} & \frac{a}{ad-bc} \end{array}\right]$$

$$\to \left[\begin{array}{cc|cc} 1 & 0 & \frac{d}{D} & \frac{-b}{D} \\ 0 & 1 & \frac{-c}{D} & \frac{a}{D} \end{array}\right]$$

Thus, $A^{-1} = \begin{bmatrix} \frac{d}{D} & \frac{-b}{D} \\ \frac{-c}{D} & \frac{a}{D} \end{bmatrix} = \frac{1}{D}\begin{bmatrix} d & -b \\ -c & a \end{bmatrix}$

where $D = ad - bc$.

76. Answers will vary.

Section 10.5

1. True

2. True

3. $3x^4 + 6x^3 + 3x^2 = 3x^2\left(x^2 + 2x + 1\right)$
$= 3x^2(x+1)^2$

4. True

5. The rational expression $\dfrac{x}{x^2 - 1}$ is proper, since the degree of the numerator is less than the degree of the denominator.

6. The rational expression $\dfrac{5x+2}{x^3 - 1}$ is proper, since the degree of the numerator is less than the degree of the denominator.

7. The rational expression $\dfrac{x^2+5}{x^2-4}$ is improper, so perform the division:

$$\begin{array}{r} 1 \\ x^2-4\overline{)x^2+5} \\ \underline{x^2-4} \\ 9 \end{array}$$

The proper rational expression is:

$$\frac{x^2+5}{x^2-4}=1+\frac{9}{x^2-4}$$

8. The rational expression $\dfrac{3x^2-2}{x^2-1}$ is improper, so perform the division:

$$\begin{array}{r} 3 \\ x^2-1\overline{)3x^2-2} \\ \underline{3x^2-3} \\ 1 \end{array}$$

The proper rational expression is:

$$\frac{3x^2-2}{x^2-1}=3+\frac{1}{x^2-1}$$

9. The rational expression $\dfrac{5x^3+2x-1}{x^2-4}$ is improper, so perform the division:

$$\begin{array}{r} 5x \\ x^2-4\overline{)5x^3+2x-1} \\ \underline{5x^3-20x} \\ 22x-1 \end{array}$$

The proper rational expression is:

$$\frac{5x^3+2x-1}{x^2-4}=5x+\frac{22x-1}{x^2-4}$$

10. The rational expression $\dfrac{3x^4+x^2-2}{x^3+8}$ is improper, so perform the division:

$$\begin{array}{r} 3x \\ x^3+8\overline{)3x^4+x^2-2} \\ \underline{3x^4+24x} \\ x^2-24x-2 \end{array}$$

The proper rational expression is:

$$\frac{3x^4+x^2-2}{x^3+8}=3x+\frac{x^2-24x-2}{x^3+8}$$

11. The rational expression

$$\frac{x(x-1)}{(x+4)(x-3)}=\frac{x^2-x}{x^2+x-12}$$ is improper, so perform the division:

$$\begin{array}{r} 1 \\ x^2+x-12\overline{)x^2-x+0} \\ \underline{x^2+x-12} \\ -2x+12 \end{array}$$

The proper rational expression is:

$$\frac{x(x-1)}{(x+4)(x-3)}=1+\frac{-2x+12}{x^2+x-12}$$
$$=1+\frac{-2(x-6)}{(x+4)(x-3)}$$

12. The rational expression $\dfrac{2x(x^2+4)}{x^2+1}=\dfrac{2x^3+8x}{x^2+1}$ is improper, so perform the division:

$$\begin{array}{r} 2x \\ x^2+1\overline{)2x^3+8x} \\ \underline{2x^3+2x} \\ 6x \end{array}$$

The proper rational expression is:

$$\frac{2x(x^2+4)}{x^2+1}=2x+\frac{6x}{x^2+1}$$

13. Find the partial fraction decomposition:

$$\frac{4}{x(x-1)}=\frac{A}{x}+\frac{B}{x-1}$$
$$x(x-1)\left(\frac{4}{x(x-1)}\right)=x(x-1)\left(\frac{A}{x}+\frac{B}{x-1}\right)$$
$$4=A(x-1)+Bx$$

Let $x=1$, then $4=A(0)+B$
$B=4$
Let $x=0$, then $4=A(-1)+B(0)$
$A=-4$

$$\frac{4}{x(x-1)}=\frac{-4}{x}+\frac{4}{x-1}$$

14. Find the partial fraction decomposition

$$\frac{3x}{(x+2)(x-1)}=\frac{A}{x+2}+\frac{B}{x-1}$$

Multiplying both sides by $(x+2)(x-1)$, we obtain: $3x=A(x-1)+B(x+2)$

Let $x=1$, then $3(1)=A(0)+B(3)$

$$3B=3$$
$$B=1$$

Let $x=-2$, then $3(-2)=A(-3)+B(0)$

$$-3A=-6$$
$$A=2$$

$$\frac{3x}{(x+2)(x-1)}=\frac{2}{x+2}+\frac{1}{x-1}$$

15. Find the partial fraction decomposition:

$$\frac{1}{x(x^2+1)}=\frac{A}{x}+\frac{Bx+C}{x^2+1}$$

$$x(x^2+1)\left(\frac{1}{x(x^2+1)}\right)=x(x^2+1)\left(\frac{A}{x}+\frac{Bx+C}{x^2+1}\right)$$

$$1=A(x^2+1)+(Bx+C)x$$

Let $x=0$, then $1=A(0^2+1)+(B(0)+C)(0)$

$$A=1$$

Let $x=1$, then $1=A(1^2+1)+(B(1)+C)(1)$

$$1=2A+B+C$$
$$1=2(1)+B+C$$
$$B+C=-1$$

Let $x=-1$, then

$$1=A((-1)^2+1)+(B(-1)+C)(-1)$$
$$1=A(1+1)+(-B+C)(-1)$$
$$1=2A+B-C$$
$$1=2(1)+B-C$$
$$B-C=-1$$

Solve the system of equations:

$$B+C=-1$$
$$B-C=-1$$
$$2B=-2 \qquad -1+C=-1$$
$$B=-1 \qquad C=0$$

$$\frac{1}{x(x^2+1)}=\frac{1}{x}+\frac{-x}{x^2+1}$$

16. Find the partial fraction decomposition:

$$\frac{1}{(x+1)(x^2+4)}=\frac{A}{x+1}+\frac{Bx+C}{x^2+4}$$

Multiplying both sides by $(x+1)(x^2+4)$, we obtain: $1=A(x^2+4)+(Bx+C)(x+1)$

Let $x=-1$, then $1=A(5)+(B(-1)+C)(0)$

$$5A=1$$
$$A=\frac{1}{5}$$

Let $x=1$, then $1=A(1^2+4)+(B(1)+C)(1+1)$

$$1=5A+(B+C)(2)$$
$$1=5(1/5)+2B+2C$$
$$1=1+2B+2C$$
$$0=2B+2C$$
$$0=B+C$$

Let $x=0$, then $1=A(0^2+4)+(B(0)+C)(0+1)$

$$1=4A+C$$
$$1=4(1/5)+C$$
$$1=\frac{4}{5}+C$$
$$C=\frac{1}{5}$$

Since $B+C=0$, we have that $B+\frac{1}{5}=0$

$$B=-\frac{1}{5}$$

$$\frac{1}{(x+1)(x^2+4)}=\frac{(1/5)}{x+1}+\frac{(-(1/5)x+1/5)}{x^2+4}$$

17. Find the partial fraction decomposition:

$$\frac{x}{(x-1)(x-2)}=\frac{A}{x-1}+\frac{B}{x-2}$$

Multiplying both sides by $(x-1)(x-2)$, we obtain: $x=A(x-2)+B(x-1)$

Let $x=1$, then $1=A(1-2)+B(1-1)$

$$1=-A$$
$$A=-1$$

Let $x=2$, then $2=A(2-2)+B(2-1)$

$$2=B$$

$$\frac{x}{(x-1)(x-2)}=\frac{-1}{x-1}+\frac{2}{x-2}$$

18. Find the partial fraction decomposition:

$$\frac{3x}{(x+2)(x-4)}=\frac{A}{x+2}+\frac{B}{x-4}$$

Multiplying both sides by $(x+2)(x-4)$, we obtain: $3x=A(x-4)+B(x+2)$

Let $x=-2$, then $3(-2)=A(-2-4)+B(-2+2)$

$$-6=-6A$$
$$A=1$$

Let $x=4$, then $3(4)=A(4-4)+B(4+2)$

$$12=6B$$
$$B=2$$

$$\frac{3x}{(x+2)(x-4)}=\frac{1}{x+2}+\frac{2}{x-4}$$

19. Find the partial fraction decomposition:

$$\frac{x^2}{(x-1)^2(x+1)}=\frac{A}{x-1}+\frac{B}{(x-1)^2}+\frac{C}{x+1}$$

Multiplying both sides by $(x-1)^2(x+1)$, we obtain: $x^2=A(x-1)(x+1)+B(x+1)+C(x-1)^2$

Let $x=1$, then

$$1^2=A(1-1)(1+1)+B(1+1)+C(1-1)^2$$
$$1=A(0)(2)+B(2)+C(0)^2$$
$$1=2B$$
$$B=\frac{1}{2}$$

Let $x=-1$, then

$$(-1)^2=A(-1-1)(-1+1)+B(-1+1)+C(-1-1)^2$$
$$1=A(-2)(0)+B(0)+C(-2)^2$$
$$1=4C$$
$$C=\frac{1}{4}$$

Let $x=0$, then

$$0^2=A(0-1)(0+1)+B(0+1)+C(0-1)^2$$
$$0=-A+B+C$$
$$A=B+C$$
$$A=\frac{1}{2}+\frac{1}{4}=\frac{3}{4}$$

$$\frac{x^2}{(x-1)^2(x+1)}=\frac{(3/4)}{x-1}+\frac{(1/2)}{(x-1)^2}+\frac{(1/4)}{x+1}$$

20. Find the partial fraction decomposition:

$$\frac{x+1}{x^2(x-2)}=\frac{A}{x}+\frac{B}{x^2}+\frac{C}{x-2}$$

Multiplying both sides by $x^2(x-2)$, we obtain:

$$x+1=Ax(x-2)+B(x-2)+Cx^2$$

Let $x=0$, then

$$0+1=A(0)(0-2)+B(0-2)+C(0)^2$$
$$1=-2B$$
$$B=-\frac{1}{2}$$

Let $x=2$, then

$$2+1=A(2)(2-2)+B(2-2)+C(2)^2$$
$$3=4C$$
$$C=\frac{3}{4}$$

Let $x=1$, then $2=-A-B+C$

$$A=-B+C-2$$
$$A=-\left(-\frac{1}{2}\right)+\frac{3}{4}-2=-\frac{3}{4}$$

$$\frac{x+1}{x^2(x-2)}=\frac{-(3/4)}{x}+\frac{-(1/2)}{x^2}+\frac{(3/4)}{x-2}$$

21. Find the partial fraction decomposition:

$$\frac{1}{x^3-8}=\frac{1}{(x-2)(x^2+2x+4)}$$

$$\frac{1}{(x-2)(x^2+2x+4)}=\frac{A}{x-2}+\frac{Bx+C}{x^2+2x+4}$$

Multiplying both sides by $(x-2)(x^2+2x+4)$, we obtain: $1=A(x^2+2x+4)+(Bx+C)(x-2)$

Let $x=2$, then

$$1=A\left(2^2+2(2)+4\right)+(B(2)+C)(2-2)$$
$$1=12A$$
$$A=\frac{1}{12}$$

Let $x=0$, then

$$1=A\left(0^2+2(0)+4\right)+(B(0)+C)(0-2)$$
$$1=4A-2C$$
$$1=4(1/12)-2C$$
$$-2C=\frac{2}{3}$$
$$C=-\frac{1}{3}$$

Let $x=1$, then

$$1=A\left(1^2+2(1)+4\right)+(B(1)+C)(1-2)$$
$$1=7A-B-C$$
$$1=7(1/12)-B+\frac{1}{3}$$
$$B=-\frac{1}{12}$$
$$\frac{1}{x^3-8}=\frac{(1/12)}{x-2}+\frac{-(1/12)x-1/3}{x^2+2x+4}$$
$$=\frac{(1/12)}{x-2}+\frac{-(1/12)(x+4)}{x^2+2x+4}$$

22. Find the partial fraction decomposition:

$$\frac{2x+4}{x^3-1}=\frac{2x+4}{(x-1)(x^2+x+1)}=\frac{A}{x-1}+\frac{Bx+C}{x^2+x+1}$$

Multiplying both sides by $(x-1)(x^2+x+1)$, we obtain: $2x+4=A(x^2+x+1)+(Bx+C)(x-1)$

Let $x=1$, then

$$2(1)+4=A\left(1^2+1+1\right)+\left(B(1)+C\right)(1-1)$$
$$6=3A$$
$$A=2$$

Let $x=0$, then

$$2(0)+4=A\left(0^2+0+1\right)+(B(0)+C)(0-1)$$
$$4=A-C$$
$$4=2-C$$
$$C=-2$$

Let $x=-1$, then

$$2(-1)+4=A\left((-1)^2+(-1)+1\right)+(B(-1)+C)(-1-1)$$
$$2=A+2B-2C$$
$$2=2+2B-2(-2)$$
$$2B=-4$$
$$B=-2$$
$$\frac{2x+4}{x^3-1}=\frac{2}{x-1}+\frac{-2x-2}{x^2+x+1}$$

23. Find the partial fraction decomposition:

$$\frac{x^2}{(x-1)^2(x+1)^2}=\frac{A}{x-1}+\frac{B}{(x-1)^2}+\frac{C}{x+1}+\frac{D}{(x+1)^2}$$

Multiplying both sides by $(x-1)^2(x+1)^2$, we obtain:

$$x^2=A(x-1)(x+1)^2+B(x+1)^2+C(x-1)^2(x+1)+D(x-1)^2$$

Let $x=1$, then

$$1^2=A(1-1)(1+1)^2+B(1+1)^2+C(1-1)^2(1+1)+D(1-1)^2$$
$$1=4B$$
$$B=\frac{1}{4}$$

Let $x=-1$, then

$$(-1)^2=A(-1-1)(-1+1)^2+B(-1+1)^2+C(-1-1)^2(-1+1)+D(-1-1)^2$$
$$1=4D$$
$$D=\frac{1}{4}$$

Let $x=0$, then

$$0^2=A(0-1)(0+1)^2+B(0+1)^2+C(0-1)^2(0+1)+D(0-1)^2$$
$$0=-A+B+C+D$$
$$A-C=B+D$$
$$A-C=\frac{1}{4}+\frac{1}{4}=\frac{1}{2}$$

Let $x=2$, then

$$2^2=A(2-1)(2+1)^2+B(2+1)^2+C(2-1)^2(2+1)+D(2-1)^2$$
$$4=9A+9B+3C+D$$
$$9A+3C=4-9B-D$$
$$9A+3C=4-9\left(\frac{1}{4}\right)-\frac{1}{4}=\frac{3}{2}$$
$$3A+C=\frac{1}{2}$$

Solve the system of equations:

$$A-C=\frac{1}{2}$$
$$3A+C=\frac{1}{2}$$
$$4A\quad=1$$
$$A=\frac{1}{4}$$
$$\frac{3}{4}+C=\frac{1}{2}$$
$$C=-\frac{1}{4}$$
$$\frac{x^2}{(x-1)^2(x+1)^2}=\frac{(1/4)}{x-1}+\frac{(1/4)}{(x-1)^2}+\frac{(-1/4)}{x+1}+\frac{(1/4)}{(x+1)^2}$$

24. Find the partial fraction decomposition:

$$\frac{x+1}{x^2(x-2)^2}=\frac{A}{x}+\frac{B}{x^2}+\frac{C}{x-2}+\frac{D}{(x-2)^2}$$

Multiplying both sides by $x^2(x-2)^2$, we obtain:

$$x+1=Ax(x-2)^2+B(x-2)^2+Cx^2(x-2)+Dx^2$$

Let $x=0$, then

$$0+1=A(0)(0-2)^2+B(0-2)^2+C(0)^2(0-2)+D(0)^2$$
$$1=4B$$
$$B=\frac{1}{4}$$

Let $x=2$, then

$$2+1=A(2)(2-2)^2+B(2-2)^2+C(2)^2(2-2)+D(2)^2$$
$$3=4D$$
$$D=\frac{3}{4}$$

Let $x=1$, then

$$1+1=A(1)(1-2)^2+B(1-2)^2+C(1)^2(1-2)+D(1)^2$$
$$2=A+B-C+D$$
$$A-C=2-B-D$$
$$A-C=2-\frac{1}{4}-\frac{3}{4}=1$$

Let $x=3$, then

$$3+1=A(3)(3-2)^2+B(3-2)^2+C(3)^2(3-2)+D(3)^2$$
$$4=3A+B+9C+9D$$
$$3A+9C=4-B-9D$$
$$3A+9C=4-\frac{1}{4}-9\left(\frac{3}{4}\right)=-3$$
$$A+3C=-1$$

Solve the system of equations:

$$\begin{cases} A-C=1 \\ A+3C=-1 \end{cases}$$

$$A-C=1$$
$$A=C+1$$
$$A+3C=-1$$
$$(C+1)+3C=-1$$
$$4C=-2$$
$$C=-\frac{1}{2}$$
$$A=C+1=-\frac{1}{2}+1=\frac{1}{2}$$

$$\frac{x+1}{x^2(x-2)^2}=\frac{(1/2)}{x}+\frac{(1/4)}{x^2}+\frac{(-1/2)}{x-2}+\frac{(3/4)}{(x-2)^2}$$

25. Find the partial fraction decomposition:

$$\frac{x-3}{(x+2)(x+1)^2}=\frac{A}{x+2}+\frac{B}{x+1}+\frac{C}{(x+1)^2}$$

Multiplying both sides by $(x+2)(x+1)^2$, we obtain:

$$x-3=A(x+1)^2+B(x+2)(x+1)+C(x+2)$$

Let $x=-2$, then

$$-2-3=A(-2+1)^2+B(-2+2)(-2+1)+C(-2+2)$$
$$-5=A$$
$$A=-5$$

Let $x=-1$, then

$$-1-3=A(-1+1)^2+B(-1+2)(-1+1)+C(-1+2)$$
$$-4=C$$
$$C=-4$$

Let $x=0$, then

$$0-3=A(0+1)^2+B(0+2)(0+1)+C(0+2)$$
$$-3=A+2B+2C$$
$$-3=-5+2B+2(-4)$$
$$2B=10$$
$$B=5$$

$$\frac{x-3}{(x+2)(x+1)^2}=\frac{-5}{x+2}+\frac{5}{x+1}+\frac{-4}{(x+1)^2}$$

26. Find the partial fraction decomposition:

$$\frac{x^2+x}{(x+2)(x-1)^2}=\frac{A}{x+2}+\frac{B}{x-1}+\frac{C}{(x-1)^2}$$

Multiplying both sides by $(x+2)(x-1)^2$, we obtain:

$$x^2+x=A(x-1)^2+B(x+2)(x-1)+C(x+2)$$

Let $x=-2$, then

$$(-2)^2+(-2)=A(-2-1)^2+B(-2+2)(-2-1)+C(-2+2)$$
$$2=9A$$
$$A=\frac{2}{9}$$

Let $x = 1$, then

$$1^2 + 1 = A(1-1)^2 + B(1+2)(1-1) + C(1+2)$$
$$2 = 3C$$
$$C = \frac{2}{3}$$

Let $x = 0$, then

$$0^2 + 0 = A(0-1)^2 + B(0+2)(0-1) + C(0+2)$$
$$0 = A - 2B + 2C$$
$$2B = A + 2C$$
$$2B = \frac{2}{9} + 2\left(\frac{2}{3}\right) = \frac{14}{9}$$
$$B = \frac{7}{9}$$

$$\frac{x^2+x}{(x+2)(x-1)^2} = \frac{(2/9)}{x+2} + \frac{(7/9)}{x-1} + \frac{(2/3)}{(x-1)^2}$$

27. Find the partial fraction decomposition:

$$\frac{x+4}{x^2(x^2+4)} = \frac{A}{x} + \frac{B}{x^2} + \frac{Cx+D}{x^2+4}$$

Multiplying both sides by $x^2(x^2+4)$, we obtain:

$$x + 4 = Ax(x^2+4) + B(x^2+4) + (Cx+D)x^2$$

Let $x = 0$, then

$$0 + 4 = A(0)(0^2+4) + B(0^2+4) + \big(C(0)+D\big)(0)^2$$
$$4 = 4B$$
$$B = 1$$

Let $x = 1$, then

$$1 + 4 = A(1)(1^2+4) + B(1^2+4) + (C(1)+D)(1)^2$$
$$5 = 5A + 5B + C + D$$
$$5 = 5A + 5 + C + D$$
$$5A + C + D = 0$$

Let $x = -1$, then

$$-1 + 4 = A(-1)((-1)^2+4) + B((-1)^2+4) + (C(-1)+D)(-1)^2$$
$$3 = -5A + 5B - C + D$$
$$3 = -5A + 5 - C + D$$
$$-5A - C + D = -2$$

Let $x = 2$, then

$$2 + 4 = A(2)(2^2+4) + B(2^2+4) + (C(2)+D)(2)^2$$
$$6 = 16A + 8B + 8C + 4D$$
$$6 = 16A + 8 + 8C + 4D$$
$$16A + 8C + 4D = -2$$

Solve the system of equations:

$$\begin{aligned} 5A + C + D &= 0 \\ -5A - C + D &= -2 \\ \hline 2D &= -2 \\ D &= -1 \end{aligned}$$

$$5A + C - 1 = 0$$
$$C = 1 - 5A$$
$$16A + 8(1-5A) + 4(-1) = -2$$
$$16A + 8 - 40A - 4 = -2$$
$$-24A = -6$$
$$A = \frac{1}{4}$$

$$C = 1 - 5\left(\frac{1}{4}\right)$$
$$C = 1 - \frac{5}{4} = -\frac{1}{4}$$

$$\frac{x+4}{x^2(x^2+4)} = \frac{(1/4)}{x} + \frac{1}{x^2} + \frac{\left(-(1/4)x - 1\right)}{x^2+4}$$
$$= \frac{(1/4)}{x} + \frac{1}{x^2} + \frac{-(1/4)(x+4)}{x^2+4}$$

28. Find the partial fraction decomposition:

$$\frac{10x^2+2x}{(x-1)^2(x^2+2)} = \frac{A}{x-1} + \frac{B}{(x-1)^2} + \frac{Cx+D}{x^2+2}$$

Multiply both sides by $(x-1)^2(x^2+2)$:

$$10x^2 + 2x = A(x-1)(x^2+2) + B(x^2+2) + (Cx+D)(x-1)^2$$

Let $x = 1$, then

$$10(1)^2 + 2(1) = A(1-1)(1^2+2) + B(1^2+2) + \big(C(1)+D\big)(1-1)^2$$
$$12 = 3B$$
$$B = 4$$

Let $x = 0$, then

$$10(0)^2 + 2(0) = A(0-1)(0^2+2) + B(0^2+2) + \big(C(0)+D\big)(0-1)^2$$
$$0 = -2A + 2B + D$$
$$0 = -2A + 8 + D$$
$$2A - D = 8$$
$$D = 2A - 8$$

Let $x=-1$, then

$$10(-1)^2+2(-1)=A(-1-1)((-1)^2+2)$$
$$+B((-1)^2+2)$$
$$+(C(-1)+D)(-1-1)^2$$
$$8=-6A+3B-4C+4D$$
$$8=-6A+12-4C+4D$$
$$-6A-4C+4D=-4$$

Let $x=2$, then

$$10(2)^2+2(2)=A(2-1)(2^2+2)+B(2^2+2)$$
$$+(C(2)+D)(2-1)^2$$
$$44=6A+6B+2C+D$$
$$44=6A+24+2C+D$$
$$6A+2C+D=20$$

Solve the system of equations (Substitute for D):

$$D=2A-8$$
$$-6A-4C+4D=-4$$
$$-6A-4C+4(2A-8)=-4$$
$$2A-4C=28$$
$$A-2C=14$$

$$6A+2C+D=20$$
$$6A+2C+(2A-8)=20$$
$$8A+2C=28$$

Add the equations and solve:

$$A-2C=14$$
$$\underline{8A+2C=28}$$
$$9A \quad =42$$
$$A=\frac{14}{3}$$
$$2C=A-14$$
$$2C=\frac{14}{3}-14=-\frac{28}{3}$$
$$C=-\frac{14}{3}$$
$$D=2A-8$$
$$D=2\left(\frac{14}{3}\right)-8=\frac{4}{3}$$

$$\frac{10x^2+2x}{(x-1)^2(x^2+2)}$$
$$=\frac{(14/3)}{x-1}+\frac{4}{(x-1)^2}+\frac{(-(14/3)x+4/3)}{x^2+2}$$

29. Find the partial fraction decomposition:

$$\frac{x^2+2x+3}{(x+1)(x^2+2x+4)}=\frac{A}{x+1}+\frac{Bx+C}{x^2+2x+4}$$

Multiplying both sides by $(x+1)(x^2+2x+4)$, we obtain:

$$x^2+2x+3=A(x^2+2x+4)+(Bx+C)(x+1)$$

Let $x=-1$, then

$$(-1)^2+2(-1)+3=A((-1)^2+2(-1)+4)$$
$$+(B(-1)+C)(-1+1)$$
$$2=3A$$
$$A=\frac{2}{3}$$

Let $x=0$, then

$$0^2+2(0)+3=A(0^2+2(0)+4)+(B(0)+C)(0+1)$$
$$3=4A+C$$
$$3=4(2/3)+C$$
$$C=\frac{1}{3}$$

Let $x=1$, then

$$1^2+2(1)+3=A(1^2+2(1)+4)+(B(1)+C)(1+1)$$
$$6=7A+2B+2C$$
$$6=7(2/3)+2B+2(1/3)$$
$$2B=6-\frac{14}{3}-\frac{2}{3}$$
$$2B=\frac{2}{3}$$
$$B=\frac{1}{3}$$

$$\frac{x^2+2x+3}{(x+1)(x^2+2x+4)}=\frac{(2/3)}{x+1}+\frac{((1/3)x+1/3)}{x^2+2x+4}$$
$$=\frac{(2/3)}{x+1}+\frac{(1/3)(x+1)}{x^2+2x+4}$$

30. Find the partial fraction decomposition:

$$\frac{x^2-11x-18}{x(x^2+3x+3)}=\frac{A}{x}+\frac{Bx+C}{x^2+3x+3}$$

Multiplying both sides by $x(x^2+3x+3)$ $(x+1)(x^2+2x+4)$, we obtain:

$$x^2-11x-18=A(x^2+3x+3)+(Bx+C)x$$

Let $x=0$, then

$$0^2-11(0)-18=A\left(0^2+3(0)+3\right)+\left(B(0)+C\right)(0)$$
$$-18=3A$$
$$A=-6$$

Let $x=1$, then

$$1^2-11(1)-18=A\left(1^2+3(1)+3\right)+\left(B(1)+C\right)(1)$$
$$-28=7A+B+C$$
$$-28=7(-6)+B+C$$
$$B+C=14$$

Let $x=-1$, then

$$(-1)^2-11(-1)-18=A\left((-1)^2+3(-1)+3\right)+\left(B(-1)+C\right)(-1)$$
$$-6=A+B-C$$
$$-6=-6+B-C$$
$$B-C=0$$

Add the last two equations and solve:

$$B+C=14$$
$$B-C=0$$
$$2B \quad =14$$
$$B=7$$
$$B+C=14$$
$$7+C=14$$
$$C=7$$

$$\frac{x^2-11x-18}{x(x^2+3x+3)}=\frac{-6}{x}+\frac{7x+7}{x^2+3x+3}$$

31. Find the partial fraction decomposition:

$$\frac{x}{(3x-2)(2x+1)}=\frac{A}{3x-2}+\frac{B}{2x+1}$$

Multiplying both sides by $(3x-2)(2x+1)$, we obtain: $x=A(2x+1)+B(3x-2)$

Let $x=-\frac{1}{2}$, then

$$-\frac{1}{2}=A\left(2(-1/2)+1\right)+B\left(3(-1/2)-2\right)$$
$$-\frac{1}{2}=-\frac{7}{2}B$$
$$B=\frac{1}{7}$$

Let $x=\frac{2}{3}$, then

$$\frac{2}{3}=A\left(2(2/3)+1\right)+B\left(3(2/3)-2\right)$$
$$\frac{2}{3}=\frac{7}{3}A$$
$$A=\frac{2}{7}$$

$$\frac{x}{(3x-2)(2x+1)}=\frac{(2/7)}{3x-2}+\frac{(1/7)}{2x+1}$$

32. Find the partial fraction decomposition:

$$\frac{1}{(2x+3)(4x-1)}=\frac{A}{2x+3}+\frac{B}{4x-1}$$

Multiplying both sides by $(2x+3)(4x-1)$, we obtain: $1=A(4x-1)+B(2x+3)$

Let $x=-\frac{3}{2}$, then

$$1=A\left(4(-3/2)-1\right)+B\left(2(-3/2)+3\right)$$
$$1=-7A$$
$$A=-\frac{1}{7}$$

Let $x=\frac{1}{4}$, then

$$1=A\left(4(1/4)-1\right)+B\left(2(1/4)+3\right)$$
$$1=\frac{7}{2}B$$
$$B=\frac{2}{7}$$

$$\frac{1}{(2x+3)(4x-1)}=\frac{(-1/7)}{2x+3}+\frac{(2/7)}{4x-1}$$

33. Find the partial fraction decomposition:

$$\frac{x}{x^2+2x-3}=\frac{x}{(x+3)(x-1)}=\frac{A}{x+3}+\frac{B}{x-1}$$

Multiplying both sides by $(x+3)(x-1)$, we obtain: $x=A(x-1)+B(x+3)$

Let $x=1$, then $1=A(1-1)+B(1+3)$

$$1=4B$$
$$B=\frac{1}{4}$$

Let $x=-3$, then $-3=A(-3-1)+B(-3+3)$

$$-3=-4A$$

$$A=\frac{3}{4}$$

$$\frac{x}{x^2+2x-3}=\frac{(3/4)}{x+3}+\frac{(1/4)}{x-1}$$

34. Find the partial fraction decomposition:

$$\frac{x^2-x-8}{(x+1)(x^2+5x+6)}=\frac{x^2-x-8}{(x+1)(x+2)(x+3)}$$

$$=\frac{A}{x+1}+\frac{B}{x+2}+\frac{C}{x+3}$$

Multiplying both sides by $(x+1)(x+2)(x+3)$, we obtain:

$$x^2-x-8=A(x+2)(x+3)+B(x+1)(x+3)+C(x+1)(x+2)$$

Let $x=-1$, then

$$(-1)^2-(-1)-8=A(-1+2)(-1+3)+B(-1+1)(-1+3)+C(-1+1)(-1+2)$$

$$-6=2A$$

$$A=-3$$

Let $x=-2$, then

$$(-2)^2-(-2)-8=A(-2+2)(-2+3)+B(-2+1)(-2+3)+C(-2+1)(-2+2)$$

$$-2=-B$$

$$B=2$$

Let $x=-3$, then

$$(-3)^2-(-3)-8=A(-3+2)(-3+3)+B(-3+1)(-3+3)+C(-3+1)(-3+2)$$

$$4=2C$$

$$C=2$$

$$\frac{x^2-x-8}{(x+1)(x^2+5x+6)}=\frac{-3}{x+1}+\frac{2}{x+2}+\frac{2}{x+3}$$

35. Find the partial fraction decomposition:

$$\frac{x^2+2x+3}{(x^2+4)^2}=\frac{Ax+B}{x^2+4}+\frac{Cx+D}{(x^2+4)^2}$$

Multiplying both sides by $(x^2+4)^2$, we obtain:

$$x^2+2x+3=(Ax+B)(x^2+4)+Cx+D$$

$$x^2+2x+3=Ax^3+Bx^2+4Ax+4B+Cx+D$$

$$x^2+2x+3=Ax^3+Bx^2+(4A+C)x+4B+D$$

$A=0$; $B=1$;

$$4A+C=2 \qquad 4B+D=3$$

$$4(0)+C=2 \qquad 4(1)+D=3$$

$$C=2 \qquad D=-1$$

$$\frac{x^2+2x+3}{(x^2+4)^2}=\frac{1}{x^2+4}+\frac{2x-1}{(x^2+4)^2}$$

36. Find the partial fraction decomposition:

$$\frac{x^3+1}{(x^2+16)^2}=\frac{Ax+B}{x^2+16}+\frac{Cx+D}{(x^2+16)^2}$$

Multiplying both sides by $(x^2+16)^2$, we obtain:

$$x^3+1=(Ax+B)(x^2+16)+Cx+D$$

$$x^3+1=Ax^3+Bx^2+16Ax+16B+Cx+D$$

$$x^3+1=Ax^3+Bx^2+(16A+C)x+16B+D$$

$A=1$; $B=0$;

$$16A+C=0$$

$$16(1)+C=0$$

$$C=-16$$

$$16B+D=1$$

$$16(0)+D=1$$

$$D=1$$

$$\frac{x^3+1}{(x^2+16)^2}=\frac{x}{x^2+16}+\frac{-16x+1}{(x^2+16)^2}$$

37. Find the partial fraction decomposition:

$$\frac{7x+3}{x^3-2x^2-3x}=\frac{7x+3}{x(x-3)(x+1)}$$

$$=\frac{A}{x}+\frac{B}{x-3}+\frac{C}{x+1}$$

Multiplying both sides by $x(x-3)(x+1)$, we obtain:

$$7x+3=A(x-3)(x+1)+Bx(x+1)+Cx(x-3)$$

Let $x=0$, then

$$7(0)+3=A(0-3)(0+1)+B(0)(0+1)+C(0)(0-3)$$

$$3=-3A$$

$$A=-1$$

Let $x=3$, then

$$7(3)+3=A(3-3)(3+1)+B(3)(3+1)+C(3)(3-3)$$

$$24=12B$$

$$B=2$$

Let $x=-1$, then

$$7(-1)+3=A(-1-3)(-1+1)+B(-1)(-1+1)+C(-1)(-1-3)$$

$$-4=4C$$

$$C=-1$$

$$\frac{7x+3}{x^3-2x^2-3x}=\frac{-1}{x}+\frac{2}{x-3}+\frac{-1}{x+1}$$

38. Find the partial fraction decomposition:

$$\frac{x^5+1}{x^6-x^4}=\frac{(x+1)(x^4-x^3+x^2-x+1)}{x^4(x-1)(x+1)}$$
$$=\frac{x^4-x^3+x^2-x+1}{x^4(x-1)}$$
$$=\frac{A}{x}+\frac{B}{x^2}+\frac{C}{x^3}+\frac{D}{x^4}+\frac{E}{x-1}$$

Multiplying both sides by $x^4(x-1)$, we obtain:

$$x^4-x^3+x^2-x+1=Ax^3(x-1)+Bx^2(x-1)+Cx(x-1)+D(x-1)+Ex^4$$

Let $x=0$, then

$$0^4-0^3+0^2-0+1=A\cdot 0^3(0-1)+B\cdot 0^2(0-1)+C\cdot 0(0-1)+D(0-1)+E\cdot 0^4$$

$$1=-D$$

$$D=-1$$

Let $x=1$, then

$$1^4-1^3+1^2-1+1=A\cdot 1^3(1-1)+B\cdot 1^2(1-1)+C\cdot 1(1-1)+D(1-1)+E\cdot 1^4$$

$$1=E$$

Let $x=-1$, then

$$(-1)^4-(-1)^3+(-1)^2-(-1)+1=A(-1)^3(-1-1)+B(-1)^2(-1-1)+C(-1)(-1-1)+D(-1-1)+E(-1)^4$$

$$5=2A-2B+2C-2D+E$$

$$5=2A-2B+2C-2(-1)+1$$

$$2A-2B+2C=2$$

$$A-B+C=1$$

Let $x=2$, then

$$2^4-2^3+2^2-2+1=A\cdot 2^3(2-1)+B\cdot 2^2(2-1)+C\cdot 2(2-1)+D(2-1)+E\cdot 2^4$$

$$11=8A+4B+2C+D+16E$$

$$11=8A+4B+2C+(-1)+16(1)$$

$$8A+4B+2C=-4$$

$$4A+2B+C=-2$$

Let $x=3$, then

$$3^4-3^3+3^2-3+1=A\cdot 3^3(3-1)+B\cdot 3^2(3-1)+C\cdot 3(3-1)+D(3-1)+E\cdot 3^4$$

$$61=54A+18B+6C+2D+81E$$

$$61=54A+18B+6C+2(-1)+81(1)$$

$$-18=54A+18B+6C$$

$$9A+3B+C=-3$$

Solve the system of equations by using a matrix equation:

$$\begin{cases}4A+2B+C=-2\\ A-B+C=1\\ 9A+3B+C=-3\end{cases}$$

$$\begin{bmatrix}4 & 2 & 1\\ 1 & -1 & 1\\ 9 & 3 & 1\end{bmatrix}\begin{bmatrix}A\\ B\\ C\end{bmatrix}=\begin{bmatrix}-2\\ 1\\ -3\end{bmatrix}$$

$$\begin{bmatrix}A\\ B\\ C\end{bmatrix}=\begin{bmatrix}4 & 2 & 1\\ 1 & -1 & 1\\ 9 & 3 & 1\end{bmatrix}^{-1}\begin{bmatrix}-2\\ 1\\ -3\end{bmatrix}=\begin{bmatrix}0\\ -1\\ 0\end{bmatrix}$$

So, $A=0$, $B=-1$, and $C=0$. Thus,

$$\frac{x^5+1}{x^6-x^4}=\frac{x^4-x^3+x^2-x+1}{x^4(x-1)}$$
$$=\frac{-1}{x^2}+\frac{-1}{x^4}+\frac{1}{x-1}$$

39. Perform synthetic division to find a factor:

$$\begin{array}{r|rrrr} 2 & 1 & -4 & 5 & -2\\ & & 2 & -4 & 2\\ \hline & 1 & -2 & 1 & 0\end{array}$$

$$x^3-4x^2+5x-2=(x-2)(x^2-2x+1)$$
$$=(x-2)(x-1)^2$$

Find the partial fraction decomposition:

$$\frac{x^2}{x^3-4x^2+5x-2}=\frac{x^2}{(x-2)(x-1)^2}$$
$$=\frac{A}{x-2}+\frac{B}{x-1}+\frac{C}{(x-1)^2}$$

Multiplying both sides by $(x-2)(x-1)^2$, we obtain:

$x^2=A(x-1)^2+B(x-2)(x-1)+C(x-2)$

Let $x=2$, then

$2^2=A(2-1)^2+B(2-2)(2-1)+C(2-2)$

$4=A$

Let $x=1$, then

$1^2=A(1-1)^2+B(1-2)(1-1)+C(1-2)$

$1=-C$

$C=-1$

Let $x=0$, then

$0^2=A(0-1)^2+B(0-2)(0-1)+C(0-2)$

$0=A+2B-2C$

$0=4+2B-2(-1)$

$-2B=6$

$B=-3$

$$\frac{x^2}{x^3-4x^2+5x-2}=\frac{4}{x-2}+\frac{-3}{x-1}+\frac{-1}{(x-1)^2}$$

40. Perform synthetic division to find a factor:

```
1)1  1  -5   3
     1   2  -3
  ------------
  1  2  -3   0
```

$x^3+x^2-5x+3=(x-1)(x^2+2x-3)$
$=(x+3)(x-1)^2$

Find the partial fraction decomposition:

$$\frac{x^2+1}{x^3+x^2-5x+3}=\frac{x^2+1}{(x+3)(x-1)^2}$$
$$=\frac{A}{x+3}+\frac{B}{x-1}+\frac{C}{(x-1)^2}$$

Multiplying both sides by $(x+3)(x-1)^2$, we obtain:

$x^2+1=A(x-1)^2+B(x+3)(x-1)+C(x+3)$

Let $x=-3$, then

$(-3)^2+1=A(-3-1)^2+B(-3+3)(-3-1)+C(-3+3)$

$10=16A$

$A=\frac{5}{8}$

Let $x=1$, then

$1^2+1=A(1-1)^2+B(1+3)(1-1)+C(1+3)$

$2=4C$

$C=\frac{1}{2}$

Let $x=0$, then

$0^2+1=A(0-1)^2+B(0+3)(0-1)+C(0+3)$

$1=A-3B+3C$

$1=\frac{5}{8}-3B+3\left(\frac{1}{2}\right)$

$3B=\frac{9}{8}$

$B=\frac{3}{8}$

$$\frac{x^2+1}{x^3+x^2-5x+3}=\frac{(5/8)}{x+3}+\frac{(3/8)}{x-1}+\frac{(1/2)}{(x-1)^2}$$

41. Find the partial fraction decomposition:

$$\frac{x^3}{(x^2+16)^3}=\frac{Ax+B}{x^2+16}+\frac{Cx+D}{(x^2+16)^2}+\frac{Ex+F}{(x^2+16)^3}$$

Multiplying both sides by $(x^2+16)^3$, we obtain:

$x^3=(Ax+B)(x^2+16)^2+(Cx+D)(x^2+16)+Ex+F$

$x^3=(Ax+B)(x^4+32x^2+256)+Cx^3+Dx^2+16Cx+16D+Ex+F$

$x^3=Ax^5+Bx^4+32Ax^3+32Bx^2+256Ax+256B+Cx^3+Dx^2+16Cx+16D+Ex+F$

$x^3=Ax^5+Bx^4+(32A+C)x^3+(32B+D)x^2+(256A+16C+E)x+(256B+16D+F)$

$A=0;\ B=0;$

$32A+C=1$

$32(0)+C=1$

$C=1$

$$32B + D = 0 \qquad 256A + 16C + E = 0$$
$$32(0) + D = 0 \qquad 256(0) + 16(1) + E = 0$$
$$D = 0 \qquad E = -16$$

$$256B + 16D + F = 0$$
$$256(0) + 16(0) + F = 0$$
$$F = 0$$

$$\frac{x^3}{(x^2+16)^3} = \frac{x}{(x^2+16)^2} + \frac{-16x}{(x^2+16)^3}$$

42. Find the partial fraction decomposition:

$$\frac{x^2}{(x^2+4)^3} = \frac{Ax+B}{x^2+4} + \frac{Cx+D}{(x^2+4)^2} + \frac{Ex+F}{(x^2+4)^3}$$

Multiplying both sides by $(x^2+4)^3$, we obtain:

$$x^2 = (Ax+B)(x^2+4)^2 + (Cx+D)(x^2+4) + Ex + F$$
$$x^2 = (Ax+B)(x^4+8x^2+16) + Cx^3 + Dx^2 + 4Cx + 4D + Ex + F$$
$$x^2 = Ax^5 + Bx^4 + 8Ax^3 + 8Bx^2 + 16Ax + 16B + Cx^3 + Dx^2 + 4Cx + 4D + Ex + F$$
$$x^2 = Ax^5 + Bx^4 + (8A+C)x^3 + (8B+D)x^2 + (16A+4C+E)x + (16B+4D+F)$$

$A = 0$; $B = 0$;
$$8A + C = 0$$
$$8(0) + C = 0$$
$$C = 0$$

$$8B + D = 1 \qquad 16A + 4C + E = 0$$
$$8(0) + D = 1 \qquad 16(0) + 4(0) + E = 0$$
$$D = 1 \qquad E = 0$$

$$16B + 4D + F = 0$$
$$16(0) + 4(1) + F = 0$$
$$F = -4$$

$$\frac{x^2}{(x^2+4)^3} = \frac{1}{(x^2+4)^2} + \frac{-4}{(x^2+4)^3}$$

43. Find the partial fraction decomposition:

$$\frac{4}{2x^2-5x-3} = \frac{4}{(x-3)(2x+1)} = \frac{A}{x-3} + \frac{B}{2x+1}$$

Multiplying both sides by $(x-3)(2x+1)$, we obtain: $4 = A(2x+1) + B(x-3)$

Let $x = -\frac{1}{2}$, then

$$4 = A(2(-1/2)+1) + B\left(-\frac{1}{2} - 3\right)$$
$$4 = -\frac{7}{2}B$$
$$B = -\frac{8}{7}$$

Let $x = 3$, then $4 = A(2(3)+1) + B(3-3)$
$$4 = 7A$$
$$A = \frac{4}{7}$$

$$\frac{4}{2x^2-5x-3} = \frac{(4/7)}{x-3} + \frac{(-8/7)}{2x+1}$$

44. Find the partial fraction decomposition:

$$\frac{4x}{2x^2+3x-2} = \frac{4x}{(x+2)(2x-1)} = \frac{A}{x+2} + \frac{B}{2x-1}$$

Multiplying both sides by $(x+2)(2x-1)$, we obtain:

$$4x = A(2x-1) + B(x+2)$$

Let $x = \frac{1}{2}$, then

$$4(1/2) = A(2(1/2)-1) + B\left(\frac{1}{2} + 2\right)$$
$$2 = \frac{5}{3}B$$
$$B = \frac{4}{5}$$

Let $x = -2$, then
$$4(-2) = A(2(-2)-1) + B(-2+2)$$
$$-8 = -5A$$
$$A = \frac{8}{5}$$

$$\frac{4x}{2x^2+3x-2} = \frac{(8/5)}{x+2} + \frac{(4/5)}{2x-1}$$

45. Find the partial fraction decomposition:

$$\frac{2x+3}{x^4-9x^2}=\frac{2x+3}{x^2(x-3)(x+3)}$$
$$=\frac{A}{x}+\frac{B}{x^2}+\frac{C}{x-3}+\frac{D}{x+3}$$

Multiplying both sides by $x^2(x-3)(x+3)$, we obtain:

$$2x+3=Ax(x-3)(x+3)+B(x-3)(x+3)+Cx^2(x+3)+Dx^2(x-3)$$

Let $x=0$, then

$$2\cdot 0+3=A\cdot 0(0-3)(0+3)+B(0-3)(0+3)+C\cdot 0^2(0+3)+D\cdot 0^2(0-3)$$
$$3=-9B$$
$$B=-\frac{1}{3}$$

Let $x=3$, then

$$2\cdot 3+3=A\cdot 3(3-3)(3+3)+B(3-3)(3+3)+C\cdot 3^2(3+3)+D\cdot 3^2(3-3)$$
$$9=54C$$
$$C=\frac{1}{6}$$

Let $x=-3$, then

$$2(-3)+3=A(-3)(-3-3)(-3+3)+B(-3-3)(-3+3)+C(-3)^2(-3+3)+D(-3)^2(-3-3)$$
$$-3=-54D$$
$$D=\frac{1}{18}$$

Let $x=1$, then

$$2\cdot 1+3=A\cdot 1(1-3)(1+3)+B(1-3)(1+3)+C\cdot 1^2(1+3)+D\cdot 1^2(1-3)$$
$$5=-8A-8B+4C-2D$$
$$5=-8A-8(-1/3)+4(1/6)-2(1/18)$$
$$5=-8A+\frac{8}{3}+\frac{2}{3}-\frac{1}{9}$$
$$-8A=\frac{16}{9}$$
$$A=-\frac{2}{9}$$

$$\frac{2x+3}{x^4-9x^2}=\frac{(-2/9)}{x}+\frac{(-1/3)}{x^2}+\frac{(1/6)}{x-3}+\frac{(1/18)}{x+3}$$

46. Find the partial fraction decomposition:

$$\frac{x^2+9}{x^4-2x^2-8}=\frac{x^2+9}{(x^2+2)(x-2)(x+2)}$$
$$=\frac{A}{x-2}+\frac{B}{x+2}+\frac{Cx+D}{x^2+2}$$

Multiplying both sides by $(x^2+2)(x-2)(x+2)$, we obtain:

$$x^2+9=A(x^2+2)(x+2)+B(x-2)(x^2+2)+(Cx+D)(x-2)(x+2)$$

Let $x=2$, then

$$2^2+9=A(2^2+2)(2+2)+B(2-2)(2^2+2)+(C(2)+D)(2-2)(2+2)$$
$$13=24A$$
$$A=\frac{13}{24}$$

Let $x=-2$, then

$$(-2)^2+9=A((-2)^2+2)(-2+2)+B(-2-2)((-2)^2+2)+(C(-2)+D)(-2-2)(-2+2)$$
$$13=-24B$$
$$B=-\frac{13}{24}$$

Let $x=0$, then

$$0^2+9=A(0^2+2)(0+2)+B(0-2)(0^2+2)+(C(0)+D)(0-2)(0+2)$$
$$9=4A-4B-4D$$
$$9=4\left(\frac{13}{24}\right)-4\left(-\frac{13}{24}\right)-4D$$
$$4D=-\frac{14}{3}$$
$$D=-\frac{7}{6}$$

Let $x=1$, then

$$1^2+9=A(1^2+2)(1+2)+B(1-2)(1^2+2)+(C(1)+D)(1-2)(1+2)$$
$$10=9A-3B-3C-3D$$
$$10=\frac{39}{8}+\frac{13}{8}-3C+\frac{7}{2}$$
$$3C=0$$
$$C=0$$

$$\frac{x^2+9}{x^4-2x^2-8}=\frac{(13/24)}{x-2}+\frac{(-13/24)}{x+2}+\frac{(-7/6)}{x^2+2}$$

Section 10.6

1. $y = 3x + 2$
   The graph is a line.
   x-intercept:
   $0 = 3x + 2$
   $3x = -2$
   $x = -\frac{2}{3}$
   y-intercept: $y = 3(0) + 2 = 2$

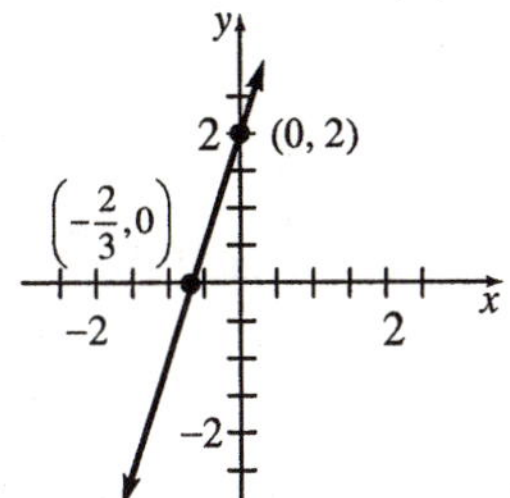

2. $y = x^2 - 4$
   The graph is a parabola.
   x-intercepts:
   $0 = x^2 - 4$
   $x^2 = 4$
   $x = -2, x = 2$
   y-intercept: $y = 0^2 - 4 = -4$
   The vertex has x-coordinate:
   $x = -\frac{b}{2a} = -\frac{0}{2(1)} = 0$.
   The y-coordinate of the vertex is
   $y = 0^2 - 4 = -4$.

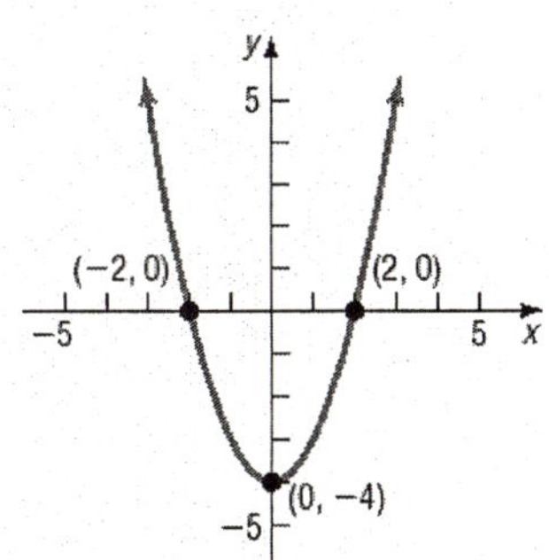

3. $y^2 = x^2 - 1$
   $x^2 - y^2 = 1$
   $\frac{x^2}{1^2} - \frac{y^2}{1^2} = 1$

The graph is a hyperbola with center (0, 0), transverse axis along the x-axis, and vertices at $(-1, 0)$ and $(1, 0)$. The asymptotes are $y = -x$ and $y = x$.

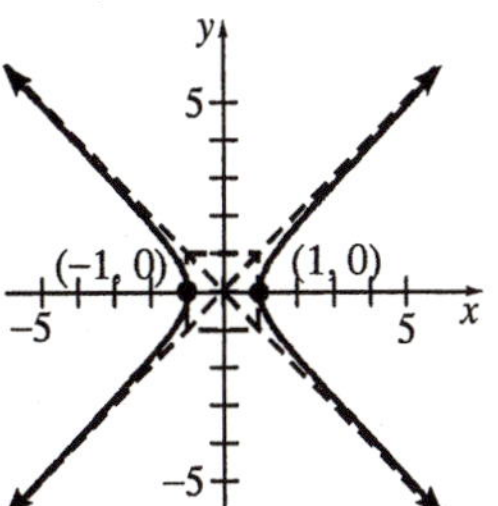

4. $x^2 + 4y^2 = 4$
   $\frac{x^2 + 4y^2}{4} = \frac{4}{4}$
   $\frac{x^2}{4} + y^2 = 1$
   $\frac{x^2}{2^2} + \frac{y^2}{1^2} = 1$

The graph is an ellipse with center $(0, 0)$, major axis along the x-axis, vertices at $(-2, 0)$ and $(2, 0)$. The graph also has y-intercepts at $(0, -1)$ and $(0, 1)$.

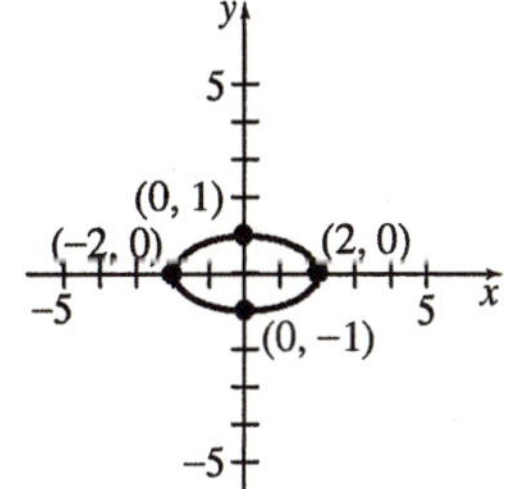

5. $\begin{cases} y = x^2 + 1 \\ y = x + 1 \end{cases}$

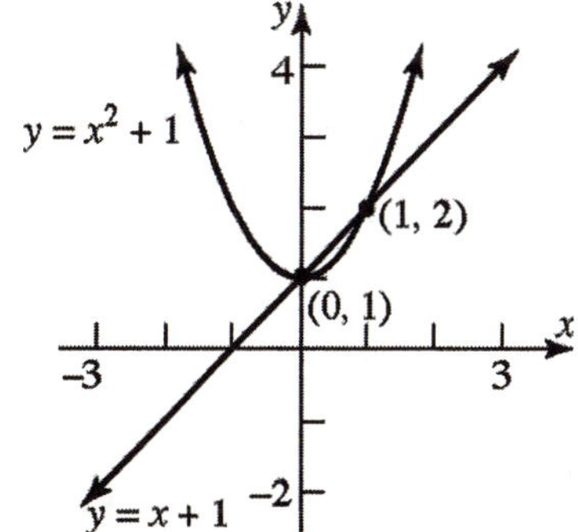

(0, 1) and (1, 2) are the intersection points.

Solve by substitution:

$$x^2+1=x+1$$
$$x^2-x=0$$
$$x(x-1)=0$$
$$x=0 \text{ or } x=1$$
$$y=1 \qquad y=2$$

Solutions: (0, 1) and (1, 2)

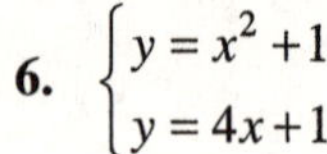

6. $\begin{cases} y=x^2+1 \\ y=4x+1 \end{cases}$

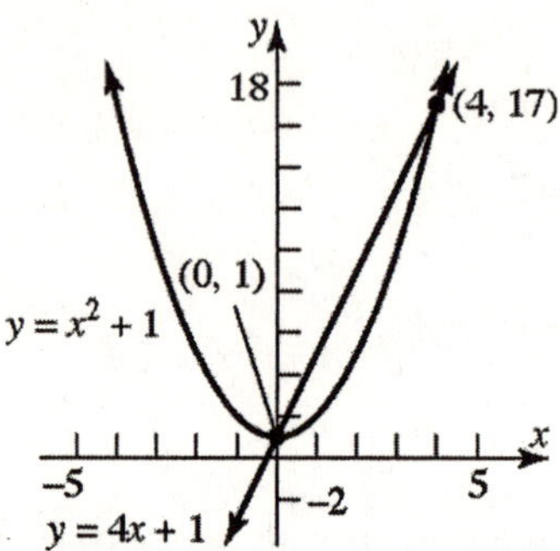

(0, 1) and (4, 17) are the intersection points.

Solve by substitution:

$$x^2+1=4x+1$$
$$x^2-4x=0$$
$$x(x-4)=0$$
$$x=0 \text{ or } x=4$$
$$y=1 \qquad y=17$$

Solutions: (0, 1) and (4, 17)

7. $\begin{cases} y=\sqrt{36-x^2} \\ y=8-x \end{cases}$

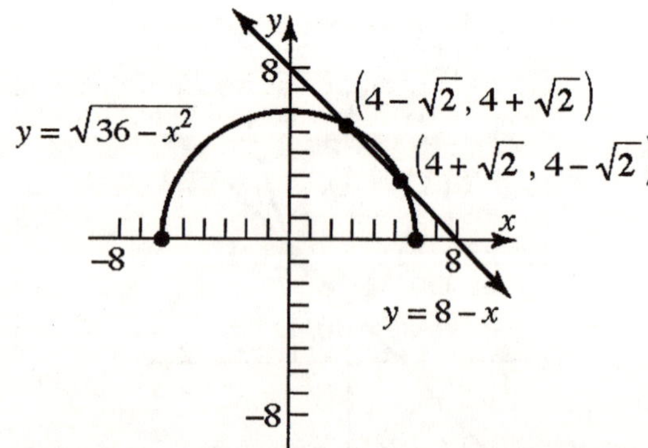

(2.59, 5.41) and (5.41, 2.59) are the intersection points.

Solve by substitution:

$$\sqrt{36-x^2}=8-x$$
$$36-x^2=64-16x+x^2$$
$$2x^2-16x+28=0$$
$$x^2-8x+14=0$$
$$x=\frac{8\pm\sqrt{64-56}}{2}$$
$$=\frac{8\pm 2\sqrt{2}}{2}$$
$$=4\pm\sqrt{2}$$

If $x=4+\sqrt{2},\ y=8-\left(4+\sqrt{2}\right)=4-\sqrt{2}$

If $x=4-\sqrt{2},\ y=8-\left(4-\sqrt{2}\right)=4+\sqrt{2}$

Solutions: $\left(4+\sqrt{2},4-\sqrt{2}\right)$ and $\left(4-\sqrt{2},4+\sqrt{2}\right)$

8. $\begin{cases} y=\sqrt{4-x^2} \\ y=2x+4 \end{cases}$

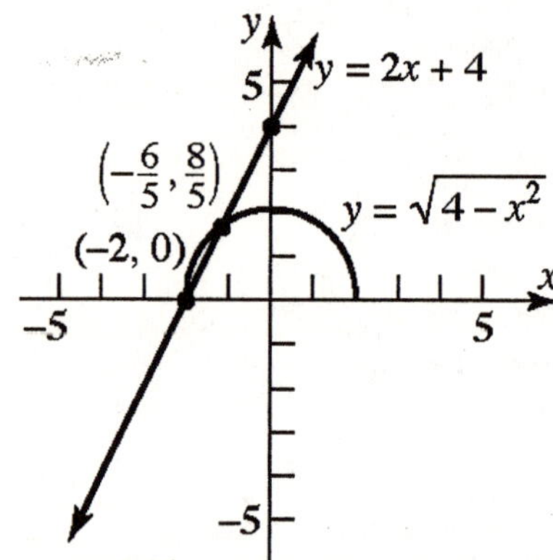

(–2, 0) and (–1.2, 1.6) are the intersection points.

Solve by substitution:

$$\sqrt{4-x^2}=2x+4$$
$$4-x^2=4x^2+16x+16$$
$$5x^2+16x+12=0$$
$$(x+2)(5x+6)=0$$
$$x=-2 \text{ or } x=-\frac{6}{5}$$
$$y=0 \quad \text{or } y=\frac{8}{5}$$

Solutions: $(-2,0)$ and $\left(-\frac{6}{5},\frac{8}{5}\right)$

9. $\begin{cases} y = \sqrt{x} \\ y = 2 - x \end{cases}$

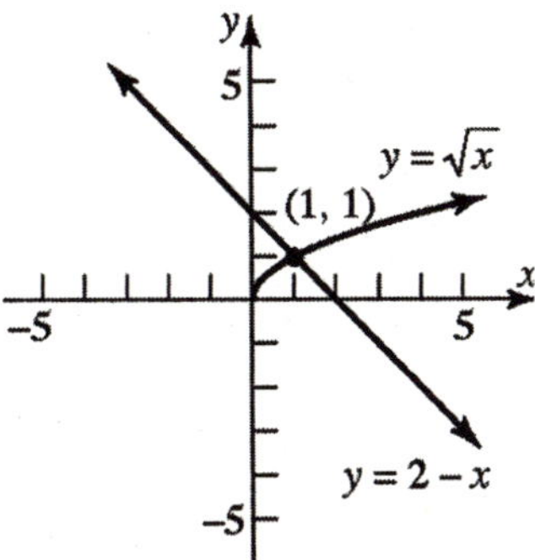

(1, 1) is the intersection point.

Solve by substitution:

$$\sqrt{x} = 2 - x$$
$$x = 4 - 4x + x^2$$
$$x^2 - 5x + 4 = 0$$
$$(x-4)(x-1) = 0$$
$$x = 4 \quad \text{or } x = 1$$
$$y = -2 \quad \text{or } y = 1$$

Eliminate (4, –2); we must have $y \geq 0$.

Solution: (1, 1)

10. $\begin{cases} y = \sqrt{x} \\ y = 6 - x \end{cases}$

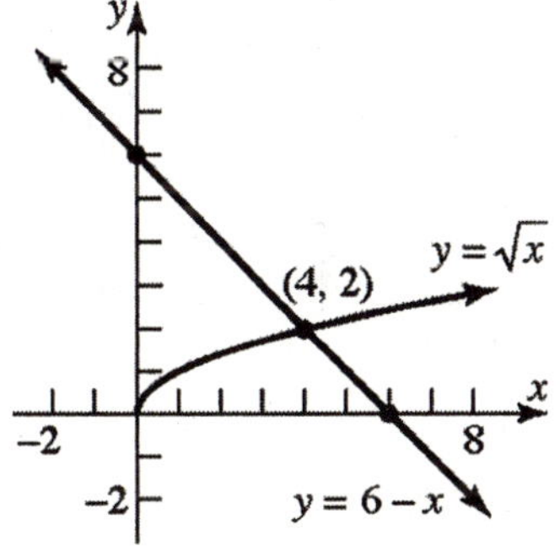

(4, 2) is the intersection point.

Solve by substitution:

$$\sqrt{x} = 6 - x$$
$$x = 36 - 12x + x^2$$
$$x^2 - 13x + 36 = 0$$
$$(x-4)(x-9) = 0$$
$$x = 4 \quad \text{or } x = 9$$
$$y = 2 \quad \text{or } y = -3$$

Eliminate (9, –3); we must have $y \geq 0$.

Solution: (4, 2)

11. $\begin{cases} x = 2y \\ x = y^2 - 2y \end{cases}$

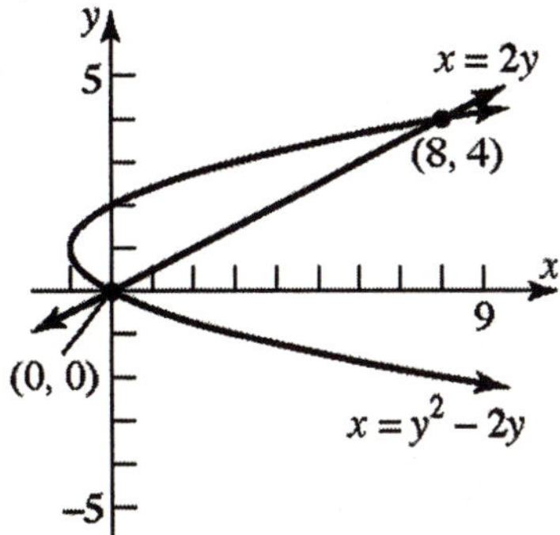

(0, 0) and (8, 4) are the intersection points.

Solve by substitution:

$$2y = y^2 - 2y$$
$$y^2 - 4y = 0$$
$$y(y-4) = 0$$
$$y = 0 \text{ or } y = 4$$
$$x = 0 \text{ or } x = 8$$

Solutions: (0, 0) and (8, 4)

12. $\begin{cases} y = x - 1 \\ y = x^2 - 6x + 9 \end{cases}$

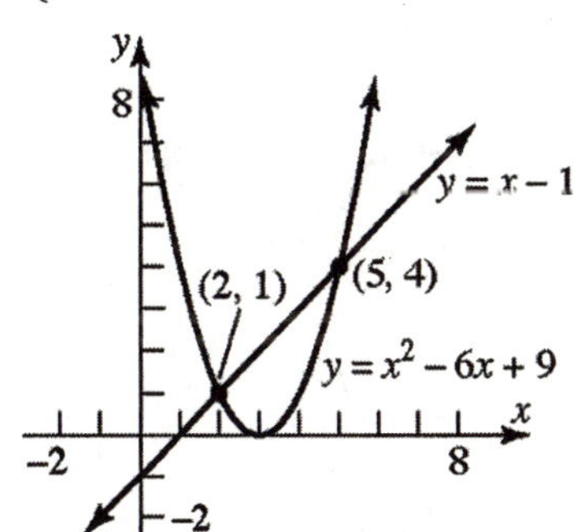

(2, 1) and (5, 4) are the intersection points.

Solve by substitution:

$$x^2 - 6x + 9 = x - 1$$
$$x^2 - 7x + 10 = 0$$
$$(x-2)(x-5) = 0$$
$$x = 2 \text{ or } x = 5$$
$$y = 1 \text{ or } y = 4$$

Solutions: (2, 1) and (5, 4)

13. $\begin{cases} x^2+y^2=4 \\ x^2+2x+y^2=0 \end{cases}$

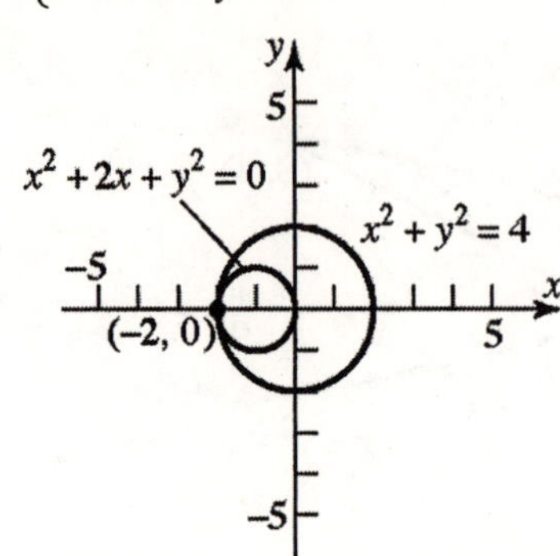

(–2, 0) is the intersection point.

Substitute 4 for x^2+y^2 in the second equation.

$$2x+4=0$$
$$2x=-4$$
$$x=-2$$
$$y=\sqrt{4-(-2)^2}=0$$

Solution: (–2, 0)

14. $\begin{cases} x^2+y^2=8 \\ x^2+y^2+4y=0 \end{cases}$

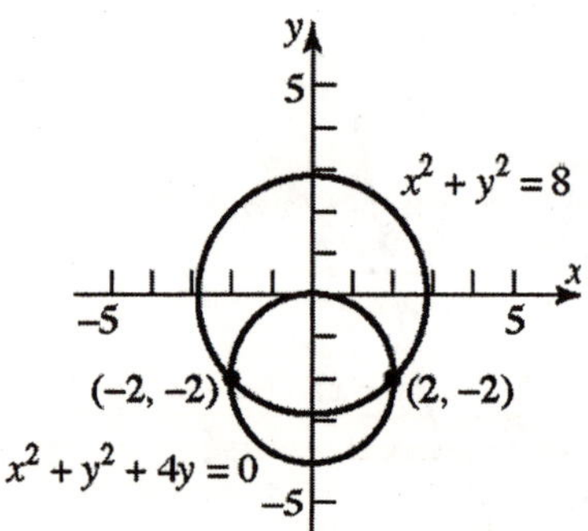

(–2, –2) and (2, –2) are the intersection points.

Substitute 8 for x^2+y^2 in the second equation.

$$8+4y=0$$
$$4y=-8$$
$$y=-2$$
$$x=\pm\sqrt{8-(-2)^2}=\pm 2$$

Solution: (–2, –2) and (2, –2)

15. $\begin{cases} y=3x-5 \\ x^2+y^2=5 \end{cases}$

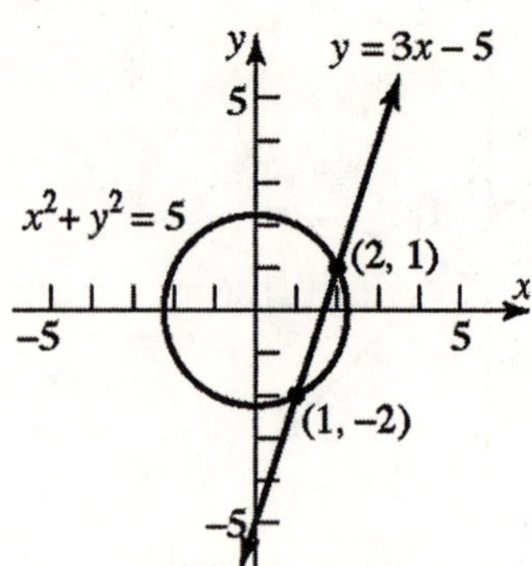

(1, –2) and (2, 1) are the intersection points.

Solve by substitution:

$$x^2+(3x-5)^2=5$$
$$x^2+9x^2-30x+25=5$$
$$10x^2-30x+20=0$$
$$x^2-3x+2=0$$
$$(x-1)(x-2)=0$$
$$x=1 \quad \text{or} \quad x=2$$
$$y=-2 \qquad y=1$$

Solutions: (1, –2) and (2, 1)

16. $\begin{cases} x^2+y^2=10 \\ y=x+2 \end{cases}$

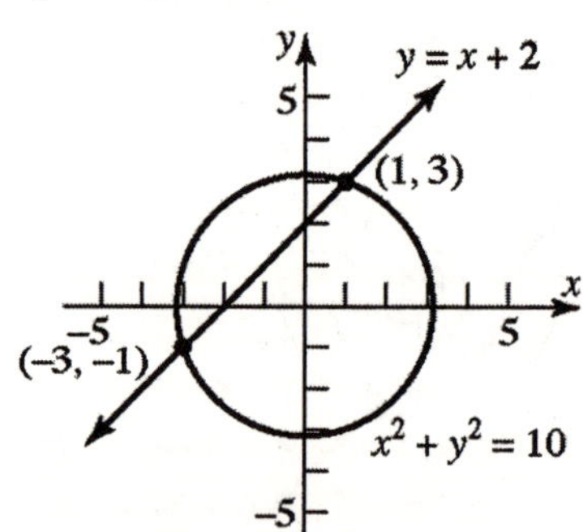

(1, 3) and (–3, –1) are the intersection points.

Solve by substitution:

$$x^2+(x+2)^2=10$$
$$x^2+x^2+4x+4=10$$
$$2x^2+4x-6=0$$
$$2(x+3)(x-1)=0$$
$$x=-3 \text{ or } x=1$$
$$y=-1 \qquad y=3$$

Solutions: (–3, –1) and (1, 3)

17. $\begin{cases} x^2 + y^2 = 4 \\ y^2 - x = 4 \end{cases}$

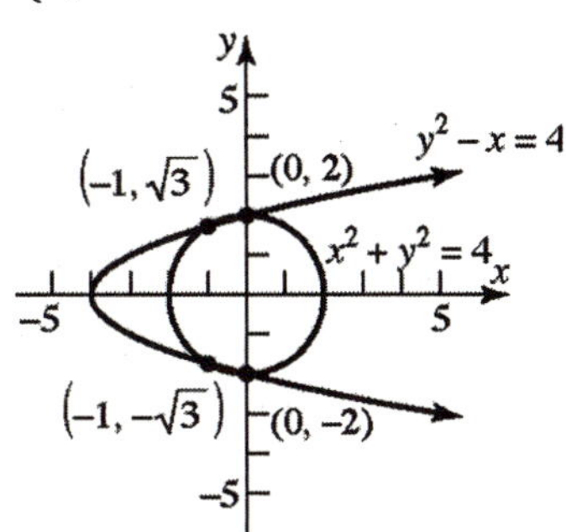

(–1, 1.73), (–1, –1.73), (0, 2), and (0, –2) are the intersection points.

Substitute $x+4$ for y^2 in the first equation:

$$x^2 + x + 4 = 4$$
$$x^2 + x = 0$$
$$x(x+1) = 0$$
$$x = 0 \quad \text{or} \quad x = -1$$
$$y^2 = 4 \qquad y^2 = 3$$
$$y = \pm 2 \qquad y = \pm\sqrt{3}$$

Solutions: $(0, -2), (0, 2), \left(-1, \sqrt{3}\right), \left(-1, -\sqrt{3}\right)$

18. $\begin{cases} x^2 + y^2 = 16 \\ x^2 - 2y = 8 \end{cases}$

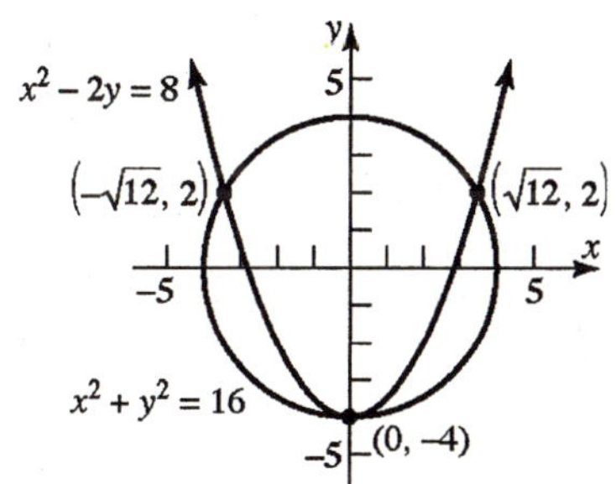

(–3.46, 2), (0, –4), and (3.46, 2) are the intersection points.

Substitute $2y+8$ for x^2 in the first equation.

$$2y + 8 + y^2 = 16$$
$$y^2 + 2y - 8 = 0$$
$$(y+4)(y-2) = 0$$
$$y = -4 \quad \text{or} \quad y = 2$$
$$x^2 = 0 \quad \text{or} \quad x^2 = 12$$
$$x = 0 \qquad x = \pm 2\sqrt{3}$$

Solutions: $(0, -4), \left(2\sqrt{3}, 2\right), \left(-2\sqrt{3}, 2\right)$

19. $\begin{cases} xy = 4 \\ x^2 + y^2 = 8 \end{cases}$

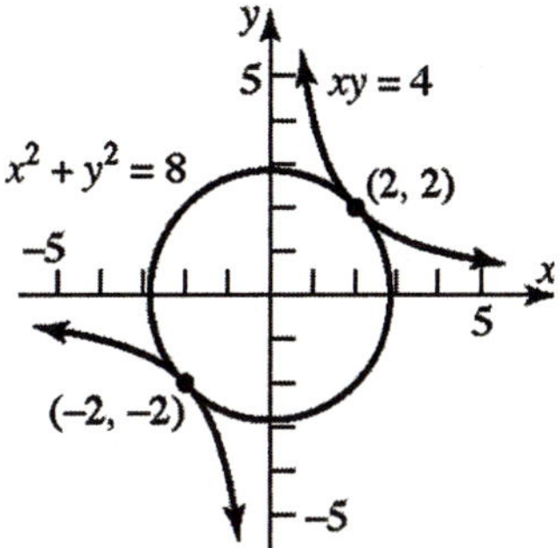

(–2, –2) and (2, 2) are the intersection points.

Solve by substitution:

$$x^2 + \left(\frac{4}{x}\right)^2 = 8$$
$$x^2 + \frac{16}{x^2} = 8$$
$$x^4 + 16 = 8x^2$$
$$x^4 - 8x^2 + 16 = 0$$
$$\left(x^2 - 4\right)^2 = 0$$
$$x^2 - 4 = 0$$
$$x^2 = 4$$
$$x = 2 \text{ or } x = -2$$
$$y = 2 \qquad y = -2$$

Solutions: (–2, –2) and (2, 2)

20. $\begin{cases} x^2 = y \\ xy = 1 \end{cases}$

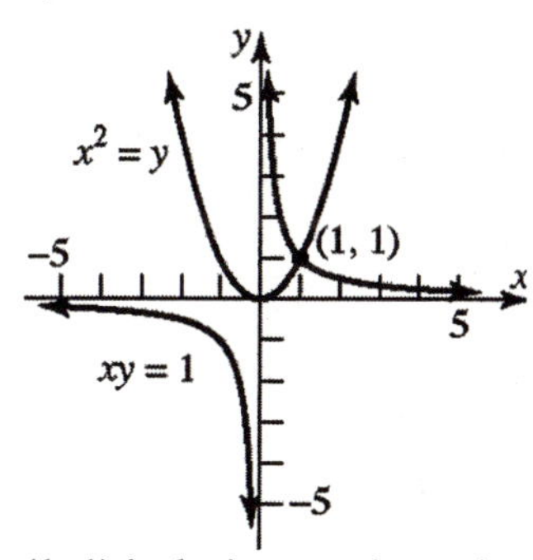

(1, 1) is the intersection point.

Solve by substitution:

$$x^2 = \frac{1}{x}$$

$$x^3 = 1$$

$$x = 1 \Rightarrow y = (1)^2 = 1$$

Solution: (1, 1)

21. $\begin{cases} x^2 + y^2 = 4 \\ y = x^2 - 9 \end{cases}$

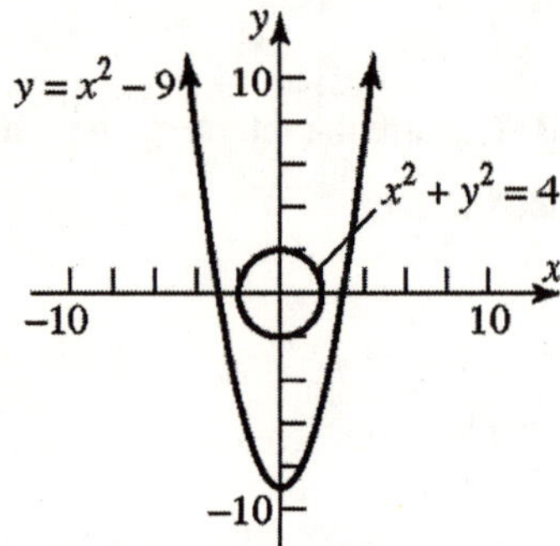

No solution; Inconsistent.

Solve by substitution:

$$x^2 + (x^2 - 9)^2 = 4$$

$$x^2 + x^4 - 18x^2 + 81 = 4$$

$$x^4 - 17x^2 + 77 = 0$$

$$x^2 = \frac{17 \pm \sqrt{289 - 4(77)}}{2}$$

$$= \frac{17 \pm \sqrt{-19}}{2}$$

There are no real solutions to this expression. Inconsistent.

22. $\begin{cases} xy = 1 \\ y = 2x + 1 \end{cases}$

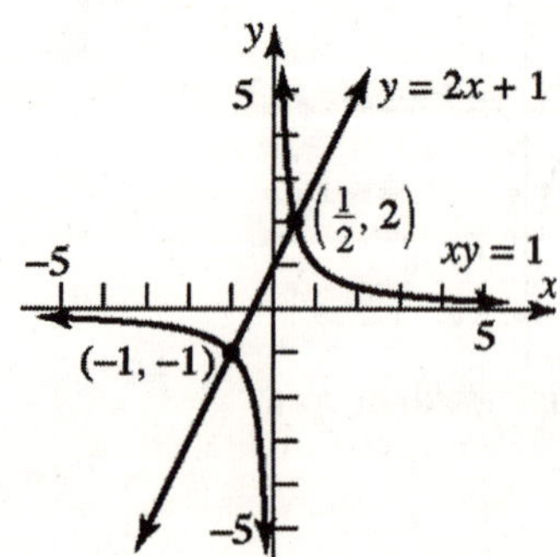

(−1, −1) and (0.5, 2) are the intersection points.

Solve by substitution:

$$x(2x + 1) = 1$$

$$2x^2 + x - 1 = 0$$

$$(x + 1)(2x - 1) = 0$$

$$x = -1 \text{ or } x = \frac{1}{2}$$

$$y = -1 \quad y = 2$$

Solutions: (−1, −1) and $\left(\frac{1}{2}, 2\right)$

23. $\begin{cases} y = x^2 - 4 \\ y = 6x - 13 \end{cases}$

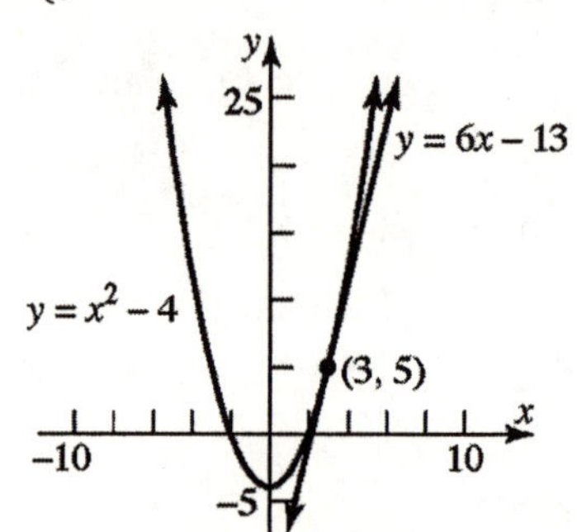

(3, 5) is the intersection point.

Solve by substitution:

$$x^2 - 4 = 6x - 13$$

$$x^2 - 6x + 9 = 0$$

$$(x - 3)^2 = 0$$

$$x - 3 = 0$$

$$x = 3 \Rightarrow y = (3)^2 - 4 = 5$$

Solution: (3,5)

24. $\begin{cases} x^2 + y^2 = 10 \\ xy = 3 \end{cases}$

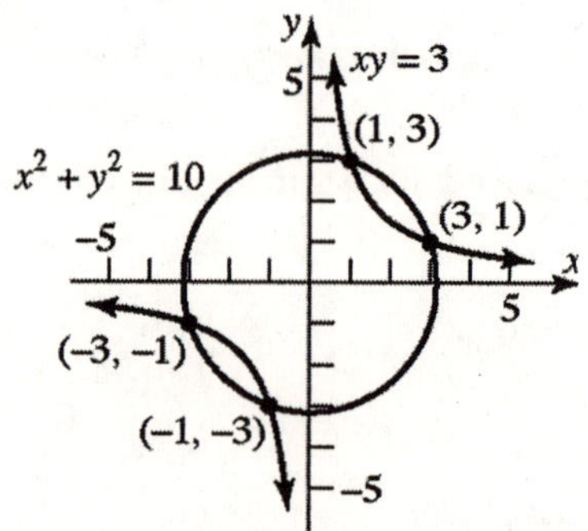

(1, 3), (3, 1), (−3, −1), and (−1, −3) are the intersection points.

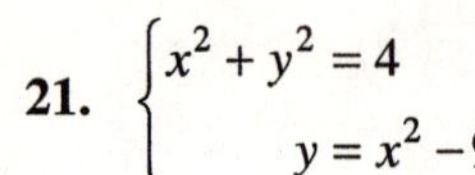

Solve by substitution:

$$x^2+\left(\tfrac{3}{x}\right)^2=10$$
$$x^2+\frac{9}{x^2}=10$$
$$x^4+9=10x^2$$
$$x^4-10x^2+9=0$$
$$(x^2-9)(x^2-1)=0$$
$$(x-3)(x+3)(x-1)(x+1)=0$$

$x=3$ or $x=-3$ or $x=1$ or $x=-1$

$y=1 \quad y=-1 \quad y=3 \quad y=-3$

Solutions: (3, 1), (–3, –1), (1, 3), (–1, –3)

25. Solve the second equation for y, substitute into the first equation and solve:

$$\begin{cases} 2x^2+y^2=18 \\ xy=4 \Rightarrow y=\dfrac{4}{x} \end{cases}$$

$$2x^2+\left(\frac{4}{x}\right)^2=18$$
$$2x^2+\frac{16}{x^2}=18$$
$$2x^4+16=18x^2$$
$$2x^4-18x^2+16=0$$
$$x^4-9x^2+8=0$$
$$(x^2-8)(x^2-1)-0$$
$$x^2=8 \Rightarrow x=\pm\sqrt{8}=\pm 2\sqrt{2}$$

or $x^2=1 \Rightarrow x=\pm 1$

If $x=2\sqrt{2}$: $\quad y=\dfrac{4}{2\sqrt{2}}=\sqrt{2}$

If $x=-2\sqrt{2}$: $\quad y=\dfrac{4}{-2\sqrt{2}}=-\sqrt{2}$

If $x=1$: $\quad y=\dfrac{4}{1}=4$

If $x=-1$: $\quad y=\dfrac{4}{-1}=-4$

Solutions:

$\left(2\sqrt{2},\sqrt{2}\right),\left(-2\sqrt{2},-\sqrt{2}\right)$, (1, 4), (−1, −4)

26. Solve the second equation for y, substitute into the first equation and solve:

$$\begin{cases} x^2-y^2=21 \\ x+y=7 \Rightarrow y=7-x \end{cases}$$

$$x^2-(7-x)^2=21$$
$$x^2-\left(49-14x+x^2\right)=21$$
$$14x=70$$
$$x=5 \Rightarrow y=7-5=2$$

Solution: (5, 2)

27. Substitute the first equation into the second equation and solve:

$$\begin{cases} y=2x+1 \\ 2x^2+y^2=1 \end{cases}$$

$$2x^2+(2x+1)^2=1$$
$$2x^2+4x^2+4x+1=1$$
$$6x^2+4x=0$$
$$2x(3x+2)=0$$
$$2x=0 \text{ or } 3x+2=0$$
$$x=0 \text{ or } x=-\frac{2}{3}$$

If $x=0$: $\quad y=2(0)+1=1$

If $x=-\dfrac{2}{3}$: $\quad y=2\left(-\dfrac{2}{3}\right)+1=-\dfrac{4}{3}+1=-\dfrac{1}{3}$

Solutions: (0, 1), $\left(-\dfrac{2}{3},-\dfrac{1}{3}\right)$

28. Solve the second equation for x and substitute into the first equation and solve:

$$\begin{cases} x^2-4y^2=16 \\ 2y-x=2 \Rightarrow x=2y-2 \end{cases}$$

$$(2y-2)^2-4y^2=16$$
$$4y^2-8y+4-4y^2=16$$
$$-8y=12$$
$$y=-\frac{3}{2} \Rightarrow x=2\left(-\frac{3}{2}\right)-2=-5$$

Solutions: $\left(-5,-\tfrac{3}{2}\right)$

29. Solve the first equation for y, substitute into the second equation and solve:

$$\begin{cases} x+y+1=0 \Rightarrow y=-x-1 \\ x^2+y^2+6y-x=-5 \end{cases}$$

$$x^2+(-x-1)^2+6(-x-1)-x=-5$$
$$x^2+x^2+2x+1-6x-6-x=-5$$
$$2x^2-5x=0$$
$$x(2x-5)=0$$
$$x=0 \text{ or } x=\frac{5}{2}$$

If $x=0$: $y=-(0)-1=-1$

If $x=\frac{5}{2}$: $y=-\frac{5}{2}-1=-\frac{7}{2}$

Solutions: $(0,-1), \left(\frac{5}{2},-\frac{7}{2}\right)$

30. Solve the second equation for y, substitute into the first equation and solve:

$$\begin{cases} 2x^2-xy+y^2=8 \\ xy=4 \Rightarrow y=\frac{4}{x} \end{cases}$$

$$2x^2-x\left(\frac{4}{x}\right)+\left(\frac{4}{x}\right)^2=8$$
$$2x^2-4+\frac{16}{x^2}=8$$
$$2x^4+16=12x^2$$
$$x^4-6x^2+8=0$$
$$\left(x^2-4\right)\left(x^2-2\right)=0$$
$$(x-2)(x+2)\left(x-\sqrt{2}\right)\left(x+\sqrt{2}\right)=0$$

$x=2$ or $x=-2$ or $x=\sqrt{2}$ or $x=-\sqrt{2}$

$y=2$ $\quad y=-2$ $\quad y=2\sqrt{2}$ $\quad y=-2\sqrt{2}$

Solutions:

$(2,2), (-2,-2), \left(\sqrt{2},2\sqrt{2}\right), \left(-\sqrt{2},-2\sqrt{2}\right)$

31. Solve the second equation for y, substitute into the first equation and solve:

$$\begin{cases} 4x^2-3xy+9y^2=15 \\ 2x+3y=5 \Rightarrow y=-\frac{2}{3}x+\frac{5}{3} \end{cases}$$

$$4x^2-3x\left(-\frac{2}{3}x+\frac{5}{3}\right)+9\left(-\frac{2}{3}x+\frac{5}{3}\right)^2=15$$
$$4x^2+2x^2-5x+4x^2-20x+25=15$$
$$10x^2-25x+10=0$$
$$2x^2-5x+2=0$$
$$(2x-1)(x-2)=0$$
$$x=\frac{1}{2} \text{ or } x=2$$

If $x=\frac{1}{2}$: $y=-\frac{2}{3}\left(\frac{1}{2}\right)+\frac{5}{3}=\frac{4}{3}$

If $x=2$: $y=-\frac{2}{3}(2)+\frac{5}{3}=\frac{1}{3}$

Solutions: $\left(\frac{1}{2},\frac{4}{3}\right), \left(2,\frac{1}{3}\right)$

32. $\begin{cases} 2y^2-3xy+6y+2x+4=0 \\ 2x-3y+4=0 \end{cases}$

Solve the second equation for x, substitute into the first equation and solve:

$$2x-3y+4=0$$
$$2x=3y-4$$
$$x=\frac{3y-4}{2}$$
$$2y^2-3\left(\frac{3y-4}{2}\right)y+6y+2\left(\frac{3y-4}{2}\right)=-4$$
$$2y^2-\frac{9}{2}y^2+6y+6y+3y-4=-4$$
$$-\frac{5}{2}y^2+15y=0$$
$$-5y^2+30y=0$$
$$-5y(y-6)=0$$

$y=0 \Rightarrow x=\frac{3(0)-4}{2}-2$

or

$y=6 \Rightarrow x=\frac{3(6)-4}{2}=7$

Solutions: $(-2,0), (7,6)$

33. Multiply each side of the second equation by 4 and add the equations to eliminate y:

$$\begin{cases} x^2-4y^2=-7 \longrightarrow x^2-4y^2=-7 \\ 3x^2+y^2=31 \xrightarrow{4} 12x^2+4y^2=124 \end{cases}$$

$$13x^2 = 117$$
$$x^2=9$$
$$x=\pm 3$$

If $x=3$: $3(3)^2+y^2=31 \Rightarrow y^2=4 \Rightarrow y=\pm 2$

If $x=-3$: $3(-3)^2+y^2=31 \Rightarrow y^2=4 \Rightarrow y=\pm 2$

Solutions: $(3,2), (3,-2), (-3,2), (-3,-2)$

34. $\begin{cases} 3x^2 - 2y^2 + 5 = 0 \\ 2x^2 - y^2 + 2 = 0 \end{cases}$

$\begin{cases} 3x^2 - 2y^2 = -5 \\ 2x^2 - y^2 = -2 \end{cases}$

Multiply each side of the second equation by –2 and add the equations to eliminate *y*:

$$\begin{aligned} 3x^2 - 2y^2 &= -5 \\ \underline{-4x^2 + 2y^2} &\underline{= 4} \\ -x^2 &= -1 \\ x^2 &= 1 \\ x &= \pm 1 \end{aligned}$$

If $x = 1$:

$2(1)^2 - y^2 = -2 \Rightarrow y^2 = 4 \Rightarrow y = \pm 2$

If $x = -1$:

$2(-1)^2 - y^2 = -2 \Rightarrow y^2 = 4 \Rightarrow y = \pm 2$

Solutions: (1, 2), (1, –2), (–1, 2), (–1, –2)

35. $\begin{cases} 7x^2 - 3y^2 + 5 = 0 \\ 3x^2 + 5y^2 = 12 \end{cases}$

$\begin{cases} 7x^2 - 3y^2 = -5 \\ 3x^2 + 5y^2 = 12 \end{cases}$

Multiply each side of the first equation by 5 and each side of the second equation by 3 and add the equations to eliminate *y*:

$$\begin{aligned} 35x^2 - 15y^2 &= -25 \\ \underline{9x^2 + 15y^2} &\underline{= 36} \\ 44x^2 &= 11 \\ x^2 &= \frac{1}{4} \\ x &= \pm\frac{1}{2} \end{aligned}$$

If $x = \frac{1}{2}$:

$3\left(\frac{1}{2}\right)^2 + 5y^2 = 12 \Rightarrow y^2 = \frac{9}{4} \Rightarrow y = \pm\frac{3}{2}$

If $x = -\frac{1}{2}$:

$3\left(-\frac{1}{2}\right)^2 + 5y^2 = 12 \Rightarrow y^2 = \frac{9}{4} \Rightarrow y = \pm\frac{3}{2}$

Solutions:

$\left(\frac{1}{2}, \frac{3}{2}\right), \left(\frac{1}{2}, -\frac{3}{2}\right), \left(-\frac{1}{2}, \frac{3}{2}\right), \left(-\frac{1}{2}, -\frac{3}{2}\right)$

36. $\begin{cases} x^2 - 3y^2 + 1 = 0 \\ 2x^2 - 7y^2 + 5 = 0 \end{cases}$

$\begin{cases} x^2 - 3y^2 = -1 \\ 2x^2 - 7y^2 = -5 \end{cases}$

Multiply each side of the first equation by –2 and add the equations to eliminate *x*:

$$\begin{aligned} -2x^2 + 6y^2 &= 2 \\ \underline{2x^2 - 7y^2} &\underline{= -5} \\ -y^2 &= -3 \\ y^2 &= 3 \\ y &= \pm\sqrt{3} \end{aligned}$$

If $y = \sqrt{3}$:

$x^2 - 3\left(\sqrt{3}\right)^2 = -1 \Rightarrow x^2 = 8 \Rightarrow x = \pm 2\sqrt{2}$

If $y = -\sqrt{3}$:

$x^2 - 3\left(-\sqrt{3}\right)^2 = -1 \Rightarrow x^2 = 8 \Rightarrow x = \pm 2\sqrt{2}$

Solutions:

$\left(2\sqrt{2}, \sqrt{3}\right), \left(2\sqrt{2}, -\sqrt{3}\right), \left(-2\sqrt{2}, \sqrt{3}\right),$
$\left(-2\sqrt{2}, -\sqrt{3}\right)$

37. Multiply each side of the second equation by 2 and add the equations to eliminate *xy*:

$$\begin{cases} x^2 + 2xy = 10 \longrightarrow & x^2 + 2xy = 10 \\ 3x^2 - \ xy = \ 2 \xrightarrow{2} & \underline{6x^2 - 2xy = 4} \end{cases}$$

$$\begin{aligned} 7x^2 \qquad &= 14 \\ x^2 &= 2 \\ x &= \pm\sqrt{2} \end{aligned}$$

If $x = \sqrt{2}$:

$3\left(\sqrt{2}\right)^2 - \sqrt{2} \cdot y = 2$

$\Rightarrow -\sqrt{2} \cdot y = -4 \Rightarrow y = \frac{4}{\sqrt{2}} \Rightarrow y = 2\sqrt{2}$

If $x = -\sqrt{2}$:

$3\left(-\sqrt{2}\right)^2 - \left(-\sqrt{2}\right)y = 2$

$\Rightarrow \sqrt{2} \cdot y = -4 \Rightarrow y = \frac{-4}{\sqrt{2}} \Rightarrow y = -2\sqrt{2}$

Solutions: $\left(\sqrt{2}, 2\sqrt{2}\right), \left(-\sqrt{2}, -2\sqrt{2}\right)$

38. $\begin{cases} 5xy+13y^2+36=0 \\ xy+7y^2=6 \end{cases}$

$\begin{cases} 5xy+13y^2=-36 \\ xy+7y^2=6 \end{cases}$

Multiply each side of the second equation by –5 and add the equations to eliminate xy:

$$\begin{aligned} 5xy+13y^2&=-36 \\ -5xy-35y^2&=-30 \\ \hline -22y^2&=-66 \\ y^2&=3 \\ y&=\pm\sqrt{3} \end{aligned}$$

If $y=\sqrt{3}$:

$x\left(\sqrt{3}\right)+7\left(\sqrt{3}\right)^2=6$

$\Rightarrow \sqrt{3}\cdot x=-15 \Rightarrow x=\dfrac{-15}{\sqrt{3}} \Rightarrow x=-5\sqrt{3}$

If $y=-\sqrt{3}$:

$x\left(-\sqrt{3}\right)+7\left(-\sqrt{3}\right)^2=6$

$\Rightarrow -\sqrt{3}\cdot x=-15 \Rightarrow x=\dfrac{15}{\sqrt{3}} \Rightarrow x=5\sqrt{3}$

Solutions: $\left(-5\sqrt{3}, \sqrt{3}\right), \left(5\sqrt{3}, -\sqrt{3}\right)$

39. $\begin{cases} 2x^2+y^2=2 \\ x^2-2y^2+8=0 \end{cases}$

$\begin{cases} 2x^2+y^2=2 \\ x^2-2y^2=-8 \end{cases}$

Multiply each side of the first equation by 2 and add the equations to eliminate y:

$$\begin{aligned} 4x^2+2y^2&=4 \\ x^2-2y^2&=-8 \\ \hline 5x^2&=-4 \\ x^2&=-\frac{4}{5} \end{aligned}$$

No real solution. The system is inconsistent.

40. $\begin{cases} y^2-x^2+4=0 \\ 2x^2+3y^2=6 \end{cases}$

$\begin{cases} x^2-y^2=4 \\ 2x^2+3y^2=6 \end{cases}$

Multiply each side of the first equation by −2 and add the equations to eliminate x:

$$\begin{aligned} -2x^2+2y^2&=-8 \\ 2x^2+3y^2&=6 \\ \hline 5y^2&=-2 \\ y^2&=-\frac{2}{5} \end{aligned}$$

No real solution. The system is inconsistent.

41. $\begin{cases} x^2+2y^2=16 \\ 4x^2-y^2=24 \end{cases}$

Multiply each side of the second equation by 2 and add the equations to eliminate y:

$$\begin{aligned} x^2+2y^2&=16 \\ 8x^2-2y^2&=48 \\ \hline 9x^2&=64 \\ x^2&=\frac{64}{9} \\ x&=\pm\frac{8}{3} \end{aligned}$$

If $x=\dfrac{8}{3}$:

$\left(\dfrac{8}{3}\right)^2+2y^2=16 \Rightarrow 2y^2=\dfrac{80}{9}$

$\Rightarrow y^2=\dfrac{40}{9} \Rightarrow y=\pm\dfrac{2\sqrt{10}}{3}$

If $x=-\dfrac{8}{3}$:

$\left(-\dfrac{8}{3}\right)^2+2y^2=16 \Rightarrow 2y^2=\dfrac{80}{9}$

$\Rightarrow y^2=\dfrac{40}{9} \Rightarrow y=\pm\dfrac{2\sqrt{10}}{3}$

Solutions:

$\left(\dfrac{8}{3}, \dfrac{2\sqrt{10}}{3}\right), \left(\dfrac{8}{3}, -\dfrac{2\sqrt{10}}{3}\right), \left(-\dfrac{8}{3}, \dfrac{2\sqrt{10}}{3}\right),$

$\left(-\dfrac{8}{3}, -\dfrac{2\sqrt{10}}{3}\right)$

42. $\begin{cases} 4x^2+3y^2=4 \\ 2x^2-6y^2=-3 \end{cases}$

Multiply each side of the first equation by 2 and add the equations to eliminate y:

$$8x^2+6y^2=8$$
$$\underline{2x^2-6y^2=-3}$$
$$10x^2=5$$
$$x^2=\frac{1}{2}$$
$$x=\pm\frac{\sqrt{2}}{2}$$

If $x=\frac{\sqrt{2}}{2}$:

$$4\left(\frac{\sqrt{2}}{2}\right)^2+3y^2=4 \Rightarrow 3y^2=2$$
$$\Rightarrow y^2=\frac{2}{3} \Rightarrow y=\pm\frac{\sqrt{6}}{3}$$

If $x=-\frac{\sqrt{2}}{2}$:

$$4\left(-\frac{\sqrt{2}}{2}\right)^2+3y^2=4 \Rightarrow 3y^2=2$$
$$\Rightarrow y^2=\frac{2}{3} \Rightarrow y=\pm\frac{\sqrt{6}}{3}$$

Solutions:

$\left(\frac{\sqrt{2}}{2},\frac{\sqrt{6}}{3}\right),\left(\frac{\sqrt{2}}{2},-\frac{\sqrt{6}}{3}\right),\left(-\frac{\sqrt{2}}{2},\frac{\sqrt{6}}{3}\right),$
$\left(-\frac{\sqrt{2}}{2},-\frac{\sqrt{6}}{3}\right)$

43. $\begin{cases} \frac{5}{x^2}-\frac{2}{y^2}+3=0 \\ \frac{3}{x^2}+\frac{1}{y^2}=7 \end{cases}$

$\begin{cases} \frac{5}{x^2}-\frac{2}{y^2}=-3 \\ \frac{3}{x^2}+\frac{1}{y^2}=7 \end{cases}$

Multiply each side of the second equation by 2 and add the equations to eliminate y:

$$\frac{5}{x^2}-\frac{2}{y^2}=-3$$
$$\underline{\frac{6}{x^2}+\frac{2}{y^2}=14}$$
$$\frac{11}{x^2}=11$$
$$x^2=1$$
$$x=\pm 1$$

If $x=1$:

$$\frac{3}{(1)^2}+\frac{1}{y^2}=7 \Rightarrow \frac{1}{y^2}=4 \Rightarrow y^2=\frac{1}{4}$$
$$\Rightarrow y=\pm\frac{1}{2}$$

If $x=-1$:

$$\frac{3}{(-1)^2}+\frac{1}{y^2}=7 \Rightarrow \frac{1}{y^2}=4 \Rightarrow y^2=\frac{1}{4}$$
$$\Rightarrow y=\pm\frac{1}{2}$$

Solutions: $\left(1,\frac{1}{2}\right),\left(1,-\frac{1}{2}\right),\left(-1,\frac{1}{2}\right),\left(-1,-\frac{1}{2}\right)$

44. $\begin{cases} \frac{2}{x^2}-\frac{3}{y^2}+1=0 \\ \frac{6}{x^2}-\frac{7}{y^2}+2=0 \end{cases}$

$\begin{cases} \frac{2}{x^2}-\frac{3}{y^2}=-1 \\ \frac{6}{x^2}-\frac{7}{y^2}=-2 \end{cases}$

Multiply each side of the first equation by –3 and add the equations to eliminate x:

$$\frac{-6}{y^2}+\frac{9}{y^2}=3$$
$$\underline{\frac{6}{x^2}-\frac{7}{y^2}=-2}$$
$$\frac{2}{y^2}=1$$
$$y^2=2$$
$$y=\pm\sqrt{2}$$

If $y=\sqrt{2}$:

$$\frac{2}{x^2}-\frac{3}{\left(\sqrt{2}\right)^2}=-1 \Rightarrow \frac{2}{x^2}=\frac{1}{2} \Rightarrow x^2=4$$

$\Rightarrow x=\pm 2$

If $y=-\sqrt{2}$:

$$\frac{2}{x^2}-\frac{3}{\left(-\sqrt{2}\right)^2}=-1 \Rightarrow \frac{2}{x^2}=\frac{1}{2} \Rightarrow x^2=4$$

$\Rightarrow x=\pm 2$

Solutions:

$\left(2,\sqrt{2}\right),\left(2,-\sqrt{2}\right),\left(-2,\sqrt{2}\right),\left(-2,-\sqrt{2}\right)$

45. $\begin{cases}\dfrac{1}{x^4}+\dfrac{6}{y^4}=6\\ \dfrac{2}{x^4}-\dfrac{2}{y^4}=19\end{cases}$

Multiply each side of the first equation by –2 and add the equations to eliminate x:

$$\frac{-2}{x^4}-\frac{12}{y^4}=-12$$
$$\frac{2}{x^4}-\frac{2}{y^4}=19$$
$$-\frac{14}{y^4}=7$$
$$y^4=-2$$

There are no real solutions. The system is inconsistent.

46. Add the equations to eliminate y:

$$\frac{1}{x^4}-\frac{1}{y^4}=1$$
$$\frac{1}{x^4}+\frac{1}{y^4}=4$$
$$\frac{2}{x^4}=5$$
$$x^4=\frac{2}{5}$$
$$x=\pm\sqrt[4]{\frac{2}{5}}$$

If $x=\sqrt[4]{\frac{2}{5}}$:

$$\frac{1}{\left(\sqrt[4]{\frac{2}{5}}\right)^4}+\frac{1}{y^4}=4 \Rightarrow \frac{1}{y^4}=\frac{3}{2} \Rightarrow y^4=\frac{2}{3}$$

$\Rightarrow y=\pm\sqrt[4]{\frac{2}{3}}$

If $x=-\sqrt[4]{\frac{2}{5}}$:

$$\frac{1}{\left(-\sqrt[4]{\frac{2}{5}}\right)^4}+\frac{1}{y^4}=4 \Rightarrow \frac{1}{y^4}=\frac{3}{2} \Rightarrow y^4=\frac{2}{3}$$

$\Rightarrow y=\pm\sqrt[4]{\frac{2}{3}}$

Solutions:

$\left(\sqrt[4]{\frac{2}{5}},\sqrt[4]{\frac{2}{3}}\right),\left(\sqrt[4]{\frac{2}{5}},-\sqrt[4]{\frac{2}{3}}\right),\left(-\sqrt[4]{\frac{2}{5}},\sqrt[4]{\frac{2}{3}}\right),$
$\left(-\sqrt[4]{\frac{2}{5}},-\sqrt[4]{\frac{2}{3}}\right)$

47. $\begin{cases}x^2-3xy+2y^2=0\\ x^2+xy=6\end{cases}$

Subtract the second equation from the first to eliminate the x^2 term.

$$-4xy+2y^2=-6$$
$$2xy-y^2=3$$

Since $y\neq 0$, we can solve for x in this equation to get

$$x=\frac{y^2+3}{2y},\quad y\neq 0$$

Now substitute for x in the second equation and solve for y.

$$x^2+xy=6$$
$$\left(\frac{y^2+3}{2y}\right)^2+\left(\frac{y^2+3}{2y}\right)y=6$$
$$\frac{y^4+6y^2+9}{4y^2}+\frac{y^2+3}{2}=6$$

$$y^4+6y^2+9+2y^4+6y^2=24y^2$$
$$3y^4-12y^2+9=0$$
$$y^4-4y^2+3=0$$
$$\left(y^2-3\right)\left(y^2-1\right)=0$$

Thus, $y=\pm\sqrt{3}$ or $y=\pm 1$.

If $y=1$: $x=2\cdot 1=2$

If $y=-1$: $x=2(-1)=-2$

If $y=\sqrt{3}$: $x=\sqrt{3}$

If $y=-\sqrt{3}$: $x=-\sqrt{3}$

Solutions:

$(2, 1), (-2, -1), \left(\sqrt{3}, \sqrt{3}\right), \left(-\sqrt{3}, -\sqrt{3}\right)$

48. $\begin{cases} x^2-xy-2y^2=0 \\ xy+\quad x+6=0 \end{cases}$

Factor the first equation, solve for *x*, substitute into the second equation and solve:

$x^2-xy-2y^2=0$

$(x-2y)(x+y)=0$

$x=2y$ or $x=-y$

Substitute $x=2y$ and solve:

$$xy+x+6=0$$
$$(2y)y+2y=-6$$
$$2y^2+2y+6=0$$
$$2(y^2+y+3)=0$$
$$y=\frac{-1\pm\sqrt{1^2-4(1)(3)}}{2(1)}$$

No real solution

Substitute $x=-y$ and solve:

$$xy+x+6=0$$
$$-y\cdot y+(-y)=-6$$
$$-y^2-y+6=0$$
$$(-y-3)(y-2)=0$$
$$y=-3 \text{ or } y=2$$

If $y=-3$: $x=3$

If $y=2$: $x=-2$

Solutions: $(3,-3), (-2, 2)$

49. $\begin{cases} y^2+y+x^2-x-2=0 \\ y+1+\dfrac{x-2}{y}=0 \end{cases}$

Multiply each side of the second equation by –y and add the equations to eliminate *y*:

$$\begin{array}{r} y^2+y+x^2-x-2=0 \\ -y^2-y\qquad -x+2=0 \\ \hline x^2-2x=0 \end{array}$$
$$x(x-2)=0$$
$$x=0 \text{ or } x=2$$

If $x=0$:

$y^2+y+0^2-0-2=0 \Rightarrow y^2+y-2=0$

$\Rightarrow (y+2)(y-1)=0 \Rightarrow y=-2$ or $y=1$

If $x=2$:

$y^2+y+2^2-2-2=0 \Rightarrow y^2+y=0$

$\Rightarrow y(y+1)=0 \Rightarrow y=0$ or $y=-1$

Note: $y\neq 0$ because of division by zero.

Solutions: (0, –2), (0, 1), (2, –1)

50. $\begin{cases} x^3-2x^2+y^2+3y-4=0 \\ x-2+\dfrac{y^2-y}{x^2}=0 \end{cases}$

Multiply each side of the second equation by $-x^2$ and add the equations to eliminate *x*:

$$\begin{array}{r} x^3-2x^2+y^2+3y-4=0 \\ -x^3+2x^2-y^2+y\qquad =0 \\ \hline 4y-4=0 \end{array}$$
$$4y=4$$
$$y=1$$

If $y=1$:

$x^3-2x^2+1^2+3\cdot 1-4=0 \Rightarrow x^3-2x^2=0$

$\Rightarrow x^2(x-2)=0 \Rightarrow x=0$ or $x=2$

Note: $x\neq 0$ because of division by zero.

Solution: (2, 1)

51. Rewrite each equation in exponential form:

$$\begin{cases} \log_x y = 3 \rightarrow \quad y = x^3 \\ \log_x(4y) = 5 \rightarrow 4y = x^5 \end{cases}$$

Substitute the first equation into the second and solve:

$$4x^3 = x^5$$
$$x^5 - 4x^3 = 0$$
$$x^3(x^2 - 4) = 0$$
$$x^3 = 0 \text{ or } x^2 = 4 \Rightarrow x = 0 \text{ or } x = \pm 2$$

The base of a logarithm must be positive, thus $x \neq 0$ and $x \neq -2$.

If $x = 2$: $\quad y = 2^3 = 8$

Solution: (2, 8)

52. Rewrite each equation in exponential form:

$$\begin{cases} \log_x(2y) = 3 \Rightarrow \quad 2y = x^3 \\ \log_x(4y) = 2 \Rightarrow \quad 4y = x^2 \end{cases}$$

Multiply the first equation by 2 then substitute the first equation into the second and solve:

$$2x^3 = x^2$$
$$2x^3 - x^2 = 0$$
$$x^2(2x - 1) = 0$$
$$x^2 = 0 \text{ or } x = \frac{1}{2} \Rightarrow x = \frac{1}{2} \text{ or } x = 0$$

The base of a logarithm must be positive, thus $x \neq 0$.

If $x = \frac{1}{2}$: $\quad 4y = \left(\frac{1}{2}\right)^2 = \frac{1}{4} \Rightarrow y = \frac{1}{16}$

Solution: $\left(\frac{1}{2}, \frac{1}{16}\right)$

53. Rewrite each equation in exponential form:

$\ln x = 4\ln y \Rightarrow \quad x = e^{4\ln y} = e^{\ln y^4} = y^4$

$\log_3 x = 2 + 2\log_3 y$

$x = 3^{2+2\log_3 y} = 3^2 \cdot 3^{2\log_3 y} = 3^2 \cdot 3^{\log_3 y^2} = 9y^2$

So we have the system $\begin{cases} x = y^4 \\ x = 9y^2 \end{cases}$

Therefore we have :

$$9y^2 = y^4 \Rightarrow 9y^2 - y^4 = 0 \Rightarrow y^2(9 - y^2) = 0$$
$$y^2(3 + y)(3 - y) = 0$$
$$y = 0 \text{ or } y = -3 \text{ or } y = 3$$

Since $\ln y$ is undefined when $y \leq 0$, the only solution is $y = 3$.

If $y = 3$: $\quad x = y^4 \Rightarrow \quad x = 3^4 = 81$

Solution: $(81, 3)$

54. Rewrite each equation in exponential form:

$\ln x = 5\ln y \Rightarrow \quad x = e^{5\ln y} = e^{\ln y^5} = y^5$

$\log_2 x = 3 + 2\log_2 y$

$x = 2^{3+2\log_2 y} = 2^3 \cdot 2^{2\log_2 y} = 2^3 \cdot 2^{\log_2 y^2} = 8y^2$

So we have the system $\begin{cases} x = y^5 \\ x = 8y^2 \end{cases}$

Therefore we have

$$8y^2 = y^5$$
$$8y^2 - y^5 = 0$$
$$y^2\left(8 - y^3\right) = 0$$
$$y = 0 \text{ or } 8 - y^3 = 0 \Rightarrow y = 2$$

Since $\ln y$ is undefined when $y \leq 0$, the only solution is $y = 2$.

If $y = 2$: $\quad x = y^5 \Rightarrow \quad x = 2^5 = 32$

Solution: $(32, 2)$

55. Solve the first equation for *x*, substitute into the second equation and solve:

$$\begin{cases} x + 2y = 0 \Rightarrow x = -2y \\ (x-1)^2 + (y-1)^2 = 5 \end{cases}$$

$$(-2y-1)^2 + (y-1)^2 = 5$$
$$4y^2 + 4y + 1 + y^2 - 2y + 1 = 5 \Rightarrow 5y^2 + 2y - 3 = 0$$
$$(5y - 3)(y + 1) = 0$$
$$y = \frac{3}{5} = 0.6 \text{ or } y = -1$$
$$x = -\frac{6}{5} = -1.2 \text{ or } x = 2$$

The points of intersection are $(-1.2, 0.6)$, $(2, -1)$.

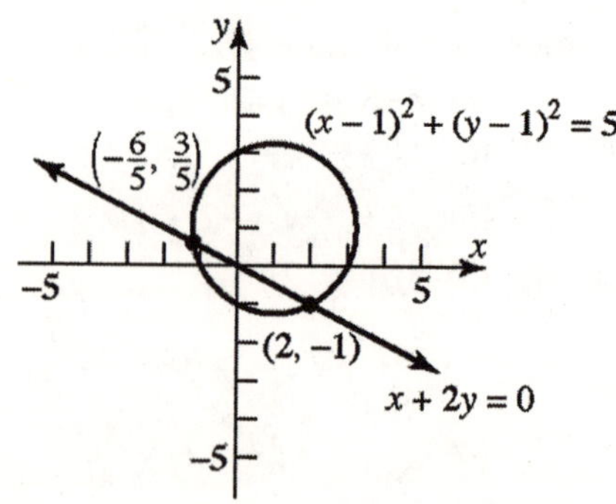

56. Solve the first equation for x, substitute into the second equation and solve:

$$\begin{cases} x+2y+6=0 \Rightarrow x=-2y-6 \\ (x+1)^2+(y+1)^2 = 5 \end{cases}$$

$$(-2y-6+1)^2+(y+1)^2=5$$
$$4y^2+20y+25+y^2+2y+1=5$$
$$5y^2+22y+21=0$$
$$(5y+7)(y+3)=0$$
$$y=-\frac{7}{5} \text{ or } y=-3$$
$$x=-\frac{16}{5} \text{ or } x=0$$

The points of intersection are $\left(-\frac{16}{5},-\frac{7}{5}\right)$, $(0,-3)$.

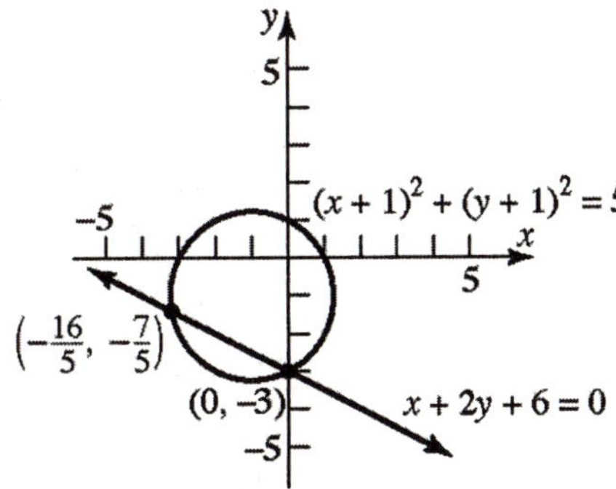

57. Complete the square on the second equation.

$$y^2+4y+4=x-1+4$$
$$(y+2)^2=x+3$$

Substitute this result into the first equation.

$$(x-1)^2+x+3=4$$
$$x^2-2x+1+x+3=4$$
$$x^2-x=0$$
$$x(x-1)=0$$
$$x=0 \text{ or } x=1$$

If $x=0$: $(y+2)^2=0+3$

$$y+2=\pm\sqrt{3} \Rightarrow y=-2\pm\sqrt{3}$$

If $x=1$: $(y+2)^2=1+3$

$$y+2=\pm 2 \Rightarrow y=-2\pm 2$$

The points of intersection are:

$\left(0,-2-\sqrt{3}\right)$, $\left(0,-2+\sqrt{3}\right)$, $(1,-4)$, $(1,0)$.

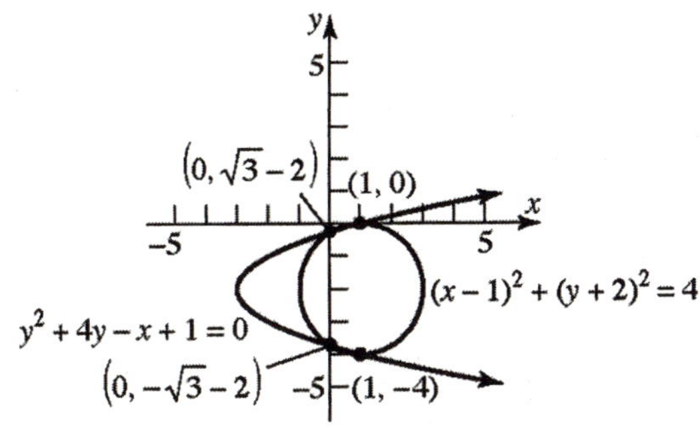

58. Complete the square on the second equation, substitute into the first equation and solve:

$$\begin{cases} (x+2)^2+(y-1)^2=4 \\ y^2-2y-x-5=0 \end{cases}$$

$$y^2-2y+1=x+5+1$$
$$(y-1)^2=x+6$$
$$(x+2)^2+x+6=4$$
$$x^2+4x+4+x+6=4$$
$$x^2+5x+6=0$$
$$(x+2)(x+3)=0$$
$$x=-2 \text{ or } x=-3$$

If $x=-2$: $(y-1)^2=-2+6 \Rightarrow y-1=\pm 2$

$$\Rightarrow y=-1 \text{ or } y=3$$

If $x=-3$: $(y-1)^2=-3+6 \Rightarrow y-1=\pm\sqrt{3}$

$$\Rightarrow y=1\pm\sqrt{3}$$

The points of intersection are:

$\left(-3,1-\sqrt{3}\right)$, $\left(-3,1+\sqrt{3}\right)$, $(-2,-1)$, $(-2,3)$.

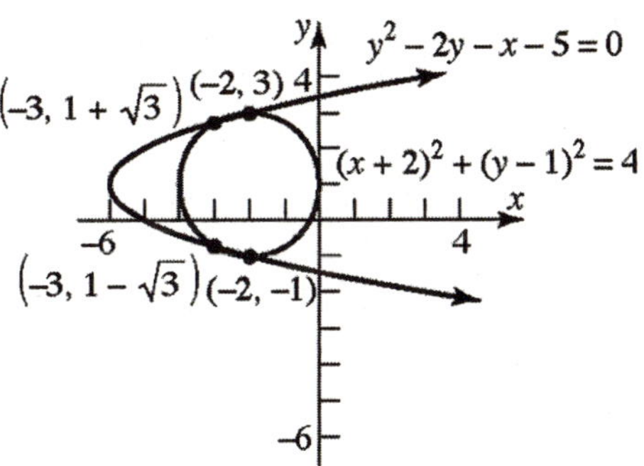

59. Solve the first equation for x, substitute into the second equation and solve:

$$\begin{cases} y = \dfrac{4}{x-3} \\ x^2 - 6x + y^2 + 1 = 0 \end{cases}$$

$$y = \frac{4}{x-3}$$

$$x - 3 = \frac{4}{y}$$

$$x = \frac{4}{y} + 3$$

$$\left(\frac{4}{y}+3\right)^2 - 6\left(\frac{4}{y}+3\right) + y^2 + 1 = 0$$

$$\frac{16}{y^2} + \frac{24}{y} + 9 - \frac{24}{y} - 18 + y^2 + 1 = 0$$

$$\frac{16}{y^2} + y^2 - 8 = 0$$

$$16 + y^4 - 8y^2 = 0$$

$$y^4 - 8y^2 + 16 = 0$$

$$(y^2 - 4)^2 = 0$$

$$y^2 - 4 = 0$$

$$y^2 = 4$$

$$y = \pm 2$$

If $y = 2$: $\quad x = \dfrac{4}{2} + 3 = 5$

If $y = -2$: $\quad x = \dfrac{4}{-2} + 3 = 1$

The points of intersection are: $(1, -2)$, $(5, 2)$.

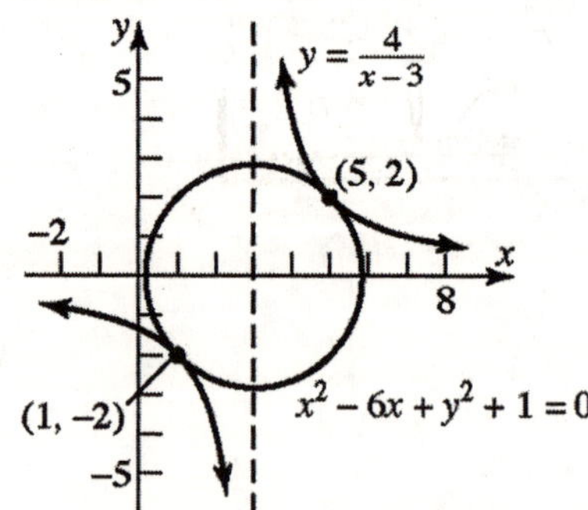

60. Substitute the first equation into the second equation and solve:

$$\begin{cases} y = \dfrac{4}{x+2} \\ x^2 + 4x + y^2 - 4 = 0 \end{cases}$$

$$x^2 + 4x + \left(\frac{4}{x+2}\right)^2 - 4 = 0$$

$$x^2 + 4x - 4 = -\left(\frac{4}{x+2}\right)^2$$

$$(x+2)^2\left(x^2 + 4x - 4\right) = -16$$

$$\left(x^2 + 4x + 4\right)\left(x^2 + 4x - 4\right) = -16$$

$$x^4 + 8x^3 + 16x^2 - 16 = -16$$

$$x^4 + 8x^3 + 16x^2 = 0$$

$$x^2\left(x^2 + 8x + 16\right) = 0$$

$$x^2(x+4)^2 = 0$$

$$x = 0 \quad \text{or} \quad x = -4$$

$$y = 2 \qquad y = -2$$

The points of intersection are: $(0, 2)$, $(-4, -2)$.

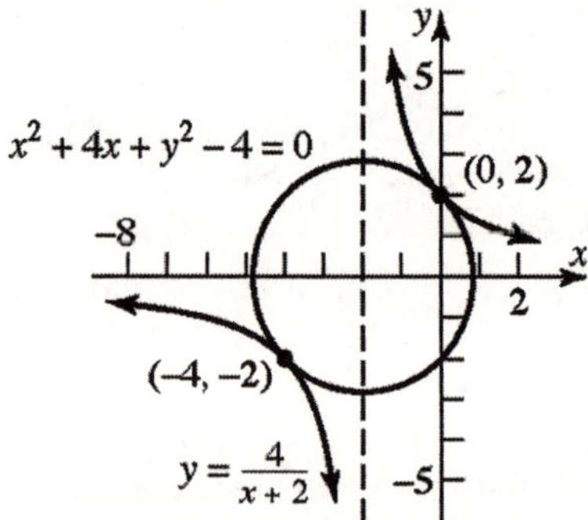

61. Graph: $y_1 = x \wedge (2/3)$; $y_2 = e \wedge (-x)$

Use INTERSECT to solve:

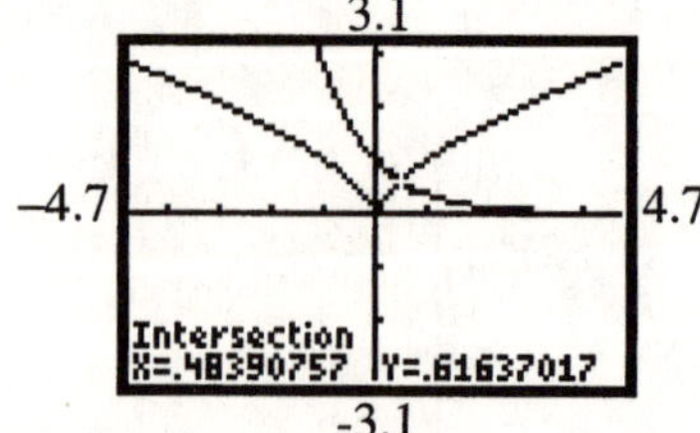

Solution: $(0.48, 0.62)$

62. Graph: $y_1 = x \wedge (3/2);\quad y_2 = e \wedge (-x)$
Use INTERSECT to solve:

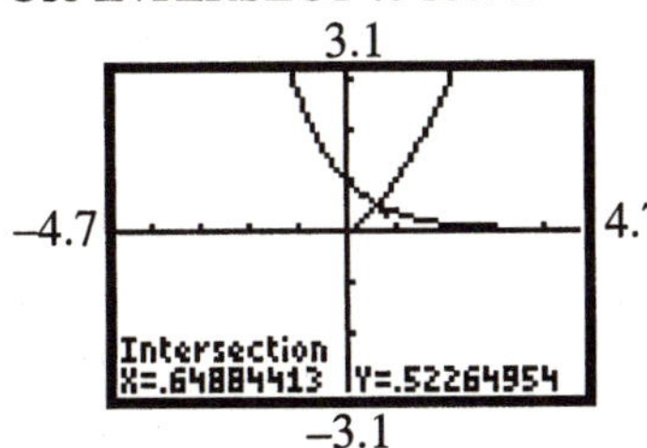

Solution: (0.65, 0.52)

63. Graph: $y_1 = \sqrt[3]{2-x^2};\quad y_2 = 4/x^3$
Use INTERSECT to solve:

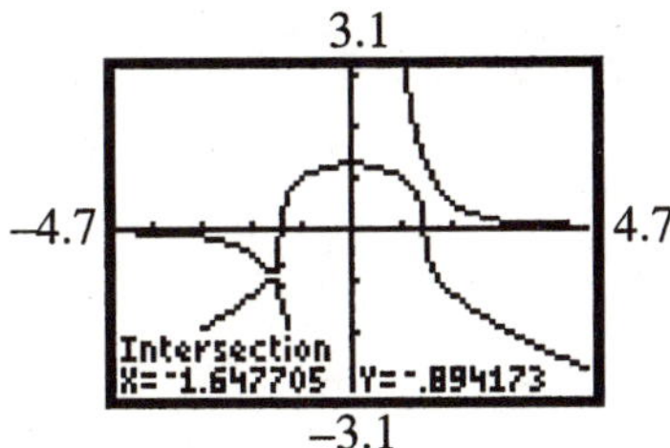

Solution: (–1.65, –0.89)

64. Graph: $y_1 = \sqrt{2-x^3};\quad y_2 = -\sqrt{2-x^3};$
$y_3 = 4/x^2$
Use INTERSECT to solve:

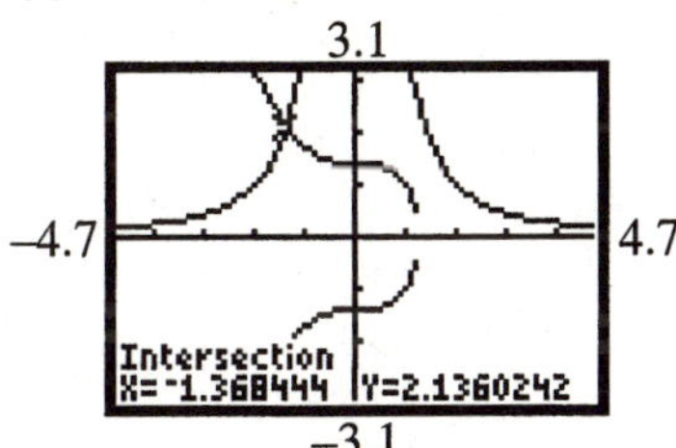

Solution: (–1.37, 2.14)

65. Graph: $y_1 = \sqrt[4]{12-x^4};\quad y_2 = -\sqrt[4]{12-x^4};$
$y_3 = \sqrt{2/x};\quad y_4 = -\sqrt{2/x}$
Use INTERSECT to solve:

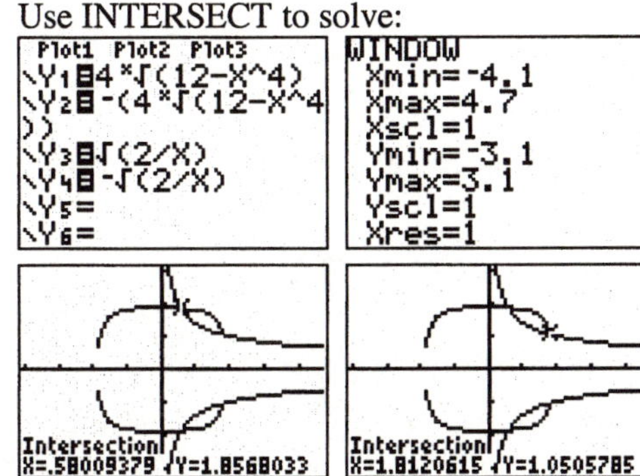

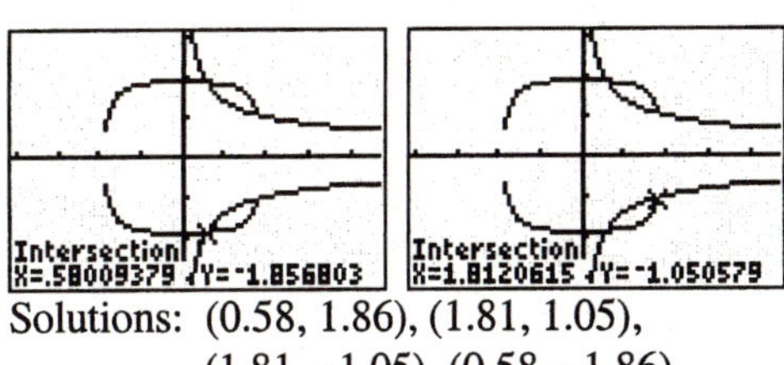

Solutions: (0.58, 1.86), (1.81, 1.05), (1.81, –1.05), (0.58, –1.86)

66. Graph: $y_1 = \sqrt[4]{6-x^4};\quad y_2 = -\sqrt[4]{6-x^4};$
$y_3 = 1/x$
Use INTERSECT to solve:

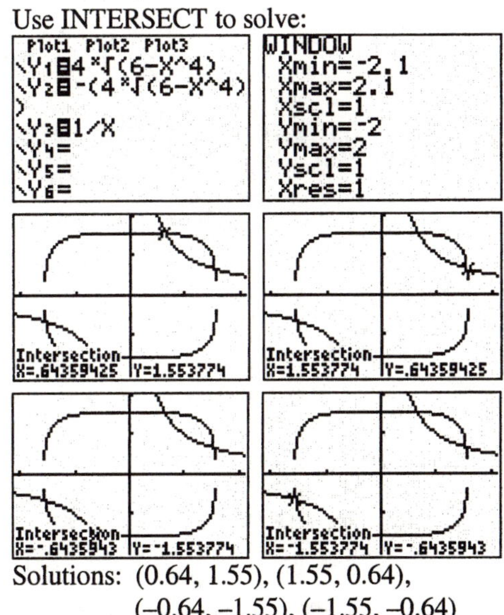

Solutions: (0.64, 1.55), (1.55, 0.64), (–0.64, –1.55), (–1.55, –0.64)

67. Graph: $y_1 = 2/x;\quad y_2 = \ln x$
Use INTERSECT to solve:

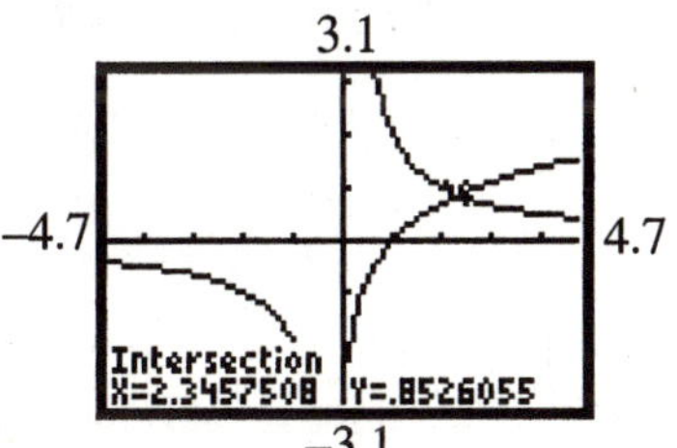

Solution: (2.35, 0.85)

68. Graph: $y_1 = \sqrt{4-x^2}$; $y_2 = -\sqrt{4-x^2}$;

$y_3 = \ln x$

Use INTERSECT to solve:

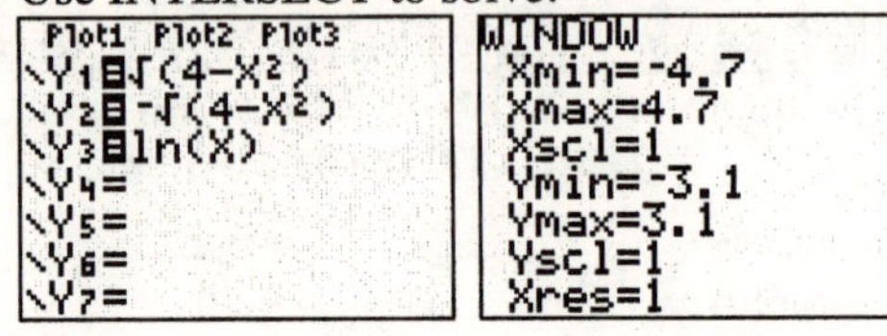

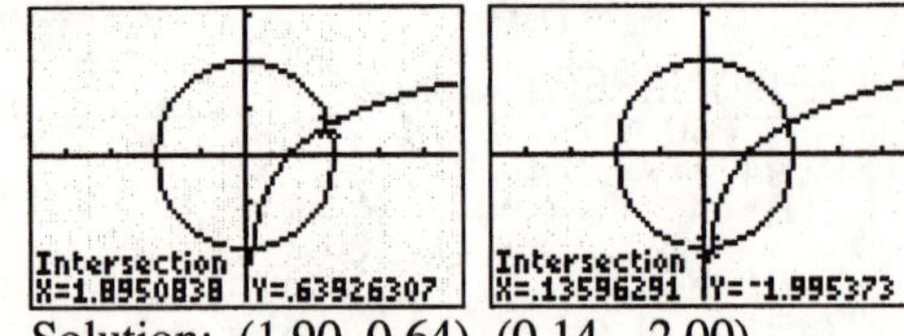

Solution: (1.90, 0.64), (0.14, –2.00)

69. Let x and y be the two numbers. The system of equations is:

$$\begin{cases} x - y = 2 \quad \Rightarrow x = y + 2 \\ x^2 + y^2 = 10 \end{cases}$$

Solve the first equation for x, substitute into the second equation and solve:

$$(y+2)^2 + y^2 = 10$$
$$y^2 + 4y + 4 + y^2 = 10$$
$$y^2 + 2y - 3 = 0$$
$$(y+3)(y-1) = 0 \Rightarrow y = -3 \text{ or } y = 1$$

If $y = -3$: $x = -3 + 2 = -1$

If $y = 1$: $x = 1 + 2 = 3$

The two numbers are 1 and 3 or –1 and –3.

70. Let x and y be the two numbers. The system of equations is:

$$\begin{cases} x + y = 7 \Rightarrow x = 7 - y \\ x^2 - y^2 = 21 \end{cases}$$

Solve the first equation for x, substitute into the second equation and solve:

$$(7-y)^2 - y^2 = 21$$
$$49 - 14y + y^2 - y^2 = 21$$
$$-14y = -28$$
$$y = 2 \Rightarrow x = 7 - 2 = 5$$

The two numbers are 2 and 5.

71. Let x and y be the two numbers. The system of equations is:

$$\begin{cases} xy = 4 \Rightarrow x = \dfrac{4}{y} \\ x^2 + y^2 = 8 \end{cases}$$

Solve the first equation for x, substitute into the second equation and solve:

$$\left(\frac{4}{y}\right)^2 + y^2 = 8$$
$$\frac{16}{y^2} + y^2 = 8$$
$$16 + y^4 = 8y^2$$
$$y^4 - 8y^2 + 16 = 0$$
$$(y^2 - 4) = 0$$
$$y^2 = 4$$
$$y = \pm 2$$

If $y = 2$: $x = \frac{4}{2} = 2$; If $y = -2$: $x = \frac{4}{-2} = -2$

The two numbers are 2 and 2 or –2 and –2.

72. Let x and y be the two numbers. The system of equations is:

$$\begin{cases} xy = 10 \Rightarrow x = \dfrac{10}{y} \\ x^2 - y^2 = 21 \end{cases}$$

Solve the first equation for x, substitute into the second equation and solve:

$$\left(\frac{10}{y}\right)^2 - y^2 = 21$$
$$\frac{100}{y^2} - y^2 = 21$$
$$100 - y^4 = 21y^2$$
$$y^4 + 21y^2 - 100 = 0$$
$$(y^2 - 4)(y^2 + 25) = 0$$

$y^2 = 4$ or $y^2 = -25$ (no real solution)

$y = \pm 2$

If $y = 2$: $x = \frac{10}{2} = 5$

If $y = -2$: $x = \frac{10}{-2} = -5$

The two numbers are 2 and 5 or –2 and –5.

73. Let x and y be the two numbers. The system of equations is:

$$\begin{cases} x-y=xy \\ \frac{1}{x}+\frac{1}{y}=5 \end{cases}$$

Solve the first equation for x, substitute into the second equation and solve:

$$x-xy=y$$

$$x(1-y)=y \quad\Rightarrow\quad x=\frac{y}{1-y}$$

$$\frac{1}{\frac{y}{1-y}}+\frac{1}{y}=5$$

$$\frac{1-y}{y}+\frac{1}{y}=5$$

$$\frac{2-y}{y}=5$$

$$2-y=5y$$

$$6y=2$$

$$y=\frac{1}{3} \quad\Rightarrow\quad x=\frac{\frac{1}{3}}{1-\frac{1}{3}}=\frac{\frac{1}{3}}{\frac{2}{3}}=\frac{1}{2}$$

The two numbers are $\frac{1}{2}$ and $\frac{1}{3}$.

74. Let x and y be the two numbers. The system of equations is:

$$\begin{cases} x+y=xy \\ \frac{1}{x}-\frac{1}{y}=3 \end{cases}$$

Solve the first equation for x, substitute into the second equation and solve:

$$xy-x=y$$

$$x(y-1)=y \quad\Rightarrow\quad x=\frac{y}{y-1}$$

$$\frac{1}{\frac{y}{y-1}}-\frac{1}{y}=3$$

$$\frac{y-1}{y}-\frac{1}{y}=3$$

$$\frac{y-2}{y}=3$$

$$y-2=3y$$

$$2y=-2$$

$$y=-1 \quad\Rightarrow\quad x=\frac{-1}{-1-1}=\frac{1}{2}$$

The two numbers are -1 and $\frac{1}{2}$.

75. $\begin{cases} \frac{a}{b}=\frac{2}{3} \\ a+b=10 \Rightarrow a=10-b \end{cases}$

Solve the second equation for a, substitute into the first equation and solve:

$$\frac{10-b}{b}=\frac{2}{3}$$

$$3(10-b)=2b$$

$$30-3b=2b$$

$$30=5b$$

$$b=6 \Rightarrow a=4$$

$$a+b=10;\ \ b-a=2$$

The ratio of $a+b$ to $b-a$ is $\frac{10}{2}=5$.

76. $\begin{cases} \frac{a}{b}=\frac{4}{3} \\ a+b=14 \Rightarrow a=14-b \end{cases}$

Solve the second equation for a, substitute into the first equation and solve:

$$\frac{14-b}{b}=\frac{4}{3}$$

$$3(14-b)=4b$$

$$42-3b=4b$$

$$42=7b$$

$$b=6 \quad\Rightarrow\quad a=8$$

$$a-b=2;\ \ a+b=14$$

The ratio of $a-b$ to $a+b$ is $\frac{2}{14}=\frac{1}{7}$.

77. Let x = the width of the rectangle.
Let y = the length of the rectangle.

$$\begin{cases} 2x+2y=16 \\ xy=15 \end{cases}$$

Solve the first equation for y, substitute into the second equation and solve.

$$2x+2y=16$$

$$2y=16-2x$$

$$y=8-x$$

$$x(8-x)=15$$

$$8x-x^2=15$$

$$x^2-8x+15=0$$

$$(x-5)(x-3)=0$$

$$x=5 \quad\text{or}\quad x=3$$

The dimensions of the rectangle are 3 inches by 5 inches.

78. Let $2x$ = the side of the first square.
Let $3x$ = the side of the second square.

$$(2x)^2+(3x)^2=52$$
$$4x^2+9x^2=52$$
$$13x^2=52$$
$$x^2=4$$
$$x=\pm 2$$

Note that we must have $x>0$.
The sides of the first square are (2)(2) = 4 feet and the sides of the second square are (3)(2) = 6 feet.

79. Let x = the radius of the first circle.
Let y = the radius of the second circle.

$$\begin{cases} 2\pi x+2\pi y=12\pi \\ \pi x^2+\pi y^2=20\pi \end{cases}$$

Solve the first equation for *y*, substitute into the second equation and solve:

$$2\pi x+2\pi y=12\pi$$
$$x+y=6$$
$$y=6-x$$

$$\pi x^2+\pi y^2=20\pi$$
$$x^2+y^2=20$$
$$x^2+(6-x)^2=20$$
$$x^2+36-12x+x^2=20 \Rightarrow 2x^2-12x+16=0$$
$$x^2-6x+8=0 \Rightarrow (x-4)(x-2)=0$$
$$x=4 \text{ or } x=2$$
$$y=2 \qquad y=4$$

The radii of the circles are 2 centimeters and 4 centimeters.

80. Let x = the length of each of the two equal sides in the isosceles triangle.
Let y = the length of the base.
The perimeter of the triangle: $x+x+y=18$
Since the altitude to the base *y* is 3, the Pythagorean theorem produces another equation.

$$\left(\frac{y}{2}\right)^2+3^2=x^2 \Rightarrow \frac{y^2}{4}+9=x^2$$

Solve the system of equations:

$$\begin{cases} 2x+y=18 \Rightarrow y=18-2x \\ \dfrac{y^2}{4}+9=x^2 \end{cases}$$

Solve the first equation for *y*, substitute into the second equation and solve.

$$\frac{(18-2x)^2}{4}+9=x^2$$
$$\frac{324-72x+4x^2}{4}+9=x^2$$
$$81-18x+x^2+9=x^2$$
$$-18x=-90$$
$$x=5 \Rightarrow y=18-2(5)=8$$

The base of the triangle is 8 centimeters.

81. The tortoise takes 9 + 3 = 12 minutes or 0.2 hour longer to complete the race than the hare.
Let r = the rate of the hare.
Let t = the time for the hare to complete the race. Then $t+0.2$ = the time for the tortoise and $r-0.5$ = the rate for the tortoise. Since the length of the race is 21 meters, the distance equations are:

$$\begin{cases} rt=21 \Rightarrow r=\dfrac{21}{t} \\ (r-0.5)(t+0.2)=21 \end{cases}$$

Solve the first equation for *r*, substitute into the second equation and solve:

$$\left(\frac{21}{t}-0.5\right)(t+0.2)=21$$
$$21+\frac{4.2}{t}-0.5t-0.1=21$$
$$10t\left(21+\frac{4.2}{t}-0.5t-0.1\right)=10t\cdot(21)$$
$$210t+42-5t^2-t=210t$$
$$5t^2+t-42=0$$
$$(5t-14)(t+3)=0$$
$$5t-14=0 \quad \text{or} \quad t+3=0$$
$$5t=14 \qquad t=-3$$
$$t=\frac{14}{5}=2.8$$

$t=-3$ makes no sense, since time cannot be negative.
Solve for *r*:

$$r=\frac{21}{2.8}=7.5$$

The average speed of the hare is 7.5 meters per hour, and the average speed for the tortoise is 7 meters per hour.

82. Let $v_1, v_2, v_3 =$ the speeds of runners 1, 2, 3.
Let $t_1, t_2, t_3 =$ the times of runners 1, 2, 3.
Then by the conditions of the problem, we have the following system:

$$\begin{cases} 5280 = v_1 t_1 \\ 5270 = v_2 t_1 \\ 5260 = v_3 t_1 \\ 5280 = v_2 t_2 \end{cases}$$

Distance between the second runner and the third runner after t_2 seconds is:

$$\begin{aligned} 5280 - v_3 t_2 &= 5280 - v_3 t_1 \left(\frac{v_2 t_2}{v_2 t_1}\right) \\ &= 5280 - 5260\left(\frac{5280}{5270}\right) \\ &\approx 10.02 \end{aligned}$$

The second place runner beats the third place runner by about 10.02 feet.

83. Let x = the width of the cardboard. Let y = the length of the cardboard. The width of the box will be $x-4$, the length of the box will be $y-4$, and the height is 2. The volume is $V = (x-4)(y-4)(2)$.
Solve the system of equations:

$$\begin{cases} xy = 216 & \Rightarrow y = \dfrac{216}{x} \\ 2(x-4)(y-4) = 224 \end{cases}$$

Solve the first equation for *y*, substitute into the second equation and solve.

$$(2x-8)\left(\frac{216}{x} - 4\right) = 224$$

$$432 - 8x - \frac{1728}{x} + 32 = 224$$

$$432x - 8x^2 - 1728 + 32x = 224x$$

$$8x^2 - 240x + 1728 = 0$$

$$x^2 - 30x + 216 = 0$$

$$(x-12)(x-18) = 0$$

$$x - 12 = 0 \quad \text{or} \quad x - 18 = 0$$

$$x = 12 \qquad x = 18$$

The cardboard should be 12 centimeters by 18 centimeters.

84. Let x = the width of the cardboard. Let y = the length of the cardboard.
The area of the cardboard is: $xy = 216$
The volume of the tube is: $V = \pi r^2 h = 224$
where $h = y$ and $2\pi r = x$ or $r = \dfrac{x}{2\pi}$.
Solve the system of equations:

$$\begin{cases} xy = 216 \Rightarrow y = \dfrac{216}{x} \\ \pi\left(\dfrac{x}{2\pi}\right)^2 y = 224 \Rightarrow \dfrac{x^2 y}{4\pi} = 224 \end{cases}$$

Solve the first equation for *y*, substitute into the second equation and solve.

$$\frac{x^2\left(\frac{216}{x}\right)}{4\pi} = 224$$

$$216x = 896\pi$$

$$x = \frac{896\pi}{216} \approx 13.03$$

$$y = \frac{216}{x} = \frac{216}{\frac{896\pi}{216}} = \frac{(216)^2}{896\pi} \approx 16.57$$

The cardboard should be about 13.03 centimeters by 16.57 centimeters.

85. Find equations relating area and perimeter:

$$\begin{cases} x^2 + y^2 = 4500 \\ 3x + 3y + (x - y) = 300 \end{cases}$$

Solve the second equation for *y*, substitute into the first equation and solve:

$$4x + 2y = 300$$

$$2y = 300 - 4x$$

$$y = 150 - 2x$$

$$x^2 + (150 - 2x)^2 = 4500$$

$$x^2 + 22{,}500 - 600x + 4x^2 = 4500$$

$$5x^2 - 600x + 18{,}000 = 0$$

$$x^2 - 120x + 3600 = 0$$

$$(x-60)^2 = 0$$

$$x - 60 = 0$$

$$x = 60$$

$$y = 150 - 2(60) = 30$$

The sides of the squares are 30 feet and 60 feet.

86. Let x = the length of a side of the square.
Let r = the radius of the circle.
The area of the square is x^2 and the area of the circle is πr^2. The perimeter of the square is $4x$ and the circumference of the circle is $2\pi r$. Find equations relating area and perimeter:

$$\begin{cases} x^2+\pi r^2=100 \\ 4x+2\pi r=60 \end{cases}$$

Solve the second equation for *x*, substitute into the first equation and solve:

$$4x+2\pi r=60$$
$$4x=60-2\pi r$$
$$x=15-\frac{1}{2}\pi r$$

$$\left(15-\frac{1}{2}\pi r\right)^2+\pi r^2=100$$
$$225-15\pi r+\frac{1}{4}\pi^2 r^2+\pi r^2=100$$
$$\left(\frac{1}{4}\pi^2+\pi\right)r^2-15\pi r+125=0$$
$$b^2-4ac=(-15\pi)^2-4\left(\frac{1}{4}\pi^2+\pi\right)(125)$$
$$=225\pi^2-500\left(\frac{1}{4}\pi^2+\pi\right)$$
$$=100\pi^2-500\pi<0$$

Since the discriminant is less than zero, it is impossible to cut the wire into two pieces whose total area equals 100 square feet.

87. Solve the system for l and w:

$$\begin{cases} 2l+2w=P \\ lw=A \end{cases}$$

Solve the first equation for l, substitute into the second equation and solve.

$$2l=P-2w$$
$$l=\frac{P}{2}-w$$
$$\left(\frac{P}{2}-w\right)w=A$$
$$\frac{P}{2}w-w^2=A$$
$$w^2-\frac{P}{2}w+A=0$$

$$w=\frac{\frac{P}{2}\pm\sqrt{\frac{P^2}{4}-4A}}{2}=\frac{\frac{P}{2}\pm\sqrt{\frac{P^2}{4}-\frac{16A}{4}}}{2}$$
$$=\frac{\frac{P}{2}\pm\frac{\sqrt{P^2-16A}}{2}}{2}=\frac{P\pm\sqrt{P^2-16A}}{4}$$

If $w=\frac{P+\sqrt{P^2-16A}}{4}$ then

$$l=\frac{P}{2}-\frac{P+\sqrt{P^2-16A}}{4}=\frac{P-\sqrt{P^2-16A}}{4}$$

If $w=\frac{P-\sqrt{P^2-16A}}{4}$ then

$$l=\frac{P}{2}-\frac{P-\sqrt{P^2-16A}}{4}=\frac{P+\sqrt{P^2-16A}}{4}$$

If it is required that length be greater than width, then the solution is:

$$w=\frac{P-\sqrt{P^2-16A}}{4} \text{ and } l=\frac{P+\sqrt{P^2-16A}}{4}$$

88. Solve the system for l and b:

$$\begin{cases} P=b+2l \Rightarrow b=P-2l \\ h^2+\frac{b^2}{4}=l^2 \end{cases}$$

Solve the first equation for b, substitute into the second equation and solve.

$$4h^2+b^2=4l^2$$
$$4h^2+(P-2l)^2=4l^2$$
$$4h^2+P^2-4Pl+4l^2=4l^2$$
$$4h^2+P^2=4Pl$$
$$l=\frac{4h^2+P^2}{4P}$$
$$b=P-\frac{4h^2+P^2}{2P}=\frac{P^2-4h^2}{2P}$$

89. Solve the equation: $m^2-4(2m-4)=0$

$$m^2-8m+16=0$$
$$(m-4)^2=0$$
$$m=4$$

Use the point-slope equation with slope 4 and the point (2, 4) to obtain the equation of the tangent line:

$$y-4=4(x-2)\Rightarrow y-4=4x-8\Rightarrow y=4x-4$$

90. Solve the system:

$$\begin{cases} x^2 + y^2 = 10 \\ y = mx + b \end{cases}$$

Solve the system by substitution:

$$x^2 + (mx+b)^2 = 10$$

$$x^2 + m^2x^2 + 2bmx + b^2 - 10 = 0$$

$$(1+m^2)x^2 + 2bmx + b^2 - 10 = 0$$

Note that the tangent line passes through (1, 3). Find the relation between m and b:

$3 = m(1) + b \Rightarrow b = 3 - m$

There is one solution to the quadratic if the discriminant is zero.

$$(2bm)^2 - 4(m^2+1)(b^2-10) = 0$$

$$4b^2m^2 - 4b^2m^2 + 40m^2 - 4b^2 + 40 = 0$$

$$40m^2 - 4b^2 + 40 = 0$$

Substitute for b and solve:

$$40m^2 - 4(3-m)^2 + 40 = 0$$

$$40m^2 - 4m^2 + 24m - 36 + 40 = 0$$

$$36m^2 + 24m + 4 = 0$$

$$9m^2 + 6m + 1 = 0$$

$$(3m+1)^2 = 0$$

$$3m = -1$$

$$m = -\frac{1}{3}$$

$$b = 3 - m = 3 - \left(-\frac{1}{3}\right) = \frac{10}{3}$$

The equation of the tangent line is

$y = -\frac{1}{3}x + \frac{10}{3}$.

91. Solve the system:

$$\begin{cases} y = x^2 + 2 \\ y = mx + b \end{cases}$$

Solve the system by substitution:

$x^2 + 2 = mx + b \Rightarrow x^2 - mx + 2 - b = 0$

Note that the tangent line passes through (1, 3). Find the relation between m and b:

$3 = m(1) + b \Rightarrow b = 3 - m$

Substitute into the quadratic to eliminate b:

$x^2 - mx + 2 - (3-m) = 0 \Rightarrow x^2 - mx + (m-1) = 0$

Find when the discriminant equals 0:

$$(-m)^2 - 4(1)(m-1) = 0$$

$$m^2 - 4m + 4 = 0$$

$$(m-2)^2 = 0$$

$$m - 2 = 0$$

$$m = 2$$

$b = 3 - m = 3 - 2 = 1$

The equation of the tangent line is $y = 2x + 1$.

92. Solve the system:

$$\begin{cases} x^2 + y = 5 \\ y = mx + b \end{cases}$$

Solve the system by substitution:

$x^2 + mx + b = 5 \Rightarrow x^2 + mx + b - 5 = 0$

Note that the tangent line passes through (–2, 1). Find the relation between m and b:

$1 = m(-2) + b \Rightarrow b = 2m + 1$

Substitute into the quadratic to eliminate b:

$x^2 + mx + 2m + 1 - 5 = 0$

$x^2 + mx + (2m - 4) = 0$

Find when the discriminant equals 0:

$$(m)^2 - 4(1)(2m-4) = 0$$

$$m^2 - 8m + 16 = 0$$

$$(m-4)^2 = 0$$

$$m - 4 = 0$$

$$m = 4$$

$b = 2m + 1 = 2(4) + 1 = 9$

The equation of the tangent line is $y = 4x + 9$.

93. Solve the system:

$$\begin{cases} 2x^2 + 3y^2 = 14 \\ y = mx + b \end{cases}$$

Solve the system by substitution:

$$2x^2 + 3(mx+b)^2 = 14$$

$$2x^2 + 3m^2x^2 + 6mbx + 3b^2 = 14$$

$$(3m^2+2)x^2 + 6mbx + 3b^2 - 14 = 0$$

Note that the tangent line passes through (1, 2). Find the relation between m and b:

$2 = m(1) + b \Rightarrow b = 2 - m$

Substitute into the quadratic to eliminate b:

$$(3m^2+2)x^2 + 6m(2-m)x + 3(2-m)^2 - 14 = 0$$

$$(3m^2+2)x^2 + (12m - 6m^2)x + (3m^2 - 12m - 2) = 0$$

Find when the discriminant equals 0:

$$(12m-6m^2)^2-4(3m^2+2)(3m^2-12m-2)=0$$
$$144m^2+96m+16=0$$
$$9m^2+6m+1=0$$
$$(3m+1)^2=0$$
$$3m+1=0$$
$$m=-\frac{1}{3}$$

$$b=2-m=2-\left(-\frac{1}{3}\right)=\frac{7}{3}$$

The equation of the tangent line is $y=-\frac{1}{3}x+\frac{7}{3}$.

94. Solve the system:
$$\begin{cases} 3x^2+y^2=7 \\ y=mx+b \end{cases}$$
Solve the system by substitution:
$$3x^2+\left(mx+b\right)^2=7$$
$$3x^2+m^2x^2+2mbx+b^2=7$$
$$\left(m^2+3\right)x^2+2mbx+b^2-7=0$$
Note that the tangent line passes through (–1, 2).
Find the relation between m and b:
$2=m(-1)+b\Rightarrow b=m+2$
There is one solution to the quadratic if the discriminant equals 0.
$$\left(2bm\right)^2-4\left(m^2+3\right)\left(b^2-7\right)=0$$
$$4b^2m^2-4b^2m^2+28m^2-12b^2+84=0$$
$$28m^2-12b^2+84=0$$
$$7m^2-3b^2+21=0$$
Substitute for b and solve:
$$7m^2-3\left(m+2\right)^2+21=0$$
$$7m^2-3m^2-12m-12+21=0$$
$$4m^2-12m+9=0$$
$$\left(2m-3\right)^2=0$$
$$2m=3$$
$$m=\frac{3}{2}$$

$$b=m+2=\frac{3}{2}+2=\frac{7}{2}$$

The equation of the tangent line is $y=\frac{3}{2}x+\frac{7}{2}$.

95. Solve the system:
$$\begin{cases} x^2-y^2=3 \\ y=mx+b \end{cases}$$
Solve the system by substitution:
$$x^2-\left(mx+b\right)^2=3$$
$$x^2-m^2x^2-2mbx-b^2=3$$
$$\left(1-m^2\right)x^2-2mbx-b^2-3=0$$
Note that the tangent line passes through (2, 1).
Find the relation between m and b:
$1=m(2)+b\Rightarrow b=1-2m$
Substitute into the quadratic to eliminate b:
$$(1-m^2)x^2-2m(1-2m)x-(1-2m)^2-3=0$$
$$(1-m^2)x^2+(-2m+4m^2)x-1+4m-4m^2-3=0$$
$$(1-m^2)x^2+(-2m+4m^2)x+(-4m^2+4m-4)=0$$
Find when the discriminant equals 0:
$$\left(-2m+4m^2\right)^2-4\left(1-m^2\right)\left(-4m^2+4m-4\right)=0$$
$$4m^2-16m^3+16m^4-16m^4+16m^3-16m+16=0$$
$$4m^2-16m+16=0$$
$$m^2-4m+4=0$$
$$\left(m-2\right)^2=0$$
$$m=2$$
The equation of the tangent line is $y=2x-3$.

96. Solve the system:
$$\begin{cases} 2y^2-x^2=14 \\ y=mx+b \end{cases}$$
Solve the system by substitution:
$$2\left(mx+b\right)^2-x^2=14$$
$$2m^2x^2+4mbx+2b^2-x^2=14$$
$$\left(2m^2-1\right)x^2+4mbx+2b^2-14=0$$
Note that the tangent line passes through (2, 3).
Find the relation between m and b:
$3=m(2)+b\Rightarrow b=3-2m$
There is one solution to the quadratic if the discriminant equals 0.
$$\left(4bm\right)^2-4\left(2m^2-1\right)\left(2b^2-14\right)=0$$
$$16b^2m^2-16b^2m^2+112m^2+8b^2-56=0$$
$$112m^2+8b^2-56=0$$
$$14m^2+b^2-7=0$$
Substitute for b and solve:

$$14m^2 + (3-2m)^2 - 7 = 0$$
$$14m^2 + 4m^2 - 12m + 9 - 7 = 0$$
$$18m^2 - 12m + 2 = 0$$
$$2(3m-1)^2 = 0$$
$$3m = 1$$
$$m = \frac{1}{3}$$
$$b = 3 - 2m = 3 - 2\left(\frac{1}{3}\right) = \frac{7}{3}$$

The equation of the tangent line is $y = \frac{1}{3}x + \frac{7}{3}$.

97. Solve for r_1 and r_2:

$$\begin{cases} r_1 + r_2 = -\frac{b}{a} \\ r_1 r_2 = \frac{c}{a} \end{cases}$$

Substitute and solve:

$$r_1 = -r_2 - \frac{b}{a}$$
$$\left(-r_2 - \frac{b}{a}\right) r_2 = \frac{c}{a}$$
$$-r_2^2 - \frac{b}{a} r_2 - \frac{c}{a} = 0$$
$$ar_2^2 + br_2 + c = 0$$
$$r_2 = \frac{-b \pm \sqrt{b^2 - 4ac}}{2a}$$
$$r_1 = -r_2 - \frac{b}{a} =$$
$$= -\left(\frac{-b \pm \sqrt{b^2 - 4ac}}{2a}\right) - \frac{b}{a}$$
$$= \frac{b \mp \sqrt{b^2 - 4ac}}{2a} - \frac{2b}{2a}$$
$$= \frac{-b \mp \sqrt{b^2 - 4ac}}{2a}$$

The solutions are:

$$\frac{-b + \sqrt{b^2 - 4ac}}{2a} \text{ and } \frac{-b - \sqrt{b^2 - 4ac}}{2a}.$$

98. Consider the circle with equation $(x-h)^2 + (y-k)^2 = r^2$ and the third degree polynomial with equation $y = ax^3 + bx^2 + cx + d$.
Substituting the second equation into the first equation yields $(x-h)^2 + (ax^3 + bx^2 + cx + d - k)^2 = r^2$.
In order to find the roots for this equation we can expand the terms on the left hand side of the equation. Notice that $(x-h)^2$ yields a 2nd degree polynomial, and $(ax^3 + bx^2 + cx + d - k)^2$ yields a 6th degree polynomial. Therefore, we need to find the roots of a 6th degree equation, and the Fundamental Theorem of Algebra states that there will be at most six real roots. Thus, the circle and the 3rd degree polynomial will intersect at most six times. Now consider the circle with equation $(x-h)^2 + (y-k)^2 = r^2$ and the polynomial of degree n with equation $y = a_0 + a_1x + a_2x^2 + a_3x^3 + \ldots + a_nx^n$.
Substituting the first equation into the first equation yields

$$(x-h)^2 + \left(a_0 + a_1x + a_2x^2 + a_3x^3 + \ldots + a_nx^n - k\right)^2 = r^2$$

In order to find the roots for this equation we can expand the terms on the left hand side of the equation.
Notice that $(x-h)^2$ yields a 2nd degree polynomial, and

$$\left(a_0 + a_1x + a_2x^2 + a_3x^3 + \ldots + a_nx^n - k\right)^2$$

yields a polynomial of degree $2n$.
Therefore, we need to find the roots of an equation of degree $2n$, and the Fundamental Theorem of Algebra states that there will be at most $2n$ real roots. Thus, the circle and the nth degree polynomial will intersect at most $2n$ times.

99. Since the area of the square piece of sheet metal is 100 square feet, the sheet's dimensions are 10 feet by 10 feet. Let $x =$ the length of the cut.

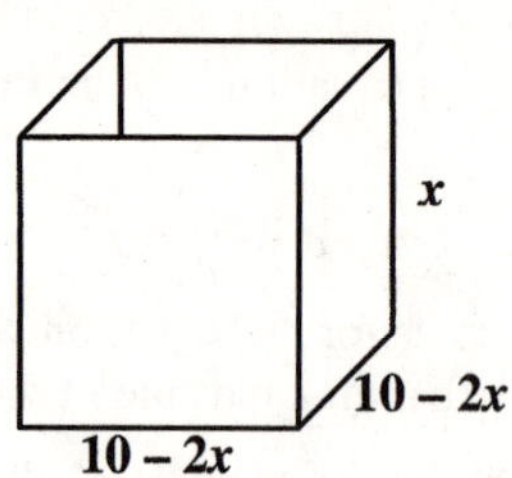

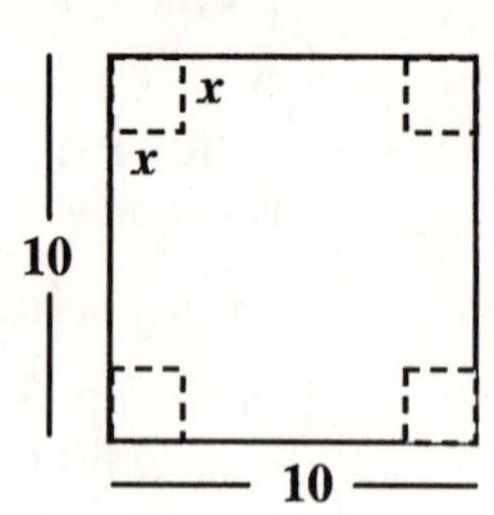

The dimensions of the box are
length $= 10 - 2x$; width $= 10 - 2x$; height $= x$
Note that each of these expressions must be positive. So we must have
$x > 0$ and $10 - 2x > 0 \Rightarrow x < 5$, that is, $0 < x < 5$.
So the volume of the box is given by

$$V = (\text{length}) \cdot (\text{width}) \cdot (\text{height})$$
$$= (10-2x)(10-2x)(x)$$
$$= (10-2x)^2 (x)$$

a. In order to get a volume equal to 9 cubic feet, we solve $(10-2x)^2(x) = 9$.

$$(10-2x)^2(x) = 9$$
$$(100 - 40x + 4x^2)x = 9$$
$$100x - 40x^2 + 4x^3 = 9$$

So we need to solve the equation
$4x^3 - 40x^2 + 100x - 9 = 0$.

Graphing the function
$y_1 = 4x^3 - 40x^2 + 100x - 9$ on a calculator yields the graph

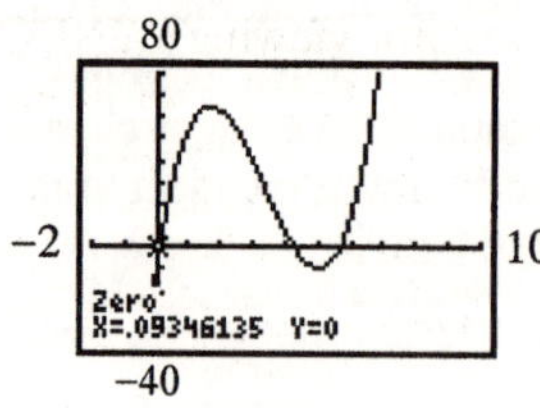

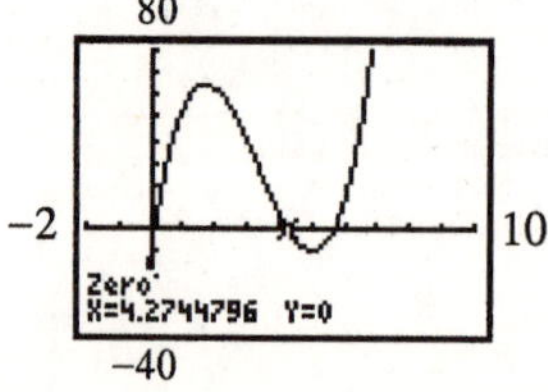

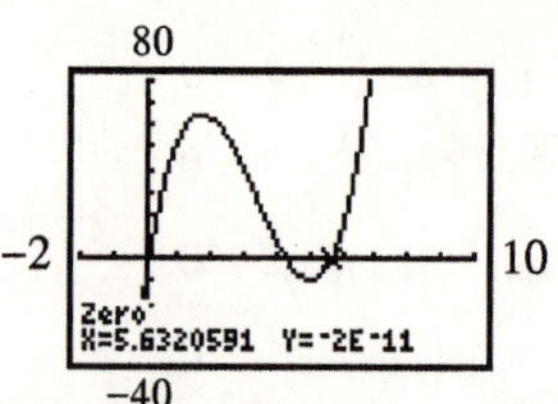

The graph indicates that there are three real zeros on the interval [0, 6].
Using the ZERO feature of a graphing calculator, we find that the three roots shown occur at $x \approx 0.093$, $x \approx 4.274$ and $x \approx 5.632$.
But we've already noted that we must have $0 < x < 5$, so the only practical values for the sides of the square base are $x \approx 0.093$ feet and $x \approx 4.274$ feet.

b. Answers will vary.

Section 10.7

1. $3x + 4 < 8 - x$

$$4x < 4$$
$$x < 1$$
$$\{x \mid x < 1\} \text{ or } (-\infty, 1)$$

2. $3x - 2y = 6$

The graph is a line.

x-intercept: $3x - 2(0) = 6$
$$3x = 6$$
$$x = 2$$

y-intercept: $3(0) - 2y = 6$
$$-2y = 6$$
$$y = -3$$

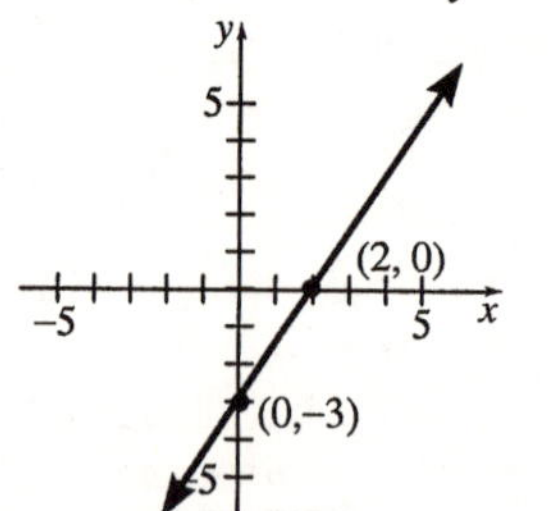

3. $x^2 + y^2 = 9$

The graph is a circle. Center: (0, 0) ; Radius: 3

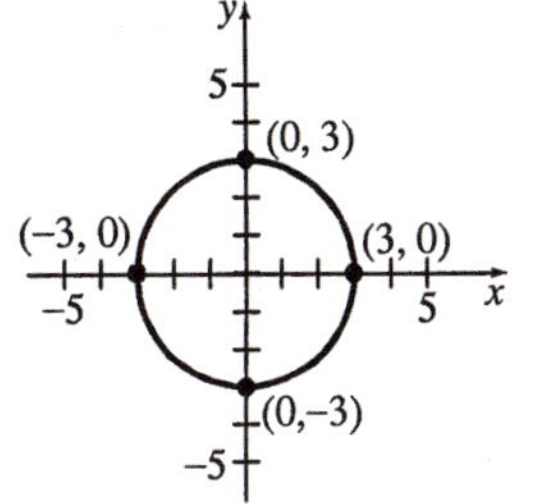

4. $y = x^2 + 4$

The graph is a parabola.

x-intercepts: $0 = x^2 + 4$

$x^2 = -4$, no x-intercepts

y-intercept: $y = 0^2 + 4 = 4$

The vertex has x-coordinate:

$x = -\frac{b}{2a} = -\frac{0}{2(1)} = 0$.

The y-coordinate of the vertex is $y = 0^2 + 4 = 4$.

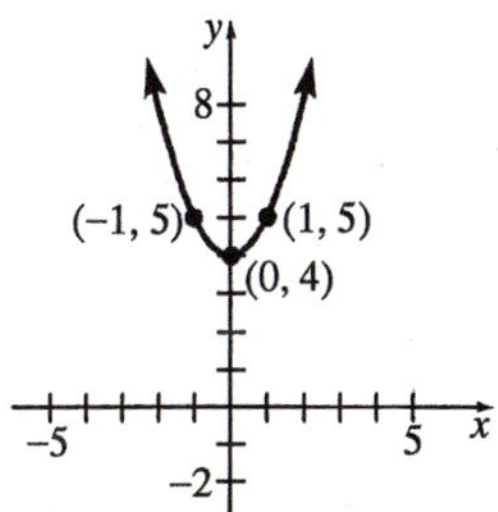

5. True

6. $y = x^2$; right; 2

7. satisfied

8. half-plane

9. False

10. True

11. a. $x \geq 0$

Graph the line $x = 0$. Use a solid line since the inequality uses $\geq$. Choose a test point not on the line, such as (2, 0). Since $2 \geq 0$ is true, shade the side of the line containing (2, 0).

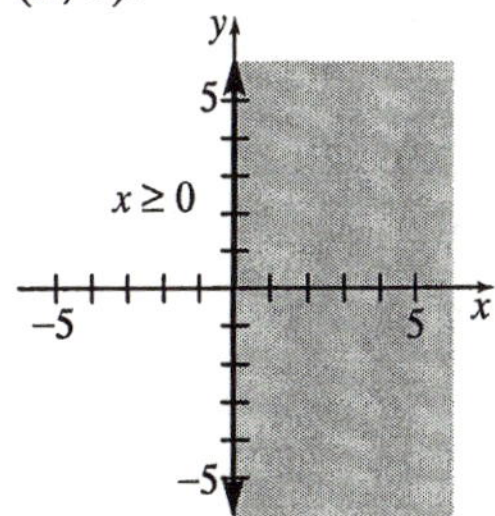

b. Set the calculator viewing WINDOW as shown below. Graph the vertical line $x = 0$ by using the SHADE commands in the calculators DRAW menu: Shade $(-5, 5, 0, 5)$

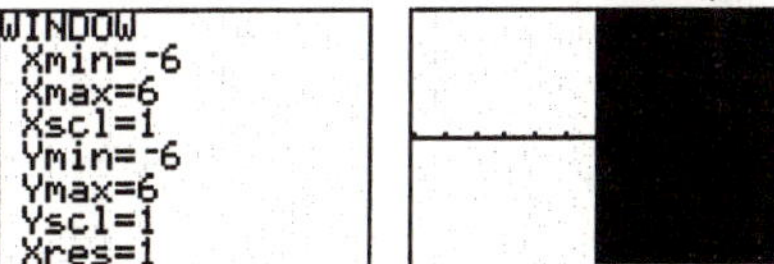

12. a. $y \geq 0$

Graph the line $y = 0$. Use a solid line since the inequality uses $\geq$. Choose a test point not on the line, such as (0, 2). Since $2 \geq 0$ is true, shade the side of the line containing (0, 2).

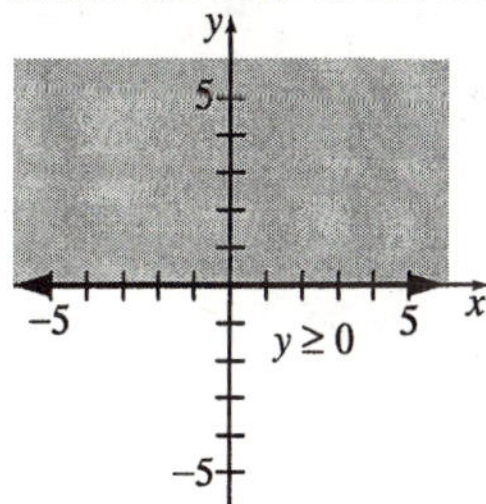

b. Set the calculator viewing WINDOW as shown below. Graph the linear inequality $y \geq 0$ by entering $Y_1 = 0$ and adjusting the setting to the left of Y_1 as shown below:

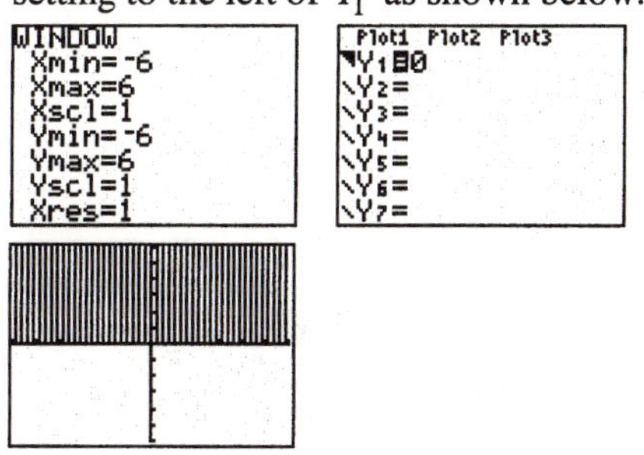

13. a. $x \geq 4$

Graph the line $x = 4$. Use a solid line since the inequality uses $\geq$. Choose a test point not on the line, such as (5, 0). Since $5 \geq 0$ is true, shade the side of the line containing (5, 0).

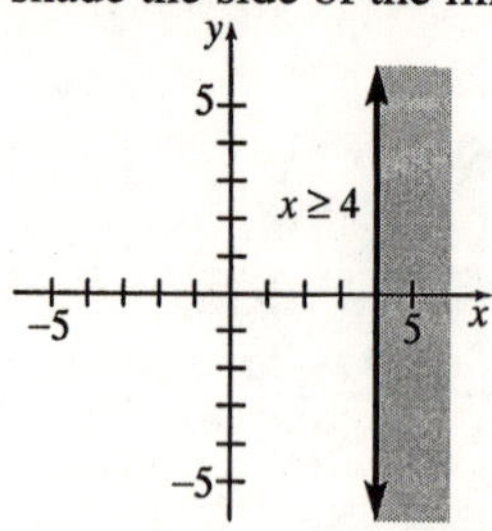

b. Set the calculator viewing WINDOW as shown below. Graph the vertical line $x = 0$ by using the SHADE commands in the calculators DRAW menu:
Shade (−5, 5, 4, 5).

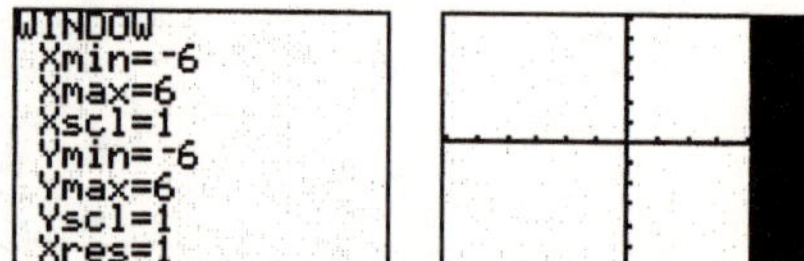

14. a. $y \leq 2$

Graph the line $y = 2$. Use a solid line since the inequality uses $\leq$. Choose a test point not on the line, such as (5, 0). Since $0 \leq 2$ is true, shade the side of the line containing (5, 0).

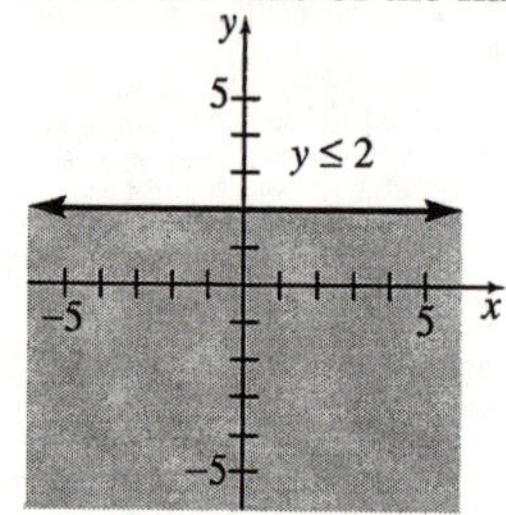

b. Set the calculator viewing WINDOW as shown below. Graph the linear inequality $y \leq 4$ by entering $Y_1 = 4$ and adjusting the setting to the left of Y_1 as shown below:

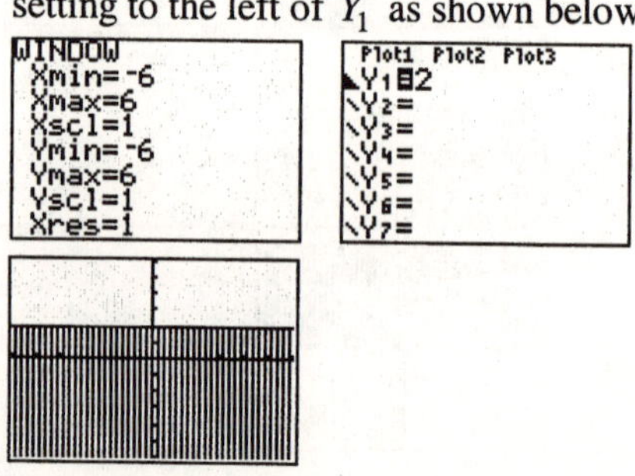

15. a. $2x + y \geq 6$

Graph the line $2x + y = 6$. Use a solid line since the inequality uses $\geq$. Choose a test point not on the line, such as (0, 0). Since $2(0) + 0 \geq 6$ is false, shade the opposite side of the line from (0, 0).

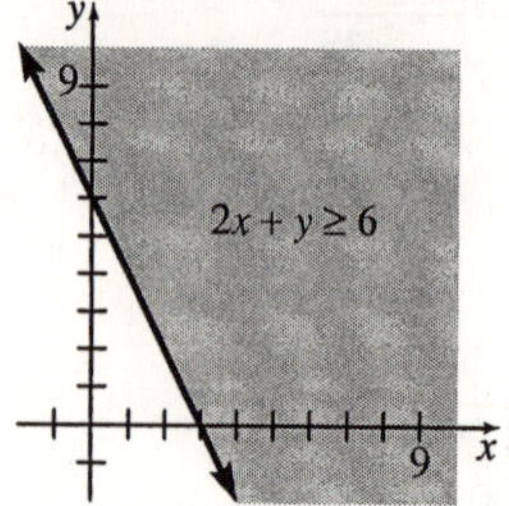

b. Set the calculator viewing WINDOW as shown below. Graph the linear inequality $2x + y \geq 6$ $(y \geq -2x + 6)$ by entering $Y_1 = -2x + 6$ and adjusting the setting to the left of Y_1 as shown below:

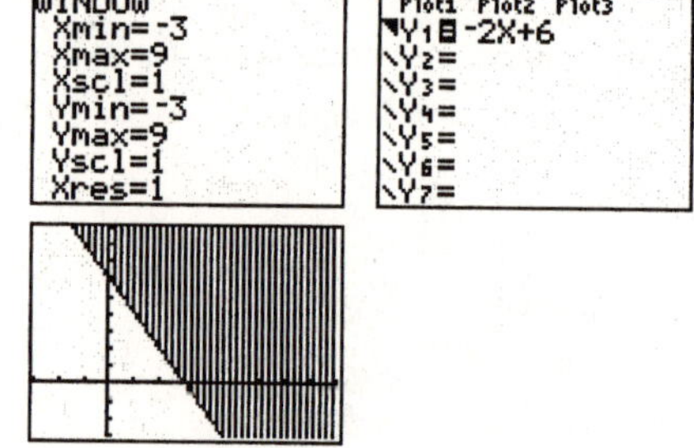

16. a. $3x + 2y \leq 6$

Graph the line $3x + 2y = 6$. Use a solid line since the inequality uses $\leq$. Choose a test point not on the line, such as (0, 0). Since $3(0) + 2(0) \leq 6$ is true, shade the side of the line containing (0, 0).

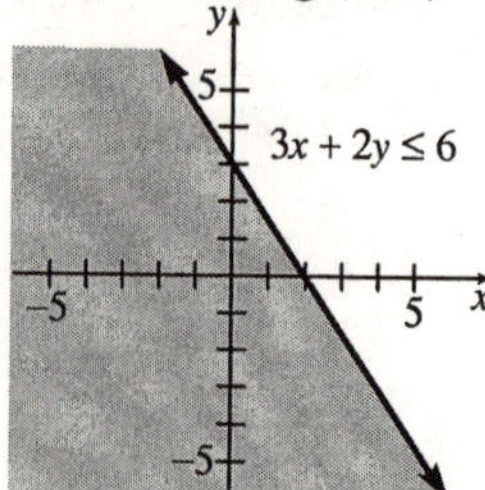

b. Set the calculator viewing WINDOW as shown below. Graph the linear inequality $3x+2y\le 6$ $\left(y\ge -\frac{3}{2}x+3\right)$ by entering $Y_1=-\frac{3}{2}x+3$ and adjusting the setting to the left of Y_1 as shown below:

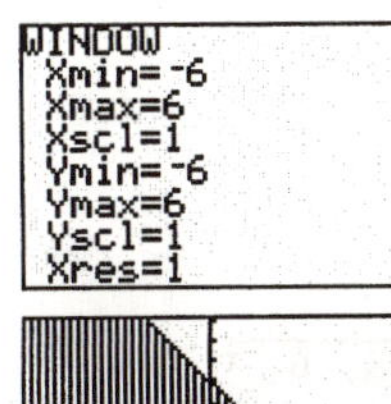

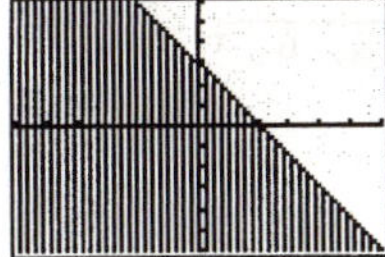

17. a. $x^2+y^2>1$

Graph the circle $x^2+y^2=1$. Use a dashed line since the inequality uses >. Choose a test point not on the circle, such as (0, 0). Since $0^2+0^2>1$ is false, shade the opposite side of the circle from (0, 0).

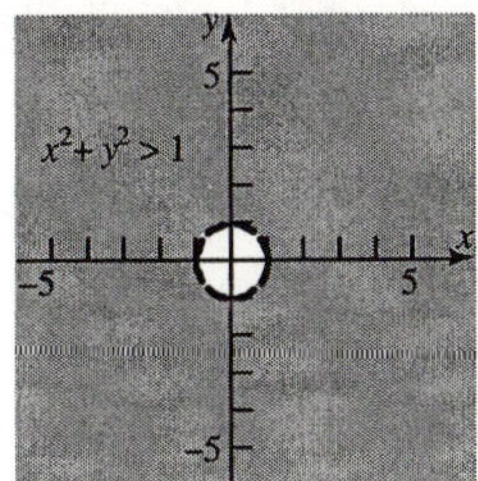

b. Solving $x^2+y^2=1$ for *y*, we obtain $y=\pm\sqrt{1-x^2}$. We need to shade above $y=\sqrt{1-x^2}$ and below $y=-\sqrt{1-x^2}$. Since these functions are only defined on the domain $-1\le x\le 1$, we must also shade to the left of $x=-1$ and to the right of $x=1$. To do so, first set the calculator viewing WINDOW as shown below. The SHADE commands shown below shade above, below, to the left, and to the right of the circle, respectively:

Shade $\left(\sqrt{1-x^2},5,-1,1\right)$

Shade $\left(-5,-\sqrt{1-x^2},-1,1\right)$

Shade $(-5,5,-5,-1)$

Shade $(-5,5,1,5)$

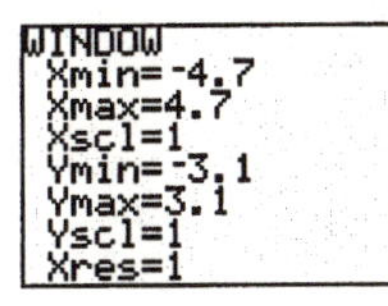

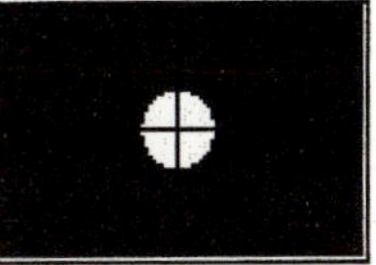

18. a. $x^2+y^2\le 9$

Graph the circle $x^2+y^2=9$. Use a solid line since the inequality uses ≤. Choose a test point not on the circle, such as (0, 0). Since $0^2+0^2\le 9$ is true, shade the same side of the circle as (0, 0).

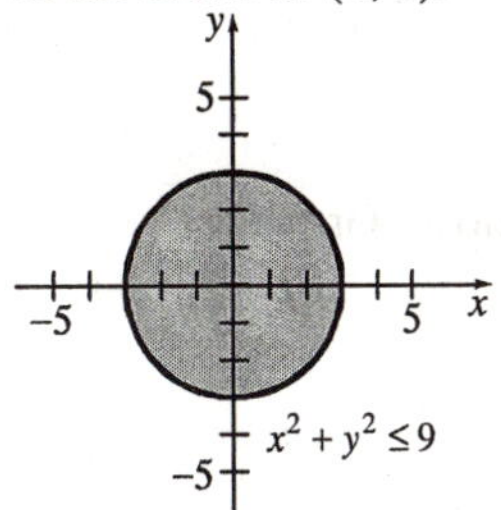

b. Solving $x^2+y^2=9$ for *y*, we obtain $y=\pm\sqrt{9-x^2}$. We need to shade above $y=-\sqrt{9-x^2}$ and below $y=\sqrt{9-x^2}$. To do so, first set the calculator viewing WINDOW as shown below. The SHADE command shown will complete the graph:

Shade $\left(-\sqrt{9-x^2},\sqrt{9-x^2},-3,3\right)$

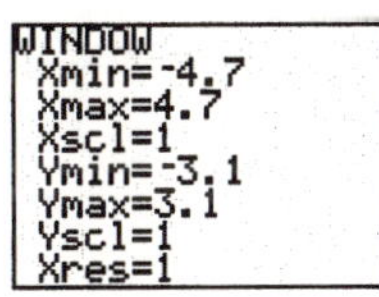

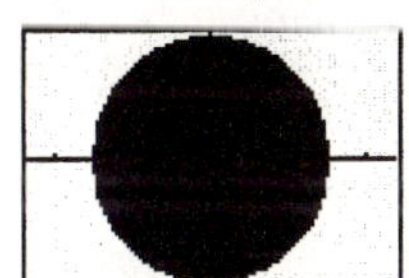

19. a. $y\le x^2-1$

Graph the parabola $y=x^2-1$. Use a solid line since the inequality uses ≤. Choose a test point not on the parabola, such as (0, 0). Since $0\le 0^2-1$ is false, shade the opposite side of the parabola from (0, 0).

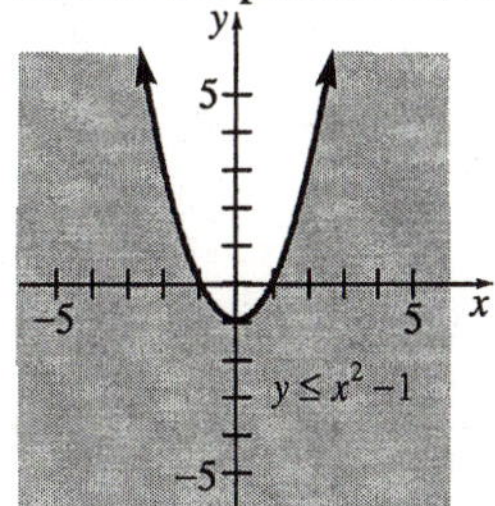

b. Set the calculator viewing WINDOW as shown below. Enter $Y_1 = x^2 - 1$ and adjust the setting to the left of Y_1 as shown below:

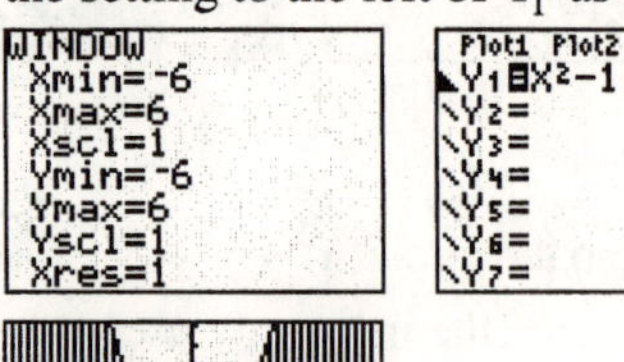

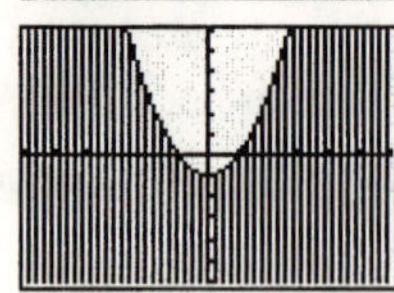

20. a. $y > x^2 + 2$

Graph the parabola $y = x^2 + 2$. Use a dashed line since the inequality uses >. Choose a test point not on the parabola, such as (0, 0). Since $0 > 0^2 + 2$ is false, shade the opposite side of the parabola from (0, 0).

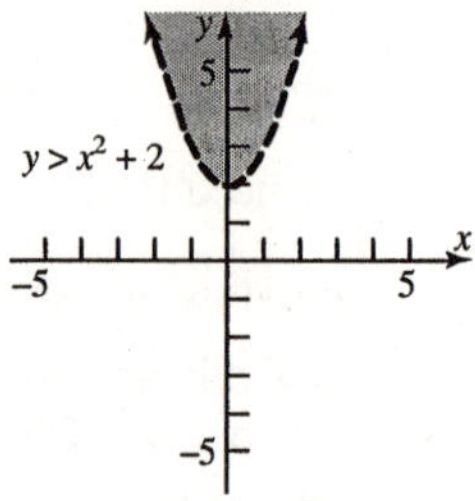

b. Set the calculator viewing WINDOW as shown below. Enter $Y_1 = x^2 + 2$ and adjust the setting to the left of Y_1 as shown below:

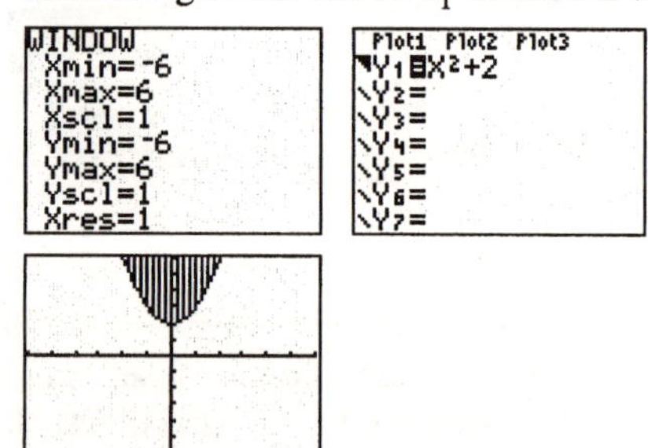

21. a. $xy \ge 4$

Graph the hyperbola $xy = 4$. Use a solid line since the inequality uses $\ge$. Choose a test point not on the hyperbola, such as (0, 0). Since $0 \cdot 0 \ge 4$ is false, shade the opposite side of the hyperbola from (0, 0).

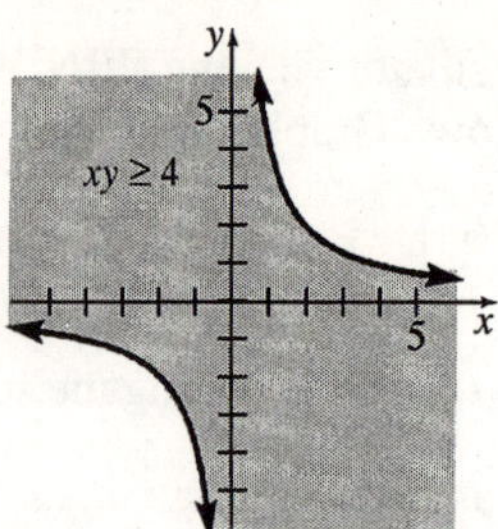

b. Solving $xy = 4$ for *y*, we obtain $y = \frac{4}{x}$. We need to shade below $y = \frac{4}{x}$ when $x < 0$ and above $y = \frac{4}{x}$ when $x > 0$. To do so, first set the calculator viewing WINDOW as shown below. The SHADE commands shown below shade the necessary area, respectively:

$$\text{Shade}\left(-6, \frac{4}{x}, -6, 0\right)$$

$$\text{Shade}\left(\frac{4}{x}, 6, 0, 6\right)$$

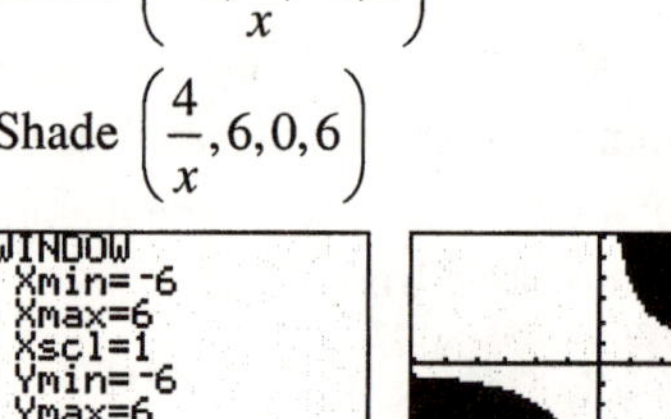

22. a. $xy \le 1$

Graph the hyperbola $xy = 1$. Use a solid line since the inequality uses $\le$. Choose a test point not on the hyperbola, such as (0, 0). Since $0 \cdot 0 \le 1$ is true, shade the same side of the hyperbola as (0, 0).

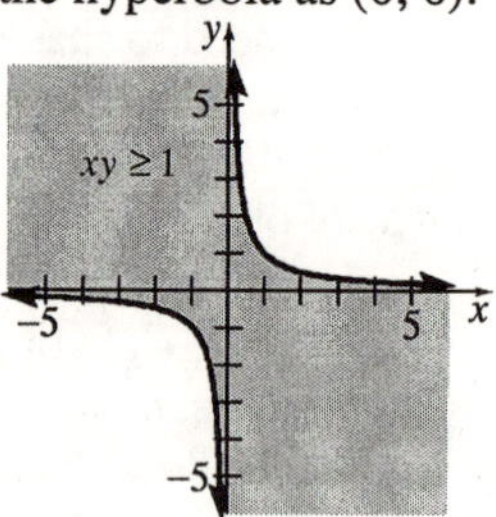

b. Solving $xy = 1$ for *y*, we obtain $y = \frac{1}{x}$. We need to shade above $y = \frac{1}{x}$ when $x < 0$ and below $y = \frac{1}{x}$ when $x > 0$. To do so, first set the calculator viewing WINDOW as shown below. The SHADE commands shown will complete the graph:

Shade $\left(\frac{1}{x}, 6, -6, 0\right)$

Shade $\left(-6, \frac{4}{x}, 0, 6\right)$

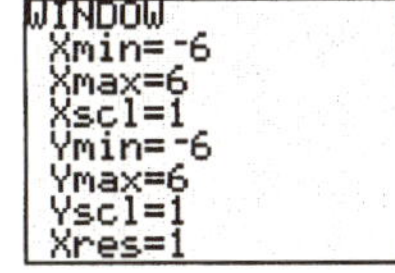

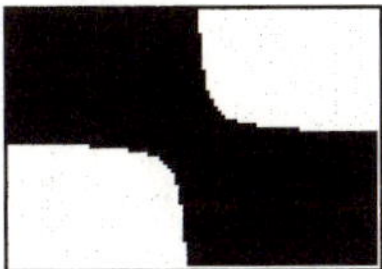

23. $\begin{cases} x+y\le 2 \\ 2x+y\ge 4 \end{cases}$

Graph the line $x+y=2$. Use a solid line since the inequality uses $\le$. Choose a test point not on the line, such as (0, 0). Since $0+0\le 2$ is true, shade the side of the line containing (0, 0). Graph the line $2x+y=4$. Use a solid line since the inequality uses $\ge$. Choose a test point not on the line, such as (0, 0). Since $2(0)+0\ge 4$ is false, shade the opposite side of the line from (0, 0). The overlapping region is the solution.

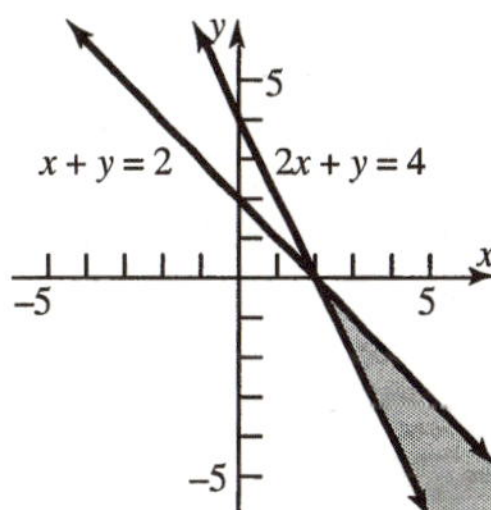

24. $\begin{cases} 3x-\ y\ge 6 \\ x+2y\le 2 \end{cases}$

Graph the line $3x-y=6$. Use a solid line since the inequality uses $\ge$. Choose a test point not on the line, such as (0, 0). Since $3(0)-0\ge 6$ is false, shade the opposite side of the line from (0, 0). Graph the line $x+2y=2$. Use a solid line since the inequality uses $\le$. Choose a test point not on the line, such as (0, 0). Since $0+2(0)\le 2$ is true, shade the side of the line containing (0, 0). The overlapping region is the solution.

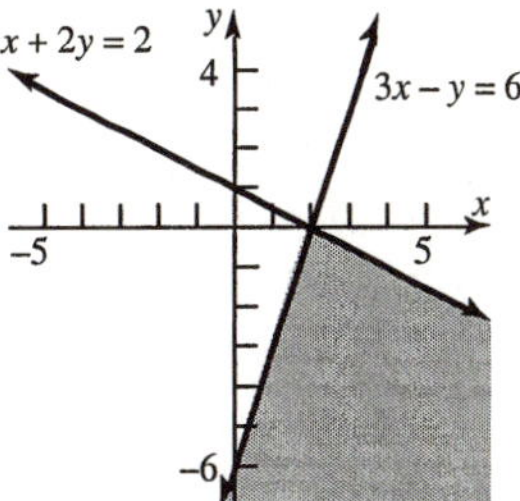

25. $\begin{cases} 2x-\ y\le\ 4 \\ 3x+2y\ge -6 \end{cases}$

Graph the line $2x-y=4$. Use a solid line since the inequality uses $\le$. Choose a test point not on the line, such as (0, 0). Since $2(0)-0\le 4$ is true, shade the side of the line containing (0, 0). Graph the line $3x+2y=-6$. Use a solid line since the inequality uses $\ge$. Choose a test point not on the line, such as (0, 0). Since $3(0)+2(0)\ge -6$ is true, shade the side of the line containing (0, 0). The overlapping region is the solution.

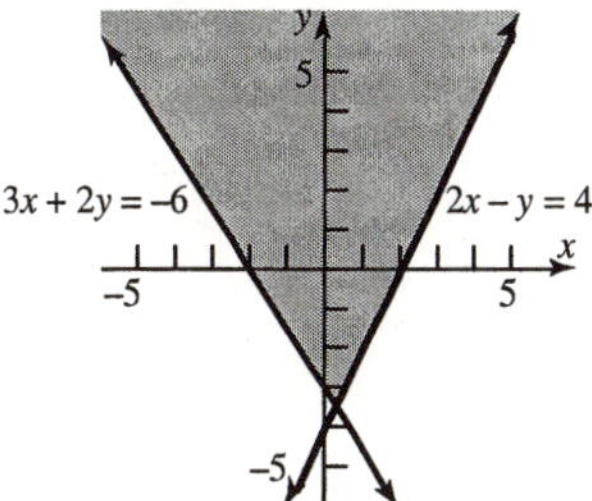

26. $\begin{cases} 4x-5y\le 0 \\ 2x-\ y\ge 2 \end{cases}$

Graph the line $4x-5y=0$. Use a solid line since the inequality uses $\le$. Choose a test point not on the line, such as (2, 0). Since $4(2)-5(0)\le 0$ is false, shade the opposite side of the line from (2, 0). Graph the line $2x-y=2$. Use a solid line since the inequality uses $\ge$. Choose a test point not on the line, such as (0, 0). Since $2(0)-0\ge 2$ is false, shade the opposite side of the line from (0, 0). The overlapping region is the solution.

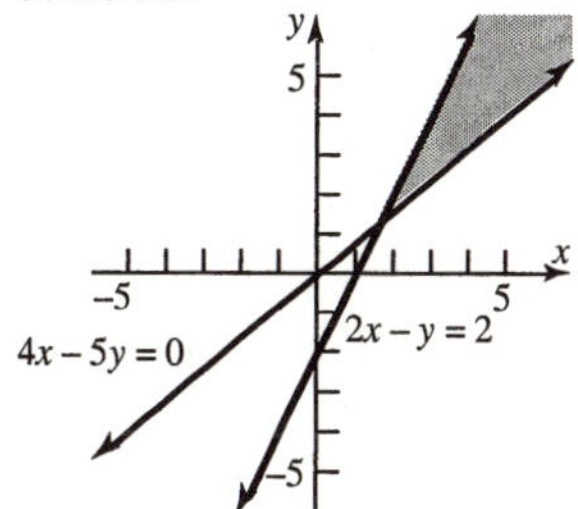

27. $\begin{cases} 2x-3y \le 0 \\ 3x+2y \le 6 \end{cases}$

Graph the line $2x-3y=0$. Use a solid line since the inequality uses $\le$. Choose a test point not on the line, such as (0, 3). Since $2(0)-3(3) \le 0$ is true, shade the side of the line containing (0, 3). Graph the line $3x+2y=6$. Use a solid line since the inequality uses $\le$. Choose a test point not on the line, such as (0, 0). Since $3(0)+2(0) \le 6$ is true, shade the side of the line containing (0, 0). The overlapping region is the solution.

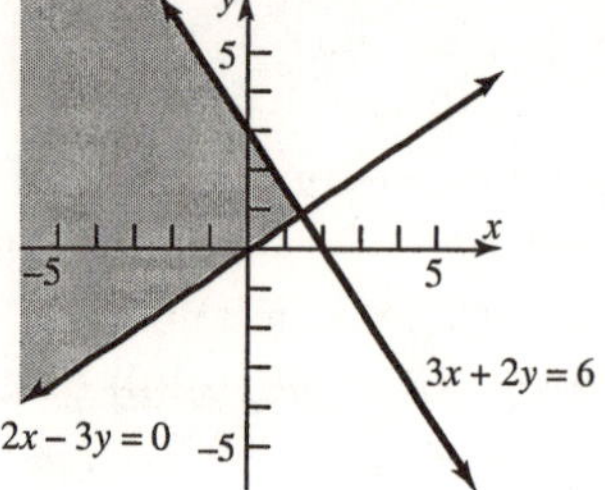

28. $\begin{cases} 4x-\ y \ge 2 \\ x+2y \ge 2 \end{cases}$

Graph the line $4x-y=2$. Use a solid line since the inequality uses $\ge$. Choose a test point not on the line, such as (0, 0). Since $4(0)-0 \ge 2$ is false, shade the opposite side of the line from (0, 0). Graph the line $x+2y=2$. Use a solid line since the inequality uses $\ge$. Choose a test point not on the line, such as (0, 0). Since $0+2(0) \ge 2$ is false, shade the opposite side of the line from (0, 0). The overlapping region is the solution.

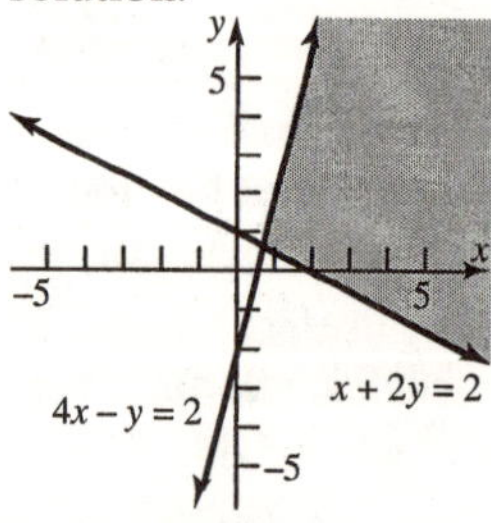

29. $\begin{cases} x-2y \le 6 \\ 2x-4y \ge 0 \end{cases}$

Graph the line $x-2y=6$. Use a solid line since the inequality uses $\le$. Choose a test point not on the line, such as (0, 0). Since $0-2(0) \le 6$ is true, shade the side of the line containing (0, 0). Graph the line $2x-4y=0$. Use a solid line since the inequality uses $\ge$. Choose a test point not on the line, such as (0, 2). Since $2(0)-4(2) \ge 0$ is false, shade the opposite side of the line from (0, 2). The overlapping region is the solution.

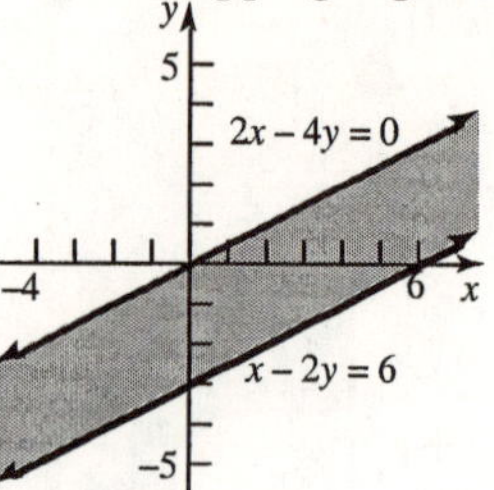

30. $\begin{cases} x+4y \le 8 \\ x+4y \ge 4 \end{cases}$

Graph the line $x+4y=8$. Use a solid line since the inequality uses $\le$. Choose a test point not on the line, such as (0, 0). Since $0+4(0) \le 8$ is true, shade the side of the line containing (0, 0). Graph the line $x+4y=4$. Use a solid line since the inequality uses $\ge$. Choose a test point not on the line, such as (0, 0). Since $0+4(0) \ge 4$ is false, shade the opposite side of the line from (0, 0). The overlapping region is the solution.

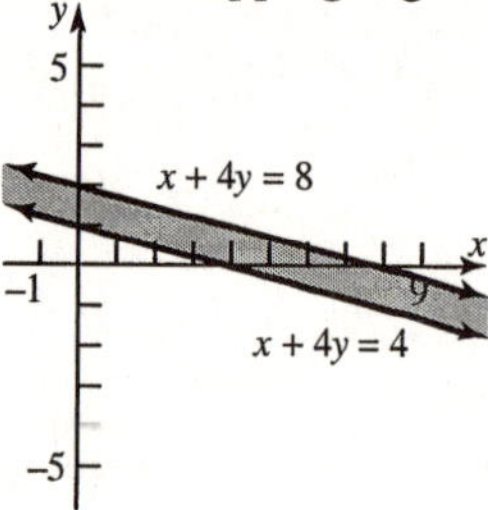

31. $\begin{cases} 2x+y \ge -2 \\ 2x+y \ge\ 2 \end{cases}$

Graph the line $2x+y=-2$. Use a solid line since the inequality uses $\ge$. Choose a test point not on the line, such as (0, 0). Since $2(0)+0 \ge -2$ is true, shade the side of the line containing (0, 0). Graph the line $2x+y=2$. Use a solid line since the inequality uses $\ge$. Choose a test point not on the line, such as (0, 0). Since $2(0)+0 \ge 2$ is false, shade the opposite side of the line from (0, 0). The overlapping region is the solution.

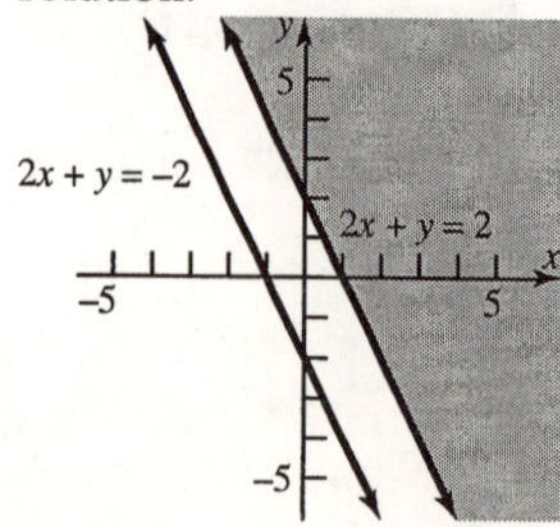

32. $\begin{cases} x-4y \le 4 \\ x-4y \ge 0 \end{cases}$

Graph the line $x-4y=4$. Use a solid line since the inequality uses $\le$. Choose a test point not on the line, such as (0, 0). Since $0-4(0) \le 4$ is true, shade the side of the line containing (0, 0). Graph the line $x-4y=0$. Use a solid line since the inequality uses $\ge$. Choose a test point not on the line, such as (1, 0). Since $1-4(0) \ge 0$ is true, shade the side of the line containing (1, 0). The overlapping region is the solution.

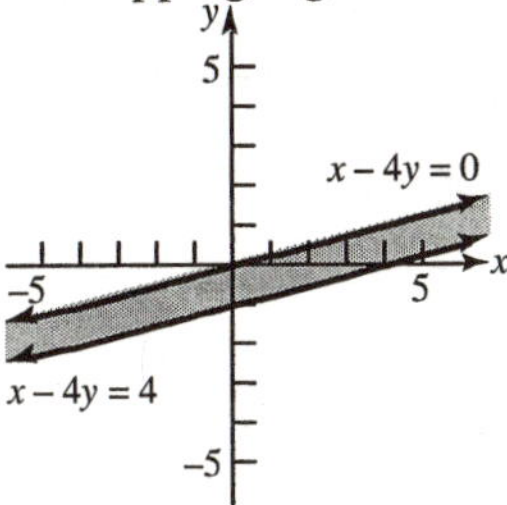

33. $\begin{cases} 2x+3y \ge 6 \\ 2x+3y \le 0 \end{cases}$

Graph the line $2x+3y=6$. Use a solid line since the inequality uses $\ge$. Choose a test point not on the line, such as (0, 0). Since $2(0)+3(0) \ge 6$ is false, shade the opposite side of the line from (0, 0). Graph the line $2x+3y=0$. Use a solid linc since the inequality uses $\le$. Choose a test point not on the line, such as (0, 2). Since $2(0)+3(2) \le 0$ is false, shade the opposite side of the line from (0, 2). Since the regions do not overlap, the solution is an empty set.

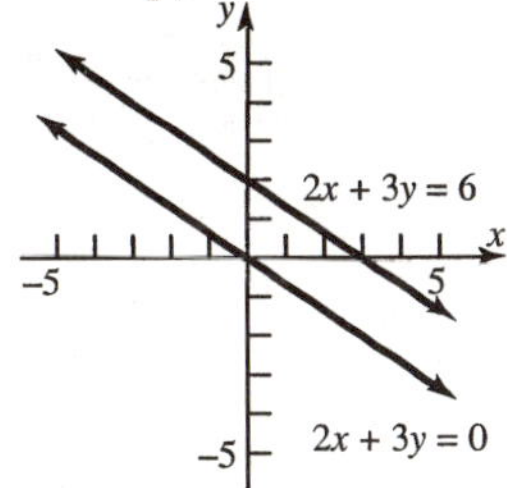

34. $\begin{cases} 2x+y \ge 0 \\ 2x+y \ge 2 \end{cases}$

Graph the line $2x+y=0$. Use a solid line since the inequality uses $\ge$. Choose a test point not on the line, such as (1, 0). Since $2(1)+0 \ge 0$ is true, shade the side of the line containing (1, 0). Graph the line $2x+y=2$. Use a solid line since the inequality uses $\ge$. Choose a test point not on the line, such as (0, 0). Since $2(0)+0 \ge 2$ is false, shade the opposite side of the line from (0, 0). The overlapping region is the solution.

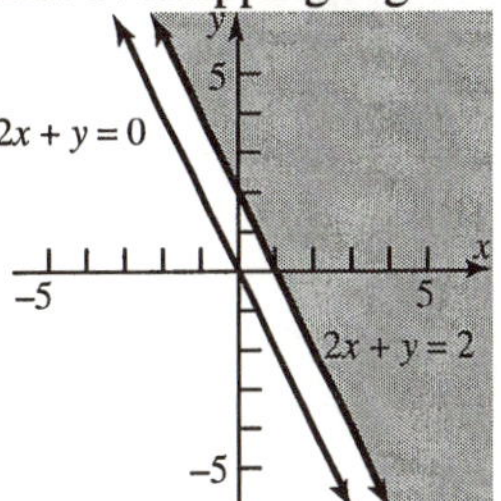

35. $\begin{cases} x^2+y^2 \le 9 \\ x+y \ge 3 \end{cases}$

Graph the circle $x^2+y^2=9$. Use a solid line since the inequality uses $\le$. Choose a test point not on the circle, such as (0, 0). Since $0^2+0^2 \le 9$ is true, shade the same side of the circle as (0, 0).
Graph the line $x+y=3$. Use a solid line since the inequality uses $\ge$. Choose a test point not on the line, such as (0, 0). Since $0+0 \ge 3$ is false, shade the opposite side of the line from (0, 0).
The overlapping region is the solution.

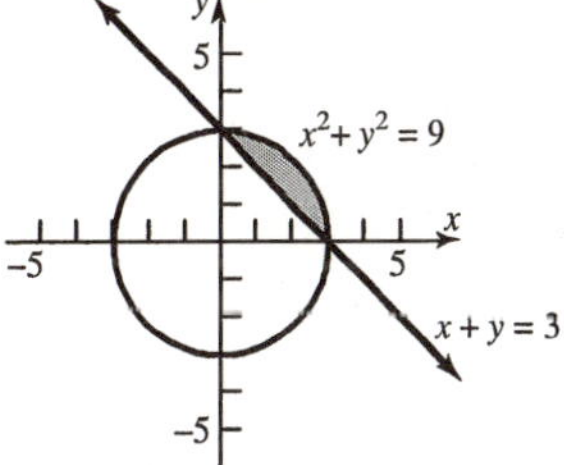

36. $\begin{cases} x^2+y^2 \ge 9 \\ x+y \le 3 \end{cases}$

Graph the circle $x^2+y^2=9$. Use a solid line since the inequality uses $\ge$. Choose a test point not on the circle, such as (0, 0). Since $0^2+0^2 \ge 9$ is false, shade the opposite side of the circle as (0, 0).
Graph the line $x+y=3$. Use a solid line since the inequality uses $\le$. Choose a test point not on the line, such as (0, 0). Since $0+0 \le 3$ is true, shade the same side of the line as (0, 0).
The overlapping region is the solution.

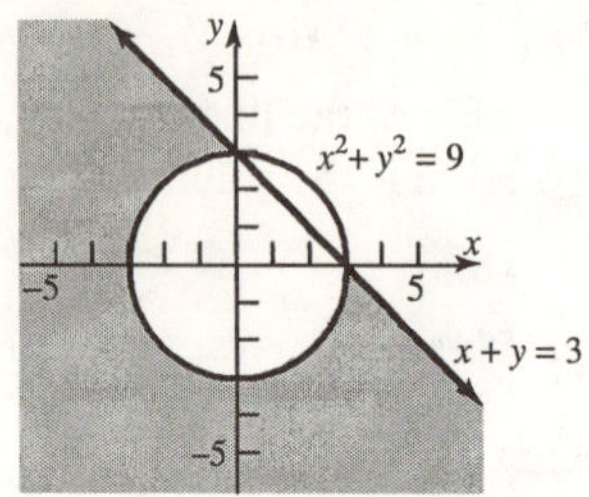

37. $\begin{cases} y \ge x^2 - 4 \\ y \le x - 2 \end{cases}$

Graph the parabola $y = x^2 - 4$. Use a solid line since the inequality uses $\ge$. Choose a test point not on the parabola, such as (0, 0). Since $0 \ge 0^2 - 4$ is true, shade the same side of the parabola as (0, 0). Graph the line $y = x - 2$. Use a solid line since the inequality uses $\le$. Choose a test point not on the line, such as (0, 0). Since $0 \le 0 - 2$ is false, shade the opposite side of the line from (0, 0). The overlapping region is the solution.

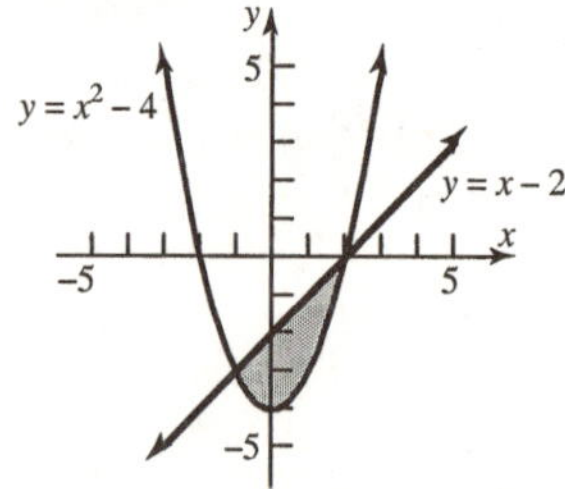

38. $\begin{cases} y^2 \le x \\ y \ge x \end{cases}$

Graph the parabola $y^2 = x$. Use a solid line since the inequality uses $\le$. Choose a test point not on the parabola, such as (1, 2). Since $2^2 \le 1$ is false, shade the opposite side of the parabola from (1, 2). Graph the line $y = x$. Use a solid line since the inequality uses $\ge$. Choose a test point not on the line, such as (1, 2). Since $2 \ge 1$ is true, shade the same side of the line as (1, 2). The overlapping region is the solution.

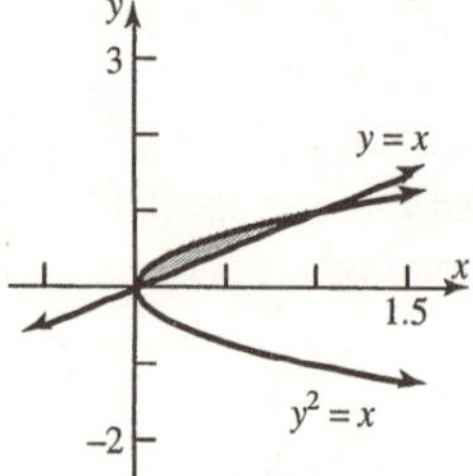

39. $\begin{cases} x^2 + y^2 \le 16 \\ y \ge x^2 - 4 \end{cases}$

Graph the circle $x^2 + y^2 = 16$. Use a sold line since the inequality is not strict. Choose a test point not on the circle, such as $(0,0)$. Since $0^2 + 0^2 \le 16$ is true, shade the side of the circle containing $(0,0)$. Graph the parabola $y = x^2 - 4$. Use a solid line since the inequality is not strict. Choose a test point not on the parabola, such as $(0,0)$. Since $0 \ge 0^2 - 4$ is true, shade the side of the parabola that contains $(0,0)$. The overlapping region is the solution.

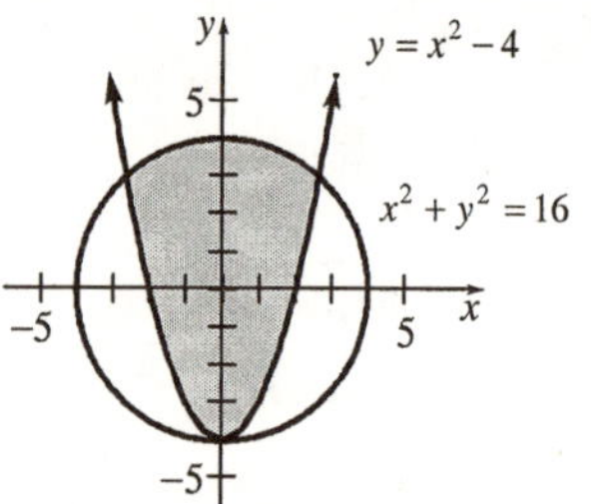

40. $\begin{cases} x^2 + y^2 \le 25 \\ y \le x^2 - 5 \end{cases}$

Graph the circle $x^2 + y^2 = 25$. Use a sold line since the inequality is not strict. Choose a test point not on the circle, such as $(0,0)$. Since $0^2 + 0^2 \le 25$ is true, shade the side of the circle containing $(0,0)$. Graph the parabola $y = x^2 - 5$. Use a solid line since the inequality is not strict. Choose a test point not on the parabola, such as $(0,0)$. Since $0 \le 0^2 - 5$ is false, shade the side of the parabola opposite that which contains the point $(0,0)$. The overlapping region is the solution.

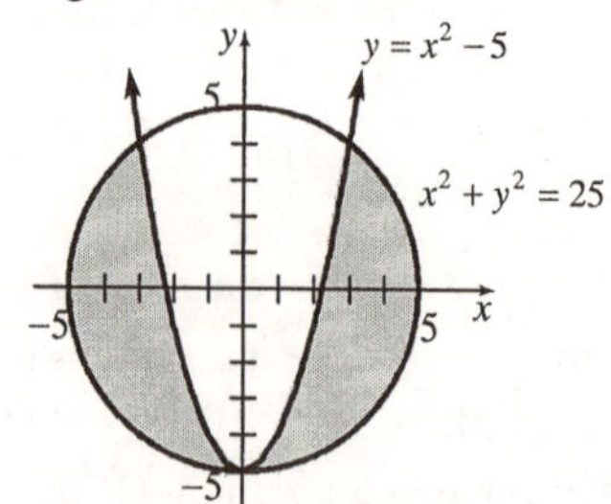

41. $\begin{cases} xy \geq 4 \\ y \geq x^2 + 1 \end{cases}$

Graph the hyperbola $xy = 4$. Use a solid line since the inequality uses $\geq$. Choose a test point not on the parabola, such as (0, 0). Since $0 \cdot 0 \geq 4$ is false, shade the opposite side of the hyperbola from (0, 0). Graph the parabola $y = x^2 + 1$. Use a solid line since the inequality uses $\geq$. Choose a test point not on the parabola, such as (0, 0). Since $0 \geq 0^2 + 1$ is false, shade the opposite side of the parabola from (0, 0).The overlapping region is the solution.

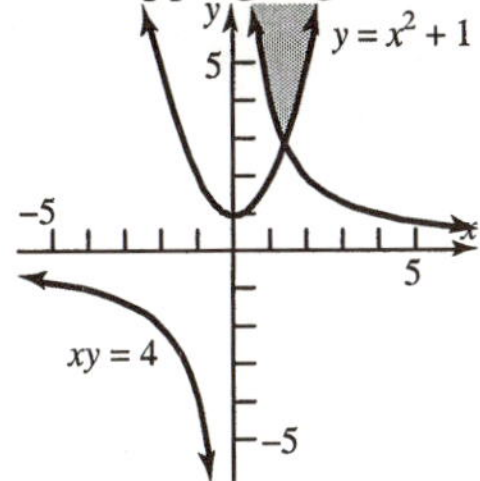

42. $\begin{cases} y + x^2 \leq 1 \\ y \geq x^2 - 1 \end{cases}$

Graph the parabola $y + x^2 = 1$. Use a solid line since the inequality uses $\leq$. Choose a test point not on the parabola, such as (0, 0). Since $0 + 0^2 \leq 1$ is true, shade the same side of the parabola as (0, 0). Graph the parabola $y = x^2 - 1$. Use a solid line since the inequality uses $\geq$. Choose a test point not on the parabola, such as (0, 0). Since $0 \geq 0^2 - 1$ is true, shade the same side of the parabola as (0, 0). The overlapping region is the solution.

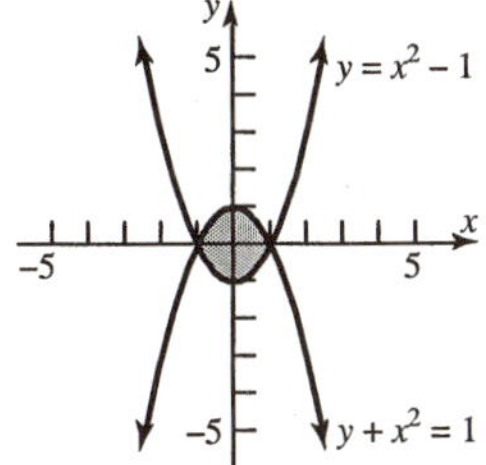

43. $\begin{cases} x \geq 0 \\ y \geq 0 \\ 2x + y \leq 6 \\ x + 2y \leq 6 \end{cases}$

Graph $x \geq 0$; $y \geq 0$. Shaded region is the first quadrant. Graph the line $2x + y = 6$. Use a solid line since the inequality uses $\leq$. Choose a test point not on the line, such as (0, 0). Since $2(0) + 0 \leq 6$ is true, shade the side of the line containing (0, 0). Graph the line $x + 2y = 6$. Use a solid line since the inequality uses $\leq$. Choose a test point not on the line, such as (0, 0). Since $0 + 2(0) \leq 6$ is true, shade the side of the line containing (0, 0). The overlapping region is the solution. The graph is bounded. Find the vertices:

The x-axis and y-axis intersect at (0, 0). The intersection of $x + 2y = 6$ and the y-axis is (0, 3). The intersection of $2x + y = 6$ and the x-axis is (3, 0). To find the intersection of $x + 2y = 6$ and $2x + y = 6$, solve the system:

$$\begin{cases} x + 2y = 6 \\ 2x + y = 6 \end{cases}$$

Solve the first equation for x: $x = 6 - 2y$.

Substitute and solve:

$$\begin{aligned} 2(6 - 2y) + y &= 6 \\ 12 - 4y + y &= 6 \\ 12 - 3y &= 6 \\ -3y &= -6 \\ y &= 2 \end{aligned}$$

$x = 6 - 2(2) = 2$

The point of intersection is (2, 2).

The four corner points are (0, 0), (0, 3), (3, 0), and (2, 2).

44. $\begin{cases} x \geq 0 \\ y \geq 0 \\ x + y \geq 4 \\ 2x + 3y \geq 6 \end{cases}$

Graph $x \geq 0$; $y \geq 0$. Shaded region is the first quadrant. Graph the line $x + y = 4$. Use a solid line since the inequality uses $\geq$. Choose a test point not on the line, such as (0, 0). Since $0 + 0 \geq 4$ is false, shade the opposite side of the line from (0, 0). Graph the line $2x + 3y = 6$. Use a solid line since the inequality uses $\geq$. Choose a test point not on the line, such as (0, 0). Since $2(0) + 3(0) \geq 6$ is false, shade the opposite side of the line from (0, 0). The overlapping region is the solution. The graph is unbounded.

Find the vertices:
The intersection of $x + y = 4$ and the y-axis is (0, 4). The intersection of $x + y = 4$ and the x-axis is (4, 0). The two corner points are (0, 4), and (4, 0).

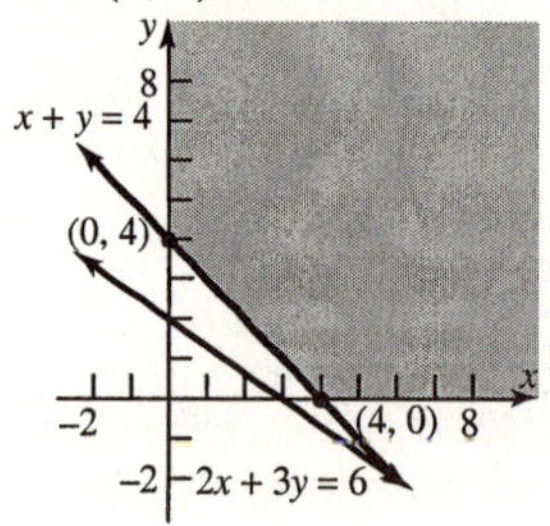

45. $\begin{cases} x \geq 0 \\ y \geq 0 \\ x + y \geq 2 \\ 2x + y \geq 4 \end{cases}$

Graph $x \geq 0$; $y \geq 0$. Shaded region is the first quadrant. Graph the line $x + y = 2$. Use a solid line since the inequality uses $\geq$. Choose a test point not on the line, such as (0, 0). Since $0 + 0 \geq 2$ is false, shade the opposite side of the line from (0, 0). Graph the line $2x + y = 4$. Use a solid line since the inequality uses $\geq$. Choose a test point not on the line, such as (0, 0). Since $2(0) + 0 \geq 4$ is false, shade the opposite side of the line from (0, 0). The overlapping region is the solution. The graph is unbounded.

Find the vertices:
The intersection of $x + y = 2$ and the x-axis is (2, 0). The intersection of $2x + y = 4$ and the y-axis is (0, 4). The two corner points are (2, 0), and (0, 4).

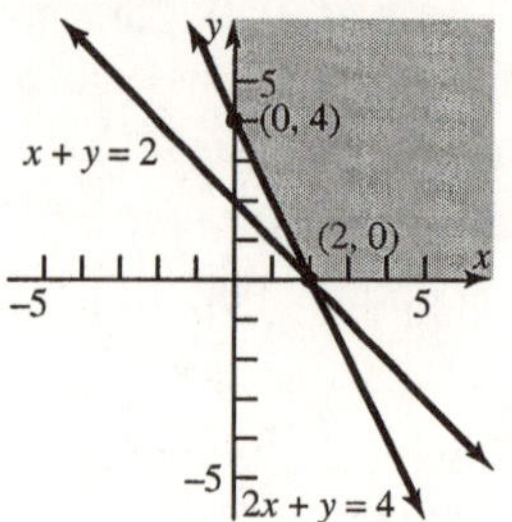

46. $\begin{cases} x \geq 0 \\ y \geq 0 \\ 3x + y \leq 6 \\ 2x + y \leq 2 \end{cases}$

Graph $x \geq 0$; $y \geq 0$. Shaded region is the first quadrant. Graph the line $3x + y = 6$. Use a solid line since the inequality uses $\leq$. Choose a test point not on the line, such as (0, 0). Since $3(0) + 0 \leq 6$ is true, shade the side of the line containing (0, 0). Graph the line $2x + y = 2$. Use a solid line since the inequality uses $\leq$. Choose a test point not on the line, such as (0, 0). Since $2(0) + 0 \leq 2$ is true, shade the side of the line containing (0, 0). The overlapping region is the solution. The graph is bounded.

Find the vertices:
The intersection of $x = 0$ and $y = 0$ is (0, 0). The intersection of $2x + y = 2$ and the x-axis is (1, 0). The intersection of $2x + y = 2$ and the y-axis is (0, 2). The three corner points are (0, 0), (1, 0), and (0, 2).

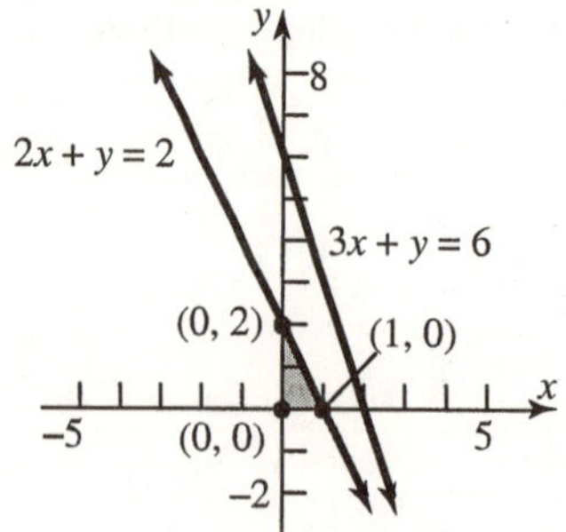

47. $\begin{cases} x \ge 0 \\ y \ge 0 \\ x + y \ge 2 \\ 2x + 3y \le 12 \\ 3x + y \le 12 \end{cases}$

Graph $x \ge 0$; $y \ge 0$. Shaded region is the first quadrant. Graph the line $x + y = 2$. Use a solid line since the inequality uses $\ge$. Choose a test point not on the line, such as (0, 0). Since $0 + 0 \ge 2$ is false, shade the opposite side of the line from (0, 0). Graph the line $2x + 3y = 12$. Use a solid line since the inequality uses $\le$. Choose a test point not on the line, such as (0, 0). Since $2(0) + 3(0) \le 12$ is true, shade the side of the line containing (0, 0). Graph the line $3x + y = 12$. Use a solid line since the inequality uses $\le$. Choose a test point not on the line, such as (0, 0). Since $3(0) + 0 \le 12$ is true, shade the side of the line containing (0, 0). The overlapping region is the solution. The graph is bounded.

Find the vertices:
The intersection of $x + y = 2$ and the y-axis is (0, 2). The intersection of $x + y = 2$ and the x-axis is (2, 0). The intersection of $2x + 3y = 12$ and the y-axis is (0, 4). The intersection of $3x + y = 12$ and the x-axis is (4, 0).

To find the intersection of $2x + 3y = 12$ and $3x + y = 12$, solve the system:

$$\begin{cases} 2x + 3y = 12 \\ 3x + y = 12 \end{cases}$$

Solve the second equation for y: $y = 12 - 3x$.
Substitute and solve:

$$2x + 3(12 - 3x) = 12$$
$$2x + 36 - 9x = 12$$
$$-7x = -24$$
$$x = \frac{24}{7}$$

$$y = 12 - 3\left(\frac{24}{7}\right) = 12 - \frac{72}{7} = \frac{12}{7}$$

The point of intersection is $\left(\frac{24}{7}, \frac{12}{7}\right)$.

The five corner points are (0, 2), (0, 4), (2, 0), (4, 0), and $\left(\frac{24}{7}, \frac{12}{7}\right)$.

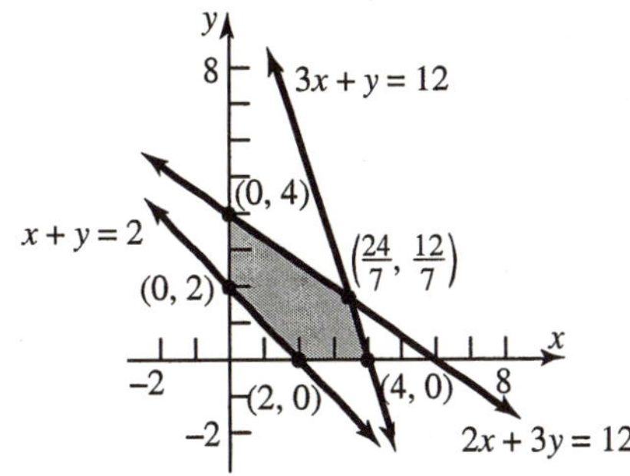

48. $\begin{cases} x \ge 0 \\ y \ge 0 \\ x + y \ge 2 \\ x + y \le 10 \\ 2x + y \le 3 \end{cases}$

Graph $x \ge 0$; $y \ge 0$. Shaded region is the first quadrant. Graph the line $x + y = 2$. Use a solid line since the inequality uses $\ge$. Choose a test point not on the line, such as (0, 0). Since $0 + 0 \ge 2$ is false, shade the opposite side of the line from (0, 0). Graph the line $x + y = 10$. Use a solid line since the inequality uses $\le$. Choose a test point not on the line, such as (0, 0). Since $0 + 0 \le 10$ is true, shade the side of the line containing (0, 0). Graph the line $2x + y = 3$. Use a solid line since the inequality uses $\le$. Choose a test point not on the line, such as (0, 0). Since $2(0) + 0 \le 3$ is true, shade the side of the line containing (0, 0). The overlapping region is the solution. The graph is bounded.

Find the vertices:
The intersection of $x + y = 2$ and the y-axis is (0, 2). The intersection of $2x + y = 3$ and the y-axis is (0, 3). To find the intersection of $2x + y = 3$ and $x + y = 2$, solve the system:

$$\begin{cases} 2x + y = 3 \\ x + y = 2 \end{cases}$$

Solve the second equation for y: $y = 2 - x$.
Substitute and solve:

$$2x + 2 - x = 3$$
$$x = 1$$
$$y = 2 - 1 = 1$$

The point of intersection is (1, 1).
The three corner points are (0, 2), (0, 3), and (1, 1).

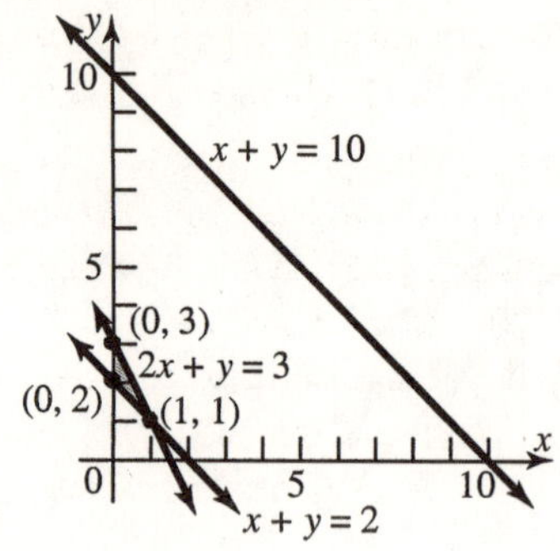

49. $\begin{cases} x \geq 0 \\ y \geq 0 \\ x+y \geq 2 \\ x+y \leq 8 \\ 2x+y \leq 10 \end{cases}$

Graph $x \geq 0$; $y \geq 0$. Shaded region is the first quadrant. Graph the line $x+y=2$. Use a solid line since the inequality uses ≥. Choose a test point not on the line, such as (0, 0). Since $0+0 \geq 2$ is false, shade the opposite side of the line from (0, 0). Graph the line $x+y=8$. Use a solid line since the inequality uses ≤. Choose a test point not on the line, such as (0, 0). Since $0+0 \leq 8$ is true, shade the side of the line containing (0, 0). Graph the line $2x+y=10$. Use a solid line since the inequality uses ≤. Choose a test point not on the line, such as (0, 0). Since $2(0)+0 \leq 10$ is true, shade the side of the line containing (0, 0). The overlapping region is the solution. The graph is bounded.

Find the vertices:
The intersection of $x+y=2$ and the *y*-axis is (0, 2). The intersection of $x+y=2$ and the *x*-axis is (2, 0). The intersection of $x+y=8$ and the *y*-axis is (0, 8). The intersection of $2x+y=10$ and the *x*-axis is (5, 0). To find the intersection of $x+y=8$ and $2x+y=10$, solve the system:

$$\begin{cases} x+y=8 \\ 2x+y=10 \end{cases}$$

Solve the first equation for *y*: $y=8-x$.
Substitute and solve:

$$2x+8-x=10$$
$$x=2$$
$$y=8-2=6$$

The point of intersection is (2, 6).
The five corner points are (0, 2), (0, 8), (2, 0), (5, 0), and (2, 6).

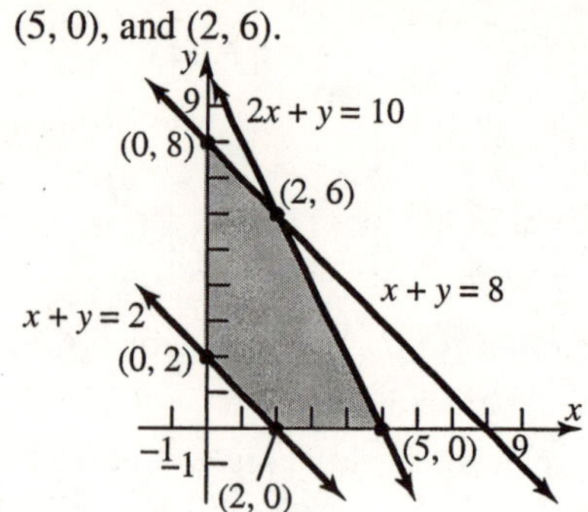

50. $\begin{cases} x \geq 0 \\ y \geq 0 \\ x+y \geq 2 \\ x+y \leq 8 \\ x+2y \geq 1 \end{cases}$

Graph $x \geq 0$; $y \geq 0$. Shaded region is the first quadrant. Graph the line $x+y=2$. Use a solid line since the inequality uses ≥. Choose a test point not on the line, such as (0, 0). Since $0+0 \geq 2$ is false, shade the opposite side of the line from (0, 0). Graph the line $x+y=8$. Use a solid line since the inequality uses ≤. Choose a test point not on the line, such as (0, 0). Since $0+0 \leq 8$ is true, shade the side of the line containing (0, 0). Graph the line $x+2y=1$. Use a solid line since the inequality uses ≥. Choose a test point not on the line, such as (0, 0). Since $0+2(0) \geq 1$ is false, shade the opposite side of the line from (0, 0). The overlapping region is the solution. The graph is bounded.

Find the vertices:
The intersection of $x+y=2$ and the *y*-axis is (0, 2). The intersection of $x+y=2$ and the *x*-axis is (2, 0). The intersection of $x+y=8$ and the *y*-axis is (0, 8). The intersection of $x+y=8$ and the *x*-axis is (8, 0). The four corner points are (0, 2), (0, 8), (2, 0), and (8, 0).

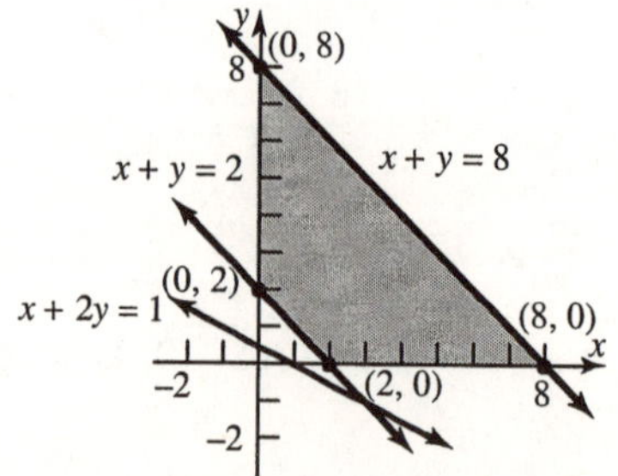

51. $\begin{cases} x \geq 0 \\ y \geq 0 \\ x+2y \geq 1 \\ x+2y \leq 10 \end{cases}$

Graph $x \geq 0$; $y \geq 0$. Shaded region is the first quadrant. Graph the line $x+2y=1$. Use a solid line since the inequality uses $\geq$. Choose a test point not on the line, such as (0, 0). Since $0+2(0) \geq 1$ is false, shade the opposite side of the line from (0, 0). Graph the line $x+2y=10$. Use a solid line since the inequality uses $\leq$. Choose a test point not on the line, such as (0, 0). Since $0+2(0) \leq 10$ is true, shade the side of the line containing (0, 0). The overlapping region is the solution. The graph is bounded.

Find the vertices:
The intersection of $x+2y=1$ and the y-axis is (0, 0.5). The intersection of $x+2y=1$ and the x-axis is (1, 0). The intersection of $x+2y=10$ and the y-axis is (0, 5). The intersection of $x+2y=10$ and the x-axis is (10, 0). The four corner points are (0, 0.5), (0, 5), (1, 0), and (10, 0).

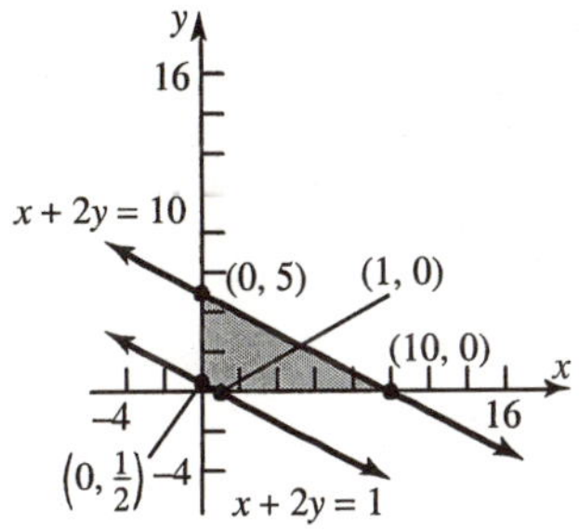

52. $\begin{cases} x \geq 0 \\ y \geq 0 \\ x+2y \geq 1 \\ x+2y \leq 10 \\ x+y \geq 2 \\ x+y \leq 8 \end{cases}$

Graph $x \geq 0$; $y \geq 0$. Shaded region is the first quadrant. Graph the line $x+2y=1$. Use a solid line since the inequality uses $\geq$. Choose a test point not on the line, such as (0, 0). Since $0+2(0) \geq 1$ is false, shade the opposite side of the line from (0, 0). Graph the line $x+2y=10$. Use a solid line since the inequality uses $\leq$. Choose a test point not on the line, such as (0, 0). Since $0+2(0) \leq 10$ is true, shade the side of the line ontaining (0, 0). Graph the line $x+y=2$. Use a solid line since the inequality uses $\geq$. Choose a test point not on the line, such as (0, 0). Since $0+0 \geq 2$ is false, shade the opposite side of the line from (0, 0). Graph the line $x+y=8$. Use a solid line since the inequality uses $\leq$. Choose a test point not on the line, such as (0, 0). Since $0+0 \leq 8$ is true, shade the side of the line containing (0, 0). The overlapping region is the solution. The graph is bounded.

Find the vertices:
The intersection of $x+y=2$ and the y-axis is (0, 2). The intersection of $x+y=2$ and the x-axis is (2, 0). The intersection of $x+2y=10$ and the y-axis is (0, 5). The intersection of $x+y=8$ and the x-axis is (8, 0). To find the intersection of $x+y=8$ and $x+2y=10$, solve the system:

$\begin{cases} x+y=8 \\ x+2y=10 \end{cases}$

Solve the first equation for x: $x=8-y$.

Substitute and solve:

$(8-y)+2y=10$

$y=2$

$x=8-2=6$

The point of intersection is (6, 2).
The five corner points are (0, 2), (0, 5), (2, 0), (8, 0), and (6, 2).

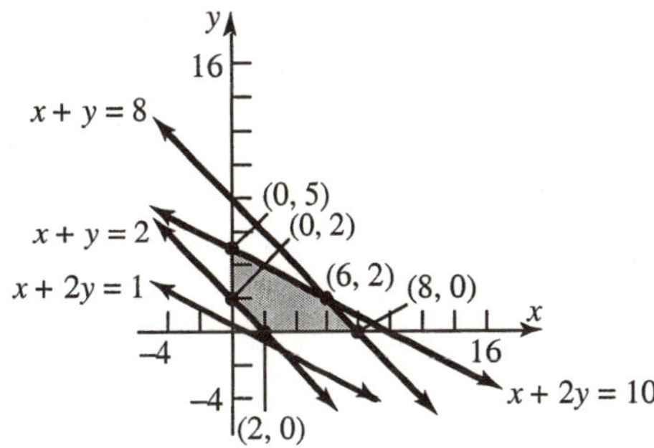

53. The system of linear inequalities is:

$\begin{cases} x \geq 0 \\ y \geq 0 \\ x \leq 4 \\ x+y \leq 6 \end{cases}$

54. The system of linear inequalities is:

$$\begin{cases} x \geq 0 \\ y \geq 0 \\ x \leq 6 \\ y \leq 5 \\ x + y \geq 2 \end{cases}$$

55. The system of linear inequalities is:

$$\begin{cases} x \geq 0 \\ y \geq 15 \\ x \leq 20 \\ x + y \leq 50 \\ x - y \leq 0 \end{cases}$$

56. The system of linear inequalities is:

$$\begin{cases} x \geq 0 \\ y \leq 6 \\ x \leq 5 \\ 3x + 4y \geq 12 \\ 2x - y \leq 8 \end{cases}$$

57. a. Let x = the amount invested in Treasury bills, and let y = the amount invested in corporate bonds.
The constraints are:
$x \geq 0,\ y \geq 0$ because a non-negative amount must be invested.
$x + y \leq 50,000$ because the total investment cannot exceed \$50,000.
$y \leq 10,000$ because the amount invested in corporate bonds must not exceed \$10,000.
$x \geq 35,000$ because the amount invested in Treasury bills must be at least \$35,000.
The system is

$$\begin{cases} x \geq 0 \\ y \geq 0 \\ x + y \leq 50,000 \\ y \leq 10,000 \\ x \geq 35,000 \end{cases}$$

b. Graph the system.

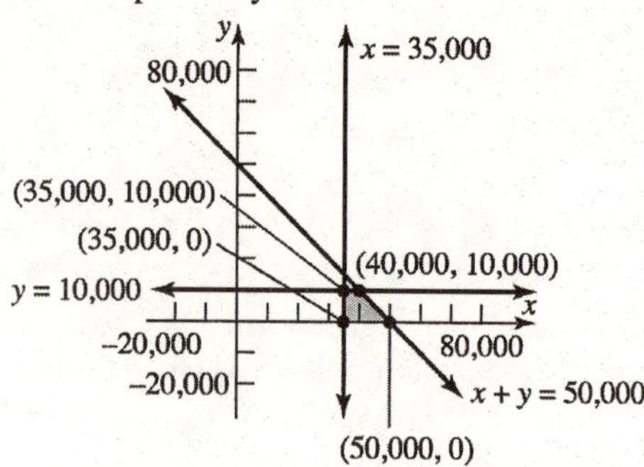

The corner points are (35,000, 0), (35,000, 10,000), (40,000, 10,000), (50,000, 0).

58. a. Let x = the # of standard model trucks, and let y = the # of deluxe model trucks.
The constraints are:
$x \geq 0, y \geq 0$ because a non-negative number of trucks must be manufactured.
$2x + 3y \leq 80$ because the total painting hours worked cannot exceed 80.
$3x + 4y \leq 120$ because the total detailing hours worked cannot exceed 120.
The system is

$$\begin{cases} x \geq 0 \\ y \geq 0 \\ 2x + 3y \leq 80 \\ 3x + 4y \leq 120 \end{cases}$$

b. Graph the system.

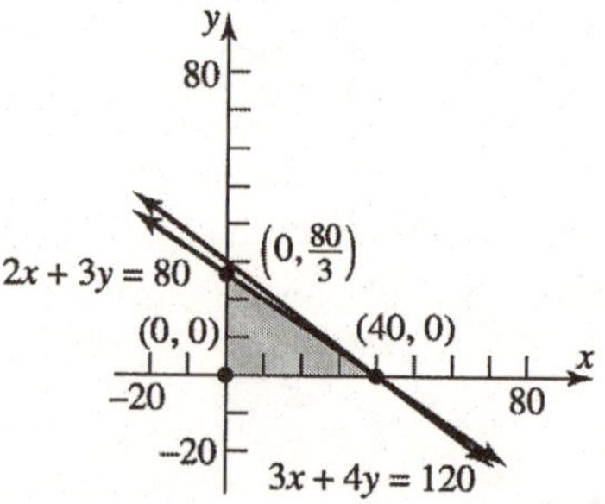

The corner points are (0, 0), (0, 80/3), (40, 0).

59. a. Let x = the # of packages of the economy blend, and let y = the # of packages of the superior blend.
The constraints are:
$x \ge 0,\ y \ge 0$ because a non-negative # of packages must be produced.
$4x+8y \le 75\cdot 16$ because the total amount of "*A* grade" coffee cannot exceed 75 pounds. (Note: 75 pounds = (75)(16) ounces.)
$12x+8y \le 120\cdot 16$ because the total amount of "*B* grade" coffee cannot exceed 120 pounds. (Note: 120 pounds = (120)(16) ounces.)
Simplifying the inequalities, we obtain:

$$4x+8y \le 75\cdot 16 \qquad 12x+8y \le 120\cdot 16$$
$$x+2y \le 75\cdot 4 \qquad 3x+2y \le 120\cdot 4$$
$$x+2y \le 300 \qquad 3x+2y \le 480$$

The system is:
$$\begin{cases} x \ge 0 \\ y \ge 0 \\ x+2y \le 300 \\ 3x+2y \le 480 \end{cases}$$

b. Graph the system.

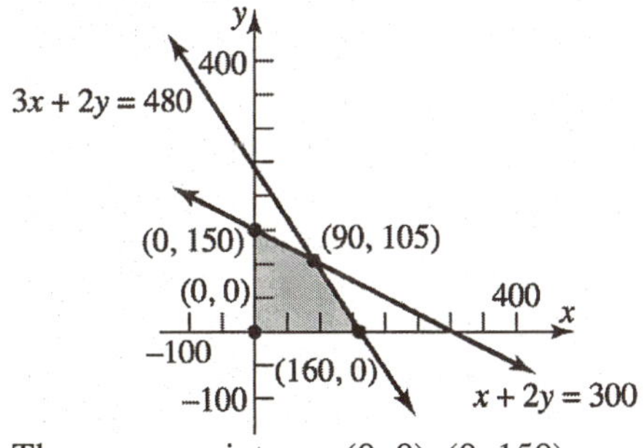

The corner points are (0, 0), (0, 150), (90, 105), (160, 0).

60. a. Let x = the # of lower-priced packages, and let y = the # of quality packages.
The constraints are:
$x \ge 0,\ y \ge 0$ because a non-negative # of packages must be produced.
$8x+6y \le 120\cdot 16$ because the total amount of peanuts cannot exceed 120 pounds. (Note: 120 pounds = (120)(16) ounces.)
$4x+6y \le 90\cdot 16$ because the total amount of cashews cannot exceed 90 pounds. (Note: 90 pounds = (90)(16) ounces.)
Simplifying the inequalities, we obtain:

$$8x+6y \le 120\cdot 16 \qquad 4x+6y \le 90\cdot 16$$
$$4x+3y \le 120\cdot 8 \qquad 2x+3y \le 90\cdot 8$$
$$4x+3y \le 960 \qquad 2x+3y \le 720$$

The system is
$$\begin{cases} x \ge 0;\ y \ge 0 \\ 4x+3y \le 960 \\ 2x+3y \le 720 \end{cases}$$

b. Graph the system.

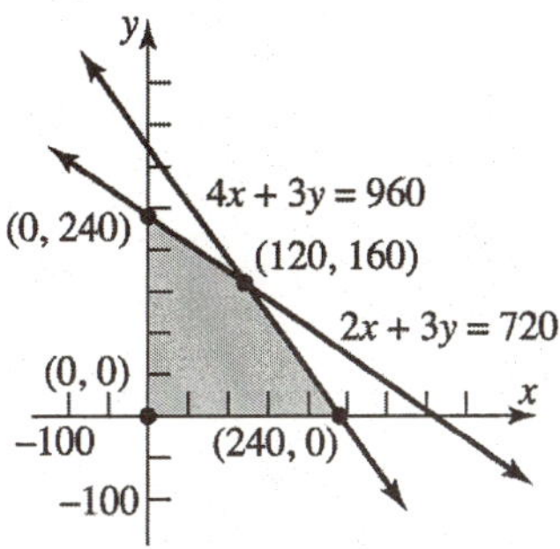

The corner points are (0, 0), (0, 240), (120, 160), (240, 0).

61. a. Let x = the # of microwaves, and let y = the # of printers.
The constraints are:
$x \ge 0,\ y \ge 0$ because a non-negative # of items must be shipped.
$30x+20y \le 1600$ because a total cargo weight cannot exceed 1600 pounds.
$2x+3y \le 150$ because the total cargo volume cannot exceed 150 cubic feet. Note that the inequality $30x+20y \le 1600$ can be simplified: $3x+2y \le 160$.
The system is:
$$\begin{cases} x \ge 0;\ y \ge 0 \\ 3x+2y \le 160 \\ 2x+3y \le 150 \end{cases}$$

b. Graph the system.

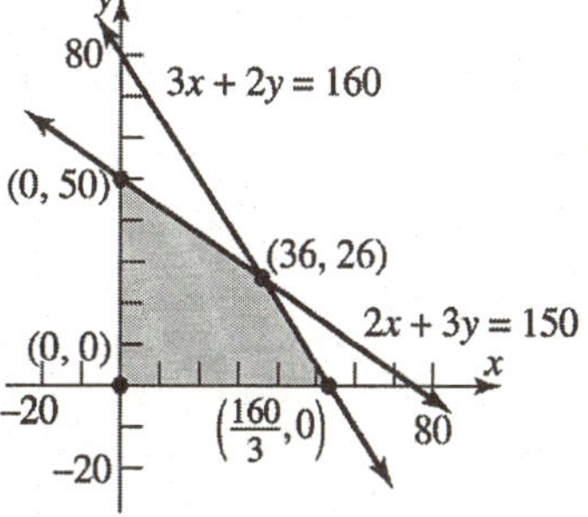

The corner points are (0, 0), (0, 50), (36, 26), (160/3, 0).

Section 10.8

1. objective function

2. True

3. $z = x + y$

Vertex	Value of $z = x + y$
(0, 3)	$z = 0 + 3 = 3$
(0, 6)	$z = 0 + 6 = 6$
(5, 6)	$z = 5 + 6 = 11$
(5, 2)	$z = 5 + 2 = 7$
(4, 0)	$z = 4 + 0 = 4$

The maximum value is 11 at (5, 6), and the minimum value is 3 at (0, 3).

4. $z = 2x + 3y$

Vertex	Value of $z = 2x + 3y$
(0, 3)	$z = 2(0) + 3(3) = 9$
(0, 6)	$z = 2(0) + 3(6) = 18$
(5, 6)	$z = 2(5) + 3(6) = 28$
(5, 2)	$z = 2(5) + 3(2) = 16$
(4, 0)	$z = 2(4) + 3(0) = 8$

The maximum value is 28 at (5, 6), and the minimum value is 8 at (4, 0).

5. $z = x + 10y$

Vertex	Value of $z = x + 10y$
(0, 3)	$z = 0 + 10(3) = 30$
(0, 6)	$z = 0 + 10(6) = 60$
(5, 6)	$z = 5 + 10(6) = 65$
(5, 2)	$z = 5 + 10(2) = 25$
(4, 0)	$z = 4 + 10(0) = 4$

The maximum value is 65 at (5, 6), and the minimum value is 4 at (4, 0).

6. $z = 10x + y$

Vertex	Value of $z = 10x + y$
(0, 3)	$z = 10(0) + 3 = 3$
(0, 6)	$z = 10(0) + 6 = 6$
(5, 6)	$z = 10(5) + 6 = 56$
(5, 2)	$z = 10(5) + 2 = 52$
(4, 0)	$z = 10(4) + 0 = 40$

The maximum value is 56 at (5, 6), and the minimum value is 3 at (0, 3).

7. $z = 5x + 7y$

Vertex	Value of $z = 5x + 7y$
(0, 3)	$z = 5(0) + 7(3) = 21$
(0, 6)	$z = 5(0) + 7(6) = 42$
(5, 6)	$z = 5(5) + 7(6) = 67$
(5, 2)	$z = 5(5) + 7(2) = 39$
(4, 0)	$z = 5(4) + 7(0) = 20$

The maximum value is 67 at (5, 6), and the minimum value is 20 at (4, 0).

8. $z = 7x + 5y$

Vertex	Value of $z = 7x + 5y$
(0, 3)	$z = 7(0) + 5(3) = 15$
(0, 6)	$z = 7(0) + 5(6) = 30$
(5, 6)	$z = 7(5) + 5(6) = 65$
(5, 2)	$z = 7(5) + 5(2) = 45$
(4, 0)	$z = 7(4) + 5(0) = 28$

The maximum value is 65 at (5, 6), and the minimum value is 15 at (0, 3).

9. Maximize $z = 2x + y$ subject to $x \geq 0$, $y \geq 0$, $x + y \leq 6$, $x + y \geq 1$. Graph the constraints.

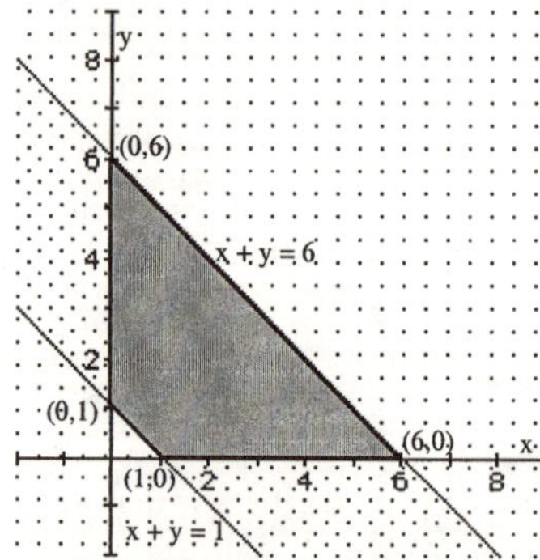

The corner points are (0, 1), (1, 0), (0, 6), (6, 0). Evaluate the objective function:

Vertex	Value of $z = 2x + y$
(0, 1)	$z = 2(0) + 1 = 1$
(0, 6)	$z = 2(0) + 6 = 6$
(1, 0)	$z = 2(1) + 0 = 2$
(6, 0)	$z = 2(6) + 0 = 12$

The maximum value is 12 at (6, 0).

10. Maximize $z = x + 3y$ subject to $x \geq 0$, $y \geq 0$, $x + y \geq 3$ $x \leq 5$, $y \leq 7$. Graph the constraints.

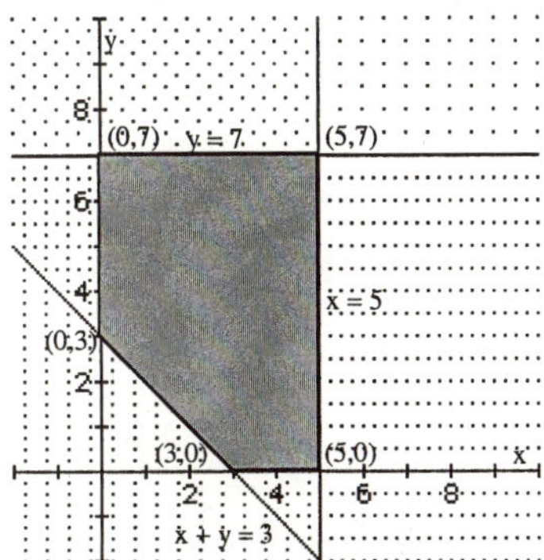

The corner points are (0, 3), (3, 0), (0, 7), (5, 0), (5, 7). Evaluate the objective function:

Vertex	Value of $z = x + 3y$
(0, 3)	$z = 0 + 3(3) = 9$
(3, 0)	$z = 3 + 3(0) = 3$
(0, 7)	$z = 0 + 3(7) = 21$
(5, 0)	$z = 5 + 3(0) = 5$
(5, 7)	$z = 5 + 3(7) = 26$

The maximum value is 26 at (5, 7).

11. Minimize $z = 2x + 5y$ subject to $x \geq 0$, $y \geq 0$, $x + y \geq 2$, $x \leq 5$, $y \leq 3$. Graph the constraints.

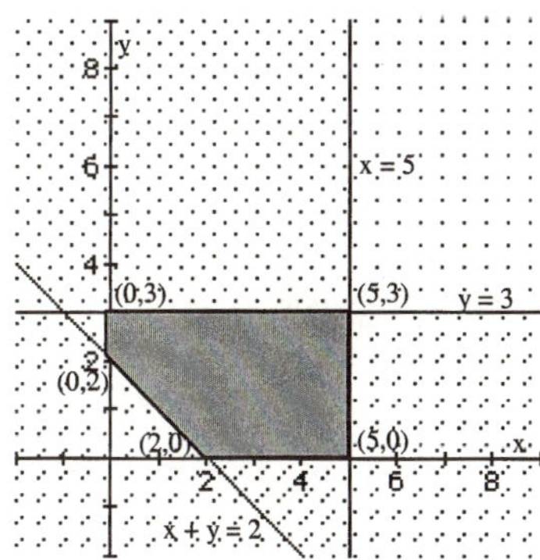

The corner points are (0, 2), (2, 0), (0, 3), (5, 0), (5, 3). Evaluate the objective function:

Vertex	Value of $z = 2x + 5y$
(0, 2)	$z = 2(0) + 5(2) = 10$
(0, 3)	$z = 2(0) + 5(3) = 15$
(2, 0)	$z = 2(2) + 5(0) = 4$
(5, 0)	$z = 2(5) + 5(0) = 10$
(5, 3)	$z = 2(5) + 5(3) = 25$

The minimum value is 4 at (2, 0).

12. Minimize $z = 3x + 4y$ subject to $x \geq 0$, $y \geq 0$, $2x + 3y \geq 6$, $x + y \leq 8$. Graph the constraints.

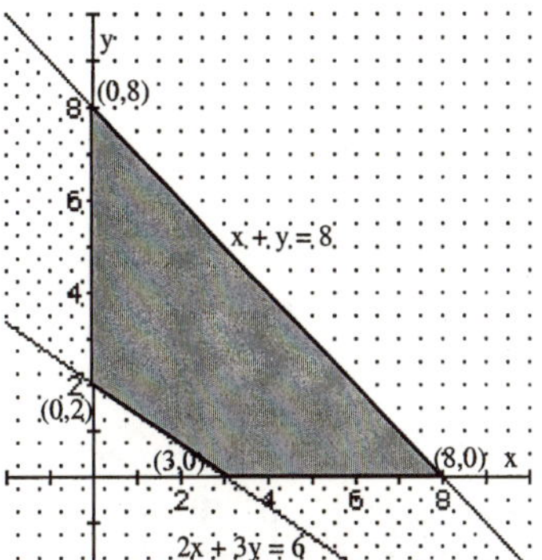

The corner points are (0, 2), (3, 0), (0, 8), (8, 0). Evaluate the objective function:

Vertex	Value of $z = 3x + 4y$
(0, 2)	$z = 3(0) + 4(2) = 8$
(3, 0)	$z = 3(3) + 4(0) = 9$
(0, 8)	$z = 3(0) + 4(8) = 32$
(8, 0)	$z = 3(8) + 4(0) = 24$

The minimum value is 8 at (0, 2).

13. Maximize $z = 3x + 5y$ subject to $x \geq 0$, $y \geq 0$, $x + y \geq 2$, $2x + 3y \leq 12$, $3x + 2y \leq 12$. Graph the constraints.

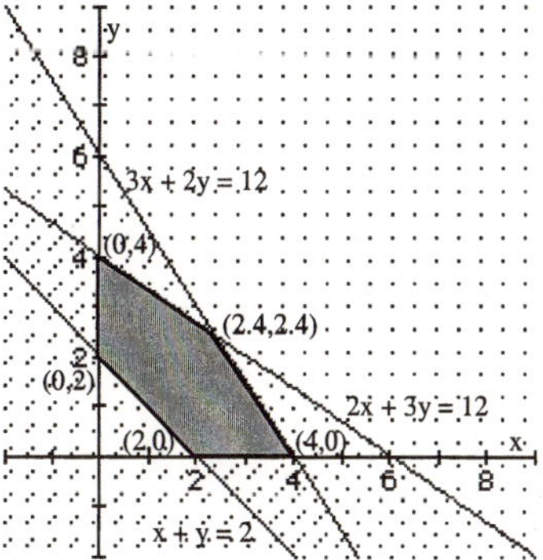

To find the intersection of $2x + 3y = 12$ and $3x + 2y = 12$, solve the system:

$$\begin{cases} 2x + 3y = 12 \\ 3x + 2y = 12 \end{cases}$$

Solve the second equation for y: $y = 6 - \frac{3}{2}x$

Substitute and solve:

$$2x+3\left(6-\frac{3}{2}x\right)=12$$
$$2x+18-\frac{9}{2}x=12$$
$$-\frac{5}{2}x=-6$$
$$x=\frac{12}{5}$$
$$y=6-\frac{3}{2}\left(\frac{12}{5}\right)=6-\frac{18}{5}=\frac{12}{5}$$

The point of intersection is $(2.4, 2.4)$.

The corner points are (0, 2), (2, 0), (0, 4), (4, 0), (2.4, 2.4). Evaluate the objective function:

Vertex	Value of $z=3x+5y$
(0, 2)	$z=3(0)+5(2)=10$
(0, 4)	$z=3(0)+5(4)=20$
(2, 0)	$z=3(2)+5(0)=6$
(4, 0)	$z=3(4)+5(0)=12$
(2.4, 2.4)	$z=3(2.4)+5(2.4)=19.2$

The maximum value is 20 at (0, 4).

14. Maximize $z=5x+3y$ subject to $x\ge 0$, $y\ge 0$, $x+y\ge 2$, $x+y\le 8$, $2x+y\le 10$. Graph the constraints.

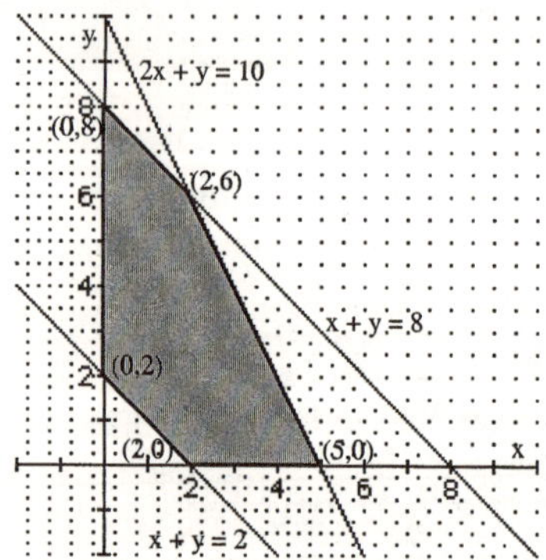

To find the intersection of $x+y=8$ and $2x+y=10$, solve the system:

$$\begin{cases} x+y=8 \\ 2x+y=10 \end{cases}$$

Solve the first equation for *y*: $y=8-x$.

Substitute and solve:

$$2x+8-x=10$$
$$x=2$$
$$y=8-2=6$$

The point of intersection is (2, 6).

The corner points are (0, 2), (2, 0), (0, 8), (5, 0), (2, 6). Evaluate the objective function:

Vertex	Value of $z=5x+3y$
(0, 2)	$z=5(0)+3(2)=6$
(0, 8)	$z=5(0)+3(8)=24$
(2, 0)	$z=5(2)+3(0)=10$
(5, 0)	$z=5(5)+3(0)=25$
(2, 6)	$z=5(2)+3(6)=28$

The maximum value is 28 at (2, 6).

15. Minimize $z=5x+4y$ subject to $x\ge 0$, $y\ge 0$, $x+y\ge 2$, $2x+3y\le 12$, $3x+y\le 12$. Graph the constraints.

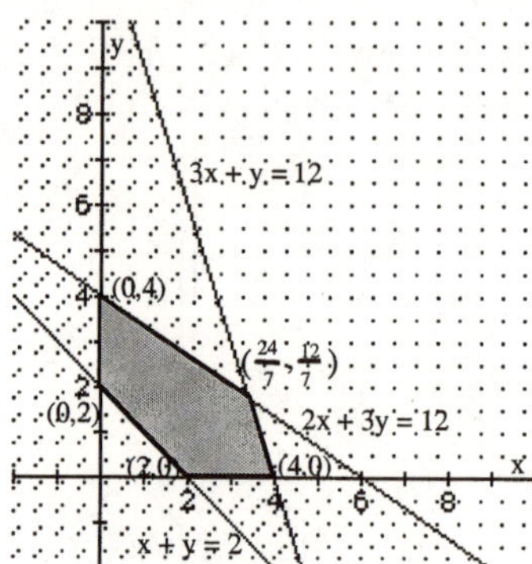

To find the intersection of $2x+3y=12$ and $3x+y=12$, solve the system:

$$\begin{cases} 2x+3y=12 \\ 3x+\ \ y=12 \end{cases}$$

Solve the second equation for *y*: $y=12-3x$

Substitute and solve:

$$2x+3(12-3x)=12$$
$$2x+36-9x=12$$
$$-7x=-24$$
$$x=\frac{24}{7}$$
$$y=12-3\left(\frac{24}{7}\right)=12-\frac{72}{7}=\frac{12}{7}$$

The point of intersection is $\left(\frac{24}{7},\frac{12}{7}\right)$.

The corner points are (0, 2), (2, 0), (0, 4), (4, 0), $\left(\frac{24}{7},\frac{12}{7}\right)$. Evaluate the objective function:

Vertex	Value of $z = 5x + 4y$
(0, 2)	$z = 5(0) + 4(2) = 8$
(0, 4)	$z = 5(0) + 4(4) = 16$
(2, 0)	$z = 5(2) + 4(0) = 10$
(4, 0)	$z = 5(4) + 4(0) = 20$
$\left(\frac{24}{7}, \frac{12}{7}\right)$	$z = 5\left(\frac{24}{7}\right) + 4\left(\frac{12}{7}\right) = 24$

The minimum value is 8 at (0, 2).

16. Minimize $z = 2x + 3y$ subject to $x \geq 0$, $y \geq 0$, $x + y \geq 3$, $x + y \leq 9$, $x + 3y \geq 6$. Graph the constraints.

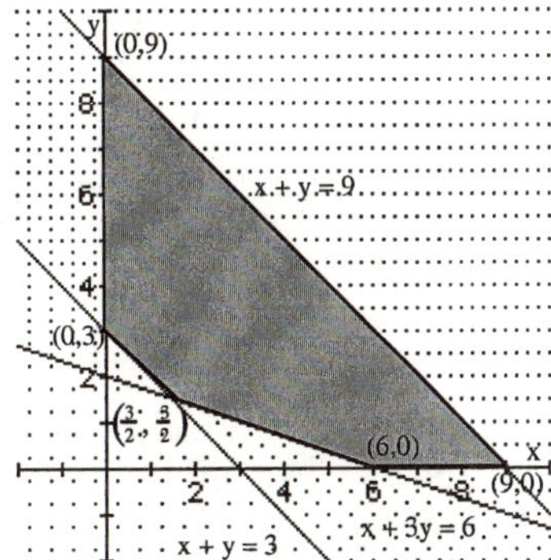

To find the intersection of $x + y = 3$ and $x + 3y = 6$, solve the system:

$$\begin{cases} x + y = 3 \\ x + 3y = 6 \end{cases}$$

Solve the first equation for *y*: $y = 3 - x$.

Substitute and solve:

$$x + 3(3 - x) = 6$$
$$x + 9 - 3x = 6$$
$$-2x = -3$$
$$x = 1.5$$
$$y = 3 - 1.5 = 1.5$$

The point of intersection is $(1.5, 1.5)$.

The corner points are (0, 3), (6, 0), (0, 9), (9, 0), (1.5, 1.5). Evaluate the objective function:

Vertex	Value of $z = 2x + 3y$
(0, 3)	$z = 2(0) + 3(3) = 9$
(0, 9)	$z = 2(0) + 3(9) = 27$
(6, 0)	$z = 2(6) + 3(0) = 12$
(9, 0)	$z = 2(9) + 3(0) = 18$
$(1.5, 1.5)$	$z = 2(1.5) + 4(1.5) = 7.5$

The minimum value is 15/2 at $(1.5, 1.5)$.

17. Maximize $z = 5x + 2y$ subject to $x \geq 0$, $y \geq 0$, $x + y \leq 10$, $2x + y \geq 10$, $x + 2y \geq 10$. Graph the constraints.

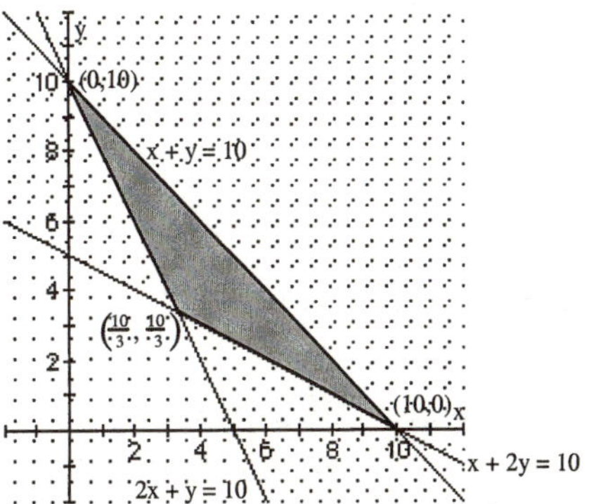

To find the intersection of $2x + y = 10$ and $x + 2y = 10$, solve the system:

$$\begin{cases} 2x + y = 10 \\ x + 2y = 10 \end{cases}$$

Solve the first equation for *y*: $y = 10 - 2x$.

Substitute and solve:

$$x + 2(10 - 2x) = 10$$
$$x + 20 - 4x = 10$$
$$-3x = -10$$
$$x = \frac{10}{3}$$

$$y = 10 - 2\left(\frac{10}{3}\right) = 10 - \frac{20}{3} = \frac{10}{3}$$

The point of intersection is (10/3 10/3).
The corner points are (0, 10), (10, 0), (10/3, 10/3).
Evaluate the objective function:

Vertex	Value of $z = 5x + 2y$
(0, 10)	$z = 5(0) + 2(10) = 20$
(10, 0)	$z = 5(10) + 2(0) = 50$
$\left(\frac{10}{3}, \frac{10}{3}\right)$	$z = 5\left(\frac{10}{3}\right) + 2\left(\frac{10}{3}\right) = \frac{70}{3} = 23\frac{1}{3}$

The maximum value is 50 at (10, 0).

18. Maximize $z = 2x + 4y$ subject to
$x \geq 0,\ y \geq 0,\ 2x + y \geq 4,\ x + y \leq 9$.
Graph the constraints.

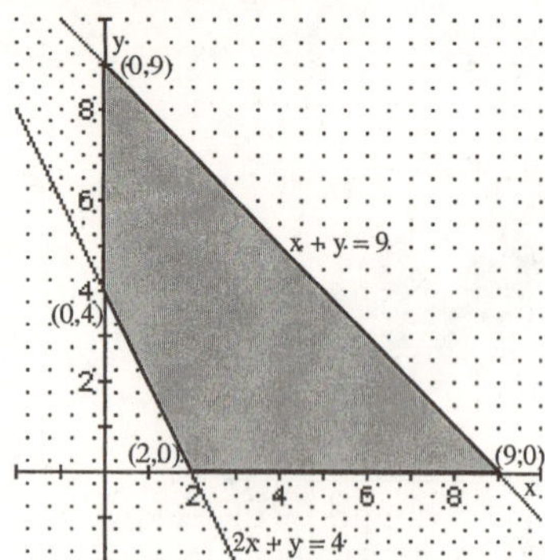

The corner points are (0, 9), (9, 0), (0, 4), (2, 0).
Evaluate the objective function:

Vertex	Value of $z = 2x + 4y$
(0, 9)	$z = 2(0) + 4(9) = 36$
(9, 0)	$z = 2(9) + 4(0) = 18$
(0, 4)	$z = 2(0) + 4(4) = 16$
(2, 0)	$z = 2(2) + 4(0) = 4$

The maximum value is 36 at (0, 9).

19. Let x = the number of downhill skis produced, and let y = the number of cross-country skis produced. The total profit is: $P = 70x + 50y$. Profit is to be maximized, so this is the objective function. The constraints are:

$x \geq 0,\ y \geq 0$ A positive number of skis must be produced.
$2x + y \leq 40$ Manufacturing time available.
$x + y \leq 32$ Finishing time available.

Graph the constraints.

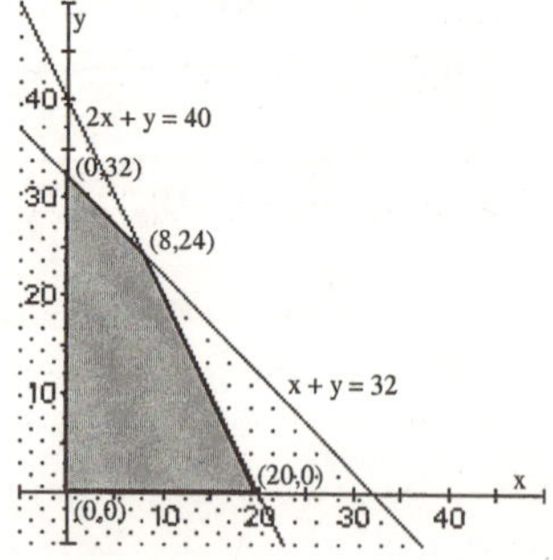

To find the intersection of $x + y = 32$ and $2x + y = 40$, solve the system:

$$\begin{cases} x + y = 32 \\ 2x + y = 40 \end{cases}$$

Solve the first equation for *y*: $y = 32 - x$.
Substitute and solve:

$$2x + (32 - x) = 40$$
$$x = 8$$
$$y = 32 - 8 = 24$$

The point of intersection is (8, 24).
The corner points are (0, 0), (0, 32), (20, 0), (8, 24). Evaluate the objective function:

Vertex	Value of $P = 70x + 50y$
(0, 0)	$P = 70(0) + 50(0) = 0$
(0, 32)	$P = 70(0) + 50(32) = 1600$
(20, 0)	$P = 70(20) + 50(0) = 1400$
(8, 24)	$P = 70(8) + 50(24) = 1760$

The maximum profit is $1760, when 8 downhill skis and 24 cross-country skis are produced.

With the increase of the manufacturing time to 48 hours, we do the following:
The constraints are:

$x \geq 0,\ y \geq 0$ A positive number of skis must be produced.
$2x + y \leq 48$ Manufacturing time available.
$x + y \leq 32$ Finishing time available.

Graph the constraints.

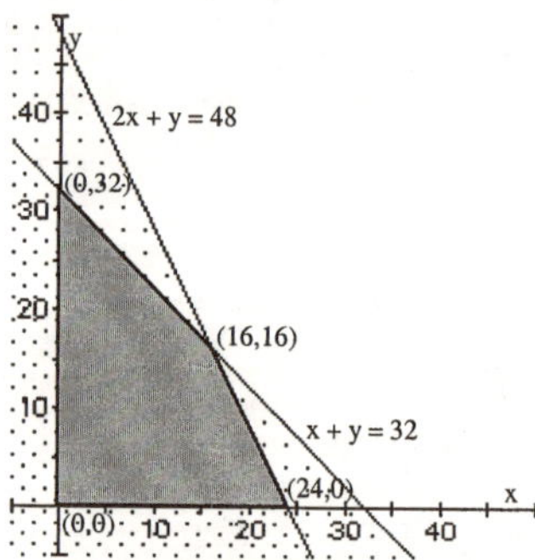

To find the intersection of $x + y = 32$ and $2x + y = 48$, solve the system:

$$\begin{cases} x + y = 32 \\ 2x + y = 48 \end{cases}$$

Solve the first equation for *y*: $y = 32 - x$.
Substitute and solve:

$$2x + (32 - x) = 48$$
$$x = 16$$
$$y = 32 - 16 = 16$$

The point of intersection is (16, 16).
The corner points are (0, 0), (0, 32), (24, 0), (16, 16). Evaluate the objective function:

Vertex	Value of $P = 70x + 50y$
(0, 0)	$P = 70(0) + 50(0) = 0$
(0, 32)	$P = 70(0) + 50(32) = 1600$
(24, 0)	$P = 70(24) + 50(0) = 1680$
(16, 16)	$P = 70(16) + 50(16) = 1920$

The maximum profit is \$1920, when 16 downhill skis and 16 cross-country skis are produced.

20. Let x = the number of acres of soybeans planted, and let y = the number of acres of wheat planted.
The total profit is: $P = 180x + 100y$. Profit is to be maximized, so this is the objective function.
The constraints are:
$x \ge 0,\ y \ge 0$ A non-negative number of acres must be planted.
$x + y \le 70$ Acres available to plant.
$60x + 30y \le 1800$ Money available for preparation.
$3x + 4y \le 120$ Workdays available.
Graph the constraints.

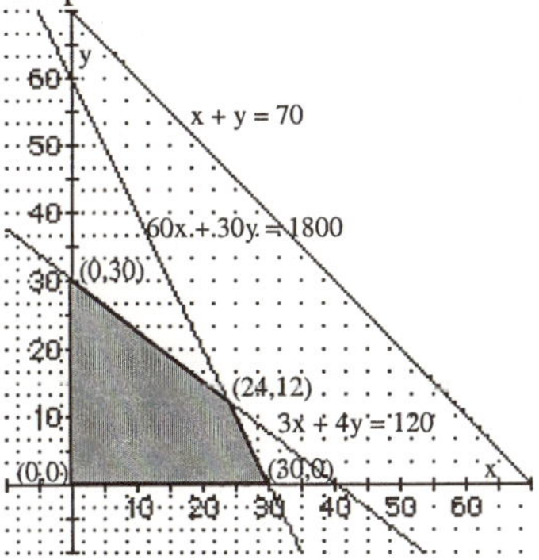

To find the intersection of $60x + 30y = 1800$ and $3x + 4y = 120$, solve the system:

$$\begin{cases} 60x + 30y = 1800 \\ 3x + 4y = 120 \end{cases}$$

Solve the first equation for y:

$$60x + 30y = 1800$$
$$2x + y = 60$$
$$y = 60 - 2x$$

Substitute and solve:

$$3x + 4(60 - 2x) = 120$$
$$3x + 240 - 8x = 120$$
$$-5x = -120$$
$$x = 24$$
$$y = 60 - 2(24) = 12$$

The point of intersection is (24, 12).
The corner points are (0, 0), (0, 30), (30, 0), (24, 12). Evaluate the objective function:

Vertex	Value of $P = 180x + 100y$
(0, 0)	$P = 180(0) + 100(0) = 0$
(0, 30)	$P = 180(0) + 100(30) = 3000$
(30, 0)	$P = 180(30) + 100(0) = 5400$
(24, 12)	$P = 180(24) + 100(12) = 5520$

The maximum profit is \$5520, when 24 acres of soybeans and 12 acres of wheat are planted.

With the increase of the preparation costs to \$2400, we do the following:
The constraints are:
$x \ge 0,\ y \ge 0$ A non-negative number of acres must be planted.
$x + y \le 70$ Acres available to plant.
$60x + 30y \le 2400$ Money available for preparation.
$3x + 4y \le 120$ Workdays available.
Graph the constraints.

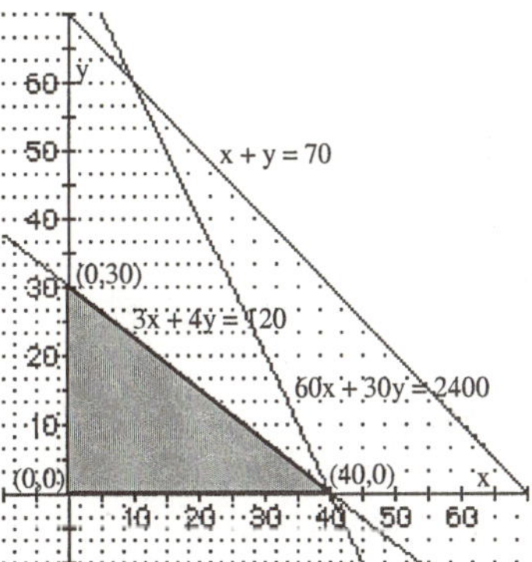

The corner points are (0, 0), (0, 30), (40, 0).
Evaluate the objective function:

Vertex	Value of $P = 180x + 100y$
(0, 0)	$P = 180(0) + 100(0) = 0$
(0, 30)	$P = 180(0) + 100(30) = 3000$
(40, 0)	$P = 180(40) + 100(0) = 7200$

The maximum profit is \$7200, when 40 acres of soybeans and 0 acres of wheat are planted.

21. Let x = the number of acres of corn planted, and let y = the number of acres of soybeans planted.
The total profit is: $P = 250x + 200y$. Profit is to be maximized, so this is the objective function.
The constraints are:
$x \ge 0,\ y \ge 0$ A non-negative number of acres must be planted.
$x + y \le 100$ Acres available to plant.
$60x + 40y \le 1800$ Money available for cultivation costs.

$60x+60y \le 2400$ Money available for labor costs.

Graph the constraints.

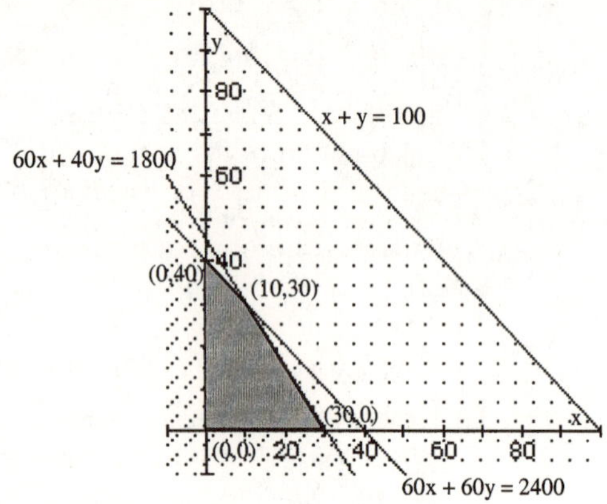

To find the intersection of $60x+40y=1800$ and $60x+60y=2400$, solve the system:

$$\begin{cases} 60x+40y=1800 \\ 60x+60y=2400 \end{cases}$$

Subtract multiply the first equation by -1 and add the result to the second equation:

$$\begin{array}{r} -60x-40y=-1800 \\ \underline{60x+60y=2400} \\ 20y=600 \\ y=30 \end{array}$$

$$\begin{aligned} 60x+40(30)&=1800 \\ 60x+1200&=1800 \\ 60x&=600 \\ x&=10 \end{aligned}$$

The point of intersection is (10, 30).
The corner points are (0, 0), (0, 40), (30, 0), (10, 30). Evaluate the objective function:

Vertex	Value of $P=250x+200y$
(0, 0)	$P=250(0)+200(0)=0$
(0, 40)	$P=250(0)+200(40)=8000$
(30, 0)	$P=250(30)+200(0)=7500$
(10, 30)	$P=250(10)+200(30)=8500$

The maximum profit is \$8500, when 10 acres of corn and 30 acres of soybeans are planted.

22. Let x = the number of ounces of Supplement A to be taken, and let y = the number of ounces of Supplement B to be taken. The total cost is: $C=1.5x+y$. Cost is to be minimized, so this is the objective function. The constraints are:

$x \ge 0,\ y \ge 0$ A non-negative number of ounces must be taken.

$5x+2y \ge 60$ Carbohydrates needed in diet.

$3x+2y \ge 45$ Protein needed in diet.

$4x+y \ge 30$ Fat needed in diet.

Graph the constraints.

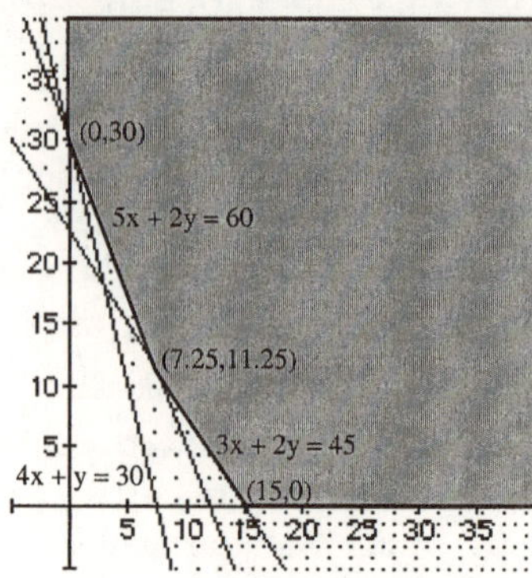

To find the intersection of $5x+2y=60$ and $3x+2y=45$, solve the system:

$$\begin{cases} 5x+2y=60 \\ 3x+2y=45 \end{cases}$$

Subtract the second equation from the first equation:

$$\begin{array}{r} 5x+2y=60 \\ \underline{-3x-2y=-45} \\ 2x \qquad =15 \\ x=7.5 \end{array}$$

Substitute and solve:

$$\begin{aligned} 5(7.5)+2y&=60 \\ 37.5+2y&=60 \\ 2y&=22.5 \\ y&=11.25 \end{aligned}$$

The point of intersection is (7.5, 11.25).
The corner points are (0, 30), (15, 0), (7.5, 11.25).
Evaluate the objective function:

Vertex	Value of $C=1.5x+y$
(0, 30)	$C=1.5(0)+30=30$
(15, 0)	$C=1.5(15)+0=22.5$
(7.5, 11.25)	$C=1.5(7.5)+11.25=22.5$

The minimum cost is \$22.50, when 15 ounces of Supplement A and 0 ounces of Supplement B are used in the diet or when 7.5 ounces of Supplement A and 11.25 ounces of Supplement B are used in the diet.

23. Let x = the number of hours that machine 1 is operated, and let y = the number of hours that machine 2 is operated. The total cost is: $C = 50x + 30y$. Cost is to be minimized, so this is the objective function.
The constraints are:
$x \geq 0,\ y \geq 0$ A positive number of hours must be used.
$x \leq 10$ Time used on machine 1.
$y \leq 10$ Time used on machine 2.
$60x + 40y \geq 240$ 8-inch pliers to be produced.
$70x + 20y \geq 140$ 6-inch pliers to be produced.
Graph the constraints.

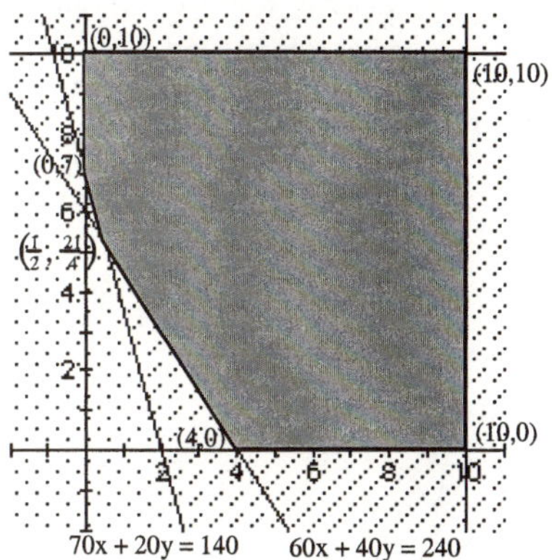

To find the intersection of $60x + 40y = 240$ and $70x + 20y = 140$, solve the system:

$$\begin{cases} 60x + 40y = 240 \\ 70x + 20y = 140 \end{cases}$$

Divide the first equation by -2 and add the result to the second equation:

$$\begin{array}{r} -30x - 20y = -120 \\ 70x + 20y = 140 \\ \hline 40x \qquad = 20 \end{array}$$

$$x = 0.5$$

Substitute and solve:

$$60(0.5) + 40y = 240$$
$$30 + 40y = 240$$
$$40y = 210$$
$$y = 5.25$$

The point of intersection is $(0.5, 5.25)$.
The corner points are (0, 7), (0, 10), (4, 0), (10, 0), (10, 10), $(0.5, 5.25)$. Evaluate the objective function:

Vertex	Value of $C = 50x + 30y$
(0, 7)	$C = 50(0) + 30(7) = 210$
(0, 10)	$C = 50(0) + 30(10) = 300$
(4, 0)	$C = 50(4) + 30(0) = 200$
(10, 0)	$C = 50(10) + 30(0) = 500$
(10, 10)	$C = 50(10) + 30(10) = 800$
(0.5, 5.25)	$P = 50(0.5) + 30(5.25) = 182.5$

The minimum cost is $182.50, when machine 1 is used for 0.5 hour and machine 2 is used for 5.25 hours.

24. Let x = the number of newer trees, and let y = the number of older trees. The total cost is: $C = 15x + 20y$. Cost is to be minimized, so this is the objective function. The constraints are:
$x \geq 0,\ y \geq 0$ A non-negative number of trees must be pruned.
$x + y \geq 25$ Must prune at least 25 trees.
$x + y \leq 50$ Only 50 trees can be in the orchard.
$x + 1.5y \geq 30$ Contract is for at least 30 hours.
Graph the constraints.

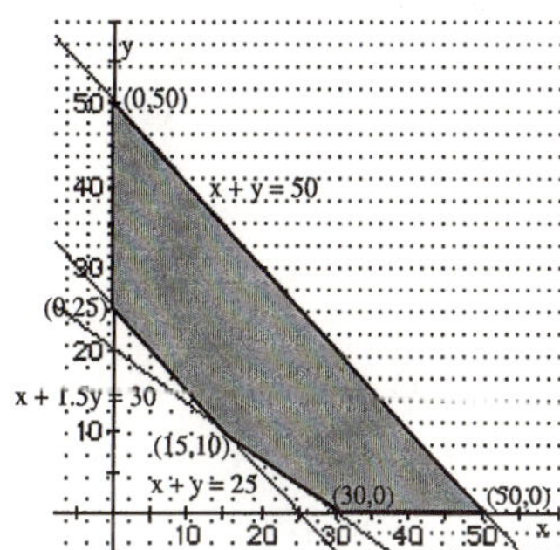

To find the intersection of $x + 1.5y = 30$ and $x + y = 25$, solve the system:

$$\begin{cases} x + y = 25 \\ x + 1.5y = 30 \end{cases}$$

Solve the second equation for x: $x = 30 - 1.5y$.
Substitute and solve:

$$(30 - 1.5y) + y = 25$$
$$-0.5y = -5$$
$$y = 10$$
$$x = 30 - 1.5(10) = 15$$

The point of intersection is (15, 10).
The corner points are (0, 25), (0, 50), (30, 0), (50, 0), (15, 10). Evaluate the objective function:

Vertex	Value of $C = 15x + 20y$
(0, 25)	$C = 15(0) + 20(25) = 500$
(0, 50)	$C = 15(0) + 20(50) = 1000$
(30, 0)	$C = 15(30) + 20(0) = 450$
(50, 0)	$C = 15(50) + 20(0) = 750$
(15, 10)	$C = 15(15) + 20(10) = 425$

The minimum cost is $425, when 15 newer trees and 10 older trees are pruned.

25. Let x = the number of pounds of ground beef, and let y = the number of pounds of ground pork. The total cost is: $C = 0.75x + 0.45y$. Cost is to be minimized, so this is the objective function. The constraints are:

$x \geq 0,\ y \geq 0$ A positive number of pounds must be used.

$x \leq 200$ Only 200 pounds of ground beef are available.

$y \geq 50$ At least 50 pounds of ground pork must be used.

$0.75x + 0.60y \geq 0.70(x + y)$ Leanness condition

(Note that the last equation will simplify to $y \leq \frac{1}{2}x$.) Graph the constraints.

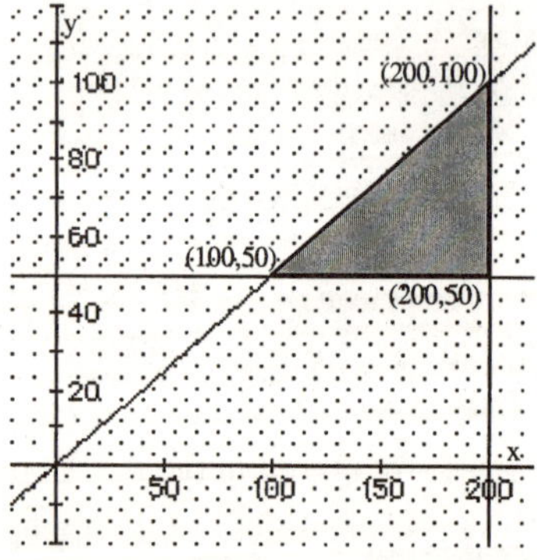

The corner points are (100, 50), (200, 50), (200, 100). Evaluate the objective function:

Vertex	Value of $C = 0.75x + 0.45y$
(100, 50)	$C = 0.75(100) + 0.45(50) = 97.50$
(200, 50)	$C = 0.75(200) + 0.45(50) = 172.50$
(200, 100)	$C = 0.75(200) + 0.45(100) = 195$

The minimum cost is $97.50, when 100 pounds of ground beef and 50 pounds of ground pork are used.

26. Let x = the amount invested in junk bonds, and let y = the amount invested in Treasury bills. The total income is: $I = 0.09x + 0.07y$. Income is to be maximized, so this is the objective function. The constraints are:

$x \geq 0,\ y \geq 0$ A non-negative amount must be invested.

$x + y \leq 20{,}000$ Total investment cannot exceed $20,000.

$x \leq 12{,}000$ Amount invested in junk bonds must not exceed $12,000.

$y \geq 8000$ Amount invested in Treasury bills must be at least $8000.

a. $y \geq x$ Amount invested in Treasury bills must be equal to or greater than the amount invested in junk bonds.

Graph the constraints.

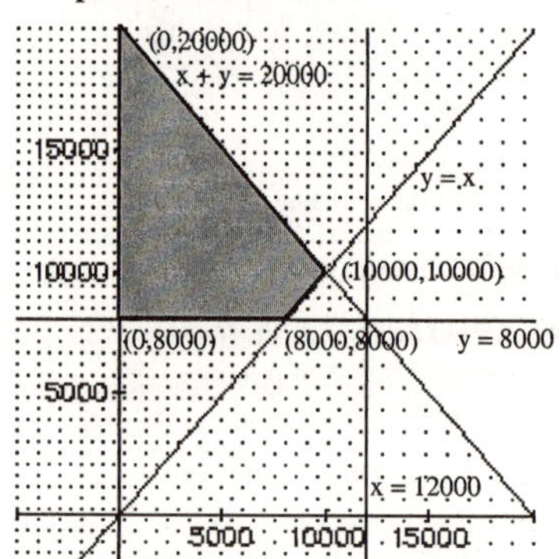

The corner points are (0, 20,000), (0, 8000), (8000, 8000), (10,000, 10,000). Evaluate the objective function:

Vertex	Value of $I = 0.09x + 0.07y$
(0, 20000)	$I = 0.09(0) + 0.07(20000) = 1400$
(0, 8000)	$I = 0.09(0) + 0.07(8000) = 560$
(8000, 8000)	$I = 0.09(8000) + 0.07(8000) = 1280$
(10000, 10000)	$I = 0.09(10000) + 0.07(10000) = 1600$

The maximum income is $1600, when $10,000 is invested in junk bonds and $10,000 is invested in Treasury bills.

b. $y \le x$ Amount invested in Treasury bills must not exceed the amount invested in junk bonds.

Graph the constraints.

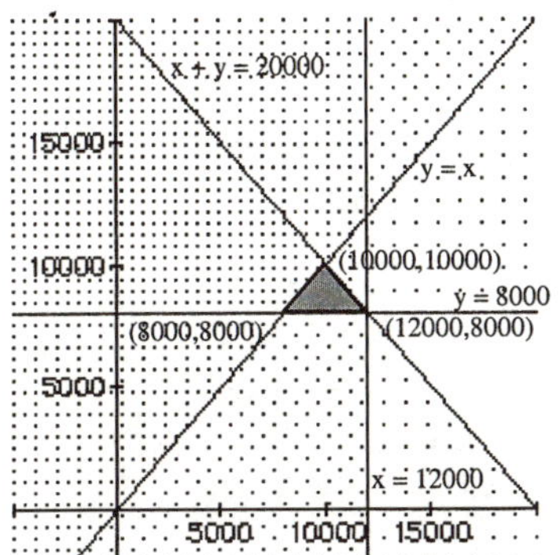

The corner points are (12,000, 8000), (8000, 8000), (10,000, 10,000).

Evaluate the objective function:

Vertex	Value of $I = 0.09x + 0.07y$
(12000, 8000)	$I = 0.09(12000) + 0.07(8000)$ $= 1640$
(8000, 8000)	$I = 0.09(8000) + 0.07(8000)$ $= 1280$
(10000, 10000)	$I = 0.09(10000) + 0.07(10000)$ $= 1600$

The maximum income is \$1640, when \$12,000 is invested in junk bonds and \$8000 is invested in Treasury bills.

27. Let x = the number of racing skates manufactured, and let y = the number of figure skates manufactured. The total profit is: $P = 10x + 12y$. Profit is to be maximized, so this is the objective function. The constraints are:

$x \ge 0,\ y \ge 0$ A positive number of skates must be manufactured.

$6x + 4y \le 120$ Only 120 hours are available for fabrication.

$x + 2y \le 40$ Only 40 hours are available for finishing.

Graph the constraints.

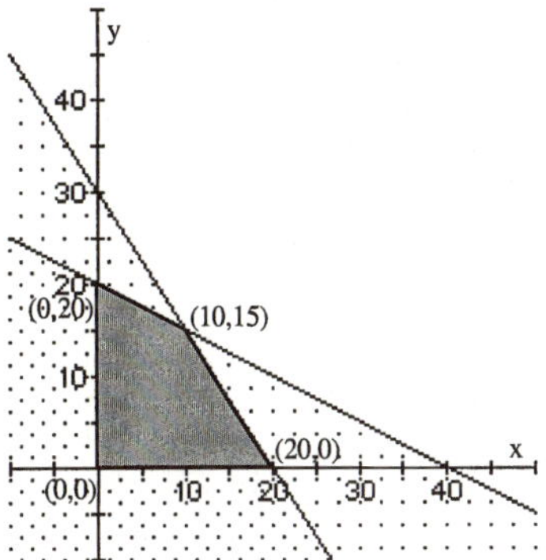

To find the intersection of $6x + 4y = 120$ and $x+2y = 40$, solve the system:

$$\begin{cases} 6x + 4y = 120 \\ x + 2y = 40 \end{cases}$$

Solve the second equation for x: $x = 40 - 2y$

Substitute and solve:

$$6(40 - 2y) + 4y = 120$$
$$240 - 12y + 4y = 120$$
$$-8y = -120$$
$$y = 15$$
$$x = 40 - 2(15) = 10$$

The point of intersection is (10, 15).

The corner points are (0, 0), (0, 20), (20, 0), (10, 15). Evaluate the objective function:

Vertex	Value of $P = 10x + 12y$
(0, 0)	$P = 10(0) + 12(0) = 0$
(0, 20)	$P = 10(0) + 12(20) = 240$
(20, 0)	$P = 10(20) + 12(0) = 200$
(10, 15)	$P = 10(10) + 12(15) = 280$

The maximum profit is \$280, when 10 racing skates and 15 figure skates are produced.

28. Let x = the amount placed in the AAA bond. Let y = the amount placed in a CD.

The total return is: $R = 0.08x + 0.04y$. Return is to be maximized, so this is the objective function. The constraints are:

$x \ge 0,\ y \ge 0$ A positive amount must be invested in each.

$x + y \le 50{,}000$ Total investment cannot exceed \$50,000.

$x \le 20{,}000$ Investment in the AAA bond cannot exceed \$20,000.

$y \ge 15{,}000$ Investment in the CD must be at least \$15,000.

$y \geq x$ Investment in the CD must exceed or equal the investment in the bond.

Graph the constraints.

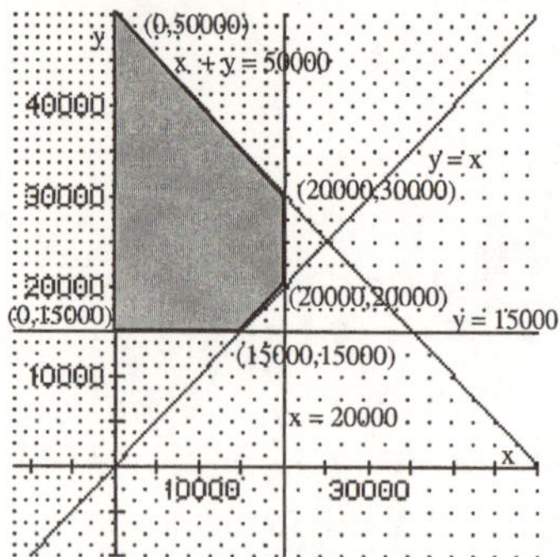

The corner points are (0, 50,000), (0, 15,000), (15,000, 15,000), (20,000, 20,000), (20,000, 30,000). Evaluate the objective function:

Vertex	Value of $R = 0.08x + 0.04y$
(0, 50000)	$R = 0.08(0) + 0.04(50000)$ $= 2000$
(0, 15000)	$R = 0.08(0) + 0.04(15000)$ $= 600$
(15000, 15000)	$R = 0.08(15000) + 0.04(15000)$ $= 1800$
(20000, 20000)	$R = 0.08(20000) + 0.04(20000)$ $= 2400$
(20000, 30000)	$R = 0.08(20000) + 0.04(30000)$ $= 2800$

The maximum return is \$2800, when \$20,000 is invested in a AAA bond and \$30,000 is invested in a CD.

29. Let x = the number of metal fasteners, and let y = the number of plastic fasteners. The total cost is: $C = 9x + 4y$. Cost is to be minimized, so this is the objective function. The constraints are:

$x \geq 2,\ y \geq 2$ At least 2 of each fastener must be made.

$x + y \geq 6$ At least 6 fasteners are needed.

$4x + 2y \leq 24$ Only 24 hours are available.

Graph the constraints.

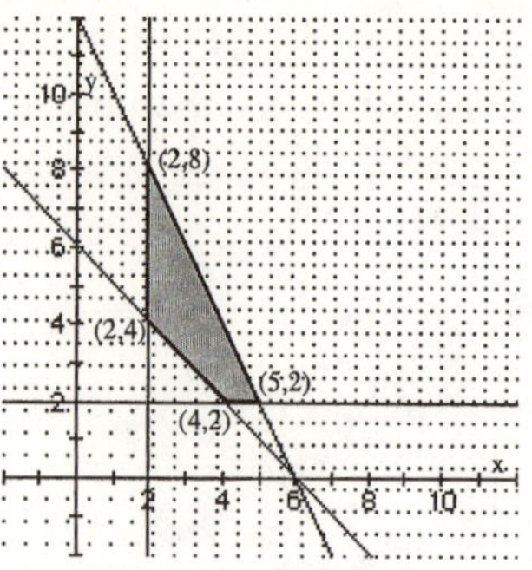

The corner points are (2, 4), (2, 8), (4, 2), (5, 2). Evaluate the objective function:

Vertex	Value of $C = 9x + 4y$
(2, 4)	$C = 9(2) + 4(4) = 34$
(2, 8)	$C = 9(2) + 4(8) = 50$
(4, 2)	$C = 9(4) + 4(2) = 44$
(5, 2)	$C = 9(5) + 4(2) = 53$

The minimum cost is \$34, when 2 metal fasteners and 4 plastic fasteners are ordered.

30. Let x = the amount of "Gourmet Dog," and let y = the amount of "Chow Hound." The total cost is: $C = 0.40x + 0.32y$. Cost is to be minimized, so this is the objective function. The constraints are:

$x \geq 0,\ y \geq 0$ A non-negative number of cans must be purchased.

$20x + 35y \geq 1175$ At least 1175 units of vitamins per month.

$75x + 50y \geq 2375$ At least 2375 calories per month.

$x + y \leq 60$ Storage space for 60 cans.

Graph the constraints.

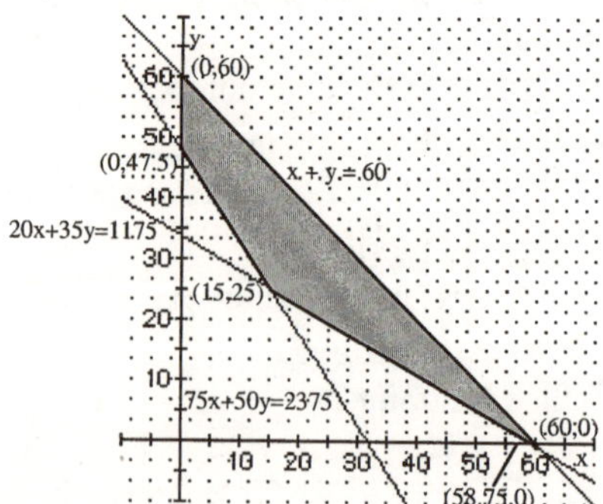

The corner points are (0, 47.5), (0, 60), (60, 0), (58.75, 0), (15, 25). Evaluate the objective function:

Vertex	Value of $C = 0.40x + 0.32y$
(0, 47.5)	$C = 0.40(0) + 0.32(47.5) = 15.20$
(0, 60)	$C = 0.40(0) + 0.32(60) = 19.20$
(60, 0)	$C = 0.40(60) + 0.32(0) = 24.00$
(58.75, 0)	$C = 0.40(58.75) + 0.32(0) = 23.50$
(15, 25)	$C = 0.40(15) + 0.32(25) = 14.00$

The minimum cost is \$14, when 15 cans of "Gourmet Dog" and 25 cans of "Chow Hound" are purchased.

31. Let x = the number of first class seats, and let y = the number of coach seats. Using the hint, the revenue from x first class seats and y coach seats is $Fx + Cy$, where $F > C > 0$. Thus, $R = Fx + Cy$ is the objective function to be maximized. The constraints are:

$8 \le x \le 16$ Restriction on first class seats.
$80 \le y \le 120$ Restriction on coach seats.

a. $\frac{x}{y} \le \frac{1}{12}$ Ratio of seats.

The constraints are:
$8 \le x \le 16$
$80 \le y \le 120$
$12x \le y$
Graph the constraints.

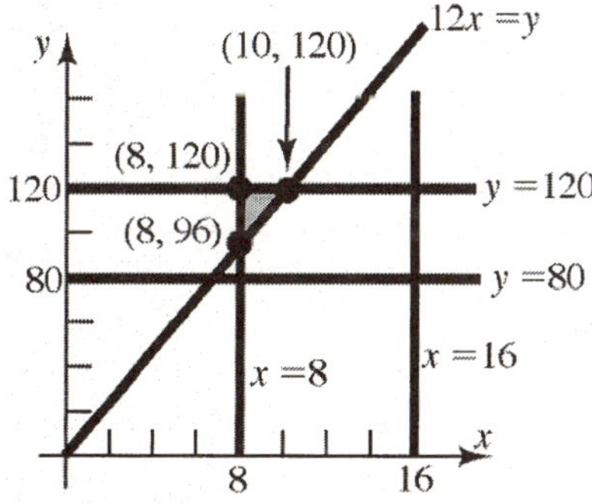

The corner points are (8, 96), (8, 120), and (10, 120). Evaluate the objective function:

Vertex	Value of $R = Fx + Cy$
(8, 96)	$R = 8F + 96C$
(8, 120)	$R = 8F + 120C$
(10, 120)	$R = 10F + 120C$

Since $C > 0$, $120C > 96C$, so
$8F + 120C > 8F + 96C$.
Since $F > 0$, $10F > 8F$, so
$10F + 120C > 8F + 120C$.
Thus, the maximum revenue occurs when the aircraft is configured with 10 first class seats and 120 coach seats.

b. $\frac{x}{y} \le \frac{1}{8}$

The constraints are:
$8 \le x \le 16$
$80 \le y \le 120$
$8x \le y$
Graph the constraints.

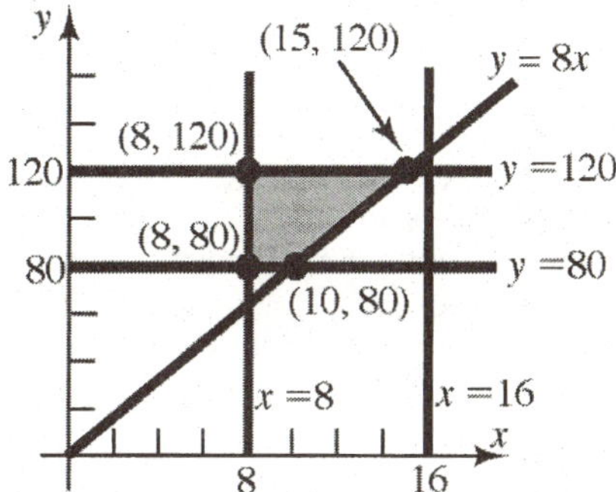

The corner points are (8, 80), (8, 120), (15, 120), and (10, 80).
Evaluate the objective function:

Vertex	Value of $R = Fx + Cy$
(8, 80)	$R = 8F + 80C$
(8, 120)	$R = 8F + 120C$
(15, 120)	$R = 15F + 120C$
(10, 80)	$R = 10F + 80C$

Since $F > 0$ and $C > 0$, $120C > 96C$, the maximum value of R occurs at (15, 120). The maximum revenue occurs when the aircraft is configured with 15 first class seats and 120 coach seats.

c. Answers will vary.

32. Let x = the number of ounces of Supplement A, and let y = the number of ounces of Supplement B. The total cost is: $C = 0.06x + 0.08y$. Cost is to be minimized; thus this is the objective function. The constraints are:

$x \ge 0,\ y \ge 0$ A non-negative number of ounces must be purchased.
$5x + 25y \ge 50$ At least 50 units of vitamin I.
$25x + 10y \ge 90$ At least 90 units of vitamin II.
$10x + 10y \ge 60$ At least 60 units of vitamin III.
$35x + 20y \ge 100$ At least 100 units of vitamin IV.

Graph the constraints.

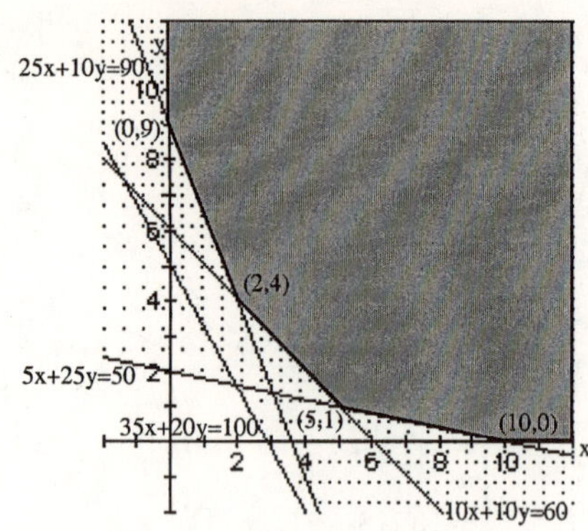

The corner points are (0, 9), (2, 4), (5, 1), (10, 0). Evaluate the objective function:

Vertex	Value of $C = 0.06x + 0.08y$
(0, 9)	$C = 0.06(0) + 0.08(9) = 0.72$
(2, 4)	$C = 0.06(2) + 0.08(4) = 0.44$
(5, 1)	$C = 0.06(5) + 0.08(1) = 0.38$
(10, 0)	$C = 0.06(10) + 0.08(0) = 0.60$

The minimum cost is \$0.38, when 5 ounces of Supplement *A*, and 1 ounce of Supplement *B* are used.

33. Answers will vary.

Chapter 10 Review

1. $\begin{cases} 2x - y = 5 \\ 5x + 2y = 8 \end{cases}$

Solve the first equation for *y*: $y = 2x - 5$.

Substitute and solve:

$$5x + 2(2x - 5) = 8$$
$$5x + 4x - 10 = 8$$
$$9x = 18$$
$$x = 2$$
$$y = 2(2) - 5 = 4 - 5 = -1$$

The solution is $x = 2,\ y = -1$.

2. $\begin{cases} 2x + 3y = 2 \\ 7x - y = 3 \end{cases}$

Solve the second equation for *y*: $y = 7x - 3$

Substitute into the first equation and solve:

$$2x + 3(7x - 3) = 2$$
$$2x + 21x - 9 = 2$$
$$23x = 11$$
$$x = \frac{11}{23}$$
$$y = 7\left(\frac{11}{23}\right) - 3 = \frac{77}{23} - \frac{69}{23} = \frac{8}{23}$$

The solution is $x = \frac{11}{23},\ y = \frac{8}{23}$.

3. $\begin{cases} 3x - 4y = 4 \\ x - 3y = \frac{1}{2} \end{cases}$

Solve the second equation for *x*: $x = 3y + \frac{1}{2}$

Substitute into the first equation and solve:

$$3\left(3y + \frac{1}{2}\right) - 4y = 4$$
$$9y + \frac{3}{2} - 4y = 4$$
$$5y = \frac{5}{2}$$
$$y = \frac{1}{2}$$
$$x = 3\left(\frac{1}{2}\right) + \frac{1}{2} = 2$$

The solution is $x = 2,\ y = \frac{1}{2}$.

4. $\begin{cases} 2x + y = 0 \\ 5x - 4y = -\frac{13}{2} \end{cases}$

Solve the first equation for *y*: $y = -2x$

Substitute into the second equation and solve:

$$5x - 4(-2x) = -\frac{13}{2}$$
$$5x + 8x = -\frac{13}{2}$$
$$13x = -\frac{13}{2}$$
$$x = -\frac{1}{2}$$
$$y = -2\left(-\frac{1}{2}\right) = 1$$

The solution is $x = -\frac{1}{2},\ y = 1$.

5. $\begin{cases} x-2y-4=0 \\ 3x+2y-4=0 \end{cases}$

Solve the first equation for x: $x=2y+4$

Substitute into the second equation and solve:

$$3(2y+4)+2y-4=0$$
$$6y+12+2y-4=0$$
$$8y=-8$$
$$y=-1$$
$$x=2(-1)+4=2$$

The solution is $x=2,\ y=-1$.

6. $\begin{cases} x-3y+5=0 \\ 2x+3y-5=0 \end{cases}$

Solve the first equation for x:
$x=3y-5$

Substitute into the second equation and solve:

$$2(3y-5)+3y-5=0$$
$$6y-10+3y-5=0$$
$$9y=15$$
$$y=\frac{5}{3}$$
$$x=3\left(\frac{5}{3}\right)-5=0$$

The solution is $x=0,\ y=\frac{5}{3}$.

7. $\begin{cases} y=2x-5 \\ x=3y+4 \end{cases}$

Substitute the first equation into the second equation and solve:

$$x=3(2x-5)+4$$
$$x=6x-15+4$$
$$-5x=-11$$
$$x=\frac{11}{5}$$
$$y=2\left(\frac{11}{5}\right)-5=-\frac{3}{5}$$

The solution is $x=\frac{11}{5},\ y=-\frac{3}{5}$.

8. $\begin{cases} x=5y+2 \\ y=5x+2 \end{cases}$

Substitute the first equation into the second equation and solve:

$$y=5(5y+2)+2$$
$$y=25y+10+2$$
$$-24y=12$$
$$y=-\frac{1}{2}$$
$$x=5\left(-\frac{1}{2}\right)+2=-\frac{5}{2}+2=-\frac{1}{2}$$

The solution is $x=-\frac{1}{2},\ y=-\frac{1}{2}$.

9. $\begin{cases} x-\ 3y+4=0 \\ \frac{1}{2}x-\frac{3}{2}y+\frac{4}{3}=0 \end{cases}$

Multiply each side of the first equation by 3 and each side of the second equation by -6 and add:

$$\begin{cases} 3x-9y+12=0 \\ -3x+9y-8=0 \end{cases}$$
$$4=0$$

There is no solution to the system. The system is inconsistent.

10. $\begin{cases} x+\frac{1}{4}y=2 \\ y+4x+2=0 \end{cases}$

Solve the second equation for y: $y=-4x-2$.

Substitute into the first equation and solve:

$$x+\frac{1}{4}(-4x-2)=2$$
$$x-x-\frac{1}{2}=2$$
$$0=\frac{5}{2}$$

There is no solution to the system. The system of equations is inconsistent.

11. $\begin{cases} 2x+3y-13=0 \\ 3x-2y \quad\quad =0 \end{cases}$

Multiply each side of the first equation by 2 and each side of the second equation by 3, and add to eliminate y:

$$\begin{cases} 4x+6y-26=0 \\ 9x-6y \quad\quad =0 \end{cases}$$
$$13x \quad -26=0$$
$$13x=26$$
$$x=2$$

Substitute and solve for y:

$$3(2)-2y=0$$
$$-2y=-6$$
$$y=3$$

The solution of the system is $x=2,\ y=3$.

12. $\begin{cases} 4x+5y=21 \\ 5x+6y=42 \end{cases}$

Multiply each side of the first equation by 5 and each side of the second equation by –4 and add to eliminate x:

$$\begin{cases} 20x+25y=\ \ 105 \\ -20x-24y=-168 \end{cases}$$
$$y=-63$$

Substitute and solve for x:

$$4x+5(-63)=21$$
$$4x-315=21$$
$$4x=336$$
$$x=84$$

The solution of the system is $x=84,\ y=-63$.

13. $\begin{cases} 3x-2y=8 \\ x-\dfrac{2}{3}y=12 \end{cases}$

Multiply each side of the second equation by –3 and add to eliminate x:

$$\begin{cases} 3x-2y=\ \ 8 \\ -3x+2y=-36 \end{cases}$$
$$0=-28$$

The system has no solution, so the system is inconsistent.

14. $\begin{cases} 2x+\ 5y=10 \\ 4x+10y=20 \end{cases}$

Multiply each side of the first equation by –2 and add to eliminate x:

$$\begin{cases} -4x-10y=-20 \\ 4x+10y=\ \ 20 \end{cases}$$
$$0=0$$

The system is dependent.

$$2x+5y=10$$
$$5y=-2x+10$$
$$y=-\frac{2}{5}x+2$$

The solution is $y=-\dfrac{2}{5}x+2$, x is any real number

15. $\begin{cases} x+2y-\ z=\ \ 6 \\ 2x-\ y+3z=-13 \\ 3x-2y+3z=-16 \end{cases}$

Multiply each side of the first equation by –2 and add to the second equation to eliminate x;

$$\begin{cases} -2x-4y+2z=-12 \\ 2x-\ y+3z=-13 \end{cases}$$
$$-5y+5z=-25$$
$$y-z=5$$

Multiply each side of the first equation by –3 and add to the third equation to eliminate x:

$$-3x-6y+3z=-18$$
$$3x-2y+3z=-16$$
$$-8y+6z=-34$$

Multiply each side of the first result by 8 and add to the second result to eliminate y:

$$8y-8z=\ \ 40$$
$$-8y+6z=-34$$
$$-2z=\ \ 6$$
$$z=-3$$

Substituting and solving for the other variables:

$$y-(-3)=5 \qquad x+2(2)-(-3)=6$$
$$y=2 \qquad x+4+3=6$$
$$x=-1$$

The solution is $x=-1,\ y=2,\ z=-3$.

16. $\begin{cases} x+5y-z=2 \\ 2x+y+z=7 \\ x-y+2z=11 \end{cases}$

Add the first equation and the second equation to eliminate z:

$$\begin{array}{l} x+5y-z=2 \\ 2x+y+z=7 \\ \hline 3x+6y=9 \\ -x-2y=-3 \end{array}$$

Multiply each side of the first equation by 2 and add to the third equation to eliminate z:

$$\begin{array}{l} 2x+10y-2z=4 \\ x-y+2z=11 \\ \hline 3x+9y=15 \\ x+3y=5 \end{array}$$

Add the two results to eliminate x:

$$\begin{array}{l} -x-2y=-3 \\ x+3y=5 \\ \hline y=2 \end{array}$$

Substituting and solving for the other variables:

$$\begin{array}{ll} x+3(2)=5 & 2(-1)+2+z=7 \\ x+6=5 & -2+2+z=7 \\ x=-1 & z=7 \end{array}$$

The solution is $x=-1,\ y=2,\ z=7$.

17. $\begin{cases} 2x-4y+z=-15 \\ x+2y-4z=27 \\ 5x-6y-2z=-3 \end{cases}$

Multiply the first equation by -1 and the second equation by 2, and then add to eliminate x:

$$\begin{array}{l} -2x+4y-z=15 \\ 2x+4y-8z=54 \\ \hline 8y-9z=69 \end{array}$$

Multiply the second equation by -5 and add to the third equation to eliminate x:

$$\begin{array}{l} -5x-10y+20z=-135 \\ 5x-6y-2z=-3 \\ \hline -16y+18z=-138 \end{array}$$

Multiply both sides of the first result by 2 and add to the second result to eliminate y:

$$\begin{array}{l} 16y-18z=138 \\ -16y+18z=-138 \\ \hline 0=0 \end{array}$$

The system is dependent.

$$-16y+18z=-138$$
$$18z+138=16y$$
$$y=\frac{9}{8}z+\frac{69}{8}$$

Substituting into the second equation and solving for x:

$$x+2\left(\frac{9}{8}z+\frac{69}{8}\right)-4z=27$$
$$x+\frac{9}{4}z+\frac{69}{4}-4z=27$$
$$x=\frac{7}{4}z+\frac{39}{4}$$

The solution is $x=\frac{7}{4}z+\frac{39}{4},\ y=\frac{9}{8}z+\frac{69}{8}$, z is any real number.

18. $\begin{cases} x-4y+3z=15 \\ -3x+y-5z=-5 \\ -7x-5y-9z=10 \end{cases}$

Multiply the first equation by 3 and then add the second equation to eliminate x:

$$\begin{array}{l} 3x-12y+9z=45 \\ -3x+y-5z=-5 \\ \hline -11y+4z=40 \end{array}$$

Multiply the first equation by 7 and add to the third equation to eliminate x:

$$\begin{array}{l} 7x-28y+21z=105 \\ -7x-5y-9z=10 \\ \hline -33y+12z=115 \\ -11y+4z=\frac{115}{3} \end{array}$$

Multiply the first result by -1 and adding it to the second result:

$$\begin{array}{l} 11y-4z=-40 \\ -11y+4z=\frac{115}{3} \\ \hline 0=-\frac{5}{3} \end{array}$$

The system has no solution. The system is inconsistent.

19. $\begin{cases} 3x+2y=8 \\ x+4y=-1 \end{cases}$

20. $\begin{cases} x+2y+5z=-2 \\ 5x-3z=8 \\ 2x-y=0 \end{cases}$

21. $A+C=\begin{bmatrix}1 & 0\\ 2 & 4\\ -1 & 2\end{bmatrix}+\begin{bmatrix}3 & -4\\ 1 & 5\\ 5 & 2\end{bmatrix}$
$=\begin{bmatrix}1+3 & 0+(-4)\\ 2+1 & 4+5\\ -1+5 & 2+2\end{bmatrix}$
$=\begin{bmatrix}4 & -4\\ 3 & 9\\ 4 & 4\end{bmatrix}$

22. $A-C=\begin{bmatrix}1 & 0\\ 2 & 4\\ -1 & 2\end{bmatrix}-\begin{bmatrix}3 & -4\\ 1 & 5\\ 5 & 2\end{bmatrix}$
$=\begin{bmatrix}1-32 & 0-(-4)\\ 2-1 & 4-5\\ -1-5 & 2-2\end{bmatrix}$
$=\begin{bmatrix}-2 & 4\\ 1 & -1\\ -6 & 0\end{bmatrix}$

23. $6A=6\cdot\begin{bmatrix}1 & 0\\ 2 & 4\\ -1 & 2\end{bmatrix}=\begin{bmatrix}6\cdot 1 & 6\cdot 0\\ 6\cdot 2 & 6\cdot 4\\ 6(-1) & 6\cdot 2\end{bmatrix}=\begin{bmatrix}6 & 0\\ 12 & 24\\ -6 & 12\end{bmatrix}$

24. $-4B=-4\cdot\begin{bmatrix}4 & -3 & 0\\ 1 & 1 & -2\end{bmatrix}$
$=\begin{bmatrix}-4\cdot 4 & -4(-3) & -4\cdot 0\\ -4\cdot 1 & -4\cdot 1 & -4(-2)\end{bmatrix}$
$=\begin{bmatrix}-16 & 12 & 0\\ -4 & -4 & 8\end{bmatrix}$

25. $AB=\begin{bmatrix}1 & 0\\ 2 & 4\\ -1 & 2\end{bmatrix}\cdot\begin{bmatrix}4 & -3 & 0\\ 1 & 1 & -2\end{bmatrix}$
$=\begin{bmatrix}1(4)+0(1) & 1(-3)+0(1) & 1(0)+2(-2)\\ 2(4)+2(1) & 2(-3)+4(1) & 2(0)+4(-2)\\ -1(4)+2(1) & -1(-3)+2(1) & -1(0)+2(-2)\end{bmatrix}$
$=\begin{bmatrix}4 & -3 & 0\\ 12 & -2 & -8\\ -2 & 5 & -4\end{bmatrix}$

26. $BA=\begin{bmatrix}4 & -3 & 0\\ 1 & 1 & -2\end{bmatrix}\cdot\begin{bmatrix}1 & 0\\ 2 & 4\\ -1 & 2\end{bmatrix}$
$=\begin{bmatrix}4(1)-3(2)+0(-1) & 4(0)-3(4)+0(-1)\\ 1(1)+1(2)-2(-1) & 1(0)+1(4)-2(2)\end{bmatrix}$
$=\begin{bmatrix}-2 & -12\\ 5 & 0\end{bmatrix}$

27. $CB=\begin{bmatrix}3 & -4\\ 1 & 5\\ 5 & 2\end{bmatrix}\cdot\begin{bmatrix}4 & -3 & 0\\ 1 & 1 & -2\end{bmatrix}$
$=\begin{bmatrix}3(4)-4(1) & 3(-3)-4(1) & 3(0)-4(-2)\\ 1(4)+5(1) & 1(-3)+5(1) & 1(0)+5(-2)\\ 5(4)+2(1) & 5(-3)+2(1) & 5(0)+2(-2)\end{bmatrix}$
$=\begin{bmatrix}8 & -13 & 8\\ 9 & 2 & -10\\ 22 & -13 & -4\end{bmatrix}$

28. $BC=\begin{bmatrix}4 & -3 & 0\\ 1 & 1 & -2\end{bmatrix}\cdot\begin{bmatrix}3 & -4\\ 1 & 5\\ 5 & 2\end{bmatrix}$
$=\begin{bmatrix}4(3)-3(1)+0(5) & 4(-4)-3(5)+0(2)\\ 1(3)+1(1)-2(5) & 1(-4)+1(5)-2(2)\end{bmatrix}$
$=\begin{bmatrix}9 & -31\\ -6 & -3\end{bmatrix}$

29. $A=\begin{bmatrix}4 & 6\\ 1 & 3\end{bmatrix}$

Augment the matrix with the identity and use row operations to find the inverse:

$\left[\begin{array}{cc|cc}4 & 6 & 1 & 0\\ 1 & 3 & 0 & 1\end{array}\right]$

$\rightarrow\left[\begin{array}{cc|cc}1 & 3 & 0 & 1\\ 4 & 6 & 1 & 0\end{array}\right]$ $\begin{pmatrix}\text{Interchange}\\ r_1 \text{ and } r_2\end{pmatrix}$

$\rightarrow\left[\begin{array}{cc|cc}1 & 3 & 0 & 1\\ 0 & -6 & 1 & -4\end{array}\right]$ $(R_2=-4r_1+r_2)$

$\rightarrow\left[\begin{array}{cc|cc}1 & 3 & 0 & 1\\ 0 & 1 & -\frac{1}{6} & \frac{2}{3}\end{array}\right]$ $(R_2=-\frac{1}{6}r_2)$

$\rightarrow\left[\begin{array}{cc|cc}1 & 0 & \frac{1}{2} & -1\\ 0 & 1 & -\frac{1}{6} & \frac{2}{3}\end{array}\right]$ $(R_1=-3r_2+r_1)$

Thus, $A^{-1}=\begin{bmatrix}\frac{1}{2} & -1\\ -\frac{1}{6} & \frac{2}{3}\end{bmatrix}$.

30. $A=\begin{bmatrix}-3 & 2\\ 1 & -2\end{bmatrix}$

Augment the matrix with the identity and use row operations to find the inverse:

$$\left[\begin{array}{cc|cc}-3 & 2 & 1 & 0\\ 1 & -2 & 0 & 1\end{array}\right] \quad \begin{pmatrix}\text{Interchange}\\ r_1 \text{ and } r_2\end{pmatrix}$$

$$\to\left[\begin{array}{cc|cc}1 & -2 & 0 & 1\\ -3 & 2 & 1 & 0\end{array}\right] \quad (R_2=3r_1+r_2)$$

$$\to\left[\begin{array}{cc|cc}1 & -2 & 0 & 1\\ 0 & -4 & 1 & 3\end{array}\right] \quad \left(R_2=-\tfrac{1}{4}r_2\right)$$

$$\to\left[\begin{array}{cc|cc}1 & -2 & 0 & 1\\ 0 & 1 & -\frac{1}{4} & -\frac{3}{4}\end{array}\right] \quad (R_1=2r_2+r_1)$$

$$\to\left[\begin{array}{cc|cc}1 & 0 & -\frac{1}{2} & -\frac{1}{2}\\ 0 & 1 & -\frac{1}{4} & -\frac{3}{4}\end{array}\right]$$

Thus, $A^{-1}=\begin{bmatrix}-\frac{1}{2} & -\frac{1}{2}\\ -\frac{1}{4} & -\frac{3}{4}\end{bmatrix}$.

31. $A=\begin{bmatrix}1 & 3 & 3\\ 1 & 2 & 1\\ 1 & -1 & 2\end{bmatrix}$

Augment the matrix with the identity and use row operations to find the inverse:

$$\left[\begin{array}{ccc|ccc}1 & 3 & 3 & 1 & 0 & 0\\ 1 & 2 & 1 & 0 & 1 & 0\\ 1 & -1 & 2 & 0 & 0 & 1\end{array}\right]$$

$$\to\left[\begin{array}{ccc|ccc}1 & 3 & 3 & 1 & 0 & 0\\ 0 & -1 & -2 & -1 & 1 & 0\\ 0 & -4 & -1 & -1 & 0 & 1\end{array}\right] \quad \begin{pmatrix}R_2=-r_1+r_2\\ R_3=-r_1+r_3\end{pmatrix}$$

$$\to\left[\begin{array}{ccc|ccc}1 & 3 & 3 & 1 & 0 & 0\\ 0 & 1 & 2 & 1 & -1 & 0\\ 0 & -4 & -1 & -1 & 0 & 1\end{array}\right] \quad (R_2=-r_2)$$

$$\to\left[\begin{array}{ccc|ccc}1 & 0 & -3 & -2 & 3 & 0\\ 0 & 1 & 2 & 1 & -1 & 0\\ 0 & 0 & 7 & 3 & -4 & 1\end{array}\right] \quad \begin{pmatrix}R_1=-3r_2+r_1\\ R_3=4r_2+r_3\end{pmatrix}$$

$$\to\left[\begin{array}{ccc|ccc}1 & 0 & -3 & -2 & 3 & 0\\ 0 & 1 & 2 & 1 & -1 & 0\\ 0 & 0 & 1 & \frac{3}{7} & -\frac{4}{7} & \frac{1}{7}\end{array}\right] \quad \left(R_3=\tfrac{1}{7}r_3\right)$$

$$\to\left[\begin{array}{ccc|ccc}1 & 0 & 0 & -\frac{5}{7} & \frac{9}{7} & \frac{3}{7}\\ 0 & 1 & 0 & \frac{1}{7} & \frac{1}{7} & -\frac{2}{7}\\ 0 & 0 & 1 & \frac{3}{7} & -\frac{4}{7} & \frac{1}{7}\end{array}\right] \quad \begin{pmatrix}R_1=3r_3+r_1\\ R_2=-2r_3+r_2\end{pmatrix}$$

Thus, $A^{-1}=\begin{bmatrix}-\frac{5}{7} & \frac{9}{7} & \frac{3}{7}\\ \frac{1}{7} & \frac{1}{7} & -\frac{2}{7}\\ \frac{3}{7} & -\frac{4}{7} & \frac{1}{7}\end{bmatrix}$.

32. $A=\begin{bmatrix}3 & 1 & 2\\ 3 & 2 & -1\\ 1 & 1 & 1\end{bmatrix}$

Augment the matrix with the identity and use row operations to find the inverse:

$$\left[\begin{array}{ccc|ccc}3 & 1 & 2 & 1 & 0 & 0\\ 3 & 2 & -1 & 0 & 1 & 0\\ 1 & 1 & 1 & 0 & 0 & 1\end{array}\right]$$

$$\to\left[\begin{array}{ccc|ccc}1 & 1 & 1 & 0 & 0 & 1\\ 3 & 2 & -1 & 0 & 1 & 0\\ 3 & 1 & 2 & 1 & 0 & 0\end{array}\right] \quad \begin{pmatrix}\text{Interchange}\\ r_1 \text{ and } r_3\end{pmatrix}$$

$$\to\left[\begin{array}{ccc|ccc}1 & 1 & 1 & 0 & 0 & 1\\ 0 & -1 & -4 & 0 & 1 & -3\\ 0 & -2 & -1 & 1 & 0 & -3\end{array}\right] \quad \begin{pmatrix}R_2=-3r_1+r_2\\ R_3=-3r_1+r_3\end{pmatrix}$$

$$\to\left[\begin{array}{ccc|ccc}1 & 1 & 1 & 0 & 0 & 1\\ 0 & 1 & 4 & 0 & -1 & 3\\ 0 & -2 & -1 & 1 & 0 & -3\end{array}\right] \quad (R_2=-r_2)$$

$$\to\left[\begin{array}{ccc|ccc}1 & 0 & -3 & 0 & 1 & -2\\ 0 & 1 & 4 & 0 & -1 & 3\\ 0 & 0 & 7 & 1 & -2 & 3\end{array}\right] \quad \begin{pmatrix}R_1=-r_2+r_1\\ R_3=2r_2+r_3\end{pmatrix}$$

$$\to\left[\begin{array}{ccc|ccc}1 & 0 & -3 & 0 & 1 & -2\\ 0 & 1 & 4 & 0 & -1 & 3\\ 0 & 0 & 1 & \frac{1}{7} & -\frac{2}{7} & \frac{3}{7}\end{array}\right] \quad \left(R_3=\tfrac{1}{7}r_3\right)$$

$$\to\left[\begin{array}{ccc|ccc}1 & 0 & 0 & \frac{3}{7} & \frac{1}{7} & -\frac{5}{7}\\ 0 & 1 & 0 & -\frac{4}{7} & \frac{1}{7} & \frac{9}{7}\\ 0 & 0 & 1 & \frac{1}{7} & -\frac{2}{7} & \frac{3}{7}\end{array}\right] \quad \begin{pmatrix}R_1=3r_3+r_1\\ R_2=-4r_3+r_2\end{pmatrix}$$

Thus, $A^{-1}=\begin{bmatrix}\frac{3}{7} & \frac{1}{7} & -\frac{5}{7}\\ -\frac{4}{7} & \frac{1}{7} & \frac{9}{7}\\ \frac{1}{7} & -\frac{2}{7} & \frac{3}{7}\end{bmatrix}$.

33. $A = \begin{bmatrix} 4 & -8 \\ -1 & 2 \end{bmatrix}$

Augment the matrix with the identity and use row operations to find the inverse:

$$\left[\begin{array}{cc|cc} 4 & -8 & 1 & 0 \\ -1 & 2 & 0 & 1 \end{array}\right]$$

$$\to \left[\begin{array}{cc|cc} -1 & 2 & 0 & 1 \\ 4 & -8 & 1 & 0 \end{array}\right] \quad \begin{pmatrix}\text{Interchange} \\ r_1 \text{ and } r_2\end{pmatrix}$$

$$\to \left[\begin{array}{cc|cc} -1 & 2 & 0 & 1 \\ 0 & 0 & 1 & 4 \end{array}\right] \quad (R_2 = 4r_1 + r_2)$$

$$\to \left[\begin{array}{cc|cc} 1 & -2 & 0 & -1 \\ 0 & 0 & 1 & 4 \end{array}\right] \quad (R_1 = -r_1)$$

There is no inverse because there is no way to obtain the identity on the left side. The matrix is singular.

34. $A = \begin{bmatrix} -3 & 1 \\ -6 & 2 \end{bmatrix}$

Augment the matrix with the identity and use row operations to find the inverse:

$$\left[\begin{array}{cc|cc} -3 & 1 & 1 & 0 \\ -6 & 2 & 0 & 1 \end{array}\right]$$

$$\to \left[\begin{array}{cc|cc} 1 & -\frac{1}{3} & -\frac{1}{3} & 0 \\ -6 & 2 & 0 & 1 \end{array}\right] \quad (R_1 = -\tfrac{1}{3}r_1)$$

$$\to \left[\begin{array}{cc|cc} 1 & -\frac{1}{3} & -\frac{1}{3} & 0 \\ 0 & 0 & -2 & 1 \end{array}\right] \quad (R_2 = 6r_1 + r_2)$$

There is no inverse because there is no way to obtain the identity on the left side. The matrix is singular.

35. $\begin{cases} 3x - 2y = 1 \\ 10x + 10y = 5 \end{cases}$

Write the augmented matrix:

$$\left[\begin{array}{cc|c} 3 & -2 & 1 \\ 10 & 10 & 5 \end{array}\right]$$

$$\to \left[\begin{array}{cc|c} 3 & -2 & 1 \\ 1 & 16 & 2 \end{array}\right] \quad (R_2 = -3r_1 + r_2)$$

$$\to \left[\begin{array}{cc|c} 1 & 16 & 2 \\ 3 & -2 & 1 \end{array}\right] \quad \begin{pmatrix}\text{Interchange} \\ r_1 \text{ and } r_2\end{pmatrix}$$

$$\to \left[\begin{array}{cc|c} 1 & 16 & 2 \\ 0 & -50 & -5 \end{array}\right] \quad (R_2 = -3r_1 + r_2)$$

$$\to \left[\begin{array}{cc|c} 1 & 16 & 2 \\ 0 & 1 & \frac{1}{10} \end{array}\right] \quad (R_2 = -\tfrac{1}{50}r_2)$$

$$\to \left[\begin{array}{cc|c} 1 & 0 & \frac{2}{5} \\ 0 & 1 & \frac{1}{10} \end{array}\right] \quad (R_1 = -16r_2 + r_1)$$

The solution is $x = \dfrac{2}{5}$, $y = \dfrac{1}{10}$.

36. $\begin{cases} 3x + 2y = 6 \\ x - y = -\dfrac{1}{2} \end{cases}$

Write the augmented matrix:

$$\left[\begin{array}{cc|c} 3 & 2 & 6 \\ 1 & -1 & -\frac{1}{2} \end{array}\right]$$

$$\to \left[\begin{array}{cc|c} 1 & -1 & -\frac{1}{2} \\ 3 & 2 & 6 \end{array}\right] \quad \begin{pmatrix}\text{Interchange} \\ r_1 \text{ and } r_2\end{pmatrix}$$

$$\to \left[\begin{array}{cc|c} 1 & -1 & -\frac{1}{2} \\ 0 & 5 & \frac{15}{2} \end{array}\right] \quad (R_2 = -3r_1 + r_2)$$

$$\to \left[\begin{array}{cc|c} 1 & -1 & -\frac{1}{2} \\ 0 & 1 & \frac{3}{2} \end{array}\right] \quad (R_2 = \tfrac{1}{5}r_2)$$

$$\to \left[\begin{array}{cc|c} 1 & 0 & 1 \\ 0 & 1 & \frac{3}{2} \end{array}\right] \quad (R_1 = r_2 + r_1)$$

The solution is $x = 1$, $y = \dfrac{3}{2}$.

37. $\begin{cases} 5x - 6y - 3z = 6 \\ 4x - 7y - 2z = -3 \\ 3x + y - 7z = 1 \end{cases}$

Write the augmented matrix:

$$\left[\begin{array}{ccc|c} 5 & -6 & -3 & 6 \\ 4 & -7 & -2 & -3 \\ 3 & 1 & -7 & 1 \end{array}\right]$$

$$\to \left[\begin{array}{ccc|c} 1 & 1 & -1 & 9 \\ 4 & -7 & -2 & -3 \\ 3 & 1 & -7 & 1 \end{array}\right] \quad (R_1 = -r_2 + r_1)$$

$$\to \left[\begin{array}{ccc|c} 1 & 1 & -1 & 9 \\ 0 & -11 & 2 & -39 \\ 0 & -2 & -4 & -26 \end{array}\right] \quad \begin{pmatrix} R_2 = -4r_1 + r_2 \\ R_3 = -3r_1 + r_3 \end{pmatrix}$$

$$\to \left[\begin{array}{ccc|c} 1 & 1 & -1 & 9 \\ 0 & 1 & -\frac{2}{11} & \frac{39}{11} \\ 0 & 1 & 2 & 13 \end{array}\right] \quad \begin{pmatrix} R_2 = -\frac{1}{11}r_2 \\ R_3 = -\frac{1}{2}r_3 \end{pmatrix}$$

$$\rightarrow \left[\begin{array}{ccc|c} 1 & 0 & -\frac{9}{11} & \frac{60}{11} \\ 0 & 1 & -\frac{2}{11} & \frac{39}{11} \\ 0 & 0 & \frac{24}{11} & \frac{104}{11} \end{array}\right] \quad \begin{pmatrix} R_1 = -r_2 + r_1 \\ R_3 = -r_2 + r_3 \end{pmatrix}$$

$$\rightarrow \left[\begin{array}{ccc|c} 1 & 0 & -\frac{9}{11} & \frac{60}{11} \\ 0 & 1 & -\frac{2}{11} & \frac{39}{11} \\ 0 & 0 & 1 & \frac{13}{3} \end{array}\right] \quad \left(R_3 = \tfrac{11}{24} r_3\right)$$

$$\rightarrow \left[\begin{array}{ccc|c} 1 & 0 & 0 & 9 \\ 0 & 1 & 0 & \frac{13}{3} \\ 0 & 0 & 1 & \frac{13}{3} \end{array}\right] \quad \begin{pmatrix} R_1 = \frac{9}{11} r_3 + r_1 \\ R_2 = \frac{2}{11} r_3 + r_2 \end{pmatrix}$$

The solution is $x = 9,\ y = \dfrac{13}{3},\ z = \dfrac{13}{3}$.

38. $\begin{cases} 2x + y + z = 5 \\ 4x - y - 3z = 1 \\ 8x + y - z = 5 \end{cases}$

Write the augmented matrix:

$$\left[\begin{array}{ccc|c} 2 & 1 & 1 & 5 \\ 4 & -1 & -3 & 1 \\ 8 & 1 & -1 & 5 \end{array}\right]$$

$$\rightarrow \left[\begin{array}{ccc|c} 2 & 1 & 1 & 5 \\ 0 & -3 & -5 & -9 \\ 0 & -3 & -5 & -15 \end{array}\right] \quad \begin{pmatrix} R_2 = -2r_1 + r_2 \\ R_3 = -4r_1 + r_3 \end{pmatrix}$$

$$\rightarrow \left[\begin{array}{ccc|c} 1 & \frac{1}{2} & \frac{1}{2} & \frac{5}{2} \\ 0 & 1 & \frac{5}{3} & 3 \\ 0 & -3 & -5 & -15 \end{array}\right] \quad \begin{pmatrix} R_1 = \frac{1}{2} r_1 \\ R_2 = -\frac{1}{3} r_2 \end{pmatrix}$$

$$\rightarrow \left[\begin{array}{ccc|c} 1 & 0 & -\frac{1}{3} & 1 \\ 0 & 1 & \frac{5}{3} & 3 \\ 0 & 0 & 0 & -6 \end{array}\right] \quad \begin{pmatrix} R_1 = -\frac{1}{2} r_2 + r_1 \\ R_3 = 3r_2 + r_3 \end{pmatrix}$$

There is no solution; the system is inconsistent.

39. $\begin{cases} x \quad\quad - 2z = 1 \\ 2x + 3y \quad\quad = -3 \\ 4x - 3y - 4z = 3 \end{cases}$

Write the augmented matrix:

$$\left[\begin{array}{ccc|c} 1 & 0 & -2 & 1 \\ 2 & 3 & 0 & -3 \\ 4 & -3 & -4 & 3 \end{array}\right]$$

$$\rightarrow \left[\begin{array}{ccc|c} 1 & 0 & -2 & 1 \\ 0 & 3 & 4 & -5 \\ 0 & -3 & 4 & -1 \end{array}\right] \quad \begin{pmatrix} R_2 = -2r_1 + r_2 \\ R_3 = -4r_1 + r_3 \end{pmatrix}$$

$$\rightarrow \left[\begin{array}{ccc|c} 1 & 0 & -2 & 1 \\ 0 & 1 & \frac{4}{3} & -\frac{5}{3} \\ 0 & -3 & 4 & -1 \end{array}\right] \quad \left(R_2 = \tfrac{1}{3} r_2\right)$$

$$\rightarrow \left[\begin{array}{ccc|c} 1 & 0 & -2 & 1 \\ 0 & 1 & \frac{4}{3} & -\frac{5}{3} \\ 0 & 0 & 8 & -6 \end{array}\right] \quad \left(R_3 = 3r_2 + r_3\right)$$

$$\rightarrow \left[\begin{array}{ccc|c} 1 & 0 & -2 & 1 \\ 0 & 1 & \frac{4}{3} & -\frac{5}{3} \\ 0 & 0 & 1 & -\frac{3}{4} \end{array}\right] \quad \left(R_3 = \tfrac{1}{8} r_3\right)$$

$$\rightarrow \left[\begin{array}{ccc|c} 1 & 0 & 0 & -\frac{1}{2} \\ 0 & 1 & 0 & -\frac{2}{3} \\ 0 & 0 & 1 & -\frac{3}{4} \end{array}\right] \quad \begin{pmatrix} R_1 = 2r_3 + r_1 \\ R_2 = -\frac{4}{3} r_3 + r_2 \end{pmatrix}$$

The solution is $x = -\dfrac{1}{2},\ y = -\dfrac{2}{3},\ z = -\dfrac{3}{4}$.

40. $\begin{cases} x + 2y - z = 2 \\ 2x - 2y + z = -1 \\ 6x + 4y + 3z = 5 \end{cases}$

Write the augmented matrix:

$$\left[\begin{array}{ccc|c} 1 & 2 & -1 & 2 \\ 2 & -2 & 1 & -1 \\ 6 & 4 & 3 & 5 \end{array}\right]$$

$$\rightarrow \left[\begin{array}{ccc|c} 1 & 2 & -1 & 2 \\ 0 & -6 & 3 & -5 \\ 0 & -8 & 9 & -7 \end{array}\right] \quad \begin{pmatrix} R_2 = -2r_1 + r_2 \\ R_3 = -6r_1 + r_3 \end{pmatrix}$$

$$\rightarrow \left[\begin{array}{ccc|c} 1 & 2 & -1 & 2 \\ 0 & 1 & -\frac{1}{2} & \frac{5}{6} \\ 0 & -8 & 9 & -7 \end{array}\right] \quad \left(R_2 = -\tfrac{1}{6} r_2\right)$$

$$\rightarrow \left[\begin{array}{ccc|c} 1 & 0 & 0 & \frac{1}{3} \\ 0 & 1 & -\frac{1}{2} & \frac{5}{6} \\ 0 & 0 & 5 & -\frac{1}{3} \end{array}\right] \quad \begin{pmatrix} R_1 = -2r_2 + r_1 \\ R_3 = 8r_2 + r_3 \end{pmatrix}$$

$$\rightarrow \left[\begin{array}{ccc|c} 1 & 0 & 0 & \frac{1}{3} \\ 0 & 1 & -\frac{1}{2} & \frac{5}{6} \\ 0 & 0 & 1 & -\frac{1}{15} \end{array}\right] \quad \left(R_3 = \tfrac{1}{5} r_3\right)$$

$$\rightarrow \left[\begin{array}{ccc|c} 1 & 0 & 0 & \frac{1}{3} \\ 0 & 1 & 0 & \frac{4}{5} \\ 0 & 0 & 1 & -\frac{1}{15} \end{array}\right] \quad \left(R_2 = \tfrac{1}{2} r_3 + r_2\right)$$

The solution is $x = \dfrac{1}{3},\ y = \dfrac{4}{5},\ z = -\dfrac{1}{15}$.

41. $\begin{cases} x-y+z=0 \\ x-y-5z=6 \\ 2x-2y+z=1 \end{cases}$

Write the augmented matrix:

$$\left[\begin{array}{ccc|c} 1 & -1 & 1 & 0 \\ 1 & -1 & -5 & 6 \\ 2 & -2 & 1 & 1 \end{array}\right]$$

$$\rightarrow \left[\begin{array}{ccc|c} 1 & -1 & 1 & 0 \\ 0 & 0 & -6 & 6 \\ 0 & 0 & -1 & 1 \end{array}\right] \quad \begin{pmatrix} R_2=-r_1+r_2 \\ R_3=-2r_1+r_3 \end{pmatrix}$$

$$\rightarrow \left[\begin{array}{ccc|c} 1 & -1 & 1 & 0 \\ 0 & 0 & 1 & -1 \\ 0 & 0 & -1 & 1 \end{array}\right] \quad \left(R_2=-\tfrac{1}{6}r_2\right)$$

$$\rightarrow \left[\begin{array}{ccc|c} 1 & -1 & 0 & 1 \\ 0 & 0 & 1 & -1 \\ 0 & 0 & 0 & 0 \end{array}\right] \quad \begin{pmatrix} R_1=-r_2+r_1 \\ R_3=r_2+r_3 \end{pmatrix}$$

The system is dependent.

$\begin{cases} x=y+1 \\ z=-1 \end{cases}$

The solution is $x=y+1,\ z=-1$, y is any real number.

42. $\begin{cases} 4x-3y+5z=0 \\ 2x+4y-3z=0 \\ 6x+2y+z=0 \end{cases}$

Write the augmented matrix:

$$\left[\begin{array}{ccc|c} 4 & -3 & 5 & 0 \\ 2 & 4 & -3 & 0 \\ 6 & 2 & 1 & 0 \end{array}\right]$$

$$\rightarrow \left[\begin{array}{ccc|c} 1 & -\frac{3}{4} & \frac{5}{4} & 0 \\ 2 & 4 & -3 & 0 \\ 6 & 2 & 1 & 0 \end{array}\right] \quad \left(R_1=\tfrac{1}{4}r_1\right)$$

$$\rightarrow \left[\begin{array}{ccc|c} 1 & -\frac{3}{4} & \frac{5}{4} & 0 \\ 0 & \frac{11}{2} & -\frac{11}{2} & 0 \\ 0 & \frac{13}{2} & -\frac{13}{2} & 0 \end{array}\right] \quad \begin{pmatrix} R_2=-2r_1+r_2 \\ R_3=-6r_1+r_3 \end{pmatrix}$$

$$\rightarrow \left[\begin{array}{ccc|c} 1 & -\frac{3}{4} & \frac{5}{4} & 0 \\ 0 & 1 & -1 & 0 \\ 0 & 1 & -1 & 0 \end{array}\right] \quad \begin{pmatrix} R_2=\frac{2}{11}r_2 \\ R_3=\frac{2}{13}r_3 \end{pmatrix}$$

$$\rightarrow \left[\begin{array}{ccc|c} 1 & 0 & \frac{1}{2} & 0 \\ 0 & 1 & -1 & 0 \\ 0 & 0 & 0 & 0 \end{array}\right] \quad \begin{pmatrix} R_1=\frac{3}{4}r_2+r_1 \\ R_3=-r_2+r_3 \end{pmatrix}$$

The solution is $x=-\dfrac{1}{2}z$, $y=z$, z is any real number.

43. $\begin{cases} x-y-z-t=1 \\ 2x+y-z+2t=3 \\ x-2y-2z-3t=0 \\ 3x-4y+z+5t=-3 \end{cases}$

Write the augmented matrix:

$$\left[\begin{array}{cccc|c} 1 & -1 & -1 & -1 & 1 \\ 2 & 1 & -1 & 2 & 3 \\ 1 & -2 & -2 & -3 & 0 \\ 3 & -4 & 1 & 5 & -3 \end{array}\right]$$

$$\rightarrow \left[\begin{array}{cccc|c} 1 & -1 & -1 & -1 & 1 \\ 0 & 3 & 1 & 4 & 1 \\ 0 & -1 & -1 & -2 & -1 \\ 0 & -1 & 4 & 8 & -6 \end{array}\right] \quad \begin{pmatrix} R_2=-2r_1+r_2 \\ R_3=-r_1+r_3 \\ R_4=-3r_1+r_4 \end{pmatrix}$$

$$\rightarrow \left[\begin{array}{cccc|c} 1 & -1 & -1 & -1 & 1 \\ 0 & -1 & -1 & -2 & -1 \\ 0 & 3 & 1 & 4 & 1 \\ 0 & -1 & 4 & 8 & -6 \end{array}\right] \quad \begin{pmatrix} \text{Interchange} \\ r_2 \text{ and } r_3 \end{pmatrix}$$

$$\rightarrow \left[\begin{array}{cccc|c} 1 & -1 & -1 & -1 & 1 \\ 0 & 1 & 1 & 2 & 1 \\ 0 & 3 & 1 & 4 & 1 \\ 0 & -1 & 4 & 8 & -6 \end{array}\right] \quad (R_2=-r_2)$$

$$\rightarrow \left[\begin{array}{cccc|c} 1 & 0 & 0 & 1 & 2 \\ 0 & 1 & 1 & 2 & 1 \\ 0 & 0 & -2 & -2 & -2 \\ 0 & 0 & 5 & 10 & -5 \end{array}\right] \quad \begin{pmatrix} R_1=r_2+r_1 \\ R_3=-3r_2+r_3 \\ R_4=r_2+r_4 \end{pmatrix}$$

$$\rightarrow \left[\begin{array}{cccc|c} 1 & 0 & 0 & 1 & 2 \\ 0 & 1 & 1 & 2 & 1 \\ 0 & 0 & 1 & 1 & 1 \\ 0 & 0 & 1 & 2 & -1 \end{array}\right] \quad \begin{pmatrix} R_3=-\frac{1}{2}r_3 \\ R_4=\frac{1}{5}r_4 \end{pmatrix}$$

$$\rightarrow \left[\begin{array}{cccc|c} 1 & 0 & 0 & 1 & 2 \\ 0 & 1 & 0 & 1 & 0 \\ 0 & 0 & 1 & 1 & 1 \\ 0 & 0 & 0 & 1 & -2 \end{array}\right] \quad \begin{pmatrix} R_2=-r_3+r_2 \\ R_4=-r_3+r_4 \end{pmatrix}$$

$$\rightarrow \left[\begin{array}{cccc|c} 1 & 0 & 0 & 0 & 4 \\ 0 & 1 & 0 & 0 & 2 \\ 0 & 0 & 1 & 0 & 3 \\ 0 & 0 & 0 & 1 & -2 \end{array}\right] \quad \begin{pmatrix} R_1=-r_4+r_1 \\ R_2=-r_4+r_2 \\ R_3=-r_4+r_3 \end{pmatrix}$$

The solution is $x=4,\ y=2,\ z=3,\ t=-2$.

44. $\begin{cases} x-3y+3z-t=4 \\ x+2y-z=-3 \\ x+3z+2t=3 \\ x+y+5z=6 \end{cases}$

Write the augmented matrix:

$$\left[\begin{array}{cccc|c} 1 & -3 & 3 & -1 & 4 \\ 1 & 2 & -1 & 0 & -3 \\ 1 & 0 & 3 & 2 & 3 \\ 1 & 1 & 5 & 0 & 6 \end{array}\right]$$

$$\rightarrow \left[\begin{array}{cccc|c} 1 & -3 & 3 & -1 & 4 \\ 0 & 5 & -4 & 1 & -7 \\ 0 & 3 & 0 & 3 & -1 \\ 0 & 4 & 2 & 1 & 2 \end{array}\right] \quad \begin{pmatrix} R_2 = -r_1 + r_2 \\ R_3 = -r_1 + r_3 \\ R_4 = -r_1 + r_4 \end{pmatrix}$$

$$\rightarrow \left[\begin{array}{cccc|c} 1 & -3 & 3 & -1 & 4 \\ 0 & 1 & -\frac{4}{5} & \frac{1}{5} & -\frac{7}{5} \\ 0 & 3 & 0 & 3 & -1 \\ 0 & 4 & 2 & 1 & 2 \end{array}\right] \quad \left(R_2 = \tfrac{1}{5} r_2\right)$$

$$\rightarrow \left[\begin{array}{cccc|c} 1 & 0 & \frac{3}{5} & -\frac{2}{5} & -\frac{1}{5} \\ 0 & 1 & -\frac{4}{5} & \frac{1}{5} & -\frac{7}{5} \\ 0 & 0 & \frac{12}{5} & \frac{12}{5} & \frac{16}{5} \\ 0 & 0 & \frac{26}{5} & \frac{1}{5} & \frac{38}{5} \end{array}\right] \quad \begin{pmatrix} R_1 = 3r_2 + r_1 \\ R_3 = -3r_2 + r_3 \\ R_4 = -4r_2 + r_4 \end{pmatrix}$$

$$\rightarrow \left[\begin{array}{cccc|c} 1 & 0 & \frac{3}{5} & -\frac{2}{5} & -\frac{1}{5} \\ 0 & 1 & -\frac{4}{5} & \frac{1}{5} & -\frac{7}{5} \\ 0 & 0 & 1 & 1 & \frac{4}{3} \\ 0 & 0 & \frac{26}{5} & \frac{1}{5} & \frac{38}{5} \end{array}\right] \quad \left(R_3 = \tfrac{5}{12} r_3\right)$$

$$\rightarrow \left[\begin{array}{cccc|c} 1 & 0 & 0 & -1 & -1 \\ 0 & 1 & 0 & 1 & -\frac{1}{3} \\ 0 & 0 & 1 & 1 & \frac{4}{3} \\ 0 & 0 & 0 & -5 & \frac{2}{3} \end{array}\right] \quad \begin{pmatrix} R_1 = -\frac{3}{5} r_3 + r_1 \\ R_2 = \frac{4}{5} r_3 + r_2 \\ R_4 = -\frac{26}{5} r_3 + r_4 \end{pmatrix}$$

$$\rightarrow \left[\begin{array}{cccc|c} 1 & 0 & 0 & -1 & -1 \\ 0 & 1 & 0 & 1 & -\frac{1}{3} \\ 0 & 0 & 1 & 1 & \frac{4}{3} \\ 0 & 0 & 0 & 1 & -\frac{2}{15} \end{array}\right] \quad \left(R_4 = -\tfrac{1}{5} r_4\right)$$

$$\rightarrow \left[\begin{array}{cccc|c} 1 & 0 & 0 & 0 & -\frac{17}{15} \\ 0 & 1 & 0 & 0 & -\frac{1}{5} \\ 0 & 0 & 1 & 0 & \frac{22}{15} \\ 0 & 0 & 0 & 1 & -\frac{2}{15} \end{array}\right] \quad \begin{pmatrix} R_1 = r_4 + r_1 \\ R_2 = -r_4 + r_2 \\ R_3 = -r_4 + r_3 \end{pmatrix}$$

The solution is $x = -\dfrac{17}{15}, y = -\dfrac{1}{5}, z = \dfrac{22}{15}, t = -\dfrac{2}{15}$.

45. $\begin{vmatrix} 3 & 4 \\ 1 & 3 \end{vmatrix} = 3(3) - 4(1) = 9 - 4 = 5$

46. $\begin{vmatrix} -4 & 0 \\ 1 & 3 \end{vmatrix} = -4(3) - 1(0) = -12 - 0 = -12$

47. $\begin{vmatrix} 1 & 4 & 0 \\ -1 & 2 & 6 \\ 4 & 1 & 3 \end{vmatrix} = 1\begin{vmatrix} 2 & 6 \\ 1 & 3 \end{vmatrix} - 4\begin{vmatrix} -1 & 6 \\ 4 & 3 \end{vmatrix} + 0\begin{vmatrix} -1 & 2 \\ 4 & 1 \end{vmatrix}$

$$\begin{aligned} &= 1(6-6) - 4(-3-24) + 0(-1-8) \\ &= 1(0) - 4(-27) + 0(-9) = 0 + 108 + 0 \\ &= 108 \end{aligned}$$

48. $\begin{vmatrix} 2 & 3 & 10 \\ 0 & 1 & 5 \\ -1 & 2 & 3 \end{vmatrix} = 2\begin{vmatrix} 1 & 5 \\ 2 & 3 \end{vmatrix} - 3\begin{vmatrix} 0 & 5 \\ -1 & 3 \end{vmatrix} + 10\begin{vmatrix} 0 & 1 \\ -1 & 2 \end{vmatrix}$

$$\begin{aligned} &= 2(3-10) - 3(0+5) + 10(0+1) \\ &= 2(-7) - 3(5) + 10(1) \\ &= -14 - 15 + 10 \\ &= -19 \end{aligned}$$

49. $\begin{vmatrix} 2 & 1 & -3 \\ 5 & 0 & 1 \\ 2 & 6 & 0 \end{vmatrix} = 2\begin{vmatrix} 0 & 1 \\ 6 & 0 \end{vmatrix} - 1\begin{vmatrix} 5 & 1 \\ 2 & 0 \end{vmatrix} + (-3)\begin{vmatrix} 5 & 0 \\ 2 & 6 \end{vmatrix}$

$$\begin{aligned} &= 2(0-6) - 1(0-2) - 3(30-0) \\ &= 2(-6) - 1(-2) - 3(30) \\ &= -12 + 2 - 90 \\ &= -100 \end{aligned}$$

50. $\begin{vmatrix} -2 & 1 & 0 \\ 1 & 2 & 3 \\ -1 & 4 & 2 \end{vmatrix} = -2\begin{vmatrix} 2 & 3 \\ 4 & 2 \end{vmatrix} - 1\begin{vmatrix} 1 & 3 \\ -1 & 2 \end{vmatrix} + 0\begin{vmatrix} 1 & 2 \\ -1 & 4 \end{vmatrix}$

$$\begin{aligned} &= -2(4-12) - 1(2+3) + 0(4+2) \\ &= -2(-8) - 1(5) + 0(6) \\ &= 16 - 5 + 0 \\ &= 11 \end{aligned}$$

51. $\begin{cases} x-2y=4 \\ 3x+2y=4 \end{cases}$

Set up and evaluate the determinants to use Cramer's Rule:

$$D=\begin{vmatrix} 1 & -2 \\ 3 & 2 \end{vmatrix}=1(2)-3(-2)=2+6=8$$

$$D_x=\begin{vmatrix} 4 & -2 \\ 4 & 2 \end{vmatrix}=4(2)-4(-2)=8+8=16$$

$$D_y=\begin{vmatrix} 1 & 4 \\ 3 & 4 \end{vmatrix}=1(4)-3(4)=4-12=-8$$

The solution is $x=\frac{D_x}{D}=\frac{16}{8}=2$,

$y=\frac{D_y}{D}=\frac{-8}{8}=-1$

52. $\begin{cases} x-3y=-5 \\ 2x+3y=5 \end{cases}$

Set up and evaluate the determinants to use Cramer's Rule:

$$D=\begin{vmatrix} 1 & -3 \\ 2 & 3 \end{vmatrix}=1(3)-2(-3)=3+6=9$$

$$D_x=\begin{vmatrix} -5 & -3 \\ 5 & 3 \end{vmatrix}=-5(3)-5(-3)=-15+15=0$$

$$D_y=\begin{vmatrix} 1 & -5 \\ 2 & 5 \end{vmatrix}=1(5)-2(-5)=5+10=15$$

The solution is $x=\frac{D_x}{D}=\frac{0}{9}=0$, $y=\frac{D_y}{D}=\frac{15}{9}=\frac{5}{3}$.

53. $\begin{cases} 2x+3y-13=0 \\ 3x-2y=0 \end{cases}$

Write the system is standard form:

$\begin{cases} 2x+3y=13 \\ 3x-2y=0 \end{cases}$

Set up and evaluate the determinants to use Cramer's Rule:

$$D=\begin{vmatrix} 2 & 3 \\ 3 & -2 \end{vmatrix}=-4-9=-13$$

$$D_x=\begin{vmatrix} 13 & 3 \\ 0 & -2 \end{vmatrix}=-26-0=-26$$

$$D_y=\begin{vmatrix} 2 & 13 \\ 3 & 0 \end{vmatrix}=0-39=-39$$

The solution is $x=\frac{D_x}{D}=\frac{-26}{-13}=2$,

$y=\frac{D_y}{D}=\frac{-39}{-13}=3$.

54. $\begin{cases} 3x-4y-12=0 \\ 5x+2y+6=0 \end{cases}$

Write the system in standard form:

$\begin{cases} 3x-4y=12 \\ 5x+2y=-6 \end{cases}$

Set up and evaluate the determinants to use Cramer's Rule:

$$D=\begin{vmatrix} 3 & -4 \\ 5 & 2 \end{vmatrix}=6+20=26$$

$$D_x=\begin{vmatrix} 12 & -4 \\ -6 & 2 \end{vmatrix}=24-24=0$$

$$D_y=\begin{vmatrix} 3 & 12 \\ 5 & -6 \end{vmatrix}=-18-60=-78$$

The solution is $x=\frac{D_x}{D}=\frac{0}{26}=0$,

$y=\frac{D_y}{D}=\frac{-78}{26}=-3$.

55. $\begin{cases} x+2y-z=6 \\ 2x-y+3z=-13 \\ 3x-2y+3z=-16 \end{cases}$

Set up and evaluate the determinants to use Cramer's Rule:

$$D=\begin{vmatrix} 1 & 2 & -1 \\ 2 & -1 & 3 \\ 3 & -2 & 3 \end{vmatrix}$$

$$=1\begin{vmatrix} -1 & 3 \\ -2 & 3 \end{vmatrix}-2\begin{vmatrix} -1 & 3 \\ -2 & 3 \end{vmatrix}+(-1)\begin{vmatrix} 2 & -1 \\ 3 & -2 \end{vmatrix}$$

$$=1(-3+6)-2(-3+6)+(-1)(-4+3)$$

$$=3+6+1=10$$

$$D_x=\begin{vmatrix} 6 & 2 & -1 \\ -13 & -1 & 3 \\ -16 & -2 & 3 \end{vmatrix}$$

$$=6\begin{vmatrix} -1 & 3 \\ -2 & 3 \end{vmatrix}-2\begin{vmatrix} -13 & 3 \\ -16 & 3 \end{vmatrix}+(-1)\begin{vmatrix} -13 & -1 \\ -16 & -2 \end{vmatrix}$$

$$=6(-3+6)-2(-39+48)+(-1)(26-16)$$

$$=18-18-10=-10$$

$$D_y=\begin{vmatrix} 1 & 6 & -1 \\ 2 & -13 & 3 \\ 3 & -16 & 3 \end{vmatrix}$$

$$=1\begin{vmatrix} -13 & 3 \\ -16 & 3 \end{vmatrix}-6\begin{vmatrix} 2 & 3 \\ 3 & 3 \end{vmatrix}+(-1)\begin{vmatrix} 2 & -13 \\ 3 & -16 \end{vmatrix}$$

$$=1(-39+48)-6(6-9)+(-1)(-32+39)$$

$$=9+18-7=20$$

$$D_z = \begin{vmatrix} 1 & 2 & 6 \\ 2 & -1 & -13 \\ 3 & -2 & -16 \end{vmatrix}$$
$$= 1\begin{vmatrix} -1 & -13 \\ -2 & -16 \end{vmatrix} - 2\begin{vmatrix} 2 & -13 \\ 3 & -16 \end{vmatrix} + 6\begin{vmatrix} 2 & -1 \\ 3 & -2 \end{vmatrix}$$
$$= 1(16-26) - 2(-32+39) + 6(-4+3)$$
$$= -10 - 14 - 6 = -30$$

The solution is $x = \frac{D_x}{D} = \frac{-10}{10} = -1$,

$y = \frac{D_y}{D} = \frac{20}{10} = 2$, $z = \frac{D_z}{D} = \frac{-30}{10} = -3$.

56. $\begin{cases} x - y + z = 8 \\ 2x + 3y - z = -2 \\ 3x - y - 9z = 9 \end{cases}$

Set up and evaluate the determinants to use Cramer's Rule:

$$D = \begin{vmatrix} 1 & -1 & 1 \\ 2 & 3 & -1 \\ 3 & -1 & -9 \end{vmatrix}$$
$$= 1\begin{vmatrix} 3 & -1 \\ -1 & -9 \end{vmatrix} - (-1)\begin{vmatrix} 2 & -1 \\ 3 & -9 \end{vmatrix} + 1\begin{vmatrix} 2 & 3 \\ 3 & -1 \end{vmatrix}$$
$$= 1(-27-1) + 1(-18+3) + 1(-2-9)$$
$$= -28 - 15 - 11$$
$$= -54$$

$$D_x = \begin{vmatrix} 8 & -1 & 1 \\ -2 & 3 & -1 \\ 9 & -1 & -9 \end{vmatrix}$$
$$= 8\begin{vmatrix} 3 & -1 \\ -1 & -9 \end{vmatrix} - (-1)\begin{vmatrix} -2 & -1 \\ 9 & -9 \end{vmatrix} + 1\begin{vmatrix} -2 & 3 \\ 9 & -1 \end{vmatrix}$$
$$= 8(-27-1) + 1(18+9) + 1(2-27)$$
$$= -224 + 27 - 25$$
$$= -222$$

$$D_y = \begin{vmatrix} 1 & 8 & 1 \\ 2 & -2 & -1 \\ 3 & 9 & -9 \end{vmatrix}$$
$$= 1\begin{vmatrix} -2 & -1 \\ 9 & -9 \end{vmatrix} - 8\begin{vmatrix} 2 & -1 \\ 3 & -9 \end{vmatrix} + 1\begin{vmatrix} 2 & -2 \\ 3 & 9 \end{vmatrix}$$
$$= 1(18+9) - 8(-18+3) + 1(18+6)$$
$$= 27 + 120 + 24$$
$$= 171$$

$$D_z = \begin{vmatrix} 1 & -1 & 8 \\ 2 & 3 & -2 \\ 3 & -1 & 9 \end{vmatrix}$$
$$= 1\begin{vmatrix} 3 & -2 \\ -1 & 9 \end{vmatrix} - (-1)\begin{vmatrix} 2 & -2 \\ 3 & 9 \end{vmatrix} + 8\begin{vmatrix} 2 & 3 \\ 3 & -1 \end{vmatrix}$$
$$= 1(27-2) + 1(18+6) + 8(-2-9)$$
$$= 25 + 24 - 88$$
$$= -39$$

The solution is $x = \frac{D_x}{D} = \frac{-222}{-54} = \frac{37}{9}$,

$y = \frac{D_y}{D} = \frac{171}{-54} = -\frac{19}{6}$, $z = \frac{D_z}{D} = \frac{-39}{-54} = \frac{13}{18}$.

57. Let $\begin{vmatrix} x & y \\ a & b \end{vmatrix} = 8$.

Then $\begin{vmatrix} 2x & y \\ 2a & b \end{vmatrix} = 2(8) = 16$ by Theorem (14).

The value of the determinant is multiplied by k when the elements of a column are multiplied by k.

58. Let $\begin{vmatrix} x & y \\ a & b \end{vmatrix} = 8$.

Then $\begin{vmatrix} y & x \\ b & a \end{vmatrix} = -8$ by Theorem (11). The value of the determinant changes sign when any 2 columns are interchanged.

59. Find the partial fraction decomposition:

$$x(x-4)\left(\frac{6}{x(x-4)}\right) = x(x-4)\left(\frac{A}{x} + \frac{B}{x-4}\right)$$
$$6 = A(x-4) + Bx$$

Let $x = 4$, then $6 = A(4-4) + B(4)$
$$4B = 6$$
$$B = \frac{3}{2}$$

Let $x = 0$, then $6 = A(0-4) + B(0)$
$$-4A = 6$$
$$A = -\frac{3}{2}$$

$$\frac{6}{x(x-4)} = \frac{(-3/2)}{x} + \frac{(3/2)}{x-4}$$

60. Find the partial fraction decomposition:

$$\frac{x}{(x+2)(x-3)} = \frac{A}{x+2} + \frac{B}{x-3}$$

Multiply both sides by $(x+2)(x-3)$.

$x = A(x-3) + B(x+2)$

Let $x = -2$, then $-2 = A(-2-3) + B(-2+2)$

$-5A = -2$

$A = \frac{2}{5}$

Let $x = 3$, then $3 = A(3-3) + B(3+2)$

$5B = 3$

$B = \frac{3}{5}$

$$\frac{x}{(x+2)(x-3)} = \frac{(2/5)}{x+2} + \frac{(3/5)}{x-3}$$

61. Find the partial fraction decomposition:

$$\frac{x-4}{x^2(x-1)} = \frac{A}{x} + \frac{B}{x^2} + \frac{C}{x-1}$$

Multiply both sides by $x^2(x-1)$

$x-4 = Ax(x-1) + B(x-1) + Cx^2$

Let $x = 1$, then

$1-4 = A(1)(1-1) + B(1-1) + C(1)^2$

$-3 = C$

$C = -3$

Let $x = 0$, then

$0-4 = A(0)(0-1) + B(0-1) + C(0)^2$

$-4 = -B$

$B = 4$

Let $x = 2$, then

$2-4 = A(2)(2-1) + B(2-1) + C(2)^2$

$-2 = 2A + B + 4C$

$2A = -2 - 4 - 4(-3)$

$2A = 6$

$A = 3$

$$\frac{x-4}{x^2(x-1)} = \frac{3}{x} + \frac{4}{x^2} + \frac{-3}{x-1}$$

62. Find the partial fraction decomposition:

$$\frac{2x-6}{(x-2)^2(x-1)} = \frac{A}{x-2} + \frac{B}{(x-2)^2} + \frac{C}{x-1}$$

Multiply both sides by $(x-2)^2(x-1)$.

$2x-6 = A(x-2)(x-1) + B(x-1) + C(x-2)^2$

Let $x = 1$, then

$2(1)-6 = A(1-2)(1-1) + B(1-1) + C(1-2)^2$

$-4 = C$

$C = -4$

Let $x = 2$, then

$2(2)-6 = A(2-2)(2-1) + B(2-1) + C(2-2)^2$

$-2 = B$

Let $x = 0$, then

$2(0)-6 = A(0-2)(0-1) + B(0-1) + C(0-2)^2$

$-6 = 2A - B + 4C$

$-6 = 2A - (-2) + 4(-4)$

$2A = 8$

$A = 4$

$$\frac{2x-6}{(x-2)^2(x-1)} = \frac{4}{x-2} + \frac{-2}{(x-2)^2} + \frac{-4}{x-1}$$

63. Find the partial fraction decomposition:

$$\frac{x}{(x^2+9)(x+1)} = \frac{A}{x+1} + \frac{Bx+C}{x^2+9}$$

Multiply both sides by $(x+1)(x^2+9)$.

$x = A(x^2+9) + (Bx+C)(x+1)$

Let $x = -1$, then

$-1 = A\left((-1)^2+9\right) + \left(B(-1)+C\right)(-1+1)$

$-1 = A(10) + (-B+C)(0)$

$-1 = 10A$

$A = -\frac{1}{10}$

Let $x = 0$, then

$0 = A\left(0^2+9\right) + \left(B(0)+C\right)(0+1)$

$0 = 9A + C$

$0 = 9\left(-\frac{1}{10}\right) + C$

$C = \frac{9}{10}$

Let $x = 1$, then $1 = A\left(1^2+9\right)+\left(B(1)+C\right)(1+1)$

$$1 = A(10)+(B+C)(2)$$
$$1 = 10A+2B+2C$$
$$1 = 10\left(-\frac{1}{10}\right)+2B+2\left(\frac{9}{10}\right)$$
$$1 = -1+2B+\frac{9}{5}$$
$$2B = \frac{1}{5}$$
$$B = \frac{1}{10}$$
$$\frac{x}{(x^2+9)(x+1)} = \frac{-\frac{1}{10}}{x+1}+\frac{\frac{1}{10}x+\frac{9}{10}}{x^2+9}$$

64. Find the partial fraction decomposition:
$$\frac{3x}{(x-2)(x^2+1)} = \frac{A}{x-2}+\frac{Bx+C}{x^2+1}$$
Multiply both sides by $(x-2)(x^2+1)$.
$$3x = A(x^2+1)+(Bx+C)(x-2)$$
Let $x = 2$, then
$$3(2) = A((2)^2+1)+(B(2)+C)(2-2)$$
$$6 = 5A$$
$$A = \frac{6}{5}$$
Let $x = 0$, then
$$3(0) = A(0^2+1)+(B(0)+C)(0-2)$$
$$0 = A-2C$$
$$0 = \frac{6}{5}-2C$$
$$2C = \frac{6}{5}$$
$$C = \frac{3}{5}$$
Let $x = 1$, then
$$3(1) = A(1^2+1)+(B(1)+C)(1-2)$$
$$3 = 2A-B-C$$
$$3 = 2\left(\frac{6}{5}\right)-B-\frac{3}{5}$$
$$B = -\frac{6}{5}$$
$$\frac{3x}{(x-2)(x^2+1)} = \frac{\frac{6}{5}}{x-2}+\frac{-\frac{6}{5}x+\frac{3}{5}}{x^2+1}$$

65. Find the partial fraction decomposition:
$$\frac{x^3}{(x^2+4)^2} = \frac{Ax+B}{x^2+4}+\frac{Cx+D}{(x^2+4)^2}$$
Multiply both sides by $(x^2+4)^2$.
$$x^3 = (Ax+B)(x^2+4)+Cx+D$$
$$x^3 = Ax^3+Bx^2+4Ax+4B+Cx+D$$
$$x^3 = Ax^3+Bx^2+(4A+C)x+4B+D$$
$$A = 1;\ B = 0$$
$$4A+C = 0$$
$$4(1)+C = 0$$
$$C = -4$$
$$4B+D = 0$$
$$4(0)+D = 0$$
$$D = 0$$
$$\frac{x^3}{(x^2+4)^2} = \frac{x}{x^2+4}+\frac{-4x}{(x^2+4)^2}$$

66. Find the partial fraction decomposition:
$$\frac{x^3+1}{(x^2+16)^2} = \frac{Ax+B}{x^2+16}+\frac{Cx+D}{(x^2+16)^2}$$
Multiply both sides by $(x^2+16)^2$.
$$x^3+1 = (Ax+B)(x^2+16)+Cx+D$$
$$x^3+1 = Ax^3+Bx^2+16Ax+16B+Cx+D$$
$$x^3+1 = Ax^3+Bx^2+(16A+C)x+16B+D$$
$$A = 1;\ B = 0$$
$$16A+C = 0$$
$$16(1)+C = 0$$
$$C = -16$$
$$16B+D = 1$$
$$16(0)+D = 1$$
$$D = 1$$
$$\frac{x^3+1}{(x^2+16)^2} = \frac{x}{x^2+16}+\frac{-16x+1}{(x^2+16)^2}$$

67. Find the partial fraction decomposition:
$$\frac{x^2}{(x^2+1)(x^2-1)} = \frac{x^2}{(x^2+1)(x-1)(x+1)}$$
$$= \frac{A}{x-1}+\frac{B}{x+1}+\frac{Cx+D}{x^2+1}$$
Multiply both sides by $(x-1)(x+1)(x^2+1)$.

$$x^2 = A(x+1)(x^2+1) + B(x-1)(x^2+1) + (Cx+D)(x-1)(x+1)$$

Let $x = 1$, then

$$1^2 = A(1+1)(1^2+1) + B(1-1)(1^2+1) + (C(1)+D)(1-1)(1+1)$$

$$1 = 4A$$

$$A = \frac{1}{4}$$

Let $x = -1$, then

$$(-1)^2 = A(-1+1)((-1)^2+1) + B(-1-1)((-1)^2+1) + (C(-1)+D)(-1-1)(-1+1)$$

$$1 = -4B$$

$$B = -\frac{1}{4}$$

Let $x = 0$, then

$$0^2 = A(0+1)(0^2+1) + B(0-1)(0^2+1) + (C(0)+D)(0-1)(0+1)$$

$$0 = A - B - D$$

$$0 = \frac{1}{4} - \left(-\frac{1}{4}\right) - D$$

$$D = \frac{1}{2}$$

Let $x = 2$, then

$$2^2 = A(2+1)(2^2+1) + B(2-1)(2^2+1) + (C(2)+D)(2-1)(2+1)$$

$$4 = 15A + 5B + 6C + 3D$$

$$4 = 15\left(\frac{1}{4}\right) + 5\left(-\frac{1}{4}\right) + 6C + 3\left(\frac{1}{2}\right)$$

$$6C = 4 - \frac{15}{4} + \frac{5}{4} - \frac{3}{2}$$

$$6C = 0$$

$$C = 0$$

$$\frac{x^2}{(x^2+1)(x^2-1)} = \frac{(1/4)}{x-1} + \frac{-(1/4)}{x+1} + \frac{(1/2)}{x^2+1}$$

68. Find the partial fraction decomposition:

$$\frac{4}{(x^2+4)(x^2-1)} = \frac{4}{(x^2+4)(x-1)(x+1)} = \frac{A}{x-1} + \frac{B}{x+1} + \frac{Cx+D}{x^2+4}$$

Multiply both sides by $(x^2+4)(x-1)(x+1)$.

$$4 = A(x+1)(x^2+4) + B(x-1)(x^2+4) + (Cx+D)(x-1)(x+1)$$

Let $x = 1$, then

$$4 = A(1+1)(1^2+4) + B(1-1)(1^2+4) + (C(1)+D)(1-1)(1+1)$$

$$4 = 10A$$

$$A = \frac{4}{10} = \frac{2}{5}$$

Let $x = -1$, then

$$4 = A(-1+1)((-1)^2+4) + B(-1-1)((-1)^2+4) + (C(-1)+D)(-1-1)(-1+1)$$

$$4 = -10B$$

$$B = -\frac{4}{10} = -\frac{2}{5}$$

Let $x = 0$, then

$$4 = A(0+1)(0^2+4) + B(0-1)(0^2+4) + (C(0)+D)(0-1)(0+1)$$

$$4 = 4A - 4B - D$$

$$4 = 4\left(\frac{2}{5}\right) - 4\left(-\frac{2}{5}\right) - D$$

$$4 = \frac{8}{5} + \frac{8}{5} - D$$

$$D = -\frac{4}{5}$$

Let $x = 2$, then

$$4 = A(2+1)(2^2+4) + B(2-1)(2^2+4) + (C(2)+D)(2-1)(2+1)$$

$$4 = 24A + 8B + 6C + 3D$$

$$4 = 24\left(\frac{2}{5}\right) + 8\left(-\frac{2}{5}\right) + 6C + 3\left(-\frac{4}{5}\right)$$

$$4 = \frac{48}{5} - \frac{16}{5} + 6C - \frac{12}{5}$$

$$6C = 0$$

$$C = 0$$

$$\frac{4}{(x^2+4)(x^2-1)} = \frac{(2/5)}{x-1} + \frac{-(2/5)}{x+1} + \frac{-(4/5)}{x^2+4}$$

69. Solve the first equation for y, substitute into the second equation and solve:

$$\begin{cases} 2x+y+3=0 \rightarrow y=-2x-3 \\ x^2+y^2=5 \end{cases}$$

$$x^2+(-2x-3)^2=5 \rightarrow x^2+4x^2+12x+9=5$$

$$5x^2+12x+4=0 \rightarrow (5x+2)(x+2)=0$$

$$\Rightarrow x=-\frac{2}{5} \text{ or } x=-2$$

$$y=-\frac{11}{5} \qquad y=1$$

Solutions: $\left(-\frac{2}{5},-\frac{11}{5}\right), (-2,1)$.

70. Add the equations to eliminate y, and solve:

$$\begin{cases} x^2+y^2=16 \\ 2x-y^2=-8 \end{cases}$$

$$x^2+2x=8$$

$$x^2+2x-8=0 \rightarrow (x+4)(x-2)=0$$

$$x=-4 \text{ or } x=2$$

If $x=-4$:

$$(-4)^2+y^2=16 \rightarrow y^2=0 \rightarrow y=0$$

If $x=2$:

$$(2)^2+y^2=16 \rightarrow y^2=12 \rightarrow y=\pm 2\sqrt{3}$$

Solutions: $(-4,0), \left(2,2\sqrt{3}\right), \left(2,-2\sqrt{3}\right)$.

71. Multiply each side of the second equation by 2 and add the equations to eliminate xy:

$$\begin{cases} 2xy+y^2=10 \longrightarrow 2xy+y^2=10 \\ -xy+3y^2=2 \xrightarrow{2} -2xy+6y^2=4 \end{cases}$$

$$7y^2=14$$

$$y^2=2$$

$$y=\pm\sqrt{2}$$

If $y=\sqrt{2}$:

$$2x\left(\sqrt{2}\right)+\left(\sqrt{2}\right)^2=10 \rightarrow 2\sqrt{2}x=8 \rightarrow x=2\sqrt{2}$$

If $y=-\sqrt{2}$:

$$2x\left(-\sqrt{2}\right)+\left(-\sqrt{2}\right)^2=10 \rightarrow -2\sqrt{2}x=8$$

$$\rightarrow x=-2\sqrt{2}$$

Solutions: $\left(2\sqrt{2},\sqrt{2}\right), \left(-2\sqrt{2},-\sqrt{2}\right)$

72. Multiply each side of the first equation by –2 and add the equations to eliminate y:

$$\begin{cases} 3x^2-y^2=1 \xrightarrow{-2} -6x^2+2y^2=-2 \\ 7x^2-2y^2=5 \longrightarrow 7x^2-2y^2=5 \end{cases}$$

$$x^2=3$$

$$x=\pm\sqrt{3}$$

If $x=\sqrt{3}$:

$$3\left(\sqrt{3}\right)^2-y^2=1 \rightarrow -y^2=-8$$

$$\rightarrow y=\pm\sqrt{8}=\pm 2\sqrt{2}$$

If $x=-\sqrt{3}$:

$$3\left(-\sqrt{3}\right)^2-y^2=1 \rightarrow -y^2=-8$$

$$\rightarrow y=\pm\sqrt{8}=\pm 2\sqrt{2}$$

Solutions:

$\left(\sqrt{3},2\sqrt{2}\right), \left(\sqrt{3},-2\sqrt{2}\right), \left(-\sqrt{3},2\sqrt{2}\right),$
$\left(-\sqrt{3},-2\sqrt{2}\right)$

73. Substitute into the second equation into the first equation and solve:

$$\begin{cases} x^2+y^2=6y \\ x^2=3y \end{cases}$$

$$3y+y^2-6y$$

$$y^2-3y=0$$

$$y(y-3)=0 \rightarrow y=0 \text{ or } y=3$$

If $y=0$: $\quad x^2=3(0) \rightarrow x^2=0 \rightarrow x=0$

If $y=3$: $\quad x^2=3(3) \rightarrow x^2=9 \rightarrow x=\pm 3$

Solutions: (0, 0), (–3, 3), (3, 3).

74. Multiply each side of the second equation by –1 and add the equations to eliminate y:

$$\begin{cases} 2x^2+y^2=9 \longrightarrow 2x^2+y^2=9 \\ x^2+y^2=9 \xrightarrow{-1} -x^2-y^2=-9 \end{cases}$$

$$x^2 = 0 \Rightarrow x=0$$

If $x=0$: $0^2+y^2=9 \rightarrow y^2=9 \rightarrow y=\pm 3$

Solutions: (0, 3), (0, –3)

75. Factor the second equation, solve for x, substitute into the first equation and solve:

$$\begin{cases} 3x^2+4xy+5y^2=8 \\ x^2+3xy+2y^2=0 \end{cases}$$

$x^2+3xy+2y^2=0$

$(x+2y)(x+y)=0 \rightarrow x=-2y \text{ or } x=-y$

Substitute $x=-2y$ and solve:

$$3x^2+4xy+5y^2=8$$
$$3(-2y)^2+4(-2y)y+5y^2=8$$
$$12y^2-8y^2+5y^2=8$$
$$9y^2=8$$
$$y^2=\frac{8}{9} \Rightarrow y=\pm\frac{2\sqrt{2}}{3}$$

Substitute $x=-y$ and solve:

$$3x^2+4xy+5y^2=8$$
$$3(-y)^2+4(-y)y+5y^2=8$$
$$3y^2-4y^2+5y^2=8$$
$$4y^2=8$$
$$y^2=2 \Rightarrow y=\pm\sqrt{2}$$

If $y=\frac{2\sqrt{2}}{3}$: $\quad x=-2\left(\frac{2\sqrt{2}}{3}\right)=\frac{-4\sqrt{2}}{3}$

If $y=\frac{-2\sqrt{2}}{3}$: $\quad x=-2\left(\frac{-2\sqrt{2}}{3}\right)=\frac{4\sqrt{2}}{3}$

If $y=\sqrt{2}$: $\quad x=-\sqrt{2}$

If $y=-\sqrt{2}$: $\quad x=\sqrt{2}$

Solutions:

$\left(\frac{-4\sqrt{2}}{3},\frac{2\sqrt{2}}{3}\right),\left(\frac{4\sqrt{2}}{3},\frac{-2\sqrt{2}}{3}\right),\left(-\sqrt{2},\sqrt{2}\right),$
$\left(\sqrt{2},-\sqrt{2}\right)$

76. $\begin{cases} 3x^2+2xy-2y^2=6 \\ xy-2y^2=-4 \end{cases}$

Multiply each side of the first equation by 2 and each side of the second equation by 3 and add to eliminate the constant:

$$6x^2+4xy-4y^2=12$$
$$3xy-6y^2=-12$$
$$\overline{6x^2+7xy-10y^2=0}$$

$(6x-5y)(x+2y)=0 \Rightarrow x=\frac{5}{6}y \text{ or } x=-2y$

Substitute $x=\frac{5}{6}y$ and solve:

$\frac{5}{6}y\cdot y-2y^2=-4$

$-\frac{7}{6}y^2=-4 \rightarrow y^2=\frac{24}{7} \Rightarrow y=\pm\frac{2\sqrt{42}}{7}$

Substitute $x=-2y$ and solve:

$$-2y\cdot y-2y^2=-4$$
$$-4y^2=-4$$
$$y^2=1 \Rightarrow y=\pm 1$$

If $y=\frac{2\sqrt{42}}{7}$: $\quad x=\frac{5}{6}\cdot\frac{2\sqrt{42}}{7}=\frac{5\sqrt{42}}{21}$

If $y=-\frac{2\sqrt{42}}{7}$: $\quad x=\frac{5}{6}\cdot\frac{-2\sqrt{42}}{7}=-\frac{5\sqrt{42}}{21}$

If $y=1$: $\quad x=-2$; If $y=-1$: $\quad x=2$

Solutions:

$\left(\frac{5\sqrt{42}}{21},\frac{2\sqrt{42}}{7}\right),\left(-\frac{5\sqrt{42}}{21},-\frac{2\sqrt{42}}{7}\right),$
$(-2,1),(2,-1)$

77. $\begin{cases} x^2-3x+y^2+y=-2 \\ \dfrac{x^2-x}{y}+y+1=0 \end{cases}$

Multiply each side of the second equation by –y and add the equations to eliminate y:

$$x^2-3x+y^2+y=-2$$
$$-x^2+x-y^2-y=0$$
$$\overline{-2x=-2 \Rightarrow x=1}$$

If $x=1$: $\quad 1^2-3(1)+y^2+y=-2$

$$y^2+y=0$$
$$y(y+1)=0$$
$$y=0 \text{ or } y=-1$$

Note that $y\neq 0$ because that would cause division by zero in the original system.

Solution: $(1,-1)$

78. Multiply each side of the second equation by –x and add the equations to eliminate x:

$$\begin{cases} x^2+x+y^2=y+2 & \Rightarrow \quad x^2+x+y^2=y+2 \\ x+1=\dfrac{2-y}{x} & \Rightarrow \quad -x^2-x \quad\; =y-2 \end{cases}$$

$$y^2=2y$$
$$y^2-2y=0$$
$$y(y-2)=0$$
$$y=0 \text{ or } y=2$$

If $y=0$:

$x^2+x+0^2=0+2 \rightarrow x^2+x-2=0$

$\rightarrow (x-1)(x+2)=0 \rightarrow x=1$ or $x=-2$

If $y=2$:

$x^2+x+2^2=2+2 \rightarrow x^2+x=0$

$\rightarrow x(x+1)=0 \rightarrow x=0$ or $x=-1$

Note that $x \neq 0$ because that would cause division by zero in the original system.

Solutions: (1, 0), (–2, 0), (–1, 2)

79. a. $3x+4y \le 12$

Graph the line $3x+4y=12$. Use a solid line since the inequality uses $\le$. Choose a test point not on the line, such as (0, 0). Since $3(0)+4(0) \le 12$ is true, shade the side of the line containing (0, 0).

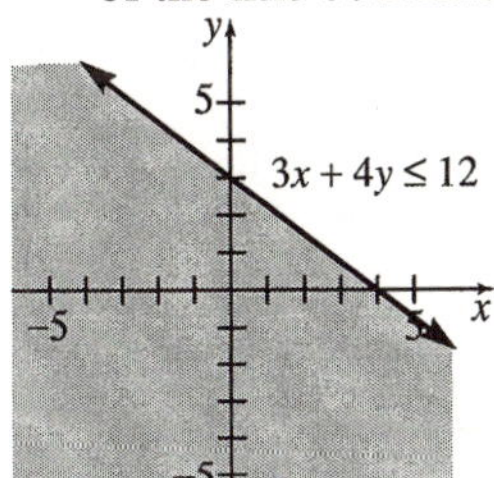

b. Solve the inequality for y:

$$3x+4y \le 12$$
$$4y \le -3x+12$$
$$y \le -\frac{3}{4}x+3$$

Set the calculator viewing WINDOW as shown below. Graph the linear inequality $y \le -\frac{3}{4}x+3$ by entering $Y_1=-\frac{3}{4}x+3$ and adjusting the setting to the left of Y_1 as shown below:

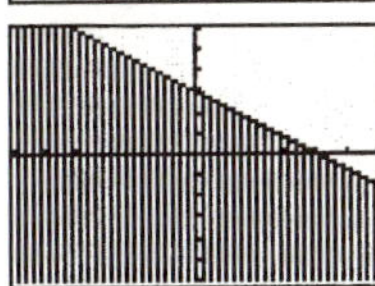

80. a. $2x-3y \ge 6$

Graph the line $2x-3y=6$. Use a solid line since the inequality uses $\ge$. Choose a test point not on the line, such as (0, 0). Since $2(0)-3(0) \ge 6$ is false, shade the side of the line opposite (0, 0).

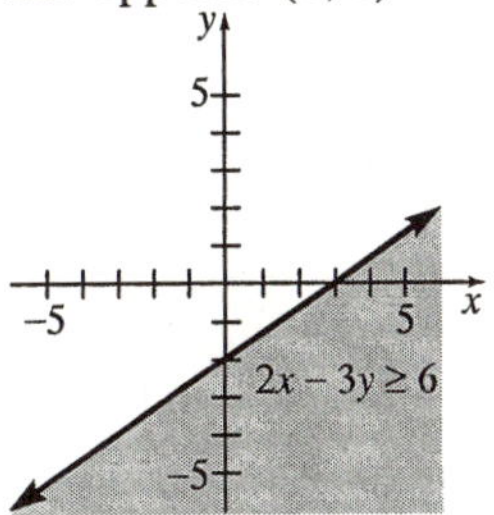

b. Solve the inequality for y:

$$2x-3y \ge 6$$
$$-3y \ge -2x+6$$
$$y \le \frac{2}{3}x-2$$

Set the calculator viewing WINDOW as shown below. Graph the linear inequality $y \le \frac{2}{3}x-2$ by entering $Y_1=-\frac{3}{4}x+3$ and adjusting the setting to the left of Y_1 as shown below:

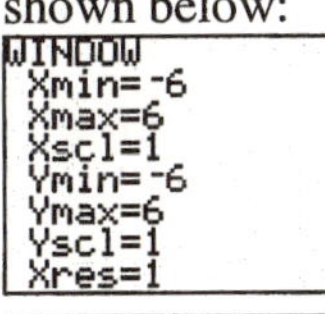

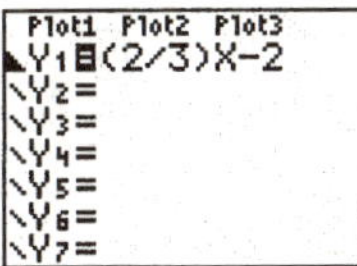

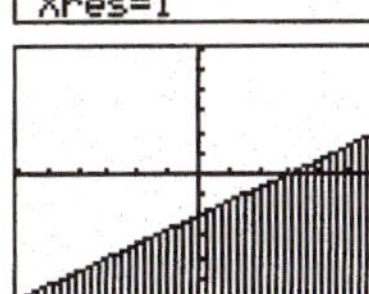

81. a. $y \le x^2$

Graph the parabola $y = x^2$. Use a solid curve since the inequality uses $\le$. Choose a test point not on the parabola, such as (0, 1). Since $0 \le 1^2$ is false, shade the opposite side of the parabola from (0, 1).

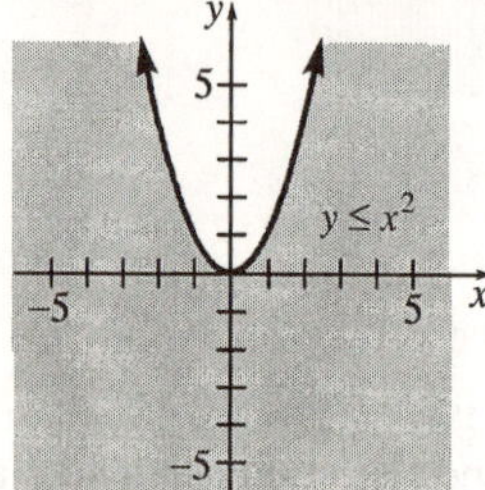

b. Set the calculator viewing WINDOW as shown below. Enter $Y_1 = x^2$ and adjust the setting to the left of Y_1 as shown below:

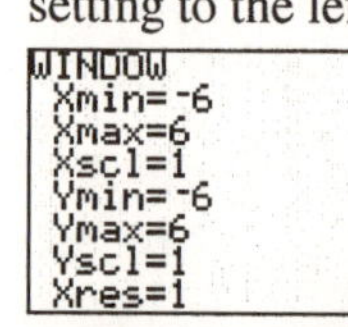

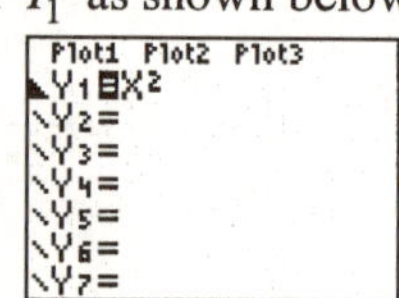

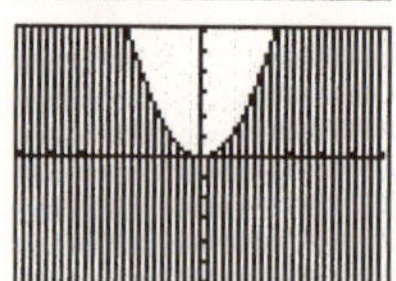

82. a. $x \ge y^2$

Graph the circle $x = y^2$. Use a solid curve since the inequality uses $\ge$. Choose a test point not on the parabola, such as (1, 0). Since $1 \ge 0^2$ is true, shade the same side of the parabola as (1, 0).

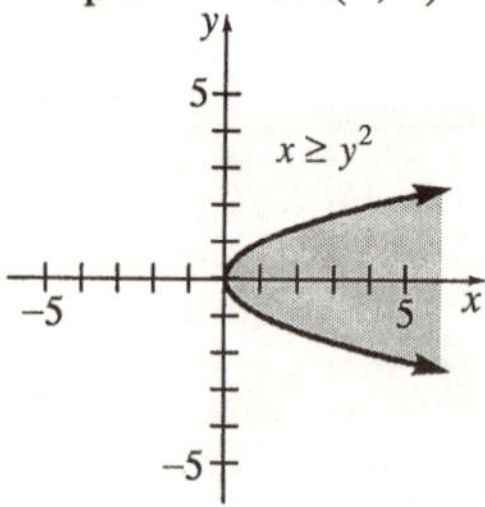

b. Solving $x = y^2$ for *y*, we obtain $y = \pm\sqrt{x}$. We need to shade above $y = -\sqrt{x}$ and below $y = \sqrt{x}$. To do so, first set the calculator viewing WINDOW as shown below. The SHADE command shown will complete the graph: Shade $\left(-\sqrt{x}, \sqrt{x}, 0, 6\right)$

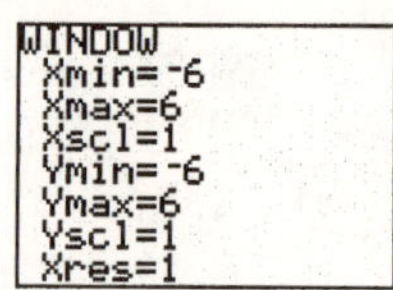

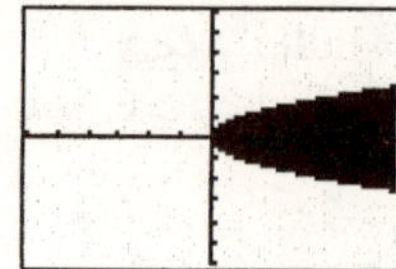

83. $\begin{cases} -2x + y \le 2 \\ x + y \ge 2 \end{cases}$

Graph the line $-2x + y = 2$. Use a solid line since the inequality uses $\le$. Choose a test point not on the line, such as (0, 0). Since $-2(0) + 0 \le 2$ is true, shade the side of the line containing (0, 0). Graph the line $x + y = 2$. Use a solid line since the inequality uses $\ge$. Choose a test point not on the line, such as (0, 0). Since $0 + 0 \ge 2$ is false, shade the opposite side of the line from (0, 0). The overlapping region is the solution.

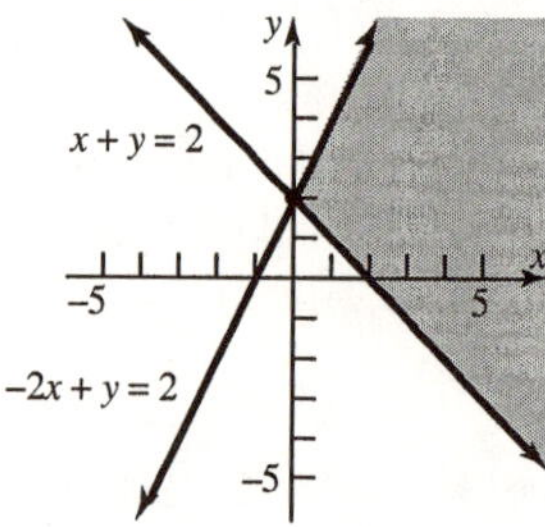

The graph is unbounded. Find the vertices:
To find the intersection of $x + y = 2$ and $-2x + y = 2$, solve the system:

$$\begin{cases} x + y = 2 \\ -2x + y = 2 \end{cases}$$

Solve the first equation for *x*: $x = 2 - y$.
Substitute and solve:

$$\begin{aligned} -2(2 - y) + y &= 2 \\ -4 + 2y + y &= 2 \\ 3y &= 6 \\ y &= 2 \end{aligned}$$

$x = 2 - 2 = 0$

The point of intersection is (0, 2).
The corner point is (0, 2).

84. $\begin{cases} x-2y\le 6 \\ 2x+\ \ y\ge 2 \end{cases}$

Graph the line $x-2y=6$. Use a solid line since the inequality uses $\le$. Choose a test point not on the line, such as (0, 0). Since $0-2(0)\le 6$ is true, shade the side of the line containing (0, 0). Graph the line $2x+y=2$. Use a solid line since the inequality uses $\ge$. Choose a test point not on the line, such as (0, 0). Since $2(0)+0\ge 2$ is false, shade the opposite side of the line from (0, 0). The overlapping region is the solution.

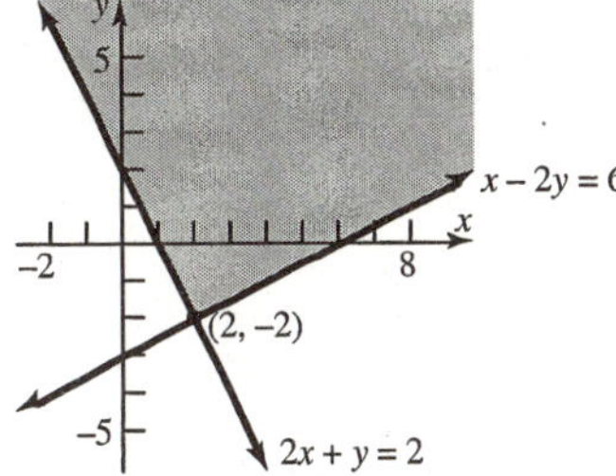

The graph is unbounded. Find the vertices: To find the intersection of $x-2y=6$ and $2x+y=2$, solve the system:

$$\begin{cases} x-2y=6 \\ 2x+y=2 \end{cases}$$

Solve the first equation for x: $x=2y+6$.

Substitute and solve:

$$2(2y+6)+y=2$$
$$4y+12+y=2$$
$$5y=-10$$
$$y=-2$$
$$x=2(-2)+6=2$$

The point of intersection is $(2,-2)$.

The corner point is $(2,-2)$.

85. $\begin{cases} x\ge 0 \\ y\ge 0 \\ x+\ \ y\le 4 \\ 2x+3y\le 6 \end{cases}$

Graph $x\ge 0$; $y\ge 0$. Shaded region is the first quadrant. Graph the line $x+y=4$. Use a solid line since the inequality uses $\le$. Choose a test point not on the line, such as (0, 0). Since $0+0\le 4$ is true, shade the side of the line containing (0, 0). Graph the line $2x+3y=6$. Use a solid line since the inequality uses $\le$. Choose a test point not on the line, such as (0, 0). Since $2(0)+3(0)\le 6$ is true, shade the side of the line containing (0, 0).

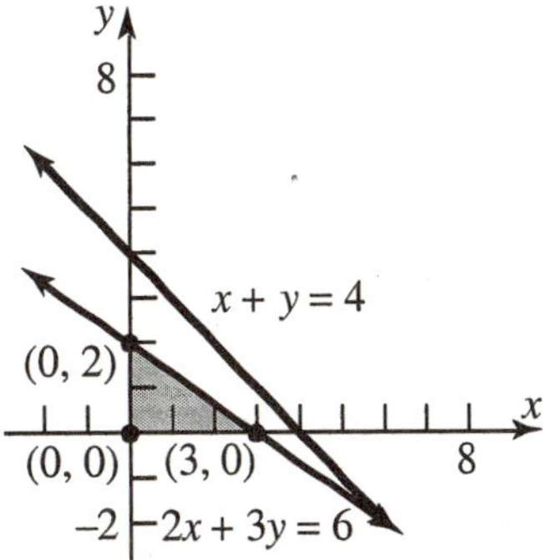

The overlapping region is the solution. The graph is bounded. Find the vertices: The x-axis and y-axis intersect at (0, 0). The intersection of $2x+3y=6$ and the y-axis is (0, 2). The intersection of $2x+3y=6$ and the x-axis is (3, 0). The three corner points are (0, 0), (0, 2), and (3, 0).

86. $\begin{cases} x\ge 0 \\ y\ge 0 \\ 3x+y\ge 6 \\ 2x+y\ge 2 \end{cases}$

Graph $x\ge 0$; $y\ge 0$. Shaded region is the first quadrant. Graph the line $3x+y=6$. Use a solid line since the inequality uses $\ge$. Choose a test point not on the line, such as (0, 0). Since $3(0)+0\ge 6$ is false, shade the opposite side of the line from (0, 0). Graph the line $2x+y=2$. Use a solid line since the inequality uses $\ge$. Choose a test point not on the line, such as (0, 0). Since $2(0)+0\ge 2$ is false, shade the opposite side of the line from (0, 0).

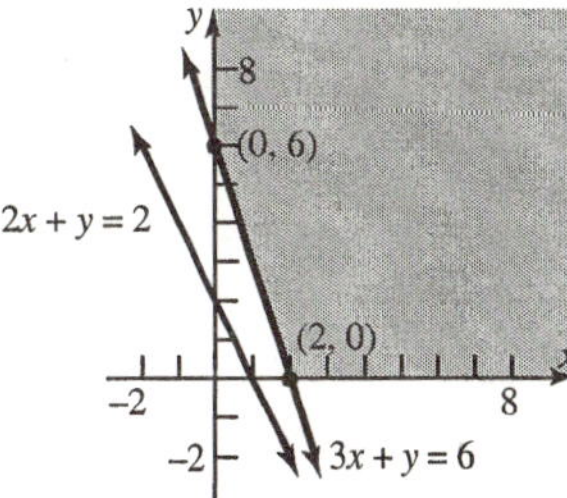

The overlapping region is the solution. The graph is unbounded. Find the vertices:
The intersection of $3x+y=6$ and the y-axis is (0, 6). The intersection of $3x+y=6$ and the x-axis is (2, 0). The two corner points are (0, 6), and (2, 0).

87. $\begin{cases} x \geq 0 \\ y \geq 0 \\ 2x + y \leq 8 \\ x + 2y \geq 2 \end{cases}$

Graph $x \geq 0$; $y \geq 0$. Shaded region is the first quadrant. Graph the line $2x + y = 8$. Use a solid line since the inequality uses $\leq$. Choose a test point not on the line, such as (0, 0). Since $2(0) + 0 \leq 8$ is true, shade the side of the line containing (0, 0). Graph the line $x + 2y = 2$. Use a solid line since the inequality uses $\geq$. Choose a test point not on the line, such as (0, 0). Since $0 + 2(0) \geq 2$ is false, shade the opposite side of the line from (0, 0).

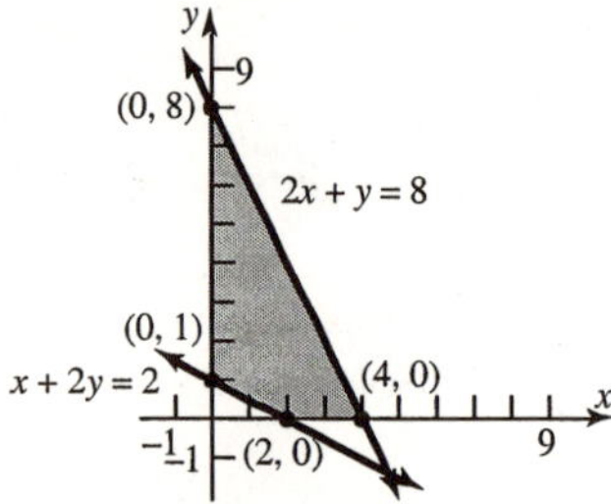

The overlapping region is the solution. The graph is bounded. Find the vertices: The intersection of $x + 2y = 2$ and the y-axis is (0, 1). The intersection of $x + 2y = 2$ and the x-axis is (2, 0). The intersection of $2x + y = 8$ and the y-axis is (0, 8). The intersection of $2x + y = 8$ and the x-axis is (4, 0). The four corner points are (0, 1), (0, 8), (2, 0), and (4, 0).

88. $\begin{cases} x \geq 0 \\ y \geq 0 \\ 3x + y \leq 9 \\ 2x + 3y \geq 6 \end{cases}$

Graph $x \geq 0$; $y \geq 0$. Shaded region is the first quadrant. Graph the line $3x + y = 9$. Use a solid line since the inequality uses $\leq$. Choose a test point not on the line, such as (0, 0). Since $3(0) + 0 \leq 9$ is true, shade the side of the line containing (0, 0).

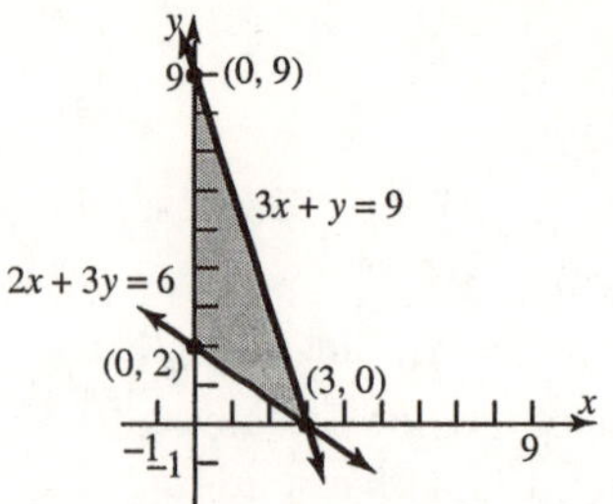

Graph the line $2x + 3y = 6$. Use a solid line since the inequality uses $\geq$. Choose a test point not on the line, such as (0, 0). Since $2(0) + 3(0) \geq 6$ is false, shade the opposite side of the line from (0, 0). The overlapping region is the solution. The graph is bounded. Find the vertices: The intersection of $2x + 3y = 6$ and the y-axis is (0, 2). The intersection of $2x + 3y = 6$ and the x-axis is (3, 0). The intersection of $3x + y = 9$ and the y-axis is (0, 9). The intersection of $3x + y = 9$ and the x-axis is (3, 0). The three corner points are (0, 2), (0, 9), and (3, 0).

89. Graph the system of inequalities:

$\begin{cases} x^2 + y^2 \leq 16 \\ x + y \geq 2 \end{cases}$

Graph the circle $x^2 + y^2 = 16$. Use a solid line since the inequality uses $\leq$. Choose a test point not on the circle, such as (0, 0). Since $0^2 + 0^2 \leq 16$ is true, shade the side of the circle containing (0, 0).

Graph the line $x + y = 2$. Use a solid line since the inequality uses $\geq$. Choose a test point not on the line, such as (0, 0). Since $0 + 0 \geq 2$ is false, shade the opposite side of the line from (0, 0). The overlapping region is the solution.

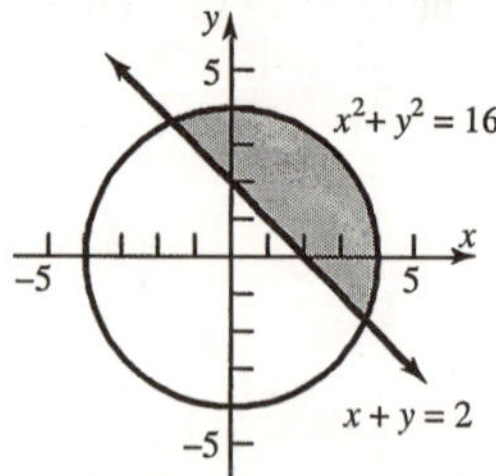

90. Graph the system of inequalities:
$$\begin{cases} y^2 \le x-1 \\ x-y \le 3 \end{cases}$$
Graph the parabola $y^2 = x-1$. Use a solid line since the inequality uses $\le$. Choose a test point not on the parabola, such as (0, 0). Since $0^2 \le 0-1$ is false, shade the opposite side of the parabola from (0, 0).
Graph the line $x-y=3$. Use a solid line since the inequality uses $\le$. Choose a test point not on the line, such as (0, 0). Since $0-0 \le 3$ is true, shade the same side of the line as (0, 0). The overlapping region is the solution.

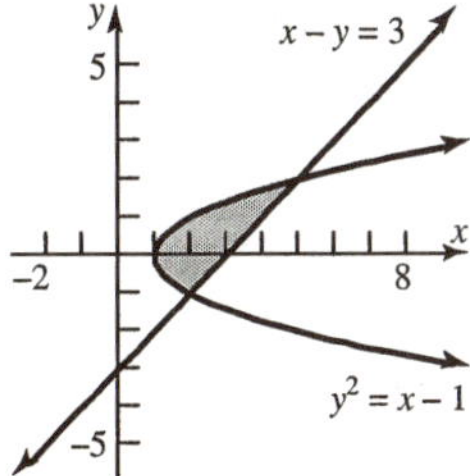

91. Graph the system of inequalities:
$$\begin{cases} y \le x^2 \\ xy \le 4 \end{cases}$$
Graph the parabola $y = x^2$. Use a solid line since the inequality uses $\le$. Choose a test point not on the parabola, such as (1, 2). Since $2 \le 1^2$ is false, shade the opposite side of the parabola from (1, 2).
Graph the hyperbola $xy = 4$. Use a solid line since the inequality uses $\le$. Choose a test point not on the hyperbola, such as (1, 2). Since $1 \cdot 2 \le 4$ is true, shade the same side of the hyperbola as (1, 2). The overlapping region is the solution.

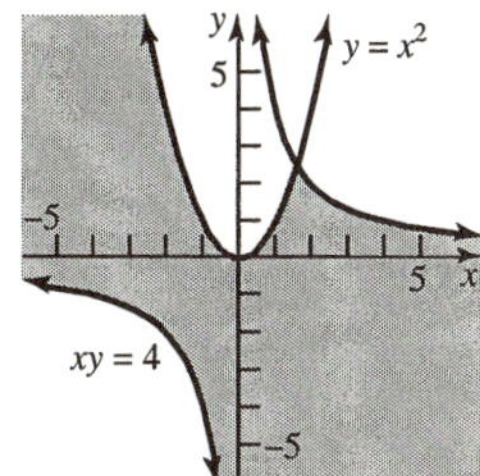

92. Graph the system of inequalities:
$$\begin{cases} x^2 + y^2 \ge 1 \\ x^2 + y^2 \le 4 \end{cases}$$
Graph the circle $x^2 + y^2 = 1$. Use a solid line since the inequality uses $\ge$. Choose a test point not on the circle, such as (0, 0). Since $0^2 + 0^2 \ge 1$ is false, shade the opposite side of the circle from (0, 0).
Graph the circle $x^2 + y^2 = 4$. Use a solid line since the inequality uses $\le$. Choose a test point not on the circle, such as (0, 0). Since $0^2 + 0^2 \le 4$ is true, shade the same side of the circle as (0, 0). The overlapping region is the solution.

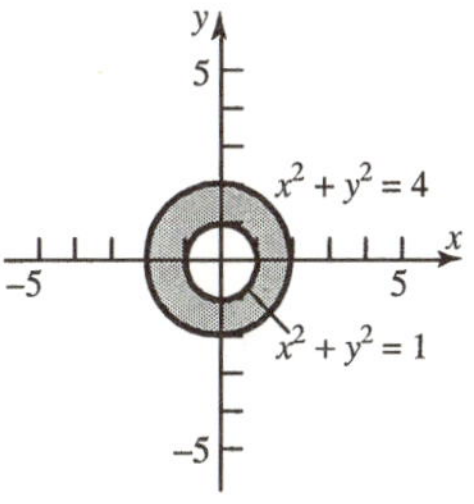

93. Maximize $z = 3x + 4y$ subject to $x \ge 0$, $y \ge 0$, $3x + 2y \ge 6$, $x + y \le 8$. Graph the constraints.

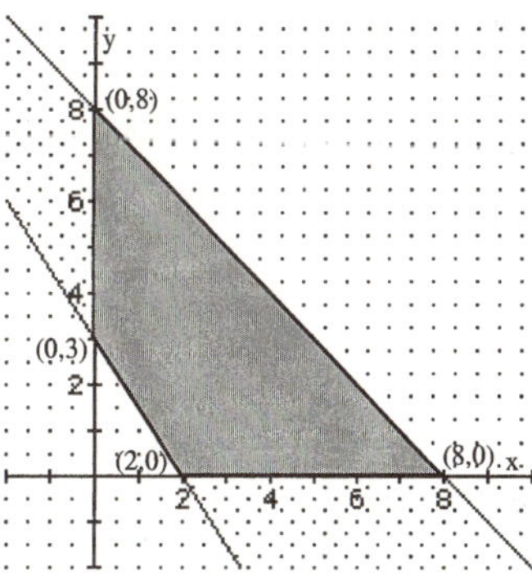

The corner points are (0, 3), (2, 0), (0, 8), (8, 0).
Evaluate the objective function:

Vertex	Value of $z = 3x + 4y$
(0, 3)	$z = 3(0) + 4(3) = 12$
(0, 8)	$z = 3(0) + 4(8) = 32$
(2, 0)	$z = 3(2) + 4(0) = 6$
(8, 0)	$z = 3(8) + 4(0) = 24$

The maximum value is 32 at (0, 8).

94. Maximize $z = 2x + 4y$ subject to $x \ge 0$, $y \ge 0$, $x + y \le 6$, $x \ge 2$. Graph the constraints.

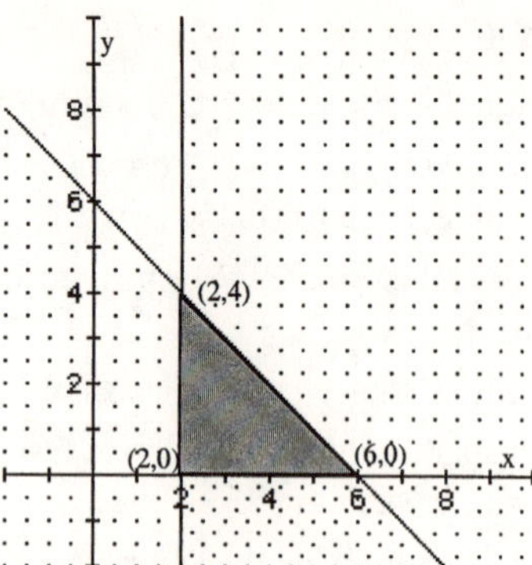

The corner points are (2, 4), (2, 0), (6, 0).
Evaluate the objective function:

Vertex	Value of $z = 2x + 4y$
(2, 4)	$z = 2(2) + 4(4) = 20$
(2, 0)	$z = 2(2) + 4(0) = 4$
(6, 0)	$z = 2(6) + 4(0) = 12$

The maximum value is 20 at (2, 4).

95. Minimize $z = 3x + 5y$ subject to $x \ge 0$, $y \ge 0$, $x + y \ge 1$, $3x + 2y \le 12$, $x + 3y \le 12$.
Graph the constraints.

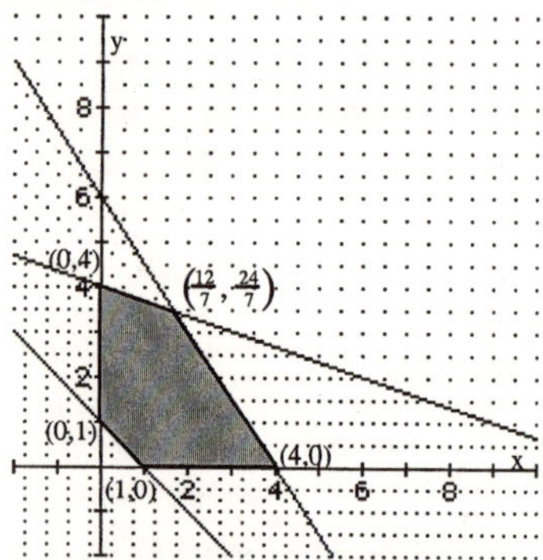

To find the intersection of $3x + 2y = 12$ and $x + 3y = 12$, solve the system:

$$\begin{cases} 3x + 2y = 12 \\ x + 3y = 12 \end{cases}$$

Solve the second equation for x: $x = 12 - 3y$
Substitute and solve:

$$3(12 - 3y) + 2y = 12$$
$$36 - 9y + 2y = 12$$
$$-7y = -24$$
$$y = \frac{24}{7}$$

$$x = 12 - 3\left(\frac{24}{7}\right) = 12 - \frac{72}{7} = \frac{12}{7}$$

The point of intersection is $\left(\frac{12}{7}, \frac{24}{7}\right)$.

The corner points are (0, 1), (1, 0), (0, 4), (4, 0), $\left(\frac{12}{7}, \frac{24}{7}\right)$.

Evaluate the objective function:

Vertex	Value of $z = 3x + 5y$
(0, 1)	$z = 3(0) + 5(1) = 5$
(0, 4)	$z = 3(0) + 5(4) = 20$
(1, 0)	$z = 3(1) + 5(0) = 3$
(4, 0)	$z = 3(4) + 5(0) = 12$
$\left(\frac{12}{7}, \frac{24}{7}\right)$	$z = 3\left(\frac{12}{7}\right) + 5\left(\frac{24}{7}\right) = \frac{156}{7}$

The minimum value is 3 at (1, 0).

96. Minimize $z = 3x + y$ subject to $x \ge 0$, $y \ge 0$, $x \le 8$, $y \le 6$, $2x + y \ge 4$. Graph the constraints.

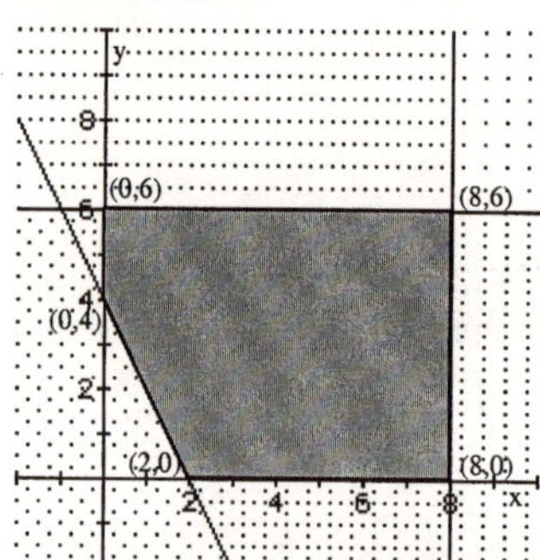

The corner points are (0, 4), (2, 0), (0, 6), (8, 0), (8, 6). Evaluate the objective function:

Vertex	Value of $z = 3x + y$
(0, 6)	$z = 3(0) + 6 = 6$
(0, 4)	$z = 3(0) + 4 = 4$
(2, 0)	$z = 3(2) + 0 = 6$
(8, 0)	$z = 3(8) + 0 = 24$
(8, 6)	$z = 3(8) + 6 = 30$

The minimum value is 4 at (0, 4).

97. $\begin{cases} 2x + 5y = 5 \\ 4x + 10y = A \end{cases}$

Multiply each side of the first equation by –2 and eliminate x:

$$\begin{cases} -4x - 10y = -10 \\ 4x + 10y = A \end{cases}$$

$$0 = A - 10$$

If there are to be infinitely many solutions, the result of elimination should be 0 = 0. Therefore, $A - 10 = 0$ or $A = 10$.

98. $\begin{cases} 2x + 5y = 5 \\ 4x + 10y = A \end{cases}$

Multiply each side of the first equation by –2 and eliminate x:

$$\begin{cases} -4x - 10y = -10 \\ 4x + 10y = A \end{cases}$$

$$0 = A - 10$$

If the system is to be inconsistent, the result of elimination should be 0 = any number except 0. Therefore, $A - 10 \neq 0$ or $A \neq 10$.

99. $y = ax^2 + bx + c$

At (0, 1) the equation becomes:

$1 = a(0)^2 + b(0) + c$

$c = 1$

At (1, 0) the equation becomes:

$0 = a(1)^2 + b(1) + c$

$0 = a + b + c$

$a + b + c = 0$

At (–2, 1) the equation becomes:

$1 = a(-2)^2 + b(-2) + c$

$1 = 4a - 2b + c$

$4a - 2b + c = 1$

The system of equations is:

$$\begin{cases} a + b + c = 0 \\ 4a - 2b + c = 1 \\ c = 1 \end{cases}$$

Substitute $c = 1$ into the first and second equations and simplify:

$a + b + 1 = 0 \qquad 4a - 2b + 1 = 1$

$a + b = -1 \qquad 4a - 2b = 0$

$a = -b - 1$

Solve the first equation for a, substitute into the second equation and solve:

$4(-b - 1) - 2b = 0$

$-4b - 4 - 2b = 0$

$-6b = 4$

$b = -\frac{2}{3}$

$a = \frac{2}{3} - 1 = -\frac{1}{3}$

The quadratic function is $y = -\frac{1}{3}x^2 - \frac{2}{3}x + 1$.

100. $x^2 + y^2 + Dx + Ey + F = 0$

At (0, 1) the equation becomes:

$0^2 + 1^2 + D(0) + E(1) + F = 0$

$E + F = -1$

At (1, 0) the equation becomes:

$1^2 + 0^2 + D(1) + E(0) + F = 0$

$D + F = -1$

At (–2, 1) the equation becomes:

$(-2)^2 + 1^2 + D(-2) + E(1) + F = 0$

$-2D + E + F = -5$

The system of equations is:

$$\begin{cases} E + F = -1 \\ D + F = -1 \\ -2D + E + F = -5 \end{cases}$$

Substitute $E + F = -1$ into the third equation and solve for D:

$-2D + (-1) = -5$

$-2D = -4$

$D = 2$

Substitute and solve:

$2 + F = -1 \qquad E + (-3) = -1$

$F = -3 \qquad E = 2$

The equation of the circle is

$x^2 + y^2 + 2x + 2y - 3 = 0$.

101. Let x = the number of pounds of coffee that costs \$3.00 per pound, and let y = the number of pounds of coffee that costs \$6.00 per pound. Then $x+y=100$ represents the total amount of coffee in the blend. The value of the blend will be represented by the equation: $3x+6y=3.90(100)$. Solve the system of equations:

$$\begin{cases} x+y=100 \\ 3x+6y=390 \end{cases}$$

Solve the first equation for y: $y=100-x$.
Solve by substitution:

$$3x+6(100-x)=390$$
$$3x+600-6x=390$$
$$-3x=-210$$
$$x=70$$
$$y=100-70=30$$

The blend is made up of 70 pounds of the \$3 per pound coffee and 30 pounds of the \$6-per pound coffee.

102. Let x = the number of acres of corn, and let y = the number of acres of soybeans. Then $x+y=1000$ represents the total acreage on the farm. The total cost will be represented by the equation: $65x+45y=54,325$. Solve the system of equations:

$$\begin{cases} x+y=1000 \\ 65x+45y=54,325 \end{cases}$$

Solve the first equation for y: $y=1000-x$
Solve by substitution:

$$65x+45(1000-x)=54,325$$
$$65x+45,000-45x=54,325$$
$$20x=9325$$
$$x=466.25$$
$$y=1000-466.25=533.75$$

Corn should be planted on 466.25 acres and soybeans should be planted on 533.75 acres.

103. Let x = the number of small boxes, let y = the number of medium boxes, and let z = the number of large boxes.
Oatmeal raisin equation: $x+2y+2z=15$
Chocolate chip equation: $x+y+2z=10$
Shortbread equation: $y+3z=11$

$$\begin{cases} x+2y+2z=15 \\ x+y+2z=10 \\ y+3z=11 \end{cases}$$

Multiply each side of the second equation by -1 and add to the first equation to eliminate x:

$$\begin{cases} x+2y+2z=15 \\ -x-y-2z=-10 \end{cases}$$
$$y+3z=11$$
$$y=5$$

Substituting and solving for the other variables:

$$5+3z=11 \qquad x+5+2(2)=10$$
$$3z=6 \qquad x+9=10$$
$$z=2 \qquad x=1$$

Thus, 1 small box, 5 medium boxes, and 2 large boxes of cookies should be purchased.

104. a. Let x = the number of lower-priced packages, and let y = the number of quality packages.

Peanut inequality: $8x+6y \le 120(16)$
$4x+3y \le 960$

Cashew inequality: $4x+6y \le 72(16)$
$2x+3y \le 576$

The system of inequalities is:

$$\begin{cases} x \ge 0 \\ y \ge 0 \\ 4x+3y \le 960 \\ 2x+3y \le 576 \end{cases}$$

b. Graphing:

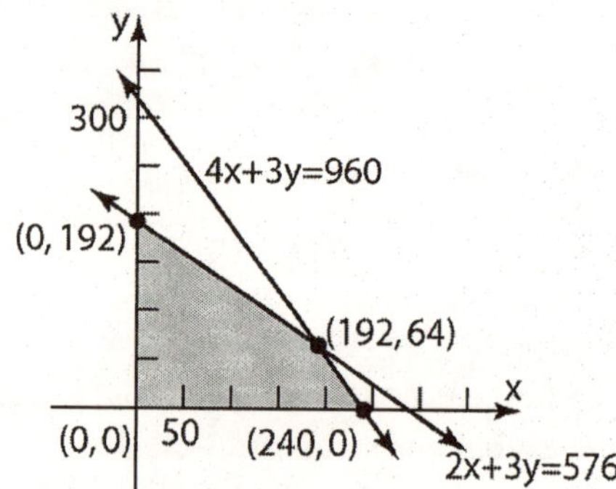

To find the intersection of $2x+3y=576$ and $4x+3y=960$, solve the system:

$$\begin{cases} 4x+3y=960 \\ 2x+3y=576 \end{cases}$$

Subtract the second equation from the first:

$$\begin{aligned} 4x+3y&=960 \\ -2x-3y&=576 \\ \hline 2x&=384 \\ x&=192 \end{aligned}$$

Substitute and solve:

$$\begin{aligned} 2(192)+3y&=576 \\ 3y&=192 \\ y&=64 \end{aligned}$$

The corner points are (0, 0), (0, 192), (240, 0), and (192, 64).

105. Let x = the speed of the boat in still water, and let y = the speed of the river current. The distance from Chiritza to the Flotel Orellana is 100 kilometers.

	Rate	Time	Distance
trip downstream	$x+y$	$5/2$	100
trip downstream	$x-y$	3	100

The system of equations is:

$$\begin{cases} \frac{5}{2}(x+y)=100 \\ 3(x-y)=100 \end{cases}$$

Multiply both sides of the first equation by 6, multiply both sides of the second equation by 5, and add the results.

$$\begin{aligned} 15x+15y&=600 \\ 15x-15y&=500 \\ \hline 30x&=1100 \\ x&=\frac{1100}{30}=\frac{110}{3} \end{aligned}$$

$$\begin{aligned} 3\left(\frac{110}{3}\right)-3y&=100 \\ 110-3y&=100 \\ 10&=3y \\ y&=\frac{10}{3} \end{aligned}$$

The speed of the boat is $110/3 \approx 36.67$ km/hr ; the speed of the current is $10/3 \approx 3.33$ km/hr .

106. Let x = the speed of the jet stream, and let d = the distance from Chicago to Ft. Lauderdale. The jet stream flows from Chicago to Ft. Lauderdale because the time is shorter in that direction.

	Rate	Time	Distance
Chicago to Ft. Lauderdale	$475+x$	$5/2$	d
Ft. Lauderdale to Chicago	$475-x$	$17/6$	d

The system of equations is:

$$\begin{cases} (475+y)(5/2)=d \\ (475-y)(17/6)=d \end{cases}$$

Simplifying the system, we obtain:

$$\begin{cases} 2d-5x=2375 \\ 6d+17x=8075 \end{cases}$$

Multiply the first equation by -3 and add the result to the second equation:

$$\begin{aligned} -6d+15x&=-7125 \\ 6d+17x&=8075 \\ \hline 32x&=950 \\ x&=\frac{950}{32}=\frac{475}{16} \end{aligned}$$

The speed of the jet stream is approximately $475/16 \approx 29.69$ miles per hour.

107. Let x = the number of hours for Bruce to do the job alone, let y = the number of hours for Bryce to do the job alone, and let z = the number of hours for Marty to do the job alone.

Then $1/x$ represents the fraction of the job that Bruce does in one hour.

$1/y$ represents the fraction of the job that Bryce does in one hour.

$1/z$ represents the fraction of the job that Marty does in one hour.

The equation representing Bruce and Bryce working together is:

$$\frac{1}{x}+\frac{1}{y}=\frac{1}{(4/3)}=\frac{3}{4}=0.75$$

The equation representing Bryce and Marty working together is:

$$\frac{1}{y}+\frac{1}{z}=\frac{1}{(8/5)}=\frac{5}{6}=0.625$$

The equation representing Bruce and Marty working together is:

$\frac{1}{x}+\frac{1}{z}=\frac{1}{(8/3)}=\frac{3}{8}=0.375$

Solve the system of equations:

$$\begin{cases} x^{-1}+y^{-1}=0.75 \\ y^{-1}+z^{-1}=0.625 \\ x^{-1}+z^{-1}=0.375 \end{cases}$$

Let $u=x^{-1},\ v=y^{-1},\ w=z^{-1}$

$$\begin{cases} u+v=0.75 \\ v+w=0.625 \\ u+w=0.375 \end{cases}$$

Solve the first equation for u: $u=0.75-v$.
Solve the second equation for w: $w=0.625-v$.
Substitute into the third equation and solve:

$$(0.75-v)+(0.625-v)=0.375$$
$$-2v=-1$$
$$v=0.5$$

$u=0.75-0.5=0.25$
$w=0.625-0.5=0.125$

Solve for x, y, and z : $x=4,\ \ y=2,\ \ z=8$ (reciprocals)

Bruce can do the job in 4 hours, Bryce in 2 hours, and Marty in 8 hours.

108. Let x = the number of dancing girls produced, and let y = the number of mermaids produced. The total profit is: $P=25x+30y$. Profit is to be maximized, so this is the objective function. The constraints are:

$x\ge 0,\ \ y\ge 0$	A non-negative number of figurines must be produced.
$3x+3y\le 90$	90 hours are available for molding.
$6x+4y\le 120$	120 hours are available for painting.
$2x+3y\le 60$	60 hours are available for glazing.

Graph the constraints.

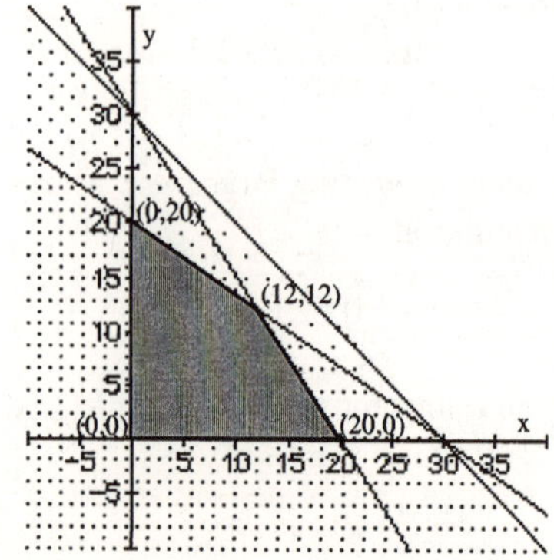

To find the intersection of $6x+4y=120$ and $2x+3y=60$, solve the system:

$$\begin{cases} 6x+4y=120 \\ 2x+3y=60 \end{cases}$$

Multiply the second equation by 3, and subtract from the first equation:

$$6x+4y=120$$
$$-6x-9y=-180$$
$$-5y=-60$$
$$y=12$$

Substitute and solve:

$$2x+3(12)=60$$
$$2x=24$$
$$x=12$$

The point of intersection is (12, 12).
The corner points are (0, 0), (0, 20), (20, 0), (12, 12).
Evaluate the objective function:

Vertex	Value of $P=25x+30y$
(0, 0)	$P=25(0)+30(0)=0$
(0, 20)	$P=25(0)+30(20)=600$
(20, 0)	$P=25(20)+30(0)=500$
(12, 12)	$P=25(12)+30(12)=660$

The maximum profit is $660, when 12 dancing girl and 12 mermaid figurines are produced each day. To determine the excess, evaluate each constraint at $x=12$ and $y=12$:

Molding: $3x+3y=3(12)+3(12)=36+36=72$

Painting: $6x+4y=6(12)+4(12)=72+48=120$

Glazing: $2x+3y=2(12)+3(12)=24+36=60$

Painting and glazing are at their capacity. Molding has 18 more hours available, since only 72 of the 90 hours are used.

109. Let x = the number of gasoline engines produced each week, and let y = the number of diesel engines produced each week. The total cost is: $C=450x+550y$. Cost is to be minimized; thus, this is the objective function. The constraints are:

$20\le x\le 60$	number of gasoline engines needed and capacity each week.
$15\le y\le 40$	number of diesel engines needed and capacity each week.
$x+y\ge 50$	number of engines produced to prevent layoffs.

Graph the constraints.

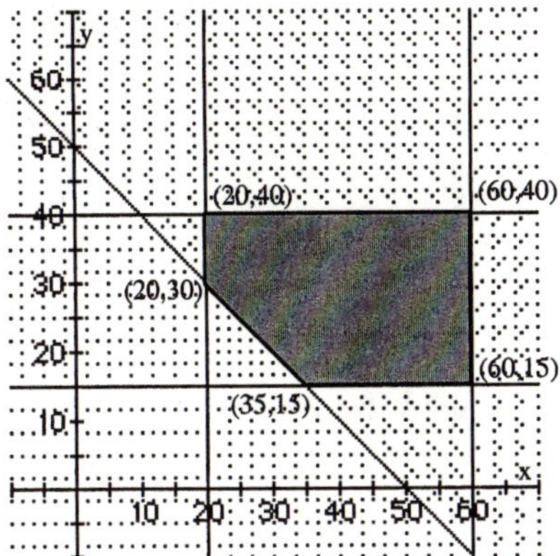

The corner points are (20, 30), (20, 40), (35, 15), (60, 15), (60, 40)

Evaluate the objective function:

Vertex	Value of $C = 450x + 550y$
(20, 30)	$C = 450(20) + 550(30) = 25{,}500$
(20, 40)	$C = 450(35) + 550(40) = 31{,}000$
(35, 15)	$C = 450(35) + 550(15) = 24{,}000$
(60, 15)	$C = 450(60) + 550(15) = 35{,}250$
(60, 40)	$C = 450(60) + 550(40) = 49{,}000$

The minimum cost is \$24,000, when 35 gasoline engines and 15 diesel engines are produced. The excess capacity is 15 gasoline engines, since only 20 gasoline engines had to be delivered.

110. Answers will vary.

Chapter 10 Test

1. $\begin{cases} -2x + y = -7 \\ 4x + 3y = 9 \end{cases}$

Substitution:

We solve the first equation for y, obtaining $y = 2x - 7$

Next we substitute this result for y in the second equation and solve for x.

$$4x + 3y = 9$$
$$4x + 3(2x - 7) = 9$$
$$4x + 6x - 21 = 9$$
$$10x = 30$$
$$x = \frac{30}{10} = 3$$

We can now obtain the value for y by letting $x = 3$ in our substitution for y.

$$y = 2x - 7$$
$$y = 2(3) - 7$$
$$= 6 - 7$$
$$= -1$$

The solution of the system is $x = 3$, $y = -1$.

Elimination:

Multiply each side of the first equation by 2 so that the coefficients of x in the two equations are negatives of each other. The result is the equivalent system

$$\begin{cases} -4x + 2y = -14 \\ 4x + 3y = 9 \end{cases}$$

We can replace the second equation of this system by the sum of the two equations. The result is the equivalent system

$$\begin{cases} -4x + 2y = -14 \\ 5y = -5 \end{cases}$$

Now we solve the second equation for y.

$$5y = -5$$
$$y = \frac{-5}{5} = -1$$

We back-substitute this value for y into the original first equation and solve for x.

$$-2x + y = -7$$
$$-2x + (-1) = -7$$
$$-2x = -6$$
$$x = \frac{-6}{-2} = 3$$

The solution of the system is $x = 3$, $y = -1$.

Check:

To check our solution graphically, we need to solve each equation for y by putting them in the form $y = mx + b$.

$$-2x + y = -7 \qquad\qquad 4x + 3y = 9$$
$$y = 2x - 7 \qquad\qquad 3y = -4x + 9$$
$$y = -\frac{4}{3}x + 3$$

Since the slopes of the two lines are not the same, we know that the two lines will intersect. We enter both equations into our calculator and find the intersection point.

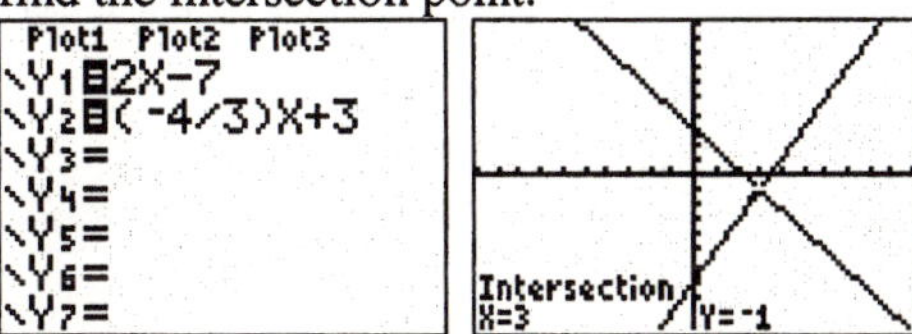

The solution of the system is $x = 3$, $y = -1$.

2. $\begin{cases} \frac{1}{3}x-2y=1 \\ 5x-30y=18 \end{cases}$

We choose to use the method of elimination and multiply the first equation by -15 to obtain the equivalent system

$\begin{cases} -5x+30y=-15 \\ 5x-30y=18 \end{cases}$

We replace the second equation by the sum of the two equations to obtain the equivalent system

$\begin{cases} -5x+30y=-15 \\ 0=3 \end{cases}$

The second equation is a contradiction and has no solution. This means that the system itself has no solution and is therefore inconsistent.

Check:
To check our solution graphically, we first solve each equation for y by putting them in the form $y=mx+b$.

$$\frac{1}{3}x-2y=1 \qquad -2y=-\frac{1}{3}x+1 \qquad y=\frac{1}{6}x-\frac{1}{2}$$

$$5x-30y=18 \qquad -30y=-5x+18 \qquad y=\frac{1}{6}x-\frac{3}{5}$$

We can see that the slopes are the same, but the y-intercepts are different. Therefore, the two lines are parallel and will not intersect. The system has no solution and is therefore inconsistent.

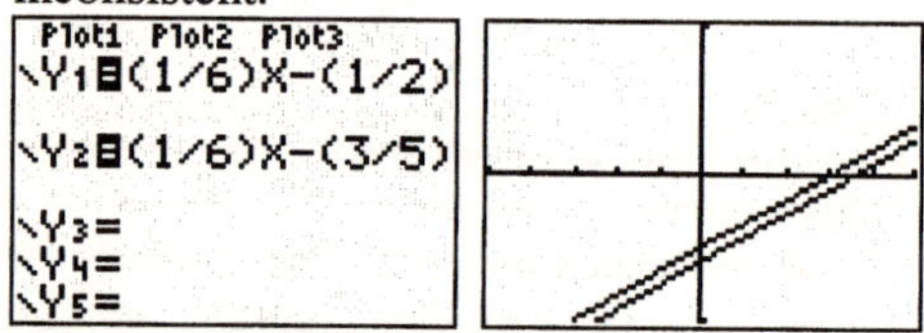

3. $\begin{cases} x-y+2z=5 & (1) \\ 3x+4y-z=-2 & (2) \\ 5x+2y+3z=8 & (3) \end{cases}$

We use the method of elimination and begin by eliminating the variable y from equation (2). Multiply each side of equation (1) by 4 and add the result to equation (2). This result becomes our new equation (2).

$$\begin{array}{ll} x-y+2z=5 & 4x-4y+8z=20 \\ 3x+4y-z=-2 & \underline{3x+4y-z=-2} \\ & 7x \qquad +7z=18 \quad (2) \end{array}$$

We now eliminate the variable y from equation (3) by multiplying each side of equation (1) by 2 and adding the result to equation (3). The result becomes our new equation (3).

$$\begin{array}{ll} x-y+2z=5 & 2x-2y+4z=10 \\ 5x+2y+3z=8 & \underline{5x+2y+3z=8} \\ & 7x \qquad +7z=18 \quad (3) \end{array}$$

Our (equivalent) system now looks like

$\begin{cases} x-y+2z=5 & (1) \\ 7x \qquad +7z=18 & (2) \\ 7x \qquad +7z=18 & (3) \end{cases}$

Treat equations (2) and (3) as a system of two equations containing two variables, and eliminate the x variable by multiplying each side of equation (2) by -1 and adding the result to equation (3). The result becomes our new equation (3).

$$\begin{array}{ll} 7x+7z=18 & -7x-7z=-18 \\ 7x+7z=18 & \underline{7x+7z=18} \\ & 0=0 \quad (3) \end{array}$$

We now have the equivalent system

$\begin{cases} x-y+2z=5 & (1) \\ 7x \qquad +7z=18 & (2) \\ 0=0 & (3) \end{cases}$

This is equivalent to a system of two equations with three variables. Since one of the equations contains three variables and one contains only two variables, the system will be dependent. There are infinitely many solutions.
We solve equation (2) for x and determine that $x=-z+\frac{18}{7}$. Substitute this expression into equation (1) to obtain y in terms of z.

$$x-y+2z=5$$
$$\left(-z+\frac{18}{7}\right)-y+2z=5$$
$$-z+\frac{18}{7}-y+2z=5$$
$$-y+z=\frac{17}{7}$$
$$y=z-\frac{17}{7}$$

We can write the solution as

$$\begin{cases} x = -z + \dfrac{18}{7} \\ y = z - \dfrac{17}{7} \end{cases}$$

where z can be any real number.

Check:
To check our solution on a graphing utility, we enter the augmented matrix and use the **rref** command.

4. $\begin{cases} 3x + 2y - 8z = -3 & (1) \\ -x - \frac{2}{3}y + z = 1 & (2) \\ 6x - 3y + 15z = 8 & (3) \end{cases}$

We start by clearing the fraction in equation (2) by multiplying both sides of the equation by 3.

$$\begin{cases} 3x + 2y - 8z = -3 & (1) \\ -3x - 2y + 3z = 3 & (2) \\ 6x - 3y + 15z = 8 & (3) \end{cases}$$

We use the method of elimination and begin by eliminating the variable x from equation (2). The coefficients on x in equations (1) and (2) are negatives of each other so we simply add the two equations together. This result becomes our new equation (2).

$$\begin{array}{r} 3x + 2y - 8z = -3 \\ \underline{-3x - 2y + 3z = 3} \\ -5z = 0 \quad (2) \end{array}$$

We now eliminate the variable x from equation (3) by multiplying each side of equation (1) by -2 and adding the result to equation (3). The result becomes our new equation (3).

$$\begin{array}{ll} 3x + 2y - 8z = -3 & -6x - 4y + 16z = 6 \\ 6x - 3y + 15z = 8 & \underline{6x - 3y + 15z = 8} \\ & -7y + 31z = 14 \quad (3) \end{array}$$

Our (equivalent) system now looks like

$$\begin{cases} 3x + 2y - 8z = -3 & (1) \\ -5z = 0 & (2) \\ -7y + 31z = 14 & (3) \end{cases}$$

We solve equation (2) for z by dividing both sides of the equation by -5.

$$-5z = 0$$
$$z = 0$$

Back-substitute $z = 0$ into equation (3) and solve for y.

$$-7y + 31z = 14$$
$$-7y + 31(0) = 14$$
$$-7y = 14$$
$$y = -2$$

Finally, back-substitute $y = -2$ and $z = 0$ into equation (1) and solve for x.

$$3x + 2y - 8z = -3$$
$$3x + 2(-2) - 8(0) = -3$$
$$3x - 4 = -3$$
$$3x = 1$$
$$x = \frac{1}{3}$$

The solution of the original system is $x = \frac{1}{3}$, $y = -2$, and $z = 0$.

Check:
To check our solution with a graphing utility, we enter the augmented matrix and use the **rref** command.

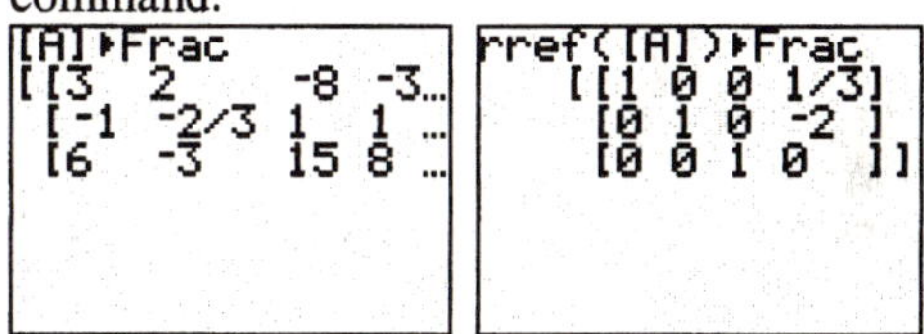

The solution of the system is $x = \frac{1}{3}$, $y = -2$, and $z = 0$.

5. We first check the equations to make sure that all variable terms are on the left side of the equation and the constants are on the right side. If a variable is missing, we put it in with a coefficient of 0. Our system can be rewritten as

$$\begin{cases} 4x - 5y + z = 0 \\ -2x - y + 0z = -25 \\ x + 5y - 5z = 10 \end{cases}$$

The augmented matrix is

$$\left[\begin{array}{ccc|c} 4 & -5 & 1 & 0 \\ -2 & -1 & 0 & -25 \\ 1 & 5 & -5 & 10 \end{array}\right]$$

6. The matrix has three rows and represents a system with three equations. The three columns to the left of the vertical bar indicate that the system has three variables. We can let *x*, *y*, and *z* denote these variables. The column to the right of the vertical bar represents the constants on the right side of the equations. The system is

$$\begin{cases} 3x+2y+4z=-6 \\ 1x+0y+8z=2 \\ -2x+1y+3z=-11 \end{cases} \text{ or } \begin{cases} 3x+2y+4z=-6 \\ x+8z=2 \\ -2x+y+3z=-11 \end{cases}$$

7. $2A+C=2\begin{bmatrix} 1 & -1 \\ 0 & -4 \\ 3 & 2 \end{bmatrix}+\begin{bmatrix} 4 & 6 \\ 1 & -3 \\ -1 & 8 \end{bmatrix}$

$$=\begin{bmatrix} 2 & -2 \\ 0 & -8 \\ 6 & 4 \end{bmatrix}+\begin{bmatrix} 4 & 6 \\ 1 & -3 \\ -1 & 8 \end{bmatrix}$$

$$=\begin{bmatrix} 6 & 4 \\ 1 & -11 \\ 5 & 12 \end{bmatrix}$$

```
2[A]+[C]
        [[6 4  ]
         [1 -11]
         [5 12 ]]
```

8. $A-3C=\begin{bmatrix} 1 & -1 \\ 0 & -4 \\ 3 & 2 \end{bmatrix}-3\begin{bmatrix} 4 & 6 \\ 1 & -3 \\ -1 & 8 \end{bmatrix}$

$$=\begin{bmatrix} 1 & -1 \\ 0 & -4 \\ 3 & 2 \end{bmatrix}-\begin{bmatrix} 12 & 18 \\ 3 & -9 \\ -3 & 24 \end{bmatrix}$$

$$=\begin{bmatrix} -11 & -19 \\ -3 & 5 \\ 6 & -22 \end{bmatrix}$$

```
[A]-3[C]
      [[-11 -19]
       [-3  5  ]
       [6   -22]]
```

9. *AC* cannot be computed because the dimensions are mismatched. To multiply two matrices, we need the number of columns in the first matrix to be the same as the number of rows in the second matrix. Matrix *A* has 2 columns, but matrix *C* has 3 rows. Therefore, the operation cannot be performed.

10. Here we are taking the product of a 2×3 matrix and a 3×2 matrix. Since the number of columns in the first matrix is the same as the number of rows in the second matrix (3 in both cases), the operation can be performed and will result in a 2×2 matrix.

$$BA=\begin{bmatrix} 1 & -2 & 5 \\ 0 & 3 & 1 \end{bmatrix}\begin{bmatrix} 1 & -1 \\ 0 & -4 \\ 3 & 2 \end{bmatrix}$$

$$=\begin{bmatrix} 1\cdot1+(-2)\cdot0+5\cdot3 & 1\cdot(-1)+(-2)\cdot(-4)+5\cdot2 \\ 0\cdot1+3\cdot0+1\cdot3 & 0\cdot(-1)+3(-4)+1\cdot2 \end{bmatrix}$$

$$=\begin{bmatrix} 16 & 17 \\ 3 & -10 \end{bmatrix}$$

```
[B][A]
     [[16 17 ]
      [3  -10]]
```

11. We first form the matrix

$$[A\,|\,I_2]=\left[\begin{array}{cc|cc} 3 & 2 & 1 & 0 \\ 5 & 4 & 0 & 1 \end{array}\right]$$

Next we use row operations to transform $[A\,|\,I_2]$ into reduced row echelon form.

$$\left[\begin{array}{cc|cc} 3 & 2 & 1 & 0 \\ 5 & 4 & 0 & 1 \end{array}\right]$$

$$\to\left[\begin{array}{cc|cc} 1 & \frac{2}{3} & \frac{1}{3} & 0 \\ 5 & 4 & 0 & 1 \end{array}\right] \quad \left(R_1=\tfrac{1}{3}r_1\right)$$

$$\to\left[\begin{array}{cc|cc} 1 & \frac{2}{3} & \frac{1}{3} & 0 \\ 0 & \frac{2}{3} & -\frac{5}{3} & 1 \end{array}\right] \quad \left(R_2=-5r_1+r_2\right)$$

$$\to\left[\begin{array}{cc|cc} 1 & \frac{2}{3} & \frac{1}{3} & 0 \\ 0 & 1 & -\frac{5}{2} & \frac{3}{2} \end{array}\right] \quad \left(R_2=\tfrac{3}{2}r_2\right)$$

$$\to\left[\begin{array}{cc|cc} 1 & 0 & 2 & -1 \\ 0 & 1 & -\frac{5}{2} & \frac{3}{2} \end{array}\right] \quad \left(R_1=-\tfrac{2}{3}r_2+r_1\right)$$

Therefore, $A^{-1}=\begin{bmatrix} 2 & -1 \\ -\frac{5}{2} & \frac{3}{2} \end{bmatrix}$.

```
[A]
        [[3 2]
         [5 4]]
```

```
[A]-1▸Frac
   [[2    -1 ]
    [-5/2 3/2]]
```

12. We first form the matrix

$$[B \mid I_3] = \left[\begin{array}{ccc|ccc} 1 & -1 & 1 & 1 & 0 & 0 \\ 2 & 5 & -1 & 0 & 1 & 0 \\ 2 & 3 & 0 & 0 & 0 & 1 \end{array}\right]$$

Next we use row operations to transform $[B \mid I_3]$ into reduced row echelon form.

$$\left[\begin{array}{ccc|ccc} 1 & -1 & 1 & 1 & 0 & 0 \\ 2 & 5 & -1 & 0 & 1 & 0 \\ 2 & 3 & 0 & 0 & 0 & 1 \end{array}\right]$$

$$\rightarrow \left[\begin{array}{ccc|ccc} 1 & -1 & 1 & 1 & 0 & 0 \\ 0 & 7 & -3 & -2 & 1 & 0 \\ 0 & 5 & -2 & -2 & 0 & 1 \end{array}\right] \quad \begin{pmatrix} R_2 = -2r_1 + r_2 \\ R_3 = -2r_1 + r_3 \end{pmatrix}$$

$$\rightarrow \left[\begin{array}{ccc|ccc} 1 & -1 & 1 & 1 & 0 & 0 \\ 0 & 1 & -\frac{3}{7} & -\frac{2}{7} & \frac{1}{7} & 0 \\ 0 & 5 & -2 & -2 & 0 & 1 \end{array}\right] \quad \left(R_2 = \tfrac{1}{7}r_2\right)$$

$$\rightarrow \left[\begin{array}{ccc|ccc} 1 & 0 & \frac{4}{7} & \frac{5}{7} & \frac{1}{7} & 0 \\ 0 & 1 & -\frac{3}{7} & -\frac{2}{7} & \frac{1}{7} & 0 \\ 0 & 0 & \frac{1}{7} & -\frac{4}{7} & -\frac{5}{7} & 1 \end{array}\right] \quad \begin{pmatrix} R_1 = r_2 + r_1 \\ R_3 = -5r_2 + r_3 \end{pmatrix}$$

$$\rightarrow \left[\begin{array}{ccc|ccc} 1 & 0 & \frac{4}{7} & \frac{5}{7} & \frac{1}{7} & 0 \\ 0 & 1 & -\frac{3}{7} & -\frac{2}{7} & \frac{1}{7} & 0 \\ 0 & 0 & 1 & -4 & -5 & 7 \end{array}\right] \quad (R_3 = 7r_3)$$

$$\rightarrow \left[\begin{array}{ccc|ccc} 1 & 0 & 0 & 3 & 3 & -4 \\ 0 & 1 & 0 & -2 & -2 & 3 \\ 0 & 0 & 1 & -4 & -5 & 7 \end{array}\right] \quad \begin{pmatrix} R_1 = -\frac{4}{7}r_3 + r_1 \\ R_2 = \frac{3}{7}r_3 + r_2 \end{pmatrix}$$

Thus, $B^{-1} = \begin{bmatrix} 3 & 3 & -4 \\ -2 & -2 & 3 \\ -4 & -5 & 7 \end{bmatrix}$

```
[B]
     [[1  -1 1 ]
      [2 5   -1]
      [2 3   0 ]]
```

```
[B]⁻¹▸Frac
     [[3   3   -4]
      [-2 -2 3 ]
      [-4 -5 7 ]]
```

13. $\begin{cases} 6x + 3y = 12 \\ 2x - y = -2 \end{cases}$

We start by writing the augmented matrix for the system.

$$\left[\begin{array}{cc|c} 6 & 3 & 12 \\ 2 & -1 & -2 \end{array}\right]$$

Next we use row operations to transform the augmented matrix into row echelon form.

$$\left[\begin{array}{cc|c} 6 & 3 & 12 \\ 2 & -1 & -2 \end{array}\right] \rightarrow \left[\begin{array}{cc|c} 2 & -1 & -2 \\ 6 & 3 & 12 \end{array}\right] \quad \begin{pmatrix} R_1 = r_2 \\ R_2 = r_1 \end{pmatrix}$$

$$\rightarrow \left[\begin{array}{cc|c} 1 & -\frac{1}{2} & -1 \\ 6 & 3 & 12 \end{array}\right] \quad \left(R_1 = \tfrac{1}{2}r_1\right)$$

$$\rightarrow \left[\begin{array}{cc|c} 1 & -\frac{1}{2} & -1 \\ 0 & 6 & 18 \end{array}\right] \quad \left(R_2 = -6r_1 + r_2\right)$$

$$\rightarrow \left[\begin{array}{cc|c} 1 & -\frac{1}{2} & -1 \\ 0 & 1 & 3 \end{array}\right] \quad \left(R_2 = \tfrac{1}{6}r_2\right)$$

$$\rightarrow \left[\begin{array}{cc|c} 1 & 0 & -\frac{1}{2} \\ 0 & 1 & 3 \end{array}\right] \quad \left(R_2 = \tfrac{1}{2}r_2 + r_1\right)$$

The solution of the system is $x = \frac{1}{2}$, $y = 3$.

```
[A]
     [[6 3   12]
      [2 -1  -2]]
```

```
rref([A])▸Frac
     [[1 0 1/2]
      [0 1 3  ]]
```

14. $\begin{cases} x + \dfrac{1}{4}y = 7 \\ 8x + 2y = 56 \end{cases}$

We start by writing the augmented matrix for the system.

$$\left[\begin{array}{cc|c} 1 & \frac{1}{4} & 7 \\ 8 & 2 & 56 \end{array}\right]$$

Next we use row operations to transform the augmented matrix into row echelon form.

$$\left[\begin{array}{cc|c} 1 & \frac{1}{4} & 7 \\ 8 & 2 & 56 \end{array}\right] R_2 = -8R_1 + r_2$$

$$\left[\begin{array}{cc|c} 1 & \frac{1}{4} & 7 \\ 0 & 0 & 0 \end{array}\right]$$

The augmented matrix is now in row echelon form. Because the bottom row consists entirely of 0's, the system actually consists of one equation in two variables. The system is dependent and therefore has an infinite number of solutions. Any ordered pair satisfying the equation $x + \dfrac{1}{4}y = 7$, or $y = -4x + 28$, is a solution to the system.

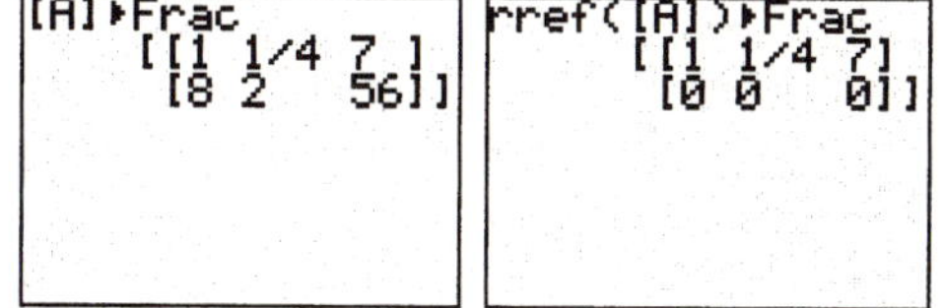

15. $\begin{cases} x+2y+4z=-3 \\ 2x+7y+15z=-12 \\ 4x+7y+13z=-10 \end{cases}$

We start by writing the augmented matrix for the system.

$$\left[\begin{array}{ccc|c} 1 & 2 & 4 & -3 \\ 2 & 7 & 15 & -12 \\ 4 & 7 & 13 & -10 \end{array}\right]$$

Next we use row operations to transform the augmented matrix into row echelon form.

$$\left[\begin{array}{ccc|c} 1 & 2 & 4 & -3 \\ 2 & 7 & 15 & -12 \\ 4 & 7 & 13 & -10 \end{array}\right]$$

$$\rightarrow \left[\begin{array}{ccc|c} 1 & 2 & 4 & -3 \\ 0 & 3 & 7 & -6 \\ 0 & -1 & -3 & 2 \end{array}\right] \quad \begin{pmatrix} R_2=-2r_1+r_2 \\ R_3=-4r_1+r_3 \end{pmatrix}$$

$$\rightarrow \left[\begin{array}{ccc|c} 1 & 2 & 4 & -3 \\ 0 & 1 & 3 & -2 \\ 0 & 3 & 7 & -6 \end{array}\right] \quad \begin{pmatrix} R_2=-r_3 \\ R_3=r_2 \end{pmatrix}$$

$$\rightarrow \left[\begin{array}{ccc|c} 1 & 2 & 4 & -3 \\ 0 & 1 & 3 & -2 \\ 0 & 0 & -2 & 0 \end{array}\right] \quad (R_3=-3r_2+r_3)$$

$$\rightarrow \left[\begin{array}{ccc|c} 1 & 2 & 4 & -3 \\ 0 & 1 & 3 & -2 \\ 0 & 0 & 1 & 0 \end{array}\right] \quad \left(R_3=-\tfrac{1}{2}r_3\right)$$

The matrix is now in row echelon form. The last row represents the equation $z=0$. Using $z=0$ we back-substitute into the equation $y+3z=-2$ (from the second row) and obtain

$$y+3z=-2$$
$$y+3(0)=-2$$
$$y=-2$$

Using $y=-2$ and $z=0$, we back-substitute into the equation $x+2y+4z=-3$ (from the first row) and obtain

$$x+2y+4z=-3$$
$$x+2(-2)+4(0)=-3$$
$$x=1$$

The solution is $x=1$, $y=-2$, $z=0$.

```
[A]
 [[1 2 4  -3 ]
  [2 7 15 -12]
  [4 7 13 -10]]
```

```
rref([A])▸Frac
 [[1 0 0 1 ]
  [0 1 0 -2]
  [0 0 1 0 ]]
```

16. $\begin{cases} 2x+2y-3z=5 \\ x-y+2z=8 \\ 3x+5y-8z=-2 \end{cases}$

We start by writing the augmented matrix for the system.

$$\left[\begin{array}{ccc|c} 2 & 2 & -3 & 5 \\ 1 & -1 & 2 & 8 \\ 3 & 5 & -8 & -2 \end{array}\right]$$

Next we use row operations to transform the augmented matrix into row echelon form.

$$\left[\begin{array}{ccc|c} 2 & 2 & -3 & 5 \\ 1 & -1 & 2 & 8 \\ 3 & 5 & -8 & -2 \end{array}\right]$$

$$= \left[\begin{array}{ccc|c} 1 & -1 & 2 & 8 \\ 2 & 2 & -3 & 5 \\ 3 & 5 & -8 & -2 \end{array}\right] \quad \begin{pmatrix} R_1=r_2 \\ R_2=r_1 \end{pmatrix}$$

$$= \left[\begin{array}{ccc|c} 1 & -1 & 2 & 8 \\ 0 & 4 & -7 & -11 \\ 0 & 8 & -14 & -26 \end{array}\right] \quad \begin{pmatrix} R_2=-2r_1+r_2 \\ R_3=-3r_1+r_3 \end{pmatrix}$$

$$= \left[\begin{array}{ccc|c} 1 & -1 & 2 & 8 \\ 0 & 1 & -\frac{7}{4} & -\frac{11}{4} \\ 0 & 8 & -14 & -26 \end{array}\right] \quad \left(R_2=\tfrac{1}{4}r_2\right)$$

$$= \left[\begin{array}{ccc|c} 1 & -1 & 2 & 8 \\ 0 & 1 & -\frac{7}{4} & -\frac{11}{4} \\ 0 & 0 & 0 & -4 \end{array}\right] \quad (R_3=-8r_2+r_3)$$

The matrix is now in row echelon form. The last row represents the equation $0=-4$ which is a contradiction. Therefore, the system has no solution and is said to be inconsistent.

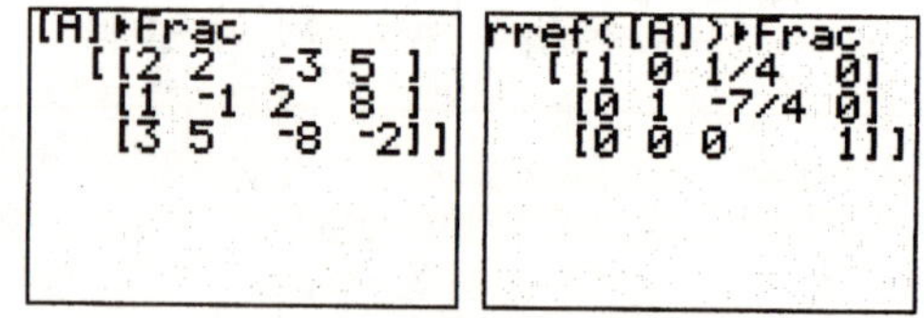

17. $\begin{vmatrix} -2 & 5 \\ 3 & 7 \end{vmatrix} = (-2)(7)-(5)(3)$

$$=-14-15$$
$$=-29$$

18. We choose to expand across row 2 because it contains a 0.

$$\begin{vmatrix} 2 & -4 & 6 \\ 1 & 4 & 0 \\ -1 & 2 & -4 \end{vmatrix}$$

$$= (-1)^{2+1}(1)\begin{vmatrix} -4 & 6 \\ 2 & -4 \end{vmatrix} + (-1)^{2+2}(4)\begin{vmatrix} 2 & 6 \\ -1 & -4 \end{vmatrix} + (-1)^{2+3}(0)\begin{vmatrix} 2 & -4 \\ -1 & 2 \end{vmatrix}$$

$$= -1\begin{vmatrix} -4 & 6 \\ 2 & -4 \end{vmatrix} + 4\begin{vmatrix} 2 & 6 \\ -1 & -4 \end{vmatrix} - 0\begin{vmatrix} 2 & -4 \\ -1 & 2 \end{vmatrix}$$

$$= -1(16-12) + 4(-8+6) - 0(4-4)$$

$$= -1(4) + 4(-2)$$

$$= -4 - 8$$

$$= -12$$

19. $\begin{cases} 4x + 3y = -23 \\ 3x - 5y = 19 \end{cases}$

The determinant D of the coefficients of the variables is

$$D = \begin{vmatrix} 4 & 3 \\ 3 & -5 \end{vmatrix} = (4)(\ 5)\ \ (3)(3) = \ 20\ \ 9 = \ 29$$

Since $D \neq 0$, Cramer's Rule can be applied.

$$D_x = \begin{vmatrix} -23 & 3 \\ 19 & -5 \end{vmatrix} = (-23)(-5) - (3)(19) = 58$$

$$D_y = \begin{vmatrix} 4 & -23 \\ 3 & 19 \end{vmatrix} = (4)(19) - (-23)(3) = 145$$

$$x = \frac{D_x}{D} = \frac{58}{-29} = -2$$

$$y = \frac{D_y}{D} = \frac{145}{-29} = -5$$

The solution of the system is $x = -2$, $y = -5$.

```
[A]
 [[4 3  -23]
  [3 -5 19 ]]

rref([A])
 [[1 0 -2]
  [0 1 -5]]
```

20. $\begin{cases} 4x - 3y + 2z = 15 \\ -2x + y - 3z = -15 \\ 5x - 5y + 2z = 18 \end{cases}$

The determinant D of the coefficients of the variables is

$$D = \begin{vmatrix} 4 & -3 & 2 \\ -2 & 1 & -3 \\ 5 & -5 & 2 \end{vmatrix}$$

$$= 4\begin{vmatrix} 1 & -3 \\ -5 & 2 \end{vmatrix} - (-3)\begin{vmatrix} -2 & -3 \\ 5 & 2 \end{vmatrix} + 2\begin{vmatrix} -2 & 1 \\ 5 & -5 \end{vmatrix}$$

$$= 4(2-15) + 3(-4+15) + 2(10-5)$$

$$= 4(-13) + 3(11) + 2(5)$$

$$= -52 + 33 + 10$$

$$= -9$$

Since $D \neq 0$, Cramer's Rule can be applied.

$$D_x = \begin{vmatrix} 15 & -3 & 2 \\ -15 & 1 & -3 \\ 18 & -5 & 2 \end{vmatrix}$$

$$= 15\begin{vmatrix} 1 & -3 \\ -5 & 2 \end{vmatrix} - (-3)\begin{vmatrix} -15 & -3 \\ 18 & 2 \end{vmatrix} + 2\begin{vmatrix} -15 & 1 \\ 18 & -5 \end{vmatrix}$$

$$= 15(2-15) + 3(-30+54) + 2(75-18)$$

$$= 15(-13) + 3(24) + 2(57)$$

$$= -9$$

$$D_y = \begin{vmatrix} 4 & 15 & 2 \\ -2 & -15 & -3 \\ 5 & 18 & 2 \end{vmatrix}$$

$$= 4\begin{vmatrix} -15 & -3 \\ 18 & 2 \end{vmatrix} - 15\begin{vmatrix} -2 & -3 \\ 5 & 2 \end{vmatrix} + 2\begin{vmatrix} -2 & -15 \\ 5 & 18 \end{vmatrix}$$

$$= 4(-30+54) - 15(-4+15) + 2(-36+75)$$

$$= 4(24) - 15(11) + 2(39)$$

$$= -9$$

$$D_z = \begin{vmatrix} 4 & -3 & 15 \\ -2 & 1 & -15 \\ 5 & -5 & 18 \end{vmatrix}$$

$$= 4\begin{vmatrix} 1 & -15 \\ -5 & 18 \end{vmatrix} - (-3)\begin{vmatrix} -2 & -15 \\ 5 & 18 \end{vmatrix} + 15\begin{vmatrix} -2 & 1 \\ 5 & -5 \end{vmatrix}$$

$$= 4(18-75) + 3(-36+75) + 15(10-5)$$

$$= 4(-57) + 3(39) + 15(5)$$

$$= -36$$

$x = \frac{D_x}{D} = \frac{-9}{-9} = 1$, $y = \frac{D_y}{D} = \frac{9}{-9} = -1$,

$z = \frac{D_z}{D} = \frac{-36}{-9} = 4$

The solution of the system is $x = 1$, $y = -1$, $z = 4$.

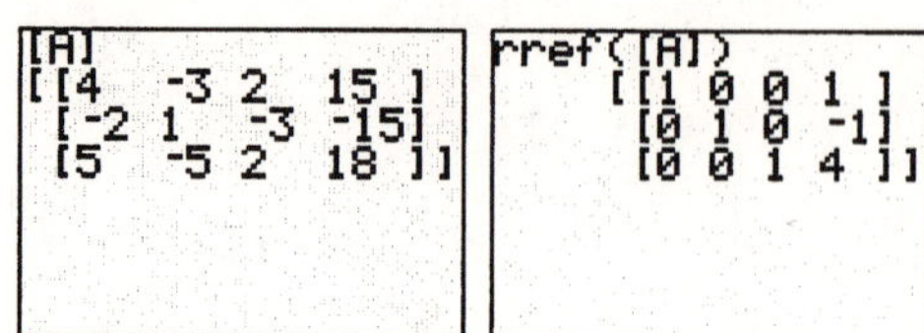

21. $\begin{cases} 3x^2 + y^2 = 12 \\ y^2 = 9x \end{cases}$

Substitute $9x$ for y^2 into the first equation and solve for x:

$$3x^2 + (9x) = 12$$
$$3x^2 + 9x - 12 = 0$$
$$x^2 + 3x - 4 = 0$$
$$(x-1)(x+4) = 0$$
$$x = 1 \text{ or } x = -4$$

Back substitute these values into the second equation to determine y:

$x = 1$: $y^2 = 9(1) = 9$

$y = \pm 3$

$x = -4$: $y^2 = 9(-4) = -36$

$y = \pm\sqrt{-36}$ (not real)

The solutions of the system are $(1, -3)$ and $(1, 3)$.

22. $\begin{cases} 2y^2 - 3x^2 = 5 \\ y - x = 1 \Rightarrow y = x + 1 \end{cases}$

Substitute $x+1$ for y into the first equation and solve for x:

$$2(x+1)^2 - 3x^2 = 5$$
$$2(x^2 + 2x + 1) - 3x^2 = 5$$
$$2x^2 + 4x + 2 - 3x^2 = 5$$
$$-x^2 + 4x - 3 = 0$$
$$x^2 - 4x + 3 = 0$$
$$(x-1)(x-3) = 0$$
$$x = 1 \text{ or } x = 3$$

Back substitute these values into the second equation to determine y:

$x = 1$: $y = 1 + 1 = 2$

$x = 3$: $y = 3 + 1 = 4$

The solutions of the system are $(1, 2)$ and $(3, 4)$.

23. $\begin{cases} x^2 + y^2 \le 100 \\ 4x - 3y \ge 0 \end{cases}$

Graph the circle $x^2 + y^2 = 100$. Use a solid curve since the inequality uses $\le$. Choose a test point not on the circle, such as (0, 0). Since $0^2 + 0^2 \le 100$ is true, shade the same side of the circle as (0, 0); that is, inside the circle.

Graph the line $4x - 3y = 0$. Use a solid line since the inequality uses $\ge$. Choose a test point not on the line, such as (0, 1). Since $4(0) - 3(1) \ge 0$ is false, shade the opposite side of the line from (0, 1). The overlapping region is the solution.

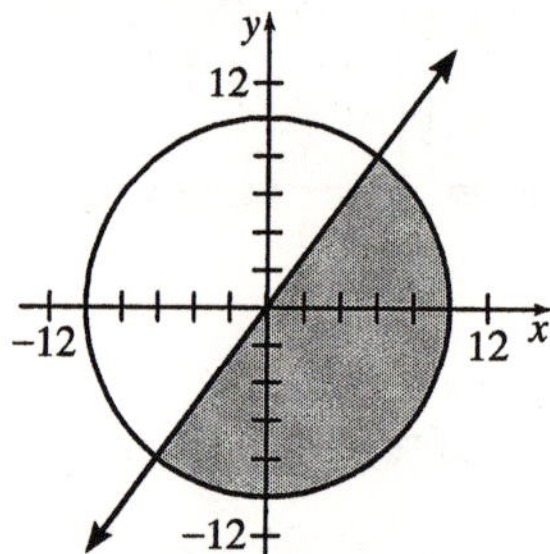

24. $\frac{3x+7}{(x+3)^2}$

The denominator contains the repeated linear factor $x+3$. Thus, the partial fraction decomposition takes on the form

$$\frac{3x+7}{(x+3)^2} = \frac{A}{x+3} + \frac{B}{(x+3)^2}$$

Clear the fractions by multiplying both sides by $(x+3)^2$. The result is the identity

$$3x + 7 = A(x+3) + B$$

or

$$3x + 7 = Ax + (3A + B)$$

We equate coefficients of like powers of x to obtain the system

$$\begin{cases} 3 = A \\ 7 = 3A + B \end{cases}$$

Therefore, we have $A = 3$. Substituting this result into the second equation gives

$7 = 3A + B$

$7 = 3(3) + B$

$-2 = B$

Thus, the partial fraction decomposition is

$$\frac{3x+7}{(x+3)^2} = \frac{3}{x+3} + \frac{-2}{(x+3)^2}.$$

25. $\dfrac{4x^2-3}{x(x^2+3)^2}$

The denominator contains the linear factor x and the repeated irreducible quadratic factor x^2+3.

The partial fraction decomposition takes on the form

$$\frac{4x^2-3}{x(x^2+3)^2} = \frac{A}{x} + \frac{Bx+C}{x^2+3} + \frac{Dx+E}{(x^2+3)^2}$$

We clear the fractions by multiplying both sides by $x(x^2+3)^2$ to obtain the identity

$$4x^2-3 = A(x^2+3)^2 + x(x^2+3)(Bx+C) + x(Dx+E)$$ C

ollecting like terms yields

$$4x^2-3 = (A+B)x^4 + Cx^3 + (6A+3B+D)x^2 + (3C+E)x + (9A)$$

Equating coefficients, we obtain the system

$$\begin{cases} A+B=0 \\ C=0 \\ 6A+3B+D=4 \\ 3C+E=0 \\ 9A=-3 \end{cases}$$

From the last equation we get $A = -\frac{1}{3}$. Substituting this value into the first equation gives $B = \frac{1}{3}$. From the second equation, we know $C = 0$. Substituting this value into the fourth equation yields $E = 0$.

Substituting $A = -\frac{1}{3}$ and $B = \frac{1}{3}$ into the third equation gives us

$6\left(-\frac{1}{3}\right) + 3\left(\frac{1}{3}\right) + D = 4$

$-2 + 1 + D = 4$

$D = 5$

Therefore, the partial fraction decomposition is

$$\frac{4x^2-3}{x(x^2+3)^2} = \frac{-\frac{1}{3}}{x} + \frac{\frac{1}{3}x}{(x^2+3)} + \frac{5x}{(x^2+3)^2}$$

26. $\begin{cases} x \geq 0 \\ y \geq 0 \\ x+2y \geq 8 \\ 2x-3y \geq 2 \end{cases}$

The inequalities $x \geq 0$ and $y \geq 0$ require that the graph be in quadrant I.

$x + 2y \geq 8$

$y \geq -\frac{1}{2}x + 4$

Test the point $(0,0)$.

$x + 2y \geq 8$

$0 + 2(0) \geq 8$?

$0 \geq 8$ false

The point $(0,0)$ is not a solution. Thus, the graph of the inequality $x + 2y \geq 8$ includes the half-plane above the line $y = -\frac{1}{2}x + 4$. Because the inequality is non-strict, the line is also part of the graph of the solution.

$2x - 3y \geq 2$

$y \leq \frac{2}{3}x - \frac{2}{3}$

Test the point $(0,0)$.

$2x - 3y \geq 2$

$2(0) - 3(0) \geq 2$?

$0 \geq 2$ false

The point $(0,0)$ is not a solution. Thus, the graph of the inequality $2x - 3y \geq 2$ includes the half-plane below the line $y = \frac{2}{3}x - \frac{2}{3}$.

Because the inequality is non-strict, the line is also part of the graph of the solution.

The overlapping shaded region (that is, the shaded region in the graph below) is the solution to the system of linear inequalities.

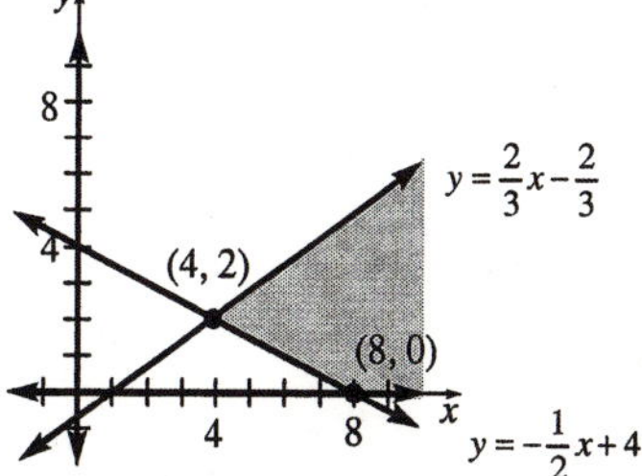

The graph is unbounded. The corner points are $(4,2)$ and $(8,0)$.

27. The objective function is $z = 5x + 8y$. We seek the largest value of z that can occur if x and y are solutions of the system of linear inequalities

$$\begin{cases} x \geq 0 \\ 2x + y \leq 8 \\ x - 3y \leq -3 \end{cases}$$

$$\begin{aligned} 2x + y &= 8 \\ y &= -2x + 8 \end{aligned} \qquad \begin{aligned} x - 3y &= -3 \\ -3y &= -x - 3 \\ y &= \frac{1}{3}x + 1 \end{aligned}$$

The graph of this system (the feasible points) is shown as the shaded region in the figure below. The corner points of the feasible region are $(0,1)$, $(3,2)$, and $(0,8)$.

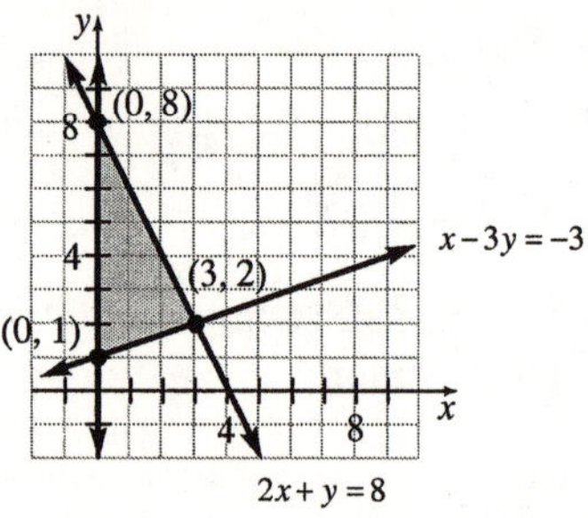

Corner point, (x, y)	Value of obj. function, z
$(0,1)$	$z = 5(0) + 8(1) = 8$
$(3,2)$	$z = 5(3) + 8(2) = 31$
$(0,8)$	$z = 5(0) + 8(8) = 64$

From the table, we can see that the maximum value of z is 64, and it occurs at the point $(0,8)$.

28. Let j = unit price for flare jeans, c = unit price for camisoles, and t = unit price for t-shirts. The given information yields a system of equations with each of the three women yielding an equation.

$$\begin{cases} 2j + 2c + 4t = 90 & \text{(Megan)} \\ j \qquad + 3t = 42.5 & \text{(Paige)} \\ j + 3c + 2t = 62 & \text{(Kara)} \end{cases}$$

We can solve this system by using matrices.

$$\left[\begin{array}{ccc|c} 2 & 2 & 4 & 90 \\ 1 & 0 & 3 & 42.5 \\ 1 & 3 & 2 & 62 \end{array}\right]$$

$$= \left[\begin{array}{ccc|c} 1 & 1 & 2 & 45 \\ 1 & 0 & 3 & 42.5 \\ 1 & 3 & 2 & 62 \end{array}\right] \quad \left(R_1 = \tfrac{1}{2}r_1\right)$$

$$= \left[\begin{array}{ccc|c} 1 & 1 & 2 & 45 \\ 0 & -1 & 1 & -2.5 \\ 0 & 2 & 0 & 17 \end{array}\right] \quad \begin{pmatrix} R_2 = -r_1 + r_2 \\ R_3 = -r_1 + r_3 \end{pmatrix}$$

$$= \left[\begin{array}{ccc|c} 1 & 1 & 2 & 45 \\ 0 & 1 & -1 & 2.5 \\ 0 & 2 & 0 & 17 \end{array}\right] \quad (R_2 = -r_2)$$

$$= \left[\begin{array}{ccc|c} 1 & 0 & 3 & 42.5 \\ 0 & 1 & -1 & 2.5 \\ 0 & 0 & 2 & 12 \end{array}\right] \quad \begin{pmatrix} R_1 = -r_2 + r_1 \\ R_3 = -2r_2 + r_3 \end{pmatrix}$$

$$= \left[\begin{array}{ccc|c} 1 & 0 & 3 & 42.5 \\ 0 & 1 & -1 & 2.5 \\ 0 & 0 & 1 & 6 \end{array}\right] \quad \left(R_3 = \tfrac{1}{2}r_3\right)$$

The last row represents the equation $z = 6$. Substituting this result into $y - z = 2.5$ (from the second row) gives

$$\begin{aligned} y - z &= 2.5 \\ y - 6 &= 2.5 \\ y &= 8.5 \end{aligned}$$

Substituting $z = 6$ into $x + 3z = 42.5$ (from the first row) gives

$$\begin{aligned} x + 3z &= 42.5 \\ x + 3(6) &= 42.5 \\ x &= 24.5 \end{aligned}$$

Thus, flare jeans cost \$24.50, camisoles cost \$8.50, and t-shirts cost \$6.00.

Chapter 10 Projects

Project 1

a.

$80\% = 0.80$	$18\% = 0.18$	$2\% = 0.02$
$40\% = 0.40$	$50\% = 0.50$	$10\% = 0.10$
$20\% = 0.20$	$60\% = 0.60$	$20\% = 0.20$

b. $$\begin{bmatrix} 0.80 & 0.18 & 0.02 \\ 0.40 & 0.50 & 0.10 \\ 0.20 & 0.60 & 0.20 \end{bmatrix}$$

c. $0.80+0.18+0.02=1.00$
$0.40+0.50+0.10=1.00$
$0.20+0.60+0.20=1.00$
The sum of each row is 1 (or 100%). These represent the three possibilities of educational achievement for a parent of a child, unless someone does not attend school at all. Since these are rounded percents, chances are the other possibilities are negligible.

d. $$P^2=\begin{bmatrix}0.8 & 0.18 & 0.02\\ 0.4 & 0.5 & 0.1\\ 0.2 & 0.6 & 0.2\end{bmatrix}^2$$
$$=\begin{bmatrix}0.716 & 0.246 & 0.038\\ 0.54 & 0.382 & 0.078\\ 0.44 & 0.456 & 0.104\end{bmatrix}$$
Grandchild of a college graduate is a college graduate: entry (1, 1): 0.716. The probability is 71.6%

e. Grandchild of a high school graduate finishes college: entry (2,1): 0.54. The probability is 54%.

f. grandchildren $\to$ k = 2.
$v^{(2)}=v^{(0)}P^2$
$$=[0.267 \quad 0.574 \quad 0.159]\begin{bmatrix}0.716 & 0.246 & 0.038\\ 0.54 & 0.382 & 0.078\\ 0.44 & 0.456 & 0.104\end{bmatrix}$$
$$=[0.475 \quad 0.43046 \quad 0.9454]$$
College: $\approx 48\%$
High School: $\approx 43\%$ 43%
Elementary: $\approx 9\%$

g. The matrix totally stops changing at
$$P^{30}=\begin{bmatrix}0.6489 & 0.2977 & 0.0544\\ 0.6489 & 0.2977 & 0.0544\\ 0.6489 & 0.2977 & 0.0544\end{bmatrix}$$

Project 2 (web)

a. $2\times2\times2\times2=16$ codewords.

b. $v=uG$
u will be the matrix representing all of the 4-digit information bit sequences.

$$u=\begin{bmatrix}0&0&0&0\\0&0&0&1\\0&0&1&0\\0&0&1&1\\0&1&0&0\\0&1&0&1\\0&1&1&0\\0&1&1&1\\1&0&0&0\\1&0&0&1\\1&0&1&0\\1&0&1&1\\1&1&0&0\\1&1&0&1\\1&1&1&0\\1&1&1&1\end{bmatrix}$$

(Remember, this is mod two. That means that you only write down the remainder when dividing by 2.)

$v=uG$

$$v=\begin{bmatrix}0&0&0&0&0&0&0\\0&0&0&1&1&1&0\\0&0&1&0&1&0&1\\0&0&1&1&0&1&1\\0&1&0&0&0&1&1\\0&1&0&1&1&0&1\\0&1&1&0&1&1&0\\0&1&1&1&0&0&0\\1&0&0&0&1&1&1\\1&0&0&1&0&0&1\\1&0&1&0&0&1&0\\1&0&1&1&1&0&0\\1&1&0&0&1&0&0\\1&1&0&1&0&1&0\\1&1&1&0&0&0&1\\1&1&1&1&1&1&1\end{bmatrix}$$

c. Answers will vary, but if we choose the 6th row and the 10th row:
0101101
<u>1001001</u>
1102102 $\to$ 1100100 (13th row)

d. $v = uG$

$VH = uGH$

$$GH = \begin{bmatrix} 0 & 0 & 0 \\ 0 & 0 & 0 \\ 0 & 0 & 0 \\ 0 & 0 & 0 \end{bmatrix}$$

e. $$rH = [0 \ 1 \ 0 \ 1 \ 0 \ 0 \ 0]\begin{bmatrix} 1 & 1 & 1 \\ 0 & 1 & 1 \\ 1 & 0 & 1 \\ 1 & 1 & 0 \\ 1 & 0 & 0 \\ 0 & 1 & 0 \\ 0 & 0 & 1 \end{bmatrix}$$

$= [1 \ 0 \ 1]$

error code: 0010 000

r : <u>0101 000</u>

0111 000

This is in the codeword list.

Project 3 (web)

a. $$A^T = \begin{bmatrix} 1 & 1 & 1 & 1 & 1 & 1 & 1 \\ 3 & 4 & 5 & 6 & 7 & 8 & 9 \end{bmatrix}$$

b. $B = (A^T A)^{-1} A^T Y$

$$B = \begin{bmatrix} -2.357 \\ 2.0357 \end{bmatrix}$$

c. $y = 2.0357x - 2.357$

d. $y = 2.0357x - 2.357$

Project 4 (web)

Answers will vary.

Cumulative Review R-10

1. $2x^2 - x = 0$

$x(2x-1) = 0$

$x = 0$ or $2x - 1 = 0$

$2x = 1$

$x = \frac{1}{2}$

The solution set is $\left\{0, \frac{1}{2}\right\}$.

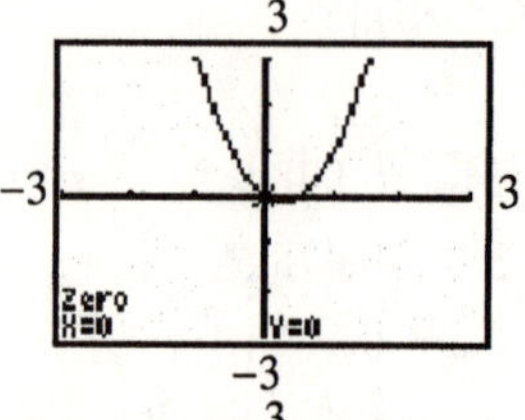

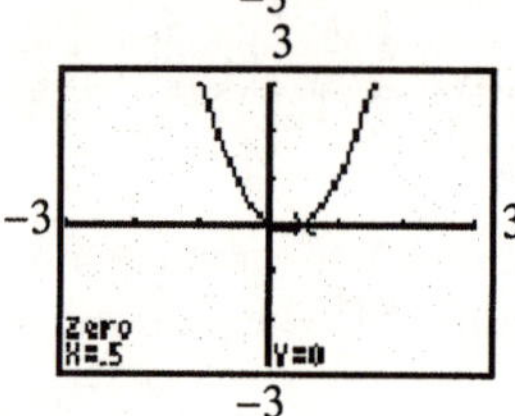

2. $\sqrt{3x+1} = 4$

$\left(\sqrt{3x+1}\right)^2 = 4^2$

$3x + 1 = 16$

$3x = 15$

$x = 5$

Check:

$\sqrt{3 \cdot 5 + 1} = 4$

$\sqrt{15 + 1} = 4$

$\sqrt{16} = 4$

$4 = 4$

The solution set is $\{5\}$.

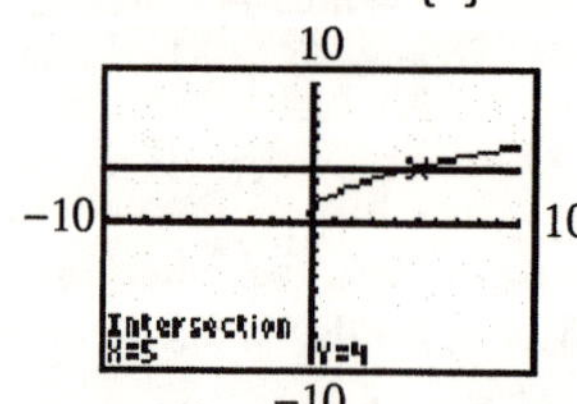

3. $2x^3 - 3x^2 - 8x - 3 = 0$

The graph of $Y_1 = 2x^3 - 3x^2 - 8x - 3$ appears to have an x-intercept at $x = 3$.

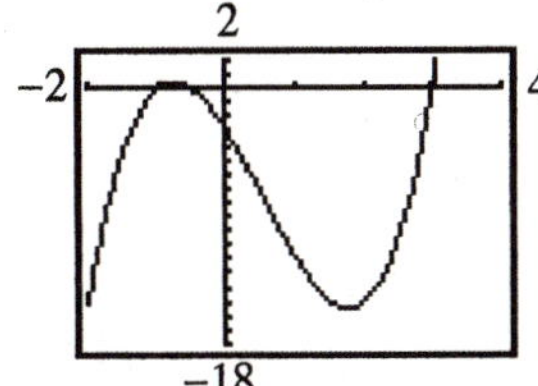

Using synthetic division:

$$\begin{array}{r|rrrr} 3 & 2 & -3 & -8 & -3 \\ & & 6 & 9 & 3 \\ \hline & 2 & 3 & 1 & 0 \end{array}$$

Therefore, $2x^3 - 3x^2 - 8x - 3 = 0$

$$(x-3)(2x^2+3x+1) = 0$$

$$(x-3)(2x+1)(x+1) = 0$$

$$x = 3 \text{ or } x = -\frac{1}{2} \text{ or } x = -1$$

The solution set is $\left\{-1, -\frac{1}{2}, 3\right\}$.

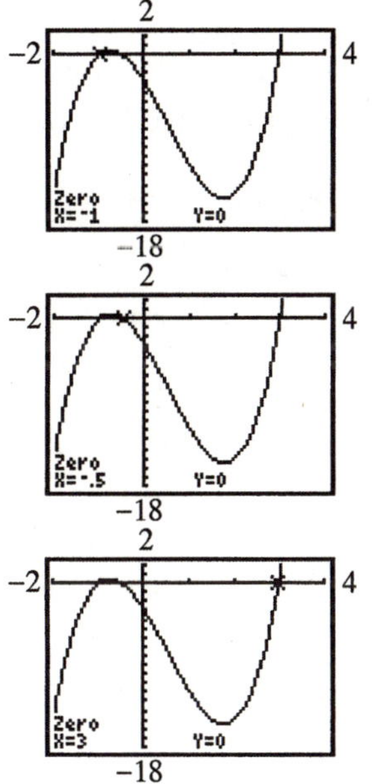

4. $3^x = 9^{x+1}$

$$3^x = \left(3^2\right)^{x+1}$$

$$3^x = 3^{2x+2}$$

$$x = 2x + 2$$

$$x = -2$$

The solution set is $\{-2\}$.

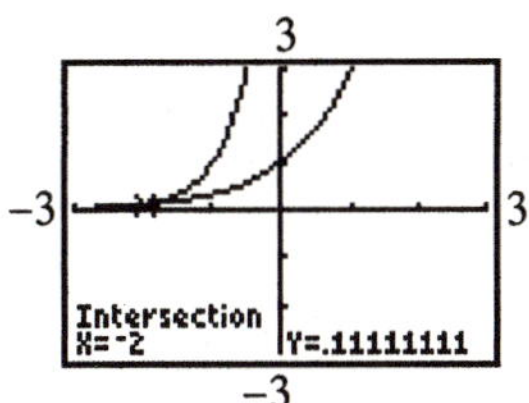

5. $\log_3(x-1) + \log_3(2x+1) = 2$

$$\log_3\left((x-1)(2x+1)\right) = 2$$

$$(x-1)(2x+1) = 3^2$$

$$2x^2 - x - 1 = 9$$

$$2x^2 - x - 10 = 0$$

$$(2x-5)(x+2) = 0$$

$$x = \frac{5}{2} \text{ or } x = -2$$

Since $x = -2$ makes the original logarithms undefined, the solution set is $\left\{\frac{5}{2}\right\}$.

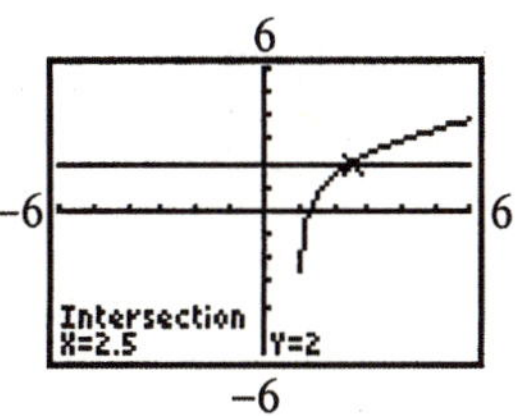

6. $3^x = e$

$$\ln\left(3^x\right) = \ln e$$

$$x \ln 3 = 1$$

$$x = \frac{1}{\ln 3} \approx 0.910$$

The solution set is $\left\{\frac{1}{\ln 3} \approx 0.910\right\}$.

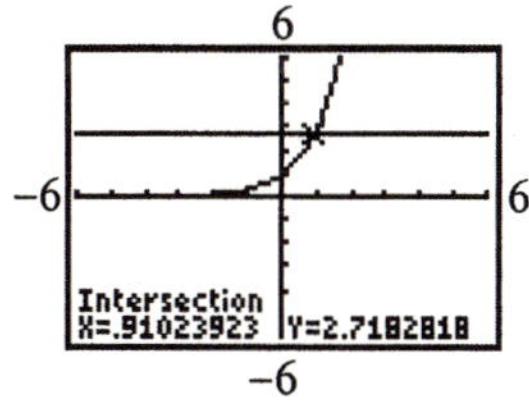

7. $g(x) = \dfrac{2x^3}{x^4+1}$

$$g(-x) = \frac{2(-x)^3}{(-x)^4+1} = \frac{-2x^3}{x^4+1} = -g(x)$$

Thus, g is an odd function and its graph is symmetric with respect to the origin.

8.
$$x^2+y^2-2x+4y-11=0$$
$$x^2-2x+y^2+4y=11$$
$$(x^2-2x+1)+(y^2+4y+4)=11+1+4$$
$$(x-1)^2+(y+2)^2=16$$
$$(x-1)^2+(y+2)^2=4^2$$

Center: (1,–2); Radius: 4

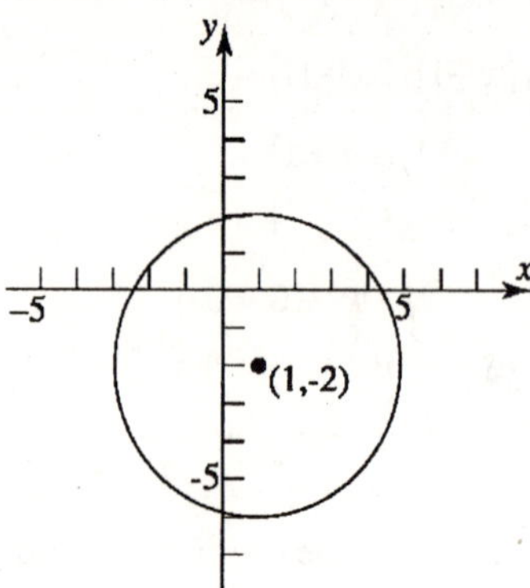

9. $f(x)=3^{x-2}+1$

Using the graph of $y=3^x$, shift the graph horizontally 2 units to the right, then shift the graph vertically upward 1 unit.

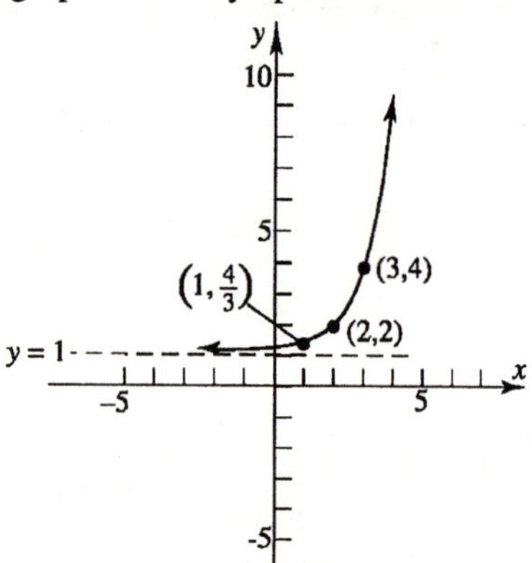

Domain: $(-\infty,\infty)$ Range: $(1,\infty)$

Horizontal Asymptote: $y=1$

10. $f(x)=\dfrac{5}{x+2}$

$$y=\frac{5}{x+2}$$
$$x=\frac{5}{y+2} \quad \text{Inverse}$$
$$x(y+2)=5$$
$$xy+2x=5$$
$$xy=5-2x$$
$$y=\frac{5-2x}{x}=\frac{5}{x}-2$$

Thus, $f^{-1}(x)=\dfrac{5}{x}-2$

Domain of f = range of f^{-1}
= all real numbers except -2.

Range of f = domain of f^{-1}
= all real numbers except 0.

11. a. $y=3x+6$

The graph is a line.

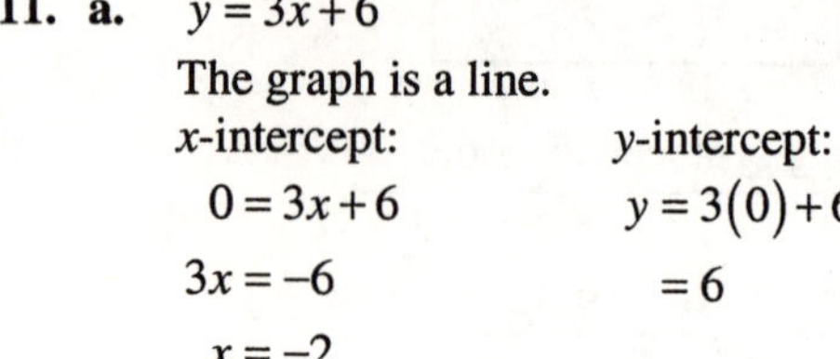

x-intercept:	*y*-intercept:
$0=3x+6$	$y=3(0)+6$
$3x=-6$	$=6$
$x=-2$	

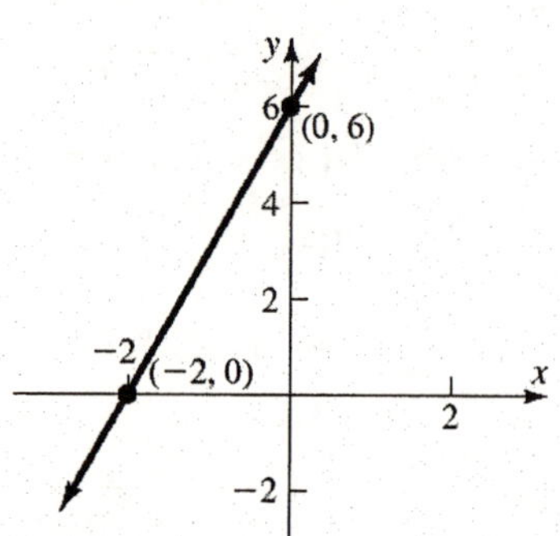

b. $x^2+y^2=4$

The graph is a circle with center (0, 0) and radius 2.

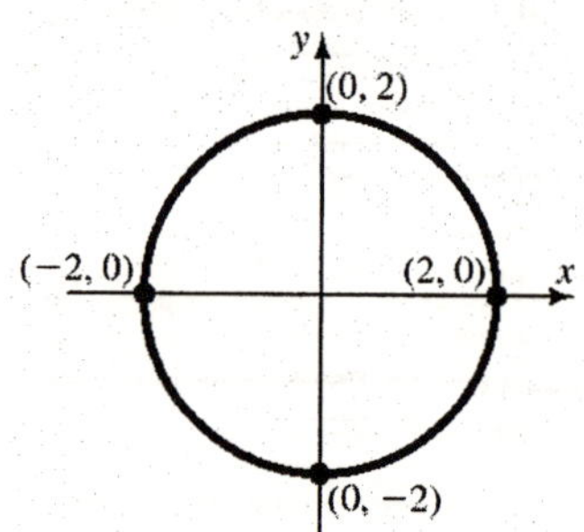

c. $y=x^3$

y
10
(2, 8)
x
–4
(0, 0)
4
(–2, –8)
–10

d. $y = \frac{1}{x}$

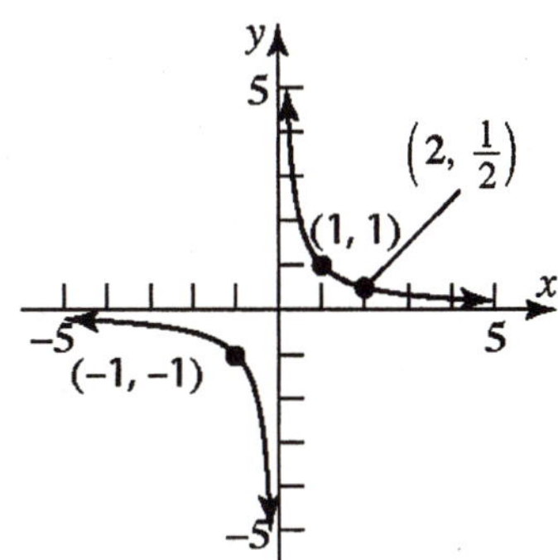

e. $y = \sqrt{x}$

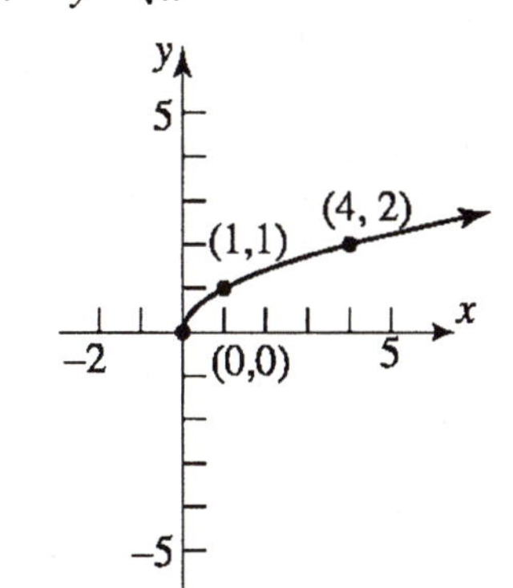

f. $y = e^x$

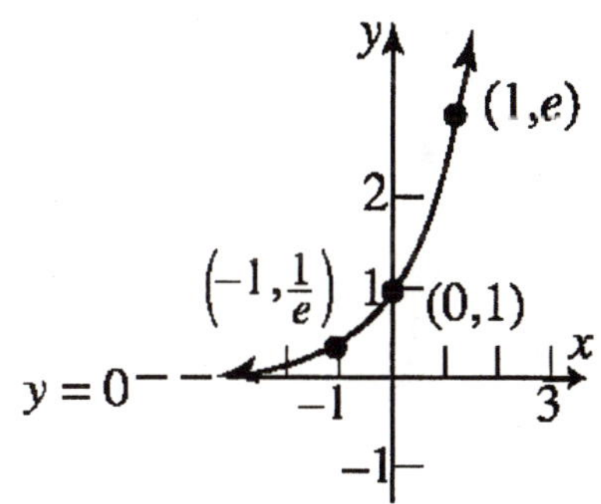

g. $y = \ln x$

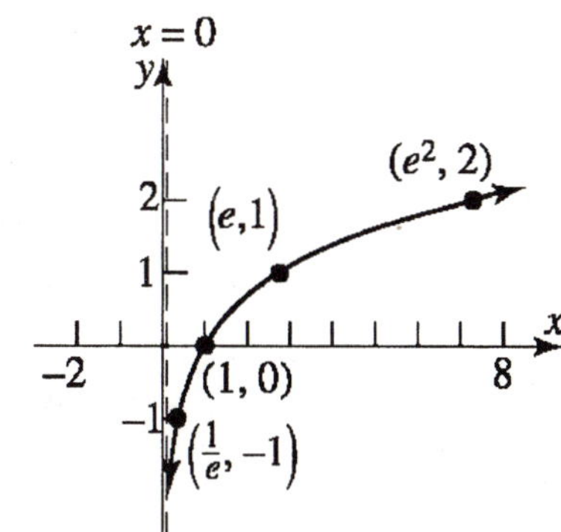

12. $f(x) = x^3 - 3x + 5$

a. Let $Y_1 = x^3 - 3x + 5$.

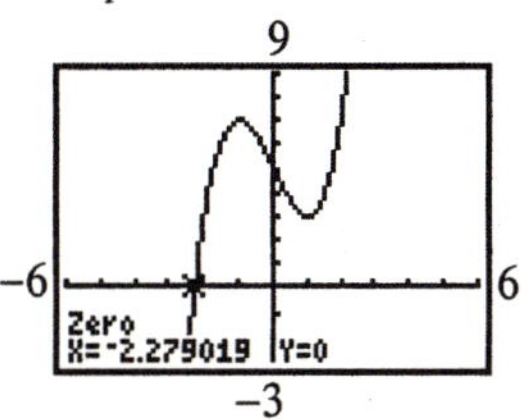

The zero of *f* is approximately -2.28.

b.

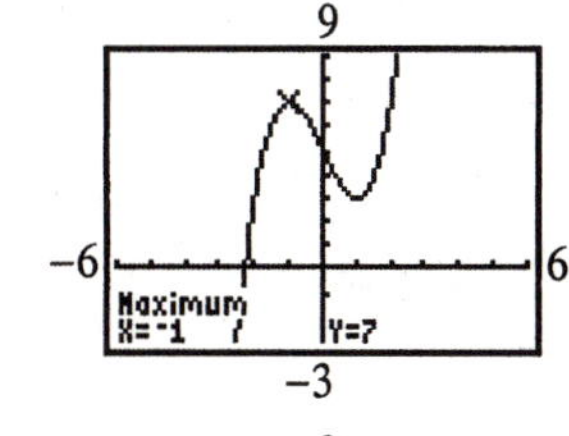

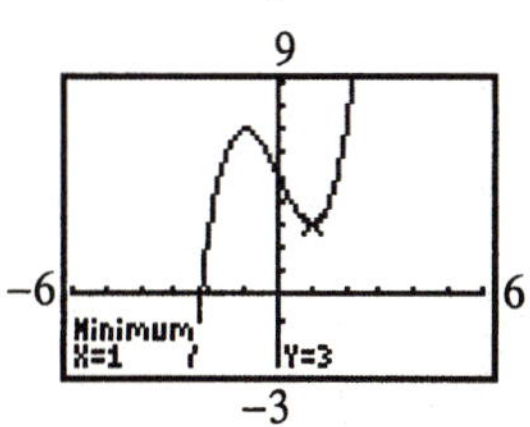

f has a local maximum of 7 at $x = -1$ and a local minimum of 3 at $x = 1$.

c. *f* is increasing on the intervals $(-\infty, -1)$ and $(1, \infty)$.

Chapter 11
Sequences; Induction; the Binomial Theorem

Section 11.1

1. $f(2)=\frac{2-1}{2}=\frac{1}{2}$; $f(3)=\frac{3-1}{3}=\frac{2}{3}$

2. True

3. $A=P\left(1+\frac{r}{n}\right)^{n\cdot t}$
$=1000\left(1+\frac{0.04}{2}\right)^{2\cdot 2}$
$=1000(1.02)^4$
$=1082.43$
After two years, the account will contain \$1082.43.

4. $A=P\left(1+\frac{r}{n}\right)^{n\cdot t}$
$10,000=P\left(1+\frac{0.05}{12}\right)^{12\cdot 1}$
$10,000=P\left(1+\frac{0.05}{12}\right)^{12}$
$10,000=P(1.051162)$
$\frac{10,000}{1.051162}=P$
$9513.28=P$
To have \$10,000 at the end of one year, you need to invest \$9513.28 now.

5. sequence

6. $s_1=4(1)-1=3$; $s_4=4(4)-1=15$

7. $\sum_{k=1}^{4}(2k)=2\sum_{k=1}^{4}k=2\cdot\frac{4(4+1)}{2}=4(5)=20$

8. True

9. True; a sequence is a function whose domain is the set of positive integers.

10. True; $\sum_{k=1}^{2}k=\frac{2(2+1)}{2}=3$

11. $10!=10\cdot 9\cdot 8\cdot 7\cdot 6\cdot 5\cdot 4\cdot 3\cdot 2\cdot 1=3,628,800$

12. $9!=9\cdot 8\cdot 7\cdot 6\cdot 5\cdot 4\cdot 3\cdot 2\cdot 1=362,880$

13. $\frac{9!}{6!}=\frac{9\cdot 8\cdot 7\cdot 6!}{6!}=9\cdot 8\cdot 7=504$

14. $\frac{12!}{10!}=\frac{12\cdot 11\cdot 10!}{10!}=12\cdot 11=132$

15. $\frac{3!\cdot 7!}{4!}=\frac{3\cdot 2\cdot 1\cdot 7\cdot 6\cdot 5\cdot 4!}{4!}$
$=3\cdot 2\cdot 1\cdot 7\cdot 6\cdot 5=1,260$

16. $\frac{5!\cdot 8!}{3!}=\frac{5\cdot 4\cdot 3!\cdot 8!}{3!}$
$=5\cdot 4\cdot 8\cdot 7\cdot 6\cdot 5\cdot 4\cdot 3\cdot 2\cdot 1$
$=806,400$

17. $a_1=1,\ a_2=2,\ a_3=3,\ a_4=4,\ a_5=5$

18. $a_1=1^2+1=2,\ a_2=2^2+1=5,\ a_3=3^2+1=10,$
$a_4=4^2+1=17,\ a_5=5^2+1=26$

19. $a_1=\frac{1}{1+2}=\frac{1}{3},\ a_2=\frac{2}{2+2}=\frac{2}{4}=\frac{1}{2},$
$a_3=\frac{3}{3+2}=\frac{3}{5},\ a_4=\frac{4}{4+2}=\frac{4}{6}=\frac{2}{3},$
$a_5=\frac{5}{5+2}=\frac{5}{7}$

20.
$a_1=\frac{2\cdot 1+1}{2\cdot 1}=\frac{3}{2},\ a_2=\frac{2\cdot 2+1}{2\cdot 2}=\frac{5}{4},$
$a_3=\frac{2\cdot 3+1}{2\cdot 3}=\frac{7}{6},\ a_4=\frac{2\cdot 4+1}{2\cdot 4}=\frac{9}{8},$
$a_5=\frac{2\cdot 5+1}{2\cdot 5}=\frac{11}{10}$

21. $a_1=(-1)^{1+1}(1^2)=1,\ a_2=(-1)^{2+1}(2^2)=-4,$
$a_3=(-1)^{3+1}(3^2)=9, a_4=(-1)^{4+1}(4^2)=-16,$
$a_5=(-1)^{5+1}(5^2)=25$

22. $a_1 = (-1)^{1-1}\left(\dfrac{1}{2\cdot 1-1}\right) = 1,$

$a_2 = (-1)^{2-1}\left(\dfrac{2}{2\cdot 2-1}\right) = -\dfrac{2}{3},$

$a_3 = (-1)^{3-1}\left(\dfrac{3}{2\cdot 3-1}\right) = \dfrac{3}{5},$

$a_4 = (-1)^{4-1}\left(\dfrac{4}{2\cdot 4-1}\right) = -\dfrac{4}{7},$

$a_5 = (-1)^{5-1}\left(\dfrac{5}{2\cdot 5-1}\right) = \dfrac{5}{9}$

23. $a_1 = \dfrac{2^1}{3^1+1} = \dfrac{2}{4} = \dfrac{1}{2},\ a_2 = \dfrac{2^2}{3^2+1} = \dfrac{4}{10} = \dfrac{2}{5},$

$a_3 = \dfrac{2^3}{3^3+1} = \dfrac{8}{28} = \dfrac{2}{7},\ a_4 = \dfrac{2^4}{3^4+1} = \dfrac{16}{82} = \dfrac{8}{41},$

$a_5 = \dfrac{2^5}{3^5+1} = \dfrac{32}{244} = \dfrac{8}{61}$

24. $a_1 = \left(\dfrac{4}{3}\right)^1 = \dfrac{4}{3},\ a_2 = \left(\dfrac{4}{3}\right)^2 = \dfrac{16}{9},$

$a_3 = \left(\dfrac{4}{3}\right)^3 = \dfrac{64}{27},\ a_4 = \left(\dfrac{4}{3}\right)^4 = \dfrac{256}{81},$

$a_5 = \left(\dfrac{4}{3}\right)^5 = \dfrac{1024}{243}$

25. $a_1 = \dfrac{(-1)^1}{(1+1)(1+2)} = \dfrac{-1}{2\cdot 3} = -\dfrac{1}{6},$

$a_2 = \dfrac{(-1)^2}{(2+1)(2+2)} = \dfrac{1}{3\cdot 4} = \dfrac{1}{12},$

$a_3 = \dfrac{(-1)^3}{(3+1)(3+2)} = \dfrac{-1}{4\cdot 5} = -\dfrac{1}{20},$

$a_4 = \dfrac{(-1)^4}{(4+1)(4+2)} = \dfrac{1}{5\cdot 6} = \dfrac{1}{30},$

$a_5 = \dfrac{(-1)^5}{(5+1)(5+2)} = \dfrac{-1}{6\cdot 7} = -\dfrac{1}{42}$

26. $a_1 = \dfrac{3^1}{1} = \dfrac{3}{1} = 3,\ a_2 = \dfrac{3^2}{2} = \dfrac{9}{2},\ a_3 = \dfrac{3^3}{3} = \dfrac{27}{3} = 9,$

$a_4 = \dfrac{3^4}{4} = \dfrac{81}{4},\ a_5 = \dfrac{3^5}{5} = \dfrac{243}{5}$

27. $a_1 = \dfrac{1}{e^1} = \dfrac{1}{e},\ a_2 = \dfrac{2}{e^2},\ a_3 = \dfrac{3}{e^3},$

$a_4 = \dfrac{4}{e^4},\ a_5 = \dfrac{5}{e^5}$

28. $a_1 = \dfrac{1^2}{2^1} = \dfrac{1}{2},\ a_2 = \dfrac{2^2}{2^2} = 1,\ a_3 = \dfrac{3^2}{2^3} = \dfrac{9}{8},$

$a_4 = \dfrac{4^2}{2^4} = \dfrac{16}{16} = 1,\ a_5 = \dfrac{5^2}{2^5} = \dfrac{25}{32}$

29. Each term is a fraction with the numerator equal to the term number and the denominator equal to one more than the term number.

$a_n = \dfrac{n}{n+1}$

30. Each term is a fraction with the numerator equal to 1 and the denominator equal to the product of the term number and one more than the term number.

$a_n = \dfrac{1}{n(n+1)}$

31. Each term is a fraction with the numerator equal to 1 and the denominator equal to a power of 2. The power is equal to one less than the term number.

$a_n = \dfrac{1}{2^{n-1}}$

32. Each term is equal to a fraction with the numerator equal to a power of 2 and the denominator equal to a power of 3. Both powers are equal to the term number. Since the powers are the same, we can use rules for exponents to write each term as a power of $\frac{2}{3}$.

$a_n = \left(\dfrac{2}{3}\right)^n$

33. The terms form an alternating sequence. Ignoring the sign, each term always contains a 1. The sign alternates by raising -1 to a power. Since the first term is positive, we use $n+1$ as the power.

$a_n = (-1)^{n+1}$

34. The terms appear to alternate between whole numbers and fractions. If we write the whole numbers as fractions (e.g. $1=\frac{1}{1}$, $3=\frac{3}{1}$, etc.), we see that each term consists of a 1 and the term number. When n is odd, the numerator is n and the denominator is 1. When n is even, the numerator is 1 and the denominator is n. This alternating behavior occurs if we have a power that alternates sign. The alternating sign is obtained by using $(-1)^{n+1}$. Thus, we get

$$a_n = n^{(-1)^{n+1}}$$

35. The terms (ignoring the sign) are equal to the term number. The alternating sign is obtained by using $(-1)^{n+1}$.

$$a_n = (-1)^{n+1}\cdot n$$

36. Here again we have alternating signs which will be taken care of by using $(-1)^{n+1}$. The rest of the term is twice the term number.

$$a_n = (-1)^{n+1}\cdot 2n$$

37. $a_1 = 2,\ a_2 = 3+2 = 5,\ a_3 = 3+5 = 8,$
$a_4 = 3+8 = 11,\ a_5 = 3+11 = 14$

38. $a_1 = 3,\ a_2 = 4-3 = 1,\ a_3 = 4-1 = 3,$
$a_4 = 4-3 = 1,\ a_5 = 4-1 = 3$

39. $a_1 = -2,\ a_2 = 2+(-2) = 0,\ a_3 = 3+0 = 3,$
$a_4 = 4+3 = 7,\ a_5 = 5+7 = 12$

40. $a_1 = 1,\ a_2 = 2-1 = 1,\ a_3 = 3-1 = 2,$
$a_4 = 4-2 = 2,\ a_5 = 5-2 = 3$

41. $a_1 = 5,\ a_2 = 2\cdot 5 = 10,\ a_3 = 2\cdot 10 = 20,$
$a_4 = 2\cdot 20 = 40,\ a_5 = 2\cdot 40 = 80$

42. $a_1 = 2,\ a_2 = -2,\ a_3 = -(-2) = 2,$
$a_4 = -2,\ a_5 = -(-2) = 2$

43. $a_1 = 3,\ a_2 = \frac{3}{2},\ a_3 = \frac{\frac{3}{2}}{3} = \frac{1}{2},$
$a_4 = \frac{\frac{1}{2}}{4} = \frac{1}{8},\ a_5 = \frac{\frac{1}{8}}{5} = \frac{1}{40}$

44. $a_1 = -2,\ a_2 = 2+3(-2) = -4,$
$a_3 = 3+3(-4) = -9,\ a_4 = 4+3(-9) = -23,$
$a_5 = 5+3(-23) = -64$

45. $a_1 = 1,\ a_2 = 2,\ a_3 = 2\cdot 1 = 2,\ a_4 = 2\cdot 2 = 4,$
$a_5 = 4\cdot 2 = 8$

46. $a_1 = -1,\ a_2 = 1,\ a_3 = -1+3\cdot 1 = 2,$
$a_4 = 1+4\cdot 2 = 9,\ a_5 = 2+5\cdot 9 = 47$

47. $a_1 = A,\ a_2 = A+d,\ a_3 = (A+d)+d = A+2d,$
$a_4 = (A+2d)+d = A+3d,$
$a_5 = (A+3d)+d = A+4d$

48. $a_1 = A,\ a_2 = rA,\ a_3 = r(rA) = r^2A,$
$a_4 = r\left(r^2A\right) = r^3A,\ a_5 = r\left(r^3A\right) = r^4A$

49. $a_1 = \sqrt{2},\ a_2 = \sqrt{2+\sqrt{2}},\ a_3 = \sqrt{2+\sqrt{2+\sqrt{2}}},$
$a_4 = \sqrt{2+\sqrt{2+\sqrt{2+\sqrt{2}}}},$
$a_5 = \sqrt{2+\sqrt{2+\sqrt{2+\sqrt{2+\sqrt{2}}}}}$

50. $a_1 = \sqrt{2},\ a_2 = \sqrt{\frac{\sqrt{2}}{2}},\ a_3 = \sqrt{\frac{\sqrt{\frac{\sqrt{2}}{2}}}{2}},$
$a_4 = \sqrt{\frac{\sqrt{\frac{\sqrt{\frac{\sqrt{2}}{2}}}{2}}}{2}},\ a_5 = \sqrt{\frac{\sqrt{\frac{\sqrt{\frac{\sqrt{\frac{\sqrt{2}}{2}}}{2}}}{2}}}{2}}$

51. $\sum_{k=1}^{n}(k+2) = 3+4+5+6+7+\cdots+(n+2)$

52. $\sum_{k=1}^{n}(2k+1) = 3+5+7+9+\cdots+(2n+1)$

53. $\sum_{k=1}^{n}\frac{k^2}{2} = \frac{1}{2}+2+\frac{9}{2}+8+\frac{25}{2}+18+\frac{49}{2}+32+\cdots+\frac{n^2}{2}$

54. $\sum_{k=1}^{n}(k+1)^2 = 4+9+16+25+36+\cdots+(n+1)^2$

55. $\sum_{k=0}^{n} \frac{1}{3^k} = 1 + \frac{1}{3} + \frac{1}{9} + \frac{1}{27} + \cdots + \frac{1}{3^n}$

56. $\sum_{k=0}^{n} \left(\frac{3}{2}\right)^k = 1 + \frac{3}{2} + \frac{9}{4} + \cdots + \left(\frac{3}{2}\right)^n$

57. $\sum_{k=0}^{n-1} \frac{1}{3^{k+1}} = \frac{1}{3} + \frac{1}{9} + \frac{1}{27} + \cdots + \frac{1}{3^n}$

58. $\sum_{k=0}^{n-1} (2k+1) = 1+3+5+7+\cdots+(2(n-1)+1)$

$= 1+3+5+7+\cdots+(2n-1)$

59. $\sum_{k=2}^{n} (-1)^k \ln k = \ln 2 - \ln 3 + \ln 4 + \cdots + (-1)^n \ln n$

60. $\sum_{k=3}^{n} (-1)^{k+1} 2^k$

$= (-1)^4 2^3 + (-1)^5 2^4 + (-1)^6 2^5 + \cdots + (-1)^{n+1} 2^n$

$= 2^3 - 2^4 + 2^5 - 2^6 + \cdots + (-1)^{n+1} 2^n$

$= 8 - 16 + 32 - 64 + \ldots + (-1)^{n+1} 2^n$

61. $1+2+3+\cdots+20 = \sum_{k=1}^{20} k$

62. $1^3 + 2^3 + 3^3 + \cdots + 8^3 = \sum_{k=1}^{8} k^3$

63. $\frac{1}{2} + \frac{2}{3} + \frac{3}{4} + \cdots + \frac{13}{13+1} = \sum_{k=1}^{13} \frac{k}{k+1}$

64. $1+3+5+7+\cdots+[2(12)-1] = \sum_{k=1}^{12} (2k-1)$

65. $1 - \frac{1}{3} + \frac{1}{9} - \frac{1}{27} + \cdots + (-1)^6\left(\frac{1}{3^6}\right) = \sum_{k=0}^{6} (-1)^k \left(\frac{1}{3^k}\right)$

66. $\frac{2}{3} - \frac{4}{9} + \frac{8}{27} + \cdots + (-1)^{11+1}\left(\frac{2}{3}\right)^{11} = \sum_{k=1}^{11} (-1)^{k+1}\left(\frac{2}{3}\right)^k$

67. $3 + \frac{3^2}{2} + \frac{3^3}{3} + \cdots + \frac{3^n}{n} = \sum_{k=1}^{n} \frac{3^k}{k}$

68. $\frac{1}{e} + \frac{2}{e^2} + \frac{3}{e^3} + \cdots + \frac{n}{e^n} = \sum_{k=1}^{n} \frac{k}{e^k}$

69. $a + (a+d) + (a+2d) + \cdots + (a+nd) = \sum_{k=0}^{n} (a+kd)$

70. $a + ar + ar^2 + \cdots + ar^{n-1} = \sum_{k=1}^{n} ar^{k-1}$

71. a. $\sum_{k=1}^{10} 5 = \underbrace{5+5+5+\cdots+5}_{10 \text{ times}} = 10(5) = 50$

b. Verify:

```
sum(seq(5,n,1,10
))
                50
```

72. a. $\sum_{k=1}^{20} 8 = \underbrace{8+8+8+\cdots+8}_{20 \text{ times}} = 20(8) = 160$

b. Verify:

```
sum(seq(8,n,1,20
))
               160
```

73. a. $\sum_{k=1}^{6} k = 1+2+3+4+5+6 = 21$

b. Verify:

```
sum(seq(n,n,1,6)
)
                21
```

74. a. $\sum_{k=1}^{4} (-k) = (-1)+(-2)+(-3)+(-4) = -10$

b. Verify:

```
sum(seq(⁻n,n,1,4
))
               -10
```

75. a. $\sum_{k=1}^{5}(5k+3)=(5\cdot1+3)+(5\cdot2+3)+(5\cdot3+3)+(5\cdot4+3)+(5\cdot5+3)$
$=8+13+18+23+28=90$

b. Verify:

```
sum(seq(5n+3,n,1
,5))
                90
```

76. a. $\sum_{k=1}^{6}(3k-7)=(3\cdot1-7)+(3\cdot2-7)+(3\cdot3-7)+(3\cdot4-7)+(3\cdot5-7)+(3\cdot6-7)$
$=-4+(-1)+2+5+8+11=21$

b. Verify:

```
sum(seq(3n-7,n,1
,6))
                21
```

77. a. $\sum_{k=1}^{3}\left(k^2+4\right)=\left(1^2+4\right)+\left(2^2+4\right)+\left(3^2+4\right)$
$=5+8+13=26$

b. Verify:

```
sum(seq(n²+4,n,1
,3))
                26
```

78. a. $\sum_{k=0}^{4}\left(k^2-4\right)=\left(0^2-4\right)+\left(1^2-4\right)+\left(2^2-4\right)+\left(3^2-4\right)+\left(4^2-4\right)$
$=-4+(-3)+0+5+12=10$

b. Verify:

```
sum(seq(n²-4,n,0
,4))
                10
```

79 a. $\sum_{k=1}^{6}(-1)^k 2^k=(-1)^1\cdot2^1+(-1)^2\cdot2^2+(-1)^3\cdot2^3+(-1)^4\cdot2^4+(-1)^5\cdot2^5+(-1)^6\cdot2^6$
$=-2+4-8+16-32+64=42$

b. Verify:

```
sum(seq((-1)^n*2
^n,n,1,6))
                42
```

80. a. $\sum_{k=1}^{4}(-1)^k 3^k=(-1)^1\cdot3^1+(-1)^2\cdot3^2+(-1)^3\cdot3^3+(-1)^4\cdot3^4$
$=-3+9-27+81=60$

b. Verify:

```
sum(seq((-1)^n*3
^n,n,1,4))
                60
```

81. a. $\sum_{k=1}^{4}\left(k^3-1\right)$
$=\left(1^3-1\right)+\left(2^3-1\right)+\left(3^3-1\right)+\left(4^3-1\right)$
$=0+7+26+63=96$

b. Verify:

```
sum(seq(n^3-1,n,
1,4))
                96
```

82. a. $\sum_{k=0}^{3}\left(k^3+2\right)$
$=\left(0^3+2\right)+\left(1^3+2\right)+\left(2^3+2\right)+\left(3^3+2\right)$
$=2+3+10+29=44$

b. Verify:

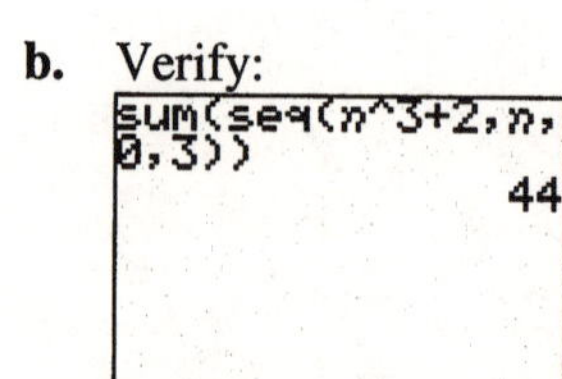

83. a. $B_1 = 1.01(3000) - 100 = \$2930$

b. Put the graphing utility in SEQuence mode. Enter Y= as follows, then examine the TABLE:

n	u(n)
9	2344.2
10	2267.6
11	2190.3
12	2112.2
13	2033.3
14	1953.7
15	1873.2

n=14

From the table we see that the balance is below \$2000 after 14 payments have been made. The balance then is \$1953.70.

c. Scrolling down the table, we find that balance is paid off in the 36th month. The last payment is \$83.78. There are 35 payments of \$100 and the last payment of \$83.78. The total amount paid is: 35(100) + 83.78(1.01) = \$3584.62. (we have to add the interest for the last month).

n	u(n)
31	470.71
32	375.42
33	279.17
34	181.96
35	83.781
36	-15.38
37	-115.5

n=36

d. The interest expense is:
3584.62 – 3000.00 = \$584.62

84. a. $B_1 = 1.005(18500) - 534.47 = \$18,058.03$

b. Put the graphing utility in SEQuence mode. Enter Y= as follows, then examine the TABLE:

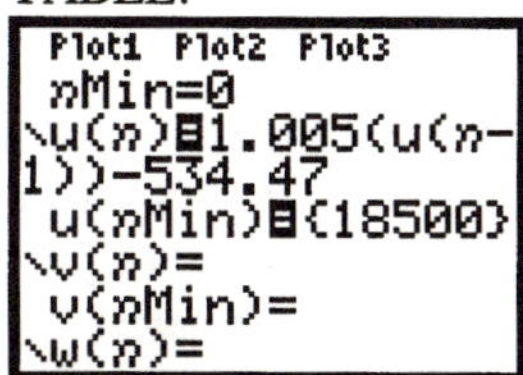

n	u(n)
15	11633
16	11157
17	10678
18	10197
19	9713.8
20	9227.9
21	8739.5

n=21

From the table we see that the balance is below \$10,000 after 19 payments have been made. The balance then is \$9713.76.

c. Scrolling down the table, we find that balance is paid off in the 39th month. The last payment is \$54.18. There are 38 payments of \$534.47 and the last payment of \$54.18 plus interest. The total amount paid is: 38(534.47) + 54.18(1.005) = \$20,364.31.

n	u(n)
34	2164.5
35	1640.9
36	1114.6
37	585.72
38	54.183
39	-480
40	-1017

n=40

d. The interest expense is:
20,364.31 – 18,500.00 = \$1864.31

85. a. $p_1 = 1.03(2000) + 20 = 2080;$

$p_2 = 1.03(2080) + 20 = 2162.4$

b. Scrolling down the table, we find the trout population exceeds 5000 at the end of the 26th month when the population is 5084.

```
Plot1 Plot2 Plot3
nMin=0
u(n)=1.03(u(n-1))+20
u(nMin)={2000}
v(n)=
v(nMin)=
w(n)=
```

n	u(n)
22	4442.9
23	4596.2
24	4754.1
25	4916.7
26	5084.2
27	5256.8
28	5434.5

n=28

86. a. $p_1 = 0.9(250) + 15 = 240;$

$p_2 = 0.9(240) + 15 = 231$

There are 240 tons of pollutants at the end of the first year, and 231 tons of pollutants at the end of the second year.

b. Scrolling down the table, we display the pollutant levels for the next 20 years.

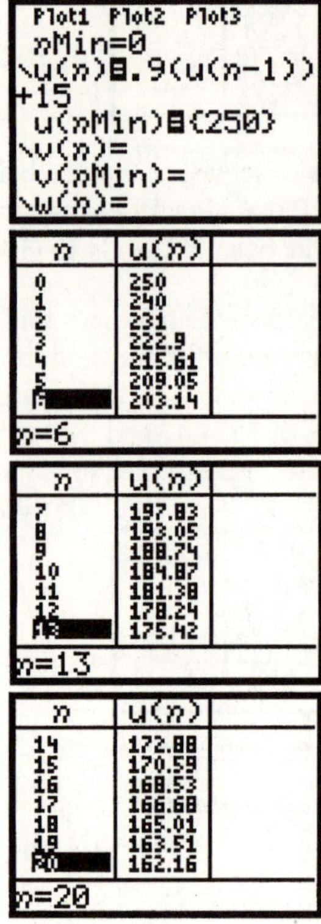

c. The equilibrium level of pollution occurs when $x = 0.9x + 15$. That is, when $x = 150$ tons.

$$x = 0.9x + 15$$
$$0.1x = 15$$
$$x = 150$$

87. a. Since the fund returns 8% compound annually, this is equivalent to a return of 2% each quarter. Defining a recursive sequence, we have:

$a_1 = 500, \quad a_n = 1.02a_{n-1} + 500$

b. Insert the formulas in your graphing utility and use the table feature to find when the value of the account will exceed $100,000:

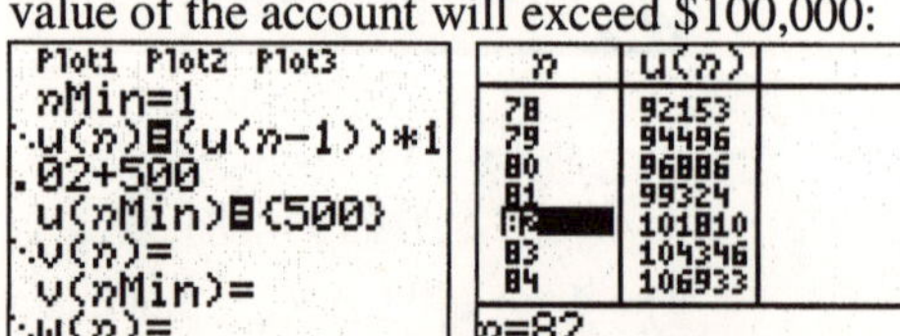

In the 82nd quarter (approximately May 2019) the value of the account will exceed $100,000 with a value of $101,810.

c. Find the value of the account in 25 years or 100 quarters:

n	u(n)
97	145670
98	149083
99	152565
100	156116
101	159738
102	163433
103	167202

n=100

The value of the account will be $156,116.

88. a. Since the fund returns 6% compounded annually, this is equivalent to a return of 0.5% each month. Defining a recursive sequence, we have:

$a_0 = 0, \quad a_n = 1.005a_{n-1} + 45$

b. Insert the formulas in your graphing utility and use the table feature to find when the value of the account will exceed $4,000:

```
Plot1 Plot2 Plot3
nMin=0
\u(n)=1.005(u(n-
1))+45
 u(nMin)={0}
\v(n)=
 v(nMin)=
\w(n)=
```

n	u(n)
71	3824.3
72	3888.4
73	3952.8
74	4017.6
75	4082.7
76	4148.1
77	4213.8

n=77

In the 74th month (February 2006) the value of the account will exceed $4,000 with a value of about $4,017.60.

c. Find the value of the account in 16 years or 192 months:

n	u(n)
188	13986
189	14101
190	14216
191	14332
192	14449
193	14566
194	14684

n=194

The value of the account will be approximately $14,449.

89. a. Since the interest rate is 6% per annum compounded monthly, this is equivalent to a rate of 0.5% each month. Defining a recursive sequence, we have:

$a_0 = 150{,}000, \quad a_n = 1.005a_{n-1} - 899.33$

b. $1.005(150,000) - 899.33 = \$149,850.67$

c. Enter the recursive formula in Y= and create the table:

d. Scroll through the table:

n	u(n)
57	140180
58	139981
59	139782
60	139581
61	139380
62	139177
63	138974

n=58

After 58 payments have been made, the balance is below \$140,000. The balance is about \$139,981.

e. Scroll through the table:

n	u(n)
354	5298.7
355	4425.9
356	3548.7
357	2667.1
358	1781.1
359	890.65
360	-4.231

n=360

The loan will be paid off at the end of 360 months or 30 years.
Total amount paid = (359)(\$899.33) + \$890.65(1.005) = \$323,754.57.

f. The total interest expense is the difference of the total of the payments and the original loan: $323,754.57 - 150,000 = \$173,754.57$

g. (a) Since the interest rate is 6% per annum compounded monthly, this is equivalent to a rate of 0.5% each month. Defining a recursive sequence, we have:

$$a_0 = 150,000, \quad a_n = 1.005a_{n-1} - 999.33$$

(b) $1.005(150,000) - 999.33 = \$149,750.67$

(c) Enter the recursive formula in Y= and create the table:

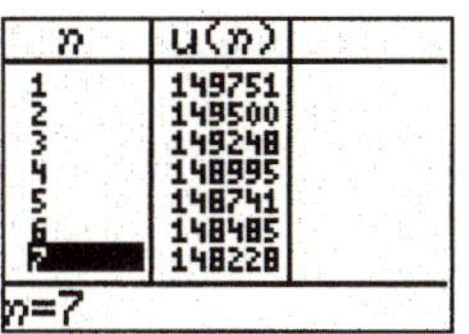

(d) Scroll through the table:

n	u(n)
36	140192
37	139894
38	139594
39	139293
40	138990
41	138685
42	138380

n=37

After 37 payments have been made, the balance is below \$140,000. The balance is \$139,894.

(e) Scroll through the table:

n	u(n)
278	353.69
279	-643.9
280	-1646
281	-2654
282	-3667
283	-4684
284	-5707

n=279

The loan will be paid off at the end of 279 months or 23 years and 3 months.
Total amount paid = (278)(\$999.33) + 353.69(1.005) = \$278,169.20

(f) The total interest expense is the difference of the total of the payments and the original loan:
$278,169.20 - 150,000 = \$128,169.20$

h. Yes, if they can afford the additional monthly payment. They would save \$44,586.07 in interest payments by paying the loan off sooner.

90. a. Since the interest rate is 6.5% per annum compounded at a rate of (6.5/12)% each month. Defining a recursive sequence, we have:

$$a_0 = 120,000,$$

$$a_n = \left(1 + \frac{0.065}{12}\right)a_{n-1} - 758.48$$

b. $\left(1 + \frac{0.065}{12}\right)(120,000) - 758.48 = \$119,891.52$

c. Enter the recursive formula in Y= and create the table:

d. Scroll through the table:

n	u(n)
125	100683
126	100470
127	100256
128	100040
129	99824
130	99606
131	99387

n=129

After 129 payments have been made, the balance is below \$100,000. The balance is about \$99,824.

e. Scroll through the table:

n	u(n)
356	2995
357	2252.8
358	1506.5
359	[illegible]
360	1.8011
361	-756.7
362	-1519

u(n)=756.1850706

The loan will be paid in the 360th month i.e. after 30 years.
Total amount paid = (359)(758.48) + 756.19(1+0.065/12) = \$273,054.60.

f. The total interest expense is the difference of the total of the payments and the original loan: $273{,}054.60 - 120{,}000 = \$153{,}054.60$

g. (a) Since the interest rate is 6.5% per annum compounded monthly, this is equivalent to a rate of (6.5/12)% each month. Defining a recursive sequence, we have:

$a_0 = 120{,}000,$

$$a_n = \left(1 + \frac{0.065}{12}\right) a_{n-1} - 858.48$$

(b) $\left(1 + \frac{0.065}{12}\right)(120{,}000) - 858.48$

$= \$119{,}791.52$

(c) Enter the recursive formula in Y= and create the table:

n	u(n)
0	120000
1	119792
2	119582
3	119371
4	119159
5	118946
6	118732

n=0

(d) Scroll through the table:

n	u(n)
74	101085
75	100774
76	100461
77	100147
78	99831
79	99513
80	99194

n=78

After 78 payments have been made, the balance is below \$100,000. The balance is \$99,831.

(e) Scroll through the table:

n	u(n)
259	2545.2
260	1700.5
261	851.23
262	-2.638
263	-861.1
264	-1724
265	-2592

n=262

The loan will be paid off at the end of 262 months or 21 years and 10 months.
Total amount paid = (261)(858.48) + (851.23)(1+0.065/12) = \$224,919.12.

(f) The total interest expense is the difference of the total of the payments and the original loan:
$224{,}919.12 - 120{,}000 = \$104{,}919.12$

h. Yes, if they can afford the additional monthly payment. They would save \$48,238.49 in interest payments by paying the loan off sooner.

91. $a_1 = 1,\ a_2 = 1,\ a_3 = 2,\ a_4 = 3,\ a_5 = 5,$
$a_6 = 8,\ a_7 = 13,\ a_8 = 21,\ a_n = a_{n-1} + a_{n-2}$
$a_8 = a_7 + a_6 = 13 + 8 = 21$
After 7 months there are 21 mature pairs of rabbits.

92. a. $$u_1 = \frac{\left(1+\sqrt{5}\right)^1 - \left(1-\sqrt{5}\right)^1}{2^1\sqrt{5}} = \frac{1+\sqrt{5}-1+\sqrt{5}}{2\sqrt{5}} = \frac{2\sqrt{5}}{2\sqrt{5}} = 1$$

$$u_2 = \frac{\left(1+\sqrt{5}\right)^2 - \left(1-\sqrt{5}\right)^2}{2^2\sqrt{5}} = \frac{1+2\sqrt{5}+5-1+2\sqrt{5}-5}{4\sqrt{5}} = \frac{4\sqrt{5}}{4\sqrt{5}} = 1$$

b. $$u_{n+1} + u_n = \frac{\left(1+\sqrt{5}\right)^{n+1} - \left(1-\sqrt{5}\right)^{n+1}}{2^{n+1}\sqrt{5}} + \frac{\left(1+\sqrt{5}\right)^n - \left(1-\sqrt{5}\right)^n}{2^n\sqrt{5}}$$

$$= \frac{\left(1+\sqrt{5}\right)^{n+1} - \left(1-\sqrt{5}\right)^{n+1} + 2\left(1+\sqrt{5}\right)^n - 2\left(1-\sqrt{5}\right)^n}{2^{n+1}\sqrt{5}}$$

$$= \frac{\left(1+\sqrt{5}\right)^n\left[1+\sqrt{5}+2\right] - \left(1-\sqrt{5}\right)^n\left[1-\sqrt{5}+2\right]}{2^{n+1}\sqrt{5}}$$

$$= \frac{\left(1+\sqrt{5}\right)^n\left[3+\sqrt{5}\right] - \left(1-\sqrt{5}\right)^n\left[3-\sqrt{5}\right]}{2^{n+1}\sqrt{5}}$$

$$= \frac{\left(1+\sqrt{5}\right)^{n+2}\dfrac{\left(3+\sqrt{5}\right)}{\left(1+\sqrt{5}\right)^2} - \left(1-\sqrt{5}\right)^{n+2}\dfrac{\left(3-\sqrt{5}\right)}{\left(1-\sqrt{5}\right)^2}}{2^{n+1}\sqrt{5}}$$

$$= \frac{\left(1+\sqrt{5}\right)^{n+2}\dfrac{\left(3+\sqrt{5}\right)}{\left(6+2\sqrt{5}\right)} - \left(1-\sqrt{5}\right)^{n+2}\dfrac{\left(3-\sqrt{5}\right)}{\left(6-2\sqrt{5}\right)}}{2^{n+1}\sqrt{5}}$$

$$= \frac{\left(1+\sqrt{5}\right)^{n+2}\frac{1}{2} - \left(1-\sqrt{5}\right)^{n+2}\frac{1}{2}}{2^{n+1}\sqrt{5}} = \frac{\left(1+\sqrt{5}\right)^{n+2} - \left(1-\sqrt{5}\right)^{n+2}}{2^{n+2}\sqrt{5}}$$

$$= u_{n+2}$$

c. Since $u_1 = 1,\ u_2 = 1,\ u_{n+2} = u_{n+1} + u_n,\ \{u_n\}$ is the Fibonacci sequence.

93. 1, 1, 2, 3, 5, 8, 13
This is the Fibonacci sequence.

94. a. $u_1 = 1,\ u_2 = 1,\ u_3 = 2,\ u_4 = 3,\ u_5 = 5,\ u_6 = 8,\ u_7 = 13,\ u_8 = 21,\ u_9 = 34,\ u_{10} = 55$

b. $\frac{u_2}{u_1} = \frac{1}{1} = 1,\ \frac{u_3}{u_2} = \frac{2}{1} = 1,\ \frac{u_4}{u_3} = \frac{3}{2} = 1.5,\ \frac{u_5}{u_4} = \frac{5}{3} = 1.67,\ \frac{u_6}{u_5} = \frac{8}{5} = 1.6,$

$\frac{u_7}{u_6} = \frac{13}{8} = 1.625,\ \frac{u_8}{u_7} = \frac{21}{13} = 1.615,\ \frac{u_9}{u_8} = \frac{34}{21} = 1.619,$

$\frac{u_{10}}{u_9} = \frac{55}{34} = 1.618,\ \frac{u_{11}}{u_{10}} = \frac{89}{55} = 1.618$

c. 1.618 (the exact value is $\frac{1+\sqrt{5}}{2}$)

d. $\frac{u_1}{u_2} = \frac{1}{1} = 1,\ \frac{u_2}{u_3} = \frac{1}{2} = 0.5,\ \frac{u_3}{u_4} = \frac{2}{3} = 0.667,\ \frac{u_4}{u_5} = \frac{3}{5} = 0.6,$

$\frac{u_5}{u_6} = \frac{5}{8} = 0.625,\ \frac{u_6}{u_7} = \frac{8}{13} = 0.615,\ \frac{u_7}{u_8} = \frac{13}{21} = 0.619,$

$\frac{u_8}{u_9} = \frac{21}{34} = 0.618,\ \frac{u_9}{u_{10}} = \frac{34}{55} = 0.618,\ \frac{u_{10}}{u_{11}} = \frac{55}{89} = 0.618$

e. 0.618 (the exact value is $\frac{2}{1+\sqrt{5}}$)

95. a. $e^{1.3} \approx \sum_{k=0}^{4} \frac{1.3^k}{k!} = \frac{1.3^0}{0!} + \frac{1.3^1}{1!} + \ldots + \frac{1.3^4}{4!}$

≈ 3.630170833

b. $e^{1.3} \approx \sum_{k=0}^{7} \frac{1.3^k}{k!} = \frac{1.3^0}{0!} + \frac{1.3^1}{1!} + \ldots + \frac{1.3^7}{7!}$

≈ 3.669060828

c. $e^{1.3} \approx 3.669296668$

```
e^(1.3)
         3.669296668
```

d. It will take $n = 12$ to approximate $e^{1.3}$ correct to 8 decimal places.

```
         3.669296614
sum(seq(1.3^n/n!
,n,0,13))
         3.669296667
sum(seq(1.3^n/n!
,n,0,12))
         3.669296662
```

96. a. $e^{-2.4} \approx \sum_{k=0}^{3} \frac{(-2.4)^k}{k!} = \frac{(-2.4)^0}{0!} + \frac{(-2.4)^1}{1!} + \frac{(-2.4)^2}{2!} + \frac{(-2.4)^3}{3!} = -0.824$

b. $e^{-2.4} \approx \sum_{k=0}^{6} \frac{(-2.4)^k}{k!} = \frac{(-2.4)^0}{0!} + \frac{(-2.4)^1}{1!} + \ldots + \frac{(-2.4)^6}{6!} = 0.1602688$

c. $e^{-2.4} \approx 0.0907179533$

e^(-2.4)
.0907179533

d. It will take $n = 17$ to approximate $e^{-2.4}$ correct to 8 decimal places.

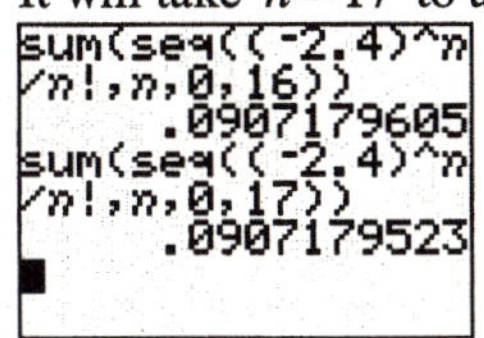

97. To show that $1+2+3+...+(n-1)+n = \dfrac{n(n+1)}{2}$

Let $S = 1 + \ 2 \ + \ 3 \ + ... + (n-1) + n,$ we can reverse the order to get

$+S = n + (n-1) + (n-2) + ... + \ 2 \ + \ 1,$ now add these two lines to get

$2S = [1+n] + [2+(n-1)] + [3+(n-2)] + ... + [(n-1)+2] + [n+1]$

So we have $2S = [1+n] + [1+n] + [1+n] + ... + [n+1] + [n+1] = n \cdot [n+1]$

$\therefore 2S = n(n+1)$

$S = \dfrac{n \cdot (n+1)}{2}$

98. Answers will vary.

Section 11.2

1. arithmetic

2. False; the sum of the first and last terms equals twice the sum of all the terms divided by the number of terms.

3. $d = a_{n+1} - a_n$
$= (n+1+4) - (n+4)$
$= n + 5 - n - 4$
$= 1$
The difference between consecutive terms is constant, therefore the sequence is arithmetic.
$a_1 = 1 + 4 = 5,\ a_2 = 2 + 4 = 6,\ a_3 = 3 + 4 = 7,$
$a_4 = 4 + 4 = 8$

4. $d = a_{n+1} - a_n$
$= (n+1-5) - (n-5)$
$= n - 4 - n + 5$
$= 1$
The difference between consecutive terms is constant, therefore the sequence is arithmetic.
$a_1 = 1 - 5 = -4,\ a_2 = 2 - 5 = -3,\ a_3 = 3 - 5 = -2,$
$a_4 = 4 - 5 = -1$

5. $d = a_{n+1} - a_n$
$= (2(n+1) - 5) - (2n - 5)$
$= 2n + 2 - 5 - 2n + 5$
$= 2$
The difference between consecutive terms is constant, therefore the sequence is arithmetic.
$a_1 = 2 \cdot 1 - 5 = -3,\ a_2 = 2 \cdot 2 - 5 = -1,$
$a_3 = 2 \cdot 3 - 5 = 1,\ a_4 = 2 \cdot 4 - 5 = 3$

6. $d = a_{n+1} - a_n$
$= (3(n+1)+1) - (3n+1)$
$= 3n + 3 + 1 - 3n - 1$
$= 3$
The difference between consecutive terms is constant, therefore the sequence is arithmetic.
$a_1 = 3 \cdot 1 + 1 = 4,\ a_2 = 3 \cdot 2 + 1 = 7,$
$a_3 = 3 \cdot 3 + 1 = 10,\ a_4 = 3 \cdot 4 + 1 = 13$

7. $d = a_{n+1} - a_n$
$= (6 - 2(n+1)) - (6 - 2n)$
$= 6 - 2n - 2 - 6 + 2n$
$= -2$
The difference between consecutive terms is constant, therefore the sequence is arithmetic.
$a_1 = 6 - 2 \cdot 1 = 4,\ a_2 = 6 - 2 \cdot 2 = 2,$
$a_3 = 6 - 2 \cdot 3 = 0,\ a_4 = 6 - 2 \cdot 4 = -2$

8. $d = a_{n+1} - a_n$
$= (4 - 2(n+1)) - (4 - 2n)$
$= 4 - 2n - 2 - 4 + 2n$
$= -2$
The difference between consecutive terms is constant, therefore the sequence is arithmetic.
$a_1 = 4 - 2 \cdot 1 = 2,\ a_2 = 4 - 2 \cdot 2 = 0,$
$a_3 = 4 - 2 \cdot 3 = -2,\ a_4 = 4 - 2 \cdot 4 = -4$

9. $d = a_{n+1} - a_n$
$= \left(\frac{1}{2} - \frac{1}{3}(n+1)\right) - \left(\frac{1}{2} - \frac{1}{3}n\right)$
$= \frac{1}{2} - \frac{1}{3}n - \frac{1}{3} - \frac{1}{2} + \frac{1}{3}n$
$= -\frac{1}{3}$
The difference between consecutive terms is constant, therefore the sequence is arithmetic.
$a_1 = \frac{1}{2} - \frac{1}{3} \cdot 1 = \frac{1}{6},\ a_2 = \frac{1}{2} - \frac{1}{3} \cdot 2 = -\frac{1}{6},$
$a_3 = \frac{1}{2} - \frac{1}{3} \cdot 3 = -\frac{1}{2},\ a_4 = \frac{1}{2} - \frac{1}{3} \cdot 4 = -\frac{5}{6}$

10. $d = a_{n+1} - a_n$
$= \left(\frac{2}{3} + \frac{1}{4}(n+1)\right) - \left(\frac{2}{3} + \frac{1}{4}n\right)$
$= \frac{2}{3} + \frac{1}{4}n + \frac{1}{4} - \frac{2}{3} - \frac{1}{4}n$
$= \frac{1}{4}$
The difference between consecutive terms is constant, therefore the sequence is arithmetic.
$a_1 = \frac{2}{3} + \frac{1}{4} \cdot 1 = \frac{11}{12},\ a_2 = \frac{2}{3} + \frac{1}{4} \cdot 2 = \frac{7}{6},$
$a_3 = \frac{2}{3} + \frac{1}{4} \cdot 3 = \frac{17}{12},\ a_4 = \frac{2}{3} + \frac{1}{4} \cdot 4 = \frac{5}{3}$

11. $d = a_{n+1} - a_n$
$= \ln\left(3^{n+1}\right) - \ln\left(3^n\right)$
$= (n+1)\ln(3) - n\ln(3)$
$= \ln 3(n+1-n)$
$= \ln(3)$
The difference between consecutive terms is constant, therefore the sequence is arithmetic.
$a_1 = \ln\left(3^1\right) = \ln(3),\ a_2 = \ln\left(3^2\right) = 2\ln(3),$
$a_3 = \ln\left(3^3\right) = 3\ln(3),\ a_4 = \ln\left(3^4\right) = 4\ln(3)$

12. $d = a_{n+1} - a_n = e^{\ln(n+1)} - e^{\ln n} = (n+1) - n = 1$
The difference between consecutive terms is constant, therefore the sequence is arithmetic.
$a_1 = e^{\ln 1} = 1,\ a_2 = e^{\ln 2} = 2,\ a_3 = e^{\ln 3} = 3,$
$a_4 = e^{\ln 4} = 4$

13. $a_n = a + (n-1)d$
$= 2 + (n-1)3$
$= 2 + 3n - 3$
$= 3n - 1$
$a_5 = 3 \cdot 5 - 1 = 14$

14. $a_n = a + (n-1)d$
$= -2 + (n-1)4$
$= -2 + 4n - 4$
$= 4n - 6$
$a_5 = 4 \cdot 5 - 6 = 14$

15. $a_n = a+(n-1)d$
$= 5+(n-1)(-3)$
$= 5-3n+3$
$= 8-3n$
$a_5 = 8-3\cdot 5 = -7$

16. $a_n = a+(n-1)d$
$= 6+(n-1)(-2)$
$= 6-2n+2$
$= 8-2n$
$a_5 = 8-2\cdot 5 = -2$

17. $a_n = a+(n-1)d$
$= 0+(n-1)\frac{1}{2}$
$= \frac{1}{2}n-\frac{1}{2}$
$= \frac{1}{2}(n-1)$
$a_5 = \frac{1}{2}\left(5-\frac{1}{2}\right) = 2$

18. $a_n = a+(n-1)d$
$= 1+(n-1)\left(-\frac{1}{3}\right)$
$= 1-\frac{1}{3}n+\frac{1}{3}$
$= \frac{4}{3}-\frac{1}{3}n$
$a_5 = \frac{4}{3}-\frac{1}{3}\cdot 5 = \frac{4}{3}-\frac{5}{3} = -\frac{1}{3}$

19. $a_n = a+(n-1)d$
$= \sqrt{2}+(n-1)\sqrt{2}$
$= \sqrt{2}+\sqrt{2}n-\sqrt{2}$
$= \sqrt{2}n$
$a_5 = 5\sqrt{2}$

20. $a_n = a+(n-1)d = 0+(n-1)\pi = (n-1)\pi$
$a_5 = 5\pi-\pi = 4\pi$

21. $a_1 = 2,\ d = 2,\ a_n = a+(n-1)d$
$a_{12} = 2+(12-1)2 = 2+11(2) = 2+22 = 24$

22. $a_1 = -1,\ d = 2,\ a_n = a+(n-1)d$
$a_8 = -1+(8-1)2 = -1+7(2) = -1+14 = 13$

23. $a_1 = 1,\ d = -2-1 = -3,\ a_n = a+(n-1)d$
$a_{10} = 1+(10-1)(-3) = 1+9(-3) = 1-27 = -26$

24. $a_1 = 5,\ d = 0-5 = -5,\ a_n = a+(n-1)d$
$a_9 = 5+(9-1)(-5) = 5+8(-5) = 5-40 = -35$

25. $a_1 = a,\ d = (a+b)-a = b,\ a_n = a+(n-1)d$
$a_8 = a+(8-1)b = a+7b$

26. $a_1 = 2\sqrt{5},\ d = 4\sqrt{5}-2\sqrt{5} = 2\sqrt{5},$
$a_n = a+(n-1)d$
$a_7 = 2\sqrt{5}+(7-1)2\sqrt{5}$
$= 2\sqrt{5}+6\left(2\sqrt{5}\right)$
$= 2\sqrt{5}+12\sqrt{5}$
$= 14\sqrt{5}$

27. $a_8 = a+7d = 8 \qquad a_{20} = a+19d = 44$
Solve the system of equations by subtracting the first equation from the second:
$12d = 36 \Rightarrow d = 3$
$a = 8-7(3) = 8-21 = -13$
Recursive formula: $a_1 = -13 \qquad a_n = a_{n-1}+3$
nth term: $a_n = a+(n-1)d$
$= -13+(n-1)(3)$
$= -13+3n-3$
$= 3n-16$

28. $a_4 = a+3d = 3 \qquad a_{20} = a+19d = 35$
Solve the system of equations by subtracting the first equation from the second:
$16d = 32 \Rightarrow d = 2$
$a = 3-3(2) = 3-6 = -3$
Recursive formula: $a_1 = -3 \qquad a_n = a_{n-1}+2$
nth term: $a_n = a+(n-1)d$
$= -3+(n-1)(2)$
$= -3+2n-2$
$= 2n-5$

29. $a_9 = a + 8d = -5 \quad a_{15} = a + 14d = 31$
Solve the system of equations by subtracting the first equation from the second:
$6d = 36 \Rightarrow d = 6$
$a = -5 - 8(6) = -5 - 48 = -53$
Recursive formula: $a_1 = -53 \quad a_n = a_{n-1} + 6$
nth term: $a_n = a + (n-1)d$
$= -53 + (n-1)(6)$
$= -53 + 6n - 6$
$= 6n - 59$

30. $a_8 = a + 7d = 4 \quad a_{18} = a + 17d = -96$
Solve the system of equations by subtracting the first equation from the second:
$10d = -100 \Rightarrow d = -10$
$a = 4 - 7(-10) = 4 + 70 = 74$
Recursive formula: $a_1 = 74 \quad a_n = a_{n-1} - 10$
nth term: $a_n = a + (n-1)d$
$= 74 + (n-1)(-10)$
$= 74 - 10n + 10$
$= 84 - 10n$

31. $a_{15} = a + 14d = 0 \quad a_{40} = a + 39d = -50$
Solve the system of equations by subtracting the first equation from the second:
$25d = -50 \Rightarrow d = -2$
$a = -14(-2) = 28$
Recursive formula: $a_1 = 28 \quad a_n = a_{n-1} - 2$
nth term: $a_n = a + (n-1)d$
$= 28 + (n-1)(-2)$
$= 28 - 2n + 2$
$= 30 - 2n$

32. $a_5 = a + 4d = -2 \quad a_{13} = a + 12d = 30$
Solve the system of equations by subtracting the first equation from the second:
$8d = 32 \Rightarrow d = 4$
$a = -2 - 4(4) = -18$
Recursive formula: $a_1 = -18 \quad a_n = a_{n-1} + 4$
nth term: $a_n = a + (n-1)d$
$= -18 + (n-1)(4)$
$= -18 + 4n - 4$
$= 4n - 22$

33. $a_{14} = a + 13d = -1 \quad a_{18} = a + 17d = -9$
Solve the system of equations by subtracting the first equation from the second:
$4d = -8 \Rightarrow d = -2$
$a = -1 - 13(-2) = -1 + 26 = 25$
Recursive formula: $a_1 = 25 \quad a_n = a_{n-1} - 2$
nth term: $a_n = a + (n-1)d$
$= 25 + (n-1)(-2)$
$= 25 - 2n + 2$
$= 27 - 2n$

34. $a_{12} = a + 11d = 4 \quad a_{18} = a + 17d = 28$
Solve the system of equations by subtracting the first equation from the second:
$6d = 24 \Rightarrow d = 4$
$a = 4 - 11(4) = 4 - 44 = -40$
Recursive formula: $a_1 = -40 \quad a_n = a_{n-1} + 4$
nth term: $a_n = a + (n-1)d$
$= -40 + (n-1)(4)$
$= -40 + 4n - 4$
$= 4n - 44$

35. $S_n = \frac{n}{2}(a + a_n) = \frac{n}{2}(1 + (2n-1)) = \frac{n}{2}(2n) = n^2$

36. $S_n = \frac{n}{2}(a + a_n) = \frac{n}{2}(2 + 2n) = n + n^2 = n(n+1)$

37. $S_n = \frac{n}{2}(a + a_n) = \frac{n}{2}(7 + (2 + 5n)) = \frac{n}{2}(9 + 5n)$

38. $S_n = \frac{n}{2}(a + a_n) = \frac{n}{2}(-1 + (4n-5))$
$= \frac{n}{2}(4n-6) = 2n^2 - 3n$
$= n(2n-3)$

39. $a_1 = 2,\ d = 4 - 2 = 2,\quad a_n = a + (n-1)d$

$$70 = 2 + (n-1)2$$
$$70 = 2 + 2n - 2$$
$$70 = 2n$$
$$n = 35$$
$$S_n = \frac{n}{2}(a + a_n) = \frac{35}{2}(2+70)$$
$$= \frac{35}{2}(72) = 35(36)$$
$$= 1260$$

40. $a_1 = 1,\ d = 3 - 1 = 2,\quad a_n = a + (n-1)d$

$$59 = 1 + (n-1)2$$
$$59 = 1 + 2n - 2$$
$$60 = 2n$$
$$n = 30$$
$$S_n = \frac{n}{2}(a + a_n) = \frac{30}{2}(1+59) = 15(60) = 900$$

41. $a_1 = 5,\ d = 9 - 5 = 4,\quad a_n = a + (n-1)d$

$$49 = 5 + (n-1)4$$
$$49 = 5 + 4n - 4$$
$$48 = 4n$$
$$n = 12$$
$$S_n = \frac{n}{2}(a + a_n) = \frac{12}{2}(5+49) = 6(54) = 324$$

42. $a_1 = 2,\ d = 5 - 2 = 3,\quad a_n = a + (n-1)d$

$$41 = 2 + (n-1)3$$
$$41 = 2 + 3n - 3$$
$$42 = 3n$$
$$n = 14$$
$$S_n = \frac{n}{2}(a + a_n) = \frac{14}{2}(2+41) = 7(43) = 301$$

43. Using the sum of the sequence feature:

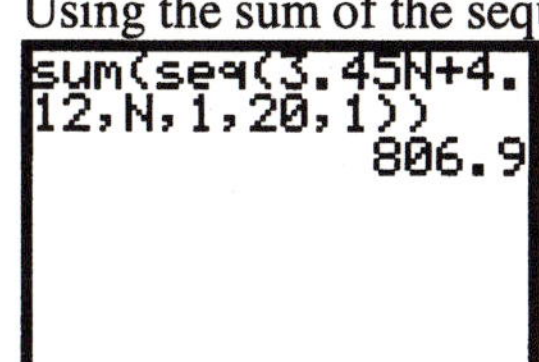

44. Using the sum of the sequence feature:

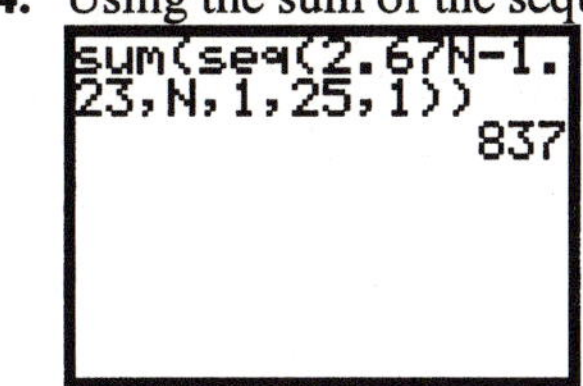

45. $d = 5.2 - 2.8 = 2.4$

$$a = 2.8$$
$$36.4 = 2.8 + (n-1)2.4$$
$$36.4 = 2.8 + 2.4n - 2.4$$
$$36 = 2.4n$$
$$n = 15$$
$$a_n = 2.8 + (n-1)2.4$$
$$= 2.8 + 2.4n - 2.4$$
$$= 2.4n + 0.4$$

```
sum(seq(2.4N+.4,
N,1,15,1))
                294
```

46. $d = 7.3 - 5.4 = 1.9$

$$a = 5.4$$
$$32 = 5.4 + (n-1)1.9$$
$$32 = 5.4 + 1.9n - 1.9$$
$$28.5 = 1.9n$$
$$n = 15$$
$$a_n = 5.4 + (n-1)1.9 = 5.4 + 1.9n - 1.9 = 1.9n + 3.5$$

```
sum(seq(1.9N+3.5
,N,1,15,1))
              280.5
```

47. $d = 7.48 - 4.9 = 2.58$

$$a = 4.9$$
$$66.82 = 4.9 + (n-1)2.58$$
$$66.82 = 4.9 + 2.58n - 2.58$$
$$64.5 = 2.58n$$
$$n = 25$$
$$a_n = 4.9 + (n-1)2.58 = 4.9 + 2.58n - 2.58$$
$$a_n = 2.58n + 2.32$$

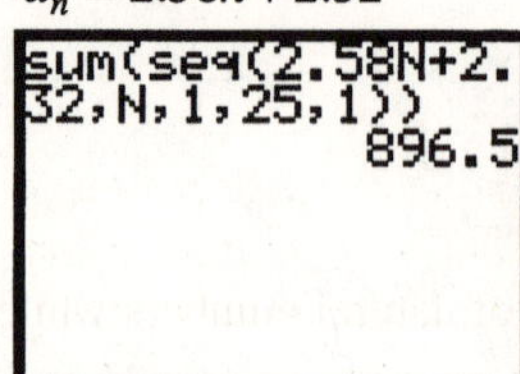

48. $d = 6.9 - 3.71 = 3.19$

$$a = 3.71$$
$$80.27 = 3.71 + (n-1)3.19$$
$$80.27 = 3.71 + 3.19n - 3.19$$
$$79.75 = 3.19n$$
$$n = 25$$
$$a_n = 3.71 + (n-1)3.19 = 3.71 + 3.19n - 3.19$$
$$a_n = 3.19n + 0.52$$

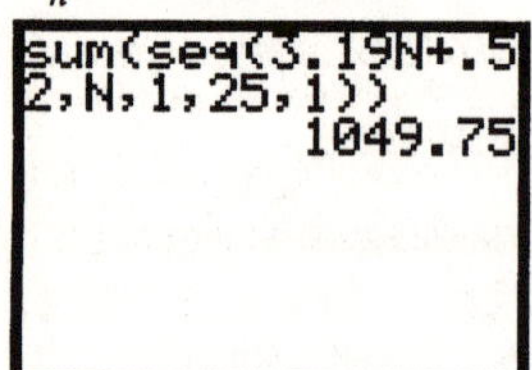

49. Find the common difference of the terms and solve the system of equations:

$$(2x+1) - (x+3) = d \Rightarrow x - 2 = d$$
$$(5x+2) - (2x+1) = d \Rightarrow 3x + 1 = d$$
$$3x + 1 = x - 2$$
$$2x = -3$$
$$x = -\frac{3}{2}$$

50. Find the common difference of the terms and solve the system of equations:

$$(3x+2) - (2x) = d \Rightarrow x + 2 = d$$
$$(5x+3) - (3x+2) = d \Rightarrow 2x + 1 = d$$
$$2x + 1 = x + 2$$
$$x = 1$$

51. The total number of seats is:

$$S = 25 + 26 + 27 + \cdots + \left(25 + 29(1)\right)$$

This is the sum of an arithmetic sequence with $d = 1,\ a = 25,$ and $n = 30$.

Find the sum of the sequence:

$$S_{30} = \frac{30}{2}[2(25) + (30-1)(1)]$$
$$= 15(50 + 29) = 15(79)$$
$$= 1185$$

There are 1185 seats in the theater.

52. The total number of seats is:

$$S = 15 + 17 + 19 + \cdots + \left(15 + \left(39(2)\right)\right)$$

This is the sum of an arithmetic sequence with $d = 2,\ a = 15,$ and $n = 40$.

Find the sum of the sequence:

$$S_{40} = \frac{40}{2}[2(15) + (40-1)(2)]$$
$$= 20(30 + 78) = 20(108)$$
$$= 2160$$

The corner section has 2160 seats.

53. The lighter colored tiles have 20 tiles in the bottom row and 1 tile in the top row. The number decreases by 1 as we move up the triangle. This is an arithmetic sequence with $a_1 = 20,\ d = -1,$ and $n = 20$. Find the sum:

$$S = \frac{20}{2}[2(20) + (20-1)(-1)]$$
$$= 10(40 - 19) = 10(21)$$
$$= 210$$

There are 210 lighter tiles.

The darker colored tiles have 19 tiles in the bottom row and 1 tile in the top row. The number decreases by 1 as we move up the triangle. This is an arithmetic sequence with $a_1 = 19,\ d = -1,$ and $n = 19$. Find the sum:

$$S = \frac{19}{2}[2(19) + (19-1)(-1)]$$
$$= \frac{19}{2}(38 - 18) = \frac{19}{2}(20) = 190$$

There are 190 darker tiles.

54. The number of bricks required decreases by 2 on each successive step. This is an arithmetic sequence with $a_1 = 100$, $d = -2$, and $n = 30$.

a. The number of bricks for the top step is:

$$a_{30} = a_1 + (n-1)d = 100 + (30-1)(-2)$$
$$= 100 + 29(-2) = 100 - 58$$
$$= 42$$

42 bricks are required for the top step.

b. The total number of bricks required is the sum of the sequence:

$$S = \frac{30}{2}[100 + 42] = 15(142) = 2130$$

2130 bricks are required to build the staircase.

55. Find n in an arithmetic sequence with $a_1 = 10$, $d = 4$, $S_n = 2040$.

$$S_n = \frac{n}{2}[2a_1 + (n-1)d]$$
$$2040 = \frac{n}{2}[2(10) + (n-1)4]$$
$$4080 = n[20 + 4n - 4]$$
$$4080 = n(4n + 16)$$
$$4080 = 4n^2 + 16n$$
$$1020 = n^2 + 4n$$
$$n^2 + 4n - 1020 = 0$$
$$(n+34)(n-30) = 0 \Rightarrow n = -34 \text{ or } n = 30$$

There are 30 rows in the corner section of the stadium.

56. The yearly salaries form an arithmetic sequence with $a_1 = 35{,}000$, $d = 1400$, $S_n = 280{,}000$. Find the number of years for the aggregate salary to equal \$280,000.

$$S_n = \frac{n}{2}[2a_1 + (n-1)d]$$
$$280{,}000 = \frac{n}{2}[2(35{,}000) + (n-1)1400]$$
$$280{,}000 = n[35{,}000 + 700n - 700]$$
$$280{,}000 = n(700n + 34{,}300)$$
$$280{,}000 = 700n^2 + 34{,}300n$$
$$400 = n^2 + 49n$$
$$n^2 + 49n - 400 = 0$$

$$n = \frac{-49 \pm \sqrt{49^2 - 4(1)(-400)}}{2(1)}$$
$$= \frac{-49 \pm \sqrt{4001}}{2} \approx \frac{-49 \pm 63.25}{2}$$

$n \approx 7.13$ or $n \approx -56.13$

It takes about 8 years to have an aggregate salary of at least \$280,000. The aggregate salary after 8 years will be \$319,200.

57. Answers will vary.

58. Answers will vary. Both increase (or decrease) at a constant rate, but the domain of an arithmetic sequence is the set of natural numbers while the domain of a linear function is the set of all real numbers.

Section 11.3

1. geometric

2. $\sum_{k-1}^{\infty} ar^{k-1} = \frac{a}{1-r}$

3. True

4. False; the common ratio can be positive or negative (or 0, but this results in a sequence of only 0s).

5. $r = \frac{3^{n+1}}{3^n} = 3^{n+1-n} = 3$

The ratio of consecutive terms is constant, therefore the sequence is geometric.

$a_1 = 3^1 = 3$, $a_2 = 3^2 = 9$,

$a_3 = 3^3 = 27$, $a_4 = 3^4 = 81$

6. $r = \frac{(-5)^{n+1}}{(-5)^n} = (-5)^{n+1-n} = -5$

The ratio of consecutive terms is constant, therefore the sequence is geometric.

$a_1 = (-5)^1 = -5$, $a_2 = (-5)^2 = 25$,

$a_3 = (-5)^3 = -125$, $a_4 = (-5)^4 = 625$

7. $r=\dfrac{-3\left(\frac{1}{2}\right)^{n+1}}{-3\left(\frac{1}{2}\right)^{n}}=\left(\dfrac{1}{2}\right)^{n+1-n}=\dfrac{1}{2}$

The ratio of consecutive terms is constant, therefore the sequence is geometric.

$a_1=-3\left(\dfrac{1}{2}\right)^1=-\dfrac{3}{2},\ a_2=-3\left(\dfrac{1}{2}\right)^2=-\dfrac{3}{4},$

$a_3=-3\left(\dfrac{1}{2}\right)^3=-\dfrac{3}{8},\ a_4=-3\left(\dfrac{1}{2}\right)^4=-\dfrac{3}{16}$

8. $r=\dfrac{\left(\frac{5}{2}\right)^{n+1}}{\left(\frac{5}{2}\right)^{n}}=\left(\dfrac{5}{2}\right)^{n+1-n}=\dfrac{5}{2}$

The ratio of consecutive terms is constant, therefore the sequence is geometric.

$a_1=\left(\dfrac{5}{2}\right)^1=\dfrac{5}{2},\ a_2=\left(\dfrac{5}{2}\right)^2=\dfrac{25}{4},$

$a_3=\left(\dfrac{5}{2}\right)^3=\dfrac{125}{8},\ a_4=\left(\dfrac{5}{2}\right)^4=\dfrac{625}{16}$

9. $r=\dfrac{\left(\frac{2^{n+1-1}}{4}\right)}{\left(\frac{2^{n-1}}{4}\right)}=\dfrac{2^n}{2^{n-1}}=2^{n-(n-1)}=2$

The ratio of consecutive terms is constant, therefore the sequence is geometric.

$a_1=\dfrac{2^{1-1}}{4}=\dfrac{2^0}{2^2}=2^{-2}=\dfrac{1}{4},$

$a_2=\dfrac{2^{2-1}}{4}=\dfrac{2^1}{2^2}=2^{-1}=\dfrac{1}{2},$

$a_3=\dfrac{2^{3-1}}{4}=\dfrac{2^2}{2^2}=1,$

$a_4=\dfrac{2^{4-1}}{4}=\dfrac{2^3}{2^2}=2$

10. $r=\dfrac{\left(\frac{3^{n+1}}{9}\right)}{\left(\frac{3^n}{9}\right)}=\dfrac{3^{n+1}}{3^n}=3^{n+1-n}=3$

The ratio of consecutive terms is constant, therefore the sequence is geometric.

$a_1=\dfrac{3^1}{9}=\dfrac{1}{3},\ a_2=\dfrac{3^2}{9}=\dfrac{9}{9}=1,$

$a_3=\dfrac{3^3}{9}=\dfrac{27}{9}=3,\ a_4=\dfrac{3^4}{9}=\dfrac{81}{9}=9$

11. $r=\dfrac{2^{\left(\frac{n+1}{3}\right)}}{2^{\left(\frac{n}{3}\right)}}=2^{\left(\frac{n+1}{3}-\frac{n}{3}\right)}=2^{1/3}$

The ratio of consecutive terms is constant, therefore the sequence is geometric.

$a_1=2^{1/3},\ a_2=2^{2/3},\ a_3=2^{3/3}=2,\ a_4=2^{4/3}$

12. $r=\dfrac{3^{2(n+1)}}{3^{2n}}=3^{2n+2-2n}=3^2=9$

The ratio of consecutive terms is constant, therefore the sequence is geometric.

$a_1=3^{2\cdot1}=9,\ a_2=3^{2\cdot2}=3^4=81,$

$a_3=3^{2\cdot3}=3^6=729,\ a_4=3^{2\cdot4}=3^8=6561$

13. $r=\dfrac{\left(\frac{3^{n+1-1}}{2^{n+1}}\right)}{\left(\frac{3^{n-1}}{2^n}\right)}=\dfrac{3^n}{3^{n-1}}\cdot\dfrac{2^n}{2^{n+1}}$

$=3^{n-(n-1)}\cdot2^{n-(n+1)}=3\cdot2^{-1}=\dfrac{3}{2}$

The ratio of consecutive terms is constant, therefore the sequence is geometric.

$a_1=\dfrac{3^{1-1}}{2^1}=\dfrac{3^0}{2}=\dfrac{1}{2},\ a_2=\dfrac{3^{2-1}}{2^2}=\dfrac{3^1}{2^2}=\dfrac{3}{4},$

$a_3=\dfrac{3^{3-1}}{2^3}=\dfrac{3^2}{2^3}=\dfrac{9}{8},\ a_4=\dfrac{3^{4-1}}{2^4}=\dfrac{3^3}{2^4}=\dfrac{27}{16}$

14. $r = \dfrac{\left(\dfrac{2^{n+1}}{3^{n+1-1}}\right)}{\left(\dfrac{2^n}{3^{n-1}}\right)} = \dfrac{3^{n-1}}{3^n} \cdot \dfrac{2^{n+1}}{2^n}$

$= 3^{n-1-n} \cdot 2^{n+1-n} = 3^{-1} \cdot 2 = \dfrac{2}{3}$

The ratio of consecutive terms is constant, therefore the sequence is geometric.

$a_1 = \dfrac{2^1}{3^{1-1}} = \dfrac{2}{3^0} = \dfrac{2}{1} = 2,\ a_2 = \dfrac{2^2}{3^{2-1}} = \dfrac{4}{3},$

$a_3 = \dfrac{2^3}{3^{3-1}} = \dfrac{8}{3^2} = \dfrac{8}{9}, a_4 = \dfrac{2^4}{3^{4-1}} = \dfrac{16}{3^3} = \dfrac{16}{27}$

15. $\{n+2\}$

$d = (n+1+2)-(n+2) = n+3-n-2 = 1$

The difference between consecutive terms is constant, therefore the sequence is arithmetic.

16. $\{2n-5\}$

$d = 2(n+1)-5-(2n-5)$

$= 2n+2-5-2n+5 = 2$

The difference between consecutive terms is constant, therefore the sequence is arithmetic.

17. $\{4n^2\}$ Examine the terms of the sequence: 4, 16, 36, 64, 100, ...

There is no common difference; there is no common ratio; neither.

18. $\{5n^2+1\}$ Examine the terms of the sequence: 6, 21, 46, 81, 126, ...

There is no common difference; there is no common ratio; neither.

19. $\left\{3-\dfrac{2}{3}n\right\}$

$d = \left(3-\dfrac{2}{3}(n+1)\right)-\left(3-\dfrac{2}{3}n\right)$

$= 3-\dfrac{2}{3}n-\dfrac{2}{3}-3+\dfrac{2}{3}n = -\dfrac{2}{3}$

The difference between consecutive terms is constant, therefore the sequence is arithmetic.

20. $\left\{8-\dfrac{3}{4}n\right\}$

$d = \left(8-\dfrac{3}{4}(n+1)\right)-\left(8-\dfrac{3}{4}n\right)$

$= 8-\dfrac{3}{4}n-\dfrac{3}{4}-8+\dfrac{3}{4}n = -\dfrac{3}{4}$

The difference between consecutive terms is constant, therefore the sequence is arithmetic.

21. 1, 3, 6, 10, ...　Neither

There is no common difference or common ratio.

22. 2, 4, 6, 8, ...

The common difference is 2.

The difference between consecutive terms is constant, therefore the sequence is arithmetic.

23. $\left\{\left(\dfrac{2}{3}\right)^n\right\}$

$r = \dfrac{\left(\dfrac{2}{3}\right)^{n+1}}{\left(\dfrac{2}{3}\right)^n} = \left(\dfrac{2}{3}\right)^{n+1-n} = \dfrac{2}{3}$

The ratio of consecutive terms is constant, therefore the sequence is geometric.

24. $\left\{\left(\dfrac{5}{4}\right)^n\right\}$

$r = \dfrac{\left(\dfrac{5}{4}\right)^{n+1}}{\left(\dfrac{5}{4}\right)^n} = \left(\dfrac{5}{4}\right)^{n+1-n} = \dfrac{5}{4}$

The ratio of consecutive terms is constant, therefore the sequence is geometric.

25. –1, –2, –4, –8, ...

$r = \dfrac{-2}{-1} = \dfrac{-4}{-2} = \dfrac{-8}{-4} = 2$

The ratio of consecutive terms is constant, therefore the sequence is geometric.

26. 1, 1, 2, 3, 5, 8, ...　Neither　There is no common difference; there is no common ratio.

27. $\left\{3^{n/2}\right\}$

$$r = \frac{3^{\left(\frac{n+1}{2}\right)}}{3^{\left(\frac{n}{2}\right)}} = 3^{\left(\frac{n+1}{2}-\frac{n}{2}\right)} = 3^{1/2}$$

The ratio of consecutive terms is constant, therefore the sequence is geometric.

28. $\left\{(-1)^n\right\}$

$$r = \frac{(-1)^{n+1}}{(-1)^n} = (-1)^{n+1-n} = -1$$

The ratio of consecutive terms is constant, therefore the sequence is geometric.

29. $a_5 = 2\cdot 3^{5-1} = 2\cdot 3^4 = 2\cdot 81 = 162$

$a_n = 2\cdot 3^{n-1}$

30. $a_5 = -2\cdot 4^{5-1} = -2\cdot 4^4 = -2\cdot 256 = -512$

$a_n = -2\cdot 4^{n-1}$

31. $a_5 = 5(-1)^{5-1} = 5(-1)^4 = 5\cdot 1 = 5$

$a_n = 5\cdot(-1)^{n-1}$

32. $a_5 = 6(-2)^{5-1} = 6(-2)^4 = 6\cdot 16 = 96$

$a_n = 6\cdot(-2)^{n-1}$

33. $a_5 = 0\cdot\left(\frac{1}{2}\right)^{5-1} = 0\cdot\left(\frac{1}{2}\right)^4 = 0$

$a_n = 0\cdot\left(\frac{1}{2}\right)^{n-1} = 0$

34. $a_5 = 1\cdot\left(-\frac{1}{3}\right)^{5-1} = 1\cdot\left(-\frac{1}{3}\right)^4 = \frac{1}{81}$

$a_n = 1\cdot\left(-\frac{1}{3}\right)^{n-1} = \left(-\frac{1}{3}\right)^{n-1}$

35. $a_5 = \sqrt{2}\cdot\left(\sqrt{2}\right)^{5-1} = \sqrt{2}\cdot\left(\sqrt{2}\right)^4 = \sqrt{2}\cdot 4 = 4\sqrt{2}$

$a_n = \sqrt{2}\cdot\left(\sqrt{2}\right)^{n-1} = \left(\sqrt{2}\right)^n$

36. $a_5 = 0\cdot\left(\frac{1}{\pi}\right)^{5-1} = 0\cdot\left(\frac{1}{\pi}\right)^4 = 0$

$a_n = 0\cdot\left(\frac{1}{\pi}\right)^{n-1} = 0$

37. $a = 1,\ r = \frac{1}{2},\ n = 7$

$a_7 = 1\cdot\left(\frac{1}{2}\right)^{7-1} = \left(\frac{1}{2}\right)^6 = \frac{1}{64}$

38. $a = 1,\ r = 3,\ n = 8$

$a_8 = 1\cdot 3^{8-1} = 3^7 = 2187$

39. $a = 1,\ r = -1,\ n = 9$

$a_9 = 1\cdot(-1)^{9-1} = (-1)^8 = 1$

40. $a = -1,\ r = -2,\ n = 10$

$a_{10} = -1\cdot(-2)^{10-1} = -1\cdot(-2)^9 = -1(-512) = 512$

41. $a = 0.4,\ r = 0.1,\ n = 8$

$a_8 = 0.4\cdot(0.1)^{8-1} = 0.4(0.1)^7 = 0.00000004$

42. $a = 0.1,\ r = 10,\ n = 7$

$a_7 = 0.1\cdot 10^{7-1} = 0.1(10)^6 = 100{,}000$

43. $a = \frac{1}{4},\ r = 2$

$$S_n = a\left(\frac{1-r^n}{1-r}\right) = \frac{1}{4}\left(\frac{1-2^n}{1-2}\right) = \frac{1}{4}\left(2^n - 1\right)$$

44. $a = \frac{3}{9} = \frac{1}{3},\ r = 3$

$$S_n = a\left(\frac{1-r^n}{1-r}\right) = \frac{1}{3}\left(\frac{1-3^n}{1-3}\right) = \frac{1}{3}\left(\frac{1-3^n}{-2}\right)$$

$$= \frac{1}{6}\left(3^n - 1\right)$$

45. $a=\frac{2}{3},\ r=\frac{2}{3}$

$$S_n=a\left(\frac{1-r^n}{1-r}\right)=\frac{2}{3}\left(\frac{1-\left(\frac{2}{3}\right)^n}{1-\frac{2}{3}}\right)$$

$$=\frac{2}{3}\left(\frac{1-\left(\frac{2}{3}\right)^n}{\frac{1}{3}}\right)=2\left(1-\left(\frac{2}{3}\right)^n\right)$$

46. $a=4,\ r=3$

$$S_n=a\left(\frac{1-r^n}{1-r}\right)=4\left(\frac{1-3^n}{1-3}\right)=4\left(\frac{1-3^n}{-2}\right)$$

$$=2\left(3^n-1\right)$$

47. $a=-1,\ r=2$

$$S_n=a\left(\frac{1-r^n}{1-r}\right)=-1\left(\frac{1-2^n}{1-2}\right)=1-2^n$$

48. $a=2,\ r=\frac{3}{5}$

$$S_n=a\left(\frac{1-r^n}{1-r}\right)=2\left(\frac{1-\left(\frac{3}{5}\right)^n}{1-\frac{3}{5}}\right)=2\left(\frac{1-\left(\frac{3}{5}\right)^n}{\left(\frac{2}{5}\right)}\right)$$

$$=5\left(1-\left(\frac{3}{5}\right)^n\right)$$

49. Using the sum of the sequence feature:

```
sum(seq(2^N/4,N,
0,14,1))
              8191.75
```

50. Using the sum of the sequence feature:

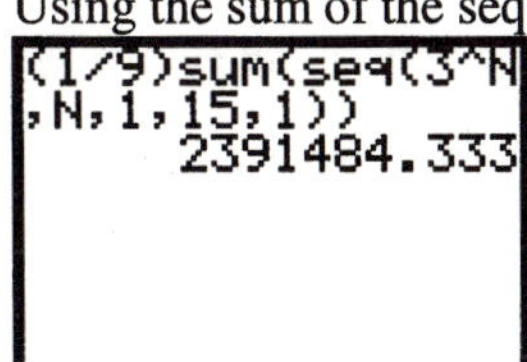

51. Using the sum of the sequence feature:

```
sum(seq((2/3)^N,
N,1,15,1))
          1.995432683
```

52. Using the sum of the sequence feature:

```
4sum(seq(3^(N-1)
,N,1,15,1))
             28697812
```

53. Using the sum of the sequence feature:

```
sum(seq(-1*2^N,N
,0,14,1))
               -32767
```

54. Using the sum of the sequence feature:

```
2sum(seq((3/5)^N
,N,0,15,1))
          4.998589445
```

55. $a=1,\ r=\frac{1}{3}$ Since $|r|<1$,

$$S_\infty=\frac{a}{1-r}=\frac{1}{\left(1-\frac{1}{3}\right)}=\frac{1}{\left(\frac{2}{3}\right)}=\frac{3}{2}$$

56. $a=2,\ r=\frac{2}{3}$ Since $|r|<1$,

$$S_\infty=\frac{a}{1-r}=\frac{2}{\left(1-\frac{2}{3}\right)}=\frac{2}{\left(\frac{1}{3}\right)}=6$$

57. $a=8,\ r=\frac{1}{2}$ Since $|r|<1$,

$$S_\infty=\frac{a}{1-r}=\frac{8}{\left(1-\frac{1}{2}\right)}=\frac{8}{\left(\frac{1}{2}\right)}=16$$

58. $a=6,\ r=\frac{1}{3}$ Since $|r|<1$,

$$S_\infty=\frac{a}{1-r}=\frac{6}{\left(1-\frac{1}{3}\right)}=\frac{6}{\left(\frac{2}{3}\right)}=9$$

59. $a=2,\ r=-\frac{1}{4}$ Since $|r|<1$,

$$S_\infty=\frac{a}{1-r}=\frac{2}{\left(1-\left(-\frac{1}{4}\right)\right)}=\frac{2}{\left(\frac{5}{4}\right)}=\frac{8}{5}$$

60. $a=1,\ r=-\frac{3}{4}$ Since $|r|<1$,

$$S_\infty=\frac{a}{1-r}=\frac{1}{\left(1-\left(-\frac{3}{4}\right)\right)}=\frac{1}{\left(\frac{7}{4}\right)}=\frac{4}{7}$$

61. $a=5,\ r=\frac{1}{4}$ Since $|r|<1$,

$$S_\infty=\frac{a}{1-r}=\frac{5}{\left(1-\frac{1}{4}\right)}=\frac{5}{\left(\frac{3}{4}\right)}=\frac{20}{3}$$

62. $a=8,\ r=\frac{1}{3}$ Since $|r|<1$,

$$S_\infty=\frac{a}{1-r}=\frac{8}{\left(1-\frac{1}{3}\right)}=\frac{8}{\left(\frac{2}{3}\right)}=12$$

63. $a=6,\ r=-\frac{2}{3}$ Since $|r|<1$,

$$S_\infty=\frac{a}{1-r}=\frac{6}{\left(1-\left(-\frac{2}{3}\right)\right)}=\frac{6}{\left(\frac{5}{3}\right)}=\frac{18}{5}$$

64. $a=4,\ r=-\frac{1}{2}$ Since $|r|<1$,

$$S_\infty=\frac{a}{1-r}=\frac{4}{\left(1-\left(-\frac{1}{2}\right)\right)}=\frac{4}{\left(\frac{3}{2}\right)}=\frac{8}{3}$$

65. Find the common ratio of the terms and solve the system of equations:

$$\frac{x+2}{x}=r;\quad \frac{x+3}{x+2}=r$$

$$\frac{x+2}{x}=\frac{x+3}{x+2}\rightarrow x^2+4x+4=x^2+3x\rightarrow x=-4$$

66. Find the common ratio of the terms and solve the system of equations:

$$\frac{x}{x-1}=r;\quad \frac{x+2}{x}=r$$

$$\frac{x+2}{x}=\frac{x}{x-1}\rightarrow x^2+x-2=x^2\rightarrow x=2$$

67. This is a geometric series with $a=\$18{,}000,\quad r=1.05,\quad n=5$. Find the 5th term:

$$a_5=18000(1.05)^{5-1}=18000(1.05)^4=\$21{,}879.11$$

68. This is a geometric series with $a=\$15{,}000,\quad r=0.85,\quad n=6$. Find the 6th term:

$$a_6=15000(0.85)^{6-1}=15000(0.85)^5=\$6655.58$$

69. a. Find the 10th term of the geometric sequence:

$a=2,\ r=0.9,\ n=10$

$$a_{10}=2(0.9)^{10-1}=2(0.9)^9=0.775 \text{ feet}$$

b. Find n when $a_n<1$:

$$2(0.9)^{n-1}<1$$
$$(0.9)^{n-1}<0.5$$
$$(n-1)\log(0.9)<\log(0.5)$$
$$n-1>\frac{\log(0.5)}{\log(0.9)}$$
$$n>\frac{\log(0.5)}{\log(0.9)}+1\approx 7.58$$

On the 8th swing the arc is less than 1 foot.

c. Find the sum of the first 15 swings:

$$S_{15}=2\left(\frac{1-(0.9)^{15}}{1-0.9}\right)=2\left(\frac{1-(0.9)^{15}}{0.1}\right)=20\left(1-(0.9)^{15}\right)=15.88 \text{ feet}$$

d. Find the infinite sum of the geometric series:

$$S_\infty=\frac{2}{1-0.9}=\frac{2}{0.1}=20 \text{ feet}$$

70. a. Find the 3rd term of the geometric sequence:
$a = 24,\ r = 0.8,\ n = 3$

$a_3 = 24(0.8)^{3-1} = 24(0.8)^2 = 15.36$ feet

b. The height after the n th bounce is:

$$a_n = 24(0.8)^{n-1} = 24(0.8)^{-1}(0.8)^n = 30(0.8)^n \text{ ft.}$$

c. Find n when $a_n < 0.5$:

$$24(0.8)^{n-1} < 0.5$$
$$(0.8)^{n-1} < 0.020833$$
$$(n-1)\log(0.8) < \log(0.020833)$$
$$n-1 > \frac{\log(0.020833)}{\log(0.8)}$$
$$n > \frac{\log(0.020833)}{\log(0.8)} + 1 \approx 18.35$$

On the 19th bounce the height is less than 0.5 feet.

d. Find the infinite sum of the geometric series:

$S_\infty = \frac{24}{1-0.8} = \frac{24}{0.2} = 120$ feet on the upward bounce.

For the downward motion of the ball:

$$S_\infty = \frac{30}{1-0.8} = \frac{30}{0.2} = 150 \text{ feet}$$

The total distance the ball travels is $120 + 150 = 270$ feet.

71. This is a geometric sequence with $a = 1,\ r = 2,\ n = 64$.
Find the sum of the geometric series:

$$S_{64} = 1\left(\frac{1-2^{64}}{1-2}\right) = \frac{1-2^{64}}{-1} = 2^{64} - 1 = 1.845\times 10^{19} \text{ grains}$$

72. This is an infinite geometric series with $a = \frac{1}{4},\ r = \frac{1}{4}$.
Find the sum of the infinite geometric series:

$$S_\infty = \frac{\left(\frac{1}{4}\right)}{\left(1-\frac{1}{4}\right)} = \frac{\left(\frac{1}{4}\right)}{\left(\frac{3}{4}\right)} = \frac{1}{3}$$

$\therefore\ \frac{1}{3}$ of the square is eventually shaded.

73. The common ratio, $r = 0.90 < 1$. The sum is:

$$S = \frac{1}{1-0.9} = \frac{1}{0.10} = 10.$$

The multiplier is 10.

74. The common ratio, $r = 0.95 < 1$. The sum is:

$$S = \frac{1}{1-0.95} = \frac{1}{0.05} = 20.$$

The multiplier is 20.

75. This is an infinite geometric series with $a = 4$, and $r = \frac{1.03}{1.09}$.

Find the sum: Price $= \dfrac{4}{\left(1-\frac{1.03}{1.09}\right)} \approx \$72.67$.

76. This is an infinite geometric series with $a = 2.5$, and $r = \frac{1.04}{1.11}$.

Find the sum: Price $= \dfrac{2.5}{\left(1-\frac{1.04}{1.11}\right)} \approx \$39.64$.

77. Given: $a = 1000,\ r = 0.9$
Find n when $a_n < 0.01$:

$$1000(0.9)^{n-1} < 0.01$$
$$(0.9)^{n-1} < 0.00001$$
$$(n-1)\log(0.9) < \log(0.00001)$$
$$n-1 > \frac{\log(0.00001)}{\log(0.9)}$$
$$n > \frac{\log(0.00001)}{\log(0.9)} + 1 \approx 110.27$$

On the 111th day or December 20, 2001, the amount will be less than \$0.01.

Find the sum of the geometric series:

$$S_{110} = a\left(\frac{1-r^n}{1-r}\right) = 1000\left(\frac{1-(0.9)^{110}}{1-0.9}\right) = 1000\left(\frac{1-(0.9)^{110}}{0.1}\right) = \$9999.91$$

78. Both options are geometric sequences:
Option A: $a = \$20,000;\ \ r = 1.06;\ \ n = 5$

$a_5 = 20,000(1.06)^{5-1} = 20,000(1.06)^4 = \$25,250$

$$S_5 = 20000\left(\frac{1-(1.06)^5}{1-1.06}\right) = \$112,742$$

Option B: $a = \$22,000;\ \ r = 1.03;\ \ n = 5$

$a_5 = 22,000(1.03)^{5-1} = 22,000(1.03)^4 = \$24,761$

$$S_5 = 22000\left(\frac{1-(1.03)^5}{1-1.03}\right) = \$116,801$$

Option A provides more money in the 5th year, while Option B provides the greatest total amount of money over the 5 year period.

79. Find the sum of each sequence:
A: Arithmetic series with:
$a = \$1000,\ \ d = -1,\ \ n = 1000$

Find the sum of the arithmetic series:

$$S_{1000} = \frac{1000}{2}(1000+1) = 500(1001) = \$500,500$$

B: This is a geometric sequence with $a = 1,\ \ r = 2,\ \ n = 19$.

Find the sum of the geometric series:

$$S_{19} = 1\left(\frac{1-2^{19}}{1-2}\right) = \frac{1-2^{19}}{-1} = 2^{19} - 1 = \$524,287$$

B results in more money.

80. Option 1:

$$\text{Total Salary} = \$2,000,000(7) + \$100,000(7) = \$14,700,000$$

Option 2: Geometric series with:
$a = \$2,000,000,\ \ r = 1.045,\ \ n = 7$
Find the sum of the geometric series:

$$S = 2,000,000\left(\frac{1-(1.045)^7}{1-1.045}\right) \approx \$16,038,304$$

Option 3: Arithmetic series with:
$a = \$2,000,000,\ \ d = \$95,000,\ \ n = 7$
Find the sum of the arithmetic series:

$$S_7 = \frac{7}{2}(2(2,000,000) + (7-1)(95,000)) = \$15,995,000$$

Option 2 provides the most money; Option 1 provides the least money.

81. Yes, a sequence can be both arithmetic and geometric. For example, the constant sequence 3,3,3,3,..... can be viewed as an arithmetic sequence with $a = 3$ and $d = 0$. Alternatively, the same sequence can be viewed as a geometric sequence with $a = 3$ and $r = 1$.

82. Answers will vary.

83. Answers will vary.

84. Answers will vary. Both increase (or decrease) exponentially, but the domain of a geometric sequence is the set of natural numbers while the domain of an exponential function is the set of all real numbers.

Section 11.4

1. I: $n = 1$: $2 \cdot 1 = 2$ and $1(1+1) = 2$

II: If $2+4+6+\cdots+2k = k(k+1)$, then

$$\begin{aligned}
&2+4+6+\cdots+2k+2(k+1)\\
&= [2+4+6+\cdots+2k] + 2(k+1)\\
&= k(k+1) + 2(k+1)\\
&= (k+1)(k+2)\\
&= (k+1)((k+1)+1)
\end{aligned}$$

Conditions I and II are satisfied; the statement is true.

2. I: $n = 1$: $4 \cdot 1 - 3 = 1$ and $1(2 \cdot 1 - 1) = 1$

II: If $1+5+9+\cdots+(4k-3) = k(2k-1)$, then

$$\begin{aligned}
&1+5+9+\cdots+(4k-3)+(4(k+1)-3)\\
&= [1+5+9+\cdots+(4k-3)] + 4k+4-3\\
&= k(2k-1) + 4k + 1\\
&= 2k^2 - k + 4k + 1\\
&= 2k^2 + 3k + 1\\
&= (k+1)(2k+1)\\
&= (k+1)(2(k+1)-1)
\end{aligned}$$

Conditions I and II are satisfied; the statement is true.

3. I: $n=1$: $1+2=3$ and $\frac{1}{2}\cdot 1(1+5)=3$

II: If $3+4+5+\cdots+(k+2)=\frac{1}{2}\cdot k(k+5)$, then

$$3+4+5+\cdots+(k+2)+[(k+1)+2]$$
$$=[3+4+5+\cdots+(k+2)]+(k+3)$$
$$=\frac{1}{2}\cdot k(k+5)+(k+3)$$
$$=\frac{1}{2}k^2+\frac{5}{2}k+k+3$$
$$=\frac{1}{2}k^2+\frac{7}{2}k+3$$
$$=\frac{1}{2}\cdot\left(k^2+7k+6\right)$$
$$=\frac{1}{2}\cdot(k+1)(k+6)$$
$$=\frac{1}{2}\cdot(k+1)\left((k+1)+5\right)$$

Conditions I and II are satisfied; the statement is true.

4. I: $n=1$: $2\cdot 1+1=3$ and $1(1+2)=3$

II: If $3+5+7+\cdots+(2k+1)=k(k+2)$, then

$$3+5+7+\cdots+(2k+1)+[2(k+1)+1]$$
$$=[3+5+7+\cdots+(2k+1)]+(2k+3)$$
$$=k(k+2)+(2k+3)$$
$$=k^2+2k+2k+3$$
$$=k^2+4k+3$$
$$=(k+1)(k+3)$$
$$=(k+1)\left((k+1)+2\right)$$

Conditions I and II are satisfied; the statement is true.

5. I: $n=1$: $3\cdot 1-1=2$ and $\frac{1}{2}\cdot 1(3\cdot 1+1)=2$

II: If $2+5+8+\cdots+(3k-1)=\frac{1}{2}\cdot k(3k+1)$,

then

$$2+5+8+\cdots+(3k-1)+[3(k+1)-1]$$
$$=[2+5+8+\cdots+(3k-1)]+(3k+2)$$
$$=\frac{1}{2}\cdot k(3k+1)+(3k+2)=\frac{3}{2}k^2+\frac{1}{2}k+3k+2$$
$$=\frac{3}{2}k^2+\frac{7}{2}k+2=\frac{1}{2}\cdot\left(3k^2+7k+4\right)$$
$$=\frac{1}{2}\cdot(k+1)(3k+4)$$
$$=\frac{1}{2}\cdot(k+1)\left(3(k+1)+1\right)$$

Conditions I and II are satisfied; the statement is true.

6. I: $n=1$: $3\cdot 1-2=1$ and $\frac{1}{2}\cdot 1(3\cdot 1-1)=1$

II: If $1+4+7+\cdots+(3k-2)=\frac{1}{2}\cdot k(3k-1)$,

then

$$1+4+7+\cdots+(3k-2)+[3(k+1)-2]$$
$$=[1+4+7+\cdots+(3k-2)]+(3k+1)$$
$$=\frac{1}{2}\cdot k(3k-1)+(3k+1)=\frac{3}{2}k^2-\frac{1}{2}k+3k+1$$
$$=\frac{3}{2}k^2+\frac{5}{2}k+1=\frac{1}{2}\cdot\left(3k^2+5k+2\right)$$
$$=\frac{1}{2}\cdot(k+1)(3k+2)$$
$$=\frac{1}{2}\cdot(k+1)\left(3(k+1)-1\right)$$

Conditions I and II are satisfied; the statement is true.

7. I: $n=1$: $2^{1-1}=1$ and $2^1-1=1$

II: If $1+2+2^2+\cdots+2^{k-1}=2^k-1$, then

$$1+2+2^2+\cdots+2^{k-1}+2^{k+1-1}$$
$$=\left[1+2+2^2+\cdots+2^{k-1}\right]+2^k$$
$$=2^k-1+2^k=2\cdot 2^k-1$$
$$=2^{k+1}-1$$

Conditions I and II are satisfied; the statement is true.

8. I: $n=1$: $3^{1-1}=1$ and $\frac{1}{2}(3^1-1)=1$

II: If $1+3+3^2+\cdots+3^{k-1}=\frac{1}{2}\cdot(3^k-1)$, then

$$1+3+3^2+\cdots+3^{k-1}+3^{k+1-1}$$
$$=\left[1+3+3^2+\cdots+3^{k-1}\right]+3^k$$
$$=\frac{1}{2}\cdot(3^k-1)+3^k=\frac{1}{2}\cdot3^k-\frac{1}{2}+3^k$$
$$=\frac{3}{2}\cdot3^k-\frac{1}{2}=\frac{1}{2}\cdot\left(3\cdot3^k-1\right)$$
$$=\frac{1}{2}\left(3^{k+1}-1\right)$$

Conditions I and II are satisfied; the statement is true.

9. I: $n=1$: $4^{1-1}=1$ and $\frac{1}{3}\cdot\left(4^1-1\right)=1$

II: If $1+4+4^2+\cdots+4^{k-1}=\frac{1}{3}\cdot\left(4^k-1\right)$, then

$$1+4+4^2+\cdots+4^{k-1}+4^{k+1-1}$$
$$=\left[1+4+4^2+\cdots+4^{k-1}\right]+4^k$$
$$=\frac{1}{3}\cdot\left(4^k-1\right)+4^k=\frac{1}{3}\cdot4^k-\frac{1}{3}+4^k$$
$$=\frac{4}{3}\cdot4^k-\frac{1}{3}=\frac{1}{3}\left(4\cdot4^k-1\right)$$
$$=\frac{1}{3}\cdot\left(4^{k+1}-1\right)$$

Conditions I and II are satisfied; the statement is true.

11. I: $n=1$: $\frac{1}{1(1+1)}=\frac{1}{2}$ and $\frac{1}{1+1}=\frac{1}{2}$

II: If $\frac{1}{1\cdot2}+\frac{1}{2\cdot3}+\frac{1}{3\cdot4}+\cdots+\frac{1}{k(k+1)}=\frac{k}{k+1}$, then

$$\frac{1}{1\cdot2}+\frac{1}{2\cdot3}+\frac{1}{3\cdot4}+\cdots+\frac{1}{k(k+1)}+\frac{1}{(k+1)(k+1+1)}=\left[\frac{1}{1\cdot2}+\frac{1}{2\cdot3}+\frac{1}{3\cdot4}+\cdots+\frac{1}{k(k+1)}\right]+\frac{1}{(k+1)(k+2)}$$
$$=\frac{k}{k+1}+\frac{1}{(k+1)(k+2)}=\frac{k}{k+1}\cdot\frac{k+2}{k+2}+\frac{1}{(k+1)(k+2)}$$
$$=\frac{k^2+2k+1}{(k+1)(k+2)}=\frac{(k+1)(k+1)}{(k+1)(k+2)}=\frac{k+1}{k+2}=\frac{k+1}{(k+1)+1}$$

Conditions I and II are satisfied; the statement is true.

10. I: $n=1$: $5^{1-1}=1$ and $\frac{1}{4}\cdot\left(5^1-1\right)=1$

II: If $1+5+5^2+\cdots+5^{k-1}=\frac{1}{4}\cdot\left(5^k-1\right)$, then

$$1+5+5^2+\cdots+5^{k-1}+5^{k+1-1}$$
$$=\left[1+5+5^2+\cdots+5^{k-1}\right]+5^k$$
$$=\frac{1}{4}\cdot\left(5^k-1\right)+5^k=\frac{1}{4}\cdot5^k-\frac{1}{4}+5^k$$
$$=\frac{5}{4}\cdot5^k-\frac{1}{4}=\frac{1}{4}\left(5\cdot5^k-1\right)$$
$$=\frac{1}{4}\cdot\left(5^{k+1}-1\right)$$

Conditions I and II are satisfied; the statement is true.

12. I: $n=1$: $\dfrac{1}{(2\cdot 1-1)(2\cdot 1+1)}=\dfrac{1}{3}$ and $\dfrac{1}{2\cdot 1+1}=\dfrac{1}{3}$

II: If $\dfrac{1}{1\cdot 3}+\dfrac{1}{3\cdot 5}+\dfrac{1}{5\cdot 7}+\cdots+\dfrac{1}{(2k-1)(2k+1)}=\dfrac{k}{2k+1}$, then

$$\frac{1}{1\cdot 3}+\frac{1}{3\cdot 5}+\frac{1}{5\cdot 7}+\cdots+\frac{1}{(2k-1)(2k+1)}+\frac{1}{(2(k+1)-1)(2(k+1)+1)}$$

$$=\left[\frac{1}{1\cdot 3}+\frac{1}{3\cdot 5}+\frac{1}{5\cdot 7}+\cdots+\frac{1}{(2k-1)(2k+1)}\right]+\frac{1}{(2k+1)(2k+3)}$$

$$=\frac{k}{2k+1}+\frac{1}{(2k+1)(2k+3)}=\frac{k}{2k+1}\cdot\frac{2k+3}{2k+3}+\frac{1}{(2k+1)(2k+3)}$$

$$=\frac{2k^2+3k+1}{(2k+1)(2k+3)}=\frac{(k+1)(2k+1)}{(2k+1)(2k+3)}=\frac{k+1}{2k+3}=\frac{k+1}{2(k+1)+1}$$

Conditions I and II are satisfied; the statement is true.

13. I: $n=1$: $1^2=1$ and $\dfrac{1}{6}\cdot 1(1+1)(2\cdot 1+1)=1$

II: If $1^2+2^2+3^2+\cdots+k^2=\dfrac{1}{6}\cdot k(k+1)(2k+1)$, then

$$1^2+2^2+3^2+\cdots+k^2+(k+1)^2=\left[1^2+2^2+3^2+\cdots+k^2\right]+(k+1)^2=\frac{1}{6}k(k+1)(2k+1)+(k+1)^2$$

$$=(k+1)\left[\frac{1}{6}k(2k+1)+k+1\right]=(k+1)\left[\frac{1}{3}k^2+\frac{1}{6}k+k+1\right]=(k+1)\left[\frac{1}{3}k^2+\frac{7}{6}k+1\right]=\frac{1}{6}(k+1)\left[2k^2+7k+6\right]$$

$$=\frac{1}{6}\cdot(k+1)(k+2)(2k+3)$$

$$=\frac{1}{6}\cdot(k+1)\left((k+1)+1\right)\left(2(k+1)+1\right)$$

Conditions I and II are satisfied; the statement is true.

14. I: $n=1$: $1^3=1$ and $\dfrac{1}{4}\cdot 1^2(1+1)^2=1$

II: If $1^3+2^3+3^3+\cdots+k^3=\dfrac{1}{4}k^2(k+1)^2$, then

$$1^3+2^3+3^3+\cdots+k^3+(k+1)^3=\left[1^3+2^3+3^3+\cdots+k^3\right]+(k+1)^3=\frac{1}{4}k^2(k+1)^2+(k+1)^3$$

$$=(k+1)^2\left[\frac{1}{4}k^2+k+1\right]=\frac{1}{4}(k+1)^2\left[k^2+4k+4\right]$$

$$=\frac{1}{4}\cdot(k+1)^2(k+2)^2$$

$$=\frac{1}{4}\cdot(k+1)^2((k+1)+1)^2$$

Conditions I and II are satisfied; the statement is true.

15. I: $n=1$: $5-1=4$ and $\frac{1}{2}\cdot 1(9-1)=4$

II: If $4+3+2+\cdots+(5-k)=\frac{1}{2}\cdot k(9-k)$, then

$$4+3+2+\cdots+(5-k)+(5-(k+1))=[4+3+2+\cdots+(5-k)]+(4-k)=\frac{1}{2}k(9-k)+(4-k)$$

$$=\frac{9}{2}k-\frac{1}{2}k^2+4-k=-\frac{1}{2}k^2+\frac{7}{2}k+4=-\frac{1}{2}\cdot\left[k^2-7k-8\right]$$

$$=-\frac{1}{2}\cdot(k+1)(k-8)=\frac{1}{2}\cdot(k+1)(8-k)=\frac{1}{2}\cdot(k+1)[9-(k+1)]$$

Conditions I and II are satisfied; the statement is true.

16. I: $n=1$: $-(1+1)=-2$ and $-\frac{1}{2}\cdot 1(1+3)=-2$

II: If $-2-3-4-\cdots-(k+1)=-\frac{1}{2}\cdot k(k+3)$, then

$$-2-3-4-\cdots-(k+1)-((k+1)+1)=[-2-3-4-\cdots-(k+1)]-(k+2)$$

$$=-\frac{1}{2}\cdot k(k+3)-(k+2)=-\frac{1}{2}k^2-\frac{3}{2}k-k-2=-\frac{1}{2}k^2-\frac{5}{2}k-2$$

$$=-\frac{1}{2}\cdot\left[k^2+5k+4\right]=-\frac{1}{2}\cdot(k+1)(k+4)$$

$$=-\frac{1}{2}\cdot(k+1)((k+1)+3)$$

Conditions I and II are satisfied; the statement is true.

17. I: $n=1$: $1(1+1)=2$ and $\frac{1}{3}\cdot 1(1+1)(1+2)=2$

II: If $1\cdot 2+2\cdot 3+3\cdot 4+\cdots+k(k+1)=\frac{1}{3}\cdot k(k+1)(k+2)$, then

$$1\cdot 2+2\cdot 3+3\cdot 4+\cdots+k(k+1)+(k+1)(k+1+1)=[1\cdot 2+2\cdot 3+3\cdot 4+\cdots+k(k+1)]+(k+1)(k+2)$$

$$=\frac{1}{3}\cdot k(k+1)(k+2)+(k+1)(k+2)=(k+1)(k+2)\left[\frac{1}{3}k+1\right]$$

$$=\frac{1}{3}\cdot(k+1)(k+2)(k+3)$$

$$=\frac{1}{3}\cdot(k+1)((k+1)+1)((k+1)+2)$$

Conditions I and II are satisfied; the statement is true.

18. I: $n=1$: $(2\cdot1-1)(2\cdot1)=2$ and $\frac{1}{3}\cdot1(1+1)(4\cdot1-1)=2$

II: If $1\cdot2+3\cdot4+5\cdot6+\cdots+(2k-1)(2k)=\frac{1}{3}\cdot k(k+1)(4k-1)$, then

$$\begin{aligned}
&1\cdot2+3\cdot4+5\cdot6+\cdots+(2k-1)(2k)+(2(k+1)-1)(2(k+1))\\
&=[1\cdot2+3\cdot4+5\cdot6+\cdots+(2k-1)(2k)]+(2k+1)(k+1)\cdot2\\
&=\frac{1}{3}k(k+1)(4k-1)+2(k+1)(2k+1)=(k+1)\left[\frac{1}{3}\cdot k(4k-1)+2(2k+1)\right]\\
&=(k+1)\left[\frac{4}{3}k^2-\frac{1}{3}k+4k+2\right]=\frac{1}{3}\cdot(k+1)\left(4k^2-k+12k+6\right)\\
&=\frac{1}{3}(k+1)\left(4k^2+11k+6\right)=\frac{1}{3}\cdot(k+1)(k+2)(4k+3)\\
&=\frac{1}{3}\cdot(k+1)((k+1)+1)(4(k+1)-1)
\end{aligned}$$

Conditions I and II are satisfied; the statement is true.

19. I: $n=1$: $1^2+1=2$ is divisible by 2

II: If k^2+k is divisible by 2, then

$$\begin{aligned}
(k+1)^2+(k+1)&=k^2+2k+1+k+1\\
&=(k^2+k)+(2k+2)
\end{aligned}$$

Since k^2+k is divisible by 2 and $2k+2$ is divisible by 2, then $(k+1)^2+(k+1)$ is divisible by 2.

Conditions I and II are satisfied; the statement is truc.

20. I: $n=1$: $1^3+2\cdot1=3$ is divisible by 3

II: If k^3+2k is divisible by 3, then

$$\begin{aligned}
&(k+1)^3+2(k+1)\\
&=k^3+3k^2+3k+1+2k+2\\
&=(k^3+2k)+(3k^2+3k+3)
\end{aligned}$$

Since k^3+2k is divisible by 3 and $3k^2+3k+3$ is divisible by 3, then $(k+1)^3+2(k+1)$ is divisible by 3.

Conditions I and II are satisfied; the statement is true.

21. I: $n=1$: $1^2-1+2=2$ is divisible by 2

II: If k^2-k+2 is divisible by 2, then

$$\begin{aligned}
(k+1)^2-(k+1)+2&=k^2+2k+1-k-1+2\\
&=(k^2-k+2)+(2k)
\end{aligned}$$

Since k^2-k+2 is divisible by 2 and $2k$ is divisible by 2, then $(k+1)^2-(k+1)+2$ is divisible by 2.

Conditions I and II are satisfied; the statement is true.

22. I: $n=1$: $1(1+1)(1+2)=6$ is divisible by 6

II: If $k(k+1)(k+2)$ is divisible by 6, then

$$\begin{aligned}
&(k+1)(k+1+1)(k+1+2)\\
&=(k+1)(k+2)(k+3)\\
&=k(k+1)(k+2)+3(k+1)(k+2).
\end{aligned}$$

Now, $k(k+1)(k+2)$ is divisible by 6; and since either $k+1$ or $k+2$ is even, $3(k+1)(k+2)$ is divisible by 6.

Thus, $(k+1)(k+2)(k+3)$
$=k(k+1)(k+2)+3(k+1)(k+2)$
is divisible by 6.

Conditions I and II are satisfied; the statement is true.

23. I: $n=1$: If $x>1$ then $x^1=x>1$.

II: Assume, for some natural number k, that if $x>1$, then $x^k>1$.
Then $x^{k+1}>1$, for $x>1$,

$$x^{k+1}=x^k\cdot x>1\cdot x=x>1$$
$$\uparrow$$
$$(x^k>1)$$

Conditions I and II are satisfied; the statement is true.

24. I: $n=1$: If $0<x<1$ then $0<x^1<1$.

II: Assume, for some natural number k, that if $0<x<1$, then $0<x^k<1$.
Then, for $0<x<1$,

$$0<x^{k+1}=x^k\cdot x<1\cdot x=x<1$$

Thus, $0<x^{k+1}<1$.

Conditions I and II are satisfied; the statement is true.

25. I: $n=1$: $a-b$ is a factor of $a^1-b^1=a-b$.

II: If $a-b$ is a factor of a^k-b^k, show that $a-b$ is a factor of $a^{k+1}-b^{k+1}$.

$$\begin{aligned}a^{k+1}-b^{k+1}&=a\cdot a^k-b\cdot b^k\\&=a\cdot a^k-a\cdot b^k+a\cdot b^k-b\cdot b^k\\&=a\left(a^k-b^k\right)+b^k(a-b)\end{aligned}$$

Since $a-b$ is a factor of a^k-b^k and $a-b$ is a factor of $a-b$, then $a-b$ is a factor of $a^{k+1}-b^{k+1}$.

Conditions I and II are satisfied; the statement is true.

26. I: $n=1$:

$a+b$ is a factor of $a^{2\cdot1+1}+b^{2\cdot1+1}=a^3+b^3$.

$$\left[a^3+b^3=(a+b)\left(a^2-ab+b^2\right)\right]$$

II: If $a+b$ is a factor of $a^{2k+1}+b^{2k+1}$, show that $a+b$ is a factor of $a^{2(k+1)+1}+b^{2(k+1)+1}$.

$$\begin{aligned}&a^{2(k+1)+1}+b^{2(k+1)+1}=a^{2k+3}+b^{2k+3}\\&=a^2\cdot a^{2k+1}+a^2\cdot b^{2k+1}-a^2\cdot b^{2k+1}+b^2\cdot b^{2k+1}\\&=a^2\left(a^{2k+1}+b^{2k+1}\right)-b^{2k+1}(a^2-b^2)\end{aligned}$$

Since $a+b$ is a factor of $a^{2k+1}+b^{2k+1}$ and $a+b$ is a factor of a^2-b^2 $[(a-b)(a+b)]$, then $a+b$ is a factor of $a^{2k+3}+b^{2k+3}$.

Conditions I and II are satisfied; the statement is true.

27. $n=1$:

$1^2-1+41=41$ is a prime number.

$n=41$:

$41^2-41+41=41^2$ is not a prime number.

28. II: If $2+4+6+\cdots+2k=k^2+k+2$, then

$$\begin{aligned}&2+4+6+\cdots+2k+2(k+1)\\&=[2+4+6+\cdots+2k]+2k+2\\&=k^2+k+2+2k+2\\&=(k^2+2k+1)+(k+1)+2\\&=(k+1)^2+(k+1)+2\end{aligned}$$

I: $n=1$: $2\cdot1=2$ and $1^2+1+2=4\neq2$

29. I: $n=1$: $ar^{1-1}=a$ and $a\left(\dfrac{1-r^1}{1-r}\right)=a$

II: If $a+ar+ar^2+\cdots+ar^{k-1}=a\left(\dfrac{1-r^k}{1-r}\right)$,

then

$$\begin{aligned}&a+ar+ar^2+\cdots+ar^{k-1}+ar^{k+1-1}\\&=\left[a+ar+ar^2+\cdots+ar^{k-1}\right]+ar^k\\&=a\left(\frac{1-r^k}{1-r}\right)+ar^k\\&=\frac{a(1-r^k)+ar^k(1-r)}{1-r}\\&=\frac{a-ar^k+ar^k-ar^{k+1}}{1-r}\\&=a\left(\frac{1-r^{k+1}}{1-r}\right)\end{aligned}$$

Conditions I and II are satisfied; the statement is true.

30. I: $n=1$:

$$a+(1-1)d=a \text{ and } 1\cdot a+d\frac{1(1-1)}{2}=a$$

II: If $a+(a+d)+(a+2d)+\cdots+\left[a+(k-1)d\right]$

$$=ka+d\frac{k(k-1)}{2}$$

then

$$a+(a+d)+(a+2d)+\cdots+\left[a+(k-1)d\right]+\left(a+kd\right)$$
$$=\left[a+(a+d)+(a+2d)+\cdots+\left[a+(k-1)d\right]\right]+(a+kd)$$
$$=ka+d\frac{k(k-1)}{2}+(a+kd)$$
$$=(k+1)a+d\left[\frac{k(k-1)}{2}+k\right]$$
$$=(k+1)a+d\left[\frac{k^2-k+2k}{2}\right]$$
$$=(k+1)a+d\left[\frac{k^2+k}{2}\right]$$
$$=(k+1)a+d\left[\frac{(k+1)k}{2}\right]$$
$$=\left(k+1\right)a+d\left[\frac{\left(k+1\right)\left(\left(k+1\right)-1\right)}{2}\right]$$

Conditions I and II are satisfied; the statement is true.

31. I: $n=4$: The number of diagonals of a quadrilateral is $\frac{1}{2}\cdot 4(4-3)=2$.

II: Assume that for any integer *k*, the number of diagonals of a convex polygon with *k* sides (*k* vertices) is $\frac{1}{2}\cdot k(k-3)$. A convex polygon with $k+1$ sides ($k+1$ vertices) consists of a convex polygon with *k* sides (*k* vertices) plus a triangle, for a total of ($k+1$) vertices. The diagonals of this $k+1$-sided convex polygon consist of the diagonals of the *k*-sided polygon plus $k-1$ additional diagonals. For example, consider the following diagrams.

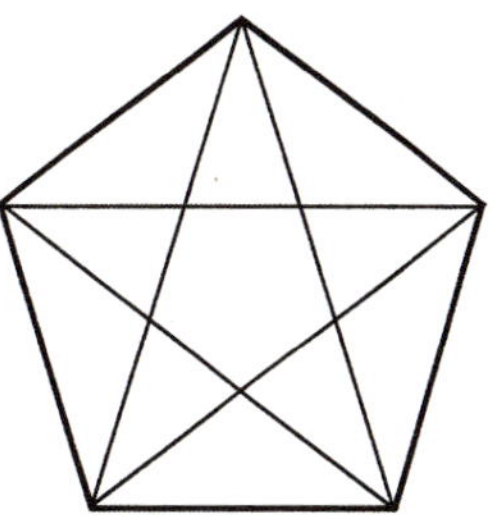

$k = 5$ sides

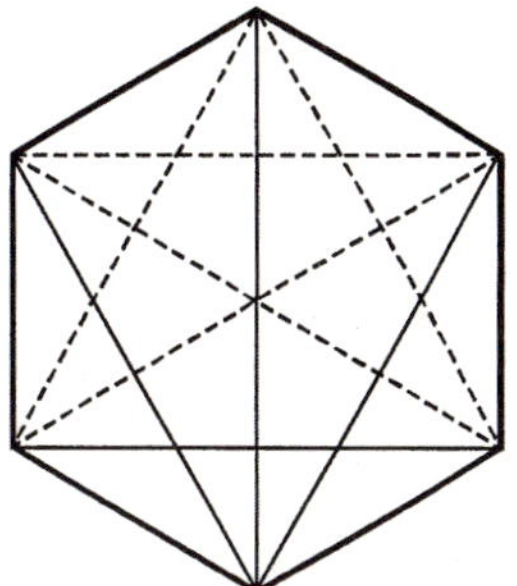

$k + 1 = 6$ sides
$k - 1 = 4$ new diagonals

Thus, we have the equation:

$$\frac{1}{2}\cdot k(k-3)+(k-1)=\frac{1}{2}k^2-\frac{3}{2}k+k-1$$
$$=\frac{1}{2}k^2-\frac{1}{2}k-1$$
$$=\frac{1}{2}\cdot\left(k^2-k-2\right)$$
$$=\frac{1}{2}\cdot(k+1)(k-2)$$
$$=\frac{1}{2}\cdot(k+1)((k+1)-3)$$

Conditions I and II are satisfied; the statement is true.

32. I: $n=3$: $(3-2)\cdot 180°=180°$ which is the sum of the angles of a triangle.

II: Assume that for any integer *k*, the sum of the angles of a convex polygon with *k* sides is $(k-2)\cdot 180°$. A convex polygon with $k+1$ sides consists of a convex polygon with *k* sides plus a triangle. Thus, the sum of the angles is

$$(k-2)\cdot 180°+180°=((k+1)-2)\cdot 180°.$$

Conditions I and II are satisfied; the statement is true.

33. Answers will vary.

Section 11.5

1. Pascal Triangle

2. $\binom{6}{2}=\frac{6!}{2!4!}=\frac{6\cdot5\cdot4\cdot3\cdot2\cdot1}{2\cdot1\cdot4\cdot3\cdot2\cdot1}=\frac{6\cdot5}{2\cdot1}=15$

3. False; $\binom{n}{j}=\frac{n!}{j!(n-j)!}$

4. Binomial Theorem

5. $\binom{5}{3}=\frac{5!}{3!2!}=\frac{5\cdot4\cdot3\cdot2\cdot1}{3\cdot2\cdot1\cdot2\cdot1}=\frac{5\cdot4}{2\cdot1}=10$

6. $\binom{7}{3}=\frac{7!}{3!4!}=\frac{7\cdot6\cdot5\cdot4\cdot3\cdot2\cdot1}{3\cdot2\cdot1\cdot4\cdot3\cdot2\cdot1}=\frac{7\cdot6\cdot5}{3\cdot2\cdot1}=35$

7. $\binom{7}{5}=\frac{7!}{5!2!}=\frac{7\cdot6\cdot5\cdot4\cdot3\cdot2\cdot1}{5\cdot4\cdot3\cdot2\cdot1\cdot2\cdot1}=\frac{7\cdot6}{2\cdot1}=21$

8. $\binom{9}{7}=\frac{9!}{7!2!}=\frac{9\cdot8\cdot7\cdot6\cdot5\cdot4\cdot3\cdot2\cdot1}{7\cdot6\cdot5\cdot4\cdot3\cdot2\cdot1\cdot2\cdot1}=\frac{9\cdot8}{2\cdot1}=36$

9. $\binom{50}{49}=\frac{50!}{49!1!}=\frac{50\cdot49!}{49!\cdot1}=\frac{50}{1}=50$

10. $\binom{100}{98}=\frac{100!}{98!2!}=\frac{100\cdot99\cdot98!}{98!\cdot2\cdot1}=\frac{100\cdot99}{2\cdot1}=4950$

11. $\binom{1000}{1000}=\frac{1000!}{1000!0!}=\frac{1}{1}=1$

12. $\binom{1000}{0}=\frac{1000!}{0!1000!}=\frac{1}{1}=1$

13. $\binom{55}{23}=\frac{55!}{23!32!}\approx1.866442159\times10^{15}$

14. $\binom{60}{20}=\frac{60!}{20!40!}\approx4.191844506\times10^{15}$

15. $\binom{47}{25}=\frac{47!}{25!22!}\approx1.483389769\times10^{13}$

16. $\binom{37}{19}=\frac{37!}{19!18!}\approx1.76726319\times10^{10}$

17. $(x+1)^5=\binom{5}{0}x^5+\binom{5}{1}x^4+\binom{5}{2}x^3+\binom{5}{3}x^2+\binom{5}{4}x^1+\binom{5}{5}x^0=x^5+5x^4+10x^3+10x^2+5x+1$

18. $(x-1)^5=\binom{5}{0}x^5+\binom{5}{1}(-1)x^4+\binom{5}{2}(-1)^2x^3+\binom{5}{3}(-1)^3x^2+\binom{5}{4}(-1)^4x^1+\binom{5}{5}(-1)^5x^0$

$=x^5-5x^4+10x^3-10x^2+5x-1$

19. $(x-2)^6=\binom{6}{0}x^6+\binom{6}{1}x^5(-2)+\binom{6}{2}x^4(-2)^2+\binom{6}{3}x^3(-2)^3+\binom{6}{4}x^2(-2)^4+\binom{6}{5}x(-2)^5+\binom{6}{6}x^0(-2)^6$

$=x^6+6x^5(-2)+15x^4\cdot4+20x^3(-8)+15x^2\cdot16+6x\cdot(-32)+64$

$=x^6-12x^5+60x^4-160x^3+240x^2-192x+64$

20. $(x+3)^5=\binom{5}{0}x^5+\binom{5}{1}x^4(3)+\binom{5}{2}x^3(3)^2+\binom{5}{3}x^2(3)^3+\binom{5}{4}x^1(3)^4+\binom{5}{5}x^0(3)^5$

$=x^5+5x^4(3)+10x^3\cdot9+10x^2(27)+5x\cdot81+243$

$=x^5+15x^4+90x^3+270x^2+405x+243$

21. $(3x+1)^4 = \binom{4}{0}(3x)^4 + \binom{4}{1}(3x)^3 + \binom{4}{2}(3x)^2 + \binom{4}{3}(3x) + \binom{4}{4}$

$= 81x^4 + 4\cdot 27x^3 + 6\cdot 9x^2 + 4\cdot 3x + 1 = 81x^4 + 108x^3 + 54x^2 + 12x + 1$

22. $(2x+3)^5 = \binom{5}{0}(2x)^5 + \binom{5}{1}(2x)^4\cdot 3 + \binom{5}{2}(2x)^3\cdot 3^2 + \binom{5}{3}(2x)^2\cdot 3^3 + \binom{5}{4}\cdot 2x\cdot 3^4 + \binom{5}{5}\cdot 3^5$

$= 32x^5 + 5\cdot 16x^4\cdot 3 + 10\cdot 8x^3\cdot 9 + 10\cdot 4x^2\cdot 27 + 5\cdot 2x\cdot 81 + 243$

$= 32x^5 + 240x^4 + 720x^3 + 1080x^2 + 810x + 243$

23. $\left(x^2+y^2\right)^5 = \binom{5}{0}\left(x^2\right)^5 + \binom{5}{1}\left(x^2\right)^4 y^2 + \binom{5}{2}\left(x^2\right)^3\left(y^2\right)^2 + \binom{5}{3}\left(x^2\right)^2\left(y^2\right)^3 + \binom{5}{4}x^2\left(y^2\right)^4 + \binom{5}{5}\left(y^2\right)^5$

$= x^{10} + 5x^8y^2 + 10x^6y^4 + 10x^4y^6 + 5x^2y^8 + y^{10}$

24. $\left(x^2-y^2\right)^6 = \binom{6}{0}\left(x^2\right)^6 + \binom{6}{1}\left(x^2\right)^5\left(-y^2\right) + \binom{6}{2}\left(x^2\right)^4\left(-y^2\right)^2 + \binom{6}{3}\left(x^2\right)^3\left(-y^2\right)^3 + \binom{6}{4}\left(x^2\right)^2\left(-y^2\right)^4$

$+\binom{6}{5}x^2\left(-y^2\right)^5 + \binom{6}{6}\left(-y^2\right)^6$

$= x^{12} - 6x^{10}y^2 + 15x^8y^4 - 20x^6y^6 + 15x^4y^8 - 6x^2y^{10} + y^{12}$

25. $\left(\sqrt{x}+\sqrt{2}\right)^6 = \binom{6}{0}\left(\sqrt{x}\right)^6 + \binom{6}{1}\left(\sqrt{x}\right)^5\left(\sqrt{2}\right)^1 + \binom{6}{2}\left(\sqrt{x}\right)^4\left(\sqrt{2}\right)^2 + \binom{6}{3}\left(\sqrt{x}\right)^3\left(\sqrt{2}\right)^3$

$+\binom{6}{4}\left(\sqrt{x}\right)^2\left(\sqrt{2}\right)^4 + \binom{6}{5}\left(\sqrt{x}\right)\left(\sqrt{2}\right)^5 + \binom{6}{6}\left(\sqrt{2}\right)^6$

$= x^3 + 6\sqrt{2}x^{5/2} + 15\cdot 2x^2 + 20\cdot 2\sqrt{2}x^{3/2} + 15\cdot 4x + 6\cdot 4\sqrt{2}x^{1/2} + 8$

$= x^3 + 6\sqrt{2}x^{5/2} + 30x^2 + 40\sqrt{2}x^{3/2} + 60x + 24\sqrt{2}x^{1/2} + 8$

26. $\left(\sqrt{x}-\sqrt{3}\right)^4 = \binom{4}{0}\left(\sqrt{x}\right)^4 + \binom{4}{1}\left(\sqrt{x}\right)^3\left(-\sqrt{3}\right)^1 + \binom{4}{2}\left(\sqrt{x}\right)^2\left(-\sqrt{3}\right)^2 + \binom{4}{3}\left(\sqrt{x}\right)\left(-\sqrt{3}\right)^3 + \binom{4}{4}\left(-\sqrt{3}\right)^4$

$= x^2 - 4\sqrt{3}x^{3/2} + 6\cdot 3x - 4\cdot 3\sqrt{3}x^{1/2} + 9$

$= x^2 - 4\sqrt{3}x^{3/2} + 18x - 12\sqrt{3}x^{1/2} + 9$

27. $(ax+by)^5 = \binom{5}{0}(ax)^5 + \binom{5}{1}(ax)^4\cdot by + \binom{5}{2}(ax)^3(by)^2 + \binom{5}{3}(ax)^2(by)^3 + \binom{5}{4}ax(by)^4 + \binom{5}{5}(by)^5$

$= a^5x^5 + 5a^4x^4by + 10a^3x^3b^2y^2 + 10a^2x^2b^3y^3 + 5axb^4y^4 + b^5y^5$

28. $(ax-by)^4 = \binom{4}{0}(ax)^4 + \binom{4}{1}(ax)^3(-by) + \binom{4}{2}(ax)^2(-by)^2 + \binom{4}{3}(ax)(-by)^3 + \binom{4}{4}(-by)^4$

$= a^4x^4 - 4a^3x^3by + 6a^2x^2b^2y^2 - 4axb^3y^3 + b^4y^4$

29. $n=10,\ j=4,\ x=x,\ a=3$

$$\binom{10}{4}x^6\cdot 3^4=\frac{10!}{4!6!}\cdot 81x^6=\frac{10\cdot 9\cdot 8\cdot 7}{4\cdot 3\cdot 2\cdot 1}\cdot 81x^6$$
$$=17{,}010x^6$$

The coefficient of x^6 is 17,010.

30. $n=10,\ j=7,\ x=x,\ a=-3$

$$\binom{10}{7}x^3\cdot(-3)^7=\frac{10!}{7!3!}\cdot(-2187)x^3$$
$$=\frac{10\cdot 9\cdot 8}{3\cdot 2\cdot 1}\cdot(-2187)x^3$$
$$=-262{,}440x^3$$

The coefficient of x^3 is $-262{,}440$.

31. $n=12,\ j=5,\ x=2x,\ a=-1$

$$\binom{12}{5}(2x)^7\cdot(-1)^5=\frac{12!}{5!7!}\cdot 128x^7(-1)$$
$$=\frac{12\cdot 11\cdot 10\cdot 9\cdot 8}{5\cdot 4\cdot 3\cdot 2\cdot 1}\cdot(-128)x^7$$
$$=-101{,}376x^7$$

The coefficient of x^7 is $-101{,}376$.

32. $n=12,\ j=9,\ x=2x,\ a=1$

$$\binom{12}{9}(2x)^3\cdot(1)^9=\frac{12!}{9!3!}\cdot 8x^3(1)$$
$$=\frac{12\cdot 11\cdot 10}{3\cdot 2\cdot 1}\cdot 8x^3$$
$$=1760x^3$$

The coefficient of x^3 is 1760.

33. $n=9,\ j=2,\ x=2x,\ a=3$

$$\binom{9}{2}(2x)^7\cdot 3^2=\frac{9!}{2!7!}\cdot 128x^7(9)$$
$$=\frac{9\cdot 8}{2\cdot 1}\cdot 128x^7\cdot 9$$
$$=41{,}472x^7$$

The coefficient of x^7 is 41,472.

34. $n=9,\ j=7,\ x=2x,\ a=-3$

$$\binom{9}{7}(2x)^2\cdot(-3)^7=\frac{9!}{7!2!}\cdot 4x^2(-2187)$$
$$=\frac{9\cdot 8}{2\cdot 1}\cdot 4x^2\cdot -2187$$
$$=-314{,}928x^2$$

The coefficient of x^2 is $-314{,}928$.

35. $n=7,\ j=4,\ x=x,\ a=3$

$$\binom{7}{4}x^3\cdot 3^4=\frac{7!}{4!3!}\cdot 81x^3=\frac{7\cdot 6\cdot 5}{3\cdot 2\cdot 1}\cdot 81x^3=2835x^3$$

36. $n=7,\ j=2,\ x=x,\ a=-3$

$$\binom{7}{2}x^5\cdot(-3)^2=\frac{7!}{2!5!}\cdot 9x^5=\frac{7\cdot 6}{2\cdot 1}\cdot 9x^5=189x^5$$

37. $n=9,\ j=2,\ x=3x,\ a=-2$

$$\binom{9}{2}(3x)^7\cdot(-2)^2=\frac{9!}{2!7!}\cdot 2187x^7\cdot 4$$
$$=\frac{9\cdot 8}{2\cdot 1}\cdot 8748x^7=314{,}928x^7$$

38. $n=8,\ j=5,\ x=3x,\ a=2$

$$\binom{8}{5}(3x)^3\cdot(2)^5=\frac{8!}{5!3!}\cdot 27x^3\cdot 32$$
$$=\frac{8\cdot 7\cdot 6}{3\cdot 2\cdot 1}\cdot 864x^3=48{,}384x^3$$

39. The x^0 term in

$$\sum_{j=0}^{12}\binom{12}{j}\left(x^2\right)^{12-j}\left(\frac{1}{x}\right)^j=\sum_{j=0}^{12}\binom{12}{j}x^{24-3j}$$

occurs when:

$$24-3j=0$$
$$24=3j$$
$$j=8$$

The coefficient is

$$\binom{12}{8}=\frac{12!}{8!4!}=\frac{12\cdot 11\cdot 10\cdot 9}{4\cdot 3\cdot 2\cdot 1}=495$$

40. The x^0 term in

$$\sum_{j=0}^{9}\binom{9}{j}(x)^{9-j}\left(-\frac{1}{x^2}\right)^j=\sum_{j=0}^{9}\binom{9}{j}(-1)^j x^{9-3j}$$

occurs when:

$$9-3j=0$$
$$9=3j$$
$$j=3$$

The coefficient is

$$\binom{9}{3}(-1)^3=-\frac{9!}{3!6!}=-\frac{9\cdot 8\cdot 7}{3\cdot 2\cdot 1}=-84$$

41. The x^4 term in

$$\sum_{j=0}^{10}\binom{10}{j}(x)^{10-j}\left(\frac{-2}{\sqrt{x}}\right)^j=\sum_{j=0}^{10}\binom{10}{j}(-2)^j x^{10-\frac{3}{2}j}$$

occurs when:

$$10-\frac{3}{2}j=4$$
$$-\frac{3}{2}j=-6$$
$$j=4$$

The coefficient is

$$\binom{10}{4}(-2)^4=\frac{10!}{6!4!}\cdot 16=\frac{10\cdot 9\cdot 8\cdot 7}{4\cdot 3\cdot 2\cdot 1}\cdot 16=3360$$

42. The x^2 term in

$$\sum_{j=0}^{8}\binom{8}{j}\left(\sqrt{x}\right)^{8-j}\left(\frac{3}{\sqrt{x}}\right)^j=\sum_{j=0}^{8}\binom{8}{j}(3)^j x^{4-j}$$

occurs when:

$$4-j=2$$
$$-j=-2$$
$$j=2$$

The coefficient is

$$\binom{8}{2}(3)^2=\frac{8!}{6!2!}\cdot 9=\frac{8\cdot 7}{2\cdot 1}\cdot 9=252$$

43. $$(1.001)^5=\left(1+10^{-3}\right)^5=\binom{5}{0}1^5+\binom{5}{1}1^4\cdot 10^{-3}+\binom{5}{2}1^3\cdot\left(10^{-3}\right)^2+\binom{5}{3}1^2\cdot\left(10^{-3}\right)^3+\cdots$$

$$=1+5(0.001)+10(0.000001)+10(0.000000001)+\cdots$$
$$=1+0.005+0.000010+0.000000010+\cdots$$
$$=1.00501 \quad \text{(correct to 5 decimal places)}$$

44. $$(0.998)^6=(1-0.002)^6=\binom{6}{0}1^6+\binom{6}{1}1^5\cdot(-0.002)+\binom{6}{2}1^4\cdot(-0.002)^2+\binom{6}{3}\cdot 1^3\cdot(-0.002)^3+\cdots$$

$$=1+6(-0.002)+15(0.000004)+20(-0.000000008)+\ldots$$
$$=1-0.012+0.00006-0.00000016$$
$$=0.98806 \quad \text{(correct to 5 decimal places)}$$

45. $$\binom{n}{n-1}=\frac{n!}{(n-1)!(n-(n-1))!}=\frac{n!}{(n-1)!(1)!}=\frac{n(n-1)!}{(n-1)!}=n$$

$$\binom{n}{n}=\frac{n!}{n!(n-n)!}=\frac{n!}{n!0!}=\frac{n!}{n!\cdot 1}=\frac{n!}{n!}=1$$

46. $\binom{n}{j}=\dfrac{n!}{j!(n-j)!}=\dfrac{n!}{(n-j)!\,j!}=\dfrac{n!}{(n-j)!(n-(n-j))!}=\binom{n}{n-j}$

47. Show that $\binom{n}{0}+\binom{n}{1}+\cdots+\binom{n}{n}=2^n$

$$\begin{aligned}2^n &= (1+1)^n\\ &=\binom{n}{0}\cdot 1^n+\binom{n}{1}\cdot 1^{n-1}\cdot 1+\binom{n}{2}\cdot 1^{n-2}\cdot 1^2+\cdots+\binom{n}{n}\cdot 1^{n-n}\cdot 1^n\\ &=\binom{n}{0}+\binom{n}{1}+\cdots+\binom{n}{n}\end{aligned}$$

48. Show that $\binom{n}{0}-\binom{n}{1}+\binom{n}{2}-\cdots+(-1)^n\binom{n}{n}=0$

$$\begin{aligned}0 &= (1-1)^n\\ &=\binom{n}{0}\cdot 1^n+\binom{n}{1}\cdot 1^{n-1}\cdot(-1)+\binom{n}{2}\cdot 1^{n-2}\cdot(-1)^2+\cdots+\binom{n}{n}\cdot 1^{n-n}\cdot(-1)^n\\ &=\binom{n}{0}-\binom{n}{1}+\binom{n}{2}-\cdots+(-1)^n\binom{n}{n}\end{aligned}$$

49. $\binom{5}{0}\left(\frac{1}{4}\right)^5+\binom{5}{1}\left(\frac{1}{4}\right)^4\left(\frac{3}{4}\right)+\binom{5}{2}\left(\frac{1}{4}\right)^3\left(\frac{3}{4}\right)^2+\binom{5}{3}\left(\frac{1}{4}\right)^2\left(\frac{3}{4}\right)^3+\binom{5}{4}\left(\frac{1}{4}\right)\left(\frac{3}{4}\right)^4+\binom{5}{5}\left(\frac{3}{4}\right)^5=\left(\frac{1}{4}+\frac{3}{4}\right)^5=(1)^5=1$

50. $12!=479{,}001{,}600=4.790016\times 10^8$

$20!\approx 2.432902008\times 10^{18}$

$25!\approx 1.551121004\times 10^{25}$

$12!\approx\sqrt{2\cdot 12\pi}\left(\dfrac{12}{e}\right)^{12}\left(1+\dfrac{1}{12\cdot 12-1}\right)\approx 479{,}013{,}972.4$

$20!\approx\sqrt{2\cdot 20\pi}\left(\dfrac{20}{e}\right)^{20}\left(1+\dfrac{1}{12\cdot 20-1}\right)\approx 2.43292403\times 10^{18}$

$25!\approx\sqrt{2\cdot 25\pi}\left(\dfrac{25}{e}\right)^{25}\left(1+\dfrac{1}{12\cdot 25-1}\right)\approx 1.551129917\times 10^{25}$

Chapter 11 Review

1. $a_1=(-1)^1\dfrac{1+3}{1+2}=-\dfrac{4}{3},\ a_2=(-1)^2\dfrac{2+3}{2+2}=\dfrac{5}{4},\ a_3=(-1)^3\dfrac{3+3}{3+2}=-\dfrac{6}{5},\ a_4=(-1)^4\dfrac{4+3}{4+2}=\dfrac{7}{6},\ a_5=(-1)^5\dfrac{5+3}{5+2}=-\dfrac{8}{7}$

2. $a_1=(-1)^{1+1}(2\cdot 1+3)=5,\ a_2=(-1)^{2+1}(2\cdot 2+3)=-7,\ a_3=(-1)^{3+1}(2\cdot 3+3)=9,$
$a_4=(-1)^{4+1}(2\cdot 4+3)=-11,\ a_5=(-1)^{5+1}(2\cdot 5+3)=13$

3. $a_1=\frac{2^1}{1^2}=\frac{2}{1}=2,\ a_2=\frac{2^2}{2^2}=\frac{4}{4}=1,\ a_3=\frac{2^3}{3^2}=\frac{8}{9},\ a_4=\frac{2^4}{4^2}=\frac{16}{16}=1,\ a_5=\frac{2^5}{5^2}=\frac{32}{25}$

4. $a_1=\frac{e^1}{1}=e,\ a_2=\frac{e^2}{2},\ a_3=\frac{e^3}{3},\ a_4=\frac{e^4}{4},\ a_5=\frac{e^5}{5}$

5. $a_1=3,\ a_2=\frac{2}{3}\cdot 3=2,\ a_3=\frac{2}{3}\cdot 2=\frac{4}{3},\ a_4=\frac{2}{3}\cdot\frac{4}{3}=\frac{8}{9},\ a_5=\frac{2}{3}\cdot\frac{8}{9}=\frac{16}{27}$

6. $a_1=4,\ a_2=-\frac{1}{4}\cdot 4=-1,\ a_3=-\frac{1}{4}\cdot -1=\frac{1}{4},\ a_4=-\frac{1}{4}\cdot\frac{1}{4}=-\frac{1}{16},\ a_5=-\frac{1}{4}\cdot -\frac{1}{16}=\frac{1}{64}$

7. $a_1=2,\ a_2=2-2=0,\ a_3=2-0=2,\ a_4=2-2=0,\ a_5=2-0=2$

8. $a_1=-3,\ a_2=4+(-3)=1,\ a_3=4+1=5,\ a_4=4+5=9,\ a_5=4+9=13$

9. $\sum_{k=1}^{4}(4k+2)=(4\cdot 1+2)+(4\cdot 2+2)+(4\cdot 3+2)+(4\cdot 4+2)=(6)+(10)+(14)+(18)=48$

10. $\sum_{k=1}^{3}(3-k^2)=(3-1^2)+(3-2^2)+(3-3^2)=(2)+(-1)+(-6)=-5$

11. $1-\frac{1}{2}+\frac{1}{3}-\frac{1}{4}+\cdots+\frac{1}{13}=\sum_{k=1}^{13}(-1)^{k+1}\left(\frac{1}{k}\right)$

12. $2+\frac{2^2}{3}+\frac{2^3}{3^2}+\frac{2^4}{3^3}+\cdots+\frac{2^{n+1}}{3^n}=\sum_{k=0}^{n}\left(\frac{2^{k+1}}{3^k}\right)$

$$=\sum_{k=1}^{n+1}\left(\frac{2^k}{3^{k-1}}\right)$$

13. $\{n+5\}$ Arithmetic

$d=(n+1+5)-(n+5)=n+6-n-5=1$

$S_n=\frac{n}{2}[6+n+5]=\frac{n}{2}(n+11)$

14. $\{4n+3\}$ Arithmetic

$d=(4(n+1)+3)-(4n+3)$
$=4n+4+3-4n-3=4$

$S_n=\frac{n}{2}[7+4n+3]$
$=\frac{n}{2}(4n+10)$
$=n(2n+5)$

15. $\{2n^3\}$ Examine the terms of the sequence: 2, 16, 54, 128, 250, ...
There is no common difference; there is no common ratio; neither.

16. $\{2n^2-1\}$ Examine the terms of the sequence: 1, 7, 17, 31, 49, ...
There is no common difference; there is no common ratio; neither.

17. $\{2^{3n}\}$ Geometric

$r=\frac{2^{3(n+1)}}{2^{3n}}=\frac{2^{3n+3}}{2^{3n}}=2^{3n+3-3n}=2^3=8$

$S_n=8\left(\frac{1-8^n}{1-8}\right)=8\left(\frac{1-8^n}{-7}\right)=\frac{8}{7}(8^n-1)$

18. $\{3^{2n}\}$ Geometric

$r=\frac{3^{2(n+1)}}{3^{2n}}=\frac{3^{2n+2}}{3^{2n}}=3^{2n+2-2n}=3^2=9$

$S_n=9\left(\frac{1-9^n}{1-9}\right)=9\left(\frac{1-9^n}{-8}\right)=\frac{9}{8}(9^n-1)$

19. 0, 4, 8, 12, ... Arithmetic $d = 4 - 0 = 4$

$$S_n = \frac{n}{2}(2(0) + (n-1)4) = \frac{n}{2}(4(n-1)) = 2n(n-1)$$

20. 1, −3, −7, −11, ... Arithmetic
$d = -3 - 1 = -4$

$$S_n = \frac{n}{2}(2(1) + (n-1)(-4)) = \frac{n}{2}(2 - 4n + 4)$$
$$= \frac{n}{2}(6 - 4n) = n(3 - 2n)$$

21. $3, \frac{3}{2}, \frac{3}{4}, \frac{3}{8}, \frac{3}{16}, \ldots$ Geometric

$$r = \frac{\left(\frac{3}{2}\right)}{3} = \frac{3}{2} \cdot \frac{1}{3} = \frac{1}{2}$$

$$S_n = 3\left(\frac{1 - \left(\frac{1}{2}\right)^n}{1 - \frac{1}{2}}\right) = 3\left(\frac{1 - \left(\frac{1}{2}\right)^n}{\left(\frac{1}{2}\right)}\right) = 6\left(1 - \left(\frac{1}{2}\right)^n\right)$$

22. $5, -\frac{5}{3}, \frac{5}{9}, -\frac{5}{27}, \frac{5}{81}, \ldots$ Geometric

$$r = \frac{\left(-\frac{5}{3}\right)}{5} = -\frac{5}{3} \cdot \frac{1}{5} = -\frac{1}{3}$$

$$S_n = 5\left(\frac{1 - \left(-\frac{1}{3}\right)^n}{1 - \left(-\frac{1}{3}\right)}\right) = 5\left(\frac{1 - \left(-\frac{1}{3}\right)^n}{\left(\frac{4}{3}\right)}\right)$$
$$= \frac{15}{4}\left(1 - \left(-\frac{1}{3}\right)^n\right)$$

23. Neither. There is no common difference or common ratio.

24. $\frac{3}{2}, \frac{5}{4}, \frac{7}{6}, \frac{9}{8}, \frac{11}{10}, \cdots$ Neither. There is no common difference or common ratio.

25. a. $\sum_{k=1}^{5}(k^2 + 12) = 13 + 16 + 21 + 28 + 37 = 115$

b. use the sum(seq) feature:

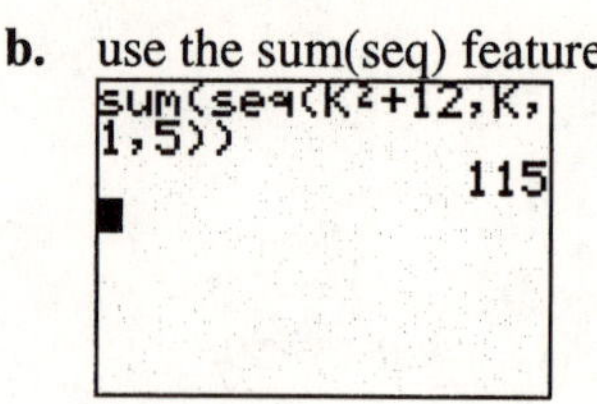

26. a. $\sum_{k=1}^{3}(k+2)^2 = 9 + 16 + 25 = 50$

b. use the sum(seq) feature:

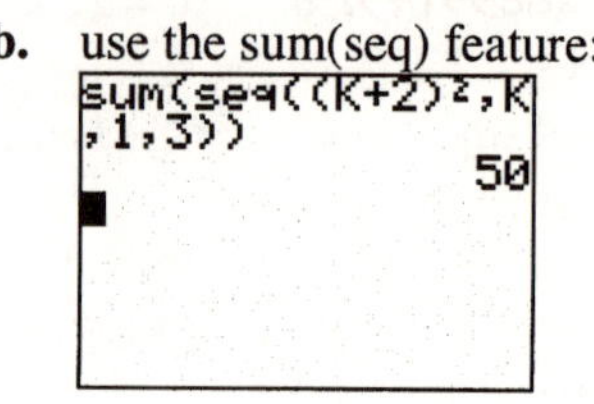

27. a. $\sum_{k=1}^{10}(3k - 9) = \sum_{k=1}^{10} 3k - \sum_{k=1}^{10} 9 = 3\sum_{k=1}^{10} k - \sum_{k=1}^{10} 9$

$$= 3\left(\frac{10(10+1)}{2}\right) - 10(9) = 165 - 90$$
$$= 75$$

b. use the sum(seq) feature:

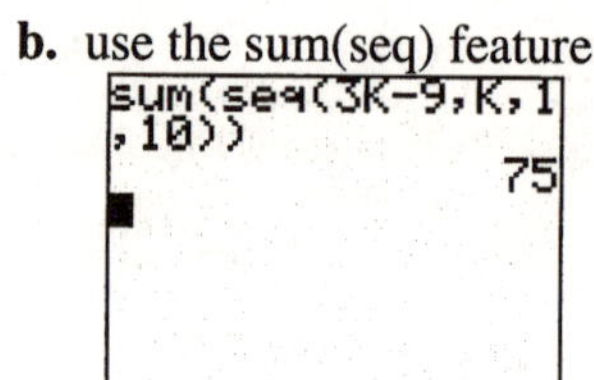

28. a. $\sum_{k=1}^{9}(-2k + 8) = \sum_{k=1}^{9} -2k + \sum_{k=1}^{9} 8$

$$= -2\sum_{k=1}^{9} k + \sum_{k=1}^{9} 8$$
$$= -2\left(\frac{9(1+9)}{2}\right) + 9(8)$$
$$= -90 + 72 = -18$$

b. use the sum(seq) feature:

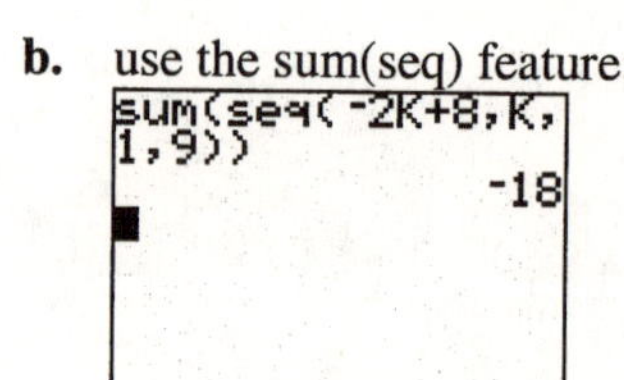

29. a. $\displaystyle\sum_{k=1}^{7}\left(\frac{1}{3}\right)^k = \frac{1}{3}\left(\frac{1-\left(\frac{1}{3}\right)^7}{1-\frac{1}{3}}\right) = \frac{1}{3}\left(\frac{1-\left(\frac{1}{3}\right)^7}{\left(\frac{2}{3}\right)}\right)$

$= \dfrac{1}{2}\left(1-\dfrac{1}{2187}\right)$

$= \dfrac{1}{2}\cdot\dfrac{2186}{2187} = \dfrac{1093}{2187} \approx 0.49977$

b. use the sum(seq) feature:

30. a. $\displaystyle\sum_{k=1}^{10}(-2)^k = -2\left(\frac{1-(-2)^{10}}{1-(-2)}\right)$

$= -2\left(\dfrac{1-1024}{3}\right) = -\dfrac{2}{3}(-1023)$

$= 682$

b. use the sum(seq) feature:

```
sum(seq((-2)^K,K
,1,10))
                 682
```

31. a. Arithmetic $a = 3,\ d = 4,\ a_n = a+(n-1)d$

$a_9 = 3+(9-1)4 = 3+8(4) = 3+32 = 35$

b. scroll to the end of the list:

```
seq(3+(A-1)*4,A,
1,9,1)→L1
…19 23 27 31 35}
```

32. a. Arithmetic $a = 1,\ d = -2,\ a_n = a+(n-1)d$

$a_8 = 1+(8-1)(-2) = 1+7(-2) = 1-14 = -13$

b. scroll to the end of the list:

```
seq(1+(A-1)*(-2)
,A,1,8,1)→L1
… -7 -9 -11 -13}
```

33. a. Geometric

$a = 1,\ r = \dfrac{1}{10},\ n = 11;\ a_n = a\,r^{n-1}$

$a_{11} = 1\cdot\left(\dfrac{1}{10}\right)^{11-1} = \left(\dfrac{1}{10}\right)^{10}$

$= \dfrac{1}{10{,}000{,}000{,}000}$

b. scroll to the end of the list:

```
seq(1*(1/10)^(A-
1),A,1,11,1)→L1
…E-8 1E-9 1E-10}
```

34. a. Geometric $a = 1,\ r = 2,\ n = 11;\ a_n = a\,r^{n-1}$

$a_{11} = 1\cdot(2)^{11-1} = (2)^{10} = 1024$

b. scroll to the end of the list:

```
seq(1*2^(A-1),A,
1,11,1)→L1
…8 256 512 1024}
```

35. a. Arithmetic

$a = \sqrt{2},\ d = \sqrt{2},\ n = 9,\ a_n = a+(n-1)d$

$a_9 = \sqrt{2}+(9-1)\sqrt{2} = \sqrt{2}+8\sqrt{2}$

$= 9\sqrt{2} \approx 12.7279$

b. scroll to the end of the list:

```
seq(√(2)+(A-1)*√
(2),A,1,9,1)→L1
…85 12.72792206}
```

36. a. Geometric

$a=\sqrt{2},\ r=\sqrt{2},\ n=9,\ a_n=ar^{n-1}$

$a_9=\sqrt{2}\left(\sqrt{2}\right)^{9-1}=\sqrt{2}\left(\sqrt{2}\right)^{8}=\sqrt{2}\cdot 16$

$=16\sqrt{2}\approx 22.6274$

b. scroll to the end of the list:

```
seq(√(2)*(√(2))^
(A-1),A,1,9,1)→L
1
…5 16 22.627417}
```

37. $a_7=a+6d=31 \quad a_{20}=a+19d=96$;

Solve the system of equations:

$a+6d=31$

$a+19d=96$

Subtract the second equation from the first equation and solve for d.

$-13d=-65$

$d=5$

$a=31-6(5)=31-30=1$

$a_n=a+(n-1)d$

$=1+(n-1)(5)$

$=1+5n-5$

$=5n-4$

General formula: $\{5n-4\}$

38. $a_8=a+7d=-20 \quad a_{17}=a+16d=-47$;

Solve the system of equations:

$a+7d=-20$

$a+16d=-47$

Subtract the second equation from the first equation and solve for d.

$-9d=27$

$d=-3$

$a=-20-7(-3)=-20+21=1$

$a_n=a+(n-1)d$

$=1+(n-1)(-3)$

$=1-3n+3$

$=-3n+4$

General formula: $\{-3n+4\}$

39. $a_{10}=a+9d=0 \quad a_{18}=a+17d=8$;

Solve the system of equations:

$a+9d=0$

$a+17d=8$

Subtract the second equation from the first equation and solve for d.

$-8d=-8$

$d=1$

$a=-9(1)=-9$

$a_n=a+(n-1)d$

$=-9+(n-1)(1)$

$=-9+n-1$

$=n-10$

General formula: $\{n-10\}$

40. $a_{12}=a+11d=30 \quad a_{22}=a+21d=50$;

Solve the system of equations:

$a+11d=30$

$a+21d=50$

Subtract the second equation from the first equation and solve for d.

$-10d=-20$

$d=2$

$a=30-11(2)=30-22=8$

$a_n=a+(n-1)d$

$=8+(n-1)(2)$

$=8+2n-2$

$=2n+6$

General formula: $\{2n+6\}$

41. $a=3,\ r=\frac{1}{3}$

Since $|r|<1$, $S_n=\frac{a}{1-r}=\frac{3}{\left(1-\frac{1}{3}\right)}=\frac{3}{\left(\frac{2}{3}\right)}=\frac{9}{2}$

42. $a=2,\ r=\frac{1}{2}$

Since $|r|<1$, $S_n=\frac{a}{1-r}=\frac{2}{\left(1-\frac{1}{2}\right)}=\frac{2}{\left(\frac{1}{2}\right)}=4$

43. $a = 2,\ r = -\frac{1}{2}$

Since $|r| < 1,\quad S_n = \frac{a}{1-r} = \frac{2}{\left(1-\left(-\frac{1}{2}\right)\right)} = \frac{2}{\left(\frac{3}{2}\right)} = \frac{4}{3}$

44. $a = 6,\ r = -\frac{2}{3}$

Since $|r| < 1,\quad S_n = \frac{a}{1-r} = \frac{6}{\left(1-\left(-\frac{2}{3}\right)\right)} = \frac{6}{\left(\frac{5}{3}\right)} = \frac{18}{5}$

45. $a = 4,\ r = \frac{1}{2}$

Since $|r| < 1,\quad S_n = \frac{a}{1-r} = \frac{4}{\left(1-\frac{1}{2}\right)} = \frac{4}{\left(\frac{1}{2}\right)} = 8$

46. $a = 3,\ r = -\frac{3}{4}$

Since $|r| < 1,\quad S_n = \frac{a}{1-r} = \frac{3}{\left(1-\left(-\frac{3}{4}\right)\right)} = \frac{3}{\left(\frac{7}{4}\right)} = \frac{12}{7}$

47. I: $n = 1$: $3\cdot 1 = 3$ and $\frac{3\cdot 1}{2}(1+1) = 3$

II: If $3+6+9+\cdots+3k = \frac{3k}{2}(k+1)$, then

$$3+6+9+\cdots+3k+3(k+1)$$
$$= [3+6+9+\cdots+3k]+3(k+1)$$
$$= \frac{3k}{2}(k+1)+3(k+1)$$
$$= (k+1)\left(\frac{3k}{2}+3\right) = \frac{3(k+1)}{2}((k+1)+1)$$

Conditions I and II are satisfied; the statement is true.

48. I: $n = 1$: $4\cdot 1-2 = 2$ and $2(1)^2 = 2$

II: If $2+6+10+\cdots+(4k-2) = 2k^2$, then

$$2+6+10+\cdots+(4k-2)+(4(k+1)-2)$$
$$= [2+6+10+\cdots+(4k-2)]+4k+2$$
$$= 2k^2+4k+2$$
$$= 2\left(k^2+2k+1\right) = 2(k+1)^2$$

Conditions I and II are satisfied; the statement is true.

49. I: $n = 1$: $2\cdot 3^{1-1} = 2$ and $3^1-1 = 2$

II: If $2+6+18+\cdots+2\cdot 3^{k-1} = 3^k-1$, then

$$2+6+18+\cdots+2\cdot 3^{k-1}+2\cdot 3^{k+1-1}$$
$$= \left[2+6+18+\cdots+2\cdot 3^{k-1}\right]+2\cdot 3^k$$
$$= 3^k-1+2\cdot 3^k = 3\cdot 3^k-1 = 3^{k+1}-1$$

Conditions I and II are satisfied; the statement is true.

50. I: $n = 1$: $3\cdot 2^{1-1} = 3$ and $3\left(2^1-1\right) = 3$

II: If $3+6+12+\cdots+3\cdot 2^{k-1} = 3\left(2^k-1\right)$, then

$$3+6+12+\cdots+3\cdot 2^{k-1}+3\cdot 2^{k+1-1}$$
$$= \left[3+6+12+\cdots+3\cdot 2^{k-1}\right]+3\cdot 2^k$$
$$= 3\left(2^k-1\right)+3\cdot 2^k$$
$$= 3\cdot\left(2^k-1+2^k\right) = 3\left(2\cdot 2^k-1\right) = 3\left(2^{k+1}-1\right)$$

Conditions I and II are satisfied; the statement is true.

51. I: $n=1$:

$(3\cdot 1-2)^2=1$ and $\frac{1}{2}\cdot 1(6\cdot 1^2-3\cdot 1-1)=1$

II: If

$1^2+4^2+\cdots+(3k-2)^2=\frac{1}{2}\cdot k\left(6k^2-3k-1\right)$,

then

$$1^2+4^2+7^2+\cdots+(3k-2)^2+\left(3(k+1)-2\right)^2$$
$$=\left[1^2+4^2+7^2+\cdots+(3k-2)^2\right]+(3k+1)^2$$
$$=\frac{1}{2}\cdot k\left(6k^2-3k-1\right)+(3k+1)^2$$
$$=\frac{1}{2}\cdot\left[6k^3-3k^2-k+18k^2+12k+2\right]$$
$$=\frac{1}{2}\cdot\left[6k^3+15k^2+11k+2\right]$$
$$=\frac{1}{2}\cdot(k+1)\left[6k^2+9k+2\right]$$
$$=\frac{1}{2}\cdot\left[6k^3+6k^2+9k^2+9k+2k+2\right]$$
$$=\frac{1}{2}\cdot\left[6k^2(k+1)+9k(k+1)+2(k+1)\right]$$
$$=\frac{1}{2}\cdot(k+1)\left[6k^2+12k+6-3k-3-1\right]$$
$$=\frac{1}{2}\cdot(k+1)\left[6(k^2+2k+1)-3(k+1)-1\right]$$
$$=\frac{1}{2}\cdot(k+1)\left[6(k+1)^2-3(k+1)-1\right]$$

Conditions I and II are satisfied; the statement is true.

52. I: $n=1$: $1(1+2)=3$ and $\frac{1}{6}\cdot(1+1)(2\cdot 1+7)=3$

II: If

$1\cdot 3+2\cdot 4+\cdots+k(k+2)=\frac{k}{6}(k+1)(2k+7)$,

then

$$1\cdot 3+2\cdot 4+\cdots+k(k+2)+(k+1)(k+1+2)$$
$$=\left[1\cdot 3+2\cdot 4+\cdots+k(k+2)\right]+(k+1)(k+3)$$
$$=\frac{k}{6}(k+1)(2k+7)+(k+1)(k+3)$$
$$=\frac{(k+1)}{6}\left(2k^2+7k+6k+18\right)$$
$$=\frac{(k+1)}{6}\left(2k^2+13k+18\right)$$
$$=\frac{(k+1)}{6}(k+2)(2(k+1)+7)$$

Conditions I and II are satisfied; the statement is true.

53. $\binom{5}{2}=\frac{5!}{2!3!}=\frac{5\cdot 4\cdot 3\cdot 2\cdot 1}{2\cdot 1\cdot 3\cdot 2\cdot 1}=\frac{5\cdot 4}{2\cdot 1}=10$

54. $\binom{8}{6}=\frac{8!}{6!2!}=\frac{8\cdot 7\cdot 6\cdot 5\cdot 4\cdot 3\cdot 2\cdot 1}{6\cdot 5\cdot 4\cdot 3\cdot 2\cdot 1\cdot 2\cdot 1}=\frac{8\cdot 7}{2\cdot 1}=28$

55. $(x+2)^5=\binom{5}{0}x^5+\binom{5}{1}x^4\cdot 2+\binom{5}{2}x^3\cdot 2^2+\binom{5}{3}x^2\cdot 2^3+\binom{5}{4}x^1\cdot 2^4+\binom{5}{5}\cdot 2^5$

$=x^5+5\cdot 2x^4+10\cdot 4x^3+10\cdot 8x^2+5\cdot 16x+1\cdot 32$

$=x^5+10x^4+40x^3+80x^2+80x+32$

56. $(x-3)^4=\binom{4}{0}x^4+\binom{4}{1}x^3(-3)+\binom{4}{2}x^2(-3)^2+\binom{4}{3}x(-3)^3+\binom{4}{4}x^0(-3)^4$

$=x^4+4(-3)x^3+6\cdot 9x^2+4(-27)x+81$

$=x^4-12x^3+54x^2-108x+81$

57. $(2x+3)^5 = \binom{5}{0}(2x)^5 + \binom{5}{1}(2x)^4 \cdot 3 + \binom{5}{2}(2x)^3 \cdot 3^2 + \binom{5}{3}(2x)^2 \cdot 3^3 + \binom{5}{4}(2x)^1 \cdot 3^4 + \binom{5}{5} \cdot 3^5$

$= 32x^5 + 5 \cdot 16x^4 \cdot 3 + 10 \cdot 8x^3 \cdot 9 + 10 \cdot 4x^2 \cdot 27 + 5 \cdot 2x \cdot 81 + 1 \cdot 243$

$= 32x^5 + 240x^4 + 720x^3 + 1080x^2 + 810x + 243$

58. $(3x-4)^4 = \binom{4}{0}(3x)^4 + \binom{4}{1}(3x)^3(-4) + \binom{4}{2}(3x)^2(-4)^2 + \binom{4}{3}(3x)(-4)^3 + \binom{4}{4}(-4)^4$

$= 81x^4 + 4 \cdot 27x^3(-4) + 6 \cdot 9x^2 \cdot 16 + 4 \cdot 3x(-64) + 1 \cdot 256$

$= 81x^4 - 432x^3 + 864x^2 - 768x + 256$

59. $n=9, \ j=2, \ x=x, \ a=2$

$\binom{9}{2}x^7 \cdot 2^2 = \frac{9!}{2!7!} \cdot 4x^7 = \frac{9 \cdot 8}{2 \cdot 1} \cdot 4x^7 = 144x^7$

The coefficient of x^7 is 144.

60. $n=8, \ j=5, \ x=x, \ a=-3$

$\binom{8}{5}x^3(-3)^5 = \frac{8!}{5!3!}(-243)x^3 = \frac{8 \cdot 7 \cdot 6}{3 \cdot 2 \cdot 1}(-243)x^3$

$= -13,608x^3$

The coefficient of x^3 is $-13,608$.

61. $n=7, \ j=5, \ x=2x, \ a=1$

$\binom{7}{5}(2x)^2 \cdot 1^5 = \frac{7!}{5!2!} \cdot 4x^2(1) = \frac{7 \cdot 6}{2 \cdot 1} \cdot 4x^2 = 84x^2$

The coefficient of x^2 is 84.

62. $n=8, \ j=2, \ x=2x, \ a=1$

$\binom{8}{2}(2x)^6 \cdot 1^2 = \frac{8!}{2!6!} \cdot 64x^6(1)$

$= \frac{8 \cdot 7}{2 \cdot 1} \cdot 64x^6 = 1792x^6$

The coefficient of x^6 is 1792.

63. This is an arithmetic sequence with $a=80, \ d=-3, \ n=25$

a. $a_{25} = 80 + (25-1)(-3) = 80 - 72 = 8$ bricks

b. $S_{25} = \frac{25}{2}(80+8) = 25(44) = 1100$ bricks

1100 bricks are needed to build the steps.

64. This is an arithmetic sequence with $a=30, \ d=-1, \ a_n=15$

$15 = 30 + (n-1)(-1)$

$-15 = -n+1$

$-16 = -n$

$n = 16$

$S_{16} = \frac{16}{2}(30+15) = 8(45) = 360$ tiles

360 tiles are required to make the trapezoid.

65. This is a geometric sequence with $a=20, \ r=\frac{3}{4}$.

a. After striking the ground the third time, the height is $20\left(\frac{3}{4}\right)^3 = \frac{135}{16} \approx 8.44$ feet.

b. After striking the ground the nth time, the height is $20\left(\frac{3}{4}\right)^n$ feet.

c If the height is less than 6 inches or 0.5 feet, then:

$0.5 \ge 20\left(\frac{3}{4}\right)^n$

$0.025 \ge \left(\frac{3}{4}\right)^n$

$\log(0.025) \ge n \log\left(\frac{3}{4}\right)$

$n \ge \frac{\log(0.025)}{\log\left(\frac{3}{4}\right)} \approx 12.82$

The height is less than 6 inches after the 13th strike.

d. Since this is a geometric sequence with $|r|<1$, the distance is the sum of the two infinite geometric series - the distances going down plus the distances going up.
Distance going down:

$$S_{down} = \frac{20}{\left(1-\frac{3}{4}\right)} = \frac{20}{\left(\frac{1}{4}\right)} = 80 \text{ feet.}$$

Distance going up:

$$S_{up} = \frac{15}{\left(1-\frac{3}{4}\right)} = \frac{15}{\left(\frac{1}{4}\right)} = 60 \text{ feet.}$$

The total distance traveled is 140 feet.

66. a. Since the interest rate is 6.75% per annum compounded monthly, this is equivalent to a rate of (6.75/12)% each month. Defining a recursive sequence, we have:

$a_0 = 190,000$

$$a_n = \left(1+\frac{0.0675}{12}\right)a_{n-1} - 1232.34$$

b. $\left(1+\frac{0.0675}{12}\right)(190,000) - 1232.34$

$= \$189,836.41$

c. Enter the recursive formula in Y= and create the table:

```
Plot1 Plot2 Plot3
nMin=0
\u(n)■(1+.0675/1
2)*(u(n-1))-1232
.34
 u(nMin)■{19000...
\v(n)=
 v(nMin)=
```

n	u(n)
0	190000
1	189836
2	189672
3	189506
4	189340
5	189173
6	189005

n=0

d. Scroll through the table:

n	u(n)
250	100874
251	100209
252	99540
253	98868
254	98192
255	97512
256	96828

n=252

At the beginning of the 252nd month, or after 251 payments have been made, the balance is below \$100,000. The balance is about \$99540.

e. Scroll through the table:

n	u(n)
357	3651.7
358	2438.9
359	[illegible]
360	-4.2
361	-1237
362	-2476
363	-3722

u(n)=1221.269892

The loan will be paid off at the end of 360 months or 30 years.
Total amount paid =
$1232.34(359) + (1221.27)(1 + 0.0675/12)$
$= \$443,638.20$

f. The total interest expense is the difference of the total of the payments and the original loan: $443,638.20 - 190,000 = \$253,638.20$.

g. (a) Since the interest rate is 6.75% per annum compounded monthly, this is equivalent to a rate of (6.75/12)% each month. Defining a recursive sequence, we have:

$a_0 = 190,000$

$$a_n = \left(1+\frac{0.0675}{12}\right)a_{n-1} - 1332.34$$

(b) $\left(1+\frac{0.0675}{12}\right)(190,000) - 1332.34$

$= \$189,736.41$

(c) Enter the recursive formula in Y= and create the table:

(d) Scroll through the table:

n	u(n)
190	100823
191	100058
192	99288
193	98515
194	97736
195	96954
196	96167

n=192

At the beginning of the 192nd month, or after 191 payments have been made, the balance is below \$100,000. The balance is about \$99,288.

(e) Scroll through the table:

n	u(n)
285	5076.2
286	3772.4
287	2461.3
288	[illegible]
289	-183.1
290	-1516
291	-2857

u(n)=1142.796503

The loan will be paid off at the end of 289 months, i.e. after 24 years 1 month.
Total amount paid =

$$1332.34(288)+(1142.80)\left(1+\frac{0.0675}{12}\right)$$
$$=\$384,863.15$$

(f) The total interest expense is the difference of the total of the payments and the original loan:
$384,863.15-190,000=\$194,863.15$.

67. a. $b_1=5,000,\quad b_n=1.015b_{n-1}-100$

$$b_2=1.015b_{(2-1)}-100=1.015b_1-100$$
$$=1.015(5000)-100=\$4975$$

b. Enter the recursive formula in Y= and draw the graph:

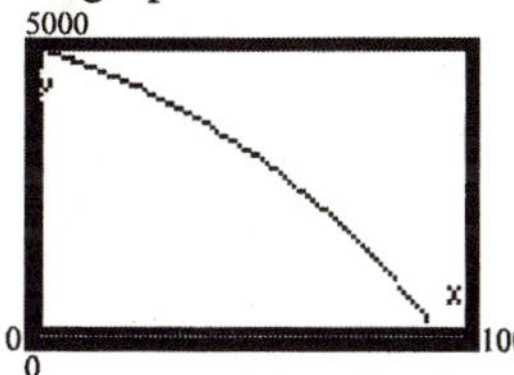

c. Scroll through the table:

n	u(n)
30	4100
31	4061.5
32	4022.5
33	3982.8
34	3942.5
35	3901.7
36	3860.2

n=33

At the beginning of the 33rd month, or after 32 payments have been made, the balance is below \$4,000. The balance is \$3982.8.

d. Scroll through the table:

n	u(n)
90	395.82
91	301.75
92	206.28
93	109.37
94	11.014
95	-88.82
96	-190.2

n=94

The balance will be paid off at the end of 94 months or 7years and 10 months.
Total payments =
$100(93)+11.01(1.015)=\$9,311.18$

e. The total interest expense is the difference between the total of the payments and the original balance:
$100(93)+11.18-5,000=\$4,311.18$

Chapter 11 Test

1. $a_n=\dfrac{n^2-1}{n+8}$

$$a_1=\frac{1^2-1}{1+8}=\frac{0}{9}=0$$
$$a_2=\frac{2^2-1}{2+8}=\frac{3}{10}$$
$$a_3=\frac{3^2-1}{3+8}=\frac{8}{11}$$
$$a_4=\frac{4^2-1}{4+8}=\frac{15}{12}=\frac{5}{4}$$
$$a_5=\frac{5^2-1}{5+8}=\frac{24}{13}$$

The first five terms of the sequence are 0, $\frac{3}{10}$, $\frac{8}{11}$, $\frac{5}{4}$, and $\frac{24}{13}$.

2. $a_1=4;\ a_n=3a_{n-1}+2$

$$a_2=3a_1+2=3(4)+2=14$$
$$a_3=3a_2+2=3(14)+2=44$$
$$a_4=3a_3+2=3(44)+2=134$$
$$a_5=3a_4+2=3(134)+2=404$$

The first five terms of the sequence are 4, 14, 44, 134, and 404.

3. $\sum_{k=1}^{3}(-1)^{k+1}\left(\frac{k+1}{k^2}\right)$

$=(-1)^{1+1}\left(\frac{1+1}{1^2}\right)+(-1)^{2+1}\left(\frac{2+1}{2^2}\right)+(-1)^{3+1}\left(\frac{3+1}{3^2}\right)$

$=(-1)^2\left(\frac{2}{1}\right)+(-1)^3\left(\frac{3}{4}\right)+(-1)^4\left(\frac{4}{9}\right)$

$=2-\frac{3}{4}+\frac{4}{9}=\frac{61}{36}$

4. $\sum_{k=1}^{4}\left[\left(\frac{2}{3}\right)^k-k\right]$

$=\left[\left(\frac{2}{3}\right)^1-1\right]+\left[\left(\frac{2}{3}\right)^2-2\right]+\left[\left(\frac{2}{3}\right)^3-3\right]+\left[\left(\frac{2}{3}\right)^4-4\right]$

$=\frac{2}{3}-1+\frac{4}{9}-2+\frac{8}{27}-3+\frac{16}{81}-4$

$=\frac{130}{81}-10=-\frac{680}{81}$

5. $-\frac{2}{5}+\frac{3}{6}-\frac{4}{7}+\ldots+\frac{11}{14}$

Notice that the signs of each term alternate, with the first term being negative. This implies that the general term will include a power of -1. Also note that the numerator is always 1 more than the term number and the denominator is 4 more than the term number. Thus, each term is in the form $(-1)^k\left(\frac{k+1}{k+4}\right)$. The last numerator is 11 which indicates that there are 10 terms.

$-\frac{2}{5}+\frac{3}{6}-\frac{4}{7}+\ldots+\frac{11}{14}=\sum_{k=1}^{10}(-1)^k\left(\frac{k+1}{k+4}\right)$

6. $\sum_{k=1}^{100}(-1)^k\left(k^2\right)=5050$

```
sum(seq((-1)^X*X
²,X,1,100))
            5050
```

7. $6, 12, 36, 144, \ldots$

$12-6=6$ and $36-12=24$

The difference between consecutive terms is not constant. Therefore, the sequence is not arithmetic.

$\frac{12}{6}=2$ and $\frac{36}{12}=3$

The ratio of consecutive terms is not constant. Therefore, the sequence is not geometric.

8. $a_n=-\frac{1}{2}\cdot 4^n$

$\frac{a_n}{a_{n-1}}=\frac{-\frac{1}{2}\cdot 4^n}{-\frac{1}{2}\cdot 4^{n-1}}=\frac{-\frac{1}{2}\cdot 4^{n-1}\cdot 4}{-\frac{1}{2}\cdot 4^{n-1}}=4$

Since the ratio of consecutive terms is constant, the sequence is geometric with common ratio $r=4$ and first term $a_1=a=-\frac{1}{2}\cdot 4^1=-2$.

The sum of the first n terms of the sequence is given by

$S_n=a\cdot\frac{1-r^n}{1-r}$

$=-2\cdot\frac{1-4^n}{1-4}$

$=\frac{2}{3}\left(1-4^n\right)$

9. $-2, -10, -18, -26, \ldots$

$-10-(-2)=-8$, $-18-(-10)=-8$,

$-26-(-18)=-8$

The difference between consecutive terms is constant. Therefore, the sequence is arithmetic with common difference $d=-8$ and first term $a_1=a=-2$.

$a_n=a+(n-1)d$

$=-2+(n-1)(-8)$

$=-2-8n+8$

$=6-8n$

The sum of the first n terms of the sequence is given by

$S_n=\frac{n}{2}(a+a_n)$

$=\frac{n}{2}(-2+6-8n)$

$=\frac{n}{2}(4-8n)$

$=n(2-4n)$

10. $a_n = -\frac{n}{2} + 7$

$$a_n - a_{n-1} = \left[-\frac{n}{2}+7\right] - \left[-\frac{(n-1)}{2}+7\right] = -\frac{n}{2}+7+\frac{n-1}{2}-7 = -\frac{1}{2}$$

The difference between consecutive terms is constant. Therefore, the sequence is arithmetic with common difference $d = -\frac{1}{2}$ and first term $a_1 = a = -\frac{1}{2} + 7 = \frac{13}{2}$.

The sum of the first n terms of the sequence is given by

$$S_n = \frac{n}{2}(a + a_n) = \frac{n}{2}\left(\frac{13}{2} + \left(-\frac{n}{2}+7\right)\right) = \frac{n}{2}\left(\frac{27}{2} - \frac{n}{2}\right) = \frac{n}{4}(27-n)$$

11. $25, 10, 4, \frac{8}{5}, \ldots$

$$\frac{10}{25} = \frac{2}{5},\ \frac{4}{10} = \frac{2}{5},\ \frac{\frac{8}{5}}{4} = \frac{8}{5}\cdot\frac{1}{4} = \frac{2}{5}$$

The ratio of consecutive terms is constant. Therefore, the sequence is geometric with common ratio $r = \frac{2}{5}$ and first term $a_1 = a = 25$.

The sum of the first n terms of the sequence is given by

$$S_n = a \cdot \frac{1-r^n}{1-r} = 25 \cdot \frac{1-\left(\frac{2}{5}\right)^n}{1-\frac{2}{5}} = 25 \cdot \frac{\left(1-\left(\frac{2}{5}\right)^n\right)}{\frac{3}{5}}$$

$$= 25 \cdot \frac{5}{3}\left(1-\left(\frac{2}{5}\right)^n\right) = \frac{125}{3}\left(1-\left(\frac{2}{5}\right)^n\right)$$

12. $a_n = \frac{2n-3}{2n+1}$

$$a_n - a_{n-1} = \frac{2n-3}{2n+1} - \frac{2(n-1)-3}{2(n-1)+1} = \frac{2n-3}{2n+1} - \frac{2n-5}{2n-1}$$

$$= \frac{(2n-3)(2n-1)-(2n-5)(2n+1)}{(2n+1)(2n-1)}$$

$$= \frac{(4n^2-8n+3)-(4n^2-8n-5)}{4n^2-1}$$

$$= \frac{8}{4n^2-1}$$

The difference of consecutive terms is not constant. Therefore, the sequence is not arithmetic.

$$\frac{a_n}{a_{n-1}} = \frac{\frac{2n-3}{2n+1}}{\frac{2(n-1)-3}{2(n-1)+1}} = \frac{2n-3}{2n+1}\cdot\frac{2n-1}{2n-5}$$

$$= \frac{(2n-3)(2n-1)}{(2n+1)(2n-5)}$$

The ratio of consecutive terms is not constant. Therefore, the sequence is not geometric.

13. For this geometric series we have $r = \frac{-64}{256} = -\frac{1}{4}$ and $a_1 = a = 256$. Since $|r| = \left|-\frac{1}{4}\right| = \frac{1}{4} < 1$, we get

$$S_\infty = \frac{a}{1-r} = \frac{256}{1-\left(-\frac{1}{4}\right)} = \frac{256}{\frac{5}{4}} = \frac{1024}{5} = 204.8$$

14. $(3m+2)^5 = \binom{5}{0}(3m)^5 + \binom{5}{1}(3m)^4(2) + \binom{5}{2}(3m)^3(2)^2 + \binom{5}{3}(3m)^2(2)^3 + \binom{5}{4}(3m)(2)^4 + \binom{5}{5}(2)^5$

$= 243m^5 + 5\cdot 81m^4\cdot 2 + 10\cdot 27m^3\cdot 4 + 10\cdot 9m^2\cdot 8 + 5\cdot 3m\cdot 16 + 32$

$= 243m^5 + 810m^4 + 1080m^3 + 720m^2 + 240m + 32$

15. First we show that the statement holds for $n = 1$.

$\left(1+\frac{1}{1}\right) = 1+1 = 2$

The equality is true for $n = 1$ so Condition I holds.

Next we assume that $\left(1+\frac{1}{1}\right)\left(1+\frac{1}{2}\right)\left(1+\frac{1}{3}\right)\ldots\left(1+\frac{1}{n}\right) = n+1$ is true for some *k*, and we determine whether the formula then holds for $k+1$. We assume that

$\left(1+\frac{1}{1}\right)\left(1+\frac{1}{2}\right)\left(1+\frac{1}{3}\right)\ldots\left(1+\frac{1}{k}\right) = k+1$

Now we need to show that

$\left(1+\frac{1}{1}\right)\left(1+\frac{1}{2}\right)\left(1+\frac{1}{3}\right)\ldots\left(1+\frac{1}{k}\right)\left(1+\frac{1}{k+1}\right) = (k+1)+1 = k+2$

We do this as follows:

$$\left(1+\frac{1}{1}\right)\left(1+\frac{1}{2}\right)\left(1+\frac{1}{3}\right)\ldots\left(1+\frac{1}{k}\right)\left(1+\frac{1}{k+1}\right) = \left[\left(1+\frac{1}{1}\right)\left(1+\frac{1}{2}\right)\left(1+\frac{1}{3}\right)\ldots\left(1+\frac{1}{k}\right)\right]\left(1+\frac{1}{k+1}\right)$$

$$= (k+1)\left(1+\frac{1}{k+1}\right) \quad \text{(using the induction assumption)}$$

$$= (k+1)\cdot 1 + (k+1)\cdot\frac{1}{k+1} = k+1+1$$

$$= k+2$$

Condition II also holds. Thus, formula holds true for all natural numbers.

16. The yearly values of the Durango form a geometric sequence with first term $a_1 = a = 31{,}000$ and common ratio $r = 0.85$ (which represents a 15% loss in value).

$a_n = 31{,}000\cdot(0.85)^{n-1}$

The *n*th term of the sequence represents the value of the Durango at the beginning of the *n*th year. Since we want to know the value *after* 10 years, we are looking for the 11th term of the sequence. That is, the value of the Durango at the beginning of the 11th year.

$a_{11} = a\cdot r^{11-1}$

$= 31{,}000\cdot(0.85)^{10}$

$= 6{,}103.11$

After 10 years, the Durango will be worth \$6,103.11.

17. The weights for each set form an arithmetic sequence with first term $a_1 = a = 100$ and common difference $d = 30$. If we imagine the weightlifter only performed one repetition per set, the total weight lifted in 5 sets would be the sum of the first five terms of the sequence.

$a_n = a + (n-1)d$

$a_5 = 100 + (5-1)(30) = 100 + 4(30) = 220$

$S_n = \frac{n}{2}(a + a_n)$

$S_5 = \frac{5}{2}(100+220) = \frac{5}{2}(320) = 800$

Since he performs 10 repetitions in each set, we multiply the sum by 10 to obtain the total weight lifted.

$10(800) = 8000$

The weightlifter will have lifted a total of 8000 pounds after 5 sets.

Chapter 11 Projects

Project 1

Data used in these solutions will be from 2001.

a.

Race	Birth rate	Death rate
White	0.0118	0.00991
African-American	0.116	0.007758
American Indian/ Alaska native	0.0137	0.00392
Asian/Pacific Islander	0.0164	0.003047
Hispanic	0.0230	0.00306
Total	0.141	0.008485

I = net immigration = 653,259
Population for 2001 = 285,545,000

b. r = 0.0141 – 0.008485 = 0.005615

c. $p_n = (1+0.005615)p_{n-1} + 653259$

$p_n = (1.005615)p_{n-1} + 653259$

$p_0 = 285545000$

d. $p_1 = (1.005615)p_0 + 653259$

$p_1 = (1.005615)(285545000) + 653259$

$p_1 = 287801594$

The population is predicted to be 287,801,594 in 2002.

e. Actual population in 2002: 288,600,000. The formula's prediction was lower.

f. It could be but one must consider trends in each of the pieces of data to find if the growth rate is increasing or decreasing over time. The same thing must be examined with respect to the net immigration.

Project 2 (web)

1. 2, 4, 8, <u>16</u>, <u>32</u>, <u>64</u>

2. length n ➔ 2^n levels

This is a geometric sequence: $a_n = 2^n$

Recursive expression: $a_n = 2a_{n-1}, \quad a_0 = 1$

3. $256 = 2^n$

$2^8 = 2^n$

$n = 8$

Project 3 (web)

a. $Q_{st} = -3 + 2P_{t-1}, \quad Q_{dt} = 18 - 3P_t$

$P_0 = 2,\ b = 2,\ d = 3,\ c = 18$

$-a = -3 \rightarrow a = 3$

$$P_t = \frac{3 + 18 - 2P_{t-1}}{3} = \frac{21 - 2P_{t-1}}{3}$$

$$P_t = 7 - \frac{2}{3}P_{t-1}, \quad P_0 = 2$$

b.

```
Plot1 Plot2 Plot3
nMin=0
·u(n)■7-(2/3)u(n
-1)
 u(nMin)■{2}
·v(n)=■
 v(nMin)=
·w(n)=
```

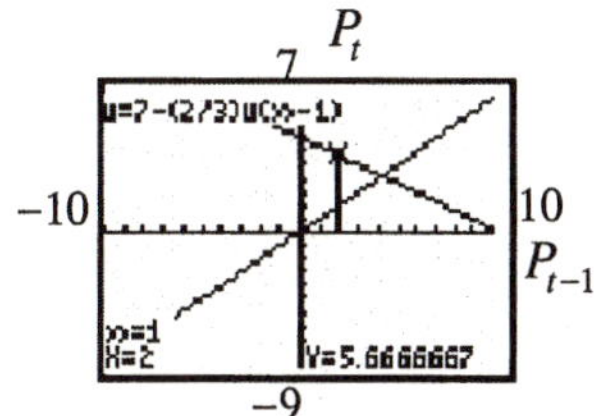

c. $P_1 = \dfrac{17}{3}$

$Q_{s1} = -3 + 2(2) \qquad Q_{d1} = 18 - 3\left(\dfrac{17}{3}\right)$

$Q_{s1} = 1 \qquad Q_{d1} = 1$

$P_2 = \dfrac{29}{9}$

$Q_{s2} = -3 + 2\left(\dfrac{17}{3}\right) \qquad Q_{d2} = 18 - 3\left(\dfrac{29}{9}\right)$

$Q_{s2} = \dfrac{25}{3} \qquad Q_{d2} = \dfrac{25}{3}$

The market (supply and demand) are getting closer to being the same.

d. The equilibrium price is 4.20.

e. It takes 17 time periods.

f. $Q_{d17} = 18 - 3(4.20) = 5.40$

$Q_{s17} = -3 + 2(4.20) = 5.40$

The equilibrium quantity is 5.4.

Project 4 (web)

a. 1, 2, 4, 7, 11, 16, 22, 29

b. It is not arithmetic because there is no common difference. It is not geometric because there is no common ration.

c. Scatter diagram

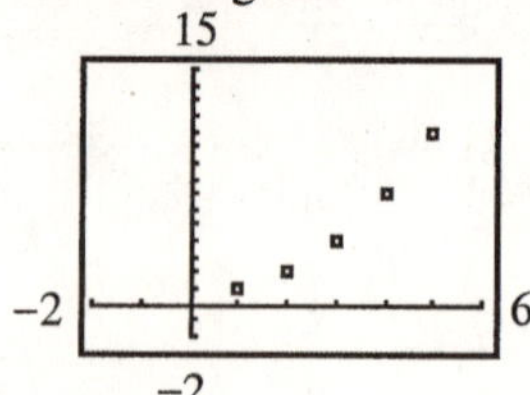

d. $y = 2.5x - 2.5$

The graph does not pass through any of the points.

$y_6 = 12.5$

$y_7 = 15$

$y_8 = 17.5$

$y_1 = 0$

$y_2 = 2.5$

$y_3 = 5$

$y_4 = 7.5$

$y_5 = 10$

$\sum_{i=1}^{5}(y_{r_i} - y_i)$

$= (0-1)+(2.5-2)+(5-4)+(7.5-7)+(10-11)$

$= -1$

This is the sum of the errors.

e. $y = 0.5x^2 - 0.5x + 1$

The graph passes through all of the points.

$y_6 = 16$

$y_7 = 22$

$y_8 = 29$

$y_1 = 1$

$y_2 = 2$

$y_3 = 4$

$y_4 = 7$

$y_5 = 11$

$\sum_{i=1}^{5}(y_{r_i} - y_i) = 0$

The sum of the errors is zero.

f. When trying to obtain the cubic and quartic polynomials of best fit, the cubic and quartic terms have coefficient zero and the polynomial of best fit is given as the quadratic in part e. For the exponential function of best fit,

$y = (0.59)(1.83)^x$.

$y_6 = 22.2 \qquad y_7 = 40.6 \qquad y_8 = 74.2$

The sum of these errors becomes quite large. This error shows that the function does not fit the data very well as x gets larger.

g. The quadratic function is best.

h. The data does not appear to be either logarithmic or sinusoidal in shape, so it does not make sense to try to fit one of those functions to the data.

Cumulative Review 1-11

1. $|x^2| = 9$

$x^2 = 9$ or $x^2 = -9$

$x = \pm 3$ or $x = \pm 3i$

2. a. graphing $x^2 + y^2 = 100$ and $y = 3x^2$.

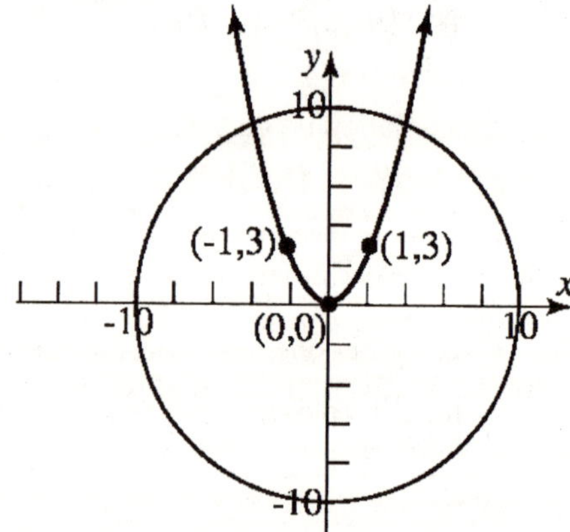

b. solving

$$\begin{cases} x^2 + y^2 = 100 \xrightarrow{3} 3x^2 + 3y^2 = 300 \\ y = 3x^2 \longrightarrow \underline{-3x^2 + y = 0} \end{cases}$$

$$3y^2 + y = 300$$

$3y^2 + y = 300 \Rightarrow 3y^2 + y - 300 = 0$

$$y = \frac{-1 \pm \sqrt{1^2 - 4(3)(-300)}}{2(3)} = \frac{-1 \pm \sqrt{3601}}{6}$$

Substitute and solve for x:

$$y=\frac{-1+\sqrt{3601}}{6}\Rightarrow\frac{-1+\sqrt{3601}}{6}=3x^2$$
$$\frac{-1+\sqrt{3601}}{18}=x^2\Rightarrow x=\pm\sqrt{\frac{-1+\sqrt{3601}}{18}}$$
or
$$y=\frac{-1-\sqrt{3601}}{6}\Rightarrow\frac{-1-\sqrt{3601}}{6}=3x^2$$
$$\frac{-1-\sqrt{3601}}{18}=x^2\Rightarrow x=\pm\sqrt{\frac{-1-\sqrt{3601}}{18}}$$
undefined since $\frac{-1-\sqrt{3601}}{18}<0$

Therefore, the system has solutions
$$x=\sqrt{\frac{-1+\sqrt{3601}}{18}},\ y=\frac{-1+\sqrt{3601}}{6}\ \text{and}$$
$$x=-\sqrt{\frac{-1+\sqrt{3601}}{18}},\ y=\frac{-1+\sqrt{3601}}{6}$$

c. The graphs intersect at the points
$$\left(\sqrt{\frac{-1+\sqrt{3601}}{18}},\frac{-1+\sqrt{3601}}{6}\right)\approx(1.81,9.84)$$
$$\left(-\sqrt{\frac{-1+\sqrt{3601}}{18}},\frac{-1+\sqrt{3601}}{6}\right)\approx(-1.81,9.84)$$

3. $2e^x=5$
$$e^x=\frac{5}{2}$$
$$\ln\left(e^x\right)=\ln\left(\frac{5}{2}\right)$$
$$x=\ln\left(\frac{5}{2}\right)\approx 0.916$$

4. slope $=m=5$; Since the x-intercept is 2, we know the point $(2,0)$ is on the graph of the line and is a solution to the equation $y=5x+b$.
$$y=5x+b$$
$$0=5(2)+b$$
$$0=10+b$$
$$-10=b$$
Therefore, the equation of the line with slope 5 and x-intercept 2 is $y=5x-10$.

5. Given a circle with center (–1, 2) and containing the point (3, 5), we first use the distance formula to determine the radius.
$$r=\sqrt{(3-(-1))^2+(5-2)^2}$$
$$=\sqrt{4^2+3^2}=\sqrt{16+9}$$
$$=\sqrt{25}$$
$$=5$$
Therefore, the equation of the circle is given by
$$(x-(-1))^2+(y-2)^2=5^2$$
$$(x+1)^2+(y-2)^2=5^2$$
$$x^2+2x+1+y^2-4y+4=25$$
$$x^2+y^2+2x-4y-20=0$$

6. $f(x)=\frac{3x}{x-2}$, $g(x)=2x+1$

a. $g(2)=2(2)+1=5$
$$f(5)=\frac{3(5)}{5-2}=\frac{15}{3}=5$$
$$(f\circ g)(2)=f(g(2))=f(5)=5$$

b. $f(4)=\frac{3(4)}{4-2}=\frac{12}{2}=6$
$$g(6)=2(6)+1=13$$
$$(g\circ f)(4)=g(f(4))=g(6)=13$$

c. $(f\circ g)(x)=f(g(x))$
$$=\frac{3(2x+1)}{(2x+1)-2}$$
$$=\frac{6x+3}{2x-1}$$

d. To determine the domain of the composition $(f\circ g)(x)$, we start with the domain of g and exclude any values in the domain of g that make the composition undefined. $g(x)$ is defined for all real numbers and $(f\circ g)(x)$ is defined for all real numbers except $x=\frac{1}{2}$. Therefore, the domain of the composite $(f\circ g)(x)$ is $\left\{x\,|\,x\neq\frac{1}{2}\right\}$.

e. $(g \circ f)(x) = 2\left(\dfrac{3x}{x-2}\right)+1$

$= \dfrac{6x}{x-2}+1$

$= \dfrac{6x+x-2}{x-2}$

$= \dfrac{7x-2}{x-2}$

f. To determine the domain of the composition $(g \circ f)(x)$, we start with the domain of *f* and exclude any values in the domain of *f* that make the composition undefined. $f(x)$ is defined for all real numbers except $x=2$ and $(g \circ f)(x)$ is defined for all real numbers except $x=2$. Therefore, the domain of the composite $(g \circ f)(x)$ is $\{x \mid x \neq 2\}$.

g. $g(x) = 2x+1$

$y = 2x+1$

$x = 2y+1$

$x-1 = 2y$

$\dfrac{x-1}{2} = y$

$g^{-1}(x) = \dfrac{x-1}{2}$

The domain of $g^{-1}(x)$ is the set of all real numbers.

h. $f(x) = \dfrac{3x}{x-2}$

$y = \dfrac{3x}{x-2}$

$x = \dfrac{3y}{y-2}$

$x(y-2) = 3y$

$xy-2x = 3y$

$xy-3y = 2x$

$y(x-3) = 2x$

$y = \dfrac{2x}{x-3}$

$f^{-1}(x) = \dfrac{2x}{x-3}$

The domain of $f^{-1}(x)$ is $\{x \mid x \neq 3\}$.

7. Center: (0, 0); Focus: (0, 3); Vertex: (0, 4); Major axis is the *y*-axis; $a=4$; $c=3$.

Find *b*: $b^2 = a^2 - c^2 = 16-9 = 7 \Rightarrow b = \sqrt{7}$

Write the equation using rectangular coordinates:

$\dfrac{x^2}{7}+\dfrac{y^2}{16}=1$

Parametric equations for the ellipse are:

$x = \sqrt{7}\cos(\pi t);\quad y = 4\sin(\pi t);\quad 0 \le t \le 2$

8. The focus is $(-1,3)$ and the vertex is $(-1,2)$. Both lie on the vertical line $x=-1$. We have $a=1$ since the distance from the vertex to the focus is 1 unit, and since $(-1,3)$ is above $(-1,2)$, the parabola opens up. The equation of the parabola is:

$(x-h)^2 = 4a(y-k)$

$(x-(-1))^2 = 4\cdot 1\cdot(y-2)$

$(x+1)^2 = 4(y-2)$

9. Center point (0, 4); passing through the pole (0,4) implies that the radius = 4 using rectangular coordinates:

$(x-h)^2+(y-k)^2 = r^2$

$(x-0)^2+(y-4)^2 = 4^2$

$x^2+y^2-8y+16 = 16$

$x^2+y^2-8y = 0$

converting to polar coordinates:

$r^2 - 8r\sin\theta = 0$

$r^2 = 8r\sin\theta$

$r = 8\sin\theta$

10. $2\sin^2 x - \sin x - 3 = 0,\quad 0 \le x \le 2\pi$

$(2\sin x - 3)(\sin x + 1) = 0$

$2\sin x - 3 = 0 \Rightarrow \sin x = \dfrac{3}{2}$, which is impossible

$\sin x + 1 = 0 \Rightarrow \sin x = -1 \Rightarrow x = \dfrac{3\pi}{2}$

Solution set $\left\{\dfrac{3\pi}{2}\right\}$.

11. $\cos^{-1}(-0.5)$

We are finding the angle θ, $-\pi \le \theta \le \pi$, whose cosine equals -0.5.

$\cos\theta = -0.5 \qquad -\pi \le \theta \le \pi$

$$\theta = \frac{2\pi}{3} \Rightarrow \cos^{-1}(-0.5) = \frac{2\pi}{3}$$

12. $\sin\theta = \frac{1}{4}$, θ is in Quadrant II

a. θ is in Quadrant II $\Rightarrow \cos\theta < 0$

$$\cos\theta = -\sqrt{1-\sin^2\theta} = -\sqrt{1-\left(\frac{1}{4}\right)^2}$$

$$= -\sqrt{1-\frac{1}{16}} = -\sqrt{\frac{15}{16}} = -\frac{\sqrt{15}}{4}$$

b. $$\tan\theta = \frac{\sin\theta}{\cos\theta} = \frac{\left(\frac{1}{4}\right)}{\left(-\frac{\sqrt{15}}{4}\right)} = \left(\frac{1}{4}\right)\left(-\frac{4}{\sqrt{15}}\right)$$

$$= -\frac{1}{\sqrt{15}} \cdot \frac{\sqrt{15}}{\sqrt{15}} = -\frac{\sqrt{15}}{15}$$

c. $\sin(2\theta) = 2\sin\theta\cos\theta$

$$= 2\left(\frac{1}{4}\right)\left(-\frac{\sqrt{15}}{4}\right)$$

$$= -\frac{\sqrt{15}}{8}$$

d. $\cos(2\theta) = \cos^2\theta - \sin^2\theta$

$$= \left(-\frac{\sqrt{15}}{4}\right)^2 - \left(\frac{1}{4}\right)^2$$

$$= \frac{15}{16} - \frac{1}{16} = \frac{14}{16} = \frac{7}{8}$$

e. $\frac{\pi}{2} < \theta < \pi \Rightarrow \frac{\pi}{4} < \frac{\theta}{2} < \frac{\pi}{2}$

$\Rightarrow \frac{\theta}{2}$ is in Quadrant I $\Rightarrow \sin\left(\frac{\theta}{2}\right) > 0$

$$\sin\left(\frac{\theta}{2}\right) = \sqrt{\frac{1-\cos\theta}{2}} = \sqrt{\frac{1-\left(-\frac{\sqrt{15}}{4}\right)}{2}}$$

$$= \sqrt{\frac{\left(\frac{4+\sqrt{15}}{4}\right)}{2}} = \sqrt{\frac{4+\sqrt{15}}{8}}$$

Chapter 12
Counting and Probability

Section 12.1

1. union

2. intersection

3. True; the union of two sets includes those elements that are in one *or* both of the sets. The intersection consists of the elements that are in *both* sets. Thus, the intersection is a subset of the union.

4. True; every element in the universal set is either in the set A or the complement of A.

5. $A \cup B = \{1, 3, 5, 7, 9\} \cup \{1, 5, 6, 7\}$
$= \{1, 3, 5, 6, 7, 9\}$

6. $A \cup C = \{1, 3, 5, 7, 9\} \cup \{1, 2, 4, 6, 8, 9\}$
$= \{1, 2, 3, 4, 5, 6, 7, 8, 9\}$

7. $A \cap B = \{1, 3, 5, 7, 9\} \cap \{1, 5, 6, 7\} = \{1, 5, 7\}$

8. $A \cap C = \{1, 3, 5, 7, 9\} \cap \{1, 2, 4, 6, 8, 9\} = \{1, 9\}$

9. $(A \cup B) \cap C$
$= (\{1, 3, 5, 7, 9\} \cup \{1, 5, 6, 7\}) \cap \{1, 2, 4, 6, 8, 9\}$
$= \{1, 3, 5, 6, 7, 9\} \cap \{1, 2, 4, 6, 8, 9\}$
$= \{1, 6, 9\}$

10. $(A \cap C) \cup (B \cap C)$
$= (\{1, 3, 5, 7, 9\} \cap \{1, 2, 4, 6, 8, 9\})$
$\cup (\{1, 5, 6, 7\} \cap \{1, 2, 4, 6, 8, 9\})$
$= \{1, 9\} \cup \{1, 6\} = \{1, 6, 9\}$

11. $(A \cap B) \cup C$
$= (\{1, 3, 5, 7, 9\} \cap \{1, 5, 6, 7\}) \cup \{1, 2, 4, 6, 8, 9\}$
$= \{1, 5, 7\} \cup \{1, 2, 4, 6, 8, 9\}$
$= \{1, 2, 4, 5, 6, 7, 8, 9\}$

12. $(A \cup B) \cup C$
$= (\{1, 3, 5, 7, 9\} \cup \{1, 5, 6, 7\}) \cup \{1, 2, 4, 6, 8, 9\}$
$= \{1, 3, 5, 6, 7, 9\} \cup \{1, 2, 4, 6, 8, 9\}$
$= \{1, 2, 3, 4, 5, 6, 7, 8, 9\}$

13. $(A \cup C) \cap (B \cup C)$
$= (\{1, 3, 5, 7, 9\} \cup \{1, 2, 4, 6, 8, 9\})$
$\cap (\{1, 5, 6, 7\} \cup \{1, 2, 4, 6, 8, 9\})$
$= \{1, 2, 3, 4, 5, 6, 7, 8, 9\} \cap \{1, 2, 4, 5, 6, 7, 8, 9\}$
$= \{1, 2, 4, 5, 6, 7, 8, 9\}$

14. $(A \cap B) \cap C$
$= (\{1, 3, 5, 7, 9\} \cap \{1, 5, 6, 7\}) \cap \{1, 2, 4, 6, 8, 9\}$
$= \{1, 5, 7\} \cap \{1, 2, 4, 6, 8, 9\}$
$= \{1\}$

15. $\overline{A} = \{0, 2, 6, 7, 8\}$

16. $\overline{C} = \{0, 2, 5, 7, 8, 9\}$

17. $\overline{A \cap B} = \overline{\{1, 3, 4, 5, 9\} \cap \{2, 4, 6, 7, 8\}}$
$= \overline{\{4\}} = \{0, 1, 2, 3, 5, 6, 7, 8, 9\}$

18. $\overline{B \cup C} = \overline{\{2, 4, 6, 7, 8\} \cup \{1, 3, 4, 6\}}$
$= \overline{\{1, 2, 3, 4, 6, 7, 8\}} = \{0, 5, 9\}$

19. $\overline{A} \cup \overline{B} = \{0, 2, 6, 7, 8\} \cup \{0, 1, 3, 5, 9\}$
$= \{0, 1, 2, 3, 5, 6, 7, 8, 9\}$

20. $\overline{B} \cap \overline{C} = \{0, 1, 3, 5, 9\} \cap \{0, 2, 5, 7, 8, 9\}$
$= \{0, 5, 9\}$

21. $\overline{A \cap \overline{C}} = \overline{\{1, 3, 4, 5, 9\} \cap \{0, 2, 5, 7, 8, 9\}}$
$= \overline{\{5, 9\}} = \{0, 1, 2, 3, 4, 6, 7, 8\}$

22. $\overline{B} \cup C = \overline{\{0,1,3,5,9\} \cup \{1,3,4,6\}}$
$= \overline{\{0,1,3,4,5,6,9\}} = \{2,7,8\}$

23. $\overline{A \cup B \cup C}$
$= \overline{\{1,3,4,5,9\} \cup \{2,4,6,7,8\} \cup \{1,3,4,6\}}$
$= \overline{\{1,2,3,4,5,6,7,8,9\}}$
$= \{0\}$

24. $\overline{A \cap B \cap C}$
$= \overline{\{1,3,4,5,9\} \cap \{2,4,6,7,8\} \cap \{1,3,4,6\}}$
$= \overline{\{4\}}$
$= \{0,1,2,3,5,6,7,8,9\}$

25. $\{a\},\{b\},\{c\},\{d\},\{a,b\},\{a,c\},\{a,d\},$
$\{b,c\},\{b,d\},\{c,d\},\{a,b,c\},\{a,b,d\},$
$\{a,c,d\},\{b,c,d\},\{a,b,c,d\},\varnothing$

26. $\{a\},\{b\},\{c\},\{d\},\{e\},\{a,b\},\{a,c\},$
$\{a,d\},\{a,e\},\{b,c\},\{b,d\},\{b,e\},$
$\{c,d\},\{c,e\},\{d,e\},\{a,b,c\},\{a,b,d\},$
$\{a,b,e\},\{a,c,d\},\{a,c,e\},\{a,d,e\},$
$\{b,c,d\},\{b,c,e\},\{b,d,e\},\{c,d,e\},$
$\{a,b,c,d\},\{a,b,c,e\},\{a,c,d,e\},$
$\{a,b,d,e\},\{b,c,d,e\},\{a,b,c,d,e\},\varnothing$

27. $n(A)=15, n(B)=20, n(A \cap B)=10$
$n(A \cup B) = n(A) + n(B) - n(A \cap B)$
$= 15 + 20 - 10 = 25$

28. $n(A)=30, n(B)=40, n(A \cup B)=45$
$n(A \cup B) = n(A) + n(B) - n(A \cap B)$
$45 = 30 + 40 - n(A \cap B)$
$n(A \cap B) = 30 + 40 - 45 = 25$

29. $n(A \cup B)=50, n(A \cap B)=10, n(B)=20$
$n(A \cup B) = n(A) + n(B) - n(A \cap B)$
$50 = n(A) + 20 - 10$
$40 = n(A)$

30. $n(A \cup B)=60, n(A \cap B)=40, n(A)=n(B)$
$n(A \cup B) = n(A) + n(B) - n(A \cap B)$
$60 = n(A) + n(A) - 40$
$100 = 2n(A)$
$n(A) = 50$

31. From the figure: $n(A) = 15+3+5+2 = 25$

32. From the figure: $n(B) = 10+3+5+2 = 20$

33. From the figure:
$n(A \text{ or } B) = n(A \cup B)$
$= 15+2+5+3+10+2 = 37$

34. From the figure:
$n(A \text{ and } B) = n(A \cap B) = 3+5 = 8$

35. From the figure:
$n(A \text{ but not } C) = n(A) - n(A \cap C) = 25 - 7 = 18$

36. From the figure: $n(\overline{A}) = 10+2+15+4 = 31$

37. From the figure:
$n(A \text{ and } B \text{ and } C) = n(A \cap B \cap C) = 5$

38. From the figure:
$n(A \text{ or } B \text{ or } C) = n(A \cup B \cup C)$
$= 15+3+5+2+10+2+15 = 52$

39. Let $A = \{\text{those who will purchase a major appliance}\}$ and
$B = \{\text{those who will buy a car}\}$
$n(U) = 500, \; n(A) = 200,$
$n(B) = 150, \; n(A \cap B) = 25$
$n(A \cup B) = n(A) + n(B) - n(A \cap B)$
$= 200 + 150 - 25 = 325$
$n(\text{purchase neither}) = n(U) - n(A \cup B)$
$= -500 - 325 = 175$
$n(\text{purchase only a car}) = n(B) - n(A \cup B)$
$= 150 - 25 = 125$

40. Let $A = \{\text{those who will attend Summer Session I}\}$ and $B = \{\text{those who will attend Summer Session II}\}$

$n(A) = 200,\ n(B) = 150,\ n(A \cap B) = 75,$

$n\left(\overline{A \cup B}\right) = 275$

$$n(A \cup B) = n(A) + n(B) - n(A \cap B) = 200 + 150 - 75 = 275$$

$$n(U) = n(A \cup B) + n\left(\overline{A \cup B}\right) = 275 + 275 = 550$$

550 students participated in the survey.

41. Construct a Venn diagram:

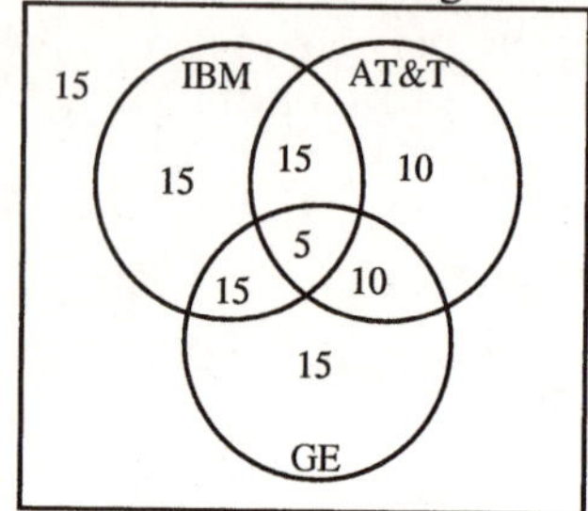

(a) 15 (b) 15

(c) 15 (d) 25

(e) 40

42. Construct a Venn diagram:

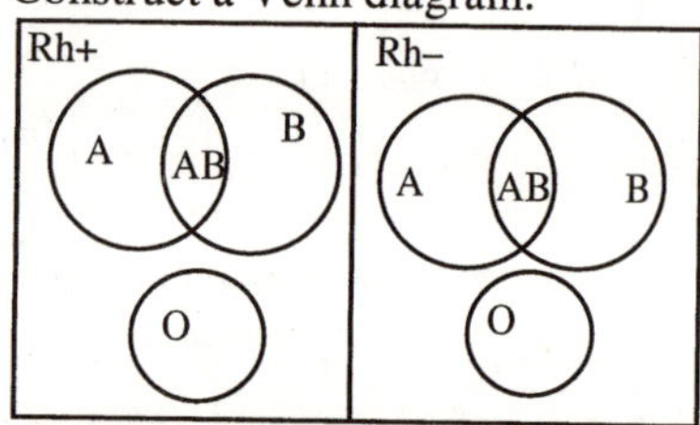

There are 8 different kinds of blood: A-Rh+, B-Rh+, AB-Rh+, O-Rh+, A-Rh–, B-Rh–, AB-Rh–, O-Rh–

43. a. $n(\text{widowed or divorced}) = n(\text{widowed}) + n(\text{divorced}) = 2,632 + 8,659 = 11,291$

There were 11,291 thousand males 18 years old and older who were widowed or divorced.

b. $n(\text{married, widowed or divorced})$

$= n(\text{married}) + n(\text{widowed}) + n(\text{divorced})$

$= 61,212 + 2,632 + 8,659 = 72,503$

There were 72,503 thousand males 18 years old and older who were married, widowed, or divorced.

44. a. $n(\text{widowed or divorced}) = n(\text{widowed}) + n(\text{divorced}) = 11,404 + 12,236 = 23,640$

There were 23,640 thousand females 18 years old and older who were widowed or divorced.

b. $n(\text{married, widowed or divorced})$

$= n(\text{married}) + n(\text{widowed}) + n(\text{divorced})$

$= 62,037 + 11,404 + 12,236$

$= 85,677$

There were 85,677 thousand females 18 years old and older who were married, widowed, or divorced.

45. Answers will vary.

46. Answers will vary.

Section 12.2

1. 1; 1

2. False; $n! = \dfrac{(n+1)!}{n+1}$

3. permutation

4. combination

5. True

6. True

7. $P(6, 2) = \dfrac{6!}{(6-2)!} = \dfrac{6!}{4!} = \dfrac{6 \cdot 5 \cdot 4!}{4!} = 30$

8. $P(7, 2) = \dfrac{7!}{(7-2)!} = \dfrac{7!}{5!} = \dfrac{7 \cdot 6 \cdot 5!}{5!} = 42$

9. $P(4, 4) = \dfrac{4!}{(4-4)!} = \dfrac{4!}{0!} = \dfrac{4 \cdot 3 \cdot 2 \cdot 1}{1} = 24$

10. $P(8, 8) = \dfrac{8!}{(8-8)!} = \dfrac{8!}{0!} = \dfrac{8 \cdot 7 \cdot 6 \cdot 5 \cdot 4 \cdot 3 \cdot 2 \cdot 1}{1} = 40,320$

11. $P(7,0)=\frac{7!}{(7-0)!}=\frac{7!}{7!}=1$

12. $P(9,0)=\frac{9!}{(9-0)!}=\frac{9!}{9!}=1$

13. $P(8,4)=\frac{8!}{(8-4)!}=\frac{8!}{4!}=\frac{8\cdot 7\cdot 6\cdot 5\cdot 4!}{4!}=1680$

14. $P(8,3)=\frac{8!}{(8-3)!}=\frac{8!}{5!}=\frac{8\cdot 7\cdot 6\cdot 5!}{5!}=336$

15. $C(8,2)=\frac{8!}{(8-2)!2!}=\frac{8!}{6!2!}=\frac{8\cdot 7\cdot 6!}{6!\cdot 2\cdot 1}=28$

16. $C(8,6)=\frac{8!}{(8-6)!6!}=\frac{8!}{2!6!}=\frac{8\cdot 7\cdot 6!}{6!\cdot 2\cdot 1}=28$

17. $C(7,4)=\frac{7!}{(7-4)!4!}=\frac{7!}{3!4!}=\frac{7\cdot 6\cdot 5\cdot 4!}{4!\cdot 3\cdot 2\cdot 1}=35$

18. $C(6,2)=\frac{6!}{(6-2)!2!}=\frac{6!}{4!2!}=\frac{6\cdot 5\cdot 4!}{4!\cdot 2\cdot 1}=15$

19. $C(15,15)=\frac{15!}{(15-15)!15!}=\frac{15!}{0!15!}=\frac{15!}{15!\cdot 1}=1$

20. $C(18,1)=\frac{18!}{(18-1)!1!}=\frac{18!}{17!1!}=\frac{18\cdot 17!}{17!\cdot 1}=18$

21. $C(26,13)=\frac{26!}{(26-13)!13!}=\frac{26!}{13!13!}=10,400,600$

22. $C(18,9)=\frac{18!}{(18-9)!9!}=\frac{18!}{9!9!}=48,620$

23. $\{abc, abd, abe, acb, acd, ace, adb, adc, ade, aeb, aec, aed, bac, bad, bae, bca, bcd, bce, bda, bdc, bde, bea, bec, bed, cab, cad, cae, cba, cbd, cbe, cda, cdb, cde, cea, ceb, ced, dab, dac, dae, dba, dbc, dbe, dca, dcb, dce, dea, deb, dec, eab, eac, ead, eba, ebc, ebd, eca, ecb, ecd, eda, edb, edc\}$

$P(5,3)=\frac{5!}{(5-3)!}=\frac{5!}{2!}=\frac{5\cdot 4\cdot 3\cdot 2!}{2!}=60$

24. $\{ab, ac, ad, ae, ba, bc, bd, be, ca, cb, cd, ce, da, db, dc, de, ea, eb, ec, ed\}$

$P(5,2)=\frac{5!}{(5-2)!}=\frac{5!}{3!}=\frac{5\cdot 4\cdot 3!}{3!}=20$

25. {123, 124, 132, 134, 142, 143, 213, 214, 231, 234, 241, 243, 312, 314, 321, 324, 341, 342, 412, 413, 421, 423, 431, 432}

$P(4,3)=\frac{4!}{(4-3)!}=\frac{4!}{1!}=\frac{4\cdot 3\cdot 2\cdot 1}{1}=24$

26. {123, 124, 125, 126, 132, 134, 135, 136, 142, 143, 145, 146, 152, 153, 154, 156, 162, 163, 164, 165, 213, 214, 215, 216, 231, 234, 235, 236, 241, 243, 245, 246, 251, 253, 254, 256, 261, 263, 264, 265, 312, 314, 315, 316, 321, 324, 325, 326, 341, 342, 345, 346, 351, 352, 354, 356, 361, 362, 364, 365, 412, 413, 415, 416, 421, 423, 425, 426, 431, 432, 435, 436, 451, 452, 453, 456, 461, 462, 463, 465, 512, 513, 514, 516, 521, 523, 524, 526, 531, 532, 534, 536, 541, 542, 543, 546, 561, 562, 563, 564, 612, 613, 614, 615, 621, 623, 624, 625, 631, 632, 634, 635, 641, 642, 643, 645, 651, 652, 653, 654}

$P(6,3)=\frac{6!}{(6-3)!}=\frac{6!}{3!}=\frac{6\cdot 5\cdot 4\cdot 3!}{3!}=120$

27. $\{abc, abd, abe, acd, ace, ade, bcd, bce, bde, cde\}$

$C(5,3)=\frac{5!}{(5-3)!3!}=\frac{5\cdot 4\cdot 3!}{2\cdot 1\cdot 3!}=10$

28. $\{ab, ac, ad, ae, bc, bd, be, cd, ce, de\}$

$C(5,2)=\frac{5!}{(5-2)!2!}=\frac{5\cdot 4\cdot 3!}{3!\cdot 2\cdot 1}=10$

29. {123, 124, 134, 234}

$C(4,3)=\frac{4!}{(4-3)!3!}=\frac{4\cdot 3!}{1!3!}=4$

30. {123, 124, 125, 126, 134, 135, 136, 145, 146, 156, 234, 235, 236, 245, 246, 256, 345, 346, 356, 456}

$$C(6,3)=\frac{6!}{(6-3)!3!}=\frac{6\cdot 5\cdot 4\cdot 3!}{3\cdot 2\cdot 1\cdot 3!}=20$$

31. There are 5 choices of shirts and 3 choices of ties; there are (5)(3) = 15 combinations.

32. There are 3 choices of blouses and 5 choices of skirts; there are (3)(5) = 15 different outfits.

33. There are 4 choices for the first letter in the code and 4 choices for the second letter in the code; there are (4)(4) = 16 possible two-letter codes.

34. There are 5 choices for the first letter in the code and 5 choices for the second letter in the code; there are (5)(5) = 25 possible two-letter codes.

35. There are two choices for each of three positions; there are (2)(2)(2) = 8 possible three-digit numbers.

36. There are ten choices for each of three positions; there are (10)(10)(10) = 1000 possible three-digit numbers.

37. To line up the four people, there are 4 choices for the first position, 3 choices for the second position, 2 choices for the third position, and 1 choice for the fourth position. Thus there are (4)(3)(2)(1) = 24 possible ways four people can be lined up.

38. To stack the five boxes, there are 5 choices for the first position, 4 choices for the second position, 3 choices for the third position, 2 choices for the fourth position, and 1 choice for the fifth position. Thus, there are (5)(4)(3)(2)(1) = 120 possible ways five boxes can be stacked.

39. Since no letter can be repeated, there are 5 choices for the first letter, 4 choices for the second letter, and 3 choices for the third letter. Thus, there are (5)(4)(3) = 60 possible three-letter codes.

40. Since no letter can be repeated, there are 6 choices for the first letter, 5 choices for the second letter, 4 choices for the third letter, and 3 choices for the fourth letter. Thus, there are (6)(5)(4)(3) = 360 possible three-letter codes.

41. There are 26 possible one-letter names. There are (26)(26) = 676 possible two-letter names. There are (26)(26)(26) = 17,576 possible three-letter names. Thus, there are
26 + 676 + 17,576 = 18,278
possible companies that can be listed on the New York Stock Exchange.

42. There are (26)(26)(26)(26) = 456,976 possible four-letter names. There are (26)(26)(26)(26)(26) = 11,881,376 possible five-letter names. Thus, there are
456,976 + 11,881,376 = 12,338,352
possible companies that can be listed on the NASDAQ.

43. A committee of 4 from a total of 7 students is given by:

$$C(7,4)=\frac{7!}{(7-4)!4!}=\frac{7!}{3!4!}=\frac{7\cdot 6\cdot 5\cdot 4!}{3\cdot 2\cdot 1\cdot 4!}=35$$

35 committees are possible.

44. A committee of 3 from a total of 8 professors is given by:

$$C(8,3)=\frac{8!}{(8-3)!3!}=\frac{8!}{5!3!}=\frac{8\cdot 7\cdot 6\cdot 5!}{3\cdot 2\cdot 1\cdot 5!}=56$$

56 committees are possible.

45. There are 2 possible answers for each question. Therefore, there are $2^{10}=1024$ different possible arrangements of the answers.

46. There are 4 possible answers for each question. Therefore, there are $4^5=1024$ different possible arrangements of the answers.

47. There are 9 choices for the first digit, and 10 choices for each of the other three digits. Thus, there are (9)(10)(10)(10) = 9000 possible four-digit numbers.

48. There are 8 choices for the first digit, and 10 choices for each of the other four digits. Thus, there are (8)(10)(10)(10)(10) = 80,000 possible five-digit numbers.

49. There are 5 choices for the first position, 4 choices for the second position, 3 choices for the third position, 2 choices for the fourth position, and 1 choice for the fifth position. Thus, there are (5)(4)(3)(2)(1) = 120 possible arrangements of the books.

50. a. There are 26 choices for each of the first two positions, and 10 choices for each of the next four positions. Thus, there are (26)(26)(10)(10)(10)(10) = 6,760,000 possible license plates.

b. There are 26 choices for each of the first two positions, 10 choices for the first digit, 9 choices for the second digit, 8 choices for the third digit, and 7 choices for the fourth digit. Thus, there are (26)(26)(10)(9)(8)(7) = 3,407,040 possible license plates.

c. There are 26 choices for the first letter, 25 choices for the second letter, 10 choices for the first digit, 9 choices for the second digit, 8 choices for the third digit, and 7 choices for the fourth digit. Thus, there are (26)(25)(10)(9)(8)(7) = 3,276,000 possible license plates.

51. There are 8 choices for the DOW stocks, 15 choices for the NASDAQ stocks, and 4 choices for the global stocks. Thus, there are (8)(15)(4) = 480 different portfolios.

52. a. If numbers can be repeated, there are (50)(50)(50) = 125,000 different lock combinations. If no number can be repeated, then there are $50\cdot 49\cdot 48 = 117{,}600$ different lock combinations.

b. Answers will vary.
Typical combination locks require two full clockwise rotations to the first number, followed by a full counter-clockwise rotation past the first number to the second number, followed by a clockwise rotation to the third number (not past the second). This is not clear from the given directions.

53. The 1st person can have any of 365 days, the 2nd person can have any of the remaining 364 days. Thus, there are (365)(364) = 132,860 possible ways two people can have different birthdays.

54. The first person can have any of 365 days, the second person can have any of the remaining 364 days, the third person can have any of the remaining 363 days, the fourth person can have any of the remaining 362 days, and the fifth person can have any of the remaining 361 days. Thus, there are

$(365)(364)(363)(362)(361) \approx 6.302555019\times 10^{12}$

possible ways five people can have different birthdays.

55. Choosing 2 boys from the 4 boys can be done $C(4,2)$ ways, and choosing 3 girls from the 8 girls can be done in $C(8,3)$ ways. Thus, there are a total of:

$$C(4,2)\cdot C(8,3) = \frac{4!}{(4-2)!2!}\cdot\frac{8!}{(8-3)!3!}$$
$$= \frac{4!}{2!2!}\cdot\frac{8!}{5!3!}$$
$$= \frac{4\cdot 3\cdot 2!}{2\cdot 1\cdot 2!}\cdot\frac{8\cdot 7\cdot 6\cdot 5!}{5!3!}$$
$$= 6\cdot 56 = 336$$

56. The committee is made up of 2 of 4 administrators, 3 of 8 faculty members, and 5 of 20 students. The number of possible committees is:

$$C(4,2)\cdot C(8,3)\cdot C(20,5)$$
$$= \frac{4!}{(4-2)!2!}\cdot\frac{8!}{(8-3)!3!}\cdot\frac{20!}{(20-5)!5!}$$
$$= \frac{4!}{2!2!}\cdot\frac{8!}{5!3!}\cdot\frac{20!}{15!5!}$$
$$= \frac{4!}{2\cdot 1\cdot 2\cdot 1}\cdot\frac{8\cdot 7\cdot 6\cdot 5!}{5\cdot 4!3!}\cdot\frac{20\cdot 19\cdot 18\cdot 17\cdot 16\cdot 15!}{15!5!}$$
$$= 5{,}209{,}344 \text{ possible committees}$$

57. This is a permutation with repetition. There are $\frac{9!}{2!2!} = 90{,}720$ different words.

58. This is a permutation with repetition. There are $\frac{11!}{2!2!2!} = 4{,}989{,}600$ different words.

59. a. $C(7,2)\cdot C(3,1) = 21\cdot 3 = 63$

b. $C(7,3)\cdot C(3,0) = 35\cdot 1 = 35$

c. $C(3,3)\cdot C(7,0) = 1\cdot 1 = 1$

60. a. $C(15,5)\cdot C(10,0) = 3003\cdot 1 = 3003$

b. $C(15,3)\cdot C(10,2) = 455\cdot 45 = 20{,}475$

c. $C(15,4)\cdot C(10,1) + C(15,5)\cdot C(10,0)$
$= 1365\cdot 10 + 3003\cdot 1$
$= 13{,}650 + 3003 = 16{,}653$

61. There are $C(100, 22)$ ways to form the first committee. There are 78 senators left, so there are $C(78, 13)$ ways to form the second committee. There are $C(65, 10)$ ways to form the third committee. There are $C(55, 5)$ ways to form the fourth committee. There are $C(50, 16)$ ways to form the fifth committee. There are C(34, 17) ways to form the sixth committee. There are $C(17, 17)$ ways to form the seventh committee.
The total number of committees
$= C(100,22)\cdot C(78,13)\cdot C(65,10)\cdot C(55,5)$
$\cdot C(50,16)\cdot C(34,17)\cdot C(17,17)$
$\approx 1.1568\times 10^{76}$

62. The team is made up of 5 of 10 linemen, 3 of 10 linebackers, and 3 of 5 safeties. The number of possible teams is:
$C(10,5)\cdot C(10,3)\cdot C(5,3) = 252\cdot 120\cdot 10$
$= 302,400$
There are 302,400 possible defensive teams.

63. There are 9 choices for the first position, 8 choices for the second position, 7 for the third position, etc. There are $9\cdot 8\cdot 7\cdot 6\cdot 5\cdot 4\cdot 3\cdot 2\cdot 1 = 9! = 362,880$ possible batting orders.

64. There are 8 choices for the first position, 7 choices for the second position, 6 for the third position, etc. and 1 choice for the last position. There are $8\cdot 7\cdot 6\cdot 5\cdot 4\cdot 3\cdot 2\cdot 1\cdot 1 = 8!\cdot 1 = 40,320$ possible batting orders.

65. The team must have 1 pitcher and 8 position players (non-pitchers). For pitcher, choose 1 player from a group of 4 players, i.e., $C(4, 1)$. For position players, choose 8 players from a group of 11 players, i.e., $C(11, 8)$. Therefore, the number different teams possible is $C(4,1)\cdot C(11,8) = 4\cdot 165 = 660$.

66. Consider the ways that the American League can win. Then multiply by 2 to get the total for both leagues. To win the World Series, the last game must be won. There is 1 way to win in four games. To win in 5 games, three of the first four must be won, so there are $C(4, 3) = 4$ ways to win in 5 games. To win in 6 games, three of the first five must be won, so there are $C(5, 3) = 10$ ways to win in 6 games. To win in 7 games, three of the first six must be won, so there are $C(6, 3) = 20$ ways to win in 7 games. Therefore, there are $1 + 4 + 10 + 20 = 35$ ways the American League can win the World Series. There are also 35 ways the National League can win the World Series. There are a total of 70 different sequences possible.

67. Choose 2 players from a group of 6 players. Therefore, there are $C(6,2) = 15$ different teams possible.

68. Choose 1 of 2 centers, 2 of 3 guards, and 2 of 7 forwards. There are $C(2,1)\cdot C(3,2)\cdot C(7,2) = 2\cdot 3\cdot 21 = 126$ different teams possible.

69. Answers will vary.

70. Answers will vary.

71. Answers will vary.

72. A permutation is an ordered arrangement of objects while with a combination order does not matter. For example, the number of ways the 11 teams in the Big Ten can come in first, second, and third would be a permutation problem. The number of ways to pick 6 numbers in the Illinois State lottery is a combination problem because the order in which the numbers are selected is irrelevant.

Section 12.3

1. equally likely

2. complement

3. False; probability may equal 0. In such cases, the corresponding event will never happen.

4. True; in a valid probability model, all probabilities are between 0 and 1, and the sum of the probabilities is 1.

5. Probabilities must be between 0 and 1, inclusive. Thus, 0, 0.01, 0.35, and 1 could be probabilities.

6. Probabilities must be between 0 and 1, inclusive. Thus, $\frac{1}{2}, \frac{3}{4}, \frac{2}{3}$, and 0 could be probabilities.

7. All the probabilities are between 0 and 1. The sum of the probabilities is $0.2 + 0.3 + 0.1 + 0.4 = 1$. This is a probability model.

8. All the probabilities are between 0 and 1. The sum of the probabilities is $0.4 + 0.3 + 0.1 + 0.2 = 1$. This is a probability model.

9. All the probabilities are between 0 and 1. The sum of the probabilities is $0.3 + 0.2 + 0.1 + 0.3 = 0.9$. This is not a probability model.

10. One probability is not between 0 and 1. This is not a probability model.

11. The sample space is: $S = \{HH, HT, TH, TT\}$.
Each outcome is equally likely to occur; so
$P(E) = \dfrac{n(E)}{n(S)}$.
The probabilities are:
$P(HH) = \dfrac{1}{4},\ P(HT) = \dfrac{1}{4},\ P(TH) = \dfrac{1}{4},\ P(TT) = \dfrac{1}{4}$.

12. The sample space is: $S = \{HH, HT, TH, TT\}$.
Each outcome is equally likely to occur; so
$P(E) = \dfrac{n(E)}{n(S)}$.
The probabilities are:
$P(HH) = \dfrac{1}{4},\ P(HT) = \dfrac{1}{4},\ P(TH) = \dfrac{1}{4},\ P(TT) = \dfrac{1}{4}$

13. The sample space of tossing two fair coins and a fair die is:
$$S = \{HH1, HH2, HH3, HH4, HH5, HH6, HT1, HT2, HT3, HT4, HT5, HT6, TH1, TH2, TH3, TH4, TH5, TH6, TT1, TT2, TT3, TT4, TT5, TT6\}$$
There are 24 equally likely outcomes and the probability of each is $\dfrac{1}{24}$.

14. The sample space of tossing a fair coin, a fair die, and a fair coin is:
$$S = \{H1H, H2H, H3H, H4H, H5H, H6H, H1T, H2T, H3T, H4T, H5T, H6T, T1H, T2H, T3H, T4H, T5H, T6H, T1T, T2T, T3T, T4T, T5T, T6T\}$$
There are 24 equally likely outcomes and the probability of each is $\dfrac{1}{24}$.

15. The sample space for tossing three fair coins is:
$$S = \{HHH, HHT, HTH, THH, HTT, THT, TTH, TTT\}$$
There are 8 equally likely outcomes and the probability of each is $\dfrac{1}{8}$.

16. The sample space for tossing one fair coin three times is:
$$S = \{HHH, HHT, HTH, THH, HTT, THT, TTH, TTT\}$$
There are 8 equally likely outcomes and the probability of each is $\dfrac{1}{8}$.

17. The sample space is:
S = {1 Yellow, 1 Red, 1 Green, 2 Yellow, 2 Red, 2 Green, 3 Yellow, 3 Red, 3 Green, 4 Yellow, 4 Red, 4 Green}
There are 12 equally likely events and the probability of each is $\frac{1}{12}$. The probability of getting a 2 or 4 followed by a Red is
$P(\text{2 Red}) + P(\text{4 Red}) = \dfrac{1}{12} + \dfrac{1}{12} = \dfrac{1}{6}$.

18. The sample space is:
S = {Forward Yellow, Forward Red, Forward Green, Backward Yellow, Backward Red, Backward Green}
There are 6 equally likely events and the probability of each is $\frac{1}{6}$. The probability of getting Forward followed by Yellow or Green is:
$P(\text{Forward Yellow}) + P(\text{Forward Green}) = \dfrac{1}{6} + \dfrac{1}{6} = \dfrac{1}{3}$.

19. The sample space is:
S = {1 Yellow Forward, 1 Yellow Backward, 1 Red Forward, 1 Red Backward, 1 Green Forward, 1 Green Backward, 2 Yellow Forward, 2 Yellow Backward, 2 Red Forward, 2 Red Backward, 2 Green Forward, 2 Green Backward, 3 Yellow Forward, 3 Yellow Backward, 3 Red Forward, 3 Red Backward, 3 Green Forward, 3 Green Backward, 4 Yellow Forward, 4 Yellow Backward, 4 Red Forward, 4 Red Backward, 4 Green Forward, 4 Green Backward}
There are 24 equally likely events and the probability of each is $\frac{1}{24}$. The probability of getting a 1, followed by a Red or Green, followed by a Backward is

$$P(\text{1 Red Backward}) + P(\text{1 Green Backward}) = \frac{1}{24} + \frac{1}{24} = \frac{1}{12}$$

20. The sample space is:
S = {Yellow 1 Forward, Yellow 1 Backward, Red 1 Forward, Red 1 Backward, Green 1 Forward, Green 1 Backward, Yellow 2 Forward, Yellow 2 Backward, Red 2 Forward, Red 2 Backward, Green 2 Forward, Green 2 Backward, Yellow 3 Forward, Yellow 3 Backward, Red 3 Forward, Red 3 Backward, Green 3 Forward, Green 3 Backward, Yellow 4 Forward, Yellow 4 Backward, Red 4 Forward, Red 4 Backward, Green 4 Forward, Green 4 Backward}
There are 24 equally likely events and the probability of each is $\frac{1}{24}$.
The probability of getting a Yellow, followed by a 2 or 4, followed by a Forward is

$$P(\text{Yellow 2 Forward}) + P(\text{Yellow 4 Forward}) = \frac{1}{24} + \frac{1}{24} = \frac{1}{12}.$$

21. The sample space is:
S = {1 1 Yellow, 1 1 Red, 1 1 Green, 1 2 Yellow, 1 2 Red, 1 2 Green, 1 3 Yellow, 1 3 Red, 1 3 Green, 1 4 Yellow, 1 4 Red, 1 4 Green, 2 1 Yellow, 2 1 Red, 2 1 Green, 2 2 Yellow, 2 2 Red, 2 2 Green, 2 3 Yellow, 2 3 Red, 2 3 Green, 2 4 Yellow, 2 4 Red, 2 4 Green, 3 1 Yellow, 3 1 Red, 3 1 Green, 3 2 Yellow, 3 2 Red, 3 2 Green, 3 3 Yellow, 3 3 Red, 3 3 Green, 3 4 Yellow, 3 4 Red, 3 4 Green, 4 1 Yellow, 4 1 Red, 4 1 Green, 4 2 Yellow, 4 2 Red, 4 2 Green, 4 3 Yellow, 4 3 Red, 4 3 Green, 4 4 Yellow, 4 4 Red, 4 4 Green}
There are 48 equally likely events and the probability of each is $\frac{1}{48}$. The probability of getting a 2, followed by a 2 or 4, followed by a Red or Green is

$$P(\text{2 2 Red}) + P(\text{2 4 Red}) + P(\text{2 2 Green}) + P(\text{2 4 Green}) = \frac{1}{48} + \frac{1}{48} + \frac{1}{48} + \frac{1}{48} = \frac{1}{12}$$

22. The sample space is:
S = {Forward 11, Forward 12, Forward 13, Forward 14, Forward 21, Forward 22, Forward 23, Forward 24, Forward 31, Forward 32, Forward 33, Forward 34, Forward 41, Forward 42, Forward 43, Forward 44, Backward 11, Backward 12, Backward 13, Backward 14, Backward 21, Backward 22, Backward 23, Backward 24, Backward 31, Backward 32, Backward 33, Backward 34, Backward 41, Backward 42, Backward 43, Backward 44}
There are 32 equally likely events and the probability of each is $\frac{1}{32}$. The probability of getting a Forward, followed by a 1 or 3, followed by a 2 or 4 is

$$P(\text{Fwd 12}) + P(\text{Fwd 14}) + P(\text{Fwd 32}) + P(\text{Fwd 34}) = \frac{1}{32} + \frac{1}{32} + \frac{1}{32} + \frac{1}{32} = \frac{1}{8}$$

23. A, B, C, F

24. A (equally likely outcomes)

25. B

26. F

27. Let $P(\text{tails}) = x$, then $P(\text{heads}) = 4x$

$$x + 4x = 1$$
$$5x = 1$$
$$x = \frac{1}{5}$$
$$P(\text{tails}) = \frac{1}{5}, \quad P(\text{heads}) = \frac{4}{5}$$

28. Let $P(\text{heads}) = x$, then $P(\text{tails}) = 2x$

$$x + 2x = 1$$
$$3x = 1$$
$$x = \frac{1}{3}$$
$$P(\text{heads}) = \frac{1}{3}, \quad P(\text{tails}) = \frac{2}{3}$$

29. $P(2) = P(4) = P(6) = x$

$$P(1) = P(3) = P(5) = 2x$$
$$P(1) + P(2) + P(3) + P(4) + P(5) + P(6) = 1$$
$$2x + x + 2x + x + 2x + x = 1$$
$$9x = 1$$
$$x = \frac{1}{9}$$
$$P(2) = P(4) = P(6) = \frac{1}{9}$$
$$P(1) = P(3) = P(5) = \frac{2}{9}$$

30. $P(1) = P(2) = P(3) = P(4) = P(5) = x; \quad P(6) = 0$

$$P(1) + P(2) + P(3) + P(4) + P(5) + P(6) = 1$$
$$x + x + x + x + x + 0 = 1$$
$$5x = 1$$
$$x = \frac{1}{5}$$
$$P(1) = P(2) = P(3) = P(4) = P(5) = \frac{1}{5}; \quad P(6) = 0$$

31. $P(E) = \dfrac{n(E)}{n(S)} = \dfrac{n\{1,2,3\}}{10} = \dfrac{3}{10}$

32. $P(F) = \dfrac{n(F)}{n(S)} = \dfrac{n\{3,5,9,10\}}{10} = \dfrac{4}{10} = \dfrac{2}{5}$

33. $P(E) = \dfrac{n(E)}{n(S)} = \dfrac{n\{2,4,6,8,10\}}{10} = \dfrac{5}{10} = \dfrac{1}{2}$

34. $P(F) = \dfrac{n(F)}{n(S)} = \dfrac{n\{1,3,5,7,9\}}{10} = \dfrac{5}{10} = \dfrac{1}{2}$

35. $P(\text{white}) = \dfrac{n(\text{white})}{n(S)} = \dfrac{5}{5+10+8+7} = \dfrac{5}{30} = \dfrac{1}{6}$

36. $P(\text{black}) = \dfrac{n(\text{black})}{n(S)} = \dfrac{7}{5+10+8+7} = \dfrac{7}{30}$

37. The sample space is: $S = \{$BBB, BBG, BGB, GBB, BGG, GBG, GGB, GGG$\}$

$$P(3\text{ boys}) = \frac{n(3\text{ boys})}{n(S)} = \frac{1}{8}$$

38. The sample space is: $S = \{$BBB, BBG, BGB, GBB, BGG, GBG, GGB, GGG$\}$

$$P(3\text{ girls}) = \frac{n(3\text{ girls})}{n(S)} = \frac{1}{8}$$

39. The sample space is:
$S = \{$BBBB, BBBG, BBGB, BGBB, GBBB, BBGG, BGBG, GBBG, BGGB, GBGB, GGBB, BGGG, GBGG, GGBG, GGGB, GGGG$\}$

$$P(1\text{ girl, }3\text{ boys}) = \frac{n(1\text{ girl, }3\text{ boys})}{n(S)} = \frac{4}{16} = \frac{1}{4}$$

40. The sample space is:
$S = \{$BBBB, BBBG, BBGB, BGBB, GBBB, BBGG, BGBG, GBBG, BGGB, GBGB, GGBB, BGGG, GBGG, GGBG, GGGB, GGGG$\}$

$$P(2\text{ girls, }2\text{ boys}) = \frac{n(2\text{ girls, }2\text{ boys})}{n(S)} = \frac{6}{16} = \frac{3}{8}$$

41. $P(\text{sum of two dice is 7})$

$$= \frac{n(\text{sum of two dice is 7})}{n(S)}$$
$$= \frac{n\{1{,}6 \text{ or } 2{,}5 \text{ or } 3{,}4 \text{ or } 4{,}3 \text{ or } 5{,}2 \text{ or } 6{,}1\}}{n(S)}$$
$$= \frac{6}{36} = \frac{1}{6}$$

42. $P(\text{sum of two dice is 11}) = \dfrac{n(\text{sum of two dice is 11})}{n(S)}$

$= \dfrac{n\{5,6 \text{ or } 6,5\}}{n(S)}$

$= \dfrac{2}{36} = \dfrac{1}{18}$

43. $P(\text{sum of two dice is 3}) = \dfrac{n(\text{sum of two dice is 3})}{n(S)}$

$= \dfrac{n\{1,2 \text{ or } 2,1\}}{n(S)} = \dfrac{2}{36} = \dfrac{1}{18}$

44. $P(\text{sum of two dice is 12}) = \dfrac{n(\text{sum of two dice is 12})}{n(S)}$

$= \dfrac{n\{6,6\}}{n(S)} = \dfrac{1}{36}$

45. $P(A \cup B) = P(A) + P(B) - P(A \cap B)$
$= 0.25 + 0.45 - 0.15 = 0.55$

46. $P(A \cap B) = P(A) + P(B) - P(A \cup B)$
$= 0.25 + 0.45 - 0.6 = 0.1$

47. $P(A \cup B) = P(A) + P(B) = 0.25 + 0.45 = 0.70$

48. $P(A \cap B) = 0$

49. $P(A \cup B) = P(A) + P(B) - P(A \cap B)$
$0.85 = 0.60 + P(B) - 0.05$
$P(B) = 0.85 - 0.60 + 0.05 = 0.30$

50. $P(A \cup B) = P(A) + P(B) - P(A \cap B)$
$0.65 = P(A) + 0.30 - 0.15$
$P(A) = 0.65 - 0.30 + 0.15 = 0.50$

51. $P(\text{not victim}) = 1 - P(\text{victim}) = 1 - 0.265 = 0.735$

52. $P(\text{not victim}) = 1 - P(\text{victim}) = 1 - 0.039 = 0.961$

53. $P(\text{not in 70's}) = 1 - P(\text{in 70's}) = 1 - 0.3 = 0.7$

54. $P(\text{not in 30's}) = 1 - P(\text{in 30's}) = 1 - 0.04 = 0.96$

55. $P(\text{white or green}) = P(\text{white}) + P(\text{green})$

$= \dfrac{n(\text{white}) + n(\text{green})}{n(S)}$

$= \dfrac{9+8}{9+8+3} = \dfrac{17}{20}$

56. $P(\text{white or orange}) = P(\text{white}) + P(\text{orange})$

$= \dfrac{n(\text{white}) + n(\text{orange})}{n(S)}$

$= \dfrac{9+3}{9+8+3} = \dfrac{12}{20}$

$= \dfrac{3}{5}$

57. $P(\text{not white}) = 1 - P(\text{white})$

$= 1 - \dfrac{n(\text{white})}{n(S)}$

$= 1 - \dfrac{9}{20} = \dfrac{11}{20}$

58. $P(\text{not green}) = 1 - P(\text{green})$

$= 1 - \dfrac{n(\text{green})}{n(S)}$

$= 1 - \dfrac{8}{20}$

$= \dfrac{12}{20} = \dfrac{3}{5}$

59. $P(\text{strike or one}) = P(\text{strike}) + P(\text{one})$

$= \dfrac{n(\text{strike}) + n(\text{one})}{n(S)}$

$= \dfrac{3+1}{8} = \dfrac{4}{8}$

$= \dfrac{1}{2}$

60. $P(\text{100 or 30}) = P(100) + P(30)$

$= \dfrac{n(100) + n(30)}{n(S)}$

$= \dfrac{1+1}{20} = \dfrac{2}{20}$

$= \dfrac{1}{10}$

61. There are 30 households out of 100 with an income of $30,000 or more.

$$P(E)=\frac{n(E)}{n(S)}=\frac{n(30{,}000\text{ or more})}{n(\text{total households})}$$
$$=\frac{30}{100}=\frac{3}{10}=0.30$$

62. There are 65 households out of 100 with an income between $10,000 and $29,999.

$$P(E)=\frac{n(E)}{n(S)}=\frac{n(10{,}000\text{ to }29{,}999)}{n(\text{total households})}$$
$$=\frac{65}{100}=\frac{13}{20}=0.65$$

63. There are 40 households out of 100 with an income of less than $20,000.

$$P(E)=\frac{n(E)}{n(S)}=\frac{n(\text{less than }\$20{,}000)}{n(\text{total households})}$$
$$=\frac{40}{100}=\frac{2}{5}=0.40$$

64. There are 60 households out of 100 with an income of $20,000 or more.

$$P(E)=\frac{n(E)}{n(S)}=\frac{n(\$20{,}000\text{ or more})}{n(\text{total households})}$$
$$=\frac{60}{100}=\frac{3}{5}=0.60$$

65. a. $P(1\text{ or }2)=P(1)+P(2)=0.24+0.33=0.57$

b. $P(1\text{ or more})=1-P(\text{none})$
$=1-0.05=0.95$

c. $P(3\text{ or fewer})=1-P(4\text{ or more})$
$=1-0.17=0.83$

d. $P(3\text{ or more})=P(3)+P(4\text{ or more})$
$=0.21+0.17=0.38$

e. $P(\text{fewer than }2)=P(0)+P(1)$
$=0.05+0.24=0.29$

f. $P(\text{fewer than }1)=P(0)=0.05$

g. $P(1,\ 2,\text{ or }3)=P(1)+P(2)+P(3)$
$=0.24+0.33+0.21=0.78$

h. $P(2\text{ or more})=P(2)+P(3)+P(4\text{ or more})$
$=0.33+0.21+0.17=0.71$

66. a. $P(\text{at most }2)=P(0)+P(1)+P(2)$
$=0.10+0.15+0.20=0.45$

b. $P(\text{at least }2)=P(2)+P(3)+P(4\text{ or more})$
$=0.20+0.24+0.31=0.75$

c. $P(\text{at least }1)=1-P(0)=1-0.10=0.90$

67. a. $P(\text{freshman or female})$
$=P(\text{freshman})+P(\text{female})-P(\text{freshman and female})$
$$=\frac{n(\text{freshman})+n(\text{female})-n(\text{freshman and female})}{n(S)}$$
$$=\frac{18+15-8}{33}=\frac{25}{33}$$

b. $P(\text{sophomore or male})$
$=P(\text{sophomore})+P(\text{male})-P(\text{sophomore and male})$
$$=\frac{n(\text{sophomore})+n(\text{male})-n(\text{sophomore and male})}{n(S)}$$
$$=\frac{15+18-8}{33}=\frac{25}{33}$$

68. a. $P(\text{female or under 40})$
$=P(\text{female})+P(\text{under 40})-P(\text{female and under 40})$
$$=\frac{n(\text{female})+n(\text{under 40})-n(\text{female and under 40})}{n(S)}$$
$$=\frac{4+5-2}{13}=\frac{7}{13}$$

b. $P(\text{male or over 40})$
$=P(\text{male})+P(\text{over 40})-P(\text{male and over 40})$
$$=\frac{n(\text{male})+n(\text{over 40})-n(\text{male and over 40})}{n(S)}$$
$$=\frac{9+8-6}{13}=\frac{11}{13}$$

69. $P(\text{at least 2 with same birthday})$
$=1-P(\text{none with same birthday})$
$$=1-\frac{n(\text{different birthdays})}{n(S)}$$
$$=1-\frac{365\cdot364\cdot363\cdot362\cdot361\cdot360\cdots354}{365^{12}}$$
$\approx1-0.833$
$=0.167$

70. $P(\text{at least 2 with same birthday})$
$= 1 - P(\text{none with same birthday})$
$= 1 - \dfrac{n(\text{different birthdays})}{n(S)}$
$= 1 - \dfrac{365 \cdot 364 \cdot 363 \cdot 362 \cdot 361 \cdot 360 \cdots 331}{365^{35}}$
$\approx 1 - 0.1856$
$= 0.8144$

71. The sample space for picking 5 out of 10 numbers in a particular order contains
$P(10,5) = \dfrac{10!}{(10-5)!} = \dfrac{10!}{5!} = 30{,}240$ possible outcomes.
One of these is the desired outcome. Thus, the probability of winning is:
$P(E) = \dfrac{n(E)}{n(S)} = \dfrac{n(\text{winning})}{n(\text{total possible outcomes})}$
$= \dfrac{1}{30{,}240} \approx 0.000033069$

Chapter 12 Review

1. $\{\text{Dave}\}, \{\text{Joanne}\}, \{\text{Erica}\},$
$\{\text{Dave, Joanne}\}, \{\text{Dave, Erica}\}, \{\text{Joanne, Erica}\},$
$\{\text{Dave, Joanne, Erica}\}, \varnothing$

2. $\{\text{Green}\}, \{\text{Blue}\}, \{\text{Red}\},$
$\{\text{Green, Blue}\}, \{\text{Green, Red}\}, \{\text{Blue, Red}\},$
$\{\text{Green, Blue, Red}\}, \varnothing$

3. $A \cup B = \{1,3,5,7\} \cup \{3,5,6,7,8\}$
$= \{1,3,5,6,7,8\}$

4. $B \cup C = \{3,5,6,7,8\} \cup \{2,3,7,8,9\}$
$= \{2,3,5,6,7,8,9\}$

5. $A \cap C = \{1,3,5,7\} \cap \{2,3,7,8,9\} = \{3,7\}$

6. $A \cap B = \{1,3,5,7\} \cap \{3,5,6,7,8\} = \{3,5,7\}$

7. $\overline{A} \cup \overline{B} = \overline{\{1,3,5,7\}} \cup \overline{\{3,5,6,7,8\}}$
$= \{2,4,6,8,9\} \cup \{1,2,4,9\}$
$= \{1,2,4,6,8,9\}$

8. $\overline{B} \cap \overline{C} = \overline{\{3,5,6,7,8\}} \cap \overline{\{2,3,7,8,9\}}$
$= \{1,2,4,9\} \cap \{1,4,5,6\}$
$= \{1,4\}$

9. $\overline{B \cap C} = \overline{\{3,5,6,7,8\} \cap \{2,3,7,8,9\}}$
$= \overline{\{3,7,8\}}$
$= \{1,2,4,5,6,9\}$

10. $\overline{A \cup B} = \overline{\{1,3,5,7\} \cup \{3,5,6,7,8\}}$
$= \overline{\{1,3,5,6,7,8\}}$
$= \{2,4,9\}$

11. $n(A) = 8, n(B) = 12, n(A \cap B) = 3$
$n(A \cup B) = n(A) + n(B) - n(A \cap B)$
$= 8 + 12 - 3 = 17$

12. $n(A) = 12, n(A \cup B) = 30, n(A \cap B) = 6$
$n(A \cup B) = n(A) + n(B) - n(A \cap B)$
$30 = 12 + n(B) - 6$
$n(B) = 30 - 12 + 6 = 24$

13. From the figure: $n(A) = 20 + 2 + 6 + 1 = 29$

14. From the figure:
$n(A \text{ or } B) = 20 + 2 + 6 + 1 + 5 + 0 = 34$

15. From the figure:
$n(A \text{ and } C) = n(A \cap C) = 1 + 6 = 7$

16. From the figure:
$n(\text{not in } B) = 20 + 1 + 4 + 20 = 45$

17. From the figure:
$n(\text{neither in } A \text{ nor in } C) = n(\overline{A \cup C}) = 20 + 5 = 25$

18. From the figure: $n(\text{in } B \text{ but not in } C) = 2 + 5 = 7$

19. $P(8,3) = \dfrac{8!}{(8-3)!} = \dfrac{8!}{5!} = \dfrac{8 \cdot 7 \cdot 6 \cdot 5!}{5!} = 336$

20. $P(7,3)=\dfrac{7!}{(7-3)!}=\dfrac{7!}{4!}=\dfrac{7\cdot 6\cdot 5\cdot 4!}{4!}=210$

21. $C(8,3)=\dfrac{8!}{(8-3)!3!}=\dfrac{8!}{5!3!}=\dfrac{8\cdot 7\cdot 6\cdot 5!}{5!\cdot 3\cdot 2\cdot 1}=56$

22. $C(7,3)=\dfrac{7!}{(7-3)!3!}=\dfrac{7!}{4!3!}=\dfrac{7\cdot 6\cdot 5\cdot 4!}{4!\cdot 3\cdot 2\cdot 1}=35$

23. There are 2 choices of material, 3 choices of color, and 10 choices of size. The complete assortment would have: $2\cdot 3\cdot 10=60$ suits.

24. This is a permutation of 5 items taken 5 at a time. There are

$P(5,5)=\dfrac{5!}{(5-5)!}=\dfrac{5!}{0!}=5!=120$ possible

wirings.

25. There are two possible outcomes for each game or $2\cdot 2\cdot 2\cdot 2\cdot 2\cdot 2\cdot 2=2^7=128$ outcomes for 7 games.

26. There are two possible outcomes for each game or $2\cdot 2\cdot 2\cdot 2\cdot 2\cdot 2=2^6=64$ outcomes for 6 games.

27. Since order is significant, this is a permutation.

$P(9,4)=\dfrac{9!}{(9-4)!}=\dfrac{9!}{5!}=\dfrac{9\cdot 8\cdot 7\cdot 6\cdot 5!}{5!}=3024$

ways to seat 4 people in 9 seats.

28. Since order is significant, this is a permutation.

$P(4,4)=\dfrac{4!}{(4-4)!}=\dfrac{4!}{0!}=\dfrac{4\cdot 3\cdot 2\cdot 1}{1}=24$

arrangements of the letters in ROSE.

29. Choose 4 runners –order is significant:

$P(8,4)=\dfrac{8!}{(8-4)!}=\dfrac{8!}{4!}=\dfrac{8\cdot 7\cdot 6\cdot 5\cdot 4!}{4!}=1680$

ways a squad can be chosen.

30. Choose 3 problems –order is not significant:

$C(10,3)=\dfrac{10!}{(10-3)!3!}=\dfrac{10!}{7!3!}=\dfrac{10\cdot 9\cdot 8\cdot 7!}{3\cdot 2\cdot 1\cdot 7!}=120$

different tests are possible.

31. Choose 2 teams from 14–order is not significant:

$C(14,2)=\dfrac{14!}{(14-2)!2!}=\dfrac{14!}{12!2!}=\dfrac{14\cdot 13\cdot 12!}{12!\cdot 2\cdot 1}=91$

ways to choose 2 teams.

32. a. Since order is important, this is a permutation:

$P(5,5)\cdot P(5,5)=\dfrac{5!}{(5-5)!}\cdot\dfrac{5!}{(5-5!}$

$=5!\cdot 5!=120\cdot 120=14,400$

different arrangements.

b. There would be $5\cdot 5\cdot 4\cdot 4\cdot 3\cdot 3\cdot 2\cdot 2\cdot 1\cdot 1=14,400$ different arrangements.

33. There are $8\cdot 10\cdot 10\cdot 10\cdot 10\cdot 10\cdot 2=1,600,000$ possible phone numbers.

34. There are $5\cdot 3\cdot 4=60$ different types of homes that can be built.

35. There are $24\cdot 9\cdot 10\cdot 10\cdot 10=216,000$ possible license plates.

36. There are two choices for each digit, so there are $2^8=256$ different numbers.

37. Since there are repeated letters:

$\dfrac{7!}{2!\cdot 2!}=\dfrac{7\cdot 6\cdot 5\cdot 4\cdot 3\cdot 2\cdot 1}{2\cdot 1\cdot 2\cdot 1}=1260$ different words

can be formed.

38. Since there are repeated colors:

$\dfrac{10!}{4!\cdot 3!\cdot 2!\cdot 1!}=\dfrac{10\cdot 9\cdot 8\cdot 7\cdot 6\cdot 5\cdot 4\cdot 3\cdot 2\cdot 1}{4\cdot 3\cdot 2\cdot 1\cdot 3\cdot 2\cdot 1\cdot 2\cdot 1\cdot 1}=12,600$

different vertical arrangements.

39. a. $C(9,4)\cdot C(9,3)\cdot C(9,2)=126\cdot 84\cdot 36$

$=381,024$

committees can be formed.

b. $C(9,4)\cdot C(5,3)\cdot C(2,2)=126\cdot 10\cdot 1=1260$

committees can be formed.

40. a. $C(5,1)\cdot C(8,3) = \dfrac{5!}{(5-1)!1!}\cdot\dfrac{8!}{(8-3)!3!}$
$= 5\cdot 56 = 280$
committees containing exactly 1 man.

b. $C(5,2)\cdot C(8,2) = 10\cdot 28 = 280$ committees containing exactly 2 women.

c. $C(5,1)\cdot C(8,3) + C(5,2)\cdot C(8,2)$
$+C(5,3)\cdot C(8,1) + C(5,4)$
$= 280 + 280 + 10\cdot 8 + 5 = 645$
committees containing at least 1 man.

41. a. $365\cdot 364\cdot 363\cdots 348 = 8.634628387\times 10^{45}$

b. P(no one has same birthday)
$= \dfrac{365\cdot 364\cdot 363\cdots 348}{365^{18}} \approx 0.6531$

c. P(at least 2 have same birthday)
$= 1 - P$(no one has same birthday)
$= 1 - 0.6531 = 0.3469$

42. a. P(heart disease) $= 0.29$

b. P(not heart disease) $= 1 - P$(heart disease)
$= 1 - 0.29 = 0.71$

43. a. P(unemployed) $= 0.058$

b. P(not unemployed) $= 1 - P$(unemployed)
$= 1 - 0.058 = 0.942$

44. $P(40 \text{ watt}) = \dfrac{n(40\text{ watt})}{n(\text{bulbs})} = \dfrac{3}{20} = 0.15$

$P(\text{not } 75 \text{ watt}) = 1 - P(75\text{ watt}) = 1 - \dfrac{n(75\text{ watt})}{n(\text{bulbs})}$
$= 1 - \dfrac{11}{20} = \dfrac{9}{20} = 0.45$

45. $P(\$1\text{ bill}) = \dfrac{n(\$1\text{ bill})}{n(S)} = \dfrac{4}{9}$

46. $P(\text{ROSE}) = \dfrac{1}{4}\cdot\dfrac{1}{3}\cdot\dfrac{1}{2}\cdot\dfrac{1}{1} = \dfrac{1}{24}$

47. Let S be all possible selections, let D be a card that is divisible by 5, and let PN be a card that is 1 or a prime number.
$n(S) = 100$
$n(D) = 20$ (There are 20 numbers divisible by 5 between 1 and 100.)
$n(PN) = 26$ (There are 25 prime numbers less than or equal to 100.)
$P(D) = \dfrac{n(D)}{n(S)} = \dfrac{20}{100} = \dfrac{1}{5} = 0.2$
$P(PN) = \dfrac{n(PN)}{n(S)} = \dfrac{26}{100} = \dfrac{13}{50} = 0.26$

48. Let S be all possible selections, let T be a car that needs a tune-up, and let B be a car that needs a brake job.

a. P(a car requires a tune-up or a brake job)
$= P(T \cup B)$
$= P(T) + P(B) - P(T \cap B)$
$= 0.6 + 0.1 - 0.02$
$= 0.68$

b. $P(\text{a car requires a tune-up but not a brake job})$
$= P(\text{a car requires a tune-up}) - P(\text{a car requires a tune-up and a brake job})$
$= P(T) - P(T \cap B)$
$= 0.6 - 0.02$
$= 0.58$

c. $P(\text{a car requires neither a tune-up nor a brake job})$
$= 1 - \left(P(\text{a car requires a tune-up or a brake job})\right)$
$= 1 - \left(P(T) + P(B) - P(T \cap B)\right)$
$= 1 - (0.6 + 0.1 - 0.02)$
$= 0.32$

Chapter 12 Test

1. $A \cap B = \{0,1,4,9\} \cap \{2,4,6,8\} = \{4\}$

2. $A \cup C = \{0,1,4,9\} \cup \{1,3,5,7,9\}$
$= \{0,1,3,4,5,7,9\}$

3. $(A \cup B) \cap C$
$= (\{0,1,4,9\} \cup \{2,4,6,8\}) \cap \{1,3,5,7,9\}$
$= \{0,1,2,4,6,8,9\} \cap \{1,3,5,7,9\}$
$= \{1,9\}$

4. $A \cup B = \{0,1,4,9\} \cup \{2,4,6,8\}$
$= \{0,1,2,4,6,8,9\}$
Since the universal set is $\{0,1,2,3,4,5,6,7,8,9\}$,
$\overline{A \cup B} = \{3,5,7\}$.

5. Since the universal set is $\{0,1,2,3,4,5,6,7,8,9\}$,
$\overline{C} = \{0,2,4,6,8\}$.

6. $\overline{C}$ can be obtained from the previous problem.
$A \cap \overline{C} = \{0,1,4,9\} \cap \{0,2,4,6,8\} = \{0,4\}$
Therefore,
$\overline{A \cap \overline{C}} = \{1,2,3,5,6,7,8,9\}$.

7. From the figure:
$n(\text{physics}) = 4 + 2 + 7 + 9 = 22$

8. From the figure:
$n(\text{biology } \textit{or} \text{ chemistry } \textit{or} \text{ physics})$
$= 22 + 8 + 2 + 4 + 9 + 7 + 15$
$= 67$
Therefore,
$n(\text{none of the three}) = 70 - 67 = 3$

9. From the figure:
$n(\text{only biology } \textit{and} \text{ chemistry})$
$= n(\text{biol. } \textit{and} \text{ chem.}) - n(\text{biol. } \textit{and} \text{ chem. } \textit{and} \text{ phys.})$
$= (8 + 2) - 2$
$= 8$

10. From the figure:
$n(\text{physics } \textit{or} \text{ chemistry}) = 4 + 2 + 7 + 9 + 15 + 8$
$= 45$

11. $7! = 7 \cdot 6 \cdot 5 \cdot 4 \cdot 3 \cdot 2 \cdot 1 = 5040$

12. $P(10,6) = \dfrac{10!}{(10-6)!} = \dfrac{10!}{4!}$
$= \dfrac{10 \cdot 9 \cdot 8 \cdot 7 \cdot 6 \cdot 5 \cdot 4!}{4!}$
$= 10 \cdot 9 \cdot 8 \cdot 7 \cdot 6 \cdot 5$
$= 151{,}200$

13. $$C(11,5)=\frac{11!}{5!(11-5)!}=\frac{11!}{5!6!}=\frac{11\cdot10\cdot9\cdot8\cdot7\cdot6!}{5\cdot4\cdot3\cdot2\cdot1\cdot6!}=\frac{11\cdot10\cdot9\cdot8\cdot7}{5\cdot4\cdot3\cdot2\cdot1}=462$$

14. Since the order in which the colors are selected doesn't matter, this is a combination problem. We have $n = 21$ colors and we wish to select $r = 6$ of them.

$$C(21,6)=\frac{21!}{6!(21-6)!}=\frac{21!}{6!15!}=\frac{21\cdot20\cdot19\cdot18\cdot17\cdot16\cdot15!}{6!15!}=\frac{21\cdot20\cdot19\cdot18\cdot17\cdot16}{6\cdot5\cdot4\cdot3\cdot2\cdot1}=54,264$$

There are 54,264 ways to choose 6 colors from the 21 available colors.

15. Because the letters are not distinct and order matters, we use the permutation formula for non-distinct objects. We have four different letters, two of which are repeated (E four times and D two times).

$$\frac{n!}{n_1!n_2!n_3!n_4!}=\frac{8!}{4!2!1!1!}=\frac{8\cdot7\cdot6\cdot5\cdot4!}{4!\cdot2\cdot1}=\frac{8\cdot7\cdot6\cdot5}{2}=4\cdot7\cdot6\cdot5=840$$

There are 840 distinct arrangements of the letters in the word REDEEMED.

16. Since the order of the horses matters and all the horses are distinct, we use the permutation formula for distinct objects.

$$P(8,2)=\frac{8!}{(8-2)!}=\frac{8!}{6!}=\frac{8\cdot7\cdot6!}{6!}=8\cdot7=56$$

There are 56 different exacta bets for an 8-horse race.

17. We are choosing 3 letters from 26 distinct letters and 4 digits from 10 distinct digits. The letters and numbers are placed in order following the format LLL DDDD with repetitions being allowed. Using the Multiplication Principle, we get

$$26\cdot26\cdot23\cdot10\cdot10\cdot10\cdot10=155,480,000$$

Note that there are only 23 possibilities for the third letter.

There are 155,480,000 possible license plates using the new format.

18. Let A = Kiersten accepted at USC, and B = Kiersten accepted at FSU. Then, we get $P(A)=0.6$, $P(B)=0.7$, and $P(A\cap B)=0.35$.

a. Here we need to use the Addition Rule.

$$P(A\cup B)=P(A)+P(B)-P(A\cap B)=0.6+0.7-0.35=0.95$$

Kiersten has a 95% chance of being admitted to at least one of the universities.

b. Here we need the Complement of an event.

$$P(\overline{B})=1-P(B)=1-0.7=0.3$$

Kiersten has a 30% chance of not being admitted to FSU.

19. a. Since the bottle is chosen at random, all bottles are equally likely to be selected. Thus,

$$P(\text{Coke})=\frac{5}{8+5+4+3}=\frac{5}{20}=\frac{1}{4}=0.25$$

There is a 25% chance that the selected bottle contains Coke.

b. $$P(\text{Pepsi}\cup\text{IBC})=\frac{8+3}{8+5+4+3}=\frac{11}{20}=0.55$$

There is a 55% chance that the selected bottle contains either Pepsi or IBC.

20. Since the ages cover all possibilities and the age groups are mutually exclusive, the sum of all the probabilities must equal 1.

$$0.03+0.23+0.29+0.25+0.01=0.81$$
$$1-0.81=0.19$$

The given probabilities sum to 0.81. This means the missing probability (for 18-20) must be 0.19.

21. The number of different selections of 6 numbers is the number of ways we can choose 5 white balls and 1 red ball, where the order of the white balls is not important. This requires the use of the Multiplication Principle and the combination formula. Thus, the total number of distinct ways to pick the 6 numbers is given by

$n(\text{white balls}) \cdot n(\text{red ball})$
$= C(53,5) \cdot C(42,1)$
$= \dfrac{53!}{5! \cdot (53-5)!} \cdot \dfrac{42!}{1! \cdot (42-1)!}$
$= \dfrac{53!}{5! \cdot 48!} \cdot \dfrac{42!}{1! \cdot 41!}$
$= \left(\dfrac{53 \cdot 52 \cdot 51 \cdot 50 \cdot 49}{5 \cdot 4 \cdot 3 \cdot 2 \cdot 1}\right)\left(\dfrac{42 \cdot 41!}{41!}\right)$
$= \dfrac{53 \cdot 52 \cdot 51 \cdot 50 \cdot 49 \cdot 42}{5 \cdot 4 \cdot 3 \cdot 2}$
$= 120{,}526{,}770$

Since each possible combination is equally likely, the probability of winning on a \$1 play is

$P(\text{win on \$1 play}) = \dfrac{1}{120{,}526{,}770}$
≈ 0.0000000083

22. The number of elements in the sample space can be obtained by using the Multiplication Principle:

$6 \cdot 6 \cdot 6 \cdot 6 \cdot 6 = 7{,}776$

Consider the rolls as a sequence of 5 slots. The number of ways to position 2 fours in 5 slots is $C(5,2)$. The remaining three slots can be filled with any of the five remaining numbers from the die. Repetitions are allowed so this can be done in $5 \cdot 5 \cdot 5 = 125$ different ways.

Therefore, the total number of ways to get exactly 2 fours is

$C(5,2) \cdot 125 = \dfrac{5!}{2! \cdot 3!} \cdot 125 = \dfrac{5 \cdot 4 \cdot 125}{2} = 1250$

The probability of getting exactly 2 fours on 5 rolls of a die is given by

$P(\text{exactly 2 fours}) = \dfrac{1250}{7776} \approx 0.1608$.

Chapter 12 Projects

Project 1

One simulation might be:

Woman has	Woman told you about	Probability
Boy-Boy	Older boy	$\frac{1}{4}$
Boy-Boy	Younger boy	$\frac{1}{4}$
Boy-Girl	Younger boy	$\frac{1}{4}$
Girl-Boy	Older boy	$\frac{1}{4}$

We leave out the combinations where she would have to tell you about a girl. Thus, the probability that she has 2 boys is $\frac{1}{4} + \frac{1}{4} = \frac{1}{2}$.

Man has	Man told you about	Probability
Boy-Boy	Older boy	$\frac{1}{2}$
Girl-Boy	Older boy	$\frac{1}{2}$

Thus the probability he has two boys is ½. The probabilities are the same.

Project 2 (web)

1. 0 bit errors: 1011

1 bit errors: 0011
1111
1001
1010

2 bit errors: 0111
0001
0010
1101
1110
1000

3 bit errors: 0110
0101
0000
1100

4 bit errors: 0100

2. $P(\text{symbol received correctly}) = \left(\frac{2}{3}\right)^4 = \frac{16}{81}$

3. # of received symbols with 2 bit errors:
$C(8,2) = 28$

$P(\text{symbol received correctly}) = \left(\frac{2}{3}\right)^8 = \frac{256}{6561}$

$P(\text{symbol received incorrectly})$
$= 1 - P(\text{received correctly})$
$= \frac{6305}{6561}$

4. Let k = # of errors, n = 8 = length of symbol.
Probability of k errors :
$P(n,k) = \binom{n}{k}(p)^k(1-p)^{n-k}$
$P(8,k) = \binom{8}{k}\left(\frac{1}{3}\right)^k\left(\frac{2}{3}\right)^{8-k}$
Since this parity code only detects odd numbers of errors,
$P(\text{error detected})$
$= P(8,1) + P(8,3) + P(8,5) + P(8,7)$
$= \binom{8}{1}\left(\frac{1}{3}\right)^1\left(\frac{2}{3}\right)^7 + \binom{8}{3}\left(\frac{1}{3}\right)^3\left(\frac{2}{3}\right)^5$
$+ \binom{8}{5}\left(\frac{1}{3}\right)^5\left(\frac{2}{3}\right)^3 + \binom{8}{7}\left(\frac{1}{3}\right)^7\left(\frac{2}{3}\right)^1$
$= 0.156464 + 0.272992 + 0.068044 + 0.002423$
$= 0.499923$
To find the probability that an error occurred but is not detected, we need to assume that an even number of errors occurred:

$P(\text{error occured, but not detected})$
$= P(8,2) + P(8,4) + P(8,6) + P(8,8)$
$= \binom{8}{2}\left(\frac{1}{3}\right)^2\left(\frac{2}{3}\right)^6 + \binom{8}{4}\left(\frac{1}{3}\right)^4\left(\frac{2}{3}\right)^4$
$+ \binom{8}{6}\left(\frac{1}{3}\right)^6\left(\frac{2}{3}\right)^2 + \binom{8}{8}\left(\frac{1}{3}\right)^8\left(\frac{2}{3}\right)^0$
$= 0.273402 + 0.170364 + 0.016985 + 0.000151$
$= 0.460951$

Project 3 (web)

Answers will vary.

Project 4 (web)

e. Answers will vary, depending on the L_2 generated by the calculator.

f. The data accumulates around y = 0.5.

Cumulative Review 1-12

1. $3x^2 - 2x = -1 \Rightarrow 3x^2 - 2x + 1 = 0$
$x = \frac{-b \pm \sqrt{b^2 - 4ac}}{2a} = \frac{-(-2) \pm \sqrt{(-2)^2 - 4(3)(1)}}{2(3)}$
$= \frac{2 \pm \sqrt{4-12}}{6} = \frac{2 \pm \sqrt{-8}}{6} = \frac{2 \pm 2\sqrt{2}i}{6}$
$= \frac{1 \pm \sqrt{2}i}{3}$
The solution set is $\left\{\frac{1-\sqrt{2}i}{3}, \frac{1+\sqrt{2}i}{3}\right\}$.

2. $f(x) = x^2 + 4x - 5$
$a = 1, b = 4, c = -5$. Since $a = 1 > 0$, the graph opens up.
The x-coordinate of the vertex is
$x = -\frac{b}{2a} = -\frac{4}{2(1)} = -\frac{4}{2} = -2$.
The y-coordinate of the vertex is
$f\left(-\frac{b}{2a}\right) = f(-2) = (-2)^2 + 4(-2) - 5$.
$= 4 - 8 - 5 = -9$
Thus, the vertex is $(-2, -9)$. The axis of symmetry is the line $x = -2$. The discriminant is:
$b^2 - 4ac = (4)^2 - 4(1)(-5) = 16 + 20 = 36 > 0$.
So the graph has two x-intercepts.
The x-intercepts are found by solving:
$x^2 + 4x - 5 = 0$
$(x+5)(x-1) = 0$
$x = -5, x = 1$
The x-intercepts are -5 and 1.

The y-intercept is $f(0) = (0)^2 + 4 \cdot (0) - 5 = -5$.

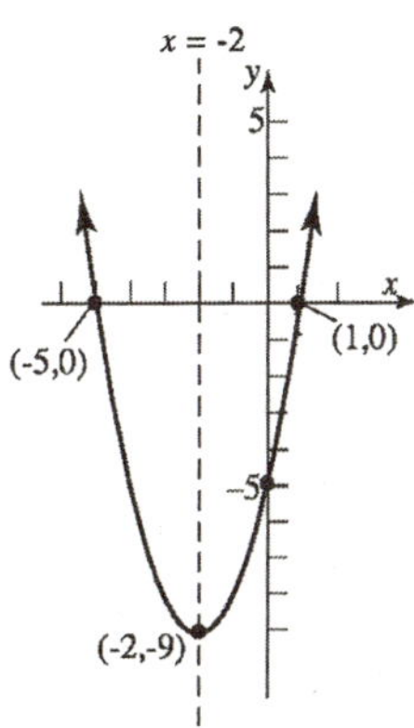

3. $y = 2(x+1)^2 - 4$

Using the graph of $y = x^2$, horizontally shift to the left 1 unit, vertically stretch by a factor of 2, and vertically shift down 4 units.

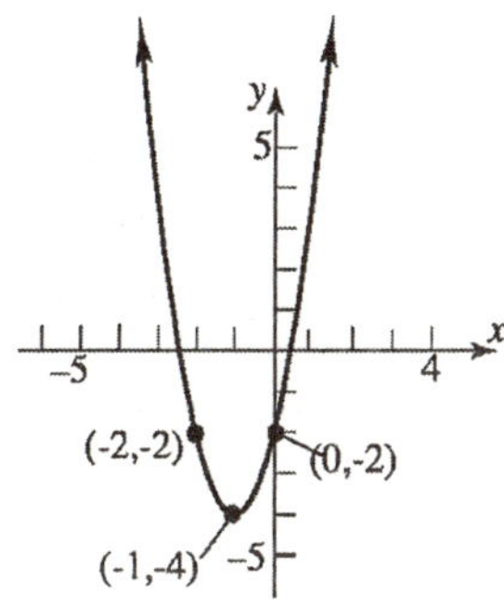

4. $|x-4| \le 0.01$

$$-0.01 \le x - 4 \le 0.01$$
$$-0.01 + 4 \le x \le 0.01 + 4$$
$$3.99 \le x \le 4.01$$

The solution set is

$\{x | 3.99 \le x \le 4.01\}$ or $[3.99, 4.01]$

5. $f(x) = 5x^4 - 9x^3 - 7x^2 - 31x - 6$

Step 1: $f(x)$ has at most 4 real zeros.

Step 2: Possible rational zeros:

$p = \pm 1, \pm 2, \pm 3, \pm 6;\quad q = \pm 1, \pm 5;$

$$\frac{p}{q} = \pm 1, \pm\frac{1}{5}, \pm 2, \pm\frac{2}{5}, \pm 3, \pm\frac{3}{5}, \pm 6, \pm\frac{6}{5}$$

Step 3: Using the Bounds on Zeros Theorem:

$f(x) = 5\left(x^4 - 1.8x^3 - 1.4x^2 - 6.2x - 1.2\right)$

$a_3 = -1.8,\ a_2 = -1.4,\ a_1 = -6.2,\ a_0 = -1.2$

$\text{Max}\ \{1, |-1.2| + |-6.2| + |-1.4| + |-1.8|\}$

$= \text{Max}\ \{1, 10.6\} = 10.6$

$1 + \text{Max}\ \{|-1.2|, |-6.2|, |-1.4|, |-1.8|\}$

$= 1 + 6.2 = 7.2$

The smaller of the two numbers is 7.2. Thus, every zero of f lies between –7.2 and 7.2.

Graphing using the bounds: (Second graph has a better window.)

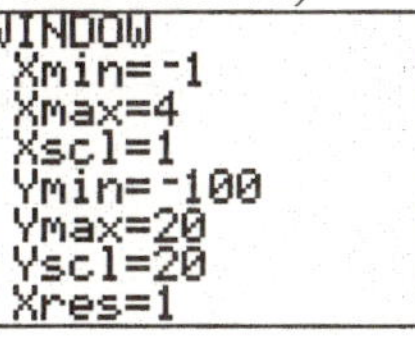

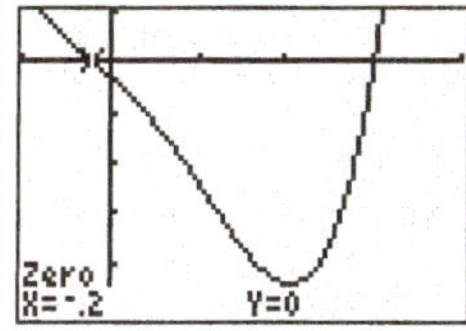

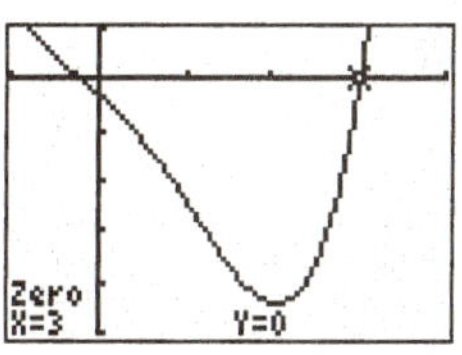

Step 4: From the graph we see that there are x-intercepts at -0.2 and 3. Using synthetic division with 3:

$$\begin{array}{r|rrrrr} 3 & 5 & -9 & -7 & -31 & -6 \\ & & 15 & 18 & 33 & 6 \\ \hline & 5 & 6 & 11 & 2 & 0 \end{array}$$

Since the remainder is 0, $x - 3$ is a factor. The other factor is the quotient: $5x^3 + 6x^2 + 11x + 2$.

Using synthetic division with 2 on the quotient:

$$\begin{array}{r|rrrr} -0.2 & 5 & 6 & 11 & 2 \\ & & -1 & -1 & -2 \\ \hline & 5 & 5 & 10 & 0 \end{array}$$

Since the remainder is 0, $x - (-0.2) = x + 0.2$ is a factor. The other factor is the quotient:

$5x^2 + 5x + 10 = 5\left(x^2 + x + 2\right)$.

Factoring, $f(x) = 5(x^2 + x + 2)(x-3)(x+0.2)$

The real zeros are 3 and -0.2.

The complex zeros come from solving

$x^2 + x + 2 = 0$.

$$x=\frac{-b\pm\sqrt{b^2-4ac}}{2a}=\frac{-1\pm\sqrt{1^2-4(1)(2)}}{2(1)}$$
$$=\frac{-1\pm\sqrt{1-8}}{2}=\frac{-1\pm\sqrt{-7}}{2}$$
$$=\frac{-1\pm\sqrt{7}i}{2}$$

Therefore, over the set of complex numbers, $f(x)=5x^4-9x^3-7x^2-31x-6$ has zeros $\left\{-\frac{1}{2}+\frac{\sqrt{7}}{2}i,-\frac{1}{2}-\frac{\sqrt{7}}{2}i,-\frac{1}{5},3\right\}$.

6. $g(x)=3^{x-1}+5$

Using the graph of $y=3^x$, shift the graph horizontally 1 unit to the right, then shift the graph vertically 5 units upward.

Domain: $(-\infty,\infty)$

Range: $(5,\infty)$

Horizontal Asymptote: $y=5$

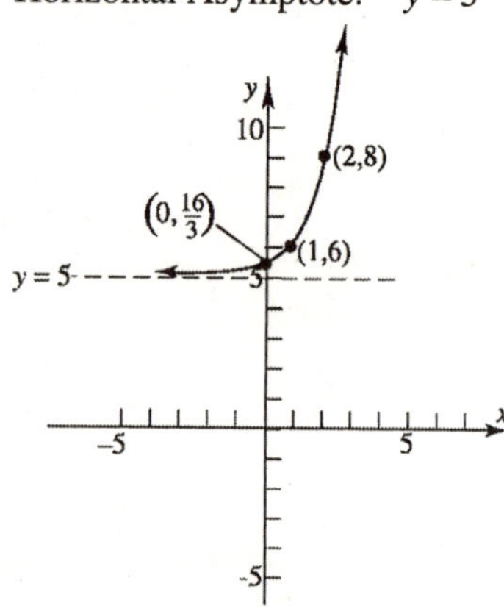

7. $\log_3(9)=\log_3(3^2)=2$

8. $\log_2(3x-2)+\log_2(x)=4$
$$\log_2((3x-2)(x))=4$$
$$(3x-2)(x)=2^4$$
$$3x^2-2x=16$$
$$3x^2-2x-16=0$$
$$(3x-8)(x+2)=0\Rightarrow x=\frac{8}{3}\text{ or } x=-2$$

Since $x=-2$ makes the original logarithms undefined, the solution set is $\left\{\frac{8}{3}\right\}$.

9. Multiply each side of the first equation by –3 and add to the second equation to eliminate x; multiply each side of the first equation by 2 and add to the third equation to eliminate x:
$$\begin{cases}x-2y+z=15\\3x+y-3z=-8\\-2x+4y-z=-27\end{cases}$$
$$\begin{aligned}-3x+6y-3z&=-45\\3x+y-3z&=-8\\\hline 7y-6z&=-53\end{aligned}$$
$$\begin{aligned}x-2y+z&=15\\-2x+4y-z&=-27\end{aligned}\qquad\begin{aligned}2x-4y+2z&=30\\-2x+4y-z&=-27\\\hline z&=3\end{aligned}$$

Substituting and solving for the other variables:
$$z=3\Rightarrow 7y-6(3)=-53$$
$$7y=-35$$
$$y=-5$$
$$z=3, y=-5\Rightarrow x-2(-5)+3=15$$
$$x+10+3=15\Rightarrow x=2$$

The solution is $x=2,\ y=-5,\ z=3$.

10. –3, 1, 5, 9, ... is an arithmetic sequence with $a=-3,\ d=4$.

Using $a_n=a+(n-1)d$,
$$a_{33}=-3+(33-1)\cdot 4$$
$$=-3+32\cdot 4$$
$$=-3+128$$
$$=125$$

To compute the sum of the first 20 terms, we use $S_{20}=\frac{20}{2}(a+a_{20})$.
$$a_{20}=-3+(20-1)\cdot 4$$
$$=-3+19\cdot 4$$
$$=-3+76$$
$$=73$$

Therefore,
$$S_{20}=\frac{20}{2}(a+a_{20})$$
$$=\frac{20}{2}(-3+73)$$
$$=10\cdot 70$$
$$=700.$$

11. $y = 3\sin(2x+\pi) = 3\sin\left(2\left(x+\frac{\pi}{2}\right)\right)$

Amplitude: $|A| = |3| = 3$

Period: $T = \frac{2\pi}{2} = \pi$

Phase Shift: $\frac{\phi}{\omega} = \frac{-\pi}{2} = -\frac{\pi}{2}$

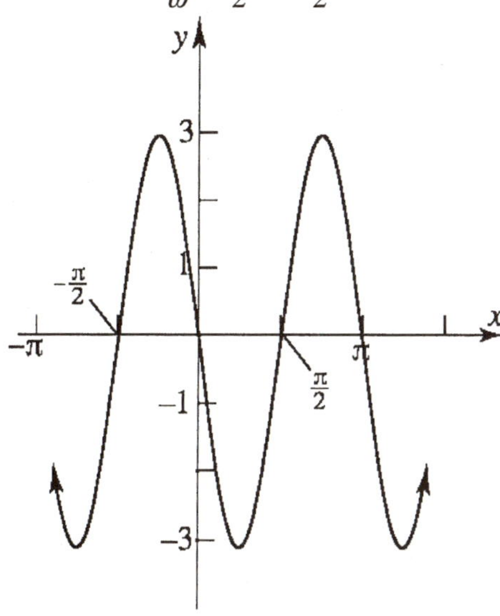

12. Use the Law of Cosines:

$a^2 = b^2 + c^2 - 2bc\cos\alpha$

$a^2 = 5^2 + 9^2 - 2\cdot 5\cdot 9\cos 40^\circ$

$= 106 - 90\cos 40^\circ$

$a \approx 6.09$

$b^2 = a^2 + c^2 - 2ac\cos\beta$

$\cos\beta = \frac{a^2+c^2-b^2}{2ac} \approx \frac{6.09^2+9^2-5^2}{2(6.09)(9)}$

$= \frac{93.0881}{109.62}$

$\beta \approx \cos^{-1}\left(\frac{93.0881}{109.62}\right) \approx 31.9^\circ$

$\gamma = 180^\circ - \alpha - \beta \approx 180^\circ - 40^\circ - 31.9^\circ = 108.1^\circ$

Area of the triangle = $\frac{1}{2}\cdot 5\cdot 9\cdot \sin(40^\circ) \approx 14.46$ square units.

Chapter 13
A Preview of Calculus: The Limit, Derivative, and Integral of a Function

Section 13.1

1. $f(x) = \begin{cases} 3x-2 & \text{if } x \neq 2 \\ 3 & \text{if } x = 2 \end{cases}$

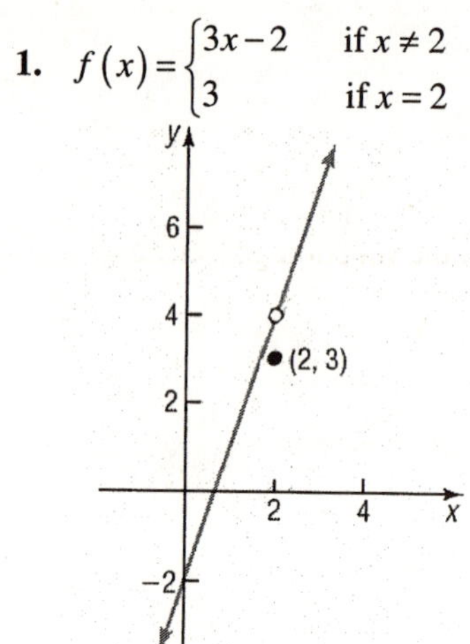

2. $f(0) = 0$

3. $\lim_{x \to c} f(x)$

4. does not exist

5. True

6. False; we are only concerned with the behavior of the function *near c*. The function does not need to be defined *at c* for the limit to exist; nor does the existence of $f(c)$ guarantee that the limit exists at *c*.

7. $\lim_{x \to 2} (4x^3)$

X	Y1
1.99	31.522
1.999	31.952
1.9999	31.995
2.0001	32.005
2.001	32.048
2.01	32.482

Y1=4X^3

$\lim_{x \to 2} (4x^3) = 32$

8. $\lim_{x \to 3} (2x^2 + 1)$

X	Y1
2.99	18.88
2.999	18.988
2.9999	18.999
3.0001	19.001
3.001	19.012
3.01	19.12

Y1=2X²+1

$\lim_{x \to 3} (2x^2 + 1) = 19$

9. $\lim_{x \to 0} \left(\frac{x+1}{x^2+1} \right)$

X	Y1
-.01	.9899
-.001	.999
-1E-4	.9999
1E-4	1.0001
.001	1.001
.01	1.0099

Y1=(X+1)/(X²+1)

$\lim_{x \to 0} \left(\frac{x+1}{x^2+1} \right) = 1$

10. $\lim_{x \to 0} \left(\frac{2-x}{x^2+4} \right)$

X	Y1
-.01	.50249
-.001	.50025
-1E-4	.50002
1E-4	.49997
.001	.49975
.01	.49749

Y1=(2-X)/(X²+4)

$\lim_{x \to 0} \left(\frac{2-x}{x^2+4} \right) = \frac{1}{2}$

11. $\lim_{x \to 4} \left(\frac{x^2-4x}{x-4} \right)$

X	Y1
3.99	3.99
3.999	3.999
3.9999	3.9999
4.0001	4.0001
4.001	4.001
4.01	4.01

Y1=(X²-4X)/(X-4)

$\lim_{x \to 4} \left(\frac{x^2-4x}{x-4} \right) = 4$

12. $\lim\limits_{x\to 3}\left(\dfrac{x^2-9}{x^2-3x}\right)$

X	Y1	
2.99	2.0033	
2.999	2.0003	
2.9999	2	
3.0001	2	
3.001	1.9997	
3.01	1.9967	

Y1■(X²-9)/(X²-3...

$\lim\limits_{x\to 3}\left(\dfrac{x^2-9}{x^2-3x}\right)=2$

13. $\lim\limits_{x\to 0}\left(e^x+1\right)$

X	Y1	
-.01	1.99	
-.001	1.999	
-1E-4	1.9999	
1E-4	2.0001	
.001	2.001	
.01	2.0101	

Y1■e^(X)+1

$\lim\limits_{x\to 0}\left(e^x+1\right)=2$

14. $\lim\limits_{x\to 0}\left(\dfrac{e^x-e^{-x}}{2}\right)$

X	Y1	
-.01	-.01	
-.001	-.001	
-1E-4	-1E-4	
1E-4	1E-4	
.001	.001	
.01	.01	

Y1■(e^(X)-e^(-X...

$\lim\limits_{x\to 0}\dfrac{e^x-e^{-x}}{2}=0$

15. $\lim\limits_{x\to 0}\left(\dfrac{\cos x-1}{x}\right)$

X	Y1	
-.01	.005	
-.001	5E-4	
-1E-4	5E-5	
1E-4	-5E-5	
.001	-5E-4	
.01	-.005	

Y1■(cos(X)-1)/X

$\lim\limits_{x\to 0}\left(\dfrac{\cos x-1}{x}\right)=0$

16. $\lim\limits_{x\to 0}\left(\dfrac{\tan x}{x}\right)$

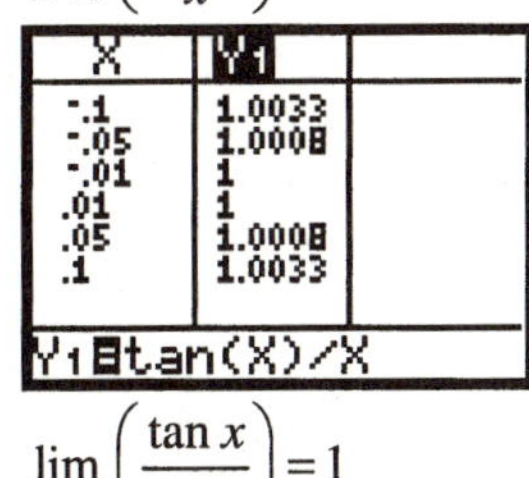

X	Y1	
-.1	1.0033	
-.05	1.0008	
-.01	1	
.01	1	
.05	1.0008	
.1	1.0033	

Y1■tan(X)/X

$\lim\limits_{x\to 0}\left(\dfrac{\tan x}{x}\right)=1$

17. $\lim\limits_{x\to 2} f(x)=3$

The value of the function gets close to 3 as x gets close to 2.

18. $\lim\limits_{x\to 4} f(x)=-3$

The value of the function gets close to -3 as x gets close to 4.

19. $\lim\limits_{x\to 2} f(x)=4$

The value of the function gets close to 4 as x gets close to 2.

20. $\lim\limits_{x\to 2} f(x)=2$

The value of the function gets close to 2 as x gets close to 2.

21. $\lim\limits_{x\to 3} f(x)$ does not exist because as x gets close to 3, but is less than 3, $f(x)$ gets close to 3. However, as x gets close to 3, but is greater than 3, $f(x)$ gets close to 6.

22. $\lim\limits_{x\to 4} f(x)$ does not exist because as x gets close to 4, but is less than 4, $f(x)$ gets close to 4. However, as x gets close to 4, but is greater than 4, $f(x)$ gets close to 2.

23. $f(x)=3x+1$

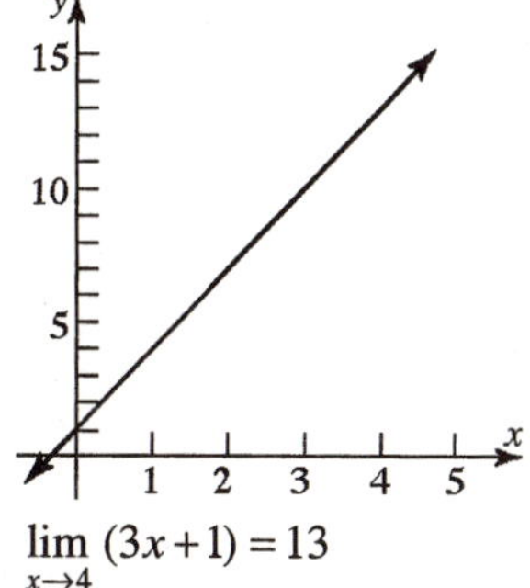

$\lim\limits_{x\to 4}(3x+1)=13$

The value of the function gets close to 13 as x gets close to 4.

24. $f(x)=2x-1$

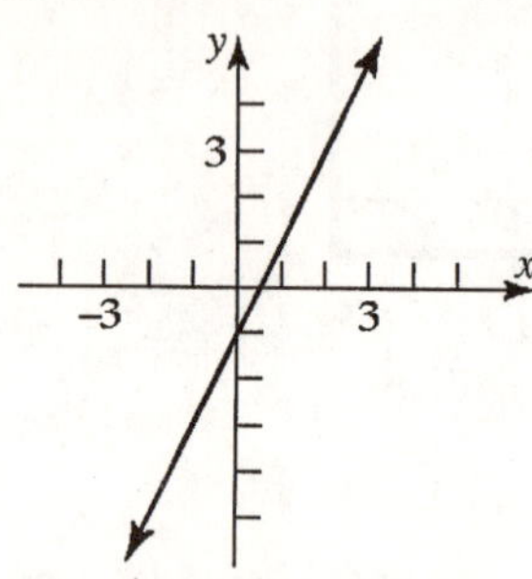

$\lim_{x\to -1}(2x-1)=-3$

The value of the function gets close to -3 as x gets close to -1.

25. $f(x)=1-x^2$

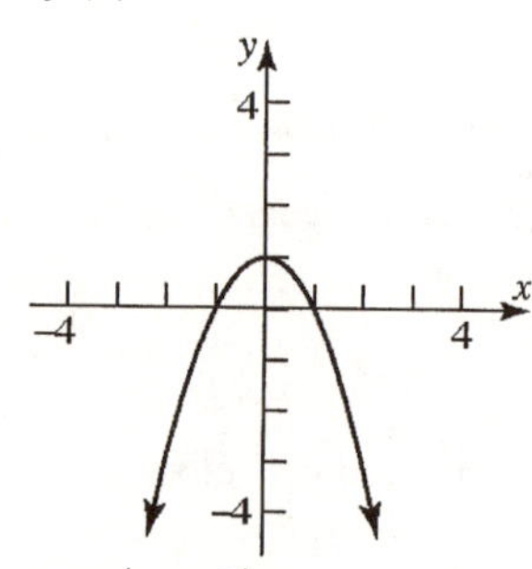

$\lim_{x\to 2}(1-x^2)=-3$

The value of the function gets close to -3 as x gets close to 2.

26. $f(x)=x^3-1$

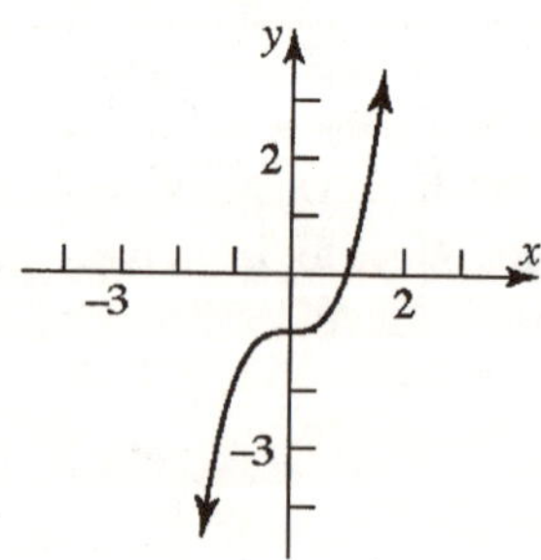

$\lim_{x\to -1}(x^3-1)=-2$

The value of the function gets close to -2 as x gets close to -1.

27. $f(x)=|2x|$

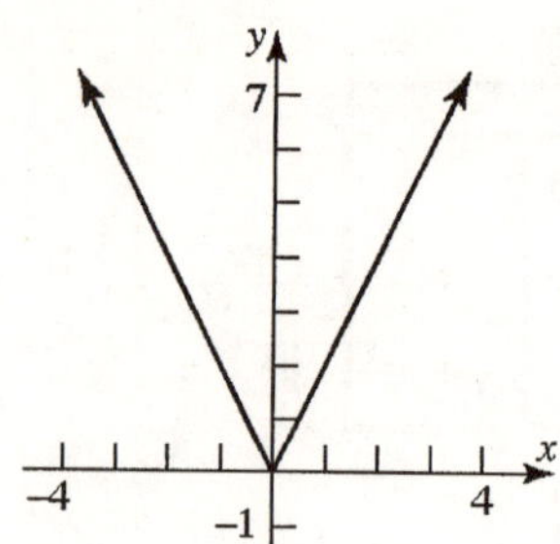

$\lim_{x\to -3}|2x|=6$

The value of the function gets close to 6 as x gets close to -3.

28. $f(x)=3\sqrt{x}$

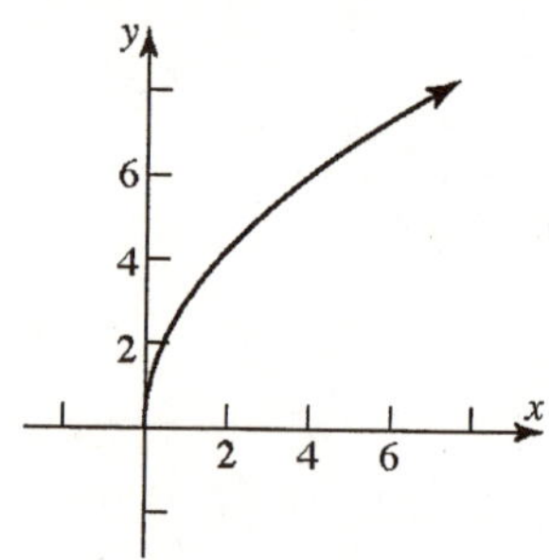

$\lim_{x\to 4}(3\sqrt{x})=6$

The value of the function gets close to 6 as x gets close to 4.

29. $f(x)=\sin x$

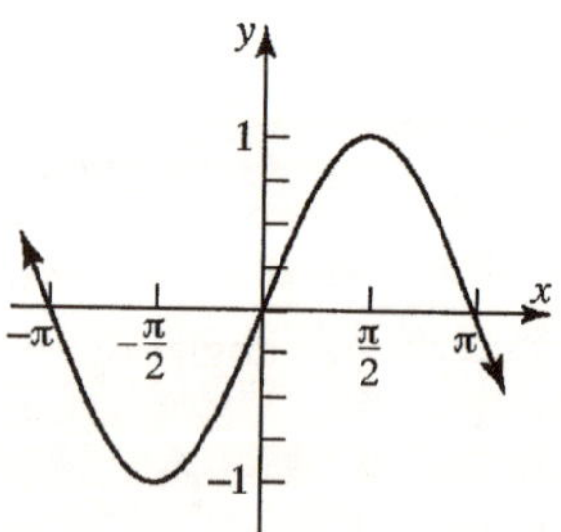

$\lim_{x\to \frac{\pi}{2}}(\sin x)=1$

The value of the function gets close to 1 as x gets close to $\frac{\pi}{2}$.

30. $f(x) = \cos x$

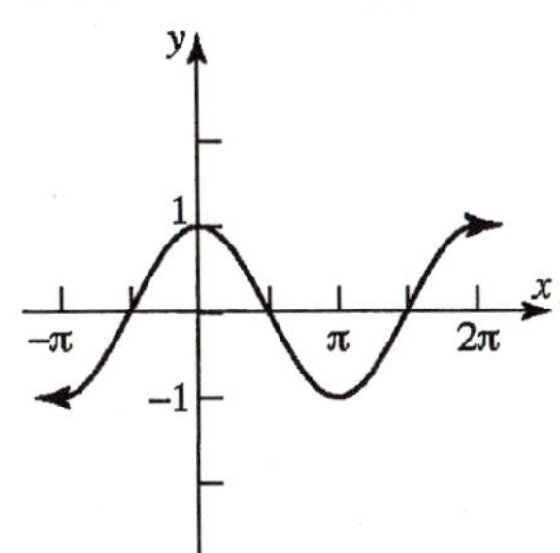

$\lim_{x \to \pi} (\cos x) = -1$

The value of the function gets close to -1 as x gets close to π.

31. $f(x) = e^x$

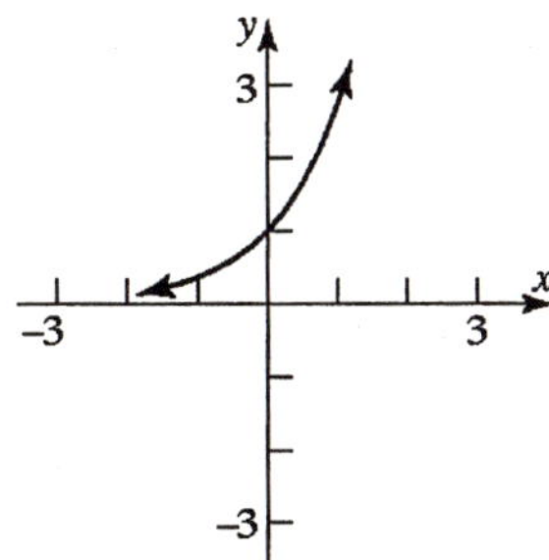

$\lim_{x \to 0} (e^x) = 1$

The value of the function gets close to 1 as x gets close to 0.

32. $f(x) = \ln x$

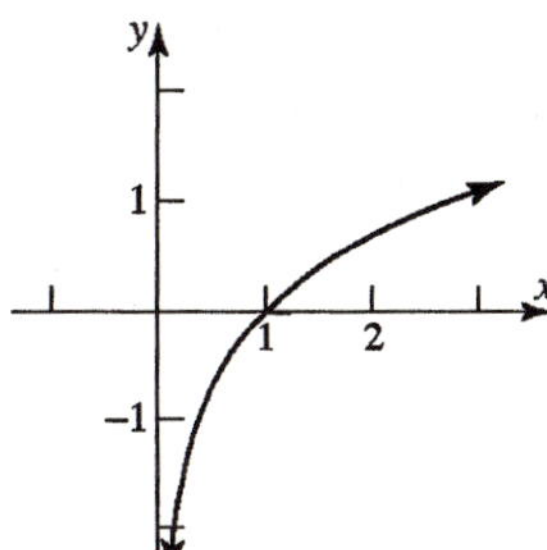

$\lim_{x \to 1} (\ln x) = 0$

The value of the function gets close to 0 as x gets close to 1.

33. $f(x) = \frac{1}{x}$

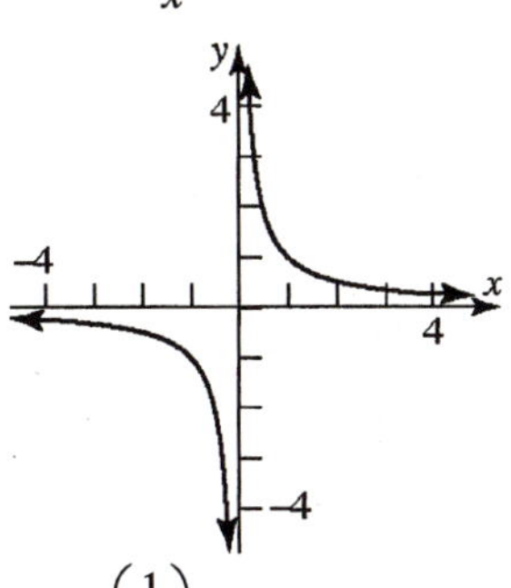

$\lim_{x \to -1} \left(\frac{1}{x}\right) = -1$

The value of the function gets close to -1 as x gets close to -1.

34. $f(x) = \frac{1}{x^2}$

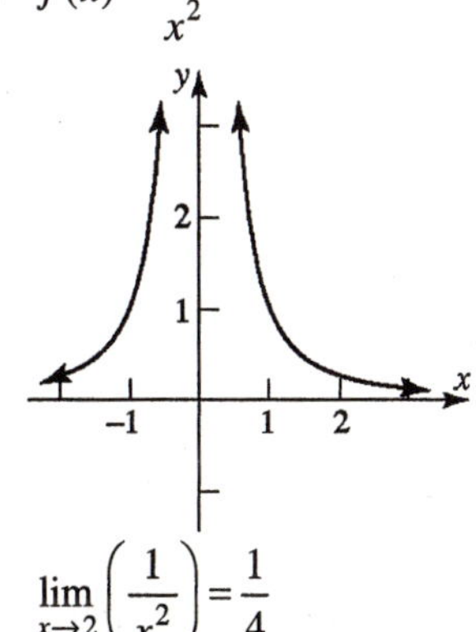

$\lim_{x \to 2} \left(\frac{1}{x^2}\right) = \frac{1}{4}$

The value of the function gets close to $\frac{1}{4}$ as x gets close to 2.

35. $f(x) = \begin{cases} x^2 & x \ge 0 \\ 2x & x < 0 \end{cases}$

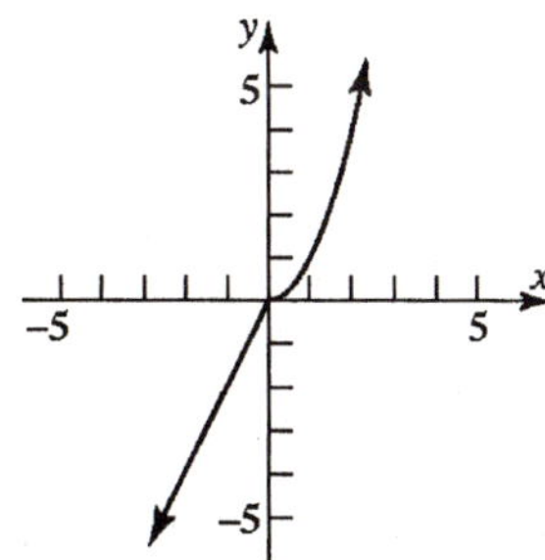

$\lim_{x \to 0} f(x) = 0$

The value of the function gets close to 0 as x gets close to 0.

36. $f(x)=\begin{cases} x-1 & x<0 \\ 3x-1 & x\geq 0 \end{cases}$

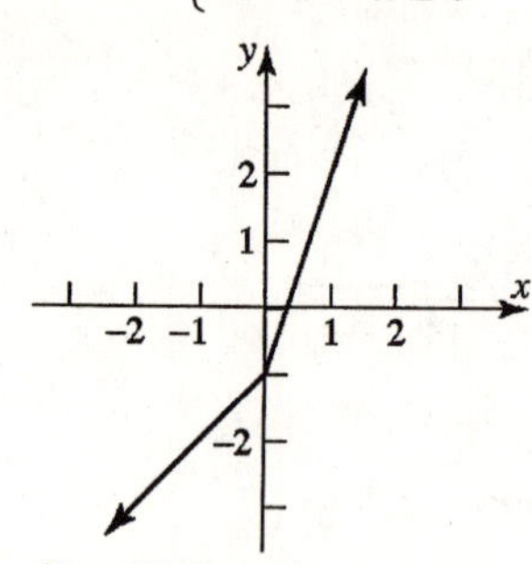

$\lim_{x\to 0} f(x) = -1$

The value of the function gets close to -1 as x gets close to 0.

37. $f(x)=\begin{cases} 3x & x\leq 1 \\ x+1 & x>1 \end{cases}$

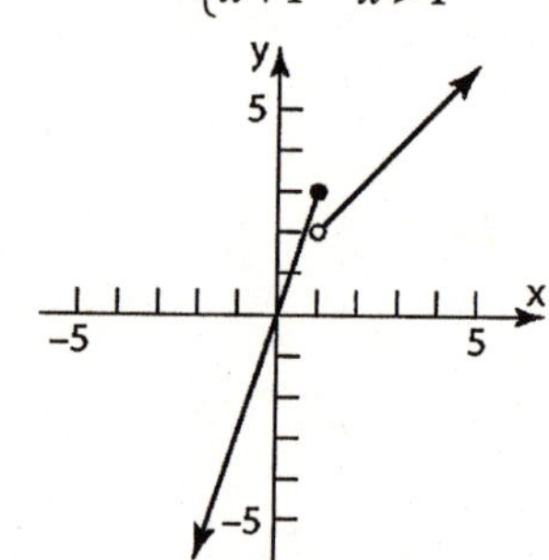

$\lim_{x\to 1} f(x)$ does not exist

The value of the function does not approach a single value as x approaches 1. For $x<1$, the function approaches the value 3, while for $x>1$ the function approaches the value 2.

38. $f(x)=\begin{cases} x^2 & x\leq 2 \\ 2x-1 & x>2 \end{cases}$

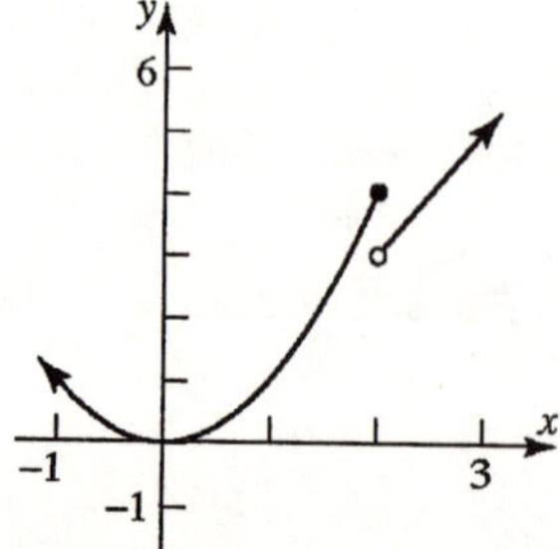

$\lim_{x\to 2} f(x)$ does not exist

The value of the function does not approach a single value as x approaches 2. For $x<2$, the function approaches the value 4, while for $x>2$ the function approaches the value 3.

39. $f(x)=\begin{cases} x & x<0 \\ 1 & x=0 \\ 3x & x>0 \end{cases}$

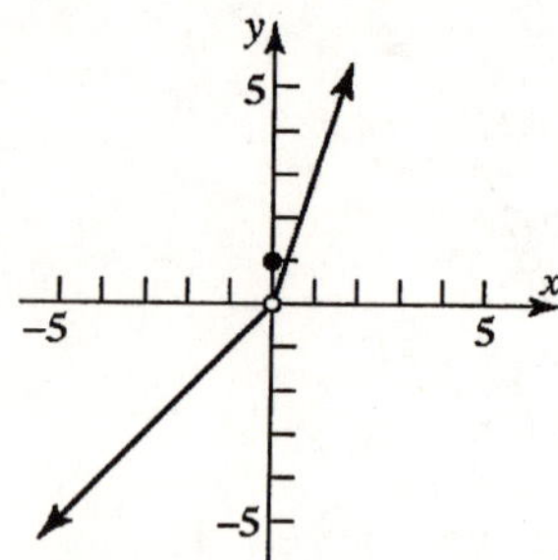

$\lim_{x\to 0} f(x) = 0$

The value of the function gets close to 0 as x gets close to 0.

40. $f(x)=\begin{cases} 1 & x<0 \\ -1 & x>0 \end{cases}$

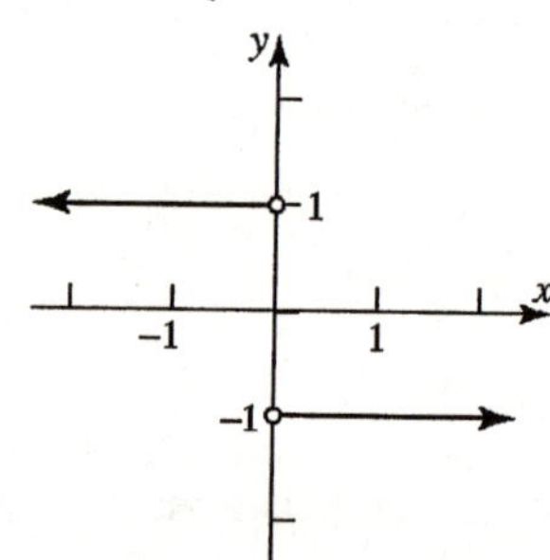

$\lim_{x\to 0} f(x)$ does not exist

The value of the function does not approach a single value as x approaches 0. For $x<0$, the function value approaches 1, while for $x>0$ the function value approaches -1.

41. $f(x)=\begin{cases} \sin x & x\leq 0 \\ x^2 & x>0 \end{cases}$

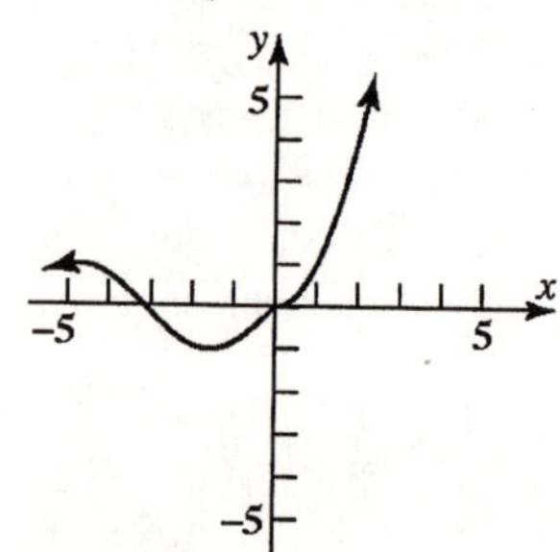

$\lim_{x\to 0} f(x) = 0$

The value of the function gets close to 0 as x gets close to 0.

42. $f(x)=\begin{cases} e^x & x>0 \\ 1-x & x\le 0 \end{cases}$

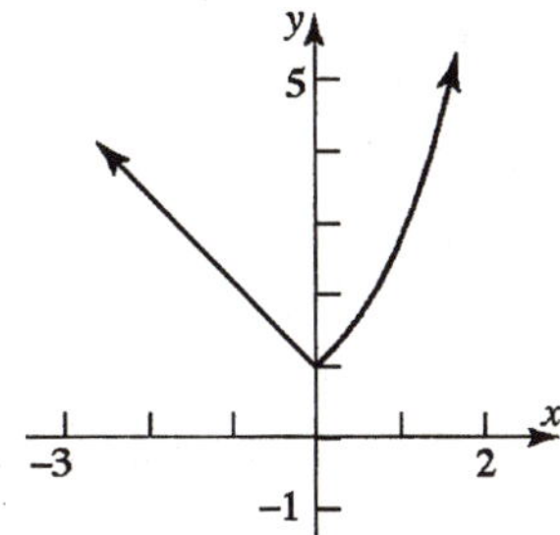

$\lim_{x\to 0} f(x)=1$

The value of the function gets close to 1 as x gets close to 0.

43. $\lim_{x\to 1}\left(\dfrac{x^3-x^2+x-1}{x^4-x^3+2x-2}\right)$

X	Y1
.99	.66663
.999	.66667
.9999	.66667
1	ERROR
1.0001	.66667
1.001	.66667
1.01	.66663

Y1=(X^3–X²+X–1)...

$\lim_{x\to 1}\left(\dfrac{x^3-x^2+x-1}{x^4-x^3+2x-2}\right)\approx 0.67$

44. $\lim_{x\to -1}\left(\dfrac{x^3+x^2+3x+3}{x^4+x^3+2x+2}\right)$

X	Y1
-1.1	6.293
-1.01	4.1457
-1.001	4.0141
-.999	3.9861
-.99	3.8653
-.9	2.9976

Y1=(X^3+X²+3X+3...

$\lim_{x\to -1}\left(\dfrac{x^3+x^2+3x+3}{x^4+x^3+2x+2}\right)=4.00$

45. $\lim_{x\to 2}\left(\dfrac{x^3-2x^2+4x-8}{x^2+x-6}\right)$

X	Y1
1.99	1.5952
1.999	1.5995
1.9999	1.6
2	ERROR
2.0001	1.6
2.001	1.6005
2.01	1.6048

Y1=(X^3–2X²+4X–...

$\lim_{x\to 2}\left(\dfrac{x^3-2x^2+4x-8}{x^2+x-6}\right)=1.60$

46. $\lim_{x\to 1}\left(\dfrac{x^3-x^2+3x-3}{x^2+3x-4}\right)$

X	Y1
.9	.77755
.99	.79762
.999	.79976
1.001	.80024
1.01	.80242
1.1	.82549

Y1=(X^3–X²+3X–3...

$\lim_{x\to 1}\left(\dfrac{x^3-x^2+3x-3}{x^2+3x-4}\right)=0.80$

47. $\lim_{x\to -1}\left(\dfrac{x^3+2x^2+x}{x^4+x^3+2x+2}\right)$

X	Y1
-1.01	.01042
-1.001	.001
-1	ERROR
-.9999	-1E-4
-.999	-1E-3
-.99	-.0096

Y1=(X^3+2X²+X)/...

$\lim_{x\to -1}\left(\dfrac{x^3+2x^2+x}{x^4+x^3+2x+2}\right)=0.00$

48. $\lim_{x\to 3}\left(\dfrac{x^3-3x^2+4x-12}{x^4-3x^3+x-3}\right)$

X	Y1
2.99	.46663
2.999	.46452
2.9999	.46431
3.0001	.46426
3.001	.46405
3.01	.46196
3.1	.44201

Y1=(X^3–3X²+4X–...

$\lim_{x\to 3}\left(\dfrac{x^3-3x^2+4x-12}{x^4-3x^3+x-3}\right)\approx 0.46$

Section 13.2

1. product
2. 5
3. 1
4. True
5. False; the function may not be defined at 5.
6. False; if the limit of the denominator equals 0, then the quotient rule for limits does not apply.

7. $\lim_{x\to 1}(5)=5$

8. $\lim_{x\to 1}(-3)=-3$

9. $\lim_{x\to 4}(x)=4$

10. $\lim_{x\to -3}(x)=-3$

11. $\lim_{x\to 2}(3x+2)=3(2)+2=8$

12. $\lim_{x\to 3}(2-5x)=2-5(3)=-13$

13. $\lim_{x\to -1}(3x^2-5x)=3(-1)^2-5(-1)=8$

14. $\lim_{x\to 2}(8x^2-4)=8(2)^2-4=28$

15. $\lim_{x\to 1}(5x^4-3x^2+6x-9)=5(1)^4-3(1)^2+6(1)-9$
$$=5-3+6-9$$
$$=-1$$

16. $\lim_{x\to -1}(8x^5-7x^3+8x^2+x-4)$
$$=8(-1)^5-7(-1)^3+8(-1)^2+(-1)-4$$
$$=-8+7+8-1-4=2$$

17. $\lim_{x\to 1}(x^2+1)^3=\left(\lim_{x\to 1}(x^2+1)\right)^3=\left(1^2+1\right)^3=2^3=8$

18. $\lim_{x\to 2}(3x-4)^2=\left(\lim_{x\to 2}(3x-4)\right)^2$
$$=\left(3(2)-4\right)^2$$
$$=2^2=4$$

19. $\lim_{x\to 1}\sqrt{5x+4}=\sqrt{\lim_{x\to 1}(5x+4)}$
$$=\sqrt{5(1)+4}$$
$$=\sqrt{9}=3$$

20. $\lim_{x\to 0}\sqrt{1-2x}=\sqrt{\lim_{x\to 0}(1-2x)}=\sqrt{1-2(0)}=\sqrt{1}=1$

21. $\lim_{x\to 0}\left(\dfrac{x^2-4}{x^2+4}\right)=\dfrac{\lim_{x\to 0}(x^2-4)}{\lim_{x\to 0}(x^2+4)}=\dfrac{0^2-4}{0^2+4}=\dfrac{-4}{4}=-1$

22. $\lim_{x\to 2}\left(\dfrac{3x+4}{x^2+x}\right)=\dfrac{\lim_{x\to 2}(3x+4)}{\lim_{x\to 2}(x^2+x)}=\dfrac{3(2)+4}{2^2+2}=\dfrac{10}{6}=\dfrac{5}{3}$

23. $\lim_{x\to 2}(3x-2)^{5/2}=\left(\lim_{x\to 2}(3x-2)\right)^{5/2}$
$$=\left(3(2)-2\right)^{5/2}$$
$$=4^{5/2}=32$$

24. $\lim_{x\to -1}(2x+1)^{5/3}=\left(\lim_{x\to -1}(2x+1)\right)^{5/3}$
$$=\left(2(-1)+1\right)^{5/3}$$
$$=(-1)^{5/3}=-1$$

25. $\lim_{x\to 2}\left(\dfrac{x^2-4}{x^2-2x}\right)=\lim_{x\to 2}\left(\dfrac{(x-2)(x+2)}{x(x-2)}\right)$
$$=\lim_{x\to 2}\left(\frac{x+2}{x}\right)$$
$$=\frac{2+2}{2}=\frac{4}{2}=2$$

26. $\lim_{x\to -1}\left(\dfrac{x^2+x}{x^2-1}\right)=\lim_{x\to -1}\left(\dfrac{x(x+1)}{(x+1)(x-1)}\right)$
$$=\lim_{x\to -1}\left(\frac{x}{x-1}\right)$$
$$=\frac{-1}{-1-1}=\frac{-1}{-2}=\frac{1}{2}$$

27. $\lim_{x\to -3}\left(\dfrac{x^2-x-12}{x^2-9}\right)=\lim_{x\to -3}\left(\dfrac{(x-4)(x+3)}{(x-3)(x+3)}\right)$
$$=\lim_{x\to -3}\left(\frac{x-4}{x-3}\right)$$
$$=\frac{-3-4}{-3-3}=\frac{-7}{-6}=\frac{7}{6}$$

28. $\lim_{x\to -3}\left(\dfrac{x^2+x-6}{x^2+2x-3}\right)=\lim_{x\to -3}\left(\dfrac{(x-2)(x+3)}{(x-1)(x+3)}\right)$
$$=\lim_{x\to -3}\left(\frac{x-2}{x-1}\right)$$
$$=\frac{-3-2}{-3-1}=\frac{-5}{-4}=\frac{5}{4}$$

29. $\lim\limits_{x\to 1}\left(\dfrac{x^3-1}{x-1}\right)=\lim\limits_{x\to 1}\left(\dfrac{(x-1)(x^2+x+1)}{x-1}\right)$
$=\lim\limits_{x\to 1}\left(x^2+x+1\right)$
$=1^2+1+1=3$

30. $\lim\limits_{x\to 1}\left(\dfrac{x^4-1}{x-1}\right)=\lim\limits_{x\to 1}\left(\dfrac{(x^2-1)(x^2+1)}{x-1}\right)$
$=\lim\limits_{x\to 1}\left(\dfrac{(x-1)(x+1)(x^2+1)}{x-1}\right)$
$=\lim\limits_{x\to 1}\left((x+1)(x^2+1)\right)$
$=(1+1)(1^2+1)=2(2)=4$

31. $\lim\limits_{x\to -1}\left(\dfrac{(x+1)^2}{x^2-1}\right)=\lim\limits_{x\to -1}\left(\dfrac{(x+1)^2}{(x-1)(x+1)}\right)$
$=\lim\limits_{x\to -1}\left(\dfrac{x+1}{x-1}\right)$
$=\dfrac{-1+1}{-1-1}=\dfrac{0}{-2}=0$

32. $\lim\limits_{x\to 2}\left(\dfrac{x^3-8}{x^2-4}\right)=\lim\limits_{x\to 2}\left(\dfrac{(x-2)(x^2+2x+4)}{(x-2)(x+2)}\right)$
$=\lim\limits_{x\to 2}\left(\dfrac{x^2+2x+4}{x+2}\right)$
$=\dfrac{2^2+2(2)+4}{2+2}=\dfrac{12}{4}=3$

33. $\lim\limits_{x\to 1}\left(\dfrac{x^3-x^2+x-1}{x^4-x^3+2x-2}\right)$
$=\lim\limits_{x\to 1}\left(\dfrac{x^2(x-1)+1(x-1)}{x^3(x-1)+2(x-1)}\right)$
$=\lim\limits_{x\to 1}\left(\dfrac{(x-1)(x^2+1)}{(x-1)(x^3+2)}\right)=\lim\limits_{x\to 1}\left(\dfrac{x^2+1}{x^3+2}\right)$
$=\dfrac{1^2+1}{1^3+2}=\dfrac{2}{3}$

34. $\lim\limits_{x\to -1}\left(\dfrac{x^3+x^2+3x+3}{x^4+x^3+2x+2}\right)$
$=\lim\limits_{x\to -1}\left(\dfrac{x^2(x+1)+3(x+1)}{x^3(x+1)+2(x+1)}\right)$
$=\lim\limits_{x\to -1}\left(\dfrac{(x+1)(x^2+3)}{(x+1)(x^3+2)}\right)=\lim\limits_{x\to -1}\left(\dfrac{x^2+3}{x^3+2}\right)$
$=\dfrac{(-1)^2+3}{(-1)^3+2}=\dfrac{4}{1}=4$

35. $\lim\limits_{x\to 2}\left(\dfrac{x^3-2x^2+4x-8}{x^2+x-6}\right)$
$=\lim\limits_{x\to 2}\left(\dfrac{x^2(x-2)+4(x-2)}{(x+3)(x-2)}\right)$
$=\lim\limits_{x\to 2}\left(\dfrac{(x-2)(x^2+4)}{(x+3)(x-2)}\right)=\lim\limits_{x\to 2}\left(\dfrac{x^2+4}{x+3}\right)$
$=\dfrac{2^2+4}{2+3}=\dfrac{8}{5}$

36. $\lim\limits_{x\to 1}\left(\dfrac{x^3-x^2+3x-3}{x^2+3x-4}\right)$
$=\lim\limits_{x\to 1}\left(\dfrac{x^2(x-1)+3(x-1)}{(x+4)(x-1)}\right)$
$=\lim\limits_{x\to 1}\left(\dfrac{(x-1)(x^2+3)}{(x+4)(x-1)}\right)=\lim\limits_{x\to 1}\left(\dfrac{x^2+3}{x+4}\right)$
$=\dfrac{1^2+3}{1+4}=\dfrac{4}{5}$

37.
$\lim\limits_{x\to -1}\left(\dfrac{x^3+2x^2+x}{x^4+x^3+2x+2}\right)$
$=\lim\limits_{x\to -1}\left(\dfrac{x(x^2+2x+1)}{x^3(x+1)+2(x+1)}\right)$
$=\lim\limits_{x\to -1}\left(\dfrac{x(x+1)^2}{(x+1)(x^3+2)}\right)$
$=\lim\limits_{x\to -1}\left(\dfrac{x(x+1)}{x^3+2}\right)=\dfrac{-1(-1+1)}{(-1)^3+2}$
$=\dfrac{-1(0)}{-1+2}=\dfrac{0}{1}=0$

38.

$$\lim_{x\to3}\left(\frac{x^3-3x^2+4x-12}{x^4-3x^3+x-3}\right)$$
$$=\lim_{x\to3}\left(\frac{x^2(x-3)+4(x-3)}{x^3(x-3)+1(x-3)}\right)$$
$$=\lim_{x\to3}\left(\frac{(x-3)(x^2+4)}{(x-3)(x^3+1)}\right)$$
$$=\lim_{x\to3}\left(\frac{x^2+4}{x^3+1}\right)=\frac{3^2+4}{3^3+1}$$
$$=\frac{9+4}{27+1}=\frac{13}{28}$$

39. $\lim_{x\to2}\left(\frac{f(x)-f(2)}{x-2}\right)=\lim_{x\to2}\left(\frac{(5x-3)-7}{x-2}\right)$
$$=\lim_{x\to2}\left(\frac{5x-10}{x-2}\right)$$
$$=\lim_{x\to2}\left(\frac{5(x-2)}{x-2}\right)$$
$$=\lim_{x\to2}(5)=5$$

40. $\lim_{x\to-2}\left(\frac{f(x)-f(-2)}{x-(-2)}\right)=\lim_{x\to-2}\left(\frac{(4-3x)-10}{x+2}\right)$
$$=\lim_{x\to-2}\left(\frac{-3x-6}{x+2}\right)$$
$$=\lim_{x\to-2}\left(\frac{-3(x+2)}{x+2}\right)$$
$$=\lim_{x\to-2}(-3)=-3$$

41. $\lim_{x\to3}\left(\frac{f(x)-f(3)}{x-3}\right)=\lim_{x\to3}\left(\frac{x^2-9}{x-3}\right)$
$$=\lim_{x\to3}\left(\frac{(x-3)(x+3)}{x-3}\right)$$
$$=\lim_{x\to3}(x+3)$$
$$=3+3=6$$

42. $\lim_{x\to3}\left(\frac{f(x)-f(3)}{x-3}\right)=\lim_{x\to3}\left(\frac{x^3-27}{x-3}\right)$
$$=\lim_{x\to3}\left(\frac{(x-3)(x^2+3x+9)}{x-3}\right)$$
$$=\lim_{x\to3}\left(x^2+3x+9\right)$$
$$=3^2+3(3)+9=27$$

43. $\lim_{x\to-1}\left(\frac{f(x)-f(-1)}{x-(-1)}\right)=\lim_{x\to-1}\left(\frac{x^2+2x-(-1)}{x+1}\right)$
$$=\lim_{x\to-1}\left(\frac{x^2+2x+1}{x+1}\right)$$
$$=\lim_{x\to-1}\left(\frac{(x+1)^2}{x+1}\right)$$
$$=\lim_{x\to-1}(x+1)$$
$$=-1+1=0$$

44. $\lim_{x\to-1}\left(\frac{f(x)-f(-1)}{x-(-1)}\right)=\lim_{x\to-1}\left(\frac{2x^2-3x-5}{x+1}\right)$
$$=\lim_{x\to-1}\left(\frac{(2x-5)(x+1)}{x+1}\right)$$
$$=\lim_{x\to-1}(2x-5)$$
$$=2(-1)-5=-7$$

45. $\lim_{x\to0}\left(\frac{f(x)-f(0)}{x-0}\right)=\lim_{x\to0}\left(\frac{3x^3-2x^2+4-4}{x}\right)$
$$=\lim_{x\to0}\left(\frac{3x^3-2x^2}{x}\right)$$
$$=\lim_{x\to0}(3x^2-2x)=0$$

46. $\lim_{x\to0}\left(\frac{f(x)-f(0)}{x-0}\right)=\lim_{x\to0}\left(\frac{4x^3-5x+8-8}{x}\right)$
$$=\lim_{x\to0}\left(\frac{4x^3-5x}{x}\right)$$
$$=\lim_{x\to0}(4x^2-5)=-5$$

47. $\lim_{x\to1}\left(\frac{f(x)-f(1)}{x-1}\right)=\lim_{x\to1}\left(\frac{\frac{1}{x}-1}{x-1}\right)=\lim_{x\to1}\left(\frac{\frac{1-x}{x}}{x-1}\right)$

$=\lim_{x\to1}\left(\frac{-1(x-1)}{x(x-1)}\right)=\lim_{x\to1}\left(\frac{-1}{x}\right)$

$=\frac{-1}{1}=-1$

48. $\lim_{x\to1}\left(\frac{f(x)-f(1)}{x-1}\right)=\lim_{x\to1}\left(\frac{\frac{1}{x^2}-1}{x-1}\right)=\lim_{x\to1}\left(\frac{\frac{1-x^2}{x^2}}{x-1}\right)$

$=\lim_{x\to1}\left(\frac{-1(x-1)(x+1)}{x^2(x-1)}\right)$

$=\lim_{x\to1}\left(\frac{-1(x+1)}{x^2}\right)=\frac{-2}{1}=-2$

49. $\lim_{x\to0}\left(\frac{\tan x}{x}\right)=\lim_{x\to0}\left(\frac{\frac{\sin x}{\cos x}}{x}\right)=\lim_{x\to0}\left(\frac{\sin x}{x}\cdot\frac{1}{\cos x}\right)$

$=\left(\lim_{x\to0}\left(\frac{\sin x}{x}\right)\right)\cdot\left(\lim_{x\to0}\left(\frac{1}{\cos x}\right)\right)$

$=1\cdot\left(\frac{\lim_{x\to0}1}{\lim_{x\to0}(\cos x)}\right)=1\cdot\frac{1}{1}=1$

50. $\lim_{x\to0}\left(\frac{\sin(2x)}{x}\right)=\lim_{x\to0}\left(\frac{2\sin x\cos x}{x}\right)$

$=2\lim_{x\to0}\left(\left(\frac{\sin x}{x}\right)\cdot\cos x\right)$

$=2\cdot1\cdot1=2$

51. $\lim_{x\to0}\left(\frac{3\sin x+\cos x-1}{4x}\right)$

$=\lim_{x\to0}\left(\frac{3\sin x}{4x}+\frac{\cos x-1}{4x}\right)$

$=\lim_{x\to0}\left(\frac{3\sin x}{4x}\right)+\lim_{x\to0}\left(\frac{\cos x-1}{4x}\right)$

$=\frac{3}{4}\lim_{x\to0}\left(\frac{\sin x}{x}\right)+\frac{1}{4}\lim_{x\to0}\left(\frac{\cos x-1}{x}\right)$

$=\frac{3}{4}\cdot1+\frac{1}{4}\cdot0=\frac{3}{4}$

52. $\lim_{x\to0}\left(\frac{\sin^2 x+\sin x(\cos x-1)}{x^2}\right)$

$=\lim_{x\to0}\left(\frac{\sin^2 x}{x^2}+\frac{\sin x(\cos x-1)}{x^2}\right)$

$=\lim_{x\to0}\left(\left(\frac{\sin x}{x}\right)\cdot\left(\frac{\sin x}{x}\right)\right)+\lim_{x\to0}\left(\left(\frac{\sin x}{x}\right)\cdot\left(\frac{\cos x-1}{x}\right)\right)$

$=\left(\lim_{x\to0}\left(\frac{\sin x}{x}\right)\right)\cdot\left(\lim_{x\to0}\left(\frac{\sin x}{x}\right)\right)$

$+\left(\lim_{x\to0}\left(\frac{\sin x}{x}\right)\right)\cdot\left(\lim_{x\to0}\left(\frac{\cos x-1}{x}\right)\right)$

$=1\cdot1+1\cdot0=1$

Section 13.3

1. $f(0)=0^2=0$, $f(2)=5-2=3$

2. $f(x)=\ln x$
   Domain: $\{x|x>0\}$
   Range: $\{y|-\infty<y<\infty\}$

3. True

4. Secant, cosecant, tangent, cotangent

5. Truc

6. True

7. one-sided

8. $\lim_{x\to c^+} f(x)=R$

9. continuous, c

10. False; a function can have different one-sided limits at a point c. For example, $f(x)=\frac{|x|}{x}$ has different one-sided limits (-1 from the left and 1 from the right).

11. True

12. True

13. Domain: $[-8,6)$ or $(-6,4)$ or $(4,6]$

14. Range: $(-\infty,\infty)$

15. x-intercepts: –8, –5, –3

16. y-intercept: 3

17. $f(-8)=0;\ f(-4)=2$

18. $f(2)=3;\ f(6)=2$

19. $\lim_{x\to -6^-} f(x)=\infty$

20. $\lim_{x\to -6^+} f(x)=-\infty$

21. $\lim_{x\to -4^-} f(x)=2$

22. $\lim_{x\to -4^+} f(x)=-2$

23. $\lim_{x\to 2^-} f(x)=1$

24. $\lim_{x\to 2^+} f(x)=-4$

25. $\lim_{x\to 4} f(x)$ does exist.

$\lim_{x\to 4} f(x)=0$ since $\lim_{x\to 4^-} f(x)=\lim_{x\to 4^+} f(x)=0$

26. $\lim_{x\to 0} f(x)$ does exist.

$\lim_{x\to 0} f(x)=3$ since $\lim_{x\to 0^-} f(x)=\lim_{x\to 0^+} f(x)=3$

27. f is not continuous at –6 because $f(-6)$ does not exist, nor does the $\lim_{x\to -6} f(x)$.

28. f is not continuous at –4 because $\lim_{x\to -4^-} f(x)\neq \lim_{x\to -4^+} f(x)$

29. f is continuous at 0 because $f(0)=\lim_{x\to 0^-} f(x)=\lim_{x\to 0^+} f(x)=3$

30. f is not continuous at 2 because $\lim_{x\to 2^-} f(x)\neq \lim_{x\to 2^+} f(x)$

31. f is not continuous at 4 because $f(4)$ does not exist.

32. f is continuous at 5 because $f(5)=\lim_{x\to 5^-} f(x)=\lim_{x\to 5^+} f(x)=1$

33. $\lim_{x\to 1^+}(2x+3)=2(1)+3=5$

34. $\lim_{x\to 2^-}(4-2x)=4-2(2)=0$

35. $\lim_{x\to 1^-}\left(2x^3+5x\right)=2(1)^3+5(1)=2+5=7$

36. $\lim_{x\to -2^+}\left(3x^2-8\right)=3(-2)^2-8=12-8=4$

37. $\lim_{x\to \frac{\pi}{2}^+}\left(\sin x\right)=\sin\frac{\pi}{2}=1$

38. $\lim_{x\to \pi^-}(3\cos x)=3\cos\pi=3(-1)=-3$

39.
$$\lim_{x\to 2^+}\left(\frac{x^2-4}{x-2}\right)=\lim_{x\to 2^+}\left(\frac{(x+2)(x-2)}{x-2}\right)=\lim_{x\to 2^+}(x+2)=2+2=4$$

40.
$$\lim_{x\to 1^-}\left(\frac{x^3-x}{x-1}\right)=\lim_{x\to 1^-}\left(\frac{x(x+1)(x-1)}{x-1}\right)=\lim_{x\to 1^-}\left(x(x+1)\right)=1(1+1)=1(2)=2$$

41.
$$\lim_{x\to -1^-}\left(\frac{x^2-1}{x^3+1}\right)=\lim_{x\to -1^-}\left(\frac{(x+1)(x-1)}{(x+1)(x^2-x+1)}\right)=\lim_{x\to -1^-}\left(\frac{x-1}{x^2-x+1}\right)=\frac{-1-1}{(-1)^2-(-1)+1}=-\frac{2}{3}$$

42.
$$\lim_{x\to 0^+}\left(\frac{x^3-x^2}{x^4+x^2}\right)=\lim_{x\to 0^+}\left(\frac{x^2(x-1)}{x^2(x^2+1)}\right)=\lim_{x\to 0^+}\left(\frac{x-1}{x^2+1}\right)=\frac{0-1}{0^2+1}=\frac{-1}{1}=-1$$

43. $\lim_{x\to-2^+}\left(\frac{x^2+x-2}{x^2+2x}\right)=\lim_{x\to-2^+}\left(\frac{(x+2)(x-1)}{x(x+2)}\right)$
$=\lim_{x\to-2^+}\left(\frac{x-1}{x}\right)$
$=\frac{-2-1}{-2}=\frac{-3}{-2}=\frac{3}{2}$

44. $\lim_{x\to-4^-}\left(\frac{x^2+x-12}{x^2+4x}\right)=\lim_{x\to-4^-}\left(\frac{(x+4)(x-3)}{x(x+4)}\right)$
$=\lim_{x\to-4^-}\left(\frac{x-3}{x}\right)$
$=\frac{-4-3}{-4}=\frac{-7}{-4}=\frac{7}{4}$

45. $f(x)=x^3-3x^2+2x-6;\ c=2$

1. $f(2)=2^3-3\cdot2^2+2\cdot2-6=-6$
2. $\lim_{x\to2^-}f(x)=2^3-3\cdot2^2+2\cdot2-6=-6$
3. $\lim_{x\to2^+}f(x)=2^3-3\cdot2^2+2\cdot2-6=-6$

Thus, $f(x)$ is continuous at $c=2$.

46. $f(x)=3x^2-6x+5;\ c=-3$

1. $f(-3)=3(-3)^2-6(-3)+5$
   $=27+18+5=50$
2. $\lim_{x\to-3^-}f(x)=3(-3)^2-6(-3)+5$
   $=27+18+5=50$
3. $\lim_{x\to-3^+}f(x)=3(-3)^2-6(-3)+5$
   $=27+18+5=50$

Thus, $f(x)$ is continuous at $c=-3$.

47. $f(x)=\frac{x^2+5}{x-6};\ c=3$

1. $f(3)=\frac{3^2+5}{3-6}=\frac{14}{-3}=-\frac{14}{3}$
2. $\lim_{x\to3^-}f(x)=\frac{3^2+5}{3-6}=\frac{14}{-3}=-\frac{14}{3}$
3. $\lim_{x\to3^+}f(x)=\frac{3^2+5}{3-6}=\frac{14}{-3}=-\frac{14}{3}$

Thus, $f(x)$ is continuous at $c=3$.

48. $f(x)=\frac{x^3-8}{x^2+4};\ c=2$

1. $f(2)=\frac{2^3-8}{2^2+4}=\frac{0}{8}=0$
2. $\lim_{x\to2^-}f(x)=\frac{2^3-8}{2^2+4}=\frac{0}{8}=0$
3. $\lim_{x\to2^+}f(x)=\frac{2^3-8}{2^2+4}=\frac{0}{8}=0$

Thus, $f(x)$ is continuous at $c=2$.

49. $f(x)=\frac{x+3}{x-3};\ c=3$

Since $f(x)$ is not defined at $c=3$, the function is not continuous at $c=3$.

50. $f(x)=\frac{x-6}{x+6};\ c=-6$

Since $f(x)$ is not defined at $c=-6$, the function is not continuous at $c=-6$.

51. $f(x)=\frac{x^3+3x}{x^2-3x};\ c=0$

Since $f(x)$ is not defined at $c=0$, the function is not continuous at $c=0$.

52. $f(x)=\frac{x^2-6x}{x^2+6x};\ c=0$

Since $f(x)$ is not defined at $c=0$, the function is not continuous at $c=0$.

53. $f(x)=\begin{cases}\frac{x^3+3x}{x^2-3x} & \text{if } x\neq0\\ 1 & \text{if } x=0\end{cases};\quad c=0$

1. $f(0)=1$
2. $\lim_{x\to0^-}f(x)=\lim_{x\to0^-}\left(\frac{x^3+3x}{x^2-3x}\right)$
   $=\lim_{x\to0^-}\left(\frac{x(x^2+3)}{x(x-3)}\right)$
   $=\lim_{x\to0^-}\left(\frac{x^2+3}{x-3}\right)=\frac{3}{-3}=-1$

Since $\lim_{x\to0^-}f(x)\neq f(0)$, the function is not continuous at $c=0$.

54. $f(x)=\begin{cases}\dfrac{x^2-6x}{x^2+6x} & \text{if } x\neq 0\\ -2 & \text{if } x=0\end{cases}$; $c=0$

1. $f(0)=-2$

2. $$\lim_{x\to 0^-} f(x)=\lim_{x\to 0^-}\left(\frac{x^2-6x}{x^2+6x}\right)=\lim_{x\to 0^-}\left(\frac{x(x-6)}{x(x+6)}\right)=\lim_{x\to 0^-}\left(\frac{x-6}{x+6}\right)=\frac{-6}{6}=-1$$

Since $\lim_{x\to 0^-} f(x)\neq f(0)$, the function is not continuous at $c=0$.

55. $f(x)=\begin{cases}\dfrac{x^3+3x}{x^2-3x} & \text{if } x\neq 0\\ -1 & \text{if } x=0\end{cases}$; $c=0$

1. $f(0)=-1$

2. $$\lim_{x\to 0^-} f(x)=\lim_{x\to 0^-}\left(\frac{x^3+3x}{x^2-3x}\right)=\lim_{x\to 0^-}\left(\frac{x(x^2+3)}{x(x-3)}\right)=\lim_{x\to 0^-}\left(\frac{x^2+3}{x-3}\right)=\frac{3}{-3}=-1$$

3. $$\lim_{x\to 0^+} f(x)=\lim_{x\to 0^+}\left(\frac{x^3+3x}{x^2-3x}\right)=\lim_{x\to 0^+}\left(\frac{x(x^2+3)}{x(x-3)}\right)=\lim_{x\to 0^+}\left(\frac{x^2+3}{x-3}\right)=\frac{3}{-3}=-1$$

The function is continuous at $c=0$.

56. $f(x)=\begin{cases}\dfrac{x^2-6x}{x^2+6x} & \text{if } x\neq 0\\ -1 & \text{if } x=0\end{cases}$; $c=0$

1. $f(0)=-1$

2. $$\lim_{x\to 0^-} f(x)=\lim_{x\to 0^-}\left(\frac{x^2-6x}{x^2+6x}\right)=\lim_{x\to 0^-}\left(\frac{x(x-6)}{x(x+6)}\right)=\lim_{x\to 0^-}\left(\frac{x-6}{x+6}\right)=\frac{-6}{6}=-1$$

3. $$\lim_{x\to 0^+} f(x)=\lim_{x\to 0^+}\left(\frac{x^2-6x}{x^2+6x}\right)=\lim_{x\to 0^+}\left(\frac{x(x-6)}{x(x+6)}\right)=\lim_{x\to 0^+}\left(\frac{x-6}{x+6}\right)=\frac{-6}{6}=-1$$

The function is continuous at $c=0$.

57. $f(x)=\begin{cases}\dfrac{x^3-1}{x^2-1} & \text{if } x<1\\ 2 & \text{if } x=1\\ \dfrac{3}{x+1} & \text{if } x>1\end{cases}$; $c=1$

1. $f(1)=2$

2. $$\lim_{x\to 1^-} f(x)=\lim_{x\to 1^-}\left(\frac{x^3-1}{x^2-1}\right)=\lim_{x\to 1^-}\left(\frac{(x-1)(x^2+x+1)}{(x-1)(x+1)}\right)=\lim_{x\to 1^-}\left(\frac{x^2+x+1}{x+1}\right)=\frac{3}{2}$$

Since $\lim_{x\to 1^-} f(x)\neq f(1)$, the function is not continuous at $c=1$.

58. $f(x)=\begin{cases}\dfrac{x^2-2x}{x-2} & \text{if } x<2\\ 2 & \text{if } x=2\\ \dfrac{x-4}{x-1} & \text{if } x>2\end{cases}$; $c=2$

1. $f(2)=2$

2. $$\lim_{x\to 2^-} f(x)=\lim_{x\to 2^-}\left(\frac{x^2-2x}{x-2}\right)=\lim_{x\to 2^-}\left(\frac{x(x-2)}{x-2}\right)=\lim_{x\to 2^-}(x)=2$$

3. $\lim_{x\to 2^+} f(x) = \lim_{x\to 2^+}\left(\frac{x-4}{x-1}\right)$
$= \frac{2-4}{2-1} = \frac{-2}{1} = -2$

Since $\lim_{x\to 2^+} f(x) \neq f(2)$, the function is not continuous at $c = 2$.

59. $f(x) = \begin{cases} 2e^x & \text{if } x<0 \\ 2 & \text{if } x=0 \\ \dfrac{x^3+2x^2}{x^2} & \text{if } x>0 \end{cases};\quad c=0$

1. $f(0) = 2$
2. $\lim_{x\to 0^-} f(x) = \lim_{x\to 0^-}\left(2e^x\right) = 2e^0 = 2\cdot 1 = 2$
3. $\lim_{x\to 0^+} f(x) = \lim_{x\to 0^+}\left(\frac{x^3+2x^2}{x^2}\right)$
$= \lim_{x\to 0^+}\left(\frac{x^2(x+2)}{x^2}\right)$
$= \lim_{x\to 0^+}(x+2) = 0+2 = 2$

The function is continuous at $c = 0$.

60. $f(x) = \begin{cases} 3\cos x & \text{if } x<0 \\ 3 & \text{if } x=0 \\ \dfrac{x^3+3x^2}{x^2} & \text{if } x>0 \end{cases};\quad c=0$

1. $f(0) = 3$
2. $\lim_{x\to 0^-} f(x) = \lim_{x\to 0^-}\left(3\cos(0)\right) = 3\cdot 1 = 3$
3. $\lim_{x\to 0^+} f(x) = \lim_{x\to 0^+}\left(\frac{x^3+3x^2}{x^2}\right)$
$= \lim_{x\to 0^+}\left(\frac{x^2(x+3)}{x^2}\right)$
$= \lim_{x\to 0^+}(x+3) = 0+3 = 3$

The function is continuous at $c = 0$.

61. The domain of $f(x) = 2x+3$ is all real numbers, and $f(x)$ is a polynomial function. Therefore, $f(x)$ is continuous everywhere.

62. The domain of $f(x) = 4-3x$ is all real numbers, and $f(x)$ is a polynomial function. Therefore, $f(x)$ is continuous everywhere.

63. The domain of $f(x) = 3x^2 + x$ is all real numbers, and $f(x)$ is a polynomial function. Therefore, $f(x)$ is continuous everywhere.

64. The domain of $f(x) = -3x^3 + 7$ is all real numbers, and $f(x)$ is a polynomial function. Therefore, $f(x)$ is continuous everywhere.

65. The domain of $f(x) = 4\sin x$ is all real numbers, and trigonometric functions are continuous at every point in their domains. Therefore, $f(x)$ is continuous everywhere.

66. The domain of $f(x) = -2\cos x$ is all real numbers, and trigonometric functions are continuous at every point in their domains. Therefore, $f(x)$ is continuous everywhere.

67. The domain of $f(x) = 2\tan x$ is all real numbers except odd integer multiples of $\frac{\pi}{2}$, and trigonometric functions are continuous at every point in their domains. Therefore, $f(x)$ is continuous everywhere except where $x = \frac{k\pi}{2}$ where k is an odd integer. $f(x)$ is discontinuous at $x = \frac{k\pi}{2}$ where k is an odd integer.

68. The domain of $f(x) = 4\csc x$ is all real numbers except integer multiples of π, and trigonometric functions are continuous at every point in their domains. Therefore, $f(x)$ is continuous everywhere except where $x = k\pi$ where k is an integer. $f(x)$ is discontinuous at $x = k\pi$ where k is an integer.

69. $f(x)=\dfrac{2x+5}{x^2-4}=\dfrac{2x+5}{(x-2)(x+2)}$. The domain of $f(x)$ is all real numbers except $x=2$ and $x=-2$, and $f(x)$ is a rational function. Therefore, $f(x)$ is continuous everywhere except at $x=2$ and $x=-2$. $f(x)$ is discontinuous at $x=2$ and $x=-2$.

70. $f(x)=\dfrac{x^2-4}{x^2-9}=\dfrac{(x-2)(x+2)}{(x-3)(x+3)}$. The domain of $f(x)$ is all real numbers except $x=3$ and $x=-3$, and $f(x)$ is a rational function. Therefore, $f(x)$ is continuous everywhere except at $x=3$ and $x=-3$. $f(x)$ is discontinuous at $x=3$ and $x=-3$.

71. $f(x)=\dfrac{x-3}{\ln x}$. The domain of $f(x)$ is $(0,1)$ or $(1,\infty)$. Thus, $f(x)$ is continuous on the interval $(0,\infty)$ except at $x=1$. $f(x)$ is discontinuous at $x=1$.

72. $f(x)=\dfrac{\ln x}{x-3}$. The domain of $f(x)$ is $(0,3)$ or $(3,\infty)$. Thus, $f(x)$ is continuous on the interval $(0,\infty)$ except at $x=3$. $f(x)$ is discontinuous at $x=3$.

73. $R(x)=\dfrac{x-1}{x^2-1}=\dfrac{x-1}{(x-1)(x+1)}$. The domain of R is $\{x \mid x\neq -1, x\neq 1\}$. Thus R is discontinuous at both –1 and 1.

$$\lim_{x\to -1^-} R(x)=\lim_{x\to -1^-}\left(\frac{x-1}{(x-1)(x+1)}\right)=\lim_{x\to -1^-}\left(\frac{1}{x+1}\right)=-\infty$$

since when $x<-1$, $\dfrac{1}{x+1}<0$, and as x approaches -1, $\dfrac{1}{x+1}$ becomes unbounded.

$\lim_{x\to -1^+} R(x)=\lim_{x\to -1^+}\left(\dfrac{1}{x+1}\right)=\infty$ since when $x>-1$, $\dfrac{1}{x+1}>0$, and as x approaches -1, $\dfrac{1}{x+1}$ becomes unbounded.

$\lim_{x\to 1} R(x)=\lim_{x\to 1}\left(\dfrac{1}{x+1}\right)=\dfrac{1}{2}$. Note there is a hole in the graph at $\left(1,\dfrac{1}{2}\right)$.

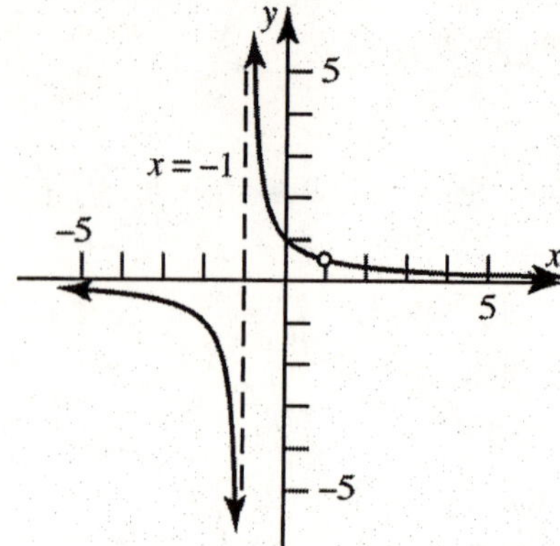

74. $R(x)=\dfrac{3x+6}{x^2-4}=\dfrac{3(x+2)}{(x-2)(x+2)}$. The domain of R is $\{x \mid x\neq -2, x\neq 2\}$. Thus R is discontinuous at both –2 and 2.

$$\lim_{x\to 2^-} R(x)=\lim_{x\to 2^-}\left(\frac{3(x+2)}{(x-2)(x+2)}\right)=\lim_{x\to 2^-}\left(\frac{3}{x-2}\right)=-\infty$$

since when $x<2$, $\dfrac{3}{x-2}<0$, and as x approaches 2, $\dfrac{3}{x-2}$ becomes unbounded.

$\lim_{x\to 2^+} R(x)=\lim_{x\to 2^+}\left(\dfrac{3}{x-2}\right)=\infty$ since when $x>2$, $\dfrac{3}{x-2}>0$, and as x approaches 2, $\dfrac{3}{x-2}$ becomes unbounded.

$\lim_{x\to -2} R(x)=\lim_{x\to -2}\left(\dfrac{3}{x-2}\right)=-\dfrac{3}{4}$. Note there is a hole in the graph at $\left(-2,-\dfrac{3}{4}\right)$.

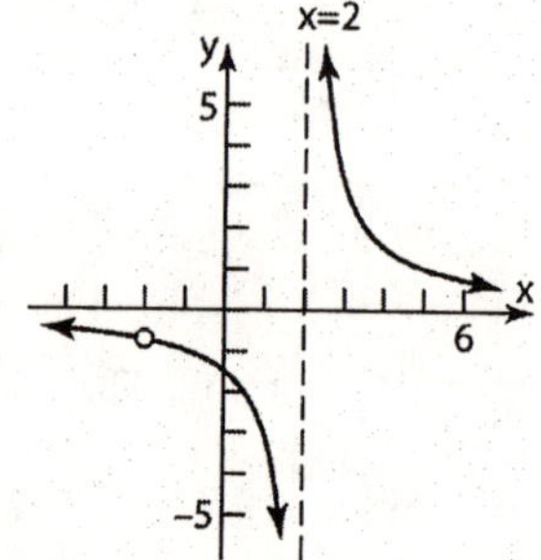

75. $R(x)=\dfrac{x^2+x}{x^2-1}=\dfrac{x(x+1)}{(x-1)(x+1)}$. The domain of R is $\{x \mid x\neq -1, x\neq 1\}$. Thus R is discontinuous at both –1 and 1.

$\lim\limits_{x\to 1^-} R(x)=\lim\limits_{x\to 1^-}\left(\dfrac{x(x+1)}{(x-1)(x+1)}\right)=\lim\limits_{x\to 1^-}\left(\dfrac{x}{x-1}\right)=-\infty$

since when

$0<x<1, \dfrac{x}{x-1}<0$, and as x approaches 1, $\dfrac{x}{x-1}$

becomes unbounded. $\lim\limits_{x\to 1^+} R(x)=\lim\limits_{x\to 1^+}\dfrac{x}{x-1}=\infty$

since when

$x>1, \dfrac{x}{x-1}>0$, and as x approaches 1, $\dfrac{x}{x-1}$

becomes unbounded.

$\lim\limits_{x\to -1} R(x)=\lim\limits_{x\to -1}\left(\dfrac{x}{x-1}\right)=\dfrac{-1}{-2}=\dfrac{1}{2}$. Note there is a hole in the graph at $\left(-1,\dfrac{1}{2}\right)$.

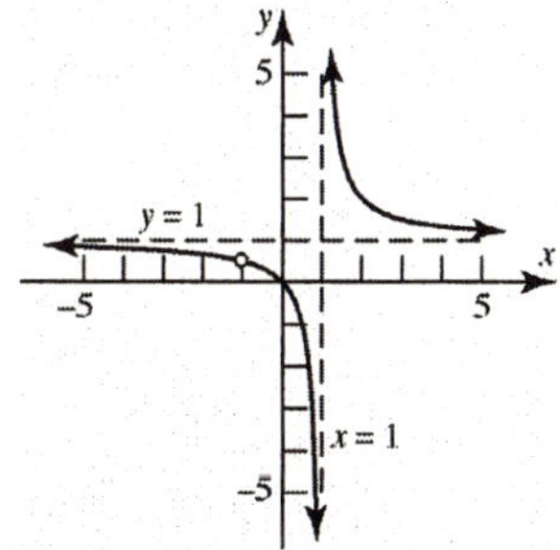

76. $R(x)=\dfrac{x^2+4x}{x^2-16}=\dfrac{x(x+4)}{(x-4)(x+4)}$. The domain of R is $\{x \mid x\neq -4, x\neq 4\}$. Thus R is discontinuous at both –4 and 4.

$\lim\limits_{x\to 4^-} R(x)=\lim\limits_{x\to 4^-}\left(\dfrac{x(x+4)}{(x-4)(x+4)}\right)$

$=\lim\limits_{x\to 4^-}\left(\dfrac{x}{x-4}\right)=-\infty$

since when

$0<x<4, \dfrac{x}{x-4}<0$, and as x approaches 4, $\dfrac{x}{x-4}$

becomes unbounded.

$\lim\limits_{x\to 4^+} R(x)=\lim\limits_{x\to 4^+}\left(\dfrac{x}{x-4}\right)=\infty$ since when

$x>4, \dfrac{x}{x-4}>0$, and as x approaches 4, $\dfrac{x}{x-4}$

becomes unbounded.

$\lim\limits_{x\to -4} R(x)=\lim\limits_{x\to -4}\left(\dfrac{x}{x-4}\right)=\dfrac{-4}{-8}=\dfrac{1}{2}$. Note there is a hole in the graph at $\left(-4,\dfrac{1}{2}\right)$.

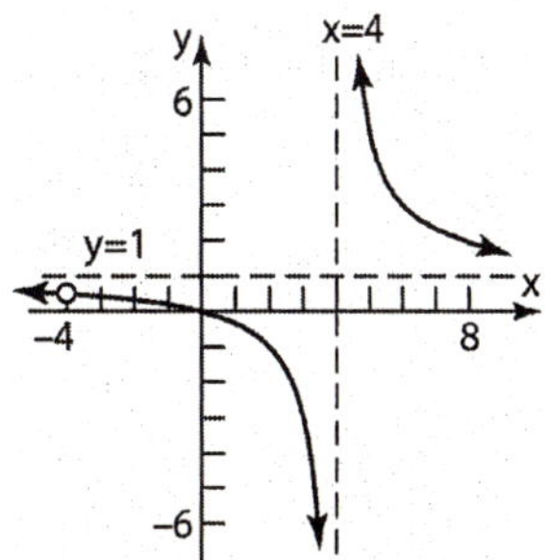

77. $R(x)=\dfrac{x^3-x^2+x-1}{x^4-x^3+2x-2}=\dfrac{x^2(x-1)+1(x-1)}{x^3(x-1)+2(x-1)}$

$=\dfrac{(x-1)(x^2+1)}{(x-1)(x^3+2)}=\dfrac{x^2+1}{x^3+2}, x\neq 1$

There is a vertical asymptote where $x^3+2=0$. $x=-\sqrt[3]{2}$ is a vertical asymptote. There is a hole in the graph at $x=1$ (at the point $\left(1,\frac{2}{3}\right)$).

78. $R(x)=\dfrac{x^3+x^2+3x+3}{x^4+x^3+2x+2}=\dfrac{x^2(x+1)+3(x+1)}{x^3(x+1)+2(x+1)}$

$=\dfrac{(x+1)(x^2+3)}{(x+1)(x^3+2)}=\dfrac{x^2+3}{x^3+2}, x\neq -1$

There is a vertical asymptote where $x^3+2=0$. $x=-\sqrt[3]{2}$ is a vertical asymptote. There is a hole in the graph at $x=-1$ (at the point $(-1,4)$).

79. $R(x)=\dfrac{x^3-2x^2+4x-8}{x^2+x-6}=\dfrac{x^2(x-2)+4(x-2)}{(x+3)(x-2)}$

$=\dfrac{(x-2)(x^2+4)}{(x+3)(x-2)}=\dfrac{x^2+4}{x+3}, x\neq 2$

There is a vertical asymptote where $x+3=0$. $x=-3$ is a vertical asymptote. There is a hole in the graph at $x=2$ (at the point $\left(2,\frac{8}{5}\right)$).

80. $R(x)=\dfrac{x^3-x^2+3x-3}{x^2+3x-4}=\dfrac{x^2(x-1)+3(x-1)}{(x+4)(x-1)}$

$=\dfrac{(x-1)(x^2+3)}{(x+4)(x-1)}=\dfrac{x^2+3}{x+4}, x\neq 1$

There is a vertical asymptote where $x+4=0$.

$x=-4$ is a vertical asymptote. There is a hole in the graph at $x=1$ (at the point $\left(1,\frac{4}{5}\right)$.

81. $R(x)=\dfrac{x^3+2x^2+x}{x^4+x^3+2x+2}=\dfrac{x(x^2+2x+1)}{x^3(x+1)+2(x+1)}$

$=\dfrac{x(x+1)^2}{(x+1)(x^3+2)}=\dfrac{x(x+1)}{x^3+2},\ x\neq -1$

There is a vertical asymptote where $x^3+2=0$. $x=-\sqrt[3]{2}$ is a vertical asymptote. There is a hole in the graph at $x=-1$ (at the point $(-1,0)$).

82. $R(x)=\dfrac{x^3-3x^2+4x-12}{x^4-3x^3+x-3}=\dfrac{x^2(x-3)+4(x-3)}{x^3(x-3)+1(x-3)}$

$=\dfrac{(x-3)(x^2+4)}{(x-3)(x^3+1)}=\dfrac{x^2+4}{x^3+1},\ x\neq 3$

There is a vertical asymptote where $x^3+1=0$. $x=-1$ is a vertical asymptote. There is a hole in the graph at $x=3$ (at the point $\left(3,\frac{13}{28}\right)$).

83. 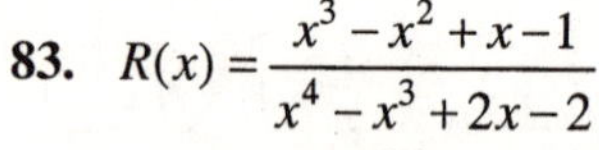
$R(x)=\dfrac{x^3-x^2+x-1}{x^4-x^3+2x-2}$

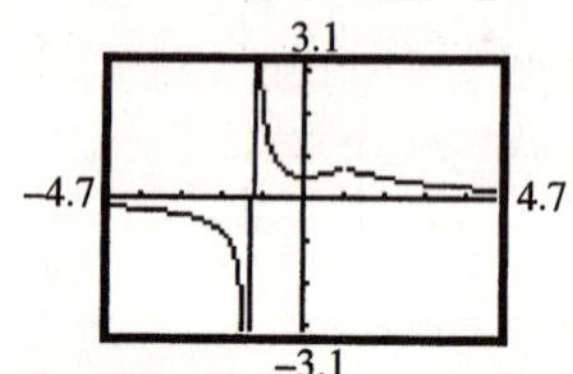

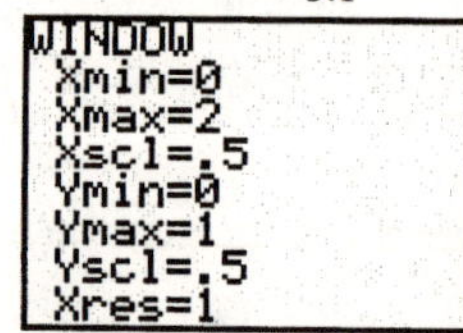

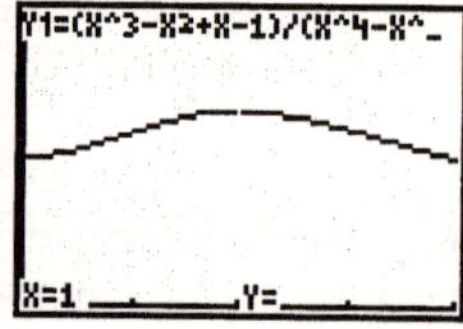

84. $R(x)=\dfrac{x^3+x^2+3x+3}{x^4+x^3+2x+2}$

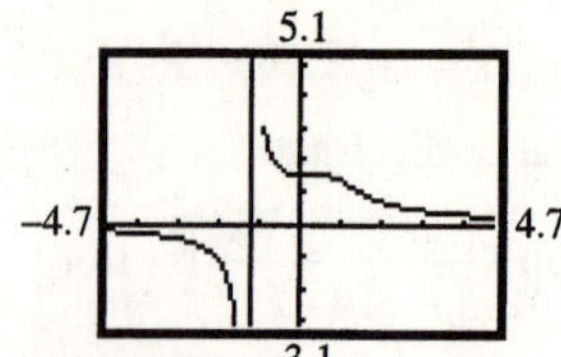

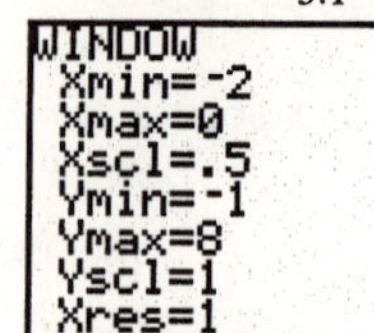

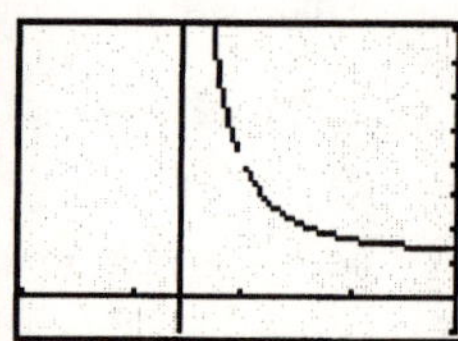

85. 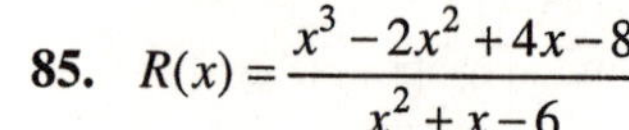
$R(x)=\dfrac{x^3-2x^2+4x-8}{x^2+x-6}$

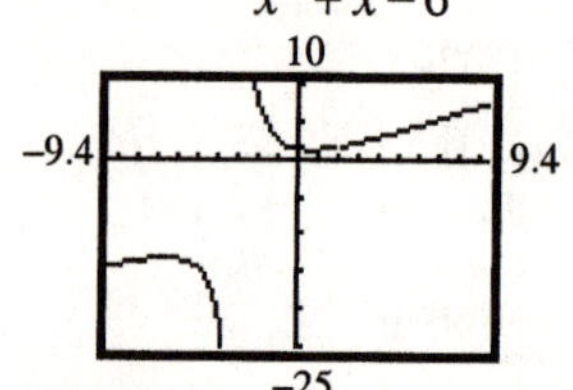

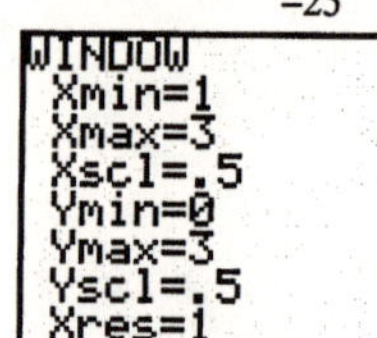

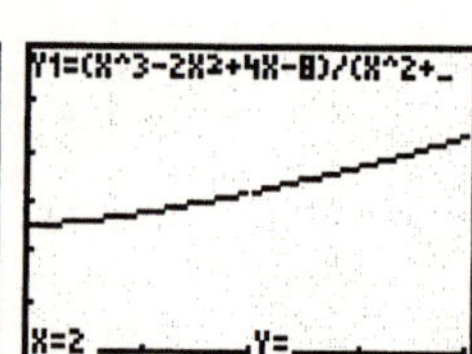

86. $R(x)=\dfrac{x^3-x^2+3x-3}{x^2+3x-4}$

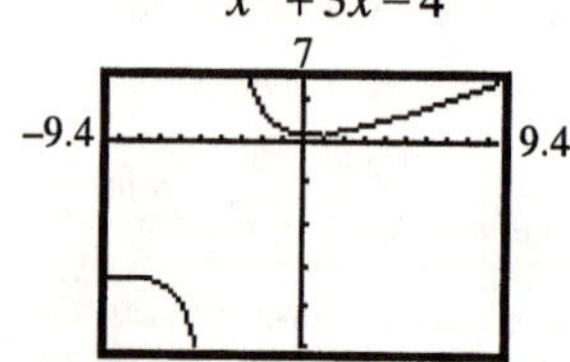

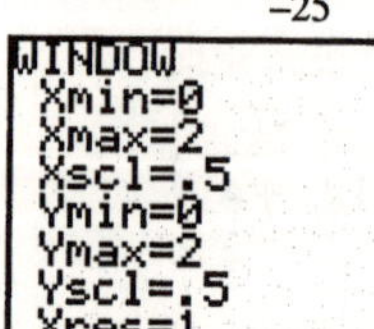

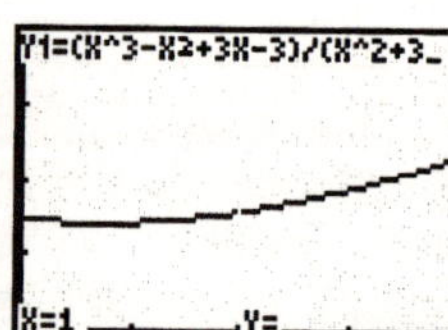

87. $R(x)=\dfrac{x^3+2x^2+x}{x^4+x^3+2x+2}$

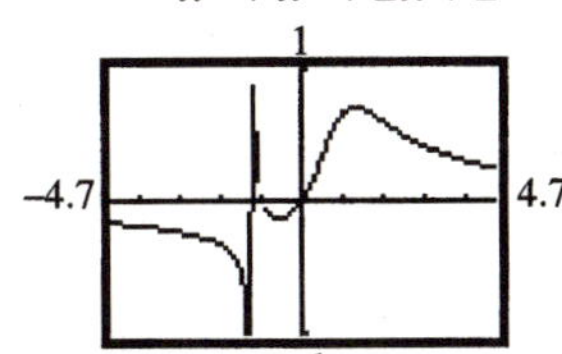

WINDOW
Xmin=-1.5
Xmax=0
Xscl=.5
Ymin=-1
Ymax=1
Yscl=.5
Xres=1

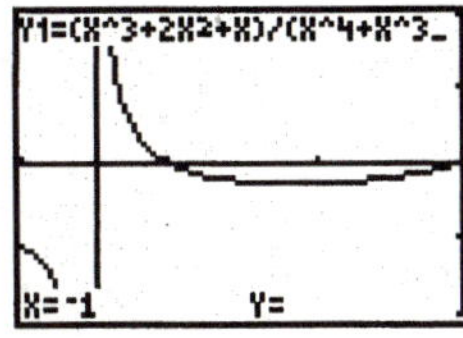

88. $R(x)=\dfrac{x^3-3x^2+4x-12}{x^4-3x^3+x-3}$

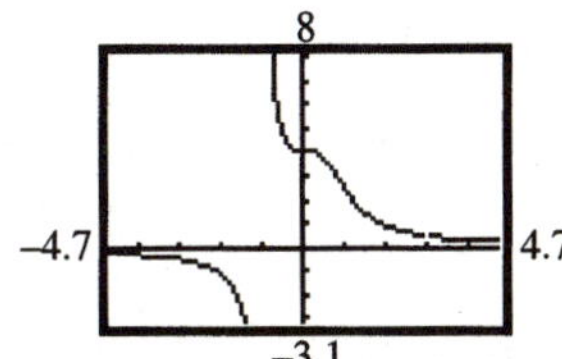

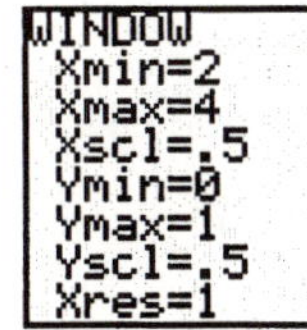

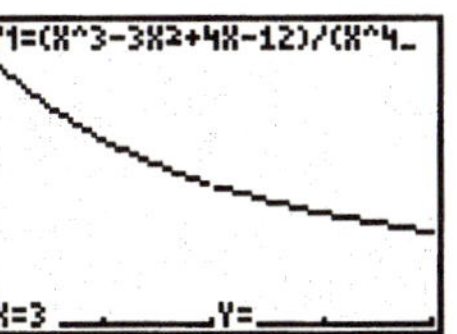

89. Answers will vary. Three possible functions are: $f(x)=x^2$; $g(x)=\sin(x)$; $h(x)=e^x$.

90. Answers will vary. One possible function is: $f(x)=\dfrac{x}{x-5}$.

Section 13.4

1. Slope 5; containing point $(2,-4)$

$$y-y_1=m(x-x_1)$$
$$y-(-4)=5(x-2)$$
$$y+4=5x-10$$
$$y=5x-14$$

2. False; it is $\dfrac{f(x)-f(c)}{x-c}$.

3. tangent line

4. derivative

5. instantaneous velocity

6. True

7. True

8. True

9. $f(x)=3x+5$ at $(1,8)$

$$m_{\tan}=\lim_{x\to1}\left(\frac{f(x)-f(1)}{x-1}\right)=\lim_{x\to1}\left(\frac{3x+5-8}{x-1}\right)$$
$$=\lim_{x\to1}\left(\frac{3x-3}{x-1}\right)=\lim_{x\to1}\left(\frac{3(x-1)}{x-1}\right)$$
$$=\lim_{x\to1}(3)=3$$

Tangent Line: $y-8=3(x-1)$
$$y-8=3x-3$$
$$y=3x+5$$

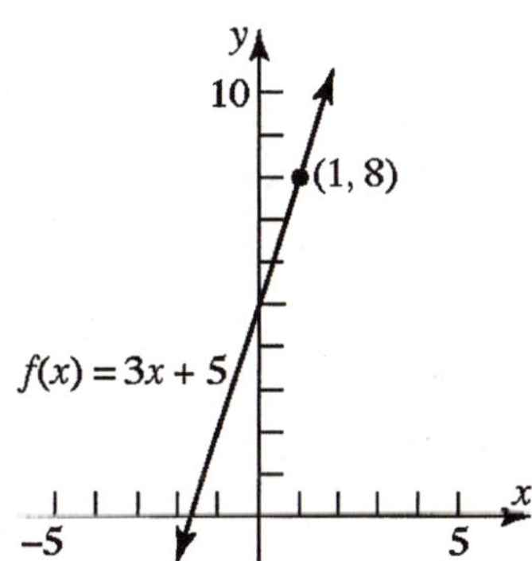

10. $f(x)=-2x+1$ at $(-1,3)$

$$m_{\tan}=\lim_{x\to-1}\left(\frac{f(x)-f(-1)}{x-(-1)}\right)$$
$$=\lim_{x\to-1}\left(\frac{-2x+1-3}{x+1}\right)$$
$$=\lim_{x\to-1}\left(\frac{-2x-2}{x+1}\right)$$
$$=\lim_{x\to-1}\left(\frac{-2(x+1)}{x+1}\right)$$
$$=\lim_{x\to-1}(-2)=-2$$

Tangent Line: $y-3=-2(x+1)$

$y-3=-2x-2$

$y=-2x+1$

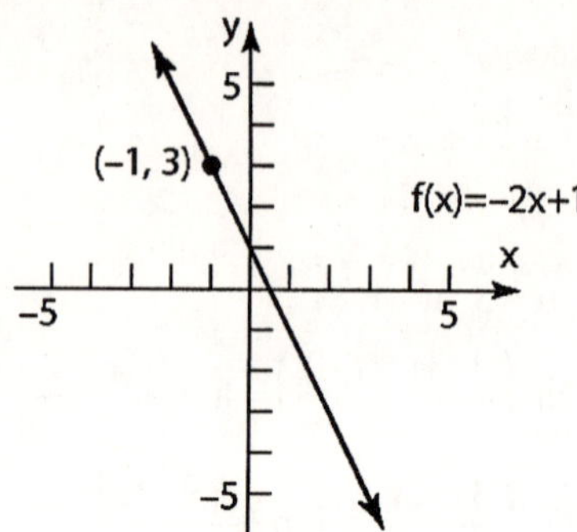

11. $f(x)=x^2+2$ at $(-1,3)$

$$m_{\tan}=\lim_{x\to-1}\left(\frac{f(x)-f(-1)}{x+1}\right)=\lim_{x\to-1}\left(\frac{x^2+2-3}{x+1}\right)$$

$$=\lim_{x\to-1}\left(\frac{x^2-1}{x+1}\right)=\lim_{x\to-1}\left(\frac{(x+1)(x-1)}{x+1}\right)$$

$$=\lim_{x\to-1}(x-1)=-1-1=-2$$

Tangent Line: $y-3=-2(x-(-1))$

$y-3=-2x-2$

$y=-2x+1$

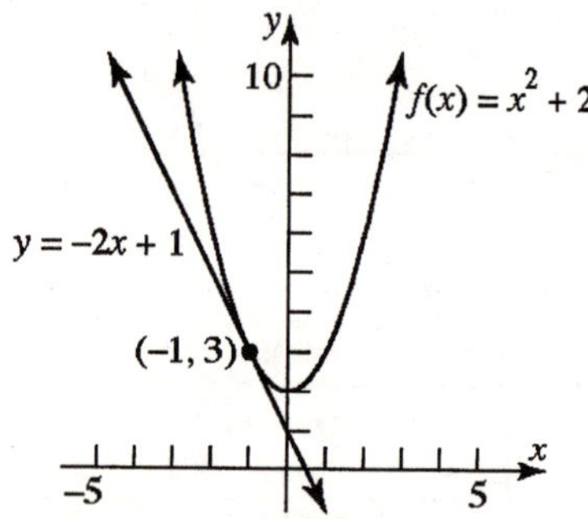

12. $f(x)=3-x^2$ at $(1,2)$

$$m_{\tan}=\lim_{x\to1}\left(\frac{f(x)-f(1)}{x-1}\right)=\lim_{x\to1}\left(\frac{3-x^2-2}{x-1}\right)$$

$$=\lim_{x\to1}\left(\frac{1-x^2}{x-1}\right)=\lim_{x\to1}\left(\frac{-1(x+1)(x-1)}{x-1}\right)$$

$$=\lim_{x\to1}\left((-1)(x+1)\right)=-1(1+1)=-2$$

Tangent Line: $y-2=-2(x-1)$

$y-2=-2x+2$

$y=-2x+4$

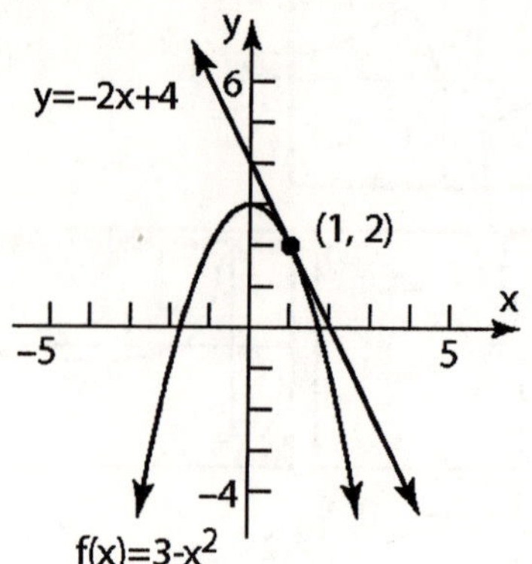

13. $f(x)=3x^2$ at $(2,12)$

$$m_{\tan}=\lim_{x\to2}\left(\frac{f(x)-f(2)}{x-2}\right)=\lim_{x\to2}\left(\frac{3x^2-12}{x-2}\right)$$

$$=\lim_{x\to2}\frac{3\left(x^2-4\right)}{x-2}=\lim_{x\to2}\left(\frac{3(x+2)(x-2)}{x-2}\right)$$

$$=\lim_{x\to2}\left(3(x+2)\right)=3(2+2)=12$$

Tangent Line: $y-12=12(x-2)$

$y-12=12x-24$

$y=12x-12$

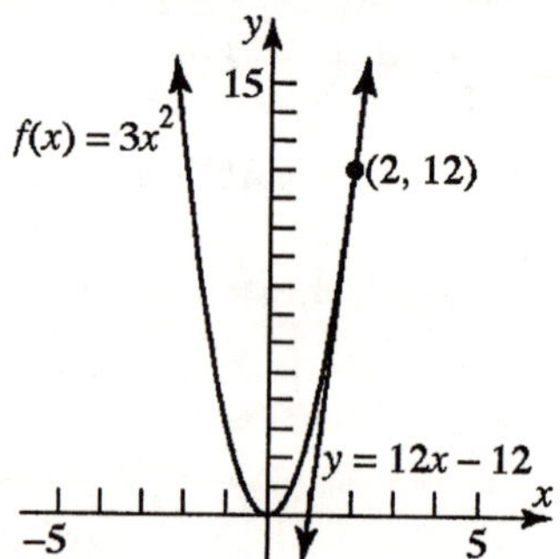

14. $f(x) = -4x^2$ at $(-2, -16)$

$$m_{\tan} = \lim_{x \to -2}\left(\frac{f(x)-f(-2)}{x-(-2)}\right)$$
$$= \lim_{x \to -2}\left(\frac{-4x^2-(-16)}{x+2}\right)$$
$$= \lim_{x \to -2}\left(\frac{-4(x^2-4)}{x+2}\right)$$
$$= \lim_{x \to -2}\left(\frac{-4(x+2)(x-2)}{x+2}\right)$$
$$= \lim_{x \to -2}\left(-4(x-2)\right)$$
$$= -4(-2-2) = 16$$

Tangent Line: $y-(-16) = 16(x-(-2))$

$$y+16 = 16x+32$$
$$y = 16x+16$$

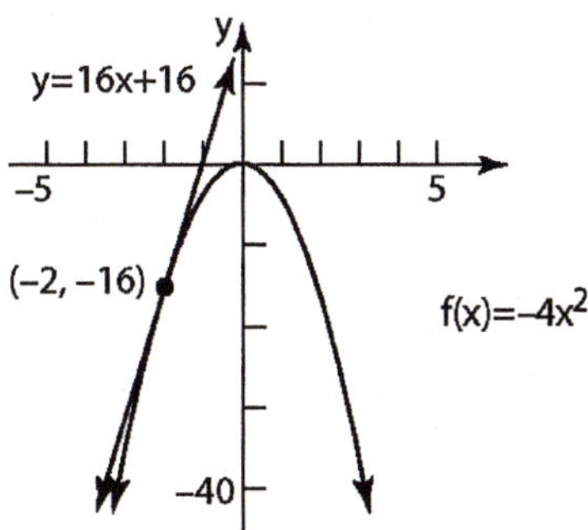

15. $f(x) = 2x^2 + x$ at $(1, 3)$

$$m_{\tan} = \lim_{x \to 1}\left(\frac{f(x)-f(1)}{x-1}\right) = \lim_{x \to 1}\left(\frac{2x^2+x-3}{x-1}\right)$$
$$= \lim_{x \to 1}\left(\frac{(2x+3)(x-1)}{x-1}\right) = \lim_{x \to 1}(2x+3)$$
$$= 2(1)+3 = 5$$

Tangent Line: $y-3 = 5(x-1)$

$$y-3 = 5x-5$$
$$y = 5x-2$$

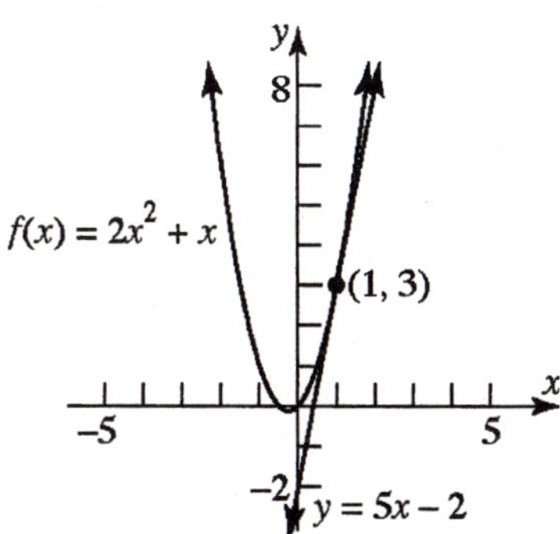

16. $f(x) = 3x^2 - x$ at $(0, 0)$

$$m_{\tan} = \lim_{x \to 0}\left(\frac{f(x)-f(0)}{x-0}\right) = \lim_{x \to 0}\left(\frac{3x^2-x-0}{x}\right)$$
$$= \lim_{x \to 0}\left(\frac{x(3x-1)}{x}\right) = \lim_{x \to 0}(3x-1)$$
$$= 3(0)-1 = -1$$

Tangent Line: $y-0 = -1(x-0)$

$$y = -x$$

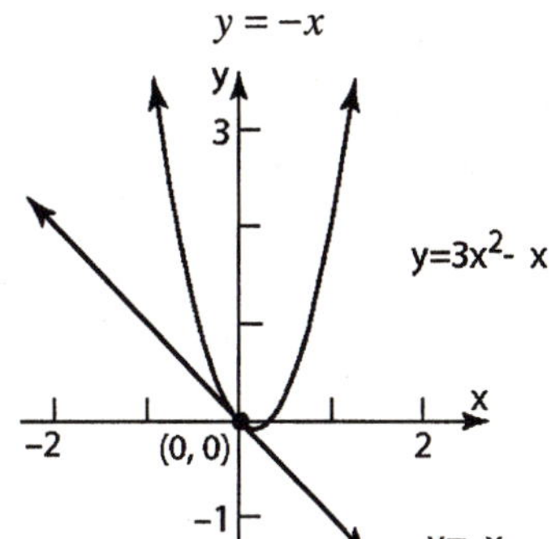

17. $f(x) = x^2 - 2x + 3$ at $(-1, 6)$

$$m_{\tan} = \lim_{x \to -1}\frac{f(x)-f(-1)}{x+1} = \lim_{x \to -1}\frac{x^2-2x+3-6}{x+1}$$
$$= \lim_{x \to -1}\frac{x^2-2x-3}{x+1} = \lim_{x \to -1}\frac{(x+1)(x-3)}{x+1}$$
$$= \lim_{x \to -1}(x-3) = -1-3 = -4$$

Tangent Line: $y-6 = -4(x-(-1))$

$$y-6 = -4x-4$$
$$y = -4x+2$$

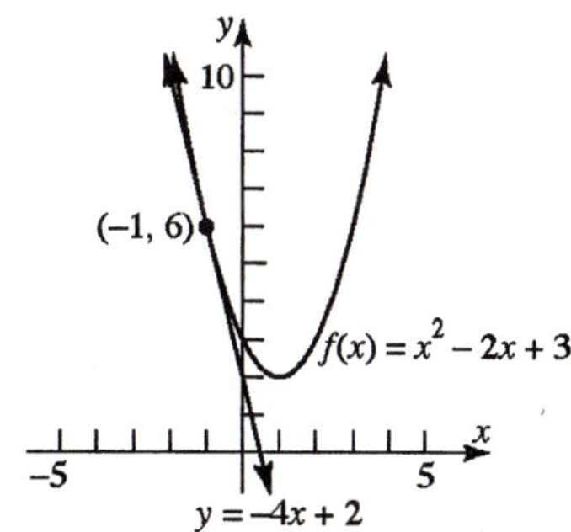

18. $f(x)=-2x^2+x-3$ at $(1,-4)$

$$m_{\tan}=\lim_{x\to 1}\left(\frac{f(x)-f(1)}{x-1}\right)$$
$$=\lim_{x\to 1}\left(\frac{-2x^2+x-3-(-4)}{x-1}\right)$$
$$=\lim_{x\to 1}\left(\frac{-2x^2+x+1}{x-1}\right)$$
$$=\lim_{x\to 1}\left(\frac{(x-1)(-2x-1)}{x-1}\right)$$
$$=\lim_{x\to 1}(-2x-1)=-2(1)-1=-3$$

Tangent Line: $y-(-4)=-3(x-1)$
$$y+4=-3x+3$$
$$y=-3x-1$$

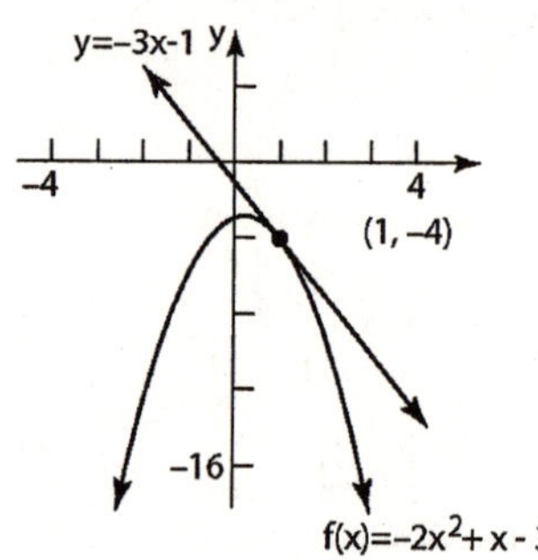

19. $f(x)=x^3+x$ at $(2,10)$

$$m_{\tan}=\lim_{x\to 2}\left(\frac{f(x)-f(2)}{x-2}\right)=\lim_{x\to 2}\left(\frac{x^3+x-10}{x-2}\right)$$
$$=\lim_{x\to 2}\left(\frac{x^3-8+x-2}{x-2}\right)$$
$$=\lim_{x\to 2}\left(\frac{(x-2)(x^2+2x+4)+(x-2)\cdot 1}{x-2}\right)$$
$$=\lim_{x\to 2}\left(\frac{(x-2)(x^2+2x+4+1)}{x-2}\right)$$
$$=\lim_{x\to 2}(x^2+2x+5)=4+4+5=13$$

Tangent Line: $y-10=13(x-2)$
$$y-10=13x-26$$
$$y=13x-16$$

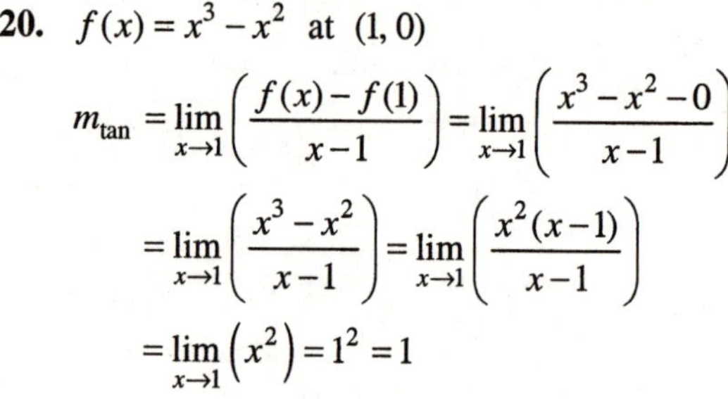

20. $f(x)=x^3-x^2$ at $(1,0)$

$$m_{\tan}=\lim_{x\to 1}\left(\frac{f(x)-f(1)}{x-1}\right)=\lim_{x\to 1}\left(\frac{x^3-x^2-0}{x-1}\right)$$
$$=\lim_{x\to 1}\left(\frac{x^3-x^2}{x-1}\right)=\lim_{x\to 1}\left(\frac{x^2(x-1)}{x-1}\right)$$
$$=\lim_{x\to 1}\left(x^2\right)=1^2=1$$

Tangent Line: $y-0=1(x-1)$
$$y=x-1$$

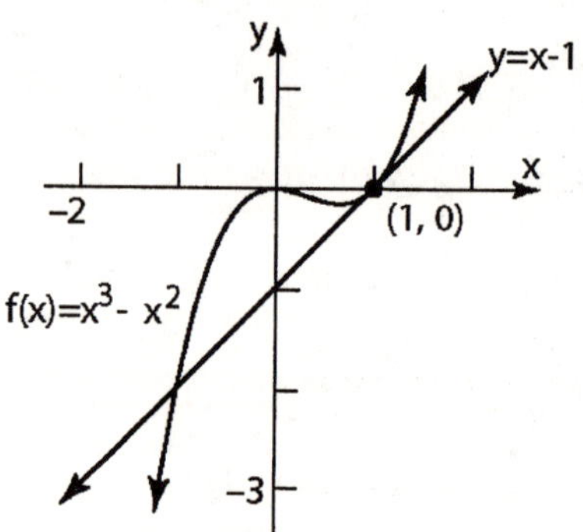

21. $f(x)=-4x+5$ at 3

$$f'(3)=\lim_{x\to 3}\left(\frac{f(x)-f(3)}{x-3}\right)$$
$$=\lim_{x\to 3}\left(\frac{-4x+5-(-7)}{x-3}\right)$$
$$=\lim_{x\to 3}\left(\frac{-4x+12}{x-3}\right)$$
$$=\lim_{x\to 3}\left(\frac{-4(x-3)}{x-3}\right)$$
$$=\lim_{x\to 3}(-4)=-4$$

22. $f(x) = -4 + 3x$ at 1

$$f'(1) = \lim_{x\to 1}\left(\frac{f(x)-f(1)}{x-1}\right) = \lim_{x\to 1}\left(\frac{-4+3x-(-1)}{x-1}\right) = \lim_{x\to 1}\left(\frac{3x-3}{x-1}\right) = \lim_{x\to 1}\left(\frac{3(x-1)}{x-1}\right) = \lim_{x\to 1}(3) = 3$$

23. $f(x) = x^2 - 3$ at 0

$$f'(0) = \lim_{x\to 0}\left(\frac{f(x)-f(0)}{x-0}\right) = \lim_{x\to 0}\left(\frac{x^2-3-(-3)}{x}\right) = \lim_{x\to 0}\left(\frac{x^2}{x}\right) = \lim_{x\to 0}(x) = 0$$

24. $f(x) = 2x^2 + 1$ at -1

$$f'(-1) = \lim_{x\to -1}\left(\frac{f(x)-f(-1)}{x-(-1)}\right) = \lim_{x\to -1}\left(\frac{2x^2+1-3}{x+1}\right) = \lim_{x\to -1}\left(\frac{2x^2-2}{x+1}\right) = \lim_{x\to -1}\left(\frac{2(x+1)(x-1)}{x+1}\right) = \lim_{x\to -1}(2(x-1)) = 2(-2) = -4$$

25. $f(x) = 2x^2 + 3x$ at 1

$$f'(1) = \lim_{x\to 1}\left(\frac{f(x)-f(1)}{x-1}\right) = \lim_{x\to 1}\left(\frac{2x^2+3x-5}{x-1}\right) = \lim_{x\to 1}\left(\frac{(2x+5)(x-1)}{x-1}\right) = \lim_{x\to 1}(2x+5) = 7$$

26. $f(x) = 3x^2 - 4x$ at 2

$$f'(2) = \lim_{x\to 2}\left(\frac{f(x)-f(2)}{x-2}\right) = \lim_{x\to 2}\left(\frac{3x^2-4x-4}{x-2}\right) = \lim_{x\to 2}\left(\frac{(3x+2)(x-2)}{x-2}\right) = \lim_{x\to 2}(3x+2) = 8$$

27. $f(x) = x^3 + 4x$ at -1

$$f'(-1) = \lim_{x\to -1}\left(\frac{f(x)-f(-1)}{x-(-1)}\right) = \lim_{x\to -1}\left(\frac{x^3+4x-(-5)}{x+1}\right) = \lim_{x\to -1}\left(\frac{x^3+1+4x+4}{x+1}\right) = \lim_{x\to -1}\left(\frac{(x+1)(x^2-x+1)+4(x+1)}{x+1}\right) = \lim_{x\to -1}\left(\frac{(x+1)(x^2-x+1+4)}{x+1}\right) = \lim_{x\to -1}(x^2-x+5) = (-1)^2-(-1)+5 = 7$$

28. $f(x) = 2x^3 - x^2$ at 2

$$f'(2) = \lim_{x\to 2}\left(\frac{f(x)-f(2)}{x-2}\right) = \lim_{x\to 2}\left(\frac{2x^3-x^2-12}{x-2}\right) = \lim_{x\to 2}\left(\frac{2x^3-4x^2+3x^2-12}{x-2}\right) = \lim_{x\to 2}\left(\frac{2x^2(x-2)+3(x+2)(x-2)}{x-2}\right) = \lim_{x\to 2}\left(\frac{(x-2)(2x^2+3x+6)}{x-2}\right) = \lim_{x\to 2}(2x^2+3x+6) = 2(2)^2+3(2)+6 = 20$$

29. $f(x) = x^3 + x^2 - 2x$ at 1

$$f'(1) = \lim_{x\to 1}\left(\frac{f(x)-f(1)}{x-1}\right)$$
$$= \lim_{x\to 1}\left(\frac{x^3+x^2-2x-0}{x-1}\right)$$
$$= \lim_{x\to 1}\left(\frac{x(x^2+x-2)}{x-1}\right)$$
$$= \lim_{x\to 1}\left(\frac{x(x+2)(x-1)}{x-1}\right)$$
$$= \lim_{x\to 1}\left(x(x+2)\right)$$
$$= 1(1+2) = 3$$

30. $f(x) = x^3 - 2x^2 + x$ at -1

$$f'(-1) = \lim_{x\to -1}\left(\frac{f(x)-f(-1)}{x-(-1)}\right)$$
$$= \lim_{x\to -1}\left(\frac{x^3-2x^2+x-(-4)}{x+1}\right)$$
$$= \lim_{x\to -1}\left(\frac{(x+1)(x^2-3x+4)}{x+1}\right)$$
$$= \lim_{x\to -1}\left(x^2-3x+4\right)$$
$$= (-1)^2 - 3(-1) + 4 = 8$$

31. $f(x) = \sin x$ at 0

$$f'(0) = \lim_{x\to 0}\left(\frac{f(x)-f(0)}{x-0}\right)$$
$$= \lim_{x\to 0}\left(\frac{\sin x - 0}{x-0}\right)$$
$$= \lim_{x\to 0}\left(\frac{\sin x}{x}\right) = 1$$

32. $f(x) = \cos x$ at 0

$$f'(0) = \lim_{x\to 0}\left(\frac{f(x)-f(0)}{x-0}\right) = \lim_{x\to 0}\left(\frac{\cos x - 1}{x}\right) = 0$$

33. Use nDeriv:

```
nDeriv(3X^3-6X²+
2,X,-2)
             60.000003
```

34. Use nDeriv:

```
nDeriv(-5X^4+6X²
-10,X,5)
          -2440.0001
```

35. Use nDeriv:

```
nDeriv((-X^3+1)/
(X²+5X+7),X,8)
        -.8587776956
```

36. Use nDeriv:

```
nDeriv((-5X^4+9X
+3)/(X^3+5X²-6),
X,-3)
         36.81251395
```

37. Use nDeriv:

```
nDeriv(Xsin(X),X
,π/3)
         1.389623659
```

38. Use nDeriv:

```
nDeriv(Xsin(X),X
,π/4)
         1.262466702
```

39. Use nDeriv:

```
nDeriv(X²sin(X),
X,π/3)
         2.362110222
```

40. Use nDeriv:

```
nDeriv(X²sin(X),
X,π/4)
         1.546899826
```

41. Use nDeriv:

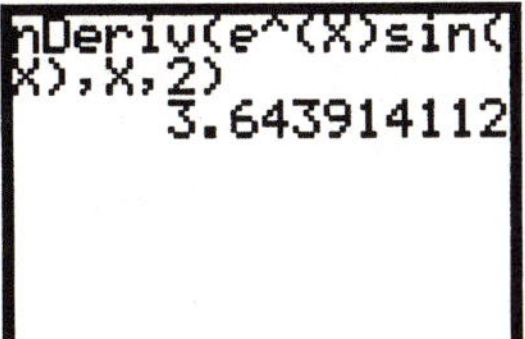

42. Use nDeriv:

43. $V(r) = 3\pi r^2$ at $r = 3$

$$\begin{aligned} V'(3) &= \lim_{r\to 3}\left(\frac{V(r)-V(3)}{r-3}\right) \\ &= \lim_{r\to 3}\left(\frac{3\pi r^2 - 27\pi}{r-3}\right) \\ &= \lim_{r\to 3}\left(\frac{3\pi(r^2-9)}{r-3}\right) \\ &= \lim_{r\to 3}\left(\frac{3\pi(r-3)(r+3)}{r-3}\right) \\ &= \lim_{r\to 3}\left(3\pi(r+3)\right) \\ &= 3\pi(3+3) = 18\pi \end{aligned}$$

At the instant $r - 3$ feet, the volume of the cylinder is increasing at a rate of 18π cubic feet per foot.

44. $S(r) = 4\pi r^2$ at $r = 2$

$$\begin{aligned} S'(2) &= \lim_{r\to 2}\left(\frac{S(r)-S(2)}{r-2}\right) \\ &= \lim_{r\to 2}\left(\frac{4\pi r^2 - 16\pi}{r-2}\right) \\ &= \lim_{r\to 2}\left(\frac{4\pi(r^2-4)}{r-2}\right) \\ &= \lim_{r\to 2}\left(\frac{4\pi(r-2)(r+2)}{r-2}\right) \\ &= \lim_{r\to 2}\left(4\pi(r+2)\right) \\ &= 4\pi(2+2) = 16\pi \end{aligned}$$

At the instant $r = 2$ feet, the surface area of the sphere is increasing at a rate of 16π square feet per foot.

45. $V(r) = \frac{4}{3}\pi r^3$ at $r = 2$

$$\begin{aligned} V'(2) &= \lim_{r\to 2}\left(\frac{V(r)-V(2)}{r-2}\right) \\ &= \lim_{r\to 2}\left(\frac{\frac{4}{3}\pi r^3 - \frac{32}{3}\pi}{r-2}\right) \\ &= \lim_{r\to 2}\left(\frac{\left(\frac{4}{3}\pi(r^3-8)\right)}{r-2}\right) \\ &= \lim_{r\to 2}\left(\frac{\left(\frac{4}{3}\pi(r-2)(r^2+2r+4)\right)}{r-2}\right) \\ &= \lim_{r\to 2}\left(\frac{4}{3}\pi(r^2+2r+4)\right) \\ &= \frac{4}{3}\pi(4+4+4) = 16\pi \end{aligned}$$

At the instant $r = 2$ feet, the volume of the sphere is increasing at a rate of 16π cubic feet per foot.

46. $V(x) = x^3$ at $x=3$

$$\begin{aligned} V'(3) &= \lim_{x\to 3}\left(\frac{V(x)-V(3)}{x-3}\right) \\ &= \lim_{x\to 3}\left(\frac{x^3-27}{x-3}\right) \\ &= \lim_{x\to 3}\left(\frac{(x-3)(x^2+3x+9)}{x-3}\right) \\ &= \lim_{x\to 3}\left(x^2+3x+9\right) \\ &= 3^2+3(3)+9 = 27 \end{aligned}$$

At the instant $x = 3$ meters, the volume of the cube is increasing at a rate of 27 cubic meters per meter.

47. a. $-16t^2 + 96t = 0$
$-16t(t-6) = 0$
$t = 0 \text{ or } t = 6$
The ball strikes the ground after 6 seconds.

b. $\frac{\Delta s}{\Delta t} = \frac{s(2)-s(0)}{2-0}$
$= \frac{-16(2)^2 + 96(2) - 0}{2}$
$= \frac{128}{2} = 64 \text{ feet/sec}$

c. $s'(t_0) = \lim_{t \to t_0}\left(\frac{s(t)-s(t_0)}{t-t_0}\right)$
$= \lim_{t \to t_0}\left(\frac{-16t^2 + 96t - \left(-16t_0^{\,2} + 96t_0\right)}{t-t_0}\right)$
$= \lim_{t \to t_0}\left(\frac{-16t^2 + 16t_0^{\,2} + 96t - 96t_0}{t-t_0}\right)$
$= \lim_{t \to t_0}\left(\frac{-16\left(t^2 - t_0^{\,2}\right) + 96(t-t_0)}{t-t_0}\right)$
$= \lim_{t \to t_0}\left(\frac{-16(t-t_0)(t+t_0) + 96(t-t_0)}{t-t_0}\right)$
$= \lim_{t \to t_0}\left(\frac{(t-t_0)(-16(t+t_0)+96)}{t-t_0}\right)$
$= \lim_{t \to t_0}\left(-16(t+t_0)+96\right)$
$= \left(-16(t_0+t_0)+96\right)$
$= -32t_0 + 96 \quad \text{ft/sec}$

d. $s'(2) = -32(2) + 96 = -64 + 96 = 32 \text{ feet/sec}$

e. $s'(t) = 0$
$-32t + 96 = 0$
$-32t = -96$
$t = 3 \text{ seconds}$

f. $s(3) = -16(3)^2 + 96(3)$
$= -144 + 288$
$= 144 \text{ feet}$

g. $s'(6) = -32(6) + 96$
$= -192 + 96$
$= -96 \text{ feet/sec}$

48. a. $-16t^2 - 48t + 160 = 0$
$-16\left(t^2 + 3t - 10\right) = 0$
$-16(t-2)(t+5) = 0$
$t = 2 \text{ or } t = -5$
The ball strikes the ground after 2 seconds in the air.

b. $\frac{\Delta s}{\Delta t} = \frac{s(1)-s(0)}{1-0}$
$= \frac{-16(1)^2 - 48(1) + 160 - 160}{1}$
$= \frac{-64}{1} = -64 \text{ feet/sec}$

c. $s'(t_0) = \lim_{t \to t_0}\left(\frac{s(t)-s(t_0)}{t-t_0}\right)$
$= \lim_{t \to t_0}\left(\frac{-16t^2 + 16t_0^{\,2} - 48t + 48t_0}{t-t_0}\right)$
$= \lim_{t \to t_0}\left(\frac{-16\left(t^2 - t_0^{\,2}\right) - 48(t-t_0)}{t-t_0}\right)$
$= \lim_{t \to t_0}\left(\frac{-16(t-t_0)(t+t_0) - 48(t-t_0)}{t-t_0}\right)$
$= \lim_{t \to t_0}\left(\frac{(t-t_0)(-16(t+t_0)-48)}{t-t_0}\right)$
$= \lim_{t \to t_0}\left(-16(t+t_0)-48\right)$
$= \left(-16(t_0+t_0)-48\right)$
$= -32t_0 - 48 \quad \text{ft/sec}$

d. $s'(1) = -32(1) - 48$
$= -32 - 48$
$= -80 \text{ feet/sec}$

e. $s'(2) = -32(2) - 48$
$= -64 - 48$
$= -112 \text{ feet/sec}$

49. a. $\frac{\Delta s}{\Delta t}=\frac{s(4)-s(1)}{4-1}$
$=\frac{917-987}{3}$
$=\frac{-70}{3}$
$=-23\frac{1}{3}$ feet/sec

b. $\frac{\Delta s}{\Delta t}=\frac{s(3)-s(1)}{3-1}$
$=\frac{945-987}{2}$
$=\frac{-42}{2}$
$=-21$ feet/sec

c. $\frac{\Delta s}{\Delta t}=\frac{s(2)-s(1)}{2-1}$
$=\frac{969-987}{1}$
$=\frac{-18}{1}=-18$ feet/sec

d. $s(t)=-2.631t^2-10.269t+999.933$

e. $s'(1)=\lim_{t\to 1}\left(\frac{s(t)-s(1)}{t-1}\right)$
$=\lim_{t\to 1}\left(\frac{-2.631t^2-10.269t+12.9}{t-1}\right)$
$=\lim_{t\to 1}\left(\frac{-2.631t^2+2.631t-12.9t+12.9}{t-1}\right)$
$=\lim_{t\to 1}\left(\frac{-2.631t(t-1)-12.9(t-1)}{t-1}\right)$
$=\lim_{t\to 1}\left(\frac{(-2.631t-12.9)(t-1)}{t-1}\right)$
$=\lim_{t\to 1}(-2.631t-12.9)$
$=-2.631(1)-12.9=-15.531$ feet/sec

At the instant when $t=1$, the instantaneous speed of the ball is –15.531 feet / sec.

50. a. $\frac{\Delta R}{\Delta x}=\frac{R(150)-R(25)}{150-25}$
$=\frac{59,160-28,000}{125}$
$=\frac{31,160}{125}=\$249.28/\text{bicycle}$

b. $\frac{\Delta R}{\Delta x}=\frac{R(102)-R(25)}{102-25}$
$=\frac{53,400-28,000}{77}$
$=\frac{25,400}{77}\approx\$329.87/\text{bicycle}$

c. $\frac{\Delta R}{\Delta x}=\frac{R(60)-R(25)}{60-25}$
$=\frac{45,000-28,000}{35}$
$=\frac{17,000}{35}\approx\$485.71/\text{bicycle}$

d. $R(x)=-1.522x^2+597.791x+7944.450$

e. $R'(25)=\lim_{x\to 25}\left(\frac{R(x)-R(25)}{x-25}\right)$
$=\lim_{x\to 25}\left(\frac{-1.522x^2+597.791x-13,993.525}{x-25}\right)$
$=\lim_{x\to 25}\left(\frac{(x-25)(-1.522x+559.741)}{x-25}\right)$
$=\lim_{x\to 25}(-1.522x+559.741)$
$=-1.522(25)+559.741$
$\approx\$521.69$ /bicycle

At the instant when $x=25$, the instantaneous rate of change of revenue is about $521.69 per bicycle.

Section 13.5

1. $A = lw$

2. $(2(1)+1)+(2(2)+1)+(2(3)+1)+(2(4)+1)=3+5+7+9=24$

3. $\int_a^b f(x)\,dx$

4. $\int_a^b f(x)\,dx$

5. $A \approx f(1)\cdot 1 + f(2)\cdot 1 = 1\cdot 1 + 2\cdot 1 = 1+2 = 3$

6. $A \approx f(2)\cdot 1 + f(3)\cdot 1 = 2\cdot 1 + 4\cdot 1 = 2+4 = 6$

7. $A \approx f(0)\cdot 2 + f(2)\cdot 2 + f(4)\cdot 2 + f(6)\cdot 2$
   $= 10\cdot 2 + 6\cdot 2 + 7\cdot 2 + 5\cdot 2$
   $= 20+12+14+10 = 56$

8. $A \approx f(2)\cdot 2 + f(4)\cdot 2 + f(6)\cdot 2 + f(8)\cdot 2$
   $= 6\cdot 2 + 7\cdot 2 + 5\cdot 2 + 1\cdot 2$
   $= 12+14+10+2 = 38$

9. **a.** Graph $f(x) = 3x$:

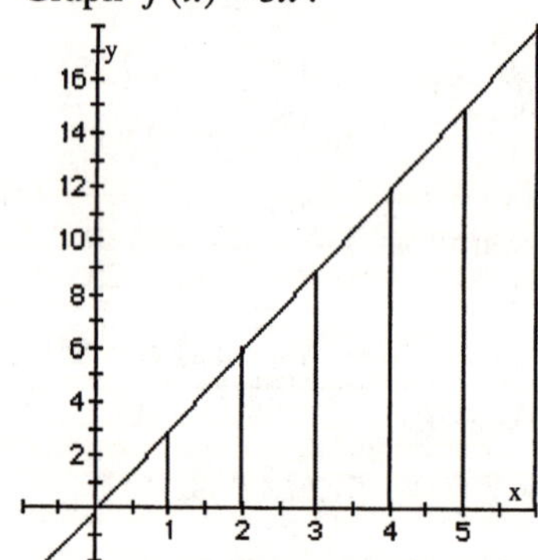

   b. $A \approx f(0)(2) + f(2)(2) + f(4)(2)$
   $= 0(2) + 6(2) + 12(2)$
   $= 0+12+24 = 36$

   c. $A \approx f(2)(2) + f(4)(2) + f(6)(2)$
   $= 6(2) + 12(2) + 18(2)$
   $= 12+24+36 = 72$

   d. $A \approx f(0)(1) + f(1)(1) + f(2)(1) + f(3)(1) + f(4)(1) + f(5)(1)$
   $= 0(1) + 3(1) + 6(1) + 9(1) + 12(1) + 15(1)$
   $= 0+3+6+9+12+15 = 45$

e. $A \approx f(1)(1) + f(2)(1) + f(3)(1) + f(4)(1) + f(5)(1) + f(6)(1)$
$= 3(1) + 6(1) + 9(1) + 12(1) + 15(1) + 18(1)$
$= 3 + 6 + 9 + 12 + 15 + 18$
$= 63$

f. The actual area is the area of a triangle: $A = \frac{1}{2}(6)(18) = 54$

10. a. Graph $f(x) = 4x$:

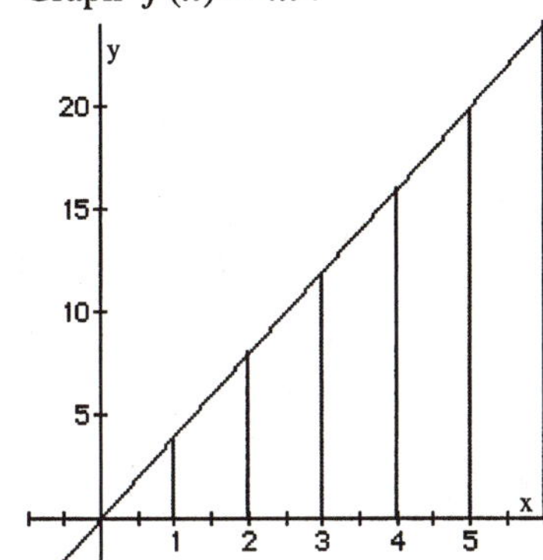

b. $A \approx f(0)(2) + f(2)(2) + f(4)(2)$
$= 0(2) + 8(2) + 16(2)$
$= 0 + 16 + 32 = 48$

c. $A \approx f(2)(2) + f(4)(2) + f(6)(2)$
$= 8(2) + 16(2) + 24(2)$
$= 16 + 32 + 48 = 96$

d. $A \approx f(0)(1) + f(1)(1) + f(2)(1) + f(3)(1) + f(4)(1) + f(5)(1)$
$= 0(1) + 4(1) + 8(1) + 12(1) + 16(1) + 20(1)$
$= 0 + 4 + 8 + 12 + 16 + 20 = 60$

e. $A \approx f(1)(1) + f(2)(1) + f(3)(1) + f(4)(1) + f(5)(1) + f(6)(1)$
$= 4(1) + 8(1) + 12(1) + 16(1) + 20(1) + 24(1)$
$= 4 + 8 + 12 + 16 + 20 + 24$
$= 84$

f. The actual area is the area of a triangle: $A = \frac{1}{2}(6)(24) = 72$

11. a. Graph $f(x) = -3x+9$:

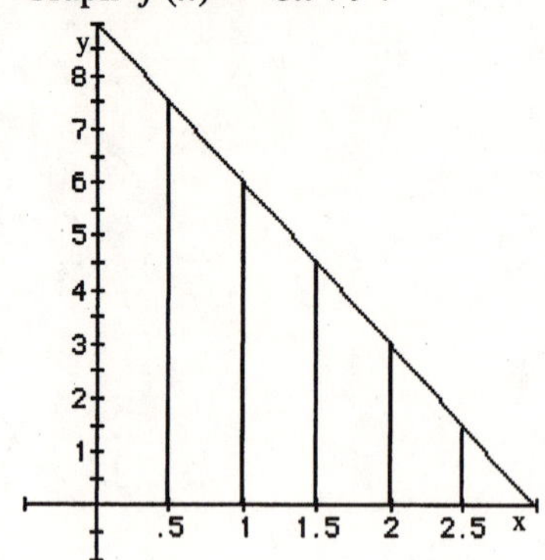

b. $A \approx f(0)(1) + f(1)(1) + f(2)(1)$
$= 9(1) + 6(1) + 3(1)$
$= 9 + 6 + 3 = 18$

c. $A \approx f(1)(1) + f(2)(1) + f(3)(1)$
$= 6(1) + 3(1) + 0(1)$
$= 6 + 3 + 0 = 9$

d. $A \approx f(0)\left(\frac{1}{2}\right) + f\left(\frac{1}{2}\right)\left(\frac{1}{2}\right) + f(1)\left(\frac{1}{2}\right) + f\left(\frac{3}{2}\right)\left(\frac{1}{2}\right) + f(2)\left(\frac{1}{2}\right) + f\left(\frac{5}{2}\right)\left(\frac{1}{2}\right)$
$= 9\left(\frac{1}{2}\right) + \frac{15}{2}\left(\frac{1}{2}\right) + 6\left(\frac{1}{2}\right) + \frac{9}{2}\left(\frac{1}{2}\right) + 3\left(\frac{1}{2}\right) + \frac{3}{2}\left(\frac{1}{2}\right)$
$= \frac{9}{2} + \frac{15}{4} + 3 + \frac{9}{4} + \frac{3}{2} + \frac{3}{4} = \frac{63}{4} = 15.75$

e. $A \approx f\left(\frac{1}{2}\right)\left(\frac{1}{2}\right) + f(1)\left(\frac{1}{2}\right) + f\left(\frac{3}{2}\right)\left(\frac{1}{2}\right) + f(2)\left(\frac{1}{2}\right) + f\left(\frac{5}{2}\right)\left(\frac{1}{2}\right) + f(3)\left(\frac{1}{2}\right)$
$= \frac{15}{2}\left(\frac{1}{2}\right) + 6\left(\frac{1}{2}\right) + \frac{9}{2}\left(\frac{1}{2}\right) + 3\left(\frac{1}{2}\right) + \frac{3}{2}\left(\frac{1}{2}\right) + 0\left(\frac{1}{2}\right)$
$= \frac{15}{4} + 3 + \frac{9}{4} + \frac{3}{2} + \frac{3}{4} + 0 = \frac{45}{4} = 11.25$

f. The actual area is the area of a triangle: $A = \frac{1}{2}(3)(9) = \frac{27}{2} = 13.5$

12. a. Graph $f(x) = -2x + 8$:

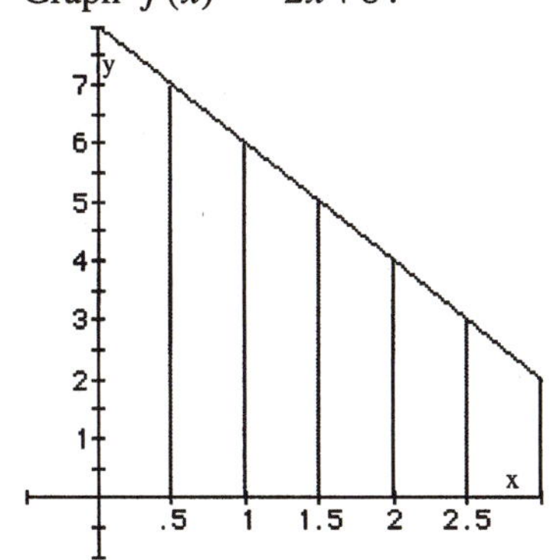

b. $A \approx f(0)(1) + f(1)(1) + f(2)(1) = 8(1) + 6(1) + 4(1) = 8 + 6 + 4 = 18$

c. $A \approx f(1)(1) + f(2)(1) + f(3)(1) = 6(1) + 4(1) + 2(1) = 6 + 4 + 2 = 12$

d.
$$A \approx f(0)\left(\frac{1}{2}\right) + f\left(\frac{1}{2}\right)\left(\frac{1}{2}\right) + f(1)\left(\frac{1}{2}\right) + f\left(\frac{3}{2}\right)\left(\frac{1}{2}\right) + f(2)\left(\frac{1}{2}\right) + f\left(\frac{5}{2}\right)\left(\frac{1}{2}\right)$$
$$= 8\left(\frac{1}{2}\right) + 7\left(\frac{1}{2}\right) + 6\left(\frac{1}{2}\right) + 5\left(\frac{1}{2}\right) + 4\left(\frac{1}{2}\right) + 3\left(\frac{1}{2}\right)$$
$$= 4 + \frac{7}{2} + 3 + \frac{5}{2} + 2 + \frac{3}{2} = \frac{33}{2} = 16.5$$

e.
$$A \approx f\left(\frac{1}{2}\right)\left(\frac{1}{2}\right) + f(1)\left(\frac{1}{2}\right) + f\left(\frac{3}{2}\right)\left(\frac{1}{2}\right) + f(2)\left(\frac{1}{2}\right) + f\left(\frac{5}{2}\right)\left(\frac{1}{2}\right) + f(3)\left(\frac{1}{2}\right)$$
$$= 7\left(\frac{1}{2}\right) + 6\left(\frac{1}{2}\right) + 5\left(\frac{1}{2}\right) + 4\left(\frac{1}{2}\right) + 3\left(\frac{1}{2}\right) + 2\left(\frac{1}{2}\right)$$
$$= \frac{7}{2} + 3 + \frac{5}{2} + 2 + \frac{3}{2} + 1 = \frac{27}{2} = 13.5$$

f. The actual area is the area of a (sideways) trapezoid: $A = \frac{1}{2}(2 + 8)(3) = \frac{30}{2} = 15$

13. a. Graph $f(x) = x^2 + 2,\ [0, 4]$:

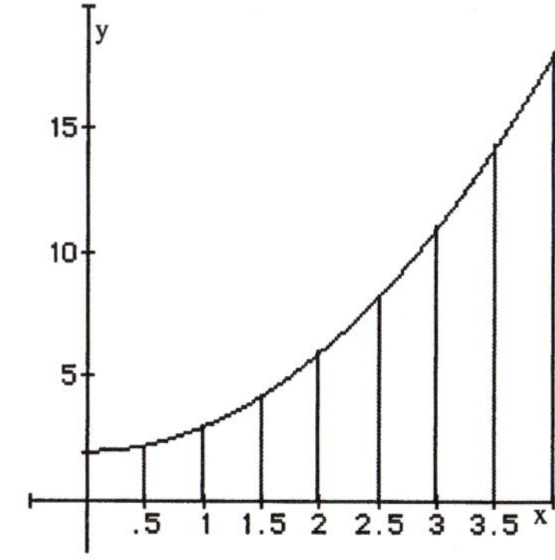

b. $A \approx f(0)(1) + f(1)(1) + f(2)(1) + f(3)(1) = 2(1) + 3(1) + 6(1) + 11(1) = 2 + 3 + 6 + 11 = 22$

c. $A \approx f(0)\left(\frac{1}{2}\right)+f\left(\frac{1}{2}\right)\left(\frac{1}{2}\right)+f(1)\left(\frac{1}{2}\right)+f\left(\frac{3}{2}\right)\left(\frac{1}{2}\right)+f(2)\left(\frac{1}{2}\right)+f\left(\frac{5}{2}\right)\left(\frac{1}{2}\right)+f(3)\left(\frac{1}{2}\right)+f\left(\frac{7}{2}\right)\left(\frac{1}{2}\right)$

$=2\left(\frac{1}{2}\right)+\frac{9}{4}\left(\frac{1}{2}\right)+3\left(\frac{1}{2}\right)+\frac{17}{4}\left(\frac{1}{2}\right)+6\left(\frac{1}{2}\right)+\frac{33}{4}\left(\frac{1}{2}\right)+11\left(\frac{1}{2}\right)+\frac{57}{4}\left(\frac{1}{2}\right)$

$=1+\frac{9}{8}+\frac{3}{2}+\frac{17}{8}+3+\frac{33}{8}+\frac{11}{2}+\frac{57}{8}=\frac{51}{2}=25.5$

d. $A=\int_0^4 (x^2+2)dx$

e. Use fnInt function:

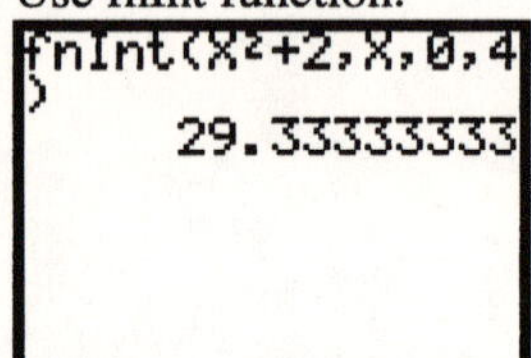

14. a. Graph $f(x)=x^2-4,\ [2,6]$:

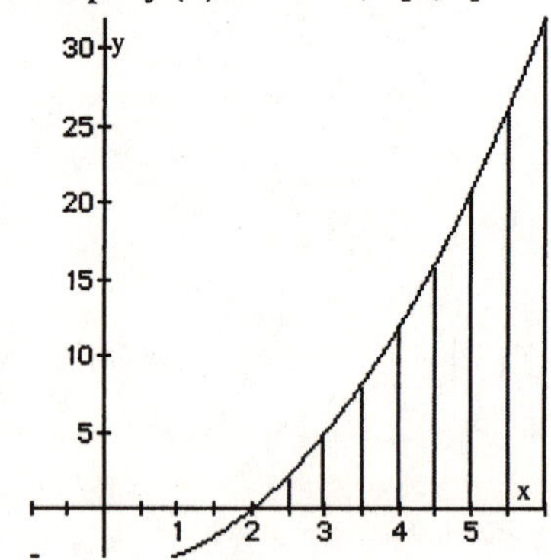

b. $A \approx f(2)(1)+f(3)(1)+f(4)(1)+f(5)(1)=0(1)+5(1)+12(1)+21(1)=0+5+12+21=38$

c. $A \approx f(2)\left(\frac{1}{2}\right)+f\left(\frac{5}{2}\right)\left(\frac{1}{2}\right)+f(3)\left(\frac{1}{2}\right)+f\left(\frac{7}{2}\right)\left(\frac{1}{2}\right)+f(4)\left(\frac{1}{2}\right)+f\left(\frac{9}{2}\right)\left(\frac{1}{2}\right)+f(5)\left(\frac{1}{2}\right)+f\left(\frac{11}{2}\right)\left(\frac{1}{2}\right)$

$=0\left(\frac{1}{2}\right)+\frac{9}{4}\left(\frac{1}{2}\right)+5\left(\frac{1}{2}\right)+\frac{33}{4}\left(\frac{1}{2}\right)+12\left(\frac{1}{2}\right)+\frac{65}{4}\left(\frac{1}{2}\right)+21\left(\frac{1}{2}\right)+\frac{105}{4}\left(\frac{1}{2}\right)$

$=0+\frac{9}{8}+\frac{5}{2}+\frac{33}{8}+6+\frac{65}{8}+\frac{21}{2}+\frac{105}{8}=\frac{91}{2}=45.5$

d. $A=\int_2^6 (x^2-4)dx$

e. Use fnInt function:

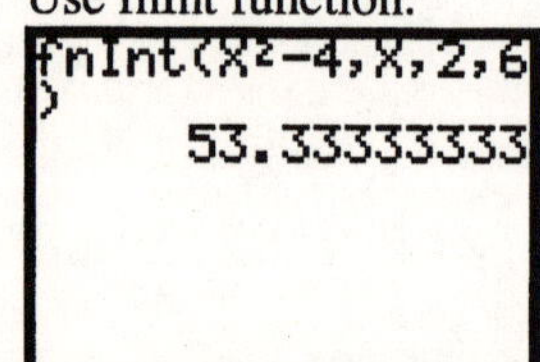

15. a. Graph $f(x) = x^3$, $[0, 4]$:

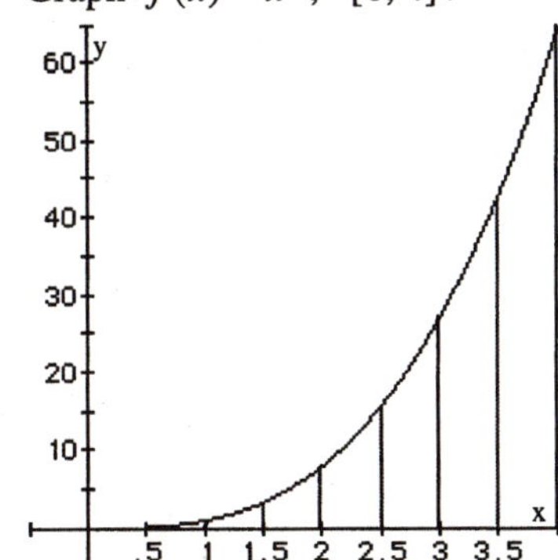

b. $A \approx f(0)(1) + f(1)(1) + f(2)(1) + f(3)(1) = 0(1) + 1(1) + 8(1) + 27(1) = 0 + 1 + 8 + 27 = 36$

c. $A \approx f(0)\left(\frac{1}{2}\right) + f\left(\frac{1}{2}\right)\left(\frac{1}{2}\right) + f(1)\left(\frac{1}{2}\right) + f\left(\frac{3}{2}\right)\left(\frac{1}{2}\right) + f(2)\left(\frac{1}{2}\right) + f\left(\frac{5}{2}\right)\left(\frac{1}{2}\right) + f(3)\left(\frac{1}{2}\right) + f\left(\frac{7}{2}\right)\left(\frac{1}{2}\right)$

$= 0\left(\frac{1}{2}\right) + \frac{1}{8}\left(\frac{1}{2}\right) + 1\left(\frac{1}{2}\right) + \frac{27}{8}\left(\frac{1}{2}\right) + 8\left(\frac{1}{2}\right) + \frac{125}{8}\left(\frac{1}{2}\right) + 27\left(\frac{1}{2}\right) + \frac{343}{8}\left(\frac{1}{2}\right)$

$= 0 + \frac{1}{16} + \frac{1}{2} + \frac{27}{16} + 4 + \frac{125}{16} + \frac{27}{2} + \frac{343}{16} = 49$

d. $A = \int_0^4 x^3\,dx$

e. Use fnInt function:

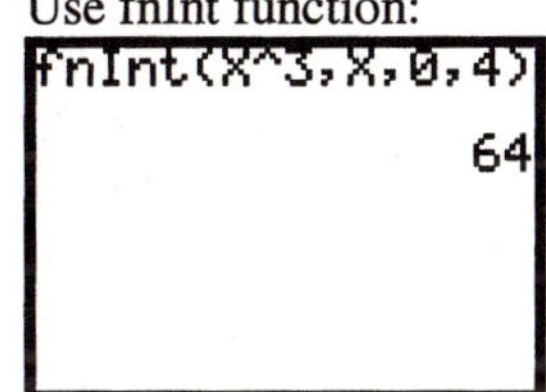

16. (a) Graph $f(x) = x^3$, $[1, 5]$:

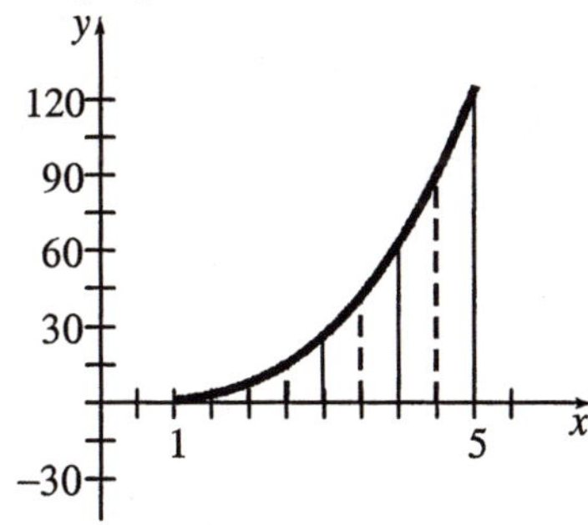

b. $A \approx f(1)(1) + f(2)(1) + f(3)(1) + f(4)(1) = 1(1) + 8(1) + 27(1) + 64(1) = 1 + 8 + 27 + 64 = 100$

c. $A \approx f(1)\left(\frac{1}{2}\right)+f\left(\frac{3}{2}\right)\left(\frac{1}{2}\right)+f(2)\left(\frac{1}{2}\right)+f\left(\frac{5}{2}\right)\left(\frac{1}{2}\right)+f(3)\left(\frac{1}{2}\right)+f\left(\frac{7}{2}\right)\left(\frac{1}{2}\right)+f(4)\left(\frac{1}{2}\right)+f\left(\frac{9}{2}\right)\left(\frac{1}{2}\right)$

$=1\left(\frac{1}{2}\right)+\frac{27}{8}\left(\frac{1}{2}\right)+8\left(\frac{1}{2}\right)+\frac{125}{8}\left(\frac{1}{2}\right)+27\left(\frac{1}{2}\right)+\frac{343}{8}\left(\frac{1}{2}\right)+64\left(\frac{1}{2}\right)+\frac{729}{8}\left(\frac{1}{2}\right)$

$=\frac{1}{2}+\frac{27}{16}+4+\frac{125}{16}+\frac{27}{2}+\frac{343}{16}+32+\frac{729}{16}=\frac{253}{2}=126.5$

d. $A=\int_1^5 x^3\,dx$

e. Use fnInt function:

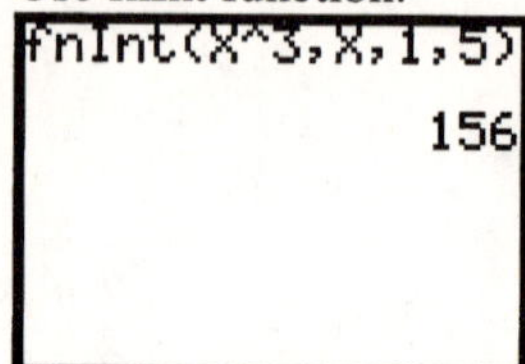

17. a. Graph $f(x)=\frac{1}{x}$, $[1, 5]$:

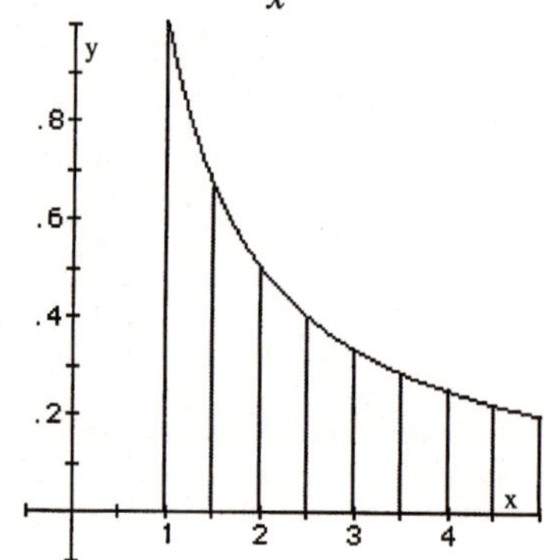

b. $A \approx f(1)(1)+f(2)(1)+f(3)(1)+f(4)(1)=1(1)+\frac{1}{2}(1)+\frac{1}{3}(1)+\frac{1}{4}(1)=1+\frac{1}{2}+\frac{1}{3}+\frac{1}{4}=\frac{25}{12}$

c. $A \approx f(1)\left(\frac{1}{2}\right)+f\left(\frac{3}{2}\right)\left(\frac{1}{2}\right)+f(2)\left(\frac{1}{2}\right)+f\left(\frac{5}{2}\right)\left(\frac{1}{2}\right)+f(3)\left(\frac{1}{2}\right)+f\left(\frac{7}{2}\right)\left(\frac{1}{2}\right)+f(4)\left(\frac{1}{2}\right)+f\left(\frac{9}{2}\right)\left(\frac{1}{2}\right)$

$=1\left(\frac{1}{2}\right)+\frac{2}{3}\left(\frac{1}{2}\right)+\frac{1}{2}\left(\frac{1}{2}\right)+\frac{2}{5}\left(\frac{1}{2}\right)+\frac{1}{3}\left(\frac{1}{2}\right)+\frac{2}{7}\left(\frac{1}{2}\right)+\frac{1}{4}\left(\frac{1}{2}\right)+\frac{2}{9}\left(\frac{1}{2}\right)$

$=\frac{1}{2}+\frac{1}{3}+\frac{1}{4}+\frac{1}{5}+\frac{1}{6}+\frac{1}{7}+\frac{1}{8}+\frac{1}{9}=\frac{4609}{2520}\approx 1.829$

d. $A=\int_1^5 \frac{1}{x}\,dx$

e. Use fnInt function:

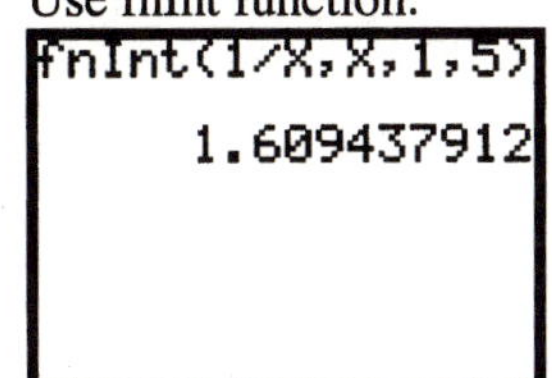

18. a. Graph $f(x)=\sqrt{x}$, $[0,4]$:

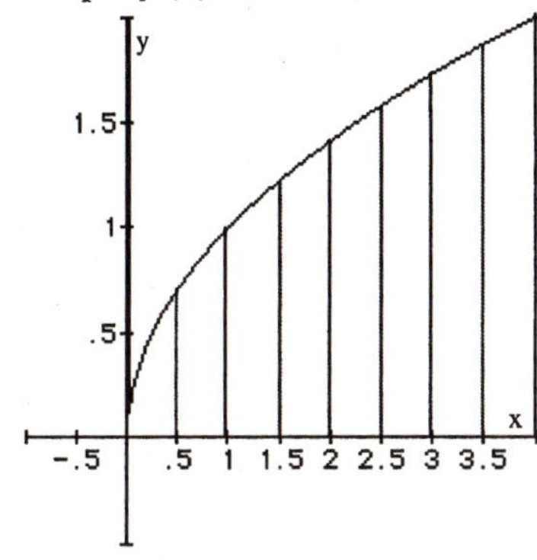

b. $A \approx f(0)(1)+f(1)(1)+f(2)(1)+f(3)(1)=0(1)+1(1)+\sqrt{2}(1)+\sqrt{3}(1)\ =1+\sqrt{2}+\sqrt{3}\approx 4.15$

c. $A \approx f(0)\left(\frac{1}{2}\right)+f\left(\frac{1}{2}\right)\left(\frac{1}{2}\right)+f(1)\left(\frac{1}{2}\right)+f\left(\frac{3}{2}\right)\left(\frac{1}{2}\right)+f(2)\left(\frac{1}{2}\right)+f\left(\frac{5}{2}\right)\left(\frac{1}{2}\right)+f(3)\left(\frac{1}{2}\right)+f\left(\frac{7}{2}\right)\left(\frac{1}{2}\right)$

$=0\left(\frac{1}{2}\right)+\sqrt{\frac{1}{2}}\left(\frac{1}{2}\right)+1\left(\frac{1}{2}\right)+\sqrt{\frac{3}{2}}\left(\frac{1}{2}\right)+\sqrt{2}\left(\frac{1}{2}\right)+\sqrt{\frac{5}{2}}\left(\frac{1}{2}\right)+\sqrt{3}\left(\frac{1}{2}\right)+\sqrt{\frac{7}{2}}\left(\frac{1}{2}\right)$

$=0+\frac{\sqrt{2}}{4}+\frac{1}{2}+\frac{\sqrt{6}}{4}+\frac{\sqrt{2}}{2}+\frac{\sqrt{10}}{4}+\frac{\sqrt{3}}{2}+\frac{\sqrt{14}}{4}\approx 4.77$

d. $A=\int_0^4 \sqrt{x}\,dx$

e. Use fnInt function:

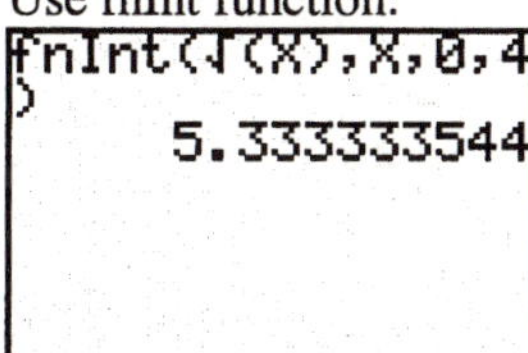

19. a. Graph $f(x) = e^x$, $[-1, 3]$:

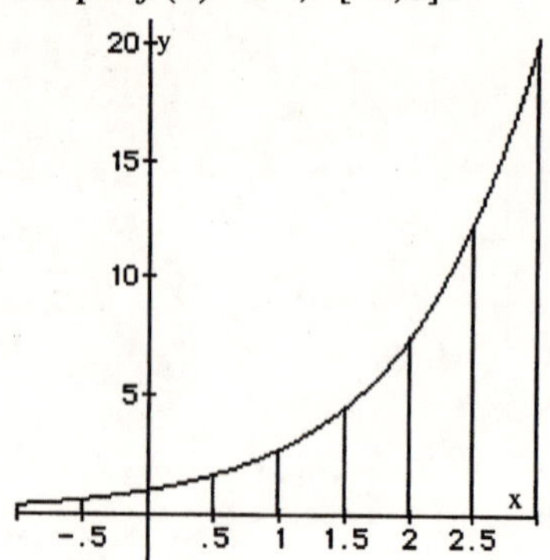

b. $A \approx \left(f(-1) + f(0) + f(1) + f(2)\right)(1) \approx (0.3679 + 1 + 2.7183 + 7.3891)(1) \approx 11.48$

c. $A \approx \left(f(-1) + f\left(-\frac{1}{2}\right) + f(0) + f\left(\frac{1}{2}\right) + f(1) + f\left(\frac{3}{2}\right) + f(2) + f\left(\frac{5}{2}\right)\right) \cdot \left(\frac{1}{2}\right)$

$\approx \left(0.3679 + 0.6065 + 1 + 1.6487 + 2.7183 + 4.4817 + 7.3891 + 12.1825\right)(0.5)$

$= 30.3947(0.5) \approx 15.20$

d. $A = \int_{-1}^{3} e^x \, dx$

e. Use fnInt function:

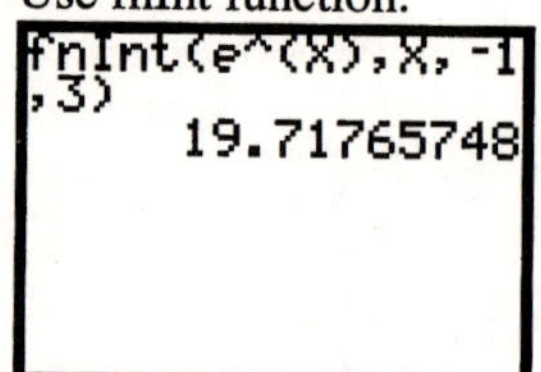

20. a. Graph $f(x) = \ln x$, $[3, 7]$:

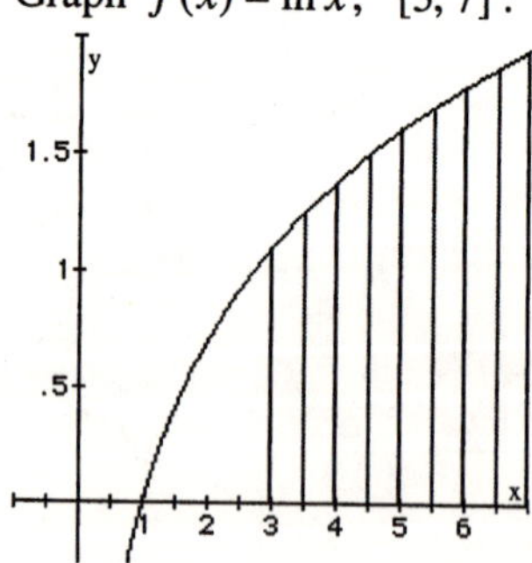

b. $A \approx \left(f(3) + f(4) + f(5) + f(6)\right)(1) = (1.0986 + 1.3863 + 1.6094 + 1.7918)(1) \approx 5.89$

c. $A \approx \left(f(3) + f\left(\frac{7}{2}\right) + f(4) + f\left(\frac{9}{2}\right) + f(5) + f\left(\frac{11}{2}\right) + f(6) + f\left(\frac{13}{2}\right)\right) \cdot \left(\frac{1}{2}\right)$

$= \left(1.0986 + 1.2528 + 1.3863 + 1.5041 + 1.6094 + 1.7047 + 1.7918 + 1.8718\right)(0.5)$

$= 12.2195(0.5) = 6.11$

d. $A = \int_3^7 \ln x \, dx$

e. Use fnInt function:

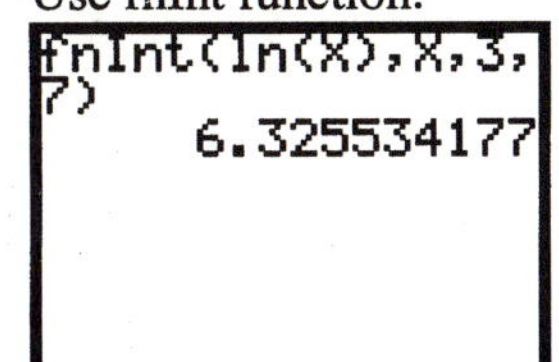

21. a. Graph $f(x) = \sin x, \quad [0, \pi]$:

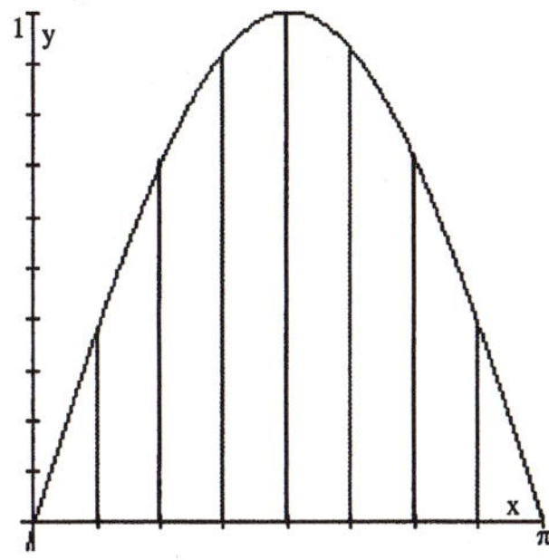

b. $A \approx \left(f(0) + f\left(\frac{\pi}{4}\right) + f\left(\frac{\pi}{2}\right) + f\left(\frac{3\pi}{4}\right)\right)\left(\frac{\pi}{4}\right) = \left(0 + \frac{\sqrt{2}}{2} + 1 + \frac{\sqrt{2}}{2}\right)\left(\frac{\pi}{4}\right)$

$= \left(1 + \sqrt{2}\right)\left(\frac{\pi}{4}\right) \approx 1.90$

c. $A \approx \left(f(0) + f\left(\frac{\pi}{8}\right) + f\left(\frac{\pi}{4}\right) + f\left(\frac{3\pi}{8}\right) + f\left(\frac{\pi}{2}\right) + f\left(\frac{5\pi}{8}\right) + f\left(\frac{3\pi}{4}\right) + f\left(\frac{7\pi}{8}\right)\right)\left(\frac{\pi}{8}\right)$

$\approx \left(0 + 0.3827 + 0.7071 + 0.9239 + 1 + 0.9239 + 0.7071 + 0.3827\right)\left(\frac{\pi}{8}\right)$

$= 5.0274\left(\frac{\pi}{8}\right) \approx 1.97$

d. $A = \int_0^{\pi} \sin x dx$

e. Use fnInt function:

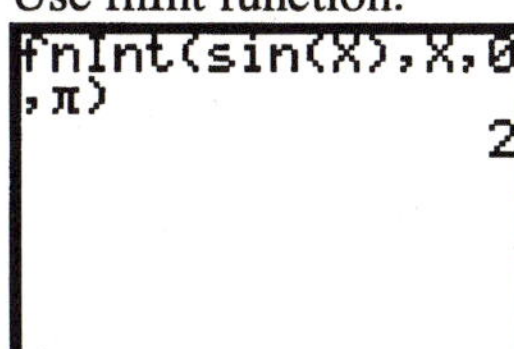

22. a. Graph $f(x)=\cos x,\ \left[0,\dfrac{\pi}{2}\right]$:

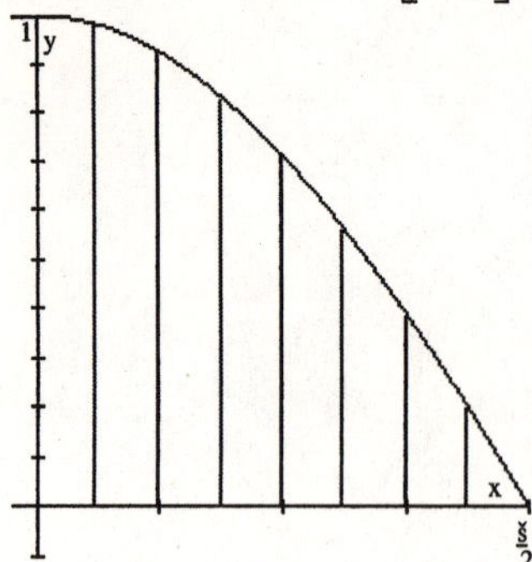

b. $A \approx \left(f(0)+f\left(\dfrac{\pi}{8}\right)+f\left(\dfrac{\pi}{4}\right)+f\left(\dfrac{3\pi}{8}\right)\right)\left(\dfrac{\pi}{8}\right)=(1+0.9239+0.7071+0.3827)\left(\dfrac{\pi}{8}\right)=(3.0137)\left(\dfrac{\pi}{8}\right)\approx 1.18$

c. $A \approx \left(f(0)+f\left(\dfrac{\pi}{16}\right)+f\left(\dfrac{\pi}{8}\right)+f\left(\dfrac{3\pi}{16}\right)+f\left(\dfrac{\pi}{4}\right)+f\left(\dfrac{5\pi}{16}\right)+f\left(\dfrac{3\pi}{8}\right)+f\left(\dfrac{7\pi}{16}\right)\right)\left(\dfrac{\pi}{16}\right)$

$\approx (1+0.9808+0.9239+0.8315+0.7071+0.5556+0.3827+0.1951)\left(\dfrac{\pi}{16}\right)$

$=5.5767\left(\dfrac{\pi}{16}\right)\approx 1.0950$

d. $A=\int_0^{\frac{\pi}{2}}\cos x\,dx$

e. Use fnInt function:

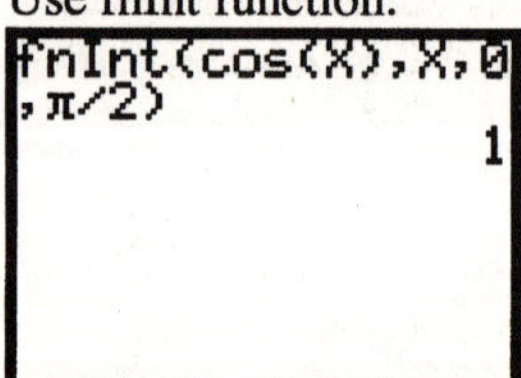

23. a. The integral represents the area under the graph of $f(x)=3x+1$ from $x=0$ to $x=4$.

b.

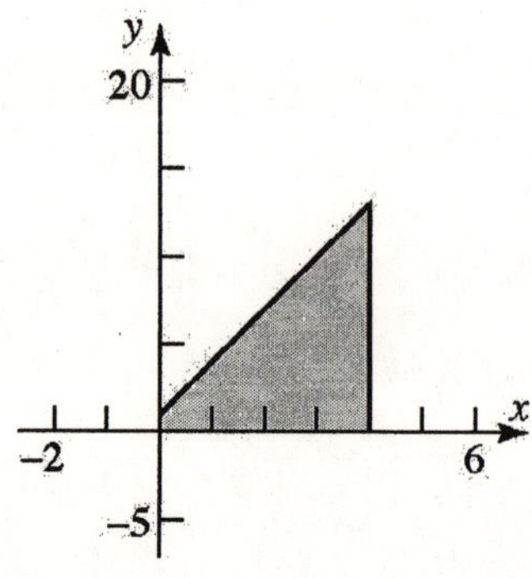

c.

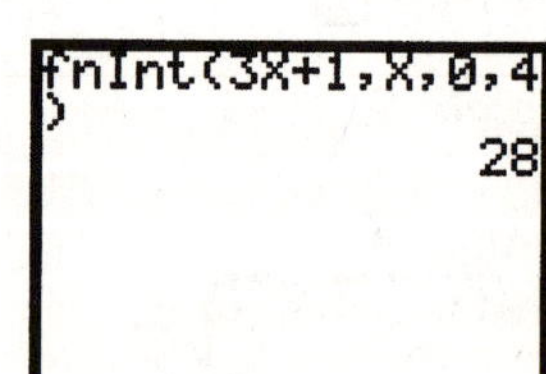

24. a. The integral represents the area under the graph of $f(x) = -2x + 7$ from $x = 1$ to $x = 3$.

b.

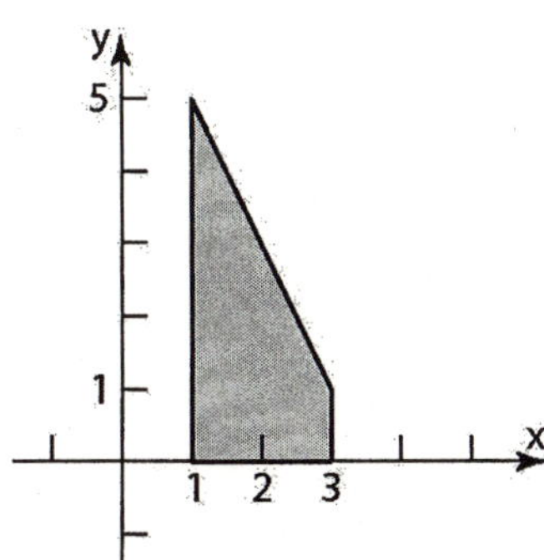

c.

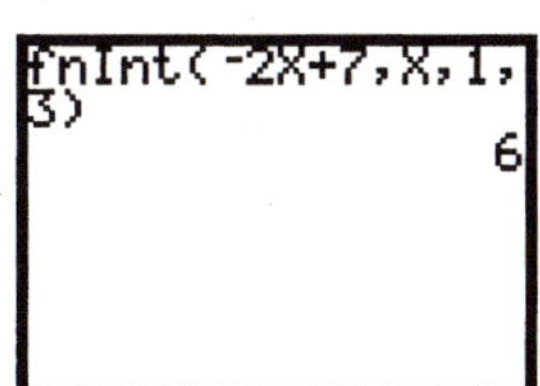

25. a. The integral represents the area under the graph of $f(x) = x^2 - 1$ from $x = 2$ to $x = 5$.

b.

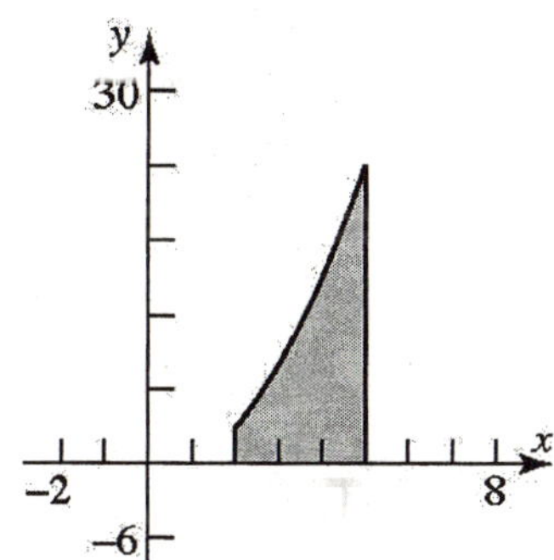

c.

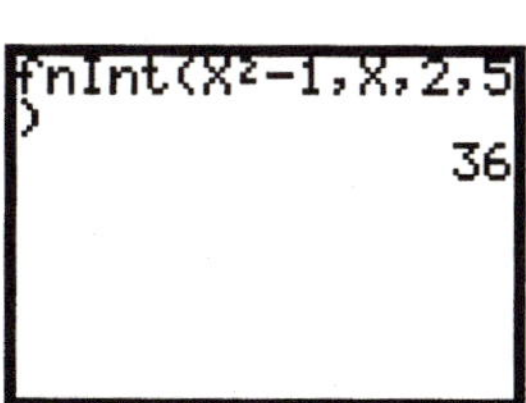

26. a. The integral represents the area under the graph of $f(x) = 16 - x^2$ from $x = 0$ to $x = 4$.

b.

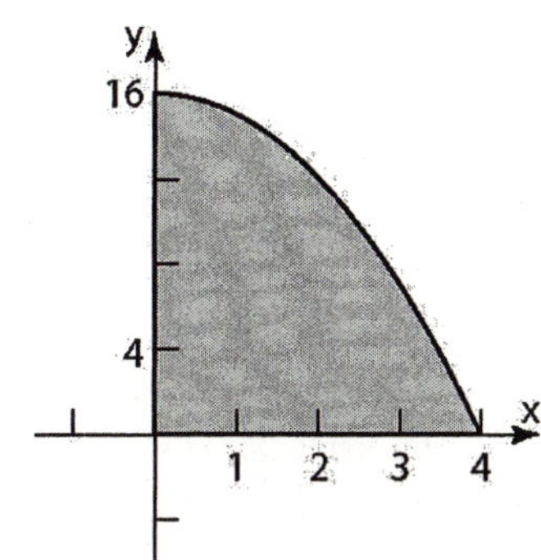

c.

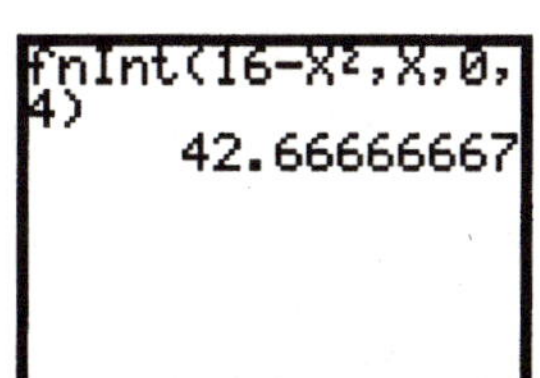

27. a. The integral represents the area under the graph of $f(x) = \sin x$ from $x = 0$ to $x = \frac{\pi}{2}$.

b.

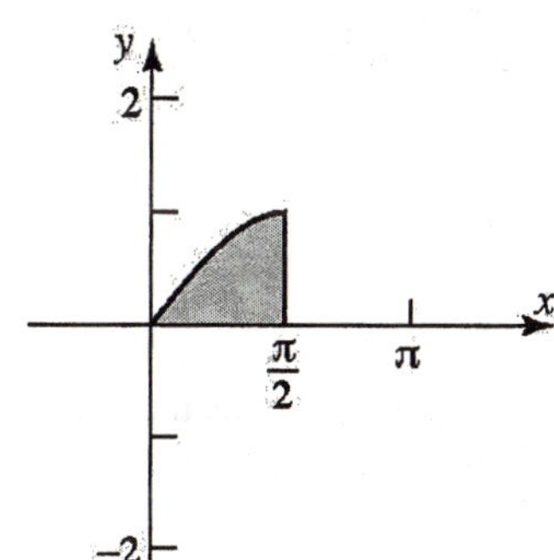

c.

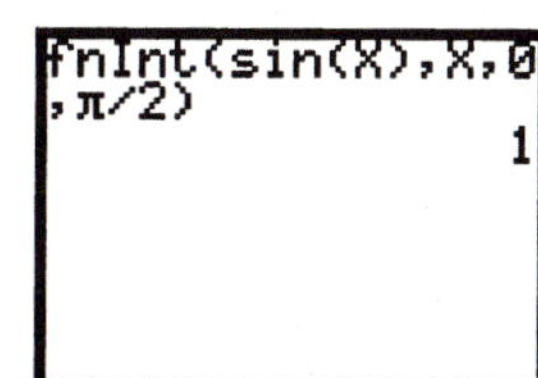

28. a. The integral represents the area under the graph of $f(x) = \cos x$ from $x = -\frac{\pi}{4}$ to $x = \frac{\pi}{4}$.

b.

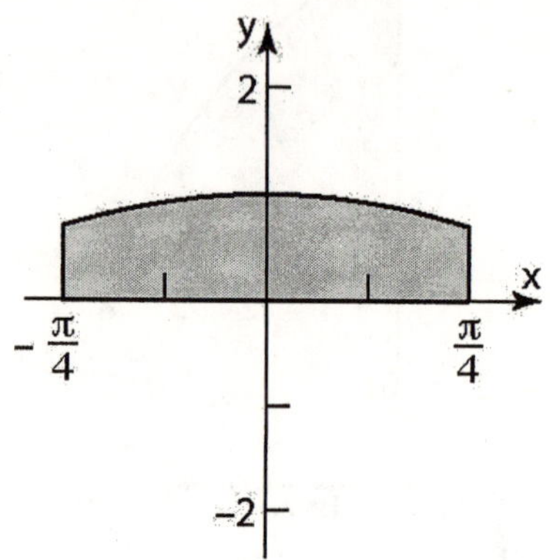

c.

fnInt(cos(X),X,-π/4,π/4)
1.414213562

29. a. The integral represents the area under the graph of $f(x) = e^x$ from $x = 0$ to $x = 2$.

b.

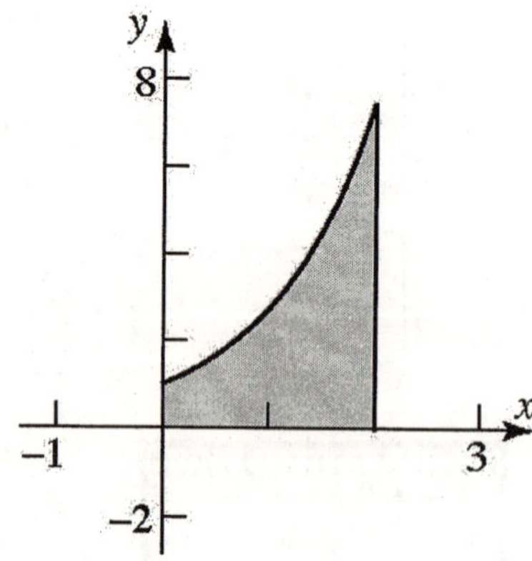

c.

fnInt(e^(X),X,0,2)
6.389056099

30. a. The integral represents the area under the graph of $f(x) = \ln x$ from $x = e$ to $x = 2e$.

b.

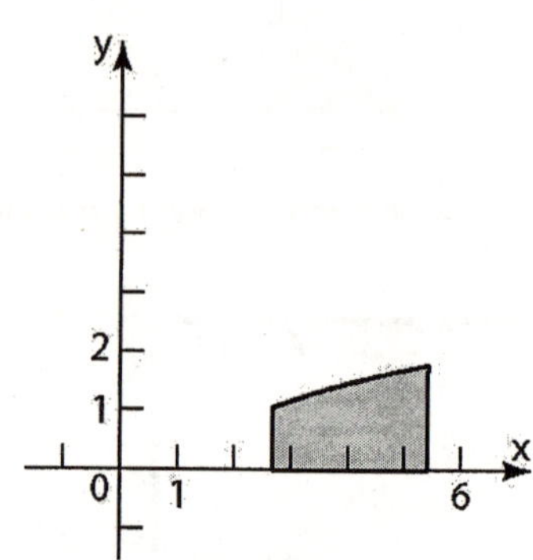

c.

fnInt(ln(X),X,e,2e)
3.768338771

31. Using left endpoints:

$n=2$: $0+0.5=0.5$

$n=4$: $0+0.125+0.25+0.375=0.75$

$n=10$: $0+0.02+0.04+0.06+\cdots+0.18=\frac{10}{2}(0+0.18)=0.9$

$n=100$: $0+0.0002+0.0004+0.0006+\cdots+0.0198=\frac{100}{2}(0+0.0198)=0.99$

Using right endpoints:

$n=2$: $0.5+1=1.5$

$n=4$: $0.125+0.25+0.375+0.5=1.25$

$n=10$: $0.02+0.04+0.06+\cdots+0.20=\frac{10}{2}(0.02+0.20)=1.1$

$n=100$: $0.0002+0.0004+0.0006+\cdots+0.02=\frac{100}{2}(0.0002+0.02)=1.01$

32. a. $f(x)=\sqrt{1-x^2}$

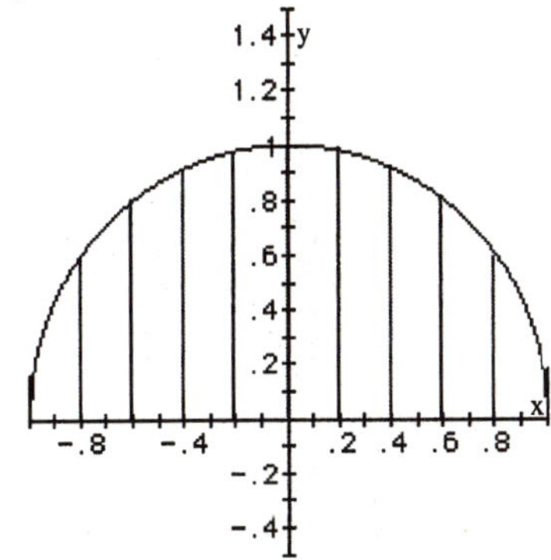

b. $A\approx\left(f(-1)+f(-0.6)+f(-0.2)+f(0.2)+f(0.6)\right)(0.4)$

$\approx(0+0.8+0.9798+0.9798+0.8)(0.4)=(3.5596)(0.4)\approx1.42$

c. $A\approx\left(f(-1)+f(-0.8)+f(-0.6)+f(-0.4)+f(-0.2)+f(0)+f(0.2)+f(0.4)+f(0.6)+f(0.8)\right)(0.2)$

$=7.5926(0.2)\approx1.52$

d. $A=\int_{-1}^{1}\sqrt{1-x^2}\,dx$

e. Use fnInt function:

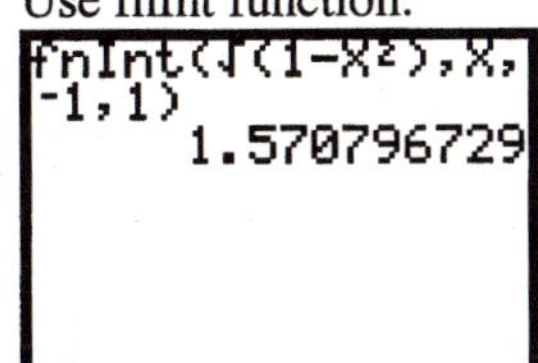

f. $A=\frac{1}{2}\pi\cdot1^2=\frac{\pi}{2}$

Chapter 13 Review

1. $\lim_{x\to 2}\left(3x^2-2x+1\right)=3(2)^2-2(2)+1$
$=12-4+1=9$

2. $\lim_{x\to 1}\left(-2x^3+x+4\right)=-2(1)^3+1+4$
$=-2+1+4=3$

3. $\lim_{x\to -2}\left(x^2+1\right)^2=\left(\lim_{x\to -2}\left(x^2+1\right)\right)^2$
$=\left((-2)^2+1\right)^2=5^2=25$

4. $\lim_{x\to -2}\left(x^3+1\right)^2=\left(\lim_{x\to -2}\left(x^3+1\right)\right)^2$
$=\left((-2)^3+1\right)^2$
$=(-7)^2=49$

5. $\lim_{x\to 3}\sqrt{x^2+7}=\sqrt{\lim_{x\to 3}(x^2+7)}$
$=\sqrt{3^2+7}$
$=\sqrt{16}=4$

6. $\lim_{x\to -2}\sqrt[3]{x+10}=\sqrt[3]{\lim_{x\to -2}(x+10)}$
$=\sqrt[3]{-2+10}$
$=\sqrt[3]{8}=2$

7. $\lim_{x\to 1^-}\sqrt{1-x^2}=\sqrt{\lim_{x\to 1^-}(1-x^2)}$
$=\sqrt{1-1^2}$
$=\sqrt{0}=0$

8. $\lim_{x\to 2^+}\sqrt{3x-2}=\sqrt{\lim_{x\to 2^+}(3x-2)}$
$=\sqrt{3(2)-2}$
$=\sqrt{4}=2$

9. $\lim_{x\to 2}(5x+6)^{3/2}=\left(\lim_{x\to 2}(5x+6)\right)^{3/2}$
$=\left(5(2)+6\right)^{3/2}$
$=16^{3/2}=64$

10. $\lim_{x\to -3}(15-3x)^{-3/2}=\left(\lim_{x\to -3}(15-3x)\right)^{-3/2}$
$=\left(15-3(-3)\right)^{-3/2}$
$=24^{-3/2}=\dfrac{1}{24^{3/2}}$
$=\dfrac{1}{\left(2\sqrt{6}\right)^3}=\dfrac{1}{48\sqrt{6}}=\dfrac{\sqrt{6}}{288}$

11. $\lim_{x\to -1}\left(\dfrac{x^2+x+2}{x^2-9}\right)=\dfrac{\lim_{x\to -1}(x^2+x+2)}{\lim_{x\to -1}(x^2-9)}$
$=\dfrac{(-1)^2+(-1)+2}{(-1)^2-9}$
$=\dfrac{2}{-8}=-\dfrac{1}{4}$

12. $\lim_{x\to 3}\left(\dfrac{3x+4}{x^2+1}\right)=\dfrac{\lim_{x\to 3}(3x+4)}{\lim_{x\to 3}(x^2+1)}$
$=\dfrac{3(3)+4}{3^2+1}=\dfrac{13}{10}$

13. $\lim_{x\to 1}\left(\dfrac{x-1}{x^3-1}\right)=\lim_{x\to 1}\left(\dfrac{x-1}{(x-1)(x^2+x+1)}\right)$
$=\lim_{x\to 1}\left(\dfrac{1}{x^2+x+1}\right)$
$=\dfrac{1}{1^2+1+1}=\dfrac{1}{3}$

14. $\lim_{x\to -1}\left(\dfrac{x^2-1}{x^2+x}\right)=\lim_{x\to -1}\left(\dfrac{(x-1)(x+1)}{x(x+1)}\right)$
$=\lim_{x\to -1}\left(\dfrac{x-1}{x}\right)$
$=\dfrac{-1-1}{-1}=\dfrac{-2}{-1}=2$

15. $\lim_{x\to -3}\left(\dfrac{x^2-9}{x^2-x-12}\right)=\lim_{x\to -3}\left(\dfrac{(x-3)(x+3)}{(x-4)(x+3)}\right)$
$=\lim_{x\to -3}\left(\dfrac{x-3}{x-4}\right)=\dfrac{-3-3}{-3-4}$
$=\dfrac{-6}{-7}=\dfrac{6}{7}$

16. $\lim_{x\to -3}\left(\frac{x^2+2x-3}{x^2-9}\right)=\lim_{x\to -3}\left(\frac{(x-1)(x+3)}{(x-3)(x+3)}\right)$

$=\lim_{x\to -3}\left(\frac{x-1}{x-3}\right)=\frac{-3-1}{-3-3}$

$=\frac{-4}{-6}=\frac{2}{3}$

17. $\lim_{x\to -1^-}\left(\frac{x^2-1}{x^3-1}\right)=\lim_{x\to -1^-}\left(\frac{(x+1)(x-1)}{(x-1)(x^2+x+1)}\right)$

$=\lim_{x\to -1^-}\left(\frac{x+1}{x^2+x+1}\right)$

$=\frac{-1+1}{(-1)^2+(-1)+1}$

$=\frac{0}{1}=0$

18. $\lim_{x\to 2^+}\left(\frac{x^2-4}{x^3-8}\right)=\lim_{x\to 2^+}\left(\frac{(x+2)(x-2)}{(x-2)(x^2+2x+4)}\right)$

$=\lim_{x\to 2^+}\left(\frac{x+2}{x^2+2x+4}\right)$

$=\frac{2+2}{2^2+2(2)+4}$

$=\frac{4}{12}=\frac{1}{3}$

19. $\lim_{x\to 2}\left(\frac{x^3-8}{x^3-2x^2+4x-8}\right)$

$=\lim_{x\to 2}\left(\frac{(x-2)(x^2+2x+4)}{x^2(x-2)+4(x-2)}\right)$

$=\lim_{x\to 2}\left(\frac{(x-2)(x^2+2x+4)}{(x-2)(x^2+4)}\right)$

$=\lim_{x\to 2}\left(\frac{x^2+2x+4}{x^2+4}\right)=\left(\frac{2^2+2(2)+4}{2^2+4}\right)$

$=\frac{12}{8}=\frac{3}{2}$

20. $\lim_{x\to 1}\left(\frac{x^3-1}{x^3-x^2+3x-3}\right)$

$=\lim_{x\to 1}\left(\frac{(x-1)(x^2+x+1)}{x^2(x-1)+3(x-1)}\right)$

$=\lim_{x\to 1}\left(\frac{(x-1)(x^2+x+1)}{(x-1)(x^2+3)}\right)$

$=\lim_{x\to 1}\left(\frac{x^2+x+1}{x^2+3}\right)$

$=\frac{1^2+1+1}{1^2+3}=\frac{3}{4}$

21. $\lim_{x\to 3}\left(\frac{x^4-3x^3+x-3}{x^3-3x^2+2x-6}\right)$

$=\lim_{x\to 3}\left(\frac{x^3(x-3)+1(x-3)}{x^2(x-3)+2(x-3)}\right)$

$=\lim_{x\to 3}\left(\frac{(x-3)(x^3+1)}{(x-3)(x^2+2)}\right)$

$=\lim_{x\to 3}\left(\frac{x^3+1}{x^2+2}\right)$

$=\frac{3^3+1}{3^2+2}=\frac{28}{11}$

22. $\lim_{x\to -1}\left(\frac{x^4+x^3+2x+2}{x^3+x^2}\right)$

$=\lim_{x\to -1}\left(\frac{x^3(x+1)+2(x+1)}{x^2(x+1)}\right)$

$=\lim_{x\to -1}\left(\frac{(x+1)(x^3+2)}{x^2(x+1)}\right)$

$=\lim_{x\to -1}\left(\frac{x^3+2}{x^2}\right)$

$=\frac{(-1)^3+2}{(-1)^2}=\frac{1}{1}=1$

23. $f(x)=3x^4-x^2+2;\ c=5$

1. $f(5)=3(5)^4-5^2+2=1852$
2. $\lim_{x\to 5^-} f(x)=3(5)^4-5^2+2=1852$
3. $\lim_{x\to 5^+} f(x)=3(5)^4-5^2+2=1852$

Thus, $f(x)$ is continuous at $c=5$.

24. $f(x)=\dfrac{x^2-9}{x+10};\ c=2$

1. $f(2)=\dfrac{2^2-9}{2+10}=-\dfrac{5}{12}$

2. $\lim\limits_{x\to 2^-} f(x)=\dfrac{2^2-9}{2+10}=-\dfrac{5}{12}$

3. $\lim\limits_{x\to 2^+} f(x)=\dfrac{2^2-9}{2+10}=-\dfrac{5}{12}$

Thus, $f(x)$ is continuous at $c=2$.

25. $f(x)=\dfrac{x^2-4}{x+2};\ c=-2$

Since $f(x)$ is not defined at $c=-2$, the function is not continuous at $c=-2$.

26. $f(x)=\dfrac{x^2+6x}{x^2-6x};\ c=0$

Since $f(x)$ is not defined at $c=0$, the function is not continuous at $c=0$.

27. $f(x)=\begin{cases}\dfrac{x^2-4}{x+2} & \text{if } x\neq -2\\ 4 & \text{if } x=-2\end{cases};\quad c=-2$

1. $f(-2)=4$

2. $\lim\limits_{x\to -2^-} f(x)=\lim\limits_{x\to -2^-}\left(\dfrac{x^2-4}{x+2}\right)$

$=\lim\limits_{x\to -2^-}\left(\dfrac{(x-2)(x+2)}{x+2}\right)$

$=\lim\limits_{x\to -2^-}(x-2)=-4$

Since $\lim\limits_{x\to -2^-} f(x)\neq f(-2)$, the function is not continuous at $c=-2$.

28. $f(x)=\begin{cases}\dfrac{x^2+6x}{x^2-6x} & \text{if } x\neq 0\\ 1 & \text{if } x=0\end{cases};\quad c=0$

1. $f(0)=1$

2. $\lim\limits_{x\to 0^-} f(x)=\lim\limits_{x\to 0^-}\left(\dfrac{x^2+6x}{x^2-6x}\right)$

$=\lim\limits_{x\to 0^-}\left(\dfrac{x(x+6)}{x(x-6)}\right)$

$=\lim\limits_{x\to 0^-}\left(\dfrac{x+6}{x-6}\right)=\dfrac{6}{-6}=-1$

Since $\lim\limits_{x\to 0^-} f(x)\neq f(0)$, the function is not continuous at $c=0$.

29. $f(x)=\begin{cases}\dfrac{x^2-4}{x+2} & \text{if } x\neq -2\\ -4 & \text{if } x=-2\end{cases};\quad c=-2$

1. $f(-2)=-4$

2. $\lim\limits_{x\to -2^-} f(x)=\lim\limits_{x\to -2^-}\left(\dfrac{x^2-4}{x+2}\right)$

$=\lim\limits_{x\to -2^-}\left(\dfrac{(x-2)(x+2)}{x+2}\right)$

$=\lim\limits_{x\to -2^-}(x-2)=-4$

3. $\lim\limits_{x\to -2^+} f(x)=\lim\limits_{x\to -2^+}\left(\dfrac{x^2-4}{x+2}\right)$

$=\lim\limits_{x\to -2^+}\left(\dfrac{(x-2)(x+2)}{x+2}\right)$

$=\lim\limits_{x\to -2^+}(x-2)=-4$

The function is continuous at $c=-2$.

30. $f(x)=\begin{cases}\dfrac{x^2+6x}{x^2-6x} & \text{if } x\neq 0\\ -1 & \text{if } x=0\end{cases};\quad c=0$

1. $f(0)=-1$

2. $\lim\limits_{x\to 0^-} f(x)=\lim\limits_{x\to 0^-}\left(\dfrac{x^2+6x}{x^2-6x}\right)$

$=\lim\limits_{x\to 0^-}\left(\dfrac{x(x+6)}{x(x-6)}\right)$

$=\lim\limits_{x\to 0^-}\left(\dfrac{x+6}{x-6}\right)=\dfrac{6}{-6}=-1$

3. $\lim\limits_{x\to 0^+} f(x)=\lim\limits_{x\to 0^+}\left(\dfrac{x^2+6x}{x^2-6x}\right)$

$=\lim\limits_{x\to 0^+}\left(\dfrac{x(x+6)}{x(x-6)}\right)$

$=\lim\limits_{x\to 0^+}\left(\dfrac{x+6}{x-6}\right)=\dfrac{6}{-6}=-1$

The function is continuous at $c=0$.

31. Domain:
$\{x\mid -6\le x<2 \text{ or } 2<x<5 \text{ or } 5<x\le 6\}$

32. Range: $(-\infty,\infty)$

33. x-intercepts: 1, 6

34. y-intercept: 4

35. $f(-6)=2;\ f(-4)=1$

36. $f(-2)=2;\ f(6)=0$

37. $\lim\limits_{x\to -4^-} f(x)=4$

38. $\lim\limits_{x\to -4^+} f(x)=-2$

39. $\lim\limits_{x\to -2^-} f(x)=-2$

40. $\lim\limits_{x\to -2^+} f(x)=2$

41. $\lim\limits_{x\to 2^-} f(x)=-\infty$

42. $\lim\limits_{x\to 2^+} f(x)=\infty$

43. $\lim\limits_{x\to 0} f(x)$ does not exist because
$\lim\limits_{x\to 0^-} f(x)=4\neq \lim\limits_{x\to 0^+} f(x)=1$

44. $\lim\limits_{x\to 2} f(x)$ does not exist because
$\lim\limits_{x\to 2^-} f(x)\neq \lim\limits_{x\to 2^+} f(x)$

45. f is not continuous at -2 because
$\lim\limits_{x\to -2^-} f(x)=-\infty\neq \lim\limits_{x\to -2^+} f(x)=\infty$

46. f is not continuous at -4 because
$\lim\limits_{x\to -4^-} f(x)=4\neq \lim\limits_{x\to -4^+} f(x)=-2$

47. f is not continuous at 0 because
$\lim\limits_{x\to 0^-} f(x)=4\neq \lim\limits_{x\to 0^+} f(x)=1$

48. f is not continuous at 2 because f is not defined at $x=2$.

49. f is continuous at 4 because
$f(4)=\lim\limits_{x\to 4^-} f(x)=\lim\limits_{x\to 4^+} f(x)$

50. f is not continuous at 5 because f is not defined at $x=5$.

51. $R(x)=\dfrac{x+4}{x^2-16}=\dfrac{x+4}{(x-4)(x+4)}$. The domain of R is $\{x\mid x\neq -4, x\neq 4\}$. Thus R is discontinuous at both -4 and 4.

$\lim\limits_{x\to 4^-} R(x)=\lim\limits_{x\to 4^-}\left(\dfrac{x+4}{(x-4)(x+4)}\right)$

$=\lim\limits_{x\to 4^-}\left(\frac{1}{x-4}\right)=-\infty$

since when $x<4$, $\dfrac{1}{x-4}<0$, and as x approaches 4, $\frac{1}{x-4}$ becomes unbounded.

$\lim\limits_{x\to 4^+} R(x)=\lim\limits_{x\to 4^+}\left(\frac{1}{x-4}\right)=\infty$ since when $x>4, \frac{1}{x-4}>0$, and as x approaches 4, $\frac{1}{x-4}$ becomes unbounded. Thus, there is a vertical asymptote at $x=4$.

$\lim\limits_{x\to -4} R(x)=\lim\limits_{x\to -4}\left(\frac{1}{x-4}\right)=-\dfrac{1}{8}$.

Thus, there is a hole in the graph at $\left(-4,-\frac{1}{8}\right)$.

52. $R(x)=\dfrac{3x^2+6x}{x^2-4}=\dfrac{3x(x+2)}{(x-2)(x+2)}$. The domain of R is $\{x \mid x \neq -2, x \neq 2\}$. Thus R is discontinuous at both –2 and 2.

$$\lim_{x\to 2^-} R(x)=\lim_{x\to 2^-}\left(\frac{3x(x+2)}{(x-2)(x+2)}\right)$$
$$=\lim_{x\to 2^-}\left(\frac{3x}{x-2}\right)=-\infty$$

since when $0<x<2$, $\dfrac{3x}{x-2}<0$, and as x approaches 2, $\dfrac{3x}{x-2}$ becomes unbounded.

$\lim\limits_{x\to 2^+} R(x)=\lim\limits_{x\to 2^+}\left(\dfrac{3x}{x-2}\right)=\infty$ since when $x>2$, $\dfrac{3x}{x-2}>0$, and as x approaches 2, $\dfrac{3x}{x-2}$ becomes unbounded. Thus, there is a vertical asymptote at $x=2$.

$$\lim_{x\to -2} R(x)=\lim_{x\to -2}\left(\frac{3x}{x-2}\right)=\frac{-6}{-4}=\frac{3}{2}.$$

Thus, there is a hole in the graph at $\left(-2,\dfrac{3}{2}\right)$.

53. $$R(x)=\frac{x^3-2x^2+4x-8}{x^2-11x+18}$$
$$=\frac{x^2(x-2)+4(x-2)}{(x-9)(x-2)}$$
$$=\frac{(x-2)(x^2+4)}{(x-9)(x-2)}$$
$$=\frac{x^2+4}{x-9},\ x\neq 2$$

Undefined at $x=2$ and $x=9$. There is a vertical asymptote where $x-9=0$. $x=9$ is a vertical asymptote. There is a hole in the graph at $x=2$, the point $\left(2,-\dfrac{8}{7}\right)$.

54. $$R(x)=\frac{x^3+3x^2-2x-6}{x^2+x-6}$$
$$=\frac{x^2(x+3)-2(x+3)}{(x+3)(x-2)}$$
$$=\frac{(x+3)(x^2-2)}{(x+3)(x-2)}$$
$$=\frac{x^2-2}{x-2},\ x\neq -3$$

Undefined at $x=-3$ and $x=2$. There is a vertical asymptote where $x-2=0$. $x=2$ is a vertical asymptote. There is a hole in the graph at $x=-3$., the point $\left(-3,-\dfrac{7}{5}\right)$.

55. $f(x)=2x^2+8x$ at $(1,10)$

$$m_{\tan}=\lim_{x\to 1}\left(\frac{f(x)-f(1)}{x-1}\right)=\lim_{x\to 1}\left(\frac{2x^2+8x-10}{x-1}\right)$$
$$=\lim_{x\to 1}\left(\frac{2(x+5)(x-1)}{x-1}\right)=\lim_{x\to 1}\left(2(x+5)\right)$$
$$=2(1+5)=12$$

Tangent Line: $y-10=12(x-1)$
$$y-10=12x-12$$
$$y=12x-2$$

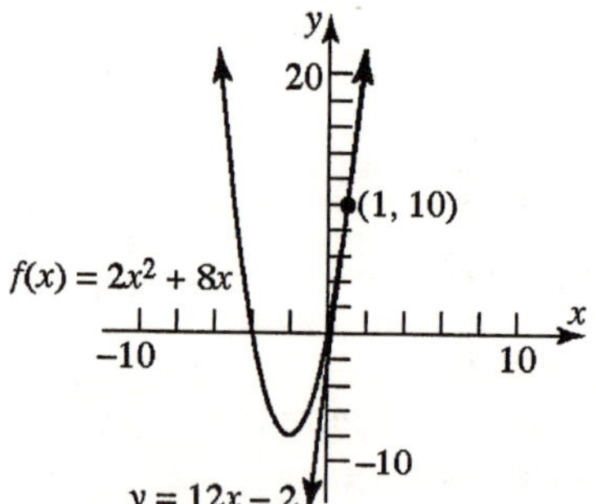

56. $f(x)=3x^2-6x$ at $(0,0)$

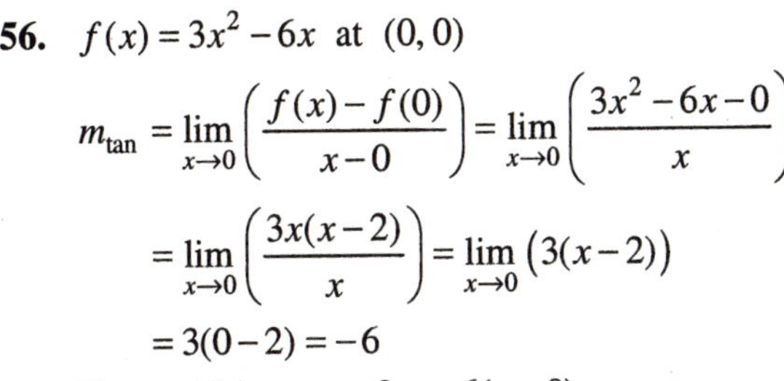

$$m_{\tan}=\lim_{x\to 0}\left(\frac{f(x)-f(0)}{x-0}\right)=\lim_{x\to 0}\left(\frac{3x^2-6x-0}{x}\right)$$

$$=\lim_{x\to 0}\left(\frac{3x(x-2)}{x}\right)=\lim_{x\to 0}\left(3(x-2)\right)$$

$$=3(0-2)=-6$$

Tangent Line: $y-0=-6(x-0)$

$$y=-6x$$

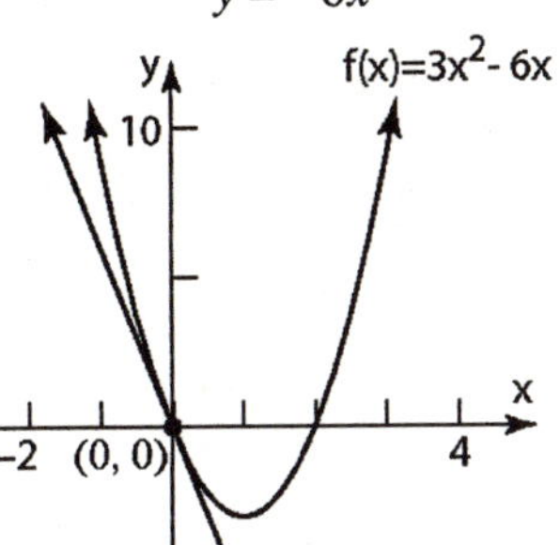

57. $f(x)=x^2+2x-3$ at $(-1,-4)$

$$m_{\tan}=\lim_{x\to -1}\left(\frac{f(x)-f(-1)}{x+1}\right)$$

$$=\lim_{x\to -1}\left(\frac{x^2+2x-3-(-4)}{x+1}\right)$$

$$=\lim_{x\to -1}\left(\frac{x^2+2x+1}{x+1}\right)$$

$$=\lim_{x\to -1}\left(\frac{(x+1)^2}{x+1}\right)$$

$$=\lim_{x\to -1}(x+1)$$

$$=-1+1=0$$

Tangent Line: $y-(-4)=0(x-(-1))$

$$y+4=0$$

$$y=-4$$

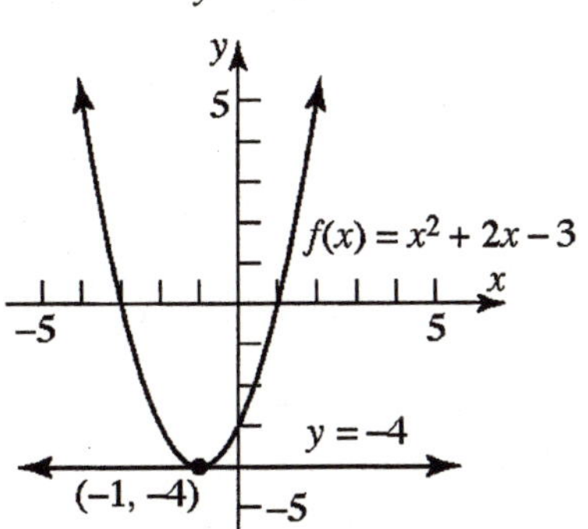

58. $f(x)=2x^2+5x-3$ at $(1,4)$

$$m_{\tan}=\lim_{x\to 1}\left(\frac{f(x)-f(1)}{x-1}\right)$$

$$=\lim_{x\to 1}\left(\frac{2x^2+5x-3-4}{x-1}\right)$$

$$=\lim_{x\to 1}\left(\frac{2x^2+5x-7}{x-1}\right)$$

$$=\lim_{x\to 1}\left(\frac{(x-1)(2x+7)}{x-1}\right)$$

$$=\lim_{x\to 1}(2x+7)$$

$$=2(1)+7=9$$

Tangent Line: $y-4=9(x-1)$

$$y-4=9x-9$$

$$y=9x-5$$

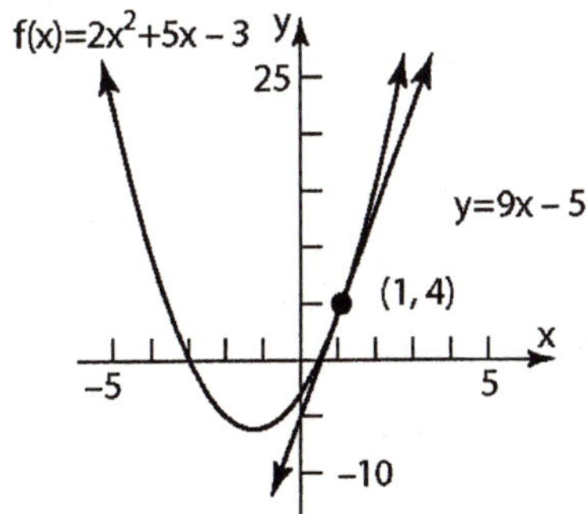

59. $f(x)=x^3+x^2$ at $(2,12)$

$$m_{\tan}=\lim_{x\to 2}\left(\frac{f(x)-f(2)}{x-2}\right)=\lim_{x\to 2}\left(\frac{x^3+x^2-12}{x-2}\right)$$

$$=\lim_{x\to 2}\left(\frac{x^3-2x^2+3x^2-12}{x-2}\right)$$

$$=\lim_{x\to 2}\left(\frac{x^2(x-2)+3(x-2)(x+2)}{x-2}\right)$$

$$=\lim_{x\to 2}\left(\frac{(x-2)(x^2+3x+6)}{x-2}\right)$$

$$=\lim_{x\to 2}(x^2+3x+6)=4+6+6=16$$

Tangent Line: $y - 12 = 16(x-2)$

$$y - 12 = 16x - 32$$
$$y = 16x - 20$$

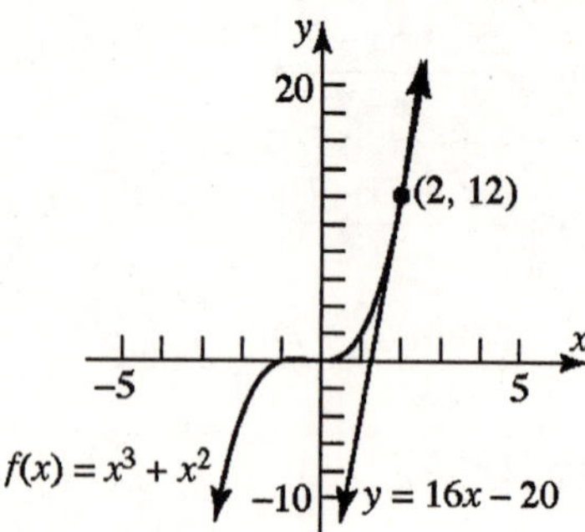

60. $f(x) = x^3 - x^2$ at $(1, 0)$

$$m_{\tan} = \lim_{x\to 1}\left(\frac{f(x)-f(1)}{x-1}\right) = \lim_{x\to 1}\left(\frac{x^3 - x^2 - 0}{x-1}\right)$$
$$= \lim_{x\to 1}\left(\frac{x^2(x-1)}{x-1}\right) = \lim_{x\to 1}(x^2) = 1^2 = 1$$

Tangent Line: $y - 0 = 1(x-1)$

$$y = x - 1$$

y
f(x)=x³ - x²
2
y=x-1
x
−2
(1, 0)
3
−3

61. $f(x) = -4x^2 + 5$ at 3

$$f'(3) = \lim_{x\to 3}\left(\frac{f(x)-f(3)}{x-3}\right)$$
$$= \lim_{x\to 3}\left(\frac{-4x^2+5-(-31)}{x-3}\right)$$
$$= \lim_{x\to 3}\left(\frac{-4x^2+36}{x-3}\right)$$
$$= \lim_{x\to 3}\left(\frac{-4(x^2-9)}{x-3}\right)$$
$$= \lim_{x\to 3}\left(\frac{-4(x-3)(x+3)}{x-3}\right)$$
$$= \lim_{x\to 3}\left((-4)(x+3)\right)$$
$$= -4(6) = -24$$

62. $f(x) = -4 + 3x^2$ at 1

$$f'(1) = \lim_{x\to 1}\left(\frac{f(x)-f(1)}{x-1}\right)$$
$$= \lim_{x\to 1}\left(\frac{-4+3x^2-(-1)}{x-1}\right)$$
$$= \lim_{x\to 1}\left(\frac{3x^2-3}{x-1}\right)$$
$$= \lim_{x\to 1}\left(\frac{3(x^2-1)}{x-1}\right)$$
$$= \lim_{x\to 1}\left(\frac{3(x-1)(x+1)}{x-1}\right)$$
$$= \lim_{x\to 1}\left(3(x+1)\right) = 3(2) = 6$$

63. $f(x) = x^2 - 3x$ at 0

$$f'(0) = \lim_{x\to 0}\left(\frac{f(x)-f(0)}{x-0}\right)$$
$$= \lim_{x\to 0}\left(\frac{x^2-3x-0}{x}\right)$$
$$= \lim_{x\to 0}\left(\frac{x(x-3)}{x}\right)$$
$$= \lim_{x\to 0}(x-3) = -3$$

64. $f(x) = 2x^2 + 4x$ at -1

$$f'(-1) = \lim_{x\to -1}\left(\frac{f(x)-f(-1)}{x-(-1)}\right)$$
$$= \lim_{x\to -1}\left(\frac{2x^2+4x-(-2)}{x+1}\right)$$
$$= \lim_{x\to -1}\left(\frac{2(x^2+2x+1)}{x+1}\right)$$
$$= \lim_{x\to -1}\left(\frac{2(x+1)^2}{x+1}\right)$$
$$= \lim_{x\to -1}\left(2(x+1)\right) = 2(0) = 0$$

65. $f(x)=2x^2+3x+2$ at 1

$$f'(1)=\lim_{x\to1}\left(\frac{f(x)-f(1)}{x-1}\right)$$
$$=\lim_{x\to1}\left(\frac{2x^2+3x+2-7}{x-1}\right)$$
$$=\lim_{x\to1}\left(\frac{2x^2+3x-5}{x-1}\right)$$
$$=\lim_{x\to1}\left(\frac{(2x+5)(x-1)}{x-1}\right)$$
$$=\lim_{x\to1}(2x+5)=7$$

66. $f(x)=3x^2-4x+1$ at 2

$$f'(2)=\lim_{x\to2}\left(\frac{f(x)-f(2)}{x-2}\right)$$
$$=\lim_{x\to2}\left(\frac{3x^2-4x+1-5}{x-2}\right)$$
$$=\lim_{x\to2}\left(\frac{3x^2-4x-4}{x-2}\right)$$
$$=\lim_{x\to2}\left(\frac{(3x+2)(x-2)}{x-2}\right)$$
$$=\lim_{x\to2}(3x+2)=8$$

67. Use nDeriv:

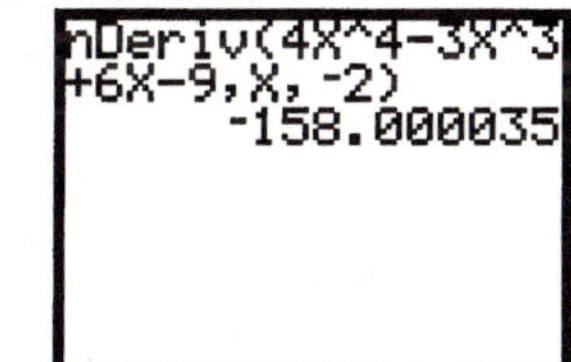

68. Use nDeriv:

69. Use nDeriv:

70. Use nDeriv:

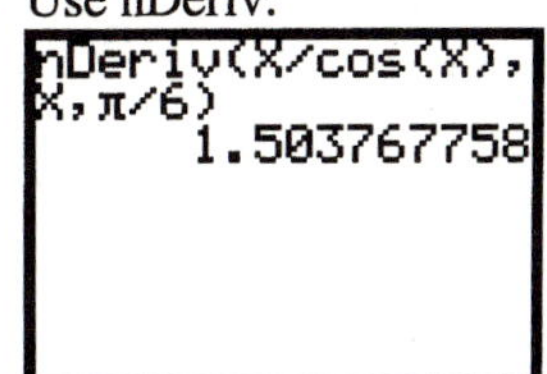

71. a. $-16t^2+96t+112=0$

$-16(t^2-6t-7)=0$

$-16(t+1)(t-7)=0$

$t=-1$ or $t=7$

The ball strikes the ground after 7 seconds in the air.

b. $-16t^2+96t+112=112$

$-16t^2+96t=0$

$-16t(t-6)=0$

$t=0$ or $t=6$

The ball passes the rooftop after 6 seconds.

c.
$$\frac{\Delta s}{\Delta t}=\frac{s(2)-s(0)}{2-0}$$
$$=\frac{-16(2)^2+96(2)+112-112}{2}$$
$$=\frac{128}{2}=64\text{ feet/sec}$$

d.
$$s'(t_0)=\lim_{t\to t_0}\left(\frac{s(t)-s(t_0)}{t-t_0}\right)$$
$$=\lim_{t\to t_0}\left(\frac{-16t^2+16t_0{}^2+96t-96t_0}{t-t_0}\right)$$
$$=\lim_{t\to t_0}\left(\frac{-16\left(t^2-t_0{}^2\right)+96(t-t_0)}{t-t_0}\right)$$
$$=\lim_{t\to t_0}\left(\frac{-16(t-t_0)(t+t_0)+96(t-t_0)}{t-t_0}\right)$$
$$=\lim_{t\to t_0}\left(\frac{(t-t_0)(-16(t+t_0)+96)}{t-t_0}\right)$$
$$=\lim_{t\to t_0}\left(-16(t+t_0)+96\right)$$
$$=\left(-16(t_0+t_0)+96\right)$$
$$=-32t_0+96\quad\text{ft/sec}$$

e. $s'(2)=-32(2)+96=-64+96=32$ feet/sec

f. $s'(t)=0$

$$-32t+96=0$$
$$-32t=-96$$
$$t=3 \text{ seconds}$$

g. $s'(6)=-32(6)+96$

$$=-192+96$$
$$=-96 \text{ feet/sec}$$

h. $s'(7)=-32(7)+96$

$$=-224+96$$
$$=-128 \text{ feet/sec}$$

72. $A'(2)=\lim_{r\to 2}\left(\dfrac{A(r)-A(2)}{r-2}\right)$

$$=\lim_{r\to 2}\left(\frac{\pi r^2-4\pi}{r-2}\right)=\lim_{r\to 2}\left(\frac{\pi(r^2-4)}{r-2}\right)$$
$$=\lim_{r\to 2}\left(\frac{\pi(r-2)(r+2)}{r-2}\right)$$
$$=\lim_{r\to 2}\left(\pi(r+2)\right)$$
$$=\pi(2+2)=4\pi \ \text{ft}^2/\text{ft}$$

At the instant when $r=2$ feet, the area is increasing at a rate of 4π square feet per foot.

$$\frac{\Delta A}{\Delta r}=\frac{A(3)-A(2)}{3-2}=\frac{9\pi-4\pi}{1}=5\pi \ \text{ft}^2/\text{ft}$$

$$\frac{\Delta A}{\Delta r}=\frac{A(2.5)-A(2)}{2.5-2}=\frac{6.25\pi-4\pi}{0.5}$$
$$=\frac{2.25\pi}{0.5}=4.5\pi \ \text{ft}^2/\text{ft}$$

$$\frac{\Delta A}{\Delta r}=\frac{A(2.1)-A(2)}{2.1-2}=\frac{4.41\pi-4\pi}{0.1}$$
$$=\frac{0.41\pi}{0.1}=4.1\pi \ \text{ft}^2/\text{ft}$$

73. a. $\dfrac{\Delta R}{\Delta x}=\dfrac{8775-2340}{130-25}=\dfrac{6435}{105}\approx \$61.29/\text{watch}$

b. $\dfrac{\Delta R}{\Delta x}=\dfrac{6975-2340}{90-25}=\dfrac{4635}{65}\approx \$71.31/\text{watch}$

c. $\dfrac{\Delta R}{\Delta x}=\dfrac{4375-2340}{50-25}=\dfrac{2035}{25}\approx \$81.40/\text{watch}$

d. $R(x)=-0.25x^2+100.01x-1.24$

e. $R'(25)=\lim_{x\to 25}\left(\dfrac{R(x)-R(25)}{x-25}\right)$

$$=\lim_{x\to 25}\left(\frac{-0.25x^2+100.014x-2344}{x-25}\right)$$
$$=\lim_{x\to 25}\left(\frac{(x-25)(-0.25x+93.76)}{x-25}\right)$$
$$=\lim_{x\to 25}\left(-0.25x+93.76\right)$$
$$=-0.25(25)+93.76=\$87.51 \text{ /watch}$$

74. a. $\dfrac{\Delta s}{\Delta t}=\dfrac{256-16}{4-1}=\dfrac{240}{3}=80 \text{ ft/sec}$

b. $\dfrac{\Delta s}{\Delta t}=\dfrac{144-16}{3-1}=\dfrac{128}{2}=64 \text{ feet/sec}$

c. $\dfrac{\Delta s}{\Delta t}=\dfrac{64-16}{2-1}=\dfrac{48}{1}=48 \text{ feet/sec}$

d. $s(t)=16t^2$

e. $s'(1)=\lim_{t\to 1}\left(\dfrac{s(t)-s(1)}{t-1}\right)$

$$=\lim_{t\to 1}\left(\frac{16t^2-16}{t-1}\right)$$
$$=\lim_{t\to 1}\left(\frac{16(t^2-1)}{t-1}\right)$$
$$=\lim_{t\to 1}\left(\frac{16(t+1)(t-1)}{t-1}\right)$$
$$=\lim_{t\to 1}\left(16(t+1)\right)$$
$$=16(2)=32 \text{ feet/sec}$$

75. a. 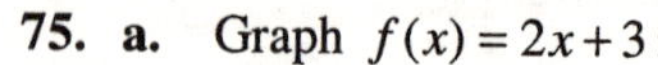
Graph $f(x)=2x+3$:

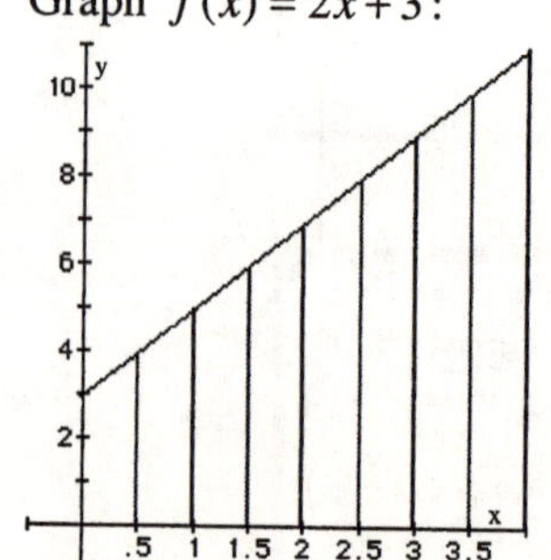

b. $A \approx \left(f(0)+f(1)+f(2)+f(3)\right)(1)$
$= (3+5+7+9)(1)$
$= 24(1) = 24$

c. $A \approx \left(f(1)+f(2)+f(3)+f(4)\right)(1)$
$= (5+7+9+11)(1)$
$= 32(1) = 32$

d. $A \approx \left(f(0)+f\left(\frac{1}{2}\right)+f(1)+f\left(\frac{3}{2}\right)+f(2) + f\left(\frac{5}{2}\right)+f(3)+f\left(\frac{7}{2}\right)\right)\cdot\left(\frac{1}{2}\right)$
$= (3+4+5+6+7+8+9+10)\left(\frac{1}{2}\right)$
$= 52\left(\frac{1}{2}\right) = 26$

e. $A \approx \left(f\left(\frac{1}{2}\right)+f(1)+f\left(\frac{3}{2}\right)+f(2)+f\left(\frac{5}{2}\right) + f(3)+f\left(\frac{7}{2}\right)+f(4)\right)\cdot\left(\frac{1}{2}\right)$
$= (4+5+6+7+8+9+10+11)\left(\frac{1}{2}\right)$
$= 60\left(\frac{1}{2}\right) = 30$

f. The actual area is the area of a trapezoid:
$A = \frac{1}{2}(3+11)(4) = \frac{56}{2} = 28$

76. a. Graph $f(x) = -2x+8$:

b. $A \approx \left(f(0)+f(1)+f(2)+f(3)\right)(1)$
$= \left(8+6+4+2\right)(1)$
$= 8+6+4+2 = 20$

c. $A \approx \left(f(1)+f(2)+f(3)+f(4)\right)(1)$
$= \left(6+4+2+0\right)(1) = 12$

d. $A \approx \left(f(0)+f\left(\frac{1}{2}\right)+f(1)+f\left(\frac{3}{2}\right)+f(2) + f\left(\frac{5}{2}\right)+f(3)+f\left(\frac{7}{2}\right)\right)\left(\frac{1}{2}\right)$
$= \left(8+7+6+5+4+3+2+1\right)\left(\frac{1}{2}\right)$
$= 36\left(\frac{1}{2}\right) = 18$

e. $A \approx \left(f\left(\frac{1}{2}\right)+f(1)+f\left(\frac{3}{2}\right)+f(2)+f\left(\frac{5}{2}\right) + f(3)+f\left(\frac{7}{2}\right)+f(4)\right)\left(\frac{1}{2}\right)$
$= \left(7+6+5+4+3+2+1+0\right)\left(\frac{1}{2}\right)$
$= 28\left(\frac{1}{2}\right) = 14$

f. The actual area is the area of a triangle:
$A = \frac{1}{2}(8)(4) = \frac{32}{2} = 16$

77. a. Graph $f(x) = 4 - x^2$, $[-1, 2]$:

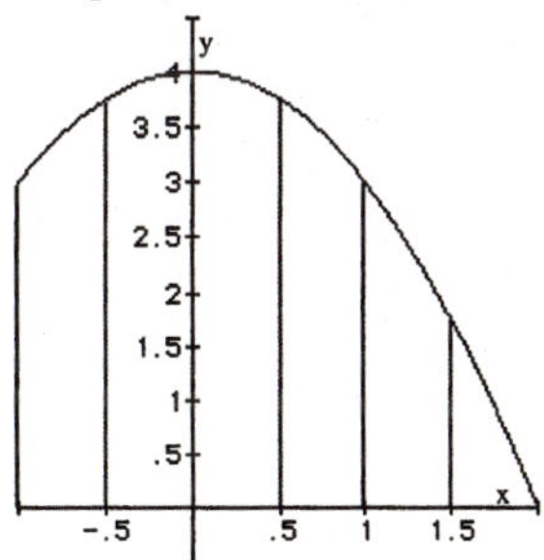

b. $A \approx \left(f(-1)+f(0)+f(1)\right)(1)$
$= (3+4+3)(1)$
$= 10(1) = 10$

c. $A \approx \left(f(-1) + f\left(-\frac{1}{2}\right) + f(0) + f\left(\frac{1}{2}\right) + f(1) + f\left(\frac{3}{2}\right) \right) \cdot \left(\frac{1}{2}\right)$

$= \left(3 + \frac{15}{4} + 4 + \frac{15}{4} + 3 + \frac{7}{4}\right)\left(\frac{1}{2}\right)$

$= \frac{77}{4}\left(\frac{1}{2}\right) = \frac{77}{8} = 9.625$

d. $A = \int_{-1}^{2} \left(4 - x^2\right) dx$

e. Use fnInt function:

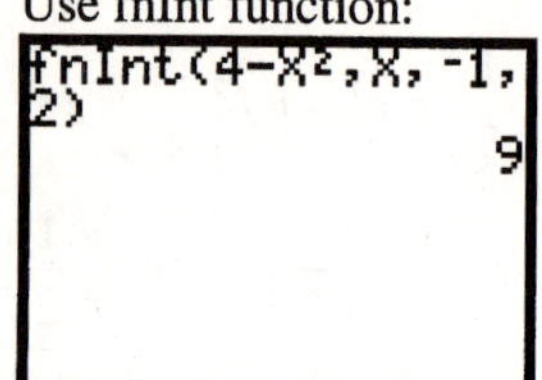

78. a. Graph $f(x) = x^2 + 3$, $[0, 6]$:

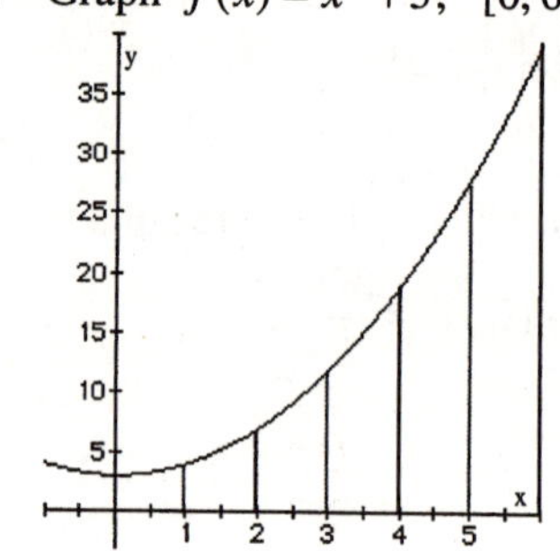

b. $A \approx \left(f(0) + f(2) + f(4)\right)(2)$

$= (3 + 7 + 19)(2)$

$= 29(2) = 58$

c. $A \approx \left(f(0) + f(1) + f(2) + f(3) + f(4) + f(5)\right) \cdot (1)$

$= (3 + 4 + 7 + 12 + 19 + 28)(1)$

$= 73(1) = 73$

d. $A = \int_{0}^{6} \left(x^2 + 3\right) dx$

e. Use fnInt function:

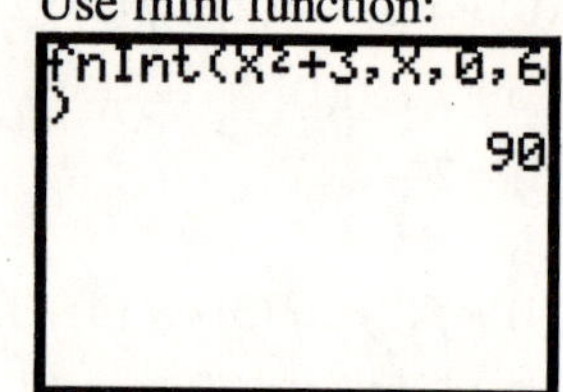

79. a. Graph $f(x) = \frac{1}{x^2}$, $[1, 4]$:

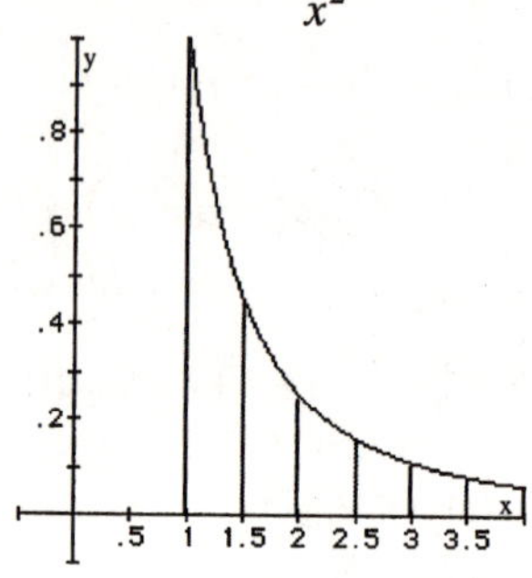

b. $A \approx \left(f(1) + f(2) + f(3)\right)(1)$

$= \left(1 + \frac{1}{4} + \frac{1}{9}\right)(1)$

$= \frac{49}{36}(1) = \frac{49}{36} \approx 1.36$

c. $A \approx \left(f(1) + f\left(\frac{3}{2}\right) + f(2) + f\left(\frac{5}{2}\right) + f(3) + f\left(\frac{7}{2}\right) \right) \cdot \left(\frac{1}{2}\right)$

$= \left(1 + \frac{4}{9} + \frac{1}{4} + \frac{4}{25} + \frac{1}{9} + \frac{4}{49}\right)\left(\frac{1}{2}\right)$

$= 1.02$

d. $A = \int_{1}^{4} \left(\frac{1}{x^2}\right) dx$

e. Use fnInt function:

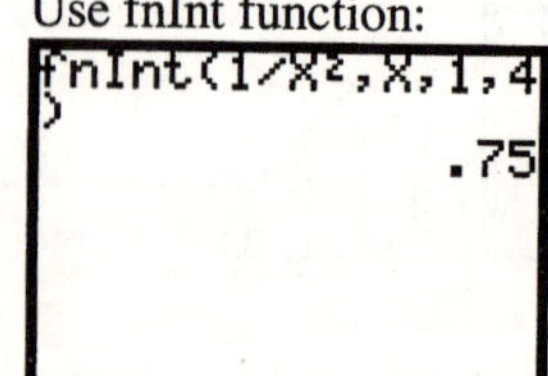

80. a. Graph $f(x) = e^x$, $[0, 6]$:

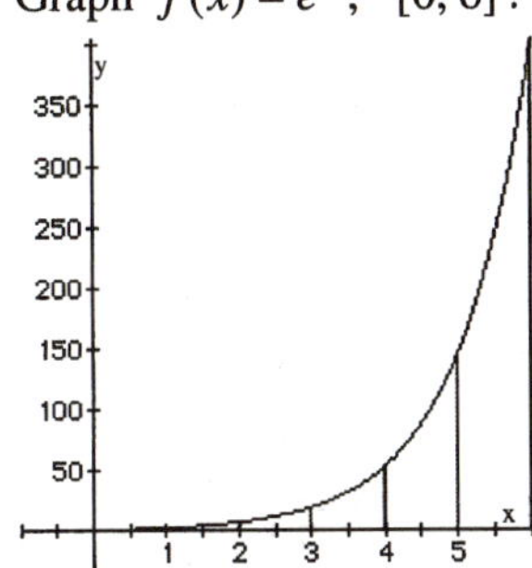

b. $A \approx (f(0) + f(2) + f(4))(2)$
$= (1 + 7.389 + 54.598)(2)$
$= 62.987(2) = 125.97$

c. $A \approx (f(0) + f(1) + f(2)$
$+ f(3) + f(4) + f(5)) \cdot (1)$
$= (1 + 2.718 + 7.389 + 20.086$
$+ 54.598 + 148.413)(1)$
$= 234.204(1) = 234.20$

d. $A = \int_0^6 (e^x)\, dx$

e. Use fnInt function:

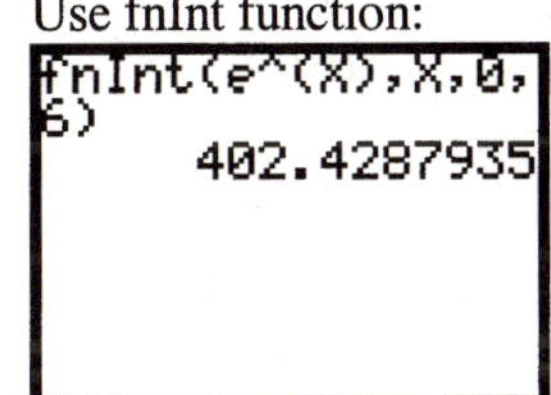

81. a. The integral represents the area under the graph of $f(x) = 9 - x^2$ from $x = -1$ to $x = 3$.

b.

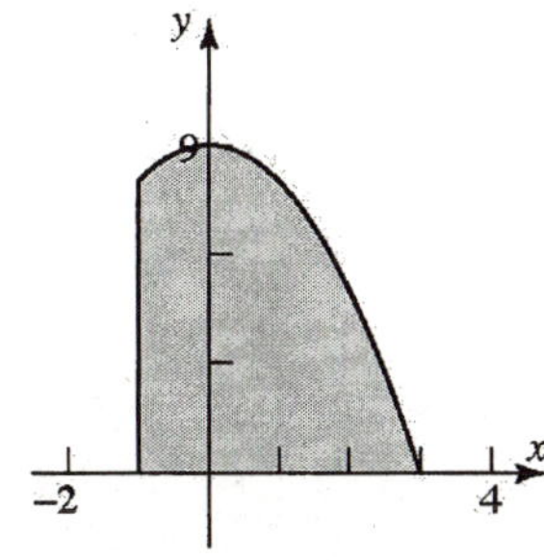

c.

82. a. The integral represents the area under the graph of $f(x) = \sqrt{x}$ from $x = 1$ to $x = 4$.

b.

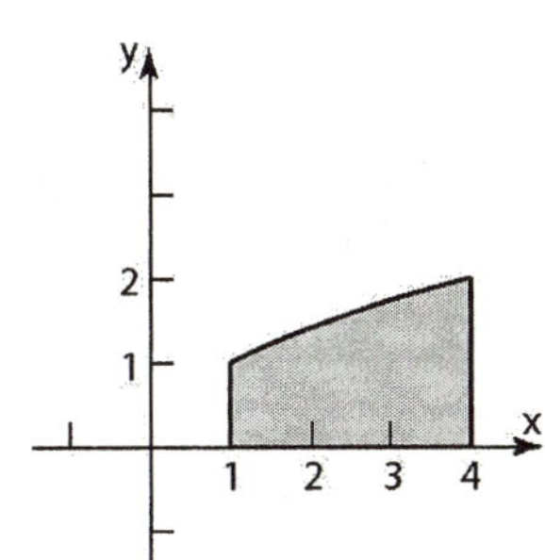

c.

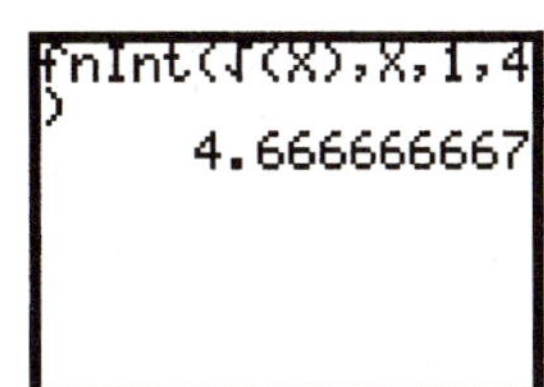

83. a. The integral represents the area under the graph of $f(x) = e^x$ from $x = -1$ to $x = 1$.

b.

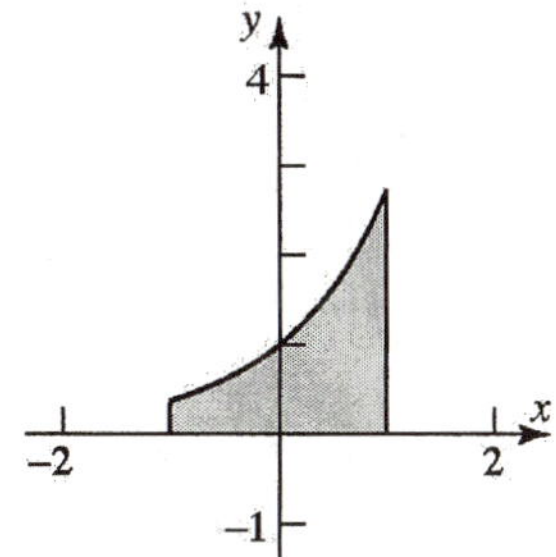

c.

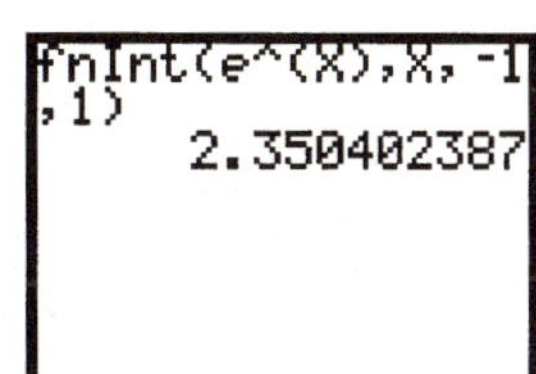

84. a. The integral represents the area under the graph of $f(x)=\sin x$ from $x=\frac{\pi}{3}$ to $x=\frac{2\pi}{3}$.

b.

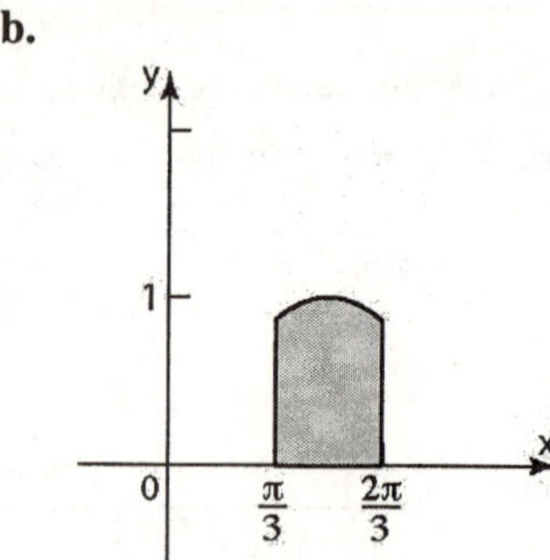

c.

Chapter 13 Test

1. Here we are taking the limit of a polynomial. Therefore, we evaluate the polynomial expression for the given value.

$$\lim_{x\to 3}\left(-x^2+3x-5\right)=-(3)^2+3(3)-5$$
$$=-9+9-5$$
$$=-5$$

```
Plot1 Plot2 Plot3
\Y1■-X²+3X-5
\Y2=
\Y3=
\Y4=
\Y5=
\Y6=
\Y7=
```

X	Y1
2.9	-4.71
2.99	-4.97
2.999	-4.997
3.001	-5.003
3.01	-5.03
3.1	-5.31

X=

2. For this problem, direct substitution does not work because it would yield the indeterminate form $\frac{0}{0}$. However, notice that this is a one-sided limit from the right. As we approach 2 from the right, we will have x values such that $x>2$. Therefore, we have $|x-2|=x-2$ and get the following:

$$\lim_{x\to 2+}\frac{|x-2|}{3x-6}=\lim_{x\to 2+}\frac{x-2}{3x-6}$$
$$=\lim_{x\to 2+}\frac{x-2}{3(x-2)}$$
$$=\lim_{x\to 2+}\frac{1}{3}$$
$$=\frac{1}{3}$$

Remember that we can cancel the common factor $(x-2)$ because we are interested in what happens *near* 2, not actually at 2.

```
Plot1 Plot2 Plot3
\Y1■abs(X-2)/(3X-6)
\Y2=
\Y3=
\Y4=
\Y5=
\Y6=
```

X	Y1
2.1	.33333
2.01	.33333
2.001	.33333
2.0001	.33333

X=

3. $\lim_{x\to -6}\sqrt{7-3x}=\sqrt{7-3(-6)}$
$$=\sqrt{7+18}$$
$$=\sqrt{25}=5$$

```
Plot1 Plot2 Plot3
\Y1■√(7-3X)
\Y2=
\Y3=
\Y4=
\Y5=
\Y6=
\Y7=
```

X	Y1
-6.1	5.0299
-6.01	5.003
-6.001	5.0003
-5.999	4.9997
-5.99	4.997
-5.9	4.9699

X=

4. Note that direct substitution will yield the indeterminate form $\frac{0}{0}$. For rational functions, this means that there is a common factor that can be cancelled before taking the limit.

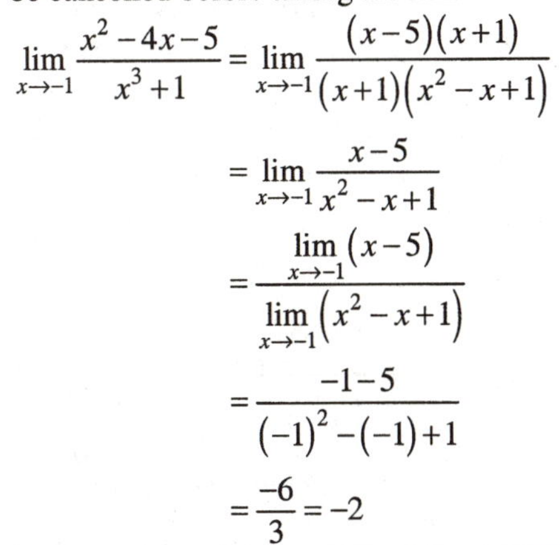

$$\lim_{x\to-1}\frac{x^2-4x-5}{x^3+1}=\lim_{x\to-1}\frac{(x-5)(x+1)}{(x+1)(x^2-x+1)}$$

$$=\lim_{x\to-1}\frac{x-5}{x^2-x+1}$$

$$=\frac{\lim\limits_{x\to-1}(x-5)}{\lim\limits_{x\to-1}(x^2-x+1)}$$

$$=\frac{-1-5}{(-1)^2-(-1)+1}$$

$$=\frac{-6}{3}=-2$$

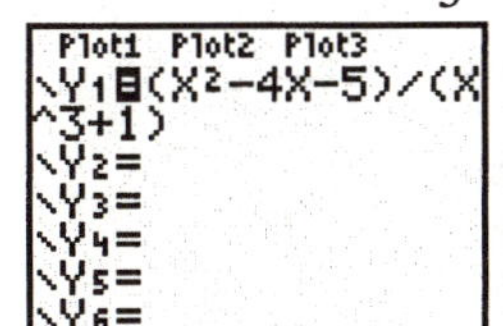

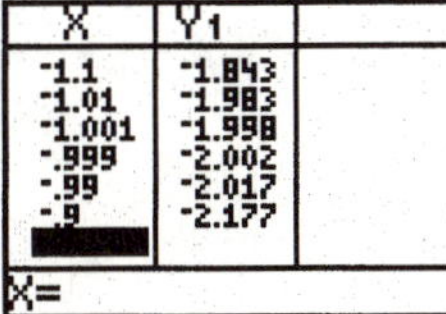

5. $\lim\limits_{x\to5}\left[(3x)(x-2)^2\right]=\lim\limits_{x\to5}(3x)\cdot\lim\limits_{x\to5}(x-2)^2$

$$=\lim_{x\to5}3\cdot\lim_{x\to5}x\cdot\left[\lim_{x\to5}(x-2)\right]^2$$

$$=3\cdot5\cdot(5-2)^2$$

$$=15(3)^2$$

$$=135$$

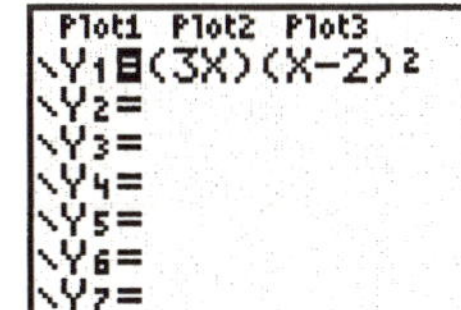

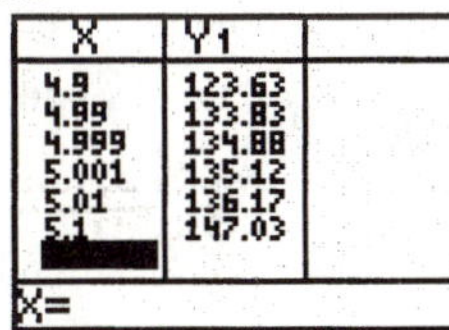

6. $\lim\limits_{x\to\frac{\pi}{4}}\frac{\tan x}{1+\cos^2 x}=\frac{\lim\limits_{x\to\frac{\pi}{4}}\tan x}{\lim\limits_{x\to\frac{\pi}{4}}1+\cos^2 x}$

$$=\frac{\tan\frac{\pi}{4}}{1+\cos^2\frac{\pi}{4}}=\frac{1}{1+\left(\frac{\sqrt{2}}{2}\right)^2}$$

$$=\frac{1}{1+\frac{1}{2}}=\frac{1}{\frac{3}{2}}$$

$$=\frac{2}{3}$$

Plot1 Plot2 Plot3
\Y1■tan(X)/(1+(cos(X))²)
\Y2=
\Y3=
\Y4=
\Y5=
\Y6=

X	Y1
.7	.53142
.78	.65714
.785	.66596
.786	.66774
.79	.6749
.8	.69317

X=

7. To be continuous at a point $x=c$, we need to show that $\lim\limits_{x\to c-}f(x)=\lim\limits_{x\to c+}f(x)=f(c)$.

$$\lim_{x\to4-}f(x)=\lim_{x\to4-}\frac{x^2-9}{x+3}=\frac{4^2-9}{4+3}=1$$

$$\lim_{x\to4+}f(x)=\lim_{x\to4+}(kx+5)=4k+5$$

$$f(4)=\frac{4^2-9}{4+3}=1$$

Therefore, we need to solve

$$4k+5=1$$

$$4k=-4$$

$$k=-1$$

8. To find the limit, we look at at the values of f when x is close to 3, but more than 3. From the graph, we conclude that $\lim\limits_{x\to3+}f(x)=-3$.

9. To find the limit, we look at at the values of f when x is close to 3, but less than 3. From the graph, we conclude that $\lim\limits_{x\to3-}f(x)=5$.

10. To find the limit, we look at at the values of f when x is close to -2 on either side. From the graph, we see that the limits from the left and right are the same and conclude that $\lim\limits_{x\to-2}f(x)=2$.

11. For a limit to exist, the limit from the left and the limit from the right must both exist and be equal. From the graph we see that
$\lim_{x\to 1-} f(x) = \lim_{x\to 1+} f(x) = 2 = \lim_{x\to 1} f(x)$
Note that the limit need not be equal to the function value at the point *c*. In fact, the function does not even need to be defined at *c*. We simply need the left and right limits to exist and be the same.

12. **a.** The graph has a hole at $x=-2$ and the function is undefined. Thus, the function is not continuous at $x=-2$.

b. The function is defined at $x=1$ and $\lim_{x\to 1} f(x)$ exists, but $\lim_{x\to 1} f(x) \neq f(1)$ so the function is not continuous at $x=1$.

c. The function is defined at $x=3$, but there is a gap in the graph. That is, $\lim_{x\to 3-} f(x) \neq \lim_{x\to 3+} f(x)$ so the two-sided limit as $x \to 3$ does not exist. Thus, the function is not continuous at $x=3$.

d. From the graph we can see that $\lim_{x\to 4-} f(x) = \lim_{x\to 4+} f(x) = f(4)$. Therefore, the function is continuous at $x=4$.

13. $R(x) = \dfrac{x^3+6x^2-4x-24}{x^2+5x-14}$

Begin by factoring the numerator and denominator, but do not cancel any common factors yet.

$$\begin{aligned} R(x) &= \frac{x^3+6x^2-4x-24}{x^2+5x-14} \\ &= \frac{x^2(x+6)-4(x+6)}{(x+7)(x-2)} \\ &= \frac{(x+6)(x^2-4)}{(x+7)(x-2)} \\ &= \frac{(x+6)(x+2)(x-2)}{(x+7)(x-2)} \end{aligned}$$

From the denominator, we can see that the function is undefined at the values $x=-7$ and $x=2$ because these values make the denominator equal 0.
To determine whether an asymptote or hole occur at these restricted values, we need to write the function in lowest terms by canceling common factors.

$$R(x) = \frac{(x+6)(x+2)(x-2)}{(x+7)(x-2)} = \frac{(x+6)(x+2)}{(x+7)}$$

(where $x \neq 2$)
Since $x=-7$ still makes the denominator equal to 0, there will be a vertical asymptote at $x=-7$. Since $x=2$ no longer makes the denominator equal to 0 when the expression is in lowest terms, there will be a hole in the graph at $x=2$.

14. **a.** $f'(2)$

$$\begin{aligned} &= \lim_{x\to 2} \frac{f(x)-f(2)}{x-2} \\ &= \lim_{x\to 2} \frac{(4x^2-11x-3)-(4(2)^2-11(2)-3)}{x-2} \\ &= \lim_{x\to 2} \frac{4x^2-11x-3-(-9)}{x-2} \\ &= \lim_{x\to 2} \frac{4x^2-11x+6}{x-2} \\ &= \lim_{x\to 2} \frac{(x-2)(4x-3)}{x-2} \\ &= \lim_{x\to 2} (4x-3) \\ &= 4(2)-3 = 5 \end{aligned}$$

b. The derivative evaluated at $x=2$ is the slope of the tangent line to the graph of *f* at $x=2$. From part (a), we have $m_{\tan}=5$. Using the slope and the given point, $(2,-9)$, we can find the equation of the tangent line.

$$\begin{aligned} y-y_1 &= m(x-x_1) \\ y-(-9) &= 5(x-2) \\ y+9 &= 5x-10 \\ y &= 5x-19 \end{aligned}$$

Therefore, the equation of the tangent line to the graph of *f* at $x=2$ is $y=5x-19$.

c. Let $Y1=4x^2-11x-3$ and $Y2=5x-19$.

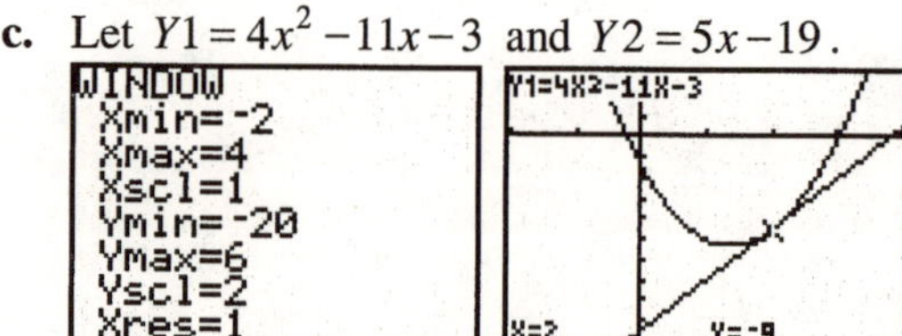

15. a.

x	$y=f(x)$	(x,y)
0	$y=\sqrt{16-0^2}=4$	$(0,4)$
2	$y=\sqrt{16-2^2}=2\sqrt{3}$	$(2,2\sqrt{3})$
3	$y=\sqrt{16-3^2}=\sqrt{7}$	$(3,\sqrt{7})$
4	$y=\sqrt{16-4^2}=0$	$(4,0)$

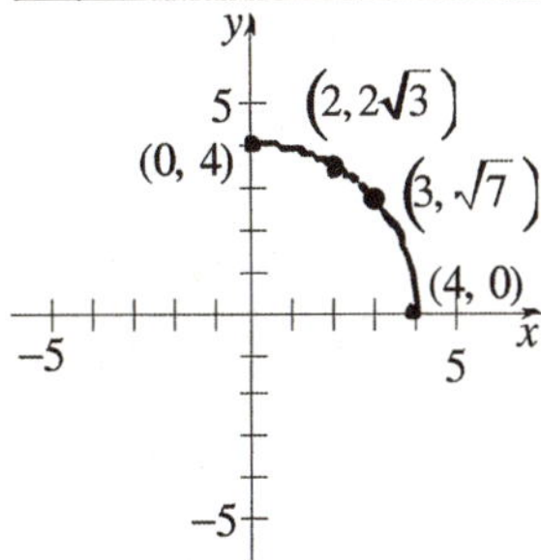

b. Each subinterval will have length

$$\frac{b-a}{8}=\frac{4-0}{8}=\frac{1}{2}$$

Since $u=$ left endpoint, we have

$$a=u_0=0,\ u_1=\frac{1}{2},\ u_2=1,\ u_3=\frac{3}{2},$$

$$u_4=2,\ u_5=\frac{5}{2},\ u_6=3,\ u_7=\frac{7}{2}$$

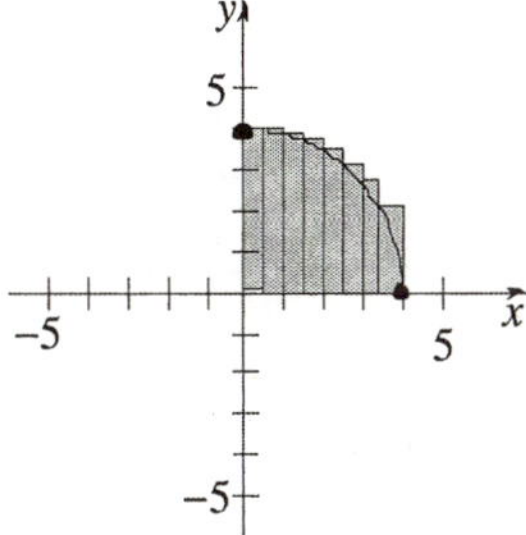

$$A\approx\frac{1}{2}[f(0)+f\left(\tfrac{1}{2}\right)+f(1)+f\left(\tfrac{3}{2}\right)+f(2)+f\left(\tfrac{5}{2}\right)+f(3)+f\left(\tfrac{7}{2}\right)]$$

$$=\frac{1}{2}[4+3.969+3.873+3.708+3.464+3.123+2.646+1.937]$$

$$=\frac{1}{2}(26.72)$$

$=13.36$ square units

c. The desired region is one quarter of a circle with radius $r=4$. The area of this region is

$$A=\frac{1}{4}\left[\pi(4)^2\right]$$

$$=\frac{1}{4}(16\pi)=4\pi\approx 12.566 \text{ square units}$$

The estimate in part (b) is slightly larger than the actual area. Since the function is decreasing over the entire interval, the largest value of the function on a subinterval always occurs at the left endpoint. Therefore, we would expect our estimate to be larger than the actual value.

16. Area $=\int_1^4\left(-x^2+5x+3\right)dx$

17. a. The average rate of change is given by

$$a.r.c.=\frac{s(6)-s(3)}{6-3}=\frac{137-31}{6-3}$$

$$=\frac{106}{3}\approx 35.33 \text{ ft. per sec.}$$

b. Use QuadReg to fit the model $y=ax^2+bx+c$.

```
QuadReg
 y=ax²+bx+c
 a=4.026515152
 b=-2.620454545
 c=.9363636364
 R²=.9975206301
```

The quadratic model of best fit is $y=4.027t^2-2.620t+0.936$.

c. For a quadratic function, the instantaneous rate of change is given by

$$s'(t_0)=\lim_{t\to t_0}\frac{\left[at^2+bt+c\right]-\left[at_0{}^2+bt_0+c\right]}{t-t_0}$$

$$=\lim_{t\to t_0}\frac{at^2+bt-at_0{}^2-bt_0}{t-t_0}$$

$$=\lim_{t\to t_0}\frac{a\left(t^2-t_0{}^2\right)+b\left(t-t_0\right)}{t-t_0}$$

$$=\lim_{t\to t_0}\frac{a(t-t_0)(t+t_0)+b(t-t_0)}{t-t_0}$$

$$=\lim_{t\to t_0}\left[a(t+t_0)+b\right]$$

$$=a(t_0+t_0)+b$$

$$=2a\cdot t_0+b$$

Therefore, the instantaneous rate of change at $t=3$ is given by

$$s'(3)=2(4.027)(3)+(-2.620)$$

$$=21.542 \text{ ft. per sec.}$$

Chapter 13 Projects

Project 1

Total Midyear Population for the World: 1950-2050

t	Year	Population
0	1950	2,555,360,972
1	1951	2,593,139,857
2	1952	2,635,192,901
3	1953	2,680,522,529
4	1954	2,728,486,476
5	1955	2,779,929,940
6	1956	2,832,880,780
7	1957	2,888,699,042
8	1958	2,945,196,478
9	1959	2,997,522,100
10	1960	3,039,585,530
11	1961	3,080,367,474
12	1962	3,136,451,432
13	1963	3,205,956,565
14	1964	3,277,024,728
15	1965	3,346,002,675
16	1966	3,416,184,968
17	1967	3,485,881,292
18	1968	3,557,690,668
19	1969	3,632,294,522
20	1970	3,707,475,887
21	1971	3,784,957,162
22	1972	3,861,537,222
23	1973	3,937,599,035
24	1974	4,013,016,398
25	1975	4,086,150,193
26	1976	4,157,827,615
27	1977	4,229,922,943
28	1978	4,301,953,661
29	1979	4,376,897,872

t	Year	Population
30	1980	4,452,584,592
31	1981	4,528,511,458
32	1982	4,608,410,617
33	1983	4,689,840,421
34	1984	4,769,886,824
35	1985	4,851,592,622
36	1986	4,934,892,988
37	1987	5,020,809,215
38	1988	5,107,404,183
39	1989	5,194,105,912
40	1990	5,281,653,820
41	1991	5,365,480,276
42	1992	5,449,369,636
43	1993	5,531,014,635
44	1994	5,611,269,983
45	1995	5,691,759,210
46	1996	5,770,701,020
47	1997	5,849,885,301
48	1998	5,927,556,529
49	1999	6,004,170,056
50	2000	6,079,603,571
51	2001	6,153,801,961
52	2002	6,226,933,918
53	2003	6,299,763,405
54	2004	6,372,797,742
55	2005	6,446,131,400

Source: U.S. Bureau of the Census, International Database.
Note: Data updated 9-30-2004
http://www.census.gov/ipc/www/worldpop.html

a.

$$y = \frac{1.13\times10^{10}}{1+3.65e^{-0.029t}}$$

b.

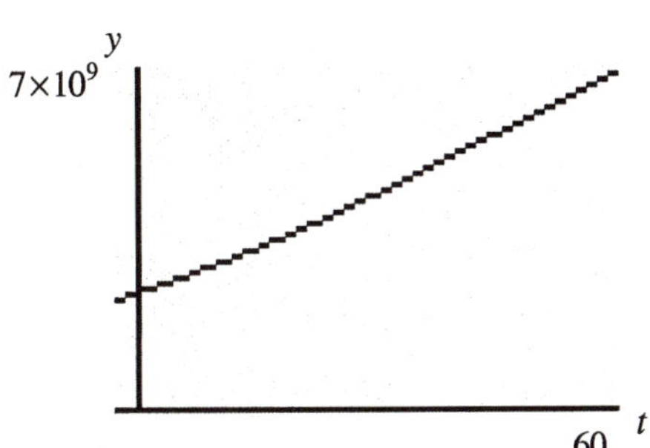

c.

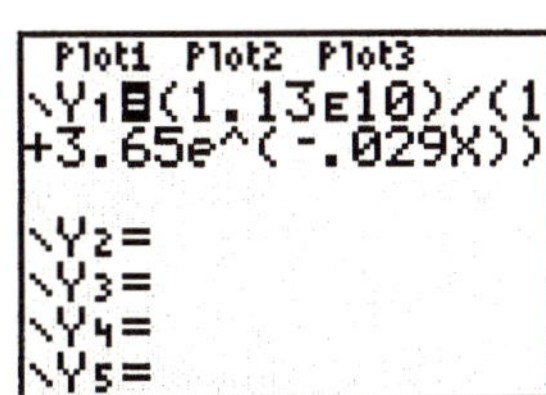

nDeriv(Y1(X),X,10)
64288854.8

d. $y(1961) \approx 3{,}039{,}585{,}530 + 64{,}288{,}855$

$= 3{,}103{,}874{,}385$

The actual population in 1961 was 3,080,367,474.

e.

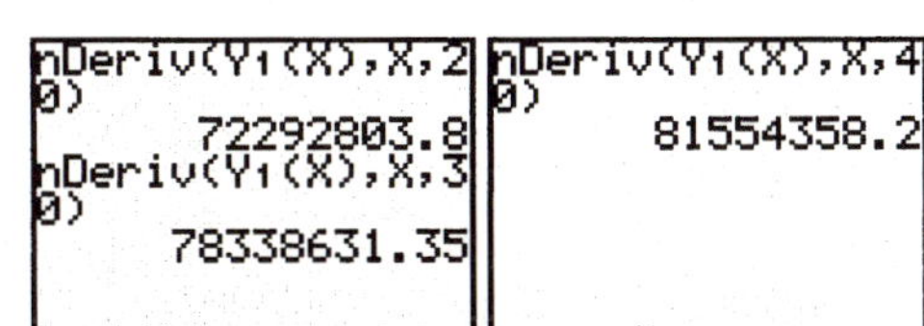

The instantaneous growth is slowing down. Thus, Malthus' contention is not true.

f.

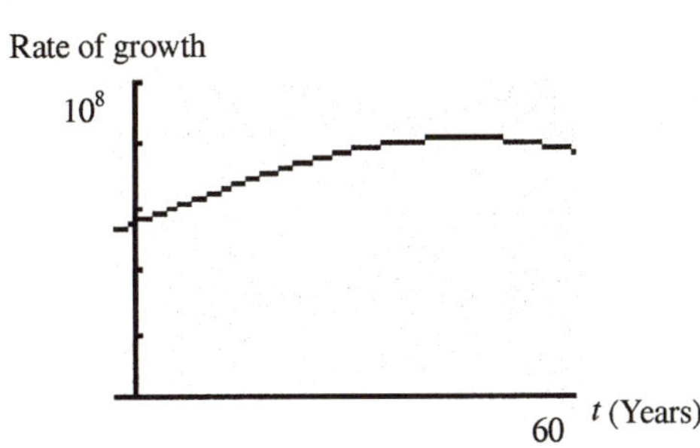

g. The growth rate is largest in 1994. Then, the growth rate begins to decrease. The y value on the graph for this time is 5.65×10^{9}.

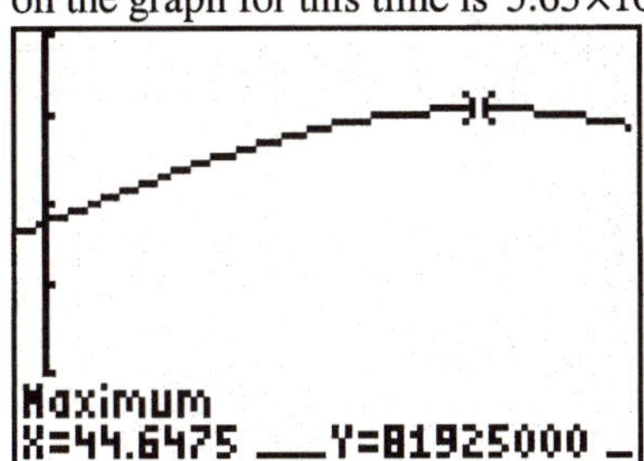

h. $\lim_{t\to\infty} f(t) = \lim_{t\to\infty}\left(\frac{1.13\times10^{10}}{1+3.65e^{-0.029t}}\right)$

$$= \frac{1.13\times10^{10}}{1+0} = 1.13\times10^{10}$$

The carrying capacity of the Earth is 1.13×10^{10} people.

i. If the population exceeds the carrying capacity, the population will begin to die off very quickly due to hunger and disease in particular. There will not be enough agricultural growth to keep up with the increase in population. Urban sprawl will cause agricultural growth to diminish since land will be taken away.

Project 2 (web)

1.

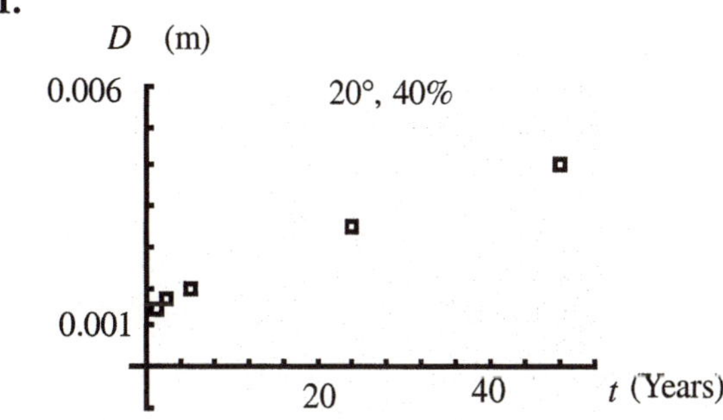

2. $t=1$ to $t=2$:

$$\frac{\Delta D}{\Delta t}=\frac{0.0014-0.0011}{2-1}=0.003 \text{ m/hr}$$

$t=2$ to $t=5$:

$$\frac{\Delta D}{\Delta t}=\frac{0.0017-0.0014}{5-2}=0.001 \text{ m/hr}$$

$t=24$ to $t=48$:

$$\frac{\Delta D}{\Delta t}=\frac{0.004-0.003}{48-24}=0.000042 \text{ m/hr}$$

3.

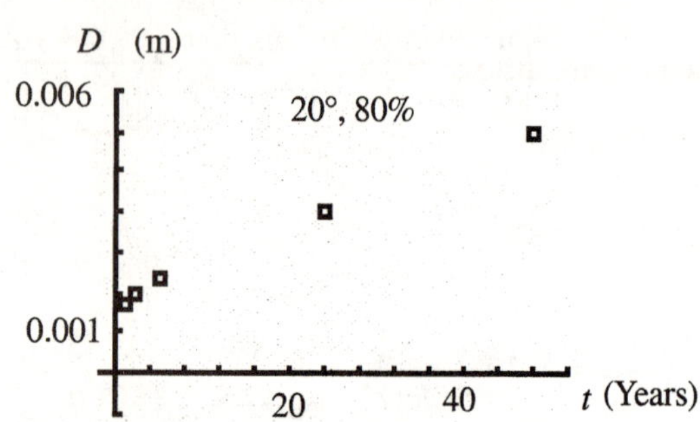

4. t = 1 to t = 2:

$$\frac{\Delta D}{\Delta t}=\frac{0.002-0.0017}{2-1}=0.0003 m/hr$$

t = 2 to t = 5:

$$\frac{\Delta D}{\Delta t}=\frac{0.0023-0.002}{5-2}=0.0001 m/hr$$

t = 24 to t = 48:

$$\frac{\Delta D}{\Delta t}=\frac{0.006-0.004}{48-24}=0.000083 m/hr$$

5. There are no differences until the time span from t = 24 to t = 48. The highest rate is for 80 % RH from t = 24 to t = 48. The lowest rate is for 20% for the time span t = 24 to t = 48.

6.

```
QuadReg
 y=ax²+bx+c
 a=-5.564938E-7
 b=1.0098218E-4
 c=.0014281645
```

$D(t)=-0.0000006t^2+0.0001t+0.0014$

7. $$D'(t)=\lim_{h\to 0}\frac{-0.0000006(t+h)^2+0.0001(t+h)+0.0014-(-0.0000006t^2+0.0001t+0.0014)}{h}$$

$$=\lim_{h\to 0}\frac{-0.0000006(t^2+2th+h^2)+0.0001(t+h)+0.0014+0.0000006t^2-0.0001t-0.0014}{h}$$

$$=\lim_{h\to 0}\frac{-0.0000012th-0.0000006h^2+0.0001h}{h}$$

$$=\lim_{h\to 0}(-0.0000012t-0.000006h+0.0001)$$

$$=-0.0000012t+0.0001$$

$D'(2)=0.0000976 \text{ m/hr} \approx 0.0001 \text{ m/hr}$

$D'(24)=0.0000712 \text{ m/hr}$

8.

```
QuadReg
 y=ax²+bx+c
 a=-2.973725E-7
 b=1.0330803E-4
 c=.0017201424
```

$D(t)=-0.0000003t^2+0.0001t+0.0017$

$$D'(t) = \lim_{h\to 0} \frac{-0.0000003(t+h)^2 + 0.0001(t+h) + 0.0017 - (-0.0000003t^2 + 0.0001t + 0.0017)}{h}$$
$$= \lim_{h\to 0} \frac{-0.0000003(t^2 + 2th + h^2) + 0.0001(t+h) + 0.0017 + 0.0000003t^2 - 0.0001t - 0.0017}{h}$$
$$= \lim_{h\to 0} \frac{-0.0000006th - 0.0000003h^2 + 0.0001h}{h}$$
$$= \lim_{h\to 0} (-0.0000006t - 0.000003h + 0.0001)$$
$$= -0.0000006t + 0.0001$$

$D'(2) = 0.0000988 m/hr \approx 0.0001 m/hr$

$D'(24) = 0.0000856 m/hr \approx 0.00009 m/hr$

The instantaneous rate of change is the same at t = 2, but the rates are different for t = 24.

Project 3 (web)

a.

X	P (total profit)
0	-24000
25	250
60	8500
102	18150
150	20160
190	19610
223	14985
249	9625

The profit-maximizing level is 150 bicycles.

b.

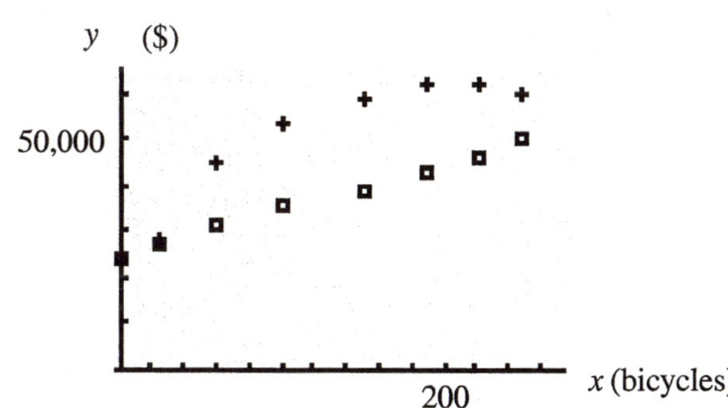

c.

CubicReg
y=ax³+bx²+cx+d
a=.0017418127
b=-.6414405444
c=156.9654523
d=24068.45549

$C(x) = 0.002x^3 - 0.6x^2 + 157x + 24068$

d.

$R(x) = -1.8x^2 + 638x + 7205$

e. $P(x) = R(x) - C(x)$

$$= (-1.8x^2 + 638x + 7205) - (0.002x^3 - 0.6x^2 + 157x + 24068)$$

$$= -0.002x^3 - 2.4x^2 + 795x + 31273$$

f.

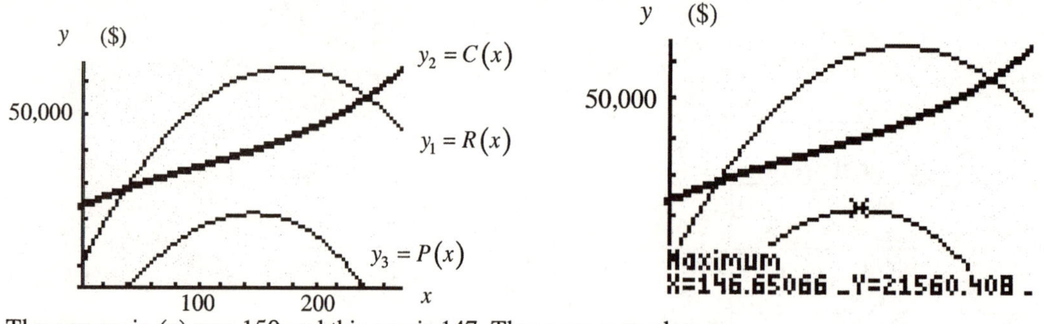

The answer in (a) was 150 and this one is 147. Those are very close.

g.

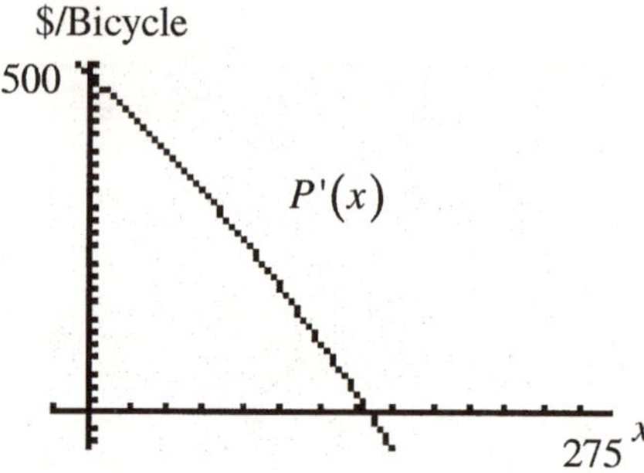

h.

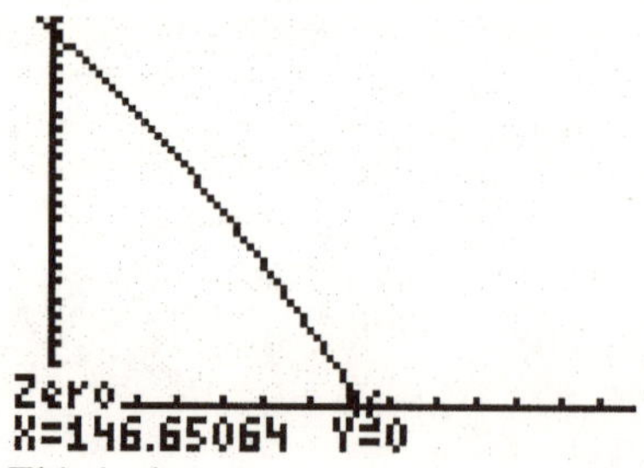

This is the same result as in part (f).

i.

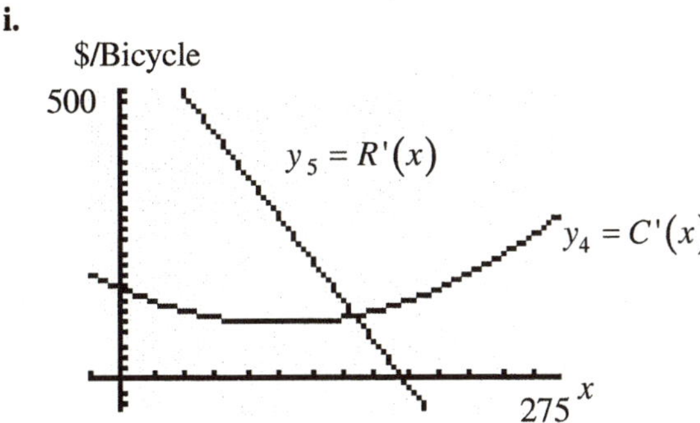

j.

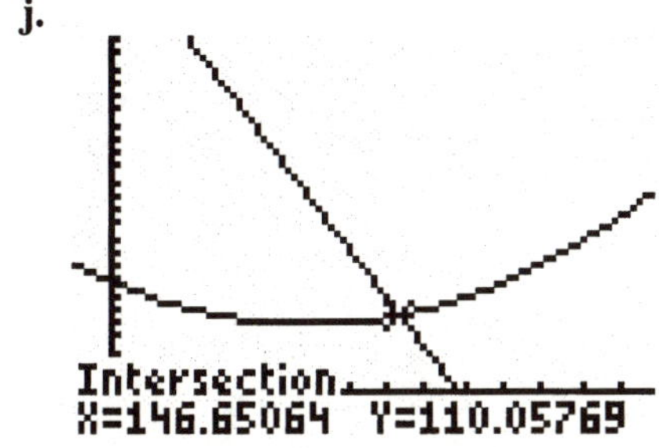

k. Marginal revenue is the rate of change of the revenue as another bicycle is produced. Marginal cost gives the change in cost that making the next bicycle will cause. When the rate of change of the revenue is the same as the rate of change of the cost, the two changes offset each other, thus maximizing the cost.

Appendix A
Review

Section A.1

1. variable

2. origin

3. strict

4. base; exponent (or power)

5. $\sqrt{(-5)^2} = \sqrt{25} = 5$;

   note that $\sqrt{(-5)^2} = |-5| = 5$.

6. False; if the points are the same, the distance between them is 0.

7. False; the absolute value of a real number is nonnegative. $|0| = 0$ which is not a positive number.

8. False; to multiply two expressions with the same base, retain the base and *add* the exponents.

9. $x + 2y = -2 + 2 \cdot 3 = -2 + 6 = 4$

10. $3x + y = 3(-2) + 3 = -6 + 3 = -3$

11. $5xy + 2 = 5(-2)(3) + 2 = -30 + 2 = -28$

12. $-2x + xy = -2(-2) + (-2)(3) = 4 - 6 = -2$

13. $\frac{2x}{x-y} = \frac{2(-2)}{-2-3} = \frac{-4}{-5} = \frac{4}{5}$

14. $\frac{x+y}{x-y} = \frac{-2+3}{-2-3} = \frac{1}{-5} = -\frac{1}{5}$

15. $\frac{3x+2y}{2+y} = \frac{3(-2)+2(3)}{2+3} = \frac{-6+6}{5} = \frac{0}{5} = 0$

16. $\frac{2x-3}{y} = \frac{2(-2)-3}{3} = \frac{-4-3}{3} = -\frac{7}{3}$

17. $\frac{x^2-1}{x}$

    Part (c) must be excluded. The value $x = 0$ must be excluded from the domain because it causes division by 0.

18. $\frac{x^2+1}{x}$

    Part (c) must be excluded. The value $x = 0$ must be excluded from the domain because it causes division by 0.

19. $\frac{x}{x^2-9} = \frac{x}{(x-3)(x+3)}$

    Part (a) must be excluded. The values $x = -3$ and $x = 3$ must be excluded from the domain because they cause division by 0.

20. $\frac{x}{x^2+9}$

    None of the given values are excluded. The domain is all real numbers.

21. $\frac{x^2}{x^2+1}$

    None of the given values are excluded. The domain is all real numbers.

22. $\frac{x^3}{x^2-1} = \frac{x^3}{(x-1)(x+1)}$

    Parts (b) and (d) must be excluded. The values $x = 1$ and $x = -1$ must be excluded from the domain because they cause division by 0.

23. $\frac{x^2+5x-10}{x^3-x} = \frac{x^2+5x-10}{x(x-1)(x+1)}$

    Parts (b), (c), and (d) must be excluded. The values $x = 0$, $x = 1$, and $x = -1$ must be excluded from the domain because they cause division by 0.

24. $\frac{-9x^2-x+1}{x^3+x} = \frac{-9x^2-x+1}{x(x^2+1)}$

    Part (c) must be excluded. The value $x = 0$ must be excluded from the domain because it causes division by 0.

25. $\dfrac{4}{x-5}$

The value $x=5$ makes the denominator equal 0, so it must be excluded from the domain.

$\text{Domain} = \{x | x \neq 5\}$

26. $\dfrac{-6}{x+4}$

The value $x=-4$ makes the denominator equal 0, so it must be excluded from the domain.

$\text{Domain} = \{x | x \neq -4\}$

27. $\dfrac{x}{x+4}$

The value $x=-4$ makes the denominator equal 0, so it must be excluded from the domain.

$\text{Domain} = \{x | x \neq -4\}$

28. $\dfrac{x-2}{x-6}$

The value $x=6$ makes the denominator equal 0, so it must be excluded from the domain.

$\text{Domain} = \{x | x \neq 6\}$

29.

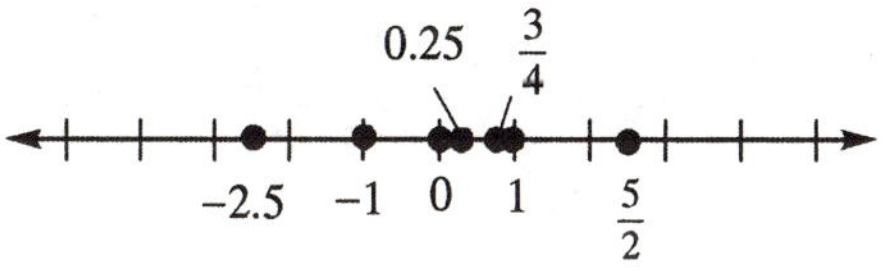

0.25 $\frac{3}{4}$

−2.5 −1 0 1 $\frac{5}{2}$

30.

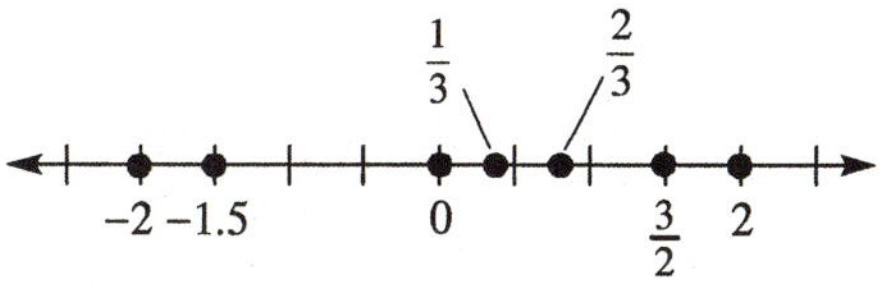

$\frac{1}{3}$ $\frac{2}{3}$

−2 −1.5 0 $\frac{3}{2}$ 2

31. $\frac{1}{2} > 0$

32. $5 < 6$

33. $-1 > -2$

34. $-3 < -\frac{5}{2}$

35. $\pi > 3.14$

36. $\sqrt{2} > 1.41$

37. $\frac{1}{2} = 0.5$

38. $\frac{1}{3} > 0.33$

39. $\frac{2}{3} < 0.67$

40. $\frac{1}{4} = 0.25$

41. $x > 0$

42. $z < 0$

43. $x < 2$

44. $y > -5$

45. $x \le 1$

46. $x \ge 2$

47. Graph on the number line: $x \ge -2$

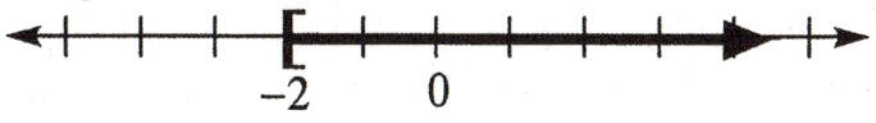

−2 0

48. Graph on the number line: $x < 4$

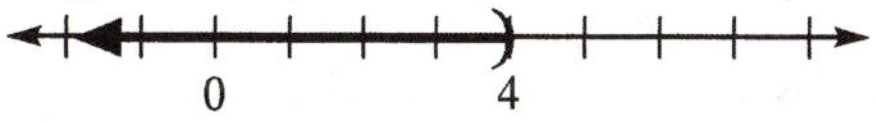

0 4

49. Graph on the number line: $x > -1$

−1 0

50. Graph on the number line: $x \le 7$

0 7

51. $|x+y| = |3+(-2)| = |1| = 1$

52. $|x-y| = |3-(-2)| = |5| = 5$

53. $|x|+|y| = |3|+|-2| = 3+2 = 5$

54. $|x|-|y| = |3|-|-2| = 3-2 = 1$

55. $\dfrac{|x|}{x} = \dfrac{|3|}{3} = \dfrac{3}{3} = 1$

56. $\dfrac{|y|}{y} = \dfrac{|-2|}{-2} = \dfrac{2}{-2} = -1$

57. $|4x-5y|=|4(3)-5(-2)|$
$=|12+10|$
$=|22|$
$=22$

58. $|3x+2y|=|3(3)+2(-2)|=|9-4|=|5|=5$

59. $||4x|-|5y||=||4(3)|-|5(-2)||$
$=||12|-|-10||$
$=|12-10|$
$=|2|$
$=2$

60. $3|x|+2|y|=3|3|+2|-2|$
$=3\cdot 3+2\cdot 2$
$=9+4$
$=13$

61. $d(C,D)=d(0,1)=|1-0|=|1|=1$

62. $d(C,A)=d(0,-3)=|-3-0|=|-3|=3$

63. $d(D,E)=d(1,3)=|3-1|=|2|=2$

64. $d(C,E)=d(0,3)=|3-0|=|3|=3$

65. $d(A,E)=d(-3,3)=|3-(-3)|=|6|=6$

66. $d(D,B)=d(1,-1)=|-1-1|=|-2|=2$

67. $(-4)^2=(-4)(-4)=16$

68. $-4^2=-(4)^2=-16$

69. $4^{-2}=\frac{1}{4^2}=\frac{1}{16}$

70. $-4^{-2}=-\frac{1}{4^2}=-\frac{1}{16}$

71. $3^{-6}\cdot 3^4=3^{-6+4}=3^{-2}=\frac{1}{3^2}=\frac{1}{9}$

72. $4^{-2}\cdot 4^3=4^{-2+3}=4^1=4$

73. $\left(3^{-2}\right)^{-1}=3^{(-2)(-1)}=3^2=9$

74. $\left(2^{-1}\right)^{-3}=2^{(-1)(-3)}=2^3=8$

75. $\sqrt{25}=\sqrt{5^2}=5$

76. $\sqrt{36}=\sqrt{6^2}=6$

77. $\sqrt{(-4)^2}=|-4|=4$

78. $\sqrt{(-3)^2}=|-3|=3$

79. $\left(8x^3\right)^{-2}=\frac{1}{\left(8x^3\right)^2}=\frac{1}{8^2\left(x^3\right)^2}=\frac{1}{64x^6}$

80. $\left(-4x^2\right)^{-1}=\frac{1}{-4x^2}=-\frac{1}{4x^2}$

81. $\left(x^2y^{-1}\right)^2=\left(x^2\right)^2\cdot\left(y^{-1}\right)^2=x^4y^{-2}=\frac{x^4}{y^2}$

82. $\left(x^{-1}y\right)^3=\left(x^{-1}\right)^3\cdot y^3=x^{-3}y^3=\frac{y^3}{x^3}$

83. $\frac{x^{-2}y^3}{xy^4}=x^{-2-1}y^{3-4}=x^{-3}y^{-1}=\frac{1}{x^3y}$

84. $\frac{x^{-2}y}{xy^2}=x^{-2-1}y^{1-2}=x^{-3}y^{-1}=\frac{1}{x^3y}$

85. $\frac{(-2)^3x^4(yz)^2}{3^2xy^3z}=\frac{-8x^4y^2z^2}{9xy^3z}$
$=\frac{-8}{9}x^{4-1}y^{2-3}z^{2-1}$
$=\frac{-8}{9}x^3y^{-1}z^1$
$=-\frac{8x^3z}{9y}$

86. $\dfrac{4x^{-2}(yz)^{-1}}{2^3x^4y}=\dfrac{4x^{-2}y^{-1}z^{-1}}{8x^4y}$
$=\dfrac{4}{8}x^{-2-4}y^{-1-1}z^{-1}$
$=\dfrac{1}{2}x^{-6}y^{-2}z^{-1}$
$=\dfrac{1}{2x^6y^2z}$

87. $\left(\dfrac{3x^{-1}}{4y^{-1}}\right)^{-2}=\left(\dfrac{3y}{4x}\right)^{-2}=\left(\dfrac{4x}{3y}\right)^{2}=\dfrac{4^2x^2}{3^2y^2}=\dfrac{16x^2}{9y^2}$

88. $\left(\dfrac{5x^{-2}}{6y^{-2}}\right)^{-3}=\left(\dfrac{5y^2}{6x^2}\right)^{-3}=\left(\dfrac{6x^2}{5y^2}\right)^{3}$
$=\dfrac{6^3\left(x^2\right)^3}{5^3\left(y^2\right)^3}=\dfrac{216x^6}{125y^6}$

89. $A=l\cdot w$

90. $P=2(l+w)$

91. $C=\pi\cdot d$

92. $A=\dfrac{1}{2}b\cdot h$

93. $A=\dfrac{\sqrt{3}}{4}\cdot x^2$

94. $P=3\cdot x$

95. $V=\dfrac{4}{3}\pi\cdot r^3$

96. $S=4\pi\cdot r^2$

97. $V=x^3$

98. $S=6\cdot x^2$

99. a. $|x-115|=|113-115|=|-2|=2\le 5$
113 volts is acceptable.

b. $|x-115|=|109-115|=|-6|=6\not\le 5$
109 volts is *not* acceptable.

100. a. $|x-220|=|214-220|=|-6|=6\le 8$
214 volts is acceptable.

b. $|x-220|=|209-220|=|-11|=11\not\le 8$
209 volts is *not* acceptable.

101. a. $|x-3|=|2.999-3|$
$=|-0.001|$
$=0.001\le 0.01$
A radius of 2.999 centimeters is acceptable.

b. $|x-3|=|2.89-3|$
$=|-0.11|$
$=0.11\not\le 0.01$
A radius of 2.89 centimeters is *not* acceptable.

102. a. $|x-98.6|=|97-98.6|$
$=|-1.6|$
$=1.6\ge 1.5$
97°F is unhealthy.

b. $|x-98.6|=|100-98.6|$
$=|1.4|$
$=1.4\not\ge 1.5$
100°F is *not* unhealthy.

103. $\dfrac{1}{3}=0.333333\ldots>0.333$
$\dfrac{1}{3}$ is larger by approximately $0.0003333\ldots$

104. $\dfrac{2}{3}=0.666666\ldots>0.666$
$\dfrac{2}{3}$ is larger by approximately $0.0006666\ldots$

105. No. For any positive number a, the value $\dfrac{a}{2}$ is smaller and therefore closer to 0.

106. We are given that $1<x^2<10$. This implies that $1<x<\sqrt{10}$. Since $x<\sqrt{10}\approx 3.162$ and $x>\pi\approx 3.142$, the number could be 3.15 or 3.16 (which are between 1 and 10 as required). The number could also be 3.14 since numbers such as 3.146 which lie between π and $\sqrt{10}$ would equal 3.14 when truncated to two decimal places.

107. Answers will vary.

Section A.2

1. right; hypotenuse

2. $A = \frac{1}{2} b \cdot h$

3. $C = 2\pi r$

4. True.

5. True. $6^2 + 8^2 = 36 + 64 = 100 = 10^2$

6. False; the volume of a sphere of radius r is given by $V = \frac{4}{3}\pi r^3$.

7. $a = 5,\ b = 12,$
$$c^2 = a^2 + b^2 = 5^2 + 12^2 = 25 + 144 = 169 \Rightarrow c = 13$$

8. $a = 6,\ b = 8,$
$$c^2 = a^2 + b^2 = 6^2 + 8^2 = 36 + 64 = 100 \Rightarrow c = 10$$

9. $a = 10,\ b = 24,$
$$c^2 = a^2 + b^2 = 10^2 + 24^2 = 100 + 576 = 676 \Rightarrow c = 26$$

10. $a = 4,\ b = 3,$
$$c^2 = a^2 + b^2 = 4^2 + 3^2 = 16 + 9 = 25 \Rightarrow c = 5$$

11. $a = 7,\ b = 24,$
$$c^2 = a^2 + b^2 = 7^2 + 24^2 = 49 + 576 = 625 \Rightarrow c = 25$$

12. $a = 14,\ b = 48,$
$$c^2 = a^2 + b^2 = 14^2 + 48^2 = 196 + 2304 = 2500 \Rightarrow c = 50$$

13. $5^2 = 3^2 + 4^2$
$25 = 9 + 16$
$25 = 25$
The given triangle is a right triangle. The hypotenuse is 5.

14. $10^2 = 6^2 + 8^2$
$100 = 36 + 64$
$100 = 100$
The given triangle is a right triangle. The hypotenuse is 10.

15. $6^2 = 4^2 + 5^2$
$36 = 16 + 25$
$36 = 41$ false
The given triangle is not a right triangle.

16. $3^2 = 2^2 + 2^2$
$9 = 4 + 4$
$9 = 8$ false
The given triangle is not a right triangle.

17. $25^2 = 7^2 + 24^2$
$625 = 49 + 576$
$625 = 625$
The given triangle is a right triangle. The hypotenuse is 25.

18. $26^2 = 10^2 + 24^2$
$676 = 100 + 576$
$676 = 676$
The given triangle is a right triangle. The hypotenuse is 26.

19. $6^2 = 3^2 + 4^2$
$36 = 9 + 16$
$36 = 25$ false
The given triangle is not a right triangle.

20. $7^2 = 5^2 + 4^2$
$49 = 25 + 16$
$49 = 41$ false
The given triangle is not a right triangle.

21. $A = l \cdot w = 4 \cdot 2 = 8 \text{ in}^2$

22. $A = l \cdot w = 9 \cdot 4 = 36 \text{ cm}^2$

23. $A = \frac{1}{2} b \cdot h = \frac{1}{2}(2)(4) = 4 \text{ in}^2$

24. $A = \frac{1}{2} b \cdot h = \frac{1}{2}(4)(9) = 18 \text{ cm}^2$

25. $A = \pi r^2 = \pi(5)^2 = 25\pi \text{ m}^2$
$C = 2\pi r = 2\pi(5) = 10\pi \text{ m}$

26. $A = \pi r^2 = \pi(2)^2 = 4\pi \text{ ft}^2$
$C = 2\pi r = 2\pi(2) = 4\pi \text{ ft}$

27. $V = lwh = 8 \cdot 4 \cdot 7 = 224 \text{ ft}^3$
$$\begin{aligned} S &= 2lw + 2lh + 2wh \\ &= 2(8)(4) + 2(8)(7) + 2(4)(7) \\ &= 64 + 112 + 56 \\ &= 232 \text{ ft}^2 \end{aligned}$$

28. $V = lwh = 9 \cdot 4 \cdot 8 = 288 \text{ in}^3$
$$\begin{aligned} S &= 2lw + 2lh + 2wh \\ &= 2(9)(4) + 2(9)(8) + 2(4)(8) \\ &= 72 + 144 + 64 \\ &= 280 \text{ in}^2 \end{aligned}$$

29. $V = \frac{4}{3}\pi r^3 = \frac{4}{3}\pi \cdot 4^3 = \frac{256}{3}\pi \text{ cm}^3$
$S = 4\pi r^2 = 4\pi \cdot 4^2 = 64\pi \text{ cm}^2$

30. $V = \frac{4}{3}\pi r^3 = \frac{4}{3}\pi \cdot 3^3 = 36\pi \text{ ft}^3$
$S = 4\pi r^2 = 4\pi \cdot 3^2 = 36\pi \text{ ft}^2$

31. $V = \pi r^2 h = \pi(9)^2(8) = 648\pi \text{ in}^3$
$$\begin{aligned} S &= 2\pi r^2 + 2\pi rh \\ &= 2\pi(9)^2 + 2\pi(9)(8) \\ &= 162\pi + 144\pi \\ &= 306\pi \approx 961.33 \text{ in}^2 \end{aligned}$$

32. $V = \pi r^2 h = \pi(8)^2(9) = 576\pi \text{ in}^3$
$$\begin{aligned} S &= 2\pi r^2 + 2\pi rh \\ &= 2\pi(8)^2 + 2\pi(8)(9) \\ &= 128\pi + 144\pi \\ &= 272\pi \approx 854.51 \text{ in}^2 \end{aligned}$$

33. The diameter of the circle is 2, so its radius is 1. $A = \pi r^2 = \pi(1)^2 = \pi$ square units

34. The diameter of the circle is 2, so its radius is 1. $A = 2^2 - \pi(1)^2 = 4 - \pi$ square units

35. The diameter of the circle is the length of the diagonal of the square.
$$\begin{aligned} d^2 &= 2^2 + 2^2 \\ &= 4 + 4 \\ &= 8 \end{aligned}$$
$d = \sqrt{8} = 2\sqrt{2}$
$r = \frac{d}{2} = \frac{2\sqrt{2}}{2} = \sqrt{2}$
The area of the circle is:
$A = \pi r^2 = \pi\left(\sqrt{2}\right)^2 = 2\pi$ square units

36. The diameter of the circle is the length of the diagonal of the square.
$$\begin{aligned} d^2 &= 2^2 + 2^2 \\ &= 4 + 4 \\ &= 8 \end{aligned}$$
$d = \sqrt{8} = 2\sqrt{2}$
$r = \frac{d}{2} = \frac{2\sqrt{2}}{2} = \sqrt{2}$
The area is:
$A = \pi\left(\sqrt{2}\right)^2 - 2^2 = 2\pi - 4$ square units

37. The total distance traveled is 4 times the circumference of the wheel.

Total Distance $= 4C = 4(\pi d) = 4\pi \cdot 16$

$= 64\pi \approx 201.1$ inches ≈ 16.8 feet

38. The distance traveled in one revolution is the circumference of the disk 4π.

The number of revolutions =

$\frac{\text{dist. traveled}}{\text{circumference}} = \frac{20}{4\pi} = \frac{5}{\pi} \approx 1.6$ revolutions

39. Area of the border = area of EFGH – area of ABCD $= 10^2 - 6^2 = 100 - 36 = 64$ ft^2

40. FG = 4 feet; BG = 4 feet and BC = 10 feet, so CG= 6 feet. The area of the triangle CGF is:

$A = \frac{1}{2} \cdot (4)(6) = 12 \text{ ft}^2$

41. Area of the window = area of the rectangle + area of the semicircle.

$A = (6)(4) + \frac{1}{2} \cdot \pi \cdot 2^2 = 24 + 2\pi \approx 30.28 \text{ ft}^2$

Perimeter of the window = 2 heights + width + one-half the circumference.

$P = 2(6) + 4 + \frac{1}{2} \cdot \pi(4) = 12 + 4 + 2\pi$

$= 16 + 2\pi \approx 22.28$ feet

42. Area of the deck = area of the pool and deck – area of the pool.

$A = \pi(13)^2 - \pi(10)^2 = 169\pi - 100\pi$

$= 69\pi \text{ ft}^2 \approx 216.77 \text{ ft}^2$

The amount of fence is the circumference of the circle with radius 13 feet.

$C = 2\pi(13) = 26\pi \text{ ft} \approx 81.68 \text{ ft}$

43. Convert 20 feet to miles, and solve the Pythagorean Theorem to find the distance:

20 feet $= 20 \text{ feet} \cdot \frac{1 \text{ mile}}{5280 \text{ feet}} = 0.003788$ miles

$d^2 = (3960 + 0.003788)^2 - 3960^2 = 30$ sq. miles

$d \approx 5.477$ miles

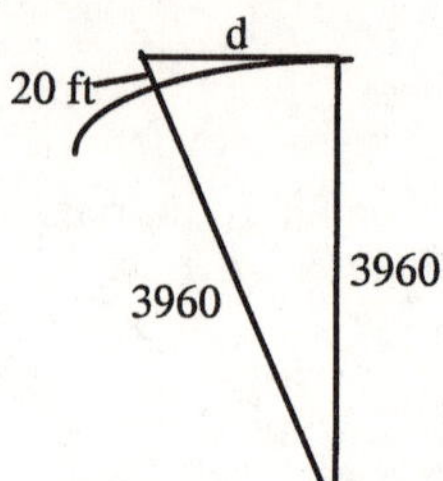

44. Convert 6 feet to miles, and solve the Pythagorean Theorem to find the distance:

6 feet $= 6 \text{ feet} \cdot \frac{1 \text{ mile}}{5280 \text{ feet}} = 0.001136$ miles

$d^2 = (3960 + 0.001136)^2 - 3960^2 = 9$ sq. miles

$d \approx 3$ miles

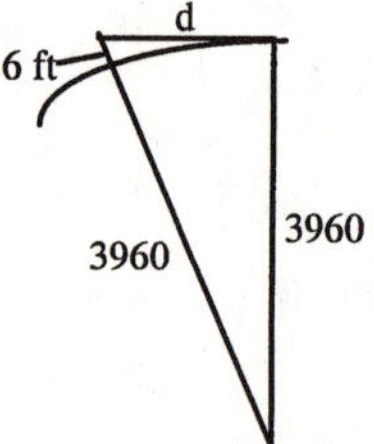

45. Convert 100 feet to miles, and solve the Pythagorean Theorem to find the distance:

100 feet $= 100 \text{ feet} \cdot \frac{1 \text{ mile}}{5280 \text{ feet}}$

$= 0.018939$ miles

$d^2 = (3960 + 0.018939)^2 - 3960^2 \approx 150$ sq. miles

$d \approx 12.247$ miles

Convert 150 feet to miles, and solve the Pythagorean Theorem to find the distance:

150 feet $= 150 \text{ feet} \cdot \frac{1 \text{ mile}}{5280 \text{ feet}}$

$= 0.028409$ miles

$d^2 = (3960 + 0.028409)^2 - 3960^2 \approx 225$ sq. miles

$d \approx 15$ miles

46. Given $m > 0$, $n > 0$ and $m > n$,
if $a = m^2 - n^2, b = 2mn$ and $c = m^2 + n^2$, then

$$a^2 + b^2 = \left(m^2 - n^2\right)^2 + (2mn)^2$$
$$= m^4 - 2m^2n^2 + n^4 + 4m^2n^2$$
$$= m^4 + 2m^2n^2 + n^4$$

and

$$c^2 = \left(m^2 + n^2\right)^2 = m^4 + 2m^2n^2 + n^4$$

$\therefore a^2 + b^2 = c^2 \rightarrow a, b$ and c represent the sides of a right triangle.

47. Let l = length of the rectangle
and w = width of the rectangle.
Notice that

$$(l + w)^2 - (l - w)^2$$
$$= [(l + w) + (l - w)][(l + w) - (l - w)]$$
$$= (2l)(2w) = 4lw = 4A$$

So $A = \frac{1}{4}[(l + w)^2 - (l - w)^2]$

Since $(l - w)^2 \geq 0$, the largest area will occur when $l - w = 0$ or $l = w$; that is, when the rectangle is a square. But

$$1000 = 2l + 2w = 2(l + w)$$
$$500 = l + w = 2l$$
$$250 = l = w$$

The largest possible rectangular area is
$250^2 = 62500$ sq ft.

A circular pool with circumference = 1000 feet yields the equation: $2\pi r = 1000 \Rightarrow r = \dfrac{500}{\pi}$

The area enclosed by the circular pool is:

$$A = \pi r^2 = \pi\left(\frac{500}{\pi}\right)^2 = \frac{500^2}{\pi} \approx 79577.47 \text{ ft}^2$$

Thus, a circular pool will enclose the most area.

48. Consider the diagram showing the lighthouse at point L, relative to the center of Earth, using the radius of Earth as 3960 miles. Let P refer to the furthest point on the horizon from which the light is visible. Note also that

362 feet $= \dfrac{362}{5280}$ miles.

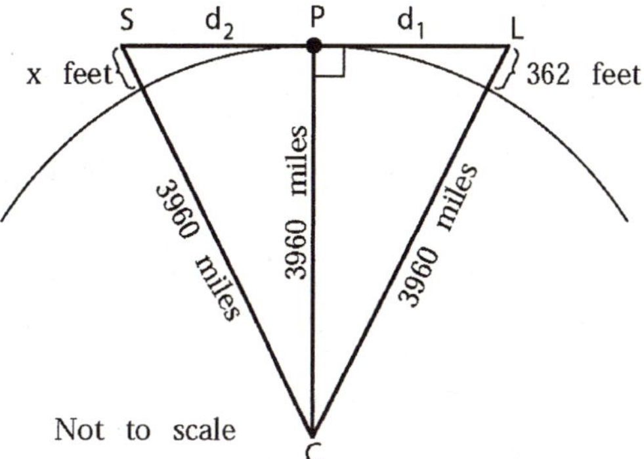

Apply the Pythagorean Theorem to ΔCPL:

$$(3960)^2 + (d_1)^2 = \left(3960 + \frac{362}{5280}\right)^2$$

$$(d_1)^2 = \left(3960 + \tfrac{362}{5280}\right)^2 - (3960)^2$$

$$d_1 = \sqrt{\left(3960 + \tfrac{362}{5280}\right)^2 - (3960)^2} \approx 23.30 \text{ mi.}$$

Therefore, the light from the lighthouse can be seen at point P on the horizon, where point P is approximately 23.30 miles away from the lighthouse. Brochure information is slightly overstated.

<u>Verify the ship information:</u>
Let S refer to the ship's location, and let x equal the height, in feet, of the ship.
We need $d_1 + d_2 \geq 40$.
Since $d_1 \approx 23.30$ miles we need
$d_2 \geq 40 - 23.30 = 16.70$ miles.

Apply the Pythagorean Theorem to ΔCPS:

$$(3960)^2 + (16.7)^2 = (3960 + x)^2$$
$$\sqrt{(3960)^2 + (16.7)^2} = 3960 + x$$
$$\sqrt{(3960)^2 + (16.7)^2} - 3960 = x$$
$$x \approx 0.035 \text{ miles}$$
$$x \approx 185.93 \text{ feet.}$$

The ship would have to be at least 186 feet tall to see the lighthouse from 40 miles away.

Verify the airplane information:

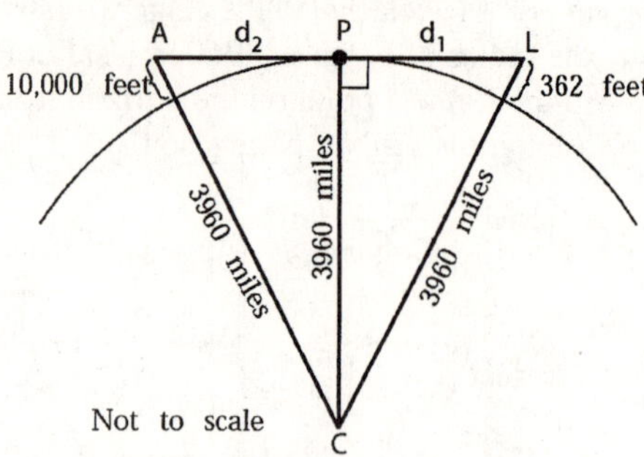

Let A refer to the airplane's location. The distance from the plane to point P is d_2.

We want to show that $d_1 + d_2 \geq 120$.

Assume the altitude of the airplane is

10,000 feet $= \dfrac{10000}{5280}$ miles.

Apply the Pythagorean Theorem to ΔCPA:

$$(3960)^2 + (d_2)^2 = \left(3960 + \frac{10000}{5280}\right)^2$$

$$(d_2)^2 = \left(3960 + \frac{10000}{5280}\right)^2 - (3960)^2$$

$$d_2 = \sqrt{\left(3960 + \frac{10000}{5280}\right)^2 - (3960)^2}$$

$$\approx 122.49 \text{ miles.}$$

Therefore,

$d_1 + d_2 \approx 23.30 + 122.49 = 145.79 \geq 120$.

The brochure information is slightly understated. Note that a plane at an altitude of 6233 feet could see the lighthouse from 120 miles away.

Section A.3

1. 4; 3

2. $x^4 - 16$

3. $x^3 - 8$

4. False; monomials cannot have negative degrees.

5. True

6. False; $x^3 + a^3 = (x+a)(x^2 - ax + a^2)$

7. $3x(x-2)(x+2)$

8. prime

9. True; $x^2 + 4$ is prime over the set of real numbers.

10. False; $3x^3 - 2x^2 - 6x + 4 = (3x-2)(x^2-2)$

11. in lowest terms

12. Least Common Multiple

13. True; $\dfrac{2x^3 - 4x}{x-2} = \dfrac{2x(x^2-2)}{x-2}$

14. False;
$$2x^3 + 6x^2 = 2x^2(x+3)$$
$$6x^4 + 4x^3 = 2x^3(3x+2)$$
$$LCM = 2x^3(x+3)(3x+2)$$

15. $(10x^5 - 8x^2) + (3x^3 - 2x^2 + 6)$
$$= 10x^5 - 8x^2 + 3x^3 - 2x^2 + 6$$
$$= 10x^5 + 3x^3 - 10x^2 + 6$$

16. $3(x^2 - 3x + 1) + 2(3x^2 + x - 4)$
$$= 3x^2 - 9x + 3 + 6x^2 + 2x - 8$$
$$= 3x^2 + 6x^2 - 9x + 2x + 3 - 8$$
$$= 9x^2 - 7x - 5$$

17. $(x+a)^2 - x^2 = x^2 + 2ax + a^2 - x^2$
$$= 2ax + a^2$$

18. $(x-a)^2 - x^2 = x^2 - 2ax + a^2 - x^2$
$$= -2ax + a^2$$

19. $(x+8)(2x+1) = 2x^2 + x + 16x + 8$
$$= 2x^2 + 17x + 8$$

20. $(2x-1)(x+2) = 2x^2 + 4x - x - 2$
$$= 2x^2 + 3x - 2$$

21. $(x^2+x-1)(x^2-x+1)$
$=(x^2+(x-1))(x^2-(x-1))$
$=(x^2)^2-(x-1)^2$
$=x^4-(x^2-2x+1)$
$=x^4-x^2+2x-1$

22. $(x^2+2x+1)(x^2-3x+4)$
$=x^4-3x^3+4x^2+2x^3-6x^2+8x+x^2-3x+4$
$=x^4-x^3-x^2+5x+4$

23. $(x+1)^3-(x-1)^3$
$=(x^3+3x^2+3x+1)-(x^3-3x^2+3x-1)$
$=x^3+3x^2+3x+1-x^3+3x^2-3x+1$
$=6x^2+2$

24. $(x+1)^3-(x+2)^3$
$=(x^3+3x^2+3x+1)-(x^3+6x^2+12x+8)$
$=x^3+3x^2+3x+1-x^3-6x^2-12x-8$
$=-3x^2-9x-7$

25. This is the difference of two squares:
$x^2-36=x^2-6^2=(x-6)(x+6)$

26. This is the difference of two squares:
$x^2-9=x^2-3^2=(x-3)(x+3)$

27. We can rewrite this expression as the difference of two squares:
$(1-4x^2)=(1^2-(2x)^2)=(1-2x)(1+2x)$

28. We can rewrite this expression as the difference of two squares:
$1-9x^2=(1^2-(3x)^2)=(1-3x)(1+3x)$

29. We are looking for two factors of 10 whose sum is 7. Since both 10 and 7 are positive, the two factors must be positive.
$10=5\cdot 2$ and $5+2=7$
$x^2+7x+10=(x+5)(x+2)$

30. We are looking for two factors of 4 whose sum is 5. Since both 4 and 5 are positive, the two factors must be positive.
$4\cdot 1=4$ and $4+1=5$
$x^2+5x+4=(x+4)(x+1)$

31. We are looking for two factors of 8 whose sum is -2. Since 8 is positive, the two factors have the same sign. Since -2 is negative, the two factors must be negative.
There are no such factors so x^2-2x+8 is prime.

32. We are looking for two factors of 5 whose sum is -4. Since 5 is positive, the two factors have the same sign. Since -4 is negative, the two factors must be negative.
There are no such factors so x^2-4x+5 is prime.

33. We are looking for two factors of 16 whose sum is 4. Since both 16 and 4 are positive, the two factors must be positive.
There are no such factors so $x^2+4x+16$ is prime.

34. $x^2+12x+36=x^2+2\cdot 6\cdot x+6^2=(x+6)^2$

35. $15+2x-x^2=(5-x)(3+x)$

36. $14+6x-x^2$ is prime

37. $3x^2-12x-36=3(x^2-4x-12)$
$=3(x-6)(x+2)$

38. $x^3+8x^2-20x=x(x^2+8x-20)$
$=x(x+10)(x-2)$

39. $y^4+11y^3+30y^2=y^2(y^2+11y+30)$
$=y^2(y+6)(y+5)$

40. $3y^3-18y^2-48y=3y(y^2-6y-16)$
$=3y(y-8)(y+2)$

41. $4x^2+12x+9=(2x)^2+2(2x)(3)+3^2$
$=(2x+3)^2$

42. $9x^2-12x+4=(3x)^2-2(3x)(2)+2^2$
$=(3x-2)^2$

43. $3x^2+4x+1=(3x+1)(x+1)$

44. $4x^2+3x-1=(4x-1)(x+1)$

45. $x^4-81=(x^2)^2-(9)^2$
$=(x^2-9)(x^2+9)$
$=(x-3)(x+3)(x^2+9)$

46. $x^4-1=(x^2)^2-1^2=(x^2-1)(x^2+1)$
$=(x-1)(x+1)(x^2+1)$

47. $x^6-2x^3+1=(x^3)^2-2(x^3)+1$
$=(x^3-1)^2$
$=\left[(x-1)(x^2+x+1)\right]^2$
$=(x-1)^2(x^2+x+1)^2$

48. $x^6+2x^3+1=(x^3)^2+2(x^3)+1$
$=(x^3+1)^2$
$=\left[(x+1)(x^2-x+1)\right]^2$
$=(x+1)^2(x^2-x+1)^2$

49. $x^7-x^5=x^5(x^2-1)=x^5(x-1)(x+1)$

50. $x^8-x^5=x^5(x^3-1)=x^5(x-1)(x^2+x+1)$

51. $5+16x-16x^2$
$a\cdot c=5(-16)=-80$
We are looking for two factors of -80 whose sum is 16.
$20\cdot(-4)=-80$ and $20+(-4)=16$
$5+16x-16x^2=5+20x-4x-16x^2$
$=5(1+4x)-4x(1+4x)$
$=(1+4x)(5-4x)$

52. $5+11x-16x^2$
$a\cdot c=5(-16)=-80$
We are looking for two factors of -80 whose sum is 11.
$16\cdot(-5)=-80$ and $16+(-5)=11$
$5+11x-16x^2=5-5x+16x-16x^2$
$=5(1-x)+16x(1-x)$
$=(1-x)(5+16x)$

53. $4y^2-16y+15$
$a\cdot c=4(15)=60$
We are looking for two factors of 60 whose sum is -16.
$(-10)(-6)=60$ and $-10+(-6)=-16$
$4y^2-16y+15=4y^2-10y-6y+15$
$=2y(2y-5)-3(2y-5)$
$=(2y-5)(2y-3)$

54. $9y^2+9y-4$
$a\cdot c=9(-4)=-36$
We are looking for two factors of -36 whose sum is 9.
$12(-3)=-36$ and $12+(-3)=9$
$9y^2+9y-4=9y^2+12y-3y-4$
$=3y(3y+4)-1(3y+4)$
$=(3y+4)(3y-1)$

55. $1-8x^2-9x^4=(1-9x^2)(1+x^2)$
$=(1-3x)(1+3x)(1+x^2)$

56. $4-14x^2-8x^4=2(2-7x^2-4x^4)$
$=2(2+x^2)(1-4x^2)$
$=2(2+x^2)(1-2x)(1+2x)$

57. $x(x+3)-6(x+3)=(x+3)(x-6)$

58. $5(3x-7)+x(3x-7)=(3x-7)(5+x)$

59. $(x+2)^2-5(x+2)=(x+2)(x+2-5)$
$=(x+2)(x-3)$

60. $(x-1)^2-2(x-1)=(x-1)(x-1-2)$
$=(x-1)(x-3)$

61. $6x(2-x)^4-9x^2(2-x)^3$
$=3x(2-x)^3(2(2-x)-3x)$
$=3x(2-x)^3(4-2x-3x)$
$=3x(2-x)^3(4-5x)$

62. $6x(1-x^2)^4-24x^3(1-x^2)^3$
$=6x(1-x^2)^3((1-x^2)-4x^2)$
$=6x[(1-x)(1+x)]^3(1-5x^2)$
$=6x(1-x)^3(1+x)^3(1-5x^2)$

63. $x^3+2x^2-x-2=(x^3+2x^2)+(-x-2)$
$=x^2(x+2)-1(x+2)$
$=(x+2)(x^2-1)$
$=(x+2)(x-1)(x+1)$

64. $x^3-3x^2-x+3=(x^3-3x^2)+(-x+3)$
$=x^2(x-3)-1(x-3)$
$=(x-3)(x^2-1)$
$=(x-3)(x-1)(x+1)$

65. $x^4-x^3+x-1=(x^4-x^3)+(x-1)$
$=x^3(x-1)+1(x-1)$
$=(x-1)(x^3+1)$
$=(x-1)(x+1)(x^2-x+1)$

66. $x^4+x^3+x+1=(x^4+x^3)+(x+1)$
$=x^3(x+1)+1(x+1)$
$=(x+1)(x^3+1)$
$=(x+1)(x+1)(x^2-x+1)$
$=(x+1)^2(x^2-x+1)$

67. $\frac{3x-6}{5x}\cdot\frac{x^2-x-6}{x^2-4}=\frac{3(x-2)(x-3)(x+2)}{5x(x+2)(x-2)}$
$=\frac{3\cancel{(x-2)}(x-3)\cancel{(x+2)}}{5x\cancel{(x+2)}\cancel{(x-2)}}$
$=\frac{3(x-3)}{5x}$

68. $\frac{9x^2-25}{2x-2}\cdot\frac{1-x^2}{6x-10}=\frac{9x^2-25}{2x-2}\cdot\frac{-(x^2-1)}{6x-10}$
$=\frac{-(3x-5)(3x+5)(x-1)(x+1)}{2(x-1)(2)(3x-5)}$
$=\frac{-\cancel{(3x-5)}(3x+5)\cancel{(x-1)}(x+1)}{2\cancel{(x-1)}(2)\cancel{(3x-5)}}$
$=-\frac{(3x+5)(x+1)}{4}$

69. $\frac{4x^2-1}{x^2-16}\cdot\frac{x^2-4x}{2x+1}=\frac{(2x-1)(2x+1)(x)(x-4)}{(x-4)(x+4)(2x+1)}$
$=\frac{(2x-1)\cancel{(2x+1)}(x)\cancel{(x-4)}}{\cancel{(x-4)}(x+4)\cancel{(2x+1)}}$
$=\frac{x(2x-1)}{x+4}$

70. $\frac{12}{x^2-x}\cdot\frac{x^2-1}{4x-2}=\frac{12(x-1)(x+1)}{x(x-1)(2)(2x-1)}$
$=\frac{{}^{6}\cancel{12}\cancel{(x-1)}(x+1)}{x\cancel{(x-1)}\cancel{(2)}(2x-1)}$
$=\frac{6(x+1)}{x(2x-1)}$

71. $\frac{x}{x^2-7x+6}-\frac{x}{x^2-2x-24}$

$=\frac{x}{(x-6)(x-1)}-\frac{x}{(x-6)(x+4)}$

$=\frac{x(x+4)}{(x-6)(x+4)(x-1)}-\frac{x(x-1)}{(x-6)(x+4)(x-1)}$

$=\frac{(x^2+4x)-(x^2-x)}{(x-6)(x+4)(x-1)}$

$=\frac{x^2+4x-x^2+x}{(x-6)(x+4)(x-1)}$

$=\frac{5x}{(x-6)(x+4)(x-1)}$

72. $\frac{x}{x-3}-\frac{x+1}{x^2+5x-24}$

$=\frac{x}{(x-3)}-\frac{(x+1)}{(x+8)(x-3)}$

$=\frac{x(x+8)}{(x+8)(x-3)}-\frac{(x+1)}{(x+8)(x-3)}$

$=\frac{x(x+8)-(x+1)}{(x+8)(x-3)}$

$=\frac{x^2+8x-x-1}{(x+8)(x-3)}$

$=\frac{x^2+7x-1}{(x+8)(x-3)}$

73. $\frac{4}{x^2-4}-\frac{2}{x^2+x-6}$

$=\frac{4}{(x-2)(x+2)}-\frac{2}{(x+3)(x-2)}$

$=\frac{4(x+3)}{(x+3)(x-2)(x+2)}-\frac{2(x+2)}{(x+3)(x-2)(x+2)}$

$=\frac{4(x+3)-2(x+2)}{(x+3)(x-2)(x+2)}$

$=\frac{4x+12-2x-4}{(x+3)(x-2)(x+2)}$

$=\frac{2x+8}{(x+3)(x-2)(x+2)}$

$=\frac{2(x+4)}{(x+3)(x-2)(x+2)}$

74. $\frac{3}{x-1}-\frac{x-4}{x^2-2x+1}$

$=\frac{3}{(x-1)}-\frac{(x-4)}{(x-1)^2}=\frac{3(x-1)}{(x-1)^2}-\frac{(x-4)}{(x-1)^2}$

$=\frac{3(x-1)-(x-4)}{(x-1)^2}=\frac{3x-3-x+4}{(x-1)^2}$

$=\frac{2x+1}{(x-1)^2}$

75. $\frac{1}{x}-\frac{2}{x^2+x}+\frac{3}{x^3-x^2}$

$=\frac{1}{x}-\frac{2}{x(x+1)}+\frac{3}{x^2(x-1)}$

$=\frac{x(x+1)(x-1)-2x(x-1)+3(x+1)}{x^2(x+1)(x-1)}$

$=\frac{x(x^2-1)-2x^2+2x+3x+3}{x^2(x+1)(x-1)}$

$=\frac{x^3-x-2x^2+5x+3}{x^2(x+1)(x-1)}$

$=\frac{x^3-2x^2+4x+3}{x^2(x+1)(x-1)}$

76. $\frac{x}{(x-1)^2}+\frac{2}{x}-\frac{x+1}{x^3-x^2}$

$=\frac{x}{(x-1)^2}+\frac{2}{x}-\frac{x+1}{x^2(x-1)}$

$=\frac{x\cdot x^2+2x(x-1)^2-(x+1)(x-1)}{x^2(x-1)^2}$

$=\frac{x^3+2x(x^2-2x+1)-(x^2-1)}{x^2(x-1)^2}$

$=\frac{x^3+2x^3-4x^2+2x-x^2+1}{x^2(x-1)^2}$

$=\frac{3x^3-5x^2+2x+1}{x^2(x-1)^2}$

77. $\frac{1}{h}\left(\frac{1}{x+h}-\frac{1}{x}\right)=\frac{1}{h}\left(\frac{1\cdot x}{(x+h)x}-\frac{1(x+h)}{x(x+h)}\right)$
$=\frac{1}{h}\left(\frac{x-x-h}{x(x+h)}\right)$
$=\frac{-h}{hx(x+h)}$
$=\frac{-1}{x(x+h)}$

78. $\frac{1}{h}\left(\frac{1}{(x+h)^2}-\frac{1}{x^2}\right)$
$=\frac{1}{h}\left(\frac{1\cdot x^2}{(x+h)^2x^2}-\frac{1(x+h)^2}{x^2(x+h)^2}\right)$
$=\frac{1}{h}\left(\frac{x^2-(x^2+2xh+h^2)}{x^2(x+h)^2}\right)$
$=\frac{-2xh-h^2}{hx^2(x+h)^2}$
$=\frac{h(-2x-h)}{hx^2(x+h)^2}$
$=\frac{-2x-h}{x^2(x+h)^2}$
$=-\frac{2x+h}{x^2(x+h)^2}$

79. $2(3x+4)^2+(2x+3)\cdot 2(3x+4)\cdot 3$
$=2(3x+4)((3x+4)+(2x+3)\cdot 3)$
$=2(3x+4)(3x+4+6x+9)$
$=2(3x+4)(9x+13)$

80. $5(2x+1)^2+(5x-6)\cdot 2(2x+1)\cdot 2$
$=(2x+1)(5(2x+1)+(5x-6)\cdot 4)$
$=(2x+1)(10x+5+20x-24)$
$=(2x+1)(30x-19)$

81. $2x(2x+5)+x^2\cdot 2=2x((2x+5)+x)$
$=2x(2x+5+x)$
$=2x(3x+5)$

82. $3x^2(8x-3)+x^3\cdot 8=x^2(3(8x-3)+8x)$
$=x^2(24x-9+8x)$
$=x^2(32x-9)$

83. $2(x+3)(x-2)^3+(x+3)^2\cdot 3(x-2)^2$
$=(x+3)(x-2)^2(2(x-2)+(x+3)\cdot 3)$
$=(x+3)(x-2)^2(2x-4+3x+9)$
$=(x+3)(x-2)^2(5x+5)$
$=5(x+3)(x-2)^2(x+1)$

84. $4(x+5)^3(x-1)^2+(x+5)^4\cdot 2(x-1)$
$=2(x+5)^3(x-1)(2(x-1)+(x+5))$
$=2(x+5)^3(x-1)(2x-2+x+5)$
$=2(x+5)^3(x-1)(3x+3)$
$=2\cdot 3(x+5)^3(x-1)(x+1)$
$=6(x+5)^3(x-1)(x+1)$

85. $(4x-3)^2+x\cdot 2(4x-3)\cdot 4$
$=(4x-3)((4x-3)+8x)$
$=(4x-3)(4x-3+8x)$
$=(4x-3)(12x-3)$
$=3(4x-3)(4x-1)$

86. $3x^2(3x+4)^2+x^3\cdot 2(3x+4)\cdot 3$
$=3x^2(3x+4)((3x+4)+2x)$
$=3x^2(3x+4)(3x+4+2x)$
$=3x^2(3x+4)(5x+4)$

87. $2(3x-5)\cdot 3(2x+1)^3+(3x-5)^2\cdot 3(2x+1)^2\cdot 2$
$=6(3x-5)(2x+1)^2((2x+1)+(3x-5))$
$=6(3x-5)(2x+1)^2(2x+1+3x-5)$
$=6(3x-5)(2x+1)^2(5x-4)$

88. $3(4x+5)^2 \cdot 4(5x+1)^2 + (4x+5)^3 \cdot 2(5x+1) \cdot 5$

$= 2(4x+5)^2(5x+1)\big(6(5x+1)+5(4x+5)\big)$

$= 2(4x+5)^2(5x+1)(30x+6+20x+25)$

$= 2(4x+5)^2(5x+1)(50x+31)$

89. $\dfrac{(2x+3)\cdot 3-(3x-5)\cdot 2}{(3x-5)^2} = \dfrac{6x+9-6x+10}{(3x-5)^2}$

$= \dfrac{19}{(3x-5)^2}$

90. $\dfrac{(4x+1)\cdot 5-(5x-2)\cdot 4}{(5x-2)^2} = \dfrac{20x+5-20x+8}{(5x-2)^2}$

$= \dfrac{13}{(5x-2)^2}$

91. $\dfrac{x\cdot 2x-(x^2+1)\cdot 1}{(x^2+1)^2} = \dfrac{2x^2-x^2-1}{(x^2+1)^2}$

$= \dfrac{x^2-1}{(x^2+1)^2}$

$= \dfrac{(x-1)(x+1)}{(x^2+1)^2}$

92. $\dfrac{x\cdot 2x-(x^2-4)\cdot 1}{(x^2-4)^2} = \dfrac{2x^2-x^2+4}{(x^2-4)^2}$

$= \dfrac{x^2+4}{(x^2-4)^2}$

93. $\dfrac{(3x+1)\cdot 2x-x^2\cdot 3}{(3x+1)^2} = \dfrac{6x^2+2x-3x^2}{(3x+1)^2}$

$= \dfrac{3x^2+2x}{(3x+1)^2}$

$= \dfrac{x(3x+2)}{(3x+1)^2}$

94. $\dfrac{(2x-5)\cdot 3x^2-x^3\cdot 2}{(2x-5)^2} = \dfrac{6x^3-15x^2-2x^3}{(2x-5)^2}$

$= \dfrac{4x^3-15x^2}{(2x-5)^2}$

$= \dfrac{x^2(4x-15)}{(2x-5)^2}$

95. $\dfrac{(x^2+1)\cdot 3-(3x+4)\cdot 2x}{(x^2+1)^2} = \dfrac{3x^2+3-6x^2-8x}{(x^2+1)^2}$

$= \dfrac{-3x^2-8x+3}{(x^2+1)^2}$

$= \dfrac{-(3x^2+8x-3)}{(x^2+1)^2}$

$= -\dfrac{(3x-1)(x+3)}{(x^2+1)^2}$

96.

$\dfrac{(x^2+9)\cdot 2-(2x-5)\cdot 2x}{(x^2+9)^2} = \dfrac{2x^2+18-4x^2+10x}{(x^2+9)^2}$

$= \dfrac{-2x^2+10x+18}{(x^2+9)^2}$

$= \dfrac{-2(x^2-5x-9)}{(x^2+9)^2}$

Section A.4

1. quotient; divisor; remainder

2. $-3\overline{)2 \quad 0 \quad -5 \quad 1}$

3. True

4. True

5.
$$\begin{array}{r} 4x^2-11x+23 \\ x+2\overline{)4x^3-3x^2+\quad x+1} \\ \underline{4x^3+8x^2} \\ -11x^2+\quad x \\ \underline{-11x^2-22x} \\ 23x+1 \\ \underline{23x+46} \\ -45 \end{array}$$

Check:

$(x+2)(4x^2-11x+23)+(-45)$
$=4x^3-11x^2+23x+8x^2-22x+46-45$
$=4x^3-3x^2+x+1$

The quotient is $4x^2-11x+23$; the remainder is –45.

6.
$$\begin{array}{r} 3x^2-7x+15 \\ x+2\overline{)3x^3-\ x^2+\quad x-2} \\ \underline{3x^3+6x^2} \\ -7x^2+\quad x \\ \underline{-7x^2-14x} \\ 15x-2 \\ \underline{15x+30} \\ -32 \end{array}$$

Check:

$(x+2)(3x^2-7x+15)+(-32)$
$=3x^3-7x^2+15x+6x^2-14x+30-32$
$=3x^3-x^2+x-2$

The quotient is $3x^2-7x+15$; the remainder is –32.

7.
$$\begin{array}{r} 4x-3 \\ x^2\overline{)4x^3-3x^2+x+1} \\ \underline{4x^3} \\ -3x^2+x+1 \\ \underline{-3x^2} \\ x+1 \end{array}$$

Check:

$(x^2)(4x-3)+(x+1)=4x^3-3x^2+x+1$

The quotient is $4x-3$; the remainder is $x+1$.

8.
$$\begin{array}{r} 3x-1 \\ x^2\overline{)3x^3-x^2+x-2} \\ \underline{3x^3} \\ -x^2+x-2 \\ \underline{-x^2} \\ x-2 \end{array}$$

Check:

$(x^2)(3x-1)+(x-2)=3x^3-x^2+x-2$

The quotient is $3x-1$; the remainder is $x-2$.

9.
$$\begin{array}{r} 5x^2-13 \\ x^2+2\overline{)5x^4+0x^3-3x^2+x+1} \\ \underline{5x^4\qquad +10x^2} \\ -13x^2+x+1 \\ \underline{-13x^2\qquad -26} \\ x+27 \end{array}$$

Check:

$\left(x^2+2\right)\left(5x^2-13\right)+\left(x+27\right)$
$=5x^4+10x^2-13x^2-26+x+27$
$=5x^4-3x^2+x+1$

The quotient is $5x^2-13$; the remainder is $x+27$.

10. $x^2+2\,\overline{\big)\,5x^4+0x^3-x^2+x-2}$, quotient $5x^2-11$

$$\begin{array}{r} 5x^4 \qquad +10x^2 \\ \hline -11x^2+x-2 \\ -11x^2 \qquad -22 \\ \hline x+20 \end{array}$$

Check:

$(x^2+2)(5x^2-11)+(x+20)$

$=5x^4+10x^2-11x^2-22+x+20$

$=5x^4-x^2+x-2$

The quotient is $5x^2-11$; the remainder is $x+20$.

11. $2x^3-1\,\overline{\big)\,4x^5+0x^4+0x^3-3x^2+x+1}$, quotient $2x^2$

$$\begin{array}{r} 4x^5 \qquad\qquad -2x^2 \\ \hline -x^2+x+1 \end{array}$$

Check:

$(2x^3-1)(2x^2)+(-x^2+x+1)$

$=4x^5-2x^2-x^2+x+1=4x^5-3x^2+x+1$

The quotient is $2x^2$; the remainder is $-x^2+x+1$.

12. $3x^3-1\,\overline{\big)\,3x^5+0x^4+0x^3-x^2+x-2}$, quotient x^2

$$\begin{array}{r} 3x^5 \qquad\qquad -x^2 \\ \hline x-2 \end{array}$$

Check:

$(3x^3-1)(x^2)+(x-2)=3x^5-x^2+x-2$

The quotient is x^2; the remainder is $x-2$.

13. $2x^2+x+1\,\overline{\big)\,2x^4-3x^3+0x^2+x+1}$, quotient $x^2-2x+\frac{1}{2}$

$$\begin{array}{r} 2x^4+x^3+x^2 \\ \hline -4x^3-x^2+x \\ -4x^3-2x^2-2x \\ \hline x^2+3x+1 \\ x^2+\frac{1}{2}x+\frac{1}{2} \\ \hline \frac{5}{2}x+\frac{1}{2} \end{array}$$

Check:

$(2x^2+x+1)(x^2-2x+\frac{1}{2})+\frac{5}{2}x+\frac{1}{2}$

$=2x^4-4x^3+x^2+x^3-2x^2+\frac{1}{2}x$

$\quad +x^2-2x+\frac{1}{2}+\frac{5}{2}x+\frac{1}{2}$

$=2x^4-3x^3+x+1$

The quotient is $x^2-2x+\frac{1}{2}$; the remainder is $\frac{5}{2}x+\frac{1}{2}$.

14. $3x^2+x+1\,\overline{\big)\,3x^4-x^3+0x^2+x-2}$, quotient $x^2-\frac{2}{3}x-\frac{1}{9}$

$$\begin{array}{r} 3x^4+x^3+x^2 \\ \hline -2x^3-x^2+x \\ -2x^3-\frac{2}{3}x^2-\frac{2}{3}x \\ \hline -\frac{1}{3}x^2+\frac{5}{3}x-2 \\ -\frac{1}{3}x^2-\frac{1}{9}x-\frac{1}{9} \\ \hline \frac{16}{9}x-\frac{17}{9} \end{array}$$

Check:

$(3x^2+x+1)(x^2-\frac{2}{3}x-\frac{1}{9})+(\frac{16}{9}x-\frac{17}{9})$

$=3x^4+x^3+x^2-2x^3-\frac{2}{3}x^2-\frac{2}{3}x$

$\quad -\frac{1}{3}x^2-\frac{1}{9}x-\frac{1}{9}+\frac{16}{9}x-\frac{17}{9}$

$=3x^4-x^3+x-2$

The quotient is $x^2-\frac{2}{3}x-\frac{1}{9}$; the remainder is $\frac{16}{9}x-\frac{17}{9}$.

15.

$$\begin{array}{r} -4x^2-3x-3 \\ x-1\overline{)-4x^3+x^2+0x-4} \\ \underline{-4x^3+4x^2} \\ -3x^2 \\ \underline{-3x^2+3x} \\ -3x-4 \\ \underline{-3x+3} \\ -7 \end{array}$$

Check:

$(x-1)(-4x^2-3x-3)+(-7)$

$=-4x^3-3x^2-3x+4x^2+3x+3-7$

$=-4x^3+x^2-4$

The quotient is $-4x^2-3x-3$; the remainder is –7.

16.

$$\begin{array}{r} -3x^3-3x^2-3x-5 \\ x-1\overline{)-3x^4+0x^3+0x^2-2x-1} \\ \underline{-3x^4+3x^3} \\ -3x^3 \\ \underline{-3x^3+3x^2} \\ -3x^2-2x \\ \underline{-3x^2+3x} \\ -5x-1 \\ \underline{-5x+5} \\ -6 \end{array}$$

Check:

$(x-1)(-3x^3-3x^2-3x-5)+(-6)$

$=-3x^4-3x^3-3x^2-5x+3x^3+3x^2$

$+3x+5-6$

$=-3x^4-2x-1$

The quotient is $-3x^3-3x^2-3x-5$; the remainder is –6.

17.

$$\begin{array}{r} x^2-x-1 \\ x^2+x+1\overline{)x^4+0x^3-x^2+0x+1} \\ \underline{x^4+x^3+x^2} \\ -x^3-2x^2 \\ \underline{-x^3-x^2-x} \\ -x^2+x+1 \\ \underline{-x^2-x-1} \\ 2x+2 \end{array}$$

Check:

$(x^2+x+1)(x^2-x-1)+2x+2$

$=x^4+x^3+x^2-x^3-x^2-x-x^2-x$

$-1+2x+2$

$=x^4-x^2+1$

The quotient is x^2-x-1; the remainder is $2x+2$.

18.

$$\begin{array}{r} x^2+x-1 \\ x^2-x+1\overline{)x^4+0x^3-x^2+0x+1} \\ \underline{x^4-x^3+x^2} \\ x^3-2x^2 \\ \underline{x^3-x^2+x} \\ -x^2-x+1 \\ \underline{-x^2+x-1} \\ -2x+2 \end{array}$$

Check:

$(x^2-x+1)(x^2+x-1)+(-2x+2)$

$=x^4+x^3-x^2-x^3-x^2+x+x^2+x$

$-1-2x+2$

$=x^4-x^2+1$

The quotient is x^2+x-1; the remainder is $-2x+2$.

19.

$$\begin{array}{r} x^2+ax+a^2 \\ x-a\overline{)x^3+0x^2+0x-a^3} \\ \underline{x^3-ax^2} \\ ax^2 \\ \underline{ax^2-a^2x} \\ a^2x-a^3 \\ \underline{a^2x-a^3} \\ 0 \end{array}$$

Check:

$(x-a)(x^2+ax+a^2)+0$

$=x^3+ax^2+a^2x-ax^2-a^2x-a^3$

$=x^3-a^3$

The quotient is x^2+ax+a^2; the remainder is 0.

20.

$$\begin{array}{r} x^4+ax^3+a^2x^2+a^3x+a^4 \\ x-a\overline{)x^5+0x^4+0x^3+0x^2+0x-a^5} \\ \underline{x^5-ax^4} \\ ax^4 \\ \underline{ax^4-a^2x^3} \\ a^2x^3 \\ \underline{a^2x^3-a^3x^2} \\ a^3x^2 \\ \underline{a^3x^2-a^4x} \\ a^4x-a^5 \\ \underline{a^4x-a^5} \\ 0 \end{array}$$

Check:

$(x-a)(x^4+ax^3+a^2x^2+a^3x+a^4)+0$

$=x^5+ax^4+a^2x^3+a^3x^2+a^4x-ax^4$

$-a^2x^3-a^3x^2-a^4x-a^5$

$=x^5-a^5$

The quotient is $x^4+ax^3+a^2x^2+a^3x+a^4$; the remainder is 0.

21.

$$\begin{array}{r|rrrr} 2 & 1 & -1 & 2 & 4 \\ & & 2 & 2 & 8 \\ \hline & 1 & 1 & 4 & 12 \end{array}$$

Quotient: x^2+x+4
Remainder: 12

22.

$$\begin{array}{r|rrrr} -1 & 1 & 2 & -3 & 1 \\ & & -1 & -1 & 4 \\ \hline & 1 & 1 & -4 & 5 \end{array}$$

Quotient: x^2+x-4
Remainder: 5

23.

$$\begin{array}{r|rrrr} 3 & 3 & 2 & -1 & 3 \\ & & 9 & 33 & 96 \\ \hline & 3 & 11 & 32 & 99 \end{array}$$

Quotient: $3x^2+11x+32$
Remainder: 99

24.

$$\begin{array}{r|rrrr} -2 & -4 & 2 & -1 & 1 \\ & & 8 & -20 & 42 \\ \hline & -4 & 10 & -21 & 43 \end{array}$$

Quotient: $-4x^2+10x-21$
Remainder: 43

25.

$$\begin{array}{r|rrrrrr} -3 & 1 & 0 & -4 & 0 & 1 & 0 \\ & & -3 & 9 & -15 & 45 & -138 \\ \hline & 1 & -3 & 5 & -15 & 46 & -138 \end{array}$$

Quotient: $x^4-3x^3+5x^2-15x+46$
Remainder: -138

26.

$$\begin{array}{r|rrrrr} 2 & 1 & 0 & 1 & 0 & 2 \\ & & 2 & 4 & 10 & 20 \\ \hline & 1 & 2 & 5 & 10 & 22 \end{array}$$

Quotient: $x^3+2x^2+5x+10$
Remainder: 22

27.

$$\begin{array}{r|rrrrrrr} 1 & 4 & 0 & -3 & 0 & 1 & 0 & 5 \\ & & 4 & 4 & 1 & 1 & 2 & 2 \\ \hline & 4 & 4 & 1 & 1 & 2 & 2 & 7 \end{array}$$

Quotient: $4x^5+4x^4+x^3+x^2+2x+2$
Remainder: 7

28. $$\begin{array}{r|rrrrrr} -1 & 1 & 0 & 5 & 0 & 0 & -10 \\ & & -1 & 1 & -6 & 6 & -6 \\ \hline & 1 & -1 & 6 & -6 & 6 & -16 \end{array}$$

Quotient: $x^4 - x^3 + 6x^2 - 6x + 6$
Remainder: -16

29. $$\begin{array}{r|rrrr} -1.1 & 0.1 & 0 & 0.2 & 0 \\ & & -0.11 & 0.121 & -0.3531 \\ \hline & 0.1 & -0.11 & 0.321 & -0.3531 \end{array}$$

Quotient: $0.1x^2 - 0.11x + 0.321$
Remainder: -0.3531

30. $$\begin{array}{r|rrr} -2.1 & 0.1 & 0 & -0.2 \\ & & -0.21 & 0.441 \\ \hline & 0.1 & -0.21 & 0.241 \end{array}$$

Quotient: $0.1x - 0.21$
Remainder: 0.241

31. $$\begin{array}{r|rrrrrr} 1 & 1 & 0 & 0 & 0 & 0 & -1 \\ & & 1 & 1 & 1 & 1 & 1 \\ \hline & 1 & 1 & 1 & 1 & 1 & 0 \end{array}$$

Quotient: $x^4 + x^3 + x^2 + x + 1$
Remainder: 0

32. $$\begin{array}{r|rrrrrr} -1 & 1 & 0 & 0 & 0 & 0 & 1 \\ & & -1 & 1 & -1 & 1 & -1 \\ \hline & 1 & -1 & 1 & -1 & 1 & 0 \end{array}$$

Quotient: $x^4 - x^3 + x^2 - x + 1$
Remainder: 0

33. $$\begin{array}{r|rrrr} 2 & 4 & -3 & -8 & 4 \\ & & 8 & 10 & 4 \\ \hline & 4 & 5 & 2 & 8 \end{array}$$

Remainder $= 8 \neq 0$; therefore $x - 2$ is not a factor of $4x^3 - 3x^2 - 8x + 4$.

34. $$\begin{array}{r|rrrr} -3 & -4 & 5 & 0 & 8 \\ & & 12 & -51 & 153 \\ \hline & -4 & 17 & -51 & 161 \end{array}$$

Remainder $= 161 \neq 0$; therefore $x + 3$ is not a factor of $-4x^3 + 5x^2 + 8$.

35. $$\begin{array}{r|rrrrr} 2 & 3 & -6 & 0 & -5 & 10 \\ & & 6 & 0 & 0 & -10 \\ \hline & 3 & 0 & 0 & -5 & 0 \end{array}$$

Remainder $= 0$; therefore $x - 2$ is a factor of $3x^4 - 6x^3 - 5x + 10$.

36. $$\begin{array}{r|rrrrr} 2 & 4 & 0 & -15 & 0 & -4 \\ & & 8 & 16 & 2 & 4 \\ \hline & 4 & 8 & 1 & 2 & 0 \end{array}$$

Remainder $= 0$; therefore $x - 2$ is a factor of $4x^4 - 15x^2 - 4$.

37. $$\begin{array}{r|rrrrrrr} -3 & 3 & 0 & 0 & 82 & 0 & 0 & 27 \\ & & -9 & 27 & -81 & -3 & 9 & -27 \\ \hline & 3 & -9 & 27 & 1 & -3 & 9 & 0 \end{array}$$

Remainder $= 0$; therefore $x + 3$ is a factor of $3x^6 + 82x^3 + 27$.

38. $$\begin{array}{r|rrrrrrr} -3 & 2 & 0 & -18 & 0 & 1 & 0 & -9 \\ & & -6 & 18 & 0 & 0 & -3 & 9 \\ \hline & 2 & -6 & 0 & 0 & 1 & -3 & 0 \end{array}$$

Remainder $= 0$; therefore $x + 3$ is a factor of $2x^6 - 18x^4 + x^2 - 9$.

39. $$\begin{array}{r|rrrrrrr} -4 & 4 & 0 & -64 & 0 & 1 & 0 & -15 \\ & & -16 & 64 & 0 & 0 & -4 & 16 \\ \hline & 4 & -16 & 0 & 0 & 1 & -4 & 1 \end{array}$$

Remainder $= 1 \neq 0$; therefore $x + 4$ is not a factor of $4x^6 - 64x^4 + x^2 - 15$.

40. $$\begin{array}{r|rrrrrrr} -4 & 1 & 0 & -16 & 0 & 1 & 0 & -16 \\ & & -4 & 16 & 0 & 0 & -4 & 16 \\ \hline & 1 & -4 & 0 & 0 & 1 & -4 & 0 \end{array}$$

Remainder $= 0$; therefore $x + 4$ is a factor $x^6 - 16x^4 + x^2 - 16$.

41. $$\begin{array}{r|rrrrr} \frac{1}{2} & 2 & -1 & 0 & 2 & -1 \\ & & 1 & 0 & 0 & 1 \\ \hline & 2 & 0 & 0 & 2 & 0 \end{array}$$

Remainder $= 0$; therefore $x - \frac{1}{2}$ is a factor of $2x^4 - x^3 + 2x - 1$.

42. $-\frac{1}{3}\overline{)3 \quad 1 \quad 0 \quad -3 \quad 1}$

$$\begin{array}{rrrrr} 3 & 1 & 0 & -3 & 1 \\ & -1 & 0 & 0 & 1 \\ \hline 3 & 0 & 0 & -3 & 2 \end{array}$$

Remainder $= 2 \neq 0$; therefore $x+\frac{1}{3}$ is not a factor of $3x^4+x^3-3x+1$.

43. $-2\overline{)1 \quad -2 \quad 3 \quad 5}$

$$\begin{array}{rrrr} 1 & -2 & 3 & 5 \\ & -2 & 8 & -22 \\ \hline 1 & -4 & 11 & -17 \end{array}$$

$$\frac{x^3-2x^2+3x+5}{x+2} = x^2-4x+11+\frac{-17}{x+2}$$

$a=1; b=-4; c=11; d=-17$

$a+b+c+d = 1-4+11-17 = -9$

44. Answers will vary.

Section A.5

1. $x^2-4=(x-2)(x+2)$

$x^2-3x+2=(x-1)(x-2)$

The least common denominator is $(x-1)(x-2)(x+2)$.

2. We are looking for two factors of $a \cdot c=(2)(-3)=-6$ whose sum is -1. Since the product is negative, the two factors have opposite signs. Since the sum is negative, the factor with the larger absolute value will be negative.

$(-3)(2)=-6$ and $-3+2=-1$

$$\begin{aligned} 2x^2-x-3 &= 2x^2+2x-3x-3 \\ &= 2x(x+1)-3(x+1) \\ &= (x+1)(2x-3) \end{aligned}$$

3. $(x-3)(3x+5)=0$

$x-3=0$ or $3x+5=0$

$x=3 \qquad 3x=-5$

$x=-\frac{5}{3}$

Solution set: $\left\{-\frac{5}{3}, 3\right\}$

4. True

5. equivalent equations

6. identity

7. False; the solution is $\frac{8}{3}$.

$3x-8=0$

$3x=8$

$x=\frac{8}{3}$

8. True; e.g. $2x+1=2x+3$.

9. add; $\frac{25}{4}$

10. discriminant; negative

11. False; a quadratic equation may also have 1 repeated real solution or no real solutions.

12. False; the equation will have two real solutions but not necessarily negatives of one another.

13. $3x=21$

$\frac{3x}{3}=\frac{21}{3}$

$x=7$

The solution set is $\{7\}$

14. $3x=-24$

$\frac{3x}{3}=\frac{-24}{3}$

$x=-8$

The solution set is $\{-8\}$

15. $5x+15=0$

$5x+15-15=0-15$

$5x=-15$

$\frac{5x}{5}=\frac{-15}{5}$

$x=-3$

The solution set is $\{-3\}$

16. $3x+18=0$

$3x+18-18=0-18$

$3x=-18$

$\frac{3x}{3}=\frac{-18}{3}$

$x=-6$

The solution set is $\{-6\}$

17.
$$\begin{aligned} 2x-3&=5\\ 2x-3+3&=5+3\\ 2x&=8\\ \frac{2x}{2}&=\frac{8}{2}\\ x&=4 \end{aligned}$$
The solution set is $\{4\}$

18.
$$\begin{aligned} 3x+4&=-8\\ 3x+4-4&=-8-4\\ 3x&=-12\\ \frac{3x}{3}&=\frac{-12}{3}\\ x&=-4 \end{aligned}$$
The solution set is $\{-4\}$

19.
$$\begin{aligned} \frac{1}{3}x&=\frac{5}{12}\\ 3\cdot\frac{1}{3}x&=3\cdot\frac{5}{12}\\ x&=\frac{5}{4} \end{aligned}$$
The solution set is $\left\{\frac{5}{4}\right\}$

20.
$$\begin{aligned} \frac{2}{3}x&=\frac{9}{2}\\ \frac{3}{2}\cdot\frac{2}{3}x&=\frac{3}{2}\cdot\frac{9}{2}\\ x&=\frac{27}{4} \end{aligned}$$
The solution set is $\left\{\frac{27}{4}\right\}$

21.
$$\begin{aligned} 6-x&=2x+9\\ 6-x-6&=2x+9-6\\ -x&=2x+3\\ -x-2x&=2x+3-2x\\ -3x&=3\\ \frac{-3x}{-3}&=\frac{3}{-3}\\ x&=-1 \end{aligned}$$
The solution set is $\{-1\}$.

22.
$$\begin{aligned} 3-2x&=2-x\\ 3-2x-3&=2-x-3\\ -2x&=-x-1\\ -2x+x&=-x-1+x\\ -x&=-1\\ -1(-x)&=-1(-1)\\ x&=1 \end{aligned}$$
The solution set is $\{1\}$.

23.
$$\begin{aligned} 2(3+2x)&=3(x-4)\\ 6+4x&=3x-12\\ 6+4x-6&=3x-12-6\\ 4x&=3x-18\\ 4x-3x&=3x-18-3x\\ x&=-18 \end{aligned}$$
The solution set is $\{-18\}$.

24.
$$\begin{aligned} 3(2-x)&=2x-1\\ 6-3x&=2x-1\\ 6-3x-6&=2x-1-6\\ -3x&=2x-7\\ -3x-2x&=2x-7-2x\\ -5x&=-7\\ \frac{-5x}{-5}&=\frac{-7}{-5}\\ x&=\frac{7}{5} \end{aligned}$$
The solution set is $\left\{\frac{7}{5}\right\}$.

25.
$$\begin{aligned} 8x-(2x+1)&=3x-10\\ 8x-2x-1&=3x-10\\ 6x-1&=3x-10\\ 6x-1+1&=3x-10+1\\ 6x&=3x-9\\ 6x-3x&=3x-9-3x\\ 3x&=-9\\ \frac{3x}{3}&=\frac{-9}{3}\\ x&=-3 \end{aligned}$$
The solution set is $\{-3\}$.

26. $5-(2x-1)=10$

$$5-2x+1=10$$
$$-2x+6=10$$
$$-2x+6-6=10-6$$
$$-2x=4$$
$$\frac{-2x}{-2}=\frac{4}{-2}$$
$$x=-2$$

The solution set is $\{-2\}$.

27. $\frac{1}{2}x-4=\frac{3}{4}x$

$$4\left(\frac{1}{2}x-4\right)=4\left(\frac{3}{4}x\right)$$
$$2x-16=3x$$
$$2x-16+16=3x+16$$
$$2x=3x+16$$
$$2x-3x=3x+16-3x$$
$$-x=16$$
$$-1(-x)=-1(16)$$
$$x=-16$$

The solution set is $\{-16\}$.

28. $1-\frac{1}{2}x=5$

$$2\left(1-\frac{1}{2}x\right)=2(5)$$
$$2-x=10$$
$$2-x-2=10-2$$
$$-x=8$$
$$-1(-x)=-1(8)$$
$$x=-8$$

The solution set is $\{-8\}$.

29. $0.9t=0.4+0.1t$

$$0.9t-0.1t=0.4+0.1t-0.1t$$
$$0.8t=0.4$$
$$\frac{0.8t}{0.8}=\frac{0.4}{0.8}$$
$$t=0.5$$

The solution set is $\{0.5\}$.

30. $0.9t=1+t$

$$0.9t-t=1+t-t$$
$$-0.1t=1$$
$$\frac{-0.1t}{-0.1}=\frac{1}{-0.1}$$
$$t=-10$$

The solution set is $\{-10\}$.

31. $\frac{2}{y}+\frac{4}{y}=3$

$$\frac{6}{y}=3$$
$$\left(\frac{6}{y}\right)^{-1}=3^{-1}$$
$$\frac{y}{6}=\frac{1}{3}$$
$$6\cdot\frac{y}{6}=6\cdot\frac{1}{3}$$
$$y=2$$

The solution set is $\{2\}$.

32. $\frac{4}{y}-5=\frac{5}{2y}$

$$2y\left(\frac{4}{y}-5\right)=2y\left(\frac{5}{2y}\right)$$
$$8-10y=5$$
$$8-10y-8=5-8$$
$$-10y=-3$$
$$\frac{-10y}{-10}=\frac{-3}{-10}$$
$$y=\frac{3}{10}$$

The solution set is $\left\{\frac{3}{10}\right\}$.

33. $(x+7)(x-1)=(x+1)^2$

$$x^2-x+7x-7=x^2+2x+1$$
$$x^2+6x-7=x^2+2x+1$$
$$x^2+6x-7-x^2=x^2+2x+1-x^2$$
$$6x-7=2x+1$$
$$6x-7-2x=2x+1-2x$$
$$4x-7=1$$
$$4x-7+7=1+7$$
$$4x=8$$
$$\frac{4x}{4}=\frac{8}{4}$$
$$x=2$$

The solution set is $\{2\}$.

34. $(x+2)(x-3)=(x-3)^2$

$$x^2+2x-3x-6=x^2-6x+9$$
$$x^2-x-6=x^2-6x+9$$
$$x^2-x-6-x^2=x^2-6x+9-x^2$$
$$-x-6=-6x+9$$
$$-x-6+6x=-6x+9+6x$$
$$5x-6=9$$
$$5x-6+6=9+6$$
$$5x=15$$
$$\frac{5x}{5}=\frac{15}{5}$$
$$x=3$$

The solution set is $\{3\}$.

35. $z(z^2+1)=3+z^3$

$$z^3+z=3+z^3$$
$$z^3+z-z^3=3+z^3-z^3$$
$$z=3$$

The solution set is $\{3\}$.

36. $w(4-w^2)=8-w^3$

$$4w-w^3=8-w^3$$
$$4w=8$$
$$w=2$$

The solution set is $\{2\}$.

37. $x^2=9x$

$$x^2-9x=0$$
$$x(x-9)=0$$
$$x=0 \text{ or } x-9=0$$
$$x=9$$

The solution set is $\{0,9\}$.

38. $x^3=x^2$

$$x^3-x^2=0$$
$$x^2(x-1)=0$$
$$x^2=0 \text{ or } x-1=0$$
$$x=0 \qquad x=1$$

The solution set is $\{0,1\}$.

39. $t^3-9t^2=0$

$$t^2(t-9)=0$$
$$t^2=0 \text{ or } t-9=0$$
$$t=0 \qquad t=9$$

The solution set is $\{0,9\}$.

40. $4z^3-8z^2=0$

$$4z^2(z-2)=0$$
$$4z^2=0 \text{ or } z-2=0$$
$$z=0 \qquad z=2$$

The solution set is $\{0,2\}$.

41. $\frac{3}{2x-3}=\frac{2}{x+5}$

$$3(x+5)=2(2x-3)$$
$$3x+15=4x-6$$
$$3x+15-3x=4x-6-3x$$
$$15=x-6$$
$$15+6=x-6+6$$
$$21=x$$

The solution set is $\{21\}$.

42. $$\frac{-2}{x+4} = \frac{-3}{x+1}$$
$$-2(x+1) = -3(x+4)$$
$$-2x-2 = -3x-12$$
$$-2x-2+3x = -3x-12+3x$$
$$x-2 = -12$$
$$x-2+2 = -12+2$$
$$x = -10$$
The solution set is $\{-10\}$.

43. $$(x+2)(3x) = (x+2)(6)$$
$$3x^2+6x = 6x+12$$
$$3x^2+6x-6x = 6x+12-6x$$
$$3x^2 = 12$$
$$x^2 = 4$$
$$x = \pm 2$$
The solution set is $\{-2, 2\}$.

44. $$(x-5)(2x) = (x-5)(4)$$
$$2x^2-10x = 4x-20$$
$$2x^2-14x+20 = 0$$
$$x^2-7x+10 = 0$$
$$(x-5)(x-2) = 0$$
$$x-5 = 0 \quad \text{or} \quad x-2 = 0$$
$$x = 5 \qquad x = 2$$
The solution set is $\{2, 5\}$.

45. $$\frac{2}{x-2} = \frac{3}{x+5} + \frac{10}{(x+5)(x-2)}$$
LCD = $(x+5)(x-2)$
$$\frac{2(x+5)}{(x+5)(x-2)} = \frac{3(x-2)}{(x+5)(x-2)} + \frac{10}{(x+5)(x-2)}$$
$$2(x+5) = 3(x-2)+10$$
$$2x+10 = 3x-6+10$$
$$2x+10 = 3x+4$$
$$10 = x+4$$
$$6 = x$$
The solution set is $\{6\}$.

46. $$\frac{1}{2x+3} + \frac{1}{x-1} = \frac{1}{(2x+3)(x-1)}$$
LCD = $(2x+3)(x-1)$
$$\frac{x-1}{(2x+3)(x-1)} + \frac{2x+3}{(2x+3)(x-1)} = \frac{1}{(2x+3)(x-1)}$$
$$x-1+2x+3 = 1$$
$$3x+2 = 1$$
$$3x = -1$$
$$x = -\frac{1}{3}$$
The solution set is $\left\{-\frac{1}{3}\right\}$.

47. $|2x| = 6$
$$2x = 6 \quad \text{or} \quad 2x = -6$$
$$x = 3 \qquad x = -3$$
The solution set is $\{-3, 3\}$.

48. $|3x| = 12$
$$3x = 12 \quad \text{or} \quad 3x = -12$$
$$x = 4 \qquad x = -4$$
The solution set is $\{-4, 4\}$.

49. $|2x+3| = 5$
$$2x+3 = 5 \quad \text{or} \quad 2x+3 = -5$$
$$2x = 2 \quad \text{or} \quad 2x = -8$$
$$x = 1 \quad \text{or} \quad x = -4$$
The solution set is $\{-4, 1\}$.

50. $|3x-1| = 2$
$$3x-1 = 2 \quad \text{or} \quad 3x-1 = -2$$
$$3x = 3 \quad \text{or} \quad 3x = -1$$
$$x = 1 \quad \text{or} \quad x = -\frac{1}{3}$$
The solution set is $\left\{-\frac{1}{3}, 1\right\}$.

51. $|1-4t| = 5$
$$1-4t = 5 \quad \text{or} \quad 1-4t = -5$$
$$-4t = 4 \quad \text{or} \quad -4t = -6$$
$$t = -1 \quad \text{or} \quad t = \frac{3}{2}$$
The solution set is $\left\{-1, \frac{3}{2}\right\}$.

52. $|1-2z|=3$

$1-2z=3$ or $1-2z=-3$

$-2z=2$ or $-2z=-4$

$z=-1$ or $z=2$

The solution set is $\{-1, 2\}$.

53. $|-2x|=8$

$-2x=8$ or $-2x=-8$

$x=-4$ or $x=4$

The solution set is $\{-4, 4\}$.

54. $|-x|=1$

$-x=1$ or $-x=-1$

$x=-1$ $x=1$

The solution set is $\{-1, 1\}$.

55. $|-2|x=4$

$2x=4$

$x=2$

The solution set is $\{2\}$.

56. $|3|x=9$

$3x=9$

$x=3$

The solution set is $\{3\}$.

57. $|x-2|=-\frac{1}{2}$

Since absolute values are never negative, this equation has no solution.

58. $|2-x|=-1$

Since absolute values are never negative, this equation has no solution.

59. $|x^2-4|=0$

$x^2-4=0$

$x^2=4$

$x=\pm 2$

The solution set is $\{-2,2\}$.

60. $|x^2-9|=0$

$x^2-9=0$

$x^2=9$

$x=\pm 3$

The solution set is $\{-3, 3\}$.

61. $|x^2-2x|=3$

$x^2-2x=3$ or $x^2-2x=-3$

$x^2-2x-3=0$ or $x^2-2x+3=0$

$(x-3)(x+1)=0$ or $x=\frac{2\pm\sqrt{4-12}}{2}$

$=\frac{2\pm\sqrt{-8}}{2}$ no real sol.

$x=3$ or $x=-1$

The solution set is $\{-1, 3\}$.

62. $|x^2+x|=12$

$x^2+x=12$ or $x^2+x=-12$

$x^2+x-12=0$ or $x^2+x+12=0$

$(x-3)(x+4)=0$ or $x=\frac{-1\pm\sqrt{1-48}}{2}$

$=\frac{1\pm\sqrt{-47}}{2}$ no real sol.

$x=3$ or $x=-4$

The solution set is $\{-4, 3\}$.

63. $|x^2+x-1|=1$

$x^2+x-1=1$ or $x^2+x-1=-1$

$x^2+x-2=0$ or $x^2+x=0$

$(x-1)(x+2)=0$ or $x(x+1)=0$

$x=1, x=-2$ or $x=0, x=-1$

The solution set is $\{-2,-1,0,1\}$.

64. $|x^2+3x-2|=2$

$x^2+3x-2=2$ or $x^2+3x-2=-2$

$x^2+3x-4=0$ or $x^2+3x=0$

$(x+4)(x-1)=0$ or $x(x+3)=0$

$x=-4,\ x=1$ or $x=0, x=-3$

The solution set is $\{-4,-3,0,1\}$.

65. $x^2 = 4x$

$x^2 - 4x = 0$

$x(x-4) = 0$

$x = 0$ or $x - 4 = 0$

$x = 4$

The solution set is $\{0, 4\}$.

66. $x^2 = -8x$

$x^2 + 8x = 0$

$x(x+8) = 0$

$x = 0$ or $x + 8 = 0$

$x = -8$

The solution set is $\{-8, 0\}$.

67. $z^2 + 4z - 12 = 0$

$(z+6)(z-2) = 0$

$z + 6 = 0$ or $z - 2 = 0$

$z = -6$ $\quad z = 2$

The solution set is $\{-6, 2\}$.

68. $v^2 + 7v + 12 = 0$

$(v+4)(v+3) = 0$

$v + 4 = 0$ or $v + 3 = 0$

$v = -4$ $\quad v = -3$

The solution set is $\{-4, -3\}$.

69. $2x^2 - 5x - 3 = 0$

$(2x+1)(x-3) = 0$

$2x + 1 = 0$ or $x - 3 = 0$

$2x = -1$ $\quad x = 3$

$x = -\dfrac{1}{2}$

The solution set is $\left\{-\dfrac{1}{2}, 3\right\}$.

70. $3x^2 + 5x + 2 = 0$

$(3x+2)(x+1) = 0$

$3x + 2 = 0$ or $x + 1 = 0$

$3x = -2$ $\quad x = -1$

$x = -\dfrac{2}{3}$

The solution set is $\left\{-1, -\dfrac{2}{3}\right\}$

71. $x(x-7) + 12 = 0$

$x^2 - 7x + 12 = 0$

$(x-4)(x-3) = 0$

$x - 4 = 0$ or $x - 3 = 0$

$x = 4$ $\quad x = 3$

The solution set is $\{3, 4\}$.

72. $x(x+1) = 12$

$x^2 + x = 12$

$x^2 + x - 12 = 0$

$(x+4)(x-3) = 0$

$x + 4 = 0$ or $x - 3 = 0$

$x = -4$ $\quad x = 3$

The solution set is $\{-4, 3\}$.

73. $4x^2 + 9 = 12x$

$4x^2 - 12x + 9 = 0$

$(2x-3)^2 = 0$

$2x - 3 = 0$

$2x = 3$

$x = \dfrac{3}{2}$

The solution set is $\left\{\dfrac{3}{2}\right\}$.

74. $25x^2 + 16 = 40x$

$25x^2 - 40x + 16 = 0$

$(5x-4)^2 = 0$

$5x - 4 = 0$

$5x = 4$

$x = \dfrac{4}{5}$

The solution set is $\left\{\dfrac{4}{5}\right\}$.

75. $$6x-5=\frac{6}{x}$$
$$x(6x-5)=x\left(\frac{6}{x}\right)$$
$$6x^2-5x=6$$
$$6x^2-5x-6=0$$
$$(3x+2)(2x-3)=0$$
$$3x+2=0 \quad \text{or} \quad 2x-3=0$$
$$3x=-2 \qquad 2x=3$$
$$x=-\frac{2}{3} \qquad x=\frac{3}{2}$$

The solution set is $\left\{-\frac{2}{3},\frac{3}{2}\right\}$.

76. $$x+\frac{12}{x}=7$$
$$x\left(x+\frac{12}{x}\right)=x(7)$$
$$x^2+12=7x$$
$$x^2-7x+12=0$$
$$(x-3)(x-4)=0$$
$$x-3=0 \quad \text{or} \quad x-4=0$$
$$x=3 \qquad x=4$$

The solution set is $\{3,4\}$.

77. $$\frac{4(x-2)}{x-3}+\frac{3}{x}=\frac{-3}{x(x-3)}$$
LCD = $x(x-3)$
$$\frac{4x(x-2)}{x(x-3)}+\frac{3(x-3)}{x(x-3)}=\frac{-3}{x(x-3)}$$
$$4x(x-2)+3(x-3)=-3$$
$$4x^2-8x+3x-9=-3$$
$$4x^2-5x-9=-3$$
$$4x^2-5x-6=0$$
$$(4x+3)(x-2)=0$$
$$4x+3=0 \quad \text{or} \quad x-2=0$$
$$4x=-3 \qquad x=2$$
$$x=-\frac{3}{4}$$

The solution set is $\left\{-\frac{3}{4},2\right\}$.

78. $$\frac{5}{x+4}=4+\frac{3}{x-2}$$
LCD = $(x-2)(x+4)$
$$\frac{5(x-2)}{(x-2)(x+4)}=\frac{4(x-2)(x+4)}{(x-2)(x+4)}+\frac{3(x+4)}{(x-2)(x+4)}$$
$$5(x-2)=4(x-2)(x+4)+3(x+4)$$
$$5x-10=4\left(x^2+2x-8\right)+3x+12$$
$$5x-10=4x^2+8x-32+3x+12$$
$$5x-10=4x^2+11x-20$$
$$0=4x^2+6x-10$$
$$0=2x^2+3x-5$$
$$0=(2x+5)(x-1)$$

$$2x+5=0 \quad \text{or} \quad x-1=0$$
$$2x=-5 \qquad x=1$$
$$x=-\frac{5}{2}$$

The solution set is $\left\{-\frac{5}{2},1\right\}$.

79. $$\frac{x}{x^2-1}-\frac{x+3}{x^2-x}=\frac{-3}{x^2+x}$$
LCD = $x(x+1)(x-1)$
$$x(x+1)(x-1)\left(\frac{x}{x^2-1}-\frac{x+3}{x^2-x}=\frac{-3}{x^2+x}\right)$$
$$x^2-(x+1)(x+3)=-3(x-1)$$
$$x^2-\left(x^2+4x+3\right)=-3x+3$$
$$x^2-x^2-4x-3=-3x+3$$
$$-4x-3=-3x+3$$
$$-4x=-3x+6$$
$$-x=6$$
$$x=-6$$

The solution set is $\{-6\}$.

80. $\frac{x+1}{x^2+2x}-\frac{x+4}{x^2+x}=\frac{-3}{x^2+3x+2}$

LCD = $x(x+2)(x+1)$

$$x(x+2)(x+1)\left(\frac{x+1}{x^2+2x}-\frac{x+4}{x^2+x}=\frac{-3}{x^2+3x+2}\right)$$

$$(x+1)(x+1)-(x+4)(x+2)=-3(x)$$
$$x^2+2x+1-\left(x^2+6x+8\right)=-3x$$
$$x^2+2x+1-x^2-6x-8=-3x$$
$$-4x-7=-3x$$
$$-4x=-3x+7$$
$$-x=7$$
$$x=-7$$

The solution set is $\{-7\}$.

81. $x^4-5x^2+4=0$

$\left(x^2-4\right)\left(x^2-1\right)=0$

$x^2-4=0$ or $x^2-1=0$

$x=\pm 2$ or $x=\pm 1$

The solution set is $\{-2,-1,1,2\}$.

82. $x^4-10x^2+25=0$

$\left(x^2-5\right)\left(x^2-5\right)=0$

$x^2-5=0 \rightarrow x=\pm\sqrt{5}$

The solution set is $\{-\sqrt{5},\sqrt{5}\}$.

83. $(x+2)^2+7(x+2)+12=0$

let $p=x+2 \rightarrow p^2=(x+2)^2$

$p^2+7p+12=0$

$(p+3)(p+4)=0$

$p+3=0$ or $p+4=0$

$p=-3 \rightarrow x+2=-3 \rightarrow x=-5$

or $p=-4 \rightarrow x+2=-4 \rightarrow x=-6$

The solution set is $\{-6,-5\}$.

84. $(2x+5)^2-(2x+5)-6=0$

let $p=2x+5 \rightarrow p^2=(2x+5)^2$

$p^2-p-6=0$

$(p-3)(p+2)=0$

$p-3=0$ *or* $p+2=0$

$p=3 \rightarrow 2x+5=3 \rightarrow x=-1$

or $p=-2 \rightarrow 2x+5=-2 \rightarrow x=-\frac{7}{2}$

The solution set is $\left\{-\frac{7}{2},-1\right\}$.

85. $2(s+1)^2-5(s+1)=3$

let $p=s+1 \rightarrow p^2=(s+1)^2$

$2p^2-5p=3$

$2p^2-5p-3=0$

$(2p+1)(p-3)=0$

$2p+1=0$ or $p-3=0$

$p=-\frac{1}{2} \rightarrow s+1=-\frac{1}{2} \rightarrow s=-\frac{3}{2}$

or $p=3 \rightarrow s+1=3 \rightarrow s=2$

The solution set is $\left\{-\frac{3}{2},2\right\}$.

86. $3(1-y)^2+5(1-y)+2=0$

let $p=1-y \rightarrow p^2=(1-y)^2$

$3p^2+5p+2=0$

$(3p+2)(p+1)=0$

$3p+2=0$ or $p+1=0$

$p=-\frac{2}{3} \rightarrow 1-y=-\frac{2}{3} \rightarrow y=\frac{5}{3}$

or $p=-1 \rightarrow 1-y=-1 \rightarrow y=2$

The solution set is $\left\{\frac{5}{3},2\right\}$.

87. $x^3+x^2-20x=0$

$x\left(x^2+x-20\right)=0$

$x(x+5)(x-4)=0$

$x=0$ or $x+5=0$ or $x-4=0$

$x=-5$ $x=4$

The solution set is $\{-5,0,4\}$.

88. $x^3+6x^2-7x=0$

$x(x^2+6x-7)=0$

$x(x+7)(x-1)=0$

$x=0$ or $x+7=0$ or $x-1=0$

$x=-7$ $\quad x=1$

The solution set is $\{-7,0,1\}$.

89. $x^3+x^2-x-1=0$

$x^2(x+1)-1(x+1)=0$

$(x+1)(x^2-1)=0$

$(x+1)(x-1)(x+1)=0$

$x+1=0$ or $x-1=0$

$x=-1$ $\quad x=1$

The solution set is $\{-1,1\}$.

90. $x^3+4x^2-x-4=0$

$x^2(x+4)-1(x+4)=0$

$(x+4)(x^2-1)=0$

$(x+4)(x-1)(x+1)=0$

$x+4=0$ or $x-1=0$ or $x+1=0$

$x=-4$ $\quad x=1$ $\quad x=-1$

The solution set is $\{-4,-1,1\}$.

91. $2x^3+4=x^2+8x$

$2x^3-x^2-8x+4=0$

$x^2(2x-1)-4(2x-1)=0$

$(2x-1)(x^2-4)=0$

$(2x-1)(x-2)(x+2)=0$

$2x-1=0$ or $x-2=0$ or $x+2=0$

$2x=1$ $\quad x=2$ $\quad x=-2$

$x=\frac{1}{2}$

The solution set is $\left\{-2,\frac{1}{2},2\right\}$.

92. $3x^3+4x^2=27x+36$

$3x^3+4x^2-27x-36=0$

$x^2(3x+4)-9(3x+4)=0$

$(3x+4)(x^2-9)=0$

$(3x+4)(x-3)(x+3)=0$

$3x+4=0$ or $x-3=0$ or $x+3=0$

$3x=-4$ $\quad x=3$ $\quad x=-3$

$x=-\frac{4}{3}$

The solution set is $\left\{-3,-\frac{4}{3},3\right\}$.

93. $x^2=25 \Rightarrow x=\pm\sqrt{25} \Rightarrow x=\pm 5$

The solution set is $\{-5,5\}$.

94. $x^2=36 \Rightarrow x=\pm\sqrt{36} \Rightarrow x=\pm 6$

The solution set is $\{-6,6\}$.

95. $(x-1)^2=4$

$x-1=\pm\sqrt{4}$

$x-1=\pm 2$

$x-1=2$ or $x-1=-2$

$\Rightarrow x=3$ or $x=-1$

The solution set is $\{-1,3\}$.

96. $(x+2)^2=1$

$x+2=\pm\sqrt{1}$

$x+2=\pm 1$

$x+2=1$ or $x+2=-1$

$\Rightarrow x=-1$ or $x=-3$

The solution set is $\{-3,-1\}$.

97. $(2x+3)^2=9$

$2x+3=\pm\sqrt{9}$

$2x+3=\pm 3$

$2x+3=3$ or $2x+3=-3$

$\Rightarrow x=0$ or $x=-3$

The solution set is $\{-3,0\}$.

98. $(3x-2)^2 = 4$

$3x-2 = \pm\sqrt{4}$

$3x-2 = \pm 2$

$3x-2 = 2$ or $3x-2 = -2$

$\Rightarrow x = \frac{4}{3}$ or $x = 0$

The solution set is $\left\{0, \frac{4}{3}\right\}$.

99. $\left(\frac{8}{2}\right)^2 = 4^2 = 16$

100. $\left(\frac{-4}{2}\right)^2 = (-2)^2 = 4$

101. $\left(\frac{1/2}{2}\right)^2 = \left(\frac{1}{4}\right)^2 = \frac{1}{16}$

102. $\left(\frac{\left(-\frac{1}{3}\right)}{2}\right)^2 = \left(-\frac{1}{6}\right)^2 = \frac{1}{36}$

103. $\left(\frac{\left(-\frac{2}{3}\right)}{2}\right)^2 = \left(-\frac{1}{3}\right)^2 = \frac{1}{9}$

104. $\left(\frac{\left(-\frac{2}{5}\right)}{2}\right)^2 = \left(-\frac{1}{5}\right)^2 = \frac{1}{25}$

105. $x^2 + 4x = 21$

$x^2 + 4x + 4 = 21 + 4$

$(x+2)^2 = 25$

$x + 2 = \pm\sqrt{25} \Rightarrow x + 2 = \pm 5$

$x = -2 \pm 5 \Rightarrow x = 3$ or $x = -7$

The solution set is $\{-7, 3\}$.

106. $x^2 - 6x = 13$

$x^2 - 6x + 9 = 13 + 9$

$(x-3)^2 = 22 \Rightarrow x - 3 = \pm\sqrt{22}$

$x = 3 \pm \sqrt{22}$

The solution set is $\left\{3-\sqrt{22}, 3+\sqrt{22}\right\}$.

107. $x^2 - \frac{1}{2}x - \frac{3}{16} = 0$

$x^2 - \frac{1}{2}x = \frac{3}{16}$

$x^2 - \frac{1}{2}x + \frac{1}{16} = \frac{3}{16} + \frac{1}{16}$

$\left(x - \frac{1}{4}\right)^2 = \frac{1}{4}$

$x - \frac{1}{4} = \pm\sqrt{\frac{1}{4}} = \pm\frac{1}{2}$

$x = \frac{1}{4} \pm \frac{1}{2} \Rightarrow x = \frac{3}{4}$ or $x = -\frac{1}{4}$

The solution set is $\left\{-\frac{1}{4}, \frac{3}{4}\right\}$.

108. $x^2 + \frac{2}{3}x - \frac{1}{3} = 0$

$x^2 + \frac{2}{3}x = \frac{1}{3}$

$x^2 + \frac{2}{3}x + \frac{1}{9} = \frac{1}{3} + \frac{1}{9}$

$\left(x + \frac{1}{3}\right)^2 = \frac{4}{9}$

$x + \frac{1}{3} = \pm\sqrt{\frac{4}{9}} = \pm\frac{2}{3}$

$x = -\frac{1}{3} \pm \frac{2}{3}$

$x = \frac{1}{3}$ or $x = -1$

The solution set is $\left\{-1, \frac{1}{3}\right\}$.

109. $3x^2+x-\frac{1}{2}=0$

$$x^2+\frac{1}{3}x-\frac{1}{6}=0$$

$$x^2+\frac{1}{3}x=\frac{1}{6}$$

$$x^2+\frac{1}{3}x+\frac{1}{36}=\frac{1}{6}+\frac{1}{36}$$

$$\left(x+\frac{1}{6}\right)^2=\frac{7}{36}$$

$$x+\frac{1}{6}=\pm\sqrt{\frac{7}{36}}$$

$$x+\frac{1}{6}=\pm\frac{\sqrt{7}}{6}$$

$$x=\frac{-1\pm\sqrt{7}}{6}$$

The solution set is $\left\{\frac{-1-\sqrt{7}}{6},\frac{-1+\sqrt{7}}{6}\right\}$.

110. $2x^2-3x-1=0$

$$x^2-\frac{3}{2}x-\frac{1}{2}=0$$

$$x^2-\frac{3}{2}x=\frac{1}{2}$$

$$x^2-\frac{3}{2}x+\frac{9}{16}=\frac{1}{2}+\frac{9}{16}$$

$$\left(x-\frac{3}{4}\right)^2=\frac{17}{16}$$

$$x-\frac{3}{4}=\pm\sqrt{\frac{17}{16}}$$

$$x-\frac{3}{4}=\pm\frac{\sqrt{17}}{4}$$

$$x=\frac{3\pm\sqrt{17}}{4}$$

The solution set is $\left\{\frac{3-\sqrt{17}}{4},\frac{3+\sqrt{17}}{4}\right\}$.

111. $x^2-4x+2=0$

$a=1,\quad b=-4,\quad c=2$

$$x=\frac{-(-4)\pm\sqrt{(-4)^2-4(1)(2)}}{2(1)}$$

$$=\frac{4\pm\sqrt{16-8}}{2}=\frac{4\pm\sqrt{8}}{2}$$

$$=\frac{4\pm2\sqrt{2}}{2}=2\pm\sqrt{2}$$

The solution set is $\{2-\sqrt{2},2+\sqrt{2}\}$.

112. $x^2+4x+2=0$

$a=1,\quad b=4,\quad c=2$

$$x=\frac{-4\pm\sqrt{4^2-4(1)(2)}}{2(1)}$$

$$=\frac{-4\pm\sqrt{16-8}}{2}=\frac{-4\pm\sqrt{8}}{2}$$

$$=\frac{-4\pm2\sqrt{2}}{2}=-2\pm\sqrt{2}$$

The solution set is $\{-2-\sqrt{2},-2+\sqrt{2}\}$.

113. $x^2-5x-1=0$

$a=1,\ b=-5,\ c=-1$

$$x=\frac{-(-5)\pm\sqrt{(-5)^2-4(1)(-1)}}{2(1)}$$

$$=\frac{5\pm\sqrt{25+4}}{2}=\frac{5\pm\sqrt{29}}{2}$$

The solution set is $\left\{\frac{5-\sqrt{29}}{2},\frac{5+\sqrt{29}}{2}\right\}$.

114. $x^2+5x+3=0$

$a=1,\ b=5,\ c=3$

$$x=\frac{-5\pm\sqrt{5^2-4(1)(3)}}{2(1)}$$

$$=\frac{-5\pm\sqrt{25-12}}{2}=\frac{-5\pm\sqrt{13}}{2}$$

The solution set is $\left\{\frac{-5-\sqrt{13}}{2},\frac{-5+\sqrt{13}}{2}\right\}$.

115. $2x^2 - 5x + 3 = 0$

$a = 2, \quad b = -5, \quad c = 3$

$$x = \frac{-(-5) \pm \sqrt{(-5)^2 - 4(2)(3)}}{2(2)}$$

$$= \frac{5 \pm \sqrt{25 - 24}}{4} = \frac{5 \pm 1}{4}$$

The solution set is $\left\{1, \frac{3}{2}\right\}$.

116. $2x^2 + 5x + 3 = 0$

$a = 2, \quad b = 5, \quad c = 3$

$$x = \frac{-5 \pm \sqrt{5^2 - 4(2)(3)}}{2(2)}$$

$$= \frac{-5 \pm \sqrt{25 - 24}}{4} = \frac{-5 \pm 1}{4}$$

The solution set is $\left\{-\frac{3}{2}, -1\right\}$.

117. $4y^2 - y + 2 = 0$

$a = 4, \quad b = -1, \quad c = 2$

$$y = \frac{-(-1) \pm \sqrt{(-1)^2 - 4(4)(2)}}{2(4)}$$

$$= \frac{1 \pm \sqrt{1 - 32}}{8} = \frac{1 \pm \sqrt{-31}}{8}$$

No real solution.

118. $4t^2 + t + 1 = 0$

$a = 4, \quad b = 1, \quad c = 1$

$$t = \frac{-1 \pm \sqrt{1^2 - 4(4)(1)}}{2(4)}$$

$$= \frac{-1 \pm \sqrt{1 - 16}}{8} = \frac{-1 \pm \sqrt{-15}}{8}$$

No real solution.

119. $4x^2 = 1 - 2x$

$4x^2 + 2x - 1 = 0$

$a = 4, \quad b = 2, \quad c = -1$

$$x = \frac{-2 \pm \sqrt{2^2 - 4(4)(-1)}}{2(4)}$$

$$= \frac{-2 \pm \sqrt{4 + 16}}{8} = \frac{-2 \pm \sqrt{20}}{8}$$

$$= \frac{-2 \pm 2\sqrt{5}}{8} = \frac{-1 \pm \sqrt{5}}{4}$$

The solution set is $\left\{\frac{-1 - \sqrt{5}}{4}, \frac{-1 + \sqrt{5}}{4}\right\}$.

120. $2x^2 = 1 - 2x$

$2x^2 + 2x - 1 = 0$

$a = 2, \quad b = 2, \quad c = -1$

$$x = \frac{-2 \pm \sqrt{2^2 - 4(2)(-1)}}{2(2)}$$

$$= \frac{-2 \pm \sqrt{4 + 8}}{4} = \frac{-2 \pm \sqrt{12}}{4}$$

$$= \frac{-2 \pm 2\sqrt{3}}{4} = \frac{-1 \pm \sqrt{3}}{2}$$

The solution set is $\left\{\frac{-1 - \sqrt{3}}{2}, \frac{-1 + \sqrt{3}}{2}\right\}$.

121. $x^2 + \sqrt{3}\,x - 3 = 0$

$a = 1, b = \sqrt{3}, c = -3$

$$x = \frac{-\sqrt{3} \pm \sqrt{\left(\sqrt{3}\right)^2 - 4(1)(-3)}}{2(1)}$$

$$= \frac{-\sqrt{3} \pm \sqrt{3 + 12}}{2} = \frac{-\sqrt{3} \pm \sqrt{15}}{2}$$

The solution set is $\left\{\frac{-\sqrt{3} - \sqrt{15}}{2}, \frac{-\sqrt{3} + \sqrt{15}}{2}\right\}$.

122. $x^2 + \sqrt{2}\,x - 2 = 0$

$a = 1, b = \sqrt{2}, c = -2$

$$x = \frac{-\sqrt{2} \pm \sqrt{\left(\sqrt{2}\right)^2 - 4(1)(-2)}}{2(1)}$$

$$= \frac{-\sqrt{2} \pm \sqrt{2 + 8}}{2} = \frac{-\sqrt{2} \pm \sqrt{10}}{2}$$

The solution set is $\left\{\frac{-\sqrt{2} - \sqrt{10}}{2}, \frac{-\sqrt{2} + \sqrt{10}}{2}\right\}$.

123. $x^2 - 5x + 7 = 0$

$a = 1, \quad b = -5, \quad c = 7$

$b^2 - 4ac = (-5)^2 - 4(1)(7)$

$= 25 - 28 = -3$

Since the discriminant < 0, we have no real solutions.

124. $x^2 + 5x + 7 = 0$

$a = 1, \quad b = 5, \quad c = 7$

$b^2 - 4ac = (5)^2 - 4(1)(7)$

$= 25 - 28 = -3$

Since the discriminant < 0, we have no real solutions.

125. $9x^2 - 30x + 25 = 0$

$a = 9, \quad b = -30, \quad c = 25$

$b^2 - 4ac = (-30)^2 - 4(9)(25)$

$= 900 - 900 = 0$

Since the discriminant = 0, we have one repeated real solution.

126. $25x^2 - 20x + 4 = 0$

$a = 25, \quad b = -20, \quad c = 4$

$b^2 - 4ac = (-20)^2 - 4(25)(4)$

$= 400 - 400 = 0$

Since the discriminant = 0, we have one repeated real solution.

127. $3x^2 + 5x - 8 = 0$

$a = 3, \quad b = 5, \quad c = -8$

$b^2 - 4ac = (5)^2 - 4(3)(-8)$

$= 25 + 96 = 121$

Since the discriminant > 0, we have two unequal real solutions.

128. $2x^2 - 3x - 4 = 0$

$a = 2, \quad b = -3, \quad c = -4$

$b^2 - 4ac = (-3)^2 - 4(2)(-4)$

$= 9 + 32 = 41$

Since the discriminant > 0, we have two unequal real solutions.

129. Solving for R:

$$\frac{1}{R} = \frac{1}{R_1} + \frac{1}{R_2}$$

$$RR_1R_2\left(\frac{1}{R}\right) = RR_1R_2\left(\frac{1}{R_1} + \frac{1}{R_2}\right)$$

$$R_1R_2 = RR_2 + RR_1$$

$$R_1R_2 = R(R_2 + R_1)$$

$$\frac{R_1R_2}{R_2 + R_1} = \frac{R(R_2 + R_1)}{R_2 + R_1}$$

$$\frac{R_1R_2}{R_2 + R_1} = R \quad \text{or} \quad R = \frac{R_1R_2}{R_1 + R_2}$$

130. Solving for r:

$$A = P(1 + rt)$$

$$A = P + Prt$$

$$A - P = Prt$$

$$\frac{A - P}{Pt} = \frac{Prt}{Pt}$$

$$r = \frac{A - P}{Pt}$$

131. Solving for R:

$$F = \frac{mv^2}{R}$$

$$RF = R\left(\frac{mv^2}{R}\right)$$

$$RF = mv^2$$

$$\frac{RF}{F} = \frac{mv^2}{F} \Rightarrow R = \frac{mv^2}{F}$$

132. Solving for T:

$$PV = nRT$$

$$\frac{PV}{nR} = \frac{nRT}{nR}$$

$$T = \frac{PV}{nR}$$

133. Solving for *r*:

$$S = \frac{a}{1-r}$$
$$(1-r)\cdot S = (1-r)\cdot\frac{a}{1-r}$$
$$(1-r)S = a$$
$$\frac{(1-r)S}{S} = \frac{a}{S}$$
$$1-r = \frac{a}{S}$$
$$1-r-1 = \frac{a}{S}-1$$
$$-r = \frac{a}{S}-1$$
$$r = 1-\frac{a}{S} \quad \text{or} \quad r = \frac{S-a}{S}$$

134. Solving for *t*:

$$v = -gt + v_0$$
$$v - v_0 = -gt + v_0 - v_0$$
$$v - v_0 = -gt$$
$$\frac{v-v_0}{-g} = \frac{-gt}{-g}$$
$$\frac{v_0 - v}{g} = t \quad \text{or} \quad t = \frac{v_0 - v}{g}$$

135. The roots of a quadratic equation are

$$x_1 = \frac{-b-\sqrt{b^2-4ac}}{2a} \text{ and}$$
$$x_2 = \frac{-b+\sqrt{b^2-4ac}}{2a}$$
$$x_1 + x_2 = \frac{-b-\sqrt{b^2-4ac}}{2a} + \frac{-b+\sqrt{b^2-4ac}}{2a}$$
$$= \frac{-b-\sqrt{b^2-4ac}-b+\sqrt{b^2-4ac}}{2a}$$
$$= \frac{-2b}{2a}$$
$$= -\frac{b}{a}$$

136. The roots of a quadratic equation are

$$x_1 = \frac{-b-\sqrt{b^2-4ac}}{2a} \text{ and}$$
$$x_2 = \frac{-b+\sqrt{b^2-4ac}}{2a}$$
$$x_1 \cdot x_2 = \left(\frac{-b-\sqrt{b^2-4ac}}{2a}\right)\left(\frac{-b+\sqrt{b^2-4ac}}{2a}\right)$$
$$= \frac{(-b)^2 - \left(\sqrt{b^2-4ac}\right)^2}{(2a)^2} = \frac{b^2-b^2+4ac}{4a^2}$$
$$= \frac{4ac}{4a^2}$$
$$= \frac{c}{a}$$

137. In order to have one repeated solution, we need the discriminant to be 0.

$a = k$, $b = 1$, $c = k$

$$b^2 - 4ac = 0$$
$$1^2 - 4(k)(k) = 0$$
$$1 - 4k^2 = 0$$
$$4k^2 = 1$$
$$k^2 = \frac{1}{4}$$
$$k = \pm\sqrt{\frac{1}{4}}$$
$$k = \frac{1}{2} \quad \text{or} \quad k = -\frac{1}{2}$$

138. In order to have one repeated solution, we need the discriminant to be 0.

$a = 1$, $b = -k$, $c = 4$

$$b^2 - 4ac = 0$$
$$(-k)^2 - 4(1)(4) = 0$$
$$k^2 - 16 = 0$$
$$(k-4)(k+4) = 0$$
$$k = 4 \quad \text{or} \quad k = -4$$

139. For $ax^2+bx+c=0$:

$$x_1=\frac{-b-\sqrt{b^2-4ac}}{2a} \text{ and}$$

$$x_2=\frac{-b+\sqrt{b^2-4ac}}{2a}$$

For $ax^2-bx+c=0$:

$$x_1^*=\frac{-(-b)-\sqrt{(-b)^2-4ac}}{2a}=-\left(\frac{-b+\sqrt{b^2-4ac}}{2a}\right)=-x_2$$

and

$$x_2^*=\frac{-(-b)+\sqrt{(-b)^2-4ac}}{2a}=-\left(\frac{-b-\sqrt{b^2-4ac}}{2a}\right)=-x_1$$

140. For $ax^2+bx+c=0$:

$$x_1=\frac{-b-\sqrt{b^2-4ac}}{2a} \text{ and}$$

$$x_2=\frac{-b+\sqrt{b^2-4ac}}{2a}$$

For $cx^2+bx+a=0$:

$$x_1^*=\frac{-b-\sqrt{b^2-4(c)(a)}}{2c}=\frac{-b-\sqrt{b^2-4ac}}{2c}$$

$$=\frac{-b-\sqrt{b^2-4ac}}{2c}\cdot\frac{-b+\sqrt{b^2-4ac}}{-b+\sqrt{b^2-4ac}}$$

$$=\frac{b^2-(b^2-4ac)}{2c\left(-b+\sqrt{b^2-4ac}\right)}=\frac{4ac}{2c\left(-b+\sqrt{b^2-4ac}\right)}$$

$$=\frac{2a}{-b+\sqrt{b^2-4ac}}$$

$$=\frac{1}{x_2}$$

and

$$x_2^*=\frac{-b+\sqrt{b^2-4(c)(a)}}{2c}=\frac{-b+\sqrt{b^2-4ac}}{2c}$$

$$=\frac{-b+\sqrt{b^2-4ac}}{2c}\cdot\frac{-b-\sqrt{b^2-4ac}}{-b-\sqrt{b^2-4ac}}$$

$$=\frac{b^2-(b^2-4ac)}{2c\left(-b-\sqrt{b^2-4ac}\right)}=\frac{4ac}{2c\left(-b-\sqrt{b^2-4ac}\right)}$$

$$=\frac{2a}{-b-\sqrt{b^2-4ac}}$$

$$=\frac{1}{x_1}$$

141. **a.** $x^2=9$ and $x=3$ are not equivalent because they do not have the same solution set. In the first equation we can also have $x=-3$.

b. $x=\sqrt{9}$ and $x=3$ are equivalent because $\sqrt{9}=3$.

c. $(x-1)(x-2)=(x-1)^2$ and $x-2=x-1$ are not equivalent because they do not have the same solution set.
The first equation has the solution set $\{1\}$ while the second equation has no solutions.

142. Answers will vary.

143. Answers will vary.

144. Answers will vary. Methods may include the quadratic formula, factoring, completing the square, graphing, etc.

145. Answers will vary. Knowing the discriminant allows us to know how many real solutions the equation will have.

146. Answers will vary. One possibility:
Two distinct: $x^2-3x-18=0$
One repeated: $x^2-14x+49=0$
No real: $x^2+x+4=0$

147. Answers will vary.

148. Answers will vary. Since absolute value is never negative, the equation $|x|=-2$ has no real solution.

Section A.6

1. True

2. $(2+i)(2-i)=2^2-i^2=4-(-1)=5$

3. False; in the complex number system, a quadratic equation always has two solutions.

4. real; imaginary; imaginary unit

5. $\{-2i,2i\}$

6. False; the conjugate of $2+5i$ is $2-5i$.

7. True; the set of real numbers is a subset of the complex numbers.

8. False; if $2-3i$ is a solution of a quadratic equation with real coefficients, then its conjugate, $2+3i$, is also a solution.

9. $(2-3i)+(6+8i)=(2+6)+(-3+8)i=8+5i$

10. $(4+5i)+(-8+2i)=(4+(-8))+(5+2)i$
$=-4+7i$

11. $(-3+2i)-(4-4i)=(-3-4)+(2-(-4))i$
$=-7+6i$

12. $(3-4i)-(-3-4i)=(3-(-3))+(-4-(-4))i$
$=6+0i=6$

13. $(2-5i)-(8+6i)=(2-8)+(-5-6)i$
$=-6-11i$

14. $(-8+4i)-(2-2i)=(-8-2)+(4-(-2))i$
$=-10+6i$

15. $3(2-6i)=6-18i$

16. $-4(2+8i)=-8-32i$

17. $2i(2-3i)=4i-6i^2=4i-6(-1)=6+4i$

18. $3i(-3+4i)=-9i+12i^2$
$=-9i+12(-1)$
$=-12-9i$

19. $(3-4i)(2+i)=6+3i-8i-4i^2$
$=6-5i-4(-1)$
$=10-5i$

20. $(5+3i)(2-i)=10-5i+6i-3i^2$
$=10+i-3(-1)$
$=13+i$

21. $(-6+i)(-6-i)=36+6i-6i-i^2$
$=36-(-1)$
$=37$

22. $(-3+i)(3+i)=-9-3i+3i+i^2$
$=-9+(-1)$
$=-10$

23. $\frac{10}{3-4i}=\frac{10}{3-4i}\cdot\frac{3+4i}{3+4i}=\frac{30+40i}{9+12i-12i-16i^2}$
$=\frac{30+40i}{9-16(-1)}=\frac{30+40i}{25}$
$=\frac{30}{25}+\frac{40}{25}i$
$=\frac{6}{5}+\frac{8}{5}i$

24. $\frac{13}{5-12i}=\frac{13}{5-12i}\cdot\frac{5+12i}{5+12i}$
$=\frac{65+156i}{25+60i-60i-144i^2}$
$=\frac{65+156i}{25-144(-1)}=\frac{65+156i}{169}$
$=\frac{65}{169}+\frac{156}{169}i$
$=\frac{5}{13}+\frac{12}{13}i$

25. $\frac{2+i}{i}=\frac{2+i}{i}\cdot\frac{-i}{-i}=\frac{-2i-i^2}{-i^2}$
$=\frac{-2i-(-1)}{-(-1)}=\frac{1-2i}{1}$
$=1-2i$

26. $\frac{2-i}{-2i}=\frac{2-i}{-2i}\cdot\frac{i}{i}=\frac{2i-i^2}{-2i^2}$
$=\frac{2i-(-1)}{-2(-1)}=\frac{1+2i}{2}=\frac{1}{2}+i$

27. $\frac{6-i}{1+i}=\frac{6-i}{1+i}\cdot\frac{1-i}{1-i}=\frac{6-6i-i+i^2}{1-i+i-i^2}$
$=\frac{6-7i+(-1)}{1-(-1)}=\frac{5-7i}{2}$
$=\frac{5}{2}-\frac{7}{2}i$

28. $\frac{2+3i}{1-i}=\frac{2+3i}{1-i}\cdot\frac{1+i}{1+i}=\frac{2+2i+3i+3i^2}{1+i-i-i^2}$
$=\frac{2+5i+3(-1)}{1-(-1)}=\frac{-1+5i}{2}$
$=-\frac{1}{2}+\frac{5}{2}i$

29. $\left(\frac{1}{2}+\frac{\sqrt{3}}{2}i\right)^2=\frac{1}{4}+2\left(\frac{1}{2}\right)\left(\frac{\sqrt{3}}{2}i\right)+\frac{3}{4}i^2$
$=\frac{1}{4}+\frac{\sqrt{3}}{2}i+\frac{3}{4}(-1)$
$=-\frac{1}{2}+\frac{\sqrt{3}}{2}i$

30. $\left(\frac{\sqrt{3}}{2}-\frac{1}{2}i\right)^2=\frac{3}{4}-2\left(\frac{\sqrt{3}}{2}\right)\left(\frac{1}{2}i\right)+\frac{1}{4}i^2$
$=\frac{3}{4}-\frac{\sqrt{3}}{2}i+\frac{1}{4}(-1)$
$=\frac{1}{2}-\frac{\sqrt{3}}{2}i$

31. $(1+i)^2=1+2i+i^2=1+2i+(-1)=2i$

32. $(1-i)^2=1-2i+i^2=1-2i+(-1)=-2i$

33. $i^{23}=i^{22+1}=i^{22}\cdot i=\left(i^2\right)^{11}\cdot i=(-1)^{11}i=-i$

34. $i^{14}=\left(i^2\right)^7=(-1)^7=-1$

35. $i^{-15}=\frac{1}{i^{15}}=\frac{1}{i^{15}}\cdot\frac{i}{i}$
$=\frac{i}{i^{16}}=\frac{i}{\left(i^4\right)^4}$
$=\frac{i}{1^4}=i$

36. $i^{-23}=\frac{1}{i^{23}}=\frac{1}{i^{23}}\cdot\frac{i}{i}$
$=\frac{i}{i^{24}}=\frac{i}{\left(i^4\right)^6}$
$=\frac{i}{1^6}=i$

37. $i^6-5=\left(i^2\right)^3-5=(-1)^3-5=-1-5=-6$

38. $4+i^3=4+i^2\cdot i=4+(-1)i=4-i$

39. $6i^3-4i^5=i^3(6-4i^2)$
$=i^2\cdot i(6-4(-1))$
$=-1\cdot i(10)$
$=-10i$

40. $4i^3-2i^2+1=4i^2\cdot i-2i^2+1$
$=4(-1)i-2(-1)+1$
$=-4i+2+1$
$=3-4i$

41. $(1+i)^3=(1+i)(1+i)(1+i)=(1+2i+i^2)(1+i)$
$=(1+2i-1)(1+i)=2i(1+i)$
$=2i+2i^2=2i+2(-1)$
$=-2+2i$

42. $(3i)^4+1=81i^4+1=81(1)+1=82$

43. $i^7(1+i^2)=i^7(1+(-1))=i^7(0)=0$

44. $2i^4(1+i^2)=2(1)(1+(-1))=2(0)=0$

45. $i^6+i^4+i^2+1=\left(i^2\right)^3+\left(i^2\right)^2+i^2+1$
$=(-1)^3+(-1)^2+(-1)+1$
$=-1+1-1+1$
$=0$

46. $i^7+i^5+i^3+i=\left(i^2\right)^3\cdot i+\left(i^2\right)^2\cdot i+i^2\cdot i+i$
$=(-1)^3\cdot i+(-1)^2\cdot i+(-1)\cdot i+i$
$=-i+i-i+i$
$=0$

47. $\sqrt{-4}=\sqrt{4(-1)}=\sqrt{4}\cdot\sqrt{-1}=2i$

48. $\sqrt{-9}=\sqrt{9(-1)}=\sqrt{9}\cdot\sqrt{-1}=3i$

49. $\sqrt{-25}=\sqrt{25(-1)}=\sqrt{25}\cdot\sqrt{-1}=5i$

50. $\sqrt{-64}=\sqrt{64(-1)}=\sqrt{64}\cdot\sqrt{-1}=8i$

51. $\sqrt{(3+4i)(4i-3)}=\sqrt{12i-9+16i^2-12i}$
$=\sqrt{-9+16(-1)}$
$=\sqrt{-25}$
$=5i$

52. $\sqrt{(4+3i)(3i-4)}=\sqrt{12i-16+9i^2-12i}$
$=\sqrt{-16+9(-1)}$
$=\sqrt{-25}$
$=5i$

53. $x^2+4=0$
$x^2=-4$
$x=\pm\sqrt{-4}$
$x=\pm 2i$
The solution set is $\{\pm 2i\}$.

54. $x^2-4=0$
$(x+2)(x-2)=0 \Rightarrow x=-2$ or $x=2$
The solution set is $\{\pm 2\}$.

55. $x^2-16=0$
$(x+4)(x-4)=0 \Rightarrow x=-4, x=4$
The solution set is $\{\pm 4\}$.

56. $x^2+25=0$
$x^2=-25 \to x=\pm\sqrt{-25}=\pm 5i$
The solution set is $\{\pm 5i\}$.

57. $x^2-6x+13=0$
$a=1, b=-6, c=13,$
$b^2-4ac=(-6)^2-4(1)(13)=36-52=-16$
$x=\dfrac{-(-6)\pm\sqrt{-16}}{2(1)}=\dfrac{6\pm 4i}{2}=3\pm 2i$
The solution set is $\{3-2i, 3+2i\}$.

58. $x^2+4x+8=0$
$a=1, b=4, c=8$
$b^2-4ac=4^2-4(1)(8)=16-32=-16$
$x=\dfrac{-4\pm\sqrt{-16}}{2(1)}=\dfrac{-4\pm 4i}{2}=-2\pm 2i$
The solution set is $\{-2-2i, -2+2i\}$.

59. $x^2-6x+10=0$
$a=1, b=-6, c=10$
$b^2-4ac=(-6)^2-4(1)(10)=36-40=-4$
$x=\dfrac{-(-6)\pm\sqrt{-4}}{2(1)}=\dfrac{6\pm 2i}{2}=3\pm i$
The solution set is $\{3-i, 3+i\}$.

60. $x^2-2x+5=0$
$a=1, b=-2, c=5$
$b^2-4ac=(-2)^2-4(1)(5)=4-20=-16$
$x=\dfrac{-(-2)\pm\sqrt{-16}}{2(1)}=\dfrac{2\pm 4i}{2}=1\pm 2i$
The solution set is $\{1-2i, 1+2i\}$.

61. $8x^2-4x+1=0$
$a=8, b=-4, c=1$
$b^2-4ac=(-4)^2-4(8)(1)=16-32=-16$
$x=\dfrac{-(-4)\pm\sqrt{-16}}{2(8)}=\dfrac{4\pm 4i}{16}=\dfrac{1}{4}\pm\dfrac{1}{4}i$
The solution set is $\left\{\dfrac{1}{4}-\dfrac{1}{4}i, \dfrac{1}{4}+\dfrac{1}{4}i\right\}$.

62. $10x^2+6x+1=0$

$a=10, b=6, c=1$

$b^2-4ac=6^2-4(10)(1)=36-40=-4$

$$x=\frac{-6\pm\sqrt{-4}}{2(10)}=\frac{-6\pm 2i}{20}=-\frac{3}{10}\pm\frac{1}{10}i$$

The solution set is $\left\{-\frac{3}{10}-\frac{1}{10}i,\ -\frac{3}{10}+\frac{1}{10}i\right\}$.

63. $5x^2+1=2x$

$5x^2-2x+1=0$

$a=5, b=-2, c=1$

$b^2-4ac=(-2)^2-4(5)(1)=4-20=-16$

$$x=\frac{-(-2)\pm\sqrt{-16}}{2(5)}=\frac{2\pm 4i}{10}=\frac{1}{5}\pm\frac{2}{5}i$$

The solution set is $\left\{\frac{1}{5}-\frac{2}{5}i,\ \frac{1}{5}+\frac{2}{5}i\right\}$.

64. $13x^2+1=6x$

$13x^2-6x+1=0$

$a=13, b=-6, c=1$

$b^2-4ac=(-6)^2-4(13)(1)=36-52=-16$

$$x=\frac{-(-6)\pm\sqrt{-16}}{2(13)}=\frac{6\pm 4i}{26}=\frac{3}{13}\pm\frac{2}{13}i$$

The solution set is $\left\{\frac{3}{13}-\frac{2}{13}i,\frac{3}{13}+\frac{2}{13}i\right\}$.

65. $x^2+x+1=0$

$a=1, b=1, c=1,$

$b^2-4ac=1^2-4(1)(1)=1-4=-3$

$$x=\frac{-1\pm\sqrt{-3}}{2(1)}=\frac{-1\pm\sqrt{3}\,i}{2}=-\frac{1}{2}\pm\frac{\sqrt{3}}{2}i$$

The solution set is $\left\{-\frac{1}{2}-\frac{\sqrt{3}}{2}i,\ -\frac{1}{2}+\frac{\sqrt{3}}{2}i\right\}$.

66. $x^2-x+1=0$

$a=1, b=-1, c=1$

$b^2-4ac=(-1)^2-4(1)(1)=1-4=-3$

$$x=\frac{-(-1)\pm\sqrt{-3}}{2(1)}=\frac{1\pm\sqrt{3}\,i}{2}=\frac{1}{2}\pm\frac{\sqrt{3}}{2}i$$

The solution set is $\left\{\frac{1}{2}-\frac{\sqrt{3}}{2}i,\ \frac{1}{2}+\frac{\sqrt{3}}{2}i\right\}$.

67. $x^3-8=0$

$(x-2)\left(x^2+2x+4\right)=0$

$x-2=0 \Rightarrow x=2$

$x^2+2x+4=0$

$a=1, b=2, c=4$

$b^2-4ac=2^2-4(1)(4)=4-16=-12$

$$x=\frac{-2\pm\sqrt{-12}}{2(1)}=\frac{-2\pm 2\sqrt{3}\,i}{2}=-1\pm\sqrt{3}i$$

The solution set is $\left\{2,\ -1-\sqrt{3}i,\ -1+\sqrt{3}i\right\}$.

68. $x^3+27=0$

$(x+3)\left(x^2-3x+9\right)=0$

$x+3=0 \Rightarrow x=-3$

or $x^2-3x+9=0$

$a=1, b=-3, c=9$

$b^2-4ac=(-3)^2-4(1)(9)=9-36=-27$

$$x=\frac{-(-3)\pm\sqrt{-27}}{2(1)}=\frac{3\pm 3\sqrt{3}\,i}{2}=\frac{3}{2}\pm\frac{3\sqrt{3}}{2}i$$

The solution set is $\left\{-3,\ \frac{3}{2}-\frac{3\sqrt{3}}{2}i,\ \frac{3}{2}+\frac{3\sqrt{3}}{2}i\right\}$.

69. $x^4=16$

$x^4-16=0$

$\left(x^2-4\right)\left(x^2+4\right)=0 \Rightarrow (x-2)(x+2)\left(x^2+4\right)=0$

$x-2=0 \Rightarrow x=2$

$x+2=0 \Rightarrow x=-2$

$x^2+4=0 \Rightarrow x=\pm 2i$

The solution set is $\{-2,\ 2,\ -2i,\ 2i\}$.

70. $x^4=1$

$x^4-1=0$

$\left(x^2-1\right)\left(x^2+1\right)=0 \Rightarrow (x-1)(x+1)\left(x^2+1\right)=0$

$x-1=0 \Rightarrow x=1$

$x+1=0 \Rightarrow x=-1$

$x^2+1=0 \Rightarrow x=\pm i$

The solution set is $\{-1,\ 1,\ -i,\ i\}$.

71. $x^4+13x^2+36=0$
$\left(x^2+9\right)\left(x^2+4\right)=0$
$x^2+9=0 \Rightarrow x=\pm 3i$
$x^2+4=0 \Rightarrow x=\pm 2i$
The solution set is $\{-3i,\ 3i,\ -2i,\ 2i\}$.

72. $x^4+3x^2-4=0$
$\left(x^2-1\right)\left(x^2+4\right)=0 \Rightarrow (x-1)(x+1)\left(x^2+4\right)=0$
$x-1=0 \Rightarrow x=1$
$x+1=0 \Rightarrow x=-1$
$x^2+4=0 \Rightarrow x=\pm 2i$
The solution set is $\{-1,\ 1,\ -2i,\ 2i\}$.

73. $3x^2-3x+4=0$
$a=3, b=-3, c=4$
$b^2-4ac=(-3)^2-4(3)(4)=9-48=-39$
The equation has two conjugate complex solutions.

74. $2x^2-4x+1=0$
$a=2, b=-4, c=1$
$b^2-4ac=(-4)^2-4(2)(1)=16-8=8$
The equation has two unequal real solutions.

75. $2x^2+3x=4$
$2x^2+3x-4=0$
$a=2, b=3, c=-4$
$b^2-4ac=3^2-4(2)(-4)=9+32=41$
The equation has two unequal real solutions.

76. $x^2+6=2x$
$x^2-2x+6=0$
$a=1, b=-2, c=6$
$b^2-4ac=(-2)^2-4(1)(6)=4-24=-20$
The equation has two conjugate complex solutions.

77. $9x^2-12x+4=0$
$a=9, b=-12, c=4$
$b^2-4ac=(-12)^2-4(9)(4)=144-144=0$
The equation has a repeated real solution.

78. $4x^2+12x+9=0$
$a=4, b=12, c=9$
$b^2-4ac=12^2-4(4)(9)=144-144=0$
The equation has a repeated real solution.

79. The other solution is the conjugate of $2+3i$, or $2-3i$.

80. The other solution is the conjugate of $4-i$, or $4+i$.

81. $z+\bar{z}=3-4i+\overline{3-4i}=3-4i+3+4i=6$

82. $w-\bar{w}=8+3i-\left(\overline{8+3i}\right)=8+3i-(8-3i)$
$=8+3i-8+3i=0+6i$
$=6i$

83. $z\cdot\bar{z}=(3-4i)(\overline{3-4i})=(3-4i)(3+4i)$
$=9+12i-12i-16i^2=9-16(-1)$
$=25$

84. $\overline{z-w}=\overline{3-4i-(8+3i)}$
$=\overline{3-4i-8-3i}$
$=\overline{-5-7i}$
$=-5+7i$

85. $z+\bar{z}=a+bi+\overline{a+bi}$
$=a+bi+a-bi$
$=2a$

$z-\bar{z}=a+bi-(\overline{a+bi})$
$=a+bi-(a-bi)$
$=a+bi-a+bi$
$=2bi$

86. $\bar{\bar{z}}=\overline{\overline{a+bi}}=\overline{a-bi}=a+bi=z$

87. $\overline{z+w}=\overline{(a+bi)+(c+di)}$
$=\overline{(a+c)+(b+d)i}$
$=(a+c)-(b+d)i$
$=(a-bi)+(c-di)$
$=\overline{a+bi}+\overline{c+di}$
$=\bar{z}+\bar{w}$

88. $\overline{z \cdot w} = \overline{(a+bi)\cdot(c+di)}$
$= \overline{ac + adi + bci + bdi^2}$
$= \overline{(ac-bd)+(ad+bc)i}$
$= (ac-bd)-(ad+bc)i$

$\overline{z}\cdot\overline{w} = \overline{a+bi}\cdot\overline{c+di}$
$= (a-bi)(c-di)$
$= ac - adi - bci + bdi^2$
$= (ac-bd)-(ad+bc)i$

89. Answers will vary.

Section A.7

1. mathematical modeling
2. interest
3. uniform motion
4. True.
5. True; this is the uniform motion formula.
6. If there are x pounds of coffee A, then there are $100-x$ pounds of coffee B.
7. Let A represent the area of the circle and r the radius. The area of a circle is the product of π times the square of the radius: $A = \pi r^2$
8. Let C represent the circumference of a circle and r the radius. The circumference of a circle is the product of π times twice the radius: $C = 2\pi r$
9. Let A represent the area of the square and s the length of a side. The area of the square is the square of the length of a side: $A = s^2$
10. Let P represent the perimeter of a square and s the length of a side. The perimeter of a square is four times the length of a side: $P = 4s$
11. Let F represent the force, m the mass, and a the acceleration. Force equals the product of the mass times the acceleration: $F = ma$
12. Let P represent the pressure, F the force, and A the area. Pressure is the force per unit area: $P = \frac{F}{A}$
13. Let W represent the work, F the force, and d the distance. Work equals force times distance: $W = Fd$
14. Let K represent the kinetic energy, m the mass, and v the velocity. Kinetic energy is one-half the product of the mass and the square of the velocity: $K = \frac{1}{2}mv^2$
15. $C =$ total variable cost in dollars, $x =$ number of dishwashers manufactured: $C = 150x$
16. $R =$ total revenue in dollars, $x =$ number of dishwashers sold: $R = 250x$
17. Let $x =$ the amount invested in CDs. Then the amount invested in bonds is expressed as $x+3000$.
$20{,}000 = x + (x+3000)$
$20{,}000 = 2x + 3000$
$17{,}000 = 2x$
$8500 = x$
$x + 3000 = 8500 + 3000 = 11{,}500$
\$11,500 will be invested in bonds and \$8500 will be invested in CDs.
18. Let $x =$ the amount given to Sean. Then the amount given to George can be expressed as $x-3000$.
$10{,}000 = x + (x-3000)$
$10{,}000 = 2x - 3000$
$13{,}000 = 2x$
$6500 = x$
$x - 3000 = 6500 - 3000 = 3500$
Sean will receive \$6500 and George will receive \$3500.
19. Let $x =$ score on the final exam.
$$\frac{80+83+71+61+95+2x}{7} = 80$$
$$\frac{390+2x}{7} = 80$$
$390 + 2x = 560$
$2x = 170$
$x = 85$
Brooke needs an 85 on the final to have an average score of 80.

20. Let x = score on the final exam.

$$\frac{1}{3}\left(\frac{86+80+84+90}{4}\right)+\frac{2}{3}x=80$$
$$\frac{1}{3}(85)+\frac{2}{3}x=80$$
$$85+2x=240$$
$$2x=155$$
$$x=77.5$$

Mike needs a 78 on the final to earn a B.

$$\frac{1}{3}(85)+\frac{2}{3}x=90$$
$$85+2x=270$$
$$2x=185$$
$$x=92.5$$

Mike needs a 93 on the final to earn an A.

21. Let w = the width. Then the length is expressed as $l=w+8$.

$$P=2l+2w$$
$$60=2(w+8)+2w$$
$$60=2w+16+2w$$
$$60=4w+16$$
$$44=4w$$
$$11=w$$
$$l=w+8=11+8=19$$

The rectangle has a length of 19 feet and a width of 11 feet.

22. Let w = the width. Then the length is expressed as $l=2w$.

$$P=2l+2w$$
$$42=2(2w)+2w$$
$$42=4w+2w$$
$$42=6w$$
$$7=w$$
$$l=2w=2(7)=14$$

The rectangle has a length of 14 meters and a width of 7 meters.

23. Let x represent the amount of money invested in bonds. Then $50,000-x$ represents the amount of money invested in CD's. Since the total interest is to be \$6,000, we have:

$$0.15x+0.07(50,000-x)=6,000$$
$$(100)(0.15x+0.07(50,000-x))=(6,000)(100)$$
$$15x+7(50,000-x)=600,000$$
$$15x+350,000-7x=600,000$$
$$8x+350,000=600,000$$
$$8x=250,000$$
$$x=31,250$$

\$31,250 should be invested in bonds at 15% and \$18,750 should be invested in CD's at 7%.

24. Let x represent the amount of money invested in bonds. Then $50,000-x$ represents the amount of money invested in CD's. Since the total interest is to be \$7,000, we have:

$$0.15x+0.07(50,000-x)=7,000$$
$$(100)(0.15x+0.07(50,000-x))=(7,000)(100)$$
$$15x+7(50,000-x)=700,000$$
$$15x+350,000-7x=700,000$$
$$8x+350,000=700,000$$
$$8x=350,000$$
$$x=43,750$$

\$43,750 should be invested in bonds at 15% and \$6,250 should be invested in CD's at 7%.

25. Let x represent the amount of money loaned at 8%. Then $12,000-x$ represents the amount of money loaned at 18%. Since the total interest was \$1,000, we have:

$$0.08x+0.18(12,000-x)=1,000$$
$$(100)(0.08x+0.18(12,000-x))=(1,000)(100)$$
$$8x+18(12,000-x)=100,000$$
$$8x+216,000-18x=100,000$$
$$-10x+216,000=100,000$$
$$-10x=-116,000$$
$$x=11,600$$

\$11,600 was loaned at 8% and \$400 at 18%.

26. Let x represent the amount of money loaned at 16%. Then $1{,}000{,}000 - x$ represents the amount of money loaned at 19%. Since the total interest is to be \$1,000,000(0.18), we have:

$$0.16x + 0.19(1{,}000{,}000 - x) = 1{,}000{,}000(0.18)$$
$$0.16x + 190{,}000 - 0.19x = 180{,}000$$
$$-0.03x + 190{,}000 = 180{,}000$$
$$-0.03x = -10{,}000$$
$$x = \frac{-10{,}000}{-0.03}$$
$$x = \$333{,}333.33$$

Wendy can lend \$333,333.33 at 16%.

27. Let x represent the number of pounds of Earl Gray tea. Then $100 - x$ represents the number of pounds of Orange Pekoe tea.

$$5x + 3(100 - x) = 4.50(100)$$
$$5x + 300 - 3x = 450$$
$$2x + 300 = 450$$
$$2x = 150$$
$$x = 75$$

75 pounds of Earl Gray tea must be blended with 25 pounds of Orange Pekoe.

28. Let x represent the number of pounds of the first kind of coffee. Then $100 - x$ represents the number of pounds of the second kind of coffee.

$$2.75x + 5(100 - x) = 3.90(100)$$
$$2.75x + 500 - 5x = 390$$
$$-2.25x + 500 = 390$$
$$-2.25x = -110$$
$$x \approx 48.9$$

48.9 pounds of the first kind of coffee must be blended with 51.1 pounds of the second kind of coffee.

29. Let x represent the number of pounds of cashews. Then $x + 60$ represents the number of pounds in the mixture.

$$4x + 1.50(60) = 2.50(x + 60)$$
$$4x + 90 = 2.50x + 150$$
$$1.5x = 60$$
$$x = 40$$

40 pounds of cashews must be added to the 60 pounds of peanuts.

30. Let x represent the number of caramels in the box. Then $30 - x$ represents the number of cremes in the box.

$$\text{Revenue} - \text{Cost} = \text{Profit}$$
$$12.50 - (0.25x + 0.45(30 - x)) = 3.00$$
$$12.50 - (0.25x + 13.5 - 0.45x) = 3.00$$
$$12.50 - (13.5 - 0.20x) = 3.00$$
$$12.50 - 13.50 + 0.20x = 3.00$$
$$-1.00 + 0.20x = 3.00$$
$$0.20x = 4.00$$
$$x = \frac{4.00}{0.20}$$
$$x = 20$$

The box should contain 20 caramels and 10 cremes.

31. Let r represent the speed of the current.

	Rate	Time	Distance
Upstream	$16 - r$	$\frac{20}{60} = \frac{1}{3}$	$\frac{16-r}{3}$
Downstream	$16 + r$	$\frac{15}{60} = \frac{1}{4}$	$\frac{16+r}{4}$

Since the distance is the same in each direction:

$$\frac{16 - r}{3} = \frac{16 + r}{4}$$
$$4(16 - r) = 3(16 + r)$$
$$64 - 4r = 48 + 3r \rightarrow 16 = 7r \rightarrow r = \frac{16}{7} \approx 2.286$$

The speed of the current is approximately 2.286 miles per hour.

32. Let r represent the speed of the motorboat.

	Rate	Time	Distance
Upstream	$r - 3$	5	$5(r - 3)$
Downstream	$r + 3$	2.5	$2.5(r + 3)$

The distance is the same in each direction:

$$5(r - 3) = 2.5(r + 3)$$
$$5r - 15 = 2.5r + 7.5$$
$$2.5r = 22.5$$
$$r = 9$$

The speed of the motorboat is 9 miles per hour.

33. Let r represent the rate of the Metra train in miles per hour.

	Metra Train	Amtrak Train
Rate	r	$r+50$
Time	3	1
Dist.	$3r$	$r+50$

The Amtrak Train has traveled 10 fewer miles than the Metra Train.

$$r+50=3r-10$$
$$60=2r$$
$$r=30$$

The Metra Train is traveling 30 mph, and the Amtrak Train is traveling $30+50=80$ mph.

34. Let r represent the rate of the slower car. Then $r+10$ represents the rate of the faster car.

	Rate	Time	Distance
Slower car	r	3.5	$3.5r$
Faster car	$r+10$	3	$3(r+10)$

$$3.5r=3(r+10)$$
$$3.5r=3r+30$$
$$0.5r=30$$
$$r=60$$

The slower car travels at a rate of 60 miles per hour. The faster car travels at a rate of 70 miles per hour. The distance is (70)(3) = 210 miles.

35. Let t represent the time it takes to do the job together.

	Time to do job	Part of job done in one minute
Trent	30	$\frac{1}{30}$
Lois	20	$\frac{1}{20}$
Together	t	$\frac{1}{t}$

$$\frac{1}{30}+\frac{1}{20}=\frac{1}{t}$$
$$2t+3t=60$$
$$5t=60$$
$$t=12$$

Working together, the job can be done in 12 minutes.

36. Let t represent the time it takes April to do the job working alone.

	Time to do job	Part of job done in one hour
Patrice	10	$\frac{1}{10}$
April	t	$\frac{1}{t}$
Together	6	$\frac{1}{6}$

$$\frac{1}{10}+\frac{1}{t}=\frac{1}{6}$$
$$3t+30=5t$$
$$2t=30$$
$$t=15$$

It will take April 15 hours to paint the four rooms.

37. Let x = length of side of original sheet in feet.
Length of box: $x-2$ feet
Width of box: $x-2$ feet
Height of box: 1 foot

$$V=l\cdot w\cdot h$$
$$4=(x-2)(x-2)(1)$$
$$4=x^2-4x+4$$
$$0=x^2-4x$$
$$0=x(x-4)$$
$$x=0 \text{ or } x=4$$

Discard $x=0$ since that is not a feasible length for the original sheet. Therefore, the original sheet should measure 4 feet on each side.

38. Let x = width of original sheet in feet.
Length of sheet: $2x$
Length of box: $2x-2$ feet
Width of box: $x-2$ feet
Height of box: 1 foot

$$V=l\cdot w\cdot h$$
$$4=(2x-2)(x-2)(1)$$
$$4=2x^2-6x+4$$
$$0=2x^2-6x$$
$$0=x^2-3x$$
$$0=x(x-3)$$
$$x=0 \text{ or } x=3$$

Discard $x=0$ since that is not a feasible length for the original sheet. Therefore, the original sheet is 3 feet wide and 6 feet long.

39. Let r represent the speed of the current.

	Rate	Time	Distance
Upstream	$15-r$	$\frac{10}{15-r}$	10
Downstream	$15+r$	$\frac{10}{15+r}$	10

Since the total time is 1.5 hours, we have:

$$\frac{10}{15-r}+\frac{10}{15+r}=1.5$$

$$10(15+r)+10(15-r)=1.5(15-r)(15+r)$$

$$150+10r+150-10r=1.5(225-r^2)$$

$$300=1.5(225-r^2)$$

$$200=225-r^2$$

$$r^2-25=0$$

$$(r-5)(r+5)=0$$

$$r=5 \text{ or } r=-5$$

The speed of the current is 5 miles per hour.

40. Let x = the width and $2x$ = the length of the patio. The height is $\frac{1}{3}$ foot and the concrete available is $8(27)=216$ cubic feet..

$$V=lwh=x(2x)\cdot\frac{1}{3}=216$$

$$\frac{2}{3}x^2=216$$

$$x^2=324$$

$$x=\pm 18$$

The dimensions of the patio are 18 feet by 36 feet.

41. Let w represent the width of window.
Then $l=w+2$ represents the length of the window.
Since the area is 143 square feet, we have:

$$w(w+2)=143$$

$$w^2+2w-143=0$$

$$(w+13)(w-11)=0$$

$$\cancel{w=-13} \text{ or } w=11$$

Discard the negative solution since width cannot be negative. The width of the rectangular window is 11 feet and the length is 13 feet.

42. Let w represent the width of window.
Then $l=w+1$ represents the length of the window.
Since the area is 306 square centimeters, we have: $w(w+1)=306$

$$w^2+w-306=0$$

$$(w+18)(w-17)=0$$

$$\cancel{w=-18} \text{ or } w=17$$

Discard the negative solution since width cannot be negative. The width of the rectangular window is 17 centimeters and the length is 18 centimeters.

43. Let l represent the length of the rectangle.
Let w represent the width of the rectangle.
The perimeter is 26 meters and the area is 40 square meters.

$$2l+2w=26$$

$$l+w=13 \quad \text{so} \quad w=13-l$$

$$lw=40$$

$$l(13-l)=40$$

$$13l-l^2=40$$

$$l^2-13l+40=0$$

$$(l-8)(l-5)=0$$

$$l=8 \text{ or } l=5$$

$$w=5 \qquad w=8$$

The dimensions are 5 meters by 8 meters.

44. Let r represent the radius of the circle.
Since the field is a square with area 1250 square feet, the length of a side of the square is $\sqrt{1250}=25\sqrt{2}$ feet. The length of the diagonal is $2r$.
Use the Pythagorean Theorem to solve for r:

$$(2r)^2=\left(25\sqrt{2}\right)^2+\left(25\sqrt{2}\right)^2$$

$$4r^2=1250+1250$$

$$4r^2=2500$$

$$r^2=625$$

$$r=25$$

The shortest radius setting for the sprinkler is 25 feet.

45. a. When the ball strikes the ground, the distance from the ground will be 0. Therefore, we solve

$$96+80t-16t^2=0$$
$$-16t^2+80t+96=0$$
$$t^2-5t-6=0$$
$$(t-6)(t+1)=0$$

$t=6$ or $t=-1$
Discard the negative solution since the time of flight must be positive. The ball will strike the ground after 6 seconds.

b. When the ball passes the top of the building, it will be 96 feet from the ground. Therefore, we solve

$$96+80t-16t^2=96$$
$$-16t^2+80t=0$$
$$t^2-5t=0$$
$$t(t-5)=0$$

$t=0$ or $t=5$
The ball is at the top of the building at time $t=0$ when it is thrown. It will pass the top of the building on the way down after 5 seconds.

46. a. To find when the object will be 15 meters above the ground, we solve

$$-4.9t^2+20t=15$$
$$-4.9t^2+20t-15=0$$

$a=-4.9,\ b=20,\ c=-15$

$$t=\frac{-20\pm\sqrt{20^2-4(-4.9)(-15)}}{2(-4.9)}$$
$$=\frac{-20\pm\sqrt{106}}{-9.8}$$
$$=\frac{20\pm\sqrt{106}}{9.8}$$

$t\approx 0.99$ or $t\approx 3.09$
The object will be 15 meters above the ground after about 0.99 seconds (on the way up) and about 3.09 seconds (on the way down).

b. The object will strike the ground when the distance from the ground is 0. Therefore, we solve

$$-4.9t^2+20t=0$$
$$t(-4.9t+20)=0$$

$t=0$ or $-4.9t+20=0$

$$-4.9t=-20$$
$$t\approx 4.08$$

The object will strike the ground after about 4.08 seconds.

c.

$$-4.9t^2+20t=100$$
$$-4.9t^2+20t-100=0$$

$a=-4.9,\ b=20,\ c=-100$

$$t=\frac{-20\pm\sqrt{20^2-4(-4.9)(-100)}}{2(-4.9)}$$
$$=\frac{-20\pm\sqrt{-1560}}{-9.8}$$

There is no real solution. The object never reaches a height of 100 meters.

d. From part (b), we see that the object will be at ground level at $t=0$ and $t\approx 4.08$ seconds. Because of the symmetry of the graph of the distance equation, we see that the maximum height will be reached after $t\approx 2.04$ seconds. Therefore, the maximum height is

$$s\approx -4.9(2.04)^2+20(2.04)\approx 20.41 \text{ meters.}$$

47. $l=$ length of the garden
$w=$ width of the garden

a. The length of the garden is to be twice its width. Thus, $l=2w$.
The dimensions of the fence are $l+4$ and $w+4$.
The perimeter is 46 feet, so:

$$2(l+4)+2(w+4)=46$$
$$2(2w+4)+2(w+4)=46$$
$$4w+8+2w+8=46$$
$$6w+16=46$$
$$6w=30$$
$$w=5$$

The dimensions of the garden are 5 feet by 10 feet.

b. Area $= l \cdot w = 5 \cdot 10 = 50$ square feet

c. If the dimensions of the garden are the same, then the length and width of the fence are also the same $(l+4)$. The perimeter is 46 feet, so:

$$\begin{aligned} 2(l+4)+2(l+4) &= 46 \\ 2l+8+2l+8 &= 46 \\ 4l+16 &= 46 \\ 4l &= 30 \\ l &= 7.5 \end{aligned}$$

The dimensions of the garden are 7.5 feet by 7.5 feet.

d. Area $= l \cdot w = 7.5(7.5) = 56.25$ square feet.

48. $l =$ length of the pond
$w =$ width of the pond

a. The pond is to be a square. Thus, $l = w$. The dimensions of the fenced area are $w+6$ on each side. The perimeter is 100 feet, so:

$$\begin{aligned} 4(w+6) &= 100 \\ 4w+24 &= 100 \\ 4w &= 76 \\ w &= 19 \end{aligned}$$

The dimensions of the pond are 19 feet by 19 feet.

b. The length of the pond is to be three times the width. Thus, $l = 3w$. The dimensions of the fenced area are $w+6$ and $l+6$. The perimeter is 100 feet, so:

$$\begin{aligned} 2(w+6)+2(l+6) &= 100 \\ 2(w+6)+2(3w+6) &= 100 \\ 2w+12+6w+12 &= 100 \\ 8w+24 &= 100 \\ 8w &= 76 \\ w &= 9.5 \\ l &= 3(9.5) = 28.5 \end{aligned}$$

The dimensions of the pond are 9.5 feet by 28.5 feet.

c. If the pond is circular, the diameter is d and the diameter of the circle with the pond and the deck is $d+6$.

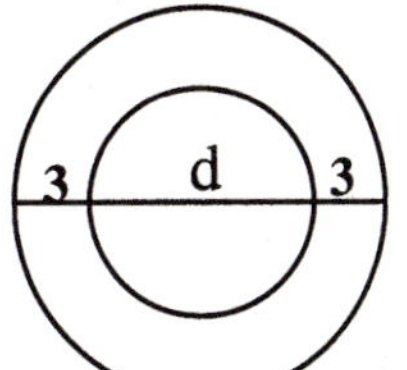

The perimeter is 100 feet, so:

$$\begin{aligned} \pi(d+6) &= 100 \\ \pi d + 6\pi &= 100 \\ \pi d &= 100 - 6\pi \\ d &= \frac{100}{\pi} - 6 \approx 25.83 \end{aligned}$$

The diameter of the pond is 25.83 feet.

d. $\text{Area}_{\text{square}} = l \cdot w = 19(19) = 361 \text{ ft}^2$.

$\text{Area}_{\text{rectangle}} = l \cdot w = 28.5(9.5) = 270.75 \text{ ft}^2$.

$\text{Area}_{\text{circle}} = \pi r^2 = \pi\left(\frac{25.83}{2}\right)^2 \approx 524 \text{ ft}^2$.

The circular pond has the largest area.

49. Let x represent the width of the border measured in feet.
The radius of the pool is 5 feet.
Then $x+5$ represents the radius of the circle, including both the pool and the border.
The total area of the pool and border is
$A_T = \pi(x+5)^2$.
The area of the pool is $A_P = \pi(5)^2 = 25\pi$.
The area of the border is
$A_B = A_T - A_P = \pi(x+5)^2 - 25\pi$.
Since the concrete is 3 inches or 0.25 feet thick, the volume of the concrete in the border is
$0.25A_B = 0.25\left(\pi(x+5)^2 - 25\pi\right)$
Solving the volume equation:

$$\begin{aligned} 0.25\left(\pi(x+5)^2 - 25\pi\right) &= 27 \\ \pi\left(x^2+10x+25-25\right) &= 108 \\ \pi x^2 + 10\pi x - 108 &= 0 \end{aligned}$$

$$x = \frac{-10\pi \pm \sqrt{(10\pi)^2 - 4(\pi)(-108)}}{2(\pi)}$$
$$= \frac{-31.42 \pm \sqrt{2344.1285}}{6.28}$$
$$= \frac{-31.42 \pm 48.42}{6.28}$$
$$= 2.71 \text{ or } -12.71$$

The width of the border is approximately 2.71 feet.

50. Let x represent the width of the border measured in feet.
The radius of the pool is 5 feet.
Then $x+5$ represents the radius of the circle, including both the pool and the border.
The total area of the pool and border is $A_T = \pi(x+5)^2$.

The area of the pool is $A_P = \pi(5)^2 = 25\pi$.
The area of the border is
$A_B = A_T - A_P = \pi(x+5)^2 - 25\pi$.

Since the concrete is 4 inches = $\frac{1}{3}$ foot thick, the volume of the concrete in the border is
$\frac{1}{3}A_B = \frac{1}{3}\left(\pi(x+5)^2 - 25\pi\right)$

Solving the volume equation:
$$\frac{1}{3}\left(\pi(x+5)^2 - 25\pi\right) = 27$$
$$\pi\left(x^2 + 10x + 25 - 25\right) = 81$$
$$\pi x^2 + 10\pi x - 81 = 0$$
$$x = \frac{-10\pi \pm \sqrt{(10\pi)^2 - 4(\pi)(-81)}}{2(\pi)}$$
$$= \frac{-31.42 \pm \sqrt{2004.8365}}{6.28} = \frac{-31.42 \pm 44.78}{6.28}$$
$$= 2.13 \text{ or } -12.13$$

Discard the negative solution. The width of the border is approximately 2.13 feet.

51. Let x represent the width of the border measured in feet.
The total area is $A_T = (6+2x)(10+2x)$.
The area of the garden is $A_G = 6 \cdot 10 = 60$.
The area of the border is
$A_B = A_T - A_G = (6+2x)(10+2x) - 60$.
Since the concrete is 3 inches or 0.25 feet thick, the volume of the concrete in the border is

$0.25A_B = 0.25\left((6+2x)(10+2x) - 60\right)$

Solving the volume equation:
$$0.25\left((6+2x)(10+2x) - 60\right) = 27$$
$$60 + 32x + 4x^2 - 60 = 108$$
$$4x^2 + 32x - 108 = 0$$
$$x^2 + 8x - 27 = 0$$
$$x = \frac{-8 \pm \sqrt{8^2 - 4(1)(-27)}}{2(1)}$$
$$= \frac{-8 \pm \sqrt{172}}{2}$$
$$= \frac{-8 \pm 13.11}{2}$$
$$= 2.56 \text{ or } -10.56$$

Discard the negative solution. The width of the border is approximately 2.56 feet.

52. $A = 2\pi r^2 + 2\pi r h$. Since $A = 188.5$ square inches and $h = 7$ inches,
$$2\pi r^2 + 2\pi r(7) = 188.5$$
$$2\pi r^2 + 14\pi r - 188.5 = 0$$
$$r = \frac{-14\pi \pm \sqrt{(14\pi)^2 - 4(2\pi)(-188.5)}}{2(2\pi)}.$$
$$= \frac{-14\pi \pm \sqrt{6671.9642}}{4\pi}$$
$r \approx 3$ or ~~$r \approx -10$~~

The radius of the coffee can is approximately 3 inches.

53. Let x represent the number of gallons of pure water. Then $x+1$ represents the number of gallons in the 60% solution.
The antifreeze equation is:
$$(\%)(\text{gallons}) + (\%)(\text{gallons}) = (\%)(\text{gallons})$$
$$0(x) + 1(1) = 0.60(x+1)$$
$$1 = 0.6x + 0.6$$
$$0.4 = 0.6x$$
$$x = \frac{4}{6} = \frac{2}{3}$$

$\frac{2}{3}$ gallon of pure water should be added.

54. Let x represent the number of liters to be drained and replaced with pure antifreeze.
The antifreeze equation is:

$$(\%)(\text{liters})+(\%)(\text{liters})=(\%)(\text{liters})$$

$$1(x)+0.40(15-x)=0.60(15)$$

$$x+6-0.40x=9$$

$$0.60x=3$$

$$x=5$$

5 liters should be drained and replaced with pure antifreeze.

55. Let x represent the number of ounces of water to be evaporated; the amount of salt remains the same. Therefore, we get

$$0.04(32)=0.06(32-x)$$

$$1.28=1.92-0.06x$$

$$0.06x=0.64$$

$$x=\frac{0.64}{0.06}=\frac{64}{6}=\frac{32}{3}=10\tfrac{2}{3}$$

$10\tfrac{2}{3}$ ounces of water need to be evaporated.

56. Let x represent the number of gallons of water to be evaporated; the amount of salt remains the same.

$$0.03(240)=0.05(240-x)$$

$$7.2=12-0.05x$$

$$0.05x=4.8$$

$$x=\frac{4.8}{0.05}=96$$

96 gallons of water need to be evaporated.

57. Let x represent the number of grams of pure gold. Then $60-x$ represents the number of grams of 12 karat gold to be used.
The gold equation is:

$$x+\frac{12}{24}(60-x)=\frac{16}{24}(60)$$

$$x+30-0.5x=40$$

$$0.5x=10$$

$$x=20$$

20 grams of pure gold should be mixed with 40 grams of 12 karat gold.

58. Let x represent the number of atoms of oxygen.
$2x$ represents the number of atoms of hydrogen.
$x+1$ represents the number of atoms of carbon.

$$x+2x+x+1=45$$

$$4x=44 \quad \rightarrow \quad x=11$$

There are 11 atoms of oxygen and 22 atoms of hydrogen in the sugar molecule.

59. Let t represent the time it takes for Mike to catch up with Dan. Since the distances are the same, we have:

$$\frac{1}{6}t=\frac{1}{9}(t+1)$$

$$3t=2t+2$$

$$t=2$$

Mike will pass Dan after 2 minutes, which is a distance of $(1/6)(2)=1/3$ mile.

60. Let t represent the time of flight with the wind. The distance is the same in each direction so we have:

$$d_{return}=d_{out}$$

$$330t=270(5-t)$$

$$330t=1350-270t$$

$$600t=1350$$

$$t=2.25$$

The distance the plane can fly and still return safely is $330(2.25)=742.5$ miles.

61. Let t represent the time the auxiliary pump needs to run. Since the two pumps are emptying one tanker, we have:

$$\frac{3}{4}+\frac{t}{9}=1$$

$$27+4t=36$$

$$4t=9$$

$$t=2.25$$

The auxiliary pump must run for 2.25 hours. It must be started at 9:45 a.m.

62. Let x represent the number of pounds of pure cement. Then $x+20$ represents the number of pounds in the 40% mixture.
The cement equation is:

$$1(x)+0.25(20)=0.40(x+20)$$

$$x+5=0.4x+8$$

$$0.6x=3 \quad \rightarrow \quad x=5$$

5 pounds of pure cement should be added.

63. Let t represent the time for the tub to fill with the faucets on and the stopper removed. Since one tub is being filled, we have:

$$\frac{t}{15}+\left(-\frac{t}{20}\right)=1$$

$$4t-3t=60$$

$$t=60$$

60 minutes is required to fill the tub.

64. Let t be the time the 5 horsepower pump needs to run to finish emptying the pool. Since the two pumps are emptying one pool, we have:

$$\frac{t+2}{5}+\frac{2}{8}=1$$

$$4(t+2)+5=20$$

$$4t+8+5=20$$

$$4t=7$$

$$t=1.75$$

The 5 horsepower pump must run for an additional 1.75 hours or 1 hour and 45 minutes to empty the pool.

65. Burke's rate is $\frac{100}{12}$ meters/sec.

In 9.99 seconds, Burke will run

$\frac{100}{12}(9.99)=83.25$ meters.

Lewis would win by 16.75 meters.

66. Let t represent the time it takes for the defensive back to catch the tight end.

	Time to run 100 yards	Time	Rate	Distance
Tight End	12 sec	t	$\frac{100}{12}=\frac{25}{3}$	$\frac{25}{3}t$
Def. Back	10 sec	t	$\frac{100}{10}=10$	$10t$

Since the defensive back has to run 5 yards farther, we have:

$$\frac{25}{3}t+5=10t$$

$$25t+15=30t$$

$$15=5t$$

$$t=3 \quad \rightarrow \quad 10t=30$$

The defensive back will catch the tight end at the 45 yard line (15 + 30 = 45).

67. Let x represent the number of highway miles traveled. Then $30{,}000-x$ represents the number of city miles traveled.

$$\frac{x}{40}+\frac{30{,}000-x}{25}=900$$

$$200\left(\frac{x}{40}+\frac{30{,}000-x}{25}\right)=200(900)$$

$$5x+240{,}000-8x=180{,}000$$

$$-3x+240{,}000=180{,}000$$

$$-3x=-60{,}000$$

$$x=20{,}000$$

Therese is allowed to claim 20,000 miles as a business expense.

68.

$$\frac{1}{2}n(n+1)=666$$

$$n(n+1)=1332$$

$$n^2+n-1332=0$$

$$(n-36)(n+37)=0$$

$$n=36 \quad \text{or} \quad n=-37$$

Since the number of consecutive integers cannot be negative, we discard the negative value. We must add 36 consecutive integers, beginning at 1, in order to get a sum of 666.

69.

$$\frac{1}{2}n(n-3)=65$$

$$n(n-3)=130$$

$$n^2-3n-130=0$$

$$(n-13)(n+10)=0$$

$$n=13 \quad \text{or} \quad n=-10$$

Since the number of sides cannot be negative, we discard the negative value. A polygon with 65 diagonals will have 13 sides.

$$\frac{1}{2}n(n-3)=80$$

$$n(n-3)=160$$

$$n^2-3n-160=0$$

$$a=1, b=-3, c=-160$$

$$n=\frac{3\pm\sqrt{(-3)^2-4(1)(-160)}}{2(1)}=\frac{3\pm\sqrt{646}}{2}$$

Neither solution is an integer, so there is no polygon that has 80 diagonals.

70. Using the Pythagorean Theorem, we get

$$(x)^2+(x+1)^2=(2x-1)^2$$
$$x^2+x^2+2x+1=4x^2-4x+1$$
$$0=2x^2-6x$$
$$0=2x(x-3)$$
$$2x=0 \text{ or } x-3=0$$
$$x=0 \qquad x=3$$
$$2x-1=2(0)-1 \text{ or } 2x-1=2(3)-1$$
$$=-1 \qquad =5$$

Since the length of the hypotenuse must be greater than 0, the length must be 5 units.

71. Let t_1 and t_2 represent the times for the two segments of the trip. Since Atlanta is halfway between Chicago and Miami, the distances are equal.

$$45t_1=55t_2$$
$$t_1=\frac{55}{45}t_2$$
$$t_1=\frac{11}{9}t_2$$

Computing the average speed:

$$\text{Avg Speed}=\frac{\text{Distance}}{\text{Time}}=\frac{45t_1+55t_2}{t_1+t_2}$$
$$=\frac{45\left(\frac{11}{9}t_2\right)+55t_2}{\frac{11}{9}t_2+t_2}=\frac{55t_2+55t_2}{\left(\frac{11t_2+9t_2}{9}\right)}$$
$$=\frac{110t_2}{\left(\frac{20t_2}{9}\right)}=\frac{990t_2}{20t_2}$$
$$=\frac{99}{2}=49.5 \text{ miles per hour}$$

The average speed for the trip from Chicago to Miami is 49.5 miles per hour.

72. The time traveled with the tail wind was:

$$t=\frac{919}{550}\approx 1.67091 \text{ hours}$$

Since they were 20 minutes $\left(\frac{1}{3}\text{ hour}\right)$ early, the time in still air would have been:

$1.67091 \text{ hrs } +20 \text{ min} \approx 2.00424 \text{ hrs}$

Thus, with no wind, the ground speed is $\frac{919}{2.00424}\approx 458.53$. Therefore, the tail wind is $550-458.53=91.47$ knots.

73. Let x be the original selling price of the shirt.

Profit = Revenue − Cost

$4=x-0.40x-20 \to 24=0.60x \to x=40$

The original price should be \$40 to ensure a profit of \$4 after the sale.

If the sale is 50% off, the profit is:

$40-0.50(40)-20=40-20-20=0$

At 50% off there will be no profit.

74. Answers will vary.

75. It is impossible to mix two solutions with a lower concentration and end up with a new solution with a higher concentration.

Algebraic Solution:

Let x = the number of liters of 25% solution.

The ethanol equation is:

$$(\%)(\text{liters})+(\%)(\text{liters})=(\%)(\text{liters})$$
$$0.25x+0.48(20)=0.58(20+x)$$
$$0.25x+9.6=10.6+0.58x$$
$$-0.33x=1$$
$$x\approx -3.03 \text{ liters}$$

(not possible)

Section A.8

1. $x\ge -2$

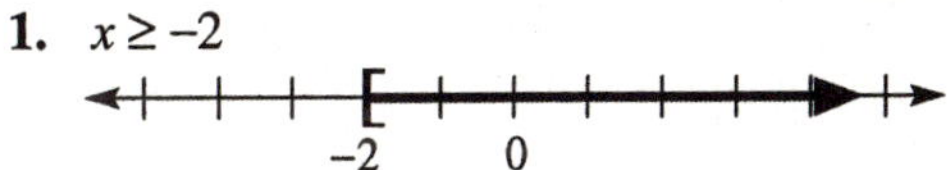

2. True; the absolute value of a number is the distance the number lies from 0 on a real number line. Negative numbers lie to the left of 0 on a real number line, but still have a positive distance from 0.

3. negative

4. closed interval

5. multiplication property for inequalities

6. True; this follows from the addition property for inequalities.

7. True; this follows from the addition property for inequalities.

8. True; this follows from the multiplication property for inequalities.

9. False; since both sides of the inequality are being divided by a negative number, the sense, or direction, of the inequality must be reversed. That is, $\frac{a}{c} > \frac{b}{c}$.

10. True

11. Interval: $[0,2]$
Inequality: $0 \le x \le 2$

12. Interval: $[2,\infty)$
Inequality: $x \ge 2$

13. Interval: $(-1,2)$
Inequality: $-1 < x < 2$

14. Interval: $(-\infty,0]$
Inequality: $x \le 0$

15. Interval: $[0,3)$
Inequality: $0 \le x < 3$

16. Interval: $(-1,1]$
Inequality: $-1 < x \le 1$

17. a.
$$\begin{gathered} 3<5 \\ 3+3<5+3 \\ 6<8 \end{gathered}$$

b.
$$\begin{gathered} 3<5 \\ 3-5<5-5 \\ -2<0 \end{gathered}$$

c.
$$\begin{gathered} 3<5 \\ 3(3)<3(5) \\ 9<15 \end{gathered}$$

d.
$$\begin{gathered} 3<5 \\ -2(3)>-2(5) \\ -6>-10 \end{gathered}$$

18. a.
$$\begin{gathered} 2>1 \\ 2+3>1+3 \\ 5>4 \end{gathered}$$

b.
$$\begin{gathered} 2>1 \\ 2-5>1-5 \\ -3>-4 \end{gathered}$$

c.
$$\begin{gathered} 2>1 \\ 3(2)>3(1) \\ 6>3 \end{gathered}$$

d.
$$\begin{gathered} 2>1 \\ -2(2)<-2(1) \\ -4<-2 \end{gathered}$$

19. a.
$$\begin{gathered} 4>-3 \\ 4+3>-3+3 \\ 7>0 \end{gathered}$$

b.
$$\begin{gathered} 4>-3 \\ 4-5>-3-5 \\ -1>-8 \end{gathered}$$

c.
$$\begin{gathered} 4>-3 \\ 3(4)>3(-3) \\ 12>-9 \end{gathered}$$

d.
$$\begin{gathered} 4>-3 \\ -2(4)<-2(-3) \\ -8<6 \end{gathered}$$

20. a.
$$\begin{gathered} -3>-5 \\ -3+3>-5+3 \\ 0>-2 \end{gathered}$$

b.
$$\begin{gathered} -3>-5 \\ -3-5>-5-5 \\ -8>-10 \end{gathered}$$

c.
$$\begin{gathered} -3>-5 \\ 3(-3)>3(-5) \\ -9>-15 \end{gathered}$$

d.
$$\begin{gathered} -3>-5 \\ -2(-3)<-2(-5) \\ 6<10 \end{gathered}$$

21. a.
$$2x+1<2$$
$$2x+1+3<2+3$$
$$2x+4<5$$

b.
$$2x+1<2$$
$$2x+1-5<2-5$$
$$2x-4<-3$$

c.
$$2x+1<2$$
$$3(2x+1)<3(2)$$
$$6x+3<6$$

d.
$$2x+1<2$$
$$-2(2x+1)>-2(2)$$
$$-4x-2>-4$$

22. a.
$$1-2x>5$$
$$1-2x+3>5+3$$
$$4-2x>8$$

b.
$$1-2x>5$$
$$1-2x-5>5-5$$
$$-2x-4>0$$

c.
$$1-2x>5$$
$$3(1-2x)>3(5)$$
$$3-6x>15$$

d.
$$1-2x>5$$
$$-2(1-2x)<-2(5)$$
$$-2+4x<-10$$

23. [0, 4]

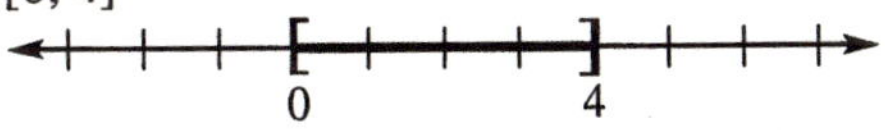

0 4

24. (–1, 5)

–1 0 5

25. [4, 6)

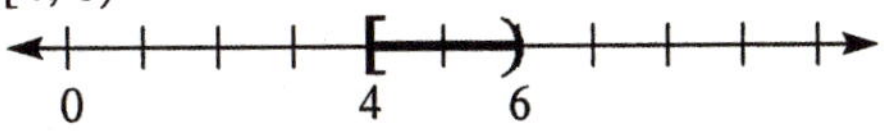

0 4 6

26. (–2, 0)

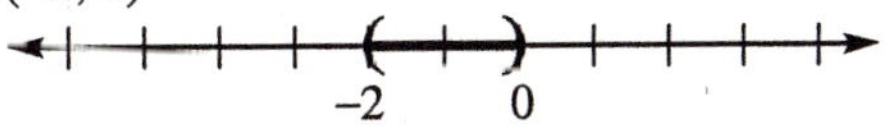

–2 0

27. $[4, \infty)$

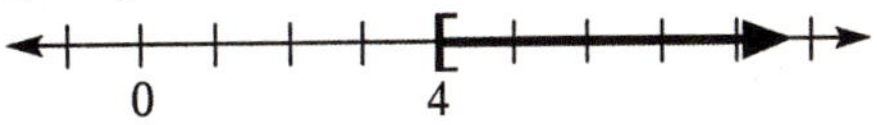

0 4

28. $(-\infty, 5]$

0 5

29. $(-\infty, -4)$

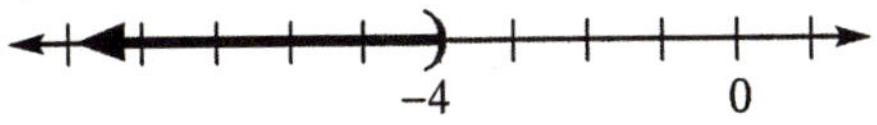

–4 0

30. $(1, \infty)$

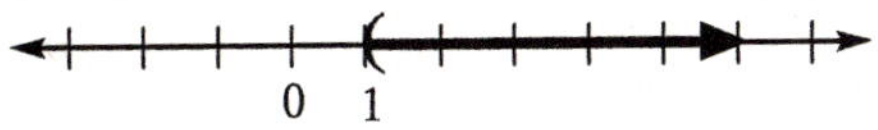

0 1

31. $2 \le x \le 5$

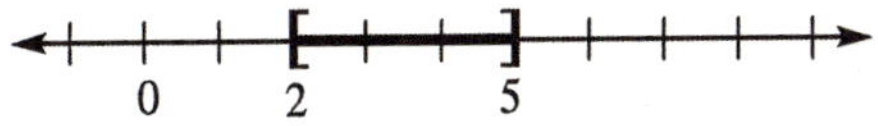

0 2 5

32. $1 < x < 2$

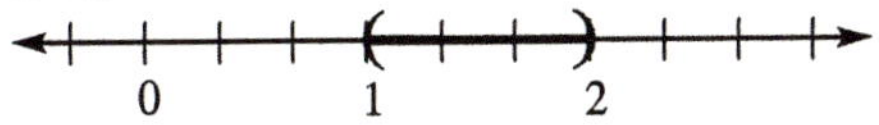

0 1 2

33. $-3 < x < -2$

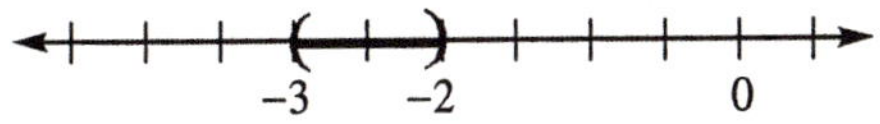

–3 –2 0

34. $0 \le x < 1$

0 1

35. $x \ge 4$

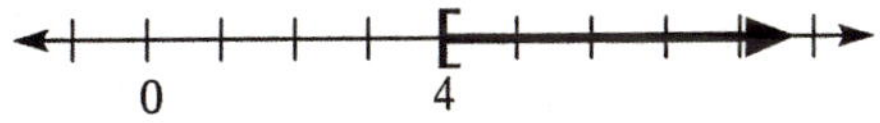

0 4

36. $x \le 2$

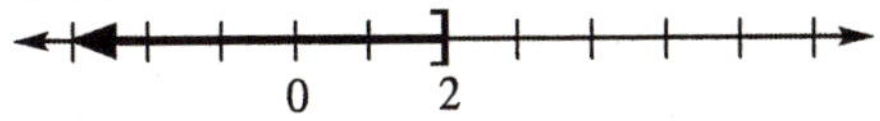

0 2

37. $x < -3$

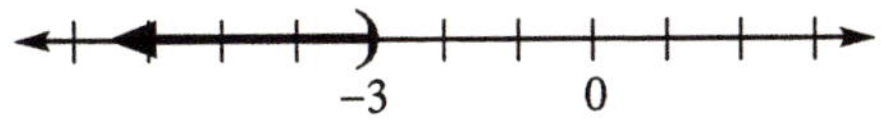

–3 0

38. $x > -8$

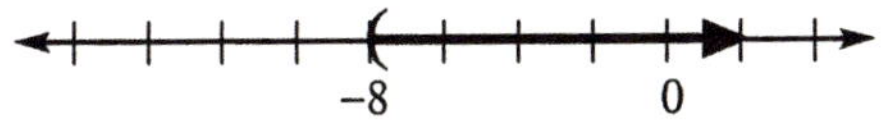

–8 0

39. If $x < 5$, then $x - 5 < 0$.

40. If $x < -4$, then $x + 4 < 0$.

41. If $x > -4$, then $x + 4 > 0$.

42. If $x > 6$, then $x - 6 > 0$.

43. If $x \ge -4$, then $3x \ge -12$.

44. If $x \le 3$, then $2x \le 6$.

45. If $x > 6$, then $-2x < -12$.

46. If $x > -2$, then $-4x < 8$.

47. If $x \ge 5$, then $-4x \le -20$.

48. If $x \le -4$, then $-3x \ge 12$.

49. If $2x < 6$, then $x < 3$.

50. If $3x \le 12$, then $x \le 4$.

51. If $-\frac{1}{2}x \le 3$, then $x \ge -6$.

52. If $-\frac{1}{4}x > 1$, then $x < -4$.

53. $x + 1 < 5$
$x + 1 - 1 < 5 - 1$
$x < 4$
$\{x \mid x < 4\}$ or $(-\infty, 4)$

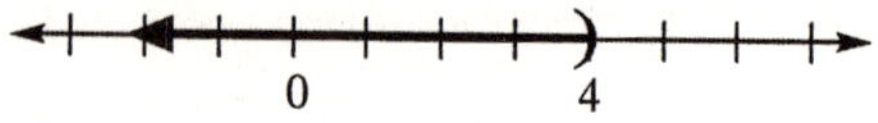

54. $x - 6 < 1$
$x - 6 + 6 < 1 + 6$
$x < 7$
$\{x \mid x < 7\}$ or $(-\infty, 7)$

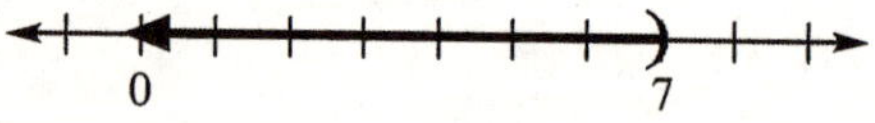

55. $1 - 2x \le 3$
$-2x \le 2$
$x \ge -1$
$\{x \mid x \ge -1\}$ or $[-1, \infty)$

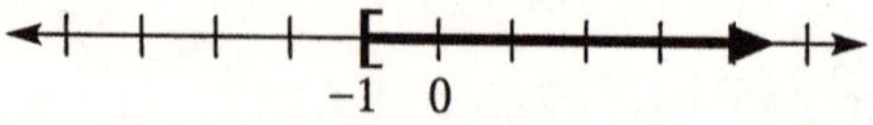

56. $2 - 3x \le 5$
$-3x \le 3$
$x \ge -1$
$\{x \mid x \ge -1\}$ or $[-1, \infty)$

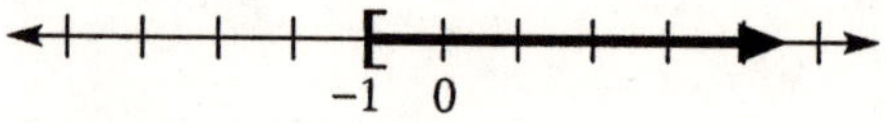

57. $3x - 7 > 2$
$3x > 9$
$x > 3$
$\{x \mid x > 3\}$ or $(3, \infty)$

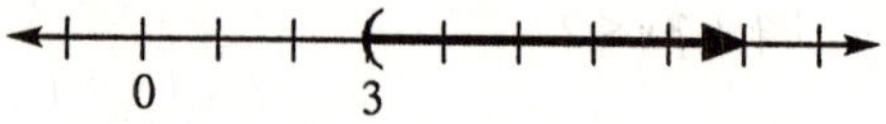

58. $2x + 5 > 1$
$2x > -4$
$x > -2$
$\{x \mid x > -2\}$ or $(-2, \infty)$

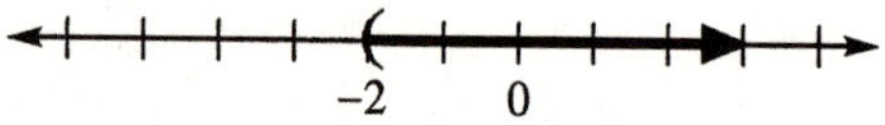

59. $3x - 1 \ge 3 + x$
$2x \ge 4$
$x \ge 2$
$\{x \mid x \ge 2\}$ or $[2, \infty)$

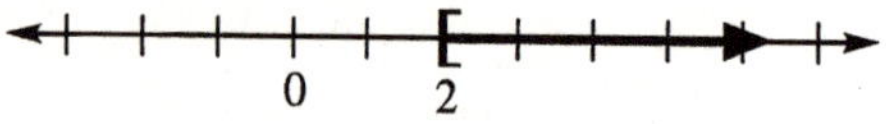

60. $2x - 2 \ge 3 + x$
$x \ge 5$
$\{x \mid x \ge 5\}$ or $[5, \infty)$

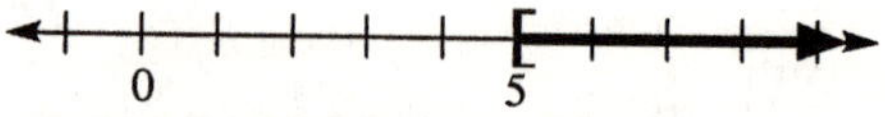

61. $-2(x + 3) < 8$
$-2x - 6 < 8$
$-2x < 14$
$x > -7$
$\{x \mid x > -7\}$ or $(-7, \infty)$

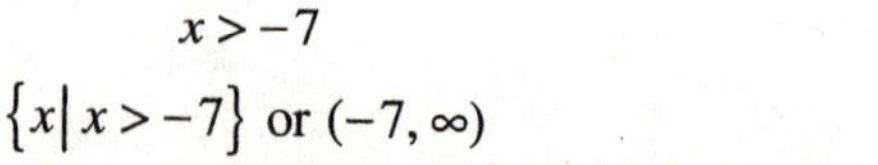

62. $-3(1-x)<12$

$-3+3x<12$

$3x<15$

$x<5$

$\{x|x<5\}$ or $(-\infty, 5)$

63. $4-3(1-x)\le 3$

$4-3+3x\le 3$

$3x+1\le 3$

$3x\le 2$

$x\le \frac{2}{3}$

$\left\{x \middle| x\le \frac{2}{3}\right\}$ or $\left(-\infty, \frac{2}{3}\right]$

64. $8-4(2-x)\le -2x$

$8-8+4x\le -2x$

$4x\le -2x$

$6x\le 0$

$x\le 0$

$\{x|x\le 0\}$ or $(-\infty, 0]$

65. $\frac{1}{2}(x-4)>x+8$

$\frac{1}{2}x-2>x+8$

$-\frac{1}{2}x>10$

$x<-20$

$\{x|x<-20\}$ or $(-\infty, -20)$

66. $3x+4>\frac{1}{3}(x-2)$

$3x+4>\frac{1}{3}x-\frac{2}{3}$

$9x+12>x-2$

$8x>-14$

$x>-\frac{7}{4}$

$\left\{x \middle| x>-\frac{7}{4}\right\}$ or $(-\frac{7}{4}, \infty)$

67. $\frac{x}{2}\ge 1-\frac{x}{4}$

$2x\ge 4-x$

$3x\ge 4$

$x\ge \frac{4}{3}$

$\left\{x \middle| x\ge \frac{4}{3}\right\}$ or $\left[\frac{4}{3}, \infty\right)$

68. $\frac{x}{3}\ge 2+\frac{x}{6}$

$2x\ge 12+x$

$x\ge 12$

$\{x|x\ge 12\}$ or $[12, \infty)$

69. $0\le 2x-6\le 4$

$6\le 2x\le 10$

$3\le x\le 5$

$\{x|3\le x\le 5\}$ or $[3, 5]$

70. $4\le 2x+2\le 10$

$2\le 2x\le 8$

$1\le x\le 4$

$\{x|1\le x\le 4\}$ or $[1, 4]$

71. $-5 \le 4-3x \le 2$

$-9 \le -3x \le -2$

$3 \ge x \ge \frac{2}{3}$

$\left\{x \middle| \frac{2}{3} \le x \le 3\right\}$ or $\left[\frac{2}{3}, 3\right]$

72. $-3 \le 3-2x \le 9$

$-6 \le -2x \le 6$

$3 \ge x \ge -3$

$\{x | -3 \le x \le 3\}$ or $[-3, 3]$

73. $-3 < \frac{2x-1}{4} < 0$

$-12 < 2x-1 < 0$

$-11 < 2x < 1$

$-\frac{11}{2} < x < \frac{1}{2}$

$\left\{x \middle| -\frac{11}{2} < x < \frac{1}{2}\right\}$ or $\left(-\frac{11}{2}, \frac{1}{2}\right)$

74. $0 < \frac{3x+2}{2} < 4$

$0 < 3x+2 < 8$

$-2 < 3x < 6$

$-\frac{2}{3} < x < 2$

$\left\{x \middle| -\frac{2}{3} < x < 2\right\}$ or $\left(-\frac{2}{3}, 2\right)$

75. $1 < 1-\frac{1}{2}x < 4$

$0 < -\frac{1}{2}x < 3$

$0 > x > -6$ or $-6 < x < 0$

$\{x | -6 < x < 0\}$ or $(-6, 0)$

76. $0 < 1-\frac{1}{3}x < 1$

$-1 < -\frac{1}{3}x < 0$

$3 > x > 0$ or $0 < x < 3$

$\{x | 0 < x < 3\}$ or $(0, 3)$

77. $(x+2)(x-3) > (x-1)(x+1)$

$x^2 - x - 6 > x^2 - 1$

$-x - 6 > -1$

$-x > 5$

$x < -5$

$\{x | x < -5\}$ or $(-\infty, -5)$

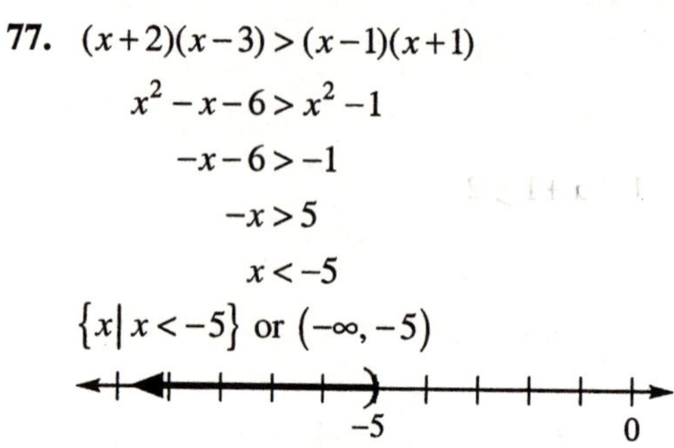

78. $(x-1)(x+1) > (x-3)(x+4)$

$x^2 - 1 > x^2 + x - 12$

$-1 > x - 12$

$-x > -11$

$x < 11$

$\{x | x < 11\}$ or $(-\infty, 11)$

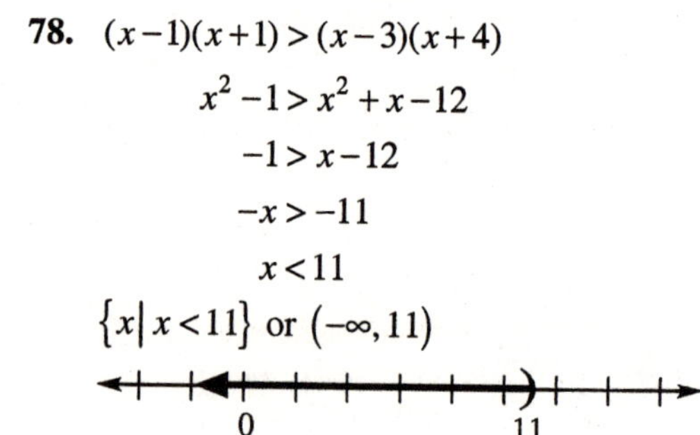

79. $x(4x+3) \le (2x+1)^2$

$4x^2 + 3x \le 4x^2 + 4x + 1$

$3x \le 4x + 1$

$-x \le 1$

$x \ge -1$

$\{x | x \ge -1\}$ or $[-1, \infty)$

80. $x(9x-5) \le (3x-1)^2$

$9x^2 - 5x \le 9x^2 - 6x + 1$

$-5x \le -6x + 1$

$x \le 1$

$\{x | x \le 1\}$ or $(-\infty, 1]$

81. $\frac{1}{2} \le \frac{x+1}{3} < \frac{3}{4}$

$6 \le 4x+4 < 9$

$2 \le 4x < 5$

$\frac{1}{2} \le x < \frac{5}{4}$

$\left\{x \middle| \frac{1}{2} \le x < \frac{5}{4}\right\}$ or $\left[\frac{1}{2}, \frac{5}{4}\right)$

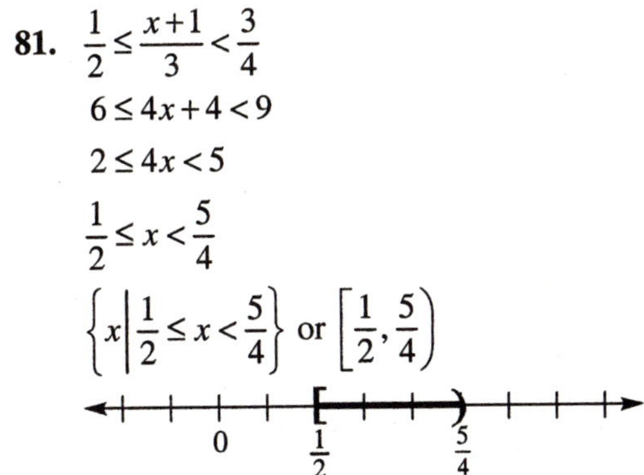

82. $\frac{1}{3} < \frac{x+1}{2} \le \frac{2}{3}$

$2 < 3x+3 \le 4$

$-1 < 3x \le 1$

$-\frac{1}{3} < x \le \frac{1}{3}$

$\left\{x \middle| -\frac{1}{3} < x \le \frac{1}{3}\right\}$ or $\left(-\frac{1}{3}, \frac{1}{3}\right]$

83. $|x| < 6$

$-6 < x < 6$

$\{x \mid -6 < x < 6\}$ or $(-6, 6)$

84. $|x| < 9$

$-9 < x < 9$

$\{x \mid -9 < x < 9\}$ or $(-9, 9)$

85. $|x| > 4$

$x < -4$ or $x > 4$

$\{x \mid x < -4 \text{ or } x > 4\}$ or $(-\infty, -4) \cup (4, \infty)$

86. $|x| > 1$

$x < -1$ or $x > 1$

$\{x \mid x < -1 \text{ or } x > 1\}$ or $(-\infty, -1) \cup (1, \infty)$

87. $|2x| < 8$

$-8 < 2x < 8$

$-4 < x < 4$

$\{x \mid -4 < x < 4\}$ or $(-4, 4)$

88. $|3x| < 15$

$-15 < 3x < 15$

$-5 < x < 5$

$\{x \mid -5 < x < 5\}$ or $(-5, 5)$

89. $|3x| > 12$

$3x < -12$ or $3x > 12$

$x < -4$ or $x > 4$

$\{x \mid x < -4 \text{ or } x > 4\}$ or $(-\infty, -4) \cup (4, \infty)$

90. $|2x| > 6$

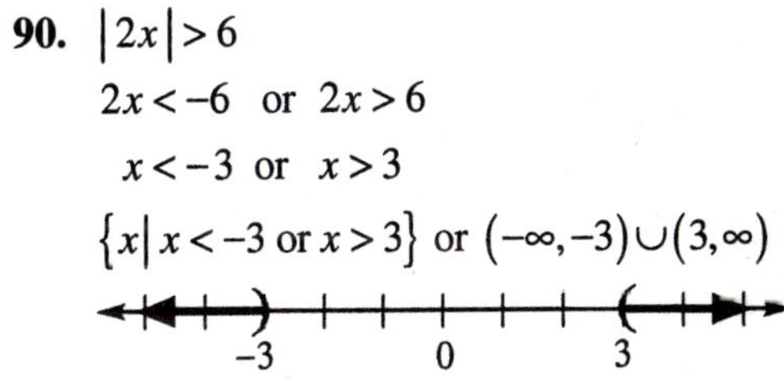

$2x < -6$ or $2x > 6$

$x < -3$ or $x > 3$

$\{x \mid x < -3 \text{ or } x > 3\}$ or $(-\infty, -3) \cup (3, \infty)$

91. $|x-2| + 2 < 3$

$|x-2| < 1$

$-1 < x-2 < 1$

$1 < x < 3$

$\{x \mid 1 < x < 3\}$ or $(1, 3)$

92. $|x+4| + 3 < 5$

$|x+4| < 2$

$-2 < x+4 < 2$

$-6 < x < -2$

$\{x \mid -6 < x < -2\}$ or $(-6, -2)$

93. $|3t-2|\le 4$

$-4\le 3t-2\le 4$

$-2\le 3t\le 6$

$-\frac{2}{3}\le t\le 2$

$\left\{t \middle| -\frac{2}{3}\le t\le 2\right\}$ or $\left[-\frac{2}{3},2\right]$

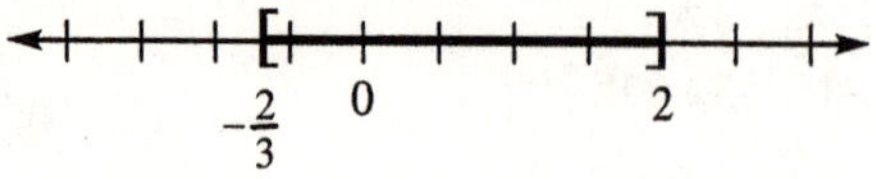

94. $|2u+5|\le 7$

$-7\le 2u+5\le 7$

$-12\le 2u\le 2$

$-6\le u\le 1$

$\{u|-6\le u\le 1\}$ or $[-6,1]$

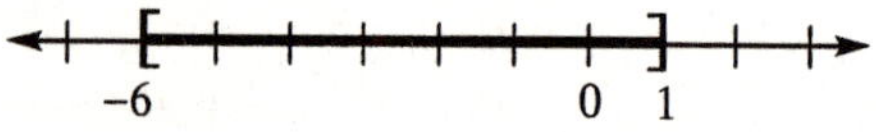

95. $|x-3|\ge 2$

$x-3\le -2$ or $x-3\ge 2$

$x\le 1$ or $x\ge 5$

$\{x|x\le 1 \text{ or } x\ge 5\}$ or $(-\infty,1]\cup[5,\infty)$

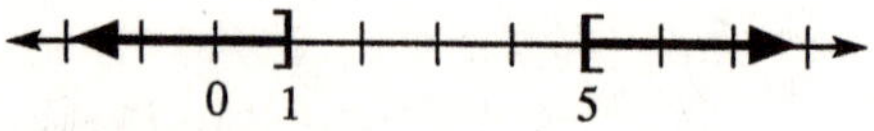

96. $|x+4|\ge 2$

$x+4\le -2$ or $x+4\ge 2$

$x\le -6$ or $x\ge -2$

$\{x|x\le -6 \text{ or } x\ge -2\}$ or $(-\infty,-6]\cup[-2,\infty)$

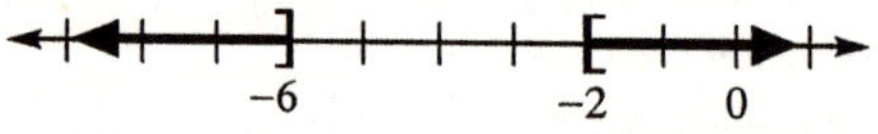

97. $|1-4x|-7<-2$

$|1-4x|<5$

$-5<1-4x<5$

$-6<-4x<4$

$\frac{-6}{-4}>x>\frac{4}{-4}$

$\frac{3}{2}>x>-1$ or $-1<x<\frac{3}{2}$

$\left\{x \middle| -1<x<\frac{3}{2}\right\}$ or $\left(-1,\frac{3}{2}\right)$

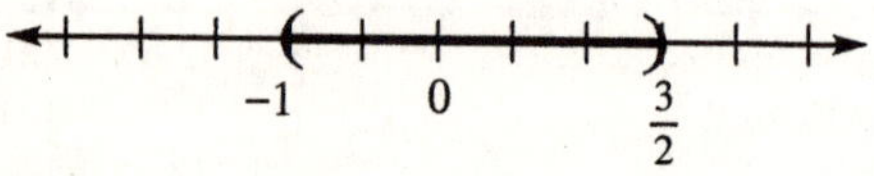

98. $|1-2x|-4<-1$

$|1-2x|<3$

$-3<1-2x<3$

$-4<-2x<2$

$\frac{-4}{-2}>x>\frac{2}{-2}$

$2>x>-1$ or $-1<x<2$

$\{x|-1<x<2\}$ or $(-1,2)$

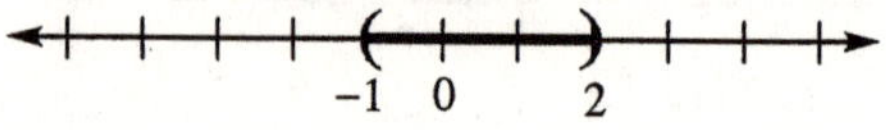

99. $|1-2x|>|-3|$

$|1-2x|>3$

$1-2x<-3$ or $1-2x>3$

$-2x<-4$ or $-2x>2$

$x>2$ or $x<-1$

$\{x|x<-1 \text{ or } x>2\}$ or $(-\infty,-1)\cup(2,\infty)$

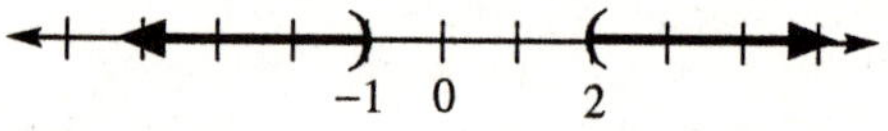

100. $|2-3x|>|-1|$

$|2-3x|>1$

$2-3x<-1$ or $2-3x>1$

$-3x<-3$ or $-3x>-1$

$x>1$ or $x<\frac{1}{3}$

$\left\{x \middle| x<\frac{1}{3} \text{ or } x>1\right\}$ or $\left(-\infty,\frac{1}{3}\right)\cup(1,\infty)$

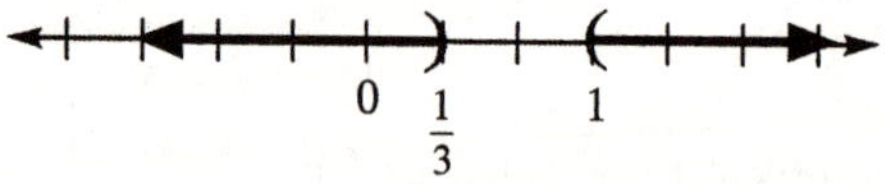

101. $|2x+1|<-1$

No solution since absolute value is always non-negative.

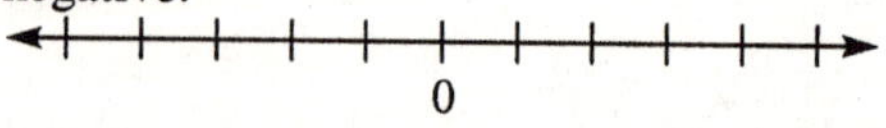

102. $|3x-4|\ge 0$

All real numbers since absolute value is always non-negative.

$\{x|-\infty<x<\infty\}$ or $(-\infty,\infty)$

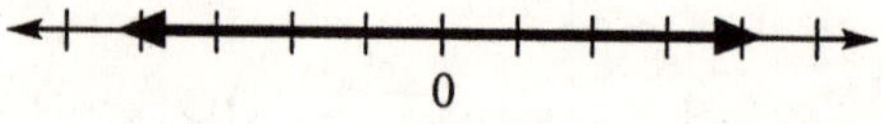

103. $-3 < x+5 < 2x$

$-3 < x+5$ and $x+5 < 2x$

$-8 < x$ $\quad$ $5 < x$

$x > -8$ $\quad$ $x > 5$

We need both $x > -8$ and $x > 5$. Therefore, $x > 5$ is sufficient to satisfy both inequalities.

$\{x \mid x > 5\}$ or $(5, \infty)$

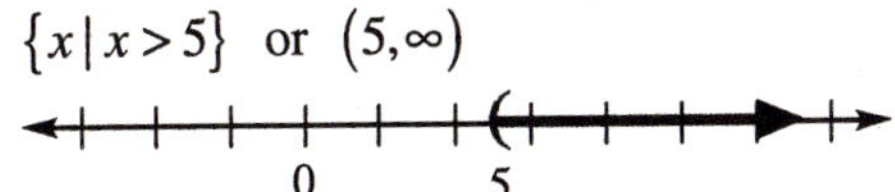

104. $2 < x-3 < 2x$

$2 < x-3$ and $x-3 < 2x$

$5 < x$ $\quad$ $-3 < x$

$x > 5$ $\quad$ $x > -3$

We need both $x > 5$ and $x > -3$. Therefore, $x > 5$ is sufficient to satisfy both inequalities.

$\{x \mid x > 5\}$ or $(5, \infty)$

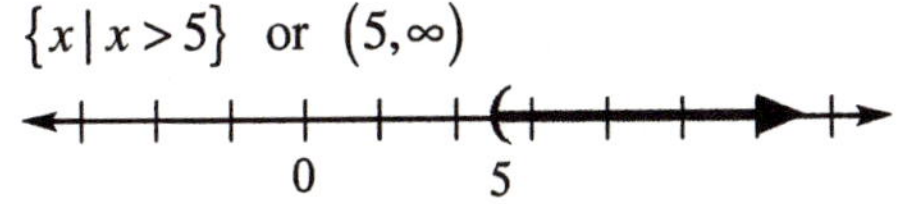

105. $x+2 < 2x-1 < 5x$

$x+2 < 2x-1$ and $2x-1 < 5x$

$2 < x-1$ $\quad$ $-3x-1 < 0$

$3 < x$ $\quad$ $-3x < 1$

$x > 3$ $\quad$ $x > -\frac{1}{3}$

We need both $x > -\frac{1}{3}$ and $x > 3$. Therefore, $x > 3$ is sufficient to satisfy both inequalities.

$\{x \mid x > 3\}$ or $(3, \infty)$.

106. $2x-1 < 3x+5 < 5x-7$

$2x-1 < 3x+5$ and $3x+5 < 5x-7$

$-x-1 < 5$ $\quad$ $-2x+5 < -7$

$-x < 6$ $\quad$ $-2x < -12$

$x > -6$ $\quad$ $x > 6$

We need both $x > -6$ and $x > 6$. Therefore, $x > 6$ is sufficient to satisfy both inequalities.

$\{x \mid x > 6\}$ or $(6, \infty)$.

107. $|x-2| < 0.5$

$-0.5 < x-2 < 0.5$

$-0.5+2 < x < 0.5+2$

$1.5 < x < 2.5$

Solution set: $\{x \mid 1.5 < x < 2.5\}$

108. $|x-(-1)| < 1$

$-1 < x-(-1) < 1$

$-1 < x+1 < 1$

$-1-1 < x < 1-1$

$-2 < x < 0$

Solution set: $\{x \mid -2 < x < 0\}$

109. $|x-(-3)| > 2$

$x-(-3) < -2$ or $x-(-3) > 2$

$x+3 < -2$ or $x+3 > 2$

$x < -5$ or $x > -1$

Solution set: $\{x \mid x < -5 \text{ or } x > -1\}$

110. $|x-2| > 3$

$x-2 < -3$ or $x-2 > 3$

$x < -1$ or $x > 5$

Solution set: $\{x \mid x < -1 \text{ or } x > 5\}$

111. 21 < young adult's age < 30

112. 40 ≤ middle-aged < 60

113. A temperature x that differs from $98.6°$ F by at least $1.5°$F.

$|x-98.6°| \geq 1.5°$

$x-98.6° \leq -1.5°$ or $x-98.6° \geq 1.5°$

$x \leq 97.1°$ or $x \geq 100.1°$

The temperatures that are considered unhealthy are those that are less than 97.1°F or greater than 100.1°F, inclusive.

114. A voltage x that differs from 115 volts by at most 5 volts.

$|x-115| \leq 5$

$-5 \leq x-115 \leq 5$

$110 \leq x \leq 120$

The actual voltage is between 110 and 120 volts, inclusive.

115. a. Let x = age at death.

$$x - 25 \geq 50.6$$
$$x \geq 75.6$$

Therefore, the average life expectancy for a 25-year-old male will be greater than or equal to 75.6 years.

b. Let x = age at death.

$$x - 25 \geq 55.4$$
$$x \geq 80.4$$

Therefore, the average life expectancy for a 25-year-old female will be greater than or equal to 80.4 years.

c. By the given information, a female can expect to live $80.4 - 75.6 = 4.8$ years longer.

116. $V = 20T$

$$353° \leq T \leq 393°$$
$$353° \leq \frac{V}{20} \leq 393°$$
$$7060 \leq V \leq 7860$$

The volume ranges from 7060 to 7860 cubic centimeters.

117. Let P represent the selling price and C represent the commission.

Calculating the commission:

$$C = 45{,}000 + 0.25(P - 900{,}000)$$
$$= 45{,}000 + 0.25P - 225{,}000$$
$$= 0.25P - 180{,}000$$

Calculate the commission range, given the price range:

$$900{,}000 \leq P \leq 1{,}100{,}000$$
$$0.25(900{,}000) \leq 0.25P \leq 0.25(1{,}100{,}000)$$
$$225{,}000 \leq 0.25P \leq 275{,}000$$
$$225{,}000 - 180{,}000 \leq 0.25P - 180{,}000 \leq 275{,}000 - 180{,}000$$
$$45{,}000 \leq C \leq 95{,}000$$

The agent's commission ranges from \$45,000 to \$95,000, inclusive.

$$\frac{45{,}000}{900{,}000} = 0.05 = 5\% \text{ to}$$
$$\frac{95{,}000}{1{,}100{,}000} = 0.086 = 8.6\% \text{, inclusive.}$$

As a percent of selling price, the commission ranges from 5% to 8.6%.

118. Let C represent the commission.

Calculate the commission range:

$$25 + 0.4(70) \leq C \leq 25 + 0.4(300)$$
$$53 \leq C \leq 145$$

The commission varies between \$53 and \$145.

119. Let W = weekly wages and T = tax withheld.

Calculating the withholding tax range, given the range of weekly wages:

$$600 \leq W \leq 700$$
$$600 - 592 \leq W - 592 \leq 700 - 592$$
$$8 \leq W - 592 \leq 108$$
$$0.25(8) \leq 0.25(W - 592) \leq 0.25(108)$$
$$2 \leq 0.25(W - 592) \leq 27$$
$$2 + 74.35 \leq 0.25(W - 592) + 74.35 \leq 27 + 74.35$$
$$76.35 \leq T \leq 101.35$$

The amount of withholding tax ranges from \$76.35 to \$101.35, inclusive.

120. Let W = weekly wages and T = tax withheld.

Calculating the withholding tax range, given the range of weekly wages:

$$800 \leq W \leq 800$$
$$800 - 592 \leq W - 592 \leq 900 - 592$$
$$208 \leq W - 592 \leq 308$$
$$0.25(208) \leq 0.25(W - 592) \leq 0.25(308)$$
$$52 \leq 0.25(W - 592) \leq 77$$
$$52 + 74.35 \leq 0.25(W - 592) + 74.35 \leq 77 + 74.35$$
$$126.35 \leq T \leq 151.35$$

The amount of withholding tax ranges from \$126.35 to \$151.35, inclusive.

121. Let K represent the monthly usage in kilowatt-hours and let C represent the monthly customer bill.

Calculating the bill:

$$C = 0.08275K + 10.07$$

Calculating the range of kilowatt-hours, given the range of bills:

$$65.96 \leq C \leq 217.02$$
$$65.96 \leq 0.08275K + 10.07 \leq 217.02$$
$$55.89 \leq 0.08275K \leq 206.95$$
$$675.41 \leq K \leq 2500.91$$

The range of usage in kilowatt-hours varied from 675.41 to 2500.91.

122. Let W represent the amount of water used (in thousands of gallons). Let C represent the customer charge (in dollars).

Calculating the charge:

$$C = 27.18 + 1.90(W - 12)$$
$$= 27.18 + 1.90W - 22.80$$
$$= 1.90W + 4.38$$

Calculating the range of water usage, given the range of charges:

$$30.03 \le C \le 43.33$$
$$30.03 \le 1.90W + 4.38 \le 43.33$$
$$25.65 \le 1.90W \le 38.95$$
$$13.5 \le W \le 20.5$$

The range of water usage varied from 13,500 to 20,500 gallons.

123. Let C represent the dealer's cost and M represent the markup over dealer's cost. If the price is \$8800, then

$$8800 = C + MC = C(1+M)$$

Solving for C yields: $C = \dfrac{8800}{1+M}$

Calculating the range of dealer costs, given the range of markups:

$$0.12 \le M \le 0.18$$
$$1.12 \le 1 + M \le 1.18$$
$$\frac{1}{1.12} \ge \frac{1}{1+M} \ge \frac{1}{1.18}$$
$$\frac{8800}{1.12} \ge \frac{8800}{1+M} \ge \frac{8800}{1.18}$$
$$7857.14 \ge C \ge 7457.63$$

The dealer's cost ranged from \$7457.63 to \$7857.14, inclusive.

124. Let T represent the test scores of the people in the top 2.5%.

$$T > 1.96(12) + 100 = 123.52$$

People in the top 2.5% will have test scores greater than 123.52.

125. Let T represent the score needed on the last test. Calculating the course grade and solving for the last test:

$$80 \le \frac{68+82+87+89+T}{5} < 90$$
$$80 \le \frac{326+T}{5} < 90$$
$$400 \le 326 + T < 450$$
$$74 \le T < 124$$

The fifth test must be greater than or equal to 74.

126. Let T represent the score needed on the last test. Calculating the course grade and solving for the last test:

$$80 \le \frac{68+82+87+89+2T}{6} < 90$$
$$80 \le \frac{326+2T}{6} < 90$$
$$80 \le \frac{163+T}{3} < 90$$
$$240 \le 163 + T < 270$$
$$77 \le T < 107$$

The fifth test must be greater than or equal to 77 to get a B.

127. Let g represent the number of gallons of gasoline in the gas tank. Since the car averages 25 miles per gallon, a trip of at least 300 miles will require at least $\dfrac{300}{25} = 12$ gallons of gas. Therefore the range of the amount of gasoline is $12 \le g \le 20$ (gallons).

128. Let g represent the number of gallons of gasoline in the gas tank. Since the car averages 25 miles per gallon, a trip of at most 250 miles will require at most $\dfrac{250}{25} = 10$ gallons of gas. Therefore the range of the amount of gasoline is $0 \le g \le 10$ (gallons).

129. Since $a < b$

$$\frac{a}{2} < \frac{b}{2} \qquad\qquad \frac{a}{2} < \frac{b}{2}$$
$$\frac{a}{2} + \frac{a}{2} < \frac{a}{2} + \frac{b}{2} \qquad\qquad \frac{a}{2} + \frac{b}{2} < \frac{b}{2} + \frac{b}{2}$$
$$a < \frac{a+b}{2} \qquad\qquad \frac{a+b}{2} < b$$

Thus, $a < \dfrac{a+b}{2} < b$.

130. $\frac{a+b}{2}-a=\frac{a+b-2a}{2}=\frac{b-a}{2}$

$b-\frac{a+b}{2}=\frac{2b-a-b}{2}=\frac{b-a}{2}$

$\therefore\ \frac{a+b}{2}$ is equidistant from a and b.

131. If $0<a<b$, then

$$ab>a^2>0 \qquad b^2>ab>0$$
$$\left(\sqrt{ab}\right)^2>a^2 \qquad b^2>\left(\sqrt{ab}\right)^2$$
$$\sqrt{ab}>a \qquad b>\sqrt{ab}$$

Thus, $a<\sqrt{ab}<b$

132. Show that $\sqrt{ab}<\frac{a+b}{2}$.

$$\frac{a+b}{2}-\sqrt{ab}=\frac{1}{2}\left(a-2\sqrt{ab}+b\right)$$
$$=\frac{1}{2}\left(\sqrt{a}-\sqrt{b}\right)^2>0, \text{ since } a\neq b.$$

Therefore, $\sqrt{ab}<\frac{a+b}{2}$.

133. For $0<a<b$, $\frac{1}{h}=\frac{1}{2}\left(\frac{1}{a}+\frac{1}{b}\right)$

$$a<b\Rightarrow\frac{1}{b}<\frac{1}{a}$$
$$\frac{1}{b}+\frac{1}{b}<\frac{1}{b}+\frac{1}{a}<\frac{1}{a}+\frac{1}{a}$$
$$\frac{2}{b}<\frac{1}{b}+\frac{1}{a}<\frac{2}{a}$$
$$\frac{1}{b}<\frac{1}{2}\left(\frac{1}{b}+\frac{1}{a}\right)<\frac{1}{a}$$
$$\frac{1}{b}<\frac{1}{h}<\frac{1}{a}$$
$$a<h<b$$

134. Show that $h=\frac{(\text{geometric mean})^2}{\text{arithmetic mean}}=\frac{\left(\sqrt{ab}\right)^2}{\left(\frac{1}{2}(a+b)\right)}$

From Problem 133 we know:

$$\frac{1}{h}=\frac{1}{2}\left(\frac{1}{a}+\frac{1}{b}\right)\Rightarrow\frac{2}{h}=\frac{1}{a}+\frac{1}{b}=\frac{b+a}{ab}$$

$$\frac{h}{2}=\frac{ab}{a+b}\Rightarrow h=2\cdot\frac{ab}{a+b}=\frac{\left(\sqrt{ab}\right)^2}{\left(\frac{1}{2}(a+b)\right)}$$

135. Answers will vary. One possibility:
No solution: $4x+6\le 2(x-5)+2x$
One solution: $3x+5\le 2(x+3)+1\le 3(x+2)-1$

136. Answers will vary.

137. Since $x^2\ge 0$, we have

$$x^2+1\ge 0+1$$
$$x^2+1\ge 1$$

Therefore, the expression x^2+1 can never be less than -5.

138. Answers will vary.

Section A.9

1. False
2. $\sqrt{(-4)^2}=|-4|=4$
3. index
4. cube root
5. True
6. False; $\sqrt[4]{(-3)^4}=|-3|=3$
7. $\sqrt[3]{27}=\sqrt[3]{3^3}=3$
8. $\sqrt[4]{16}=\sqrt[4]{2^4}=2$
9. $\sqrt[3]{-8}=\sqrt[3]{(-2)^3}=-2$
10. $\sqrt[3]{-1}=\sqrt[3]{(-1)^3}=-1$
11. $\sqrt{8}=\sqrt{4\cdot 2}=2\sqrt{2}$
12. $\sqrt[3]{54}=\sqrt[3]{27\cdot 2}=3\sqrt[3]{2}$
13. $\sqrt[3]{-8x^4}=\sqrt[3]{-8x^3\cdot x}=-2x\sqrt[3]{x}$
14. $\sqrt[4]{48x^5}=\sqrt[4]{16x^4\cdot 3x}=2x\sqrt[4]{3x}$

15. $\sqrt[4]{x^{12}y^8} = \sqrt[4]{\left(x^3\right)^4\left(y^2\right)^4} = x^3y^2$

16. $\sqrt[5]{x^{10}y^5} = \sqrt[5]{\left(x^2\right)^5 y^5} = x^2y$

17. $\sqrt[4]{\dfrac{x^9y^7}{xy^3}} = \sqrt[4]{x^8y^4} = x^2y$

18. $\sqrt[3]{\dfrac{3xy^2}{81x^4y^2}} = \sqrt[3]{\dfrac{1}{27x^3}} = \dfrac{\sqrt[3]{1}}{\sqrt[3]{27x^3}} = \dfrac{1}{3x}$

19. $\sqrt{36x} = 6\sqrt{x}$

20. $\sqrt{9x^5} = 3\sqrt{x^4 \cdot x} = 3x^2\sqrt{x}$

21. $\sqrt{3x^2}\sqrt{12x} = \sqrt{36x^2 \cdot x} = 6x\sqrt{x}$

22. $\sqrt{5x}\sqrt{20x^3} = \sqrt{100x^4} = 10x^2$

23.
$$\begin{aligned}\left(\sqrt{5}\sqrt[3]{9}\right)^2 &= \left(\sqrt{5}\right)^2\left(\sqrt[3]{9}\right)^2\\ &= 5 \cdot \sqrt[3]{9^2}\\ &= 5\sqrt[3]{81}\\ &= 5 \cdot 3\sqrt[3]{3}\\ &= 15\sqrt[3]{3}\end{aligned}$$

24.
$$\begin{aligned}\left(\sqrt[3]{3}\sqrt{10}\right)^4 &= \left(\sqrt[3]{3}\right)^4\left(\sqrt{10}\right)^4\\ &= \sqrt[3]{3^4} \cdot 10^2\\ &= 3\sqrt[3]{3} \cdot 100\\ &= 300\sqrt[3]{3}\end{aligned}$$

25. $\left(3\sqrt{6}\right)\left(2\sqrt{2}\right) = 6\sqrt{12} = 6\sqrt{4 \cdot 3} = 12\sqrt{3}$

26. $\left(5\sqrt{8}\right)\left(-3\sqrt{3}\right) = -15\sqrt{24} = -30\sqrt{6}$

27.
$$\begin{aligned}\left(\sqrt{3}+3\right)\left(\sqrt{3}-1\right) &= \left(\sqrt{3}\right)^2 + 3\sqrt{3} - \sqrt{3} - 3\\ &= 3 + 2\sqrt{3} - 3\\ &= 2\sqrt{3}\end{aligned}$$

28.
$$\begin{aligned}\left(\sqrt{5}-2\right)\left(\sqrt{5}+3\right) &= \left(\sqrt{5}\right)^2 - 2\sqrt{5} + 3\sqrt{5} - 6\\ &= 5 + \sqrt{5} - 6\\ &= \sqrt{5} - 1\end{aligned}$$

29.
$$\begin{aligned}\left(\sqrt{x}-1\right)^2 &= \left(\sqrt{x}\right)^2 - 2\sqrt{x} + 1\\ &= x - 2\sqrt{x} + 1\end{aligned}$$

30.
$$\begin{aligned}\left(\sqrt{x}+\sqrt{5}\right)^2 &= \left(\sqrt{x}\right)^2 + 2\left(\sqrt{x}\right)\left(\sqrt{5}\right) + \left(\sqrt{5}\right)^2\\ &= x + 2\sqrt{5x} + 5\end{aligned}$$

31.
$$\begin{aligned}3\sqrt{2} - 4\sqrt{8} &= 3\sqrt{2} - 4\sqrt{4 \cdot 2}\\ &= 3\sqrt{2} - 4 \cdot 2\sqrt{2}\\ &= 3\sqrt{2} - 8\sqrt{2}\\ &= (3-8)\sqrt{2}\\ &= -5\sqrt{2}\end{aligned}$$

32.
$$\begin{aligned}\sqrt[3]{-x^4} + \sqrt[3]{8x} &= \sqrt[3]{-x^3 \cdot x} + \sqrt[3]{8 \cdot x}\\ &= -x\sqrt[3]{x} + 2\sqrt[3]{x}\\ &= (-x+2)\sqrt[3]{x}\\ &= (2-x)\sqrt[3]{x}\end{aligned}$$

33.
$$\begin{aligned}\sqrt[3]{16x^4} - \sqrt[3]{2x} &= \sqrt[3]{8x^3 \cdot 2x} - \sqrt[3]{2x}\\ &= 2x\sqrt[3]{2x} - \sqrt[3]{2x}\\ &= (2x-1)\sqrt[3]{2x}\end{aligned}$$

34.
$$\begin{aligned}\sqrt[4]{32x} + \sqrt[4]{2x^5} &= \sqrt[4]{16 \cdot 2x} + \sqrt[4]{x^4 \cdot 2x}\\ &= 2\sqrt[4]{2x} + x\sqrt[4]{2x}\\ &= (2+x)\sqrt[4]{2x}\end{aligned}$$

35. $\dfrac{1}{\sqrt{2}} = \dfrac{1}{\sqrt{2}} \cdot \dfrac{\sqrt{2}}{\sqrt{2}} = \dfrac{\sqrt{2}}{2}$

36. $\dfrac{6}{\sqrt[3]{4}} = \dfrac{6}{\sqrt[3]{4}} \cdot \dfrac{\sqrt[3]{2}}{\sqrt[3]{2}} = \dfrac{6\sqrt[3]{2}}{\sqrt[3]{8}} = \dfrac{6\sqrt[3]{2}}{2} = 3\sqrt[3]{2}$

37. $\dfrac{-\sqrt{3}}{\sqrt{5}} = \dfrac{-\sqrt{3}}{\sqrt{5}} \cdot \dfrac{\sqrt{5}}{\sqrt{5}} = \dfrac{-\sqrt{15}}{5}$

38. $\dfrac{-\sqrt[3]{3}}{\sqrt{8}} = \dfrac{-\sqrt[3]{3}}{2\sqrt{2}} = \dfrac{-\sqrt[3]{3}}{2\sqrt{2}} \cdot \dfrac{\sqrt{2}}{\sqrt{2}} = \dfrac{-\sqrt[3]{3}\sqrt{2}}{2 \cdot 2} = \dfrac{-\sqrt[3]{3}\sqrt{2}}{4}$

39. $\frac{\sqrt{3}}{5-\sqrt{2}}=\frac{\sqrt{3}}{5-\sqrt{2}}\cdot\frac{5+\sqrt{2}}{5+\sqrt{2}}$
$=\frac{\sqrt{3}\left(5+\sqrt{2}\right)}{25-2}$
$=\frac{\sqrt{3}\left(5+\sqrt{2}\right)}{23}$

40. $\frac{\sqrt{2}}{\sqrt{7}+2}=\frac{\sqrt{2}}{\sqrt{7}+2}\cdot\frac{\sqrt{7}-2}{\sqrt{7}-2}$
$=\frac{\sqrt{2}\left(\sqrt{7}-2\right)}{7-4}$
$=\frac{\sqrt{2}\left(\sqrt{7}-2\right)}{3}$

41. $\frac{2-\sqrt{5}}{2+3\sqrt{5}}=\frac{2-\sqrt{5}}{2+3\sqrt{5}}\cdot\frac{2-3\sqrt{5}}{2-3\sqrt{5}}$
$=\frac{4-2\sqrt{5}-6\sqrt{5}+15}{4-45}$
$=\frac{19-8\sqrt{5}}{-41}$
$=\frac{8\sqrt{5}-19}{41}$

42. $\frac{\sqrt{3}-1}{2\sqrt{3}+3}=\frac{\sqrt{3}-1}{2\sqrt{3}+3}\cdot\frac{2\sqrt{3}-3}{2\sqrt{3}-3}$
$=\frac{6-2\sqrt{3}-3\sqrt{3}+3}{12-9}$
$=\frac{9-5\sqrt{3}}{3}$

43. $\frac{5}{\sqrt[3]{2}}=\frac{5}{\sqrt[3]{2}}\cdot\frac{\sqrt[3]{4}}{\sqrt[3]{4}}=\frac{5\sqrt[3]{4}}{2}$

44. $\frac{-2}{\sqrt[3]{9}}=\frac{-2}{\sqrt[3]{9}}\cdot\frac{\sqrt[3]{3}}{\sqrt[3]{3}}=\frac{-2\sqrt[3]{3}}{3}$

45. $\frac{\sqrt{x+h}-\sqrt{x}}{\sqrt{x+h}+\sqrt{x}}=\frac{\sqrt{x+h}-\sqrt{x}}{\sqrt{x+h}+\sqrt{x}}\cdot\frac{\sqrt{x+h}-\sqrt{x}}{\sqrt{x+h}-\sqrt{x}}$
$=\frac{(x+h)-2\sqrt{x(x+h)}+x}{(x+h)-x}$
$=\frac{x+h-2\sqrt{x^2+xh}+x}{x+h-x}$
$=\frac{2x+h-2\sqrt{x^2+xh}}{h}$

46. $\frac{\sqrt{x+h}+\sqrt{x-h}}{\sqrt{x+h}-\sqrt{x-h}}$
$=\frac{\sqrt{x+h}+\sqrt{x-h}}{\sqrt{x+h}-\sqrt{x-h}}\cdot\frac{\sqrt{x+h}+\sqrt{x-h}}{\sqrt{x+h}+\sqrt{x-h}}$
$=\frac{(x+h)+2\sqrt{(x-h)(x+h)}+(x-h)}{(x+h)-(x-h)}$
$=\frac{x+h+2\sqrt{x^2-h^2}+x-h}{x+h-x+h}$
$=\frac{2x+2\sqrt{x^2-h^2}}{2h}$
$=\frac{x+\sqrt{x^2-h^2}}{h}$

47. $\sqrt[3]{2t-1}=2$
$\left(\sqrt[3]{2t-1}\right)^3=2^3$
$2t-1=8$
$2t=9$
$t=\frac{9}{2}$
Solution set: $\left\{\frac{9}{2}\right\}$

48. $\sqrt[3]{3t+1}=-2$
$\left(\sqrt[3]{3t+1}\right)^3=(-2)^3$
$3t+1=-8$
$3t=-9$
$t=-3$
Solution set: $\{-3\}$

49. $\sqrt{15-2x} = x$

$\left(\sqrt{15-2x}\right)^2 = x^2$

$15-2x = x^2$

$x^2+2x-15=0$

$(x+5)(x-3)=0$

$x+5=0$ or $x-3=0$

$x=-5$ $\quad x=3$

Check each solution in the original equation.

$\sqrt{15-2(-5)} = \sqrt{25} = 5 \neq -5$

$\sqrt{15-2(3)} = \sqrt{9} = 3$ ✓

$x=-5$ is an extraneous solution.

Solution set: $\{3\}$

50. $\sqrt{12-x} = x$

$\left(\sqrt{12-x}\right)^2 = x^2$

$12-x = x^2$

$x^2+x-12=0$

$(x+4)(x-3)=0$

$x+4=0$ or $x-3=0$

$x=-4$ $\quad x=3$

Check each solution in the original equation.

$\sqrt{12-(-4)} = \sqrt{16} = 4 \neq -4$

$\sqrt{12-3} = \sqrt{9} = 3$ ✓

$x=-4$ is an extraneous solution.

Solution set: $\{3\}$

51. $8^{2/3} = \left(\sqrt[3]{8}\right)^2 = 2^2 = 4$

52. $4^{3/2} = \left(\sqrt{4}\right)^3 = 2^3 = 8$

53. $(-27)^{1/3} = \sqrt[3]{-27} = -3$

54. $16^{3/4} = \left(\sqrt[4]{16}\right)^3 = 2^3 = 8$

55. $16^{3/2} = \left(\sqrt{16}\right)^3 = 4^3 = 64$

56. $64^{3/2} = \left(\sqrt{64}\right)^3 = 8^3 = 512$

57. $9^{-3/2} = \dfrac{1}{9^{3/2}} = \dfrac{1}{\left(\sqrt{9}\right)^3} = \dfrac{1}{3^3} = \dfrac{1}{27}$

58. $25^{-5/2} = \dfrac{1}{25^{5/2}} = \dfrac{1}{\left(\sqrt{25}\right)^5} = \dfrac{1}{5^5} = \dfrac{1}{3125}$

59. $\left(\dfrac{9}{8}\right)^{3/2} = \left(\sqrt{\dfrac{9}{8}}\right)^3 = \left(\dfrac{3}{2\sqrt{2}}\right)^3 = \dfrac{3^3}{2^3\left(\sqrt{2}\right)^3}$

$= \dfrac{27}{8\cdot 2\sqrt{2}} = \dfrac{27}{16\sqrt{2}} = \dfrac{27}{16\sqrt{2}}\cdot\dfrac{\sqrt{2}}{\sqrt{2}}$

$= \dfrac{27\sqrt{2}}{32}$

60. $\left(\dfrac{27}{8}\right)^{2/3} = \left(\sqrt[3]{\dfrac{27}{8}}\right)^2 = \left(\dfrac{3}{2}\right)^2 = \dfrac{9}{4}$

61. $\left(\dfrac{8}{9}\right)^{-3/2} = \left(\dfrac{9}{8}\right)^{3/2} = \left(\sqrt{\dfrac{9}{8}}\right)^3 = \left(\dfrac{3}{2\sqrt{2}}\right)^3$

$= \dfrac{3^3}{2^3\left(\sqrt{2}\right)^3} = \dfrac{27}{8\cdot 2\sqrt{2}} = \dfrac{27}{16\sqrt{2}}$

$= \dfrac{27}{16\sqrt{2}}\cdot\dfrac{\sqrt{2}}{\sqrt{2}} = \dfrac{27\sqrt{2}}{32}$

62. $\left(\dfrac{8}{27}\right)^{-2/3} = \left(\dfrac{27}{8}\right)^{2/3} = \left(\sqrt[3]{\dfrac{27}{8}}\right)^2 = \left(\dfrac{3}{2}\right)^2 = \dfrac{9}{4}$

63. $x^{3/4}x^{1/3}x^{-1/2} = x^{3/4+1/3-1/2} = x^{7/12}$

64. $x^{2/3}x^{1/2}x^{-1/4} = x^{2/3+1/2-1/4} = x^{11/12}$

65. $\left(x^3y^6\right)^{1/3} = \left(x^3\right)^{1/3}\left(y^6\right)^{1/3} = xy^2$

66. $\left(x^4y^8\right)^{3/4} = \left(x^4\right)^{3/4}\left(y^8\right)^{3/4} = x^3y^6$

67. $\left(x^2y\right)^{1/3}\left(xy^2\right)^{2/3} = \left(x^2\right)^{1/3}(y)^{1/3}(x)^{2/3}\left(y^2\right)^{2/3}$

$= x^{2/3}y^{1/3}x^{2/3}y^{4/3}$

$= x^{2/3+2/3}y^{1/3+4/3}$

$= x^{4/3}y^{5/3}$

68. $\left(xy\right)^{1/4}\left(x^2y^2\right)^{1/2}=x^{1/4}y^{1/4}xy$
$=x^{1/4+1}y^{1/4+1}$
$=x^{5/4}y^{5/4}$

69. $\left(16x^2y^{-1/3}\right)^{3/4}=16^{3/4}\left(x^2\right)^{3/4}\left(y^{-1/3}\right)^{3/4}$
$=\left(\sqrt[4]{16}\right)^3x^{3/2}y^{-1/4}$
$=\dfrac{2^3x^{3/2}}{y^{1/4}}$
$=\dfrac{8x^{3/2}}{y^{1/4}}$

70. $\left(4x^{-1}y^{1/3}\right)^{3/2}=4^{3/2}\left(x^{-1}\right)^{3/2}\left(y^{1/3}\right)^{3/2}$
$=\left(\sqrt{4}\right)^3x^{-3/2}y^{1/2}$
$=\dfrac{2^3y^{1/2}}{x^{3/2}}$
$=\dfrac{8y^{1/2}}{x^{3/2}}$

71. $\dfrac{x}{(1+x)^{1/2}}+2(1+x)^{1/2}=\dfrac{x+2(1+x)^{1/2}(1+x)^{1/2}}{(1+x)^{1/2}}$
$=\dfrac{x+2(1+x)}{(1+x)^{1/2}}$
$=\dfrac{x+2+2x}{(1+x)^{1/2}}$
$=\dfrac{3x+2}{(1+x)^{1/2}}$

72. $\dfrac{1+x}{2x^{1/2}}+x^{1/2}=\dfrac{1+x+x^{1/2}\cdot 2x^{1/2}}{2x^{1/2}}$
$=\dfrac{1+x+2x}{2x^{1/2}}$
$=\dfrac{3x+1}{2x^{1/2}}$

73. $2x\left(x^2+1\right)^{1/2}+x^2\cdot\dfrac{1}{2}\left(x^2+1\right)^{-1/2}\cdot 2x$
$=2x\left(x^2+1\right)^{1/2}+\dfrac{x^3}{\left(x^2+1\right)^{1/2}}$
$=\dfrac{2x\left(x^2+1\right)^{1/2}\cdot\left(x^2+1\right)^{1/2}+x^3}{\left(x^2+1\right)^{1/2}}$
$=\dfrac{2x\left(x^2+1\right)^{1/2+1/2}+x^3}{\left(x^2+1\right)^{1/2}}=\dfrac{2x\left(x^2+1\right)^1+x^3}{\left(x^2+1\right)^{1/2}}$
$=\dfrac{2x^3+2x+x^3}{\left(x^2+1\right)^{1/2}}=\dfrac{3x^3+2x}{\left(x^2+1\right)^{1/2}}$
$=\dfrac{x\left(3x^2+2\right)}{\left(x^2+1\right)^{1/2}}$

74. $(x+1)^{1/3}+x\cdot\dfrac{1}{3}(x+1)^{-2/3}, x\neq -1$
$=(x+1)^{1/3}+\dfrac{x}{3(x+1)^{2/3}}$
$=\dfrac{3(x+1)^{2/3}(x+1)^{1/3}+x}{3(x+1)^{2/3}}$
$=\dfrac{3(x+1)^{2/3+1/3}+x}{3(x+1)^{2/3}}=\dfrac{3(x+1)^1+x}{3(x+1)^{2/3}}$
$=\dfrac{3x+3+x}{3(x+1)^{2/3}}=\dfrac{4x+3}{3(x+1)^{2/3}}$

75. $\sqrt{4x+3}\cdot\dfrac{1}{2\sqrt{x-5}}+\sqrt{x-5}\cdot\dfrac{1}{5\sqrt{4x+3}}, x>5$
$=\dfrac{\sqrt{4x+3}}{2\sqrt{x-5}}+\dfrac{\sqrt{x-5}}{5\sqrt{4x+3}}$
$=\dfrac{\sqrt{4x+3}\cdot 5\cdot\sqrt{4x+3}+\sqrt{x-5}\cdot 2\cdot\sqrt{x-5}}{10\sqrt{x-5}\sqrt{4x+3}}$
$=\dfrac{5(4x+3)+2(x-5)}{10\sqrt{(x-5)(4x+3)}}$
$=\dfrac{20x+15+2x-10}{10\sqrt{(x-5)(4x+3)}}$
$=\dfrac{22x+5}{10\sqrt{(x-5)(4x+3)}}$

76. $$\frac{\sqrt[3]{8x+1}}{3\sqrt[3]{(x-2)^2}}+\frac{\sqrt[3]{x-2}}{24\sqrt[3]{(8x+1)^2}}, x\neq 2, x\neq -\frac{1}{8}$$

$$=\frac{8\sqrt[3]{8x+1}\cdot\sqrt[3]{(8x+1)^2}+\sqrt[3]{x-2}\cdot\sqrt[3]{(x-2)^2}}{24\sqrt[3]{(x-2)^2}\cdot\sqrt[3]{(8x+1)^2}}$$

$$=\frac{8\sqrt[3]{(8x+1)^3}+\sqrt[3]{(x-2)^3}}{24\sqrt[3]{(x-2)^2}\cdot\sqrt[3]{(8x+1)^2}}$$

$$=\frac{8(8x+1)+x-2}{24\sqrt[3]{(x-2)^2(8x+1)^2}}$$

$$=\frac{64x+8+x-2}{24\sqrt[3]{(x-2)^2(8x+1)^2}}$$

$$=\frac{65x+6}{24\sqrt[3]{(x-2)^2(8x+1)^2}}$$

77. $$\frac{\left(\sqrt{1+x}-x\cdot\frac{1}{2\sqrt{1+x}}\right)}{1+x}=\frac{\left(\sqrt{1+x}-\frac{x}{2\sqrt{1+x}}\right)}{1+x}$$

$$=\frac{\left(\frac{2\sqrt{1+x}\sqrt{1+x}-x}{2\sqrt{1+x}}\right)}{1+x}$$

$$=\frac{2(1+x)-x}{2(1+x)^{1/2}}\cdot\frac{1}{1+x}$$

$$=\frac{2+x}{2(1+x)^{3/2}}$$

78. $$\frac{\left(\sqrt{x^2+1}-x\cdot\frac{2x}{2\sqrt{x^2+1}}\right)}{x^2+1}$$

$$=\frac{\left(\sqrt{x^2+1}-\frac{x^2}{\sqrt{x^2+1}}\right)}{x^2+1}$$

$$=\frac{\left(\sqrt{x^2+1}\cdot\frac{\sqrt{x^2+1}}{\sqrt{x^2+1}}-\frac{x^2}{\sqrt{x^2+1}}\right)}{x^2+1}$$

$$=\frac{\left(\frac{x^2+1-x^2}{\sqrt{x^2+1}}\right)}{x^2+1}=\frac{1}{\sqrt{x^2+1}}\cdot\frac{1}{x^2+1}$$

$$=\frac{1}{\left(x^2+1\right)^{3/2}}$$

79. $$\frac{(x+4)^{1/2}-2x(x+4)^{-1/2}}{x+4}$$

$$=\frac{\left((x+4)^{1/2}-\frac{2x}{(x+4)^{1/2}}\right)}{x+4}$$

$$=\frac{\left((x+4)^{1/2}\cdot\frac{(x+4)^{1/2}}{(x+4)^{1/2}}-\frac{2x}{(x+4)^{1/2}}\right)}{x+4}$$

$$=\frac{\left(\frac{x+4-2x}{(x+4)^{1/2}}\right)}{x+4}$$

$$=\frac{-x+4}{(x+4)^{1/2}}\cdot\frac{1}{x+4}$$

$$=\frac{-x+4}{(x+4)^{3/2}}$$

$$=\frac{4-x}{(x+4)^{3/2}}$$

80. $$\frac{\left(9-x^2\right)^{1/2}+x^2\left(9-x^2\right)^{-1/2}}{9-x^2}, -3<x<3$$

$$\frac{\left(\left(9-x^2\right)^{1/2}+\frac{x^2}{\left(9-x^2\right)^{1/2}}\right)}{9-x^2}$$

$$=\frac{\left(\frac{\left(9-x^2\right)^{1/2}\cdot\left(9-x^2\right)^{1/2}+x^2}{\left(9-x^2\right)^{1/2}}\right)}{9-x^2}$$

$$=\frac{\left(9-x^2\right)^{1/2}\cdot\left(9-x^2\right)^{1/2}+x^2}{\left(9-x^2\right)^{1/2}}\cdot\frac{1}{9-x^2}$$

$$=\frac{9-x^2+x^2}{\left(9-x^2\right)^{1/2}}\cdot\frac{1}{9-x^2}$$

$$=\frac{9}{\left(9-x^2\right)^{3/2}}$$

81. $\dfrac{\left(\dfrac{x^2}{(x^2-1)^{1/2}}-(x^2-1)^{1/2}\right)}{x^2}, x<-1 \text{ or } x>1$

$=\dfrac{\left(\dfrac{x^2-(x^2-1)^{1/2}\cdot(x^2-1)^{1/2}}{(x^2-1)^{1/2}}\right)}{x^2}$

$=\dfrac{x^2-(x^2-1)^{1/2}\cdot(x^2-1)^{1/2}}{(x^2-1)^{1/2}}\cdot\dfrac{1}{x^2}$

$=\dfrac{x^2-(x^2-1)}{(x^2-1)^{1/2}}\cdot\dfrac{1}{x^2}$

$=\dfrac{x^2-x^2+1}{(x^2-1)^{1/2}}\cdot\dfrac{1}{x^2}$

$=\dfrac{1}{x^2(x^2-1)^{1/2}}$

82. $\dfrac{(x^2+4)^{1/2}-x^2(x^2+4)^{-1/2}}{x^2+4}$

$=\dfrac{\left((x^2+4)^{1/2}-\dfrac{x^2}{(x^2+4)^{1/2}}\right)}{x^2+4}$

$=\dfrac{\left(\dfrac{(x^2+4)^{1/2}\cdot(x^2+4)^{1/2}-x^2}{(x^2+4)^{1/2}}\right)}{x^2+4}$

$=\dfrac{(x^2+4)^{1/2}\cdot(x^2+4)^{1/2}-x^2}{(x^2+4)^{1/2}}\cdot\dfrac{1}{x^2+4}$

$=\dfrac{x^2+4-x^2}{(x^2+4)^{1/2}}\cdot\dfrac{1}{x^2+4}=\dfrac{4}{(x^2+4)^{3/2}}$

83. $\dfrac{\dfrac{1+x^2}{2\sqrt{x}}-2x\sqrt{x}}{(1+x^2)^2}, x>0$

$=\dfrac{\left(\dfrac{1+x^2-(2\sqrt{x})(2x\sqrt{x})}{2\sqrt{x}}\right)}{(1+x^2)^2}$

$=\dfrac{1+x^2-(2\sqrt{x})(2x\sqrt{x})}{2\sqrt{x}}\cdot\dfrac{1}{(1+x^2)^2}$

$=\dfrac{1+x^2-4x^2}{2\sqrt{x}}\cdot\dfrac{1}{(1+x^2)^2}$

$=\dfrac{1-3x^2}{2\sqrt{x}(1+x^2)^2}$

84. $\dfrac{2x(1-x^2)^{1/3}+\frac{2}{3}x^3(1-x^2)^{-2/3}}{(1-x^2)^{2/3}}, x\neq-1, x\neq1$

$=\dfrac{\left(2x(1-x^2)^{1/3}+\dfrac{2x^3}{3(1-x^2)^{2/3}}\right)}{(1-x^2)^{2/3}}$

$=\dfrac{\left(\dfrac{2x(1-x^2)^{1/3}3(1-x^2)^{2/3}+2x^3}{3(1-x^2)^{2/3}}\right)}{(1-x^2)^{2/3}}$

$=\dfrac{6x(1-x^2)^{1/3+2/3}+2x^3}{3(1-x^2)^{2/3}}\cdot\dfrac{1}{(1-x^2)^{2/3}}$

$=\dfrac{6x(1-x^2)+2x^3}{3(1-x^2)^{2/3+2/3}}=\dfrac{6x-6x^3+2x^3}{3(1-x^2)^{4/3}}$

$=\dfrac{6x-4x^3}{3(1-x^2)^{4/3}}=\dfrac{2x(3-2x^2)}{3(1-x^2)^{4/3}}$

85. $(x+1)^{3/2} + x\cdot\frac{3}{2}(x+1)^{1/2} = (x+1)^{1/2}\left(x+1+\frac{3}{2}x\right)$

$= (x+1)^{1/2}\left(\frac{5}{2}x+1\right)$

$= \frac{1}{2}(x+1)^{1/2}(5x+2)$

86. $(x^2+4)^{4/3} + x\cdot\frac{4}{3}(x^2+4)^{1/3}\cdot 2x$

$= (x^2+4)^{1/3}\left(x^2+4+\frac{8}{3}x^2\right)$

$= (x^2+4)^{1/3}\left(\frac{11}{3}x^2+4\right)$

$= \frac{1}{3}\left(x^2+4\right)^{1/3}\left(11x^2+12\right)$

87. $6x^{1/2}\left(x^2+x\right) - 8x^{3/2} - 8x^{1/2}$

$= 2x^{1/2}\left(3(x^2+x)-4x-4\right)$

$= 2x^{1/2}\left(3x^2-x-4\right)$

$= 2x^{1/2}(3x-4)(x+1)$

88. $6x^{1/2}(2x+3) + x^{3/2}\cdot 8 = 2x^{1/2}(3(2x+3)+4x)$

$= 2x^{1/2}(10x+9)$

89. $3\left(x^2+4\right)^{4/3} + x\cdot 4\left(x^2+4\right)^{1/3}\cdot 2x$

$= \left(x^2+4\right)^{1/3}\left[3\left(x^2+4\right)+8x^2\right]$

$= \left(x^2+4\right)^{1/3}\left[3x^2+12+8x^2\right]$

$= \left(x^2+4\right)^{1/3}\left(11x^2+12\right)$

90. $2x(3x+4)^{4/3} + x^2\cdot 4(3x+4)^{1/3}$

$= 2x(3x+4)^{1/3}\left[(3x+4)+2x\right]$

$= 2x(3x+4)^{1/3}(5x+4)$

91. $4(3x+5)^{1/3}(2x+3)^{3/2} + 3(3x+5)^{4/3}(2x+3)^{1/2}$

$= (3x+5)^{1/3}(2x+3)^{1/2}\left[4(2x+3)+3(3x+5)\right]$

$= (3x+5)^{1/3}(2x+3)^{1/2}(8x+12+9x+15)$

$= (3x+5)^{1/3}(2x+3)^{1/2}(17x+27)$

where $x \ge -\frac{3}{2}$.

92. $6(6x+1)^{1/3}(4x-3)^{3/2} + 6(6x+1)^{4/3}(4x-3)^{1/2}$

$= 6(6x+1)^{1/3}(4x-3)^{1/2}\left[(4x-3)+(6x+1)\right]$

$= 6(6x+1)^{1/3}(4x-3)^{1/2}(10x-2)$

$= 6(6x+1)^{1/3}(4x-3)^{1/2}(2)(5x-1)$

$= 12(6x+1)^{1/3}(4x-3)^{1/2}(5x-1)$

where $x \ge \frac{3}{4}$.

93. $3x^{-1/2} + \frac{3}{2}x^{1/2}, x > 0$

$= \frac{3}{x^{1/2}} + \frac{3}{2}x^{1/2}$

$= \frac{3\cdot 2 + 3x^{1/2}\cdot x^{1/2}}{2x^{1/2}}$

$= \frac{6+3x}{2x^{1/2}}$

$= \frac{3(2+x)}{2x^{1/2}}$

94. $8x^{1/3} - 4x^{-2/3}, x \ne 0$

$= 8x^{1/3} - \frac{4}{x^{2/3}}$

$= \frac{8x^{1/3}\cdot x^{2/3} - 4}{x^{2/3}}$

$= \frac{8x-4}{x^{2/3}}$

$= \frac{4(2x-1)}{x^{2/3}}$

95. $x\left(\frac{1}{2}\right)\left(8-x^2\right)^{-1/2}(-2x) + \left(8-x^2\right)^{1/2}$

$= -x^2\left(8-x^2\right)^{-1/2} + \left(8-x^2\right)^{1/2}$

$= \left(8-x^2\right)^{-1/2}\left(-x^2+\left(8-x^2\right)\right)$

$= \left(8-x^2\right)^{-1/2}\left(8-2x^2\right)$

$= \frac{2\left(4-x^2\right)}{\left(8-x^2\right)^{1/2}}$

$= \frac{2(2-x)(2+x)}{\left(8-x^2\right)^{1/2}}$

96. $2x\left(1-x^2\right)^{3/2}+x^2\left(\frac{3}{2}\right)\left(1-x^2\right)^{1/2}(-2x)$

$=2x\left(1-x^2\right)^{3/2}-3x^3\left(1-x^2\right)^{1/2}$

$=x\left(1-x^2\right)^{1/2}\left(2\left(1-x^2\right)-3x^2\right)$

$=x\left(1-x^2\right)^{1/2}\left(2-2x^2-3x^2\right)$

$=x\left[(1-x)(1+x)\right]^{1/2}\left(2-5x^2\right)$

$=x(1-x)^{1/2}(1+x)^{1/2}\left(2-5x^2\right)$

97. Answers will vary.